COAL INFORMATION

2013
with 2012 data

INTERNATIONAL ENERGY AGENCY

The International Energy Agency (IEA), an autonomous agency, was established in November 1974. Its primary mandate was – and is – two-fold: to promote energy security amongst its member countries through collective response to physical disruptions in oil supply, and provide authoritative research and analysis on ways to ensure reliable, affordable and clean energy for its 28 member countries and beyond. The IEA carries out a comprehensive programme of energy co-operation among its member countries, each of which is obliged to hold oil stocks equivalent to 90 days of its net imports. The Agency's aims include the following objectives:

- Secure member countries' access to reliable and ample supplies of all forms of energy; in particular, through maintaining effective emergency response capabilities in case of oil supply disruptions.
- Promote sustainable energy policies that spur economic growth and environmental protection in a global context – particularly in terms of reducing greenhouse-gas emissions that contribute to climate change.
- Improve transparency of international markets through collection and analysis of energy data.
- Support global collaboration on energy technology to secure future energy supplies and mitigate their environmental impact, including through improved energy efficiency and development and deployment of low-carbon technologies.
- Find solutions to global energy challenges through engagement and dialogue with non-member countries, industry, international organisations and other stakeholders.

IEA member countries:

Australia
Austria
Belgium
Canada
Czech Republic
Denmark
Finland
France
Germany
Greece
Hungary
Ireland
Italy
Japan
Korea (Republic of)
Luxembourg
Netherlands
New Zealand
Norway
Poland
Portugal
Slovak Republic
Spain
Sweden
Switzerland
Turkey
United Kingdom
United States

The European Commission also participates in the work of the IEA.

International Energy Agency

International Energy Agency
9 rue de la Fédération
75739 Paris Cedex 15, France

www.iea.org

TABLE OF CONTENTS

PART I: INTRODUCTION

1. **Introduction** **I.9**
2. **Definitions** **I.11**
   Energy sources I.11
   Flows: energy balance I.14
   Coal resources and reserves I.16
   Units and conversions I.17
3. **Sources and notes** **I.19**
   General notes I.19
   Price data I.20
   Quarterly energy statistics I.21
   Data sources I.21
4. **Country notes** **I.23**
5. **Geographical coverage** **I.31**
6. **Abbreviations, conversion factors and calorific values** **I.33**

PART II: WORLD COAL OVERVIEW

1. **World coal market review** **II.3**
   Summary II.3
   Production II.4
   Coal resources and reserves II.7
   Consumption II.8
   Trade II.11
2. **CO_2 emissions** **II.17**
3. **Flexibility of coal-fuelled power plants for dynamic operation and grid stability** **II.23**

PART III: WORLD COAL OVERVIEW

1. **Production**
   Table 1.1: World coal production (kt) III.5
   Table 1.2: World coal production (ktce) III.7
   Table 1.3: World coking coal production (kt) III.9
   Table 1.4: World steam coal production (kt) III.10
   Table 1.5: World lignite production (kt) III.12
   Table 1.6: OECD production of coke oven coke (kt) III.14
2. **Consumption**
   Table 2.1: World coal consumption (kt) III.17
   Table 2.2: World coal consumption (ktce) III.20
   Table 2.3: World coking coal consumption (kt) III.23
   Table 2.4: World steam coal consumption (kt) III.25
   Table 2.5: World lignite consumption (kt) III.28
   Table 2.6: OECD consumption of coke oven coke (kt) III.30
   Table 2.7: Key energy and CO_2 emissions data for OECD countries (2011) III.31

3. Trade

Table 3.1: World and seaborne coal trade III.35
Table 3.2: World total coal trade III.35
Table 3.3: World steam coal trade III.36
Table 3.4: World coking coal trade III.36
Table 3.5: World coal imports - regional aggregates III.37
Table 3.6: World sub-bituminous coal and lignite imports - regional aggregates III.37
Table 3.7: World coal imports - selected countries III.38
Table 3.8: World sub-bituminous coal and lignite imports - selected countries III.38
Table 3.9: World coking coal imports - regional aggregates III.39
Table 3.10: World steam coal imports - regional aggregates III.39
Table 3.11: World coking coal imports - selected countries III.40
Table 3.12: World steam coal imports - selected countries III.40
Table 3.13: OECD coke oven coke imports III.41
Table 3.14: World coal exports - regional aggregates III.42
Table 3.15: World sub-bituminous coal and lignite exports - regional aggregates III.42
Table 3.16: World coal exports - selected countries III.43
Table 3.17: World sub-bituminous coal and lignite exports - selected countries III.43
Table 3.18: World coking coal exports - regional aggregates III.44
Table 3.19: World steam coal exports - regional aggregates III.44
Table 3.20: World coking coal exports - selected countries III.45
Table 3.21: World steam coal exports - selected countries III.45
Table 3.22: OECD coke oven coke exports III.46

4. Prices

Table 4.1: Japan coking coal import costs III.49
Table 4.2: EU coking coal import costs from selected countries III.49
Table 4.3: Japan steam coal import costs III.50
Table 4.4: EU steam coal import costs from selected countries III.50
Table 4.5: Steam coal export costs III.51
Table 4.6: Coking coal export costs III.51
Table 4.7: Coking coal prices for industry III.52
Table 4.8: Steam coal prices for industry III.53
Table 4.9: Steam coal prices for electricity generation III.54

5. Coal for other uses

Table 5.1: OECD coal-fired and total electricity generating capacity III.57
Table 5.2: OECD coal use for electricity production and heat sold III.58
Table 5.3: OECD coal use in coke ovens III.59
Table 5.4: World consumption of pulverised coal injection (PCI) coals III.59
Table 5.5: Coal-fired heat and electricity generation and saturation III.60
Table 5.6: World final consumption of energy by fuel III.63
Table 5.7: World use of coal for selected end uses III.64

6. Figures

Figure 1: World coal production III.67
Figure 2: World steam and coking coal trade III.67
Figure 3: Coking coal price CIF Japan and CIF EU member states (US$/t) III.68
Figure 4: Steam coal price CIF Japan and CIF EU member states (US$/t) III.68
Figure 5: Steam coal price US FOB vs. US domestic (US$/t) III.69
Figure 6: Steam coal price CIF EU member states (US$/t) and US exports (Mt) III.69
Figure 7: OECD total primary energy supply III.70
Figure 8: OECD coal consumption and indicators III.70
Figure 9: Japan steam coal supply, imports, GDP annual growth rates III.71

Figure 10: EU member states steam coal supply, imports, GDP annual growth rates III.71
Figure 11: Coal production by region III.72
Figure 12: Coal consumption by region III.72
Figure 13: Coal imports by region III.73
Figure 14: Coal exports by region III.73

PART IV: DETAILED OECD COAL DATA

Directory of part IV figures and tables IV.3

OECD Total IV.5
OECD Americas IV.16
OECD Asia Oceania IV.26
OECD Europe IV.36
IEA Total IV.47
IEA Americas IV.57
IEA Asia Oceania IV.67
IEA Europe IV.77
Australia IV.87
Austria IV.97
Belgium IV.106
Canada IV.116
Chile IV.126
Czech Republic IV.135
Denmark IV.145
Estonia IV.155
Finland IV.164
France IV.173
Germany IV.183
Greece IV.194
Hungary IV.203
Iceland IV.211
Ireland IV.219
Israel IV.228
Italy IV.236
Japan IV.245
Korea IV.254
Luxembourg IV.263
Mexico IV.271
Netherlands IV.279
New Zealand IV.290
Norway IV.300
Poland IV.309
Portugal IV.319
Slovak Republic IV.329
Slovenia IV.337
Spain IV.346
Sweden IV.356
Switzerland IV.365
Turkey IV.373
United Kingdom IV.381
United States IV.391

PART V: NON-OECD COAL DATA

General notes V.3

World coal supply and end-use statistics: 1975, 1980, 1985, 1990, 1995, 2000, 2005, 2007 to 2011, and 2012e V.5

National coal balances and trade statistics:

Brazil V.19
Bulgaria V.21
China, People's Republic of V.23
Chinese Taipei V.27
Colombia V.29
Hong Kong, China V.31
India V.33
Indonesia V.35
Kazakhstan V.37
Korea, People's Democratic Republic V.40
Malaysia V.41
Philippines V.43
Romania V.45
Russian Federation V.47
Serbia V.50
South Africa V.51
Thailand V.53
Ukraine V.55
Venezuela V.59
Vietnam V.61
Africa V.63
Asia excluding China V.64
European Union - 27 .... V.65
Non-OECD Americas V.66
Non-OECD Europe and Eurasia V.67
Non-OECD Total V.68
World V.69

PART I

INTRODUCTION

1. INTRODUCTION

IEA *Coal Information 2013* is the latest edition of an annual publication intended to provide sound market information on coal to policy and market analysts and those employed in all sectors of the coal industry.

This monitoring and reporting of historical trends and current energy market situation provides a strong foundation for policy and market analysis to better inform the policy decision process toward selecting policy instruments that are best suited to meet domestic and/or international objectives.

IEA *Coal Information 2013* brings together in one volume, the basic statistics compiled by the IEA on coal supply, consumption, trade and prices for both member and non-member countries[1]. It also includes information on coal by-products.

Part I provides important documentation that will assist the reader in correctly using the data in this publication and to understand the details of the statistical methodology and collection practices related to the coal data. In addition, calorific values used for preparing national energy balances are presented.

Part II of the publication provides a review of world coal market developments in 2011. It covers world coal production and coal reserves, coal consumption by type (total, steam, coking and lignite), steam and coking coal trade and prices.

Part III provides the reference tables to the Part II review on coal production, consumption, trade and prices.

Part IV provides, in tabular and graphic form, a more detailed and comprehensive statistical picture of coal developments in the 34 OECD member countries, by region and individually.

Part V provides for selected non-OECD countries summary statistics on coal supply and end-use statistics for about 40 countries and regions worldwide. Complete coal balances and coal trade data for selected years are presented for 20 major non-OECD coal-producing and consuming countries.

OECD data are taken from IEA/OECD databases of Energy Statistics that are based on annual, quarterly and monthly submissions from OECD Member countries to the Secretariat. The Energy Data Centre of the IEA Secretariat works closely with national administration to secure consistency in all time series with particular regard for IEA product definitions and reporting conventions. This effort is supplemented by surveys of energy industry publications, national statistics reports and other material. The finalized data provide the basis for IEA/OECD *Energy Balances of OECD Countries* and *Energy Statistics of OECD Countries*.

The non-OECD data are based upon information collected by the IEA Secretariat, official national submissions to the United Nations in Geneva and New York, and national energy publications. The resulting synthesis is published in *Energy Balances of Non-OECD Countries* and *Energy Statistics of Non-OECD Countries*. Users of this publication are directed to the Methodology Section of those publications for more detail on individual non-Member countries covered in the publication.

OECD coal balances and statistics, including itemized import and export data, are available on a CD-ROM, along with world supply data. Information on ordering the CD-ROM and other energy statistics publications is available at the end of this book and on the IEA website at *www.iea.org/stats/index.asp*.

1. This publication is without prejudice to the status of or sovereignty over any territory, to the delimitation of international frontiers and boundaries and to the name of any territory, city or area. In addition, the term "country" refers to a country or a territory, as the case may be.

Price data in Parts II, III and IV are derived from the quarterly publication *Energy Prices and Taxes*. Readers should consult this IEA/OECD publication for detailed information on methodology, data coverage and data sources. Country notes and documentation are available online in the *Energy Prices and Taxes* folder at: *http://wds.iea.org*.

In addition, a data service is available on the internet, which provides unlimited access through an annual subscription, as well as the possibility to obtain data on a pay-per-view basis. Details are available at *http://data.iea.org*.

Further information on reporting methodologies is also available on the IEA Web site.

Annual energy data are collected by the Energy Data Centre (EDC) of the IEA Secretariat, headed by Mr. Jean-Yves Garnier. The IEA would like to thank and acknowledge the dedication and professionalism of the statisticians working on energy data within national administrations. OECD coal statistics in the EDC were the responsibility of Mr. Julian Smith. Ms. Rachael Hackney, Ms. Alki Delichatsios and Mr. Raphael Vial contributed to Part II, Part III and Part V. Mr. Vladimir Kubecek had overall responsibility for this publication.

Also in the IEA Secretariat, input from the Directorate of Global Energy Economics was crucial to the compilation of this edition, while special thanks are also due to Mr. Carlos Fernandez-Alvarez, Mr. Timo Panke, Mr. Harald Hecking and Ms. Roberta Quadrelli for invaluable assistance, the non-OECD Member countries section headed by Mr. Pierre Boileau, and to the OECD Balances section headed by Ms. Karen Tréanton.

Editorial and desktop publishing support from Ms. Sharon Burghgraeve is also gratefully acknowledged.

Enquiries, comments and suggestions are most welcome and should be addressed to:

Vladimir Kubecek or Julian Smith
Energy Data Centre,
International Energy Agency
9, rue de la Fédération
75739 Paris Cedex 15, France

Telephone: (33) (1) 40 57 66 31 or 40 57 67 02
Telefax: (33) (1) 40 57 66 49
E-mail: coalaq@iea.org

What's New?

Prior to the 2012 edition, **Steam coal** consisted of anthracite, other bituminous coal, and in the case of 12 specific countries, sub-bituminous coal. Steam coal now includes sub-bituminous coal for all countries.

Definitions of hard coal and brown coal remain unchanged. This means that hard coal is no longer calculable by adding together steam coal and coking coal. To avoid confusion, trade in particular may refer to total coal, which will be the sum of hard coal and brown coal, or anthracite, coking coal, other bituminous coal, sub-bituminous coal and lignite.

Oxygen steel furnace gas is now known as **Other recovered gases**. Gases of similar nature that did not originate in a basic oxygen steel furnace were also reported here historically. The change in name makes this more apparent.

Due to reductions in budget, the IEA no longer has adequate resources to provide complete information on energy prices and taxes, so has had to suppress certain sections of *Energy Prices and Taxes* as of 1 January 2012. As a result, **some price series have not been updated** in this edition of *Coal Information*. The IEA considered it necessary to reduce the coverage of this publication in order to maintain the high quality of the remaining information in *Energy Prices and Taxes* and other publications. We remain hopeful that we may be able to restore these sections in the future if resources become available.

2. DEFINITIONS

Energy sources

Coal

Coal is a family name for a variety of solid organic fuels and refers to a whole range of combustible sedimentary rock materials spanning a continuous quality scale. For convenience, this continuous series is often divided into two main categories, which are themselves divided into two subcategories:

- Hard coal
    - Anthracite
    - Bituminous coal
        - Coking coal
        - Other bituminous coal
- Brown coal
    - Sub-bituminous coal
    - Lignite

In cases where data are presented in Mtoe or Mtce in this book and sourced to OECD/IEA *Energy Balances*, the term "Coal" includes all primary coal types listed above, peat, and coal products (patent fuel, coke oven coke, gas coke, coal tar, BKB, coke oven gas, gas works gas, blast furnace gas, and other recovered gases).

Classifying different types of coal into practical categories for use at an international level is difficult for two reasons:

Divisions between coal categories vary between classification systems, both national and international, based on calorific value, volatile matter content, fixed carbon content, caking and coking properties, or some combination of two or more of these criteria.

Although the relative value of the coals within a particular category depends on the degree of dilution by moisture and ash and contamination by sulphur, chlorine, phosphorous and certain trace elements, these factors do not affect the divisions between categories.

Coal quality can vary and it is not always possible to ensure that the available descriptive and analytical information is truly representative of the body of coal to which it refers.

The International Coal Classification of the Economic Commission for Europe (UNECE) recognises two broad categories of coal:

i) Hard coal - Coal of gross calorific value not less than 5 700 kcal/kg (23.9 GJ/t) on an ash-free but moist basis and with a mean random reflectance of vitrinite of at least 0.6.

ii) Brown coal - Non-agglomerating coal with a gross calorific value less than 5 700 kcal/kg (23.9 GJ/t) containing more than 31% volatile matter on a dry mineral matter free basis.

The IEA has adopted the basis of these definitions of hard coal and brown coal in this book and in other publications for presenting statistics relating to coal production, trade and consumption.

It should be stressed that this classification system is based on the inherent qualities of the coal in question and not on the final use of the coal. In this way the classification system attempts to be objective and simple to apply.

Primary coal used in pulverised (or granular) coal injection in blast furnaces is commonly abbreviated to PCI (or GCI) coal. (In this book PCI includes GCI). The IEA does not have a separate product classification for PCI as the term defines a particular end-use for coal. In IEA statistics, PCI is generally included in steam coal, except for Japan, Korea, the Netherlands,

Poland, the Slovak Republic and the United Kingdom where it is included with coking coal.

Data presented in this book may be different from that presented in the national publications of individual countries because the countries may have adopted a different coal classification and reporting system that better suits their particular national needs. As far as possible, national coal statistics reported by the IEA in this book and in other publications have been adjusted to be consistent with the IEA definitions noted above.

Furthermore, in order to improve the information base for coal market analysis and projections, these two categories of coal have been further sub-divided in IEA/OECD Coal Statistics from 1978 as follows:

Hard coal

Hard coal is calculated as the sum of anthracite and bituminous coals.

- ***Anthracite*** is a high-rank, hard coal used mainly for industrial and residential heat raising.
- ***Bituminous coal*** is a medium-rank coal used for gasification, industrial coking and heat raising and residential heat raising. Bituminous coal that can be used in the production of a coke capable of supporting a blast furnace charge is known as ***coking coal***.
- ***Other bituminous coal***, not included under coking coal, is also commonly known as thermal coal. Also included are recovered slurries, middlings and other low-grade, higher-rank coal products not further classified by type.

Note that for the following countries, hard coal also includes sub-bituminous coal: Australia, Belgium, Chile, Finland, France, Iceland, Japan, Korea, Mexico, New Zealand, Portugal and the United States.

Except for these 12 countries, steam coal for all other countries in the world is equivalent to the sum of anthracite and other bituminous coal.

Brown coal

Brown coal is calculated as the sum of sub-bituminous coal and lignite. Oil shale mined and combusted directly is also currently reported as lignite. (Shale oil is reported as other hydrocarbons.)

- ***Sub-bituminous coal***: non-agglomerating coals with a gross calorific value between 4 165 kcal/kg (17.4 GJ/t) and 5 700 kcal/kg (23.9 GJ/t) on an ash-free but moist basis.
- ***Lignite***: non-agglomerating coal with a gross calorific value less than 4 165 kcal/kg (17.4 GJ/t) on an ash-free but moist basis.

Note that for 12 countries (listed above), sub-bituminous coal is included in hard coal, not brown coal. Under current methodology, lignite data also includes oil shale (but not shale oil).

Steam Coal

Steam coal in this publication contains all anthracite, other bituminous coal and sub-bituminous coal.

Prior to the 2012 publication, all hard coals that were not coking coal (including the sub-bituminous coal from the excepted countries) were classed as steam coal. This also included by necessity countries (not listed) where sub-bituminous coal was unable to be separated from other bituminous coal data for reporting purposes.

For the *Coal Information* 2012 publication onwards, the definition of steam coal was adjusted to include all sub-bituminous coals. This move was done to achieve greater congruence with practical, formal and informal definitions of steam (thermal) coal in the market and coal industry at large.

The definitions of hard coal and brown coal remain unchanged and consistent with the UNECE guidelines above and InterEnerStat definitions. This means hard coal can no longer be calculated by adding steam coal data to coking coal data for all countries.

Coal products

Derived solid fuels are products resulting from the transformation from hard coal, brown coal or other primary solid fuels, sometimes with the addition of other materials.

Coke oven coke

Coke oven coke is the solid product obtained from the carbonisation of coal, principally coking coal, at high temperature. Semi-coke, the solid product obtained from the carbonisation of coal at lower temperatures is also included, along with coke and semi-coke made from lignite.

Gas coke

Gas coke is a solid by-product of coal used for the production of town gas in gas works. Gas coke is used for heating purposes.

Patent fuel

Patent fuel is a composition fuel manufactured from coal fines by shaping with the addition of a binding agent such as pitch.

Brown coal briquettes (BKB)

BKB are composition fuels manufactured from brown coal. The brown coal is crushed, dried and moulded under high pressure into an even shaped briquette without the addition of binders. Peat briquettes, dried brown coke, fines and dust and brown coal breeze are also included.

Coal tar

Coal tar is the liquid by-product of the destructive distillation of bituminous coal to make coke in the coke oven process. Coal tar can also be the result of low-temperature carbonisation of brown coal. Coal tar can be further distilled into different organic products (*e.g.* benzene, toluene, naphthalene), which normally would be reported as a feedstock to the petrochemical industry.

Quite a few countries are currently unable to report coal tar data. For these countries, coke oven transformation losses will appear larger than they actually are, and consumption data will obviously be missing from the relevant end-use sector.

Manufactured Gases

Manufactured gases created outside of refineries, sourced primarily from solid hydrocarbons are reported on the coal questionnaire. They include purpose built products like gas works gas, whose manufacture is often the main purpose of the transformation process, and products like coke oven gas and blast furnace gas which are useful energy by-products of another process.

Coke oven gas

Coke oven gas is obtained as a by-product of solid fuel carbonisation and gasification operations carried out by coke producers and iron and steel plants not connected with gas works and municipal gas plants.

Gas works gas

Gas works gas covers all types of gas produced in public utility or private plants, whose main purpose is the manufacture, transport and distribution of gas, regardless of process. It includes gas produced by carbonisation (including gas produced by coke ovens and transferred to gas works), by total gasification (with or without enrichment from oil products) and by reforming and simple mixing of gases, which may include air.

Note: Starting with the 2011 edition, gas works gas is included with coal for the years 1990 and beyond. Before 1990, gas works gas is included with natural gas.

Blast furnace gas

Blast furnace gas is obtained as a by-product in operating blast furnaces. It is recovered upon leaving the furnace and used partly within the plant and partly in other steel industry processes or used in power stations equipped to burn it. It is mainly nitrogen (N_2), with roughly equal amounts of carbon dioxide and carbon monoxide and will contain other trace gases.

Other recovered gases

Other recovered gases were previously known as oxygen steel furnace gas, which is most commonly obtained as a by-product of the production of steel in an oxygen-fired furnace; it is recovered upon leaving the furnace. The gas is also known as converter gas, LD gas or BOS gas. Other gases of similar nature are also reported in this category, hence the change of name to be intrinsically more inclusive.

Peat

A solid formed from the partial decomposition of dead vegetation under conditions of high humidity and limited air access (initial stage of coalification). It is available in two forms *for use as a fuel*, sod peat and milled peat. Milled peat is also made into briquettes for fuel use. Peat is not considered a renewable resource as its regeneration period is long.

Heat and electricity

Data for electricity and heat includes disaggregated data on inputs and outputs of 'combined heat and power' and on 'district heating'. Data on heat became available in different years for different countries and thus aggregated country data should be used with caution.

Total electricity production includes production from both main activity producers (formerly known as public) and autoproducers. Generally, the split of total electricity production between main activity producers and autoproducers is available only after 1973.

Flows: energy balance

Coal balances are presented in detail in Parts IV and V. In Part IV, Table 5 presents uses in the rows and selected years in the columns. Data are presented in millions of tonnes of coal equivalent (Mtce). One tonne of coal equivalent is 7 million kilocalories.

Each table is divided into three main parts: the first shows supply elements such as trade and production, the second shows the transformation processes and energy industries, while the third shows final consumption broken down into various end-use sectors.

Both primary fuels such as coal and peat, and derived fuels such as coke oven coke and blast furnace gas are included in the calculations. However, derived products manifest themselves as positive outputs in the relevant transformation process used to create them. Generally they should be less than the inputs, which result in a net negative entry in the transformation flow. Given that this balance is restricted to coal and associated products, inputs from other fuel types (such as pitch for patent fuels, or oil, gas and renewable inputs to blast furnaces) will not be recorded, nor will electricity generated, which differs from a full energy balance.

The energy balance flows detailed below have the following functions, and may also appear in other tables:

Supply

Production

Production is the production of primary energy, *i.e.* hard coal, brown coal, peat, shale oil, etc. Production is calculated after the removal of impurities on the bases which it is provided for sale. It is important to note that derived products such as coke oven coke and patent fuel, while included in the balances, do not appear in production as they are not primary products.

Imports and exports

Imports and exports comprise amounts having crossed the national territorial boundaries of the country, whether or not customs clearance has taken place. Coal in transit to another country should not be included here. Both primary and secondary products are reported.

Stock changes

Stock changes reflects the difference between opening stock levels on the first day of the year and closing levels on the last day of the year of stocks on national territory held by producers, importers, energy transformation industries and large consumers. A stock build is shown as a negative number and a stock draw as a positive number. It is presented this way as this is how it affects the domestic supply, as opposed to how it describes the changes in stocks.

Total primary energy supply

Total primary energy supply (TPES) consists of production + imports - exports ± stock changes as an abstract concept. Given that exports and stock builds both are represented as negative numbers, in reality TPES = production + imports + exports + stock changes. Marine and aviation bunkers also are not counted in TPES.

Statistical differences

Statistical differences includes the sum of the unexplained statistical differences for individual fuels, as they appear in the basic energy statistics. It also includes the statistical differences that arise because of the variety of conversion factors in the coal and oil columns. See the introduction to *Energy Statistics of OECD Countries* for further details.

For countries that are unable to collect stock change data, stock builds and draws will contribute to statistical differences.

Transformation processes

Transformation processes record the transformation of one kind of fuel or energy into another with both inputs and outputs being measured. This may bridge several transformation processes. For instance:

- Coking coal used to manufacture coke oven coke would be reported as a negative input to the coke oven transformation process.
- The resulting coke oven coke, coal tar and coke oven gas would be reported as a positive output to the coke oven transformation process flow.
- Energy inputs from other sources, including electricity, will not be reported in this particular instance. Therefore, numbers may not be indicative of true efficiencies, but rather map the flow of coal.

- The coke oven coke will largely be used to produce pig-iron in a blast furnace. Therefore, it will be reported where it is used – mainly as an input to the blast furnace transformation process.
- The by-product blast furnace gas will appear as an output in the blast furnace transformation flow. However, a significant amount of energy is lost in the process of making the pig-iron, so the net negative value in the blast furnace transformation flow will tend to be approximately 60% of the total energy inputs.
- The blast furnace gas (and coal tar and coke oven gas) will likewise be reported where used. Some of this will appear in the relevant consumption flows, other parts might be used to generate electricity and appear in electricity transformation.
- In a complete energy balance, the electricity generated would be converted to the appropriate energy unit and reported as a positive output in the applicable electricity transformation flow. This is not the case in the coal balance, so the number displayed in the electricity transformation flow is the fuel input, not the process efficiency loss.

The main transformation processes reported either create a derived coal product or by-product, and have been described earlier in the Energy sources section or are mentioned below.

Electricity and heat generation

Electricity and heat generation can refer to electricity plants, combined heat and power plants (CHP), or heat plants. Both main activity producer[1] and autoproducer[2] plants are included here.

Electricity plants are plants which are designed to produce electricity only. If one or more units of the plant is a CHP unit (and the inputs and outputs cannot be distinguished on a unit basis), then the whole plant is designated as a CHP plant.

Note that for autoproducer CHP plants, all fuel inputs used to generate electricity are taken into account. However, only the part of the fuel inputs used to produce the heat that is sold is shown. Fuel inputs for the production of heat that is consumed within the autoproducer's establishment are not included here but are included in the final consumption of fuels in the appropriate consumption sector.

Heat plants (including heat pumps and electric boilers) are designed to produce heat only, which is sold to a third party under the provisions of a contract. Heat pumps that are operated within the residential sector, where the heat is not sold, are not considered a transformation process and are not included here, despite the fact that equivalent electricity consumption will appear as residential use.

Other transformation

Other transformation covers non-specified transformation and transformations not shown elsewhere, such as coal liquefaction.

Energy industry own use

Energy industry own use contains the primary and secondary energy consumed by transformation industries for heating, pumping, traction and lighting purposes [ISIC[3] 05, 06, 19 and 35, Group 091 and Classes 0892 and 0721]. These quantities are shown as negative figures. Included here is, for example, own use of energy in coal mines.

Losses

Losses includes losses in gas distribution, flaring or venting of manufactured gases, electricity transmission and coal transport.

Consumption

Total final consumption (TFC) is the sum of consumption by the different end-use sectors.

Industry consumption is specified in the following sub-sectors. Note that energy used for transport by industry is not included here but is reported under transport.

Non-energy use covers those fuels that are used as raw materials in the different sectors and are not consumed as a fuel or transformed into another fuel.

1. Main activity producer generate electricity and/or heat for sale to third parties, *as their primary activity*. They may be privately or publicly owned. Note that the sale need not take place through the public grid.
2. Autoproducer undertakings generate electricity and/or heat, wholly or partly for their own use as an activity which supports their primary activity. They may be privately or publicly owned.
3. International Standard Industrial Classification of All Economic Activities, Series M, No. 4 / Rev. 4, United Nations, New York, 2008.

Iron and steel	ISIC Group 241 and Class 2431
Chemical and petrochemical industry	ISIC Divisions 20 and 21, excluding petrochemical feedstocks
Non-ferrous metals	ISIC Group 242 and Class 2432
Non-metallic minerals	ISIC Division 23, such as glass, ceramic, cement, etc.
Transport equipment	ISIC Divisions 29 and 30
Machinery	ISIC Divisions 25 to 28, comprises fabricated metal products, machinery and equipment other than transport equipment
Mining (excluding fuels) and quarrying	ISIC Divisions 07 and 08 and Group 099
Food and tobacco	ISIC Divisions 10 to 12
Paper, pulp and printing	ISIC Divisions 17 and 18
Wood and wood products	ISIC Division 16, other than pulp and paper
Construction	ISIC Divisions 41 to 43
Textile and leather	ISIC Divisions 13 to 15
Non-specified	ISIC Divisions 22, 31 and 32, any manufacturing industry not included above Note: Most countries have difficulties supplying an industrial breakdown for all fuels. In these cases, the non-specified industry row has been used. Regional aggregates of industrial consumption should therefore be used with caution.

Transport includes all fuels used for the transport [ISIC Divisions 49 to 51] of goods or persons between points of departure and destination within the national territory irrespective of the economic sector within which the activity occurs.

Other covers residential, commercial and public services [ISIC Divisions 33, 36-39, 45-47, 52, 53, 55, 56, 58-66, 68-75, 77-82, 84 (excluding Class 8422), 85-88, 90-96 and 99], agriculture/forestry [ISIC Divisions 01 and 02], fishing [ISIC Division 03] and non-specified consumption.

Coal resources and reserves

Quantifying mineable coal is based on a consideration of geological, mining and economic criteria. The amount of coal in place and, in some cases, the amount of mineable coal is influenced by national resource measurement criteria. The basis for computing these resources varies from country to country and, therefore, it must be borne in mind that for this reason, direct comparisons are sometimes not possible. During the 1990s, there was a considerable discussion on the adoption of internationally recognised standards for reporting reserves. This largely stems from the requirements of capital markets for improved transparency in reserve estimation where project financing is being sought. However, to date there has not been any adoption of international standards. There are, however, some generally recognised definitions that can be applied.

Resources

Resources refer to the amount of coal that may be present in a deposit or a coalfield. This does not take into account the feasibility of mining the coal economically. Not all resources are recoverable using current technology. Reserves constitute those resources that are recoverable.

Reserves

Reserves may be defined further in terms of proved (or measured) reserves, and probable (or indicated) reserves, based on exploration results and the degree of confidence in those results. Probable reserves have been estimated with a lower degree of confidence than proved reserves. Estimates take account of coalfields' geological characteristics, in particular the regularity, thickness and quality of seams, the spacing of exploration boreholes and other exposures, and geological discontinuities such as faults or folding, all of which affect the practical recoverability of the coal.

Proved reserves

Proved reserves are those reserves that are not only confidently considered to be recoverable, but can also be recovered economically under current market conditions. In other words, they take into account what current mining technology can achieve, as well as the economics of recovery (mining, transportation and other relevant recovery costs, such as government royalties, and coal prices). Proved reserves will, therefore, fluctuate according to economic pressures, especially price.

Units and conversions

Balance units

Most IEA/OECD publications showing inter-fuel relations and projections present such information in a common energy unit, the tonne of oil equivalent (toe). A tonne of oil equivalent is defined as 10^7 kcal (41.868 GJ), a convenient measure because it is approximately the net heat content of one ton of average crude oil. This unit is used by the IEA/OECD in the majority of its energy balances.

The change from using the original unit to tonne of oil equivalent implies choosing coefficients of equivalence between different forms and sources of energy. This problem can be approached in many different ways. For example, one could adopt a single equivalence for each major primary energy source in all countries, *e.g.* 29 307 kJ/kg (7 000 kcal/kg) for hard coal, 41 868 kJ/kg (10 000 kcal/kg) for oil.

The main objection to this method is that it results in distortions since there is a wide spread in calorific values between types of coal and individual coal products, and between calorific values of these fuels in different countries.

The Secretariat has, therefore, obtained specific calorific factors supplied by the national administrations for the main categories of each quality of coal and for each flow or use (*i.e.* production, imports, exports, electricity generation, coke ovens, blast furnaces and industry). The set of particular national calorific values that allow for the conversion of energy sources from original (physical) units to joules are presented later in Part I.

The balances are expressed in terms of net calorific value. The difference between net and gross being the latent heat of vaporisation of the water produced during combustion of the fuel. For coal and oil, net calorific value is usually around 5% less than gross and for most forms of natural and manufactured gas the difference is 9-10%. The use of net calorific value is consistent with the practice of the Statistical Offices of the European Communities and the United Nations.

Note that throughout this publication, 1 tonne means 1 metric ton or 1000 kg. Billion refers to 1 thousand million (10^9). Also, in many cases, totals shown in the tables may not be the exact sum of their components due to independent rounding.

Conversion (to toe and tce)

In this report some data are reported in terms of tonnes of coal equivalent (tce) because this unit is more widely used in the international coal industry. A tonne of coal equivalent is defined as 7 million kilocalories (29.3076 GJ). The relation between tonne of oil equivalent (toe) and tonne of coal equivalent (tce) is therefore:

1 tce = 0.7 toe

Units for gases

In the IEA/OECD publication *Energy Statistics of OECD Countries* all data on gases are expressed in terajoules, on the basis of their gross calorific value.

1 terajoule = 0.00002388 Mtoe.

To calculate the net heat content of a gas from its gross heat content, multiply the gross heat content by the appropriate following factor:

Gas	Ratio from GCV to NCV
Natural gas	0.9
Gas works gas	0.9
Coke oven gas	0.9
Blast furnace gas	1.0
Other recovered gas	1.0

Please note that this means in order to calculate gross from net, if necessary, you must divide the net value by 0.9.

Electricity

Figures for electricity production, trade, and final consumption are calculated using the energy content of the electricity, *i.e.* at a rate of 1 TWh = 86 ktoe.

Hydro-electricity production (excluding pumped storage) and electricity produced by other non-thermal means (wind, tide/wave/ocean, solar PV, etc.) are accounted for similarly using 1 TWh = 86 ktoe.

The primary energy equivalent of nuclear electricity is calculated from the gross generation by assuming a 33% efficiency, *i.e.* 1 TWh = (86 ÷ 0.33) ktoe.

In the case of electricity produced from geothermal heat, if the actual geothermal efficiency is not known, then the primary equivalent is calculated assuming an efficiency of 10%, so 1 TWh = (86 ÷ 0.1) ktoe.

For electricity produced from solar thermal heat, the primary equivalent is calculated assuming an efficiency of 33%, so 1 TWh = (86 ÷ 0.33) ktoe, unless the actual efficiency is known.

Heat

Information on heat is supplied in terajoules and 1 terajoule = 0.02388 ktoe.

In the case of heat produced in a geothermal plant, if the actual geothermal efficiency is not known, then the primary equivalent is calculated assuming an efficiency of 50%, so 1 TJ = (0.02388 ÷ 0.5) ktoe.

For heat produced (output) in a solar thermal plant, the primary equivalent is equal to the heat content (input) assuming an efficiency of 100%, *i.e.* 1 TJ = 0.02388 ktoe.

For direct use of geothermal and solar thermal heat, all the heat consumed is accounted for in production and consumption.

For autoproducers, only heat sold is reported. Fuels consumed by autoproducers for production of own-use heat are reported as consumption under the classification of their main economic activity, not as heat only, or combined heat and power (CHP) transformation.

3. SOURCES AND NOTES

General notes

Energy flows data reported for 2012 (shown as 2012e) are preliminary estimates based on the submissions received in early 2013 and on quarterly submissions to the IEA from member countries.

The fuels listed as "solar, wind, tide" also include wave and ocean, unless otherwise noted.

Statistics of non-OECD countries presented in this publication are based on available data at the time of publishing and may differ from the final non-OECD data to be published in *Energy Statistics of Non-OECD Countries*.

Additional information on methodologies and reporting conventions are included in the notes in *Energy Balances of OECD Countries* 2013 edition and *Energy Statistics of OECD Countries* 2013 edition.

Preliminary 2012 data - 2012e

Data reported for the year 2012 in this publication are preliminary and presented as 2012e. Final 2012 data on solid fuels and manufactured gases will be submitted by OECD Member countries to the Secretariat in annual questionnaires in late 2013. As a result, final data for 2012 and preliminary 2013 data will be published in the 2014 edition of *Coal Information* in July 2014.

Selected coal data for 2012 for some non-OECD countries have been estimated by the Secretariat, as is the case for three OECD countries: Belgium, Israel and Turkey.

Qualifiers

Data marked as *'e'* are estimates of the IEA Secretariat. Data marked as *'c'* mean that the data are confidential due to country specific regulations. Data marked as *'..'* mean that data are not available (either not collected or not submitted by national government). Data marked as *'x'* mean that they are not applicable or there is no meaningful explanation of a value there. For example, the price cannot be shown if the consumption in the country is forbidden or the country itself did not exist as an independent entity at a given point in time.

Treatment of blast furnace coke and PCI data

Data on coke used in and pulverised coal injected into blast furnaces (PCI), are harmonized for all OECD countries in order to ensure that blast furnace transformation data are consistently presented and that comparisons between countries for consumption are meaningful. The main effect of these revisions has been, where necessary, to revise the reported consumption of coal in the iron and steel industry and in blast furnace transformation, so discrepancies between IEA and national accounts may ensue. In effect, inputs to blast furnaces may be calibrated to be proportionate to production of blast furnace gas and some inputs to blast furnace consumption may be reported as consumption in the iron and steel industry if there are lower than normal outputs of blast furnace gas.

It should be noted that in IEA statistics of coal trade and consumption, PCI is not separately specified; rather it is included in some form of hard coal. This methodology is based on the fact that pulverised coal injection is a process, and this process, unlike for coke oven coke manufacture, is largely independent of coal type.

For Japan and Korea, PCI consumption is reported in this book as a coking coal in order to be consistent with the national practice of including imports of PCI coal with coking coal without regard to coal type. Other countries that report usage of coking coal as inputs to blast furnaces may do so for this reason, or because of the respective coal quality.

Price data

Energy prices are published quarterly in the IEA/OECD *Energy Prices and Taxes*. IEA data on coal prices are managed in two sub-systems, which vary not only in content, but also with respect to the data collection methods.

- Import and export unit values
- End-user prices

Import and export unit values

Import and export unit values are calculated quarterly (March, June, September and December) from national customs statistics import and export volumes and values. The basic data are collected from monthly national trade sources (Chile, Japan, Korea, United States, Australia and Canada) or provided monthly to the IEA by the Statistical Office of the European Communities (Eurostat).

Values recorded at the import stage are the sum of cost, insurance and freight (CIF – cost including freight/fees), but exclude import duties. Values recorded at the export stage (FOB – free on board), exclude seaborne or international transport, but include inland transport costs of the exporting country.

As far as possible, the concept of 'general imports and exports' is used. This includes coal imports for re-export with or without processing, but excludes transit trade.

The definitions of coal categories and the volume and value units used in each of the above source systems vary considerably. A certain amount of regrouping and unit conversions is necessary once the basic data are compiled.

The rules for regrouping coal categories are consistent with the definitions used in the annual IEA/OECD coal statistics. Prices are compiled for steam coal and for coking coal. Definitions and the correspondence to national and European classifications are discussed in detail in the quarterly IEA publication *Energy Prices and Taxes*. Comments in *Energy Prices and Taxes* on certain data items, as well as general background information, are developed systematically. Data comments relate mainly to calorific values of specific coal trade flows and to national coal definitions. Background information covers duties and trade regulations.

Due to reductions in budget, the IEA no longer has adequate resources to provide complete information on energy prices and taxes, so has had to suppress certain sections of *Energy Prices and Taxes* as of 1 January 2012. This included steam and coking coal import and export data for the year gone. As a result, it has not been possible to provide these price series with 2011 or 2012 data in this edition of *Coal Information*. We are hopeful that we may be able to restore these sections in the future if resources become available.

End-user prices

End-user prices are collected quarterly from national administrations and other relevant bodies and supplemented with data extracted from national publications. Although a standard approach to reporting the data has been developed, differences in definitions between countries are explained in the notes published in *Energy Prices and Taxes*.

The standard approach to reporting end-use prices can be summarised as follows:

- includes transport costs to the consumer;
- shows prices actually paid, *i.e.* net of rebates; and
- includes taxes which have to be paid by the consumer as part of the transaction and which are not refundable. This excludes value added taxes paid in many European countries by industry (including electric power stations) for all goods and services (including energy). In these cases, value added taxes are refunded to the customer, usually in the form of a tax credit. Therefore, it is not shown as part of the prices.

A standard coal quality for all international comparisons of end-use prices is not possible given the wide variety of coal qualities in domestic and international coal trade. As a result, only average prices covering a range of different qualities are collected, along with the calorific value of these averaged sales. If average prices are not available, prices of a selected coal may be chosen. Accordingly, international comparisons of coal end-use prices may be misleading if read at face value. Detailed notes concerning these price series are published in *Energy Prices and Taxes*. Also, please refer to *Energy Prices and Taxes* for the detailed description of price mechanisms in each country and country specific notes.

Derived price data

The information collected on prices is converted by the IEA Secretariat into a variety of secondary data in order to facilitate its analysis. Inter-fuel price comparisons for one country are usually made on the basis of prices per heat unit such as a tonne of coal equivalent. In the end-user price tables, the conversion factor used for converting gross calories to net calories for natural gas is 0.9.

Inter-country price comparisons are made on the basis of a standard currency unit, *e.g.*, US dollars. Prices for regional totals are calculated as the weighted average only of the available price data in the region and, therefore, prices shown should be considered as only indicative.

For coal exports and imports, customs unit values are prices reported by OECD Member countries.

Customs unit values are average values derived from customs' administrations total volume and total value data. These data indicate broad price movements as they are averages of all qualities of coal without regard to the end-use of the coal or to the contract terms and conditions under which the trade occurs.

End-user prices are those paid by end-users in the power sector and in industry and are reported by Member countries in a quarterly reporting system which the IEA's Standing Group on Long Term Co-operation initiated in 1981. Data received are published in the IEA quarterly publication *Energy Prices and Taxes*.

Unless otherwise stated, prices are reported in US dollars in the year specified (*i.e.* current US dollars).

In addition to the official price statistics presented, coal price statistics published in the industry press are used to summarise short-term spot steam and coking coal price trends. Although not "official" in that they are not provided by member countries, there is a high correlation between prices published by the industry press and national coal price statistics.

Quarterly energy statistics

Readers who are interested in more recent data should consult the OECD/IEA publication *Oil, Gas, Coal and Electricity Quarterly Statistics* which is published in January, March, June and September each year.

This book provides current, accurate and detailed statistics on quarterly production, supply and demand and trade of the major energy forms mainly in, but not limited to, the OECD area.

The information contained in this publication consists of:

Oil

- Production of crude oil and NGL for the major producers in the world.
- Refinery balances for crude oil, NGL, refinery feedstocks, and total (including inputs of origin other than crude oil and NGL);
- Complete product balances of production, trade, refinery intake and output, final consumption, stock levels and changes;
- Crude, NGL and feedstock imports from 47 origins and exports to 24 destinations; and
- Trade data for main product groups, LPG and naphtha; imports from 44 origins and exports to 30 destinations.

Natural gas

- Balances of supply and consumption of OECD member countries; and
- Imports from 28 origins and exports to 20 destinations.

Coal

- World steam and coking coal, and lignite production;
- World steam coal and coking coal trade; and
- Coking coal and steam coal imports and exports for major OECD countries.

Electricity

- Electricity production, (separately from combustible fuel, nuclear, hydro and other sources) imports, exports and (apparent) consumption in OECD member countries.

Data sources

Historical data (1960-2011)

The annual historical data in Part IV of this report are taken from the IEA/OECD databases of energy statistics which are primarily based on annual submissions from all OECD Member countries.

i) IEA/OECD coal statistics

This database of annual statistics for OECD countries covers all primary solid fuels, derived fuels and related manufactured gases. It contains detailed supply/demand balances for each fuel, as well as information on coal trade by origin and destination. The main data from this system are published annually in the IEA/OECD publication *Coal Information.*

ii) IEA/OECD electricity statistics

This database of annual statistics for OECD countries covers generating capacity and electricity production from main activity producers and autoproducers plants. It includes information on electricity production by fuel type and supply/demand balances for electricity and for heat sold to third parties from different types of power and heat plants. The main data from this system are published annually in the IEA/OECD publication *Electricity Information.*

iii) IEA/OECD oil and gas statistics

This database of annual statistics for OECD countries covers crude oil, NGL, refinery feedstocks and natural gas, as well as derived oil products. It includes detailed supply/demand balances, trade by origin and destination and stock levels and changes.

The main data from this system are published annually in the IEA/OECD publications *Oil Information* and *Natural Gas Information.*

iv) IEA/OECD renewables statistics

This database of annual statistics for OECD countries covers hydro, solid biofuels, geothermal, renewable municipal waste, wind, gas from biofuels, solar photovoltaic, solar thermal, tide/wave/ocean, non-renewable municipal waste and industrial waste. It includes detailed supply/demand balances.

The main data from this system are published annually in the IEA/OECD publication *Renewables Information.*

v) IEA/OECD energy statistics

This annual database integrates data from the four IEA/OECD statistical database systems listed above to provide a summary of energy supply and demand for each OECD country. It includes detailed statistics on production, trade and consumption for each source of energy, expressed in original units (*e.g.* metric tons, TJ, GWh).

The main data from this data system are published annually in the IEA/OECD publication *Energy Statistics of OECD Countries.* Detailed country notes referring to historical data can be found in this publication.

vi) IEA/OECD energy balances

Overall energy balances are constructed annually for all OECD countries from the basic energy statistics described above. The overall energy balance data are expressed in a common energy unit of tons of oil equivalent (toe), and presented in a standard matrix format. The balances are published annually in the IEA/OECD publication *Energy Balances of OECD Countries* in which detailed country notes referring to historical data can be found.

vii) IEA/OECD energy prices and taxes

The statistics were discussed separately above. The prices and taxes are published quarterly in IEA/OECD *Energy Prices and Taxes.*

viii) Energy statistics of non-OECD countries

The annual historical data for non-OECD countries presented in Part V of this report are taken from IEA/OECD databases of energy statistics of non-OECD countries. These databases are compiled from data submitted annually to the IEA Secretariat in questionnaires from non-OECD Member countries of the United Nations Economic Commission for Europe (UNECE), from data provided by other international organisations (the United Nations in New York, the Organización Latinoamericana de Energía (OLADE) in Quito, the Asia-Pacific Economic Cooperation forum (APEC) in Tokyo and the Food and Agriculture Organization of the United Nations (FAO) in Rome, etc.), from direct communications with national administrations, industry contacts and from published sources.

The main data from this data system are published annually in the IEA/OECD publications *Energy Statistics of Non-OECD Countries* and *Energy Balances of Non-OECD Countries.* Detailed country notes referring to historical data can be found in these publications.

4. COUNTRY NOTES

In many cases, data submitted by Member countries to the Secretariat do not conform to the standard reporting methodology or have other particular characteristics. Information set out below will assist readers to interpret data for particular countries and aid in the comparison of data among countries.

The notes given below refer to data for the years 1960 to 2011 and may also refer to 2012e preliminary data as well as the information on CD-ROM and the online data service. In general, more detailed notes are available for data starting in 1990.

Data for anthracite, coking coal, other bituminous coal, sub-bituminous coal and lignite are available separately from 1978. Prior to 1978, only data for hard coal and brown coal (lignite/sub-bituminous coal) are available. Sub-bituminous coal is included under hard coal for the specific countries as mentioned below, namely; Australia, Belgium, Chile, Finland, France, Iceland, Japan, Korea, Mexico, New Zealand, Portugal and the United States.

In 1996, the IEA Secretariat extensively revised data on coal and coke use in blast furnaces, and in the iron and steel industry (for those countries with blast furnaces), based on data provided to the OECD Steel Committee and other sources. The quantities of fuels transformed into blast furnace gas have been estimated by the IEA Secretariat based on its blast furnace model.

Australia

In the 2013 edition, data for Australia were revised back to 2003 due to the adoption of the National Greenhouse and Energy Reporting (NGER) data as the main energy consumption data source for the Australian Energy Statistics. As a result, there are breaks in the time series for many data between 2002 and 2003. The revisions have also introduced some methodological problems. The national statistics appear to have problems identifying inputs and outputs to certain transformation processes such as gas works plants, electricity plants and CHP plants. Energy industry own use and inputs to the transformation processes are sometimes not reported separately in the correct categories. More detail is given in the notes below.

All data refer to the fiscal year, (*e.g.* July 2011 to June 2012 for 2012). For the 2002 data, the Australian Administration began to use a new survey methodology which has caused shifts in the structure of industry consumption. The Australian Administration is planning to revise the historical series.

Data on blast furnace gas for electricity production by autoproducers begins in 1986. Consumption in wood and wood products is included in paper, pulp and print from 2001 onwards. The drop in BKB production in 2004 was due to a fire in the main production plant. Only anthracite for export is reported separately; the remainder that is consumed domestically is included with other bituminous coal. Reclassification of some coal types in 2013 were calculated on an energy basis and resulted in a net increase of quantities of primary coal from 2003 to 2011. Export trade in coke oven coke since 2004 exists, but data are unavailable for reasons of confidentiality. Coal tar data for 2011 include tars from other fuel sources and also some manufactured coal products in the chemical and petrochemical industry. Natural gas consumed to fuel the distribution of natural gas in natural gas networks is reported as transformation for gas works gas production instead of in energy industry own-use. Coke oven gas reported as energy industry own-use in electricity or CHP plants is used for generation purposes, while natural gas used for own-use plant support is reported in the transformation sector. Production data for all manufactured gases have been revised downwards as part of the new national methodology, leading to significant statistical differences between the revised data and consumption statistics.

For 2012 estimates, the Australian Administration provided natural gas production numbers which appeared too high. Information received after the preparation of this report suggest that a corrected TPES value for natural gas (shown in Table 1 in Part IV) should have been about 41.2 Mtce, an increase of 6.7% from 2011, not the 28.7% increase previously submitted and included in this report.

Sub-bituminous coal is included in hard coal.

Austria

Historical revisions by the Austrian Administration have resulted in some breaks in series between 1989 and 1990.

In the 2012 edition, the Austrian Administration has revised the data for consumption and transformation processes for other bituminous coal, coke oven coke and coke oven gas from 2001 onwards. Blast furnace gas was revised in selected sectors from 1990. Other bituminous coal includes hard coal briquettes. "Trockenkohle" is included with BKB because of its high calorific value. Since 1996, gas works gas is reported with natural gas because it is distributed in the same network. The amount of gas works gas is negligible and it is mostly consumed by households. The last lignite mine closed in the second quarter of 2004 and lignite use for power generation ceased in 2006.

Sub-bituminous coal is included in brown coal.

Belgium

Sub-bituminous coal data reported in *from other sources* refer to coal recuperated from coal dumps. Production of other bituminous coal ceased on 31 August 1992. The use of coke oven gas in the chemical and petrochemical industries ceased in 1996. The decrease of bituminous coal and coke oven coke in the iron and steel industry in 2002 is due to the closure of several plants.

Provisional data for 2012e are estimated by the IEA Secretariat.

Sub-bituminous coal is included in hard coal.

Canada

Revisions received by the Canadian Administration and incorporated into the 2002 edition have resulted in breaks in series between 1989 and 1990.

Due to the unavailability of data, non-energy use of coke oven coke and hard coal is included with final consumption sectors prior to 1978 and 1980, respectively. Before 1978, lignite inputs to main activity producer heat plants are included in final consumption. Starting in 1979, these inputs are included in main activity producer electricity plants.

Due to a Canadian confidentiality law, it is not possible for the Canadian Administration to submit disaggregated series for all of the coal types. Between 2002 and 2006, the IEA Secretariat has estimated some of the missing series. The data for 2007 onwards are given directly as reported. The Canadian Administration is planning to further refine its reporting.

Sub-bituminous coal is included in brown coal.

Chile

Data are available starting in 1971.

From 1990, consumption in paper and pulp includes forestry and consumption in agriculture is included in non-specified industry. In general, a new methodology has been applied for data since 1990, leading to other breaks in series between 1989 and 1990.

Other bituminous coal includes sub-bituminous coals for all years except 2012e.

Sub-bituminous coal is included in hard coal.

Czech Republic

Data are available starting in 1971.

Final consumption data were submitted by the Czech Administration starting with 1996 data. Due to economic restructuring in the consumption sectors in the late 1990s (large state enterprises subdividing and/or privatising and the utilisation of new technologies by businesses), there might be breaks in time series in these sectors. Data for 1990 to 1995 were estimated based on the Czech publication *Energy Economy Year Book*. In 1995, town gas production (included in gas works gas) ceased. Revisions by the Czech Administration have resulted in some breaks in series between 2001 and 2002. Production *from other sources* of other bituminous coal in 2004 is from coal slurries. Sub-bituminous coal is included in other bituminous coal.

Coal which had been previously classified as sub-bituminous coal until the 2008 edition is now reported under lignite for all years.

Denmark

In the 2004 edition, major revisions were made by the Danish Administration for the 1990 to 2001 data, which may cause breaks in time series between 1989 and 1990.

A large increase of steam coal imports in 2003 was related to a drought in Scandinavia. Thermal power plants were operated more intensively to replace hydro-generated electricity that was consumed in the country. Additionally, more coal-generated electricity was exported to other countries in the region.

Sub-bituminous coal is included in brown coal.

Estonia

Data for Estonia are available starting in 1990. Prior to that, they are included in Former Soviet Union in *Energy Statistics of Non-OECD Countries*.

Data reported as lignite are for oil shale. In 2013, data for lignite (oil shale) production for the period 1991 to 1997 were revised to match Estonian GHG National Inventory values. Consumption data remained unchanged. Fuels reported as coke oven coke (semi-coke) and gas works gas are by-products of oil shale liquefaction.

Sub-bituminous coal is included in brown coal.

Finland

A new survey system and a reclassification of the data lead to breaks in the time series between 1999 and 2000 for most products and sectors. The new survey system is more detailed and has better product coverage especially in electricity, CHP and heat production, as well as in industry.

A large increase of steam coal imports in 2003 is related to a drought in Scandinavia. Thermal power plants were operated more intensively to replace hydro-generated electricity that is consumed in the country. Additionally, more coal-generated electricity was exported to other countries in the region. Likewise, peat production is highly dependant upon favourable weather conditions and the pricing of other fuels. The decrease in peat and other bituminous coal usage in main activity electricity plants was due to record electricity generation from hydro plants.

The first coking plant started operation in 1987, hence imports of coking coal and production of coke oven coke and coke oven gas started in that year. Coal tars used for non-energy purposes are not reported in production or consumption. The increase of other bituminous coal inputs into main activity producer electricity plants from 1993 to 1994 was due to coal replacing imported electricity and hydro power. Production of gas works gas ceased in April 1994.

Sub-bituminous coal is included in hard coal.

France

Prior to 1985, consumption of colliery gas is included with the use of coke oven gas by autoproducers. Final consumption in industry is estimated by the Secretariat from 1986 to 2001 for some products. For 1989 to 1998, the IEA Secretariat has estimated industry consumption based on *Consommations d'Energie dans l'Industrie*, SESSI. Other manufactured gases (oxygen steel furnace gas) are included in blast furnace gas. Distinction between coke oven gas consumption, and consumption of other gases produced in the iron and steel sector is poorly implemented, resulting in jumps in time series and unusual efficiencies.

Sub-bituminous coal is included in hard coal.

Germany

German data include the new federal states of Germany from 1970 onwards.

The German Administration has changed the methodology for reporting heat. Between 2003 and 2006, autoproducer heat output was provided, but not inputs. Starting in 2007, more information is available on main activity heat plants and additional inputs started to be reported for this category. This causes breaks in series between 2006 and 2007.

Due to the reclassification of several sectors by the German Administration, breaks in series may occur between 1990 and 1992. This particularly affects BKB, lignite and coke oven coke. BKB inputs to gas works plants stopped in 1997. Breaks in time series may occur between 1998 and 2005 for coke oven gas and blast furnace gas. Up to 2003, other bituminous coal includes anthracite. Breaks in the time series for coke oven gas from 2007 are due to a change in statistical source. Consumption of non-renewable municipal waste and other solid biofuels as a reductant occurs in German blast furnaces, but is not currently quantified. Likewise, coal tar is a by-product of coke ovens, but not currently reported.

Sub-bituminous coal is included in brown coal.

Greece

Electricity production using hard coal ceased in 1989. A new main activity producer electricity plant using imported bituminous coal was brought on-line in 1991. Production of gas works gas ceased in 1997. Lignite has been used in main activity producer CHP plants since 1997. Production of BKB/peat briquettes ceased in 2009.

Sub-bituminous coal is included in brown coal.

Hungary

Data are available from 1965.

From 1992, the production of sub-bituminous coal has been included with lignite due to the low quality of the coal. For 1990 to 1999, the use of this domestic coal in main activity producer electricity and CHP plants has also been reclassified to lignite. Autoproducer heat and power plants using coke oven gas and blast furnace gas were reclassified in 1998 as main activity power plants.

Sub-bituminous coal is included in brown coal.

Iceland

Prior to 1970, final consumption includes inputs and outputs to heat production. The industrial classifications used by the Icelandic Administration were changed in 1987. Final consumption increased in 2000 due to a new iron and steel plant coming on-line.

Sub-bituminous coal is included in hard coal.

Ireland

Production data for peat briquettes (BKB) are available from 1975. Low production of peat in 1985 was due to a poor "harvest". The production of gas works gas ceased in 1987 due to fuel switching to natural gas. Other bituminous coal inputs to main activity producer electricity plants increased from 1986 due to three new generating units at Moneypoint coming on-line. A reclassification causes a break in the time series for peat consumption in the energy industry own use in BKB plants from 1989 to 1990. Due to confidentiality reasons, inputs of anthracite, other bituminous coal and BKB/peat briquettes into patent fuel transformation are reported with residential consumption.

Sub-bituminous coal is included in brown coal.

Israel

Data are available starting in 1971.

Data reported as lignite are oil shale.

Regrettably, due to resource constraints, Israel was incapable of providing data for 2012e. These data have been estimated by the IEA Secretariat.

Sub-bituminous coal is included in brown coal.

Italy

A change in methodology leads to breaks in series for industry and transformation between 2003 and 2004.

From 1986 onwards, figures from lignite are given using the same methodology as in the *Bilancio Energetico Nazionale*. In 1991, all industrial activities were reclassified on the basis of ISTAT/NACE 91. This has implied some transfer of activities which may result in some anomalies between 1991 and earlier years. Due to a change in the survey system, breaks in time series may occur between 1997 and 1998 for final consumption. Prior to 2009, sub-bituminous coal used in main activity electricity plants was included with other bituminous coal.

Sub-bituminous coal is included in brown coal.

Japan

Between 2004 and 2007, the IEA received a series of revisions from the Japanese Administration. The first set of revisions received in 2004 increased the 1990 supply by 5% for coal, 2% for natural gas and 0.7% for oil compared to the previous data. This led to an increase of 2.5% in 1990 CO_2 emissions calculated using the Reference Approach while the Sectoral Approach remained fairly constant. For the 2006 edition, the IEA received revisions to the coal and oil data which had a significant impact on both the energy data and the CO_2 emissions. The most significant revisions occurred for coke oven coke, naphtha, blast furnace gas and petroleum coke. These revisions affected consumption rather than supply in the years concerned. As a result, the Sectoral Approach CO_2 emissions increased for all the years, however at different rates. For example, the Sectoral Approach CO_2 emissions for 1990 were 4.6% higher than those calculated for the 2005 edition, while the 2003 emissions were 1.1% higher than those of the previous edition.

Due to the impact these successive revisions have had on the final energy balance, as well as on CO_2 emissions, the IEA was in close contact with the Japanese Administration to better understand the reasons behind these changes. These changes were mainly due to the Government of Japan's efforts to improve the input-output balances in the production of oil products and coal products in response to inquiries from the UNFCCC Secretariat. To cope with this issue, the Japanese Administration established a working group in March 2004. The working group completed its work in April 2006. Many of its conclusions were incorporated in the 2006 edition, but some further revisions to the time series (especially in industry and *other*) were submitted for the 2007 edition.

Starting in 1990, data are reported on a fiscal year basis (*e.g.* April 2011 to March 2012 for 2011).

From 1982, residential use of coke oven coke is included in commercial/public services sector. Other recovered gas data are available from 1982. The inputs of coke oven coke to blast furnaces, as well as the final consumption of coke oven coke in iron and steel, have been estimated by the IEA Secretariat starting in 1990. From 1998, inputs of coke oven gas, blast furnace gas and other recovered gases into autoproducer electricity plants include the amount used to produce electricity with TRT technology (Top pressure Recovery Turbines) which was previously included in industry. The net calorific values for coal have been recalculated by the IEA Secretariat based upon gross values submitted by Japan.

Coal injected in blast furnaces (PCI) is classified as coking coal in order to be consistent with Japanese trade statistics. With the 2008 edition, Japan has reclassified part of the coal inputs to coke ovens as inputs to blast furnaces.

Statistical differences in hard coal include stock changes since 2001. Large positive differences for several years since 2004 are partly due to purpose driven stock build by final consumers.

Sub-bituminous coal is included in hard coal.

Korea

Data are available from 1971.

Data for 2002 onwards have been reported on a different basis, causing breaks in series between 2001 and 2002, especially for inputs and outputs to electricity generation and consumption in the iron and steel industry. The Korean Administration is planning to revise the historical series as time and resources permit.

Data for coal and coal products from 1971 to 2001 are based on information provided by the Korean Administration, as well as information from the *Yearbook of Energy Statistics 2002*, the *Yearbook of Coal Statistics 2001* (both from the Ministry of Commerce, Industry and Energy), and *Statistics of Electric Power in Korea 2001* (from the Korea Electric Power Corporation). During this period, import data by coal type were estimated by the IEA Secretariat, based on statistics of the exporting countries.

Consumption of imported coke oven coke starting in 2002 is reported under non-specified industry. Consumption of manufactured gases in the iron and steel industry starting in 2002 includes the consumption in blast furnaces, basic oxygen steel furnaces and other iron and steel processing plants. Blast furnace gas used for energy purposes in blast furnaces prior to 2007 are reported in the iron and steel industry. Coal tar production prior to 2007 is not available at this time.

Sub-bituminous coal is included in hard coal.

Luxembourg

Steel production from blast furnaces ceased at the end of 1997. For the 2011 edition, the Luxembourgian Administration revised the time series from 2000 for most coal and coal products. Time series for brown coal briquettes consumption were revised from 1990.

Sub-bituminous coal is included in hard coal.

Mexico

Data are available starting in 1971 and are partly estimated based on the publication Balance *Nacional - Energía*. The Mexican Administration submitted data directly by questionnaire for the first time with 1992 data. As a result, some breaks in time series may occur between 1991 and 1992.

The time series for blast furnace gas and inputs of coke oven coke to blast furnaces start in 1991. Significant statistical differences are currently included in stock changes for some products.

Sub-bituminous coal is included in hard coal.

Netherlands

In the national statistical system of the Netherlands, use of fuel in manufacturing industries for CHP production is considered to be consumption in the transformation sector. However, in IEA statistics, this own use for heat production (autoproduced heat) is reported under the relevant industry sub-sector, based on estimates provided by the Central Bureau of Statistics.

For 1984 to 1986, production *from other sources* of other bituminous coal represents a stock of "smalls" washed for re-use. Prior to 1989, non-energy use is included with industry consumption.

Coal exports primarily consist of re-exported volumes after blending. International trade into and through the hub ports of Amsterdam and Rotterdam is complicated by the capacity to purchase coal directly at these points. The majority of coal traded is purely in transit, where the Netherlands is neither the country of origin or destination and these data have been removed where possible. Prior to 2011, stock changes for primary coal types were estimated by the Dutch Administration, based on trade and consumption data. In the 2013 edition, non-specified exports for 2011 are estimated by the Central Bureau of Statistics due to a lack of information from key market players.

Sub-bituminous coal is included in brown coal.

New Zealand

Where data refer to the fiscal year, April 2010 to March 2011 is shown as 2010. In the 2011 edition, the New Zealand Administration has revised some of the coal, natural gas, oil, renewable and electricity time series back to 1990.

A reorganisation of government departments during 1987 leading to the cessation of certain data series has resulted in several breaks in time series between 1987 and 1988.

Production of gas works gas ceased in 1988. Peat, although produced in New Zealand, is not used as a fuel. It is used for agricultural purposes only. In final consumption, some industry data are reported in non-specified industry for confidentiality reasons. Prior to 2009, mining and quarrying is included in agriculture. Prior to 2010, construction is included with commercial/public services.

Sub-bituminous coal inputs into coke ovens refers to coal that is merged with iron sands and limestone to form the inputs for the multi-hearth-furnaces, kilns and melters that produce direct reduced iron (Glenbrook Steel Site), with off-gases and supplemental and natural gas driving CHP plants. This method, while not a typical iron and steel process, produces similar by-products. The sub-bituminous coal inputs are reported under coke oven coke transformation and the resulting off-gases are reported as production of coke oven gas and blast furnace gas. Blast furnace gas production and distribution losses prior to 1998 are IEA Secretariat estimates. Portions of this gas will have been used for energy purposes in the multi-hearth furnaces or elsewhere in the plant. Some transformation efficiencies will appear higher than normal due to non-reporting of certain inputs, including some confidential data.

Detailed breakdown of exports of coking coal by country of destination since 2001 are estimated by the IEA, based on secondary sources and partner data.

Sub-bituminous coal is included in hard coal.

Norway

Production of coking coal, coke oven coke and coke oven gas ceased in the late 1980s. Other bituminous coal includes lignite. The decrease of bituminous coal production in 2005 is due to a fire in one of the coal mines; this entailed a break in the production for a large part of the year.

Sub-bituminous coal is included in brown coal.

Poland

Prior to 2010, own use in coal mines included worker's take home allowance which should be included in residential consumption.

Sub-bituminous coal is included in brown coal.

Portugal

Between 1997 and 2001, gas works gas was gradually replaced by natural gas in the commercial/public services and residential sectors. The production of pig iron ceased in the first quarter of 2001, leading to decreases in supply and consumption of coking coal, coke oven coke, coke oven gas and blast furnace gas in 2001.

Sub-bituminous coal is included in hard coal.

Slovak Republic

Data are available starting in 1971.

There are some breaks in series between 1992 and 1993. A new survey system in 2001 leads to major breaks in series for most products. Commercial/public services includes statistical differences for other bituminous coal, lignite, patent fuel and coke oven coke from 1980 onwards and BKB from 1989 onwards.

Sub-bituminous coal is included in brown coal.

Slovenia

Data for Slovenia are available starting in 1990. Prior to that, they are included in Energy *Statistics of Non-OECD Countries* in Former Yugoslavia. A new energy data collection system was implemented in January 2001, causing some breaks in time series between 1999 and 2000.

Sub-bituminous coal is included in brown coal.

Spain

Lignite mining was halted indefinitely in 2008. For 1999-2003, anthracite is included in other bituminous coal. The transformation efficiency for coke ovens is over 100% for 2009 and 2010, resulting in positive losses for these two years. Data associated with the coke oven coke transformation process are under review by Spain and revised data are expected next cycle.

Sub-bituminous coal is included in brown coal.

Sweden

Other bituminous coal production is coal recovered during the quarrying of clay. Autoproducer inputs to waste heat production that is sold are reported in the respective final consumption sectors and not in transformation. Some mixture of LNG with air to form a lower calorie product is reported as gas works gas production. However, the LNG inputs are reported in natural gas consumption, resulting in some minor double counting and unrealistic transformation efficiencies for gas works gas.

Sub-bituminous coal is included in brown coal.

Switzerland

From 1999, data on consumption result from a new survey and are not comparable with data of previous years.

From 1985, industrial consumption of gas works gas is reported in non-specified industry to prevent the disclosure of commercially confidential data. Allocation of consumption data between certain coal types is estimated by the Swiss Administration, as are the net calorific values for these coals.

Sub-bituminous coal is included in brown coal.

Turkey

Production of gas works gas declined in 1989 due to plant closures; the last plant closed in 1994. Use of gas coke and gas works gas ceased in 1994. Due to government regulations in industry and residential, in particular, there has been a shift from the use of domestically produced coal to imported coal and natural gas. The privatisation of state owned coke ovens in recent years results in incomplete information on coke oven gas distribution. Data from 2008 are provided from the results of an improved questionnaire. Therefore, significant changes occur in consumption patterns within the iron and steel industry, coal mining as well as across industry, residential and commercial/ public services for other bituminous coal. Some coal used in cement kilns is reported under construction instead of non-metallic minerals.

Lack of availability of national data for 2012 has resulted in all 2012e data being estimated by the IEA Secretariat.

Sub-bituminous coal is included in brown coal.

United Kingdom

Consumption shown for the commercial/public services includes consumption of some of non-specified *other*. Prior to 1994, the consumption of substitute natural gas is included with natural gas, while its production is included with gas works gas.

Sub-bituminous coal is included in brown coal.

United States

Due to problems in reporting, there are numerous breaks in series for the US data, particularly in 1992, 1999, 2001 and 2002. Care should be taken when evaluating consumption by sector since inputs of fuel to autoproducers are included in final consumption for some years. No data are available for most energy products in the construction and mining and quarrying industries.

In 2002, the United States reported "synfuel" production as patent fuel for the first time. Prior to 2002, the consumption of this fuel was reported with other bituminous coal. Production ceased in 2007 for economic reasons. Since the Energy Information Administration (EIA) and the US Department of Commerce do not collect separate data on patent fuel exports by country, total exports of patent fuel are included in the exports of other bituminous coal. Coal tar as a by-product of coke ovens is not currently reported.

Sub-bituminous coal is included in hard coal.

5. GEOGRAPHICAL COVERAGE

Australia excludes the overseas territories.

Denmark excludes the Danish Faroes and Greenland.

France includes Monaco, but excludes the following overseas departments and territories: Guadeloupe, Guyana, Martinique, New Caledonia, French Polynesia, Reunion, and St.-Pierre and Miquelon.

Germany includes the new federal states of Germany from 1970 onwards.

The statistical data for **Israel** are supplied by and under the responsibility of the relevant Israeli authorities. The use of such data by the OECD is without prejudice to the status of the Golan Heights, East Jerusalem and Israeli settlements in the West Bank under the terms of international law.

Italy includes San Marino and the Vatican.

Japan includes Okinawa.

The Netherlands excludes Suriname and the Netherlands Antilles.

Portugal includes the Açores and Madeira.

Spain includes the Canary Islands.

Switzerland does not include Liechtenstein.

United States includes 50 States and the District of Columbia.

OECD Total comprises Australia, Austria, Belgium, Canada, *Chile*, the Czech Republic, Denmark, *Estonia*[1], Finland, France, Germany, Greece, Hungary, *Iceland*, Ireland, *Israel*, Italy, Japan, Korea, Luxembourg, *Mexico*, Netherlands, New Zealand, Norway, Poland, Portugal, the Slovak Republic, *Slovenia*[1], Spain, Sweden, Switzerland, Turkey, the United Kingdom and the United States,

With the exception of Chile, Estonia, Iceland, Israel, Mexico, and Slovenia, all 34 OECD member countries are also member countries of the IEA. IEA regional totals include only IEA Member countries and, therefore, exclude non-IEA Members shown in italics below.

The following countries are included in the regional aggregates. OECD Total is the sum of the three regional aggregates.

OECD Americas comprises Canada, *Chile*, *Mexico* and the United States.

OECD Asia Oceania comprises Australia, *Israel*, Japan, Korea and New Zealand.

OECD Europe comprises Austria, Belgium, Czech Republic, Denmark, *Estonia*, Finland, France, Germany, Greece, Hungary, *Iceland*, Ireland, Italy, Luxembourg, Netherlands, Norway, Poland, Portugal, the Slovak Republic, *Slovenia*, Spain, Sweden, Switzerland, Turkey and the United Kingdom.

The **European Union - 27 (EU-27)** includes Austria, Belgium, Bulgaria, Cyprus[2], the Czech Republic, Denmark, Estonia, Finland, France, Germany, Greece, Hungary, Ireland, Italy, Latvia, Lithuania, Luxembourg,

1. Estonia and Slovenia are included starting in 1990. Prior to 1990, data for Estonia are included in Former Soviet Union and data for Slovenia in Former Yugoslavia in the publication *Energy Statistics of Non-OECD Countries*.

2. **Note by Turkey:**
The information in this document with reference to "Cyprus" relates to the southern part of the Island. There is no single authority representing both Turkish and Greek Cypriot people on the Island. Turkey recognises the Turkish Republic of Northern Cyprus (TRNC). Until a lasting and equitable solution is found within the context of United Nations, Turkey shall preserve its position concerning the "Cyprus issue".
Note by all the European Union Member States of the OECD and the European Union:
The Republic of Cyprus is recognised by all members of the United Nations with the exception of Turkey. The information in this report relates to the area under the effective control of the Government of the Republic of Cyprus.

Malta, the Netherlands, Poland, Portugal, Romania, the Slovak Republic, Slovenia, Spain, Sweden and the United Kingdom. Please note that in the interest of having comparable data, all these countries are included since 1990 despite different entry dates into the European Union.

Non-OECD Europe and Eurasia includes Albania, Armenia, Azerbaijan, Belarus[2], Bosnia and Herzegovina, Bulgaria, Croatia, Cyprus, Gibraltar, Kazakhstan, Kosovo[3] (from 2000), Kyrgyzstan, Latvia, Lithuania, the Former Yugoslav Republic of Macedonia, Malta, the Republic of Moldova, Montenegro[3] (from 2005), Romania, the Russian Federation, Serbia[3], the Former Soviet Union[4], Tajikistan, Turkmenistan, Ukraine, Uzbekistan and Former Yugoslavia[4].

Africa includes Algeria, Angola, Benin, Botswana (from 1981), Cameroon, Congo, the Democratic Republic of Congo, Côte d'Ivoire, Egypt, Eritrea, Ethiopia, Gabon, Ghana, Kenya, Libya, Morocco, Mozambique, Namibia (from 1991), Nigeria, Senegal, South Africa, Sudan (including the Republic of South Sudan[5]), the United Republic of Tanzania, Togo, Tunisia, Zambia, Zimbabwe and Other Africa.

Other Africa includes Botswana (until 1980), Burkina Faso, Burundi, Cape Verde, the Central African Republic, Chad, Comoros, Djibouti, Equatorial Guinea, Gambia, Guinea, Guinea-Bissau, Lesotho, Liberia, Madagascar, Malawi, Mali, Mauritania, Mauritius, Namibia (until 1990), Niger, Reunion, Rwanda, Sao Tome and Principe, Seychelles, Sierra Leone, Somalia, Swaziland, Uganda and Western Sahara (from 1990).

Non-OECD Americas includes Argentina, Bolivia, Brazil, Colombia, Costa Rica, Cuba, the Dominican Republic, Ecuador, El Salvador, Guatemala, Haiti, Honduras, Jamaica, the Netherlands Antilles,[6] Nicaragua, Panama, Paraguay, Peru, Trinidad and Tobago, Uruguay, Venezuela and Other non-OECD Americas.

3. Data for Serbia include data for Montenegro until 2004 and data for Kosovo until 1999.

4. Data are shown only as Former Soviet Union and Former Yugoslavia until 1989. From 1990 onward, data are shown by country only.

5. South Sudan became an independent state on 9 July 2011. Because only aggregated data were available until 2011, the data for Sudan also include South Sudan.

6. The Netherlands Antilles was dissolved on 10 October 2010, resulting in two new "constituent countries" of the Netherlands (Curaçao and Saint Maarten), with the other islands joining the Netherlands as "special municipalities". However, due to a lack of detailed data, the Secretariat's data and estimates under the "Netherlands Antilles" still refer to the whole territory of the Netherlands Antilles as it was known prior to 10 October 2010.

Other non-OECD Americas includes Antigua and Barbuda, Aruba, Bahamas, Barbados, Belize, Bermuda, the British Virgin Islands, the Cayman Islands, Dominica, the Falkland Islands, French Guyana, Grenada, Guadeloupe, Guyana, Martinique, Montserrat, Puerto Rico[7] (for natural gas and electricity), St. Kitts and Nevis, Saint Lucia, Saint Pierre et Miquelon, St. Vincent and the Grenadines, Suriname, and Turks and Caicos Islands.

Asia includes Bangladesh, Brunei Darussalam, Cambodia (from 1995), Chinese Taipei, India, Indonesia, the DPR of Korea, Malaysia, Mongolia (from 1985), Myanmar, Nepal, Pakistan, the Philippines, Singapore, Sri Lanka, Thailand, Vietnam and Other Asia.

Other Asia includes Afghanistan, Bhutan, Cambodia (until 1994), Cook Islands, East Timor, Fiji, French Polynesia, Kiribati, Laos, Macau, Maldives, Mongolia (until 1984), New Caledonia, Palau (from 1994), Papua New Guinea, Samoa, the Solomon Islands, Tonga and Vanuatu.

China includes the People's Republic of China and Hong Kong (China).

Former Yugoslavia[4] includes Bosnia and Herzegovina, Croatia, the Former Yugoslav Republic of Macedonia, Serbia[3] (including Kosovo and Montenegro) and Slovenia.

Former Soviet Union[4] includes Armenia, Azerbaijan, Belarus, Estonia, Georgia, Kazakhstan, Kyrgyzstan, Latvia, Lithuania, the Republic of Moldova, the Russian Federation, Tajikistan, Turkmenistan, Ukraine, and Uzbekistan.

Middle East includes Bahrain, the Islamic Republic of Iran, Iraq, Jordan, Kuwait, Lebanon, Oman, Qatar, Saudi Arabia, Syria, the United Arab Emirates and Yemen.

Please note that the following countries have not been considered due to lack of data:

- **Non-OECD Europe and Eurasia**: Liechtenstein[8] (except for oil data);
- **Africa:** Saint Helena;
- **America:** Anguilla; and
- **Asia** and **Oceania:** Christmas Island, Nauru and Niue.

7. Oil statistics as well as coal trade statistics for Puerto Rico are included under the United States.

8. Oil data for Liechtenstein are included under Switzerland.

6. ABBREVIATIONS, CONVERSION FACTORS AND CALORIFIC VALUES

Units and technical abbreviations

t	:	metric ton = tonne = 1000 kg
kt	:	thousand tonnes
Mt	:	million tonnes
toe	:	tonne of oil equivalent
Mtoe	:	million tonnes of oil equivalent
tce	:	tonne of coal equivalent (= 0.7 toe)
Mtce	:	million tonnes of coal equivalent
kcal	:	kilocalories (10^3 calories)
MBtu	:	million British thermal units
GWh	:	million kilowatt hours
USD	:	US dollars
CIF	:	cost, insurance and freight
FAS	:	free alongside ship
FOB	:	free on board
GDP	:	Gross Domestic Product
GCV	:	gross calorific value
PCI	:	coals for pulverised injection
TPES	:	Total primary energy supply
EU	:	European Union
FSU	:	Former Union of Soviet Socialist Republics/Soviet Union
OECD	:	Organisation for Economic Co-operation and Development
UN	:	United Nations
UNECE	:	United Nations Economic Commission for Europe
0 or 0.0	:	negligible
c	:	confidential
e	:	estimated
..	:	not available
-	:	nil
x	:	not applicable

General conversion factors for energy

To:	TJ	Gcal	Mtoe	MBtu	GWh
From:	*multiply by:*				
TJ	1	238.8	2.388×10^{-5}	947.8	0.2778
Gcal	4.1868×10^{-3}	1	10^{-7}	3.968	1.163×10^{-3}
Mtoe	4.1868×10^{4}	10^{7}	1	3.968×10^{7}	11630
MBtu	1.0551×10^{-3}	0.252	2.52×10^{-8}	1	2.931×10^{-4}
GWh	3.6	860	8.6×10^{-5}	3412	1

Conversion factors for mass

To:	kg	t	lt	st	lb
From:	*multiply by:*				
kilogramme (kg)	1	0.001	9.84×10^{-4}	1.102×10^{-3}	2.2046
tonne (t)	1000	1	0.984	1.1023	2204.6
long ton (lt)	1016	1.016	1	1.120	2240
short ton (st)	907.2	0.9072	0.893	1	2000
pound (lb)	0.454	4.54×10^{-4}	4.46×10^{-4}	5.0×10^{-4}	1

Conversion factors for volume

To:	gal U.S.	gal U.K.	bbl	ft^3	l	m^3
From:	*multiply by:*					
U.S. gallon (gal)	1	0.8327	0.02381	0.1337	3.785	0.0038
U.K. gallon (gal)	1.201	1	0.02859	0.1605	4.546	0.0045
Barrel (bbl)	42.0	34.97	1	5.615	159.0	0.159
Cubic foot (ft^3)	7.48	6.229	0.1781	1	28.3	0.0283
Litre (l)	0.2642	0.220	0.0063	0.0353	1	0.001
Cubic metre (m^3)	264.2	220.0	6.289	35.3147	1000.0	1

Decimal prefixes

10^{1}	deca (da)	10^{-1}	deci (d)
10^{2}	hecto (h)	10^{-2}	centi (c)
10^{3}	kilo (k)	10^{-3}	milli (m)
10^{6}	mega (M)	10^{-6}	micro (μ)
10^{9}	giga (G)	10^{-9}	nano (n)
10^{12}	tera (T)	10^{-12}	pico (p)
10^{15}	peta (P)	10^{-15}	femto (f)
10^{18}	exa (E)	10^{-18}	atto (a)

The conversion factors shown above are available online with greater precision at: *www.iea.org/stats/unit.asp.*

2011 COUNTRY SPECIFIC AVERAGE NET CALORIFIC VALUES [kJ/kg]

	Anthracite	Coking coal	Other bituminous coal	Sub-bituminous coal	Lignite	Peat	Patent fuels	Coke oven coke	Coal tar	BKB
Australia	-	28 500	25 700	18 478	9 800	-	-	25 650	35 714	20 995
Austria	30 140	29 070	27 239	21 383	8 096	8 800	-	29 000	41 800	19 300
Belgium	25 184	29 308	25 781	-	8 370	-	29 308	29 308	-	20 000
Canada	26 381	27 237	24 219	18 238	14 286	-	-	27 457	-	-
Chile	-	28 638	22 686	-	-	-	-	28 638	41 366	-
Czech Republic	29 646	29 213	25 134	-	12 203	-	-	27 994	36 968	21 558
Denmark	-	-	24 393	-	-	-	-	29 300	-	18 300
Estonia	-	-	27 160	-	8 900	10 254	-	28 500	-	16 000
Finland	-	29 300	24 910	-	-	10 190	-	29 300	37 000	-
France	-	30 500	26 000	-	17 000	-	32 000	28 000	38 000	-
Germany	28 982	29 000	25 877	-	9 038	-	31 400	28 650	-	21 317
Greece	-	-	23 907	-	5 354	-	-	-	-	-
Hungary	-	31 840	24 430	17 681	7 209	-	22 000	29 699	38 000	23 000
Iceland	28 050	-	28 050	-	-	-	-	26 670	-	-
Ireland	27 928	-	25 642	-	19 816	8 586	-	-	-	18 548
Israel	-	-	25 424	20 358	3 007	-	-	-	-	-
Italy	-	30 984	26 587	18 832	10 468	-	-	29 000	-	-
Japan	26 362	28 130	24 801	-	-	-	-	29 400	35 393	-
Korea	24 198	28 261	24 911	20 934	-	-	19 259	29 308	37 000	-
Luxembourg	26 700	-	24 400	-	-	-	-	28 500	-	22 200
Mexico	-	29 434	-	20 358	14 100	-	-	26 521	-	-
Netherlands	29 300	28 671	24 608	-	20 000	-	-	28 500	41 900	-
New Zealand	-	29 839	28 063	20 630	14 543	-	-	29 500	-	-
Norway	-	-	28 100	-	-	-	-	28 500	-	-
Poland	-	29 604	22 411	-	8 362	-	23 189	27 502	37 720	17 477
Portugal	30 672	-	24 955	-	-	-	-	29 557	-	-
Slovak Republic	27 384	29 250	25 535	-	11 425	-	28 000	28 260	-	17 000
Slovenia	-	-	25 341	19 336	11 221	-	-	29 498	-	-
Spain	19 960	29 443	22 611	13 372	-	-	-	26 795	38 519	-
Sweden	-	30 000	27 400	-	-	12 645	-	28 080	-	-
Switzerland	28 100	-	28 100	-	20 100	-	-	28 100	-	-
Turkey	-	27 216	26 742	18 074	9 170	-	-	27 571	-	-
United Kingdom	-	30 379	24 791	-	-	-	30 970	28 310	38 519	-
United States	28 801	29 668	26 173	18 971	13 950	-	-	28 842	-	-

Source: IEA/OECD Coal Statistics

Coal classification

The IEA collects statistics on coal production, trade and consumption according to a technically precise classification based on the quality of coal as follows:

- Anthracite is a high rank, non-agglomerating coal with a gross calorific value not less than 23 865 kJ/kg (5 700 kcal/kg) on an ash-free but moist basis;
- Coking coal is hard coal suitable for the production of coke that can support a blast furnace charge;
- Other bituminous coal is an agglomerating coal with a gross calorific value not less than 23 865 kJ/kg (5 700 kcal/kg) on an ash-free but moist basis;
- Sub-bituminous coal is a non-agglomerating coal with a gross calorific value between 23 865 kJ/kg (5 700 kcal/kg) and 17 435 kJ/kg (4 165 kcal/kg) containing more than 31% volatile matter on a dry mineral matter-free basis; and
- Lignite is a non-agglomerating coal with a gross calorific value less than 17 435 kJ/kg (4 165 kcal/kg) and greater than 31% volatile matter on a dry mineral matter-free basis. Oil shale is included in lignite.

However, when publishing these data, the IEA sometimes adopts a simplified classification of hard coal, steam coal and brown coal. The correspondence is as follows:

- Total coal is the sum of hard coal and brown coal;
- Hard coal is the sum of coking coal, anthracite and other bituminous coal for all countries, plus sub-bituminous coal for Australia, Belgium, Chile, Finland, France, Iceland, Japan, Korea, Mexico, New Zealand, Portugal and the United States;
- Brown coal contains lignite for all countries, and sub-bituminous coal for all countries not listed above; and
- Steam coal consists of anthracite, other bituminous coal and sub-bituminous coal.

The term *total coal* also refers to the sum of hard coal and brown coal after conversion to a common energy unit (tonne of coal equivalent - tce). The conversion is done by multiplying the calorific value of the coal in question (the conversion factors are submitted by national administrations to the IEA Secretariat each year) by the total volume of hard and brown coal used, measured in physical units, *i.e.* in tonnes. One tce has an energy content of 29.3 Gigajoules (GJ) or 7 000 kcal and corresponds to 0.7 tonnes of oil equivalent (toe).

Defining coal consumption

Energy statistics are compiled and presented to take account of the complexity in the way fuels are used and to avoid double counting. Misunderstandings can arise when statistics on coal consumption are used because of the particular terminology used by energy statisticians.

Coal is used in four possible ways:

- As a *primary input* to produce electricity or a secondary/tertiary fuel that is used elsewhere or sold - this is referred to as use in *transformation processes*;
  e.g. coking coal used to produce coke in a coke oven, steam coal used to *produce* electricity.
- As a *fuel* used to *support* a transformation process - this is referred to as *energy industry own use*;
  e.g. coke oven gas used to heat the coke oven, steam coal used to operate the power plant.
- As a *fuel* consumed in manufacturing industry, mining and construction, in transport, in agriculture, in commercial and public services and in households - this is referred to as use in the *final consumption* sectors;
  e.g. steam coal used to produce heat in cement kilns, steam coal used to produce industrial process steam.
- As a *raw material* - this is referred to as non-energy use;
  e.g. coal used to produce carbon electrodes for the aluminium industry.

In the wider community, the term "consumption" is commonly understood to include all of the above end-uses. In Parts IV and V of this book, the term "consumption" refers only to use in the *final consumption* sectors (*i.e.* in the third item above). In Parts II and III, "consumption", unless otherwise specified, refers to Total Primary Energy Supply as defined in the section in *Flows: energy balance* in Part I Definitions.

PART II

WORLD COAL OVERVIEW

1. WORLD COAL MARKET REVIEW

Summary

In 2012, total global coal production increased by 2.9% to 7 830.8 Mt (Table II.1), making it the 13th year in a row of uninterrupted growth. With 1999 being the last year when production declined on an annual basis, global coal production in 2012 was 74% higher in comparison to then. This expansion was primarily the result of growth in steam coal production (86%) and coking coal production (105%) increasing market share, while there was more sedentary growth experienced in lignite production, which only increased by 9%.

The pronounced divergence of coal production patterns between OECD and non-OECD economies in recent years continued in 2012 as national economies reacted to the price of oil in different ways. The United States decreased production and consumption of steam coal by 68.7 Mt and 95.4 Mt respectively, and increased exports to a record 114.1 Mt including an increase of 16.6 Mt of steam coal as the relative oversupply of unconventional gas continued to impact upon domestic prospects where coal and gas can directly compete.

These circumstances partly arose from a legislated lack of international markets for shale gas, but equally this apparent overproduction is being driven almost entirely by demand for associated light tight oil and use it or lose it leases. This was combined with a slight downturn in energy requirements as electricity generation decreased by 1% and TPES itself shrank by 2.7% or 83.9 Mtce, outweighing both the coal consumption decrease and the gas increase.

The consequent overabundance of fuels – both gas and coal – for electricity generation could not have occurred if there had not been surplus generating capacity. This overabundance of choice has resulted in or coincided with reduction in electricity production by other technologies, reduced coal prices, and also provided a cost opportunity in Europe in particular, where European Union imports from the United States rose by 8.9 Mt to 47.0 Mt and U.S. hard coal exports to Europe as a whole passed 54.1 Mt.

In Canada, ex situ production of Athabascan oil sands and similar deposits have the capacity to be reported on the coal questionnaire if the above ground retort production is taken to be liquefaction of mined oil shale. Currently this is instead interpreted to be part of the oil extraction process to generate a saleable product, rather than a transformation process between fuel types, which is also a valid interpretation. It continues to be however, a pertinent example of activities influenced by current oil prices. Of more relevance to coal markets are the billions of dollars being invested in coal to liquids infrastructure or syngas and methanation plants, particularly in the People's Republic of China and elsewhere in Asia, which will begin coming online over the next few years.

Led by the discussed decline in production by the United States, the OECD countries' 2012 production dropped to the lowest level since 2000, while over the same period, non-OECD countries increased their annual production by 3 178.8 Mt (121%). This divergence is also evident in the 2011 to 2012e growth rates, where OECD coal production decreased by 2.4%, while non-OECD total coal production increased by 4.9%.

Gross electricity production in 2012 in the OECD (not including generation from pumped storage plants) was 10 771 TWh, a decrease of 0.3%, while electricity generated from coal-fired plants in OECD countries decreased by 4.3% to 3 453 TWh. This 32.1% share of total electricity generation is a decrease from 32.9% in 2011 and 33.7% in 2010 (Part III, Table 5.5).

The People's Republic of China continues to drive much of the world's coal markets despite emerging signs of possible slowing of growth. Coal consumption directly facilitated the generation of 3 723 TWh of electricity in 2011, up from 3 250 TWh in 2010. In 2012 there was indigenous production of 407.3 Mt of coke oven coke, 709 Mt of crude steel, 654 Mt of pig iron and over 2.1 Gt of cement[1], and preliminary data for Chinese coal indicated that:

- Consumption (TPES)[2] grew to 2 794.8 Mtce (or 50.5% of world consumption); and
- production grew by 3.8% or 130.3 Mt; and
- net imports rose by 118.0 Mt to 278.3 Mt as exports dropped by 51.7% and imports grew by 58.7%, although it is possible that 2011 figures may increase in the future.

Total global coal consumption increased by 2.5% or 133.2 Mtce in 2012, as OECD consumption declined by 56.4 Mtce and growth in non-OECD countries was 189.6 Mtce (Table II.8). As a consequence, the OECD's share of global consumption declined to its lowest value ever (26.4%).

Global coal exports grew by 111.2 Mt to record levels of 1 255.3 Mt in 2012, as steam coal exports grew by 12.3% and coking coal exports grew by 2.5% (Table II.13). Global exports also reached a record level of total consumption (21.1%) as shown over time in Figure II.1.

Steam coal import spot prices in both the Atlantic and Pacific markets stayed relatively constant during the first three quarters of 2011. However, there was then a decline in prices, led by Europe, both in timing and the magnitude of drop, which has persisted throughout 2012 and the first 6 months of 2013 (Figure II.4).

Global CO_2 emissions from fuel combustion grew in 2010 by a record 1.3 Gt to reach 30.3 Gt – another record. This 4.6% relative growth rate was the largest since 2004 and followed on from the 1.8% decrease in emissions experienced in 2009. Coal remained the largest source of anthropogenic CO_2 emissions at 13.1 Gt, and the 608 Mt annual increase of emissions was 45.7% of the world's 2010 increase and over twice that of oil (284 Mt). In 2011, OECD coal emissions decreased from 4.18 Gt to 4.07 Gt, in no small part as coal emissions in the United States decreased from 1.94 Gt to 1.83 Gt (Part III, Table 2.7).

1. Iron and steel data are provided by the World Steel Association and cement data by the Federación Interamericana del Cemento.

2. Consumption (TPES) when discussed with regard to production, imports, exports and stock changes, differs to consumption calculated from sectoral end-use consumption data by the statistical difference. The statistical difference reflects unexplained differences caused by different data collection methodologies, application of conversion factors and so forth. The magninitude and volatility of the statistical difference over the years may lead to issues when calculating some specific growth rates for the People's Republic of China. (For more information, see the Coal balance in Part V.) See definitions in Part I for more information.

Production

Total world coal production

Total world coal production reached a record level of 7 830.8 Mt in 2012, increasing by 2.9% over 2011 (Table II.1), achieving an average annual growth rate of 4.3% since 1999.

Table II.1 Total world coal production* [Mt]

	2010	2011	2012e
Steam coal	5 437.3	5 726.6	5 942.4
Coking coal	911.6	970.0	983.8
Lignite	861.0	911.0	904.6
Peat	17.5	16.5	..
Total coal	**7 209.9**	**7 607.6**	**7 830.8**

* Production includes recovered slurries. Total coal excludes peat.

The People's Republic of China was once again the world's leading producer, as it has been since 1984 (Table II.2), with 3 549.1 Mt of total coal production, followed by the United States (934.9 Mt), India (595.0 Mt), Indonesia (442.8 Mt) and Australia (420.7 Mt). What appears to be a 22.9% rise in production (82.5 Mt) for Indonesia in provisional data should be tempered as it also will include significant producer stock changes (draw downs) in this instance. Stock changes are currently not available for Indonesia for any years adding apparent volatility to the production time series.

Table II.2 Major coal producers* [Mt]

	2010	2011	2012e
PR of China	3 140.2	3 418.8	3 549.1
United States	996.1	1 005.9	934.9
India	570.4	582.3	595.0
Indonesia	325.0	360.3	442.8
Australia	424.4	402.2	420.7
Russian Federation	321.7	321.7	353.9
South Africa	254.5	252.8	259.3
Germany	182.3	188.6	197.0
Poland	133.2	139.3	144.1
Kazakhstan	110.9	116.5	126.0
Colombia	74.4	85.8	89.5
Ukraine	57.7	69.7	71.6
Turkey	73.4	76.1	69.5
Canada	67.9	67.1	66.6
Greece	56.5	58.7	61.8
Czech Republic	55.2	57.9	55.0
Other	*366.1*	*404.1*	*393.9*
World	**7 209.9**	**7607.6**	**7 830.8**

* Production includes recovered slurries.

See Table 1.1 in Part III for historic data on selected countries.

The largest OECD coal-producing region continued to be OECD Americas, with 50.1% of the OECD production, despite production in the United States declining by 71.0 Mt. OECD Europe was responsible for three quarters of OECD lignite production and 28.9% of OECD total coal production.

Steam coal production

In 2012, world steam coal production (anthracite, other bituminous coal and sub-bituminous coal) increased by 3.8%, compared to 5.3% in 2011 (Table II.3), and was the 13th straight year of increased production. This continued to be driven by growth in the non-OECD countries who profited from an increase of 5.9% (or 269.7 Mt) in 2012. Among the major producing non-OECD countries, production increased noticeably in Indonesia (23.1%), Kazakhstan (12.4%), the Russian Federation (12.0%) and Ukraine (8.1%). The data for the People's Republic of China currently includes lignite within steam coal and annual production increased by 129.3 Mt or 4.4%.

OECD steam coal production was 1 133.0 Mt in 2012, down from the 2011 production of 1 186.8 Mt, reaching the lowest level since 1993 despite an increase in production in Australia of 8.5% or 15.7 Mt.

Table II.3 Major steam coal producers* [Mt]

	2010	2011	2012e
PR of China	2 680.7	2 909.3	3 038.6
United States	856.5	850.7	782.0
India	491.3	495.6	504.3
Indonesia	322.8	357.5	440.0
South Africa	252.7	251.1	258.5
Russian Federation	178.7	179.9	201.5
Australia	189.4	184.5	200.3
Kazakhstan	91.7	95.7	107.6
Colombia	70.5	81.8	85.5
Poland	65.1	65.0	68.1
Ukraine	40.0	49.9	53.9
Vietnam	44.8	44.5	42.1
DPR of Korea	32.0	39.1	39.2
Canada	29.5	27.9	26.0
United Kingdom	18.1	18.2	16.5
Mexico	11.2	13.7	13.0
Other	*62.3*	*62.1*	*65.6*
World	**5 437.3**	**5 726.6**	**5 942.4**

* Production includes recovered slurries and middlings.

See Table 1.4 in Part III for historic data on selected countries.

In 1978, the OECD accounted for 42.8% of the world steam coal production and this metric remained above 37% until the year 2000. However, since then its share has generally declined as non-OECD countries have increased their steam coal production at a faster rate, predominantly led by the expansion of the Chinese coal industry since 2001. In 2012, the OECD's share (19.1%) was less than half its 1978 share despite its production being 227.6 Mt higher.

The largest OECD steam coal-producing region is OECD Americas, with almost three quarters of the OECD production. Of the remainder, 18% of the OECD production is in the Asia Oceania region and the remaining 9% is in the OECD Europe region.

Steam coal production in the non-OECD countries increased by 5.9% to 4 809.4 Mt. Of note, over 99% of the non-OECD steam coal production currently is accounted for by the 10 largest producing countries. The Philippines comes in at number 11, with production of 8.0 Mt in 2012.

The People's Republic of China (including lignite) now accounts for 51.1% of world steam coal production and 63.2% of non-OECD production.

As a major coal exporter, the People's Republic of China moved from the second largest coal exporter in 2001 to be 12th in the world in 2012. In addition to this trend, they also became the largest coal-importing country in 2011, surpassing Japan's imports by 7.9 Mt and shifting their trade balance from net exports of 4.9 Mt in 2008, to net imports of 278.3 Mt in 2012, up from 160.3 Mt in 2011. Their imports grew dramatically from 40.3 Mt in 2008 to 288.8 Mt (Table II.15), while exports were reduced to less than a quarter of their prior quantity, from 45.3 Mt to 10.5 Mt (Table II.14), over this four-year period.

India was the second largest non-OECD coal producer and third largest in the world in 2012. Their steam coal production increased by 1.8% to 504.3 Mt. Although it dominates South Asian production, it contains a very high ash quotient and is almost exclusively consumed domestically and may need supplementing with import quality coals. Projections also tend to foresee higher reliance upon imports as time goes on as projected demand outstrips the speed at which production capacity can be augmented and delivered. Similarly to China, rail logistics in India with internal coal distribution from mine mouth to end-user are a key challenge, particularly if the large mid- to long-term increases in domestic production envisaged are to be realised. July 2012 was also the month where the Northern States of India experienced a widespread blackout as generation was unable to keep up with the combination of legal and illegal load demand. There is rapidly increasing demand which is a good sign with respect to economic growth, fuel poverty and expanded availability of cleaner domestic energy sources, but many challenges lie ahead.

Indonesia was the 4th largest steam coal producer in 2012 with coal production calculated to reach 440.0 Mt, up 23.1% from its 2011 level, which in turn was up 10.8% from 2010. As mentioned in Part I and Part III, 2012 provisional data for non-OECD countries, generally do not have stock builds shown. Normally this is not of too much concern until unusual events occur, large scale year-on-year trends happen or there is a systematic bias somewhere in the data. Indonesia is in the unusual position where we use data sources to estimate production, rather than consumption. The inability to obtain stock numbers for Indonesia may in this case, where large quantities of stocks were finally dispatched after delays, appear to inflate production and growth rate numbers slightly.

Coking coal production

World coking coal production increased by 1.4% in 2012 (Table II.4) and was the tenth straight year of increased production. Although it was not particularly robust in 2012, growth continued to be driven by non-OECD countries with an increase of 13.1 Mt in 2012 as production in OECD countries only grew by 0.2% or 667 kt.

Table II.4 Major coking coal producers [Mt]

	2010	2011	2012e
PR of China	459.5	509.5	510.4
Australia	162.9	146.7	146.9
United States	68.6	81.7	81.3
Russian Federation	66.9	65.4	74.6
India	41.4	44.3	47.2
Canada	28.2	29.5	31.1
Mongolia	15.2	21.1	20.9
Ukraine	17.7	19.8	17.8
Kazakhstan	11.9	12.4	12.9
Poland	11.7	11.4	11.7
Germany	7.1	7.3	6.3
Czech Republic	6.0	5.2	5.0
Colombia	3.8	4.0	4.0
Indonesia	2.2	2.8	2.8
Mozambique	-	0.3	2.7
Mexico	1.6	2.0	2.2
Other	*5.3*	*5.3*	*5.8*
World	**911.6**	**970.0**	**983.8**

* Significant proportions of production in some countries may be designated for thermal usage.

See Table 1.3 in Part III for historic data on selected countries.

Australia is the world's second largest producer of coking coal and the largest exporter, but its production declined by 16.2 Mt due to adverse weather conditions in 2011, and this similarly reduced level of production also occurred in 2012. There was taken to be very little increase in indigenous production for coking coal for China, as there was a large increase in imports of coking coal which took up almost all of the increase in expected demand. This does not take into account the possibility that some of this coal may have been intended for use in the ferrosilicon industry, PCI, foundries or other uses for high grade coal. Countries including India, Germany and Mexico use a significant proportion of their indigenously sourced coking coal (Table II.4) for thermal purposes.

Lignite production

Worldwide, lignite production decreased by 0.7% to 904.6 Mt in 2012 (Table II.5). OECD lignite production rose by 3.6 Mt to 611.0 Mt, led by increases in Germany (8.9 Mt) as German electricity production from lignite increased by 5.9% in 2012. Greece (3.1 Mt) and Australia (2.5 Mt) also increased production. Non-OECD lignite production decreased by 10.0 Mt, despite increases in Mongolia (1.7 Mt), the Russian Federation (1.5 Mt) and India (1.2 Mt). It should be noted that lignite production in the People's Republic of China is still currently being included in general coal on their submission which is significant as they are the world's second largest producer and consumer of lignite behind Germany. Turkey's production was estimated in 2012, based upon reported electricity generation from lignite fired plants.

Table II.5 Major lignite* producers [Mt]

	2010	2011	2012e
Germany	169.4	176.5	185.4
Russian Federation	76.1	76.4	77.9
Australia	72.1	71.0	73.5
United States	71.0	73.6	71.6
Turkey	69.7	72.6	66.0
Poland	56.5	62.8	64.3
Greece	56.5	58.7	61.8
Czech Republic	43.8	46.6	43.5
India	37.7	42.3	43.5
Serbia	38.0	41.1	38.0
Romania	31.1	35.5	34.0
Bulgaria	29.4	37.1	32.5
Estonia	17.9	18.7	18.8
Thailand	18.3	21.3	18.3
Mongolia	8.5	8.3	10.0
Canada	10.3	9.7	9.5
Other	*54.6*	*58.7*	*56.0*
World	**861.0**	**911.0**	**904.6**

* Lignite also includes oil shale.

See Table 1.5 in Part III for historic data on selected countries.

Following its 1989 peak of 1 210.9 Mt, world lignite production declined as a general trend through to 1999, largely as a result of contractions of demand and supply in Central and Eastern Europe. Production stabilised from 2000 onwards in countries such as Australia, Canada, Germany, Greece, Romania, the Russian Federation and Serbia.

Table II.6 Proved recoverable reserves in 2011 and 2012 production [Mt]

Region	Proved recoverable reserves 2011			Production 2012e			
	Hard coal(2)	Soft brown coal(1)	Total	Steam coal	Coking coal	Lignite	Total
OECD Europe	18 474	56 094	74 568	106.3	24.6	455.6	586.5
OECD Americas	231 669	32 956	264 655	821.7	114.5	81.1	1 017.4
OECD Asia Oceania	59 029	50 979	110 008	204.9	149.0	74.3	428.2
OECD Total	**309 202**	**140 029**	**449 231**	**1 133.0**	**288.1**	**611.0**	**2 032.1**
Transition Economies	122 884	106 776	229 660	370.2	105.3	217.2	692.6
of which: Russian Federation	*68 944*	*91 184*	*160 128*	*201.5*	*74.6*	*77.9*	*353.9*
People's Republic of China	180 600	11 000	191 600	3 038.6	510.4	0	3 549.1
Asia excluding China	96 524	20 009	116 533	1 044.5	70.9	73.4	1 188.8
of which: India	*77 197*	*4 847*	*82 044*	*504.3*	*47.2*	*43.5*	*595.0*
Non OECD Americas	7 762	5 073	12 835	92.1	4.0	3.0	99.1
of which: Colombia	*4 881*	-	*4 881*	*85.5*	*4.0*	-	*89.5*
Africa and Middle East	37 443	9	37 452	264.0	5.0	-	269.0
World	**754 595**	**282 957**	**1 037 552**	**5 942.4**	**983.8**	**904.6**	**7 830.8**

(1) Soft brown coal is lignite with a calorific value of less that 16.5 MJ/kg, which differs from the IEA definition of lignite.
(2) "Hard" coal for this purpose is all other coals.

Sources: Reserves - Bundesanstalt für Geowissenschaften und Rohstoffe (BGR) http://www.bgr.bund.de
Production - International Energy Agency, *Coal Information* 2013

Coal resources and reserves

Coal reserve estimates

In its most recent study, the German Federal Institute for Geosciences and Natural Resources (BGR) estimates the current proved economically recoverable global coal reserves to be 1 037.6 billion tonnes (Table II.6), which is an increase of 33.8 billion tonnes. These proved reserves represent 132.5 years of production at current levels, up from the 131.9 years calculated last year, but significantly down from 138.8 years supply, two years ago and 145.8 years, the year before that. The decreasing number of years available at current production levels (Table II.7) is increasingly evident when looking over wider timescales and is primarily due to the rapid expansion year-over-year of annual production in recent times. It is also evident that rising coal demand and higher prices do affect quantities of proven reserves, as seen by the demand downturn experienced in 2009 and 2010 overriding positive development trends and correlation with price in other years.

Another estimate published by the World Energy Council (WEC) in 1978 set proved coal reserves at 636.4 billion tonnes. Thus, world proved coal reserves are 63% larger than they were 33 years before, even accounting for the fact that production from 1978 to 2012 amounted to a further 175.8 billion tonnes being physically removed from proven resources. Losses and other factors would compound this. Better understanding of resources along with technological and process improvements have been responsible for much of the growth in proven reserves in addition to price trends.

Table II.7 Changes in production at current levels over time. World and China (Gt)

Coal Edition	Proved reserves World	Proved reserves *China*	Production World	Production *China*	Years World	Years *China*
2003	984.5	*114.5*	4.92	*1.54*	200.0	*74.6*
2004	907.3	*114.5*	5.27	*1.81*	172.1	*63.2*
2005	909.1	*114.5*	5.67	*2.08*	160.4	*54.9*
2006	989.5	*114.5*	6.01	*2.30*	164.6	*49.8*
2007	934.9	*133.2*	6.33	*2.48*	147.7	*53.8*
2008	1 019.3	*192.0*	6.56	*2.63*	155.4	*72.9*
2009	989.9	*192.0*	6.74	*2.73*	146.8	*70.2*
2010	997.2	*191.6*	6.84	*2.90*	145.8	*66.2*
2011	1 000.5	*191.6*	7.21	*3.14*	138.8	*61.0*
2012	1 003.8	*191.6*	7.61	*3.42*	131.9	*56.0*
2013	1 037.6	*191.6*	7.83	*3.55*	132.5	54.0

Reserves data from 2003 to 2005 are from the World Energy Council, while other data are provided by BGR.

Reserve data are as submitted in initial publications. Production data are inclusive of current country revisions.

It is also important to note that world hard coal total resources are estimated to be 17.2 trillion tonnes or nearly 17 times current proved reserves.[3] Taking the example of the People's Republic of China from Table II.7, while the estimated number of years production at current rates have been dramatically dropping, proven reserves currently constitute just 3.5% of their total hard coal resources (5.0 Tt), which provides

3. *Reserves, Resources and Availability of Energy Resources*, Federal Institute for Geosciences and Natural Resources, Hanover, Germany 2010.

powerful incentive for development of technologies including in situ and ex situ coal gasification, coal-to-liquids and clean coal technologies.

Another estimate published by the WEC on reserves at the end of 2002 set proved coal reserves at 909.1 billion tonnes. Even though it is difficult to compare the estimates from two different sources, the increase observed between the end-2002 coal reserves and the 2010 coal reserves suggests that, as the prices increased, a part of the resources were reallocated to recoverable reserves as they became economical to mine.

Geographic location of proved coal reserves

Although coal resources are widely distributed around the world[4], proved coal reserves tend to be concentrated in the countries which rely on coal for domestic energy or export revenue. OECD Americas and the Transition Economy countries controlled 25.5% and 22.1% of the proved total coal reserves, respectively. The People's Republic of China accounts for 18.5% of proved reserves and 11.2% are in Asia (excluding the People's Republic of China). Thus, over 75% of proved coal reserves are concentrated in just four (rather large) regions. When solely concentrating on hard coal reserves, OECD Americas contains 30.7% of world reserves and the same four regions possess 83.7% of all coal reserves.

Consumption

Total coal consumption[5]

In 2012, 76.8% of world coal consumption was accounted for by just five countries: the People's Republic of China (excluding Hong Kong), the United States, India, the Russian Federation and Japan (Table II.8). This is an increase from 76.6% in 2011. If a second set of five countries is added to the 2012 list, South Africa, Germany, Korea, Poland and Australia, then 86.0% of the world coal consumption is accounted for.

Table II.8 Total coal consumption [Mtce] (selected countries)

	2010	2011	2012e
OECD Countries			
United States	718.2	683.9	608.4
Japan	163.8	153.8	161.7
Korea	104.3	114.2	111.2
Germany	106.6	107.9	108.8
Poland	86.0	84.6	78.0
Australia	71.5	68.8	70.3
United Kingdom	43.7	44.8	55.4
Turkey	45.6	48.2	49.9
Canada	31.2	28.0	27.4
Czech Republic	25.9	25.7	24.5
Italy	20.4	23.0	22.5
Spain	11.5	18.0	21.9
France	16.1	13.5	15.7
Mexico	13.3	14.0	13.8
Israel	10.6	11.1	13.2
Greece	11.2	11.3	11.7
Netherlands	10.6	10.4	11.4
Chile	6.6	7.6	8.0
Estonia	5.5	5.8	5.8
Other OECD	41.3	40.0	38.5
Non-OECD Countries			
PR of China	2 430.1	2 665.5	2 794.8
India	442.2	463.7	492.9
Russian Federation	165.8	167.5	189.5
South Africa	144.2	140.7	140.2
Ukraine	56.0	61.1	65.8
Kazakhstan	48.8	53.3	59.0
Chinese Taipei	58.1	58.4	57.8
Indonesia	43.5	45.0	45.3
DPR of Korea	22.3	22.7	23.0
Thailand	23.1	25.9	22.7
Malaysia	20.9	22.3	22.2
Brazil	18.9	19.9	21.1
Vietnam	20.9	22.2	19.5
Philippines	10.9	11.8	15.0
Hong Kong (China)	9.1	11.0	11.1
Serbia	10.3	11.9	10.7
Romania	9.0	10.7	10.3
Bulgaria	9.8	11.5	9.6
Other Non-OECD	53.4	57.0	61.3
European Union	*395.4*	*404.5*	*411.5*
Total IEA	*1 505.9*	*1 473.9*	*1 415.1*
Total OECD	*1 544.0*	*1 514.4*	*1 458.0*
Total Non-OECD	*3 597.3*	*3 882.0*	*4 071.6*
World	**5 141.3**	**5 396.4**	**5 529.6**

See Table 2.2 Part III for historical data.

The term "total coal" refers to the sum of anthracite, other bituminous coal, coking coal, sub-bituminous coal and lignite after conversion to a common energy unit (tonne of coal equivalent - tce). The conversion is done by multiplying the calorific value of the coal in question by the total volume of the coal consumed, measured in physical units, *i.e.* in tonnes. The energy content of one tonne of coal equivalent is 29.3 Gigajoules (GJ) or 7000 kcal and corresponds to 0.7 tonnes of oil equivalent (toe).

Consumption refers to Total Primary Energy Supply.

4. The range of definitions of coal resources and reserves and an overview of the national classification systems is provided in an Appendix to *Major Coalfields of the World*, IEA Coal Research, London, June 2000.

5. Total coal refers to the sum of anthracite, other bituminous coal, coking coal, sub-bituminous coal and lignite, converted to a common energy unit, million tonnes of coal equivalent (Mtce).

OECD consumption

As OECD countries initially rebounded from the onset of the economic crisis in 2009, so did their coal consumption increasing from 1 468.6 Mtce (the lowest value since 1995) to 1 544.0 Mtce in 2010. This has been followed by two years of declines, firstly 29.6 Mtce in 2011 and then 56.4 Mtce in 2012. The main reason behind this change is fuel switching for electricity generation in the United States, where declines of coal consumption already amounted to 34.3 Mtce in 2011 and jumped to 75.5 Mtce in 2012.

Two years of reduced consumption in the OECD in combination with two years of growth in non-OECD countries ensured that the OECD share of global consumption dropped below 30% for the first time ever – decreasing to 28.1% in 2011 and 26.4% in 2012.

In OECD Europe, coal consumption is reported to have reached 440.3 Mtce in 2012, a 2.5% increase from the level of consumption in 2011, while total coal consumption in the OECD Asia Oceania region was 357.8 Mtce, an increase of 7.7 Mtce (Part III, Table 2.2). Increases in consumption in OECD Europe were led by the United Kingdom with an extra 10.6 Mtce (23.8%) and Spain with an extra 3.9 Mtce (21.5%). Increases in consumption in OECD Asia Oceania were led by Japan with an additional 8.0 Mtce of consumption (5.2%).

Non-OECD consumption

Total coal consumption in the non-OECD countries continued with strong growth, increasing by 4.9% to 4 071.6 Mtce. This was mainly due to the People's Republic of China, which further strengthened its position as the world's largest coal-consuming country in 2012. Consumption in China was calculated to increase to 2 794.8 Mtce, or 50.5% of total world coal consumption on a calorific basis. This is the first time that the 50% benchmark has been passed when looking at domestic supply.

Consumption also increased in other major coal consuming countries, including India (6.3%), the Russian Federation (13.1%) and Ukraine (7.5%).

Steam coal consumption

World steam coal consumption was up 2.4% in 2012, to 5 813.6 Mt (Table II.9). Steam coal consumption in the OECD decreased by 66.7 Mt to 1 370.0 Mt. Viewed on a regional basis, steam coal consumption sharply decreased by 10.8% in OECD Americas, while it increased by 4.2% in OECD Asia Oceania and 6.0% in OECD Europe. Given the predominant position of the United States in this respect, it is of no surprise that overall, OECD consumption decreased by 4.6%. The countries with the largest increases in 2012 were: the United Kingdom (13.2 Mt), Japan (11.2 Mt) and Spain (4.9 Mt). These increases were overshadowed by the decrease in the United States (95.4 Mt), while Poland also decreased by 7.6 Mt.

Table II.9 Major steam coal consumers* [Mt]

	2010	2011	2012e
PR of China	2 698.6	2 960.1	3 086.5
United States	862.3	826.1	730.7
India	565.0	588.4	625.9
South Africa	186.3	182.8	185.0
Japan	127.7	120.3	131.6
Russian Federation	97.8	93.6	116.9
Korea	92.8	98.3	95.7
Kazakhstan	61.1	65.6	76.3
Poland	72.5	71.6	64.1
Indonesia	57.8	59.7	60.2
Australia	61.9	58.2	59.7
United Kingdom	45.1	45.2	58.5
Chinese Taipei	57.8	60.6	56.1
Ukraine	39.7	45.4	51.9
Germany	43.5	41.5	40.9
DPR of Korea	27.4	27.9	28.3
Other	*314.0*	*331.2*	*345.5*
World	**5 411.3**	**5 676.7**	**5 813.6**

* Steam coal comprises anthracite, other bituminous coal and sub-bituminous for all countries.

See Table 2.4 in Part III for historic data on selected countries.

Non-OECD steam coal consumption increased by 4.8% to 4 443.6 Mt in 2012. Increases were once again led by the People's Republic of China, who increased steam coal consumption (including exceptionally, lignite) from 2 960.1 Mt in 2011 to 3 086.5 Mt in 2012. In addition, India (37.4 Mt), the Russian Federation (23.3 Mt) and Kazakhstan (10.7 Mt) also achieved prominent increases.

Coking coal consumption

Coking coal consumption in the OECD (Table II.10) decreased by 3.8% to 185.6 Mt in 2012 to remain 6.2% below the pre-economic crisis level in 2008.

Coking coal consumption has generally been declining in the OECD since the mid-1980s. Several factors have combined to weaken coking coal consumption. First, there is the growing use of pulverised coal injection (PCI) and heavy fuel oil injection into blast furnaces. Coal used for PCI can be classified as being of steam coal quality, so steam coal is displacing coke oven coke (usually manufactured from coking coal) as a carbon source. Second, the steel industries in the OECD countries are experiencing technological evolution where traditional integrated steel production, which relies on coke oven coke, is losing ground to electric arc furnace technology, which does not require coke. Third, there has been a general migration of integrated steel manufacturing from the OECD to the non-OECD countries.

Table II.10 Major coking coal consumers [Mt]

	2010	2011	2012e
PR of China	502.0	541.1	579.4
India	74.3	79.8	83.8
Russian Federation	49.7	53.7	56.3
Japan	57.7	53.8	52.2
Korea	27.2	32.6	31.6
Ukraine	26.4	27.5	27.1
United States	19.2	19.4	19.0
Germany	15.1	16.5	15.3
Kazakhstan	11.6	12.1	12.6
Poland	12.3	11.9	11.6
Brazil	11.0	11.4	10.9
Turkey	7.5	8.0	8.7
Chinese Taipei	5.6	5.9	8.4
United Kingdom	6.4	6.3	5.5
Canada	3.7	5.6	4.7
France	4.5	3.8	4.6
Netherlands	4.0	4.2	4.2
Other	*42.5*	*42.7*	*40.3*
World	**880.7**	**936.2**	**976.2**

See Table 2.3 in Part III for historic data on selected countries.

World consumption of coking coal in 2012 is estimated to have increased by 4.3% to 976.2 Mt with the People's Republic of China accounting for 59.3% of global coking coal consumption. The next five major coking coal consumers are: India, the Russian Federation, Japan, Korea and Ukraine. Together they account for another 25.7% of world annual coking coal consumption.

The OECD countries mostly used their coking coal for the production of coke oven coke. In 2012, they produced 124.0 Mt of coke oven coke and consumed 122.6 Mt (Part III, Tables 1.6 and 2.6).

The latest available data for 2011 shows that total PCI consumption increased by 10.3% or 4.1 Mt (Table II.11). The top five PCI consumers reported (Japan, Korea, the Russian Federation, Germany and India) accounted for 76.3% of all PCI consumption. It would however be remiss to not point out that many emerging economies are using PCI techniques and injecting metallurgical quality coal, but are not yet providing statistics in this area. For example, reasonable estimates for the magnitude of PCI use in the People's Republic of China would outweigh the entire table below in its current form by a factor of more than two on an annual basis.

Table II.11 PCI Used in blast furnaces (major consumers in thousand tonnes)

	2010	2011
Japan	11 644	12 593
Korea	7 548	9 310
Russian Federation	3 820	4 604
Germany	3 959	3 772
India	2 743	2 877
France	1 878	1 808
United States	1 284	1 479
Chinese Taipei	1 109	1 379
Netherlands	1 018	1 266
United Kingdom	898	895
Italy	905	862
Spain	562	636
Australia	697	620
Slovak Republic	356	421
Turkey	464	401
Sweden	312	318
Belgium	172	147
Poland	32	80
World	**39 401**	**43 468**

See Table 5.4 in Part III for other countries and historical data. World totals are Secretariat estimates. Includes granular coal injection for some countries. Data for PR China are not available.

Lignite consumption

Consumption of lignite on a global basis decreased by 6.7 Mt or 0.7% in 2012 (Table II.12). Germany remained the largest producer and consumer of lignite, using 185.2 Mt, an increase of 8.5 Mt, ahead of Russia (77.8 Mt), Australia (73.5 Mt) and the United States (72.2 Mt).

Table II.12 Major lignite* consumers [Mt]

	2010	2011	2012e
Germany	169.5	176.7	185.2
Russian Federation	76.3	77.6	77.8
Australia	72.1	71.0	73.5
United States	68.3	74.8	72.2
Turkey	69.2	73.9	66.0
Poland	56.6	62.7	64.1
Greece	57.7	60.0	63.7
India	37.7	41.9	43.5
Czech Republic	43.7	44.1	42.6
Serbia	37.5	42.1	38.6
Romania	30.8	37.1	33.7
Bulgaria	29.4	37.0	32.5
Estonia	17.9	18.7	18.8
Thailand	18.0	20.3	18.3
Mongolia	6.9	6.8	9.7
Hungary	8.9	9.7	9.6
Canada	10.1	9.6	9.4
Other	*44.2*	*49.9*	*47.8*
World	**855.1**	**913.8**	**907.1**

* Lignite also includes oil shale.

See Table 2.5 in Part III for historic data on selected countries.

Currently, production and consumption of lignite are reported as sub-bituminous coal in Indonesia, and under other bituminous coal (along with sub-bituminous coal) in the People's Republic of China.

Trade

World coal trade

Export trade of all types of coal in the world rose by 9.7% in 2012, to reach a record level of 1 255.3 Mt (Table II.13). This was composed of 962.7 Mt of steam coal exports, 290.1 Mt of coking coal exports (also both records) and 2.5 Mt of lignite exports. These increases followed on from a 6.3% increase in 2011.

Table II.13 World coal trade [Mt]

	2010	2011	2012e
Steam coal exports	789.2	857.5	962.7
Coking coal exports	282.0	282.9	290.1
Lignite exports	4.8	3.7	2.5
Steam coal imports	806.1	867.3	989.3
Coking coal imports	258.4	261.7	283.9
Lignite imports	3.1	5.0	2.8
Total exports	1 076.1	1 144.1	1 255.3
Total imports	1 067.6	1 134.0	1 276.0
Balancing item	*-8.5*	*-10.1*	*20.7*

Note: The balancing item is the difference between total coal imports and total coal exports. This is primarily due to the different coal classification methodologies used by the importing and exporting countries, which does not hold on a global basis. It also occurs because of coal in-transit, coal that is unaccounted for, and reporting discrepancies by importing and exporting countries.

The breakdown between coking coal, steam coal, and lignite statistics in world coal trade is affected by the practice of some countries. For example, Australia includes some types of coal (soft or semi-soft coking coal, used for blending or as PCI) in their coking coal statistics; however, other countries may include this trade in their steam coal import statistics.

Global trade has been growing faster that global consumption on a relatively consistent basis, as evidenced in Figure II.1, which compares regional trade data as a portion of corresponding consumption. Global exports reached a record share of total consumption (21.1%), up from 11.0% in 1980 and 16.5% in 1990, and in one sense it tells a story of optimised globalisation.

Given that the consumption data are consistent; this chart can also be used to visually compare ratios of imports to exports for OECD and non-OECD regions. It is also interesting to see non-OECD imports almost keeping pace with the large increases in production and consumption in recent years, and the change of OECD consumption patterns, moving from coal being 79.1% domestically sourced in 2002 to 65.8% in 2012.

Figure II.1: Steam and coking coal trade as a percentage of consumption

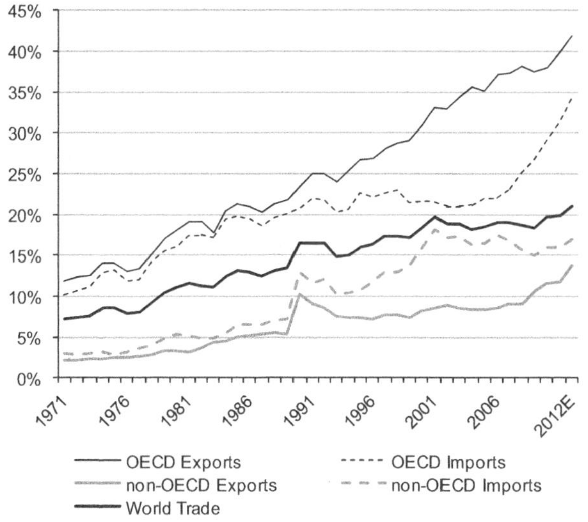

Some sub-bituminous coal prior to 1978 is unintentionally excluded. Calculations are based on energy, not tonnage.

Exports

Indonesia became the world's leading coal exporter in 2011 with total exports reaching 300.7 Mt compared to 284.5 Mt for Australia. In 2012 Australian exports increased by 17.0 Mt to 301.5Mt, and the United States exports grew by 16.8 Mt (17.2%) with strong momentum for the second year in a row, however, in 2012, Indonesian exports increased by 27.3% or 81.9 Mt to 382.6 Mt although 2011 data may be revised upwards in the 2014 edition. Also of note, exports from the Russian Federation increased by 7.7% and South African exports increased by 8.1%.

Table II.14 Major coal exporters [Mt]

	2010	2011	2012e
Indonesia	267.2	300.7	382.6
Australia	292.6	284.5	301.5
Russian Federation	132.8	124.6	134.2
United States	74.1	97.3	114.1
Colombia	68.1	79.3	82.2
South Africa	66.4	68.8	74.3
Canada	33.4	33.7	34.8
Kazakhstan	31.3	30.4	31.9
Mongolia	16.7	23.7	22.1
Vietnam	19.7	17.7	19.1
DPR of Korea	4.6	11.2	12.0
PR of China	21.3	21.7	10.5
Other	*47.6*	*50.6*	*36.1*
World	**1 076.1**	**1 144.1**	**1 255.3**

Statistics for hard coal exports presented in this table are shown for selected countries with historic data in Part III, Table 3.16.

The People's Republic of China's ranking dropped to twelfth in terms of largest coal exporters – exporting 10.5 Mt. This is less than a quarter of the amount they exported in 2008 (45.3 Mt) and is one ninth of what they exported in 2003. Moreover, the People's Republic of China now imports over 24 times more than it exports.

Imports

Table II.15 Major coal importers [Mt]

	2010	2011	2012e
PR of China	163.1	182.0	288.8
Japan	185.4	174.1	183.8
India	115.7	132.1	159.6
Korea	118.6	129.2	125.5
Chinese Taipei	63.2	66.6	64.5
Germany	45.7	47.8	45.2
United Kingdom	26.5	32.5	44.8
Russian Federation	25.5	27.6	31.4
Turkey	21.3	23.7	28.7
Italy	22.1	23.5	24.3
Spain	12.8	16.2	22.4
Malaysia	20.7	21.9	22.0
Brazil	15.9	18.0	18.2
Thailand	16.8	19.5	16.8
France	17.6	14.5	15.9
Other	*196.6*	*204.8*	*184.1*
OECD Americas	*44.8*	*38.6*	*35.7*
OECD Asia Oceania	*316.5*	*315.9*	*324.0*
OECD Europe	*218.8*	*239.7*	*239.1*
OECD Total	***580.2***	***594.2***	***598.8***
Africa + Mid. East	*12.2*	*13.3*	*14.2*
Other Asia Oceania	*408.3*	*453.5*	*587.1*
Oth. Europe + Eurasia	*47.2*	*50.9*	*54.1*
Other Americas	*19.8*	*22.1*	*21.8*
Non-OECD Total	***487.4***	***539.8***	***677.2***
World	**1 067.6**	**1 134.0**	**1 276.0**

Statistics for coal imports presented in this table are shown for selected countries with historic data in Part III, Tables 3.5 and 3.7.

Total world coal imports were 1 276.0 Mt in 2012, a 12.5% increase over 2011 (Table II.15). Of this 142.1 Mt of additional trade, 137.5 Mt had a non-OECD country destination, with just 4.6 Mt of import growth in OECD countries.

Looking at the OECD Asia Oceania and non-OECD Asia Oceania regions, we can see that total imports reached 911.0 Mt, or 71.4% of all imports, with the top five individual importers being from this region. The following six countries were from Europe or Eurasia. However their combined imports of 196.8 Mt were 92.0 Mt less than the People's Republic of China's alone.

The largest growth in coal imports occurred in the People's Republic of China (106.8 Mt), India (27.5 Mt), the United Kingdom (12.3 Mt) and Japan (9.6 Mt).

Steam coal trade

Total world steam coal exports increased by 105.2 Mt or 12.3% in 2012 (Table II.16). Ranked in terms of change in tonnage, Indonesia (81.9 Mt), the United States (16.6 Mt) and Australia (15.1 Mt) all provided substantial increases in exports.

Table II.16 Major steam coal exporters [Mt]

	2010	2011	2012e
Indonesia	265.0	297.8	379.8
Australia	135.4	144.1	159.2
Russian Federation	114.2	109.6	115.9
Colombia	66.9	77.9	81.7
South Africa	66.4	68.4	73.6
United States	23.0	34.1	50.6
Kazakhstan	29.1	29.8	31.3
Vietnam	19.7	17.7	19.1
DPR of Korea	4.6	11.2	12.0
PR of China	20.2	18.1	8.8
Poland	8.2	5.3	5.4
Ukraine	5.9	6.7	5.3
Other	*30.6*	*36.9*	*20.1*
World	**789.2**	**857.5**	**962.7**

Statistics for steam coal exports presented in this table are shown for selected countries with historic data in Part III, Table 3.21.

Indonesia further increased their world steam coal market share from 34.7% to 39.5%. Australia, the Russian Federation, Colombia, South Africa, and the United States followed with respective shares of 16.5%, 12.0%, 8.5%, 7.6% and 5.3%, as all six of these countries also produced record levels of steam coal exports in 2012.

In 2012, steam coal imports in the Asia-Oceania market increased by 113.3 Mt to reach 711.7 Mt (Table II.17), 240.5 Mt of which was to OECD countries. Asia-Oceania imports represented 71.9% of total world steam coal trade in 2012, up from 69.0% in the previous year. In recent years, Asia Oceania's share has been steadily increasing.

Despite the conscious move away from nuclear power in 2011 (as discussed in a feature article in the 2012 *Coal Information* book), Japan did not initially experience a surge in coal demand as a result of potential fuel switching. This occurred because of deliberate load management, naturally depressed demand and supply chain disruptions, amongst other factors. As a consequence, steam coal imports decreased by 7.4 Mt in 2011,, however, in 2012 this was followed by an 11.2 Mt increase in imports.

The People's Republic of China's steam coal imports grew by 58.8% to 218.1 Mt in 2012 as international trade continued to be an appealing option. Other major importers in the region were India (123.0 Mt – up 24.9 Mt), Korea (94.3 Mt – down 2.6 Mt) and Chinese Taipei (56.1 Mt – down 4.4 Mt). In 2012, the major steam coal suppliers to this Asia Oceania market were Indonesia, Australia, South Africa, the Russian Federation and Vietnam.

Table II.17 Major steam coal importers [Mt]

	2010	2011	2012e
PR of China	116.0	137.3	218.1
Japan	127.7	120.3	131.6
India	81.3	98.2	123.0
Korea	90.4	96.9	94.3
Chinese Taipei	57.6	60.6	56.1
United Kingdom	19.9	26.6	40.0
Germany	37.9	39.1	35.9
Russian Federation	24.0	23.1	31.4
Malaysia	20.7	21.9	22.0
Turkey	16.2	16.9	21.1
Spain	10.0	13.7	20.2
Italy	17.0	17.9	19.8
Thailand	16.8	19.5	16.8
Israel	12.3	12.5	14.7
Hong Kong (China)	10.3	12.5	12.6
Other	*147.8*	*150.5*	*131.7*
OECD Americas	*38.2*	*31.0*	*28.7*
OECD Asia Oceania	*230.7*	*229.9*	*240.5*
OECD Europe	*164.7*	*185.7*	*186.6*
OECD Total	***433.6***	***446.7***	***455.9***
Africa + Mid. East	*7.9*	*8.8*	*10.0*
Other Asia Oceania	*320.8*	*368.5*	*471.2*
Oth. Europe + Eurasia	*35.5*	*33.6*	*41.9*
Other Americas	*8.3*	*9.7*	*10.4*
Non-OECD Total	***372.5***	***420.7***	***533.4***
World	**806.1**	**867.3**	**989.3**

Statistics for steam coal imports presented in this table are shown for selected countries with historic data in Part III, Tables 3.10 and 3.12.

Steam coal imports in the European market were 228.5 Mt in 2012, 4.2% higher than in 2011. This market now represents 23.1% of total world steam coal trade. Within the region, the major Europe/ Eurasian importers were the United Kingdom (40.0 Mt), Germany (35.9 Mt), the Russian Federation (31.4 Mt), Turkey (21.1 Mt), Spain (20.2 Mt) and Italy (19.8 Mt). The 2012 main steam coal suppliers to this market were the Russian Federation, Colombia, the United States and South Africa.

In the OECD Americas, steam coal imports decreased by 7.4% to 28.7 Mt in 2012. The largest supplier to the North American market was Colombia, followed by trade between Canada and the United States.

In 2012, 10.4 Mt of steam coal was imported by non-OECD American countries, 70.5% of which went to Brazil. The largest supplier to the non-OECD Americas was Colombia, followed by the United States and South Africa.

The annual cost of steam coal imported by OECD countries was unavailable for 2011and 2012 due to budgetary restraints. We remain hopeful that this regretable situation will only be a temporary inconvenience.

Coking coal trade

Total world coking coal export trade increased by 2.5% to 290.1 Mt in 2012 (Table II.18). Australia remained by far the largest exporter of coking coal at 142.4 Mt, accounting for 49.1% of coking coal trade in 2012. The United States remained as the second-ranked coking coal exporter with a volume of 63.4 Mt, while third-ranked Canada exported 30.7 Mt of coking coal, an 11.1% increase over 2011 levels.

Table II.18 Major coking coal exporters [Mt]

	2010	2011	2012e
Australia	157.3	140.5	142.4
United States	50.9	63.1	63.4
Canada	27.5	27.7	30.7
Mongolia	15.1	21.0	19.1
Russian Federation	18.0	14.2	18.3
Czech Republic	3.5	3.0	2.9
Indonesia	2.2	2.8	2.8
Mozambique	-	0.3	2.7
New Zealand	2.3	2.1	2.2
PR of China	1.1	3.6	1.7
Poland	1.8	1.7	1.6
Other	*2.2*	*3.1*	*2.4*
World	**282.0**	**282.9**	**290.1**

Statistics for coking coal exports presented in this table are shown for selected countries with historic data in Part III, Table 3.21.

Coking coal exports from Mongolia fell slightly to 19.1 Mt in 2012 as issues were experienced for part of the year with the financing of transportation to market. Similar problems are not forecast for 2013. 2012 was also the first full year of coking coal production in Mozambique, which again has certain logistical and infrastructure issues to address as it ramps up production.

Coking coal imports reported by Asia Oceania countries increased by 28.3 Mt to 199.3 Mt in 2012. Japan accounted for 18.4% (52.2 Mt) of the world market in 2011, but was surpassed by the People's Republic of China importing 24.9% or 70.6 Mt in 2012 (Table II.19).

Other major importers in the region were India (36.6 Mt) and Korea (31.3 Mt).The major coking coal supplier to this regional market in 2012 was Australia, with over half of the supply, followed by Mongolia, Canada and the United States.

Table II.19 Major coking coal importers [Mt]

	2010	2011	2012e
PR of China	47.1	44.7	70.6
Japan	57.7	53.8	52.2
India	34.4	33.9	36.6
Korea	28.2	32.2	31.3
Brazil	10.8	11.7	10.9
Ukraine	7.7	9.9	9.6
Germany	7.8	8.8	9.3
Chinese Taipei	5.5	6.0	8.4
Turkey	5.1	6.8	7.6
United Kingdom	6.6	5.9	4.8
France	4.6	3.8	4.7
Italy	5.1	5.6	4.6
Canada	3.1	3.8	4.4
Netherlands	4.6	4.3	4.1
Slovak Republic	2.5	2.5	2.6
Other	*27.7*	*27.9*	*22.3*
OECD Americas	*6.5*	*7.4*	*6.8*
OECD Asia Oceania	*85.8*	*86.1*	*83.5*
OECD Europe	*52.9*	*52.4*	*50.6*
OECD Total	***145.2***	***145.9***	***140.9***
Africa + Mid. East	*4.3*	*4.5*	*4.3*
Other Asia Oceania	*87.5*	*85.0*	*115.9*
Oth. Europe + Eurasia	*10.0*	*13.9*	*11.5*
Other Americas	*11.5*	*12.4*	*11.5*
Non-OECD Total	***113.3***	***115.8***	***143.1***
World	**258.4**	**261.7**	**283.9**

Statistics for coking coal imports presented in this table are shown for selected countries with historic data in Part III, Tables 3.09 and 3.11.

Coking coal imports to the Europe/Eurasian market were 62.1 Mt in 2012, a decrease of 4.3 Mt from 2011. Within the region, the major importers were: Ukraine (9.6 Mt), Germany (9.3 Mt), Turkey (7.6 Mt), the United Kingdom (4.8 Mt), France (4.7 Mt), Italy (4.6 Mt) and the Netherlands (4.1 Mt). The main coking coal suppliers to this market in 2011 were the United States, Australia, the Russian Federation, Canada, the Czech Republic and Poland.

Within OECD Americas, coking coal imports are primarily between Canada and the United States. About 6.8 Mt of coking coal was traded in the region in 2012, down from 7.4 Mt in 2011.

In 2012, non-OECD Americas had imports of 11.5 Mt of coking coal, 10.9 Mt of which went to Brazil. The main suppliers to this regional market were the United States, Australia and Canada.

Coke oven coke trade

The OECD countries' imports of coke oven coke decreased by 8.8% (1.1 Mt) between 2011 and 2012 as imports fell to 11.3 Mt (Table 3.13, Part III). Germany accounted for 28.8% of the OECD coke imports in 2012 despite a decline of 151 kt. Austria remained the OECD's second highest importer in 2012, importing 1.34 Mt, mainly from Poland, the Czech Republic and Hungary.

Exports of coke oven coke from the OECD countries increased from 11.5 Mt in 2011 to 12.4 Mt in 2012, an increase of 7.6% (Table 3.22, Part III). Poland (6.4 Mt), Japan (1.5 Mt) and the United States (0.9 Mt) were the top exporters, with more than half of all OECD coke oven coke exports coming from Poland.

Putting these numbers in some context, the largest consumer of coke oven coke in the OECD – Japan – consumed over 39.5 Mt in 2012, and OECD consumption of coke oven coke was 122.6 Mt, down 2.8 Mt from 2011, 25.4 Mt from 2000 and 56.2 Mt from 1990 (Table 2.6, Part III).

Coke oven coke statistics for non-OECD countries are not available for 2012. However, in 2011 OECD countries only contributed 20% of world coking coal consumption as total global consumption was reported as being 624.8 Mt. Consumption within the People's Republic of China contributed 63% of global consumption and 79% of non-OECD consumption, using 393.8 Mt in 2011. This is 11.7 times more than the second largest non-OECD consumer, the Russian Federation and 10.1 times larger than Japan's 2011 consumption.

Also in 2011, non-OECD imports declined relative to 2010, decreasing 6.7% to 10.6 Mt, while exports increased by 1.9% to 11.5 Mt. Within the non-OECD countries, India (2.4 Mt), Brazil (2.1 Mt), the Islamic Republic of Iran (1.3 Mt) and Kazakhstan and Romania (both 0.8 Mt) were the main importers, while China (3.3 Mt), the Russian Federation (2.5 Mt), Ukraine (2.0 Mt) and Colombia (1.5 Mt) were the major exporters, with these four countries providing 81% of non-OECD exports.

Figure II.2: OECD international trade values for steam coal and oil (USD/tce)

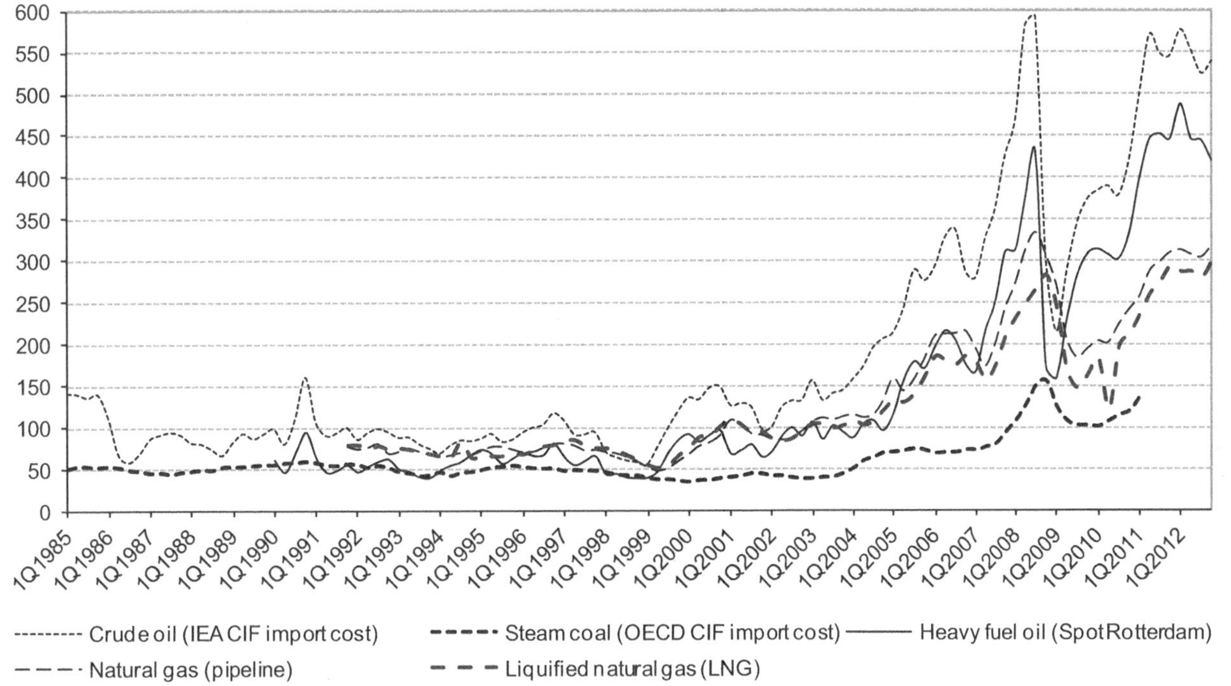

Notes: Spot prices for heavy fuel oil are not directly comparable to customs unit values for steam coal. They are, however, closely correlated with average CIF crude oil prices, which are, by definition comparable to customs unit values for steam coal. As a consequence, it is not unreasonable to compare customs unit values for steam coal with spot prices of heavy fuel oil. Steam coal and crude oil are IEA average and CIF import values. Steam coal excludes intra-EU trade. Heavy fuel oil is Rotterdam spot market value, 3.5% sulphur.

Source: *IEA/OECD Energy Prices and Taxes.*

Figure II.3: Steam coal import and export value comparison (USD/t)

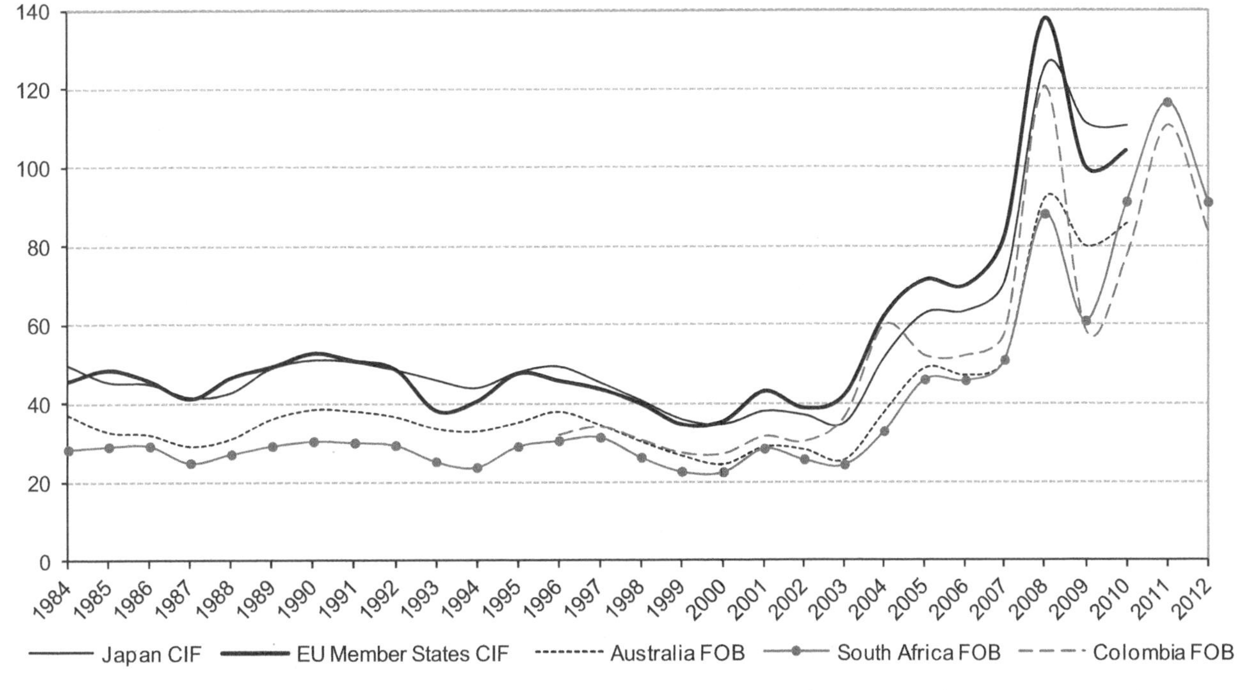

Coal data for some countries and regions for Figures II.2 and II.3 are currently unavailable for 2011 and 2012 due to resource constraints.

Source: *IEA/OECD Energy Prices and Taxes.*

Figure II.4: Delivered steam coal prices in Europe and Asia (USD/t CIF)

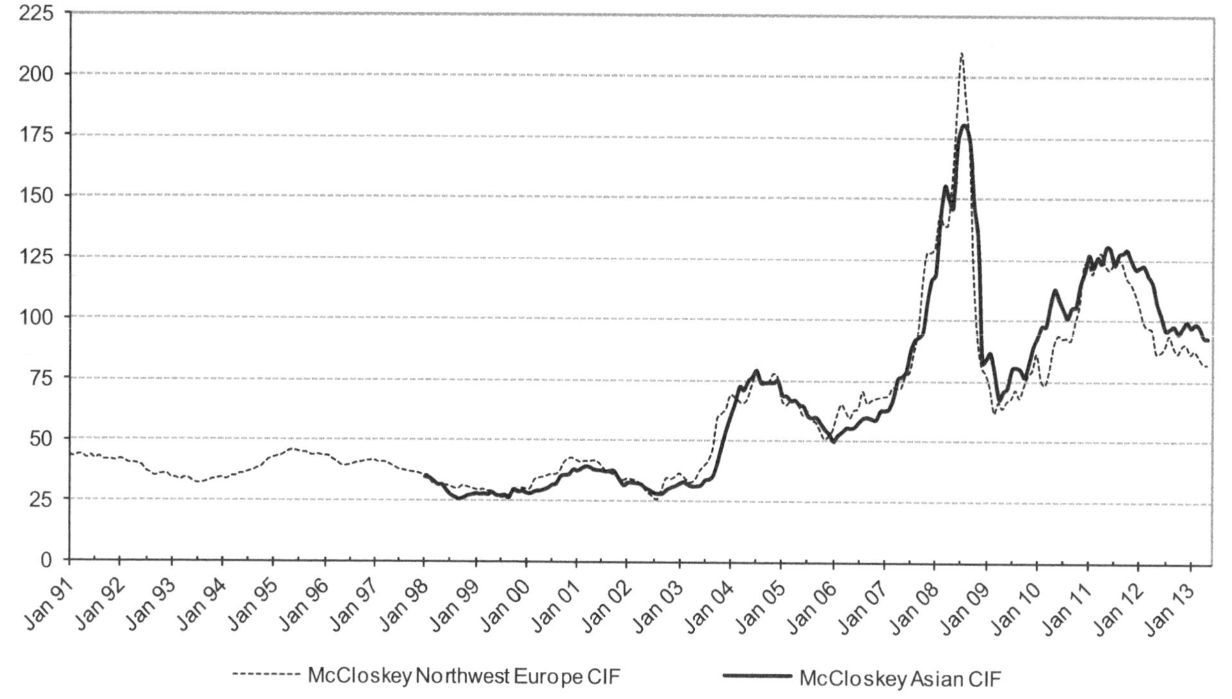

Source: IHS McCloskey, *McCloskey's Coal Report*

Figure II.5: FOB port steam coal prices in South Africa and Australia (USD/t FOB)

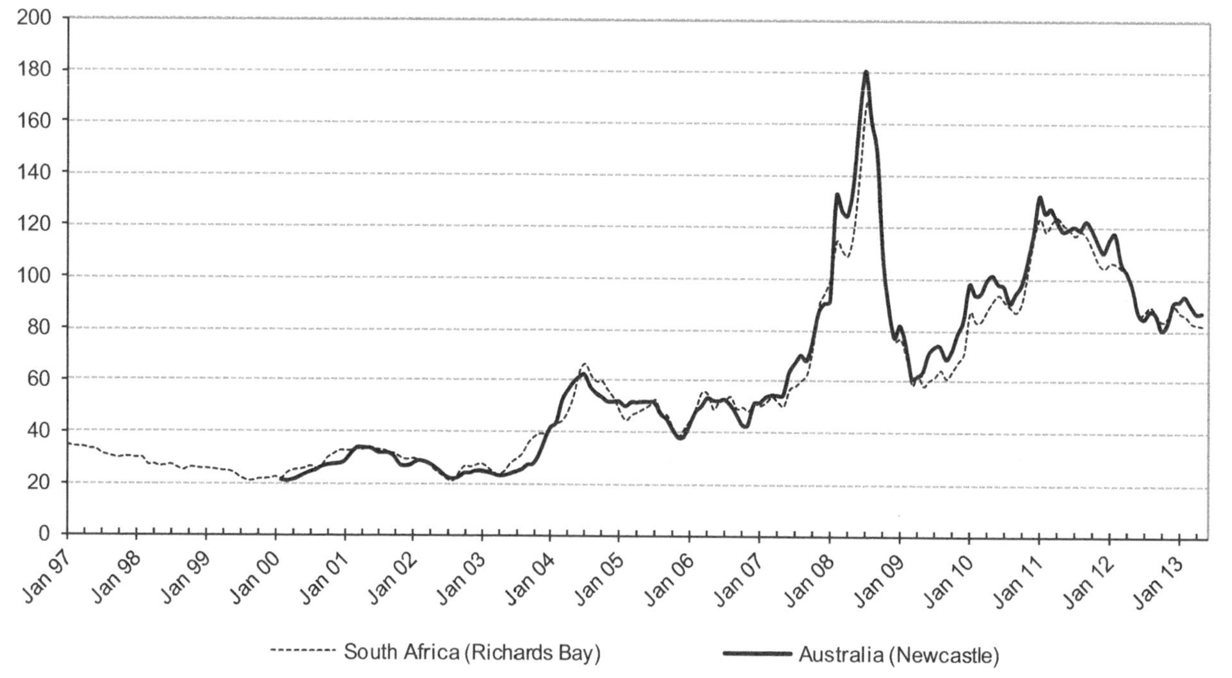

Source: IHS McCloskey, *McCloskey's Coal Report*

2. CO_2 EMISSIONS (2010)

World CO_2 emissions from combustion of fossil fuels reached a record level of 30.3 Gt in 2010 after increasing by 1.3 Gt – also a record. As a consequence, the 1.8% emissions decrease experienced in 2009, largely due to the global economic downturn was more than countered by the subsequent 4.6% increase, ensuring 2010 levels were 0.8 Gt greater than the previous record in 2008. This was the 6th instance in the eight years between 2003 and 2010 that global emissions increased by more than 800 Mt[1]. Examining this trend over a longer period, we find that emissions increased 115% from 1971 where emissions were 14.1 Gt, and that they increased 44% from the 21.0 Gt emitted in the Kyoto baseline year of 1990 (Figure II.6).

Figure II.6: World CO_2 emissions by region

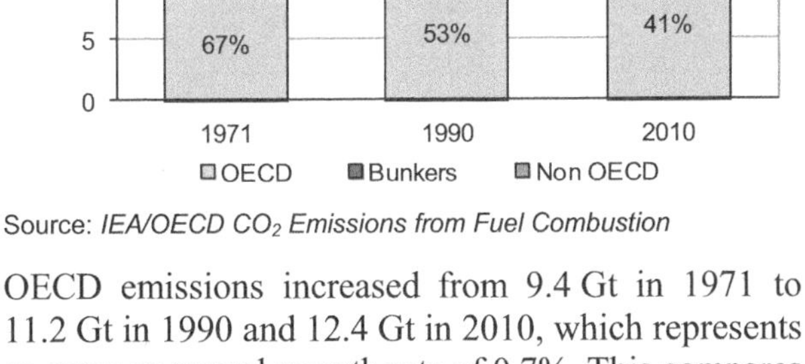

Source: *IEA/OECD CO_2 Emissions from Fuel Combustion*

OECD emissions increased from 9.4 Gt in 1971 to 11.2 Gt in 1990 and 12.4 Gt in 2010, which represents an average annual growth rate of 0.7%. This compares to a global rate of 2.0%. As a result, the share for the OECD countries in total world CO_2 emissions declined from 67% in 1971 to 53% in 1990 and 41% in 2010. This occurred as the share of the non-OECD countries in total world CO_2 emissions rose from 30% in 1971 to 44% in 1990 and 55% in 2010.

1. The only other two occasions where the annual increase in CO_2 emissions from fuel combustion exceeded 800 Mt were in 1973 and 1976.

In 2010, global coal emissions increased by 608 Mt, while natural gas emissions increased by 411 Mt and those from oil increased by 284 Mt (Figure II.7). The OECD contribution to global emissions grew in 2010 by 417 Mt, reversing two years of reduction, while non-OECD sources of CO_2 emissions rose by 843 Mt. Finally, emissions from international marine and aviation bunkers rose by 70 Mt or 6.8%.

Figure II.7. Global change in CO_2 emissions (2009-2010)

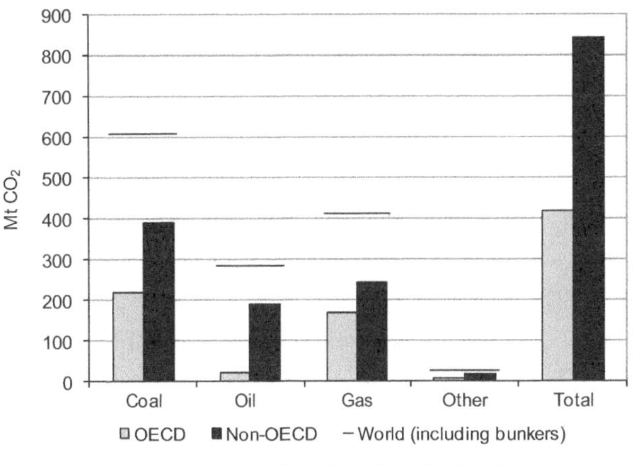

Source: *IEA/OECD CO_2 Emissions from Fuel Combustion*

Since 1971, coal-related emissions have increased by 152% from 5.2 Gt to 13.1 Gt in 2010 – passing 8.3 Gt in 1990, to now account for 43.2% of CO_2 emissions (Figure II.8). Coal, since 2005, is the leading source of global anthropogenic CO_2 emissions, outstripping those from oil and natural gas and other sources. Emissions of CO_2 from coal have grown by 4.7% on an average annual basis (or 500 Mt/y) since 2002. A large majority (83.6%[2]) of which has come from rapid growth in the People's Republic of China.

2. Excluding the People's Republic of China, countries with growth in emissions over this period, increased emissions by 1 044 Mt, while emissions in the remaining countries decreased by 350 Mt. Thus, overall emissions increased by 4.00 Gt, of which the People's Republic of China's contribution was 3.35 Gt or 83.6% of net growth, and 76.9% of all growth.

Figure II.8: World CO_2 emissions by fuel

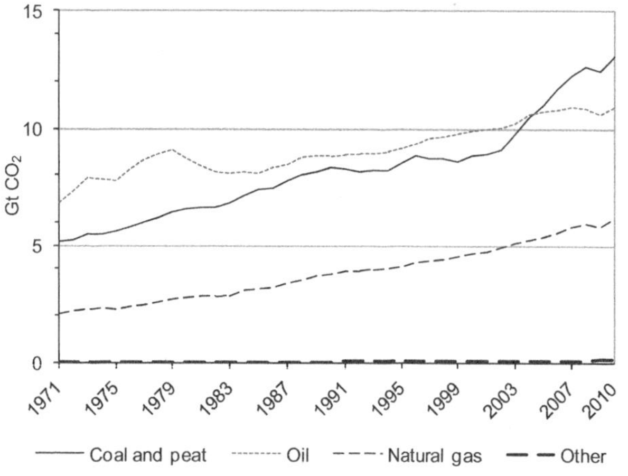

Source: *IEA/OECD CO_2 Emissions from Fuel Combustion*

The annual growth in emissions from fossil fuels averaged 2.1% between 1971 and 1990, before averaging 1.1% between 1990 and 2000 and 2.6% between 2000 and 2010. This lower pace of growth in the 1990s is in part related to the collapse of the formerly centrally-planned economies of Central and Eastern Europe. Other important factors include the use of more efficient technologies, fuel switching and renewable energy sources. The most prominent feature since 2000 has been the rapid expansion of non-OECD economies which has caused their emissions to increase on average by 5.2% between 2000 and 2010, with non-OECD coal emissions increasing by 7.0% per annum over this period.

The economic integration of large non-OECD countries, in particular Brazil, the Russian Federation, India, the People's Republic of China and South Africa, the so-called BRICS countries is having growing significance. In 2010, these five countries represented 33.3% of global energy use and 36.9% of CO_2 emissions from fuel combustion (Figure II.9). The People's Republic of China, the Russian Federation and India constitute three of the four countries that produced the most CO_2 emissions in absolute terms and their shares are likely to (and are planned to) continue to rise further in upcoming years.

Iran, Saudi Arabia and Indonesia are also becoming increasingly prominent, placed 9th, 11th and 13th in rankings for CO_2 emissions respectively in 2010. Emissions are linked to varying degrees to economic growth. For example, in the People's Republic of China, where robustly increasing electricity production as well as high demand for cement, iron and steel and other energy intensive commodities have been the main drivers for increases in CO_2 emissions.

Figure II.9: The growing importance of the BRICS countries

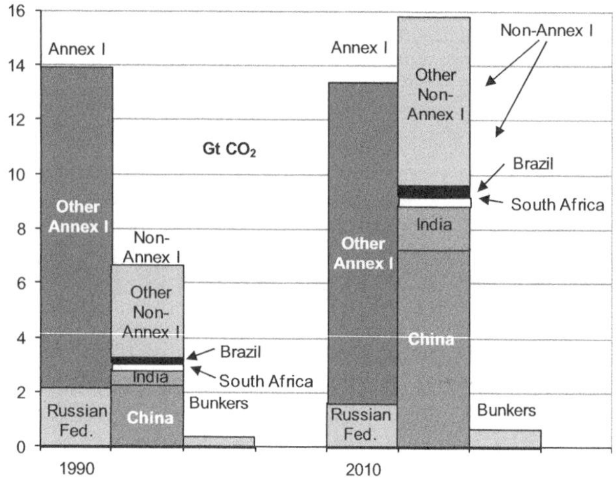

Source: *IEA/OECD CO_2 Emissions from Fuel Combustion*

Annex I countries are: OECD nations (with the exception of Chile, Estonia, Israel, Korea, Mexico and Slovenia), the European Union (with the exception of Cyprus[3]), Belarus, Croatia, Liechtenstein, Monaco, the Russian Federation and Ukraine.

While many developing countries are actively pursuing aggressive and ambitious electrification programs, none have been able to do at the same build rate as the People's Republic of China, where coal consumption for electricity generation increased by 164% between 2000 and 2010, or 494% when compared to 1990. Therefore, the OECD share of world CO_2 emissions from coal since 1990 reduced from 50% to 32%, as non-OECD coal-related emissions increased from 4.1 Gt to 8.9 Gt over the same period. In 2010, China contributed 6.0 Gt of those 8.9 Gt, or 67% of non-OECD countries' coal emissions (and 46% of global coal emissions).

Figure II.10: Coal-related CO_2 emissions by region

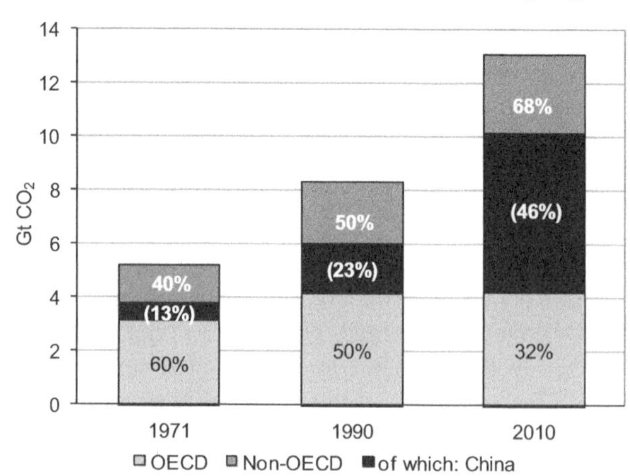

Source: *IEA/OECD CO_2 Emissions from Fuel Combustion*

3. See the Geographical coverage section in Part I.

Coal-related emissions in the non-OECD countries became responsible for the majority of coal-related annual global emissions from 1991 onward. The non-OECD countries more than quadrupled their coal-related emissions from 2.0 Gt in 1971 to 8.9 Gt in 2010, whereas the increase in OECD countries was 33.4% from 3.1 Gt in 1971 to 4.2 Gt, down from a peak of 4.5 Gt in 2007.

World coal-related CO_2 emissions due to electricity and heat generation more than tripled between 1971 and 2010, from 2.4 Gt in 1971 to 8.9 Gt in 2010 (Figure II.11). During the same period, coal emissions due to manufacturing industries and construction increase more than doubled, from 1.6 Gt in 1971 to 3.3 Gt. Emissions from the other sectors were cut by 32% between 1971 and 2010, falling from 1.2 Gt in 1971 to 0.8 Gt in 2010. As a result, the electricity and heat generation share of world coal-related CO_2 emissions increased from 46% in 1971 to 68% in 2010. On the other hand, the share of CO_2 emissions from manufacturing industries and construction fell from 31% in 1971 to 25% in 2010 in spite of its 107% increase from 1971 levels, while the share from other sector emissions fell from 23% to 6%.

Figure II.11: World coal-related CO_2 emissions by sector

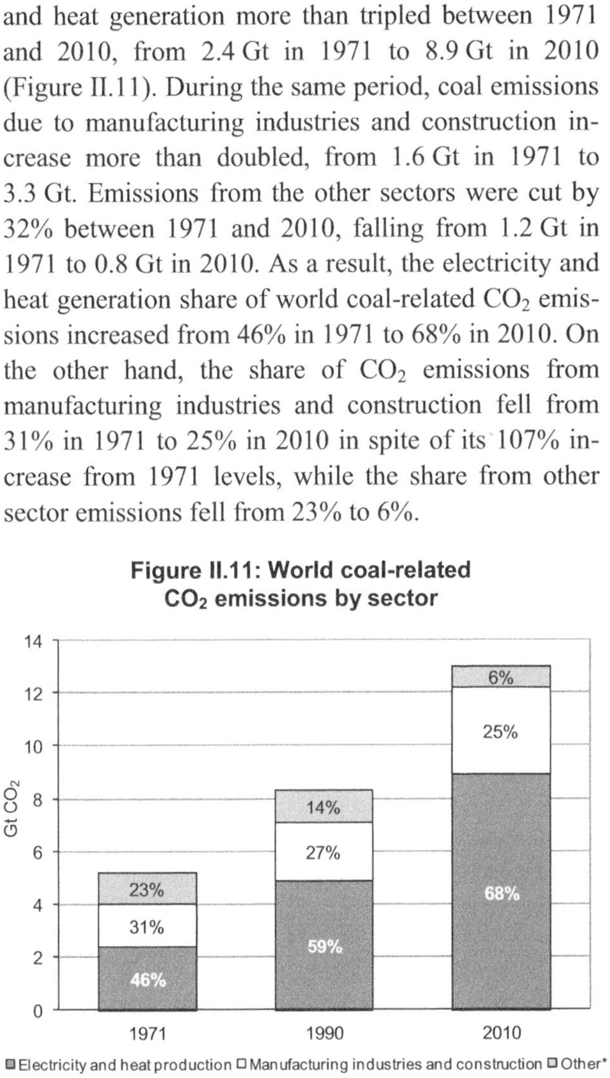

* Other includes residential, transport, commercial/public services, agriculture, other energy industries and non-specified other sectors.

Source: *IEA/OECD CO_2 Emissions from Fuel Combustion*

World CO_2 emissions have also been influenced by changes in international energy prices. Oil had been the largest contributor to world emissions until 2005, even though the upward trend was interrupted by the oil shocks of 1974 and 1980. Even after the decline in oil prices after 1985, the growth in emissions was partially mitigated as consuming countries adopted more energy-efficient technologies and behaviour, and some fuel switching occurred. In fact, as OECD countries started diversifying their energy consumption after the two oil shocks, their share of oil contributing to CO_2 emissions gradually declined from a peak of 53.4% in 1978 to under 45% in 1985. Since then oil's contribution has largely continued to lessen in relative terms, despite 2010 OECD road transport emissions having increased in real terms to 3.0 Gt, or by 27% since 1990 and 108% since 1971.

Oil, however, remains the main contributor to OECD emissions and it generated 41.1% of the CO_2 emissions in 2010, down from 42.3% in 2009, as coal's share of emissions rose to 33.6% from 33.0%. Natural gas increased its share of emissions from 24.0% to 24.5%; and both consumption of and emissions from the combustion of non-renewable waste continued to increase (Figure II.12).

Figure II.12: OECD CO_2 emission shares by fuel

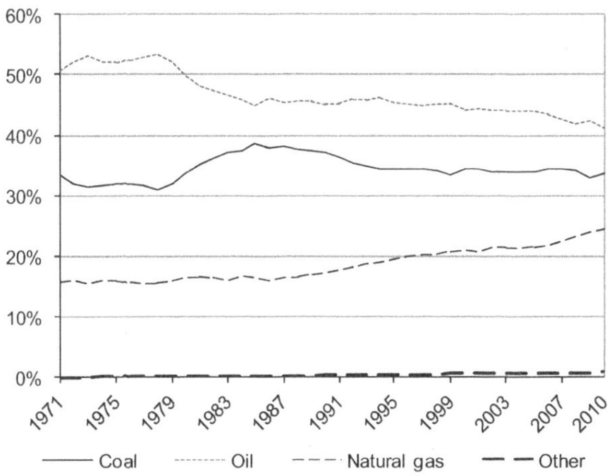

Source: *IEA/OECD CO_2 Emissions from Fuel Combustion*

Within the OECD, coal is the second largest source of CO_2 emissions. Between 1971 and 1985, coal was used as the main substitute for oil and its share of emissions increased, especially between 1978 and 1985, from its lowest point (31%) to its highest point (39%). After 1985, the share of emissions from coal declined consistently and after 1994, coal's share of emissions remained between 33% and 35% of toal OECD emissions.

Sectoral and reference approaches

The *CO_2 Emissions from Fuel Combustion* publication is produced in October each year. The 2010 data shown in the previous section were obtained from the 2012 edition. As mentioned in Part I, statistics of non-OECD countries presented in this publication are based on available data at the time of publishing so may differ from the final non-OECD data to be published in *Energy Statistics of Non-OECD Countries*. Hence non-OECD data for 2011 which will be released in the 2013 edition may be subject to change between what is shown here and what will finally appear in the CO_2 publication. However, bearing this caveat in mind, it is possible to look at global data for 2011 as it currently stands.

Data in the previous section were obtained using a bottom-up, sectoral approach which can take advantage of activity specific information to make more precise calculations. It is however also possible to calculate emissions based upon trade and production, and do so not only with 2011 data, but also with preliminary data for 2012. More information concerning distinctions between the two approaches is available in the free, *CO_2 Emissions from Fuel Combustion* highlights pdf downloadable from *www.iea.org*.

Figure II.13: Coal CO_2 emissions - reference vs. sectoral approaches

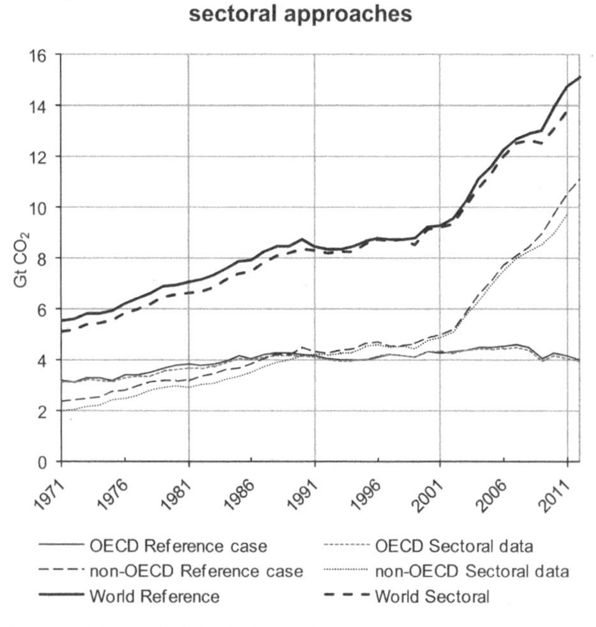

Source: *IEA/OECD CO_2 Emissions from Fuel Combustion*

As a general rule, the reference case does return higher values, however the year-on-year increases are a good guide as to what has been occurring for the last two years. OECD coal emissions decreased in both 2011 and 2012 to be the lowest since 1984 (Figure II.13). Substantial increases in coal consumption in the United Kingdom, Spain and Japan, could not offset decreases in consumption by the United States, where coal-based emissions apparently fell by roughly 100 Mt in 2011 and over 200 Mt in 2012.

This is in stong contrast to what is occurring to coal emissions in non-OECD countries, where emissions seemingly have increased by over 1.3 Gt in this two year period. The noticeable divergence, particularly in recent years between the reference and sectoral data are mostly due to statistical differences provided in the energy balances by the People's Republic of China, where divergence between domestic supply and reported consumption amounted to 0.4 Gt between 2009 and 2011 alone. Larger than reported stock builds may also result higher emission calculations than what actually occurred.

Atmospheric CO_2 concentrations

There is in the order of 900 Gt more CO_2 currently in the atmosphere than there was in preindustrial times, and this is currently increasing by around a further 15 Gt every year. The reasons behind this are both diverse and complex, but the key factor is the rate of anthropogenic CO_2 emissions outweighing the capacity of carbon sinks to redress the evident and still growing equilibrium peturbation.

There are real concerns that the major sink, the ocean's solubility pump will slow with warming and as concentrations of carbonic acid increase. The ocean's uptake of carbon dioxide loses efficacy at warmer temperatures, which enables the cooler deep ocean to draw down and absorb more carbon, driving the pump. However, rapid warming may cause the oceans to stratify thermally, which will not only strongly impact on their ability to absorb extra CO_2 from the atmosphere, but also impair their heat absorbtion capacity, causing the climate to warm at a quicker rate than otherwise expected. Ocean acidification is also of concern, as is the increased probability of increased outgassing activity.

The chart below shows several series of CO_2 metrics, displayed in terms of atmospheric parts per million (volume) (Figure II.14). The fossil-fuel emissions data are obtained from the *CO_2 Emissions from Fuel Combustion* 2012 edition, while other emissions include CO_2 from other anthropogenic activities which categorically do not have an immediate replenishment mechanism. Both are zero based, so display as they are. Other emissions + short cycle emissions also include CO_2 from anthropogenic activities where the regeneration (biomass recovery) is expected to be complete or near complete in the very short term. These short cycle data are added on top of the other emissions, so are only slightly larger than the other emissions, as opposed to more than twice the size. The short-cycle data are not expected to contribute to increases in CO_2.on a medium-term basis. Both datasets were obtained from the European Commission's, Joint Research Centre's, Emission Database for Global Atmospheric Research (EDGAR) database.

Some examples of non short-cycle anthropogenic CO_2 other emissions shown here include:

- Some fugitive emissions, deliberately or unintentionally omitted; and
- Some emissions produced by industry from non-fuel sources, such as limestone; and
- Basic omissions from national statistics; and
- Changes originating from forest fires, and decomposition of forests after fires; and
- Emissions from drainage of peat bogs and other land use change.

Finally there are the data showing the increase in global annual mean concentrations. These data fluctuate annually with strong correlation to El Nino and La Nina cycles, but have an underlying trend in recent decades as evidenced by the 10 year average of this series. Both series are collected by the United States', National Oceanic and Atmospheric Adminstration's (NOAA), Earth System Reserach Laboratory (ESRL).

Atmospheric concentrations of CO_2, have increased each year on an annual basis since measurements began. To achieve the stabilisation of atmospheric concentrations at safe levels, the net carbon flux needs to begin to approach or even fall below zero. There currently is an undeniable increasing trend alongside the increase in anthropogenic emissions, where the annual 10 year moving average increased from 0.82 ppmv/y in the 1960s to over 2 ppmv/y since 2001.

It is also important to note that this upward trend is occurring despite a statistically significant and possibly temporary increase in uptake of terrestrial carbon over the last two decades. The prospect of potentially decreasing efficacy of sinks in addition to release of significant quantities of long-life greenhouse gases through natural feedback cycles are sobering uncertainties that need to be considered when planning the scale of any effective emissions mitigation program.

Figure II.14: Atmospheric CO_2 concentration, rate of annual increase.

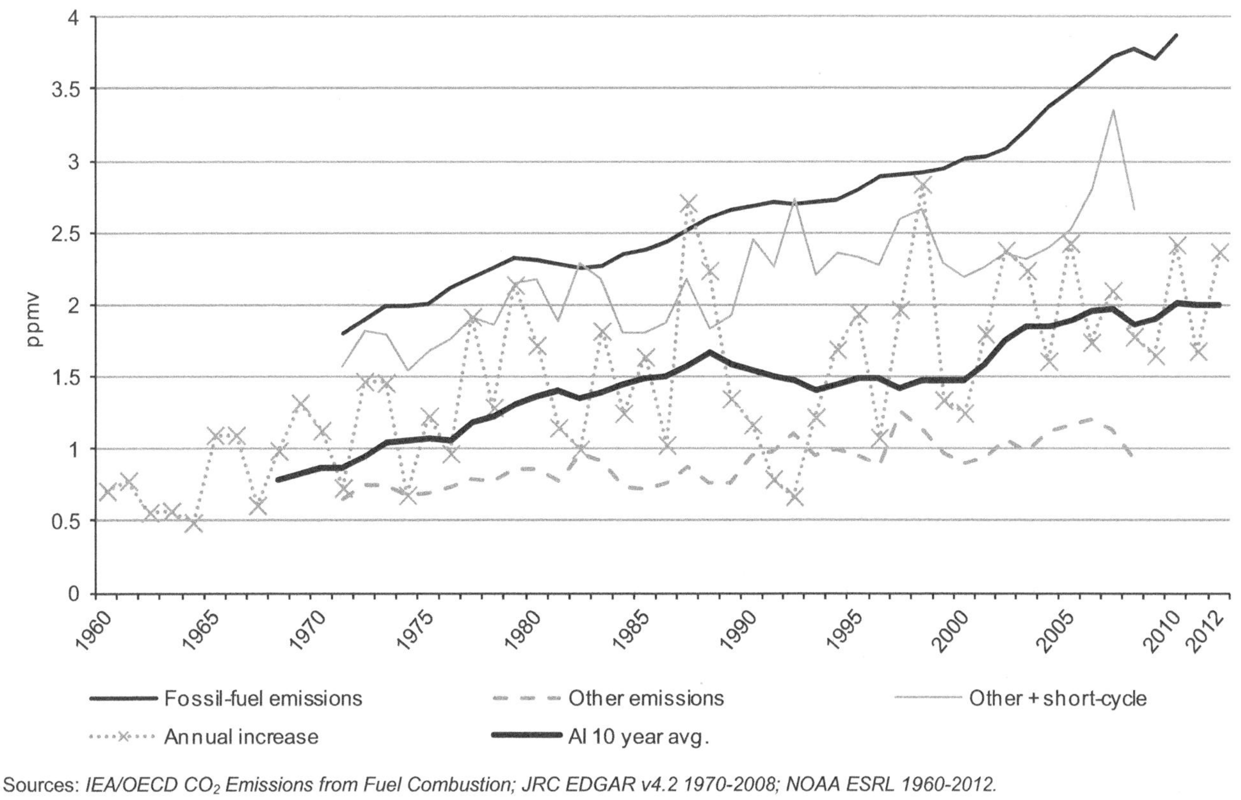

Sources: *IEA/OECD CO_2 Emissions from Fuel Combustion; JRC EDGAR v4.2 1970-2008; NOAA ESRL 1960-2012.*

3. FLEXIBILITY OF COAL-FUELLED POWER PLANTS FOR DYNAMIC OPERATION AND GRID STABILITY

The following feature article is provided to acquaint the reader with some of the ongoing analytical work at the IEA, including that of the Coal Industry Advisory Board (CIAB), established by the IEA in 1979.

The CIAB's *21st Century Coal* is a publication in the International Energy Agency's Insights series. It is a summary of *Advanced Technology and Global Energy Solutions* from a coal-based perspective and this article is a condensed excerpt from chapter 4. Some portions have been edited to add context that was not immediately apparent in the excerpt, but was provided elsewhere in the report. The views expressed in this paper do not necessarily reflect the views or policy of the International Energy Agency (IEA) Secretariat or of its individual member countries. This paper is a work in progress and/or is produced in parallel with or contributing to other IEA work or formal publication. Comments are welcome, directed to carlos.fernandezalvarez@iea.org.

Introduction

In 2010, renewable energy accounted for 4207 TWh of the world's electricity generation, or 19.7% of the total. The majority (82%) of this low-carbon electricity came from hydro, but the amount of both renewable energy and non-hydro renewables is projected to grow significantly as civilization begins to address its current emissions conundrum.

In the IEA's, *World Energy Outlook* 2012, middle-ground New Policy Scenario (NPS); from 2010 to 2035, 47% of the increase in electricity generation comes from renewables as shown in Figure II.15. Mainly driven by government policies, generation from non-hydro renewables is projected to increase from 4% of the total in 2010 to over 15% in 2035. It should also be noted that the measures in the NPS are incompatible with an emissions trajectory which successfully limits anthropogenic warming to 2°C.

The increasing contribution of renewables to meeting rising electricity demand has important implications on other generating units on the electricity grid – both existing and future ones – because of the variable nature or intermittency of the output of several renewables-based technologies, such as wind and solar. The increasing penetration of intermittent renewable generation generally requires non-renewable units to operate more flexibly to ensure grid stability. Grid instabilities can cause blackouts (a worst-case scenario of this happened in India in late July 2012, when over 600 million people went without power for several days) and incur significant repair cost.

The ability of coal power plants to be more flexible in response to the changing electricity market is therefore a key consideration. While most coal power plants are capable of some dynamic operation and are designed to be able to cycle with moderate ramp rates and potentially even handle two-shift operation (where the plant is started up and shutdown daily), the increased need for flexibility will impact costs, maintenance, and reliability. Most notably, higher cycling will increase wear and tear while the number of operating hours decreases, resulting in an increase of specific maintenance costs per MWh over time. Moreover, as coal power plants add more complex environmental control systems such as CCS in the future, their ability to operate dynamically may be reduced (IEACCC, 2011).

Figure II.15: Incremental global renewables-based electricity generation relative to 2010

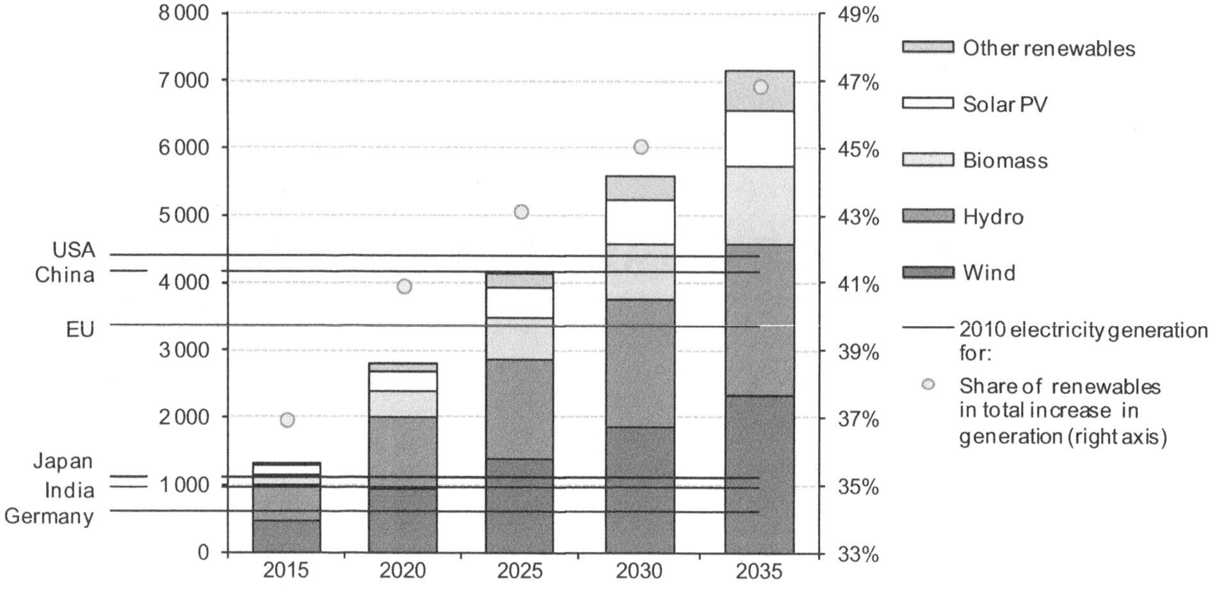

Cycling of coal-fuelled plant

Coal assets worldwide will remain a critical element of the generation portfolio mix for decades. Reliable and dispatchable generation will be required to balance increased deployment of must-take or preferential variable renewable energy resources.

Traditional coal plant design focused on economical base-load operation, which has led in some markets to an evolution in unit size and main steam conditions that challenge the ability of many existing plants to operate flexibly in the future. With demands to extend operating life of these current assets to beyond forty years, affordable strategies must be found for improving the cycling capability of existing coal plants. Lessons learned from these efforts to operate current plants more flexibly, together with advances in materials and control schemes, can form a basis for improved design of future plants with greater focus on meeting a load balancing role.

Challenges for coal assets with increased renewables penetration

As more renewable energy sources come on-line, particularly intermittent ones, the role of coal- and gas-fuelled generation will likely need to change in response. A reverse correlation exists between the daily and monthly trends in renewable generation and the demand trends. Morning and evening demand peaks do not generally align with peak output of renewable energy resources. Optimal load balancing using coal and gas will likely evolve with coal supplying an increasing level of spinning reserve and load following, and gas supplying power during peak demand. Regional coal-gas price dynamics will influence this trend in a particular market, for example more gas plants supplying bulk power in essentially base-load operation. Studies indicate that spinning reserve requirements will increase with increasing renewable penetration, which could be primarily supplied by coal (GE Energy, 2010).

Even with continued uncertainty in fuel prices and regulatory constraints, it is evident that many coal assets must improve performance in the following key areas: 1) increased turndown, 2) faster and less damaging startups, 3) faster load changes, and 4) reserve shutdown at minimal cost. These desired operational attributes can conflict with the traditional design objectives of existing optimised coal plants, which favored high thermal efficiency base-load operation. Operational constraints on NO_X and SO_X control equipment, depending on how they are integrated into the plant, can also result in reduced flexibility of coal assets.

In regions like North America, much of the flexibility of the existing coal fleet is supplied by relatively small, subcritical units operating without environmental control equipment. With the long-term future of these units in question due to their lower efficiency, and capital spending required for emissions compliance, the emerging challenge to coal in these markets will be to attain flexibility with the remaining larger units.

Existing power plants currently in Germany are all designed to cater to flexible operation and these requirements are equally met by new natural gas combined cycle (NGCC) plants and new coal-fuelled power plants. Many of the conventional power plants available in Germany today were built in the 1980s and 1990s before expansion targets for wind and photovoltaic plants had been adopted. In many plants, measures to allow greater flexibility have been implemented subsequently, so that power plants can meet increased requirements for market load adjustments. As a result, there are very few dedicated base-load power plants in Germany that do not allow for flexible operation.

Example: High fluctuation of wind power feed-in in January 2012 – how flexible operation of coal- and gas-fuelled power plants was used to balance dynamic grid demand.

On Sunday, 1 January 2012, power demand was relatively low due to low industrial demand and mild temperatures of approximately 8°C (46°F). Around the evening peak, a temporary daily maximum consumption of 56 GW was reached in the German power grid, after which demand decreased to a minimum value of less than 41 GW until late at night. At the same time, the amount of wind feed-in reached a level of more than 16 GW. Further feed-ins that day came from other renewable energy sources, including run-of-the-river hydro and biomass power plants, which also benefit from feed-in priority. The feed-in from those plants continuously amounted to about 5 GW. The power generation from photovoltaic plants was negligible due to the season as well as the cloudy weather conditions over that weekend.

On Sunday night, after renewable energy feed-ins, only a residual load of 21 GW had to be temporarily covered by other power plants available according to schedule.

At 4 a.m. on Monday, power consumption markedly increased and reached a demand level of approximately 73 GW at around noon. This corresponds to an increase of 32 GW within 8 hours. At the same time the feed-in from wind power plants decreased in the early hours of the morning due to declining wind speeds and amounted to 4 GW around noon, which helped contribute to a decrease in feed-in of about 12 GW. Thus, overall, an additional power output of nearly 45 GW had to be provided by the conventional power plant portfolio within those 8 hours.

Power generation from German nuclear power plants contributed almost without interruption a supply of about 12 GW. There is a degree of flexibility available from the German nuclear power stations, although their low variable power generation costs ensures that this is only used once the load adaptability of the fossil-fuelled power plants is exhausted. As seen in Figure II.16, the necessary load adjustment of about 50 GW on Monday morning was almost completely provided by the coal- and gas-fuelled power plants.

Figure II.16: Power consumption and deployment of power plants in Germany on 1st and 2nd January 2012

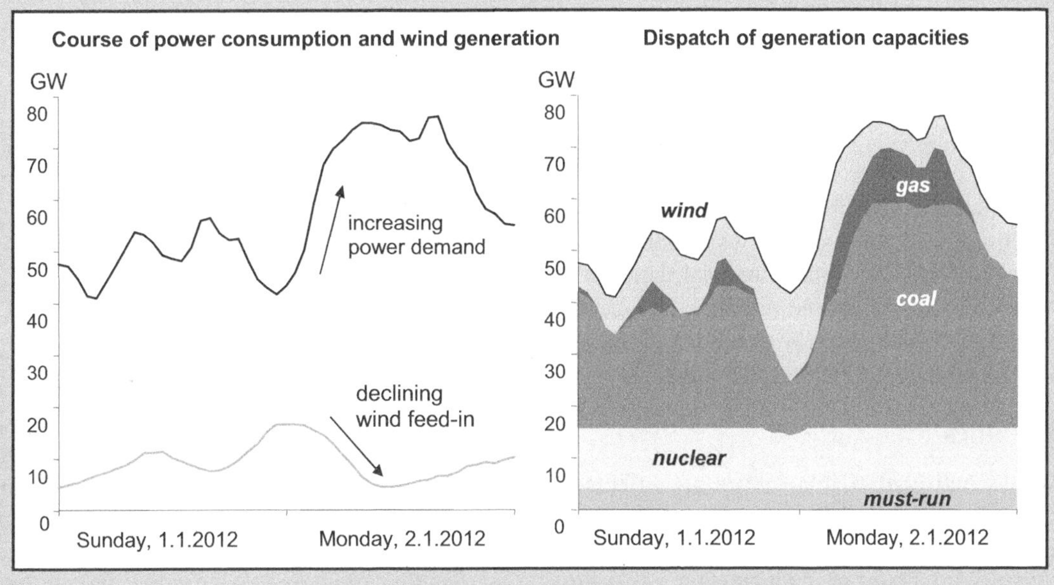

Figure II.16 displays on the left-hand side the parallel development of power consumption and intermittent wind feed-in, requiring a high degree of load adjustment from the conventional power generation portfolio.

Cycling impacts on existing coal assets

Increased rates of component life consumption due to thermal fatigue, mechanical fatigue, and wear caused by differential thermal expansion comprise the bulk of cycling impacts on thermal plants. Ultimately, these will manifest themselves on the levelised cost of electricity of the system. Although estimating the cost addition is difficult as the impacts are cumulative and complex, the following sections summarise the cycling impacts on major plant systems.

Boilers and Piping

Boiler thermal fatigue in both drum and supercritical units primarily affects the superheat, reheat, and economiser sections. In each case, the source of fatigue is temperature mismatch between the fluid and the pressure boundary wall. Supercritical boiler waterwalls are also subjected to thermal fatigue due to temperature stratification. In addition to thermal cycles associated with starts and load following, header fatigue cracking can be caused by repeated quenching of header internal surfaces due to inadequate condensate draining (EPRI, 2010). The extent of thermal fatigue in cycling units can be reduced through increased efforts to control steam temperature, attemperator sprays, and condensate draining.

Corrosion fatigue damage, which can occur on riser tubes in drum units (EPRI, 2012) and waterwalls (EPRI, 2009), increases the operational challenges associated with maintaining boiler cycle chemistry under cycling operation. Increased levels of dissolved oxygen in feedwater during cycling can be the result of condenser leaks, aggravated by more frequent shutdowns. Other factors affecting cycle chemistry include increased need for make-up water and interruption in operation of the condensate polishers and deaerators (EPRI, 2009a). These operational factors can be controlled or eliminated with increased attention to plant material condition and staff expertise.

Steam Turbines

Steam turbines and control/stop valves are affected by a number of damage mechanisms due to flexible operation (EPRI, 2004). The most costly damage is increased thermal fatigue of thick-section pressure boundaries (casings and valve bodies), as well as high-pressure turbine (HPT) and intermediate-pressure turbine (IPT) rotor forgings. Valve bodies tend to cool more rapidly during shutdowns, resulting in more warm-starts and cold-starts as opposed to less-damaging hot-starts. Thermal fatigue cracking of casings is common, even prior to undergoing increased cycling operation. The risk associated with increased cycling is the potential for more rapid fatigue crack propagation. Spare casings difficult to procure, so casing life consumption must be managed through a program of inspections and weld repair.

HPT and IPT rotor forgings under cycling operation are subjected to thermal and mechanical low-cycle fatigue. Allowing HPTs and IPTs to retain as much heat as possible during brief shutdowns will reduce low-cycle fatigue damage due to frequent starts. Plants with rotors that have already experienced 20-30 years of operation may have bore defects that require monitoring more frequently using ultrasonics.

Cycling impacts on the low-pressure turbine (LPT) include greater risk of corrosion-related crack formation, and risk of blade vibration and fatigue associated with reduced minimum load operation. A reduction in steam reheat temperature associated with low-load operation results in an upstream shift in the steam phase transition zone in the LPT. The associated deposition and concentration of corrosive salts over a larger extent of the steam path can contribute to increased risk of disk rim stress corrosion cracking and corrosion fatigue of blading. Greater attention to maintaining steam purity under dynamic operation and more frequent LPT steam path inspections will be required to manage risk in cycling operation.

Electrical Generators

The primary cycling-related damage mechanism in generators is associated with repeated differential thermal expansion of the copper-insulation system, and the surrounding rotor steel and/or core iron (EPRI, 2004). The three materials involved (copper, insulation, iron/steel) each have different coefficients of thermal expansion, resulting in strain at the interface of bonded connections, or rubbing at the interface of tight sliding connections. In both cases, accelerated mechanical wear of the insulation material takes place as a result of even small temperature variations if occurring continuously. The repeated relative motion at these material interfaces can reduce the design life of the windings from a nominal twenty years under base-load operation. Periodic monitoring of winding condition is possible and is the subject of ongoing research (EPRI, 2006, 2006a). Consequences of undetected insulation damage include winding short circuits and ground faults.

Generator cores consist of a compressed laminate of iron plates separated by thin dielectric material. Continuous mechanical movements of the core assembly associated with frequent starts, and generator load (torque) changes, will accelerate dielectric wear and reduce the required time interval between core restacking.

Selective Catalytic Reduction Systems

Operational parameters that are critical to control of SCR operation are: 1) maintaining flue gas temperature above a minimum required for the catalyst to function, and 2) achieving optimum proportions and mixing of ammonia and flue gas.

SO_3 and H_2SO_4 present in the flue gas can combine with the injected ammonia to form ammonium bisulphate (ABS), $(NH_4)HSO_4$, on the catalyst surface. This occurs primarily when the flue gas temperature drops below the ABS formation temperature of approximately 285–305°C (545–580°F) (EPRI, 2010b). Liquid ABS fouls the catalyst pores and reduces its effectiveness in NO_X removal. From an operational perspective it is imperative that economiser exit temperature remains above the ABS formation temperature for that unit. The impact of these temperature restrictions for units without economiser bypass is to limit minimum load level for long-term operation to a range of approximately 40–60% of unit maximum continuous rating. Large fleets of units with NO_X control equipment will therefore have limitations on turndown due to the need to avoid costly ABS fouling of the catalyst. Ongoing research is exploring ways to incrementally reduce the ABS formation temperature on a unit-specific basis, thus allowing greater unit turndown (EPRI, 2011b).

Flue Gas Desulphurisation Systems

Wet FGD systems involve complex chemical reactions in which control of the reagent process flow is critical to achieving SO_X removal. Typically the process is tuned to designed load and fuel characteristics, so unit cycling and load-following necessarily requires more advanced process control functionality. Cycling impacts FGD systems in two key areas: water balance and oxidation-reduction potential.

Prolonged low-load operation affects the overall water balance by lowering the amount of evaporative losses. This reduces the fresh water available for the mist eliminator and filter cake washing. If the makeup water is not adjusted at low load, it can result in lower recycle solids, potentially changing gypsum crystal formation, resulting in lower-quality gypsum.

A more serious cycling impact is the resulting change in the oxygen/sulphur ratio and resultant oxidation air requirements. The consequence of variations in oxygen/sulphur ratio in units not designed to adjust oxidation air with load is to change the process chemistry by increasing oxidation of other species such as mercury, selenium, or manganese and the formation of higher oxidation states of sulphur. This effect is indicated by a high oxidation-reduction potential (ORP) measured in the scrubber liquid. High ORP operation keeps more mercury in solution phase and increases potential to re-emit mercury. High ORP also oxidises selenite to selenate which is more difficult to remove in downstream wastewater treatment systems. Another detrimental effect of high ORP operation is the formation of MnO_2 scale and potential crevice corrosion attack of the FGD vessel. The formation of higher oxidised states of sulphur compounds make it more difficult to reduce the ORP level when the unit increases load, continuing these other detrimental effects discussed.

Overall Plant Operational Impacts

Although the boiler and turbine-generator experience the greatest component damage in cycling operation, other operational impacts deserve mention. Some examples are highlighted below:

Cycling units require increased attention to human factors issues that affect operators. Alarm management, human-machine interface, and documented procedures need to be in place. New distributed control systems provide the capability to significantly improve human-machine interfaces and reduce cycling impacts.

Uncertainty in the future dispatch scheduling may result in coal bunkers being filled at the start of an economic reserve shutdown. For some fuels with high volatility, this represents a significant fire hazard, resulting in the need to empty the bunker (an example of this occurred at Xcel Energy's Black Dog Plant [Nelson & Dunbar, 2010]).

Two-phase flow-assisted corrosion (FAC) is a potentially dangerous condition in which pipes and vessels experience wall thinning over time. Prolonged part-load operation can cause FAC to appear in areas of the plant that have not previously been affected.

Increased maintenance required on boiler ignitors, precipitator fouling due to increased oil firing, increased condensation/corrosion in air pre-heaters, and reliability issues with ash-handling systems caused by intermittent operation are also key considerations.

Future plant design considerations

Designers will continue to face the requirement to balance high thermal efficiency with improved flexibility in future coal-fuelled plants. It is essential that operational flexibility be considered early in the design phase so that material selection and features such as bypass systems can be incorporated into the plant layout.

Lessons-learned from the existing fleet, combined with new research on advanced materials (Viswanathan, *et al.*, 2008), leads to the following list of recommended considerations for achieving future plant flexibility:

- Use of advanced creep strength enhanced ferritic steels such as P92, and new nickel alloys such as Inconel 740 and Haynes 282. These materials for piping, headers, and casing components provide good creep-life characteristics at the higher temperatures. Additional advantages include reduced wall thickness and a reduction in the level of thermal stresses under cycling operation.
- Design boiler headers, piping systems, valve casing, and turbine rotors/shells with both creep life and thermal cycling as a requirement. Existing commercial analysis tools can assist in this process (Payten, Bendiech & Snowden, 2009).
- Consideration of a broader range of unit load when designing the furnace, burner systems, and controls and seeking to attain a lower minimum load below which oil firing is necessary for stable furnace operation.
- Installation of a main steam turbine bypass to improve temperature matching during startups
- Installation of an economiser bypass to extend the load range over which the SCR system will operate without catalyst fouling.
- Incorporation of partial-arc admission for the steam turbine to improve part-load efficiency
- Improve measures to retain heat in the boiler-turbine systems during brief shutdowns, particularly in the turbine valve bodies and steam chests which typically lose heat faster.
- Consideration of long-term cost advantages of employing variable speed motors as prime movers for large pumps and fans.
- Installation of adequate condensate drains in all boiler header and piping systems, with instrumentation to detect presence of condensate. Enable remote actuation of these drains in the control room.
- Incorporate the latest advanced human-machine interface technology in the control room. Operator situational awareness is critical in cycling operation and is greatly influenced by screen layouts, trending, and strict adherence to alarm management standards.

Cost-effective cycling impact mitigation strategies

Many plants are successfully instituting cycling impact mitigation efforts by initially employing low-cost operational and controls strategies rather than costly whole-component upgrades. One European power producer has used a "toolbox" approach to reduce cycling impacts, which includes the following elements (Powermag 2011):

- Flexible operation studies on those key components that either limit flexibility or accrue damage during cycling. This includes additional instrumentation to monitor key thermal transients followed by repeated operational trials using this new data to reduce startup times and equipment damage.
- Identification and elimination of any hot surface quenching caused by inadequate drain capacity/control as well as attemperator spray control
- Continuous operator coaching using simple component damage algorithms to instill a focus on reducing cycling related equipment impacts. One current example of commercial cycling damage assessment tools is REMLIFE, now in use by Australian power plants (Payten, Bendiech & Snowden, 2009).
- Institute a more aggressive equipment inspection regime to seek and proactively address new component degradation modes. Plants need to adopt a formal maintenance basis, and continuously adapt this basis with input from observed component degradation.
- Where design changes are cost-justified, seek to "design-out" the failure modes rather than simply replace in-kind.

Improved controls strategies have emerged as perhaps the most effective overall approach to improving coal plant flexibility. The capability of new distributed control systems to optimise over wider operating ranges is generally under-utilised; however, some successful case studies have been documented recently (Powermag, 2011a).

Protection of the boiler-turbine water circuit during layup periods of indeterminate duration has emerged as a major challenge to the industry. An effective new strategy to provide corrosion protection in these situations is the use of filming amines in the feedwater, which continuously coat the wetted surfaces during operation and provide a barrier against oxidation when exposed to air during shutdown. This new strategy has the potential to significantly reduce the cost and protection effectiveness over traditional layup practices.

Plant efficiency considerations under cycling operations

To reduce throttling losses in the turbine at part-load operation, partial-admission control stage designs should be incorporated. The efficiency gains at part-load can more than offset the increased losses at design-point operation.

Recent studies have outlined cost-effective strategies to incrementally reduce the roughly 10-20% part load heat-rate penalty (relative to full load). These strategies include the implementation of sliding pressure operation, in which the turbine inlet pressure is set using feedpump controls rather than through throttling of the turbine control valves. Sliding pressure has been accepted by many plants as the most cost-beneficial approach to reducing the heat rate penalty at low load. Additional actions to improve performance of cycling units include (EPRI, 2010a):

- Boiler combustion and air quality control system tuning over a range of unit conditions rather than just at design operating point (referred to as gain scheduling)
- Replacement of fixed-speed motors with variable-speed drives.
- Use of heat rate monitoring systems that trend process parameters and apply algorithms to calculate recoverable losses in real-time.
- Increased attention to improving cycle isolation, which improves heat rate across the entire load range
- Improved schemes for operation of condenser cooling system (pumps, fans) at part load operation to reduce auxiliary loads and avoid wasteful condensate sub-cooling
- Improved schemes for operation of boiler draft fan systems during startup and part load operation to reduce auxiliary loads.

A new area of research on cycling impact is focusing on the effect of continuous load following on heat rate. Units that are never stabilised at constant load experience thermal hysteresis effects on even minor load swings that result in additional heat rate penalties due simply to the non-steady operation. Controlled testing has been conducted on commercial coal plants (EPRI, 2011).

CCS impacts on plant flexibility

Much of the inherent flexibility of CCS plants will not be fully understood until such time as full-scale capture plants are demonstrated and new operators begin to become familiar with the potential bottlenecks and restrictions associated with startup, shutdown and varying plant loads. Currently, modeling activities are the tools to best understand the potential limitations of these evolving technologies.

Several issues could limit the flexibility of advanced coal plants with CCS. In general, increased dynamic operation implies lower capacity factors, and with the capital intensity of advanced coal plants with CCS, lower capacity factors could pose a large cost barrier to running at lower loads. If the plant is capturing CO_2 for enhanced oil recovery (EOR), disruption or change in the supply of CO_2 might be detrimental for oil-field production, particularly if there is an existing contract for CO_2. Such plants may be required to be base-loaded as a result at least while EOR is occurring. The low-carbon nature of CCS based applications, however, will also make base-load operating schedules more acceptable.

Oxy-combustion flexibility

Essential to oxy-combustion startup is the requirement for its air separation unit (ASU) to be up and running for nearly 48 hours in order to achieve acceptable oxygen purity from ambient conditions. This long startup time for the ASU is the rate limiting step for the entire oxy-combustion system startup.

Once the ASU is close to achieving acceptable oxygen purity, the boiler startup can be initiated, which is essentially the same as for air firing. At about 40% load on air, the transition to oxygen firing and flue gas recycling will begin.

A full-scale oxy-combustion plant is envisaged to operate similarly to a conventional PC plant. The transition from the air firing to the oxy-fuel firing during startup and shutdown has been shown (at pilot scale) to be relatively simple and automatic. Load changing has been found to be essentially the same as for air firing with the addition of oxygen control.

With oxy-combustion there is no steam (heat source) required to be diverted from the plant's steam turbines. This has implications on its options in terms of flexibility.

One element of oxy-combustion flexibility that could be considered to reduce costs would be the storing of its oxygen. In this scenario, the ASU could come down in load and reduce the overall parasitic power loss of the oxy-combustion plant for peak periods yet still maintain the required oxygen flow to the boiler. Such flexibility would require pre-investments on an appropriate oxygen storage tank, but may prove cost effective in a competitive electricity bidding market. The disadvantage would be that it also requires some element of cycling of the ASU equipment, which may be deleterious for existing ASU systems that are historically not designed to cycle.

Likewise if the CO_2 compressors were to be turned off on the oxy-combustion plant and CO_2 vented for a period, this might allow an opportunity where the plant could fire at full MW capacity with a considerably lower parasitic load (albeit, venting its CO_2 for this short period).

Post-combustion capture flexibility

Unlike their oxy-combustion counterparts, amine capture systems require large amounts of heat to regenerate the solvent used in the CO_2 removal process (as opposed to large amounts of MW to power an ASU for oxygen supply in the oxy-combustion case). The heat source is commonly supplied from the steam generated in the PC plant for both new-build and retrofit PCC plants. The steam is required to be at a certain condition as dictated by the capture solvent. The crossover steam between the IPT and LPT is the common place for extraction.

Obtaining appropriate source of heat for post-combustion-capture systems

Studies on advanced amine PCC with commercial systems show that to achieve 90% CO_2 capture for an entire plant, significant amounts (40-50%) of the steam produced is required to be extracted from the steam turbine for solvent regeneration (EPRI, 2011a). This significant extraction has implications on the IPTs and LPTs in both normal operation and load following scenarios.

An approach intended to avoid major modifications to the IPT design is to add a pressure control valve downstream of the crossover extraction location as shown in Figure II.17 (EPRI, 2011a). This valve will maintain IPT exhaust pressure at its normal value, and thus the IPT is unaffected. Controls will be implemented such that this valve modulates IPT exhaust pressure to the value appropriate to the loading of the IPT.

Understanding the potentiale effects/limits on plant turndown operation

Extracting the large amount of low-pressure steam required for the PCC plant will cause a substantial reduction in LPT power output. Since the IPT exhaust pressure is maintained by the pressure control valve, the steam pressure available to the PCC plant will be much higher than required. A let-down back-pressure turbine (BPT) generator can be installed upstream of the PCC plant to recover some of the power. This turbine accepts steam from the crossover line and exhausts at the pressure required for the PCC plant. Because the IPT exhaust pressure, which is also roughly the PCC extraction pressure, decreases with turndown, eventually the difference between the PCC extraction pressure and the required PCC reboil er pressure will be too small to generate power efficiently from the BPT. Hence, a bypass line with shut- off valve is required around the BPT to allow the plant to operate with PCC during low turndown rate.Therefore two major criteria that limit PCC turndown are: Firstly maintaining the minimum PCC extraction pressure for solvent regeneration. This is usually around 4.14 bar (75 psig) for 30% monoethanolamine (MEA) systems, but different for other solvents systems; and also maintaining the steam flow through the LPTs for stable turbine operation, which should not be less than the pre-PCC design flow.

Figure II.17: Turbine and feedwater heater circuit for a pulverised coal plant with CO_2 capture added

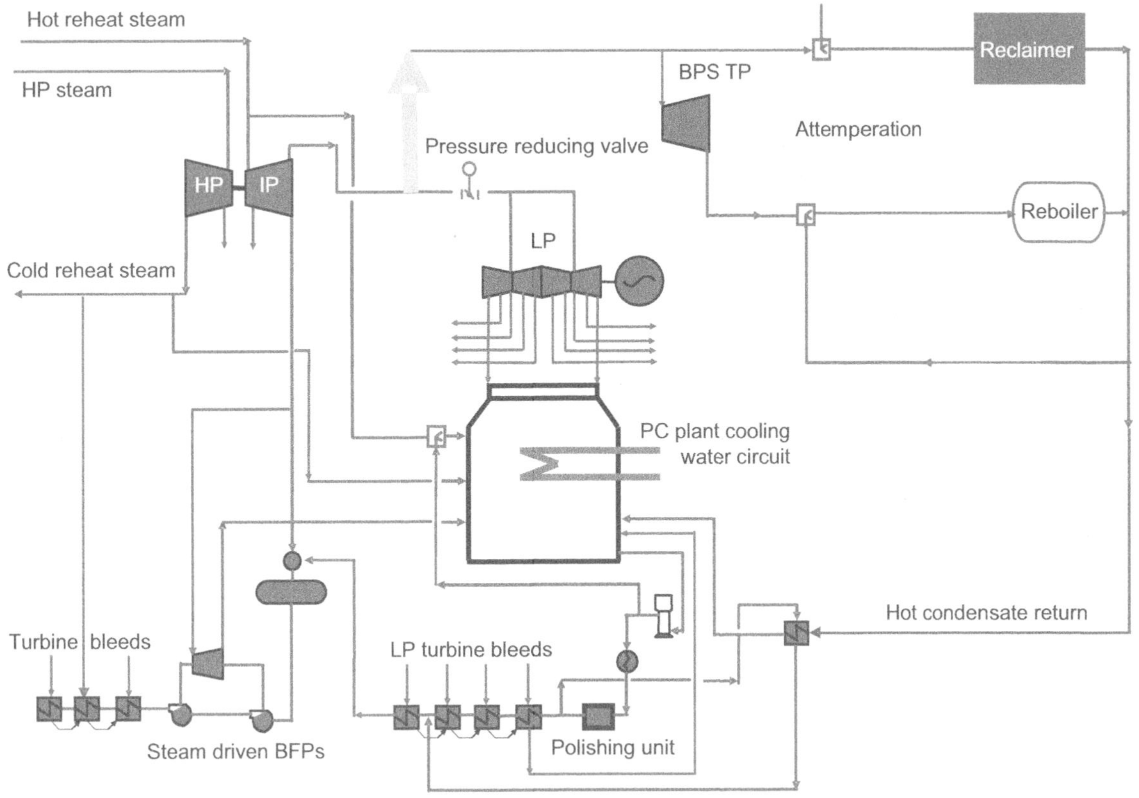

Flexibility of IGCC plants with capture

In general, integrated gasification combined cycle (IGCC) plants have been designed to operate at base-load and do not have a great deal of experience in cycling. The plants tend to be highly integrated and complex, especially those designed for carbon capture, so turndown of the unit is slow, making load-following difficult. In general, they have longer startup times than a PC unit or NGCC, so single or two-shifting a unit becomes impractical.

However, several gasifiers do have experience operating in a load-following manner: The Nuon gasification project at Buggenum, the Netherlands, operated in load dispatch from 1999 to 2003. However, the plant reverted to base-load operation in 2004. The plant demonstrated that in an integrated mode it could vary load at better than 3% per minute. Also, Nippon Petroleum Refining's Negishi IGCC plant reportedly operates at 75% load at night and weekends, but can ramp up to full load over a 30-minute time period.

One advantage of IGCC over other technologies when it comes to cycling is that the technology has the ability to "poly-generate" or can make multiple products. During times of reduced electricity demand, the syngas produced by the IGCC could be used to make chemicals, which are typically stored in liquid form. This would allow the gasification portion of the process to operate at a constant load, while the combined cycle is operating at a reduced load (or even shutdown). Two IGCC projects currently in development in the U.S. plan to use this operating strategy (Summit Power's TCEP and SCS Energy's HECA).

Flexibility of advanced cycles

Chapter 2 of the report presented several future advanced cycles for coal-powered electricity generation, which are unlikely to be in commercial use until after 2020. Similarly to CCS, much of the detail will not be fully understood until full scale commercial plants are in operation, however, the status of the electricity grid and the impacts of intermittent resources by then will in all probability increase the need for flexible operation.

Advanced fuel cells

The dynamic control of fuel cells through startup, shutdown, emergency, and load-following has been identified as a critical issue primarily due to the complexity of the system (NETL, 2012). Designing fuel cells for flexible operation is in its infancy, but one study has identified the need of a DC/DC converter for a solid oxide fuel cell (SOFC) to be used for load following applications (IEEE, 2007).

Chemical looping combustion

CLC is a complex system that requires balancing of multiple sub-systems each of which may have different startup and shutdown sequences and times as well as different operating characteristics at part-load operation. One study showed that CLC had a promising efficiency at part-load operation down to 60% of full load; however, it required an advanced control strategy. A combination of control strategies is required for startup and shutdown and for part load when airflow reduction is not practically possible with the current generation of compressors (Energy, 2007).

Closed Brayton power cycles

Closed Brayton power cycles are likely to have excellent flexibility. Supercritical CO_2 closed Brayton cycles have smaller weight and volume, lower thermal mass, and less complex power blocks versus Rankine cycles due to the higher density of the fluid and simpler cycle design. The lower thermal mass makes startup and load change faster for frequent startup/shutdown operations and load adaption than a steam-based system (NREL, 2011).

Pressurised oxy-combustion

The flexibility of a pressurised oxy-combustion plant is not envisaged to be significantly different from an atmospheric oxy-combustion plant discussed earlier in this chapter.

Conclusion

The increasing amount of intermittent renewables means that the generation portfolio must be increasingly flexible to ensure grid adequacy with an acceptably high probability.

Coal assets must improve performance in the following key areas related to flexibility: 1) increased turndown, 2) faster and less damaging startups, 3) faster load changes, and 4) reserve shutdown at minimal cost. These can conflict to varying extent with the desire for high thermal efficiency base-load operation for coal. Operational constraints on emission control equipment also impact upon flexibility.

Currently much of the flexibility of the existing coal fleet is supplied by small, subcritical units operating without environmental control equipment. With the long-term future of these units in question due to their lower efficiency, and capital spending required for emissions compliance, the challenge to coal will be to attain flexibility with the remaining larger units. Improved controls strategies have emerged as perhaps the most effective overall approach to improving coal plant flexibility. For new coal plants, it is essential that operational flexibility be considered early in the design phase.

Advanced high-temperature materials and plant controls can be employed to improve flexibility. Conducting rigorous operational trials to optimise details of plant process changes during transient operation has also proven to be cost effective. The change in flexibility of advanced coal plants with the addition of CCS will not be fully understood until demonstrated and the resulting experience is accrued. However, it is not envisaged that adding CCS will substantially change the base plant's flexibility.

For advanced cycles such as CLC and fuel cells, the complexity of the system could impact its flexibility, requiring advanced control strategies. Pressurised oxy-combustion should behave similarly to atmospheric oxy-combustion. Closed Brayton cycles are likely to have excellent flexibility arising from their smaller size, lower thermal mass, and less complex power block.

Coal will continue to play an essential role in the global energy mix for decades to come. The ability of coal power plants – both existing and future – to be more flexible in response to the changing electricity market has been questioned largely due to the perception that coal plants are designed to be base-loaded units that have a limited ability to respond dynamically. As shown, while considerable challenges exist and significant investment is required, this loading constraint does not need to apply and there is rapidly mounting evidence that 21st century coal can contribute to our 21st century energy needs in both a meaningful and sustainable manner.

References

Energy (2007), "Part-load Analysis of a Chemical Looping Combustion (CLC) Combined Cycle with CO_2 Capture", retrieved from SciVerse - Science Direct: *www.sciencedirect.com/science/article/pii/S0360544206002003.*

EPRI (2004), *Effects of Flexible Operation on Turbines and Generators*, Palo Alto, CA.

EPRI. (2006), *Steam Turbines-Generators and Auxiliary Systems - Program 65*, Palo Alto, CA.

EPRI (2006a), *Main Generator Rotor Maintenance*, Palo Alto, CA.

EPRI (2009), *Circumferential Cracking in Supercritical Waterwalls*, Palo Alto, CA.

EPRI (2009a), *Cycling, Startup, Shutdown, and Layup Fossil Plant Cycle Chemistry Guidelines for Operators and Chemists*, Palo Alto, CA.

EPRI (2010), *Boiler Condition Assessment Guideline*, retrieved from Electric Power Research Institute: *http://176.9.24.71/book/1447300.*

EPRI (2010a), *Efficiency Improvement for Cycling Service*, Palo Alto, CA.

EPRI (2010b), *Summary of Selective Catalytic Reduction System Operational Issues at Low Load*, Palo Alto, CA.

EPRI (2011), *Cycling and Load-Following Effects on Heat Rate*, Palo Alto, CA.

EPRI (2011a), *Engineering and Economic Assessment of Post-Combustion CO2 Capture for 1100°F Ultra-Supercritical Pulverized Coal Power Plant Applications*, Palo Alto, CA.

EPRI (2011b), *Investigation of Catalyst Deactivation from Operation below the Minimum Operating Temperature: Laboratory Bench-Scale Testing*, Palo Alto, CA.

EPRI (2012), *Laboratory Characterization of the Factors Affecting Corrosion Fatigue Damage in Supply and Riser Tubes*. Palo Alto, CA.

GE Energy (2010), "Western Wind and Solar Integration Study: Executive Summary", retrieved from the National Renewable Energy Laboratory: *www.nrel.gov/docs/fy10osti/47781.pdf.*

IEACCC (2011), *Integrating intermittent renewable energy technologies with coal-fired power plant*, IEA/OECD, Paris. Retrieved from IEA Clean Coal Centre: *www.ieacoal.org.uk/documents/82824/8303/Integrating-intermittent-renewable-energy technologies-with-coal-fired-power-plant,- CCC/189.*

IEEE (2007), *SOFC-based Fuel Cells for Load Following Stationary Applications*, retrieved from Institute of Electrical and Electronics Engineers: *http://ieeexplore.ieee.org/xpl/login.jsp?tp=&arnumber=4201369&url=http%3A%2F%2Fieeexplore.ieee.org%2Fxpls%2Fabs_all.jsp%3Farnumber%3D4201369.*

Nelson, T., & Dunbar, E. (2010), "Black Dog coal plant shut down after explosion", retrieved from MPR News: *http://minnesota.publicradio.org/display/web/2010/09/21/black-dog-plantexplosion.*

NETL (2012), "Solid State Energy Converson Alliance (SECA)", retrieved from National Energy Technology Laboratory: *www.netl.doe.gov/technologies/coalpower/fuelcells/seca/.*

NREL (2011), *Advanced Supercritical Carbon Dioxide Power Cycle Configurations for Use in Concentrating Solar Power Systems*, retrieved from National Renewable Energy Laboratory: *www.nrel.gov/docs/fy11osti/50787.pdf.*

Payten, W. M., Bendiech, P., & Snowden, K. (2009), "High temperature remaining life cost assessment", Ninth International Conference on Operating Pressure Equipment (OPE9). AINDT Biennial Conference . Surfers Paradise, Queensland, Australia.

Powermag (2011), "Increasing Generation Ramp Rate at Morgantown Generating Station's Coal-Fired Units", retrieved from PowerMag.com: *www.powermag.com/o_and_m/Increasing-Generation-Ramp-Rate-at-Morgantown-Generating-Stations-Coal-Fired-Units_3388.html.*

Powermag (2011a), "Mitigating the Effects of Flexible Operation on Coal-Fired Power Plants", retrieved from PowerMag: *www.powermag.com/gas/3886.html.*

Viswanathan, R., Purgert, R., Goodstine, S., Tanzosh, J., Stanko, G., Shingledecker, J. P., *et al.* (2008), *United States Program on Materials Technology for Ultrasupercritical Coal-Fired Boilers*, retrieved from ASM International: *www.asminternational.org/content/ASM/StoreFiles/05226G_Sample.pdf.*

PART III

WORLD COAL OVERVIEW

1. PRODUCTION

Table 1.1: World coal[(1)] production

(thousand tonnes)

	1973	1980	1985	1990	1995	2000	2005	2009	2010	2011	2012e
Australia	79604	104504	160726	204562	241807	306722	370590	407898	424402	402237	420737
Austria	3328	2865	3081	2448	1298	1249	-	-	-	-	-
Belgium	10362	8018	7666	2357	637	375	109	-	-	-	-
Canada	20472	36688	60853	68332	74981	69163	65345	62938	67896	67114	66565
Chile	1435	1165	1326	2183	1038	366	544	636	619	654	717
Czech Republic	103745	116807	121037	101398	74901	65162	62026	56417	55209	57904	54973
Estonia	x	x	x	22486	13310	11727	14591	14939	17933	18734	18796
France	29114	22750	18894	13532	9896	4100	617	147	261	149	290
Germany	470816	484218	522886	434021	251614	205067	205925	183623	182303	188561	196990
Greece	13301	23198	35888	51896	57662	63887	69398	64893	56520	58666	61789
Hungary	27111	26025	24092	17830	14772	14033	9570	8986	9113	9555	9290
Ireland	64	60	57	25	1	-	-	-	-	-	-
Israel	-	-	-	303	470	390	429	444	432	416	400
Italy	1190	1286	1892	1014	172	14	95	72	101	92	80
Japan	25190	18054	16381	7979	6317	2964	-	-	-	-	-
Korea	13571	18625	22543	17217	5720	8300	2832	2519	2084	2084	2092
Mexico	2578	3089	5193	6933	9320	11344	10755	10548	12833	15761	15189
Netherlands	1829	-	101	-	-	-	-	-	-	-	-
New Zealand	2468	2138	2526	2578	3577	3459	5267	4564	5331	4945	4928
Norway	415	288	507	303	292	632	1471	2641	1935	1386	1229
Poland	195845	229987	249388	215320	200713	162815	159540	135172	133238	139289	144093
Portugal	221	177	237	281	-	-	-	-	-	-	-
Slovak Republic	5804	5796	5731	4766	3759	3648	2511	2573	2378	2376	2292
Slovenia	x	x	x	5583	4884	4480	4540	4429	4430	4501	4278
Spain	12994	28292	39663	35682	28305	23471	19481	9445	8430	6621	6145
Sweden	12	18	13	11	-	-	-	-	-	-	-
Turkey	12396	18625	39997	47428	55073	63268	58340	79499	73399	76116	69510
United Kingdom	131985	130097	94111	92762	53037	31198	20498	17874	18416	18627	16788
United States	543012	752961	801636	933561	937098	971591	1038591	987552	996107	1005921	934927
IEA Total	**1704849**	**2031477**	**2229906**	**2255303**	**2021632**	**2001118**	**2092206**	**2026813**	**2037123**	**2041643**	**1992718**
OECD Total	**1708862**	**2035731**	**2236425**	**2292791**	**2050654**	**2029425**	**2123065**	**2057809**	**2073370**	**2081709**	**2032098**
Algeria	333	3	23	-	-	-	-	-	-	-	-
Botswana	-	-	437	794	898	947	985	738	988	788	740
Dem. Rep. of Congo	130	138	121	126	93	96	120	135	139	143	132
Egypt	-	-	-	-	-	58	25	-	-	-	-
Morocco	565	680	775	526	650	31	12	-	-	-	-
Mozambique	394	207	35	40	38	16	3	38	38	648	3768
Nigeria	327	176	140	90	20	3	8	34	38	32	32
South Africa	62352	115120	173500	174800	206211	224200	244986	249489	254522	252757	259302
United Rep. of Tanzania	-	1	15	4	44	79	31	-	47	45	96
Zambia	940	570	511	377	152	196	150	1	-	-	-
Zimbabwe	2806	2768	3104	5345	4693	4260	3621	2669	2926	3199	2996
Other Africa	160	196	317	314	342	585	678	726	781	818	800

Table 1.1: World coal[(1)] production (continued)

(thousand tonnes)

	1973	1980	1985	1990	1995	2000	2005	2009	2010	2011	2012e
Argentina	451	390	400	276	305	259	25	82	65	90	80
Brazil	2339	5242	7712	4595	5199	6806	6255	5709	5415	5505	6303
Colombia	2834	4164	8766	21375	25651	38242	59064	72807	74350	85803	89451
Peru	33	41	127	97	51	17	43	322	88	163	149
Venezuela	50	42	40	2189	4064	7885	7195	3282	2730	2100	3120
Bangladesh	-	-	-	-	-	-	77	888	770	900	1000
India	79908	116110	158508	225258	290426	335675	437267	566113	570427	582282	595011
Indonesia	149	304	1908	10230	41828	79377	170541	291247	325000	360336	442808
DPR of Korea	30198	44106	52000	46353	31300	29743	34610	31556	31994	39081	39174
Malaysia	-	-	-	111	135	384	788	2138	2397	2916	3061
Mongolia	-	-	6523	7157	5019	5185	7511	14539	25282	31958	33610
Myanmar	10	38	86	78	35	580	574	620	686	693	1128
Nepal	-	-	-	-	-	17	12	15	16	17	16
Pakistan	1143	1569	2238	2746	3637	3094	4871	3481	3451	3613	3091
Philippines	39	326	1256	1232	1293	1357	2880	4687	6650	6881	8000
Chinese Taipei	3327	2574	1858	472	235	83	-	-	-	-	-
Thailand	361	1525	5188	12421	18421	17708	20878	17786	18344	21327	18326
Vietnam	2990	5200	5594	4638	8350	11609	34093	44078	44835	44493	42096
Other Asia	2323	4496	151	108	97	463	740	1426	1754	1876	1496
PR of China	417000	620150	837272	1039820	1338746	1353801	2299696	2895345	3140153	3418766	3549061
Albania	811	1420	2150	2071	80	30	64	14	14	9	20
Bosnia and Herzegovina	x	x	x	19670	1640	7439	9119	11469	10985	12628	12627
Bulgaria	26810	30213	30880	31675	30830	26432	24695	27279	29427	37128	32519
Croatia	x	x	x	174	82	-	-	-	-	-	-
Georgia	x	x	x	1103	34	7	5	152	105	140	254
Kazakhstan	x	x	x	131443	84494	77444	87197	100854	110929	116450	126024
Kosovo	x	x	x	..	..	4989	6554	8591	8649	8293	8684
Kyrgyzstan	x	x	x	3635	463	423	354	582	576	790	1166
F.Y.R. of Macedonia	x	x	x	6644	7249	7516	6881	7426	6724	8209	7470
Montenegro	x	x	x	..	..	..	1297	957	1938	1973	2000
Romania	24851	35164	46581	38183	41121	29285	31106	33961	31127	35512	34031
Russian Federation	x	x	x	371899	245728	240324	282881	275991	321701	321663	353940
Serbia	x	x	x	45937	40595	37094	35100	38499	37976	41105	38022
Tajikistan	x	x	x	925	41	22	99	202	200	236	200
Ukraine	x	x	x	152763	76298	62403	60361	59634	57659	69664	71636
Uzbekistan	x	x	x	6400	3054	2569	3076	3654	3630	3844	4036
Former Soviet Union	667600	716000	726000	x	x	x	x	x	x	x	x
Former Yugoslavia	32450	41301	68472	x	x	x	x	x	x	x	x
Islam. Rep. of Iran	903	925	1106	835	1084	1148	1556	1152	1025	1036	1176
Non-OECD Total	**1364587**	**1751159**	**2143794**	**2374929**	**2520726**	**2619881**	**3888084**	**4780368**	**5136551**	**5525910**	**5798652**
World	**3073449**	**3786890**	**4380219**	**4667720**	**4571380**	**4649306**	**6011149**	**6838177**	**7209921**	**7607619**	**7830750**

(1) Coal comprises all coals from anthracite through lignite, however excludes peat and manufacture of all coal products. For further information, see notes and definitions in Part I.

Source: *IEA/OECD Energy Statistics of OECD Countries* , *IEA/OECD Energy Statistics of Non-OECD Countries*

Table 1.2: World coal[1] production

(kilotonnes of coal equivalent)

	1973	1980	1985	1990	1995	2000	2005	2009	2010	2011	2012e
Australia	57501	74140	119913	151576	182605	235112	287969	319047	337722	317961	333272
Austria	1455	1203	1324	911	483	418	-	-	-	-	-
Belgium	9178	6726	6745	1687	436	294	82	-	-	-	-
Canada	16714	28924	48205	54182	58352	49153	45522	43891	48286	48083	47958
Chile	1369	1115	1273	2073	986	348	385	337	362	351	413
Czech Republic	54023	57642	58983	51877	39390	35784	33671	29790	29614	29851	28679
Estonia	x	x	x	6905	4175	3696	4416	4588	5507	5689	5708
France	25766	19108	15573	11772	8594	3546	547	130	232	132	257
Germany	202003	204491	207193	173842	112722	86571	80660	65291	64464	66471	68747
Greece	2419	4219	6911	10169	10729	11745	12197	11680	10451	10721	11292
Hungary	8647	9059	8202	6034	4665	4133	2497	2223	2276	2351	2285
Ireland	65	60	51	22	0.9	-	-	-	-	-	-
Israel	-	-	-	30	47	39	43	44	43	43	41
Italy	425	459	676	393	61	5.0	86	65	92	84	73
Japan	25576	15577	14144	6379	5050	2196	-	-	-	-	-
Korea	9500	11707	14170	10822	3521	5197	1801	1655	1370	1370	1375
Mexico	2165	2475	3894	4839	6179	7735	7370	7592	9035	11128	10788
Netherlands	1623	-	93	-	-	-	-	-	-	-	-
New Zealand	1842	1632	1868	2033	3086	2962	4520	3791	4482	4134	4104
Norway	415	288	507	291	280	606	1410	2532	1855	1329	1178
Poland	143895	171922	175068	141385	130101	101855	98367	80597	79116	79659	82739
Portugal	189	104	139	165	-	-	-	-	-	-	-
Slovak Republic	2428	2424	2397	1995	1452	1455	910	933	876	859	829
Slovenia	x	x	x	1929	1703	1517	1691	1657	1709	1715	1630
Spain	9251	14035	19098	16779	14505	11380	8950	5444	4709	3784	3500
Sweden	12	8.7	6.3	5.3	-	-	-	-	-	-	-
Turkey	7445	8790	15248	17671	17261	17836	15438	24861	25034	25485	23423
United Kingdom	108416	105652	76771	76593	45820	26654	16998	14793	15358	15533	13998
United States	476228	639884	664373	774742	756835	766936	807545	757333	759782	766241	710899
IEA Total	**1165013**	**1378054**	**1457658**	**1511323**	**1395950**	**1363837**	**1419171**	**1364057**	**1385719**	**1374046**	**1334608**
OECD Total	**1168547**	**1381644**	**1462825**	**1527099**	**1409039**	**1377173**	**1433077**	**1378274**	**1402374**	**1392971**	**1353188**
Algeria	299	2.9	22	-	-	-	-	-	-	-	-
Botswana	-	-	385	699	791	834	867	650	870	694	651
Dem. Rep. of Congo	117	119	104	109	80	83	103	116	120	123	114
Egypt	-	-	-	-	-	51	22	-	-	-	-
Morocco	507	599	620	421	520	25	9.6	-	-	-	-
Mozambique	336	177	30	34	32	14	2.6	32	32	583	3508
Nigeria	293	155	123	79	18	2.6	7.0	30	34	28	28
South Africa	50203	95366	142506	143090	168597	181323	197664	201297	205393	203921	208990
United Rep. of Tanzania	-	0.9	13	3.6	39	70	27	-	41	40	85
Zambia	792	481	431	318	128	165	126	0.8	-	-	-
Zimbabwe	2585	2550	2859	4923	4323	3924	3335	2459	2695	2947	2760
Other Africa	144	173	279	276	301	515	597	639	688	720	704

Table 1.2: World coal[(1)] production (continued)
(kilotonnes of coal equivalent)

	1973	1980	1985	1990	1995	2000	2005	2009	2010	2011	2012e
Argentina	380	329	337	233	257	218	21	69	55	76	67
Brazil	1257	3560	5059	2759	2931	3759	3548	3199	2991	3030	3361
Colombia	2632	3867	8140	19848	23819	35510	54845	67606	69039	79674	83061
Peru	33	41	127	97	51	17	43	322	88	163	149
Venezuela	52	44	42	2283	4238	8223	7503	3423	2847	2190	3254
Bangladesh	-	-	-	-	-	-	55	634	550	643	714
India	53817	76613	105232	149560	186880	208907	268237	350617	353324	360238	369054
Indonesia	134	248	1557	8353	34133	64935	140329	238289	266183	295547	356287
DPR of Korea	25018	36338	42618	38026	25424	24153	28383	25922	26440	32842	32926
Malaysia	-	-	-	100	122	346	710	1926	2159	2626	2757
Mongolia	-	-	3438	3805	3091	2582	4687	10562	20342	26874	27677
Myanmar	9.0	21	55	51	20	458	472	512	585	588	993
Nepal	-	-	-	-	-	15	10	13	14	15	14
Pakistan	731	896	1278	1568	2077	1767	2652	1951	1955	2022	1709
Philippines	17	246	946	928	974	1022	2171	3534	5014	5188	6032
Chinese Taipei	2994	2280	1646	418	208	74	-	-	-	-	-
Thailand	119	588	2149	5146	7634	7336	8650	7369	7600	8836	7592
Vietnam	2392	4160	4475	3710	6680	9287	27148	35262	35868	35594	33677
Other Asia	1189	2408	133	95	85	322	527	1074	1349	1443	1163
PR of China	295410	443884	584359	740559	962323	1020431	1745400	2196893	2383591	2604845	2701266
Albania	406	710	1075	696	27	10	22	4.7	4.7	3.0	6.7
Bosnia and Herzegovina	x	x	x	5969	498	3512	4210	5295	5001	5801	5801
Bulgaria	6646	7412	7546	7690	7564	6172	5969	6569	7059	8871	7767
Croatia	x	x	x	144	68	-	-	-	-	-	-
Georgia	x	x	x	944	20	4.2	3.0	91	63	84	153
Kazakhstan	x	x	x	82873	53068	48757	54692	63260	69353	72708	79159
Kosovo	x	x	x	..	..	1328	1745	2287	2303	2208	2312
Kyrgyzstan	x	x	x	2018	256	225	190	305	302	410	606
F.Y.R. of Macedonia	x	x	x	1735	1893	1732	1755	1848	1706	2083	1895
Montenegro	x	x	x	..	..	..	410	301	609	620	629
Romania	8644	11575	14698	12355	11266	8002	8276	9376	8433	9517	9123
Russian Federation	x	x	x	273224	185302	182946	224334	219176	256839	256716	284580
Serbia	x	x	x	14531	12643	11931	10658	10473	10327	11178	10340
Tajikistan	x	x	x	526	22	12	61	125	123	146	127
Ukraine	x	x	x	121864	60584	51760	49302	49757	48021	57461	58465
Uzbekistan	x	x	x	3227	1537	1294	1548	1841	1824	1930	2049
Former Soviet Union	451456	476897	439907	x	x	x	x	x	x	x	x
Former Yugoslavia	10926	13748	22226	x	x	x	x	x	x	x	x
Islam. Rep. of Iran	810	883	1056	797	1029	1087	1446	1100	978	989	1123
Non-OECD Total	**920346**	**1186369**	**1395470**	**1656085**	**1771551**	**1895138**	**2862771**	**3526208**	**3802808**	**4102211**	**4312725**
World	**2088893**	**2568013**	**2858295**	**3183184**	**3180590**	**3272310**	**4295848**	**4904482**	**5205183**	**5495182**	**5665913**

(1) Coal comprises all coals from anthracite through lignite, however excludes peat and manufacture of all coal products.
For further information, see notes and definitions in Part I.

Source: *IEA/OECD Energy Statistics of OECD Countries* , *IEA/OECD Energy Statistics of Non-OECD Countries*

Table 1.3: World coking coal production

(thousand tonnes)

	1978	1985	1990	1995	2000	2005	2008	2009	2010	2011	2012e
Australia	37668	54835	64631	79375	103750	128358	143991	129810	162929	146712	146944
Belgium	3805	3484	-	-	-	-	-	-	-	-	-
Canada	13780	24371	27660	28624	28164	26613	28345	22984	28156	29452	31086
Czech Republic	18553	16465	14383	10824	8136	7136	7512	5900	6023	5183	5049
France	4664	3388	1821	439	-	-	-	-	-	-	-
Germany	52239	51401	44577	31686	18862	15171	10554	9064	7147	7304	6325
Hungary	861	592	169	-	-	-	-	-	-	-	-
Japan	8659	3921	-	-	-	-	-	-	-	-	-
Mexico	3085	3713	2963	1645	2214	1792	1841	1793	1587	2043	2158
New Zealand	11	319	578	1474	1310	2446	2284	1903	2341	2120	2075
Norway	227	174	-	-	-	-	-	-	-	-	-
Poland	40845	31143	28793	28714	17222	14071	12024	8540	11658	11436	11738
Spain	1800	907	279	-	-	-	-	-	-	-	-
Turkey	2964	2247	1824	929	735	648	858	1562	1088	1181	1123
United Kingdom	15110	2608	1600	599	255	274	307	246	270	383	338
United States	92201	89463	93259	77166	54287	46444	57367	46559	68645	81656	81300
IEA Total	**293387**	**285318**	**279574**	**259830**	**232721**	**241161**	**263242**	**226568**	**288257**	**285427**	**285978**
OECD Total	**296472**	**289031**	**282537**	**261475**	**234935**	**242953**	**265083**	**228361**	**289844**	**287470**	**288136**
Algeria	-	23	-	-	-	-	-	-	-	-	-
Mozambique	-	-	-	-	-	-	-	-	-	275	2689
South Africa	9718	11142	9308	10165	3204	1640	2207	1668	1841	1639	845
United Rep. of Tanzania	-	-	1	1	-	-	-	-	-	-	-
Zimbabwe	946	269	612	621	813	690	305	323	354	386	386
Brazil	1317	1407	499	106	15	210	260	-	-	-	-
Colombia	1313	1489	1721	1840	1818	1451	1303	1962	3837	3990	3990
India	13938	25847	36088	28803	22088	23584	26735	36144	41432	44328	47224
Indonesia	-	10	29	241	616	1222	1922	2049	2201	2826	2826
DPR of Korea	3415	2500	2543	-	-	-	-	-	-	-	-
Mongolia	-	-	-	-	-	1954	3891	4800	15222	21077	20868
Chinese Taipei	-	107	3	-	-	-	-	-	-	-	-
PR of China	52604	68370	85657	147206	123028	296648	384962	416458	459492	509493	510447
Georgia	x	x	574	-	-	-	-	-	-	-	-
Kazakhstan	x	x	29983	12756	10687	10981	10661	11001	11906	12416	12926
Romania	2134	3825	1482	349	13	-	-	-	-	-	-
Russian Federation	x	x	85458	55645	51035	55505	54402	59854	66884	65362	74599
Tajikistan	x	x	-	7	-	-	-	-	-	-	-
Ukraine	x	x	62283	25783	27844	23166	20733	19244	17688	19809	17764
Former Soviet Union	139250	134985	x	x	x	x	x	x	x	x	x
Islam. Rep. of Iran	678	1006	760	914	931	930	1266	1048	926	936	1076
Non-OECD Total	**225313**	**250980**	**317001**	**284437**	**242092**	**417981**	**508647**	**554551**	**621783**	**682537**	**695640**
World	**521785**	**540011**	**599538**	**545912**	**477027**	**660934**	**773730**	**782912**	**911627**	**970007**	**983776**

Source: *IEA/OECD Energy Statistics of OECD Countries , IEA/OECD Energy Statistics of Non-OECD Countries*

Table 1.4: World steam coal[1] production

(thousand tonnes)

	1973[2]	1980	1985	1990	1995	2000	2005	2009	2010	2011	2012e
Australia[3]	55483	31524	67511	93941	111680	135679	171699	206051	189383	184525	200255
Austria	-	-	-	-	1	-	-	-	-	-	-
Belgium[3]	10362	3982	4182	2357	637	375	109	-	-	-	-
Canada	12337	16566	26810	31265	35618	29809	27715	29404	29476	27931	25983
Chile[3]	1374	1125	1291	2183	1038	366	544	636	619	654	717
Czech Republic	27780	10318	9936	8032	6914	6719	6118	5101	5412	6082	6391
France[3]	26350	16076	13667	9378	8056	3804	617	147	261	149	290
Germany	104407	38456	37448	31976	27172	18514	12847	4702	5753	4755	5233
Hungary	1186	2498	2088	329	-	-	-	-	-	-	-
Ireland	64	60	57	25	1	-	-	-	-	-	-
Italy	-	-	-	58	-	-	95	72	101	92	80
Japan[3]	25090	11084	12460	7979	6317	2964	-	-	-	-	-
Korea[3]	13571	18625	22543	17217	5720	8300	2832	2519	2084	2084	2092
Mexico[3]	2578	-	1480	3970	7675	9130	8963	8755	11246	13718	13031
Netherlands	1829	-	101	-	-	-	-	-	-	-	-
New Zealand[3]	2325	1757	1960	1841	1860	1936	2575	2401	2695	2505	2527
Norway	415	128	333	303	292	632	1471	2641	1935	1386	1229
Poland	156630	161626	160499	118943	108452	86109	83833	69524	65070	65012	68075
Portugal[3]	221	177	237	281	-	-	-	-	-	-	-
Spain	9991	15544	21464	19030	17529	14947	11894	9445	8430	6621	6145
Sweden	12	18	13	11	-	-	-	-	-	-	-
Turkey	4642	1721	1881	1197	1386	1679	2410	2360	2613	2385	2437
United Kingdom	131985	120047	91503	91162	52438	30943	20224	17628	18146	18244	16450
United States[3]	530064	592462	646472	760388	781461	839685	915996	875242	856492	850691	782017
IEA Total	**1114744**	**1042669**	**1121165**	**1195713**	**1165534**	**1182095**	**1260435**	**1227237**	**1187851**	**1172462**	**1119204**
OECD Total	**1118696**	**1043794**	**1123936**	**1201866**	**1174247**	**1191591**	**1269942**	**1236628**	**1199716**	**1186834**	**1132952**
Algeria	333	-	-	-	-	-	-	-	-	-	-
Botswana	-	-	437	794	898	947	985	738	988	788	740
Dem. Rep. of Congo	130	138	121	126	93	96	120	135	139	143	132
Egypt	-	-	-	-	-	58	25	-	-	-	-
Morocco	565	680	775	526	650	31	12	-	-	-	-
Mozambique	394	207	35	40	38	16	3	38	38	373	1079
Nigeria	327	176	140	90	20	3	8	34	38	32	32
South Africa	62352	104515	162358	165492	196046	220996	243346	247821	252681	251118	258457
United Rep. of Tanzania	-	1	15	3	43	79	31	-	47	45	96
Zambia	940	570	511	377	152	196	150	1	-	-	-
Zimbabwe	2806	1777	2835	4733	4072	3447	2931	2346	2572	2813	2610
Other Africa	160	196	317	314	342	585	678	726	781	818	800
Argentina	451	390	400	276	305	259	25	82	65	90	80
Brazil	1015	2570	4318	1935	2673	4061	3542	3660	3320	3369	3264
Colombia	2834	2781	7277	19654	23811	36424	57613	70845	70513	81813	85461
Peru	33	41	127	97	51	17	43	322	88	163	149
Venezuela	50	42	40	2189	4064	7885	7195	3282	2730	2100	3120

Table 1.4: World steam coal[1] production (continued)

(thousand tonnes)

	1973[2]	1980	1985	1990	1995	2000	2005	2009	2010	2011	2012e
Bangladesh	-	-	-	-	-	-	77	888	770	900	1000
India	76588	92945	124621	175096	239477	289340	383455	495898	491262	495622	504296
Indonesia[4]	149	304	1898	10201	41587	78761	169319	289198	322799	357510	439982
DPR of Korea[4]	23198	40270	49500	43810	31300	29743	34610	31556	31994	39081	39174
Malaysia	-	-	-	111	135	384	788	2138	2397	2916	3061
Mongolia	-	-	480	595	1290	70	156	2390	1551	2573	2758
Myanmar	10	11	43	40	12	468	504	548	646	646	1128
Nepal	-	-	-	-	-	17	12	15	16	17	16
Pakistan	1143	1098	1567	1922	2546	2166	2861	2281	2350	2355	1924
Philippines[4]	-	326	1252	1229	1290	1354	2880	4687	6650	6881	8000
Chinese Taipei	3327	2574	1751	469	235	83	-	-	-	-	-
Thailand	-	-	-	-	5	-	-	-	-	-	-
Vietnam	2990	5200	5594	4638	8350	11609	33771	44078	44835	44493	42096
Other Asia	117	511	151	108	97	243	420	960	1252	1339	1100
PR of China[4]	417000	551928	768902	954163	1191540	1230773	2003048	2478887	2680661	2909273	3038614
Bosnia and Herzegovina	x	x	x	-	-	4038	4643	5841	5367	6339	6339
Bulgaria	351	267	223	143	3381	118	9	131	48	18	8
Croatia	x	x	x	156	75	-	-	-	-	-	-
Georgia	x	x	x	529	34	7	5	152	105	140	254
Kazakhstan	x	x	x	98017	67998	64199	71807	84769	91740	95666	107574
Kyrgyzstan	x	x	x	1495	183	97	93	106	106	112	168
Romania	7172	6232	4832	2964	799	268	32	11	4	35	40
Russian Federation	x	x	x	152056	106766	101503	153708	147126	178696	179949	201486
Serbia	x	x	x	137	55	176	24	8	-	-	-
Tajikistan	x	x	x	475	34	22	99	202	200	236	200
Ukraine[4]	x	x	x	81200	48219	33757	36841	40390	39971	49855	53872
Uzbekistan	x	x	x	200	74	69	73	101	65	60	232
Former Soviet Union	510600	409466	434015	x	x	x	x	x	x	x	x
Former Yugoslavia	576	388	400	x	x	x	x	x	x	x	x
Islam. Rep. of Iran	903	83	100	75	170	217	626	104	99	100	100
Non-OECD Total	**1116514**	**1225687**	**1575035**	**1726475**	**1978910**	**2124582**	**3216568**	**3962495**	**4237584**	**4539781**	**4809442**
World	**2235210**	**2269481**	**2698971**	**2928341**	**3153157**	**3316173**	**4486510**	**5199123**	**5437300**	**5726615**	**5942394**

(1) Steam coal is also commonly known as thermal coal. From 1978 onwards it comprises anthracite, bituminous coal and sub-bituminous coal. For further information, see notes and definitions in Part I.

(2) Data prior to 1978 are hard coal. Hard coal comprises anthracite, coking coal, other bituminous coal and for certain countries sub-bituminous coal.

(3) Includes sub-bituminous coal prior to 1978.

(4) Data includes lignite for at least some years.

Source: *IEA/OECD Energy Statistics of OECD Countries* , *IEA/OECD Energy Statistics of Non-OECD Countries*

Table 1.5: World lignite[1] production

(thousand tonnes)

	1973[2]	1980	1985	1990	1995	2000	2005	2009	2010	2011	2012e
Australia[3]	24121	32894	38380	45990	50752	67293	70533	72037	72090	71000	73538
Austria	3328	2865	3081	2448	1297	1249	-	-	-	-	-
Canada	8135	5971	9672	9407	10739	11190	11017	10550	10264	9731	9496
Chile[3]	61	40	35	-	-	-	-	-	-	-	-
Czech Republic	75965	89086	94636	78983	57163	50307	48772	45416	43774	46639	43533
Estonia	x	x	x	22486	13310	11727	14591	14939	17933	18734	18796
France[3]	2764	2560	1839	2333	1401	296	-	-	-	-	-
Germany	366409	389726	434037	357468	192756	167691	177907	169857	169403	176502	185432
Greece	13301	23198	35888	51896	57662	63887	69398	64893	56520	58666	61789
Hungary	25925	22644	21412	17332	14772	14033	9570	8986	9113	9555	9290
Israel	-	-	-	303	470	390	429	444	432	416	400
Italy	1190	1286	1892	956	172	14	-	-	-	-	-
Japan[3]	100	27	-	-	-	-	-	-	-	-	-
New Zealand[3]	143	208	247	159	243	213	246	260	295	320	326
Poland	39215	36866	57746	67584	63547	59484	61636	57108	56510	62841	64280
Slovak Republic	5804	5796	5731	4766	3759	3648	2511	2573	2378	2376	2292
Slovenia	x	x	x	5583	4884	4480	4540	4429	4430	4501	4278
Spain	3003	11415	17292	16373	10776	8524	7587	-	-	-	-
Turkey	7754	14469	35869	44407	52758	60854	55282	75577	69698	72550	65950
United States[3]	12948	42783	65701	79914	78471	77619	76151	65751	70970	73574	71610
IEA Total	**590105**	**681794**	**823423**	**780016**	**596268**	**586302**	**590610**	**573008**	**561015**	**583754**	**587536**
OECD Total	**590166**	**681834**	**823458**	**808388**	**614932**	**602899**	**610170**	**592820**	**583810**	**607405**	**611010**

Table 1.5: World lignite[1] production (continued)

(thousand tonnes)

	1973[2]	1980	1985	1990	1995	2000	2005	2009	2010	2011	2012e
Brazil	1324	1127	1987	2161	2420	2730	2503	2049	2095	2136	3039
India	3320	5110	8040	14074	22146	24247	30228	34071	37733	42332	43491
DPR of Korea[4]	7000	-	-	-	-	-	-	-	-	-	-
Mongolia	-	-	6043	6562	3729	5115	5401	7349	8509	8308	9984
Myanmar	-	27	43	38	23	112	70	72	40	47	-
Pakistan	-	471	671	824	1091	928	2010	1200	1101	1258	1167
Philippines[4]	39	-	4	3	3	3	-	-	-	-	-
Thailand	361	1525	5188	12421	18416	17708	20878	17786	18344	21327	18326
Vietnam	-	-	-	-	-	-	322	-	-	-	-
Other Asia	2206	3985	-	-	-	220	320	466	502	537	396
Albania	811	1420	2150	2071	80	30	64	14	14	9	20
Bosnia and Herzegovina	x	x	x	19670	1640	3401	4476	5628	5618	6289	6288
Bulgaria	26459	29946	30657	31532	27449	26314	24686	27148	29379	37110	32511
Croatia	x	x	x	18	7	-	-	-	-	-	-
Kazakhstan	x	x	x	3443	3740	2558	4409	5084	7283	8368	5524
Kosovo	x	x	x	..	..	4989	6554	8591	8649	8293	8684
Kyrgyzstan	x	x	x	2140	280	326	261	476	470	678	998
F.Y.R. of Macedonia	x	x	x	6644	7249	7516	6881	7426	6724	8209	7470
Montenegro	x	x	x	..	..	..	1297	957	1938	1973	2000
Romania	17679	27104	37924	33737	39973	29004	31074	33950	31123	35477	33991
Russian Federation	x	x	x	134385	83317	87786	73668	69011	76121	76352	77855
Serbia	x	x	x	45800	40540	36918	35076	38491	37976	41105	38022
Tajikistan	x	x	x	450	-	-	-	-	-	-	-
Ukraine[4]	x	x	x	9280	2296	802	354	-	-	-	-
Uzbekistan	x	x	x	6200	2980	2500	3003	3553	3565	3784	3804
Former Soviet Union	157000	163000	157000	x	x	x	x	x	x	x	x
Former Yugoslavia	31874	40913	68072	x	x	x	x	x	x	x	x
Non-OECD Total	**248073**	**274628**	**317779**	**331453**	**257379**	**253207**	**253535**	**263322**	**277184**	**303592**	**293570**
World	**838239**	**956462**	**1141237**	**1139841**	**872311**	**856106**	**863705**	**856142**	**860994**	**910997**	**904580**

(1) Lignite includes oil shale. Some countries, most notably the People's Republic of China and Indonesia, produce and consume lignite, however these data are reported under other coal types included in steam coal and are not shown here.

(2) Data before 1978 are brown coal, which may include sub-bituminous coal.

(3) Brown coal data excludes sub-bituminous coal.

(4) Data are reported as other coal types for at least some years.

For further information, see notes and definitions in Part I.

Source: *IEA/OECD Energy Statistics of OECD Countries , IEA/OECD Energy Statistics of Non-OECD Countries*

Table 1.6: OECD production of coke oven coke[(1)]

(thousand tonnes)

	1973	1980	1985	1990	1995	2000	2005	2009	2010	2011	2012e
Australia	4983	4991	3603	4495	4617	3739	3494	2563	2795	3212	2807
Austria	1719	1729	1751	1725	1448	1385	1388	1281	1391	1316	1308
Belgium	7774	6048	5964	5420	3696	3104	2856	1574	1935	1923	1918
Canada	5370	5250	4684	3708	3283	3242	3305	2190	2720	2804	2949
Chile	315	303	301	336	457	475	493	431	361	549	588
Czech Republic	9383	8725	8328	7125	4963	3411	3412	2295	2548	2586	2467
Denmark	-	-	-	-	-	-	-	-	-	-	-
Estonia	x	x	x	41	40	23	37	18	22	24	25
Finland	-	-	-	487	920	910	894	738	827	852	881
France	11881	11120	8691	7197	5566	5234	4445	3222	3151	2958	3205
Germany	41614	35492	30171	21926	11102	9115	8397	6771	8150	7990	8050
Greece	400	247	-	-	-	-	-	-	-	-	-
Hungary	1082	975	607	672	1033	937	614	746	1018	1049	1026
Iceland	-	-	-	-	-	-	-	-	-	-	-
Ireland	-	-	-	-	-	-	-	-	-	-	-
Israel	-	-	-	-	-	-	-	-	-	-	-
Italy	7665	8264	7410	6356	5185	4504	4574	2755	4110	4788	4034
Japan	52300	47463	48621	51444	45077	41795	42917	38086	42212	39263	40084
Korea	318	2965	5253	8800	10593	12288	8935	9632	13549	15349	15150
Luxembourg	-	-	-	-	-	-	-	-	-	-	-
Mexico	2021	2447	2924	2384	2148	2100	2002	1763	2209	2122	2085
Netherlands	2655	2455	2958	2736	2895	2127	2250	1690	2031	2015	1883
New Zealand	20	13	5	373	369	349	418	414	458	478	486
Norway	282	349	313	-	-	-	-	-	-	-	-
Poland	16938	19598	15828	13516	11578	8972	8404	7091	9844	9377	8903
Portugal	269	216	275	230	331	371	-	-	-	-	-
Slovak Republic	1500	1598	1909	2340	1861	1706	1846	1573	1658	1620	1561
Slovenia	x	x	x	-	-	-	-	-	-	-	-
Spain	4475	3900	3440	3211	2438	2470	2662	1917	2051	1974	1740
Sweden	533	1188	1203	1084	1149	1146	1411	987	1197	1223	1115
Switzerland	-	-	-	-	-	-	-	-	-	-	-
Turkey	1251	1937	2711	3158	3131	2925	2992	3437	4274	3903	4253
United Kingdom	17776	10060	9276	8350	6274	6206	4364	3693	4023	4053	3725
United States	62803	41850	25992	25053	21545	18876	15168	10108	13628	13989	13764
IEA Americas	68173	47100	30676	28761	24828	22118	18473	12298	16348	16793	16713
IEA Asia Oceania	57621	55432	57482	65112	60656	58171	55764	50695	59014	58302	58527
IEA Europe	127197	113901	100835	85533	63570	54523	50509	39770	48208	47627	46069
OECD Americas	70509	49850	33901	31481	27433	24693	20968	14492	18918	19464	19386
OECD Asia Oceania	57621	55432	57482	65112	60656	58171	55764	50695	59014	58302	58527
OECD Europe	127197	113901	100835	85574	63610	54546	50546	39788	48230	47651	46094
IEA Total	**252991**	**216433**	**188993**	**179406**	**149054**	**134812**	**124746**	**102763**	**123570**	**122722**	**121309**
OECD Total	**255327**	**219183**	**192218**	**182167**	**151699**	**137410**	**127278**	**104975**	**126162**	**125417**	**124007**

(1) Solid product obtained from carbonization of coal, principally coking coal, used mainly in the iron and steel industry.
Also includes coke and semi-coke made from lignite.
For further information, see notes and definitions in Part I.

Source: *IEA/OECD Energy Statistics of OECD Countries*

2. CONSUMPTION

Table 2.1: World coal[1] consumption

(thousand tonnes)

	1973	1980	1985	1990	1995	2000	2005	2009	2010	2011	2012e[2]
Australia	51413	67316	78360	95289	102491	128123	139601	143867	138145	133650	137290
Austria	6184	6145	7046	6663	5135	5105	5325	3343	3818	3870	3478
Belgium	17169	17099	15851	16429	12589	11266	7757	4855	5164	4691	4691
Canada	25614	37272	48175	49146	52634	62968	60211	49253	48172	43025	41762
Chile	1772	1718	1827	3720	3391	4590	4571	6817	8351	9788	10420
Czech Republic	103745	103759	105884	91832	66480	61089	56843	50995	51544	52120	49886
Denmark	3146	9669	11935	9992	11003	6641	6293	6778	6496	5521	4176
Estonia	x	x	x	26336	15523	13319	14860	13856	17949	18809	18859
Finland	3035	5692	5318	5648	6540	5193	4598	5413	6982	5477	4633
France	43064	50650	38925	30885	24127	22156	21178	16373	17385	14629	16899
Germany	478298	488138	525987	451015	269035	238905	241844	221862	228108	234697	241391
Greece	13651	23237	37964	53433	58442	65685	70659	65550	58318	60358	63823
Hungary	29033	27544	25626	20305	16931	15173	11588	10676	10989	11614	11405
Iceland	1	12	69	65	65	101	117	98	106	105	104
Ireland	822	1066	1586	3236	2734	2973	3039	1869	2023	2073	2450
Israel	-	-	2927	4023	7038	10981	12553	12291	12742	13162	15568
Italy	12902	18409	23935	22416	17642	18043	24248	19949	21767	24445	24138
Japan	81790	87726	109391	114960	133564	153190	176972	163716	185390	174139	183768
Korea	16329	27790	42505	44776	44634	71799	82272	106861	120048	130894	127256
Luxembourg	305	374	199	197	217	172	122	104	102	89	82
Mexico	2894	3973	5317	6881	10245	12294	17559	16367	17692	18631	18366
Netherlands	4814	6129	10379	14101	14314	12786	13059	11899	11903	11712	12833
New Zealand	2460	1976	2074	2243	2138	2096	4286	2986	2645	2933	2853
Norway	772	951	1118	749	1018	999	795	538	706	780	796
Poland	156379	199086	214135	187622	171019	142859	142027	132814	141381	146235	139732
Portugal	805	604	1050	4397	5708	6154	5476	4677	2702	3697	4873
Slovak Republic	18618	21412	21593	18360	12551	8869	8290	7593	7214	7209	6911
Slovenia	x	x	x	6090	5239	4925	5192	4942	4914	5029	5057
Spain	16322	31222	48440	46823	42542	45654	44498	19678	14661	24186	28866
Sweden	1060	2138	4158	3709	3444	2861	3070	2433	2859	3043	2776
Switzerland	258	315	640	494	253	179	217	221	228	213	188
Turkey	12237	20431	41490	54324	61019	79932	76736	98694	95608	101170	98192
United Kingdom	133527	123610	105980	106722	75916	59839	61779	48786	51447	51500	63984
United States	505515	650167	744671	815949	863552	966391	1029721	918493	949702	920300	821925
IEA Americas	531129	687439	792846	865095	916186	1029359	1089932	967746	997874	963325	863687
IEA Asia Oceania	151992	184808	232330	257268	282827	355208	403131	417430	446228	441616	451167
IEA Europe	1056146	1157680	1249239	1149352	878659	812533	809441	735100	741405	769329	786203
OECD Americas	535795	693130	799990	875696	929822	1046243	1112062	990930	1023917	991744	892473
OECD Asia Oceania	151992	184808	235257	261291	289865	366189	415684	429721	458970	454778	466735
OECD Europe	1056147	1157692	1249308	1181843	899486	830878	829610	753996	764374	793272	810223
IEA Total	**1739267**	**2029927**	**2274415**	**2271715**	**2077672**	**2197100**	**2302504**	**2120276**	**2185507**	**2174270**	**2101057**
OECD Total	**1743934**	**2035630**	**2284555**	**2318830**	**2119173**	**2243310**	**2357356**	**2174647**	**2247261**	**2239794**	**2169431**

Table 2.1: World coal[1] consumption (continued)

(thousand tonnes)

	1973	1980	1985	1990	1995	2000	2005	2009	2010	2011	2012e[2]
Algeria	63	93	1226	1005	632	689	819	290	-	-	-
Botswana	-	-	466	815	912	1040	1002	742	992	792	740
Dem. Rep. of Congo	170	167	156	169	136	132	162	198	212	226	220
Egypt	487	894	1192	1340	1540	1820	1810	1791	2158	2097	2191
Kenya	70	16	90	151	156	107	145	154	268	380	480
Morocco	582	635	1110	1774	2665	4018	4938	4100	4230	4511	4577
Mozambique	587	288	106	58	56	-	-	10	10	26	73
Namibia	-	-	-	-	16	3	20	49	35	13	12
Nigeria	289	151	94	55	20	3	8	34	38	32	32
Senegal	-	-	-	-	-	-	152	254	264	271	403
South Africa	60408	86961	125870	124900	147205	157135	175403	192164	190235	186340	187178
United Rep. of Tanzania	-	1	15	4	44	79	31	-	47	45	96
Tunisia	33	21	21	15	-	-	-	-	-	-	-
Zambia	941	618	471	375	148	130	140	1	-	-	-
Zimbabwe	2758	2614	3026	5355	4494	4272	3673	2710	2971	3248	3053
Other Africa	222	278	379	436	442	1020	1219	1616	1791	1876	1867
Argentina	1072	1425	1247	1367	1439	1058	1383	1458	1504	1716	1521
Brazil	4122	9142	16861	15436	17120	20270	20003	17420	21707	22868	24851
Colombia	2859	2801	3142	4825	5608	4231	4173	5399	6202	6530	7279
Costa Rica	1	1	1	-	-	1	2	4	1	9	-
Cuba	-	95	126	153	77	22	22	40	36	37	-
Dominican Republic	-	-	224	17	80	93	476	747	709	882	476
Guatemala	-	22	-	-	-	215	409	294	492	469	666
Haiti	-	-	61	12	-	-	-	-	-	-	-
Honduras	-	-	-	-	-	135	183	103	107	161	50
Jamaica	-	-	-	52	55	53	58	51	54	56	123
Panama	13	-	32	32	51	60	-	-	-	237	312
Peru	86	74	107	149	389	708	1075	1142	1182	910	688
Uruguay	32	4	-	1	-	1	1	3	4	2	-
Venezuela	53	42	42	355	7	181	51	325	273	281	300
Bangladesh	243	235	98	563	642	660	899	1725	1808	1824	2000
Cambodia	-	-	-	-	-	-	-	16	17	19	195
Hong Kong (China)	12	3	5523	8928	9109	6058	10824	12331	10324	12529	12599
India	77172	107796	156229	220707	294875	357009	463510	650149	676895	710100	753150
Indonesia	129	236	925	6320	11892	22720	42031	57885	57854	59716	60198
DPR of Korea	30580	44456	54200	48453	31940	29383	31806	28576	27441	27903	28252
Malaysia	13	84	574	2150	2558	3661	10926	16730	23161	24717	24633
Mongolia	-	-	6167	6649	5204	5212	5472	6522	7029	6931	9847
Myanmar	74	248	266	118	38	580	574	620	686	693	1139
Nepal	78	83	17	81	123	430	413	322	505	484	496
Pakistan	1270	1667	2954	4246	4722	4044	7714	8139	7718	7670	6362
Philippines	40	558	2419	2576	3004	8603	9909	10060	13319	14639	19007
Singapore	1	1	2	2	-	-	1	-	-	-	-
Sri Lanka	-	-	1	8	5	-	93	93	95	514	711
Chinese Taipei	3572	5956	11085	17230	26229	46780	59716	60018	63415	66501	64530
Thailand	362	1619	5344	12707	20801	21270	29525	33318	35419	39725	35123
Vietnam	2770	4052	4990	3951	5917	7808	14812	22397	26146	27728	24317
Other Asia	2426	4857	345	303	244	575	727	1274	1886	2018	3817
PR of China	414180	626010	803907	1049632	1312632	1337525	2219618	2935245	3200683	3501294	3665893

Table 2.1: World coal[1] consumption (continued)

(thousand tonnes)

	1973	1980	1985	1990	1995	2000	2005	2009	2010	2011	2012e[2]
Albania	899	1580	2370	2145	80	73	77	226	248	307	54
Armenia	x	x	x	552	3	-	-	-	1	-	-
Azerbaijan	x	x	x	200	6	-	-	-	-	-	-
Belarus	x	x	x	2389	1125	504	168	79	79	91	120
Bosnia and Herzegovina	x	x	x	19670	1640	7437	9457	11547	12090	13889	13788
Bulgaria	32447	36703	38934	37824	34316	29223	29231	30186	32601	40179	34664
Croatia	x	x	x	1893	331	703	1140	848	1171	1197	1310
Cyprus*	-	-	74	97	20	49	53	22	27	13	13
Georgia	x	x	x	1323	44	27	18	226	113	147	262
Kazakhstan	x	x	x	89249	64825	44090	63767	72239	78029	85725	94131
Kosovo	x	x	x	..	..	5163	6619	8489	8903	8608	8684
Kyrgyzstan	x	x	x	6154	792	1127	1193	1262	1677	1776	2159
Latvia	x	x	x	920	252	97	120	130	167	187	312
Lithuania	x	x	x	1303	372	131	287	242	301	333	317
F.Y.R. of Macedonia	x	x	x	6937	7435	7769	7473	7480	6939	8025	7638
Malta	-	-	192	300	52	-	-	-	-	-	-
Republic of Moldova	x	x	x	4510	1315	181	167	181	186	198	155
Montenegro	x	x	x	..	..	..	1261	937	1869	1929	2000
Romania	26180	39373	53109	46223	45700	31962	36002	34483	31606	38059	35106
Russian Federation	x	x	x	374080	245331	230479	214594	186662	223820	224843	251120
Serbia	x	x	x	45937	40605	37324	35391	39085	37679	42273	38682
Tajikistan	x	x	x	1494	41	29	105	213	212	248	200
Turkmenistan	x	x	x	670	-	-	-	-	-	-	-
Ukraine	x	x	x	147423	89898	66680	64023	61377	66095	72929	78997
Uzbekistan	x	x	x	8940	3028	3542	3185	3817	3713	3905	4136
Former Soviet Union	647358	692140	700445	x	x	x	x	x	x	x	x
Former Yugoslavia	33896	44306	73306	x	x	x	x	x	x	x	x
Islam. Rep. of Iran	948	1783	1444	1061	1546	1781	2074	1137	952	740	1105
Lebanon	-	1	-	-	180	200	200	200	225	250	313
Syrian Arab Republic	1	1	-	-	-	-	-	-	-	-	-
United Arab Emirates	-	-	-	-	-	-	236	894	1162	2029	2750
Non-OECD Total	**1349499**	**1720091**	**2080985**	**2344769**	**2452164**	**2518365**	**3602769**	**4538481**	**4899788**	**5286900**	**5527473**
World	**3093433**	**3755721**	**4365540**	**4663599**	**4571337**	**4761675**	**5960125**	**6713128**	**7147049**	**7526694**	**7696904**

(1) Coal comprises all coals from anthracite through lignite, however excludes peat and manufacture of all coal products. For further information, see notes and definitions in Part I.

(2) Consumption data for 2012e are supplied by OECD member countries. Non-OECD country data are calculated from production and net trade data from varied sources. Stock changes are not accounted for, with the exceptions of the People's Republic of China, where stock changes have also been sourced independently; Mozambique, where producer stock builds were able to be estimated; and Brazil, Mongolia, Romania, Sri Lanka and Ukraine, where data was provided by official sources.

* Please refer to the Geographical notes in Part I.

Source: *IEA/OECD Energy Statistics of OECD Countries* , *IEA/OECD Energy Statistics of Non-OECD Countries*

Table 2.2: World coal[(1)] consumption

(thousand tonnes of coal equivalent)

	1973	1980	1985	1990	1995	2000	2005	2009	2010	2011	2012e[(2)]
Australia	32250	38985	43074	51013	54309	69020	73542	75461	71545	68846	70345
Austria	4311	4078	4742	4990	3959	4103	4419	3240	3624	3683	3329
Belgium	15207	15705	14798	15192	11781	10428	7075	4394	4711	4255	4258
Canada	21724	29148	36417	34957	35796	45055	41132	31629	31195	28000	27386
Chile	1695	1651	1759	3532	3219	4358	3826	5032	6558	7604	7970
Czech Republic	54023	50086	50247	46396	33766	31202	29155	24812	25925	25663	24518
Denmark	2697	8288	10334	8653	9236	5654	5272	5693	5417	4595	3476
Estonia	x	x	x	8234	4866	4192	4527	4227	5546	5756	5767
Finland	2732	5123	4786	5083	5886	4685	4182	4781	6152	4847	4115
France	38773	44811	34140	28230	22222	20551	19732	15048	16099	13529	15674
Germany	205035	206438	211078	185039	127428	115708	113443	100063	106633	107911	108819
Greece	2992	4649	8616	11481	11972	12910	12813	12035	11232	11267	11720
Hungary	9790	10441	9705	7874	6627	5645	4288	3809	4202	4320	4182
Iceland	0.9	11	62	64	62	97	112	94	101	100	100
Ireland	828	1072	1619	2945	2538	2589	2687	1636	1760	1796	2159
Israel	-	-	2639	3265	5758	9249	10581	10134	10581	11071	13170
Italy	12067	17541	21980	20845	17073	17406	22837	18478	20411	22999	22456
Japan	83146	86700	107448	110805	125826	138503	156498	144741	163760	153773	161744
Korea	11575	19148	32567	36512	38016	59932	70463	92379	104265	114216	111183
Luxembourg	298	359	199	164	181	145	104	88	87	75	69
Mexico	2478	3336	4072	4822	7110	8803	13434	12762	13299	13958	13789
Netherlands	4292	5477	9808	13100	13187	11233	11758	10520	10603	10421	11355
New Zealand	1836	1465	1430	1692	1624	1581	3135	2106	1874	2052	1991
Norway	772	951	1118	718	976	958	762	516	677	748	763
Poland	109504	144229	141978	116130	103602	83489	83174	77644	86032	84601	77993
Portugal	690	510	975	3929	5178	5482	4780	4086	2364	3154	4151
Slovak Republic	10095	11394	11701	10549	7600	6003	5991	5426	5202	5097	4946
Slovenia	x	x	x	2180	1913	1794	2140	2005	2041	2060	2083
Spain	12000	17736	27493	27179	26472	30447	29969	14947	11544	18006	21869
Sweden	1015	2019	3760	3566	3332	2832	3034	2402	2839	2990	2727
Switzerland	247	302	613	469	240	170	197	194	202	185	167
Turkey	7356	9968	17133	24047	23667	32055	32127	42450	45591	48171	49919
United Kingdom	109682	100725	88643	90091	67264	52262	53274	42179	43736	44754	55385
United States	443017	541605	607139	657253	676462	760150	796374	693557	718216	683928	608418
IEA Americas	464741	570753	643556	692210	712258	805205	837505	725187	749411	711928	635804
IEA Asia Oceania	128807	146298	184519	200022	219776	269036	303638	314687	341445	338888	345262
IEA Europe	604406	661900	675467	626671	504188	455958	451075	394440	415041	423067	434048
OECD Americas	468914	575740	649386	700564	722587	818365	854766	742980	769268	733490	657563
OECD Asia Oceania	128807	146298	187157	203287	225534	278284	314218	324821	352025	349959	358432
OECD Europe	604407	661911	675529	637149	511030	462040	457853	400765	422729	430984	441998
IEA Total	**1197954**	**1378952**	**1503541**	**1518902**	**1436222**	**1530198**	**1592218**	**1434314**	**1505896**	**1473883**	**1415114**
OECD Total	**1202128**	**1383949**	**1512073**	**1541000**	**1459152**	**1558690**	**1626837**	**1468567**	**1544023**	**1514433**	**1457993**

Table 2.2: World coal(1) consumption (continued)

(thousand tonnes of coal equivalent)

	1973	1980	1985	1990	1995	2000	2005	2009	2010	2011	2012e(2)
Algeria	57	89	1180	967	608	663	788	279	-	-	-
Botswana	-	-	410	717	803	916	882	653	873	697	651
Dem. Rep. of Congo	151	144	134	145	117	114	139	170	182	195	189
Egypt	437	860	1147	1289	1482	1751	1742	1723	2076	2018	2108
Kenya	63	14	79	133	137	94	128	136	236	335	423
Morocco	522	559	947	1593	2411	3784	4654	3866	3988	4253	4315
Mozambique	501	246	90	49	48	-	-	8.5	8.5	22	62
Namibia	-	-	-	-	12	2.3	15	38	27	10	9.3
Nigeria	259	133	83	48	18	2.6	7.0	30	33	28	28
Senegal	-	-	-	-	-	-	134	225	233	240	356
South Africa	48346	68117	96486	95056	111618	117142	131339	146932	144206	140681	140231
United Rep. of Tanzania	-	0.9	13	3.6	39	70	27	-	41	40	85
Tunisia	30	18	18	13	-	-	-	-	-	-	-
Zambia	793	521	397	316	125	110	118	0.8	-	-	-
Zimbabwe	2540	2408	2787	4933	4140	3935	3383	2496	2737	2992	2812
Other Africa	199	245	334	384	389	898	1073	1423	1577	1651	1644
Argentina	1048	1363	1193	1381	1427	1040	1433	1490	1538	1754	1556
Brazil	3151	8026	14337	13881	15248	17074	17043	15244	18949	19852	21055
Colombia	2655	2601	2918	4480	5207	3929	3875	5013	5759	6064	6759
Costa Rica	0.9	0.9	0.9	-	-	0.9	1.8	3.5	0.9	7.9	-
Cuba	-	86	114	139	70	20	20	36	32	33	-
Dominican Republic	-	-	197	15	70	82	419	658	624	776	419
Guatemala	-	19	-	-	-	189	360	259	433	413	586
Haiti	-	-	54	11	-	-	-	-	-	-	-
Honduras	-	-	-	-	-	119	161	91	94	142	44
Jamaica	-	-	-	46	48	47	51	45	48	49	108
Panama	12	-	28	28	45	53	-	-	-	209	275
Peru	86	74	107	149	389	708	1075	1142	1182	910	688
Uruguay	29	3.5	-	0.9	-	0.9	0.9	2.6	3.5	1.8	-
Venezuela	55	44	44	370	7.3	189	53	339	285	293	313
Bangladesh	174	168	70	402	458	471	642	1232	1291	1302	1428
Cambodia	-	-	-	-	-	-	-	11	12	13	144
Hong Kong (China)	11	2.6	4862	7859	8019	5333	9529	10855	9088	11030	11091
India	51756	70639	104438	147543	193105	228329	294832	422951	442229	463717	492865
Indonesia	116	193	735	5024	9043	17156	31610	43590	43547	44966	45318
DPR of Korea	25360	36683	44759	40087	26072	23836	25828	23207	22289	22659	22966
Malaysia	10.0	76	517	1936	2304	3297	9840	15067	20858	22260	22184
Mongolia	-	-	3263	3556	3182	2595	2689	3250	3512	3460	4891
Myanmar	66	206	214	86	23	457	472	512	585	588	1003
Nepal	67	71	15	69	105	369	354	276	433	415	425
Pakistan	821	988	1951	2799	3097	2660	5332	6344	5979	5848	4793
Philippines	17	407	1755	1869	2436	7121	8101	8319	10861	11831	14957
Singapore	0.3	0.4	0.8	0.8	-	-	0.9	-	-	-	-
Sri Lanka	-	-	0.9	7.0	4.5	0.1	93	94	97	543	741
Chinese Taipei	3220	5537	10298	16137	24573	42368	54438	54436	58053	58447	57831
Thailand	120	673	2317	5386	9737	10601	16350	21320	23115	25914	22710
Vietnam	2216	3242	3992	3161	4734	6246	11723	17918	20917	22182	19454
Other Asia	1282	2642	304	267	215	489	603	1087	1624	1738	3206
PR of China	292827	446724	559358	742865	937586	998018	1673945	2224571	2430114	2665464	2794788

Table 2.2: World coal[(1)] consumption (continued)
(thousand tonnes of coal equivalent)

	1973	1980	1985	1990	1995	2000	2005	2009	2010	2011	2012e[(2)]
Albania	488	859	1279	899	27	25	26	76	83	103	18
Armenia	x	x	x	350	1.9	-	-	-	0.6	-	-
Azerbaijan	x	x	x	127	3.8	-	-	-	-	-	-
Belarus	x	x	x	2009	980	439	138	65	63	78	102
Bosnia and Herzegovina	x	x	x	5969	498	3517	4725	5808	6226	7203	7025
Bulgaria	11487	13028	14334	12761	10691	9222	9792	9098	9803	11518	9629
Croatia	x	x	x	1455	218	598	954	698	948	976	1089
Cyprus*	-	-	65	92	19	46	51	21	24	11	11
Georgia	x	x	x	1280	30	19	14	159	70	90	159
Kazakhstan	x	x	x	56122	40613	27626	39866	45169	48761	53274	58977
Kosovo	x	x	x	..	..	1380	1770	2264	2392	2318	2312
Kyrgyzstan	x	x	x	3615	464	668	716	733	993	1032	1235
Latvia	x	x	x	892	245	94	107	116	149	167	279
Lithuania	x	x	x	1117	319	102	222	196	245	283	271
F.Y.R. of Macedonia	x	x	x	1897	2036	1835	2041	1910	1842	2080	2027
Malta	-	-	169	264	46	-	-	-	-	-	-
Republic of Moldova	x	x	x	2859	834	115	106	128	133	140	115
Montenegro	x	x	x	..	..	..	401	295	587	606	629
Romania	9862	15252	20352	17588	15559	10296	11983	10194	9024	10740	10294
Russian Federation	x	x	x	269903	183716	171792	163117	137290	165775	167495	189460
Serbia	x	x	x	14531	12649	12086	10886	10798	10313	11877	10735
Tajikistan	x	x	x	887	22	17	65	132	131	153	127
Turkmenistan	x	x	x	425	-	-	-	-	-	-	-
Ukraine	x	x	x	120835	72331	55805	53039	51966	55965	61143	65753
Uzbekistan	x	x	x	4837	1525	1780	1602	1922	1865	1961	2099
Former Soviet Union	436639	459831	423673	x	x	x	x	x	x	x	x
Former Yugoslavia	12816	17229	27067	x	x	x	x	x	x	x	x
Islam. Rep. of Iran	851	1709	1381	1015	1475	1701	1947	1086	908	705	1055
Lebanon	-	0.9	-	-	170	189	189	189	212	236	296
Syrian Arab Republic	0.9	0.9	-	-	-	-	-	-	-	-	-
United Arab Emirates	-	-	-	-	-	-	208	787	1023	1786	2421
Non-OECD Total	**911140**	**1161733**	**1350267**	**1626931**	**1713622**	**1801633**	**2619272**	**3318437**	**3597309**	**3882016**	**4071632**
World	**2113269**	**2545682**	**2862339**	**3167931**	**3172773**	**3360323**	**4246109**	**4787004**	**5141332**	**5396448**	**5529625**

(1) Coal comprises all coals from anthracite through lignite, however excludes peat and manufacture of all coal products. For further information, see notes and definitions in Part I.

(2) Consumption data for 2012e are supplied by OECD member countries. Non-OECD country data are calculated from production and net trade data from varied sources. Stock changes are not accounted for, with the exceptions of the People's Republic of China, where stock changes have also been sourced independently; Mozambique, where producer stock builds were able to be estimated; and Brazil, Mongolia, Romania, Sri Lanka and Ukraine, where data was provided by official sources.

* Please refer to the Geographical notes in Part I.

Source: *IEA/OECD Energy Statistics of OECD Countries* , *IEA/OECD Energy Statistics of Non-OECD Countries*

Table 2.3: World coking coal consumption
(thousand tonnes)

	1978	1985	1990	1995	2000	2005	2008	2009	2010	2011	2012e[1]
Australia	7216	5476	5932	5874	4799	4462	4547	3482	4146	4420	4053
Austria	2006	2391	2337	1908	1877	1899	1863	1693	1838	1785	1798
Belgium	7524	8086	7157	4737	4045	3263	2912	1958	2627	2579	2600
Canada	6776	6483	5023	4417	4461	4289	5084	3672	3720	5556	4743
Chile	385	427	492	715	714	703	675	591	509	642	725
Czech Republic	12567	11365	9941	6648	4972	4334	4301	2932	3369	3345	3163
Denmark	-	6	-	-	-	-	-	-	-	-	-
Finland	-	-	711	1650	1284	1401	1077	915	1206	1279	1184
France	12980	11332	9669	7739	6543	6222	7218	3504	4504	3753	4592
Germany	41881	47825	42216	33993	24462	22193	20013	15472	15113	16479	15346
Greece	213	-	-	-	-	-	-	-	-	-	-
Hungary	1740	1043	971	1402	1280	808	1432	1022	1415	1465	1428
Iceland	-	20	-	-	-	-	-	-	-	-	-
Ireland	-	7	14	-	-	-	-	-	-	-	-
Italy	9909	10165	8633	6966	6658	5630	5925	3585	5145	5870	4104
Japan	58724	73456	64934	59796	57081	56528	57366	52334	57679	53817	52199
Korea	2009	6959	11735	16305	19415	20883	23835	21182	27210	32579	31590
Mexico	4053	3837	2911	2693	2724	2445	2854	2440	2696	2864	2673
Netherlands	2953	4061	4391	4893	4054	4682	4419	3423	3953	4245	4177
New Zealand	-	6	243	140	1	115	1	-	67	1	1
Norway	414	394	-	-	-	-	-	-	-	-	-
Poland	25847	17745	18127	17405	13332	11157	13012	9907	12336	11903	11563
Portugal	423	370	313	455	497	-	-	-	-	-	-
Slovak Republic	2180	2790	3136	2745	2597	2738	2617	2391	2490	2503	2623
Slovenia	x	x	1	-	-	-	-	-	-	-	-
Spain	4863	4908	4456	3312	3556	3463	3453	2502	2498	2539	2350
Sweden	1227	1620	1515	1646	1772	1847	1586	1440	1868	1633	1480
Turkey	3429	4825	5335	4658	7042	5585	6878	7151	7515	7997	8700
United Kingdom	14988	11122	10517	8487	8824	6569	6931	5763	6372	6271	5511
United States	68885	37246	35269	29934	25963	20893	19989	14042	19152	19403	19002
IEA Americas	75661	43729	40292	34351	30424	25182	25073	17714	22872	24959	23745
IEA Asia Oceania	67949	85897	82844	82115	81296	81988	85749	76998	89102	90817	87843
IEA Europe	145144	140055	129439	108644	92795	81791	83637	63658	72249	73646	70619
OECD Americas	80099	47993	43695	37759	33862	28330	28602	20745	26077	28465	27143
OECD Asia Oceania	67949	85897	82844	82115	81296	81988	85749	76998	89102	90817	87843
OECD Europe excl Estonia	145144	140075	129440	108644	92795	81791	83637	63658	72249	73646	70619
IEA Total	**288754**	**269681**	**252575**	**225110**	**204515**	**188961**	**194459**	**158370**	**184223**	**189422**	**182207**
OECD Total	**293192**	**273965**	**255979**	**228518**	**207953**	**192109**	**197988**	**161401**	**187428**	**192928**	**185605**

Table 2.3: World coking coal consumption (continued)

(thousand tonnes)

	1978	1985	1990	1995	2000	2005	2008	2009	2010	2011	2012e(1)
Algeria	50	1226	1005	632	689	819	856	290	-	-	-
Egypt	975	1192	1339	1540	1820	1810	2003	1791	2158	2097	2191
South Africa	7018	6000	5675	4220	2569	2975	3346	3028	3950	3573	2166
United Rep. of Tanzania	-	-	1	1	-	-	-	-	-	-	-
Zimbabwe	946	269	612	621	813	690	305	323	354	386	386
Argentina	990	826	1121	589	558	789	1018	994	686	686	629
Brazil	4869	9602	10489	11093	9936	9742	10345	9146	10993	11381	10852
Colombia	705	745	775	681	587	514	541	1198	2621	2569	3472
Peru	40	50	37	51	44	-	-	-	-	-	-
India	15757	28149	39491	39409	35852	39041	52245	62385	74256	79795	83801
Indonesia	-	-	-	-	140	98	107	69	55	42	-
DPR of Korea	3857	5000	5143	1040	-	-	-	-	-	-	-
Mongolia	-	-	-	-	-	-	155	96	123	115	109
Pakistan	16	716	1102	1085	950	565	1200	431	429	275	220
Chinese Taipei	1386	2562	4150	4208	5236	4919	4597	4175	5642	5879	8412
PR of China	52304	62873	80140	139861	119038	298365	387528	438568	502049	541147	579368
Albania	25	33	62	-	-	-	-	-	-	-	-
Bosnia and Herzegovina	x	x	-	-	-	592	708	859	1264	1276	1172
Bulgaria	1921	1553	1854	1693	1325	1051	434	-	-	-	-
Croatia	x	x	747	-	-	-	-	-	-	-	-
Georgia	x	x	686	6	-	-	-	-	-	-	-
Kazakhstan	x	x	29983	10853	10343	10734	10332	10718	11612	12115	12625
F.Y.R. of Macedonia	x	x	117	70	67	-	-	-	-	-	-
Republic of Moldova	x	x	-	-	-	-	-	2	-	-	-
Romania	5734	7825	5082	5047	2257	2963	1951	337	131	125	746
Russian Federation	x	x	53885	50669	43938	44991	42200	46807	49701	53665	56348
Tajikistan	x	x	-	7	2	2	2	2	2	2	-
Ukraine	x	x	54508	31642	30581	29821	27722	24771	26369	27487	27130
Former Soviet Union	129250	123985	x	x	x	x	x	x	x	x	x
Former Yugoslavia	1447	4689	x	x	x	x	x	x	x	x	x
Islam. Rep. of Iran	1107	1344	986	1388	1626	1484	1467	1033	856	649	1005
Non-OECD Total	**228397**	**258639**	**298990**	**306406**	**268371**	**451965**	**549062**	**607023**	**693251**	**743264**	**790632**
World	**521589**	**532604**	**554969**	**534924**	**476324**	**644074**	**747050**	**768424**	**880679**	**936192**	**976237**

(1) Consumption data for 2012e are supplied by OECD member countries. Non-OECD country data are calculated from production and net trade data from varied sources. Stock changes are not accounted for, with the exceptions of the People's Republic of China, where stock changes have also been sourced independently; Mozambique, where producer stock builds were able to be estimated; and Brazil, Mongolia, Romania, Sri Lanka and Ukraine, where data was provided by official sources.

Source: *IEA/OECD Energy Statistics of OECD Countries* , *IEA/OECD Energy Statistics of Non-OECD Countries*

Table 2.4: World steam coal[1] consumption

(thousand tonnes)

	1973[2]	1980	1985	1990	1995	2000	2005	2009	2010	2011	2012e[3]
Australia[4]	27292	27373	34504	43367	45865	56031	64606	68348	61909	58230	59699
Austria	2856	500	787	1822	1484	1885	2215	1643	1966	2068	1679
Belgium[4]	17169	8994	7490	8996	7657	7001	4288	2724	2293	1820	1799
Canada	17556	24269	31904	34765	37318	47299	44782	35158	34316	27860	27622
Chile[4]	1711	1224	1365	3228	2676	3876	3868	6226	7842	9146	9695
Czech Republic	27780	11771	10665	10119	7532	5745	4887	3825	4443	4674	4147
Denmark	3146	9662	11929	9992	11003	6641	6293	6778	6496	5521	4176
Estonia	x	x	x	382	85	87	56	87	60	70	65
Finland[4]	3035	5692	5318	4937	4890	3909	3197	4498	5776	4198	3449
France[4]	40289	33510	25175	19122	14872	15258	14920	12818	12829	10794	12231
Germany	105801	46092	45641	44749	40231	44501	41759	36544	43475	41541	40874
Greece	651	161	1750	1380	1480	1121	563	337	614	380	97
Hungary	2137	3249	3456	1435	287	390	1269	757	653	487	411
Iceland[4]	1	12	49	65	65	101	117	98	106	105	104
Ireland	822	1066	1579	3184	2689	2938	2988	1845	2002	2051	2422
Israel	-	-	2927	3720	6568	10591	12124	11847	12310	12746	15168
Italy	11603	5787	11729	12694	10480	11355	18610	16357	16616	18570	20029
Japan[4]	81690	17496	35935	50026	73768	96109	120444	111382	127711	120322	131569
Korea[4]	16329	23803	35546	33041	28329	52384	61389	85679	92838	98315	95666
Luxembourg	305	346	199	197	217	172	122	104	102	89	82
Mexico[4]	2894	-	1480	3970	7550	9566	15111	13923	14992	15764	15690
Netherlands	4794	2452	6215	9641	9388	8686	8327	8448	7920	7429	8615
New Zealand[4]	2317	1760	1804	1841	1755	1882	3925	2726	2286	2613	2525
Norway	772	519	724	749	1018	999	795	538	706	780	796
Poland	122097	138506	138825	102104	90418	70039	69281	65823	72452	71624	64055
Portugal[4]	805	241	680	4084	5253	5657	5476	4677	2702	3697	4873
Slovak Republic	5834	3037	3019	2743	2585	2059	2245	1936	1673	1555	1294
Slovenia	x	x	x	262	328	446	612	492	495	472	488
Spain	13260	14715	26058	25788	28696	33695	33471	17176	12163	21647	26516
Sweden	1060	484	2538	2194	1798	1089	1223	993	991	1410	1296
Switzerland	258	315	640	481	245	173	178	155	166	144	138
Turkey	4595	1781	1887	3098	3956	8506	14574	15902	18854	19240	23542
United Kingdom	133527	111982	94858	96205	67429	51015	55210	43023	45075	45229	58473
United States[4]	492567	546581	645362	701657	752854	866163	932692	839346	862251	826136	730720
IEA Americas	510123	570850	677266	736422	790172	913462	977474	874504	896567	853996	758342
IEA Asia Oceania	127628	70432	107789	128275	149717	206406	250364	268135	284744	279480	289459
IEA Europe	502596	400862	401162	365714	313608	282834	291891	246901	259967	264948	280994
OECD Americas	514728	572074	680111	743620	800398	926904	996453	894653	919401	878906	783727
OECD Asia Oceania	127628	70432	110716	131995	156285	216997	262488	279982	297054	292226	304627
OECD Europe	502597	400874	401211	366423	314086	283468	292676	247578	260628	265595	281651
IEA Total	**1140347**	**1042144**	**1186217**	**1230411**	**1253497**	**1402702**	**1519729**	**1389540**	**1441278**	**1398424**	**1328795**
OECD Total	**1144953**	**1043380**	**1192038**	**1242038**	**1270769**	**1427369**	**1551617**	**1422213**	**1477083**	**1436727**	**1370005**

Table 2.4: World steam coal[1] consumption (continued)

(thousand tonnes)

	1973[2]	1980	1985	1990	1995	2000	2005	2009	2010	2011	2012e[3]
Algeria	63	-	-	-	-	-	-	-	-	-	-
Botswana	-	-	466	815	912	1040	1002	742	992	792	740
Dem. Rep. of Congo	170	167	156	169	136	132	162	198	212	226	220
Egypt	487	-	-	1	-	-	-	-	-	-	-
Kenya	70	16	90	151	156	107	145	154	268	380	480
Morocco	582	635	1110	1774	2665	4018	4938	4100	4230	4511	4577
Mozambique	587	288	106	58	56	-	-	10	10	26	73
Namibia	-	-	-	-	16	3	20	49	35	13	12
Nigeria	289	151	94	55	20	3	8	34	38	32	32
Senegal	-	-	-	-	-	-	152	254	264	271	403
South Africa	60408	79803	119870	119225	142985	154566	172428	189136	186285	182767	185012
United Rep. of Tanzania	-	1	15	3	43	79	31	-	47	45	96
Tunisia	33	21	21	15	-	-	-	-	-	-	-
Zambia	941	618	471	375	148	130	140	1	-	-	-
Zimbabwe	2758	1623	2757	4743	3873	3459	2983	2387	2617	2862	2667
Other Africa	222	278	379	436	442	1020	1219	1616	1791	1876	1867
Argentina	1072	624	421	246	850	500	594	464	818	1030	892
Brazil	2842	2066	5212	2703	3375	7463	7686	6792	9238	9351	11143
Colombia	2859	2085	2397	4050	4927	3644	3659	4201	3581	3961	3807
Costa Rica	1	1	1	-	-	1	2	4	1	9	-
Cuba	-	95	126	153	77	22	22	40	36	37	-
Dominican Republic	-	-	224	17	80	93	476	747	709	882	476
Guatemala	-	22	-	-	-	215	409	294	492	469	666
Haiti	-	-	61	12	-	-	-	-	-	-	-
Honduras	-	-	-	-	-	135	183	103	107	161	50
Jamaica	-	-	-	52	55	53	58	51	54	56	123
Panama	13	-	32	32	51	60	-	-	-	237	312
Peru	86	25	57	112	338	664	1075	1142	1182	910	688
Uruguay	32	4	-	1	-	1	1	3	4	2	-
Venezuela	53	42	42	355	7	181	51	325	273	281	300
Bangladesh	243	235	98	563	642	660	899	1725	1808	1824	2000
Cambodia	-	-	-	-	-	-	-	16	17	19	195
Hong Kong (China)	12	3	5523	8928	9109	6058	10824	12331	10324	12529	12599
India	73410	87296	120167	166231	233168	296333	394230	553355	564951	588414	625858
Indonesia[5]	129	236	925	6320	11892	22580	41933	57816	57799	59674	60198
DPR of Korea[5]	23580	40170	49200	43310	30900	29383	31806	28576	27441	27903	28252
Malaysia	13	84	574	2150	2558	3661	10926	16730	23161	24717	24633
Mongolia	-	-	480	595	1290	70	-	-	-	-	-
Myanmar	74	221	223	80	15	468	504	548	646	646	1139
Nepal	78	83	17	81	123	430	413	322	505	484	496
Pakistan	1270	1098	1567	2320	2546	2166	5139	6508	6188	6137	4975
Philippines[5]	-	558	2415	2573	3001	8600	9909	10060	13319	14639	19007
Singapore	-	-	-	-	-	-	1	-	-	-	-
Sri Lanka	-	-	1	8	5	-	93	93	95	514	711
Chinese Taipei	3572	4443	8523	13080	22021	41544	54797	55843	57773	60622	56118
Thailand	1	94	212	250	2305	3684	8479	15475	17378	19468	16797
Vietnam	2770	4052	4990	3951	5917	7808	14490	22397	26146	27728	24317
Other Asia	220	657	345	303	244	530	632	1186	1793	1918	3421
PR of China[5]	414180	559188	741034	969492	1172771	1218487	1921253	2496677	2698634	2960147	3086525

Table 2.4: World steam coal(1) consumption (continued)

(thousand tonnes)

	1973(2)	1980	1985	1990	1995	2000	2005	2009	2010	2011	2012e(3)
Albania	89	135	187	240	-	-	-	-	-	-	-
Armenia	x	x	x	552	3	-	-	-	1	-	-
Azerbaijan	x	x	x	200	6	-	-	-	-	-	-
Belarus	x	x	x	2389	1125	504	168	79	79	91	120
Bosnia and Herzegovina	x	x	x	-	-	4057	4556	5382	5224	6509	6341
Bulgaria	6136	5073	6724	4192	4944	2054	3310	3015	3156	3155	2153
Croatia	x	x	x	403	143	623	1057	799	1112	1121	1306
Cyprus*	-	-	74	97	20	49	52	21	26	12	13
Georgia	x	x	x	637	38	27	18	226	113	147	262
Kazakhstan	x	x	x	55823	50342	31309	48836	56812	61120	65581	76267
Kosovo	x	x	x	..	..	9	12	7	34	40	-
Kyrgyzstan	x	x	x	4014	505	781	892	761	1150	1077	1161
Latvia	x	x	x	917	252	97	120	130	167	187	312
Lithuania	x	x	x	1303	372	130	284	242	300	332	316
F.Y.R. of Macedonia	x	x	x	12	72	-	97	110	155	210	162
Malta	-	-	192	300	52	-	-	-	-	-	-
Republic of Moldova	x	x	x	4510	1315	181	167	179	186	198	155
Romania	8490	6781	6880	4269	843	392	715	788	645	869	694
Russian Federation	x	x	x	186148	112597	98284	96447	69630	97843	93609	116929
Serbia	x	x	x	137	55	306	227	170	147	196	60
Tajikistan	x	x	x	1044	34	27	103	211	210	246	200
Turkmenistan	x	x	x	670	-	-	-	-	-	-	-
Ukraine(5)	x	x	x	84932	55256	35306	33793	36606	39726	45442	51867
Uzbekistan	x	x	x	2740	81	69	73	101	65	60	232
Former Soviet Union	490220	397576	419445	x	x	x	x	x	x	x	x
Former Yugoslavia	2540	388	400	x	x	x	x	x	x	x	x
Islam. Rep. of Iran	948	83	100	75	158	155	590	104	96	91	100
Lebanon	-	1	-	-	180	200	200	200	225	250	313
Syrian Arab Republic	1	1	-	-	-	-	-	-	-	-	-
United Arab Emirates	-	-	-	-	-	-	236	894	1162	2029	2750
Non-OECD Total	**1101544**	**1197021**	**1504404**	**1711392**	**1888112**	**1994611**	**2895725**	**3668972**	**3934204**	**4240022**	**4443590**
World	**2246497**	**2240401**	**2696442**	**2953430**	**3158881**	**3421980**	**4447342**	**5091185**	**5411287**	**5676749**	**5813595**

(1) Steam coal is also commonly known as thermal coal. From 1978 onwards it comprises anthracite, bituminous coal and sub-bituminous coal. For further information, see notes and definitions in Part I.

(2) Data prior to 1978 are hard coal. Hard coal comprises anthracite, coking coal, other bituminous coal and for certain countries sub-bituminous coal.

(3) Consumption data for 2012e are supplied by OECD member countries. Non-OECD country data are calculated from production and net trade data from varied sources. Stock changes are not accounted for, with the exceptions of the People's Republic of China, where stock changes have also been sourced independently; Mozambique, where producer stock builds were able to be estimated; and Brazil, Mongolia, Romania, Sri Lanka and Ukraine, where data was provided by official sources.

(4) Includes sub-bituminous coal prior to 1978.

(5) Data includes lignite for at least some years.

* Please refer to the Geographical notes in Part I.

Source: *IEA/OECD Energy Statistics of OECD Countries* , *IEA/OECD Energy Statistics of Non-OECD Countries*

Table 2.5: World lignite[(1)] consumption

(thousand tonnes)

	1973[(2)]	1980	1985	1990	1995	2000	2005	2009	2010	2011	2012e[(3)]
Australia[(4)]	24121	32894	38380	45990	50752	67293	70533	72037	72090	71000	73538
Austria	3328	3274	3868	2504	1743	1343	1211	7	14	17	1
Belgium[(4)]	-	95	275	276	195	220	206	173	244	292	292
Canada	8058	5682	9788	9358	10899	11208	11140	10423	10136	9609	9397
Chile[(4)]	61	40	35	-	-	-	-	-	-	-	-
Czech Republic	75965	79822	83854	71772	52300	50372	47622	44238	43732	44101	42576
Estonia	x	x	x	25954	15438	13232	14804	13769	17889	18739	18794
France[(4)]	2775	2570	2418	2094	1516	355	36	51	52	82	76
Germany	372497	391753	432521	364050	194811	169942	177892	169846	169520	176677	185171
Greece	13000	22692	36214	52053	56962	64564	70096	65213	57704	59978	63726
Hungary	26896	22600	21127	17899	15242	13503	9511	8897	8921	9662	9566
Ireland	-	-	-	38	45	35	51	24	21	22	28
Israel	-	-	-	303	470	390	429	444	432	416	400
Italy	1299	1385	2041	1089	196	30	8	7	6	5	5
Japan[(4)]	100	27	-	-	-	-	-	-	-	-	-
Luxembourg	-	28	-	-	-	-	-	-	-	-	-
Mexico[(4)]	-	-	-	-	2	4	3	4	4	3	3
Netherlands	20	156	103	69	33	46	50	28	30	38	41
New Zealand[(4)]	143	208	264	159	243	213	246	260	292	319	327
Poland	34282	35308	57565	67391	63196	59488	61589	57084	56593	62708	64114
Slovak Republic	12784	16039	15784	12481	7221	4213	3307	3266	3051	3151	2994
Slovenia	x	x	x	5827	4911	4479	4580	4450	4419	4557	4569
Spain	3062	11100	17474	16579	10534	8403	7564	-	-	-	-
Switzerland	-	-	-	13	8	6	39	66	62	69	50
Turkey	7642	15243	34778	45891	52405	64384	56577	75641	69239	73933	65950
United States[(4)]	12948	42129	62063	79023	80764	74265	76136	65105	68299	74761	72203
IEA Americas	21006	47811	71851	88381	91663	85473	87276	75528	78435	84370	81600
IEA Asia Oceania	24364	33129	38644	46149	50995	67506	70779	72297	72382	71319	73865
IEA Europe	553550	602065	708022	654199	456407	436904	435759	424541	409189	430735	434590
OECD Americas	21067	47851	71886	88381	91665	85477	87279	75532	78439	84373	81603
OECD Asia Oceania	24364	33129	38644	46452	51465	67896	71208	72741	72814	71735	74265
OECD Europe	553550	602065	708022	685980	476756	454615	455143	442760	431497	454031	457953
IEA Total	**598920**	**683005**	**818517**	**788729**	**599065**	**589883**	**593814**	**572366**	**560006**	**586424**	**590055**
OECD Total	**598981**	**683045**	**818552**	**820813**	**619886**	**607988**	**613630**	**591033**	**582750**	**610139**	**613821**

Table 2.5: World lignite[1] consumption (continued)

(thousand tonnes)

	1973[2]	1980	1985	1990	1995	2000	2005	2009	2010	2011	2012e[3]
Brazil	1280	922	2047	2244	2652	2871	2575	1482	1476	2136	2856
India	3762	5059	7913	14985	22298	24824	30239	34409	37688	41891	43491
DPR of Korea[5]	7000	-	-	-	-	-	-	-	-	-	-
Mongolia	-	-	5687	6054	3914	5142	5472	6426	6906	6816	9738
Myanmar	-	27	43	38	23	112	70	72	40	47	-
Pakistan	-	471	671	824	1091	928	2010	1200	1101	1258	1167
Philippines[5]	40	-	4	3	3	3	-	-	-	-	-
Singapore	1	1	2	2	-	-	-	-	-	-	-
Thailand	361	1525	5132	12457	18496	17586	21046	17843	18041	20257	18326
Vietnam	-	-	-	-	-	-	322	-	-	-	-
Other Asia	2206	4200	-	-	-	45	95	88	93	100	396
Albania	810	1420	2150	1843	80	73	77	226	248	307	54
Bosnia and Herzegovina	x	x	x	19670	1640	3380	4309	5306	5602	6104	6275
Bulgaria	26311	29704	30657	31778	27679	25844	24870	27171	29445	37024	32511
Croatia	x	x	x	743	188	80	83	49	59	76	4
Cyprus*	-	-	-	-	-	-	1	1	1	1	-
Kazakhstan	x	x	x	3443	3630	2438	4197	4709	5297	8029	5239
Kosovo	x	x	x	..	..	5154	6607	8482	8869	8568	8684
Kyrgyzstan	x	x	x	2140	287	346	301	501	527	699	998
Latvia	x	x	x	3	-	-	-	-	-	-	-
Lithuania	x	x	x	-	-	1	3	-	1	1	1
F.Y.R. of Macedonia	x	x	x	6808	7293	7702	7376	7370	6784	7815	7476
Montenegro	x	x	x	..	..	..	1261	937	1869	1929	2000
Romania	17690	27364	38404	36872	39810	29313	32324	33358	30830	37065	33666
Russian Federation	x	x	x	134047	82065	88257	73156	70225	76276	77569	77843
Serbia	x	x	x	45800	40550	37018	35164	38915	37532	42077	38622
Tajikistan	x	x	x	450	-	-	-	-	-	-	-
Ukraine[5]	x	x	x	7983	3000	793	409	-	-	-	-
Uzbekistan	x	x	x	6200	2947	3473	3112	3716	3648	3845	3904
Former Soviet Union	157138	163030	157015	x	x	x	x	x	x	x	x
Former Yugoslavia	31356	40516	68217	x	x	x	x	x	x	x	x
Non-OECD Total	**247955**	**274239**	**317942**	**334387**	**257646**	**255383**	**255079**	**262486**	**272333**	**303614**	**293251**
World	**846936**	**957284**	**1136494**	**1155200**	**877532**	**863371**	**868709**	**853519**	**855083**	**913753**	**907072**

(1) Lignite includes oil shale. Some countries, most notably the People's Republic of China and Indonesia, produce and consume lignite, however these data are reported under other coal types included in steam coal and not shown here.

(2) Data before 1978 are brown coal, which may include sub-bituminous coal.

(3) Consumption data for 2012e are supplied by OECD member countries. Non-OECD country data are calculated from production and net trade data from varied sources. Stock changes are not accounted for, with the exceptions of the People's Republic of China, where stock changes have also been sourced independently; Mozambique, where producer stock builds were able to be estimated; and Brazil, Mongolia, Romania, Sri Lanka and Ukraine, where data was provided by official sources.

(4) Brown coal data excludes sub-bituminous coal.

(5) Data are reported as other coal types for at least some years.

* Please refer to the Geographical notes in Part I.

For further information, see notes and definitions in Part I.

Source: *IEA/OECD Energy Statistics of OECD Countries , IEA/OECD Energy Statistics of Non-OECD Countries*

Table 2.6: OECD consumption of coke oven coke[(1)]

(thousand tonnes)

	1973	1980	1985	1990	1995	2000	2005	2009	2010	2011	2012e
Australia	4983	5077	3520	3300	3952	3456	2729	2348	2705	3175	2144
Austria	2681	2674	2993	2403	2354	2436	2733	2128	2567	2577	2587
Belgium	8429	6542	5789	5265	4287	3928	2951	1415	1750	1784	1538
Canada	5454	5519	5212	3412	3682	3429	3361	1873	3453	2808	2884
Chile	333	399	334	373	590	506	540	431	319	582	588
Czech Republic	5913	6244	6472	5713	3857	3144	2949	2290	2610	2743	2433
Denmark	96	116	68	40	43	40	34	26	23	24	20
Estonia	x	x	x	20	1	2	-	-	-	-	-
Finland	877	1224	1218	1272	1192	1429	1416	1034	1270	1245	1163
France	14783	13180	10353	7714	6219	6119	5043	4102	4234	4074	3847
Germany	37220	32691	29657	21127	14879	14982	12209	9770	12243	11198	11086
Greece	414	261	55	42	12	1	4	1	1	-	-
Hungary	2317	2243	2000	1183	990	804	751	644	754	756	726
Iceland	-	16	34	30	18	47	33	31	30	32	43
Ireland	13	8	22	29	6	-	-	-	-	-	-
Israel	2	1	1	-	-	-	-	-	-	-	-
Italy	7099	7379	7258	6413	5656	5041	5271	2487	3938	4515	3762
Japan	51800	45795	44859	49890	41939	41604	43191	37964	42572	38838	39508
Korea	356	3086	5359	8800	10593	12288	9276	9888	14178	15755	15518
Luxembourg	3235	2282	1854	1447	521	1	1	1	1	1	1
Mexico	2173	2494	3022	2528	2951	3591	3166	1644	2516	2507	2476
Netherlands	2657	2363	2539	2339	2454	2027	2082	1758	2146	2162	2093
New Zealand	20	13	6	373	369	349	418	414	458	478	486
Norway	826	849	881	529	500	559	356	294	425	474	461
Poland	14212	17849	14671	9854	8294	5762	3399	2693	3165	2977	2803
Portugal	306	317	401	240	296	323	5	3	3	3	1
Slovak Republic	2422	1642	2126	2833	1964	1815	1984	1752	2078	1877	1636
Slovenia	x	x	x	70	57	72	60	30	33	35	35
Spain	5349	3917	3758	3365	3131	1923	2027	1732	1841	1714	1573
Sweden	1900	1618	1494	1422	1496	1495	1701	840	1424	1323	1163
Switzerland	185	131	82	43	31	27	20	22	18	20	21
Turkey	1251	1954	2803	3256	3201	3592	3428	3500	4454	4212	4439
United Kingdom	17034	7296	9209	8244	6517	6114	5001	3735	3425	3085	3486
United States	64235	37446	26553	25230	22180	21085	16547	9364	13469	14356	14037
IEA Americas	69689	42965	31765	28642	25862	24514	19908	11237	16922	17164	16921
IEA Asia Oceania	57159	53971	53744	62363	56853	57697	55614	50614	59913	58246	57656
IEA Europe	129219	112780	105703	84773	67900	61562	53365	40227	48370	46764	44839
OECD Americas	72195	45858	35121	31543	29403	28611	23614	13312	19757	20253	19985
OECD Asia Oceania	57161	53972	53745	62363	56853	57697	55614	50614	59913	58246	57656
OECD Europe	129219	112796	105737	84893	67976	61683	53458	40288	48433	46831	44917
IEA Total	**256067**	**209716**	**191212**	**175778**	**150615**	**143773**	**128887**	**102078**	**125205**	**122174**	**119416**
OECD Total	**258575**	**212626**	**194603**	**178799**	**154232**	**147991**	**132686**	**104214**	**128103**	**125330**	**122558**

(1) Solid product obtained from carbonization of coal, principally coking coal, used mainly in the iron and steel industry.
Also includes coke and semi-coke made from lignite.
For further information, see notes and definitions in Part I.

Source: *IEA/OECD Energy Statistics of OECD Countries*

Table 2.7: Key energy and CO_2 emissions data for OECD countries (2011)

	TPES	Energy-related CO_2 emissions					% Coal in total energy-related CO_2 emissions	Energy-related CO_2/TPES	Energy-related CO_2/GDP	Energy-related CO_2/population	Share in total OECD TPES	Share in total OECD energy-related CO_2 emissions
		Coal	Oil	Gas	Other*	Total						
	(Mtce)	(Million tonnes CO_2)					(%)	(t CO_2 / tce)	(kg CO_2/ 2005 USD)	(t CO_2 / capita)	(%)	(%)
Australia	175.55	198.82	123.73	73.76	0.46	396.77	50.1	4.61	0.44	17.43	2.3	3.2
Austria	47.17	14.98	31.85	17.86	3.81	68.49	21.9	2.96	0.20	8.13	0.6	0.6
Belgium	84.42	10.63	50.59	34.92	12.45	108.59	9.8	2.63	0.27	9.89	1.1	0.9
Canada	359.78	84.16	253.14	191.80	0.73	529.84	15.9	3.01	0.43	15.37	4.7	4.3
Chile	47.96	20.95	44.50	10.85	0.01	76.31	27.5	3.25	0.49	4.42	0.6	0.6
Czech Republic	62.04	73.79	22.15	15.74	0.99	112.68	65.5	3.71	0.75	10.73	0.8	0.9
Denmark	25.71	12.65	18.45	8.70	1.88	41.68	30.3	3.31	0.16	7.48	0.3	0.3
Estonia	8.00	15.09	3.03	1.18	0.00	19.30	78.2	4.92	1.27	14.40	0.1	0.2
Finland	49.64	22.91	24.39	7.65	0.66	55.61	41.2	2.29	0.27	10.32	0.7	0.5
France	361.18	35.14	204.60	82.92	5.65	328.31	10.7	1.86	0.15	5.04	4.8	2.7
Germany	445.39	311.10	255.85	163.03	17.60	747.58	41.6	3.43	0.25	9.14	5.9	6.1
Greece	38.18	33.52	41.23	8.76	0.13	83.64	40.1	4.47	0.37	7.40	0.5	0.7
Hungary	35.66	10.49	15.36	21.09	0.46	47.39	22.1	2.71	0.43	4.75	0.5	0.4
Iceland	8.19	0.37	1.48	0.00	0.00	1.85	20.2	0.46	0.11	5.81	0.1	0.0
Ireland	18.88	7.98	17.43	9.46	0.06	34.93	22.8	3.78	0.17	7.63	0.2	0.3
Israel	33.22	29.79	28.02	9.42	0.00	67.24	44.3	4.13	0.39	8.66	0.4	0.5
Italy	239.17	58.38	181.47	147.97	5.15	392.97	14.9	3.35	0.22	6.47	3.2	3.2
Japan	659.24	400.00	528.69	249.38	7.97	1 186.04	33.7	3.67	0.26	9.28	8.7	9.6
Korea	372.06	295.98	182.25	97.81	11.68	587.73	50.4	3.22	0.56	11.81	4.9	4.8
Luxembourg	5.96	0.23	7.65	2.41	0.14	10.43	2.2	3.57	0.25	20.10	0.1	0.1
Mexico	265.96	40.78	261.00	130.53	0.00	432.30	9.4	3.32	0.45	3.96	3.5	3.5
Netherlands	110.60	27.32	65.19	78.50	3.45	174.47	15.7	3.22	0.25	10.45	1.5	1.4
New Zealand	25.95	5.20	17.49	7.62	0.00	30.31	17.2	2.38	0.25	6.87	0.3	0.2
Norway	40.20	2.84	22.74	11.33	1.19	38.10	7.5	1.93	0.12	7.69	0.5	0.3
Poland	144.73	201.96	66.47	28.11	3.45	300.00	67.3	4.23	0.75	7.79	1.9	2.4
Portugal	32.98	8.65	28.22	10.56	0.65	48.08	18.0	2.98	0.25	4.51	0.4	0.4
Slovak Republic	24.78	13.87	9.71	10.12	0.17	33.86	40.9	2.79	0.55	6.22	0.3	0.3
Slovenia	10.36	6.00	7.41	1.72	0.13	15.26	39.3	3.01	0.39	7.43	0.1	0.1
Spain	179.39	49.98	152.65	66.90	0.79	270.32	18.5	3.08	0.23	5.86	2.4	2.2
Sweden	70.06	8.31	31.63	2.63	2.33	44.90	18.5	1.31	0.11	4.75	0.9	0.4
Switzerland	36.25	0.57	29.51	6.22	3.56	39.86	1.4	2.24	0.09	5.07	0.5	0.3
Turkey	160.66	125.10	74.72	85.75	0.16	285.73	43.8	3.63	0.46	3.86	2.1	2.3
United Kingdom	268.68	116.33	163.96	160.77	1.95	443.01	26.3	3.37	0.19	7.06	3.5	3.6
United States	3 130.28	1 830.26	2 111.25	1 314.74	30.93	5 287.18	34.6	3.45	0.40	16.94	41.3	42.8
IEA Total	**7 204.57**	**3 961.17**	**4 732.37**	**2 916.52**	**118.46**	**11 728.52**	**33.8**	**3.32**	**0.32**	**10.64**	**95.1**	**95.0**
OECD Total	**7 578.26**	**4 074.16**	**5 077.80**	**3 070.22**	**118.60**	**12 340.78**	**33.0**	**3.32**	**0.32**	**9.95**	**100.0**	**100.0**
EU27	**2 362.87**	**1 107.89**	**1 453.00**	**919.53**	**62.31**	**3 542.73**	**31.27**	**2.14**	**0.24**	**7.04**	**x**	**x**

Note: Energy-related CO_2 emissions are calculated using the Tier 1 IPCC Sectoral Approach. Emissions from the combustion of biomass-derived fuels are not included in accordance with the IPCC greenhouse gas inventory methodology.

* Other includes industrial wastes and non-renewable municipal wastes.

Sources: IEA, *Energy Balances of OECD Countries, 2010-2011* (Paris: OECD, 2013) and CO_2 *Emissions from Fuel Combustion* (Paris: OECD, 2013).

3. TRADE

Table 3.1: World and seaborne coal trade

(million tonnes)

	Steam coal		Coking coal		Total coal[1]	
	Total	**Seaborne**	**Total**	**Seaborne**	**Total**	**Seaborne**
1989	298.5	166.1	190.9	152.0	489.5	318.1
1990	308.3	183.0	199.1	153.8	509.6	337.1
1991	297.3	201.0	193.3	164.3	492.4	365.9
1992	296.2	209.9	172.8	153.9	480.6	364.3
1993	261.5	197.0	186.5	160.6	450.1	358.2
1994	273.8	216.7	182.9	163.2	458.2	380.5
1995	303.7	246.4	193.9	172.1	498.3	418.8
1996	323.3	270.9	195.0	174.7	520.8	445.8
1997	344.6	284.5	200.7	181.4	545.8	466.0
1998	365.7	292.3	186.2	171.3	554.0	464.3
1999	360.9	302.1	185.0	172.4	547.7	475.0
2000	424.1	355.6	183.3	167.7	609.2	523.5
2001	457.0	382.5	199.1	181.3	657.8	563.9
2002	466.0	405.6	187.6	172.4	654.0	578.1
2003	514.1	442.5	205.9	190.4	720.3	633.0
2004	557.6	486.0	197.3	179.9	755.7	666.4
2005	597.2	524.8	206.4	186.8	806.1	711.8
2006	648.7	566.1	222.8	204.4	876.7	773.6
2007	681.5	602.8	246.8	225.5	935.1	832.9
2008	704.2	613.1	234.2	211.6	948.5	832.4
2009	749.4	670.4	210.2	190.4	962.9	861.5
2010	784.6	699.3	281.2	247.1	1069.7	946.9
2011	853.5	774.5	281.2	243.0	1136.8	1017.8
2012	967.4	887.3	288.1	249.7	1258.4	1137.4

(1) Total coal is steam coal + coking coal + lignite.

Table 3.2: World total coal trade

(million tonnes)

	Japan		Other Asia		OECD Europe		Oth. Eur. + Eurasia		Africa + Mid. East		North America		Latin America		Balancing item		World	
Exporters	**2011**	**2012**	**2011**	**2012**	**2011**	**2012**	**2011**	**2012**	**2011**	**2012**	**2011**	**2012**	**2011**	**2012**	**2011**	**2012**	**2011**	**2012**
Australia	108.9	115.4	145.7	161.5	18.8	17.4	0.1	-	1.5	1.6	3.4	3.6	4.6	3.8	1.6	-1.9	284.5	301.5
Canada	7.9	8.0	14.6	18.2	5.3	4.1	-	-	0.1	-	1.7	1.2	2.5	2.1	1.6	1.1	33.6	34.7
Poland	-	-	0.0	-	7.9	7.2	0.2	0.4	-	-	-	-	-	-	-1.1	-0.7	7.0	7.0
United States	5.3	5.2	18.9	24.4	38.8	50.6	7.4	4.6	3.7	4.1	9.3	11.2	10.6	10.4	3.2	3.5	97.1	114.0
Other OECD	0.4	0.4	1.5	1.5	11.3	9.7	0.2	0.1	0.3	0.3	0.3	-	0.1	0.1	11.3	0.5	25.4	12.5
Total OECD	**122.4**	**129.0**	**180.7**	**205.6**	**82.0**	**89.0**	**7.8**	**5.1**	**5.5**	**6.0**	**14.7**	**16.0**	**17.9**	**16.3**	**16.6**	**2.6**	**447.6**	**469.7**
PR of China	4.4	4.3	7.7	7.3	0.1	0.1	-	-	0.0	-	0.0	0.1	0.0	0.1	9.4	-1.4	21.7	10.5
Colombia	0.2	0.2	8.2	9.7	52.9	50.8	1.2	0.5	7.8	5.9	10.9	7.8	10.0	11.1	-15.5	-3.7	75.5	82.2
Indonesia	33.5	35.2	266.5	333.8	10.4	9.9	0.9	0.2	0.2	0.2	0.9	0.5	0.6	0.2	-12.4	2.4	300.7	382.5
Kazakhstan	-	-	-	-	0.6	0.3	24.8	32.3	-	-	-	-	-	-	4.7	-1.0	30.1	31.6
Russian Fed.	11.8	13.1	28.9	41.0	59.5	61.2	17.0	12.8	2.9	3.5	0.1	0.1	1.2	1.2	1.9	1.3	123.3	134.2
South Africa	0.7	0.8	37.0	52.6	19.2	15.0	0.1	-	7.8	10.0	1.1	0.3	1.9	1.7	0.9	-5.2	68.7	75.1
Oth. non-OECD	1.2	1.1	61.1	62.5	13.5	10.9	4.2	3.0	1.8	1.3	1.0	0.7	2.2	1.4	-17.9	-11.3	67.2	69.7
Tot. Imp./Exp.	**174.1**	**183.8**	**592.4**	**729.8**	**238.2**	**237.2**	**55.0**	**53.4**	**24.5**	**26.6**	**28.8**	**25.4**	**37.0**	**34.5**	**-15.1**	**-35.1**	**1134.7**	**1255.5**

Notes: The data in this table come from a variety of sources. The columns for OECD Europe, Japan and North America hold import statistics from those regions and countries. The data in the rows from Australia to Total OECD, except the data in the columns mentioned above, are export statistics from these countries and regions. The data in the World column are based on export statistics. Other data are based on national and international sources and estimates, with this itemised trade data stored in an independent database. Trade aggregates may differ from data reported elsewhere. There are additional uncertainties in the regional breakdown of the different types of coal. The Balancing item is used to account for this. In addition, the Balancing item accounts for regional differences in national methodologies countries use to classify their coal imports and exports, coal in-transit, coal that is unaccounted for, confidentiality, and reporting discrepancies by importing and exporting countries.

Source: IEA/OECD Coal Statistics and Secretariat sources.

Table 3.3: World steam coal trade

(million tonnes)

	Japan		Other Asia		OECD Europe		Oth. Eur. + Eurasia		Africa + Mid. East		North America		Latin America		Balancing Item		World	
Exporters	**2011**	**2012**	**2011**	**2012**	**2011**	**2012**	**2011**	**2012**	**2011**	**2012**	**2011**	**2012**	**2011**	**2012**	**2011**	**2012**	**2011**	**2012**
Australia	79.3	86.8	74.2	85.3	3.3	2.4	-	-	0.5	0.5	2.7	3.4	0.7	0.4	-16.7	-19.6	144.1	159.2
Canada	2.6	2.8	3.5	1.7	1.3	1.1	-	-	-	-	0.3	0.2	-	-	-1.7	-1.9	5.9	4.0
Poland	-	-	0.0	-	5.7	4.9	0.1	0.1	-	-	-	-	-	-	-0.5	0.4	5.3	5.4
United States	0.7	0.8	6.7	9.1	17.6	29.6	0.6	0.5	2.4	2.9	4.8	6.7	2.4	2.4	-1.1	-1.3	34.1	50.6
Other OECD	-	-	0.0	0.1	8.1	6.6	0.1	0.1	0.1	0.0	0.3	-	0.1	0.0	11.3	0.0	20.0	6.8
Total OECD	**82.6**	**90.3**	**84.5**	**96.2**	**36.1**	**44.5**	**0.8**	**0.7**	**3.0**	**3.5**	**8.1**	**10.2**	**3.2**	**2.9**	**-8.8**	**-22.4**	**209.5**	**226.0**
PR of China	3.9	3.8	5.3	5.2	0.1	0.1	-	-	0.0	-	0.0	0.1	0.0	0.0	8.8	-0.4	18.1	8.7
Colombia	0.2	0.2	8.1	9.6	52.0	50.1	1.2	0.5	7.8	5.9	10.8	7.7	10.0	11.0	-14.5	-3.3	75.4	81.7
Indonesia	22.0	24.1	266.0	333.8	10.2	9.8	0.9	0.2	0.2	0.2	0.9	0.5	0.6	0.2	-2.9	13.6	297.8	382.4
Kazakhstan	-	-	-	-	0.6	0.3	24.8	32.3	-	-	-	-	-	-	4.4	-1.0	29.8	31.6
Russian Fed.	9.9	11.2	23.6	34.0	55.8	58.0	11.5	5.7	2.8	3.4	0.1	0.1	1.2	1.2	4.3	2.2	109.4	115.9
South Africa	0.7	0.8	37.0	52.6	18.7	14.6	0.1	-	7.8	10.0	1.1	0.1	1.9	1.7	0.1	-6.1	67.4	73.6
Oth. non-OECD	1.2	1.1	40.7	41.0	12.3	9.2	3.1	1.9	1.8	1.3	1.0	0.6	1.5	1.3	-15.4	-8.9	46.2	47.5
Tot. Imp./Exp.	**120.3**	**131.6**	**471.5**	**582.1**	**185.7**	**186.6**	**42.5**	**41.3**	**22.0**	**24.3**	**22.0**	**19.3**	**21.8**	**22.2**	**-32.3**	**-40.0**	**853.5**	**967.4**

Notes: The data in this table come from a variety of sources. The columns for OECD Europe, Japan and North America hold import statistics from those regions and countries. The data in the rows from Australia to Total OECD, except the data in the columns mentioned above, are export statistics from these countries and regions. The data in the World column are based on export statistics. Other data are based on national and international sources and estimates, with this itemised trade data stored in an independent database. Trade aggregates may differ from data reported elsewhere. There are additional uncertainties in the regional breakdown of the different types of coal. The Balancing item is used to account for this. In addition, the Balancing item accounts for regional differences in national methodologies countries use to classify their coal imports and exports, coal in-transit, coal that is unaccounted for, confidentiality, and reporting discrepancies by importing and exporting countries.

Source: IEA/OECD Coal Statistics and Secretariat sources.

Table 3.4: World coking coal trade

(million tonnes)

	Japan		Other Asia		OECD Europe		Oth. Eur. + Eurasia		Africa + Mid. East		North America		Latin America		Balancing Item		World	
Exporters	**2011**	**2012**	**2011**	**2012**	**2011**	**2012**	**2011**	**2012**	**2011**	**2012**	**2011**	**2012**	**2011**	**2012**	**2011**	**2012**	**2011**	**2012**
Australia(1)	29.6	28.7	71.5	76.2	15.5	15.0	0.1	-	1.0	1.1	0.7	0.2	3.9	3.5	18.3	17.7	140.5	142.4
Canada	5.3	5.2	11.0	16.5	4.0	3.0	-	-	0.1	-	1.5	1.0	2.5	2.1	3.3	3.0	27.7	30.7
Poland	-	-	-	-	2.2	2.4	0.1	0.3	-	-	-	-	-	-	-0.5	-1.1	1.7	1.6
United States	4.6	4.5	12.2	15.3	21.1	21.0	6.8	4.1	1.3	1.2	4.5	4.6	8.3	7.9	4.3	4.8	63.1	63.4
Other OECD	0.4	0.4	1.5	1.5	3.1	3.1	0.1	0.1	0.2	0.2	-	-	0.0	0.0	0.0	0.5	5.3	5.7
Total OECD	**39.8**	**38.6**	**96.2**	**109.5**	**45.9**	**44.5**	**7.1**	**4.4**	**2.5**	**2.5**	**6.6**	**5.8**	**14.7**	**13.5**	**25.3**	**24.9**	**238.2**	**243.8**
PR of China	0.5	0.5	2.4	2.1	-	-	-	-	-	-	-	0.0	-	0.1	0.7	-1.0	3.6	1.7
Colombia	0.0	0.0	0.0	0.1	0.9	0.7	-	-	-	-	0.1	0.0	-	0.1	-1.0	-0.5	0.1	0.5
Indonesia	11.5	11.2	0.6	0.0	0.2	0.0	-	-	-	-	-	-	-	-	-9.4	-11.2	2.8	0.0
Kazakhstan	-	-	-	-	-	-	-	-	-	-	-	-	-	-	0.3	-	0.3	-
Russian Fed.	1.9	1.9	5.2	7.0	3.7	3.2	5.5	7.1	0.1	0.0	-	-	-	-	-2.5	-1.0	13.9	18.3
South Africa	-	-	-	-	0.5	0.5	-	-	-	-	-	0.1	-	-	0.8	0.9	1.3	1.5
Oth. non-OECD	0.0	0.0	20.4	21.6	1.2	1.7	1.1	1.1	-	-	0.1	0.2	0.7	0.1	-2.5	-2.4	21.0	22.2
Tot. Imp./Exp.	**53.8**	**52.2**	**120.8**	**147.7**	**52.4**	**50.6**	**12.5**	**12.0**	**2.4**	**2.3**	**6.8**	**6.1**	**15.2**	**12.3**	**17.2**	**4.9**	**281.2**	**288.1**

Notes: The data in this table come from a variety of sources. The columns for OECD Europe, Japan and North America hold import statistics from those regions and countries. The data in the rows from Australia to Total OECD, except the data in the columns mentioned above, are export statistics from these countries and regions. The data in the World column are based on export statistics. Other data are based on national and international sources and estimates, with this itemised trade data stored in an independent database. Trade aggregates may differ from data reported elsewhere. There are additional uncertainties in the regional breakdown of the different types of coal. The Balancing item is used to account for this. In addition, the Balancing item accounts for regional differences in national methodologies countries use to classify their coal imports and exports, coal in-transit, coal that is unaccounted for, confidentiality, and reporting discrepancies by importing and exporting countries.

(1) Includes exports of soft and semi-soft coking coal used for pulverised coal injection (PCI).

Source: IEA/OECD Coal Statistics and Secretariat sources.

Table 3.5: World coal imports - regional aggregates

(thousand tonnes)

	World	OECD	Non OECD	OECD			Non-OECD			
				Americas	Asia Oceania	Europe	Africa M. East	Americas	Asia	Europe Eurasia
1989	391757	303439	88318	18618	129421	155400	4681	10973	41738	30926
1990	510425	327069	183356	18506	134646	173917	4123	11687	44576	122970
1991	507784	350059	157725	17157	145691	187211	4722	12261	42832	97910
1992	487447	344342	143105	17967	146844	179531	4649	11927	48359	78170
1993	456130	330293	125837	16270	154815	159208	4756	12645	53799	54637
1994	471793	348744	123049	18467	168607	161670	5199	13171	54715	49964
1995	502675	369555	133120	19297	180275	169983	5415	13467	61833	52405
1996	514504	381567	132937	23204	183602	174761	6650	14448	65862	45977
1997	542397	405939	136458	27064	197018	181857	6936	14348	72265	42909
1998	549222	411774	137448	32408	193266	186100	8268	14773	72125	42282
1999	551546	418396	133150	35561	202782	180053	7892	14646	80531	30081
2000	627131	468538	158593	41420	225168	201950	9459	15264	91115	42755
2001	669981	500404	169577	48655	232446	219303	10591	14988	99187	44811
2002	684254	501735	182519	47103	245834	208798	11645	15639	115803	39432
2003	729319	530500	198819	55434	252539	222527	11309	16893	121543	49074
2004	787222	566522	220700	52453	273443	240626	11473	17319	146611	45297
2005	805327	565934	239393	59654	267502	238778	11871	17235	166853	43434
2006	869640	600638	269002	66236	272818	261584	12610	16480	190803	49109
2007	921863	617523	304340	63600	289483	264440	12507	18378	221759	51696
2008	933897	616010	317887	63236	297048	255726	12669	18726	228168	58324
2009	951187	545323	405864	45709	279245	220369	11088	16110	338537	40129
2010	1067620	580185	487435	44801	316546	218838	12171	19793	408309	47162
2011	1133963	594200	539763	38595	315938	239667	13307	22135	453467	50854
2012e	*1276031*	*598793*	*677238*	*35706*	*323975*	*239112*	*14217*	*21840*	*587057*	*54124*

Table 3.6: World sub-bituminous and lignite imports - regional aggregates

(thousand tonnes)

	World	OECD	Non OECD	OECD			Non-OECD			
				Americas	Asia Oceania	Europe	Africa M. East	Americas	Asia	Europe/ Eurasia
1989	16881	13086	3795	-	716	12370	-	-	1163	2632
1990	19862	14580	5282	-	882	13698	-	-	825	4457
1991	20447	17631	2816	24	1038	16569	-	-	2	2814
1992	16616	14916	1700	14	1510	13392	-	-	1	1699
1993	14777	14010	767	149	1511	12350	-	-	1	766
1994	12780	12393	387	317	1947	10129	-	-	1	386
1995	13453	11982	1471	431	2438	9113	-	-	271	1200
1996	15126	11860	3266	1277	2518	8065	-	-	779	2487
1997	15523	12887	2636	1749	3491	7647	-	-	1514	1122
1998	16711	13731	2980	3111	4600	6020	-	-	2284	696
1999	18801	15616	3185	5347	4150	6119	-	-	2796	389
2000	23649	16175	7474	5205	5469	5501	-	-	5832	1642
2001	27928	18175	9753	6927	5907	5341	-	-	7769	1984
2002	29640	21205	8435	13973	3751	3481	-	-	6860	1575
2003	31298	21342	9956	16085	2070	3187	-	-	7280	2676
2004	27614	17353	10261	10655	3432	3266	-	-	7984	2277
2005	28689	20322	8367	13758	3884	2680	-	-	5722	2645
2006	31386	20830	10556	13074	4888	2868	-	-	7383	3173
2007	33354	21402	11952	14476	4095	2831	-	-	7962	3990
2008	43884	22503	21381	15401	3669	3433	-	-	18307	3074
2009	47911	18732	29179	11601	4201	2930	-	-	26910	2269
2010	64719	19230	45489	11467	5289	2474	-	257	42534	2698
2011	86786	14863	71923	7836	4518	2509	-	378	67269	4276
2012e	*105244*	*15061*	*90183*	*9049*	*3002*	*3010*	-	-	*88723*	*1460*

Note: Prior to 1990 excludes trade within the Former Soviet Union and Former Yugoslavia.

Table 3.7: World coal imports - selected countries

(thousand tonnes)

	Belgium	France	Germany	Spain	UK	USA	Canada	Russian Fed.	India	PR of China	Chinese Taipei	Japan	Korea
1989	12925	16000	13061	10569	12137	2587	14522	x	4410	2290	16852	101509	24216
1990	15037	19458	15663	10456	14783	2449	14169	53210	6046	2003	18527	106918	23729
1991	14697	21794	18851	12988	19611	3075	12567	46911	5272	1368	18444	112675	29094
1992	14258	21991	19092	14279	20339	3450	13003	39723	6495	1630	22139	110630	30781
1993	12120	14231	16081	12726	18400	6631	8468	28200	7330	1428	25345	112053	37381
1994	12871	12238	18106	11769	15088	6880	9366	27203	10556	1209	26721	121477	41009
1995	14294	13243	17184	13889	15896	6533	9735	22734	12512	1635	28757	127354	45831
1996	13041	15800	18298	12130	17799	6464	12207	20081	13175	3217	31148	130372	46074
1997	13015	13641	22182	11340	19757	6792	14469	20804	16440	2013	36219	136217	51997
1998	12830	18456	24193	14554	21244	7914	18657	21822	16535	1586	37093	130044	53586
1999	10995	17878	24731	20098	20293	8245	19759	16047	19700	1673	41104	138615	54569
2000	11566	19032	29744	21649	23446	11351	23231	25528	20930	2178	45409	150340	64895
2001	12888	16019	35508	18916	35542	18751	23627	28062	20548	2661	49091	154616	66381
2002	10111	18165	33455	24514	28687	16181	22061	20866	23260	11258	51814	161304	71708
2003	9578	16810	34952	21552	31891	22721	22568	25344	21683	11098	54670	166418	73405
2004	9980	19500	39553	24473	36153	24749	19223	22429	28950	18614	60483	180803	78963
2005	9010	19887	37114	24756	43968	27634	20745	22643	38586	26172	60252	176975	76758
2006	8345	20428	42185	23704	50528	32882	20716	26083	43081	38106	62311	179118	79707
2007	7732	19009	46314	24439	43364	32973	18432	23711	49794	51016	65232	186986	88285
2008	7713	21355	45455	20967	43876	31032	20549	31267	65201	40340	63840	184091	99584
2009	4978	15459	38485	17038	38167	20538	12912	24146	96161	125840	58635	163718	102982
2010	5772	17631	45725	12817	26540	17556	12611	25540	115717	163065	63155	185394	118591
2011	5483	14509	47845	16168	32528	11873	9472	27584	132121	181976	66589	174141	129150
2012e	*4804*	*15932*	*45195*	*22414*	*44797*	*8310*	*9817*	*31381*	*159619*	*288787*	*64530*	*183771*	*125535*

Table 3.8: World sub-bituminous and lignite imports - selected countries

(thousand tonnes)

	Belgium	France	Germany	Spain	UK	USA	Canada	Russian Fed.	India	PR of China	Chinese Taipei	Japan	Korea
1989	267	57	1990	-	-	-	-	x	-	-	-	-	716
1990	276	69	2083	1	-	-	-	-	-	-	-	-	882
1991	276	2	3428	-	-	24	-	-	-	-	-	-	1038
1992	244	2	3640	-	-	13	-	-	-	-	-	-	1510
1993	226	-	2991	-	-	147	-	-	-	-	-	-	1511
1994	212	48	2623	265	-	316	-	5	-	-	-	-	1947
1995	195	53	2132	481	-	429	-	-	-	-	59	-	2438
1996	227	42	1950	136	-	808	467	-	-	-	241	-	2518
1997	219	37	2151	-	-	782	965	89	-	-	442	-	3491
1998	196	32	1944	-	-	946	2163	22	-	-	2010	-	4600
1999	159	37	2053	-	-	580	3630	3	-	-	2435	-	4150
2000	219	52	1796	-	-	124	4441	10	-	-	4701	-	5469
2001	207	59	1997	-	-	880	3486	242	-	-	6017	-	5906
2002	205	23	848	-	-	994	8074	-	-	-	5244	-	3751
2003	188	42	35	-	-	932	9117	127	-	-	5141	-	1741
2004	190	40	17	-	-	1461	6426	170	-	-	5726	-	2610
2005	206	36	9	-	-	1625	6641	253	-	-	4797	-	2862
2006	289	37	53	-	-	2264	4848	341	-	-	5250	-	3706
2007	317	51	27	-	-	2483	8033	270	-	-	5888	-	3486
2008	282	67	28	-	-	3179	9523	275	10591	-	5821	-	3160
2009	172	51	10	-	-	1595	4544	338	20841	-	5800	-	3562
2010	245	52	-	-	-	1503	3794	681	35245	-	4244	-	5090
2011	291	82	-	-	-	524	1330	2048	51033	-	12945	-	4215
2012e	*300*	*76*	-	-	-	*387*	*954*	*10*	*73822*	-	*8772*	-	*2834*

Table 3.9: World coking coal imports - regional aggregates

(thousand tonnes)

	World	OECD	Non OECD	OECD			Non-OECD			
				Americas	Asia Oceania	Europe	Africa M. East	Americas	Asia	Europe Eurasia
1989	177410	141289	36121	6377	80340	54572	3001	10845	13129	9146
1990	171129	136967	34162	5211	76223	55533	2478	11304	13841	6539
1991	169042	137117	31925	5384	77412	54321	2918	11619	13501	3887
1992	166887	132099	34788	6087	74282	51730	3070	11233	13958	6527
1993	163737	130160	33577	5436	74425	50299	3210	11366	14301	4700
1994	174224	133447	40777	5425	77203	50819	3421	11517	16646	9193
1995	180393	135328	45065	5662	76950	52716	2960	11572	15893	14640
1996	182018	138420	43598	6493	77705	54222	3305	11932	15559	12802
1997	184485	138756	45729	6167	76493	56096	3409	11401	19292	11627
1998	178201	137674	40527	7546	71628	58500	4116	11062	16798	8551
1999	169797	132135	37662	7196	72107	52832	3771	10336	17587	5968
2000	180074	141066	39008	8353	76656	56057	4217	10446	17650	6695
2001	172492	134339	38153	7673	74526	52140	4113	10363	17704	5973
2002	176286	135458	40828	8184	78437	48837	5071	10029	19616	6112
2003	177222	132187	45035	6751	78058	47378	4979	10338	21690	8028
2004	199358	142417	56941	7490	82692	52235	5102	10359	30202	11278
2005	195799	138378	57421	8296	77155	52927	5063	10184	29717	12457
2006	197205	141927	55278	8129	77753	56045	5262	9947	27834	12235
2007	208873	146695	62178	7053	80731	58911	4976	11059	33265	12878
2008	214761	147868	66893	7438	81450	58980	5434	11457	38285	11717
2009	208755	117865	90890	4268	72993	40604	3940	10140	70181	6629
2010	258431	145176	113255	6458	85840	52878	4309	11458	87514	9974
2011	261680	145908	115772	7421	86054	52433	4499	12387	84950	13936
2012e	*283933*	*140864*	*143069*	*6813*	*83456*	*50595*	*4258*	*11482*	*115853*	*11476*

Table 3.10: World steam coal imports - regional aggregates

(thousand tonnes)

	World	OECD	Non OECD	OECD			Non-OECD			
				Americas	Asia Oceania	Europe	Africa M. East	Americas	Asia	Europe Eurasia
1989	200796	151309	49487	12241	49081	89987	1680	128	28531	19148
1990	322313	177526	144787	13295	58423	105808	1645	383	30660	112099
1991	321245	198261	122984	11773	68279	118209	1804	642	29329	91209
1992	306230	199613	106617	11879	72562	115172	1579	694	34400	69944
1993	279956	188463	91493	10832	80390	97241	1546	1279	39497	49171
1994	288020	206135	81885	13041	91404	101690	1778	1654	38068	40385
1995	312952	226285	86667	13633	103325	109327	2455	1895	45728	36589
1996	322919	236091	86828	16709	105897	113485	3345	2516	50279	30688
1997	349987	260252	89735	20895	120525	118832	3527	2947	52873	30388
1998	364925	268737	96188	24814	121638	122285	4152	3711	55289	33036
1999	376112	281042	95070	28319	130675	122048	4121	4310	62914	23725
2000	440809	322790	118019	32983	148512	141295	5242	4818	73422	34537
2001	490903	361370	129533	40897	157920	162553	6478	4625	81473	36957
2002	503860	363529	140331	38794	167397	157338	6574	5610	96187	31960
2003	547325	395916	151409	48570	174481	172865	6330	6555	99853	38671
2004	583930	421755	162175	44844	190751	186160	6371	6960	116406	32438
2005	606542	426058	180484	51226	190347	184485	6808	7051	137136	29489
2006	668879	456959	211920	57913	195065	203981	7348	6533	162969	35070
2007	709565	469117	240448	56436	208752	203929	7531	7319	188490	37108
2008	715865	466173	249692	55675	215598	194900	7235	7269	189882	45306
2009	739440	425820	313620	41305	206252	178263	7148	5970	268356	32146
2010	806128	433620	372508	38200	230706	164714	7862	8335	320795	35516
2011	867316	446657	420659	31035	229884	185738	8808	9748	368517	33586
2012e	*989269*	*455858*	*533411*	*28746*	*240519*	*186593*	*9959*	*10358*	*471204*	*41890*

Notes: Steam coal comprises anthracite, other bituminous coal and sub-bituminous coal.

Prior to 1990, Europe/Eurasia excludes trade within the Former Soviet Union and Former Yugoslavia.

Table 3.11: World coking coal imports - selected countries

(thousand tonnes)

	Belgium	France	Germany	Spain	UK	USA	Canada	Russian Fed.	India	PR of China	Chinese Taipei	Japan	Korea
1989	7064	7769	2281	4000	7998	-	5917	x	4410	303	4921	68689	11651
1990	7132	7848	1706	4169	8614	-	4491	-	5854	250	4237	64935	11287
1991	6426	7660	1057	4652	9196	-	4744	-	5272	408	4350	61839	15572
1992	5795	7457	1357	4343	8385	-	4848	-	6325	400	3748	57900	16381
1993	4748	6904	987	4572	8602	-	4681	-	6936	-	3915	57104	17321
1994	4444	6747	1093	3893	8065	-	4467	3237	9874	-	4032	60301	16902
1995	5267	7300	1427	3244	7754	-	4129	2215	9378	-	4390	59799	17151
1996	5325	7387	2189	3318	8245	-	4833	1708	10617	-	3946	59479	18226
1997	4328	7235	2536	3745	8072	-	4301	1274	11745	398	6033	59098	17395
1998	4195	7052	4299	3905	8646	1050	4597	292	10023	103	5636	53649	17979
1999	3953	6950	3519	3548	8020	1065	4041	2	10992	263	5375	54880	17227
2000	3818	6543	4608	3755	8462	1547	4296	200	11063	339	5158	57081	19575
2001	4169	6942	3984	3424	7723	2091	3987	-	11107	277	5373	56627	17899
2002	3363	6405	5174	3425	6315	2207	4315	-	12947	256	5272	58434	20003
2003	3220	5577	5504	3321	6474	1556	3272	-	12992	2605	5274	57743	20315
2004	3577	6798	6875	4043	6345	1987	3429	-	16925	6830	5078	60884	21808
2005	3533	6255	7152	3571	6551	1603	4228	866	16891	7195	4968	56528	20627
2006	3490	5995	8692	3622	6774	1533	4253	167	17877	4663	4872	57671	20081
2007	3247	6191	9627	3682	7481	1515	3322	-	22029	6219	4483	58198	22532
2008	2993	7239	9255	3371	6349	1580	3286	1450	25364	6857	4757	57366	24083
2009	1666	3581	6448	2058	5264	947	2219	225	31145	34417	4119	52334	20659
2010	2801	4615	7793	2777	6634	1385	3092	847	34424	47082	5524	57679	28160
2011	2704	3799	8778	2505	5908	1446	3770	2485	33943	44654	6036	53817	32234
2012e	*2600*	*4744*	*9256*	*2260*	*4835*	*1015*	*4382*	*-*	*36577*	*70644*	*8412*	*52199*	*31256*

Table 3.12: World steam coal imports - selected countries

(thousand tonnes)

	Belgium	France	Germany	Spain	UK	USA	Canada	Russian Fed.	India	PR of China	Chinese Taipei	Japan	Korea
1989	5594	8174	8790	6569	4139	2587	8605	x	-	1987	11931	32820	12565
1990	7629	11541	11874	6287	6169	2449	9678	53210	192	1753	14290	41983	12442
1991	7995	14132	14366	8336	10415	3075	7823	46911	-	960	14094	50836	13522
1992	8219	14532	14095	9936	11954	3450	8155	39723	170	1230	18391	52730	14400
1993	7146	7327	12103	8154	9798	6631	3787	28200	394	1428	21430	54949	20060
1994	8215	5443	14390	7876	7023	6880	4899	23961	682	1209	22689	61176	24107
1995	8832	5890	13625	10645	8142	6533	5606	20519	3134	1635	24367	67555	28680
1996	7489	8371	14159	8812	9554	6464	7374	18373	2558	3217	27202	70893	27848
1997	8468	6369	17495	7595	11685	6792	10168	19441	4695	1615	30186	77119	34602
1998	8439	11372	17950	10649	12598	6818	14060	21508	6512	1483	31457	76395	35607
1999	6883	10891	19159	16550	12273	7139	15718	16042	8708	1410	35729	83735	37342
2000	7529	12437	23340	17894	14984	9724	18935	25318	9867	1839	40251	93259	45320
2001	8512	9018	29527	15492	27819	16581	19636	27820	9441	2384	43718	97989	48482
2002	6543	11737	27433	21089	22372	13853	17744	20866	10313	11002	46542	102870	51705
2003	6170	11191	29413	18231	25417	21058	19294	25217	8691	8493	49396	108675	53090
2004	6213	12662	32661	20430	29808	22648	15793	22259	12025	11784	55405	119919	57155
2005	5271	13596	29953	21185	37417	25903	16516	21524	21695	18977	55284	120447	56131
2006	4566	14396	33440	20082	43754	31158	16463	25575	25204	33443	57439	121447	59626
2007	4168	12767	36660	20757	35883	31350	15110	23441	27765	44797	60749	128788	65753
2008	4438	14049	36172	17596	37527	29333	17262	29542	39837	33483	59083	126725	75501
2009	3140	11827	32027	14980	32903	19461	10691	23583	65016	91423	54516	111384	82323
2010	2726	12964	37932	10040	19906	16036	9515	24012	81293	115983	57631	127715	90431
2011	2488	10628	39067	13663	26620	10298	5695	23051	98178	137322	60553	120324	96916
2012e	*1904*	*11112*	*35939*	*20154*	*39962*	*7162*	*5424*	*31371*	*123042*	*218143*	*56118*	*131572*	*94279*

Notes: Steam coal comprises anthracite, other bituminous coal and sub-bituminous coal for all countries.

Table 3.13: OECD coke oven coke[1] imports

(thousand tonnes)

	1973	1980	1985	1990	1995	2000	2005	2009	2010	2011	2012e
Australia	-	-	-	-	-	-	22	6	14	55	37
Austria	1055	986	1274	815	718	981	1402	813	1252	1324	1342
Belgium	1110	1261	676	822	1259	1108	152	179	166	149	355
Canada	357	267	340	339	818	631	88	346	750	403	340
Chile	11	148	53	44	51	27	68	-	1	33	-
Czech Republic	-	-	-	-	320	602	510	536	885	566	431
Denmark	129	112	75	41	45	47	36	21	22	21	20
Estonia	x	x	x	12	-	1	-	-	-	-	-
Finland	832	1229	1231	786	274	505	505	246	441	444	320
France	3646	3000	2147	1109	885	1448	1525	1048	1270	1209	931
Germany	4497	4202	2693	2001	2601	5987	3793	3168	4310	3405	3254
Greece	45	35	53	32	11	1	4	1	1	-	-
Hungary	1207	1468	1486	478	351	5	205	6	10	9	7
Iceland	-	16	34	30	18	47	33	21	25	23	43
Ireland	13	8	22	29	6	-	-	-	-	-	-
Israel	2	1	1	-	-	-	-	-	-	-	-
Italy	82	101	134	115	745	506	848	-	18	33	1
Japan	-	-	-	261	493	2467	2500	413	879	679	815
Korea	38	121	106	-	-	-	341	256	629	406	368
Luxembourg	3247	2292	1854	1447	521	1	1	1	1	1	1
Mexico	150	122	121	124	437	631	390	208	391	336	391
Netherlands	677	832	405	360	649	515	341	78	243	159	109
New Zealand	-	-	-	-	-	-	-	-	-	-	-
Norway	577	531	700	534	522	543	382	279	434	455	469
Poland	-	-	-	-	34	16	117	55	137	147	138
Portugal	38	102	157	13	34	-	6	1	3	3	1
Slovak Republic	922	130	268	456	178	155	260	275	610	468	194
Slovenia	x	x	x	68	57	72	56	25	29	31	30
Spain	459	553	209	172	846	137	136	121	204	165	168
Sweden	1495	453	386	318	484	328	393	146	247	214	92
Switzerland	158	127	77	27	24	27	20	22	18	20	21
Turkey	-	-	40	-	182	722	414	274	173	309	186
United Kingdom	53	-	374	304	640	483	909	178	113	26	192
United States	978	598	524	694	1648	3430	3202	315	1101	1286	1029
IEA Americas	1335	865	864	1033	2466	4061	3290	661	1851	1689	1369
IEA Asia Oceania	38	121	106	261	493	2467	2863	675	1522	1140	1220
IEA Europe	20242	17422	14261	9859	11329	14117	11959	7448	10558	9127	8232
OECD Americas	1496	1135	1038	1201	2954	4719	3748	869	2243	2058	1760
OECD Asia Oceania	40	122	107	261	493	2467	2863	675	1522	1140	1220
OECD Europe	20242	17438	14295	9969	11404	14237	12048	7494	10612	9181	8305
IEA Total	**21615**	**18408**	**15231**	**11153**	**14288**	**20645**	**18112**	**8784**	**13931**	**11956**	**10821**
OECD Total	**21778**	**18695**	**15440**	**11431**	**14851**	**21423**	**18659**	**9038**	**14377**	**12379**	**11285**

(1) Solid product obtained from carbonization of coal, principally coking coal, used mainly in the iron and steel industry.
Also includes coke and semi-coke made from lignite.
For further information, see notes and definitions in Part I.

Source: *IEA/OECD Energy Statistics of OECD Countries*

Table 3.14: World coal exports - regional aggregates

(thousand tonnes)

	World	OECD	Non OECD	OECD			Non-OECD			
				Americas	Asia Oceania	Europe	Africa M. East	Americas	Asia	Europe Eurasia
1989	401435	280559	120876	124324	99155	57080	47091	14176	19809	39800
1990	519996	285886	234110	126997	103733	55156	50066	15371	23875	144798
1991	505976	302209	203767	132968	120793	48448	47515	18590	29229	108433
1992	497704	292290	205414	121175	127015	44100	52149	16937	41212	95116
1993	449384	272214	177170	95910	132542	43762	51959	21441	38841	64929
1994	457357	277759	179598	96483	132247	49029	55115	22613	47710	54160
1995	504052	305496	198556	114322	138038	53136	59896	22553	63917	52190
1996	525105	307504	217601	116550	142448	48506	60410	28398	77419	51374
1997	550881	319986	230895	112320	158802	48864	64672	32685	81454	52084
1998	557908	324730	233178	106576	167899	50255	61833	35969	84367	51009
1999	553917	303523	250394	86656	173316	43551	66798	36760	98202	48634
2000	624385	318399	305986	85147	188516	44736	70446	43321	116996	75223
2001	678430	320550	357880	73853	193897	52800	69924	46428	163249	78279
2002	667021	313166	353855	62854	206267	44045	69896	43854	165303	74802
2003	724886	317320	407566	67405	210961	38954	72087	52392	190809	92278
2004	762550	331617	430933	69407	221253	40957	68608	57650	206675	98000
2005	819535	345906	473629	73514	234664	37728	71977	60834	225502	115316
2006	894067	347768	546299	73282	235187	39299	69284	69699	282593	124723
2007	929239	370169	559070	84467	246408	39294	67486	71665	290580	129339
2008	936434	393441	542993	106989	254753	31699	58441	73299	274707	136546
2009	931986	372912	559074	82216	263856	26840	52516	69764	295952	140842
2010	1076051	430976	645075	107548	295045	28383	67036	70726	336178	171135
2011	1144110	449250	694860	131040	286697	31513	70217	81366	380442	162835
2012e	*1255341*	*471686*	*783655*	*149095*	*303728*	*18863*	*77569*	*85068*	*449235*	*171783*

Table 3.15: World sub-bituminous and lignite exports - regional aggregates

(thousand tonnes)

	World	OECD	Non OECD	OECD			Non-OECD			
				Americas	Asia Oceania	Europe	Africa M. East	Americas	Asia	Europe/ Eurasia
1989	12453	10793	1660	10	-	10783	-	-	1660	-
1990	17772	10234	7538	79	-	10155	-	-	2137	5401
1991	17707	11223	6484	55	-	11168	-	-	2682	3802
1992	18624	9888	8736	55	-	9833	-	-	5548	3188
1993	18790	11490	7300	1960	-	9530	-	-	5953	1347
1994	16845	9220	7625	2221	-	6999	-	-	7127	498
1995	23814	9970	13844	3086	-	6884	-	-	11273	2571
1996	25041	9146	15895	3334	-	5812	-	-	13094	2801
1997	22212	5988	16224	1100	-	4888	-	-	15017	1207
1998	24698	5695	19003	1881	-	3814	-	-	17226	1777
1999	28777	6781	21996	3484	-	3297	-	-	20266	1730
2000	22538	4019	18519	1134	-	2885	-	-	16646	1873
2001	33370	12203	21167	9155	-	3048	-	-	19443	1724
2002	33496	9414	24082	7373	-	2041	-	-	22853	1229
2003	40833	9771	31062	8567	-	1204	-	-	28905	2157
2004	37843	8090	29753	6819	-	1271	-	-	28588	1165
2005	44965	8819	36146	7180	-	1639	-	-	34949	1197
2006	57305	8183	49122	6245	-	1938	-	-	47852	1270
2007	67277	9689	57588	8094	-	1595	-	-	55947	1641
2008	86264	10888	75376	9177	-	1711	-	-	73177	2199
2009	95279	7193	88086	5793	73	1327	-	-	86427	1659
2010	114158	8353	105805	7166	-	1187	-	-	102631	3174
2011	122996	7559	115437	6229	9	1321	-	-	113524	1913
2012e	*212928*	*10145*	*202783*	*8345*	-	*1800*	-	-	*202439*	*344*

Note: Prior to 1990 excludes trade within the Former Soviet Union and Former Yugoslavia.

Table 3.16: World coal exports - selected countries

(thousand tonnes)

	Poland	Canada	USA	Aus-tralia	Co-lombia	Vene-zuela	Russian Fed.	Kazakh-stan	Ukraine	PR of China	India	Indo-nesia	South Africa
1989	28943	32827	91457	98669	12487	1689	x	x	x	15338	160	2489	46931
1990	28258	31009	95982	103396	13505	1834	59185	51480	23650	17290	100	4660	49900
1991	23739	34113	98854	120183	16379	2196	40381	51963	13070	20001	110	7242	47357
1992	23605	28174	93001	126242	14614	2309	43396	42459	7342	23363	130	15407	52059
1993	23877	28303	67602	131752	17616	3825	27202	33821	3101	19815	100	16878	51711
1994	28414	31746	64736	131201	18437	4135	23332	26266	4106	24194	673	20214	54838
1995	32236	33993	80329	136702	18274	4242	28434	20970	2400	28617	651	31308	59676
1996	28965	34459	82076	140856	24781	3617	27400	21072	2290	36485	478	36370	60224
1997	29503	36530	75790	157557	27580	5105	24376	25080	2374	35331	540	41714	64200
1998	28078	34183	72391	166796	30061	5908	25662	23403	1881	32297	823	47600	61300
1999	24115	33539	53048	171861	29932	6828	29313	17098	2138	37437	1156	55750	66235
2000	23254	32082	53061	186962	35391	7930	38329	34428	2329	55057	1292	56797	69910
2001	23044	29696	44149	192178	38868	7560	42798	31567	3729	90125	1903	66344	69210
2002	22664	26924	35927	204334	36510	7344	44280	27062	3095	83887	1517	72981	69231
2003	20155	28389	39016	208750	45644	6748	55942	32756	2914	93986	1627	87888	71531
2004	19711	25863	43543	219343	50902	6748	68931	24537	3885	86613	1374	105121	67946
2005	19377	28204	45306	232330	53607	7143	86558	24658	3666	71676	1989	128608	71442
2006	16735	28240	45039	232465	61968	7589	91930	28853	3457	63211	1554	183188	68747
2007	11900	30786	53673	244390	65263	6355	98638	26178	3621	53111	1627	194885	66963
2008	8462	33029	73953	252189	68523	4729	98119	32912	4794	45288	1655	199947	57891
2009	8464	28599	53612	261747	66756	2957	106445	28604	5290	22348	2450	233431	51977
2010	10080	33411	74132	292621	68148	2457	132801	31296	6194	21308	1875	267201	66396
2011	7152	33728	97303	284536	79273	1819	124593	30350	6991	21692	2014	300662	68807
2012e	*7123*	*34806*	*114073*	*301515*	*82172*	*2836*	*134201*	*31893*	*5443*	*10475*	*1480*	*382610*	*74349*

Table 3.17: World sub-bituminous and lignite exports - selected countries

(thousand tonnes)

	Poland	Canada	USA	Aus-tralia	Co-lombia	Vene-zuela	Russian Fed.	Kazakh-stan	Ukraine	PR of China	India	Indo-nesia	South Africa
1989	-	10	-	-	-	-	x	x	x	-	-	884	-
1990	193	9	70	-	-	-	3134	-	1433	-	-	1647	-
1991	1275	10	45	-	-	-	2979	-	348	-	-	2561	-
1992	1063	9	46	-	-	-	2896	-	-	-	-	5460	-
1993	909	29	1931	-	-	-	1321	-	-	-	-	5953	-
1994	719	51	2170	-	-	-	232	266	-	-	-	7127	-
1995	368	-	3086	-	-	-	2171	202	-	-	-	11271	-
1996	45	11	3323	-	-	-	2059	232	1	-	-	13093	-
1997	37	-	1100	-	-	-	883	223	-	-	-	15017	-
1998	23	-	1881	-	-	-	1620	109	-	-	-	17136	-
1999	13	-	3484	-	-	-	1604	73	-	-	-	20070	-
2000	9	-	1134	-	-	-	1592	147	9	-	-	16471	-
2001	15	78	9077	-	-	-	1245	299	-	-	-	19240	-
2002	41	120	7253	-	-	-	783	87	3	-	-	22624	-
2003	36	109	8458	-	-	-	1317	184	2	-	-	28734	-
2004	27	117	6702	-	-	-	315	211	2	-	-	28383	-
2005	8	140	7040	-	-	-	552	212	-	-	-	34724	-
2006	-	251	5994	-	-	-	539	248	-	-	-	47629	-
2007	-	364	7730	-	-	-	584	225	-	-	-	54568	-
2008	1	1516	7661	-	-	-	649	867	-	-	-	71981	-
2009	68	146	5647	-	-	-	893	309	-	-	-	84035	-
2010	115	139	7027	-	-	-	526	1924	-	-	-	98050	-
2011	145	176	6053	-	-	-	831	286	-	-	-	110199	-
2012e	*134*	*158*	*8187*	-	-	-	*22*	*285*	-	-	-	*201151*	-

Table 3.18: World coking coal exports - regional aggregates

(thousand tonnes)

	World	OECD	Non OECD	OECD			Non-OECD			
				Americas	Asia Oceania	Europe	Africa M. East	Americas	Asia	Europe Eurasia
1989	192758	164800	27958	87800	55714	21286	4161	892	2905	20000
1990	209721	162292	47429	84425	58086	19781	3633	946	3502	39348
1991	202492	169976	32516	87432	66059	16485	3523	968	3889	24136
1992	203262	161704	41558	77034	68331	16339	5088	1017	3789	31664
1993	185535	161119	24416	68970	74729	17420	4843	1041	4453	14079
1994	195475	158391	37084	70008	72540	15843	5764	1102	5370	24848
1995	195133	169662	25471	75819	75625	18218	6305	1159	7314	10693
1996	193529	170157	23372	76773	78558	14826	6133	1183	7978	8078
1997	197040	175094	21946	77406	84706	12982	5650	1224	5207	9865
1998	184410	165764	18646	71077	84414	10273	5167	1231	5525	6723
1999	179219	162067	17152	58161	93450	10456	2517	1179	6855	6601
2000	186980	168629	18351	58170	100712	9747	1744	1231	7710	7666
2001	195070	164921	30149	49975	106645	8301	1086	1252	13064	14747
2002	182403	156607	25796	42506	106458	7643	759	1283	14284	9470
2003	186472	160526	25946	43755	110004	6767	584	1333	14207	9822
2004	190792	168594	22198	48197	113640	6757	917	1624	7056	12601
2005	207306	186624	20682	52715	127246	6663	524	937	8482	10739
2006	201915	181443	20472	50152	123199	8092	672	729	8245	10826
2007	216595	197093	19502	55880	133979	7234	910	688	7438	10466
2008	236130	210738	25392	65249	139482	6007	1266	762	9224	14140
2009	211616	188527	23089	55339	127272	5916	638	764	7659	14028
2010	282007	243496	38511	78439	159566	5491	112	1216	18550	18633
2011	282925	238188	44737	90753	142568	4867	1030	1421	27479	14807
2012e	*290134*	*243757*	*46377*	*94333*	*144573*	*4851*	*3506*	*519*	*23608*	*18744*

Table 3.19: World steam coal exports - regional aggregates

(thousand tonnes)

	World	OECD	Non OECD	OECD			Non-OECD			
				Americas	Asia Oceania	Europe	Africa M. East	Americas	Asia	Europe Eurasia
1989	197191	105049	92142	36514	43441	25094	42930	13284	16128	19800
1990	294744	113417	181327	42493	45647	25277	46433	14425	19883	100586
1991	288735	121102	167633	45481	54734	20887	43992	17622	25219	80800
1992	281414	120712	160702	44086	58684	17942	47061	15920	37335	60386
1993	252980	101555	151425	26911	57813	16831	47116	20400	34388	49521
1994	254422	112406	142016	26424	59707	26275	49351	21511	42340	28814
1995	299476	128952	170524	38503	62413	28036	53591	21394	56601	38938
1996	323051	131623	191428	39766	63890	27967	54277	27215	69440	40496
1997	347750	140006	207744	34912	74096	30998	59022	31461	76247	41014
1998	367529	154863	212666	35206	83485	36172	56666	34738	78752	42510
1999	369429	138113	231316	28448	79866	29799	64281	35581	91151	40303
2000	432491	146891	285600	26922	87804	32165	68702	42090	109111	65697
2001	478267	152458	325809	23717	87252	41489	68838	45176	149982	61813
2002	480958	154312	326646	20107	99809	34396	69137	42571	150790	64148
2003	535024	155409	379615	23423	100957	31029	71503	51059	176431	80622
2004	569147	161544	407603	20906	107613	33025	67691	56026	199414	84472
2005	608996	157350	451646	20460	107418	29472	71453	59897	216795	103501
2006	688777	164151	524626	22841	111988	29322	68612	68970	274125	112919
2007	708806	171124	537682	28163	112429	30532	66576	70977	282562	117567
2008	696108	180674	515434	41356	115271	24047	57175	72537	265180	120542
2009	716818	182697	534121	26515	136584	19598	51878	69000	287895	125348
2010	789220	185959	603261	28774	135479	21706	66924	69510	317146	149681
2011	857457	209450	648007	39990	144129	25331	69187	79945	352374	146501
2012e	*962678*	*225971*	*736707*	*54600*	*159155*	*12216*	*74063*	*84549*	*425381*	*152714*

Notes: Steam coal comprises anthracite, other bituminous coal and sub-bituminous coal.

Prior to 1990, Europe/Eurasia excludes trade within the Former Soviet Union and Former Yugoslavia.

Table 3.20: World coking coal exports - selected countries

(thousand tonnes)

	Poland	Canada	USA	Aus-tralia	Co-lombia	Vene-zuela	Russian Fed.	Kazakh-stan	Ukraine	PR of China	India	Indo-nesia	South Africa
1989	11467	28677	59083	55228	892	-	x	x	x	2896	-	9	4161
1990	11226	26851	57568	57750	946	-	31573	-	7775	3473	-	29	3633
1991	9767	28786	58645	65450	968	-	15633	-	8503	3840	-	49	3523
1992	10781	23124	53910	67561	1017	-	24322	-	7342	3700	-	89	5088
1993	12829	23921	45044	73941	1041	-	6729	4247	3101	4283	-	170	4843
1994	10694	27066	42941	71496	1102	-	17367	3375	4106	4906	254	210	5764
1995	12296	28564	47255	74291	1159	-	8693	1999	-	6744	329	241	6305
1996	9886	28722	48036	76968	1183	-	6538	1509	31	7487	188	303	6133
1997	9138	30092	47314	83462	1224	-	8450	1371	44	4601	272	334	5650
1998	6506	28353	42722	83312	1231	-	6440	271	12	4855	385	285	5167
1999	6635	28946	29146	91996	1179	-	6400	184	17	5246	774	835	2517
2000	5290	28386	29780	99161	1231	-	7300	344	22	6470	624	616	1744
2001	3813	26914	23053	104935	1252	-	14431	316	-	11445	879	740	970
2002	3521	22964	19539	104526	1283	-	9196	271	3	13295	163	826	759
2003	2710	23716	20039	107794	1333	-	9470	328	24	13135	158	914	584
2004	3036	23847	24349	111732	1624	-	11935	245	417	5757	240	1059	917
2005	3151	26710	26001	124915	937	-	9983	247	509	5260	46	1222	524
2006	3601	25203	24946	120479	729	-	10007	289	530	4344	107	1550	672
2007	2363	26674	29198	131965	688	-	10019	262	118	2543	36	1736	910
2008	1683	26643	38599	136921	762	-	13614	329	197	3457	109	1922	1266
2009	1725	21531	33803	125238	764	-	13276	283	453	636	270	2049	616
2010	1815	27528	50906	157265	1216	-	18030	294	261	1139	111	2201	-
2011	1670	27666	63078	140455	1421	-	14182	301	286	3594	97	2826	456
2012e	*1587*	*30725*	*63392*	*142363*	*518*	-	*18251*	*301*	*192*	*1723*	-	*2826*	*746*

Table 3.21: World steam coal exports - selected countries

(thousand tonnes)

	Poland	Canada	USA	Aus-tralia	Co-lombia	Vene-zuela	Russian Fed.	Kazakh-stan	Ukraine	PR of China	India	Indo-nesia	South Africa
1989	17476	4140	32374	43441	11595	1689	x	x	x	12442	160	2480	42770
1990	16839	4149	38344	45646	12559	1834	24478	51480	14442	13817	100	4631	46267
1991	12697	5317	40164	54733	15411	2196	21769	51963	4219	16161	110	7193	43834
1992	11761	5041	39045	58681	13597	2309	16178	42459	-	19663	130	15318	46971
1993	10139	4353	22558	57811	16575	3825	19152	29574	-	15532	100	16708	46868
1994	17001	4629	21795	59705	17335	4135	5733	22625	-	19288	419	20004	49074
1995	19572	5429	33074	62411	17115	4242	17570	18769	2400	21873	322	31067	53371
1996	19034	5726	34040	63888	23598	3617	18803	19331	2258	28998	290	36067	54091
1997	20328	6438	28474	74095	26356	5105	15043	23486	2330	30730	268	41380	58550
1998	21549	5830	29376	83484	28830	5908	17602	23023	1869	27442	438	47315	56133
1999	17467	4593	23855	79865	28753	6828	21309	16841	2121	32191	382	54915	63718
2000	17955	3696	23226	87801	34160	7930	29437	33937	2298	48587	668	56181	68166
2001	19216	2704	21013	87243	37616	7560	27122	30952	3729	78680	1024	65604	68240
2002	19102	3841	16266	99808	35227	7344	34301	26704	3089	70592	1354	72155	68472
2003	17409	4566	18857	100956	44311	6748	45155	32244	2888	80851	1469	86974	70947
2004	16648	1904	19002	107611	49278	6748	56681	24081	3466	80856	1134	104062	67029
2005	16218	1366	19094	107415	52670	7143	76023	24199	3157	66416	1943	127386	70918
2006	13134	2927	19914	111986	61239	7589	81384	28316	2927	58867	1447	181638	68075
2007	9537	4002	24161	112425	64575	6355	88035	25691	3503	50568	1591	193149	66053
2008	6778	6267	35089	115268	67761	4729	83856	31716	4597	41831	1546	198025	56625
2009	6671	6939	19576	136509	65992	2957	92276	28012	4837	21712	2180	231382	51361
2010	8150	5751	23023	135356	66932	2457	114245	29078	5933	20169	1764	265000	66396
2011	5337	5933	34057	144081	77852	1819	109580	29763	6705	18098	1917	297836	68351
2012e	*5402*	*3971*	*50629*	*159152*	*81654*	*2836*	*115928*	*31307*	*5251*	*8752*	*1480*	*379784*	*73603*

Notes: Steam coal comprises anthracite, other bituminous coal and sub-bituminous coal for all countries.

Table 3.22: OECD coke oven coke[1] exports

(thousand tonnes)

	1973	1980	1985	1990	1995	2000	2005	2009	2010	2011	2012e
Australia	-	122	11	574	334	22	c	c	c	c	-
Austria	82	2	5	1	1	1	4	-	3	-	6
Belgium	469	768	847	915	557	293	20	405	446	364	735
Canada	368	80	22	193	334	300	154	229	82	353	420
Chile	-	45	34	-	19	41	39	-	-	-	-
Czech Republic	3467	2450	1927	1451	1409	948	913	562	875	524	465
Denmark	-	-	1	-	-	-	-	-	-	-	-
Estonia	x	x	x	35	39	20	37	26	21	25	23
Finland	24	5	13	1	2	-	-	3	5	6	46
France	1012	870	553	383	307	711	643	617	122	86	72
Germany	10197	7692	6038	2945	287	75	75	105	189	177	218
Greece	31	-	-	-	-	-	-	-	-	-	-
Hungary	103	-	-	-	420	183	53	154	300	303	309
Iceland	-	-	-	-	-	-	-	-	-	-	-
Ireland	-	-	-	-	-	-	-	-	-	-	-
Israel	-	-	-	-	-	-	-	-	-	-	-
Italy	628	747	451	201	130	123	229	276	303	273	254
Japan	600	2068	4055	1880	3428	2593	1674	935	652	979	1476
Korea	-	-	-	-	-	-	-	-	-	-	-
Luxembourg	-	-	-	-	-	-	-	-	-	-	-
Mexico	-	79	95	4	1	2	2	-	1	-	-
Netherlands	680	702	880	829	1025	562	358	135	148	42	92
New Zealand	-	-	-	-	-	-	-	-	-	-	-
Norway	51	36	122	5	-	2	2	-	4	-	-
Poland	2780	1770	1639	3662	3331	3691	4624	4813	6347	6492	6361
Portugal	16	-	-	11	74	80	-	-	-	-	-
Slovak Republic	-	13	23	-	58	59	138	164	324	176	61
Slovenia	x	x	x	-	-	-	-	-	-	-	-
Spain	2	3	11	42	81	744	610	199	370	374	461
Sweden	11	114	77	41	67	29	35	274	33	25	22
Switzerland	-	-	-	4	-	-	-	-	-	-	-
Turkey	-	-	-	-	-	-	-	-	-	-	-
United Kingdom	665	1265	1163	435	264	381	119	146	483	467	520
United States	1266	1879	1018	519	680	1039	1585	1186	1327	880	882
IEA Americas	1634	1959	1040	712	1014	1339	1739	1415	1409	1233	1302
IEA Asia Oceania	600	2190	4066	2454	3762	2615	1674	935	652	979	1476
IEA Europe	20218	16437	13750	10926	8013	7882	7823	7853	9952	9309	9622
OECD Americas	1634	2083	1169	716	1034	1382	1780	1415	1410	1233	1302
OECD Asia Oceania	600	2190	4066	2454	3762	2615	1674	935	652	979	1476
OECD Europe	20218	16437	13750	10961	8052	7902	7860	7879	9973	9334	9645
IEA Total	**22452**	**20586**	**18856**	**14092**	**12789**	**11836**	**11236**	**10203**	**12013**	**11521**	**12400**
OECD Total	**22452**	**20710**	**18985**	**14131**	**12848**	**11899**	**11314**	**10229**	**12035**	**11546**	**12423**

(1) Solid product obtained from carbonization of coal, principally coking coal, used mainly in the iron and steel industry.
Also includes coke and semi-coke made from lignite.
For further information, see notes and definitions in Part I.

Source: *IEA/OECD Energy Statistics of OECD Countries*

4. PRICES

Table 4.1: Japan coking coal import costs

(average unit value, CIF, USD/tonne)

	Total[(1)] (all sources)	Australia	Canada	USA	South Africa	Russian Federation	PR of China
1992	57.86	53.55	70.98	63.68	52.36	55.80	51.66
1993	55.39	51.49	67.12	61.82	46.78	54.37	50.52
1994	51.91	47.89	64.65	59.34	46.93	51.80	47.03
1995	55.03	51.15	64.49	61.37	49.54	54.81	49.49
1996	56.39	54.07	64.44	61.05	50.94	57.21	51.88
1997	55.19	52.73	64.84	61.24	49.64	57.09	49.98
1998	50.98	49.47	59.73	59.53	47.06	54.63	46.78
1999	42.95	41.83	51.05	55.79	39.74	45.17	40.01
2000	39.46	39.01	45.46	52.69	39.99	43.62	37.12
2001	41.13	40.96	47.30	47.81	47.09	45.52	39.35
2002	42.14	43.32	50.50	52.07	x	45.59	38.97
2003	41.73	43.56	51.64	42.93	x	46.05	39.09
2004	61.40	56.85	62.04	163.61	x	67.98	72.57
2005	88.80	96.44	106.05	159.01	x	114.96	100.45
2006	93.10	106.20	125.68	159.35	x	116.69	91.98
2007	88.43	96.03	112.61	516.32	x	105.45	100.91
2008	184.13	206.71	234.34	308.56	x	250.90	256.42
2009	163.82	191.58	220.94	241.65	x	190.54	138.99
2010	151.45	170.10	191.97	208.24	x	185.00	154.71
2011[(2)]	..	..	..	..	..	..	..
2012[(2)]	..	..	..	..	..	..	..

Note: It should be noted that as a result of the import coal classification system used by Japanese customs authorities, most imports of Indonesian coal are recorded by customs as coking coal even though most of the coal is not used in the metallurgical industry. As this coal has a lower unit value than coking coal reported in other categories, the data presented in the Total column in this table, from 1991 onwards, tend to understate the total average unit value of coals imported into Japan for metallurgical use. Prior to 1991, the volume of imports reported by customs in this manner was not so large as to significantly affect the total averages.

Table 4.2: EU coking coal import costs from selected countries

(Average Unit Value, CIF, USD/tonne)

	Total[(1)] (all sources)	Australia	Canada	USA	South Africa	Russian Federation	Poland
1992	60.95	60.76	61.51	61.22	50.54	47.95	59.64
1993	55.82	55.67	55.40	57.13	48.21	36.22	54.09
1994	55.97	55.56	52.42	58.22	44.13	46.11	53.55
1995	58.49	57.03	57.02	61.20	49.65	56.00	59.26
1996	59.74	59.72	58.52	62.03	47.44	59.88	61.11
1997	57.99	57.97	57.61	60.08	43.31	54.93	61.02
1998	54.53	55.00	54.18	58.53	35.83	51.08	52.52
1999	48.97	47.33	45.58	54.14	36.21	57.96	47.48
2000	47.88	45.45	45.92	52.91	39.09	42.02	50.43
2001	53.56	51.24	54.59	58.54	42.10	58.89	51.58
2002	56.63	55.52	56.90	61.48	38.69	58.11	50.15
2003	59.61	58.03	61.31	64.18	39.20	47.45	62.24
2004	78.12	73.63	76.64	84.75	61.31	86.41	108.71
2005	109.61	114.89	113.81	110.91	71.77	86.05	138.92
2006	125.86	135.52	137.66	123.44	66.18	81.34	118.82
2007	125.73	127.74	129.45	126.95	96.82	92.15	139.47
2008	197.84	220.54	220.91	175.35	141.18	147.83	245.85
2009	187.29	243.76	223.39	160.22	81.99	108.78	137.24
2010	194.02	213.42	205.20	191.43	95.76	128.15	194.36
2011[(2)]	..	..	..	..	..	..	..
2012[(2)]	..	..	..	..	..	..	..

(1) Weighted average based only on imports for which prices are available. Calculated average prices may not be comparable from one year to the next due to differing components.

(2) Coking coal import prices for 2011 and 2012 are unavailable due to resource constraints.

Source: *IEA/OECD Energy Prices & Taxes* (Table 21 and Table 19 in Part I) [For editions prior to 2011]

Table 4.3: Japan steam coal import costs

(average unit value, CIF, USD/tonne)

	Total[(1)] (all sources)	Australia	USA	South Africa	Russian Federation	PR of China	Canada	Indonesia
1992	48.47	49.29	51.22	46.93	43.73	46.04	46.30	44.49
1993	45.92	46.31	51.05	44.34	40.24	43.55	45.76	42.38
1994	43.88	44.92	49.91	42.47	38.68	39.05	42.88	39.89
1995	47.85	48.87	52.65	48.27	43.45	44.48	44.20	43.79
1996	49.29	50.24	54.32	49.86	44.63	46.52	45.91	45.49
1997	45.26	45.59	49.86	46.58	42.57	44.73	42.01	41.63
1998	40.68	40.80	47.52	41.50	38.46	39.96	38.25	36.58
1999	35.87	36.12	45.37	37.34	30.24	34.89	33.50	32.40
2000	34.59	34.59	45.49	35.82	30.68	33.69	34.72	31.85
2001	37.95	38.32	45.99	38.05	37.61	36.95	36.94	35.17
2002	36.95	37.39	48.63	40.30	34.35	36.25	37.34	34.52
2003	34.93	35.13	x	35.21	34.28	35.28	33.82	32.72
2004	51.48	50.20	186.48	x	59.02	55.06	52.60	46.98
2005	62.73	61.90	x	80.64	66.10	65.65	63.79	59.65
2006	63.33	63.90	596.68	x	62.20	63.88	56.32	60.31
2007	70.92	71.03	553.99	77.46	72.28	73.60	66.04	66.45
2008	125.42	127.23	151.26	105.49	121.34	124.10	125.52	116.08
2009	111.12	111.87	47.12	87.62	101.62	118.18	103.84	112.99
2010	110.40	111.12	84.43	107.07	106.42	112.29	107.80	108.20
2011[(2)]	..	..	..	..	..	..	..	..
2012[(2)]	..	..	..	..	..	..	..	..

Table 4.4: EU steam coal import costs from selected countries

(average unit value, CIF, USD/tonne)

	Total[(1)] (all sources)	Australia	USA	South Africa	Russian Federation	PR of China	Poland	Colombia
1992	48.59	50.07	50.69	42.48	39.50	45.46	54.11	49.65
1993	37.99	38.85	46.96	32.86	24.10	45.65	36.60	35.39
1994	40.35	40.74	46.05	36.18	36.67	41.78	41.44	39.68
1995	47.51	45.06	49.12	43.76	42.55	45.29	46.78	43.42
1996	45.62	48.25	48.03	42.23	42.48	46.18	46.50	41.93
1997	43.52	46.38	48.47	40.33	39.84	52.87	42.62	42.29
1998	39.72	40.03	46.99	36.30	37.23	52.85	39.61	37.56
1999	34.43	36.68	40.67	31.78	30.96	32.58	35.03	32.69
2000	35.22	39.04	41.07	33.83	33.59	31.45	35.30	34.22
2001	42.96	44.13	46.52	42.32	42.16	40.90	44.41	42.32
2002	38.69	43.28	43.48	36.65	36.79	44.30	40.75	36.99
2003	41.94	46.83	47.37	39.10	42.22	51.38	43.26	41.65
2004	61.91	69.05	61.50	58.00	65.14	60.61	68.95	61.54
2005	71.27	106.40	86.75	67.64	68.46	93.41	78.34	67.98
2006	69.80	109.71	82.08	66.24	67.93	150.20	75.65	66.07
2007	82.21	103.73	97.50	80.25	79.03	73.02	94.13	78.16
2008	137.79	184.75	138.40	142.07	131.62	161.12	156.01	138.32
2009	100.28	149.29	107.28	95.66	92.41	x	114.60	95.99
2010	104.10	197.74	117.46	98.50	97.66	x	104.06	95.84
2011[(2)]	..	..	..	..	..	..	..	..
2012[(2)]	..	..	..	..	..	..	..	..

(1) Weighted average based only on imports for which prices are available. Calculated average prices may not be comparable from one year to the next due to differing components.

(2) Steam coal import prices for 2011 and 2012 are unavailable due to resource constraints.

Source: *IEA/OECD Energy Prices & Taxes* (Table 16 and Table 14 in Part I) [For editions prior to 2011]

Table 4.5: Steam coal export costs

(Average Unit Value, USD/tonne)

Exported from:	Australia (FOB)			Canada	United States (FAS)			Colombia[2]	Indonesia	S. Africa
To:	Total	Japan	EU[1]		Total	Japan	EU[1]			
1992	36.49	38.85	32.47	38.15	39.53	39.66	40.90	..	41.22	29.53
1993	33.45	35.49	30.02	34.61	39.78	39.08	41.60	..	38.52	25.38
1994	32.75	33.72	29.09	31.51	38.22	37.13	40.23	..	36.40	23.89
1995	34.99	36.85	31.05	32.90	38.94	38.61	40.91	..	42.71	29.20
1996	37.77	38.89	33.19	35.67	36.94	40.02	36.80	31.81	42.71	30.63
1997	34.35	35.40	31.76	32.44	34.84	38.55	35.88	33.84	39.90	31.47
1998	30.10	31.50	24.54	30.24	32.24	37.00	33.05	30.51	35.82	26.37
1999	26.49	27.02	21.63	25.72	32.02	35.40	29.39	27.25	31.00	22.77
2000	24.27	24.72	20.86	24.99	31.84	34.90	28.26	26.99	29.60	22.63
2001	28.71	29.69	25.86	26.01	34.51	37.70	34.69	31.36	32.07	28.59
2002	28.01	29.20	24.42	28.11	37.70	40.41	33.09	30.09	29.98	25.82
2003	25.41	26.00	25.18	35.75	37.00	166.40	32.18	36.41	25.00	24.47
2004	37.65	38.45	34.44	44.75	53.93	91.47	42.06	59.55	38.00	32.77
2005	48.86	49.42	49.46	53.23	67.09	105.21	76.79	51.89	44.50	46.46
2006	47.05	50.59	49.29	45.24	55.66	166.53	64.17	51.92	47.25	45.90
2007	51.11	54.61	55.42	50.51	52.61	198.41	66.30	57.76	60.91	51.06
2008	92.23	103.23	102.85	97.37	62.76	143.02	81.46	120.27	115.48	88.19
2009	80.03	88.54	72.97	84.45	81.33	113.97	95.04	58.26	55.29	60.99
2010	85.82	89.03	120.56	95.97	71.63	83.85	82.21	77.31	71.01	91.26
2011[3]	..	..	..	..	..	..	..	110.38	80.87	116.52
2012[3]	..	..	..	..	..	..	..	83.30	68.30	91.13

Table 4.6: Coking coal export costs

(Average Unit Value, USD/tonne)

Exported from:	Australia (FOB)			Canada	United States (FAS)			Colombia[2]	Indonesia	S. Africa
To:	Total	Japan	EU[1]		Total	Japan	EU[1]			
1992	46.78	46.44	46.53	57.08	50.01	49.58	51.72	36.50	..	40.35
1993	44.28	43.85	44.06	52.73	48.63	47.99	50.71	..	..	41.16
1994	41.36	40.24	41.41	48.03	47.11	44.72	49.82	..	..	37.48
1995	44.23	43.37	44.45	48.76	48.47	45.35	51.13	..	..	39.66
1996	47.56	46.11	48.56	51.52	50.05	46.51	53.12	..	..	39.68
1997	47.37	45.52	47.73	50.04	49.99	45.94	52.87	..	..	37.72
1998	43.97	42.47	44.10	46.15	48.55	45.98	51.67	31.97	..	31.39
1999	35.99	34.60	37.59	38.19	46.19	44.46	49.27	31.26	..	34.38
2000	32.85	31.64	33.66	34.01	42.98	41.99	45.58	33.15	..	32.09
2001	36.93	34.46	39.99	43.03	45.88	x	50.10	..	..	37.81
2002	40.16	36.79	44.34	43.90	50.06	x	53.14	..	..	32.43
2003	39.75	36.18	43.14	44.59	49.11	56.01	52.19	..	..	27.91
2004	48.84	45.51	51.43	51.37	70.14	98.14	63.60	..	..	43.15
2005	88.94	83.74	100.84	96.26	89.91	100.61	89.53	..	..	53.00
2006	98.27	94.04	106.29	109.29	111.72	89.80	102.37	..	..	54.94
2007	84.16	79.50	93.67	98.44	98.10	x	99.84	..	..	67.93
2008	194.87	193.79	215.74	209.51	148.39	157.59	132.75	..	..	114.58
2009	143.83	165.07	168.94	194.92	129.77	181.25	124.64	..	..	95.04
2010	171.76	154.53	193.93	212.01	160.32	166.81	157.76	..	..	..
2011[3]	..	..	..	..	..	..	..	..	..	..
2012[3]	..	..	..	..	..	..	..	..	..	..

(1) Weighted average based only on imports for which prices are available. Calculated average prices may not be comparable from one year to the next due to differing components.

(2) Low ash bituminous; injection grade to Japanese steel mills

(3) OECD steam and coking coal export values are unavailable for 2011 and 2012 due to resource constraints.

Sources: *IEA/OECD Energy Prices & Taxes* (Tables 23 to 27) for Australia, Canada and the US, prior to 2011; *Latin America Coal & Power* for Colombia prior to 2003, *Indonesia Mineral and Coal Statistics* for Indonesia prior to 2003; *Coal Americas,* IHS Energy Publishing Inc. for Colombia and Indonesia 2003 till 2011, IHS McCloskey, *McCloskey's Coal Report* from 2011 onwards; Republic of S. Africa Minerals Bureau for South Africa prior to 2004, Xavier Provost, private consultant since 2004 and *South African Coal Report,* IHS Energy Publishing Inc

Table 4.7: Coking coal prices for industry

(USD/tce)

	1978	1985	1990	1995	1997	1998	1999	2000	2001	2002
Australia	29.69	24.85	..	..	..	..	..	..	..	..
Austria	..	..	..	..	..	..	..	..	..	..
Belgium	63.67	60.40	61.72	57.55	57.58	55.48	49.29	47.21	54.03	..
Canada	62.70	54.01	..	..	..	..	..	..	..	..
Chile	..	..	..	..	..	..	..	..	..	55.47
Czech Republic	..	44.30	52.62	64.92	56.28	58.08	56.79	50.73	55.32	65.73
Denmark	..	..	..	..	..	..	..	..	..	..
Estonia	..	..	..	..	..	..	..	..	..	..
Finland	..	..	..	101.01	106.56	107.06	105.93	99.78	105.14	109.88
France	51.84	50.31	62.62	59.55	55.72	53.43	47.44	45.43	50.42	58.03
Germany	62.91	63.40	63.02	60.34	60.05	56.13	..	..	..	..
Greece	..	..	..	..	..	..	..	..	..	..
Hungary	..	..	..	..	..	..	..	..	..	..
Ireland	..	..	..	..	..	..	..	..	..	..
Israel	..	..	..	..	..	..	..	..	..	..
Italy	69.79	62.38	64.88	60.70	61.12	59.26	52.70	53.26	58.22	60.99
Japan	72.47	66.24	68.03	61.13	62.65	58.61	49.64	45.29	46.93	48.88
Korea	..	..	..	..	..	..	..	..	..	..
Luxembourg	..	..	..	..	..	..	..	..	..	..
Mexico	..	..	..	..	..	..	..	..	..	..
Netherlands	105.55	97.88	154.34	..	..	..	..	..	..	..
New Zealand	..	..	..	..	..	..	..	..	..	..
Norway	..	72.51	93.32	89.69	..	..	..	..	..	..
Poland	..	..	..	60.05	49.94	47.08	38.32	39.10	46.73	47.46
Portugal	175.90	68.37	52.51	43.92	42.79	37.82	31.74	31.67	38.85	32.64
Slovak Republic	27.50	41.02	56.55	..	..	..	..	..	..	..
Slovenia	..	..	..	..	..	..	..	..	..	..
Spain	68.57	63.42	..	..	..	..	..	..	..	..
Sweden	64.11	64.56	..	..	..	..	..	..	..	..
Switzerland	..	102.86	..	..	..	..	..	..	..	..
Turkey	70.23	57.26	82.07	86.05	110.65	94.11	70.90	80.19	71.75	86.26
United Kingdom	..	..	..	..	..	..	..	..	..	..
United States	48.94	59.10	52.01	51.52	51.58	50.15	49.91	48.41	50.67	55.73

	2003	2004	2005	2006	2007	2008	2009	2010	2011	2012
Australia	..	..	..	..	..	..	..	..	..	..
Austria	..	..	..	..	..	..	..	..	..	..
Belgium	..	..	..	..	..	..	..	..	..	..
Canada	..	..	..	..	..	..	..	..	..	..
Chile	45.45	66.19	99.06	116.90	106.42	144.74	196.62	130.57	269.88	222.89
Czech Republic	..	..	..	..	..	..	..	..	..	..
Denmark	..	..	..	..	..	..	..	..	..	..
Estonia	..	..	..	..	..	..	..	..	..	..
Finland	126.04	158.63	196.65	203.29	225.89	382.62	287.47	288.81	479.99	398.61
France	59.67	72.00	106.19	127.95	119.55	207.93	197.67	209.31	278.39	..
Germany	..	..	..	..	..	..	..	..	..	..
Greece	..	..	..	..	..	..	..	..	..	..
Hungary	..	..	..	..	..	..	..	..	..	..
Ireland	..	..	..	..	..	..	..	..	..	..
Israel	..	..	..	..	..	..	..	..	..	..
Italy	63.48	81.00	108.96	117.95	124.19	199.90	152.00	185.72	254.80	..
Japan	48.81	69.33	104.02	112.09	108.07	220.66	203.29	186.84	251.65	212.40
Korea	..	72.66	116.53	120.23	110.12	247.02	169.43	..	..	..
Luxembourg	..	..	..	..	..	..	..	..	..	..
Mexico	..	..	..	..	..	..	..	..	..	..
Netherlands	..	..	..	..	..	..	..	..	..	..
New Zealand	..	..	..	..	..	..	..	..	..	..
Norway	..	..	..	..	..	..	..	..	..	..
Poland	51.92	95.31	114.57	94.24	111.17	221.50	129.55	187.88	253.68	184.15
Portugal	38.73	..	..	..	..	..	..	321.28	489.93	428.98
Slovak Republic	..	..	..	..	..	..	..	..	..	..
Slovenia	..	..	..	..	..	..	..	..	..	..
Spain	..	..	..	..	..	..	..	..	..	..
Sweden	..	..	..	..	..	..	..	..	..	..
Switzerland	..	..	..	..	..	..	..	..	..	..
Turkey	92.36	118.78	148.07	161.57	200.52	258.70	269.73	272.13	254.23	268.97
United Kingdom	..	..	..	..	..	..	..	..	..	..
United States	55.15	66.99	91.26	101.49	103.44	128.62	156.19	158.80	197.97	200.07

Source: *IEA/OECD Energy Prices & Taxes*

Table 4.8: Steam coal prices for industry

(USD/tce)

	1978	1980	1985	1990	1997	1998	1999	2000	2001	2002
Australia	..	..	27.23	..	..	..	..	..	..	..
Austria	90.20	..	65.63	78.84	64.91	64.72	56.93	54.73	56.42	75.33
Belgium	29.62	..	54.28	55.41	..	..	..	..	..	..
Canada	43.98	..	50.08	..	..	..	..	..	..	..
Chile	..	..	..	..	..	..	..	..	..	..
Czech Republic	..	..	19.06	24.44	27.73	29.43	28.33	26.96	28.43	33.87
Denmark	76.39	..	81.93	119.94	..	..	..	..	..	..
Estonia	..	..	..	..	..	..	..	..	..	..
Finland	51.74	..	57.34	72.35	98.95	97.05	97.39	89.47	97.40	96.68
France	61.82	..	89.61	128.20	125.17	126.61	121.78	105.48	107.49	109.32
Germany	98.52	..	98.18	185.76	..	..	..	..	..	..
Greece	93.42	..	..	..	..	..	..	..	..	..
Hungary	..	..	57.76	100.92	62.66	66.26	..	..	..	..
Ireland	..	..	..	..	..	..	..	..	..	..
Israel	..	..	..	..	..	..	..	..	..	..
Italy	70.05	..	56.50	58.84	53.73	46.24	37.90	43.46	50.90	47.01
Japan	71.23	..	66.49	70.53	59.78	53.76	47.47	45.68	49.69	48.29
Korea	..	..	40.62	61.03	45.46	33.38	49.56	58.36	51.13	52.96
Luxembourg	..	..	..	..	..	..	..	..	..	..
Mexico	..	..	..	..	..	..	..	..	..	..
Netherlands	57.76	..	60.33	62.74	..	..	..	..	..	..
New Zealand	36.23	..	..	..	..	..	..	..	..	..
Norway	39.79	..	45.67	53.13	..	..	..	..	..	..
Poland	..	..	..	..	50.53	57.85	49.23	48.83	54.39	50.23
Portugal	..	..	..	..	47.43	35.69	32.14	37.98	46.43	39.74
Slovak Republic	7.88	..	11.70	15.25	35.95	35.36	31.85	..	..	..
Slovenia	..	..	..	..	..	..	..	..	..	..
Spain	..	..	..	..	..	..	..	..	..	..
Sweden	77.55	..	69.90	92.56	..	..	..	..	..	..
Switzerland	68.81	..	70.26	72.45	59.54	58.75	55.12	60.13	68.67	61.64
Turkey	..	..	51.63	79.67	63.09	61.75	57.96	53.26	52.41	69.46
United Kingdom	49.74	..	72.98	88.51	62.89	64.98	62.93	59.32	63.69	65.08
United States	36.26	..	44.31	40.00	38.59	38.42	37.60	37.77	39.02	39.97

	2003	2004	2005	2006	2007	2008	2009	2010	2011	2012
Australia	..	..	..	..	..	..	..	..	..	..
Austria	87.64	171.25	178.69	181.33	203.12	249.48	243.64	208.76	249.91	233.09
Belgium	..	..	..	..	..	..	..	..	..	..
Canada	..	..	..	..	..	..	..	..	..	..
Chile	..	..	..	..	..	..	..	..	..	..
Czech Republic	..	..	..	..	..	..	..	..	..	..
Denmark	..	..	..	..	..	..	..	..	..	..
Estonia	..	..	..	..	..	..	..	..	..	..
Finland	113.53	140.81	146.63	149.81	166.59	249.09	192.19	193.76	362.38	322.66
France	130.49	..	..	..	..	..	..	..	..	..
Germany	..	..	..	..	..	..	..	..	..	..
Greece	..	..	..	..	..	..	..	..	..	..
Hungary	..	..	..	..	..	..	..	..	..	..
Ireland	..	..	..	..	..	..	..	..	..	..
Israel	..	..	..	..	..	..	..	..	..	..
Italy	48.09	72.13	83.23	78.59	97.48	163.22	129.51	127.19	159.34	..
Japan	45.29	67.16	81.25	87.10	97.98	167.62	152.90	151.91	193.10	189.16
Korea	58.42	57.88	66.77	61.12	75.29	124.80	94.12	..	..	..
Luxembourg	..	..	..	..	..	..	..	..	..	..
Mexico	..	..	..	..	..	..	..	..	..	..
Netherlands	..	..	..	..	..	..	..	..	..	..
New Zealand	..	..	..	..	..	..	..	..	..	..
Norway	..	..	..	..	..	..	..	..	..	..
Poland	56.63	65.40	71.39	81.68	96.83	135.19	122.71	125.20	142.48	142.17
Portugal	47.00	..	..	..	..	..	..	140.85	192.01	189.83
Slovak Republic	..	..	..	..	..	..	..	..	..	..
Slovenia	..	..	..	..	..	..	..	..	..	..
Spain	..	..	..	..	..	..	..	..	..	..
Sweden	..	..	..	..	..	..	..	..	..	..
Switzerland	75.26	110.08	110.03	111.25	145.85	252.48	160.86	181.26	233.62	177.96
Turkey	73.39	67.16	78.71	79.99	115.02	152.51	139.04	137.83	142.56	161.45
United Kingdom	70.29	86.76	99.65	99.17	113.43	130.55	111.10	130.90	166.67	165.78
United States	40.73	46.79	56.31	61.57	64.85	75.60	77.28	76.20	85.14	87.90

Source: *IEA/OECD Energy Prices & Taxes*

Table 4.9: Steam coal prices for electricity generation

(USD/tce)

	1978	1980	1985	1990	1997	1998	1999	2000	2001	2002
Australia	10.83	..	19.11	29.05	..	..	..	..	..	..
Austria	173.19	..	75.95	80.98	..	..	..	56.94	48.41	55.78
Belgium	53.08	..	65.42	59.60	55.95	44.82	39.63	42.95	49.46	45.21
Canada	37.88	..	..	61.02	..	..	..	20.54	21.27	22.02
Chile	..	..	..	..	..	..	..	32.41	36.66	33.56
Czech Republic	10.29	..	15.70	20.49	24.42	26.02	26.43	21.88	22.10	23.40
Denmark	..	..	56.88	..	..	..	..	..	..	..
Estonia	..	..	..	..	..	..	..	..	..	..
Finland	51.74	..	57.34	72.35	59.94	49.45	45.31	44.40	53.62	50.59
France	..	..	43.58	57.35	53.83	46.68	43.11	41.98	51.04	48.35
Germany	83.97	..	82.45	141.95	49.61	47.17	42.45	42.94	52.51	46.27
Greece	..	..	..	..	..	..	..	..	..	..
Hungary	..	..	38.53	90.60	..	..	..	..	..	..
Ireland	..	..	84.91	63.11	51.66	41.07	34.88	34.26	39.99	42.37
Israel	..	..	..	..	..	..	..	..	..	..
Italy	41.32	..	57.38	67.72	63.14	56.89	..	..	..	..
Japan	90.58	..	77.86	101.84	74.99	63.46	55.83	51.42	53.65	49.77
Korea	..	..	..	..	..	..	..	..	..	..
Luxembourg	..	..	..	..	..	..	..	..	..	..
Mexico	..	..	42.39	48.20	41.14	39.66	44.02	48.69	51.94	51.56
Netherlands	40.99	..	65.72	71.37	..	..	..	..	..	..
New Zealand	..	..	..	..	..	..	..	..	..	..
Norway	..	..	..	..	..	..	..	..	..	..
Poland	..	..	16.93	17.03	40.84	44.20	39.56	38.86	43.31	44.13
Portugal	31.26	..	65.23	58.90	48.20	41.25	36.10	34.67	44.23	37.06
Slovak Republic	4.30	..	6.55	8.61	24.33	23.73	20.47	..	..	..
Slovenia	..	..	..	..	..	..	..	..	..	..
Spain	42.60	..	50.82	..	..	..	..	..	..	..
Sweden	50.00	..	50.43	..	..	..	..	..	..	..
Switzerland	..	..	..	..	..	..	..	..	..	..
Turkey	..	..	33.65	31.85	54.68	55.64	54.01	50.45	35.92	53.19
United Kingdom	52.18	..	72.71	96.27	68.44	61.89	58.20	55.02	57.50	55.10
United States	29.96	..	43.67	38.52	33.40	32.86	31.92	31.51	32.33	32.91

	2003	2004	2005	2006	2007	2008	2009	2010	2011	2012
Australia	..	..	..	..	..	..	..	..	..	..
Austria	68.29	86.09	92.70	98.23	101.93	129.35	141.37	115.56	135.53	126.89
Belgium	47.13	95.02	105.36	82.93	99.07	171.19	100.77	..	..	..
Canada	24.15	23.34	28.16	30.02	32.70	33.19	28.67	35.65	33.13	..
Chile	34.56	51.39	58.26	57.37	69.81	116.66	95.43	83.62	106.83	96.33
Czech Republic	..	..	..	..	..	..	..	..	..	..
Denmark	..	..	..	..	..	..	..	..	..	..
Estonia	..	..	..	..	..	..	..	..	..	..
Finland	55.49	76.99	82.82	85.36	96.21	164.25	111.56	116.89	157.56	133.45
France	47.85	71.64	84.41	82.15	94.53	153.44	128.44	122.05	144.71	..
Germany	50.64	70.87	80.73	78.92	91.38	154.47	111.47	119.27	155.16	128.02
Greece	..	..	..	..	..	..	..	..	..	..
Hungary	..	..	..	..	..	..	..	..	..	..
Ireland	40.16	75.99	79.21	69.22	94.31	112.40	128.53	94.21	141.76	99.60
Israel	..	..	..	..	..	169.74	165.17	130.46	189.72	179.45
Italy	43.38	70.11	83.45	79.24	100.83	169.04	122.09	127.81	..	..
Japan	..	..	..	..	..	..	..	..	..	..
Korea	..	50.51	58.44	54.73	63.96	96.49	87.54	..	..	..
Luxembourg	..	..	..	..	..	..	..	..	..	..
Mexico	49.13	56.83	62.69	64.79	70.26	80.19	78.20	83.49	85.90	87.11
Netherlands	..	..	..	..	..	..	..	..	..	..
New Zealand	..	..	..	..	..	..	..	..	..	..
Norway	..	..	..	..	..	..	..	..	..	..
Poland	49.33	54.89	65.19	69.75	79.21	110.19	109.84	107.21	116.06	113.53
Portugal	44.00	65.92	77.51	66.84	87.61	162.01	93.32	101.20	129.94	103.39
Slovak Republic	..	..	..	..	..	..	..	..	..	..
Slovenia	..	..	..	..	..	..	..	..	..	..
Spain	..	..	..	..	..	..	..	..	..	..
Sweden	..	..	..	..	..	..	..	..	..	..
Switzerland	..	..	..	..	..	..	..	..	..	..
Turkey	66.45	90.65	88.29	86.87	97.12	113.22	112.43	127.32	131.08	135.91
United Kingdom	56.85	73.98	81.19	86.73	102.00	148.68	105.08	119.15	158.99	130.13
United States	33.33	35.48	40.50	44.47	46.45	54.32	57.91	59.42	62.80	63.12

Source: *IEA/OECD Energy Prices & Taxes*

5. COAL FOR OTHER USES

Table 5.1: OECD coal-fired and total electricity generating capacity, 1990 to 2011

(GW)

	Coal					Total				
	1990	2000	2005	2010	2011	1990	2000	2005	2010	2011
Australia	24.92	28.65	29.89	30.69	30.87	38.45	46.20	50.14	59.86	61.12
Austria	2.10	2.24	2.11	2.02	2.16	16.69	17.80	19.10	21.41	22.79
Belgium	4.90	2.47	4.09	..	..	14.14	15.69	16.10	18.32	20.10
Canada	19.24	..	..	..	..	104.14	111.32	122.84	137.11	138.69
Chile	0.89	1.93	2.01	2.53	3.33	5.10	9.89	12.59	16.23	17.57
Czech Republic	12.11	11.47	11.46	11.79	11.89	15.28	15.32	17.41	19.83	20.18
Denmark	7.54	5.60	6.46	5.89	5.86	9.14	12.32	13.04	13.44	13.58
Estonia	..	2.79	2.51	2.63	2.36	..	2.80	2.56	2.75	2.83
Finland	5.79	7.71	7.76	7.79	7.78	13.22	16.26	16.47	16.69	16.69
France	14.23	..	..	..	7.94	103.41	114.68	115.79	124.55	131.35
Germany	42.73	51.59	..	..	..	97.75	118.38	125.03	156.98	163.26
Greece	3.89	4.49	4.81	4.79	4.79	8.51	10.90	13.31	15.12	16.52
Hungary	2.24	2.02	1.51	1.52	1.37	7.18	8.28	8.59	8.99	9.65
Iceland	..	..	..	..	..	0.94	1.38	1.54	2.58	2.67
Ireland	0.87	0.87	0.89	0.85	0.85	3.81	4.71	6.27	8.53	8.79
Israel	..	4.29	4.84	4.84	4.84	5.07	9.13	10.61	14.35	14.04
Italy	9.03	12.56	9.99	11.19	11.20	56.56	75.51	85.50	106.49	118.44
Japan	40.47	51.78	59.18	47.23	48.29	194.73	260.49	277.32	287.03	292.06
Korea	..	14.44	21.18	29.41	28.80	2.34	53.69	66.54	84.70	84.65
Luxembourg	0.09	..	..	..	..	1.24	1.22	1.68	1.72	1.74
Mexico	1.61	5.11	5.13	5.90	5.88	27.37	40.35	52.50	62.29	61.60
Netherlands	3.77	4.18	4.18	..	..	17.56	21.06	21.80	26.69	28.05
New Zealand	1.09	1.11	1.16	1.16	1.16	7.18	8.39	8.85	9.63	9.63
Norway	0.05	0.08	0.08	0.01	0.01	27.13	28.42	29.12	31.70	31.70
Poland	25.99	27.80	28.45	28.41	28.78	27.88	30.56	32.26	33.36	34.55
Portugal	1.47	1.99	2.04	2.38	2.33	7.40	10.90	13.39	18.92	19.94
Slovak Republic	..	..	1.77	1.52	1.46	..	7.45	8.26	7.87	8.06
Slovenia	..	1.03	0.95	0.85	0.85	..	2.61	2.99	3.19	3.27
Spain	10.41	11.36	..	..	..	43.42	53.92	76.57	101.79	102.80
Sweden	..	..	2.76	..	..	34.19	33.72	33.39	36.45	35.23
Switzerland	0.01	..	..	0.22	0.30	15.11	17.25	17.47	19.94	20.03
Turkey	5.58	7.40	9.59	12.40	13.03	16.32	27.26	38.84	49.52	52.91
United Kingdom	41.17	33.37	30.51	30.48	29.37	73.21	78.39	82.38	93.73	93.82
United States	307.96	321.06	315.44	318.96	319.92	733.59	812.63	978.54	1041.00	1055.28

Notes: Includes multi-fired units.
Includes autoproducers for all countries except Japan.
For further information, see notes and definitions in Part I.

Source: *IEA/OECD Electricity Information*

Table 5.2: OECD coal and peat use for electricity production and heat sold

(Mtce)

	1973	1980	1985	1990	1995	2000	2005	2008	2009	2010	2011
Australia	19.7	28.8	33.7	41.3	46.3	60.0	67.7	68.8	72.0	67.0	64.0
Austria	1.2	1.2	1.7	2.4	2.0	2.0	2.6	2.1	1.6	2.0	2.2
Belgium	3.9	5.5	5.3	6.5	6.3	4.7	3.5	2.7	2.1	1.9	1.7
Canada	11.9	21.5	28.3	28.5	29.7	38.5	34.3	33.4	26.7	27.1	25.2
Chile	0.4	0.7	0.7	2.5	2.1	3.1	2.4	5.0	4.8	5.5	6.8
Czech Republic	11.9	15.5	16.5	20.5	21.3	21.3	20.6	21.1	19.6	20.7	20.8
Denmark	2.3	8.0	9.7	8.1	8.7	5.2	4.9	5.5	5.6	5.4	4.4
Estonia	..	..	..	6.8	3.4	3.2	3.7	4.0	3.4	4.6	4.8
Finland	1.7	4.9	4.5	4.7	5.9	5.0	4.9	5.8	6.2	7.9	6.1
France	14.8	24.4	15.0	12.0	9.6	10.8	10.3	9.6	8.3	8.3	5.5
Germany	104.2	115.9	123.3	115.6	103.2	98.8	97.3	95.5	87.3	89.6	90.3
Greece	2.0	3.8	6.9	9.8	11.1	11.8	12.4	11.7	11.8	10.9	11.2
Hungary	5.8	6.9	6.0	4.5	4.0	4.1	3.1	3.0	2.6	2.7	2.7
Iceland	-	-	-	-	-	-	-	-	-	-	-
Ireland	0.9	0.8	1.2	2.6	3.0	2.7	2.7	2.2	1.9	1.9	2.0
Israel	-	-	2.6	3.4	5.9	9.2	10.6	10.9	10.5	10.6	11.0
Italy	1.9	6.0	9.9	11.4	8.8	9.6	16.5	16.6	14.4	14.5	16.2
Japan	17.1	15.0	30.0	36.3	51.5	69.1	89.7	86.6	83.1	88.1	83.5
Korea	0.5	1.2	6.9	8.5	16.9	38.7	51.9	61.6	71.1	74.9	79.4
Luxembourg	0.4	0.2	0.2	0.2	0.1	-	-	-	-	-	-
Mexico	0.1	-	1.0	2.6	5.3	6.8	11.4	7.6	10.2	11.3	11.8
Netherlands	1.1	2.6	5.3	8.3	9.1	7.9	8.0	7.8	7.7	7.5	7.1
New Zealand	0.5	0.2	0.3	0.2	0.3	0.5	2.0	1.6	1.1	0.7	0.8
Norway	-	-	-	-	-	-	-	-	-	-	-
Poland	52.9	77.0	82.9	74.9	60.0	56.2	56.4	54.9	52.8	54.7	54.7
Portugal	0.3	0.1	0.3	2.9	4.2	4.6	4.8	3.5	4.0	2.3	3.2
Slovak Republic	2.7	4.1	4.1	3.3	3.1	2.4	2.5	2.1	1.8	1.8	1.8
Slovenia	..	..	..	1.9	1.8	1.8	2.0	2.0	1.9	2.0	2.0
Spain	4.3	11.2	17.2	20.3	21.9	26.7	25.7	16.1	12.2	8.7	15.3
Sweden	0.1	0.1	1.9	1.6	1.4	1.1	1.2	1.2	1.0	1.2	1.0
Switzerland	-	-	-	-	-	-	-	-	-	-	-
Turkey	1.8	2.6	6.0	7.8	10.5	14.1	13.5	20.5	19.3	19.8	23.5
United Kingdom	65.0	74.3	61.4	68.3	49.7	41.1	46.0	42.2	35.2	36.6	36.8
United States	309.9	417.2	501.8	565.7	636.7	716.5	718.7	707.3	628.2	661.1	622.6
IEA Americas	321.7	438.7	530.1	594.3	666.4	755.1	753.0	740.7	654.9	688.2	647.7
IEA Asia Oceania	37.8	45.2	70.9	86.3	115.0	168.3	211.3	218.6	227.2	230.7	227.6
IEA Europe	279.3	365.1	379.1	386.0	344.0	330.2	336.9	324.1	295.4	298.6	306.6
OECD Americas	322.2	439.3	531.7	599.3	673.8	765.0	766.8	753.3	669.9	704.9	666.4
OECD Asia Oceania	37.8	45.2	73.6	89.7	120.9	177.5	221.9	229.5	237.7	241.3	238.6
OECD Europe	279.3	365.1	379.1	394.7	349.2	335.3	342.6	330.1	300.7	305.1	313.4
IEA Total	**638.9**	**849.0**	**980.1**	**1066.6**	**1125.4**	**1253.6**	**1301.1**	**1283.4**	**1177.5**	**1217.4**	**1181.9**
OECD Total	**639.4**	**849.6**	**984.4**	**1083.8**	**1144.0**	**1277.7**	**1331.4**	**1312.9**	**1208.4**	**1251.3**	**1218.4**

Note: "Coal" refers to all coal types, primary (including hard and brown coal), and derived fuels (including patent fuel, coke oven coke, gas coke, coal tar, BKB, coke oven gas, blast furnace gas and oxygen steel furnace gas) as well as peat, peat briquettes and oil shale. Quantities have been converted to Mtce units using calorific values largely submitted in annual questionnaires to the IEA Secretariat by OECD member countries.

For further information, see notes and definitions in Part I.

Source: *IEA/OECD Energy Balances of OECD Countries*

Table 5.3: OECD coal[1] use in coke ovens

(million tonnes)

	1973	1980	1985	1990	1995	2000	2005	2008	2009	2010	2011
Australia	9.5	7.0	5.5	5.9	5.9	4.8	4.5	4.5	3.5	4.1	4.4
Austria	2.3	2.4	2.4	2.3	1.9	1.9	1.9	1.9	1.7	1.8	1.8
Belgium	9.9	7.9	7.8	7.2	4.7	3.9	3.4	2.8	2.0	2.6	2.6
Canada	7.7	7.3	6.3	5.0	4.2	4.2	4.3	4.3	3.0	4.0	4.2
Chile	0.5	0.5	0.4	0.5	0.7	0.7	0.7	0.7	0.6	0.5	0.7
Czech Republic	18.6	12.2	10.6	8.5	6.6	4.6	4.3	4.3	3.0	3.2	3.3
Estonia	x	x	x	0.2	0.2	0.1	0.1	0.2	0.1	0.1	0.1
Finland	-	-	-	0.7	1.7	1.3	1.3	1.2	1.1	1.2	1.2
France	15.3	14.6	11.3	9.5	7.7	6.5	5.8	6.0	4.4	4.3	4.1
Germany	46.3	38.7	31.5	24.1	13.7	11.4	10.2	10.3	8.4	10.6	10.4
Hungary	1.6	1.4	0.9	1.0	1.4	1.3	0.8	1.4	1.0	1.4	1.5
Italy	10.4	11.2	10.1	8.6	6.7	6.4	5.1	5.6	3.4	4.9	5.6
Japan	62.3	65.7	67.6	65.2	60.8	57.4	59.4	56.6	53.5	57.9	54.8
Korea	0.4	4.0	7.0	11.7	14.1	16.4	15.4	16.8	14.8	19.5	22.9
Mexico	2.8	4.0	3.8	2.9	2.7	2.7	2.4	2.9	2.4	2.7	2.9
Netherlands	3.4	3.7	4.1	3.8	4.0	3.0	3.2	3.1	2.5	2.9	3.0
New Zealand	-	-	-	0.6	0.7	0.7	0.8	0.7	0.8	0.8	0.8
Norway	0.4	0.4	0.4	-	-	-	-	-	-	-	-
Poland	21.6	25.3	20.5	18.2	15.4	12.4	11.4	13.3	9.4	13.0	12.4
Portugal	0.4	0.3	0.4	0.3	0.5	0.5	-	-	-	-	-
Slovak Republic	2.8	2.1	2.5	2.9	2.4	2.2	2.4	2.2	2.0	2.1	2.1
Spain	6.4	5.2	4.9	4.5	3.3	3.6	3.6	3.5	2.3	2.6	2.7
Sweden	0.7	1.7	1.6	1.5	1.6	1.8	1.8	2.0	1.5	1.8	1.6
Turkey	2.0	2.6	3.7	4.7	4.2	4.2	4.2	5.5	4.9	5.3	5.2
United Kingdom	24.0	13.5	9.7	10.1	8.1	8.2	5.6	5.9	4.9	5.7	5.4
United States	87.3	60.5	37.2	35.3	29.9	26.0	18.4	20.0	13.9	19.1	19.4
Other OECD	0.5	0.4	-	-	-	-	-	-	-	-	-
IEA Total	**333.5**	**288.1**	**246.0**	**231.8**	**199.7**	**182.6**	**167.7**	**171.9**	**142.0**	**169.2**	**169.3**
OECD Total	**336.8**	**292.5**	**250.2**	**235.4**	**203.2**	**186.1**	**171.0**	**175.6**	**145.1**	**172.5**	**172.9**

(1) Primary coal only. Coal products such as briquettes are not included. For further information, see notes and definitions in Part I.

Source: *IEA/OECD Energy Statistics of OECD Countries*

Table 5.4: World* consumption of pulverised coal injection (PCI) coals

(thousand tonnes)

	1990	1995	2000	2004	2005	2006	2007	2008	2009	2010	2011
Australia	..	..	..	807	881	790	808	825	647	697	760
Belgium	513	859	1458	642	522	514	450	478	46	172	147
France	1310	2305	2462	2534	2936	3193	3110	2636	1343	2410	2309
Germany	1730	1843	2530	2641	2770	2975	3115	3071	1833	3959	3772
Italy	170	1230	1198	1628	1678	1818	1030	1102	942	1459	1328
Japan	5252	8286	10936	10416	10440	10670	11594	9705	8625	11646	12596
Korea	..	2181	3031	5065e	5481	5852	6284	6997	6381	7695	9414
Netherlands	680	845	1040	1468	1522	1324	1593	1353	938	1018	1284
Poland	..	..	..	..	..	..	..	..	..	32	80
Slovak Republic	219	313	390	412	388	470	468	406	383	356	421
Spain	..	..	681	649	717	511	769	661	269	681	708
Sweden	210	245	277	548	506	493	532	515	209	413	438
Turkey	..	..	..	..	..	..	293	444	462	553	459
United Kingdom	767	585	456	895	1039	1121	1242	1170	852	978	995
United States	200	2300	2924e	2405	1519	1814	1959	2482	1501	1964	2046
Chinese Taipei	130	472	801	1292	1156	1409	1445	1113	678	1109	1379
India	..	..	2111	2059	2160	2266	2377	2493	2615	2743	2877
Russian Federation	..	..	319	410	2682	2840	3499	3995	3314	3820	4604
Total World	**11181**	**21464**	**30614**	**33871**	**36397**	**38060**	**40568**	**39446**	**31038**	**41705**	**45617**

* Data are not present for many countries - most significantly, the People's Republic of China.

Source: *IEA/OECD Energy Statistics of OECD Countries, IEA/OECD Energy Statistics of Non-OECD Countries*

Table 5.5: Coal-fired heat and electricity generation efficiency and share

	2011				Share of total generation from coal					
	Coal[(1)] (Mtce)	Electricity (TWh)	Heat (PJ)	Efficiency	1971	1980	1990	2000	2010	2011
Australia	44.77	173.29	-	33.3%	71.0%	73.7%	78.8%	83.0%	71.8%	68.6%
Austria	1.52	7.32	3.02	46.0%	13.1%	6.7%	14.7%	10.3%	8.2%	9.7%
Belgium	1.18	5.38	-	39.0%	26.9%	27.8%	27.8%	18.0%	5.7%	5.4%
Canada	17.63	76.36	-	37.2%	18.8%	15.5%	16.8%	19.1%	13.7%	11.9%
Chile	4.79	19.62	-	35.2%	16.1%	16.1%	35.5%	21.1%	27.9%	29.9%
Czech Republic	14.57	49.89	84.38	43.3%	84.8%	71.1%	77.1%	71.2%	61.9%	60.6%
Denmark	3.11	13.97	30.96	62.5%	18.6%	83.2%	74.1%	39.7%	33.5%	31.4%
Estonia	3.33	11.34	4.52	32.5%	x	x	59.3%	62.3%	64.8%	65.4%
Finland	2.56	10.31	31.17	63.7%	14.9%	31.6%	25.6%	15.8%	17.5%	15.2%
France	3.85	17.31	13.38	46.9%	28.9%	27.1%	8.4%	6.3%	4.9%	3.5%
Germany	63.24	271.87	153.22	42.8%	73.6%	62.8%	59.8%	53.1%	41.9%	42.9%
Greece	7.87	31.06	2.21	34.6%	43.2%	44.8%	72.4%	64.4%	54.1%	53.0%
Hungary	1.86	6.57	7.08	39.4%	60.3%	43.0%	31.1%	27.4%	15.9%	17.2%
Ireland	0.91	4.70	-	44.3%	1.2%	0.7%	41.6%	28.8%	14.5%	17.0%
Israel	7.67	35.19	-	39.4%	-	-	50.1%	68.8%	58.5%	59.0%
Italy	11.31	50.14	2.93	38.8%	4.7%	9.9%	16.8%	11.3%	12.7%	14.1%
Japan	58.46	281.14	-	41.4%	11.9%	9.6%	13.9%	22.0%	26.7%	26.8%
Korea	55.57	224.52	58.54	37.3%	6.9%	6.7%	16.8%	36.8%	42.6%	41.8%
Luxembourg	-	-	-	-	63.2%	51.6%	76.4%	-	-	-
Mexico	8.27	34.13	-	35.5%	0.5%	-	6.7%	9.5%	12.0%	11.5%
Netherlands	4.99	24.74	18.44	51.5%	12.4%	13.7%	38.2%	23.3%	19.7%	19.6%
New Zealand	0.53	2.17	-	34.9%	4.8%	1.9%	2.0%	3.9%	4.6%	4.8%
Norway	0.03	0.12	0.26	59.0%	0.0%	0.0%	0.1%	0.1%	0.1%	0.1%
Poland	38.27	141.44	265.73	48.4%	90.6%	91.2%	93.2%	95.2%	87.8%	86.4%
Portugal	2.21	9.85	-	38.4%	3.6%	2.3%	31.7%	32.7%	11.9%	17.1%
Slovak Republic	1.26	4.04	11.31	49.1%	37.2%	27.5%	37.9%	22.7%	17.9%	17.8%
Slovenia	1.40	5.31	5.61	42.3%	x	x	36.5%	37.8%	36.7%	36.9%
Spain	10.71	44.93	-	36.1%	21.7%	30.0%	40.1%	36.6%	8.8%	15.5%
Sweden	0.44	1.39	10.87	86.2%	0.3%	0.7%	5.3%	2.5%	2.5%	2.2%
Switzerland	-	-	-	-	-	0.1%	0.2%	-	-	-
Turkey	16.46	66.22	1.04	34.7%	30.5%	25.6%	35.1%	29.7%	24.5%	27.3%
United Kingdom	25.79	109.62	8.36	37.3%	63.9%	72.8%	65.0%	31.7%	28.2%	29.4%
United States	435.79	1875.41	96.13	37.5%	44.8%	51.2%	52.8%	52.3%	45.0%	42.6%
IEA Total	**824.89**	**3503.74**	**799.03**	**38.8%**	**42.1%**	**43.2%**	**42.6%**	**39.3%**	**34.2%**	**33.3%**
IEA Americas	*453.42*	*1951.78*	*96.13*	*37.5%*	*41.8%*	*46.3%*	*48.0%*	*48.0%*	*41.3%*	*38.7%*
IEA Asia Oceania	*159.33*	*681.12*	*58.54*	*37.6%*	*18.3%*	*17.7%*	*22.8%*	*32.4%*	*36.4%*	*36.3%*
IEA Europe	*212.13*	*870.85*	*644.36*	*42.6%*	*49.1%*	*47.3%*	*43.2%*	*31.2%*	*24.6%*	*25.3%*
OECD Total	**850.34**	**3609.33**	**809.16**	**38.8%**	**41.6%**	**42.6%**	**42.1%**	**38.7%**	**33.7%**	**32.9%**
OECD Americas	*466.47*	*2005.53*	*96.13*	*37.5%*	*41.0%*	*45.1%*	*46.7%*	*46.2%*	*39.7%*	*37.1%*
OECD Asia Oceania	*167.00*	*716.31*	*58.54*	*37.7%*	*18.0%*	*17.4%*	*23.3%*	*33.3%*	*37.1%*	*36.9%*
OECD Europe	*216.86*	*887.49*	*654.48*	*42.4%*	*49.1%*	*47.2%*	*43.4%*	*31.2%*	*24.7%*	*25.4%*

Table 5.5: Coal-fired heat and electricity generation efficiency and share (continued)

	2011				Share of total generation from coal					
	Coal[(1)] (Mtce)	Electricity / Heat (TWh)	(PJ)	Effici-ency	1971	1980	1990	2000	2010	2011
Africa	**63.13**	**262.12**	**-**	**35.7%**	**62.2%**	**54.3%**	**52.2%**	**47.3%**	**38.7%**	**37.9%**
Botswana	0.17	0.37	-	18.7%	..	..	88.1%	97.1%	100.0%	100.0%
Morocco	2.96	11.68	-	34.0%	13.4%	19.5%	23.0%	68.3%	46.2%	47.0%
Mozambique	-	-	-	-	..	17.5%	13.9%	-	-	-
Namibia	0.01	0.02	-	24.4%	..	..	..	0.4%	4.2%	1.4%
Nigeria	-	-	-	-	..	-	0.1%	-	-	-
South Africa	58.11	243.41	-	36.0%	99.8%	99.0%	94.3%	93.1%	94.2%	93.8%
United Rep. of Tanzania	0.02	0.06	-	31.0%	..	-	-	2.7%	1.2%	1.1%
Zambia	-	-	-	-	20.3%	0.7%	0.5%	0.2%	-	-
Zimbabwe	0.81	2.26	-	24.0%	32.7%	11.7%	53.3%	53.4%	25.3%	25.3%
Other Africa	1.06	4.32	-	34.9%	..	-	10.6%	16.1%	22.4%	22.4%
Non-OECD Americas	**5.31**	**22.20**	**-**	**35.9%**	**3.0%**	**2.0%**	**1.9%**	**2.1%**	**2.0%**	**2.0%**
Argentina	0.82	3.24	-	34.1%	3.1%	2.1%	1.3%	2.0%	2.2%	2.5%
Brazil	2.73	12.38	-	38.9%	3.3%	2.5%	2.1%	3.1%	2.2%	2.3%
Colombia	0.60	2.14	-	30.9%	13.1%	7.9%	10.2%	5.1%	6.6%	3.5%
Dominican Republic	0.49	1.99	-	35.0%	..	-	1.2%	2.7%	15.0%	15.4%
Guatemala	0.29	1.18	-	35.0%	..	-	-	8.9%	13.9%	14.4%
Honduras	0.03	0.13	-	35.9%	..	-	-	-	-	1.8%
Panama	0.15	0.51	-	30.0%	..	-	-	-	-	6.5%
Peru	0.21	0.64	-	26.0%	..	-	-	1.7%	2.5%	1.6%
Asia excluding China	**304.66**	**1072.24**	**36.21**	**30.5%**	**31.2%**	**29.7%**	**42.5%**	**46.0%**	**46.9%**	**49.0%**
Bangladesh	0.20	0.78	-	32.7%	..	-	-	-	1.7%	1.8%
Cambodia	0.01	0.03	-	32.4%	..	..	..	-	3.1%	3.2%
Hong Kong (China)	6.39	27.78	-	37.4%	..	-	98.2%	60.4%	62.1%	71.2%
India	216.39	714.95	-	28.4%	49.1%	51.5%	66.2%	68.9%	67.1%	67.9%
Indonesia	21.88	81.00	-	31.8%	..	-	29.9%	36.4%	40.6%	44.4%
DPR of Korea	2.44	7.85	-	27.7%	36.0%	48.0%	40.1%	43.3%	35.5%	36.3%
Malaysia	13.01	52.98	-	35.0%	..	-	12.7%	11.1%	34.3%	40.7%
Mongolia	1.86	4.52	36.21	67.4%	..	..	95.6%	98.4%	98.5%	98.2%
Myanmar	0.18	0.56	-	26.9%	3.9%	2.0%	1.6%	-	8.9%	7.6%
Pakistan	0.06	0.10	-	13.6%	1.2%	0.2%	0.1%	0.4%	0.1%	0.1%
Philippines	6.19	25.34	-	35.2%	0.1%	1.0%	7.3%	36.8%	34.4%	36.6%
Sri Lanka	0.28	1.04	-	32.3%	..	-	-	-	-	8.9%
Chinese Taipei	28.16	127.03	-	38.8%	12.3%	14.0%	27.7%	48.9%	51.4%	51.0%
Thailand	8.76	34.81	-	34.1%	6.1%	9.8%	25.0%	18.5%	18.8%	22.3%
Vietnam	5.14	20.92	-	35.0%	73.3%	39.9%	23.1%	11.8%	20.7%	21.1%
Other Asia	0.08	0.33	-	34.1%	..	-	-	1.2%	1.5%	1.5%
PR of China	**987.73**	**3723.24**	**2923.25**	**39.5%**	**70.5%**	**64.7%**	**72.9%**	**80.1%**	**79.5%**	**81.0%**

Table 5.5: Coal-fired heat and electricity generation efficiency and share (continued)

	2011				Share of total generation from coal					
	Coal[1] (Mtce)	Electricity (TWh)	Heat (PJ)	Efficiency	1971	1980	1990	2000	2010	2011
Non-OECD Europe and Eurasia	**145.06**	**417.93**	**1819.08**	**54.7%**	**43.4%**	**32.2%**	**22.5%**	**24.9%**	**23.2%**	**23.3%**
Albania	-	-	-	-	..	-	-	0.5%	-	-
Belarus	0.06	0.01	1.73	69.6%	x	x	2.5%	2.2%	0.6%	0.5%
Bosnia and Herzegovina	3.84	10.81	3.52	26.4%	x	x	69.9%	49.1%	52.4%	69.1%
Bulgaria	7.60	27.54	22.84	38.3%	82.6%	51.7%	29.4%	43.6%	45.9%	51.1%
Croatia	0.56	2.58	-	39.8%	x	x	5.3%	11.2%	13.7%	18.4%
Georgia	-	-	-	-	x	x	4.5%	-	-	-
Kazakhstan	18.96	70.22	401.50	82.4%	x	x	89.2%	83.9%	91.1%	91.2%
Kosovo	1.54	5.68	-	31.7%	x	x	..	95.0%	95.4%	96.9%
Kyrgyzstan	0.19	0.48	2.94	59.8%	x	x	17.4%	11.0%	10.1%	7.7%
Latvia	0.01	0.00	0.29	56.3%	x	x	6.7%	0.8%	0.7%	0.6%
Lithuania	0.00	-	0.11	74.4%	x	x	0.7%	0.3%	0.2%	0.2%
F.Y.R. of Macedonia	1.32	5.29	0.08	34.5%	x	x	75.7%	60.8%	58.0%	69.6%
Malta	-	-	-	-	..	-	55.9%	-	-	-
Republic of Moldova	0.00	-	0.05	76.7%	x	x	24.8%	2.7%	0.2%	0.1%
Montenegro	0.42	1.45	-	29.9%	x	x	..	..	31.6%	54.7%
Romania	7.20	24.80	25.41	38.1%	30.5%	42.3%	37.9%	27.4%	31.5%	35.6%
Russian Federation	72.86	163.66	1268.77	60.9%	x	x	20.6%	25.0%	18.7%	18.3%
Serbia	7.32	28.79	10.01	37.1%	x	x	73.8%	56.1%	57.8%	63.7%
Ukraine	22.25	74.49	77.09	37.1%	x	x	15.2%	14.2%	26.1%	26.4%
Uzbekistan	0.93	2.14	4.76	31.9%	x	x	8.1%	4.1%	4.4%	4.3%
Former Soviet Union	x	x	x	x	36.9%	34.2%	x	x	x	x
Former Yugoslavia	x	x	x	x	43.3%	42.8%	x	x	x	x
Middle East	**0.15**	**0.43**	**-**	**25.0%**	**0.1%**	**0.1%**	**0.0%**	**0.1%**	**0.0%**	**0.1%**
Islam. Rep. of Iran	0.15	0.43	-	25.0%	0.3%	0.5%	0.1%	0.4%	0.2%	0.2%
Non-OECD Total	**1512.43**	**5525.94**	**4778.55**	**39.0%**	**41.7%**	**31.4%**	**28.6%**	**37.6%**	**45.5%**	**47.4%**
World	**2362.77**	**9135.27**	**5587.70**	**38.9%**	**40.0%**	**37.8%**	**35.5%**	**38.2%**	**40.0%**	**40.9%**

(1) "Coal" refers to all types of coal, primary (anthracite, coking coal, other bituminous coal, sub-bituminous coal, lignite/brown coal) and derived fuels (including patent fuel, BKB, coke oven coke, gas coke, coal tar, coke oven gas, gas works gas, blast furnace gas and other recovered gases), as well as peat, peat briquettes and oil shale.

Quantities have been converted to Mtce using calorific values reported by the respective countries.

Please refer to notes and definitions in Part I.

Source: *IEA/OECD Energy Balances of OECD Countries, IEA/OECD Energy Balances of non-OECD Countries*

WORLD

Table 5.6: Final consumption of energy by fuel*

(Mtce)

	1973	1980	1990	2000	2009	2010	2011	Average annual percent change 73-90	90-11
Total final consumption[(1)]	**6677.2**	**7696.9**	**8989.8**	**10133.8**	**11904.8**	**12531.1**	**12739.3**	**1.76**	**1.67**
Coal and peat	914.3	1017.1	1100.4	825.0	1188.5	1235.2	1290.9	1.10	0.76
Oil	3213.6	3493.1	3726.0	4465.5	4946.2	5171.7	5190.4	0.87	1.59
Natural Gas	931.8	1165.5	1348.8	1598.4	1803.0	1936.6	1972.1	2.20	1.83
Biofuels and wastes	880.8	1011.0	1140.2	1321.9	1529.2	1572.0	1588.2	1.53	1.59
Geothermal	0.5	0.7	2.8	6.2	9.8	10.5	11.1	10.65	6.71
Solar, wind, tide	-	0.0	2.7	7.0	18.4	20.8	25.1	-	11.11
Electricity	626.6	837.1	1190.1	1556.8	2048.4	2195.3	2260.2	3.85	3.10
Heat	109.5	172.4	478.8	353.1	361.3	389.2	401.4	9.07	-0.84
of which:									
Total industry	**2207.6**	**2545.2**	**2591.4**	**2721.6**	**3264.0**	**3539.9**	**3652.5**	**0.95**	**1.65**
Coal and peat	517.0	606.0	679.0	615.8	928.3	981.3	1041.3	1.62	2.06
Oil	640.9	679.1	468.7	468.6	455.8	496.7	461.6	-1.82	-0.07
Natural Gas	509.9	596.9	512.1	592.7	635.1	693.9	723.4	0.03	1.66
Biofuels and wastes	130.7	154.7	169.5	244.4	266.7	279.6	283.1	1.54	2.47
Geothermal	0.0	0.0	0.2	0.5	0.6	0.6	0.7	10.47	5.08
Solar, wind, tide	-	-	0.0	0.1	0.0	0.0	0.0	-	3.71
Electricity	334.8	425.6	543.9	657.4	822.6	914.7	962.5	2.89	2.76
Heat	74.2	82.8	218.0	142.1	154.8	173.1	179.8	6.55	-0.91
Total transport	**1543.7**	**1781.7**	**2258.0**	**2798.1**	**3280.3**	**3443.6**	**3492.8**	**2.26**	**2.10**
Coal and peat	47.1	35.9	18.4	6.3	4.9	4.9	4.9	-5.39	-6.13
Oil	1455.8	1698.1	2120.9	2668.5	3059.4	3196.3	3236.0	2.24	2.03
Natural Gas	25.3	25.7	80.2	82.2	110.7	127.4	132.2	7.02	2.41
Biofuels and wastes	0.3	2.1	8.4	14.2	72.9	80.9	83.7	20.67	11.55
Electricity	15.1	19.8	30.1	26.9	32.5	34.2	35.9	4.13	0.84
Residential	**1562.5**	**1826.6**	**2190.4**	**2598.8**	**2876.5**	**2962.9**	**2961.2**	**2.01**	**1.45**
Coal and peat	176.9	180.4	222.6	111.5	112.5	113.6	113.0	1.36	-3.18
Oil	329.5	282.6	280.0	325.4	301.9	299.8	294.3	-0.95	0.24
Natural Gas	211.9	303.3	391.6	525.1	579.9	605.4	593.4	3.68	2.00
Biofuels and wastes	683.0	772.1	914.6	1032.4	1151.2	1172.3	1180.2	1.73	1.22
Geothermal	0.4	0.6	1.4	3.4	5.8	6.4	6.7	7.38	7.74
Solar, wind, tide	-	0.0	2.5	5.1	6.8	7.1	7.6	-	5.38
Electricity	144.3	211.1	309.8	439.3	578.2	608.9	612.2	4.60	3.30
Heat	16.4	76.4	67.8	156.6	140.2	149.4	153.8	8.72	3.98
Comm & public services	**524.9**	**598.2**	**654.4**	**807.0**	**998.7**	**1025.7**	**1021.8**	**1.31**	**2.14**
Coal and peat	113.6	127.8	69.7	22.1	36.2	33.3	35.5	-2.83	-3.17
Oil	205.4	168.3	142.3	155.0	144.8	146.3	144.8	-2.14	0.08
Natural Gas	100.1	136.4	169.0	210.0	249.5	259.3	254.5	3.13	1.97
Biofuels and waste	9.0	10.6	13.2	17.4	24.1	24.9	25.4	2.32	3.17
Geothermal	0.0	0.0	0.3	1.6	2.4	2.4	2.6	17.86	10.79
Solar, wind, tide	-	-	0.1	0.1	0.4	0.4	0.5	-	7.54
Electricity	95.6	148.3	246.7	367.7	492.6	511.4	514.6	5.73	3.56
Heat	1.2	6.7	13.1	33.1	48.7	47.7	43.9	14.99	5.94
Non-energy use	**408.8**	**505.6**	**682.5**	**886.9**	**1072.9**	**1128.6**	**1170.6**	**3.06**	**2.60**
Coal and peat	8.6	14.1	35.6	39.2	52.4	54.9	56.0	8.73	2.18
Oil	374.0	430.7	515.8	686.3	810.9	848.8	869.8	1.91	2.52
Natural Gas	26.2	60.9	131.1	161.4	209.6	224.9	244.8	9.92	3.02

(1) Total Final Consumption (TFC) excludes use in transformation processes and energy industry own use.

* Please refer to notes and definitions in Part I. Agriculture, Forestry, Fishing and "Non-specified other" sectors are not reported here individually, but are accounted for in Total final consumption.

Source: *IEA/OECD Energy Balances of OECD Countries, IEA/OECD Energy Balances of non-OECD Countries*

WORLD

Table 5.7: Use of coal for selected end-uses*

(million tonnes)

	1978	1990	2000	2005	2009	2010	2011	Average annual percent change 78-90	90-11
Steam coal	**2108.3**	**2953.4**	**3422.0**	**4447.3**	**5091.2**	**5411.3**	**5676.8**	**2.85**	**3.16**
Total electricity and heat	1058.8	1814.4	2508.1	3152.6	3707.4	3892.4	4105.2	4.59	3.96
Main activity producers	*986.3*	*1726.7*	*2420.7*	*3053.6*	*3586.2*	*3776.0*	*3981.8*	*4.78*	*4.06*
Autoproducers	*72.6*	*87.7*	*87.4*	*99.0*	*121.2*	*116.4*	*123.4*	*1.59*	*1.64*
Patent fuel/BKB plants	26.3	32.6	13.1	137.7	15.7	15.1	15.9	1.79	-3.34
Coke ovens/Liquefaction(1)	3.8	70.8	101.9	101.7	82.2	60.4	75.7	27.73	0.32
Blast furnace inputs	0.0	3.9	12.9	15.7	12.7	17.9	18.9	74.13	7.83
Gas manufacture	4.9	6.5	10.9	18.8	16.2	16.4	14.5	2.33	3.92
Industry	505.8	524.4	536.0	692.8	807.2	845.2	890.0	0.30	2.55
Iron and steel	*18.5*	*42.0*	*62.4*	*93.4*	*125.3*	*150.3*	*172.3*	*7.09*	*6.95*
Chemical	*23.2*	*64.3*	*65.6*	*75.5*	*84.2*	*86.7*	*92.3*	*8.86*	*1.73*
Non-metallic minerals	*57.4*	*165.6*	*206.1*	*271.6*	*314.0*	*311.6*	*333.9*	*9.22*	*3.40*
Paper, pulp and print	*11.1*	*33.1*	*28.5*	*36.9*	*37.9*	*38.4*	*36.6*	*9.50*	*0.48*
Other industry	*395.5*	*219.4*	*173.3*	*215.4*	*245.8*	*258.1*	*255.0*	*-4.79*	*0.72*
Other sectors(2)	387.6	382.9	179.2	210.5	243.9	227.8	219.9	-0.10	-2.60
Non-energy use	10.5	36.5	36.8	42.6	45.6	46.0	47.0	10.95	1.22
Coking coal	**521.6**	**555.0**	**476.3**	**644.1**	**768.4**	**880.7**	**936.2**	**0.52**	**2.52**
Total electricity and heat	7.1	22.0	22.0	23.2	32.1	34.6	43.9	9.84	3.33
Main activity producers	*6.2*	*18.1*	*18.8*	*21.2*	*29.0*	*33.1*	*40.6*	*9.34*	*3.92*
Autoproducers	*0.9*	*3.9*	*3.2*	*2.0*	*3.1*	*1.5*	*3.3*	*12.59*	*-0.86*
Patent fuel/BKB plants	-	-	-	2.8	-	-	-	-	-
Coke ovens/Liquefaction(1)	503.8	465.6	407.4	567.4	666.3	759.2	807.9	-0.66	2.66
Blast furnace inputs	-	6.7	15.7	18.7	16.8	21.5	24.6	-	6.38
Gas manufacture	6.9	2.1	3.8	4.8	6.0	4.5	3.4	-9.50	2.31
Industry	1.9	55.2	20.2	25.2	36.4	40.9	40.6	32.33	-1.45
Iron and steel	*1.7*	*6.4*	*6.5*	*10.6*	*14.8*	*14.1*	*15.5*	*11.50*	*4.31*
Chemical	*0.0*	*3.4*	*1.8*	*3.0*	*4.0*	*3.5*	*3.8*	*62.43*	*0.54*
Non-metallic minerals	*0.0*	*1.7*	*1.8*	*3.6*	*4.9*	*4.8*	*4.3*	*69.39*	*4.55*
Paper, pulp and print	-	*0.8*	*0.1*	*0.3*	*0.3*	*0.3*	*0.3*	-	*-4.89*
Other industry	*0.2*	*42.9*	*10.0*	*7.6*	*12.4*	*18.3*	*16.8*	*58.10*	*-4.38*
Other sectors(2)	0.3	0.8	0.5	0.1	0.4	0.2	0.3	8.99	-4.97
Non-energy use	-	0.2	0.8	1.3	1.6	1.5	1.5	-	9.83
Lignite	**915.1**	**1155.2**	**863.4**	**868.7**	**853.5**	**855.1**	**913.8**	**1.96**	**-1.11**
Total electricity and heat	648.3	878.7	790.7	796.1	782.5	782.1	831.0	2.57	-0.27
Main activity producers	*631.3*	*841.1*	*762.6*	*767.2*	*759.4*	*753.2*	*803.8*	*2.42*	*-0.22*
Autoproducers	*17.0*	*37.6*	*28.1*	*28.9*	*23.1*	*28.9*	*27.2*	*6.84*	*-1.53*
Patent fuel/BKB plants	141.3	125.4	18.8	19.1	18.1	19.2	20.2	-0.99	-8.32
Coke ovens/Liquefaction(1)	-	0.9	1.4	2.0	2.8	3.1	3.1	-	6.17
Gas manufacture	6.4	9.1	7.6	7.7	7.8	8.2	8.0	2.99	-0.60
Industry	61.6	61.7	28.7	28.2	23.3	26.7	30.9	0.01	-3.24
Iron and steel	*1.1*	*1.8*	*0.3*	*0.6*	*0.2*	*0.3*	*0.3*	*3.98*	*-7.69*
Chemical	*14.7*	*12.7*	*4.9*	*4.8*	*4.2*	*5.1*	*4.4*	*-1.21*	*-4.96*
Non-metallic minerals	*1.9*	*4.7*	*5.0*	*7.6*	*5.4*	*5.4*	*7.5*	*7.60*	*2.25*
Paper, pulp and print	*4.5*	*2.2*	*0.7*	*0.7*	*2.3*	*2.9*	*1.1*	*-5.95*	*-3.15*
Other industry	*39.4*	*40.4*	*17.8*	*14.5*	*11.2*	*13.0*	*17.6*	*0.21*	*-3.87*
Other sectors(2)	43.8	63.5	15.2	11.9	15.2	14.9	16.7	3.15	-6.15
Non-energy use	0.8	0.7	0.7	0.3	0.1	0.1	0.0	-1.78	-11.79

(1) Liquefaction primarily refers to direct distillation processes. Indirect liquefaction may be reported here or under gas manufacture.

(2) Other sectors are Residential, Commercial and public services, Agriculture/Forestry, Fishing and Non-specified other.

* Coal totals may not match the sum of the individual rows due to statistical differences. Please refer to notes and definitions in Part I.

Source: *IEA/OECD Energy Balances of OECD Countries, IEA/OECD Energy Balances of non-OECD Countries*

6. CHARTS

Figure 1: World coal production (million tonnes)

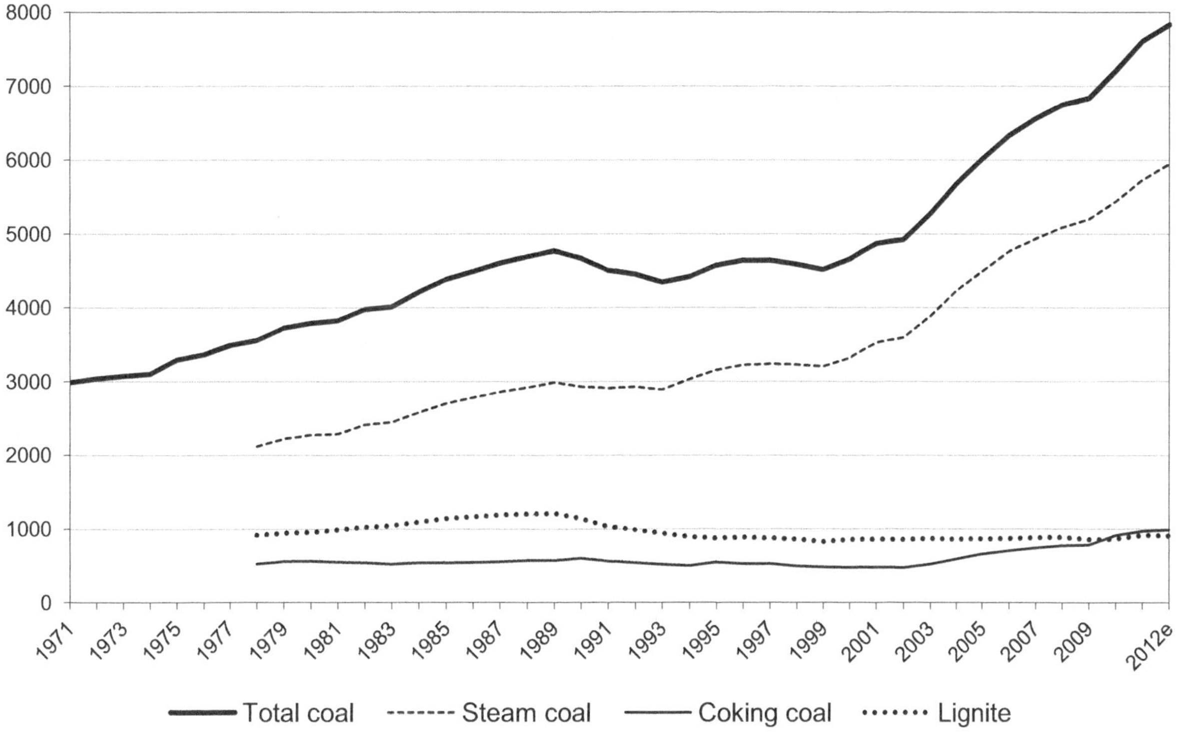

Figure 2: World steam and coking coal trade (million tonnes)

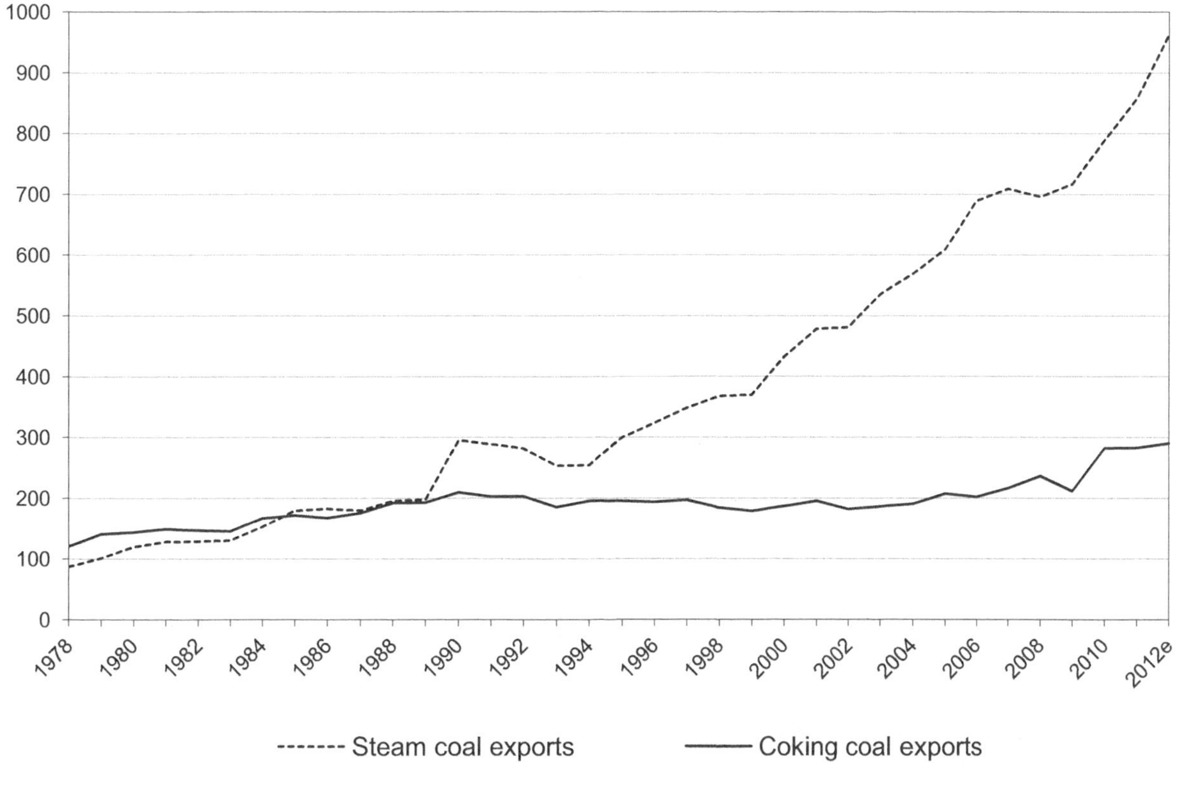

Figure 3: Coking coal price
CIF Japan and CIF EU member states (USD/tonne)

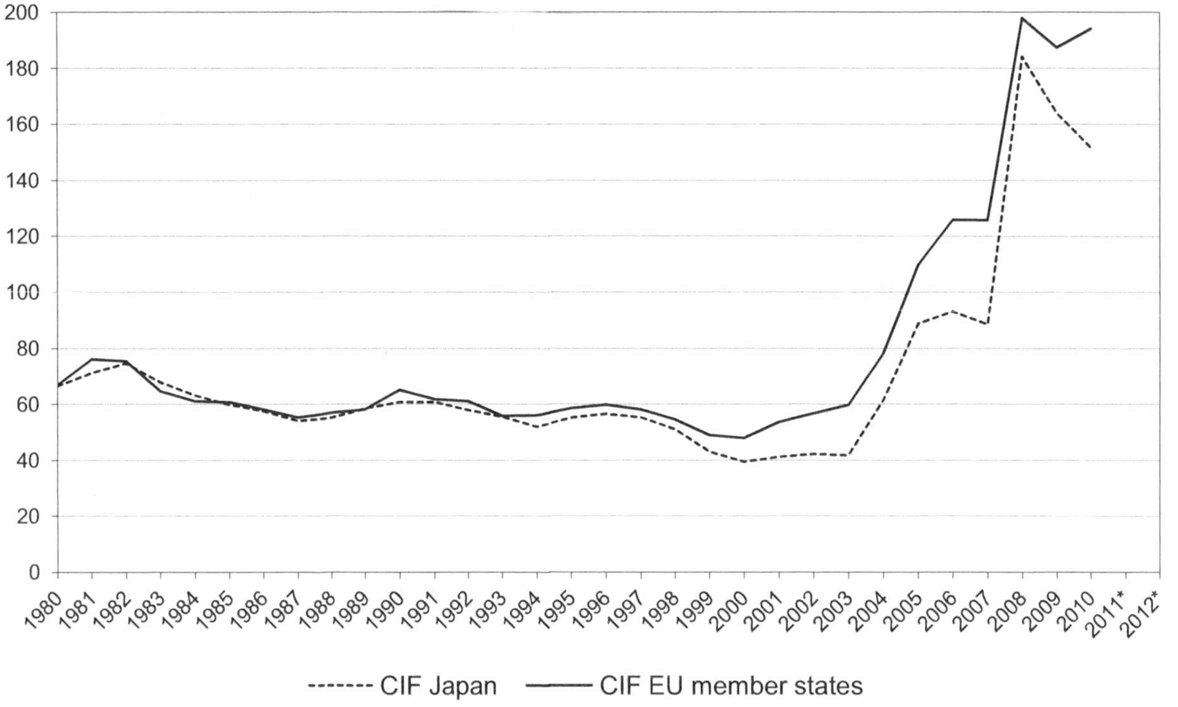

Figure 4: Steam coal price
CIF Japan and CIF EU member states (USD/tonne)

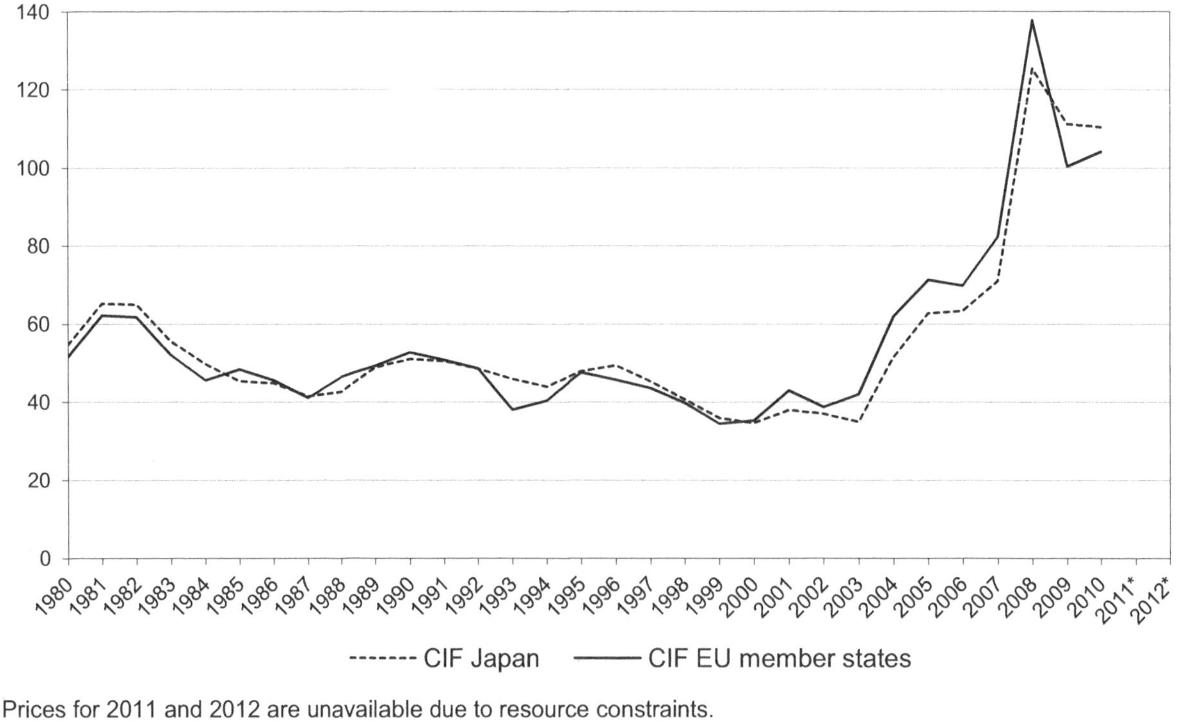

* Prices for 2011 and 2012 are unavailable due to resource constraints.

Figure 5: Steam coal price
US FOB vs. US domestic (USD/tonne)

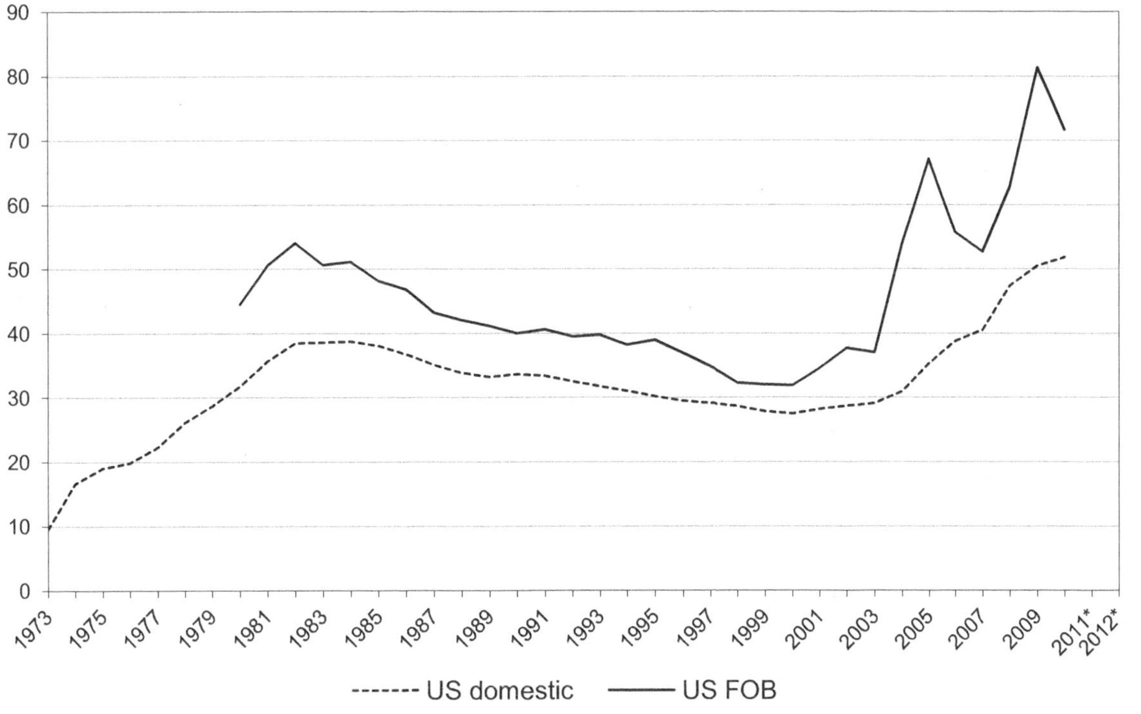

Figure 6: Steam coal price CIF EU member states (USD/tonne)
and US exports (million tonnes)

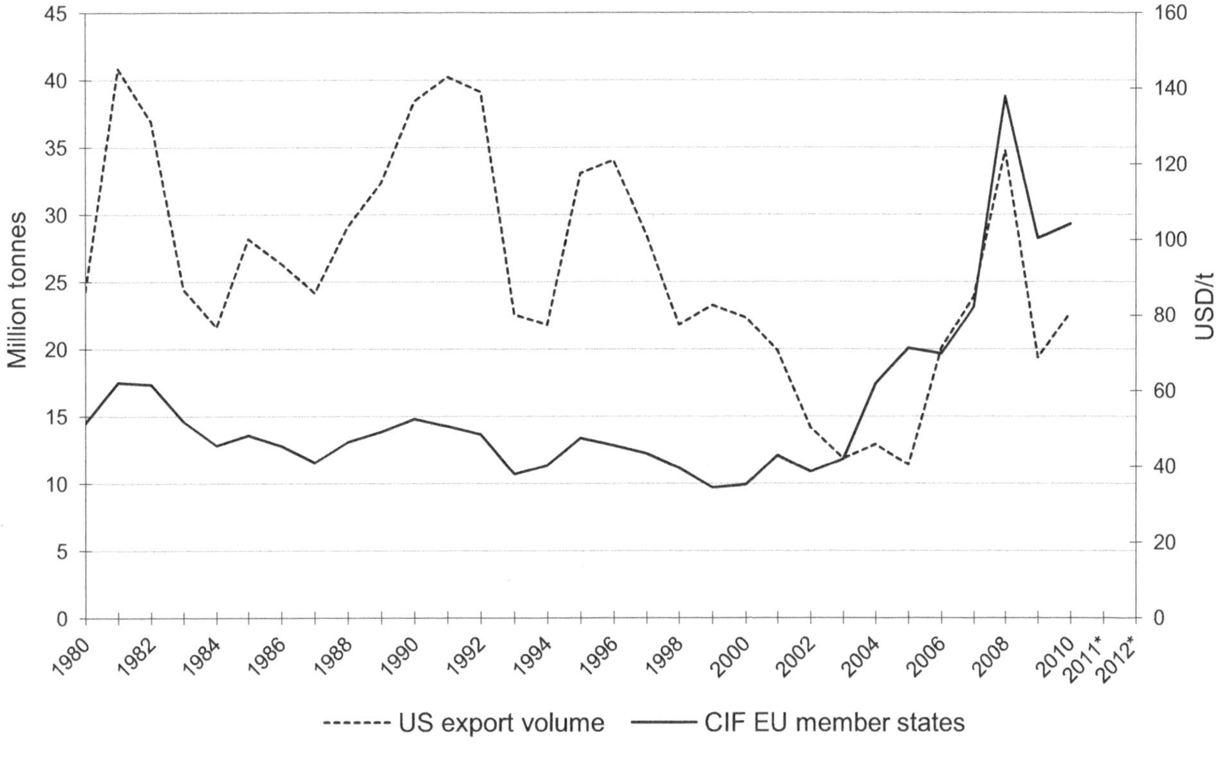

* Prices for 2011 and 2012 are unavailable due to resource constraints.

Figure 7: OECD total primary energy supply (Mtce)

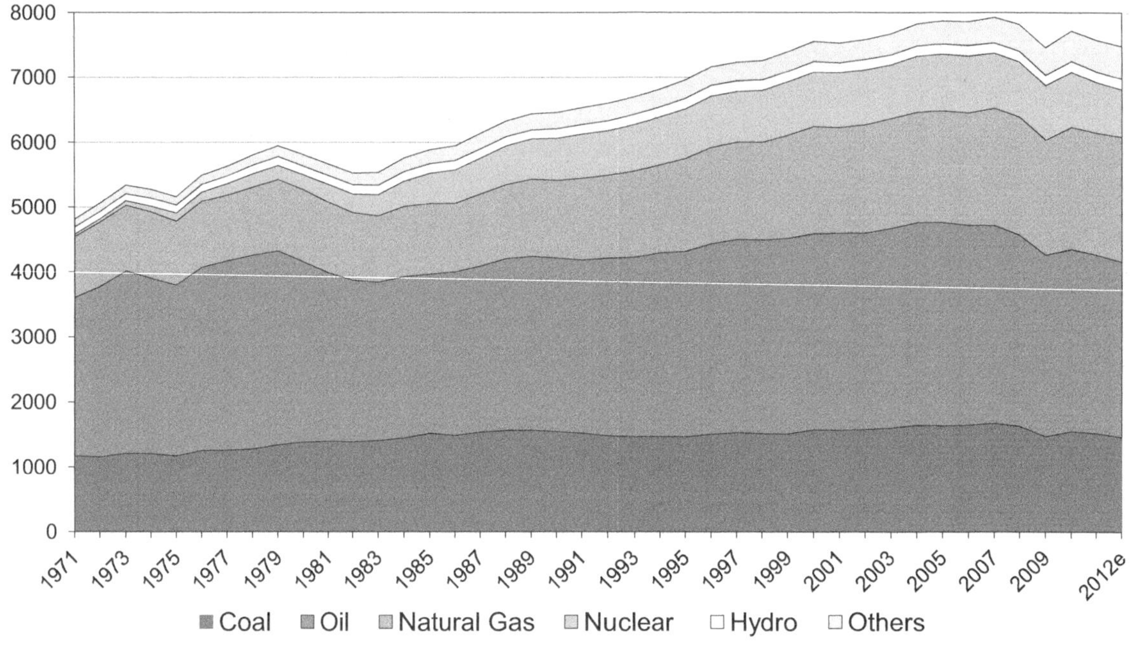

Figure 8: OECD coal consumption and indicators (2005=100)

Figure 9: Japan steam coal supply, Imports, GDP annual growth rates

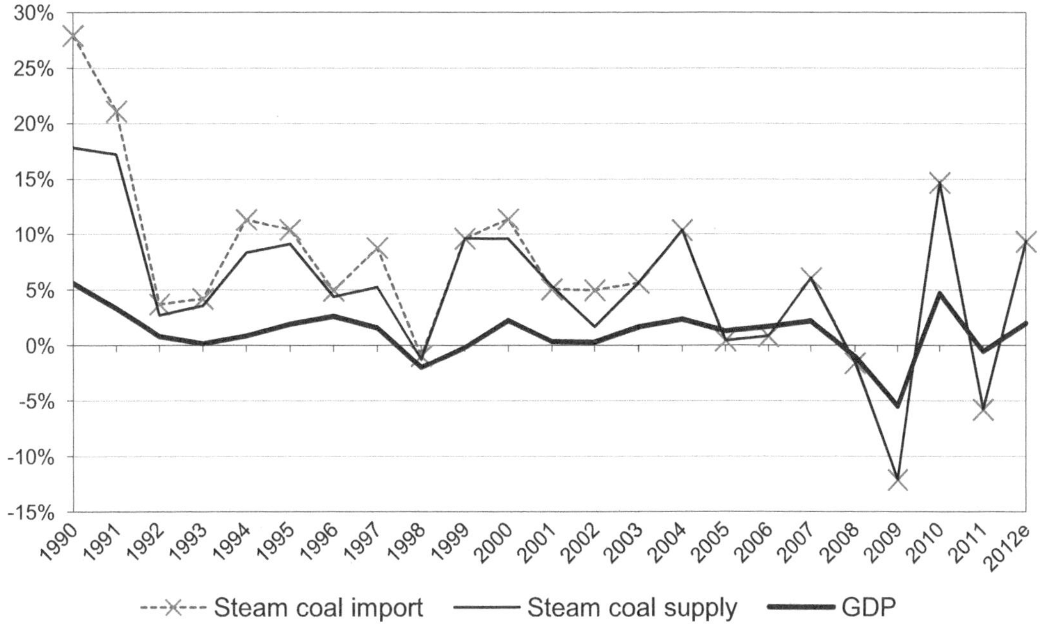

Figure 10: EU member states steam coal supply, Imports, GDP annual growth rates

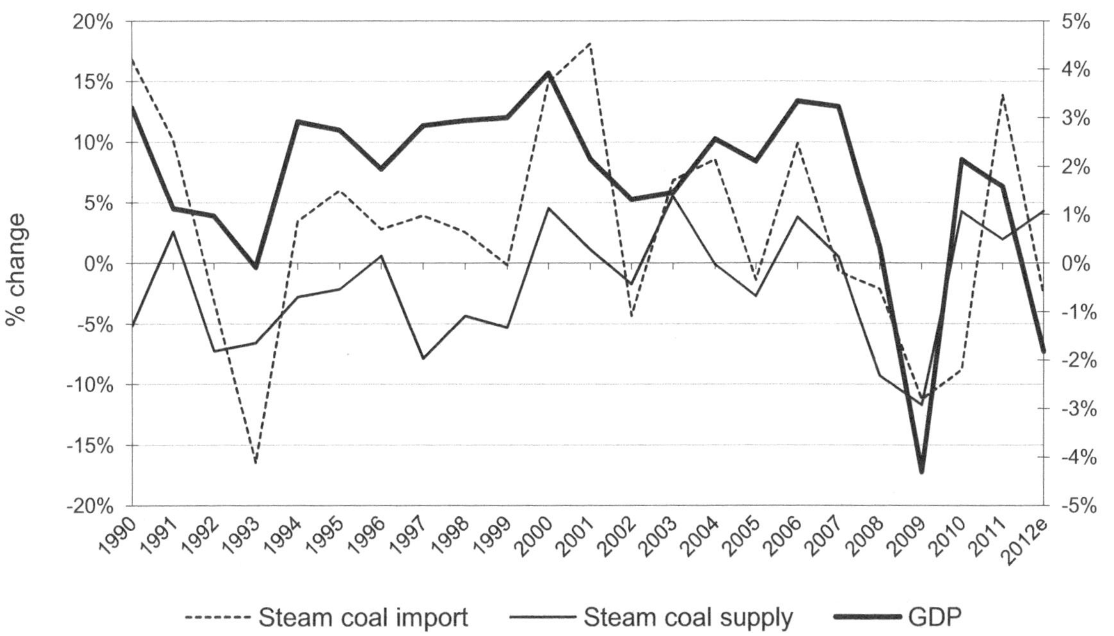

Figure 11: Coal production by region
(million tonnes)

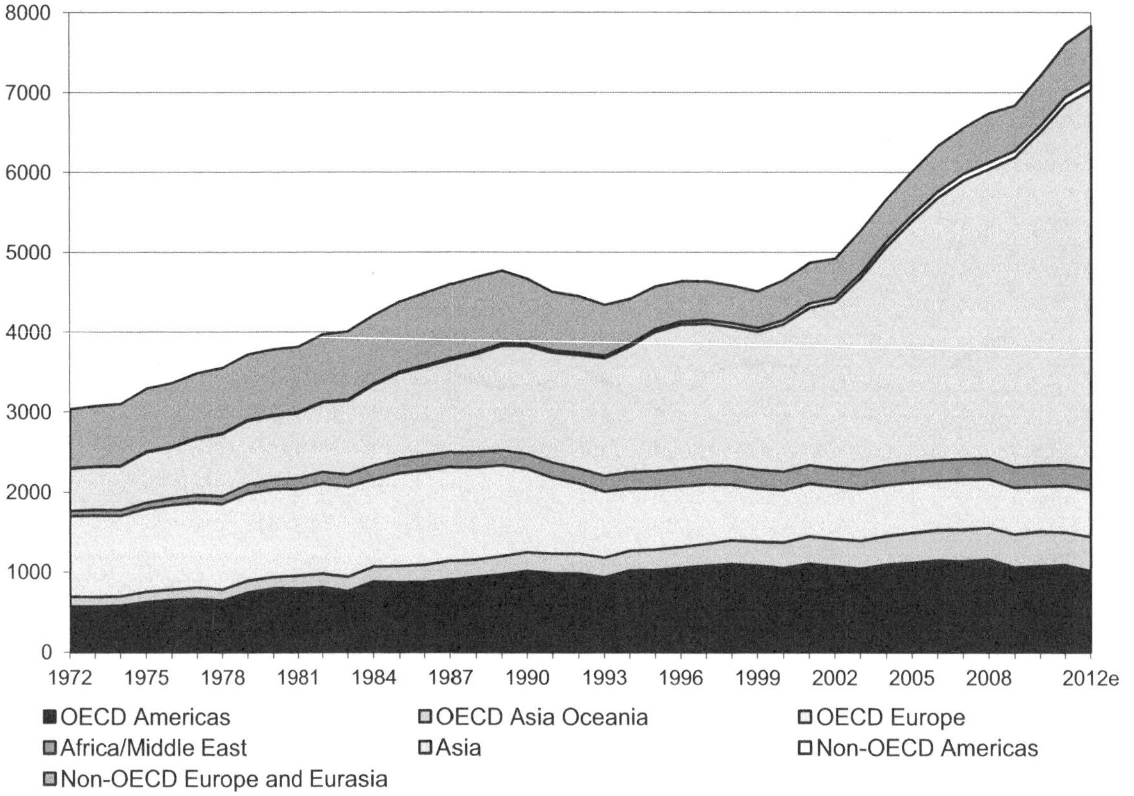

Figure 12: Coal consumption by region
(million tonnes)

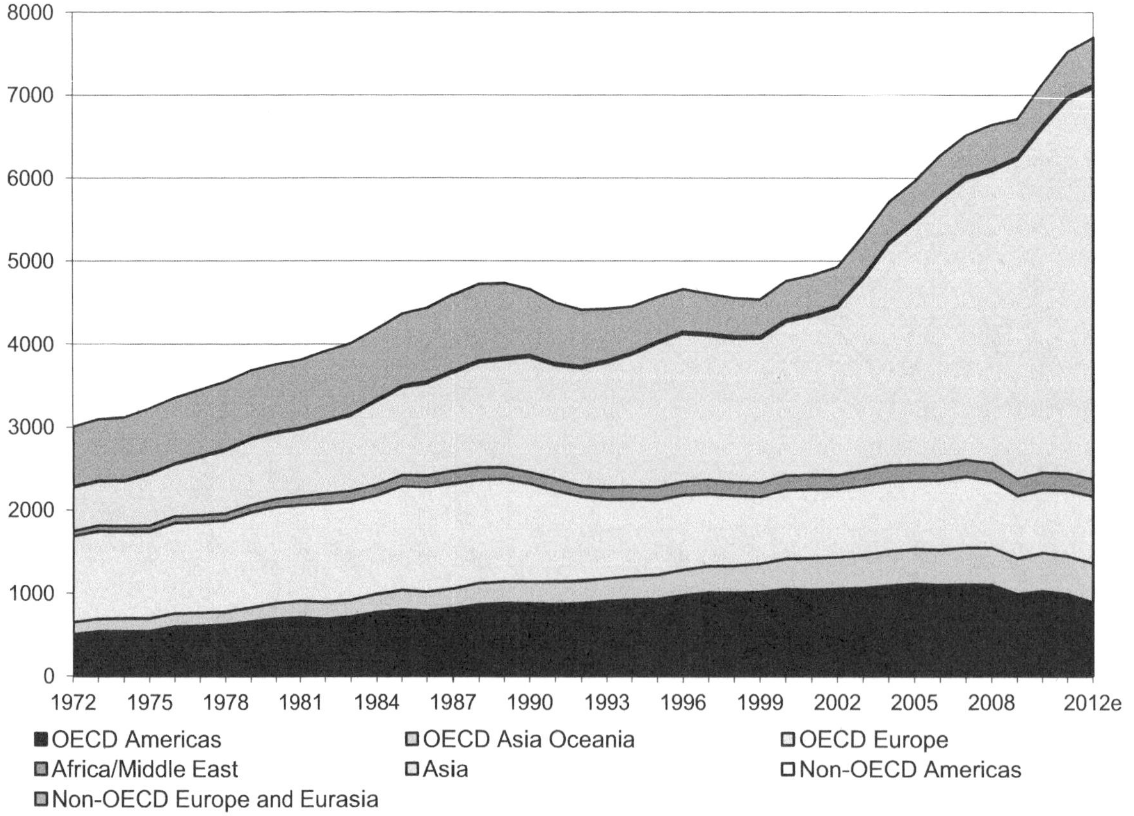

Figure 13: Coal imports by region
(million tonnes)

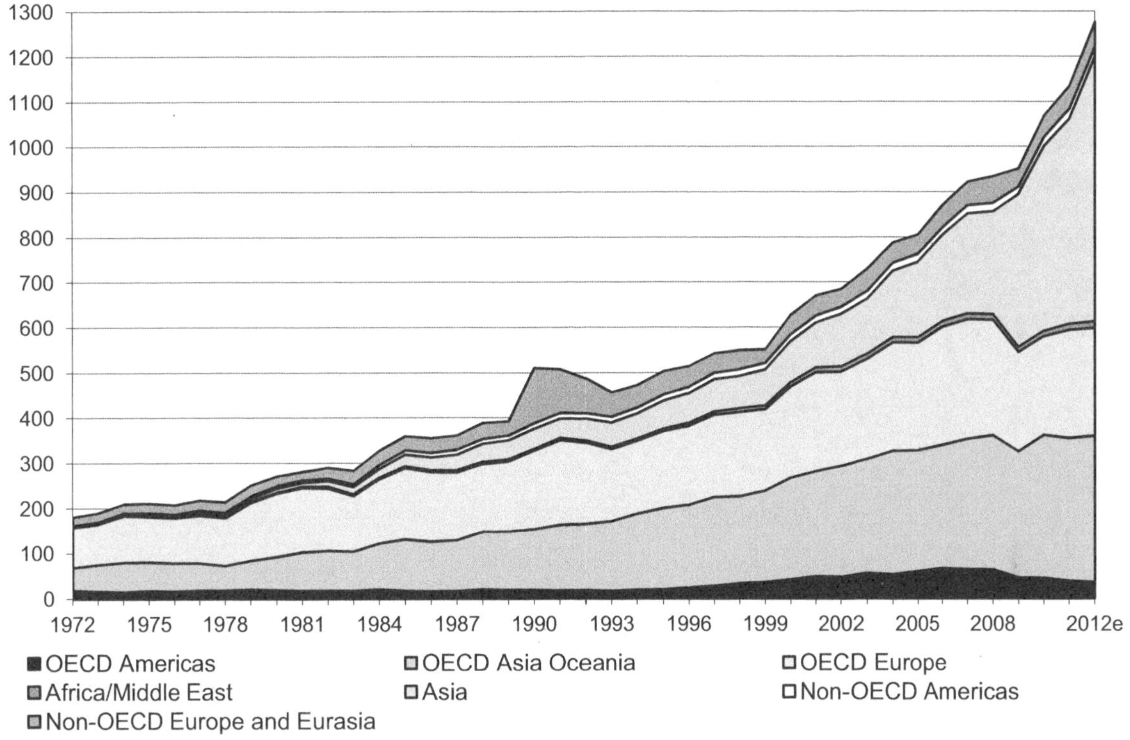

Figure 14: Coal exports by region
(million tonnes)

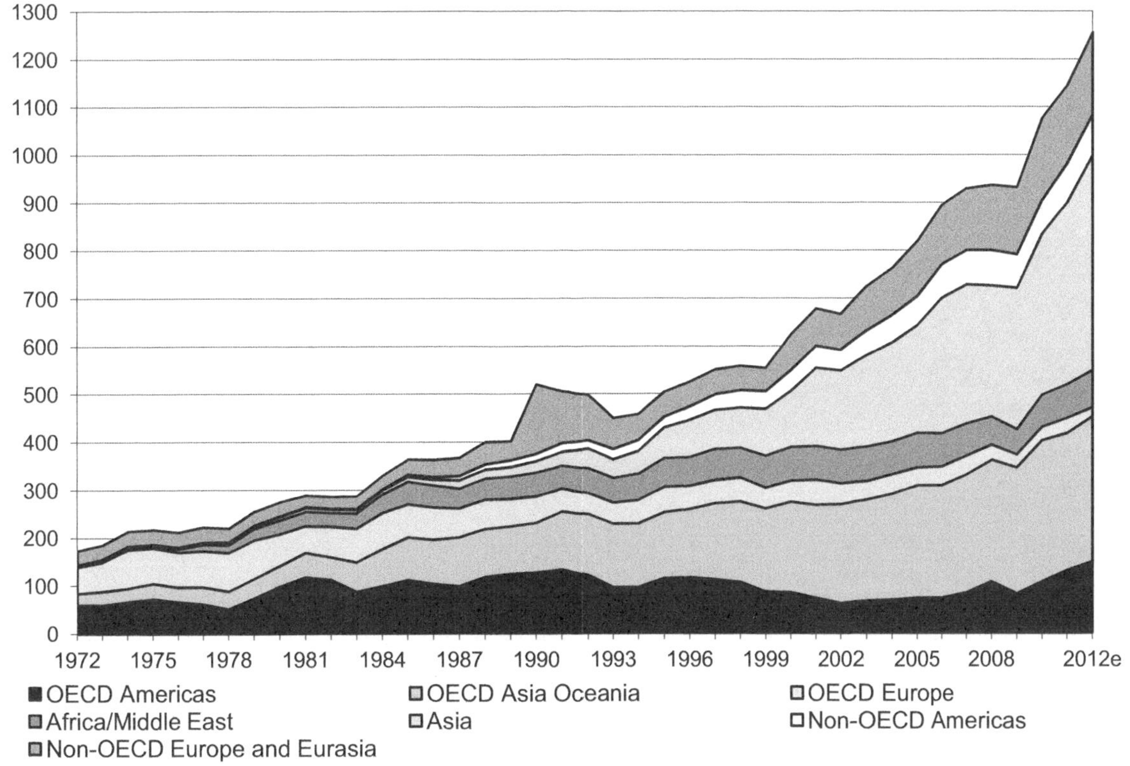

PART IV

DETAILED OECD COAL DATA

DIRECTORY OF PART IV FIGURES AND TABLES

Part IV of *Coal Information* contains detailed statistical information on coal for the 34 Member countries of the OECD and for regional aggregates (OECD Total, OECD Americas, OECD Asia Oceania, OECD Europe, IEA Total, IEA Americas, IEA Asia Oceania, and IEA Europe). The figures and tables of regional aggregates are presented before the country tables which are set out in alphabetical order.

Data for each OECD and IEA region and country are illustrated in figures at the beginning of each section. This is followed by detailed statistical information presented in several tables for each region and country.

Interpreting energy data and comparing statistics between countries is made difficult by differences in definitions used by countries in the collection and reporting of data. In Part I, such differences are explained. The conventions used by the Secretariat in presenting energy data are also reported. In particular, readers are strongly advised to read the country notes. Conversion factors are also included for reference.

Figures

1. Total primary energy supply by fuel
2. Coal supply indicators
3. Primary coal supply
4. Coal consumption
5. Electricity generation by fuel
6. CO_2 emissions by fuel
7. Electricity generation by fuel share
8. CO_2 emissions by fuel share

Tables

Where present, tables presented are numbered as follows:

1. Total primary energy supply (TPES) by fuel
2. Energy supply, GDP and population
3. Coal and peat production by type
4. Final consumption of energy by fuel
5. Coal balance
6. Use of coal for selected end-uses
7. Fuel prices to end users
8. Coal and peat trade by type
9. Total coal imports by origin
10. Coking coal exports by destination
11. Steam coal exports by destination
12. Coal import values by origin
13. Coal export values by destination

It should be noted that not all tables are shown for all countries. For example, in the case where a country has no or very few coal exports, the related tables (on volumes and values) are omitted.

Data for 2012 are provisional with the exception of Tables 7, 12 and 13 where data for 2012 are final. USD refers to dollars used in the United States of America. Prices for regional totals, weighted by national consumption, are calculated as an average of available price data in the region and therefore prices shown should be considered only indicative. Data are converted from unit prices to tonnes of coal equivalent (tce) using country specific calorific values.

OECD TOTAL*

Figure 1: TPES by fuel (Mtce)

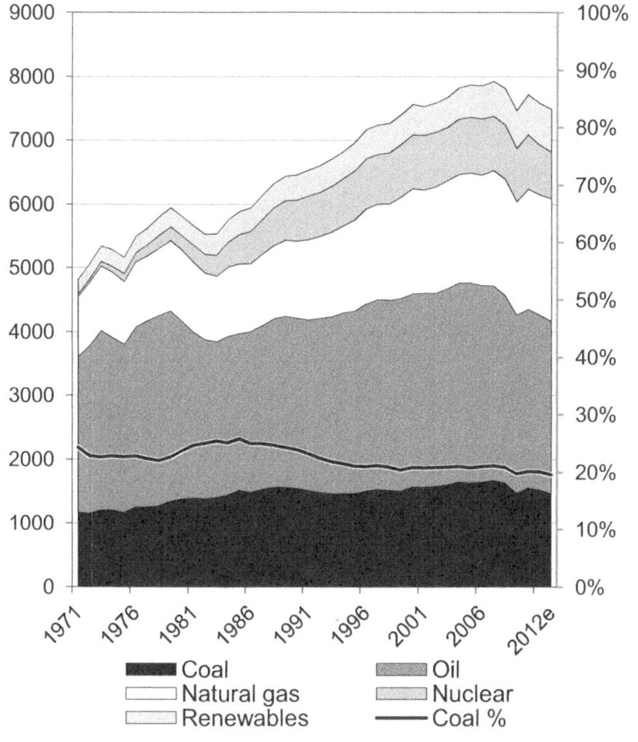

Figure 2: Coal supply indicators (1971 = 100)

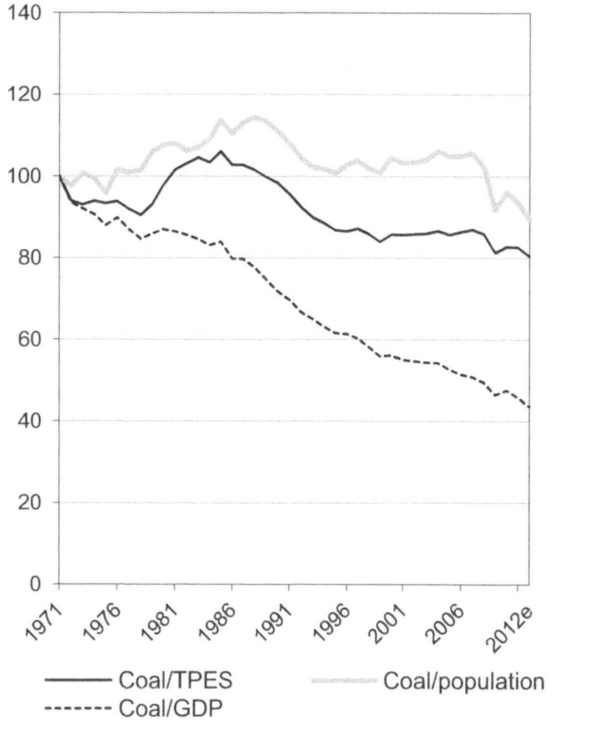

Figure 3: Primary coal supply (Mtce)

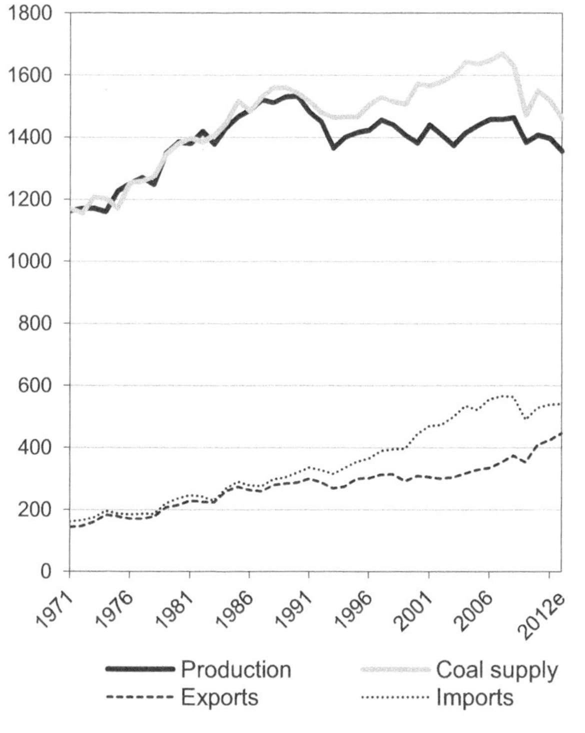

Figure 4: Coal consumption (Mtce)

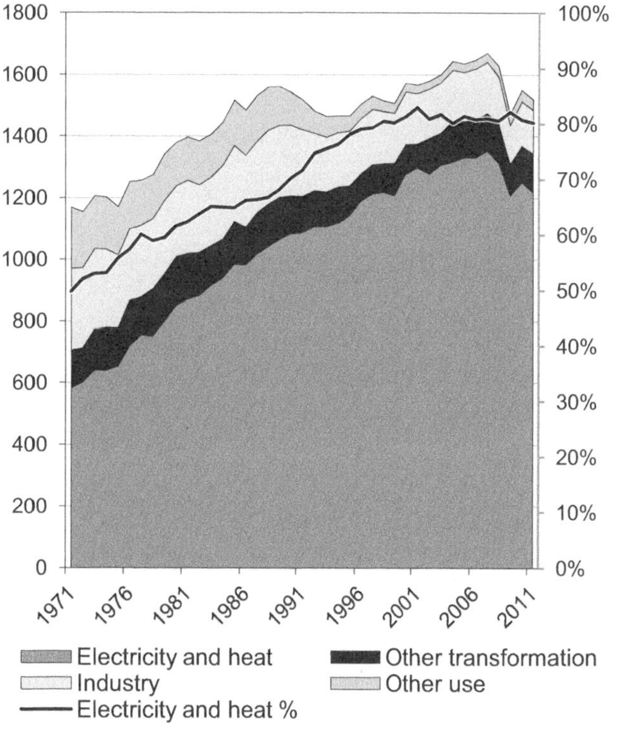

*Please refer to notes and definitions in Part I. Peat is included in coal for all figures.

Source: *IEA/OECD Coal Information, IEA/OECD Energy Balances of OECD Countries, OECD Main Economic Indicators*

OECD TOTAL*

Figure 5: Electricity generation by fuel (TWh)

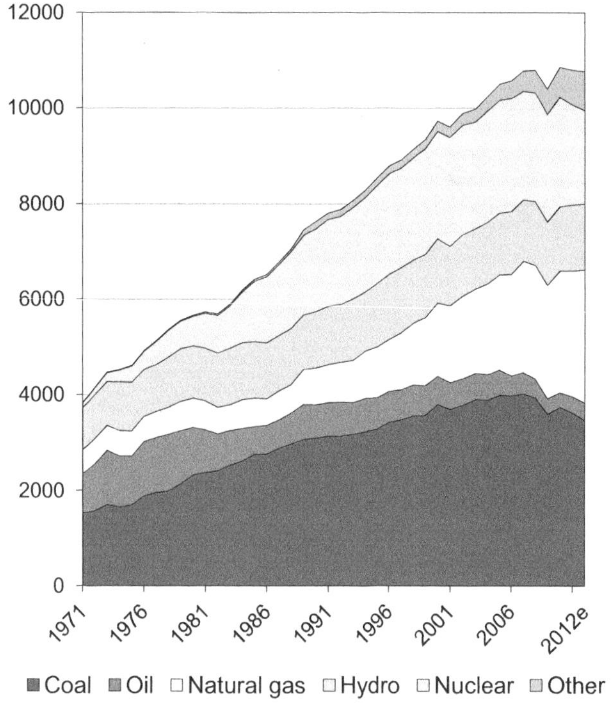

Figure 6: CO_2 emissions by fuel (Mt CO_2)

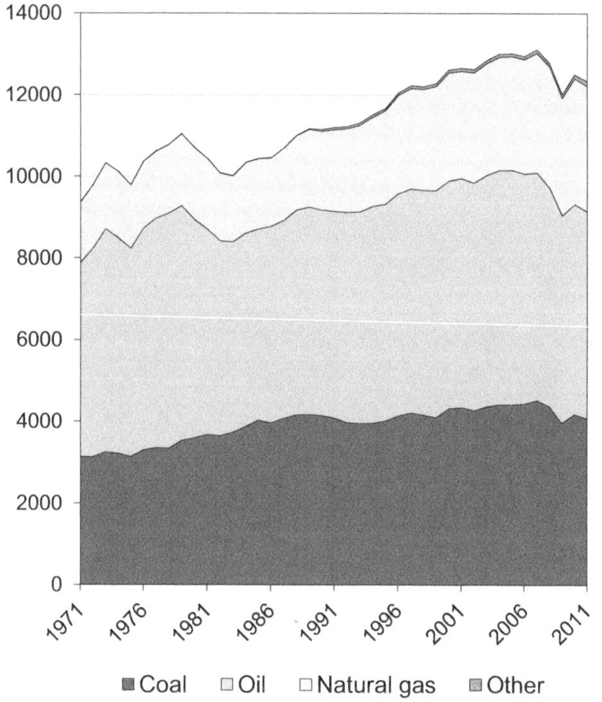

Figure 7: Electricity generation by fuel share

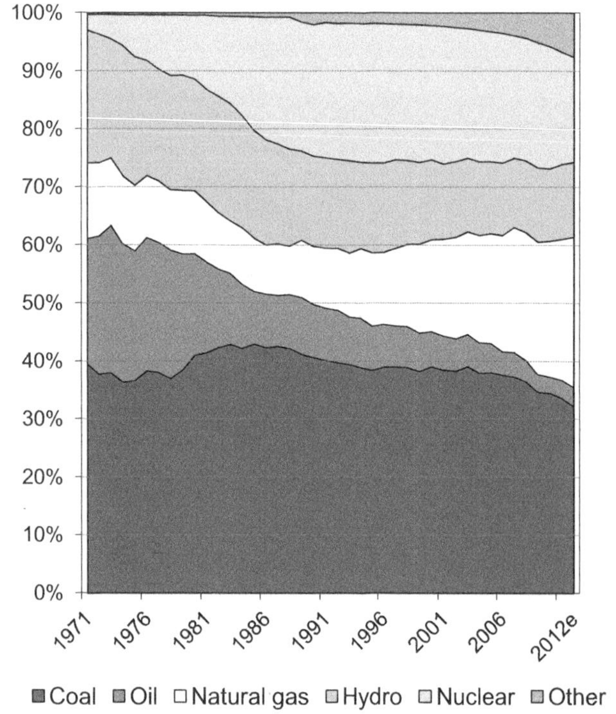

Figure 8: CO_2 emissions by fuel share

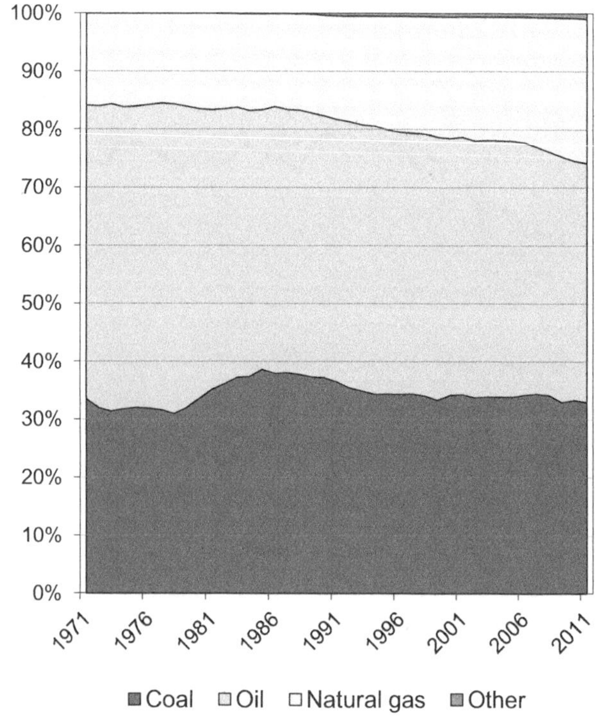

*Please refer to notes and definitions in Part I. Peat is included in coal for all figures.

Source: *IEA/OECD Electricity Information, IEA/OECD CO2 Emissions from Fuel Combustion*

OECD TOTAL

1. Total primary energy supply (TPES) by fuel*

	1973	1980	1990	2000	2010	2011	2012e	Average annual percent change 73-90	90-11
TPES (Mtce)	**5343.64**	**5810.88**	**6460.75**	**7560.93**	**7723.20**	**7578.26**	**7482.78**	**1.12**	**0.76**
Coal and peat	1206.57	1379.43	1543.40	1572.44	1550.51	1518.62	1460.59	1.46	-0.08
Oil	2810.63	2779.30	2671.13	3020.54	2801.28	2744.76	2701.30	-0.30	0.13
Natural Gas	1009.03	1111.54	1203.89	1651.39	1883.62	1881.77	1925.39	1.04	2.15
Biofuels and waste	124.81	158.03	209.88	260.87	377.10	380.97	386.32	3.10	2.88
Nuclear	70.31	231.79	644.59	837.40	852.13	777.17	726.52	13.92	0.89
Hydro	112.77	134.23	145.17	164.87	166.07	170.54	170.06	1.50	0.77
Geothermal	8.68	14.55	37.56	43.51	45.17	46.65	47.73	9.00	1.04
Solar, wind, tide	0.07	0.10	3.20	9.13	45.01	56.66	66.14	25.36	14.66
Net electricity trade(1)	0.77	1.91	1.84	0.07	1.35	0.22	-2.22	5.28	-9.61
Heat(2)	-	-	0.08	0.70	0.96	0.90	0.94	-	12.06

(1) Net trade between OECD and non-OECD countries.

(2) Ambient heat from heat pumps used in the transformation sector.

Source: *IEA/OECD Energy Balances of OECD Countries*

2. Energy supply, GDP and population*

	1973	1980	1990	2000	2010	2011	2012e	Average annual percent change 73-90	90-11
GDP (2005 bil. USD)	15117	18269	24793	32279	37595	38239	38672	2.95	2.08
TPES/GDP(1)	0.35	0.32	0.26	0.23	0.21	0.20	0.19	-1.78	-1.30
Population (millions)	915.7	980.0	1064.0	1151.8	1233.6	1240.5	1246.6	0.89	0.73
TPES/population(1)	5.84	5.93	6.07	6.56	6.26	6.11	6.00	0.23	0.03
TPES/GDP(2)	161.0	144.8	118.7	106.7	93.5	90.2	88.1	-1.78	-1.30
Coal and peat supply/GDP(2)	174.9	165.5	136.4	106.8	90.4	87.0	82.8	-1.45	-2.12
Elec. consumption/GDP(2)	98.8	103.1	102.4	102.3	99.0	96.7	..	0.21	-0.28
Elec. generation (TWh)	4472	5668	7629	9728	10855	10802	10771	3.19	1.67
Industrial production(2)	..	57.4	72.8	94.7	99.8	102.6	103.5	..	1.65

(1) TPES/GDP in units of tce/2005 thousand USD, TPES/population in units of tce/capita.

(2) As index, 2005 (not shown) = 100

Source: *IEA/OECD Energy Balances of OECD Countries* , *OECD Main Economic Indicators*

3. Coal and peat production by type*

	1978(1)	1990	2000	2005	2010	2011	2012e	Average annual percent change 78-90	90-11
Mtce:									
Coking coal	290.52	281.52	228.73	233.65	278.01	275.22	275.67	-0.26	-0.11
Steam coal	748.82	985.64	952.10	1003.41	930.54	918.05	877.82	2.32	-0.34
Lignite	205.58	259.93	196.34	196.01	193.82	199.70	199.70	1.97	-1.25
Peat	2.30	5.52	3.32	4.68	4.45	3.92	2.24	7.57	-1.61
Mt:									
Coking coal	296.47	282.54	234.94	242.95	289.84	287.47	288.14	-0.40	0.08
Steam coal	905.35	1201.87	1191.59	1269.94	1199.72	1186.83	1132.95	2.39	-0.06
Lignite	652.37	808.39	602.90	610.17	583.81	607.41	611.01	1.80	-1.35
Peat	8.05	16.41	10.32	13.95	13.57	11.70	6.52	6.11	-1.60

(1) Earliest year for which split by coal type is available.

* Please refer to notes and definitions in Part I.

Source: *IEA/OECD Energy Balances of OECD Countries, IEA/OECD Energy Statistics of OECD Countries*

OECD TOTAL

4. Final consumption of energy by fuel*

(Mtce)

	1973	1980	1990	2000	2009	2010	2011	Average annual percent change 73-90	90-11
Total final consumption[(1)]	**4022.28**	**4202.24**	**4441.20**	**5207.65**	**5087.68**	**5307.62**	**5216.33**	**0.58**	**0.77**
Coal and peat	433.30	369.81	336.02	199.23	161.76	181.34	174.97	-1.48	-3.06
Oil	2262.88	2247.91	2275.01	2644.26	2495.77	2542.46	2493.71	0.03	0.44
Natural Gas	712.11	798.19	841.53	1066.04	998.46	1069.28	1043.01	0.99	1.03
Biofuels and wastes	120.43	149.81	134.01	193.38	237.73	255.81	259.47	0.63	3.20
Geothermal	0.51	0.73	2.83	3.95	4.90	5.15	5.42	10.65	3.14
Solar, wind, tide	-	0.04	2.57	5.24	6.87	7.24	7.82	-	5.43
Electricity	463.47	584.81	788.38	1024.96	1098.63	1155.85	1147.49	3.17	1.80
Heat	29.59	50.93	60.86	70.60	83.56	90.49	84.44	4.33	1.57
of which:									
Total industry	**1368.86**	**1345.17**	**1184.12**	**1302.70**	**1095.11**	**1194.02**	**1194.80**	**-0.85**	**0.04**
Coal and peat	261.16	229.01	228.55	170.49	126.19	143.37	140.73	-0.78	-2.28
Oil	447.56	398.14	241.93	207.68	163.09	165.68	159.96	-3.55	-1.95
Natural Gas	357.77	358.53	322.54	402.72	334.95	374.13	380.34	-0.61	0.79
Biofuels and wastes	60.36	70.87	52.29	100.50	95.76	102.24	103.26	-0.84	3.29
Geothermal	0.04	0.04	0.23	0.45	0.41	0.41	0.41	10.47	2.85
Solar, wind, tide	-	-	0.01	0.14	0.01	0.02	0.03	-	3.41
Electricity	226.45	266.54	317.39	396.82	340.38	373.89	375.93	2.01	0.81
Heat	15.52	22.05	21.19	23.89	34.32	34.26	34.14	1.85	2.30
Total transport	**993.31**	**1117.56**	**1343.45**	**1637.14**	**1668.26**	**1702.53**	**1688.66**	**1.79**	**1.09**
Coal and peat	10.48	3.71	0.41	0.14	0.27	0.26	0.19	-17.34	-3.66
Oil	950.97	1080.95	1305.24	1588.26	1574.06	1599.80	1581.93	1.88	0.92
Natural Gas	24.28	24.22	26.71	29.86	30.57	33.04	33.66	0.56	1.11
Biofuels and wastes	0.00	0.00	0.01	5.76	50.19	56.11	59.64	5.41	x
Electricity	7.58	8.67	11.08	13.12	13.17	13.31	13.24	2.26	0.85
Residential	**784.51**	**794.90**	**855.89**	**994.72**	**1005.23**	**1051.26**	**998.13**	**0.51**	**0.73**
Coal and peat	109.53	91.24	67.94	19.74	22.91	25.77	22.81	-2.77	-5.06
Oil	283.83	214.22	172.24	183.28	138.47	137.93	125.57	-2.90	-1.49
Natural Gas	205.91	248.46	283.80	368.02	369.42	386.62	362.06	1.91	1.17
Biofuels and wastes	40.41	42.99	57.83	79.32	82.27	87.52	85.10	2.13	1.86
Geothermal	0.42	0.64	1.40	2.24	3.20	3.55	3.72	7.38	4.77
Solar, wind, tide	-	0.04	2.44	4.88	6.39	6.70	7.19	-	5.29
Electricity	133.34	177.88	242.41	307.96	354.35	371.10	362.84	3.58	1.94
Heat	11.07	19.42	27.83	29.28	28.22	32.06	28.84	5.57	0.17
Comm & public services	**406.66**	**420.03**	**504.14**	**623.10**	**690.72**	**703.24**	**686.78**	**1.27**	**1.48**
Coal and peat	28.49	28.68	23.14	4.40	6.82	5.74	5.44	-1.22	-6.66
Oil	186.43	151.20	120.54	119.18	90.16	87.36	83.06	-2.53	-1.76
Natural Gas	99.94	111.41	143.55	187.83	212.90	219.23	212.07	2.15	1.88
Biofuels and waste	1.16	1.54	1.03	3.90	6.70	6.87	6.91	-0.67	9.47
Geothermal	0.02	0.02	0.30	0.90	0.92	0.85	0.88	17.86	5.28
Solar, wind, tide	-	-	0.10	0.13	0.36	0.38	0.43	-	7.08
Electricity	89.41	123.98	206.95	295.77	358.00	365.75	363.60	5.06	2.72
Heat	1.21	3.21	8.53	10.99	14.87	17.07	14.38	12.20	2.52
Non-energy use	**315.18**	**348.59**	**415.01**	**531.32**	**500.71**	**522.95**	**515.23**	**1.63**	**1.04**
Coal and peat	4.42	3.27	3.51	2.44	3.55	3.88	3.80	-1.36	0.39
Oil	302.79	320.84	360.36	467.29	456.89	474.97	467.48	1.03	1.25
Natural Gas	7.97	24.48	51.15	61.59	40.27	44.09	43.94	11.56	-0.72

(1) Total Final Consumption (TFC) excludes use in transformation processes and energy industry own use.

* Please refer to notes and definitions in Part I. Agriculture, Forestry, Fishing and "Non-specified other" sectors are not reported here individually, but are accounted for in Total final consumption.

Source: *IEA/OECD Energy Balances of OECD Countries*

OECD TOTAL

5. Coal balance*

(Mtce)

	1973	1980	1990	2000	2005	2010	2011	2012e	Average annual percent change 73-90	90-11
Production	1170.4	1384.3	1532.6	1380.5	1437.8	1406.8	1396.9	1355.4	1.6	-0.4
Imports	174.2	236.3	318.9	442.5	521.1	528.2	537.8	540.1	3.6	2.5
Exports	-158.7	-214.9	-286.7	-308.1	-327.7	-408.6	-423.8	-445.2	3.5	1.9
Stock changes	20.7	-26.3	-21.4	57.5	4.9	24.1	7.7	10.3		
Primary supply	**1206.6**	**1379.4**	**1543.4**	**1572.4**	**1636.0**	**1550.5**	**1518.6**	**1460.6**	**1.5**	**-0.1**
Statistical differences	**21.2**	**-19.9**	**1.2**	**23.9**	**-14.1**	**-8.9**	**-13.0**	..		
Total transformation	**-754.0**	**-959.5**	**-1185.8**	**-1376.6**	**-1414.9**	**-1338.8**	**-1309.3**	..	**2.7**	**0.5**
Electricity and heat gen.	-639.4	-849.6	-1083.8	-1277.7	-1331.4	-1251.3	-1218.4	..	3.2	0.6
Main activity producers (1)	*-618.2*	*-786.0*	*-1015.7*	*-1224.1*	*-1285.8*	*-1208.3*	*-1175.8*	..	*3.0*	*0.7*
Autoproducers	*-21.1*	*-63.6*	*-68.1*	*-53.7*	*-45.6*	*-43.1*	*-42.5*	..	*7.1*	*-2.2*
Gas works	15.7	7.8	-0.5	-2.7	-2.6	-3.0	-3.0	..	-	8.7
Coal transformation(2)	-130.3	-117.7	-101.2	-95.6	-80.2	-83.4	-86.8	..	-1.5	-0.7
BKB plants	*2.2*	*1.4*	*-1.6*	*-0.2*	*-0.3*	*-0.0*	*-0.1*	..	-	*-13.6*
Blast furnaces	*-93.6*	*-80.0*	*-78.3*	*-81.2*	*-74.4*	*-72.6*	*-74.3*	..	*-1.0*	*-0.3*
Coke ovens	*-38.2*	*-39.2*	*-17.1*	*-13.5*	*-12.7*	*-10.8*	*-12.5*	..	*-4.6*	*-1.5*
Patent fuel plants	*-0.8*	*0.1*	*-4.1*	*-0.8*	*7.2*	*0.0*	*0.0*	..	*10.5*	-
Other transformation(3)	-	-	-0.3	-0.5	-0.7	-1.1	-1.1	..	-	7.0
Energy ind. own use	**-35.0**	**-27.7**	**-21.5**	**-19.4**	**-18.1**	**-20.2**	**-20.1**	..	**-2.8**	**-0.3**
Losses	**-5.4**	**-2.4**	**-1.4**	**-1.2**	**-1.3**	**-1.3**	**-1.3**	..		
Final consumption(4)	**433.3**	**369.8**	**336.0**	**199.2**	**187.7**	**181.3**	**175.0**	..	**-1.5**	**-3.1**
Industry(5)	261.2	229.0	228.5	170.5	159.2	143.4	140.7	..	-0.8	-2.3
Iron and steel	*134.1*	*106.3*	*81.6*	*61.8*	*57.3*	*55.3*	*57.1*	..	*-2.9*	*-1.7*
Chemical	*26.6*	*24.9*	*28.2*	*18.2*	*16.3*	*16.1*	*15.5*	..	*0.3*	*-2.8*
Non-metallic minerals	*21.2*	*32.7*	*41.8*	*36.1*	*31.0*	*25.2*	*26.0*	..	*4.1*	*-2.2*
Paper, pulp and print	*11.6*	*11.2*	*16.2*	*8.3*	*10.9*	*9.6*	*9.1*	..	*2.0*	*-2.7*
Other industry (6)	*67.6*	*53.9*	*60.7*	*46.1*	*43.8*	*37.1*	*33.0*	..	*-0.6*	*-2.9*
Transport(7)	10.5	3.7	0.4	0.1	0.3	0.3	0.2	..	-17.3	-3.7
Other	157.2	133.8	103.6	26.2	25.1	33.8	30.2	..	-2.4	-5.7
Comm. and pub. services	*28.5*	*28.7*	*23.1*	*4.4*	*5.8*	*5.7*	*5.4*	..	*-1.2*	*-6.7*
Residential	*109.5*	*91.2*	*67.9*	*19.7*	*16.9*	*25.8*	*22.8*	..	*-2.8*	*-5.1*
Other sectors (8)	*19.2*	*13.9*	*12.5*	*2.0*	*2.3*	*2.3*	*2.0*	..	*-2.5*	*-8.3*
Non-energy use	4.4	3.3	3.5	2.4	3.1	3.9	3.8	..	-1.4	0.4

(1) Main activity electricity and heat generation includes district heating.

(2) Coal transformation refers to the transformation of coal and peat to secondary and tertiary products (mainly cokes, briquettes, coal tar and off-gases). Gas works and Liquefaction are shown separately to match the *Energy Balances of OECD Countries.*

(3) Other transformation includes Liquefaction and Non-specified transformations.

(4) Final Consumption includes non-energy use and energy use (Industry, Transport and Other).

(5) Please refer to notes and definitions in Part I for detailed explanation.

(6) Other industry includes Non-ferrous metals, Transport equipment, Machinery, Mining and quarrying, Food and tobacco, Wood and wood products, Construction, Textile and leather, and Non-specified industry.

(7) Transport includes Rail and Inland waterways.

(8) Other sectors includes Agriculture/Forestry, Fishing and Non-specified other.

* "Coal" refers to all types of coal, primary (anthracite, coking coal, other bituminous coal, sub-bituminous coal, lignite/brown coal) and derived fuels (including patent fuel, BKB, coke oven coke, gas coke, coal tar, coke oven gas, gas works gas, blast furnace gas and other recovered gases), as well as peat, peat briquettes and oil shale. Quantities have been converted to Mtce using calorific values reported by the respective countries. Please refer to notes and definitions in Part I.

Source: *IEA/OECD Energy Balances of OECD Countries*

OECD TOTAL

6. Use of coal for selected end-uses*

(million tonnes)

	1978	1990	2000	2005	2009	2010	2011	Average annual percent change 78-90	90-11
Steam coal	**929.13**	**1242.04**	**1427.37**	**1551.62**	**1422.21**	**1477.08**	**1436.73**	**2.45**	**0.70**
Total electricity and heat	714.31	1041.72	1288.16	1280.04	1285.70	1334.07	1290.38	3.19	1.02
Main activity producers	*664.54*	*991.90*	*1255.07*	*1253.19*	*1262.16*	*1308.32*	*1263.45*	*3.39*	*1.16*
Autoproducers	*49.77*	*49.82*	*33.10*	*26.86*	*23.54*	*25.74*	*26.93*	*0.01*	*-2.89*
Patent fuel/BKB plants	25.54	23.75	3.24	122.22	2.29	2.21	2.16	-0.60	-10.79
Coke ovens/Liquefaction(1)	2.40	6.61	12.13	12.41	10.20	13.62	14.48	8.80	3.80
Blast furnace inputs	0.01	3.76	9.71	9.69	6.06	10.23	10.04	73.63	4.80
Gas manufacture	1.85	0.41	-	-	-	-	-	-11.86	-
Industry	93.80	119.83	103.02	96.67	77.60	83.34	81.75	2.06	-1.80
Iron and steel	*9.78*	*8.07*	*8.97*	*10.33*	*7.42*	*8.92*	*10.82*	*-1.59*	*1.41*
Chemical	*13.86*	*22.59*	*17.49*	*14.74*	*14.23*	*15.52*	*14.97*	*4.16*	*-1.94*
Non-metallic minerals	*27.37*	*42.02*	*36.85*	*31.58*	*24.90*	*25.72*	*26.59*	*3.64*	*-2.16*
Paper, pulp and print	*9.02*	*15.83*	*8.64*	*11.38*	*10.27*	*10.43*	*9.77*	*4.80*	*-2.27*
Other industry	*33.76*	*31.32*	*31.07*	*28.65*	*20.77*	*22.75*	*19.60*	*-0.62*	*-2.21*
Other sectors(2)	71.43	47.92	20.32	20.64	25.85	27.93	24.54	-3.27	-3.14
Non-energy use	0.08	0.33	0.45	0.77	1.36	1.30	1.28	12.80	6.66
Coking coal	**293.19**	**255.98**	**207.95**	**192.11**	**161.40**	**187.43**	**192.93**	**-1.12**	**-1.34**
Total electricity and heat	7.14	18.23	13.37	12.70	7.70	5.14	6.72	8.12	-4.64
Main activity producers	*6.19*	*14.32*	*10.17*	*10.73*	*6.71*	*4.21*	*5.60*	*7.23*	*-4.37*
Autoproducers	*0.95*	*3.91*	*3.20*	*1.97*	*0.98*	*0.93*	*1.12*	*12.52*	*-5.77*
Patent fuel/BKB plants	-	-	-	2.81	-	-	-	-	-
Coke ovens/Liquefaction(1)	275.86	228.57	173.91	158.45	134.83	158.77	158.38	-1.55	-1.73
Blast furnace inputs	-	6.70	15.75	18.72	16.76	21.50	24.57	-	6.38
Gas manufacture	6.91	0.26	-	-	-	-	-	-23.92	-
Industry	0.26	1.97	2.64	0.79	1.86	1.75	2.28	18.42	0.69
Iron and steel	*0.07*	*1.63*	*0.24*	*0.24*	*0.90*	*1.12*	*1.33*	*30.02*	*-0.99*
Chemical	*0.01*	-	*0.00*	-	*0.02*	-	-	-	-
Non-metallic minerals	*0.00*	-	*0.00*	-	*0.00*	*0.00*	-	-	-
Paper, pulp and print	-	-	-	-	-	-	-	-	-
Other industry	*0.18*	*0.34*	*2.39*	*0.55*	*0.94*	*0.63*	*0.95*	*5.54*	*5.06*
Other sectors(2)	0.28	0.14	0.10	0.13	0.28	0.20	0.24	-5.70	2.75
Non-energy use	-	-	0.23	0.22	0.14	0.18	0.22	-	-
Lignite	**652.00**	**820.81**	**607.99**	**613.63**	**591.03**	**582.75**	**610.14**	**1.94**	**-1.40**
Total electricity and heat	415.48	612.04	562.94	570.93	545.16	536.22	559.87	3.28	-0.42
Main activity producers	*398.48*	*583.34*	*554.17*	*562.66*	*539.26*	*526.22*	*552.60*	*3.23*	*-0.26*
Autoproducers	*16.99*	*28.70*	*8.77*	*8.27*	*5.90*	*10.00*	*7.27*	*4.46*	*-6.33*
Patent fuel/BKB plants	136.86	110.15	13.11	13.43	12.82	13.81	14.98	-1.79	-9.06
Coke ovens/Liquefaction(1)	-	0.88	1.40	1.98	2.80	3.10	3.08	-	6.17
Gas manufacture	6.40	9.12	7.64	7.71	7.84	8.16	8.03	2.99	-0.60
Industry	50.71	43.49	15.74	9.44	10.57	12.30	11.08	-1.27	-6.30
Iron and steel	*0.93*	*1.37*	*0.22*	*0.09*	*0.06*	*0.20*	*0.16*	*3.32*	*-9.77*
Chemical	*14.35*	*10.93*	*3.37*	*3.23*	*3.12*	*3.49*	*2.93*	*-2.24*	*-6.08*
Non-metallic minerals	*1.57*	*1.63*	*0.72*	*0.55*	*0.59*	*0.72*	*0.73*	*0.31*	*-3.77*
Paper, pulp and print	*4.51*	*2.11*	*0.57*	*0.49*	*0.42*	*0.28*	*0.40*	*-6.14*	*-7.62*
Other industry	*29.36*	*27.45*	*10.87*	*5.09*	*6.38*	*7.61*	*6.87*	*-0.56*	*-6.38*
Other sectors(2)	32.90	41.36	9.08	7.31	8.76	8.70	9.61	1.93	-6.72
Non-energy use	0.83	0.67	0.31	0.29	0.09	0.09	0.04	-1.78	-12.25

(1) Liquefaction primarily refers to direct distillation processes. Indirect liquefaction may be reported here or under gas manufacture.

(2) Other sectors are Residential, Commercial and public services, Agriculture/Forestry, Fishing and Non-specified other.

* Coal totals may not match the sum of the individual rows due to statistical differences. Please refer to notes and definitions in Part I.

Source: *IEA/OECD Energy Statistics of OECD Countries*

OECD TOTAL

7. Fuel prices to end users*

	1978(1)	1990	1995	2000	2005	2009	2010	2011	2012
				(US dollars / tce)					
For electricity generation									
Steam coal	..	..	..	..	..	..	..	..	..
Heavy fuel oil	..	..	..	..	..	..	..	..	..
Natural gas	..	..	..	..	..	..	..	..	..
For industry									
Steam coal	..	..	..	..	..	..	..	..	..
Coking coal	..	..	..	..	..	..	..	..	..
High sulphur fuel oil	..	..	..	..	..	..	..	..	..
Low sulphur fuel oil	..	..	..	..	..	..	..	..	..
Natural gas	..	..	..	..	..	..	..	..	..
				(US dollars / unit)(2)					
For electricity generation									
Steam coal	..	..	..	..	..	..	..	..	..
Heavy fuel oil	..	..	..	..	..	..	..	..	..
Natural gas	..	..	..	..	..	..	..	..	..
For industry									
Steam coal	..	..	..	..	..	..	..	..	..
Coking coal	..	..	..	..	..	..	..	..	..
High sulphur fuel oil	..	..	..	..	..	..	..	..	..
Low sulphur fuel oil	..	..	..	..	..	..	..	..	..
Natural gas	..	..	..	..	..	..	..	..	..

(1) Earliest year for which split by coal type is available.

(2) For steam coal, coking coal and heavy fuel oil per metric tonne; for natural gas per 10^7 kilocalories GCV.

* Please refer to notes and definitions in Part I. Steam coal prices relate to historical definitions.

Source: *IEA/OECD Energy Prices & Taxes*

8. Coal and peat trade by type of coal*

(Mtce)

	1978	1990	1995	2000	2005	2009	2010	2011	2012e
Total imports	**185.87**	**318.88**	**354.42**	**442.50**	**521.09**	**488.99**	**528.20**	**537.83**	**540.09**
Bituminous coal(1)	58.79	156.85	195.05	273.54	351.58	350.87	357.57	371.31	379.76
Coking coal	99.77	141.34	138.11	137.93	135.72	115.00	141.80	142.65	137.57
Sub-bituminous coal	0.60	1.40	2.58	7.50	13.70	12.43	12.60	9.57	9.34
Lignite	4.69	5.13	3.24	1.93	0.64	0.76	0.65	0.72	0.96
Peat	-	0.10	0.13	0.10	0.14	0.23	0.16	0.20	0.19
Coal products(2)	22.01	14.06	15.31	21.51	19.31	9.70	15.42	13.39	12.26
Total exports	**176.71**	**286.65**	**300.48**	**308.08**	**327.68**	**352.47**	**408.59**	**423.81**	**445.21**
Bituminous coal(1)	44.50	103.69	113.76	130.89	132.56	156.72	158.04	179.93	193.82
Coking coal	105.43	161.25	167.92	162.61	177.47	179.74	231.87	226.17	231.21
Sub-bituminous coal	-	0.04	2.01	0.77	4.43	3.54	4.40	3.84	5.29
Lignite	4.38	4.37	2.96	1.34	0.82	0.86	0.79	0.86	1.04
Peat	0.01	0.12	0.07	0.08	0.06	0.02	0.04	0.01	0.01
Coal products(2)	22.40	17.19	13.76	12.39	12.34	11.60	13.46	12.99	13.85

(1) Bituminous coal includes anthracite.

(2) Coal products includes products derived from coal, for example: coke, coal tar, briquettes, patent fuels.

* Please refer to notes and definitions in Part I.

Source: *IEA/OECD Energy Balances of OECD Countries, IEA/OECD Energy Statistics of OECD Countries*

OECD TOTAL

9. Total coal imports by origin*

(thousand tonnes)

	1978(1)	1985	1990	1995	2000	2005	2010	2011	2012e
Total coal(2)	**178767**	**289141**	**327069**	**369555**	**468538**	**565934**	**580185**	**594200**	**598793**
Coking coal	**99172**	**136741**	**136967**	**135328**	**141066**	**138378**	**145176**	**145908**	**140864**
Australia	29796	43475	49601	52923	68229	62162	66840	62369	60102
Canada	11598	20623	16250	18001	18510	17912	16893	17596	15789
Czech Republic	909	695	774	2985	3388	3366	3704	2990	2698
Germany	10948	5219	3141	37	2	289	1	15	46
Poland	6619	5025	2570	6573	3118	3246	1821	2168	2384
United Kingdom	79	87	52	-	-	6	1	-	-
United States	25075	44013	43205	32737	24099	20449	31140	35433	35078
Other OECD	116	686	305	682	391	932	476	500	724
China, People's Rep.	420	1211	2803	5339	9558	7313	2621	2437	2368
Colombia	-	50	177	137	140	313	882	1034	743
Indonesia	-	103	714	4843	6327	15374	13157	11698	11183
South Africa	2566	5342	3213	3643	991	295	574	515	598
Former Soviet Union(3)	5249	4216	x	x	x	x	x	x	x
Russian Federation (4)	*x*	*x*	*8569*	*3267*	*4891*	*5202*	*6210*	*7868*	*7284*
Other FSU (4)	*x*	*x*	*-*	*-*	*26*	*520*	*80*	*661*	*121*
Venezuela	-	-	-	990	510	872	187	406	553
Vietnam	-	-	-	-	80	-	37	-	-
Non-specified/other	5797	5996	5593	3171	806	127	552	218	1178
Steam coal	**66833**	**138684**	**177526**	**226285**	**322790**	**426058**	**433620**	**446657**	**455858**
Australia	4556	30433	34664	51101	78080	109146	118594	114572	120612
Canada	970	4051	9885	13108	10916	2577	8830	12031	11975
Czech Republic	243	462	327	1685	2443	1416	2889	3958	2460
Germany	6716	3526	2123	835	470	641	627	564	327
Poland	16292	15429	13084	15832	18894	15564	10510	5746	4864
United Kingdom	2285	2394	2441	713	593	319	311	729	265
United States	9296	26912	35894	33335	26287	16558	22070	27508	41774
Other OECD	986	2720	3653	2081	4481	3210	3934	3193	3528
China, People's Rep.	534	2663	6287	15372	35873	40190	11918	7287	7126
Colombia	-	2083	10450	14281	29428	47929	61796	72988	69114
Indonesia	-	306	1000	12444	24124	44492	74124	74068	73515
South Africa	11967	32671	33156	39460	46469	53042	25940	26376	23229
Former Soviet Union(3)	3012	2915	x	x	x	x	x	x	x
Russian Federation (4)	*x*	*x*	*9866*	*10565*	*21013*	*62284*	*72972*	*78056*	*82315*
Other FSU (4)	*x*	*x*	*-*	*647*	*845*	*3019*	*2351*	*3682*	*2452*
Venezuela	-	-	1752	3506	5096	5169	1382	1639	879
Vietnam	-	157	150	1860	1639	2699	3477	2833	2678
Non-specified/other	9976	11962	12794	9460	16139	17803	11895	11424	8745
Lignite	**12762**	**13716**	**12576**	**7942**	**4682**	**1498**	**1389**	**1635**	**2071**

(1) Earliest year for which split by coal type is available.

(2) Total coal consists of steam coal, coking coal and lignite.

(3) For years prior to 1990, or if no details for individual Former Soviet Union republics.

(4) Data for former republics are not available separately for some OECD member countries.

* Please refer to notes and definitions in Part I.

Source: IEA/OECD Coal Statistics

OECD TOTAL

10. Coking coal exports by destination*

(thousand tonnes)

	1978(1)	1985	1990	1995	2000	2005	2010	2011	2012e
World	**107072**	**152492**	**162292**	**169662**	**168629**	**186624**	**243496**	**238188**	**243757**
Total OECD	**90313**	**123858**	**122941**	**128766**	**128525**	**132846**	**139897**	**142942**	**138888**
Australia	-	-	-	125	122	49	-	-	-
Austria	1283	1582	1351	1897	1844	1753	1724	1302	2711
Belgium	3860	5031	7450	5890	4551	3429	2787	2716	2800
Canada	5410	6360	4018	4039	3501	4034	3091	3772	4379
Chile	32	345	492	733	1088	1190	661	1024	1182
Czech Republic	-	-	-	871	214	523	720	1017	848
Denmark	-	-	50	9	-	-	-	-	-
Estonia	-	-	-	-	-	-	-	-	-
Finland	-	-	330	1249	1005	1181	754	797	903
France	8972	8609	8494	6177	6629	5535	4480	4359	5100
Germany	714	950	877	999	4891	4373	4040	3990	4421
Greece	245	7	-	1	-	-	-	-	-
Hungary	-	-	-	1287	1075	449	542	521	391
Iceland	-	13	28	39	48	57	59	58	62
Ireland	-	62	3	3	4	15	-	138	9
Israel	-	-	50	127	56	129	55	-	-
Italy	8393	10228	8734	7824	7381	6494	4961	6688	5947
Japan	43380	59198	55410	53979	52798	54226	60321	55965	52832
Korea	2503	7014	7852	15654	15305	17954	23873	27554	25815
Luxembourg	286	-	-	10	-	-	77	-	-
Mexico	10	498	3	517	1406	1582	1338	1264	1181
Netherlands	3347	5206	5273	5766	4745	8055	8978	10989	10739
New Zealand	-	-	-	-	-	-	-	-	-
Norway	193	232	99	239	95	18	75	81	79
Poland	-	-	-	896	538	592	2882	1944	1024
Portugal	387	369	805	434	198	-	-	93	134
Slovak Republic	5126	4183	3681	2657	1570	1909	2092	1802	1797
Slovenia	-	-	-	-	-	163	223	663	364
Spain	3257	4283	4499	3267	4163	4598	2531	2947	2648
Sweden	840	1564	1568	2006	2128	1690	1887	1703	1515
Switzerland	18	5	3	3	46	-	37	-	1
Turkey	498	2232	2869	3681	3954	3599	3901	3833	5445
United Kingdom	1402	5777	8230	7612	8626	7366	6425	6367	5686
United States	157	110	772	775	544	1883	1383	1355	875
Total non-OECD	**8808**	**22793**	**30067**	**40203**	**36647**	**49645**	**99015**	**91877**	**102913**
Brazil	2121	7958	8867	11064	10695	8049	12996	12731	11242
China(2)	-	192	860	317	265	5286	35649	23305	37729
Chinese Taipei	1186	2772	3155	5451	7713	8365	6221	9119	9589
Egypt	218	769	1009	1672	1211	1414	1366	629	375
India	232	2080	5179	9831	10795	19117	34749	35785	35606
Romania	1348	1646	3915	2455	505	593	812	937	607
Oth. Africa & Mid. East	521	1564	1068	947	1825	3485	1840	1863	2201
Oth. non-OECD Americas	919	1484	2129	784	781	991	929	914	1046
Other Asia & Oceania	69	1041	963	1366	1051	841	781	467	688
Other non-OECD Europe and Eurasia	2194	3287	2922	6316	1806	1504	3672	6127	3830
Non-specified/Other	**7951**	**5841**	**9284**	**693**	**3457**	**4133**	**4584**	**3369**	**1956**

(1) Earliest year for which split by coal type is available.

(2) China includes the People's Republic of China and Hong Kong, China.

* Please refer to notes and definitions in Part I.

Source: IEA/OECD Coal Statistics

OECD TOTAL

11. Steam coal exports by destination*

(thousand tonnes)

	1978(1)	1985	1990	1995	2000	2005	2010	2011	2012e
World	**49931**	**105530**	**113417**	**128952**	**146891**	**157350**	**185959**	**209450**	**225971**
Total OECD	**40497**	**83710**	**95379**	**108337**	**127163**	**132704**	**140805**	**149802**	**158921**
Australia	-	2	2	12	-	-	105	325	158
Austria	275	1001	1216	1050	1643	2182	1447	1634	1321
Belgium	2660	2856	3778	3480	1467	1292	974	1728	1186
Canada	8782	8508	10083	4541	13524	13644	7291	2476	2163
Chile	-	99	514	1230	1349	668	1406	2298	2023
Czech Republic	274	1378	2282	1874	864	719	685	844	676
Denmark	4620	8297	5857	6195	2594	1220	873	404	273
Estonia	-	-	-	-	3	-	-	-	-
Finland	4095	3669	2815	3089	1228	963	356	197	129
France	6755	3461	3859	2049	3413	3612	3055	2984	1580
Germany	4157	5228	5451	7364	16478	13612	9780	10158	11179
Greece	1	1278	-	1	112	144	47	66	32
Hungary	-	-	-	374	322	321	158	202	38
Iceland	-	28	33	20	7	44	56	54	6
Ireland	540	1622	2248	1636	1208	1164	604	563	602
Israel	-	998	1058	1257	2623	1170	516	501	511
Italy	1552	5070	5051	4122	1430	422	1785	1850	5858
Japan	1732	18173	30637	40735	51874	58536	68593	69611	72495
Korea	356	7942	4352	13462	13730	18719	29033	35384	34228
Luxembourg	52	107	3	50	164	38	50	109	-
Mexico	-	44	188	-	373	4579	4487	3623	6179
Netherlands	1285	6539	9559	5194	3556	2240	2197	4155	2590
New Zealand	-	-	1	-	16	56	59	58	24
Norway	167	429	401	336	677	495	461	434	297
Poland	-	-	1	304	117	127	1451	1500	1260
Portugal	15	805	1572	1977	348	580	776	988	1023
Slovak Republic	237	79	198	404	1217	722	449	602	491
Slovenia	-	-	-	2	12	10	186	-	199
Spain	21	1630	762	2350	2327	729	426	614	1167
Sweden	244	1896	1040	861	255	443	659	471	417
Switzerland	90	295	81	11	20	8	32	176	30
Turkey	79	275	15	216	110	178	438	439	1663
United Kingdom	1081	1846	2097	3901	3866	3674	2117	4964	9100
United States	1427	155	225	240	236	393	253	390	23
Total non-OECD	**384**	**13445**	**13032**	**19718**	**19134**	**24439**	**44627**	**52430**	**65771**
Brazil	11	91	345	190	22	726	138	395	125
China(2)	-	2695	2554	3420	1440	2121	18130	20050	33156
Chinese Taipei	76	6336	6866	7953	10034	14332	19554	20124	17744
Egypt	-	-	1	1	2	4	168	179	22
India	-	93	48	1	2469	1679	783	1132	2998
Romania	-	-	49	381	-	844	-	-	26
Oth. Africa & Mid. East	32	253	1057	1099	828	499	1075	2369	2940
Oth. non-OECD Americas	82	113	128	-	89	13	130	509	723
Other Asia & Oceania	129	2229	1621	1672	3303	4019	4374	6912	7369
Other non-OECD Europe and Eurasia	54	1635	363	5001	947	202	275	760	668
Non-specified/Other	**9050**	**8375**	**5006**	**897**	**594**	**207**	**527**	**7218**	**1279**

(1) Earliest year for which split by coal type is available.

(2) China includes the People's Republic of China and Hong Kong, China.

* Please refer to notes and definitions in Part I. Steam coal includes all sub-bituminous coal.

Source: IEA/OECD Coal Statistics

OECD TOTAL

12. Coal import values by origin*

(average unit value, CIF, USD/tonne)

	1990	1995	2000	2005	2007	2008	2009	2010	2011/12[1]
Coking coal[2]	**62.27**	**56.28**	**42.45**	**97.30**	**103.65**	**189.81**	**168.33**	**168.32**	**..**
Imports from:									
Australia	57.64	52.80	40.80	102.50	105.33	210.45	194.80	180.60	..
Canada	70.37	62.94	45.22	109.46	116.80	220.71	200.49	192.44	..
Czech Republic	67.03	82.00	x	242.07	133.32	235.16	160.08	210.67	..
Poland	63.98	59.26	50.43	138.92	139.47	245.85	137.24	194.36	..
United States	64.21	61.25	53.13	108.46	126.15	191.98	165.30	182.88	..
China	54.50	49.58	38.32	109.89	110.31	263.49	147.32	182.33	..
Colombia	60.89	50.11	36.28	105.35	92.53	..	87.36	137.52	..
Indonesia	51.82	48.34	32.48	53.97	61.42	105.88	97.23	96.95	..
South Africa	50.58	49.57	39.35	71.77	96.82	141.18	81.99	95.76	..
Former Soviet Union[3]	57.45	54.95	43.26	104.12	98.26	211.14	143.56	155.68	..
Other bituminous coal[4]	**51.81**	**47.14**	**33.65**	**63.71**	**72.80**	**120.27**	**97.47**	**99.14**	**..**
Imports from:									
Australia	52.17	48.23	34.76	64.10	71.58	122.96	107.12	111.02	..
Canada	43.38	40.99	33.74	77.47	70.15	109.39	90.72	96.47	..
Czech Republic	65.54	63.84	40.98	113.59	112.36	158.65	140.77	140.68	..
Poland	61.05	46.78	35.30	78.34	94.13	155.92	114.49	104.05	..
United States	54.63	49.60	40.96	49.31	94.49	135.17	103.06	106.06	..
China	48.34	44.63	30.62	62.75	71.25	117.72	110.69	107.60	..
Colombia	50.88	41.58	32.77	59.23	65.80	..	86.48	87.54	..
Indonesia	55.74	41.80	29.84	53.90	59.58	93.10	81.07	81.51	..
South Africa	45.38	44.19	33.80	67.68	80.16	140.90	95.58	98.40	..
Former Soviet Union[3]	45.48	42.87	32.14	67.35	77.44	129.53	92.97	98.31	..

Note: On occasion, shipment of extremely small quantities of high valued coal results in high import costs.

(1) Import data for steam and coking coals are currently unavailable for 2011 and 2012 due to resource constraints.

(2) Weighted average of individual countries using import volumes as weights. Prices exclude intra-EU trade.

(3) Former Soviet Union until 1991, Russian Federation starting in 1992.

(4) Bituminous steam coal only. Weighted average of trades. Prices exclude intra-EU trade.

* Please refer to notes and definitions in Part I.

Source: IEA/OECD Coal Statistics

OECD AMERICAS*

Figure 1: TPES by fuel (Mtce)

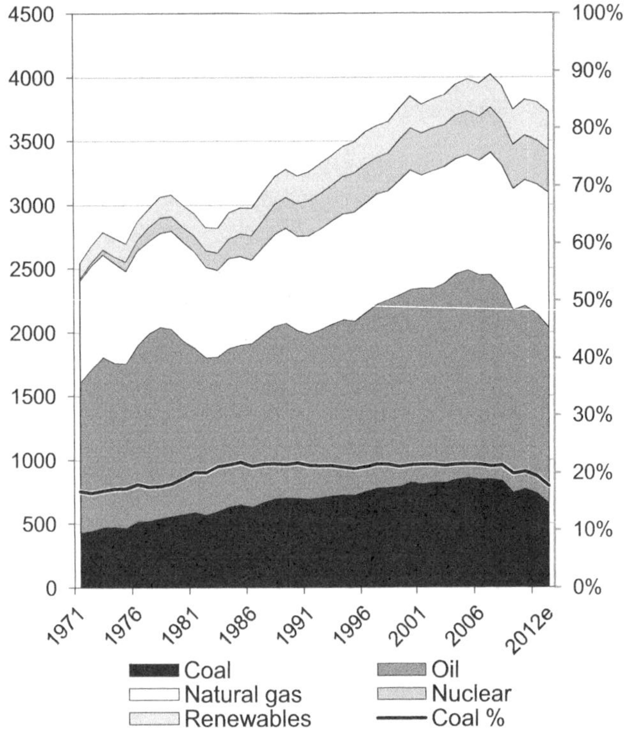

Figure 2: Coal supply indicators (1971 = 100)

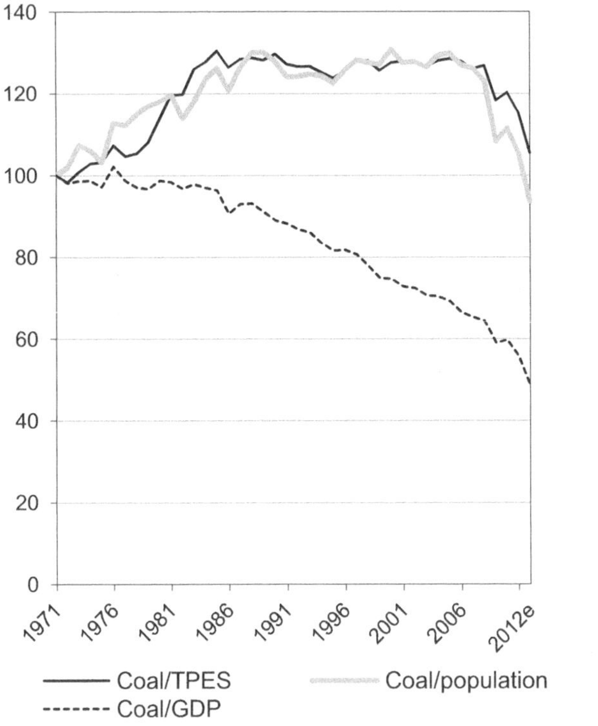

Figure 3: Primary coal supply (Mtce)

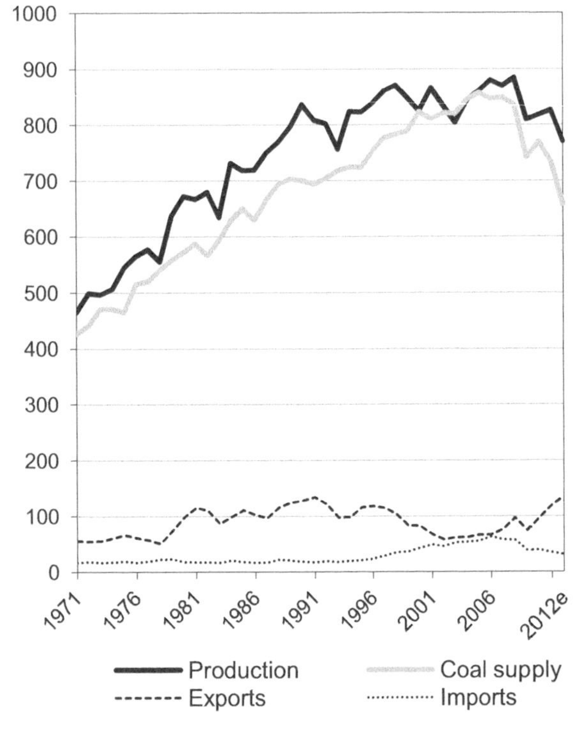

Figure 4: Coal consumption (Mtce)

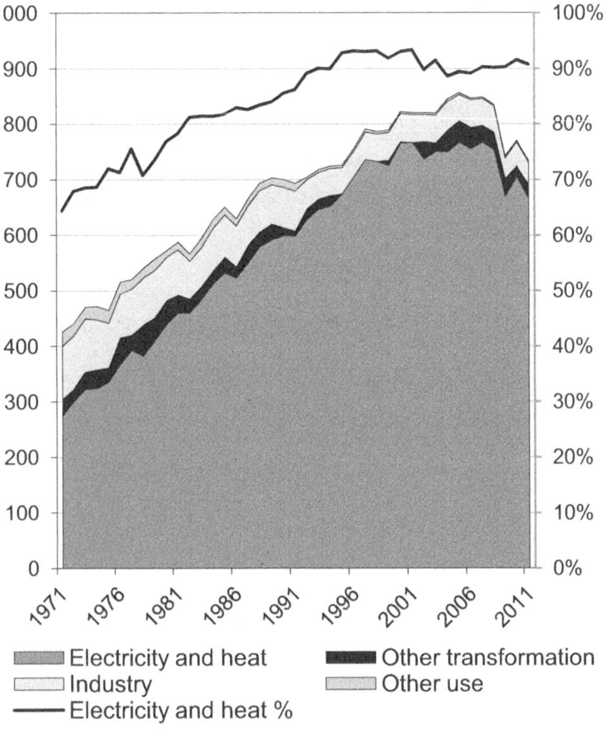

*Please refer to notes and definitions in Part I. Peat is included in coal for all figures.

Source: *IEA/OECD Coal Information, IEA/OECD Energy Balances of OECD Countries, OECD Main Economic Indicators*

OECD AMERICAS*

Figure 5: Electricity generation by fuel (TWh)

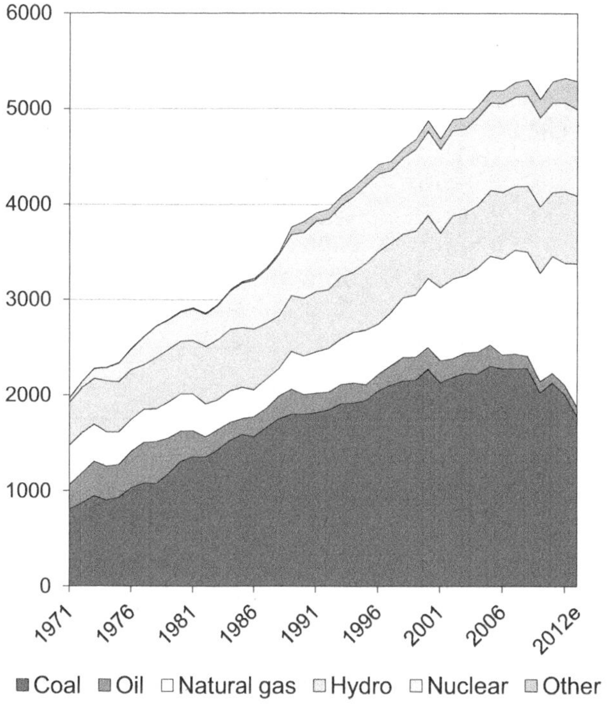

Figure 6: CO_2 emissions by fuel (Mt CO_2)

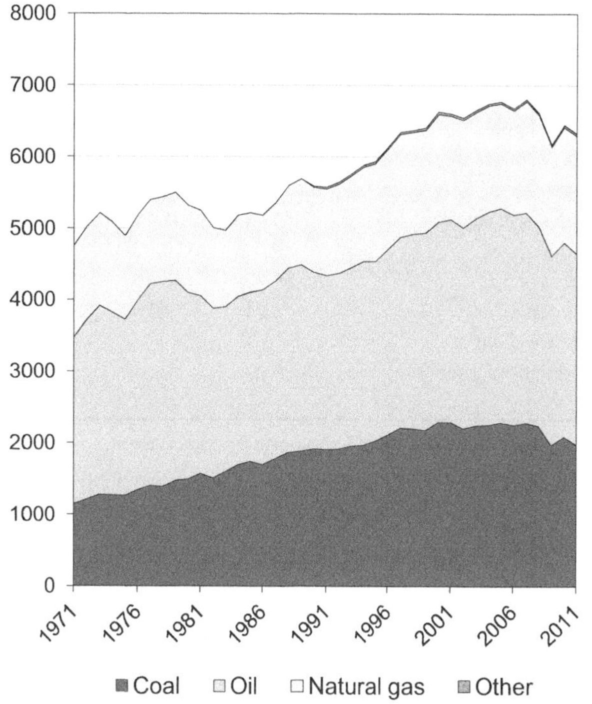

Figure 7: Electricity generation by fuel share

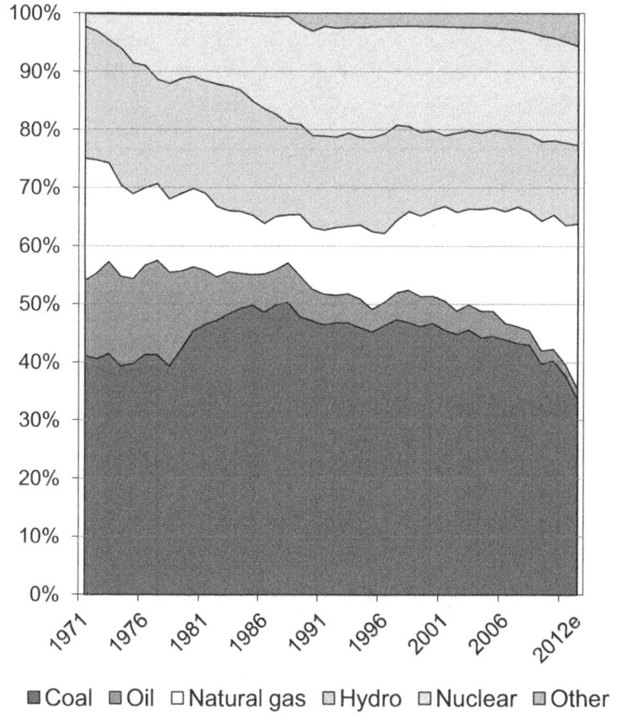

Figure 8: CO_2 emissions by fuel share

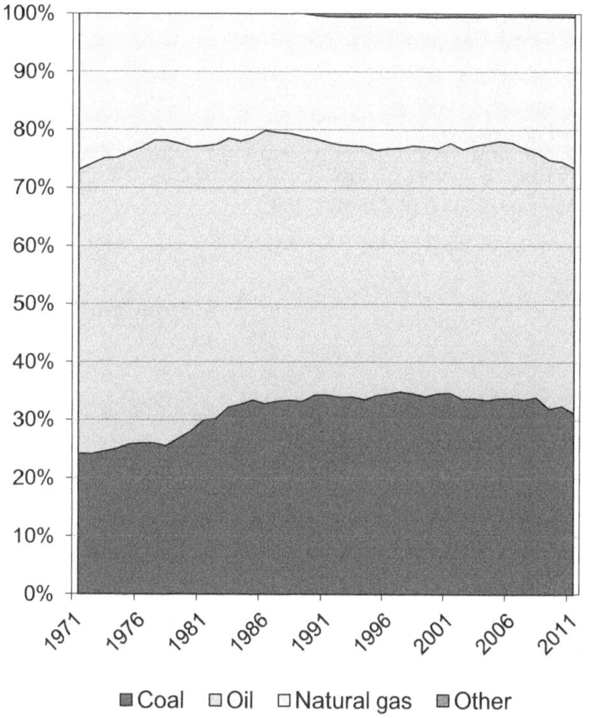

*Please refer to notes and definitions in Part I. Peat is included in coal for all figures.

Source: *IEA/OECD Electricity Information, IEA/OECD CO2 Emissions from Fuel Combustion*

OECD AMERICAS

1. Total primary energy supply (TPES) by fuel*

	1973	1980	1990	2000	2010	2011	2012e	Average annual percent change 73-90	90-11
TPES (Mtce)	**2786.22**	**3001.76**	**3228.66**	**3850.49**	**3823.34**	**3803.98**	**3728.20**	**0.87**	**0.78**
Coal and peat	470.50	572.00	700.63	822.09	770.03	734.24	658.13	2.37	0.22
Oil	1334.74	1364.24	1314.42	1512.79	1433.69	1405.57	1377.52	-0.09	0.32
Natural Gas	804.01	774.55	738.89	937.06	989.04	1018.68	1057.97	-0.50	1.54
Biofuels and waste	75.49	101.15	117.28	140.84	163.74	168.48	166.59	2.63	1.74
Nuclear	39.01	113.96	256.50	327.15	348.27	344.40	336.90	11.71	1.41
Hydro	59.20	68.15	74.00	81.50	82.62	92.72	87.37	1.32	1.08
Geothermal	3.21	7.70	26.44	25.94	20.14	20.21	20.84	13.21	-1.27
Solar, wind, tide	-	-	0.49	3.07	15.78	19.57	23.86	-	19.23
Net electricity trade(1)	0.07	0.01	0.03	0.04	0.03	0.11	-0.97	-4.38	5.89
Heat(2)	-	-	-	-	-	-	-	-	-

(1) Net trade between OECD and non-OECD countries.

(2) Ambient heat from heat pumps used in the transformation sector.

Source: *IEA/OECD Energy Balances of OECD Countries*

2. Energy supply, GDP and population*

	1973	1980	1990	2000	2010	2011	2012e	Average annual percent change 73-90	90-11
GDP (2005 bil. USD)	5645	6859	9313	13027	15265	15575	15936	2.99	2.48
TPES/GDP(1)	0.49	0.44	0.35	0.30	0.25	0.24	0.23	-2.06	-1.65
Population (millions)	297.8	329.1	372.3	426.8	469.3	473.0	477.5	1.32	1.15
TPES/population(1)	9.36	9.12	8.67	9.02	8.15	8.04	7.81	-0.45	-0.36
TPES/GDP(2)	181.7	161.1	127.6	108.8	92.2	89.9	86.1	-2.06	-1.65
Coal and peat supply/GDP(2)	142.6	142.7	128.7	108.0	86.3	80.7	70.7	-0.60	-2.20
Elec. consumption/GDP(2)	112.1	114.8	111.7	105.4	98.4	96.9	..	-0.02	-0.68
Elec. generation (TWh)	2281	2879	3819	4876	5288	5325	5292	3.08	1.60
Industrial production(2)	..	..	..	..	..	..	..	..	..

(1) TPES/GDP in units of tce/2005 thousand USD, TPES/population in units of tce/capita.

(2) As index, 2005 (not shown) = 100

Source: *IEA/OECD Energy Balances of OECD Countries , OECD Main Economic Indicators*

3. Coal and peat production by type*

	1978(1)	1990	2000	2005	2010	2011	2012e	Average annual percent change 78-90	90-11
Mtce:									
Coking coal	109.25	125.74	81.69	69.25	91.30	105.16	106.31	1.18	-0.85
Steam coal	427.40	668.18	700.52	750.65	687.52	680.88	625.03	3.79	0.09
Lignite	18.09	41.91	41.95	40.91	38.65	39.76	38.71	7.25	-0.25
Peat	-	-	-	-	-	-	-	-	-
Mt:									
Coking coal	109.07	123.88	84.67	74.85	98.39	113.15	114.54	1.07	-0.43
Steam coal	497.37	797.81	878.99	953.22	897.83	892.99	821.75	4.02	0.54
Lignite	36.26	89.32	88.81	87.17	81.23	83.31	81.11	7.80	-0.33
Peat	-	-	-	-	-	-	-	-	-

(1) Earliest year for which split by coal type is available.

* Please refer to notes and definitions in Part I.

Source: *IEA/OECD Energy Balances of OECD Countries, IEA/OECD Energy Statistics of OECD Countries*

OECD AMERICAS

4. Final consumption of energy by fuel*

(Mtce)

	1973	1980	1990	2000	2009	2010	2011	Average annual percent change 73-90	90-11
Total final consumption(1)	**2132.94**	**2199.38**	**2211.24**	**2649.33**	**2547.11**	**2650.62**	**2641.33**	**0.21**	**0.85**
Coal and peat	116.54	89.52	87.33	54.60	40.49	46.21	42.53	-1.68	-3.37
Oil	1135.96	1161.21	1155.40	1350.07	1294.78	1337.56	1320.20	0.10	0.64
Natural Gas	568.57	552.24	516.22	613.62	538.78	563.47	569.00	-0.57	0.46
Biofuels and wastes	75.08	100.62	58.67	105.92	123.61	129.21	132.53	-1.44	3.96
Geothermal	-	-	0.48	0.74	0.31	0.32	0.32	-	-1.96
Solar, wind, tide	-	-	0.02	2.12	2.19	2.15	2.26	-	23.96
Electricity	236.66	294.32	389.14	513.55	535.97	561.53	564.24	2.97	1.78
Heat	0.14	1.47	3.98	8.70	10.96	10.18	10.27	21.94	4.62
of which:									
Total industry	**651.32**	**660.64**	**515.43**	**602.59**	**492.02**	**543.00**	**546.20**	**-1.37**	**0.28**
Coal and peat	95.67	77.70	72.84	50.93	37.07	43.05	39.69	-1.59	-2.85
Oil	129.96	154.80	85.46	59.49	61.77	65.58	65.42	-2.44	-1.26
Natural Gas	279.20	254.56	196.52	238.38	192.36	216.35	219.26	-2.04	0.52
Biofuels and wastes	51.20	59.31	24.81	65.84	55.89	58.17	57.75	-4.17	4.11
Geothermal	-	-	-	0.16	0.15	0.15	0.15	-	-
Solar, wind, tide	-	-	0.00	0.00	0.01	0.01	0.01	-	10.31
Electricity	95.15	112.84	134.90	180.68	135.84	151.50	155.65	2.07	0.68
Heat	0.14	1.43	0.90	7.11	8.93	8.19	8.27	11.73	11.15
Total transport	**660.16**	**706.40**	**803.23**	**974.58**	**985.61**	**1018.78**	**1010.73**	**1.16**	**1.10**
Coal and peat	0.36	0.07	-	-	-	-	-	-	-
Oil	634.74	682.18	776.04	940.07	926.43	953.89	942.77	1.19	0.93
Natural Gas	24.06	23.41	26.16	28.50	24.76	26.85	27.22	0.49	0.19
Biofuels and wastes	-	-	-	4.74	32.80	36.43	39.12	-	-
Electricity	0.99	0.74	1.03	1.26	1.62	1.61	1.61	0.24	2.14
Residential	**390.93**	**367.46**	**367.81**	**454.10**	**453.30**	**460.28**	**457.07**	**-0.36**	**1.04**
Coal and peat	5.16	2.41	2.27	1.98	0.06	0.06	0.06	-4.71	-16.01
Oil	125.22	81.86	54.79	62.68	46.74	45.33	42.74	-4.75	-1.18
Natural Gas	170.56	171.05	164.07	186.93	180.44	178.97	178.57	-0.23	0.40
Biofuels and wastes	11.24	12.11	14.15	31.42	31.20	30.93	32.26	1.36	4.00
Geothermal	-	-	-	0.31	0.05	0.05	0.05	-	-
Solar, wind, tide	-	-	0.01	2.09	2.10	2.08	2.17	-	27.10
Electricity	78.76	100.02	132.51	168.68	192.72	202.87	201.23	3.11	2.01
Heat	-	-	0.00	-	0.00	-	-	-	-
Comm & public services	**237.74**	**232.80**	**261.36**	**320.75**	**338.64**	**337.44**	**338.85**	**0.56**	**1.24**
Coal and peat	4.91	2.66	3.47	1.30	2.29	2.17	1.87	-2.03	-2.90
Oil	79.66	54.63	37.21	34.75	28.77	28.41	27.71	-4.38	-1.39
Natural Gas	94.02	96.77	99.42	120.31	118.51	116.86	119.04	0.33	0.86
Biofuels and waste	-	-	0.38	1.78	3.24	3.15	2.90	-	10.21
Geothermal	-	-	-	0.27	0.12	0.12	0.12	-	-
Solar, wind, tide	-	-	0.01	0.02	0.09	0.06	0.08	-	10.44
Electricity	59.15	78.70	117.80	160.72	183.59	184.68	185.14	4.14	2.18
Heat	-	0.04	3.08	1.59	2.04	1.99	1.99	-	-2.06
Non-energy use	**140.55**	**169.41**	**205.80**	**264.67**	**226.35**	**238.03**	**236.52**	**2.27**	**0.66**
Coal and peat	-	0.18	0.50	0.38	1.07	0.93	0.91	-	2.87
Oil	139.82	163.17	175.96	225.62	203.32	213.48	211.73	1.36	0.89
Natural Gas	0.73	6.06	29.34	38.67	21.96	23.63	23.88	24.30	-0.97

(1) Total Final Consumption (TFC) excludes use in transformation processes and energy industry own use.

* Please refer to notes and definitions in Part I. Agriculture, Forestry, Fishing and "Non-specified other" sectors are not reported here individually, but are accounted for in Total final consumption.

Source: *IEA/OECD Energy Balances of OECD Countries*

OECD AMERICAS

5. Coal balance*

(Mtce)

	1973	1980	1990	2000	2005	2010	2011	2012e	Average annual percent change 73-90	90-11
Production	496.5	672.4	835.8	824.2	860.8	817.5	825.8	770.1	3.1	-0.1
Imports	16.7	18.2	18.6	42.8	55.2	39.2	34.6	31.5	0.7	3.0
Exports	-55.3	-98.0	-127.8	-82.3	-66.2	-97.1	-119.2	-135.6	5.1	-0.3
Stock changes	12.6	-20.6	-26.1	37.4	7.5	10.4	-6.9	-7.8		
Primary supply	**470.5**	**572.0**	**700.6**	**822.1**	**857.3**	**770.0**	**734.2**	**658.1**	**2.4**	**0.2**
Statistical differences	**18.6**	**-10.2**	**9.5**	**21.5**	**-24.1**	**-1.4**	**-6.5**	..		
Total transformation	**-363.7**	**-468.5**	**-620.7**	**-786.7**	**-777.2**	**-719.9**	**-682.4**	..	**3.2**	**0.5**
Electricity and heat gen.	-322.2	-439.3	-599.3	-765.0	-766.8	-704.9	-666.4	..	3.7	0.5
Main activity producers (1)	*-322.2*	*-439.3*	*-592.3*	*-747.5*	*-758.1*	*-698.4*	*-660.7*	..	*3.6*	*0.5*
Autoproducers	*-0.0*	*-0.0*	*-7.0*	*-17.5*	*-8.7*	*-6.6*	*-5.7*	..	*51.8*	*-1.0*
Gas works	0.5	0.0	-2.6	-2.6	-2.6	-2.7	-2.7	..	-	0.3
Coal transformation(2)	-42.0	-29.2	-18.8	-19.1	-7.9	-12.2	-13.2	..	-4.6	-1.7
BKB plants	-	-	-	-	-	-	-	..	-	-
Blast furnaces	*-27.8*	*-18.5*	*-13.2*	*-13.3*	*-10.8*	*-7.5*	*-8.5*	..	*-4.3*	*-2.0*
Coke ovens	*-14.2*	*-10.7*	*-5.6*	*-5.8*	*-4.1*	*-4.7*	*-4.7*	..	*-5.3*	*-0.8*
Patent fuel plants	-	-	-	-	*7.0*	-	-	..	-	-
Other transformation(3)	-	-	-	-	-	-	-	..	-	-
Energy ind. own use	**-8.8**	**-3.7**	**-2.1**	**-2.3**	**-3.2**	**-2.5**	**-2.8**	..	**-8.2**	**1.4**
Losses	**-0.1**	**-0.1**	**-0.0**	**-0.1**	**-0.0**	**-0.0**	**-0.0**	..		
Final consumption(4)	**116.5**	**89.5**	**87.3**	**54.6**	**52.7**	**46.2**	**42.5**	..	**-1.7**	**-3.4**
Industry(5)	95.7	77.7	72.8	50.9	48.4	43.0	39.7	..	-1.6	-2.8
Iron and steel	*54.4*	*36.2*	*22.7*	*15.8*	*10.4*	*11.5*	*11.0*	..	*-5.0*	*-3.4*
Chemical	*11.9*	*11.2*	*12.5*	*9.6*	*7.1*	*6.1*	*5.7*	..	*0.3*	*-3.7*
Non-metallic minerals	*6.0*	*10.5*	*11.9*	*12.7*	*12.0*	*8.3*	*7.9*	..	*4.1*	*-2.0*
Paper, pulp and print	*7.1*	*7.4*	*10.5*	*3.8*	*6.6*	*5.8*	*5.3*	..	*2.3*	*-3.2*
Other industry (6)	*16.1*	*12.4*	*15.2*	*9.1*	*12.4*	*11.4*	*9.8*	..	*-0.3*	*-2.1*
Transport(7)	0.4	0.1	-	-	-	-	-	..	-	-
Other	20.5	11.6	14.0	3.3	3.7	2.2	1.9	..	-2.2	-9.0
Comm. and pub. services	*4.9*	*2.7*	*3.5*	*1.3*	*3.0*	*2.2*	*1.9*	..	*-2.0*	*-2.9*
Residential	*5.2*	*2.4*	*2.3*	*2.0*	*0.1*	*0.1*	*0.1*	..	*-4.7*	*-16.0*
Other sectors (8)	*10.4*	*6.5*	*8.3*	*0.0*	*0.6*	*0.0*	*0.0*	..	*-1.4*	*-30.6*
Non-energy use	-	0.2	0.5	0.4	0.6	0.9	0.9	..	-	2.9

(1) Main activity electricity and heat generation includes district heating.

(2) Coal transformation refers to the transformation of coal and peat to secondary and tertiary products (mainly cokes, briquettes, coal tar and off-gases). Gas works and Liquefaction are shown separately to match the *Energy Balances of OECD Countries.*

(3) Other transformation includes Liquefaction and Non-specified transformations.

(4) Final Consumption includes non-energy use and energy use (Industry, Transport and Other).

(5) Please refer to notes and definitions in Part I for detailed explanation.

(6) Other industry includes Non-ferrous metals, Transport equipment, Machinery, Mining and quarrying, Food and tobacco, Wood and wood products, Construction, Textile and leather, and Non-specified industry.

(7) Transport includes Rail and Inland waterways.

(8) Other sectors includes Agriculture/Forestry, Fishing and Non-specified other.

* "Coal" refers to all types of coal, primary (anthracite, coking coal, other bituminous coal, sub-bituminous coal, lignite/brown coal) and derived fuels (including patent fuel, BKB, coke oven coke, gas coke, coal tar, coke oven gas, gas works gas, blast furnace gas and other recovered gases), as well as peat, peat briquettes and oil shale. Quantities have been converted to Mtce using calorific values reported by the respective countries. Please refer to notes and definitions in Part I.

Source: *IEA/OECD Energy Balances of OECD Countries*

OECD AMERICAS

6. Use of coal for selected end-uses*

(million tonnes)

	1978	1990	2000	2005	2009	2010	2011	Average annual percent change 78-90	90-11
Steam coal	**505.43**	**743.62**	**926.90**	**996.45**	**894.65**	**919.40**	**878.91**	**3.27**	**0.80**
Total electricity and heat	426.81	676.52	893.20	822.84	846.15	886.14	840.10	3.91	1.04
Main activity producers	*426.79*	*669.44*	*877.71*	*816.32*	*841.59*	*880.97*	*834.82*	*3.82*	*1.06*
Autoproducers	*0.03*	*7.08*	*15.49*	*6.52*	*4.56*	*5.16*	*5.28*	*60.08*	*-1.38*
Patent fuel/BKB plants	-	-	-	119.79	-	-	-	-	-
Coke ovens/Liquefaction(1)	-	-	-	-	-	-	-	-	-
Blast furnace inputs	-	0.17	2.39	1.20	1.03	1.28	1.48	-	10.73
Gas manufacture	0.03	-	-	-	-	-	-	-	-
Industry	46.43	54.07	38.34	38.38	32.14	34.92	31.81	1.28	-2.49
Iron and steel	*3.73*	*1.71*	*1.46*	*0.32*	*0.47*	*0.68*	*0.57*	*-6.32*	*-5.11*
Chemical	*10.34*	*14.26*	*10.51*	*7.23*	*6.68*	*6.84*	*6.48*	*2.71*	*-3.69*
Non-metallic minerals	*12.00*	*12.85*	*13.50*	*12.94*	*8.57*	*9.19*	*8.74*	*0.57*	*-1.82*
Paper, pulp and print	*7.82*	*11.38*	*4.14*	*6.84*	*6.22*	*6.50*	*5.98*	*3.17*	*-3.02*
Other industry	*12.53*	*13.88*	*8.74*	*11.04*	*10.20*	*11.72*	*10.05*	*0.86*	*-1.53*
Other sectors(2)	19.19	15.52	3.75	4.31	2.64	2.52	2.22	-1.75	-8.84
Non-energy use	-	0.26	0.33	0.53	1.22	1.07	1.05	-	6.90
Coking coal	**80.10**	**43.70**	**33.86**	**28.33**	**20.75**	**26.08**	**28.47**	**-4.92**	**-2.02**
Total electricity and heat	-	-	-	-	-	-	-	-	-
Main activity producers	-	-	-	-	-	-	-	-	-
Autoproducers	-	-	-	-	-	-	-	-	-
Patent fuel/BKB plants	-	-	-	2.81	-	-	-	-	-
Coke ovens/Liquefaction(1)	76.10	43.65	33.63	25.88	19.98	26.36	27.18	-4.53	-2.23
Blast furnace inputs	-	-	-	-	-	-	-	-	-
Gas manufacture	-	-	-	-	-	-	-	-	-
Industry	-	-	-	-	-	-	-	-	-
Iron and steel	-	-	-	-	-	-	-	-	-
Chemical	-	-	-	-	-	-	-	-	-
Non-metallic minerals	-	-	-	-	-	-	-	-	-
Paper, pulp and print	-	-	-	-	-	-	-	-	-
Other industry	-	-	-	-	-	-	-	-	-
Other sectors(2)	-	-	-	-	-	-	-	-	-
Non-energy use	-	-	-	-	-	-	-	-	-
Lignite	**35.78**	**88.38**	**85.48**	**87.28**	**75.53**	**78.44**	**84.37**	**7.83**	**-0.22**
Total electricity and heat	33.12	81.12	80.87	81.10	68.12	72.02	76.57	7.75	-0.27
Main activity producers	*33.12*	*80.03*	*79.62*	*79.04*	*68.12*	*69.44*	*76.26*	*7.63*	*-0.23*
Autoproducers	-	*1.10*	*1.25*	*2.06*	-	*2.57*	*0.31*	-	*-5.82*
Patent fuel/BKB plants	-	-	-	-	-	-	-	-	-
Coke ovens/Liquefaction(1)	-	-	0.00	0.00	0.00	0.00	0.00	-	-
Gas manufacture	-	5.64	5.67	5.39	5.43	5.57	5.41	-	-0.20
Industry	2.46	1.41	1.59	0.39	0.29	0.30	0.08	-4.53	-12.73
Iron and steel	-	-	-	-	-	-	-	-	-
Chemical	-	*0.23*	*0.24*	*0.24*	*0.26*	*0.26*	*0.05*	-	*-7.00*
Non-metallic minerals	-	*0.00*	*0.00*	*0.01*	-	-	-	-	-
Paper, pulp and print	*0.17*	*0.10*	*0.10*	*0.06*	-	-	-	*-4.60*	-
Other industry	*2.30*	*1.08*	*1.25*	*0.08*	*0.03*	*0.04*	*0.03*	*-6.09*	*-15.69*
Other sectors(2)	0.20	0.12	0.09	0.06	0.09	0.09	0.10	-4.34	-0.88
Non-energy use	-	0.09	0.14	0.16	-	-	-	-	-

(1) Liquefaction primarily refers to direct distillation processes. Indirect liquefaction may be reported here or under gas manufacture.

(2) Other sectors are Residential, Commercial and public services, Agriculture/Forestry, Fishing and Non-specified other.

* Coal totals may not match the sum of the individual rows due to statistical differences. Please refer to notes and definitions in Part I.

Source: *IEA/OECD Energy Statistics of OECD Countries*

OECD AMERICAS

7. Fuel prices to end users*

	1978(1)	1990	1995	2000	2005	2009	2010	2011	2012
				(US dollars / tce)					
For electricity generation									
Steam coal	..	..	..	..	..	..	..	..	..
Heavy fuel oil	..	..	..	..	..	..	..	..	..
Natural gas	..	..	..	..	..	..	..	..	..
For industry									
Steam coal	..	..	..	..	..	..	..	..	..
Coking coal	..	..	..	..	..	..	..	..	..
High sulphur fuel oil	..	..	..	..	..	..	..	..	..
Low sulphur fuel oil	..	..	..	..	..	..	..	..	..
Natural gas	..	..	..	..	..	..	..	..	..
				(US dollars / unit)(2)					
For electricity generation									
Steam coal	..	..	..	..	..	..	..	..	..
Heavy fuel oil	..	..	..	..	..	..	..	..	..
Natural gas	..	..	..	..	..	..	..	..	..
For industry									
Steam coal	..	..	..	..	..	..	..	..	..
Coking coal	..	..	..	..	..	..	..	..	..
High sulphur fuel oil	..	..	..	..	..	..	..	..	..
Low sulphur fuel oil	..	..	..	..	..	..	..	..	..
Natural gas	..	..	..	..	..	..	..	..	..

(1) Earliest year for which split by coal type is available.

(2) For steam coal, coking coal and heavy fuel oil per metric tonne; for natural gas per 10^7 kilocalories GCV.

* Please refer to notes and definitions in Part I. Steam coal prices relate to historical definitions.

Source: *IEA/OECD Energy Prices & Taxes*

8. Coal and peat trade by type of coal*

(Mtce)

	1978	1990	1995	2000	2005	2009	2010	2011	2012e
Total imports	**22.37**	**18.62**	**21.03**	**42.81**	**55.19**	**38.84**	**39.24**	**34.58**	**31.45**
Bituminous coal(1)	10.87	12.55	12.53	26.77	33.33	25.26	22.68	19.59	16.56
Coking coal	6.04	4.91	5.39	8.24	8.09	4.16	6.31	7.25	6.63
Sub-bituminous coal	-	-	0.28	3.20	10.05	8.54	8.04	5.69	6.50
Lignite	-	-	0.00	0.04	0.06	0.06	0.07	0.07	0.07
Peat	-	-	-	-	-	-	-	-	-
Coal products(2)	5.45	1.15	2.83	4.56	3.65	0.82	2.14	1.98	1.69
Total exports	**50.81**	**127.76**	**115.19**	**82.31**	**66.21**	**74.03**	**97.09**	**119.23**	**135.61**
Bituminous coal(1)	9.63	41.27	35.39	25.38	12.75	18.98	19.98	31.33	43.16
Coking coal	40.44	85.76	76.80	54.83	47.14	50.01	71.16	82.73	85.82
Sub-bituminous coal	-	-	2.01	0.73	4.40	3.48	4.40	3.83	5.28
Lignite	0.04	0.04	-	0.03	0.17	0.17	0.16	0.14	0.08
Peat	-	-	-	-	-	-	-	-	-
Coal products(2)	0.70	0.69	1.00	1.34	1.74	1.38	1.38	1.20	1.26

(1) Bituminous coal includes anthracite.

(2) Coal products includes products derived from coal, for example: coke, coal tar, briquettes, patent fuels.

* Please refer to notes and definitions in Part I.

Source: *IEA/OECD Energy Balances of OECD Countries, IEA/OECD Energy Statistics of OECD Countries*

OECD AMERICAS

9. Total coal imports by origin*

(thousand tonnes)

	1978[1]	1985	1990	1995	2000	2005	2010	2011	2012e
Total coal[2]	**17684**	**17312**	**18506**	**19297**	**41420**	**59654**	**44801**	**38595**	**35706**
Coking coal	**6209**	**7150**	**5211**	**5662**	**8353**	**8296**	**6458**	**7421**	**6813**
Australia	-	-	-	-	1074	948	701	1053	627
Canada	-	195	122	818	2049	2246	1900	1703	1252
Czech Republic	-	-	-	-	-	-	-	-	-
Germany	-	-	-	-	-	-	-	-	-
Poland	-	-	-	-	-	-	-	-	-
United Kingdom	-	-	-	-	-	-	-	-	-
United States	5924	6491	4497	4129	4357	4697	3762	4488	4585
Other OECD	-	-	-	-	-	222	-	-	-
China, People's Rep.	-	-	-	-	1	3	3	-	4
Colombia	-	-	-	-	-	120	92	89	35
Indonesia	-	-	-	-	-	-	-	-	-
South Africa	-	-	-	-	-	-	-	-	146
Former Soviet Union[3]	-	-	x	x	x	x	x	x	x
Russian Federation [4]	*x*	*x*	-	-	*66*	-	-	-	-
Other FSU [4]	*x*	*x*	-	-	-	-	-	-	*24*
Venezuela	-	-	-	-	92	-	-	88	140
Vietnam	-	-	-	-	-	-	-	-	-
Non-specified/other	285	464	592	715	714	60	-	-	-
Steam coal	**11475**	**10162**	**13295**	**13633**	**32983**	**51226**	**38200**	**31035**	**28746**
Australia	933	35	22	192	1482	5436	3413	3437	3602
Canada	49	367	883	1197	960	576	361	253	185
Czech Republic	-	-	-	-	-	-	-	-	-
Germany	-	-	-	-	-	-	-	3	-
Poland	645	-	-	-	-	70	-	-	-
United Kingdom	-	-	5	-	-	22	4	302	-
United States	8796	8391	9639	5095	19167	13862	10023	8374	10575
Other OECD	-	6	-	39	164	293	84	7	-
China, People's Rep.	-	-	-	-	235	69	47	19	59
Colombia	-	539	1296	2994	6928	21333	19711	14960	12817
Indonesia	-	-	-	924	1282	3505	2474	1451	630
South Africa	996	824	-	-	182	70	1275	1131	139
Former Soviet Union[3]	-	-	x	x	x	x	x	x	x
Russian Federation [4]	*x*	*x*	-	-	*26*	*707*	*12*	*145*	*96*
Other FSU [4]	*x*	*x*	-	-	-	*141*	*191*	*221*	*390*
Venezuela	-	-	277	1675	2015	4012	560	679	176
Vietnam	-	-	-	-	-	85	-	-	-
Non-specified/other	56	-	1173	1517	542	1045	45	50	77
Lignite	**-**	**-**	**-**	**2**	**84**	**132**	**143**	**139**	**147**

(1) Earliest year for which split by coal type is available.

(2) Total coal consists of steam coal, coking coal and lignite.

(3) For years prior to 1990, or if no details for individual Former Soviet Union republics.

(4) Data for former republics are not available separately for some OECD member countries.

* Please refer to notes and definitions in Part I.

Source: IEA/OECD Coal Statistics

OECD AMERICAS

10. Coking coal exports by destination*

(thousand tonnes)

	1978[1]	1985	1990	1995	2000	2005	2010	2011	2012e
World	**40450**	**77198**	**84425**	**75819**	**58170**	**52715**	**78439**	**90753**	**94333**
Total OECD	**36030**	**65898**	**67629**	**61804**	**45508**	**41668**	**52379**	**60299**	**57236**
Australia	-	-	-	-	-	-	-	-	-
Austria	-	-	-	-	-	239	412	330	1670
Belgium	1103	3443	5538	3807	2717	1528	1737	1371	1392
Canada	5410	6360	3988	4039	3501	4034	3091	3772	4379
Chile	-	281	292	267	312	368	215	611	639
Czech Republic	-	-	-	-	-	-	-	-	-
Denmark	-	-	50	-	-	-	-	-	-
Estonia	-	-	-	-	-	-	-	-	-
Finland	-	-	127	715	288	776	678	729	569
France	1468	3823	4880	3233	2782	1610	1958	1797	2647
Germany	560	709	708	433	1211	2230	2728	2565	3340
Greece	-	7	-	-	-	-	-	-	-
Hungary	-	-	-	-	65	-	40	26	-
Iceland	-	13	28	36	48	57	59	58	62
Ireland	-	54	-	-	-	-	-	136	-
Israel	-	-	50	127	56	-	-	-	-
Italy	3069	5734	6536	5054	4467	3672	3402	5067	4196
Japan	19925	29639	26588	22688	12937	8446	11387	12927	12164
Korea	1173	3950	2908	6615	4947	5497	8009	11111	9417
Luxembourg	-	-	-	-	-	-	77	-	-
Mexico	-	498	3	517	740	994	1002	1010	758
Netherlands	929	2975	3975	4068	2143	2355	5636	7152	5710
New Zealand	-	-	-	-	-	-	-	-	-
Norway	68	98	99	83	42	18	75	81	79
Poland	-	-	-	-	-	-	2149	1258	465
Portugal	265	244	753	188	198	-	-	93	134
Slovak Republic	-	-	-	-	-	-	260	470	462
Slovenia	-	-	-	-	-	163	223	663	292
Spain	838	2460	3156	2313	2331	2029	1462	1226	1160
Sweden	453	1310	866	1161	642	464	401	480	535
Switzerland	-	-	-	-	-	-	37	-	-
Turkey	409	1716	1957	1895	2403	2667	2914	3097	3931
United Kingdom	360	2474	4355	3831	3134	2915	3044	2914	2360
United States	-	110	772	734	544	1606	1383	1355	875
Total non-OECD	**4420**	**11300**	**11592**	**13631**	**9437**	**11047**	**25991**	**30454**	**37097**
Brazil	1942	6139	6327	6826	5564	4831	8762	9849	8904
China[2]	-	-	300	-	-	956	8149	7265	16091
Chinese Taipei	205	782	357	1600	1440	1274	864	1069	1004
Egypt	218	457	586	1165	682	706	1042	629	375
India	200	38	-	-	22	1078	2299	3728	5181
Romania	673	1051	1559	1528	443	547	812	937	607
Oth. Africa & Mid. East	1	872	614	350	269	377	230	683	879
Oth. non-OECD Americas	914	588	580	400	184	207	326	306	476
Other Asia & Oceania	24	353	229	548	-	104	109	110	127
Other non-OECD Europe and Eurasia	243	1020	1040	1214	833	967	3398	5878	3453
Non-specified/Other	**-**	**-**	**5204**	**384**	**3225**	**-**	**69**	**-**	**-**

(1) Earliest year for which split by coal type is available.

(2) China includes the People's Republic of China and Hong Kong, China.

* Please refer to notes and definitions in Part I.

Source: IEA/OECD Coal Statistics

OECD AMERICAS

11. Steam coal exports by destination*

(thousand tonnes)

	1978(1)	1985	1990	1995	2000	2005	2010	2011	2012e
World	**10808**	**34228**	**42493**	**38503**	**26922**	**20460**	**28774**	**39990**	**54600**
Total OECD	**10341**	**30501**	**36299**	**34885**	**25445**	**18309**	**24131**	**33386**	**43792**
Australia	-	1	1	-	-	-	105	302	152
Austria	-	-	-	-	-	-	-	-	-
Belgium	27	712	2178	950	429	411	367	1470	1013
Canada	8782	8508	10083	4541	13524	13625	7245	2431	2161
Chile	-	-	394	207	48	256	1097	1607	1635
Czech Republic	-	-	-	-	-	-	-	-	-
Denmark	309	2286	3321	1905	70	66	73	146	-
Estonia	-	-	-	-	-	-	-	-	-
Finland	-	1208	-	530	-	-	166	145	-
France	38	978	1758	621	564	28	1080	2078	1146
Germany	528	595	384	1648	522	133	935	2394	7402
Greece	-	1138	-	-	-	-	47	-	-
Hungary	-	-	-	-	-	-	-	-	-
Iceland	-	7	15	-	-	-	-	-	-
Ireland	-	335	1322	829	456	-	-	86	208
Israel	-	421	530	562	-	-	-	-	17
Italy	22	3633	4451	3449	79	23	613	1004	4318
Japan	243	2847	4007	6179	4425	962	2180	2650	2678
Korea	356	2489	719	3800	2275	749	4193	7111	5303
Luxembourg	-	-	-	-	-	-	-	-	-
Mexico	-	44	188	-	373	341	1378	1920	2584
Netherlands	27	2793	3982	3015	643	829	1700	3992	2532
New Zealand	-	-	1	-	-	-	-	-	-
Norway	-	143	62	25	74	-	-	11	17
Poland	-	-	-	-	-	-	65	129	245
Portugal	-	626	1386	1562	343	143	531	798	992
Slovak Republic	-	-	-	-	-	-	-	-	-
Slovenia	-	-	-	-	-	-	182	-	197
Spain	-	786	282	2111	441	-	374	521	816
Sweden	-	230	21	8	-	71	275	153	138
Switzerland	-	8	-	-	-	-	-	169	-
Turkey	5	275	15	191	55	67	220	422	1515
United Kingdom	-	283	1005	2705	1016	361	1219	3800	8700
United States	4	155	194	47	108	244	86	47	23
Total non-OECD	**95**	**3727**	**4967**	**3616**	**923**	**1948**	**4643**	**6603**	**9564**
Brazil	11	78	177	153	22	693	118	310	125
China(2)	-	506	108	-	9	-	2938	2182	3188
Chinese Taipei	-	2575	3820	2033	-	3	1	-	228
Egypt	-	-	-	1	-	-	146	144	-
India	-	-	-	-	-	217	171	635	1829
Romania	-	-	-	271	-	844	-	-	-
Oth. Africa & Mid. East	1	176	682	1099	825	63	1044	2292	2913
Oth. non-OECD Americas	82	110	128	-	1	13	115	463	677
Other Asia & Oceania	1	2	5	-	-	-	77	1	103
Other non-OECD Europe and Eurasia	-	280	47	59	66	115	33	576	501
Non-specified/Other	**372**	**-**	**1227**	**2**	**554**	**203**	**-**	**1**	**1244**

(1) Earliest year for which split by coal type is available.

(2) China includes the People's Republic of China and Hong Kong, China.

* Please refer to notes and definitions in Part I. Steam coal includes all sub-bituminous coal.

Source: IEA/OECD Coal Statistics

OECD ASIA OCEANIA*

Figure 1: TPES by fuel (Mtce)

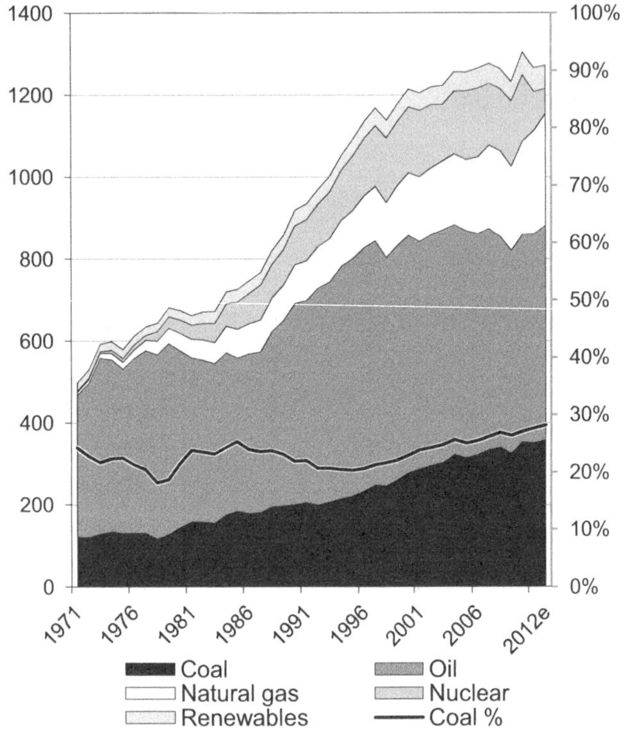

Figure 2: Coal supply indicators (1971 = 100)

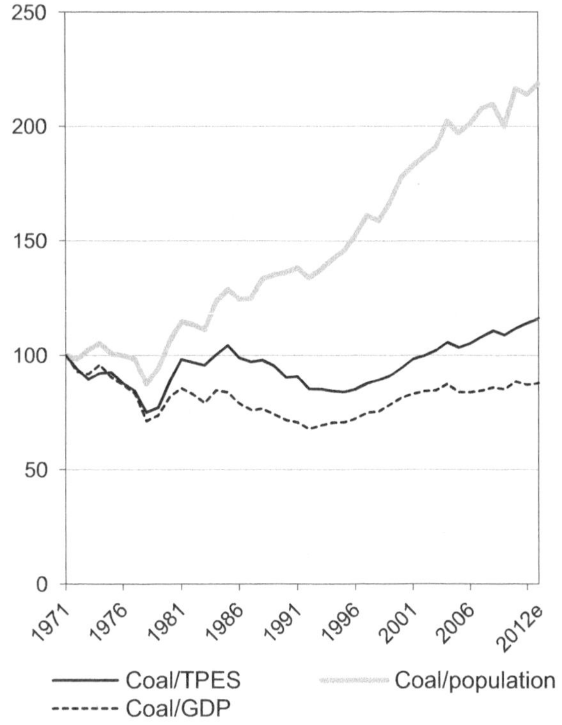

Figure 3: Primary coal supply (Mtce)

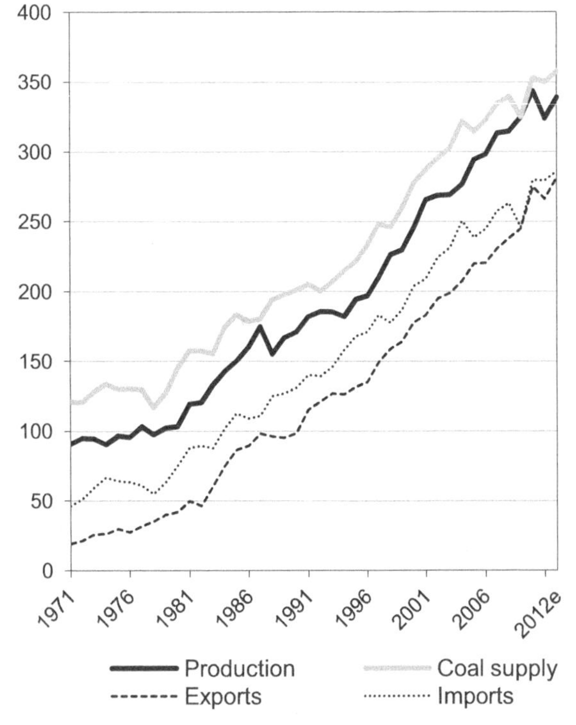

Figure 4: Coal consumption (Mtce)

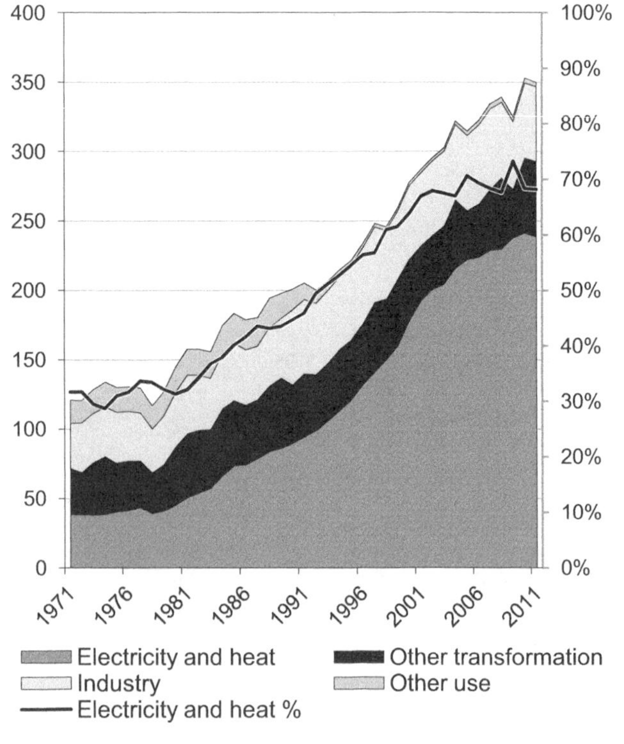

*Please refer to notes and definitions in Part I. Peat is included in coal for all figures.

Source: *IEA/OECD Coal Information, IEA/OECD Energy Balances of OECD Countries, OECD Main Economic Indicators*

OECD ASIA OCEANIA*

Figure 5: Electricity generation by fuel (TWh)

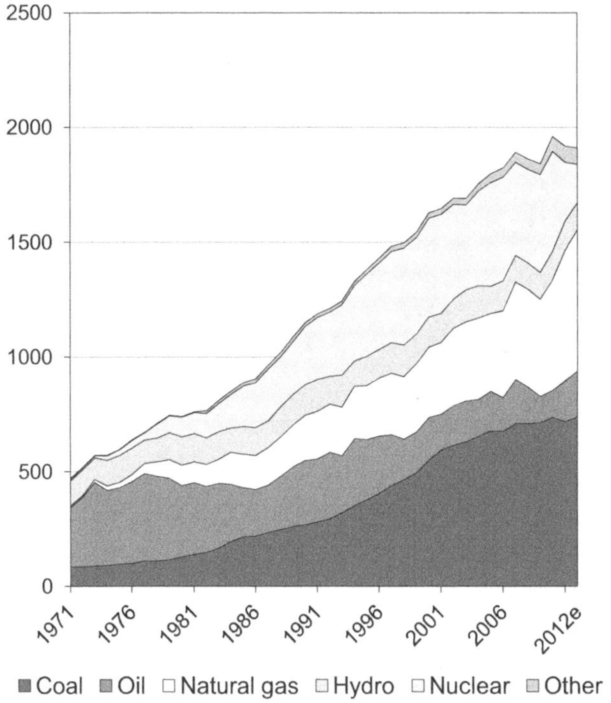

Figure 6: CO_2 emissions by fuel (Mt CO_2)

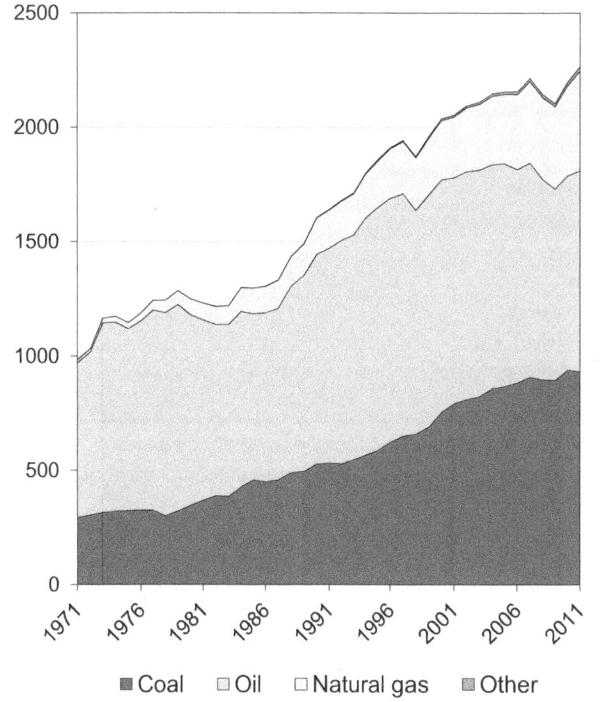

Figure 7: Electricity generation by fuel share

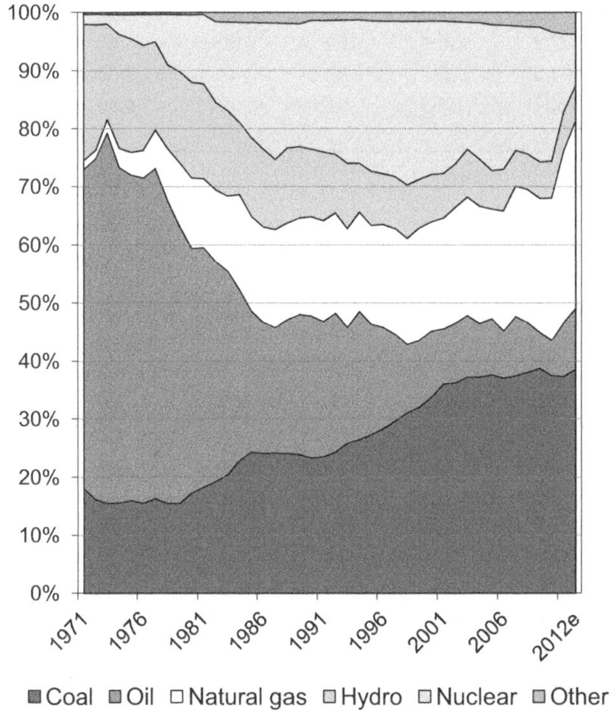

Figure 8: CO_2 emissions by fuel share

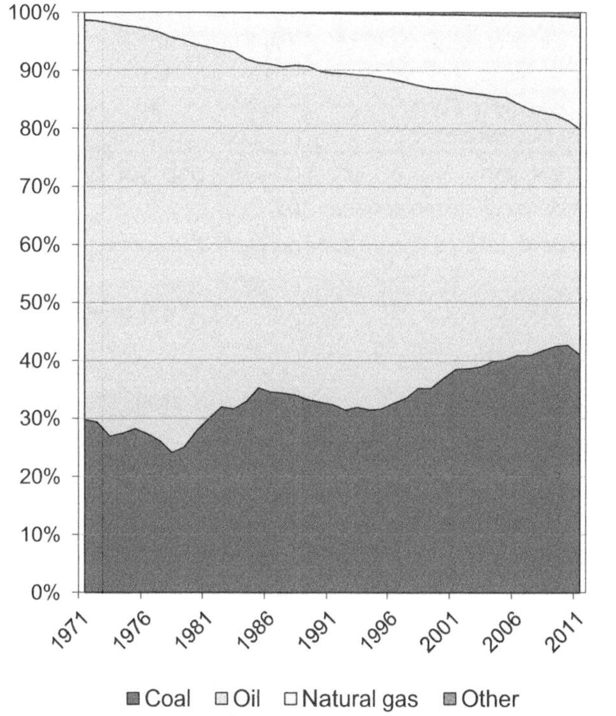

*Please refer to notes and definitions in Part I. Peat is included in coal for all figures.

Source: *IEA/OECD Electricity Information, IEA/OECD CO2 Emissions from Fuel Combustion*

OECD ASIA OCEANIA

1. Total primary energy supply (TPES) by fuel*

	1973	1980	1990	2000	2010	2011	2012e	Average annual percent change 73-90	90-11
TPES (Mtce)	**592.50**	**674.49**	**918.53**	**1215.04**	**1304.36**	**1266.02**	**1272.61**	**2.61**	**1.54**
Coal and peat	128.34	144.83	200.89	277.93	353.08	350.03	357.77	2.67	2.68
Oil	429.59	431.60	491.05	579.08	506.45	511.34	523.90	0.79	0.19
Natural Gas	12.54	42.55	93.67	152.84	227.11	251.51	272.52	12.55	4.82
Biofuels and waste	5.04	5.91	14.92	19.14	26.78	27.90	28.37	6.59	3.03
Nuclear	3.61	32.04	94.99	160.46	162.63	95.49	61.85	21.20	0.02
Hydro	11.52	15.01	16.34	16.23	15.25	15.93	14.53	2.08	-0.12
Geothermal	1.86	2.56	4.36	7.21	8.74	9.04	8.85	5.15	3.53
Solar, wind, tide	-	0.03	2.31	2.27	4.63	5.12	5.17	-	3.86
Net electricity trade(1)	-0.01	-0.02	-0.06	-0.18	-0.49	-0.52	-0.52	13.01	11.18
Heat(2)	-	-	0.05	0.05	0.18	0.18	0.18	-	5.82

(1) Net trade between OECD and non-OECD countries.

(2) Ambient heat from heat pumps used in the transformation sector.

Source: *IEA/OECD Energy Balances of OECD Countries*

2. Energy supply, GDP and population*

	1973	1980	1990	2000	2010	2011	2012e	Average annual percent change 73-90	90-11
GDP (2005 bil. USD)	2392	3029	4801	5840	6821	6870	6962	4.18	1.72
TPES/GDP(1)	0.25	0.22	0.19	0.21	0.19	0.18	0.18	-1.51	-0.18
Population (millions)	162.9	177.0	191.7	203.4	211.9	212.6	212.7	0.96	0.49
TPES/population(1)	3.64	3.81	4.79	5.97	6.16	5.96	5.98	1.63	1.04
TPES/GDP(2)	126.6	113.8	97.8	106.4	97.8	94.2	93.5	-1.51	-0.18
Coal and peat supply/GDP(2)	109.6	97.7	85.5	97.2	105.7	104.1	105.0	-1.45	0.94
Elec. consumption/GDP(2)	84.3	86.6	85.0	99.5	102.4	99.5	..	0.05	0.75
Elec. generation (TWh)	572	740	1148	1629	1961	1920	1911	4.19	2.48
Industrial production(2)	..	..	..	..	..	..	..	..	..

(1) TPES/GDP in units of tce/2005 thousand USD, TPES/population in units of tce/capita.

(2) As index, 2005 (not shown) = 100

Source: *IEA/OECD Energy Balances of OECD Countries* , *OECD Main Economic Indicators*

3. Coal and peat production by type*

	1978(1)	1990	2000	2005	2010	2011	2012e	Average annual percent change 78-90	90-11
Mtce:									
Coking coal	44.87	63.09	102.24	127.32	160.81	144.83	145.01	2.88	4.04
Steam coal	42.82	93.04	122.17	143.26	158.51	154.74	168.99	6.68	2.45
Lignite	9.73	14.72	21.10	23.76	24.30	23.94	24.79	3.51	2.34
Peat	-	-	-	-	-	-	-	-	-
Mt:									
Coking coal	46.34	65.21	105.06	130.80	165.27	148.83	149.02	2.89	4.01
Steam coal	60.93	120.98	148.88	177.11	194.16	189.11	204.87	5.88	2.15
Lignite	30.66	46.45	67.90	71.21	72.82	71.74	74.26	3.52	2.09
Peat	-	-	-	-	-	-	-	-	-

(1) Earliest year for which split by coal type is available.

* Please refer to notes and definitions in Part I.

Source: *IEA/OECD Energy Balances of OECD Countries, IEA/OECD Energy Statistics of OECD Countries*

OECD ASIA OCEANIA

4. Final consumption of energy by fuel*

(Mtce)

	1973	1980	1990	2000	2009	2010	2011	Average annual percent change 73-90	90-11
Total final consumption[1]	**429.33**	**459.12**	**625.10**	**806.94**	**803.22**	**835.40**	**828.71**	**2.23**	**1.35**
Coal and peat	52.37	57.60	69.35	55.74	52.17	57.85	57.20	1.67	-0.91
Oil	302.31	298.95	379.55	482.13	435.17	439.11	434.04	1.35	0.64
Natural Gas	7.70	16.21	37.69	69.18	92.22	99.59	102.49	9.80	4.88
Biofuels and wastes	4.99	5.64	10.33	13.21	12.95	14.09	14.64	4.37	1.67
Geothermal	-	-	0.35	0.59	0.67	0.62	0.64	-	2.87
Solar, wind, tide	-	0.03	2.31	2.18	2.45	2.59	2.58	-	0.54
Electricity	61.92	80.56	125.26	178.44	200.46	214.51	209.91	4.23	2.49
Heat	0.04	0.15	0.27	5.47	7.14	7.04	7.20	11.86	17.01
of which:									
Total industry	**185.18**	**175.36**	**208.17**	**236.82**	**213.77**	**231.77**	**229.02**	**0.69**	**0.46**
Coal and peat	35.20	39.16	54.44	53.22	48.71	54.39	53.90	2.60	-0.05
Oil	102.88	76.29	73.85	71.58	48.74	48.94	47.05	-1.93	-2.12
Natural Gas	4.26	8.31	15.09	23.49	30.29	33.35	36.15	7.72	4.25
Biofuels and wastes	2.13	3.17	6.80	9.92	9.33	10.16	10.43	7.06	2.05
Geothermal	-	-	0.16	0.19	0.22	0.21	0.21	-	1.44
Solar, wind, tide	-	-	-	-	0.00	0.00	0.00	-	-
Electricity	40.71	48.43	57.84	75.37	72.62	81.39	77.80	2.09	1.42
Heat	-	-	-	3.04	3.86	3.34	3.47	-	-
Total transport	**84.68**	**113.22**	**160.86**	**210.42**	**204.11**	**206.70**	**205.05**	**3.85**	**1.16**
Coal and peat	0.33	0.01	0.11	0.13	0.25	0.25	0.17	-6.54	2.44
Oil	82.62	111.17	158.24	207.03	198.19	200.45	198.74	3.90	1.09
Natural Gas	-	0.01	0.10	0.44	1.91	2.10	2.19	-	16.10
Biofuels and wastes	-	-	-	-	0.65	0.85	0.90	-	-
Electricity	1.73	2.03	2.42	2.83	3.11	3.07	3.05	2.00	1.11
Residential	**47.65**	**60.63**	**87.07**	**109.29**	**114.97**	**120.96**	**118.55**	**3.61**	**1.48**
Coal and peat	15.17	16.84	12.64	0.87	1.31	1.25	1.23	-1.07	-10.51
Oil	13.98	15.88	23.27	31.00	24.40	25.80	24.95	3.04	0.33
Natural Gas	2.24	5.62	13.97	25.60	29.25	30.48	30.53	11.37	3.79
Biofuels and wastes	2.86	2.43	3.45	3.05	2.26	2.25	2.27	1.11	-1.97
Geothermal	-	-	0.01	0.01	0.01	0.01	0.02	-	2.23
Solar, wind, tide	-	0.03	2.24	2.14	2.41	2.54	2.53	-	0.58
Electricity	13.40	19.83	31.44	45.03	53.08	56.08	54.41	5.15	2.65
Heat	-	-	0.04	1.57	2.25	2.56	2.62	-	21.51
Comm & public services	**32.48**	**37.53**	**84.88**	**122.94**	**132.22**	**134.96**	**135.22**	**5.81**	**2.24**
Coal and peat	1.56	1.59	1.58	1.00	0.85	0.84	0.81	0.06	-3.10
Oil	24.66	24.45	43.47	50.54	31.74	29.49	29.13	3.39	-1.89
Natural Gas	0.85	1.61	6.93	16.90	28.60	31.26	31.29	13.16	7.45
Biofuels and waste	-	0.03	0.08	0.24	0.70	0.83	1.00	-	13.10
Geothermal	-	-	0.13	0.24	0.29	0.27	0.27	-	3.68
Solar, wind, tide	-	-	0.07	0.04	0.04	0.05	0.05	-	-1.32
Electricity	5.38	9.70	32.41	53.12	68.97	71.08	71.55	11.15	3.84
Heat	0.04	0.15	0.22	0.86	1.03	1.14	1.11	10.68	7.94
Non-energy use	**58.67**	**50.34**	**66.55**	**106.31**	**119.88**	**122.13**	**121.67**	**0.74**	**2.92**
Coal and peat	-	-	0.57	0.50	1.01	1.07	1.00	-	2.68
Oil	58.35	49.71	64.44	103.15	116.77	118.72	118.41	0.59	2.94
Natural Gas	0.32	0.63	1.54	2.66	2.10	2.34	2.26	9.66	1.84

(1) Total Final Consumption (TFC) excludes use in transformation processes and energy industry own use.

* Please refer to notes and definitions in Part I. Agriculture, Forestry, Fishing and "Non-specified other" sectors are not reported here individually, but are accounted for in Total final consumption.

Source: *IEA/OECD Energy Balances of OECD Countries*

OECD ASIA OCEANIA

5. Coal balance*

(Mtce)

	1973	1980	1990	2000	2005	2010	2011	2012e	Average annual percent change 73-90	90-11
Production	94.4	103.1	170.8	245.5	294.3	343.6	323.5	338.8	3.5	3.1
Imports	59.6	75.0	130.8	203.6	238.1	279.7	279.1	286.0	4.7	3.7
Exports	-26.0	-41.9	-98.3	-177.7	-219.8	-274.8	-266.2	-281.8	8.1	4.9
Stock changes	0.3	8.7	-2.4	6.6	1.7	4.5	13.6	14.8		
Primary supply	**128.3**	**144.8**	**200.9**	**277.9**	**314.5**	**353.1**	**350.0**	**357.8**	**2.7**	**2.7**
Statistical differences	**-2.5**	**-2.3**	**3.4**	**1.7**	**10.6**	**-4.2**	**-1.4**	..		
Total transformation	**-68.7**	**-79.1**	**-128.7**	**-217.1**	**-262.2**	**-282.3**	**-282.4**	..	**3.8**	**3.8**
Electricity and heat gen.	-37.8	-45.2	-89.7	-177.5	-221.9	-241.3	-238.6	..	5.2	4.8
Main activity producers (1)	*-37.8*	*-40.9*	*-78.7*	*-161.2*	*-203.7*	*-221.2*	*-217.8*	..	*4.4*	*5.0*
Autoproducers	*-0.1*	*-4.3*	*-11.0*	*-16.3*	*-18.3*	*-20.1*	*-20.8*	..	*36.1*	*3.1*
Gas works	3.7	5.2	-0.3	-0.2	0.1	0.0	0.0	..	-	-
Coal transformation(2)	-34.7	-39.1	-38.6	-39.4	-40.3	-41.0	-43.8	..	0.6	0.6
BKB plants	*-0.2*	*-0.2*	*-0.1*	*-0.0*	*-0.1*	*-0.0*	*-0.0*	..	*-5.3*	*-3.7*
Blast furnaces	*-27.1*	*-23.7*	*-32.2*	*-37.8*	*-36.0*	*-39.6*	*-40.9*	..	*1.0*	*1.2*
Coke ovens	*-5.8*	*-14.5*	*-1.6*	*-0.6*	*-4.3*	*-1.4*	*-2.9*	..	*-7.3*	*2.8*
Patent fuel plants	*-1.5*	*-0.6*	*-4.8*	*-0.9*	*0.1*	-	-	..	*6.9*	-
Other transformation(3)	-	-	-	-	-	-	-	..	-	-
Energy ind. own use	**-4.6**	**-5.5**	**-6.1**	**-6.8**	**-5.5**	**-8.8**	**-9.0**	..	**1.7**	**1.9**
Losses	**-0.1**	**-0.3**	**-0.2**	**-0.0**	**-0.0**	**-0.0**	**-0.0**	..		
Final consumption(4)	**52.4**	**57.6**	**69.3**	**55.7**	**57.3**	**57.9**	**57.2**	..	**1.7**	**-0.9**
Industry(5)	35.2	39.2	54.4	53.2	54.1	54.4	53.9	..	2.6	-0.0
Iron and steel	*26.9*	*27.3*	*19.8*	*18.7*	*21.6*	*23.2*	*24.1*	..	*-1.8*	*0.9*
Chemical	*0.4*	*0.8*	*4.3*	*4.4*	*4.5*	*4.7*	*5.0*	..	*15.3*	*0.6*
Non-metallic minerals	*0.9*	*5.9*	*11.9*	*12.1*	*11.1*	*10.0*	*10.2*	..	*16.4*	*-0.7*
Paper, pulp and print	*0.3*	*0.5*	*1.9*	*2.3*	*2.3*	*2.2*	*2.2*	..	*10.8*	*0.6*
Other industry (6)	*6.7*	*4.7*	*16.5*	*15.8*	*14.6*	*14.3*	*12.4*	..	*5.5*	*-1.4*
Transport(7)	0.3	0.0	0.1	0.1	0.3	0.2	0.2	..	-6.5	2.4
Other	16.8	18.4	14.2	1.9	2.4	2.2	2.1	..	-1.0	-8.6
Comm. and pub. services	*1.6*	*1.6*	*1.6*	*1.0*	*0.9*	*0.8*	*0.8*	..	*0.1*	*-3.1*
Residential	*15.2*	*16.8*	*12.6*	*0.9*	*1.4*	*1.2*	*1.2*	..	*-1.1*	*-10.5*
Other sectors (8)	*0.1*	-	*0.0*	*0.0*	*0.0*	*0.1*	*0.1*	..	*-11.4*	*9.5*
Non-energy use	-	-	0.6	0.5	0.5	1.1	1.0	..	-	2.7

(1) Main activity electricity and heat generation includes district heating.

(2) Coal transformation refers to the transformation of coal and peat to secondary and tertiary products (mainly cokes, briquettes, coal tar and off-gases). Gas works and Liquefaction are shown separately to match the *Energy Balances of OECD Countries.*

(3) Other transformation includes Liquefaction and Non-specified transformations.

(4) Final Consumption includes non-energy use and energy use (Industry, Transport and Other).

(5) Please refer to notes and definitions in Part I for detailed explanation.

(6) Other industry includes Non-ferrous metals, Transport equipment, Machinery, Mining and quarrying, Food and tobacco, Wood and wood products, Construction, Textile and leather, and Non-specified industry.

(7) Transport includes Rail and Inland waterways.

(8) Other sectors includes Agriculture/Forestry, Fishing and Non-specified other.

* "Coal" refers to all types of coal, primary (anthracite, coking coal, other bituminous coal, sub-bituminous coal, lignite/brown coal) and derived fuels (including patent fuel, BKB, coke oven coke, gas coke, coal tar, coke oven gas, gas works gas, blast furnace gas and other recovered gases), as well as peat, peat briquettes and oil shale. Quantities have been converted to Mtce using calorific values reported by the respective countries.
Please refer to notes and definitions in Part I.

Source: *IEA/OECD Energy Balances of OECD Countries*

OECD ASIA OCEANIA

6. Use of coal for selected end-uses*

(million tonnes)

	1978	1990	2000	2005	2009	2010	2011	Average annual percent change 78-90	90-11
Steam coal	**57.24**	**132.00**	**217.00**	**262.49**	**279.98**	**297.05**	**292.23**	**7.21**	**3.86**
Total electricity and heat	30.57	80.92	165.65	219.30	240.88	243.16	239.94	8.45	5.31
Main activity producers	*30.21*	*75.93*	*157.93*	*208.01*	*228.39*	*229.86*	*225.69*	*7.98*	*5.32*
Autoproducers	*0.36*	*4.99*	*7.72*	*11.29*	*12.49*	*13.30*	*14.25*	*24.49*	*5.12*
Patent fuel/BKB plants	18.64	20.81	2.41	2.01	1.94	1.86	1.82	0.92	-10.95
Coke ovens/Liquefaction(1)	0.13	6.53	12.13	12.41	10.20	13.36	14.36	38.86	3.83
Blast furnace inputs	-	-	-	0.80	0.55	0.70	0.62	-	-
Gas manufacture	0.19	-	-	-	-	-	-	-	-
Industry	7.37	25.17	32.46	28.52	26.01	26.96	28.00	10.78	0.51
Iron and steel	*1.46*	*2.05*	*3.42*	*4.81*	*4.05*	*4.59*	*5.57*	*2.86*	*4.88*
Chemical	*0.34*	*2.63*	*3.92*	*4.21*	*4.19*	*4.54*	*4.86*	*18.50*	*2.97*
Non-metallic minerals	*1.88*	*12.63*	*13.30*	*12.29*	*10.90*	*11.02*	*11.44*	*17.23*	*-0.47*
Paper, pulp and print	*0.35*	*2.25*	*2.64*	*2.83*	*2.71*	*2.66*	*2.62*	*16.65*	*0.73*
Other industry	*3.34*	*5.62*	*9.18*	*4.37*	*4.16*	*4.16*	*3.52*	*4.45*	*-2.20*
Other sectors(2)	1.49	1.04	1.05	1.01	0.94	0.94	0.97	-2.95	-0.37
Non-energy use	-	0.02	-	-	-	-	-	-	-
Coking coal	**67.95**	**82.84**	**81.30**	**81.99**	**77.00**	**89.10**	**90.82**	**1.67**	**0.44**
Total electricity and heat	-	-	-	-	-	-	-	-	-
Main activity producers	-	-	-	-	-	-	-	-	-
Autoproducers	-	-	-	-	-	-	-	-	-
Patent fuel/BKB plants	-	-	-	-	-	-	-	-	-
Coke ovens/Liquefaction(1)	64.35	77.00	67.15	67.67	62.33	69.01	68.56	1.51	-0.55
Blast furnace inputs	-	5.25	13.97	15.92	14.75	19.19	21.90	-	7.04
Gas manufacture	4.45	-	-	-	-	-	-	-	-
Industry	0.12	0.00	0.00	0.00	0.32	0.23	0.12	-32.71	25.61
Iron and steel	*0.06*	*0.00*	*0.00*	*0.00*	*0.26*	*0.15*	*0.11*	*-29.20*	*24.92*
Chemical	*0.01*	-	-	-	-	-	-	-	-
Non-metallic minerals	-	-	-	-	*0.00*	*0.00*	-	-	-
Paper, pulp and print	-	-	-	-	-	-	-	-	-
Other industry	*0.04*	-	-	-	*0.06*	*0.07*	*0.01*	-	-
Other sectors(2)	0.01	-	-	-	-	0.01	-	-	-
Non-energy use	-	-	-	-	-	-	-	-	-
Lignite	**30.64**	**46.45**	**67.90**	**71.21**	**72.74**	**72.81**	**71.74**	**3.53**	**2.09**
Total electricity and heat	27.32	44.45	66.81	68.54	72.65	72.01	70.59	4.14	2.23
Main activity producers	*26.32*	*43.37*	*65.52*	*68.35*	*72.19*	*71.56*	*70.16*	*4.25*	*2.32*
Autoproducers	*0.99*	*1.08*	*1.29*	*0.19*	*0.46*	*0.45*	*0.43*	*0.72*	*-4.28*
Patent fuel/BKB plants	2.85	1.78	0.96	1.00	0.60	0.48	0.54	-3.84	-5.55
Coke ovens/Liquefaction(1)	-	-	-	-	-	-	-	-	-
Gas manufacture	-	-	-	-	-	-	-	-	-
Industry	0.42	0.21	0.17	0.21	0.21	0.20	0.24	-5.76	0.68
Iron and steel	-	-	-	-	-	-	-	-	-
Chemical	*0.03*	*0.03*	-	-	-	-	-	*-1.23*	-
Non-metallic minerals	*0.00*	*0.00*	-	*0.00*	*0.00*	*0.00*	*0.00*	*-3.32*	*0.00*
Paper, pulp and print	*0.30*	-	-	-	-	-	-	-	-
Other industry	*0.10*	*0.18*	*0.17*	*0.21*	*0.21*	*0.20*	*0.24*	*5.52*	*1.31*
Other sectors(2)	0.05	0.06	0.08	0.06	0.03	0.06	0.07	0.90	0.46
Non-energy use	-	-	-	-	-	-	-	-	-

(1) Liquefaction primarily refers to direct distillation processes. Indirect liquefaction may be reported here or under gas manufacture.

(2) Other sectors are Residential, Commercial and public services, Agriculture/Forestry, Fishing and Non-specified other.

* Coal totals may not match the sum of the individual rows due to statistical differences. Please refer to notes and definitions in Part I.

Source: *IEA/OECD Energy Statistics of OECD Countries*

OECD ASIA OCEANIA

7. Fuel prices to end users*

	1978[(1)]	1990	1995	2000	2005	2009	2010	2011	2012
				(US dollars / tce)					
For electricity generation									
Steam coal	..	..	..	..	..	..	..	..	..
Heavy fuel oil	..	..	..	..	..	..	..	..	..
Natural gas	..	..	..	..	..	..	..	..	..
For industry									
Steam coal	..	..	..	..	..	..	..	..	..
Coking coal	..	..	..	..	..	..	..	..	..
High sulphur fuel oil	..	..	..	..	..	..	..	..	..
Low sulphur fuel oil	..	..	..	..	..	..	..	..	..
Natural gas	..	..	..	..	..	..	..	..	..
				(US dollars / unit)[(2)]					
For electricity generation									
Steam coal	..	..	..	..	..	..	..	..	..
Heavy fuel oil	..	..	..	..	..	..	..	..	..
Natural gas	..	..	..	..	..	..	..	..	..
For industry									
Steam coal	..	..	..	..	..	..	..	..	..
Coking coal	..	..	..	..	..	..	..	..	..
High sulphur fuel oil	..	..	..	..	..	..	..	..	..
Low sulphur fuel oil	..	..	..	..	..	..	..	..	..
Natural gas	..	..	..	..	..	..	..	..	..

(1) Earliest year for which split by coal type is available.

(2) For steam coal, coking coal and heavy fuel oil per metric tonne; for natural gas per 10^7 kilocalories GCV.

* Please refer to notes and definitions in Part I. Steam coal prices relate to historical definitions.

Source: *IEA/OECD Energy Prices & Taxes*

8. Coal and peat trade by type of coal*

(Mtce)

	1978	1990	1995	2000	2005	2009	2010	2011	2012e
Total imports	**54.88**	**130.83**	**167.86**	**203.55**	**238.15**	**245.66**	**279.73**	**279.11**	**285.96**
Bituminous coal[(1)]	1.59	50.21	86.52	124.27	158.56	171.82	191.78	191.88	202.19
Coking coal	53.19	79.50	79.14	73.02	73.71	70.15	82.52	82.74	80.24
Sub-bituminous coal	-	0.60	1.59	3.67	2.78	3.00	3.78	3.22	2.14
Lignite	-	-	-	-	-	-	-	-	-
Peat	-	-	-	-	-	-	-	-	-
Coal products[(2)]	0.10	0.51	0.61	2.60	3.10	0.69	1.65	1.27	1.39
Total exports	**35.25**	**98.34**	**131.47**	**177.72**	**219.77**	**244.68**	**274.75**	**266.15**	**281.80**
Bituminous coal[(1)]	4.25	39.96	54.64	77.02	94.23	119.71	118.82	126.41	139.59
Coking coal	29.83	55.92	72.93	98.02	123.85	123.85	155.26	138.74	140.69
Sub-bituminous coal	-	-	-	-	-	0.05	-	0.01	-
Lignite	-	-	-	-	-	-	-	-	-
Peat	-	-	-	-	-	-	-	-	-
Coal products[(2)]	1.17	2.46	3.91	2.68	1.68	1.07	0.67	1.00	1.52

(1) Bituminous coal includes anthracite.

(2) Coal products includes products derived from coal, for example: coke, coal tar, briquettes, patent fuels.

* Please refer to notes and definitions in Part I.

Source: *IEA/OECD Energy Balances of OECD Countries, IEA/OECD Energy Statistics of OECD Countries*

OECD ASIA OCEANIA

9. Total coal imports by origin*

(thousand tonnes)

	1978(1)	1985	1990	1995	2000	2005	2010	2011	2012e
Total coal(2)	**55041**	**113643**	**134646**	**180275**	**225168**	**267502**	**316546**	**315938**	**323975**
Coking coal	**52885**	**77367**	**76223**	**76950**	**76656**	**77155**	**85840**	**86054**	**83456**
Australia	25299	33741	39558	39880	45398	40884	48994	45837	44454
Canada	11304	19321	13316	13007	9389	8923	11261	11891	11532
Czech Republic	-	-	-	-	-	-	-	-	-
Germany	398	-	-	-	-	-	-	-	-
Poland	429	-	-	-	-	-	-	-	-
United Kingdom	-	-	-	-	-	-	-	-	-
United States	10406	14729	10275	7541	2841	1838	5859	9802	9505
Other OECD	11	401	225	298	367	530	384	364	354
China, People's Rep.	420	1211	2802	5327	9555	7031	2615	2437	2364
Colombia	-	-	153	52	-	-	24	24	23
Indonesia	-	103	668	4629	5886	15374	13157	11497	11151
South Africa	2360	5086	3055	3437	603	-	-	-	-
Former Soviet Union(3)	2244	2775	x	x	x	x	x	x	x
Russian Federation (4)	*x*	*x*	*6150*	*2741*	*2498*	*2534*	*3499*	*4184*	*4057*
Other FSU (4)	*x*	*x*	-	-	-	-	-	-	-
Venezuela	-	-	-	-	-	-	-	-	-
Vietnam	-	-	-	-	27	-	-	-	-
Non-specified/other	14	-	21	38	92	41	47	18	1
Steam coal	**2156**	**36276**	**58423**	**103325**	**148512**	**190347**	**230706**	**229884**	**240519**
Australia	668	19238	24852	43431	67182	96282	111677	107820	114585
Canada	105	2249	8170	11604	9944	1194	7610	10521	10706
Czech Republic	-	-	-	-	-	-	-	-	-
Germany	-	-	-	1	-	-	-	-	-
Poland	-	-	-	-	-	-	-	-	-
United Kingdom	-	-	-	-	-	-	-	-	-
United States	1	1980	5751	7780	2396	365	492	1521	1615
Other OECD	-	14	138	319	637	56	-	-	-
China, People's Rep.	513	2427	3504	13065	33751	38226	11592	7134	7008
Colombia	-	-	94	56	-	-	4372	6061	6195
Indonesia	-	306	806	8323	14185	26174	61492	62376	63051
South Africa	157	5508	7057	6493	3448	97	5836	7061	8510
Former Soviet Union(3)	149	1020	x	x	x	x	x	x	x
Russian Federation (4)	*x*	*x*	*3402*	*3189*	*5497*	*12369*	*19002*	*22095*	*24236*
Other FSU (4)	*x*	*x*	-	-	-	*21*	-	-	-
Venezuela	-	-	-	-	-	15	-	-	-
Vietnam	-	157	150	1496	1079	2401	3454	2815	2678
Non-specified/other	563	3377	4499	7568	10393	13147	5179	2480	1935
Lignite	**-**	**-**	**-**	**-**	**-**	**-**	**-**	**-**	**-**

(1) Earliest year for which split by coal type is available.

(2) Total coal consists of steam coal, coking coal and lignite.

(3) For years prior to 1990, or if no details for individual Former Soviet Union republics.

(4) Data for former republics are not available separately for some OECD member countries.

* Please refer to notes and definitions in Part I.

Source: IEA/OECD Coal Statistics

OECD ASIA OCEANIA

10. Coking coal exports by destination*

(thousand tonnes)

	1978(1)	1985	1990	1995	2000	2005	2010	2011	2012e
World	**31085**	**49698**	**58086**	**75625**	**100712**	**127246**	**159566**	**142568**	**144573**
Total OECD	**29188**	**42067**	**44287**	**54979**	**74911**	**85006**	**82157**	**77956**	**77181**
Australia	-	-	-	123	122	49	-	-	-
Austria	-	-	-	-	-	-	-	-	-
Belgium	374	16	1054	1524	1833	1881	1034	1314	1276
Canada	-	-	30	-	-	-	-	-	-
Chile	32	64	200	466	776	822	446	413	543
Czech Republic	-	-	-	-	-	-	-	-	-
Denmark	-	-	-	-	-	-	-	-	-
Estonia	-	-	-	-	-	-	-	-	-
Finland	-	-	-	-	-	392	76	68	314
France	1173	2106	1917	2310	3739	3895	2506	2560	2453
Germany	-	-	25	204	2619	1746	1160	1275	908
Greece	191	-	-	-	-	-	-	-	-
Hungary	-	-	-	-	-	-	-	-	-
Iceland	-	-	-	-	-	-	-	-	-
Ireland	-	-	-	-	-	14	-	-	-
Israel	-	-	-	-	-	129	55	-	-
Italy	1319	2055	1181	2752	2914	2527	1559	1621	1751
Japan	22651	29559	28822	31291	39861	45780	48934	43038	40668
Korea	1330	3064	4944	9039	10358	12457	15864	16443	16398
Luxembourg	-	-	-	-	-	-	-	-	-
Mexico	10	-	-	-	666	588	336	254	423
Netherlands	992	1380	725	703	2279	5652	3342	3837	5029
New Zealand	-	-	-	-	-	-	-	-	-
Norway	-	-	-	-	-	-	-	-	-
Poland	-	-	-	-	-	-	-	-	40
Portugal	-	-	-	129	-	-	-	-	-
Slovak Republic	-	-	-	-	-	-	-	-	-
Slovenia	-	-	-	-	-	-	-	-	72
Spain	593	1186	694	896	1767	2470	1069	1721	1486
Sweden	-	-	702	535	992	1226	1486	1223	980
Switzerland	-	-	-	-	46	-	-	-	-
Turkey	-	516	912	1608	1451	720	909	736	1514
United Kingdom	469	2121	3081	3360	5488	4451	3381	3453	3326
United States	54	-	-	39	-	207	-	-	-
Total non-OECD	**1897**	**7631**	**13799**	**20415**	**25569**	**38109**	**72894**	**61243**	**65438**
Brazil	164	958	1291	3448	4988	3218	4234	2882	2338
China(2)	-	192	560	317	265	4330	27500	16040	21638
Chinese Taipei	981	1990	2798	3851	6273	7091	5357	8050	8585
Egypt	-	312	323	459	-	353	324	-	-
India	32	2042	4895	9831	10773	18039	32450	32057	30425
Romania	675	595	2256	710	-	46	-	-	-
Oth. Africa & Mid. East	-	688	454	597	1554	3108	1610	1180	1322
Oth. non-OECD Americas	-	35	300	384	595	784	603	608	570
Other Asia & Oceania	45	688	734	818	1051	737	671	356	560
Other non-OECD Europe and Eurasia	-	131	188	-	70	403	145	70	-
Non-specified/Other	**-**	**-**	**-**	**231**	**232**	**4131**	**4515**	**3369**	**1954**

(1) Earliest year for which split by coal type is available.

(2) China includes the People's Republic of China and Hong Kong, China.

* Please refer to notes and definitions in Part I.

Source: IEA/OECD Coal Statistics

OECD ASIA OCEANIA

11. Steam coal exports by destination*

(thousand tonnes)

	1978(1)	1985	1990	1995	2000	2005	2010	2011	2012e
World	**5533**	**39339**	**45647**	**62413**	**87804**	**107418**	**135479**	**144129**	**159155**
Total OECD	**4809**	**29593**	**38304**	**51291**	**70501**	**85460**	**95692**	**98650**	**103280**
Australia	-	-	-	-	-	-	-	-	-
Austria	-	-	-	-	-	-	-	-	-
Belgium	-	771	129	1116	428	17	-	-	-
Canada	-	-	-	-	-	-	-	-	-
Chile	-	99	120	1023	1301	412	309	691	388
Czech Republic	-	-	-	-	-	-	-	-	-
Denmark	150	2557	1149	1076	142	-	-	-	-
Estonia	-	-	-	-	-	-	-	-	-
Finland	-	-	-	-	-	-	-	-	-
France	682	508	934	276	434	469	66	-	-
Germany	458	591	125	61	72	115	-	-	-
Greece	-	125	-	-	110	-	-	-	-
Hungary	-	-	-	-	-	-	-	-	-
Iceland	-	-	-	-	-	-	-	-	-
Ireland	-	-	133	121	284	568	-	-	-
Israel	-	577	528	695	2623	1165	516	501	494
Italy	-	226	-	366	428	141	-	-	-
Japan	1489	15326	26569	34556	47449	57574	66413	66961	69817
Korea	-	5453	3633	9662	11455	17970	24840	28273	28925
Luxembourg	-	-	-	-	-	-	-	-	-
Mexico	-	-	-	-	-	4238	3109	1703	3595
Netherlands	320	2464	4236	1562	2550	760	127	-	-
New Zealand	-	-	-	-	16	56	59	58	22
Norway	-	-	-	-	-	-	-	-	-
Poland	-	-	-	-	-	-	-	-	-
Portugal	-	-	-	264	-	-	-	-	-
Slovak Republic	-	-	-	-	-	-	-	-	-
Slovenia	-	-	-	-	-	-	-	-	-
Spain	-	455	205	110	1445	671	-	-	-
Sweden	-	-	155	44	83	164	73	145	39
Switzerland	-	-	29	-	-	-	-	-	-
Turkey	-	-	-	25	55	45	-	-	-
United Kingdom	932	441	328	142	1499	993	13	-	-
United States	778	-	31	192	127	102	167	318	-
Total non-OECD	**204**	**8082**	**7343**	**11122**	**17303**	**21958**	**39668**	**45432**	**55875**
Brazil	-	-	158	-	-	33	20	41	-
China(2)	-	2189	2443	3420	1429	2121	15192	17867	29968
Chinese Taipei	76	3761	3046	5920	10034	14329	19553	20124	17516
Egypt	-	-	-	-	-	-	-	-	-
India	-	93	47	-	2469	1461	610	495	1165
Romania	-	-	33	110	-	-	-	-	-
Oth. Africa & Mid. East	-	-	-	-	-	-	-	2	-
Oth. non-OECD Americas	-	-	-	-	72	-	-	-	-
Other Asia & Oceania	128	2039	1616	1672	3299	4014	4293	6903	7226
Other non-OECD Europe and Eurasia	-	-	-	-	-	-	-	-	-
Non-specified/Other	**520**	**1664**	**-**	**-**	**-**	**-**	**119**	**47**	**-**

(1) Earliest year for which split by coal type is available.

(2) China includes the People's Republic of China and Hong Kong, China.

* Please refer to notes and definitions in Part I. Steam coal includes all sub-bituminous coal.

Source: IEA/OECD Coal Statistics

OECD EUROPE*

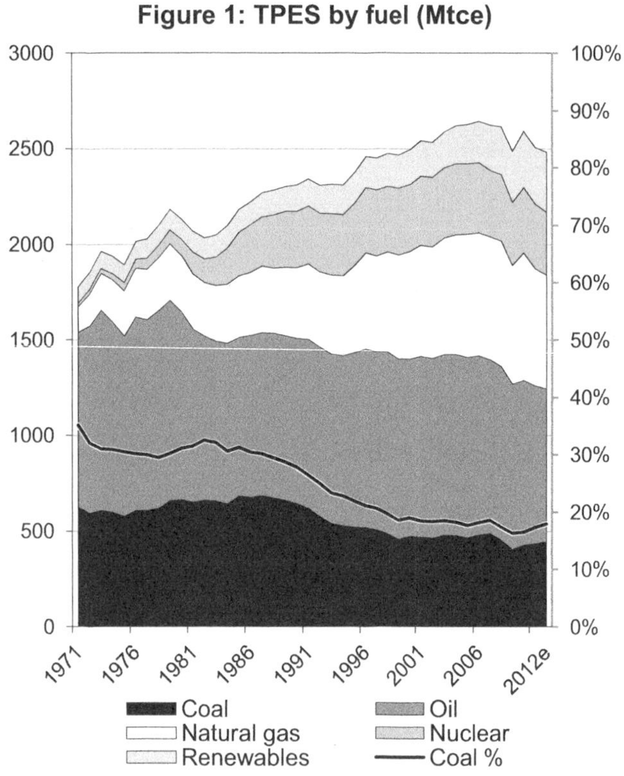

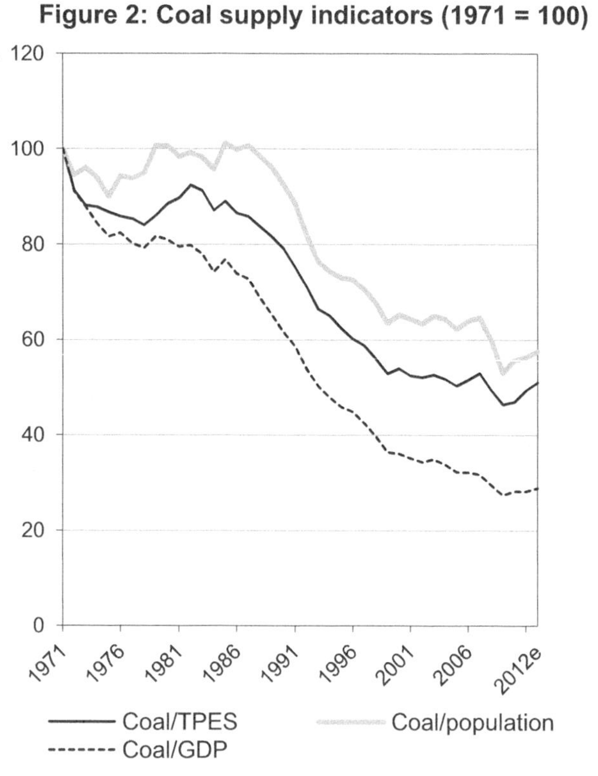

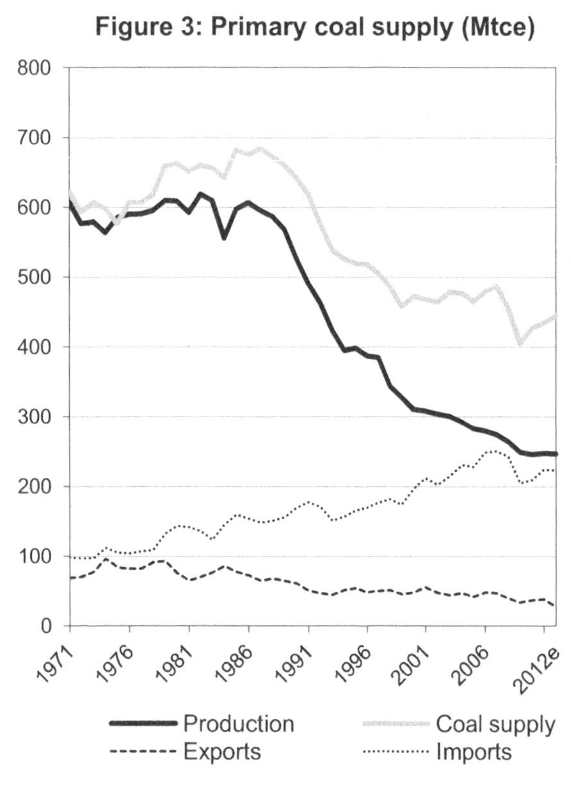

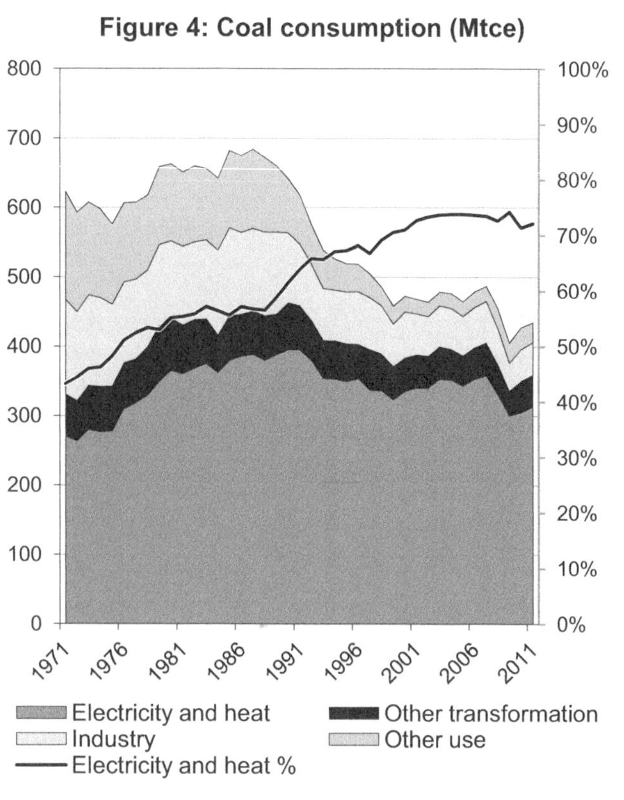

*Please refer to notes and definitions in Part I. Peat is included in coal for all figures.

Source: *IEA/OECD Coal Information, IEA/OECD Energy Balances of OECD Countries, OECD Main Economic Indicators*

OECD EUROPE*

Figure 5: Electricity generation by fuel (TWh)

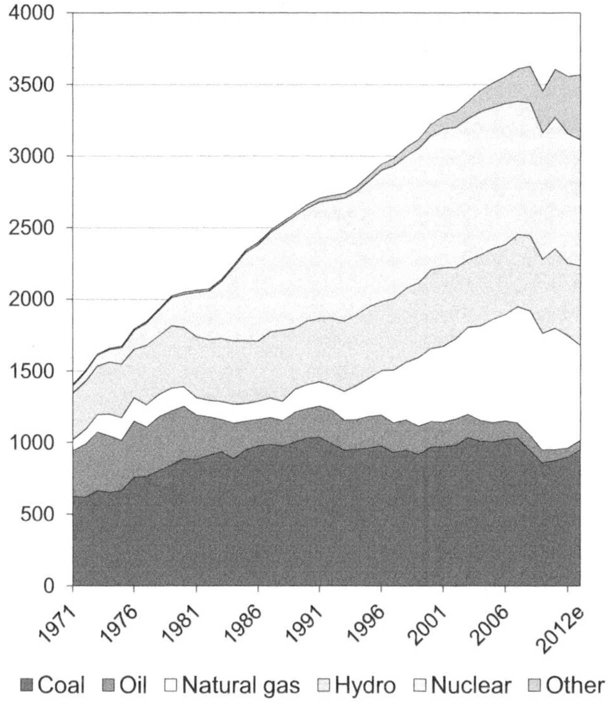

Figure 6: CO_2 emissions by fuel (Mt CO_2)

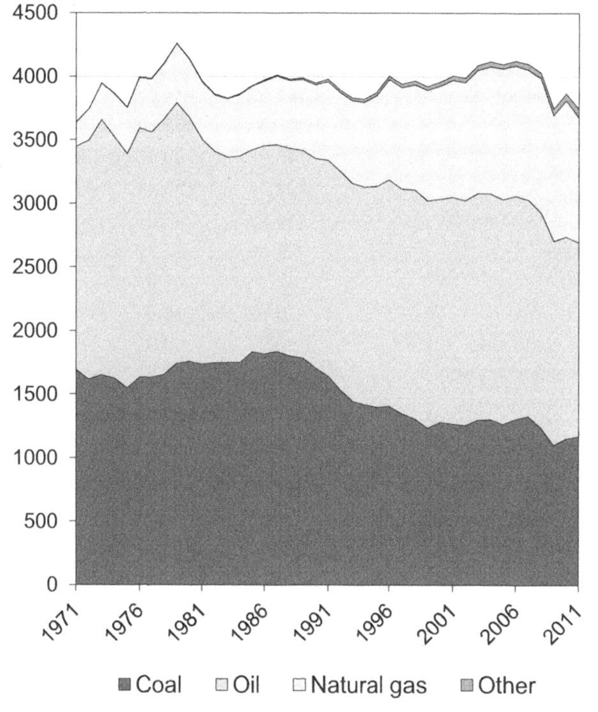

Figure 7: Electricity generation by fuel share

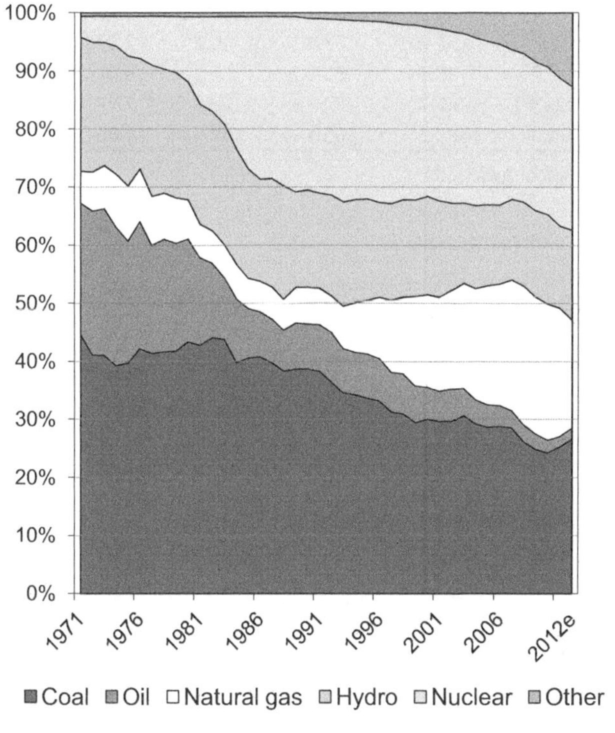

Figure 8: CO_2 emissions by fuel share

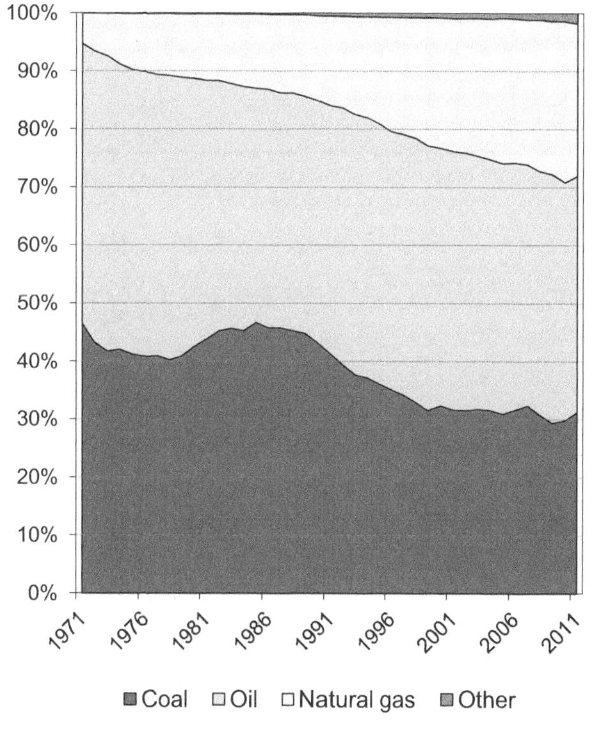

*Please refer to notes and definitions in Part I. Peat is included in coal for all figures.

Source: *IEA/OECD Electricity Information, IEA/OECD CO2 Emissions from Fuel Combustion*

OECD EUROPE

1. Total primary energy supply (TPES) by fuel*

	1973	1980	1990	2000	2010	2011	2012e	Average annual percent change 73-90	90-11
TPES (Mtce)	**1964.92**	**2134.63**	**2313.56**	**2495.41**	**2595.50**	**2508.26**	**2481.96**	**0.97**	**0.39**
Coal and peat	607.73	662.60	641.89	472.42	427.39	434.35	444.70	0.32	-1.84
Oil	1046.31	983.46	865.67	928.68	861.14	827.84	799.88	-1.11	-0.21
Natural Gas	192.47	294.44	371.33	561.48	667.47	611.57	594.90	3.94	2.40
Biofuels and waste	44.27	50.97	77.68	100.89	186.58	184.59	191.36	3.36	4.21
Nuclear	27.69	85.79	293.10	349.79	341.23	337.28	327.77	14.89	0.67
Hydro	42.06	51.07	54.84	67.14	68.21	61.89	68.17	1.57	0.58
Geothermal	3.62	4.30	6.76	10.35	16.30	17.40	18.04	3.74	4.60
Solar, wind, tide	0.07	0.07	0.41	3.80	24.60	31.98	37.11	11.02	23.11
Net electricity trade(1)	0.71	1.92	1.86	0.21	1.80	0.63	-0.73	5.88	-5.02
Heat(2)	-	-	0.03	0.65	0.78	0.72	0.76	-	16.82

(1) Net trade between OECD and non-OECD countries.

(2) Ambient heat from heat pumps used in the transformation sector.

Source: *IEA/OECD Energy Balances of OECD Countries*

2. Energy supply, GDP and population*

	1973	1980	1990	2000	2010	2011	2012e	Average annual percent change 73-90	90-11
GDP (2005 bil. USD)	7079	8381	10679	13412	15509	15795	15775	2.45	1.88
TPES/GDP(1)	0.28	0.25	0.22	0.19	0.17	0.16	0.16	-1.45	-1.47
Population (millions)	455.0	473.8	500.0	521.7	552.4	555.0	556.4	0.56	0.50
TPES/population(1)	4.32	4.51	4.63	4.78	4.70	4.52	4.46	0.41	-0.11
TPES/GDP(2)	155.7	142.9	121.5	104.4	93.9	89.1	88.2	-1.45	-1.47
Coal and peat supply/GDP(2)	272.9	251.4	191.1	112.0	87.6	87.4	89.6	-2.07	-3.66
Elec. consumption/GDP(2)	93.6	100.7	103.1	99.8	98.4	95.0	..	0.57	-0.39
Elec. generation (TWh)	1618	2049	2662	3223	3607	3558	3568	2.97	1.39
Industrial production(2)	57.4	64.6	78.4	94.5	99.6	103.1	101.5	1.85	1.31

(1) TPES/GDP in units of tce/2005 thousand USD, TPES/population in units of tce/capita.

(2) As index, 2005 (not shown) = 100

Source: *IEA/OECD Energy Balances of OECD Countries* , *OECD Main Economic Indicators*

3. Coal and peat production by type*

	1978(1)	1990	2000	2005	2010	2011	2012e	Average annual percent change 78-90	90-11
Mtce:									
Coking coal	136.40	92.69	44.80	37.08	25.90	25.24	24.35	-3.17	-6.01
Steam coal	278.59	224.43	129.41	109.50	84.51	82.43	83.80	-1.79	-4.66
Lignite	177.75	203.31	133.28	131.34	130.88	135.99	136.19	1.13	-1.90
Peat	2.30	5.52	3.32	4.68	4.45	3.92	2.24	7.57	-1.61
Mt:									
Coking coal	141.07	93.45	45.21	37.30	26.19	25.49	24.57	-3.37	-6.00
Steam coal	347.05	283.08	163.72	139.62	107.72	104.73	106.33	-1.68	-4.62
Lignite	585.45	672.62	446.19	451.79	429.76	452.36	455.64	1.16	-1.87
Peat	8.05	16.41	10.32	13.95	13.57	11.70	6.52	6.11	-1.60

(1) Earliest year for which split by coal type is available.

* Please refer to notes and definitions in Part I.

Source: *IEA/OECD Energy Balances of OECD Countries, IEA/OECD Energy Statistics of OECD Countries*

OECD EUROPE

4. Final consumption of energy by fuel*

(Mtce)

	1973	1980	1990	2000	2009	2010	2011	Average annual percent change 73-90	90-11
Total final consumption(1)	**1460.02**	**1543.73**	**1604.85**	**1751.38**	**1737.35**	**1821.60**	**1746.29**	**0.56**	**0.40**
Coal and peat	264.40	222.69	179.34	88.89	69.10	77.28	75.24	-2.26	-4.05
Oil	824.60	787.76	740.06	812.06	765.83	765.79	739.47	-0.63	-0.00
Natural Gas	135.85	229.74	287.63	383.24	367.46	406.22	371.52	4.51	1.23
Biofuels and wastes	40.36	43.56	65.01	74.25	101.16	112.51	112.30	2.84	2.64
Geothermal	0.51	0.73	2.00	2.61	3.92	4.21	4.46	8.40	3.90
Solar, wind, tide	-	0.01	0.24	0.94	2.22	2.50	2.98	-	12.77
Electricity	164.90	209.93	273.98	332.97	362.19	379.81	373.35	3.03	1.48
Heat	29.41	49.31	56.61	56.43	65.46	73.27	66.98	3.93	0.80
of which:									
Total industry	**532.35**	**509.18**	**460.52**	**463.29**	**389.32**	**419.25**	**419.58**	**-0.85**	**-0.44**
Coal and peat	130.29	112.14	101.27	66.33	40.40	45.94	47.13	-1.47	-3.58
Oil	214.72	167.06	82.62	76.61	52.57	51.16	47.49	-5.46	-2.60
Natural Gas	74.30	95.66	110.93	140.84	112.30	124.43	124.93	2.39	0.57
Biofuels and wastes	7.03	8.39	20.68	24.75	30.55	33.92	35.08	6.55	2.55
Geothermal	0.04	0.04	0.07	0.09	0.04	0.05	0.05	3.16	-1.69
Solar, wind, tide	-	-	0.01	0.14	0.00	0.01	0.02	-	1.48
Electricity	90.59	105.28	124.65	140.77	131.92	141.01	142.49	1.90	0.64
Heat	15.38	20.62	20.29	13.75	21.53	22.74	22.39	1.64	0.47
Total transport	**248.48**	**297.93**	**379.35**	**452.14**	**478.53**	**477.04**	**472.87**	**2.52**	**1.05**
Coal and peat	9.79	3.63	0.31	0.01	0.02	0.02	0.01	-18.44	-13.96
Oil	233.60	287.60	370.95	441.15	449.44	445.47	440.42	2.76	0.82
Natural Gas	0.22	0.81	0.45	0.93	3.90	4.09	4.24	4.25	11.23
Biofuels and wastes	0.00	0.00	0.01	1.01	16.73	18.83	19.61	5.41	44.37
Electricity	4.86	5.89	7.63	9.03	8.44	8.64	8.58	2.69	0.56
Residential	**345.93**	**366.81**	**401.01**	**431.34**	**436.95**	**470.03**	**422.50**	**0.87**	**0.25**
Coal and peat	89.20	71.99	53.02	16.89	21.54	24.47	21.52	-3.01	-4.20
Oil	144.63	116.48	94.19	89.60	67.34	66.80	57.89	-2.49	-2.29
Natural Gas	33.12	71.78	105.76	155.49	159.74	177.17	152.96	7.07	1.77
Biofuels and wastes	26.32	28.45	40.24	44.85	48.81	54.35	50.56	2.53	1.09
Geothermal	0.42	0.64	1.39	1.92	3.14	3.49	3.66	7.34	4.72
Solar, wind, tide	-	0.01	0.18	0.64	1.88	2.08	2.49	-	13.26
Electricity	41.18	58.03	78.45	94.24	108.54	112.15	107.21	3.86	1.50
Heat	11.07	19.42	27.79	27.70	25.97	29.51	26.22	5.56	-0.28
Comm & public services	**136.43**	**149.71**	**157.91**	**179.41**	**219.87**	**230.84**	**212.71**	**0.86**	**1.43**
Coal and peat	22.02	24.44	18.10	2.10	3.68	2.73	2.76	-1.15	-8.57
Oil	82.12	72.12	39.86	33.89	29.65	29.45	26.22	-4.16	-1.97
Natural Gas	5.07	13.02	37.21	50.62	65.79	71.10	61.74	12.44	2.44
Biofuels and waste	1.16	1.50	0.58	1.89	2.76	2.90	3.01	-3.98	8.14
Geothermal	0.02	0.02	0.17	0.39	0.52	0.47	0.49	14.06	5.12
Solar, wind, tide	-	-	0.02	0.07	0.23	0.26	0.30	-	12.93
Electricity	24.88	35.58	56.74	81.93	105.44	110.00	106.91	4.97	3.06
Heat	1.17	3.03	5.23	8.53	11.80	13.93	11.29	9.23	3.73
Non-energy use	**115.96**	**128.83**	**142.67**	**160.34**	**154.48**	**162.78**	**157.04**	**1.23**	**0.46**
Coal and peat	4.42	3.08	2.44	1.57	1.47	1.89	1.90	-3.45	-1.17
Oil	104.62	107.96	119.96	138.52	136.80	142.77	137.34	0.81	0.65
Natural Gas	6.92	17.79	20.28	20.25	16.21	18.12	17.80	6.53	-0.62

(1) Total Final Consumption (TFC) excludes use in transformation processes and energy industry own use.

* Please refer to notes and definitions in Part I. Agriculture, Forestry, Fishing and "Non-specified other" sectors are not reported here individually, but are accounted for in Total final consumption.

Source: *IEA/OECD Energy Balances of OECD Countries*

OECD EUROPE

5. Coal balance*

(Mtce)

	1973	1980	1990	2000	2005	2010	2011	2012e	Average annual percent change 73-90	90-11
Production	579.5	608.9	525.9	310.8	282.6	245.7	247.6	246.6	-0.6	-3.5
Imports	97.9	143.1	169.4	196.1	227.8	209.2	224.1	222.7	3.3	1.3
Exports	-77.5	-75.1	-60.5	-48.0	-41.7	-36.7	-38.4	-27.8	-1.4	-2.1
Stock changes	7.9	-14.3	7.1	13.5	-4.4	9.2	1.0	3.2		
Primary supply	**607.7**	**662.6**	**641.9**	**472.4**	**464.3**	**427.4**	**434.4**	**444.7**	**0.3**	**-1.8**
Statistical differences	**5.1**	**-7.5**	**-11.6**	**0.7**	**-0.5**	**-3.3**	**-5.0**	..		
Total transformation	**-321.5**	**-411.9**	**-436.4**	**-372.8**	**-375.4**	**-336.7**	**-344.5**	..	**1.8**	**-1.1**
Electricity and heat gen.	-279.3	-365.1	-394.7	-335.3	-342.6	-305.1	-313.4	..	2.1	-1.1
Main activity producers (1)	*-258.3*	*-305.8*	*-344.7*	*-315.4*	*-324.0*	*-288.7*	*-297.3*	..	*1.7*	*-0.7*
Autoproducers	*-21.1*	*-59.3*	*-50.1*	*-19.9*	*-18.6*	*-16.4*	*-16.1*	..	*5.2*	*-5.3*
Gas works	11.5	2.5	2.4	0.1	-0.2	-0.3	-0.3	..	-8.9	-
Coal transformation(2)	-53.7	-49.4	-43.8	-37.2	-31.9	-30.2	-29.8	..	-1.2	-1.8
BKB plants	*2.4*	*1.6*	*-1.5*	*-0.2*	*-0.2*	*-0.0*	*-0.0*	..	-	*-15.3*
Blast furnaces	*-38.6*	*-37.7*	*-33.0*	*-30.1*	*-27.6*	*-25.6*	*-24.9*	..	*-0.9*	*-1.3*
Coke ovens	*-18.2*	*-14.0*	*-9.9*	*-7.0*	*-4.3*	*-4.6*	*-4.9*	..	*-3.5*	*-3.3*
Patent fuel plants	*0.8*	*0.7*	*0.7*	*0.2*	*0.1*	*0.0*	*0.0*	..	*-1.0*	*-14.0*
Other transformation(3)	-	-	-0.3	-0.5	-0.7	-1.1	-1.1	..	-	7.0
Energy ind. own use	**-21.7**	**-18.5**	**-13.4**	**-10.3**	**-9.4**	**-8.8**	**-8.3**	..	**-2.8**	**-2.2**
Losses	**-5.2**	**-2.0**	**-1.2**	**-1.1**	**-1.3**	**-1.3**	**-1.3**	..		
Final consumption(4)	**264.4**	**222.7**	**179.3**	**88.9**	**77.7**	**77.3**	**75.2**	..	**-2.3**	**-4.1**
Industry(5)	130.3	112.1	101.3	66.3	56.7	45.9	47.1	..	-1.5	-3.6
Iron and steel	*52.7*	*42.8*	*39.2*	*27.3*	*25.3*	*20.7*	*22.0*	..	*-1.7*	*-2.7*
Chemical	*14.3*	*12.9*	*11.4*	*4.3*	*4.7*	*5.3*	*4.9*	..	*-1.3*	*-3.9*
Non-metallic minerals	*14.3*	*16.3*	*17.9*	*11.3*	*7.9*	*7.0*	*7.9*	..	*1.4*	*-3.8*
Paper, pulp and print	*4.2*	*3.4*	*3.8*	*2.3*	*2.0*	*1.6*	*1.6*	..	*-0.5*	*-4.1*
Other industry (6)	*44.9*	*36.8*	*29.0*	*21.2*	*16.8*	*11.4*	*10.7*	..	*-2.5*	*-4.6*
Transport(7)	9.8	3.6	0.3	0.0	0.0	0.0	0.0	..	-18.4	-14.0
Other	119.9	103.8	75.3	21.0	18.9	29.4	26.2	..	-2.7	-4.9
Comm. and pub. services	*22.0*	*24.4*	*18.1*	*2.1*	*1.8*	*2.7*	*2.8*	..	*-1.1*	*-8.6*
Residential	*89.2*	*72.0*	*53.0*	*16.9*	*15.4*	*24.5*	*21.5*	..	*-3.0*	*-4.2*
Other sectors (8)	*8.7*	*7.4*	*4.2*	*2.0*	*1.7*	*2.2*	*1.9*	..	*-4.2*	*-3.7*
Non-energy use	4.4	3.1	2.4	1.6	2.0	1.9	1.9	..	-3.4	-1.2

(1) Main activity electricity and heat generation includes district heating.

(2) Coal transformation refers to the transformation of coal and peat to secondary and tertiary products (mainly cokes, briquettes, coal tar and off-gases). Gas works and Liquefaction are shown separately to match the *Energy Balances of OECD Countries.*

(3) Other transformation includes Liquefaction and Non-specified transformations.

(4) Final Consumption includes non-energy use and energy use (Industry, Transport and Other).

(5) Please refer to notes and definitions in Part I for detailed explanation.

(6) Other industry includes Non-ferrous metals, Transport equipment, Machinery, Mining and quarrying, Food and tobacco, Wood and wood products, Construction, Textile and leather, and Non-specified industry.

(7) Transport includes Rail and Inland waterways.

(8) Other sectors includes Agriculture/Forestry, Fishing and Non-specified other.

* "Coal" refers to all types of coal, primary (anthracite, coking coal, other bituminous coal, sub-bituminous coal, lignite/brown coal) and derived fuels (including patent fuel, BKB, coke oven coke, gas coke, coal tar, coke oven gas, gas works gas, blast furnace gas and other recovered gases), as well as peat, peat briquettes and oil shale. Quantities have been converted to Mtce using calorific values reported by the respective countries. Please refer to notes and definitions in Part I.

Source: *IEA/OECD Energy Balances of OECD Countries*

OECD EUROPE

6. Use of coal for selected end-uses*

(million tonnes)

	1978	1990	2000	2005	2009	2010	2011	Average annual percent change 78-90	90-11
Steam coal	**366.47**	**366.42**	**283.47**	**292.68**	**247.58**	**260.63**	**265.60**	**-0.00**	**-1.52**
Total electricity and heat	256.93	284.28	229.31	237.91	198.67	204.77	210.35	0.85	-1.42
Main activity producers	*207.55*	*246.53*	*219.42*	*228.86*	*192.18*	*197.49*	*202.95*	*1.44*	*-0.92*
Autoproducers	*49.38*	*37.75*	*9.88*	*9.05*	*6.49*	*7.29*	*7.40*	*-2.21*	*-7.47*
Patent fuel/BKB plants	6.90	2.94	0.84	0.42	0.35	0.35	0.34	-6.86	-9.83
Coke ovens/Liquefaction(1)	2.28	0.09	-	-	-	0.25	0.12	-23.82	1.38
Blast furnace inputs	0.01	3.58	7.32	7.69	4.48	8.25	7.94	72.95	3.87
Gas manufacture	1.63	0.41	-	-	-	-	-	-10.93	-
Industry	40.00	40.60	32.22	29.78	19.45	21.46	21.93	0.12	-2.89
Iron and steel	*4.59*	*4.32*	*4.10*	*5.20*	*2.91*	*3.65*	*4.68*	*-0.52*	*0.39*
Chemical	*3.18*	*5.71*	*3.06*	*3.29*	*3.37*	*4.15*	*3.63*	*5.00*	*-2.13*
Non-metallic minerals	*13.49*	*16.55*	*10.05*	*6.35*	*5.43*	*5.51*	*6.42*	*1.72*	*-4.41*
Paper, pulp and print	*0.84*	*2.21*	*1.85*	*1.71*	*1.33*	*1.27*	*1.18*	*8.39*	*-2.96*
Other industry	*17.90*	*11.82*	*13.16*	*13.23*	*6.41*	*6.87*	*6.03*	*-3.40*	*-3.15*
Other sectors(2)	50.74	31.36	15.53	15.32	22.26	24.47	21.35	-3.93	-1.81
Non-energy use	0.08	0.05	0.12	0.25	0.14	0.22	0.23	-3.32	7.34
Coking coal	**145.14**	**129.44**	**92.80**	**81.79**	**63.66**	**72.25**	**73.65**	**-0.95**	**-2.65**
Total electricity and heat	7.14	18.23	13.37	12.70	7.70	5.14	6.72	8.12	-4.64
Main activity producers	*6.19*	*14.32*	*10.17*	*10.73*	*6.71*	*4.21*	*5.60*	*7.23*	*-4.37*
Autoproducers	*0.95*	*3.91*	*3.20*	*1.97*	*0.98*	*0.93*	*1.12*	*12.52*	*-5.77*
Patent fuel/BKB plants	-	-	-	-	-	-	-	-	-
Coke ovens/Liquefaction(1)	135.41	107.92	73.13	64.90	52.53	63.40	62.64	-1.87	-2.56
Blast furnace inputs	-	1.45	1.78	2.81	2.01	2.30	2.66	-	2.93
Gas manufacture	2.46	0.26	-	-	-	-	-	-17.08	-
Industry	0.14	1.97	2.64	0.79	1.54	1.52	2.16	24.43	0.43
Iron and steel	*0.01*	*1.63*	*0.24*	*0.23*	*0.64*	*0.97*	*1.22*	*57.52*	*-1.38*
Chemical	*0.00*	-	*0.00*	-	*0.02*	-	-	-	-
Non-metallic minerals	*0.00*	-	*0.00*	-	-	-	-	-	-
Paper, pulp and print	-	-	-	-	-	-	-	-	-
Other industry	*0.13*	*0.34*	*2.39*	*0.55*	*0.88*	*0.55*	*0.94*	*8.10*	*4.99*
Other sectors(2)	0.27	0.14	0.10	0.13	0.28	0.18	0.24	-5.56	2.75
Non-energy use	-	-	0.23	0.22	0.14	0.18	0.22	-	-
Lignite	**585.58**	**685.98**	**454.62**	**455.14**	**442.76**	**431.50**	**454.03**	**1.33**	**-1.95**
Total electricity and heat	355.04	486.46	415.27	421.29	404.39	392.20	412.71	2.66	-0.78
Main activity producers	*339.04*	*459.94*	*409.03*	*415.26*	*398.95*	*385.21*	*406.18*	*2.57*	*-0.59*
Autoproducers	*16.00*	*26.53*	*6.24*	*6.03*	*5.44*	*6.98*	*6.53*	*4.30*	*-6.46*
Patent fuel/BKB plants	134.01	108.37	12.15	12.42	12.22	13.34	14.44	-1.75	-9.15
Coke ovens/Liquefaction(1)	-	0.88	1.39	1.98	2.79	3.09	3.08	-	6.16
Gas manufacture	6.40	3.48	1.97	2.32	2.42	2.59	2.63	-4.95	-1.34
Industry	47.82	41.87	13.98	8.85	10.07	11.81	10.76	-1.10	-6.27
Iron and steel	*0.93*	*1.37*	*0.22*	*0.09*	*0.06*	*0.20*	*0.16*	*3.32*	*-9.77*
Chemical	*14.32*	*10.67*	*3.13*	*2.99*	*2.86*	*3.23*	*2.88*	*-2.42*	*-6.05*
Non-metallic minerals	*1.57*	*1.63*	*0.72*	*0.55*	*0.59*	*0.72*	*0.73*	*0.30*	*-3.78*
Paper, pulp and print	*4.04*	*2.01*	*0.47*	*0.43*	*0.42*	*0.28*	*0.40*	*-5.65*	*-7.42*
Other industry	*26.97*	*26.19*	*9.45*	*4.80*	*6.14*	*7.38*	*6.60*	*-0.24*	*-6.35*
Other sectors(2)	32.65	41.19	8.91	7.20	8.63	8.55	9.44	1.96	-6.77
Non-energy use	0.83	0.58	0.17	0.12	0.09	0.09	0.04	-2.94	-11.65

(1) Liquefaction primarily refers to direct distillation processes. Indirect liquefaction may be reported here or under gas manufacture.

(2) Other sectors are Residential, Commercial and public services, Agriculture/Forestry, Fishing and Non-specified other.

* Coal totals may not match the sum of the individual rows due to statistical differences. Please refer to notes and definitions in Part I.

Source: *IEA/OECD Energy Statistics of OECD Countries*

OECD EUROPE

7. Fuel prices to end users*

	1978(1)	1990	1995	2000	2005	2009	2010	2011	2012
			(US dollars / tce)						
For electricity generation									
Steam coal	..	..	..	..	..	..	..	..	..
Heavy fuel oil	..	..	..	..	..	..	..	..	..
Natural gas	..	..	..	..	..	..	..	..	..
For industry									
Steam coal	..	..	..	..	..	..	..	..	..
Coking coal	..	..	..	..	..	..	..	..	..
High sulphur fuel oil	..	..	..	..	..	..	..	..	..
Low sulphur fuel oil	..	..	..	..	..	..	..	..	..
Natural gas	..	..	..	..	..	..	..	..	..
			(US dollars / unit)(2)						
For electricity generation									
Steam coal	..	..	..	..	..	..	..	..	..
Heavy fuel oil	..	..	..	..	..	..	..	..	..
Natural gas	..	..	..	..	..	..	..	..	..
For industry									
Steam coal	..	..	..	..	..	..	..	..	..
Coking coal	..	..	..	..	..	..	..	..	..
High sulphur fuel oil	..	..	..	..	..	..	..	..	..
Low sulphur fuel oil	..	..	..	..	..	..	..	..	..
Natural gas	..	..	..	..	..	..	..	..	..

(1) Earliest year for which split by coal type is available.

(2) For steam coal, coking coal and heavy fuel oil per metric tonne; for natural gas per 10^7 kilocalories GCV.

* Please refer to notes and definitions in Part I. Steam coal prices relate to historical definitions.

Source: *IEA/OECD Energy Prices & Taxes*

8. Coal and peat trade by type of coal*

(Mtce)

	1978	1990	1995	2000	2005	2009	2010	2011	2012e
Total imports	**108.62**	**169.43**	**165.54**	**196.14**	**227.75**	**204.48**	**209.24**	**224.15**	**222.68**
Bituminous coal(1)	46.33	94.09	96.00	122.50	159.69	153.78	143.11	159.84	161.01
Coking coal	40.54	56.93	53.59	56.67	53.92	40.70	52.97	52.66	50.70
Sub-bituminous coal	0.60	0.79	0.70	0.63	0.86	0.89	0.79	0.65	0.70
Lignite	4.69	5.13	3.24	1.89	0.57	0.70	0.58	0.65	0.89
Peat	-	0.10	0.13	0.10	0.14	0.23	0.16	0.20	0.19
Coal products(2)	16.47	12.39	11.87	14.35	12.57	8.18	11.63	10.14	9.19
Total exports	**90.65**	**60.54**	**53.81**	**48.05**	**41.70**	**33.76**	**36.75**	**38.42**	**27.80**
Bituminous coal(1)	30.61	22.45	23.74	28.49	25.58	18.02	19.24	22.19	11.07
Coking coal	35.17	19.56	18.19	9.76	6.47	5.88	5.44	4.70	4.69
Sub-bituminous coal	-	0.04	0.00	0.04	0.03	0.00	0.00	0.00	0.00
Lignite	4.33	4.33	2.96	1.31	0.65	0.69	0.63	0.72	0.97
Peat	0.01	0.12	0.07	0.08	0.06	0.02	0.04	0.01	0.01
Coal products(2)	20.53	14.03	8.86	8.37	8.92	9.15	11.40	10.79	11.06

(1) Bituminous coal includes anthracite.

(2) Coal products includes products derived from coal, for example: coke, coal tar, briquettes, patent fuels.

* Please refer to notes and definitions in Part I.

Source: *IEA/OECD Energy Balances of OECD Countries, IEA/OECD Energy Statistics of OECD Countries*

OECD EUROPE

9. Total coal imports by origin*

(thousand tonnes)

	1978(1)	1985	1990	1995	2000	2005	2010	2011	2012e
Total coal(2)	**106042**	**158186**	**173917**	**169983**	**201950**	**238778**	**218838**	**239667**	**239112**
Coking coal	**40078**	**52224**	**55533**	**52716**	**56057**	**52927**	**52878**	**52433**	**50595**
Australia	4497	9734	10043	13043	21757	20330	17145	15479	15021
Canada	294	1107	2812	4176	7072	6743	3732	4002	3005
Czech Republic	909	695	774	2985	3388	3366	3704	2990	2698
Germany	10550	5219	3141	37	2	289	1	15	46
Poland	6190	5025	2570	6573	3118	3246	1821	2168	2384
United Kingdom	79	87	52	-	-	6	1	-	-
United States	8745	22793	28433	21067	16901	13914	21519	21143	20988
Other OECD	105	285	80	384	24	180	92	136	370
China, People's Rep.	-	-	1	12	2	279	3	-	-
Colombia	-	50	24	85	140	193	766	921	685
Indonesia	-	-	46	214	441	-	-	201	32
South Africa	206	256	158	206	388	295	574	515	452
Former Soviet Union(3)	3005	1441	x	x	x	x	x	x	x
Russian Federation (4)	*x*	*x*	*2419*	*526*	*2327*	*2668*	*2711*	*3684*	*3227*
Other FSU (4)	*x*	*x*	*-*	*-*	*26*	*520*	*80*	*661*	*97*
Venezuela	-	-	-	990	418	872	187	318	413
Vietnam	-	-	-	-	53	-	37	-	-
Non-specified/other	5498	5532	4980	2418	-	26	505	200	1177
Steam coal	**53202**	**92246**	**105808**	**109327**	**141295**	**184485**	**164714**	**185738**	**186593**
Australia	2955	11160	9790	7478	9416	7428	3504	3315	2425
Canada	816	1435	832	307	12	807	859	1257	1084
Czech Republic	243	462	327	1685	2443	1416	2889	3958	2460
Germany	6716	3526	2123	834	470	641	627	561	327
Poland	15647	15429	13084	15832	18894	15494	10510	5746	4864
United Kingdom	2285	2394	2436	713	593	297	307	427	265
United States	499	16541	20504	20460	4724	2331	11555	17613	29584
Other OECD	986	2700	3515	1723	3680	2861	3850	3186	3528
China, People's Rep.	21	236	2783	2307	1887	1895	279	134	59
Colombia	-	1544	9060	11231	22500	26596	37713	51967	50102
Indonesia	-	-	194	3197	8657	14813	10158	10241	9834
South Africa	10814	26339	26099	32967	42839	52875	18829	18184	14580
Former Soviet Union(3)	2863	1895	x	x	x	x	x	x	x
Russian Federation (4)	*x*	*x*	*6464*	*7376*	*15490*	*49208*	*53958*	*55816*	*57983*
Other FSU (4)	*x*	*x*	*-*	*647*	*845*	*2857*	*2160*	*3461*	*2062*
Venezuela	-	-	1475	1831	3081	1142	822	960	703
Vietnam	-	-	-	364	560	213	23	18	-
Non-specified/other	9357	8585	7122	375	5204	3611	6671	8894	6733
Lignite	**12762**	**13716**	**12576**	**7940**	**4598**	**1366**	**1246**	**1496**	**1924**

(1) Earliest year for which split by coal type is available.

(2) Total coal consists of steam coal, coking coal and lignite.

(3) For years prior to 1990, or if no details for individual Former Soviet Union republics.

(4) Data for former republics are not available separately for some OECD member countries.

* Please refer to notes and definitions in Part I.

Source: IEA/OECD Coal Statistics

OECD EUROPE

10. Coking coal exports by destination*

(thousand tonnes)

	1978[1]	1985	1990	1995	2000	2005	2010	2011	2012e
World	**35537**	**25596**	**19781**	**18218**	**9747**	**6663**	**5491**	**4867**	**4851**
Total OECD	**25095**	**15893**	**11025**	**11983**	**8106**	**6172**	**5361**	**4687**	**4471**
Australia	-	-	-	2	-	-	-	-	-
Austria	1283	1582	1351	1897	1844	1514	1312	972	1041
Belgium	2383	1572	858	559	1	20	16	31	132
Canada	-	-	-	-	-	-	-	-	-
Chile	-	-	-	-	-	-	-	-	-
Czech Republic	-	-	-	871	214	523	720	1017	848
Denmark	-	-	-	9	-	-	-	-	-
Estonia	-	-	-	-	-	-	-	-	-
Finland	-	-	203	534	717	13	-	-	20
France	6331	2680	1697	634	108	30	16	2	-
Germany	154	241	144	362	1061	397	152	150	173
Greece	54	-	-	1	-	-	-	-	-
Hungary	-	-	-	1287	1010	449	502	495	391
Iceland	-	-	-	3	-	-	-	-	-
Ireland	-	8	3	3	4	1	-	2	9
Israel	-	-	-	-	-	-	-	-	-
Italy	4005	2439	1017	18	-	295	-	-	-
Japan	804	-	-	-	-	-	-	-	-
Korea	-	-	-	-	-	-	-	-	-
Luxembourg	286	-	-	10	-	-	-	-	-
Mexico	-	-	-	-	-	-	-	-	-
Netherlands	1426	851	573	995	323	48	-	-	-
New Zealand	-	-	-	-	-	-	-	-	-
Norway	125	134	-	156	53	-	-	-	-
Poland	-	-	-	896	538	592	733	686	519
Portugal	122	125	52	117	-	-	-	-	-
Slovak Republic	5126	4183	3681	2657	1570	1909	1832	1332	1335
Slovenia	-	-	-	-	-	-	-	-	-
Spain	1826	637	649	58	65	99	-	-	2
Sweden	387	254	-	310	494	-	-	-	-
Switzerland	18	5	3	3	-	-	-	-	1
Turkey	89	-	-	178	100	212	78	-	-
United Kingdom	573	1182	794	421	4	-	-	-	-
United States	103	-	-	2	-	70	-	-	-
Total non-OECD	**2491**	**3862**	**4676**	**6157**	**1641**	**489**	**130**	**180**	**378**
Brazil	15	861	1249	790	143	-	-	-	-
China[2]	-	-	-	-	-	-	-	-	-
Chinese Taipei	-	-	-	-	-	-	-	-	-
Egypt	-	-	100	48	529	355	-	-	-
India	-	-	284	-	-	-	-	-	-
Romania	-	-	100	217	62	-	-	-	-
Oth. Africa & Mid. East	520	4	-	-	2	-	-	-	-
Oth. non-OECD Americas	5	861	1249	-	2	-	-	-	-
Other Asia & Oceania	-	-	-	-	-	-	1	1	1
Other non-OECD Europe and Eurasia	1951	2136	1694	5102	903	134	129	179	377
Non-specified/Other	**7951**	**5841**	**4080**	**78**	**-**	**2**	**-**	**-**	**2**

(1) Earliest year for which split by coal type is available.
(2) China includes the People's Republic of China and Hong Kong, China.
* Please refer to notes and definitions in Part I.
Source: IEA/OECD Coal Statistics

OECD EUROPE

11. Steam coal exports by destination*

(thousand tonnes)

	1978(1)	1985	1990	1995	2000	2005	2010	2011	2012e
World	**33590**	**31963**	**25277**	**28036**	**32165**	**29472**	**21706**	**25331**	**12216**
Total OECD	**25347**	**23616**	**20776**	**22161**	**31217**	**28935**	**20982**	**17766**	**11849**
Australia	-	1	1	12	-	-	-	23	6
Austria	275	1001	1216	1050	1643	2182	1447	1634	1321
Belgium	2633	1373	1471	1414	610	864	607	258	173
Canada	-	-	-	-	-	19	46	45	2
Chile	-	-	-	-	-	-	-	-	-
Czech Republic	274	1378	2282	1874	864	719	685	844	676
Denmark	4161	3454	1387	3214	2382	1154	800	258	273
Estonia	-	-	-	-	3	-	-	-	-
Finland	4095	2461	2815	2559	1228	963	190	52	129
France	6035	1975	1167	1152	2415	3115	1909	906	434
Germany	3171	4042	4942	5655	15884	13364	8845	7764	3777
Greece	1	15	-	1	2	144	-	66	32
Hungary	-	-	-	374	322	321	158	202	38
Iceland	-	21	18	20	7	44	56	54	6
Ireland	540	1287	793	686	468	596	604	477	394
Israel	-	-	-	-	-	5	-	-	-
Italy	1530	1211	600	307	923	258	1172	846	1540
Japan	-	-	61	-	-	-	-	-	-
Korea	-	-	-	-	-	-	-	-	-
Luxembourg	52	107	3	50	164	38	50	109	-
Mexico	-	-	-	-	-	-	-	-	-
Netherlands	938	1282	1341	617	363	651	370	163	58
New Zealand	-	-	-	-	-	-	-	-	2
Norway	167	286	339	311	603	495	461	423	280
Poland	-	-	1	304	117	127	1386	1371	1015
Portugal	15	179	186	151	5	437	245	190	31
Slovak Republic	237	79	198	404	1217	722	449	602	491
Slovenia	-	-	-	2	12	10	4	-	2
Spain	21	389	275	129	441	58	52	93	351
Sweden	244	1666	864	809	172	208	311	173	240
Switzerland	90	287	52	11	20	8	32	7	30
Turkey	74	-	-	-	-	66	218	17	148
United Kingdom	149	1122	764	1054	1351	2320	885	1164	400
United States	645	-	-	1	1	47	-	25	-
Total non-OECD	**85**	**1636**	**722**	**4980**	**908**	**533**	**316**	**395**	**332**
Brazil	-	13	10	37	-	-	-	44	-
China(2)	-	-	3	-	2	-	-	1	-
Chinese Taipei	-	-	-	-	-	-	-	-	-
Egypt	-	-	1	-	2	4	22	35	22
India	-	-	1	1	-	1	2	2	4
Romania	-	-	16	-	-	-	-	-	26
Oth. Africa & Mid. East	31	77	375	-	3	436	31	75	27
Oth. non-OECD Americas	-	3	-	-	16	-	15	46	46
Other Asia & Oceania	-	188	-	-	4	5	4	8	40
Other non-OECD Europe and Eurasia	54	1355	316	4942	881	87	242	184	167
Non-specified/Other	**8158**	**6711**	**3779**	**895**	**40**	**4**	**408**	**7170**	**35**

(1) Earliest year for which split by coal type is available.

(2) China includes the People's Republic of China and Hong Kong, China.

* Please refer to notes and definitions in Part I. Steam coal includes all sub-bituminous coal.

Source: IEA/OECD Coal Statistics

OECD EUROPE

12. Coal import values by origin*

(average unit value, CIF, USD/tonne)

	1990	1995	2000	2005	2007	2008	2009	2010	2011/12(1)
Coking coal(2)	**65.01**	**58.49**	**47.88**	**109.61**	**126.10**	**197.62**	**187.35**	**194.02**	**..**
Imports from:									
Australia	67.29	57.03	45.45	114.89	127.97	219.74	243.76	213.42	..
Canada	64.53	57.02	45.92	113.81	129.45	218.81	223.39	205.20	..
Czech Republic	67.03	82.00	..	242.07	133.32	235.16	160.08	210.67	..
Poland	63.98	59.26	50.43	138.92	139.47	245.85	137.24	194.36	..
United States	63.07	61.20	52.91	110.91	126.16	174.66	159.91	191.43	..
China	63.56	51.72	43.24	219.14	238.11	276.79	..	253.86	..
Colombia	68.65	59.04	35.62	104.26	92.53	..	86.35	136.20	..
Indonesia	65.10	49.63	34.74	67.23	81.67	141.10	76.85	109.86	..
South Africa	58.81	49.65	39.09	71.77	96.82	141.18	81.99	95.76	..
Former Soviet Union(3)	57.44	56.00	42.02	86.05	86.75	142.20	109.45	128.19	..
Other bituminous coal(4)	**52.64**	**47.51**	**35.22**	**71.27**	**82.28**	**137.61**	**100.24**	**104.08**	**..**
Imports from:									
Australia	52.03	45.06	39.04	106.40	103.93	185.19	149.29	197.74	..
Canada	52.72	44.71	37.33	119.03	112.33	180.46	167.55	214.66	..
Czech Republic	65.54	63.84	40.97	113.59	112.36	158.65	140.77	140.68	..
Poland	61.05	46.78	35.30	78.34	94.13	155.92	114.49	104.05	..
United States	54.69	49.12	41.07	86.75	97.50	138.02	107.28	117.53	..
China	49.36	45.29	31.45	93.41	73.02	161.11	552.50	368.71	..
Colombia	52.60	43.42	34.22	67.98	78.16	..	95.79	95.82	..
Indonesia	56.70	40.67	30.05	61.27	71.95	113.71	93.70	97.55	..
South Africa	45.18	43.76	33.83	67.64	80.25	142.17	95.51	98.45	..
Former Soviet Union(3)	44.69	42.55	33.59	68.46	79.09	132.66	92.18	97.47	..

Note: On occasion, shipment of extremely small quantities of high valued coal results in high import costs.

(1) Import data for steam and coking coals are currently unavailable for 2011 and 2012 due to resource constraints.

(2) Weighted average of individual countries using import volumes as weights. Prices exclude intra-EU trade.

(3) Former Soviet Union until 1991, Russian Federation starting in 1992.

(4) Bituminous steam coal only. Weighted average of trades. Prices exclude intra-EU trade.

* Please refer to notes and definitions in Part I.

Source: IEA/OECD Coal Statistics

IEA TOTAL*

Figure 1: TPES by fuel (Mtce)

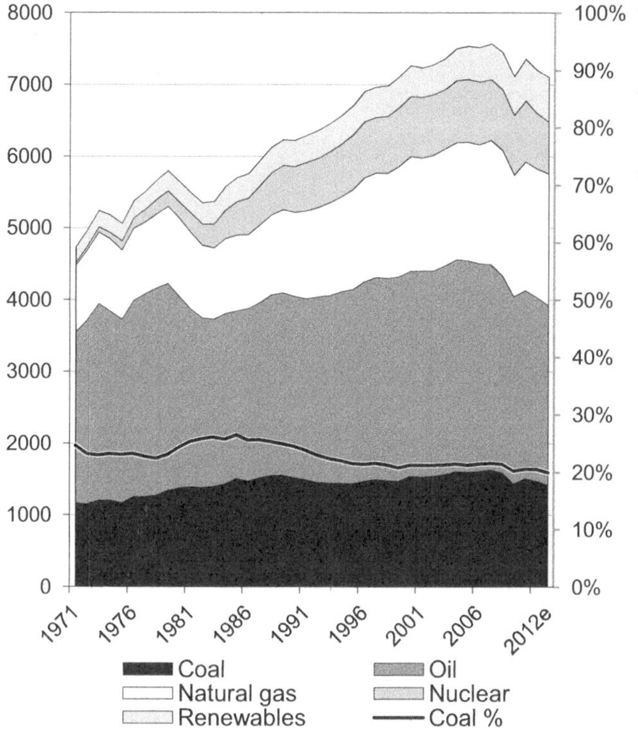

Figure 2: Coal supply indicators (1971 = 100)

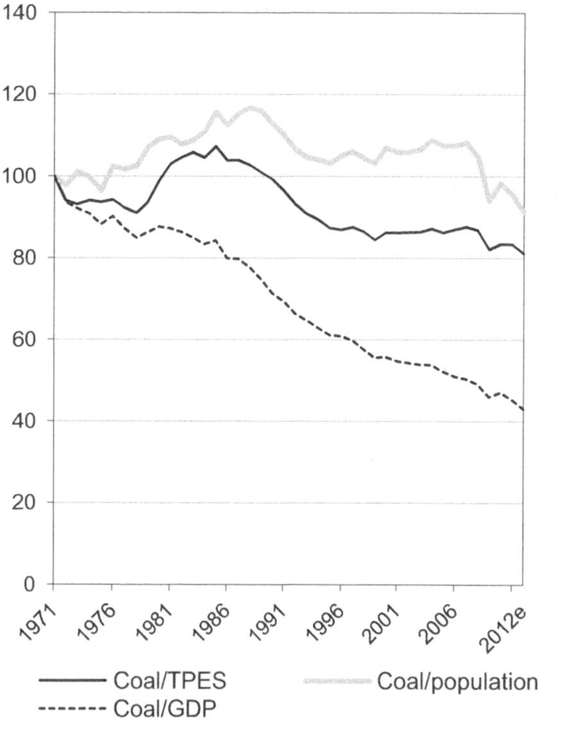

Figure 3: Primary coal supply (Mtce)

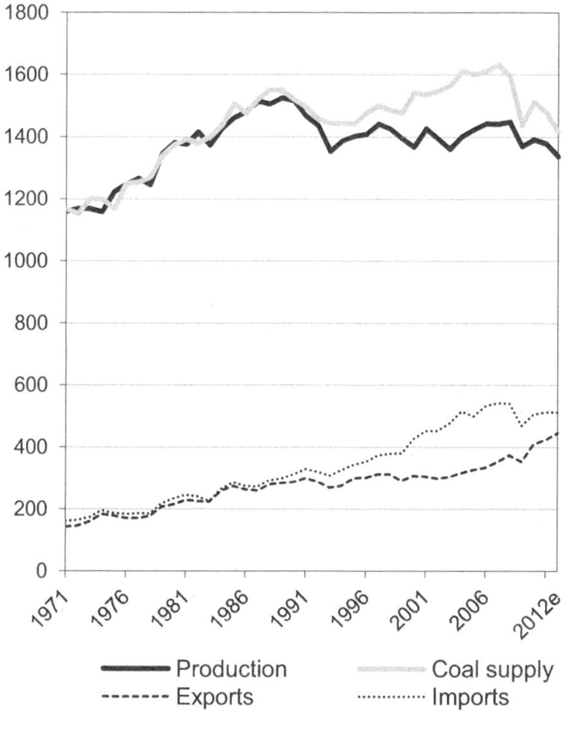

Figure 4: Coal consumption (Mtce)

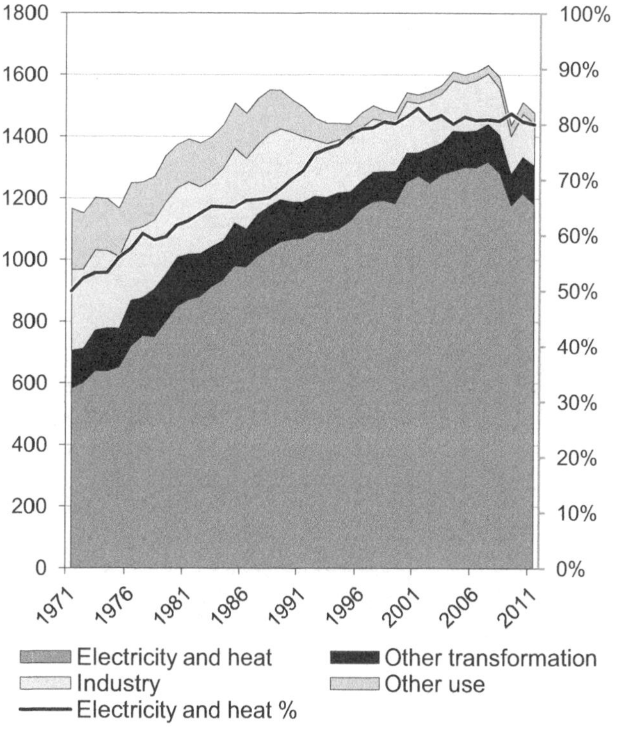

*Please refer to notes and definitions in Part I. Peat is included in coal for all figures.

Source: *IEA/OECD Coal Information, IEA/OECD Energy Balances of OECD Countries, OECD Main Economic Indicators*

IEA TOTAL*

Figure 5: Electricity generation by fuel (TWh)

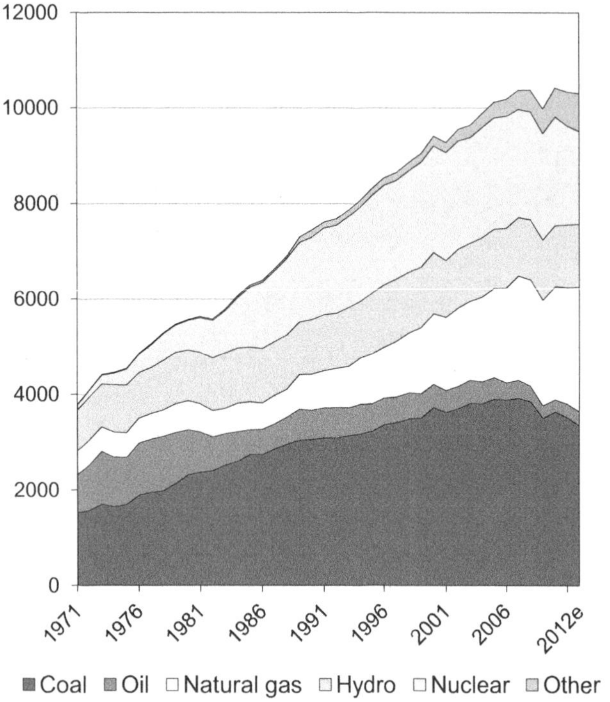

Figure 6: CO_2 emissions by fuel (Mt CO_2)

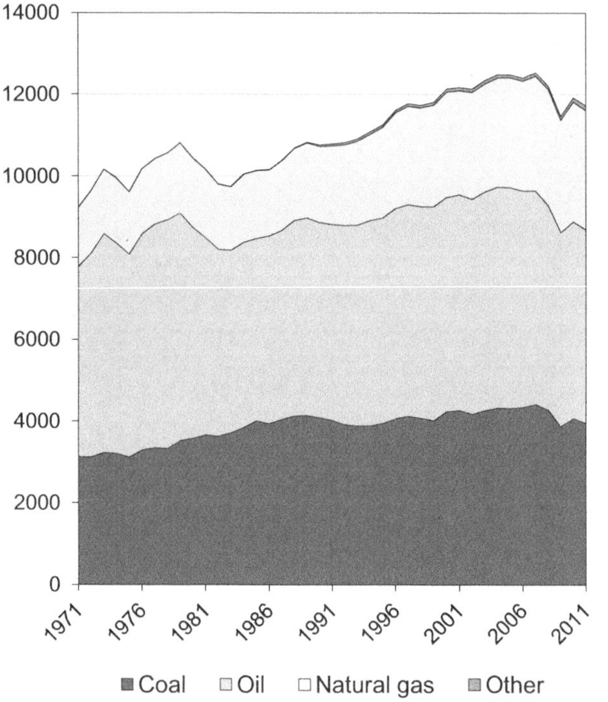

Figure 7: Electricity generation by fuel share

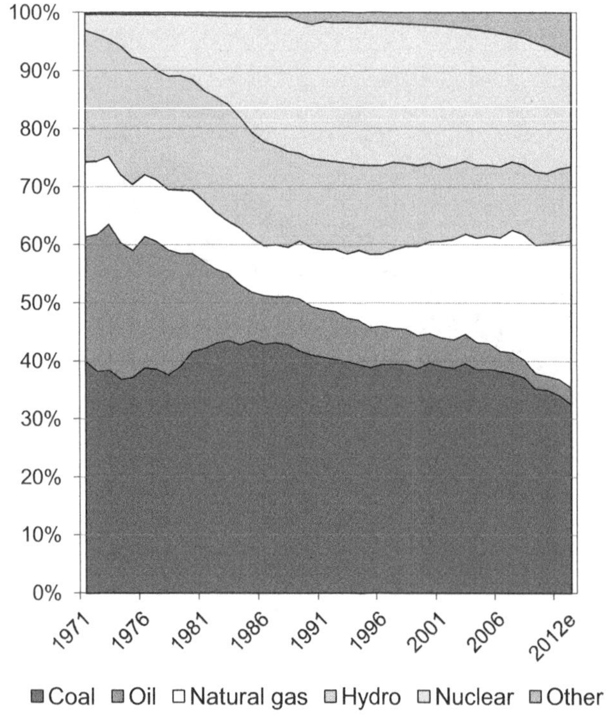

Figure 8: CO_2 emissions by fuel share

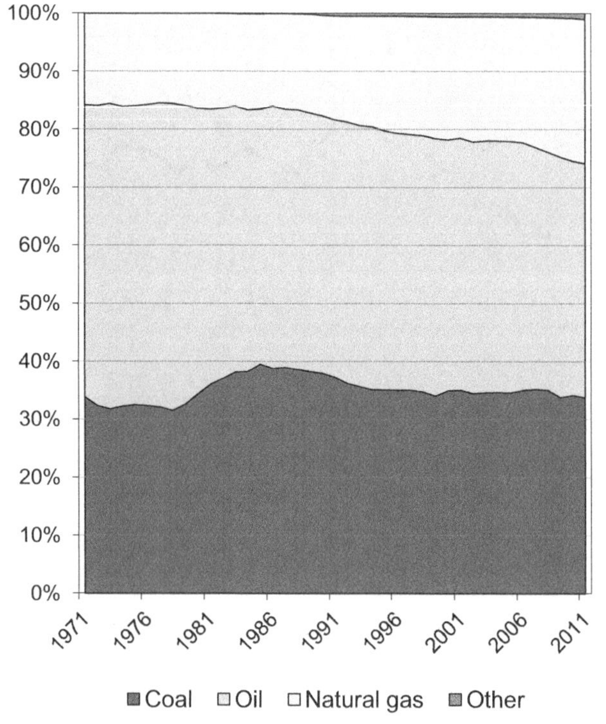

*Please refer to notes and definitions in Part I. Peat is included in coal for all figures.

Source: *IEA/OECD Electricity Information, IEA/OECD CO2 Emissions from Fuel Combustion*

IEA TOTAL

1. Total primary energy supply (TPES) by fuel*

	1973	1980	1990	2000	2010	2011	2012e	Average annual percent change 73-90	90-11
TPES (Mtce)	**5243.71**	**5648.15**	**6224.07**	**7270.90**	**7364.33**	**7204.57**	**7100.48**	**1.01**	**0.70**
Coal and peat	1202.23	1374.28	1520.51	1542.40	1512.03	1477.58	1417.28	1.39	-0.14
Oil	2745.29	2668.16	2527.01	2855.27	2620.25	2558.37	2507.44	-0.49	0.06
Natural Gas	993.22	1083.00	1166.35	1600.50	1792.82	1787.31	1831.54	0.95	2.05
Biofuels and waste	114.04	145.64	192.54	239.95	356.01	358.66	363.31	3.13	3.01
Nuclear	70.31	231.79	641.77	832.57	847.84	771.10	721.19	13.89	0.88
Hydro	109.83	130.81	140.31	157.27	156.74	161.52	161.79	1.45	0.67
Geothermal	7.98	12.52	29.78	33.75	32.22	33.28	34.76	8.05	0.53
Solar, wind, tide	0.07	0.10	2.67	8.22	42.98	54.55	63.74	24.02	15.46
Net electricity trade(1)	0.74	1.86	3.05	0.27	2.49	1.31	-1.51	8.72	-3.93
Heat(2)	-	-	0.08	0.70	0.96	0.90	0.94	-	12.06

(1) Net trade between OECD and non-OECD countries.

(2) Ambient heat from heat pumps used in the transformation sector.

Source: *IEA/OECD Energy Balances of OECD Countries*

2. Energy supply, GDP and population*

	1973	1980	1990	2000	2010	2011	2012e	Average annual percent change 73-90	90-11
GDP (2005 bil. USD)	14750	17720	24078	31237	36291	36881	37262	2.92	2.05
TPES/GDP(1)	0.36	0.32	0.26	0.23	0.20	0.20	0.19	-1.86	-1.32
Population (millions)	848.8	898.9	961.1	1028.2	1096.8	1102.6	1107.3	0.73	0.66
TPES/population(1)	6.18	6.28	6.48	7.07	6.71	6.53	6.41	0.28	0.04
TPES/GDP(2)	163.5	146.6	118.9	107.1	93.3	89.8	87.6	-1.86	-1.32
Coal and peat supply/GDP(2)	176.7	168.1	136.9	107.0	90.3	86.8	82.4	-1.49	-2.14
Elec. consumption/GDP(2)	100.0	104.6	103.0	102.5	98.9	96.4	..	0.18	-0.32
Elec. generation (TWh)	4415	5574	7440	9412	10419	10335	10301	3.12	1.58
Industrial production(2)	..	..	..	..	..	..	..	..	..

(1) TPES/GDP in units of tce/2005 thousand USD, TPES/population in units of tce/capita.

(2) As index, 2005 (not shown) = 100

Source: *IEA/OECD Energy Balances of OECD Countries , OECD Main Economic Indicators*

3. Coal and peat production by type*

	1978(1)	1990	2000	2005	2010	2011	2012e	Average annual percent change 78-90	90-11
Mtce:									
Coking coal	288.05	279.15	226.96	232.22	276.43	273.18	273.51	-0.26	-0.10
Steam coal	747.72	981.11	945.79	997.09	922.73	908.62	868.78	2.29	-0.36
Lignite	205.56	251.07	191.09	189.86	186.56	192.25	192.32	1.68	-1.26
Peat	2.30	4.96	3.21	4.56	4.32	3.81	2.18	6.61	-1.24
Mt:									
Coking coal	293.39	279.57	232.72	241.16	288.26	285.43	285.98	-0.40	0.10
Steam coal	904.22	1195.71	1182.10	1260.44	1187.85	1172.46	1119.20	2.36	-0.09
Lignite	652.34	780.02	586.30	590.61	561.02	583.75	587.54	1.50	-1.37
Peat	8.05	14.68	9.96	13.57	13.21	11.38	6.35	5.13	-1.20

(1) Earliest year for which split by coal type is available.

* Please refer to notes and definitions in Part I.

Source: *IEA/OECD Energy Balances of OECD Countries, IEA/OECD Energy Statistics of OECD Countries*

IEA TOTAL

4. Final consumption of energy by fuel*

(Mtce)

	1973	1980	1990	2000	2009	2010	2011	Average annual percent change 73-90	90-11
Total final consumption[1]	**3949.58**	**4089.37**	**4278.74**	**5007.68**	**4862.79**	**5074.38**	**4978.93**	**0.47**	**0.72**
Coal and peat	430.34	366.68	331.15	195.68	159.28	177.85	171.62	-1.53	-3.08
Oil	2220.77	2179.77	2181.05	2526.11	2356.77	2400.00	2351.10	-0.11	0.36
Natural Gas	701.61	779.52	818.32	1041.45	978.06	1046.33	1019.43	0.91	1.05
Biofuels and wastes	109.69	137.48	117.22	174.58	219.41	237.78	240.55	0.39	3.48
Geothermal	0.07	0.10	2.05	3.16	4.30	4.51	4.77	21.83	4.11
Solar, wind, tide	-	0.04	2.04	4.33	5.14	5.46	6.03	-	5.31
Electricity	457.50	574.89	769.45	993.02	1057.57	1113.34	1102.27	3.11	1.73
Heat	29.59	50.88	57.46	69.35	82.26	89.12	83.15	3.98	1.78
of which:									
Total industry	**1343.45**	**1307.22**	**1132.87**	**1247.51**	**1039.64**	**1134.73**	**1132.86**	**-1.00**	**-0.00**
Coal and peat	258.53	226.05	224.51	167.11	123.79	139.96	137.43	-0.83	-2.31
Oil	439.42	386.25	226.55	192.77	148.62	151.34	146.17	-3.82	-2.07
Natural Gas	348.67	343.58	305.98	388.02	321.93	358.74	363.85	-0.77	0.83
Biofuels and wastes	58.14	67.91	48.45	96.69	91.92	98.94	99.18	-1.07	3.47
Geothermal	-	-	0.17	0.36	0.39	0.39	0.40	-	4.25
Solar, wind, tide	-	-	0.01	0.14	0.00	0.01	0.02	-	1.73
Electricity	223.17	261.37	307.52	378.61	318.82	351.22	351.81	1.90	0.64
Heat	15.52	22.05	19.68	23.79	34.18	34.12	34.00	1.41	2.64
Total transport	**971.14**	**1079.79**	**1292.43**	**1569.73**	**1576.95**	**1608.46**	**1594.32**	**1.70**	**1.00**
Coal and peat	10.30	3.64	0.40	0.14	0.27	0.26	0.19	-17.35	-3.58
Oil	929.04	1043.34	1254.41	1521.08	1483.05	1506.08	1487.91	1.78	0.82
Natural Gas	24.28	24.22	26.70	29.84	30.52	33.00	33.62	0.56	1.10
Biofuels and wastes	0.00	0.00	0.01	5.76	50.14	56.04	59.59	5.41	x
Electricity	7.51	8.59	10.91	12.92	12.96	13.08	13.02	2.22	0.85
Residential	**768.16**	**774.51**	**823.43**	**954.67**	**964.04**	**1009.33**	**955.72**	**0.41**	**0.71**
Coal and peat	109.47	91.18	67.45	19.66	22.88	25.75	22.78	-2.81	-5.04
Oil	278.51	206.54	160.42	169.04	126.05	125.37	113.03	-3.19	-1.65
Natural Gas	205.31	247.68	282.34	366.79	367.69	384.77	360.24	1.89	1.17
Biofuels and wastes	31.90	33.61	45.43	64.35	67.88	72.90	70.35	2.10	2.11
Geothermal	0.07	0.10	0.77	1.68	2.78	3.13	3.28	15.03	7.14
Solar, wind, tide	-	0.04	1.91	3.99	4.76	4.99	5.49	-	5.16
Electricity	131.82	175.98	238.26	300.77	344.63	361.25	352.52	3.54	1.88
Heat	11.07	19.38	26.87	28.40	27.38	31.17	28.00	5.35	0.20
Comm & public services	**405.47**	**416.59**	**499.71**	**614.54**	**680.46**	**692.46**	**675.45**	**1.24**	**1.45**
Coal and peat	28.47	28.66	23.05	4.37	6.80	5.73	5.43	-1.23	-6.66
Oil	185.43	149.36	119.53	116.05	87.30	84.25	79.72	-2.55	-1.91
Natural Gas	99.93	111.38	143.35	187.46	212.35	218.65	211.48	2.15	1.87
Biofuels and waste	1.16	1.54	0.51	3.89	6.68	6.84	6.88	-4.77	13.24
Geothermal	-	-	0.27	0.84	0.81	0.73	0.75	-	5.07
Solar, wind, tide	-	-	0.09	0.11	0.27	0.31	0.35	-	6.58
Electricity	89.28	122.44	204.71	291.09	351.65	359.18	356.73	5.00	2.68
Heat	1.21	3.21	8.20	10.73	14.60	16.77	14.11	11.94	2.62
Non-energy use	**311.38**	**340.49**	**399.21**	**510.46**	**485.21**	**507.01**	**499.82**	**1.47**	**1.08**
Coal and peat	4.42	3.27	3.47	2.40	3.53	3.86	3.79	-1.41	0.42
Oil	299.79	315.66	349.55	454.74	446.44	464.14	456.72	0.91	1.28
Natural Gas	7.18	21.56	46.19	53.33	35.24	39.01	39.30	11.58	-0.77

(1) Total Final Consumption (TFC) excludes use in transformation processes and energy industry own use.

* Please refer to notes and definitions in Part I. Agriculture, Forestry, Fishing and "Non-specified other" sectors are not reported here individually, but are accounted for in Total final consumption.

Source: *IEA/OECD Energy Balances of OECD Countries*

IEA TOTAL

5. Coal balance*

(Mtce)

	1973	1980	1990	2000	2005	2010	2011	2012e	Average annual percent change 73-90	90-11
Production	1166.8	1380.7	1516.3	1367.0	1423.7	1390.0	1377.9	1336.8	1.6	-0.5
Imports	173.5	234.4	312.1	425.9	499.0	505.0	512.2	512.6	3.5	2.4
Exports	-158.7	-214.8	-286.6	-308.0	-327.6	-408.5	-423.7	-444.9	3.5	1.9
Stock changes	20.6	-26.0	-21.3	57.4	5.0	25.5	11.3	12.9		
Primary supply	**1202.2**	**1374.3**	**1520.5**	**1542.4**	**1600.2**	**1512.0**	**1477.6**	**1417.3**	**1.4**	**-0.1**
Statistical differences	**21.2**	**-19.8**	**1.2**	**24.2**	**-13.7**	**-10.2**	**-14.5**	..		
Total transformation	**-752.8**	**-957.9**	**-1168.1**	**-1350.6**	**-1383.1**	**-1302.7**	**-1270.4**	..	**2.6**	**0.4**
Electricity and heat gen.	-638.9	-849.0	-1066.6	-1253.6	-1301.1	-1217.4	-1181.9	..	3.1	0.5
Main activity producers (1)	*-617.7*	*-785.4*	*-999.4*	*-1200.6*	*-1255.8*	*-1174.7*	*-1139.7*	..	*2.9*	*0.6*
Autoproducers	*-21.1*	*-63.6*	*-67.3*	*-53.1*	*-45.3*	*-42.7*	*-42.2*	..	*7.0*	*-2.2*
Gas works	15.7	7.8	-0.5	-2.7	-2.6	-2.8	-2.9	..	-	8.5
Coal transformation (2)	-129.7	-116.6	-100.9	-94.3	-79.4	-82.5	-85.6	..	-1.5	-0.8
BKB plants	*2.2*	*1.4*	*-1.5*	*-0.2*	*-0.3*	*-0.0*	*-0.1*	..	-	*-13.5*
Blast furnaces	*-93.5*	*-79.8*	*-78.1*	*-80.2*	*-73.8*	*-72.2*	*-73.8*	..	*-1.1*	*-0.3*
Coke ovens	*-37.6*	*-38.3*	*-17.2*	*-13.1*	*-12.5*	*-10.3*	*-11.8*	..	*-4.5*	*-1.8*
Patent fuel plants	*-0.8*	*0.1*	*-4.1*	*-0.8*	*7.2*	*0.0*	*0.0*	..	*10.5*	-
Other transformation (3)	-	-	-	0.0	0.0	0.0	0.0	..	-	-
Energy ind. own use	**-34.9**	**-27.7**	**-21.2**	**-19.2**	**-17.8**	**-19.9**	**-19.8**	..	**-2.9**	**-0.3**
Losses	**-5.4**	**-2.3**	**-1.3**	**-1.1**	**-1.3**	**-1.3**	**-1.3**	..		
Final consumption (4)	**430.3**	**366.7**	**331.2**	**195.7**	**184.3**	**177.9**	**171.6**	..	**-1.5**	**-3.1**
Industry (5)	258.5	226.1	224.5	167.1	156.1	140.0	137.4	..	-0.8	-2.3
Iron and steel	*131.9*	*103.8*	*79.2*	*59.6*	*55.1*	*52.8*	*54.5*	..	*-3.0*	*-1.8*
Chemical	*26.6*	*24.9*	*28.2*	*18.2*	*16.3*	*16.1*	*15.5*	..	*0.3*	*-2.8*
Non-metallic minerals	*21.0*	*32.5*	*41.5*	*35.6*	*30.5*	*24.8*	*25.7*	..	*4.1*	*-2.3*
Paper, pulp and print	*11.6*	*11.2*	*16.2*	*8.3*	*10.9*	*9.6*	*9.1*	..	*2.0*	*-2.7*
Other industry (6)	*67.3*	*53.6*	*59.4*	*45.4*	*43.3*	*36.6*	*32.6*	..	*-0.7*	*-2.8*
Transport (7)	10.3	3.6	0.4	0.1	0.3	0.3	0.2	..	-17.3	-3.6
Other	157.1	133.7	102.8	26.0	25.0	33.8	30.2	..	-2.5	-5.7
Comm. and pub. services	*28.5*	*28.7*	*23.1*	*4.4*	*5.8*	*5.7*	*5.4*	..	*-1.2*	*-6.7*
Residential	*109.5*	*91.2*	*67.4*	*19.7*	*16.8*	*25.7*	*22.8*	..	*-2.8*	*-5.0*
Other sectors (8)	*19.1*	*13.9*	*12.3*	*2.0*	*2.3*	*2.3*	*2.0*	..	*-2.6*	*-8.3*
Non-energy use	4.4	3.3	3.5	2.4	3.0	3.9	3.8	..	-1.4	0.4

(1) Main activity electricity and heat generation includes district heating.

(2) Coal transformation refers to the transformation of coal and peat to secondary and tertiary products (mainly cokes, briquettes, coal tar and off-gases). Gas works and Liquefaction are shown separately to match the *Energy Balances of OECD Countries.*

(3) Other transformation includes Liquefaction and Non-specified transformations.

(4) Final Consumption includes non-energy use and energy use (Industry, Transport and Other).

(5) Please refer to notes and definitions in Part I for detailed explanation.

(6) Other industry includes Non-ferrous metals, Transport equipment, Machinery, Mining and quarrying, Food and tobacco, Wood and wood products, Construction, Textile and leather, and Non-specified industry.

(7) Transport includes Rail and Inland waterways.

(8) Other sectors includes Agriculture/Forestry, Fishing and Non-specified other.

* "Coal" refers to all types of coal, primary (anthracite, coking coal, other bituminous coal, sub-bituminous coal, lignite/brown coal) and derived fuels (including patent fuel, BKB, coke oven coke, gas coke, coal tar, coke oven gas, gas works gas, blast furnace gas and other recovered gases), as well as peat, peat briquettes and oil shale. Quantities have been converted to Mtce using calorific values reported by the respective countries. Please refer to notes and definitions in Part I.

Source: *IEA/OECD Energy Balances of OECD Countries*

IEA TOTAL

6. Use of coal for selected end-uses*

(million tonnes)

	1978	1990	2000	2005	2009	2010	2011	Average annual percent change 78-90	90-11
Steam coal	**928.15**	**1230.41**	**1402.70**	**1519.73**	**1389.54**	**1441.28**	**1398.42**	**2.38**	**0.61**
Total electricity and heat	713.89	1031.16	1264.87	1248.74	1253.41	1299.13	1253.19	3.11	0.93
Main activity producers	*664.13*	*982.13*	*1231.81*	*1221.91*	*1229.92*	*1273.49*	*1226.35*	*3.31*	*1.06*
Autoproducers	*49.76*	*49.03*	*33.06*	*26.83*	*23.49*	*25.64*	*26.85*	*-0.12*	*-2.83*
Patent fuel/BKB plants	25.54	23.75	3.24	122.22	2.29	2.21	2.16	-0.60	-10.79
Coke ovens/Liquefaction(1)	2.40	6.61	12.13	12.41	10.20	13.62	14.48	8.80	3.80
Blast furnace inputs	0.01	3.76	9.71	9.69	6.06	10.23	10.04	73.63	4.80
Gas manufacture	1.83	0.41	-	-	-	-	-	-11.75	-
Industry	93.42	118.97	102.05	95.60	76.95	82.42	81.05	2.04	-1.81
Iron and steel	*9.78*	*8.00*	*8.89*	*10.23*	*7.34*	*8.82*	*10.72*	*-1.66*	*1.40*
Chemical	*13.86*	*22.59*	*17.49*	*14.74*	*14.23*	*15.52*	*14.97*	*4.15*	*-1.94*
Non-metallic minerals	*27.24*	*41.82*	*36.56*	*31.12*	*24.53*	*25.35*	*26.35*	*3.64*	*-2.18*
Paper, pulp and print	*9.02*	*15.83*	*8.53*	*11.30*	*10.22*	*10.39*	*9.74*	*4.80*	*-2.29*
Other industry	*33.52*	*30.73*	*30.58*	*28.20*	*20.64*	*22.33*	*19.28*	*-0.72*	*-2.20*
Other sectors(2)	71.37	47.71	20.25	20.57	25.82	27.92	24.53	-3.30	-3.12
Non-energy use	0.08	0.33	0.45	0.76	1.36	1.30	1.28	12.80	6.66
Coking coal	**288.75**	**252.58**	**204.52**	**188.96**	**158.37**	**184.22**	**189.42**	**-1.11**	**-1.36**
Total electricity and heat	7.14	18.23	13.37	12.70	7.70	5.14	6.72	8.12	-4.64
Main activity producers	*6.19*	*14.32*	*10.17*	*10.73*	*6.71*	*4.21*	*5.60*	*7.23*	*-4.37*
Autoproducers	*0.95*	*3.91*	*3.20*	*1.97*	*0.98*	*0.93*	*1.12*	*12.52*	*-5.77*
Patent fuel/BKB plants	-	-	-	2.81	-	-	-	-	-
Coke ovens/Liquefaction(1)	271.50	225.18	170.47	155.30	131.80	155.56	154.84	-1.55	-1.77
Blast furnace inputs	-	6.70	15.75	18.72	16.76	21.50	24.57	-	6.38
Gas manufacture	6.91	0.26	-	-	-	-	-	-23.92	-
Industry	0.26	1.97	2.64	0.79	1.86	1.75	2.28	18.42	0.69
Iron and steel	*0.07*	*1.63*	*0.24*	*0.24*	*0.90*	*1.12*	*1.33*	*30.02*	*-0.99*
Chemical	*0.01*	-	*0.00*	-	*0.02*	-	-	-	-
Non-metallic minerals	*0.00*	-	*0.00*	-	*0.00*	*0.00*	-	-	-
Paper, pulp and print	-	-	-	-	-	-	-	-	-
Other industry	*0.18*	*0.34*	*2.39*	*0.55*	*0.94*	*0.63*	*0.95*	*5.51*	*5.08*
Other sectors(2)	0.28	0.14	0.10	0.13	0.28	0.20	0.24	-5.70	2.75
Non-energy use	-	-	0.23	0.22	0.14	0.18	0.22	-	-
Lignite	**651.97**	**788.73**	**589.88**	**593.81**	**572.37**	**560.01**	**586.42**	**1.60**	**-1.40**
Total electricity and heat	415.45	584.10	547.17	554.30	530.42	517.84	540.53	2.88	-0.37
Main activity producers	*398.45*	*555.48*	*538.60*	*546.21*	*525.00*	*508.29*	*533.70*	*2.81*	*-0.19*
Autoproducers	*16.99*	*28.62*	*8.57*	*8.09*	*5.42*	*9.55*	*6.83*	*4.44*	*-6.59*
Patent fuel/BKB plants	136.86	110.15	13.11	13.43	12.82	13.81	14.98	-1.79	-9.06
Coke ovens/Liquefaction(1)	-	-	-	-	-	-	-	-	-
Gas manufacture	6.40	8.47	7.02	6.88	6.94	7.13	7.05	2.36	-0.87
Industry	50.71	41.83	15.51	9.17	10.41	12.13	10.79	-1.59	-6.25
Iron and steel	*0.93*	*1.35*	*0.22*	*0.09*	*0.06*	*0.20*	*0.16*	*3.17*	*-9.70*
Chemical	*14.35*	*10.93*	*3.37*	*3.23*	*3.12*	*3.49*	*2.93*	*-2.24*	*-6.08*
Non-metallic minerals	*1.57*	*1.63*	*0.50*	*0.28*	*0.44*	*0.57*	*0.47*	*0.31*	*-5.79*
Paper, pulp and print	*4.51*	*2.03*	*0.57*	*0.49*	*0.42*	*0.26*	*0.37*	*-6.45*	*-7.79*
Other industry	*29.36*	*25.89*	*10.86*	*5.09*	*6.38*	*7.61*	*6.87*	*-1.04*	*-6.12*
Other sectors(2)	32.90	40.91	9.07	7.31	8.76	8.70	9.61	1.83	-6.67
Non-energy use	0.83	0.67	0.16	0.21	0.03	0.03	0.03	-1.78	-13.35

(1) Liquefaction primarily refers to direct distillation processes. Indirect liquefaction may be reported here or under gas manufacture.

(2) Other sectors are Residential, Commercial and public services, Agriculture/Forestry, Fishing and Non-specified other.

* Coal totals may not match the sum of the individual rows due to statistical differences. Please refer to notes and definitions in Part I.

Source: *IEA/OECD Energy Statistics of OECD Countries*

IEA TOTAL

7. Fuel prices to end users*

	1978[1]	1990	1995	2000	2005	2009	2010	2011	2012
				(US dollars / tce)					
For electricity generation									
Steam coal	..	..	..	..	..	..	..	..	..
Heavy fuel oil	..	..	..	..	..	..	..	..	..
Natural gas	..	..	..	..	..	..	..	..	..
For industry									
Steam coal	..	..	..	..	..	..	..	..	..
Coking coal	..	..	..	..	..	..	..	..	..
High sulphur fuel oil	..	..	..	..	..	..	..	..	..
Low sulphur fuel oil	..	..	..	..	..	..	..	..	..
Natural gas	..	..	..	..	..	..	..	..	..
				(US dollars / unit)[2]					
For electricity generation									
Steam coal	..	..	..	..	..	..	..	..	..
Heavy fuel oil	..	..	..	..	..	..	..	..	..
Natural gas	..	..	..	..	..	..	..	..	..
For industry									
Steam coal	..	..	..	..	..	..	..	..	..
Coking coal	..	..	..	..	..	..	..	..	..
High sulphur fuel oil	..	..	..	..	..	..	..	..	..
Low sulphur fuel oil	..	..	..	..	..	..	..	..	..
Natural gas	..	..	..	..	..	..	..	..	..

(1) Earliest year for which split by coal type is available.

(2) For steam coal, coking coal and heavy fuel oil per metric tonne; for natural gas per 10^7 kilocalories GCV.

* Please refer to notes and definitions in Part I. Steam coal prices relate to historical definitions.

Source: *IEA/OECD Energy Prices & Taxes*

8. Coal and peat trade by type of coal*

(Mtce)

	1978	1990	1995	2000	2005	2009	2010	2011	2012e
Total imports	**184.95**	**312.09**	**344.04**	**425.92**	**498.98**	**468.16**	**504.99**	**512.17**	**512.56**
Bituminous coal[1]	58.79	151.86	187.31	261.21	337.60	336.40	341.84	353.27	360.30
Coking coal	99.04	140.65	136.60	135.44	133.25	113.91	139.81	140.44	136.17
Sub-bituminous coal	0.60	1.26	2.39	6.81	8.58	7.46	7.55	4.58	3.18
Lignite	4.69	4.43	2.80	1.55	0.59	0.69	0.61	0.68	0.89
Peat	-	0.10	0.13	0.10	0.14	0.23	0.16	0.20	0.19
Coal products[2]	21.83	13.80	14.80	20.80	18.81	9.46	15.02	13.00	11.84
Total exports	**176.68**	**286.56**	**300.38**	**307.96**	**327.56**	**352.42**	**408.50**	**423.72**	**444.94**
Bituminous coal[1]	44.50	103.69	113.76	130.89	132.56	156.71	158.04	179.93	193.81
Coking coal	105.43	161.24	167.92	162.61	177.47	179.73	231.86	226.17	230.99
Sub-bituminous coal	-	0.04	2.01	0.77	4.42	3.54	4.40	3.84	5.29
Lignite	4.38	4.32	2.95	1.33	0.82	0.86	0.79	0.86	1.04
Peat	0.01	0.12	0.07	0.06	0.05	0.02	0.01	0.00	0.00
Coal products[2]	22.37	17.15	13.67	12.30	12.24	11.56	13.40	12.93	13.80

(1) Bituminous coal includes anthracite.

(2) Coal products includes products derived from coal, for example: coke, coal tar, briquettes, patent fuels.

* Please refer to notes and definitions in Part I.

Source: *IEA/OECD Energy Balances of OECD Countries, IEA/OECD Energy Statistics of OECD Countries*

IEA TOTAL

9. Total coal imports by origin*

(thousand tonnes)

	1978(1)	1985	1990	1995	2000	2005	2010	2011	2012e
Total coal(2)	**178012**	**285047**	**317988**	**357408**	**449765**	**541030**	**552489**	**563770**	**565722**
Coking coal	**98417**	**135759**	**136246**	**133795**	**138556**	**135913**	**143195**	**143703**	**139448**
Australia	29796	43471	49601	52923	67155	61358	66139	61316	59475
Canada	11598	20428	16128	17183	17972	17124	16378	17251	15412
Czech Republic	909	695	774	2985	3388	3366	3704	2990	2698
Germany	10948	5219	3141	37	2	289	1	15	46
Poland	6619	5025	2570	6573	3118	3246	1821	2168	2384
United Kingdom	79	71	52	-	-	6	1	-	-
United States	24605	43710	43199	32737	24038	19912	30470	34715	34875
Other OECD	116	686	305	682	391	710	476	500	724
China, People's Rep.	420	1211	2803	5339	9557	7310	2618	2437	2364
Colombia	-	50	177	137	140	202	790	945	708
Indonesia	-	103	714	4843	6327	15374	13157	11698	11183
South Africa	2566	5342	3213	3643	991	295	574	515	452
Former Soviet Union(3)	5249	4216	x	x	x	x	x	x	x
Russian Federation (4)	*x*	*x*	*8569*	*3267*	*4825*	*5202*	*6210*	*7868*	*7284*
Other FSU (4)	*x*	*x*	*-*	*-*	*26*	*520*	*80*	*661*	*97*
Venezuela	-	-	-	990	454	872	187	406	553
Vietnam	-	-	-	-	80	-	37	-	-
Non-specified/other	5512	5532	5000	2456	92	127	552	218	1178
Steam coal	**66833**	**135572**	**171669**	**217221**	**307899**	**403802**	**407984**	**418520**	**424355**
Australia	4556	30433	34664	51101	76750	103776	115526	111191	116516
Canada	970	4051	9885	13108	10111	2245	8555	11824	11813
Czech Republic	243	462	327	1685	2435	1394	2880	3929	2406
Germany	6716	3523	2123	835	464	601	611	533	293
Poland	16292	15429	13084	15821	18884	15545	10510	5746	4864
United Kingdom	2285	2394	2428	713	589	319	307	729	265
United States	9296	26870	35842	33298	25603	15542	19274	22386	34957
Other OECD	986	2718	3653	2064	4297	2883	3796	3126	3461
China, People's Rep.	534	2663	6287	15372	35811	40189	11918	7287	7126
Colombia	-	2081	10450	14281	29418	47927	53125	62855	58235
Indonesia	-	306	1000	12137	23066	42725	72944	73038	72979
South Africa	11967	32671	33156	39460	46348	53042	21520	21908	18338
Former Soviet Union(3)	3012	2915	x	x	x	x	x	x	x
Russian Federation (4)	*x*	*x*	*9493*	*10478*	*20915*	*62189*	*70963*	*76273*	*79826*
Other FSU (4)	*x*	*x*	*-*	*647*	*845*	*3019*	*2351*	*3682*	*2449*
Venezuela	-	-	1752	3506	4985	5169	1371	1630	823
Vietnam	-	157	150	1860	1639	2699	3477	2833	2678
Non-specified/other	9976	8899	7375	855	5739	4538	8856	9547	7326
Lignite	**12762**	**13716**	**10073**	**6392**	**3310**	**1315**	**1310**	**1547**	**1919**

(1) Earliest year for which split by coal type is available.

(2) Total coal consists of steam coal, coking coal and lignite.

(3) For years prior to 1990, or if no details for individual Former Soviet Union republics.

(4) Data for former republics are not available separately for some OECD member countries.

* Please refer to notes and definitions in Part I.

Source: IEA/OECD Coal Statistics

IEA TOTAL

10. Coking coal exports by destination*

(thousand tonnes)

	1978(1)	1985	1990	1995	2000	2005	2010	2011	2012e
World	**107072**	**152492**	**162286**	**169662**	**168625**	**186620**	**243491**	**238179**	**243541**
Total OECD	**90313**	**123858**	**122941**	**128766**	**128521**	**132842**	**139897**	**142937**	**138677**
Australia	-	-	-	125	122	49	-	-	-
Austria	1283	1582	1351	1897	1844	1753	1724	1302	2711
Belgium	3860	5031	7450	5890	4551	3429	2787	2716	2800
Canada	5410	6360	4018	4039	3501	4034	3091	3772	4379
Chile	32	345	492	733	1088	1190	661	1024	1182
Czech Republic	-	-	-	871	214	523	720	1017	848
Denmark	-	-	50	9	-	-	-	-	-
Estonia	-	-	-	-	-	-	-	-	-
Finland	-	-	330	1249	1005	1181	754	797	903
France	8972	8609	8494	6177	6629	5535	4480	4359	5100
Germany	714	950	877	999	4891	4373	4040	3990	4421
Greece	245	7	-	1	-	-	-	-	-
Hungary	-	-	-	1287	1075	449	542	521	391
Iceland	-	13	28	39	48	57	59	58	62
Ireland	-	62	3	3	4	15	-	138	9
Israel	-	-	50	127	56	129	55	-	-
Italy	8393	10228	8734	7824	7381	6494	4961	6688	5947
Japan	43380	59198	55410	53979	52798	54226	60321	55960	52729
Korea	2503	7014	7852	15654	15305	17954	23873	27554	25815
Luxembourg	286	-	-	10	-	-	77	-	-
Mexico	10	498	3	517	1406	1582	1338	1264	1181
Netherlands	3347	5206	5273	5766	4745	8055	8978	10989	10739
New Zealand	-	-	-	-	-	-	-	-	-
Norway	193	232	99	239	95	18	75	81	79
Poland	-	-	-	896	538	592	2882	1944	1024
Portugal	387	369	805	434	198	-	-	93	134
Slovak Republic	5126	4183	3681	2657	1570	1909	2092	1802	1797
Slovenia	-	-	-	-	-	163	223	663	364
Spain	3257	4283	4499	3267	4163	4598	2531	2947	2648
Sweden	840	1564	1568	2006	2128	1690	1887	1703	1515
Switzerland	18	5	3	3	46	-	37	-	1
Turkey	498	2232	2869	3681	3954	3599	3901	3833	5337
United Kingdom	1402	5777	8230	7612	8626	7366	6425	6367	5686
United States	157	110	772	775	540	1879	1383	1355	875
Total non-OECD	**8808**	**22793**	**30067**	**40203**	**36647**	**49645**	**99010**	**91873**	**102908**
Brazil	2121	7958	8867	11064	10695	8049	12996	12727	11242
China(2)	-	192	860	317	265	5286	35649	23305	37729
Chinese Taipei	1186	2772	3155	5451	7713	8365	6221	9119	9589
Egypt	218	769	1009	1672	1211	1414	1366	629	375
India	232	2080	5179	9831	10795	19117	34749	35785	35606
Romania	1348	1646	3915	2455	505	593	812	937	607
Oth. Africa & Mid. East	521	1564	1068	947	1825	3485	1840	1863	2201
Oth. non-OECD Americas	919	1484	2129	784	781	991	924	914	1041
Other Asia & Oceania	69	1041	963	1366	1051	841	781	467	688
Other non-OECD Europe and Eurasia	2194	3287	2922	6316	1806	1504	3672	6127	3830
Non-specified/Other	**7951**	**5841**	**9278**	**693**	**3457**	**4133**	**4584**	**3369**	**1956**

(1) Earliest year for which split by coal type is available.

(2) China includes the People's Republic of China and Hong Kong, China.

* Please refer to notes and definitions in Part I.

Source: IEA/OECD Coal Statistics

IEA TOTAL

11. Steam coal exports by destination*

(thousand tonnes)

	1978(1)	1985	1990	1995	2000	2005	2010	2011	2012e
World	**49931**	**105530**	**113417**	**128952**	**146891**	**157330**	**185959**	**209448**	**225970**
Total OECD	**40497**	**83710**	**95379**	**108337**	**127163**	**132691**	**140805**	**149802**	**158921**
Australia	-	2	2	12	-	-	105	325	158
Austria	275	1001	1216	1050	1643	2182	1447	1634	1321
Belgium	2660	2856	3778	3480	1467	1292	974	1728	1186
Canada	8782	8508	10083	4541	13524	13644	7291	2476	2163
Chile	-	99	514	1230	1349	668	1406	2298	2023
Czech Republic	274	1378	2282	1874	864	719	685	844	676
Denmark	4620	8297	5857	6195	2594	1220	873	404	273
Estonia	-	-	-	-	3	-	-	-	-
Finland	4095	3669	2815	3089	1228	963	356	197	129
France	6755	3461	3859	2049	3413	3612	3055	2984	1580
Germany	4157	5228	5451	7364	16478	13612	9780	10158	11179
Greece	1	1278	-	1	112	144	47	66	32
Hungary	-	-	-	374	322	321	158	202	38
Iceland	-	28	33	20	7	44	56	54	6
Ireland	540	1622	2248	1636	1208	1164	604	563	602
Israel	-	998	1058	1257	2623	1170	516	501	511
Italy	1552	5070	5051	4122	1430	409	1785	1850	5858
Japan	1732	18173	30637	40735	51874	58536	68593	69611	72495
Korea	356	7942	4352	13462	13730	18719	29033	35384	34228
Luxembourg	52	107	3	50	164	38	50	109	-
Mexico	-	44	188	-	373	4579	4487	3623	6179
Netherlands	1285	6539	9559	5194	3556	2240	2197	4155	2590
New Zealand	-	-	1	-	16	56	59	58	24
Norway	167	429	401	336	677	495	461	434	297
Poland	-	-	1	304	117	127	1451	1500	1260
Portugal	15	805	1572	1977	348	580	776	988	1023
Slovak Republic	237	79	198	404	1217	722	449	602	491
Slovenia	-	-	-	2	12	10	186	-	199
Spain	21	1630	762	2350	2327	729	426	614	1167
Sweden	244	1896	1040	861	255	443	659	471	417
Switzerland	90	295	81	11	20	8	32	176	30
Turkey	79	275	15	216	110	178	438	439	1663
United Kingdom	1081	1846	2097	3901	3866	3674	2117	4964	9100
United States	1427	155	225	240	236	393	253	390	23
Total non-OECD	**384**	**13445**	**13032**	**19718**	**19134**	**24432**	**44627**	**52429**	**65771**
Brazil	11	91	345	190	22	726	138	395	125
China(2)	-	2695	2554	3420	1440	2121	18130	20050	33156
Chinese Taipei	76	6336	6866	7953	10034	14332	19554	20124	17744
Egypt	-	-	1	1	2	4	168	179	22
India	-	93	48	1	2469	1679	783	1132	2998
Romania	-	-	49	381	-	844	-	-	26
Oth. Africa & Mid. East	32	253	1057	1099	828	499	1075	2369	2940
Oth. non-OECD Americas	82	113	128	-	89	13	130	509	723
Other Asia & Oceania	129	2229	1621	1672	3303	4019	4374	6912	7369
Other non-OECD Europe and Eurasia	54	1635	363	5001	947	195	275	759	668
Non-specified/Other	**9050**	**8375**	**5006**	**897**	**594**	**207**	**527**	**7217**	**1278**

(1) Earliest year for which split by coal type is available.

(2) China includes the People's Republic of China and Hong Kong, China.

* Please refer to notes and definitions in Part I. Steam coal includes all sub-bituminous coal.

Source: IEA/OECD Coal Statistics

IEA AMERICAS*

Figure 1: TPES by fuel (Mtce)

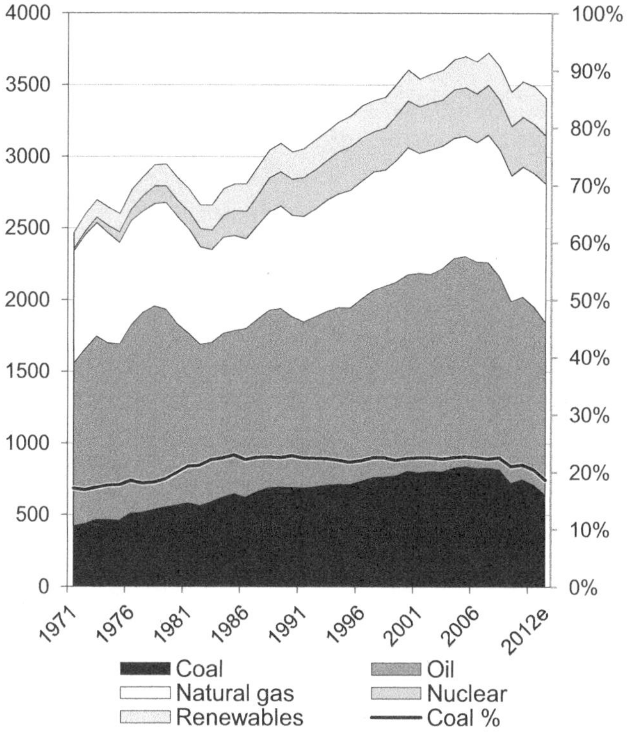

Figure 2: Coal supply indicators (1971 = 100)

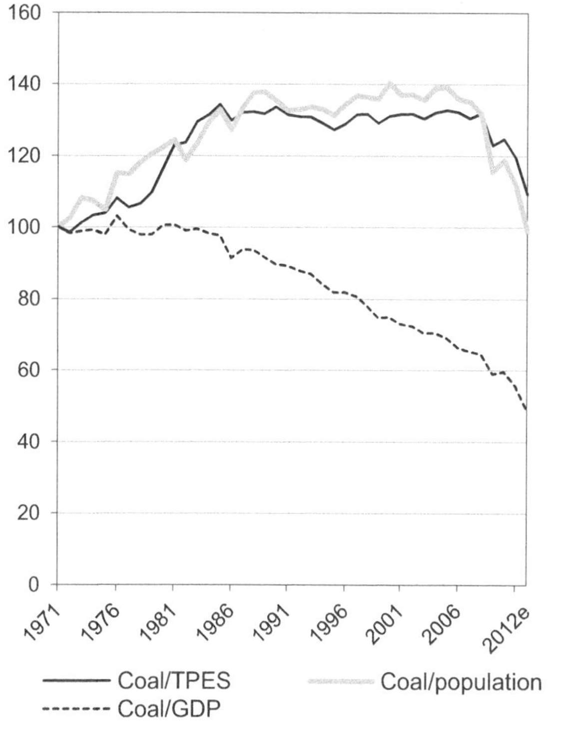

Figure 3: Primary coal supply (Mtce)

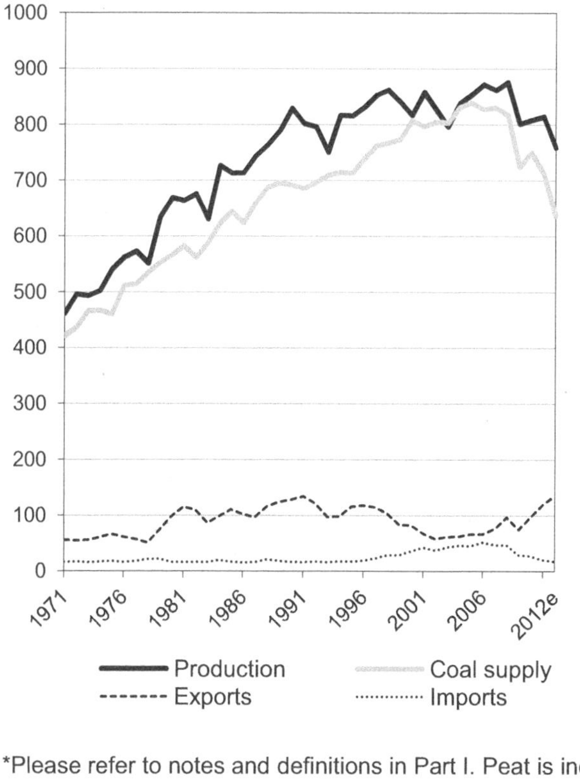

Figure 4: Coal consumption (Mtce)

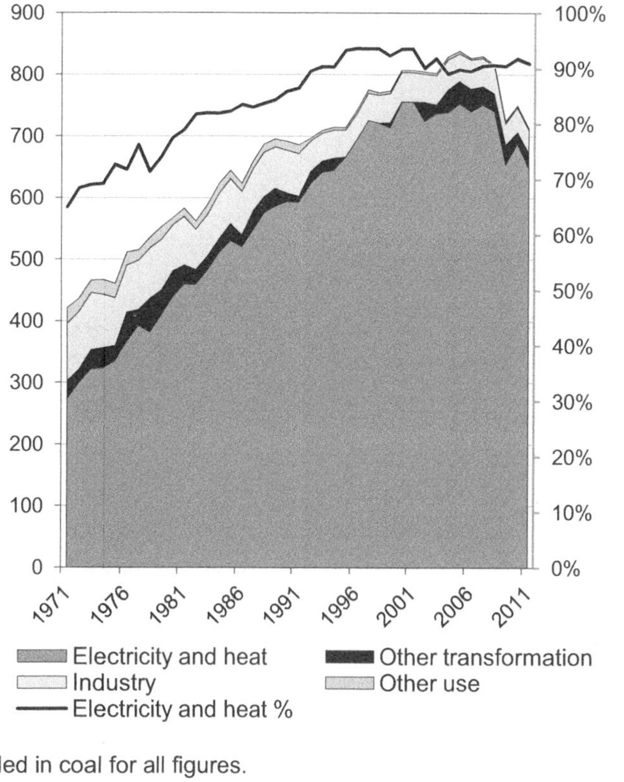

*Please refer to notes and definitions in Part I. Peat is included in coal for all figures.

Source: *IEA/OECD Coal Information, IEA/OECD Energy Balances of OECD Countries, OECD Main Economic Indicators*

IEA AMERICAS*

Figure 5: Electricity generation by fuel (TWh)

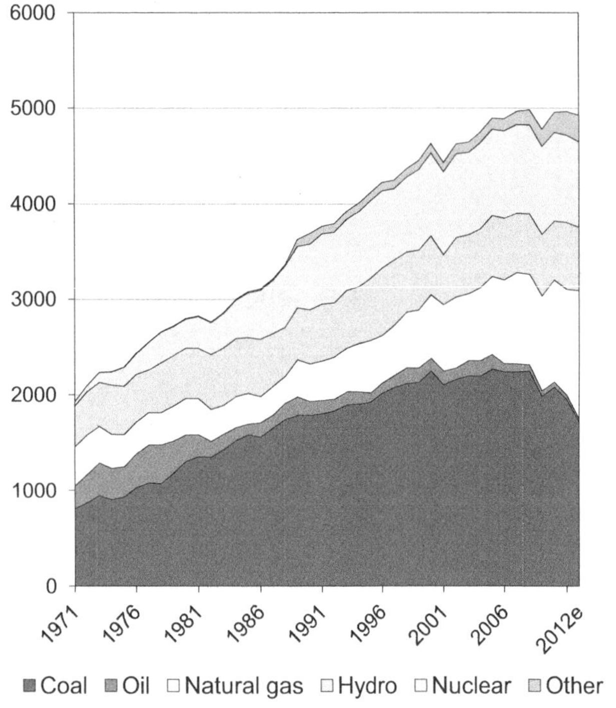

Figure 6: CO_2 emissions by fuel (Mt CO_2)

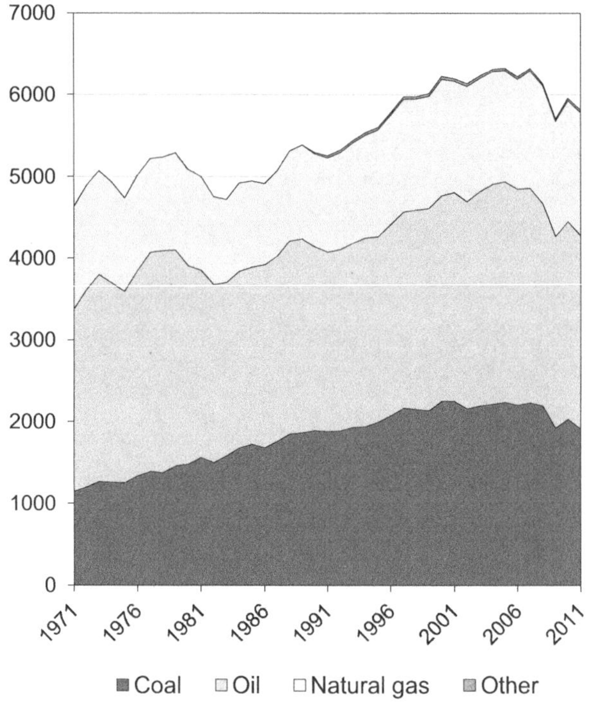

Figure 7: Electricity generation by fuel share

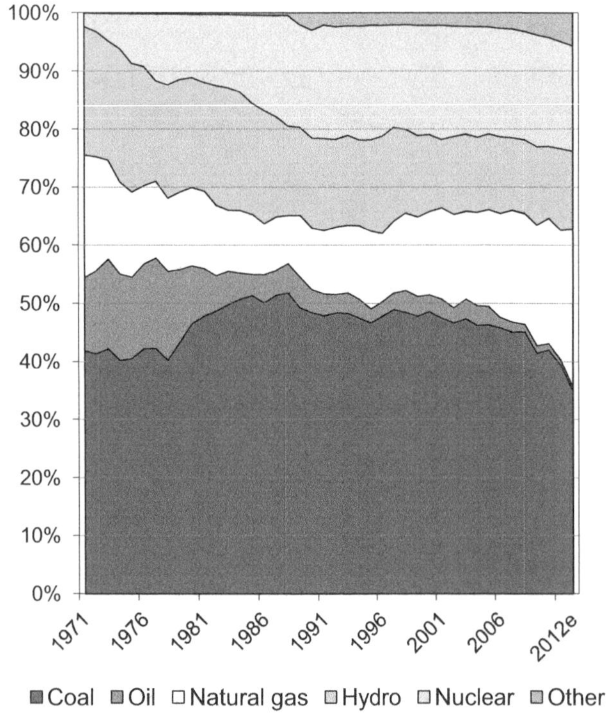

Figure 8: CO_2 emissions by fuel share

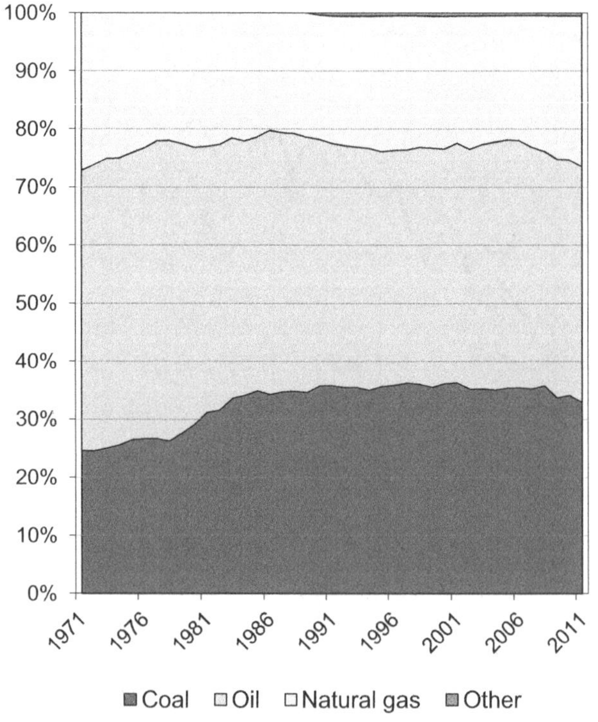

*Please refer to notes and definitions in Part I. Peat is included in coal for all figures.

Source: *IEA/OECD Electricity Information, IEA/OECD CO2 Emissions from Fuel Combustion*

IEA AMERICAS

1. Total primary energy supply (TPES) by fuel*

	1973	1980	1990	2000	2010	2011	2012e	Average annual percent change 73-90	90-11
TPES (Mtce)	**2698.98**	**2852.34**	**3033.66**	**3606.83**	**3523.57**	**3490.06**	**3407.28**	**0.69**	**0.67**
Coal and peat	466.16	566.88	692.11	807.55	749.94	712.29	636.01	2.35	0.14
Oil	1281.25	1264.93	1190.51	1368.92	1273.34	1239.94	1206.30	-0.43	0.19
Natural Gas	788.27	746.19	704.23	888.31	906.57	931.85	968.91	-0.66	1.34
Biofuels and waste	64.73	88.77	100.59	121.31	144.79	148.22	145.64	2.63	1.86
Nuclear	39.01	113.96	255.40	324.09	346.08	340.65	333.63	11.69	1.38
Hydro	56.52	65.11	70.01	75.15	75.39	85.68	81.09	1.27	0.97
Geothermal	3.01	6.57	20.14	18.70	12.01	12.22	13.70	11.83	-2.35
Solar, wind, tide	-	-	0.46	3.00	15.41	19.12	23.18	-	19.40
Net electricity trade(1)	0.03	-0.06	0.20	-0.21	0.03	0.09	-1.17	11.90	-3.90
Heat(2)	-	-	-	-	-	-	-	-	-

(1) Net trade between OECD and non-OECD countries.

(2) Ambient heat from heat pumps used in the transformation sector.

Source: *IEA/OECD Energy Balances of OECD Countries*

2. Energy supply, GDP and population*

	1973	1980	1990	2000	2010	2011	2012e	Average annual percent change 73-90	90-11
GDP (2005 bil. USD)	5320	6365	8712	12158	14196	14461	14775	2.94	2.44
TPES/GDP(1)	0.51	0.45	0.35	0.30	0.25	0.24	0.23	-2.19	-1.73
Population (millions)	234.4	252.2	277.9	313.1	343.9	346.5	349.9	1.01	1.06
TPES/population(1)	11.51	11.31	10.92	11.52	10.25	10.07	9.74	-0.31	-0.38
TPES/GDP(2)	187.7	165.8	128.9	109.8	91.9	89.3	85.3	-2.19	-1.73
Coal and peat supply/GDP(2)	143.1	145.4	129.7	108.5	86.3	80.4	70.3	-0.57	-2.25
Elec. consumption/GDP(2)	114.9	118.5	113.5	106.1	98.3	96.5	..	-0.07	-0.77
Elec. generation (TWh)	2236	2801	3685	4631	4956	4964	4927	2.98	1.43
Industrial production(2)	..	..	..	..	..	..	..	..	..

(1) TPES/GDP in units of tce/2005 thousand USD, TPES/population in units of tce/capita.

(2) As index, 2005 (not shown) = 100

Source: *IEA/OECD Energy Balances of OECD Countries* , *OECD Main Economic Indicators*

3. Coal and peat production by type*

	1978(1)	1990	2000	2005	2010	2011	2012e	Average annual percent change 78-90	90-11
Mtce:									
Coking coal	106.78	123.37	79.92	67.82	89.72	103.11	104.15	1.21	-0.85
Steam coal	426.31	663.65	694.21	744.34	679.71	671.45	615.99	3.76	0.06
Lignite	18.07	41.91	41.95	40.91	38.65	39.76	38.71	7.26	-0.25
Peat	-	-	-	-	-	-	-	-	-
Mt:									
Coking coal	105.98	120.92	82.45	73.06	96.80	111.11	112.39	1.10	-0.40
Steam coal	496.24	791.65	869.49	943.71	885.97	878.62	808.00	3.97	0.50
Lignite	36.23	89.32	88.81	87.17	81.23	83.31	81.11	7.81	-0.33
Peat	-	-	-	-	-	-	-	-	-

(1) Earliest year for which split by coal type is available.

* Please refer to notes and definitions in Part I.

Source: *IEA/OECD Energy Balances of OECD Countries, IEA/OECD Energy Statistics of OECD Countries*

IEA AMERICAS

4. Final consumption of energy by fuel*

(Mtce)

	1973	1980	1990	2000	2009	2010	2011	Average annual percent change 73-90	90-11
Total final consumption[(1)]	**2066.85**	**2094.79**	**2074.94**	**2479.80**	**2357.48**	**2454.26**	**2439.55**	**0.02**	**0.77**
Coal and peat	113.57	86.42	83.92	51.56	38.38	43.06	39.56	-1.76	-3.52
Oil	1098.78	1098.76	1074.40	1248.86	1174.55	1214.76	1196.16	-0.13	0.51
Natural Gas	558.13	533.76	494.70	590.42	519.86	541.91	546.72	-0.71	0.48
Biofuels and wastes	64.35	88.28	42.53	88.35	106.83	112.81	115.10	-2.41	4.85
Geothermal	-	-	0.48	0.74	0.31	0.32	0.32	-	-1.96
Solar, wind, tide	-	-	-	2.06	1.96	1.98	2.06	-	-
Electricity	231.88	286.10	374.92	489.11	504.62	529.25	529.36	2.87	1.66
Heat	0.14	1.47	3.98	8.70	10.96	10.18	10.27	21.94	4.62
of which:									
Total industry	**627.85**	**624.92**	**472.63**	**553.21**	**443.21**	**491.16**	**490.69**	**-1.66**	**0.18**
Coal and peat	93.05	74.78	69.61	47.96	35.00	39.93	36.75	-1.69	-3.00
Oil	123.24	144.40	72.91	45.88	48.41	52.89	52.18	-3.04	-1.58
Natural Gas	270.11	239.61	181.05	224.56	180.37	201.90	203.71	-2.33	0.56
Biofuels and wastes	48.97	56.35	21.06	62.25	52.28	55.15	53.94	-4.84	4.58
Geothermal	-	-	-	0.16	0.15	0.15	0.15	-	-
Solar, wind, tide	-	-	-	-	-	-	-	-	-
Electricity	92.35	108.34	127.09	165.29	118.08	132.95	135.68	1.90	0.31
Heat	0.14	1.43	0.90	7.11	8.93	8.19	8.27	11.73	11.15
Total transport	**639.81**	**670.84**	**758.12**	**914.83**	**903.66**	**934.51**	**925.97**	**1.00**	**0.96**
Coal and peat	0.18	-	-	-	-	-	-	-	-
Oil	614.65	646.77	731.07	880.51	844.71	869.86	858.26	1.03	0.77
Natural Gas	24.06	23.41	26.15	28.48	24.72	26.81	27.18	0.49	0.18
Biofuels and wastes	-	-	-	4.74	32.80	36.43	39.12	-	-
Electricity	0.92	0.66	0.91	1.10	1.43	1.41	1.42	-0.10	2.13
Residential	**375.63**	**348.39**	**340.85**	**421.31**	**420.74**	**427.22**	**423.39**	**-0.57**	**1.04**
Coal and peat	5.11	2.35	2.19	1.95	0.04	0.05	0.05	-4.86	-16.65
Oil	120.29	74.47	44.32	50.22	36.16	34.59	32.00	-5.70	-1.54
Natural Gas	169.96	170.28	162.73	185.84	178.93	177.36	176.99	-0.26	0.40
Biofuels and wastes	2.73	2.73	2.13	17.45	18.02	17.55	18.64	-1.44	10.88
Geothermal	-	-	-	0.31	0.05	0.05	0.05	-	-
Solar, wind, tide	-	-	-	2.06	1.96	1.98	2.06	-	-
Electricity	77.55	98.57	129.47	163.48	185.58	195.65	193.60	3.06	1.93
Heat	-	-	0.00	-	0.00	-	-	-	-
Comm & public services	**236.58**	**229.78**	**258.56**	**314.96**	**331.90**	**330.38**	**331.25**	**0.52**	**1.19**
Coal and peat	4.89	2.63	3.45	1.27	2.27	2.16	1.86	-2.03	-2.91
Oil	78.65	52.80	36.26	32.05	26.23	25.60	24.63	-4.45	-1.83
Natural Gas	94.01	96.74	99.39	119.98	118.03	116.36	118.55	0.33	0.84
Biofuels and waste	-	-	-	1.77	3.24	3.15	2.90	-	-
Geothermal	-	-	-	0.27	0.12	0.12	0.12	-	-
Solar, wind, tide	-	-	-	-	-	-	-	-	-
Electricity	59.03	77.57	116.38	158.01	179.98	181.01	181.19	4.07	2.13
Heat	-	0.04	3.08	1.59	2.04	1.99	1.99	-	-2.06
Non-energy use	**137.18**	**162.28**	**191.52**	**247.17**	**213.65**	**224.91**	**224.06**	**1.98**	**0.75**
Coal and peat	-	0.18	0.47	0.38	1.07	0.93	0.91	-	3.22
Oil	137.18	158.76	166.39	216.07	195.50	205.32	203.90	1.14	0.97
Natural Gas	-	3.33	24.67	30.73	17.08	18.66	19.25	-	-1.17

(1) Total Final Consumption (TFC) excludes use in transformation processes and energy industry own use.

* Please refer to notes and definitions in Part I. Agriculture, Forestry, Fishing and "Non-specified other" sectors are not reported here individually, but are accounted for in Total final consumption.

Source: *IEA/OECD Energy Balances of OECD Countries*

IEA AMERICAS

5. Coal balance*

(Mtce)

	1973	1980	1990	2000	2005	2010	2011	2012e	Average annual percent change 73-90	90-11
Production	492.9	668.8	828.9	816.1	853.1	808.1	814.3	758.9	3.1	-0.1
Imports	16.0	16.4	16.7	35.8	44.8	27.2	20.3	17.3	0.2	0.9
Exports	-55.3	-97.9	-127.8	-82.3	-66.2	-97.1	-119.2	-135.4	5.1	-0.3
Stock changes	12.5	-20.4	-25.7	37.9	7.2	11.8	-3.1	-4.7		
Primary supply	**466.2**	**566.9**	**692.1**	**807.6**	**838.9**	**749.9**	**712.3**	**636.0**	**2.4**	**0.1**
Statistical differences	**18.7**	**-10.0**	**9.2**	**21.7**	**-23.4**	**-2.4**	**-7.7**	..		
Total transformation	**-362.6**	**-466.8**	**-615.5**	**-775.6**	**-762.7**	**-702.2**	**-662.5**	..	**3.2**	**0.4**
Electricity and heat gen.	-321.7	-438.7	-594.3	-755.1	-753.0	-688.2	-647.7	..	3.7	0.4
Main activity producers (1)	*-321.7*	*-438.7*	*-588.0*	*-738.1*	*-744.5*	*-681.9*	*-642.3*	..	*3.6*	*0.4*
Autoproducers	-	*-0.0*	*-6.2*	*-17.0*	*-8.4*	*-6.3*	*-5.5*	..	-	*-0.6*
Gas works	0.5	-	-2.6	-2.7	-2.6	-2.7	-2.8	..	-	0.2
Coal transformation(2)	-41.3	-28.2	-18.6	-17.8	-7.1	-11.3	-12.0	..	-4.6	-2.1
BKB plants	-	-	-	-	-	-	-	..	-	-
Blast furnaces	*-27.7*	*-18.4*	*-12.9*	*-12.3*	*-10.3*	*-7.1*	*-8.0*	..	*-4.4*	*-2.3*
Coke ovens	*-13.6*	*-9.8*	*-5.7*	*-5.5*	*-3.9*	*-4.2*	*-4.0*	..	*-5.0*	*-1.6*
Patent fuel plants	-	-	-	-	*7.0*	-	-	..	-	-
Other transformation(3)	-	-	-	-	-	-	-	..	-	-
Energy ind. own use	**-8.7**	**-3.6**	**-1.9**	**-2.1**	**-3.0**	**-2.3**	**-2.5**	..	**-8.5**	**1.2**
Losses	-	-	-	-	-	-	-	..		
Final consumption(4)	**113.6**	**86.4**	**83.9**	**51.6**	**49.8**	**43.1**	**39.6**	..	**-1.8**	**-3.5**
Industry(5)	93.1	74.8	69.6	48.0	45.6	39.9	36.8	..	-1.7	-3.0
Iron and steel	*52.3*	*33.7*	*20.4*	*13.8*	*8.4*	*9.1*	*8.5*	..	*-5.4*	*-4.1*
Chemical	*11.9*	*11.2*	*12.5*	*9.6*	*7.1*	*6.1*	*5.7*	..	*0.3*	*-3.7*
Non-metallic minerals	*5.9*	*10.3*	*11.7*	*12.4*	*11.6*	*8.0*	*7.7*	..	*4.2*	*-2.0*
Paper, pulp and print	*7.1*	*7.4*	*10.5*	*3.7*	*6.6*	*5.8*	*5.3*	..	*2.3*	*-3.2*
Other industry (6)	*15.8*	*12.2*	*14.5*	*8.4*	*11.9*	*10.9*	*9.5*	..	*-0.5*	*-2.0*
Transport(7)	0.2	-	-	-	-	-	-	..	-	-
Other	20.3	11.5	13.8	3.2	3.7	2.2	1.9	..	-2.2	-9.0
Comm. and pub. services	*4.9*	*2.6*	*3.5*	*1.3*	*3.0*	*2.2*	*1.9*	..	*-2.0*	*-2.9*
Residential	*5.1*	*2.4*	*2.2*	*1.9*	*0.1*	*0.0*	*0.0*	..	*-4.9*	*-16.7*
Other sectors (8)	*10.3*	*6.5*	*8.2*	-	*0.6*	-	-	..	*-1.4*	-
Non-energy use	-	0.2	0.5	0.4	0.6	0.9	0.9	..	-	3.2

(1) Main activity electricity and heat generation includes district heating.

(2) Coal transformation refers to the transformation of coal and peat to secondary and tertiary products (mainly cokes, briquettes, coal tar and off-gases). Gas works and Liquefaction are shown separately to match the *Energy Balances of OECD Countries.*

(3) Other transformation includes Liquefaction and Non-specified transformations.

(4) Final Consumption includes non-energy use and energy use (Industry, Transport and Other).

(5) Please refer to notes and definitions in Part I for detailed explanation.

(6) Other industry includes Non-ferrous metals, Transport equipment, Machinery, Mining and quarrying, Food and tobacco, Wood and wood products, Construction, Textile and leather, and Non-specified industry.

(7) Transport includes Rail and Inland waterways.

(8) Other sectors includes Agriculture/Forestry, Fishing and Non-specified other.

* "Coal" refers to all types of coal, primary (anthracite, coking coal, other bituminous coal, sub-bituminous coal, lignite/brown coal) and derived fuels (including patent fuel, BKB, coke oven coke, gas coke, coal tar, coke oven gas, gas works gas, blast furnace gas and other recovered gases), as well as peat, peat briquettes and oil shale. Quantities have been converted to Mtce using calorific values reported by the respective countries. Please refer to notes and definitions in Part I.

Source: *IEA/OECD Energy Balances of OECD Countries*

IEA AMERICAS

6. Use of coal for selected end-uses*

(million tonnes)

	1978	1990	2000	2005	2009	2010	2011	Average annual percent change 78-90	90-11
Steam coal	**504.44**	**736.42**	**913.46**	**977.47**	**874.50**	**896.57**	**854.00**	**3.20**	**0.71**
Total electricity and heat	426.39	670.07	880.51	804.72	826.59	863.94	815.94	3.84	0.94
Main activity producers	*426.38*	*663.78*	*865.03*	*798.21*	*822.06*	*858.86*	*810.73*	*3.76*	*0.96*
Autoproducers	*0.01*	*6.29*	*15.48*	*6.51*	*4.53*	*5.09*	*5.21*	*66.36*	*-0.89*
Patent fuel/BKB plants	-	-	-	119.79	-	-	-	-	-
Coke ovens/Liquefaction(1)	-	-	-	-	-	-	-	-	-
Blast furnace inputs	-	0.17	2.39	1.20	1.03	1.28	1.48	-	10.73
Gas manufacture	-	-	-	-	-	-	-	-	-
Industry	46.05	53.40	37.62	37.54	31.74	34.21	31.33	1.24	-2.51
Iron and steel	*3.73*	*1.71*	*1.46*	*0.32*	*0.47*	*0.68*	*0.57*	*-6.32*	*-5.11*
Chemical	*10.34*	*14.26*	*10.51*	*7.23*	*6.68*	*6.84*	*6.48*	*2.71*	*-3.69*
Non-metallic minerals	*11.88*	*12.65*	*13.26*	*12.55*	*8.31*	*8.89*	*8.57*	*0.53*	*-1.84*
Paper, pulp and print	*7.82*	*11.37*	*4.10*	*6.84*	*6.22*	*6.50*	*5.98*	*3.17*	*-3.02*
Other industry	*12.28*	*13.42*	*8.29*	*10.60*	*10.06*	*11.31*	*9.74*	*0.74*	*-1.52*
Other sectors(2)	19.13	15.43	3.73	4.28	2.63	2.51	2.22	-1.78	-8.82
Non-energy use	-	0.26	0.33	0.53	1.22	1.07	1.05	-	6.90
Coking coal	**75.66**	**40.29**	**30.42**	**25.18**	**17.71**	**22.87**	**24.96**	**-5.12**	**-2.25**
Total electricity and heat	-	-	-	-	-	-	-	-	-
Main activity producers	-	-	-	-	-	-	-	-	-
Autoproducers	-	-	-	-	-	-	-	-	-
Patent fuel/BKB plants	-	-	-	2.81	-	-	-	-	-
Coke ovens/Liquefaction(1)	71.74	40.27	30.20	22.73	16.95	23.15	23.64	-4.70	-2.50
Blast furnace inputs	-	-	-	-	-	-	-	-	-
Gas manufacture	-	-	-	-	-	-	-	-	-
Industry	-	-	-	-	-	-	-	-	-
Iron and steel	-	-	-	-	-	-	-	-	-
Chemical	-	-	-	-	-	-	-	-	-
Non-metallic minerals	-	-	-	-	-	-	-	-	-
Paper, pulp and print	-	-	-	-	-	-	-	-	-
Other industry	-	-	-	-	-	-	-	-	-
Other sectors(2)	-	-	-	-	-	-	-	-	-
Non-energy use	-	-	-	-	-	-	-	-	-
Lignite	**35.75**	**88.38**	**85.47**	**87.28**	**75.53**	**78.44**	**84.37**	**7.84**	**-0.22**
Total electricity and heat	33.09	81.12	80.87	81.10	68.12	72.02	76.57	7.76	-0.27
Main activity producers	*33.09*	*80.03*	*79.62*	*79.04*	*68.12*	*69.44*	*76.26*	*7.64*	*-0.23*
Autoproducers	-	*1.10*	*1.25*	*2.06*	-	*2.57*	*0.31*	-	*-5.82*
Patent fuel/BKB plants	-	-	-	-	-	-	-	-	-
Coke ovens/Liquefaction(1)	-	-	-	-	-	-	-	-	-
Gas manufacture	-	5.64	5.67	5.39	5.43	5.57	5.41	-	-0.20
Industry	2.46	1.41	1.59	0.39	0.29	0.30	0.08	-4.53	-12.73
Iron and steel	-	-	-	-	-	-	-	-	-
Chemical	-	*0.23*	*0.24*	*0.24*	*0.26*	*0.26*	*0.05*	-	*-7.00*
Non-metallic minerals	-	*0.00*	*0.00*	*0.01*	-	-	-	-	-
Paper, pulp and print	*0.17*	*0.10*	*0.10*	*0.06*	-	-	-	*-4.60*	-
Other industry	*2.30*	*1.08*	*1.25*	*0.08*	*0.03*	*0.04*	*0.03*	*-6.09*	*-15.69*
Other sectors(2)	0.20	0.12	0.09	0.06	0.09	0.09	0.10	-4.34	-0.88
Non-energy use	-	0.09	0.14	0.16	-	-	-	-	-

(1) Liquefaction primarily refers to direct distillation processes. Indirect liquefaction may be reported here or under gas manufacture.

(2) Other sectors are Residential, Commercial and public services, Agriculture/Forestry, Fishing and Non-specified other.

* Coal totals may not match the sum of the individual rows due to statistical differences. Please refer to notes and definitions in Part I.

Source: *IEA/OECD Energy Statistics of OECD Countries*

IEA AMERICAS

7. Fuel prices to end users*

	1978[1]	1990	1995	2000	2005	2009	2010	2011	2012
				(US dollars / tce)					
For electricity generation									
Steam coal	..	..	..	..	..	..	..	..	..
Heavy fuel oil	..	..	..	..	..	..	..	..	..
Natural gas	..	..	..	..	..	..	..	..	..
For industry									
Steam coal	..	..	..	..	..	..	..	..	..
Coking coal	..	..	..	..	..	..	..	..	..
High sulphur fuel oil	..	..	..	..	..	..	..	..	..
Low sulphur fuel oil	..	..	..	..	..	..	..	..	..
Natural gas	..	..	..	..	..	..	..	..	..
				(US dollars / unit)[2]					
For electricity generation									
Steam coal	..	..	..	..	..	..	..	..	..
Heavy fuel oil	..	..	..	..	..	..	..	..	..
Natural gas	..	..	..	..	..	..	..	..	..
For industry									
Steam coal	..	..	..	..	..	..	..	..	..
Coking coal	..	..	..	..	..	..	..	..	..
High sulphur fuel oil	..	..	..	..	..	..	..	..	..
Low sulphur fuel oil	..	..	..	..	..	..	..	..	..
Natural gas	..	..	..	..	..	..	..	..	..

(1) Earliest year for which split by coal type is available.

(2) For steam coal, coking coal and heavy fuel oil per metric tonne; for natural gas per 10^7 kilocalories GCV.

* Please refer to notes and definitions in Part I. Steam coal prices relate to historical definitions.

Source: *IEA/OECD Energy Prices & Taxes*

8. Coal and peat trade by type of coal*

(Mtce)

	1978	1990	1995	2000	2005	2009	2010	2011	2012e
Total imports	**21.44**	**16.67**	**17.66**	**35.80**	**44.82**	**28.63**	**27.16**	**20.27**	**17.26**
Bituminous coal[1]	10.87	11.44	11.12	23.27	30.58	21.05	17.68	12.42	9.87
Coking coal	5.30	4.22	3.88	5.75	5.63	3.06	4.33	5.04	5.22
Sub-bituminous coal	-	-	0.28	2.79	5.32	3.83	3.29	1.09	0.76
Lignite	-	-	-	0.04	0.06	0.06	0.07	0.06	0.07
Peat	-	-	-	-	-	-	-	-	-
Coal products[2]	5.26	1.00	2.39	3.97	3.23	0.63	1.79	1.64	1.33
Total exports	**50.78**	**127.76**	**115.17**	**82.26**	**66.16**	**74.02**	**97.08**	**119.22**	**135.39**
Bituminous coal[1]	9.63	41.27	35.39	25.38	12.75	18.98	19.98	31.33	43.16
Coking coal	40.44	85.76	76.80	54.83	47.14	50.00	71.16	82.72	85.61
Sub-bituminous coal	-	-	2.01	0.73	4.40	3.48	4.40	3.83	5.28
Lignite	0.04	0.04	-	0.03	0.17	0.17	0.16	0.14	0.08
Peat	-	-	-	-	-	-	-	-	-
Coal products[2]	0.67	0.69	0.98	1.30	1.70	1.38	1.38	1.20	1.26

(1) Bituminous coal includes anthracite.

(2) Coal products includes products derived from coal, for example: coke, coal tar, briquettes, patent fuels.

* Please refer to notes and definitions in Part I.

Source: *IEA/OECD Energy Balances of OECD Countries, IEA/OECD Energy Statistics of OECD Countries*

IEA AMERICAS

9. Total coal imports by origin*

(thousand tonnes)

	1978(1)	1985	1990	1995	2000	2005	2010	2011	2012e
Total coal(2)	**16929**	**16350**	**16618**	**16268**	**34582**	**48379**	**30167**	**21345**	**18127**
Coking coal	**5454**	**6188**	**4491**	**4129**	**5843**	**5831**	**4477**	**5216**	**5397**
Australia	-	-	-	-	-	144	-	-	-
Canada	-	-	-	-	1511	1458	1385	1358	875
Czech Republic	-	-	-	-	-	-	-	-	-
Germany	-	-	-	-	-	-	-	-	-
Poland	-	-	-	-	-	-	-	-	-
United Kingdom	-	-	-	-	-	-	-	-	-
United States	5454	6188	4491	4129	4296	4160	3092	3770	4382
Other OECD	-	-	-	-	-	-	-	-	-
China, People's Rep.	-	-	-	-	-	-	-	-	-
Colombia	-	-	-	-	-	9	-	-	-
Indonesia	-	-	-	-	-	-	-	-	-
South Africa	-	-	-	-	-	-	-	-	-
Former Soviet Union(3)	-	-	x	x	x	x	x	x	x
Russian Federation (4)	*x*	*x*	-	-	-	-	-	-	-
Other FSU (4)	*x*	*x*	-	-	-	-	-	-	-
Venezuela	-	-	-	-	36	-	-	88	140
Vietnam	-	-	-	-	-	-	-	-	-
Non-specified/other	-	-	-	-	-	60	-	-	-
Steam coal	**11475**	**10162**	**12127**	**12139**	**28659**	**42419**	**25551**	**15993**	**12586**
Australia	933	35	22	192	152	66	345	56	-
Canada	49	367	883	1197	155	244	86	46	23
Czech Republic	-	-	-	-	-	-	-	-	-
Germany	-	-	-	-	-	-	-	3	-
Poland	645	-	-	-	-	70	-	-	-
United Kingdom	-	-	5	-	-	22	4	302	-
United States	8796	8391	9639	5095	18531	12855	7247	3266	3777
Other OECD	-	6	-	39	6	39	27	7	-
China, People's Rep.	-	-	-	-	173	69	47	19	59
Colombia	-	539	1296	2994	6928	21333	15274	10422	7650
Indonesia	-	-	-	924	651	2239	1728	777	470
South Africa	996	824	-	-	61	70	-	-	-
Former Soviet Union(3)	-	-	x	x	x	x	x	x	x
Russian Federation (4)	*x*	*x*	-	-	*26*	*707*	*12*	*145*	*96*
Other FSU (4)	*x*	*x*	-	-	-	*141*	*191*	*221*	*390*
Venezuela	-	-	277	1675	1904	4012	560	679	121
Vietnam	-	-	-	-	-	85	-	-	-
Non-specified/other	56	-	5	23	72	467	30	47	-
Lignite	**-**	**-**	**-**	**-**	**80**	**129**	**139**	**136**	**144**

(1) Earliest year for which split by coal type is available.

(2) Total coal consists of steam coal, coking coal and lignite.

(3) For years prior to 1990, or if no details for individual Former Soviet Union republics.

(4) Data for former republics are not available separately for some OECD member countries.

* Please refer to notes and definitions in Part I.

Source: IEA/OECD Coal Statistics

IEA AMERICAS

10. Coking coal exports by destination*

(thousand tonnes)

	1978[(1)]	1985	1990	1995	2000	2005	2010	2011	2012e
World	**40450**	**77198**	**84419**	**75819**	**58166**	**52711**	**78434**	**90744**	**94117**
Total OECD	**36030**	**65898**	**67629**	**61804**	**45504**	**41664**	**52379**	**60294**	**57025**
Australia	-	-	-	-	-	-	-	-	-
Austria	-	-	-	-	-	239	412	330	1670
Belgium	1103	3443	5538	3807	2717	1528	1737	1371	1392
Canada	5410	6360	3988	4039	3501	4034	3091	3772	4379
Chile	-	281	292	267	312	368	215	611	639
Czech Republic	-	-	-	-	-	-	-	-	-
Denmark	-	-	50	-	-	-	-	-	-
Estonia	-	-	-	-	-	-	-	-	-
Finland	-	-	127	715	288	776	678	729	569
France	1468	3823	4880	3233	2782	1610	1958	1797	2647
Germany	560	709	708	433	1211	2230	2728	2565	3340
Greece	-	7	-	-	-	-	-	-	-
Hungary	-	-	-	-	65	-	40	26	-
Iceland	-	13	28	36	48	57	59	58	62
Ireland	-	54	-	-	-	-	-	136	-
Israel	-	-	50	127	56	-	-	-	-
Italy	3069	5734	6536	5054	4467	3672	3402	5067	4196
Japan	19925	29639	26588	22688	12937	8446	11387	12922	12061
Korea	1173	3950	2908	6615	4947	5497	8009	11111	9417
Luxembourg	-	-	-	-	-	-	77	-	-
Mexico	-	498	3	517	740	994	1002	1010	758
Netherlands	929	2975	3975	4068	2143	2355	5636	7152	5710
New Zealand	-	-	-	-	-	-	-	-	-
Norway	68	98	99	83	42	18	75	81	79
Poland	-	-	-	-	-	-	2149	1258	465
Portugal	265	244	753	188	198	-	-	93	134
Slovak Republic	-	-	-	-	-	-	260	470	462
Slovenia	-	-	-	-	-	163	223	663	292
Spain	838	2460	3156	2313	2331	2029	1462	1226	1160
Sweden	453	1310	866	1161	642	464	401	480	535
Switzerland	-	-	-	-	-	-	37	-	-
Turkey	409	1716	1957	1895	2403	2667	2914	3097	3823
United Kingdom	360	2474	4355	3831	3134	2915	3044	2914	2360
United States	-	110	772	734	540	1602	1383	1355	875
Total non-OECD	**4420**	**11300**	**11592**	**13631**	**9437**	**11047**	**25986**	**30450**	**37092**
Brazil	1942	6139	6327	6826	5564	4831	8762	9845	8904
China[(2)]	-	-	300	-	-	956	8149	7265	16091
Chinese Taipei	205	782	357	1600	1440	1274	864	1069	1004
Egypt	218	457	586	1165	682	706	1042	629	375
India	200	38	-	-	22	1078	2299	3728	5181
Romania	673	1051	1559	1528	443	547	812	937	607
Oth. Africa & Mid. East	1	872	614	350	269	377	230	683	879
Oth. non-OECD Americas	914	588	580	400	184	207	321	306	471
Other Asia & Oceania	24	353	229	548	-	104	109	110	127
Other non-OECD Europe and Eurasia	243	1020	1040	1214	833	967	3398	5878	3453
Non-specified/Other	**-**	**-**	**5198**	**384**	**3225**	**-**	**69**	**-**	**-**

(1) Earliest year for which split by coal type is available.

(2) China includes the People's Republic of China and Hong Kong, China.

* Please refer to notes and definitions in Part I.

Source: IEA/OECD Coal Statistics

IEA AMERICAS

11. Steam coal exports by destination*

(thousand tonnes)

	1978(1)	1985	1990	1995	2000	2005	2010	2011	2012e
World	**10808**	**34228**	**42493**	**38503**	**26922**	**20460**	**28774**	**39990**	**54600**
Total OECD	**10341**	**30501**	**36299**	**34885**	**25445**	**18309**	**24131**	**33386**	**43792**
Australia	-	1	1	-	-	-	105	302	152
Austria	-	-	-	-	-	-	-	-	-
Belgium	27	712	2178	950	429	411	367	1470	1013
Canada	8782	8508	10083	4541	13524	13625	7245	2431	2161
Chile	-	-	394	207	48	256	1097	1607	1635
Czech Republic	-	-	-	-	-	-	-	-	-
Denmark	309	2286	3321	1905	70	66	73	146	-
Estonia	-	-	-	-	-	-	-	-	-
Finland	-	1208	-	530	-	-	166	145	-
France	38	978	1758	621	564	28	1080	2078	1146
Germany	528	595	384	1648	522	133	935	2394	7402
Greece	-	1138	-	-	-	-	47	-	-
Hungary	-	-	-	-	-	-	-	-	-
Iceland	-	7	15	-	-	-	-	-	-
Ireland	-	335	1322	829	456	-	-	86	208
Israel	-	421	530	562	-	-	-	-	17
Italy	22	3633	4451	3449	79	23	613	1004	4318
Japan	243	2847	4007	6179	4425	962	2180	2650	2678
Korea	356	2489	719	3800	2275	749	4193	7111	5303
Luxembourg	-	-	-	-	-	-	-	-	-
Mexico	-	44	188	-	373	341	1378	1920	2584
Netherlands	27	2793	3982	3015	643	829	1700	3992	2532
New Zealand	-	-	1	-	-	-	-	-	-
Norway	-	143	62	25	74	-	-	11	17
Poland	-	-	-	-	-	-	65	129	245
Portugal	-	626	1386	1562	343	143	531	798	992
Slovak Republic	-	-	-	-	-	-	-	-	-
Slovenia	-	-	-	-	-	-	182	-	197
Spain	-	786	282	2111	441	-	374	521	816
Sweden	-	230	21	8	-	71	275	153	138
Switzerland	-	8	-	-	-	-	-	169	-
Turkey	5	275	15	191	55	67	220	422	1515
United Kingdom	-	283	1005	2705	1016	361	1219	3800	8700
United States	4	155	194	47	108	244	86	47	23
Total non-OECD	**95**	**3727**	**4967**	**3616**	**923**	**1948**	**4643**	**6603**	**9564**
Brazil	11	78	177	153	22	693	118	310	125
China(2)	-	506	108	-	9	-	2938	2182	3188
Chinese Taipei	-	2575	3820	2033	-	3	1	-	228
Egypt	-	-	-	1	-	-	146	144	-
India	-	-	-	-	-	217	171	635	1829
Romania	-	-	-	271	-	844	-	-	-
Oth. Africa & Mid. East	1	176	682	1099	825	63	1044	2292	2913
Oth. non-OECD Americas	82	110	128	-	1	13	115	463	677
Other Asia & Oceania	1	2	5	-	-	-	77	1	103
Other non-OECD Europe and Eurasia	-	280	47	59	66	115	33	576	501
Non-specified/Other	**372**	**-**	**1227**	**2**	**554**	**203**	**-**	**1**	**1244**

(1) Earliest year for which split by coal type is available.

(2) China includes the People's Republic of China and Hong Kong, China.

* Please refer to notes and definitions in Part I. Steam coal includes all sub-bituminous coal.

Source: IEA/OECD Coal Statistics

IEA ASIA OCEANIA*

Figure 1: TPES by fuel (Mtce)

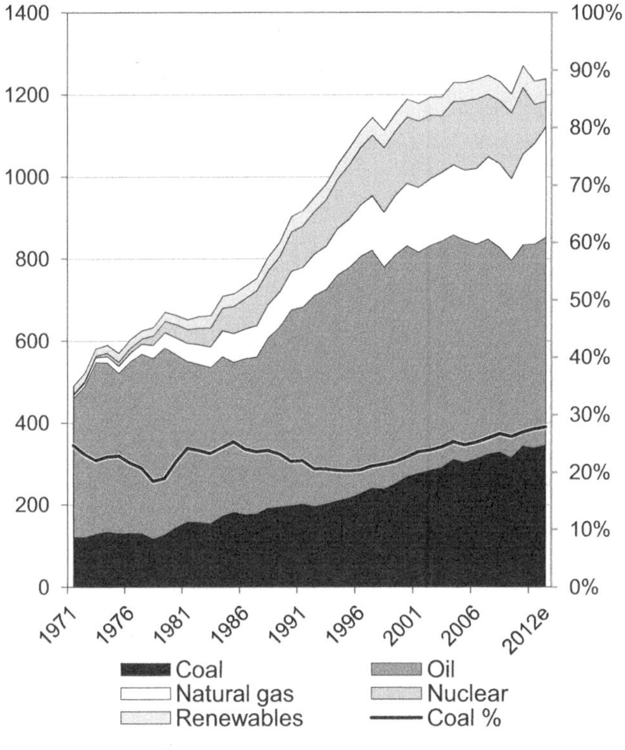

Figure 2: Coal supply indicators (1971 = 100)

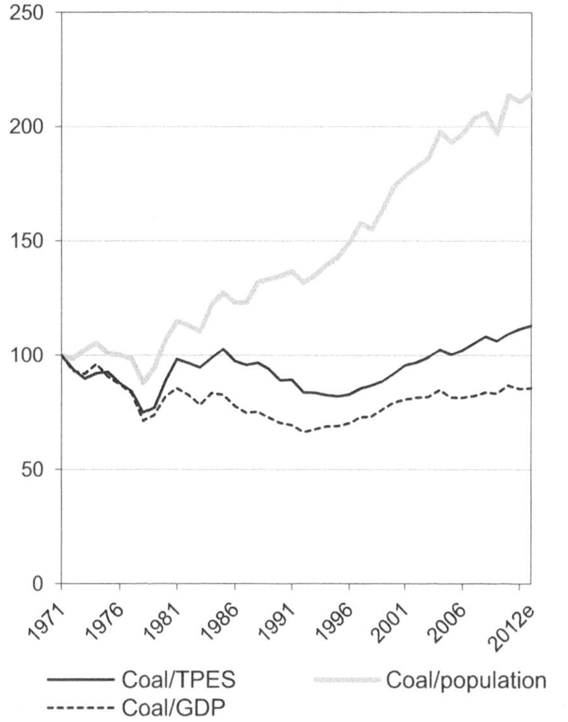

Figure 3: Primary coal supply (Mtce)

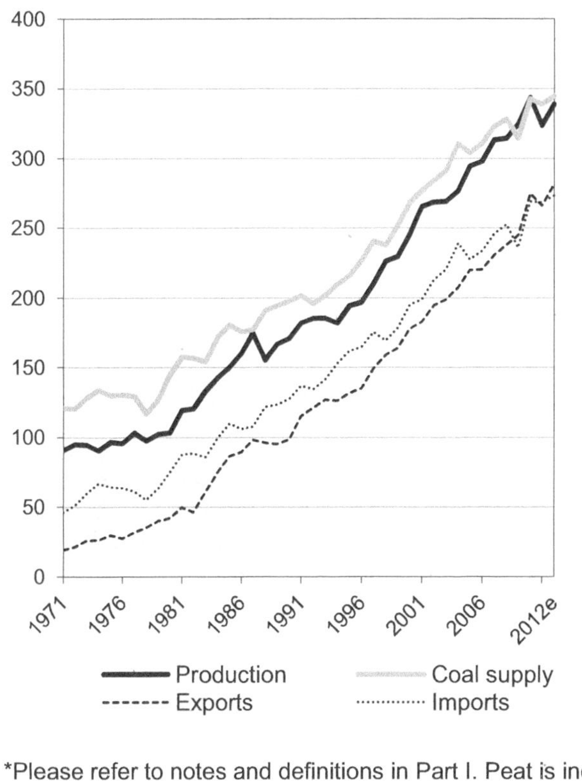

Figure 4: Coal consumption (Mtce)

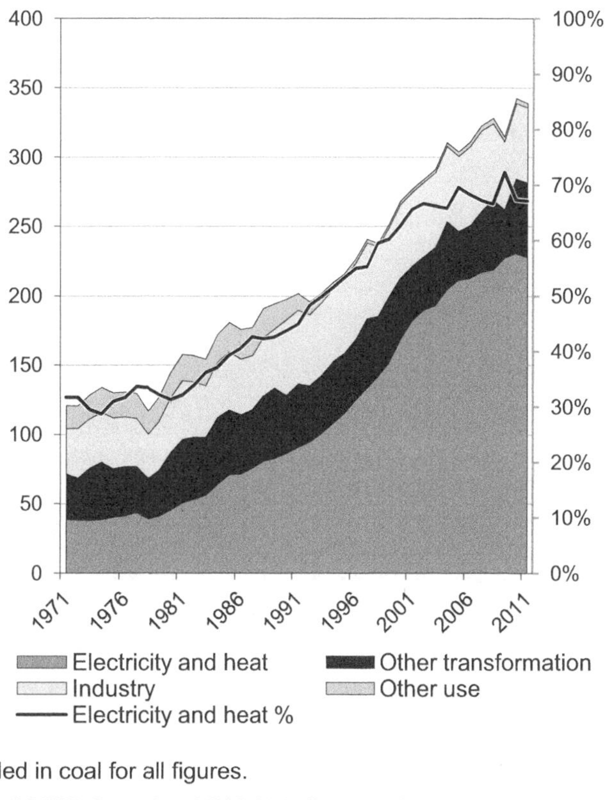

*Please refer to notes and definitions in Part I. Peat is included in coal for all figures.

Source: *IEA/OECD Coal Information, IEA/OECD Energy Balances of OECD Countries, OECD Main Economic Indicators*

IEA ASIA OCEANIA*

Figure 5: Electricity generation by fuel (TWh)

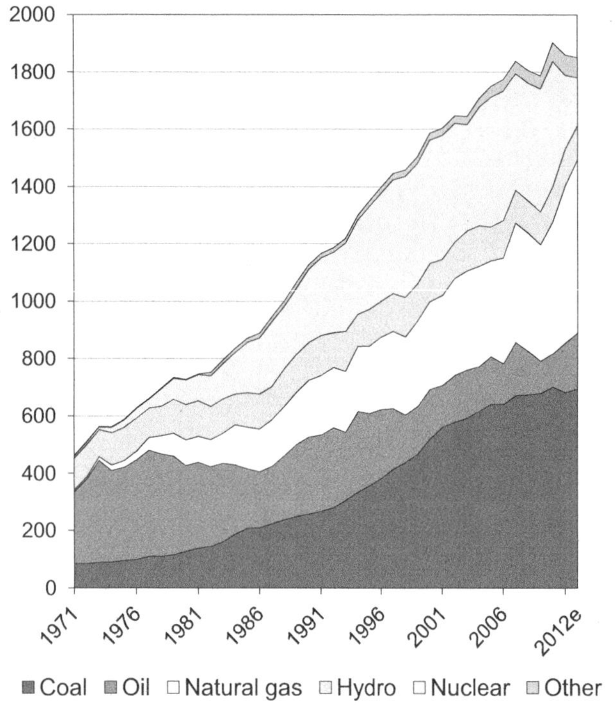

Figure 6: CO_2 emissions by fuel (Mt CO_2)

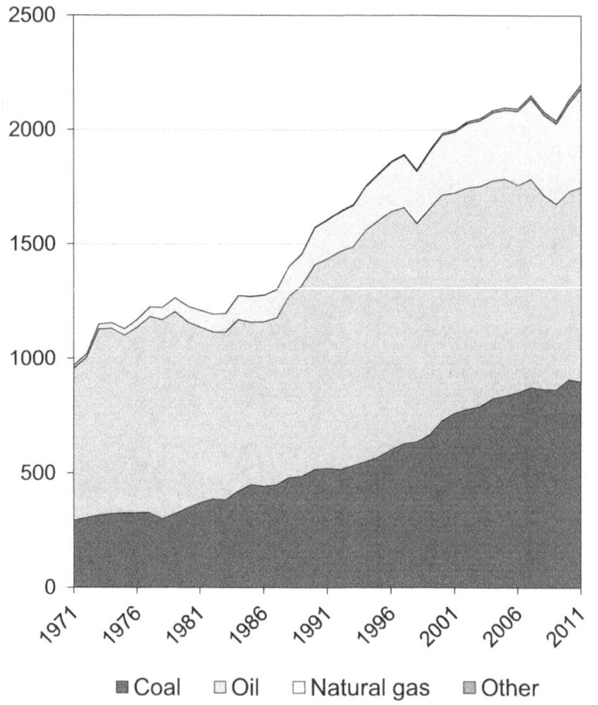

Figure 7: Electricity generation by fuel share

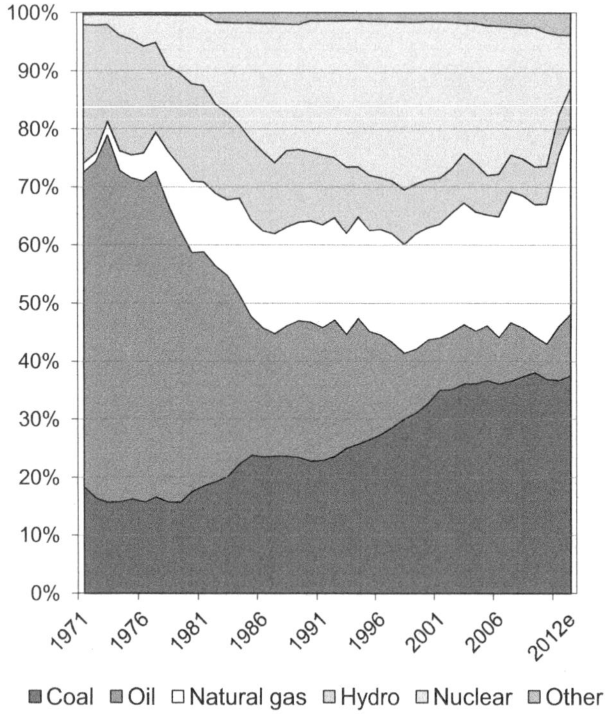

Figure 8: CO_2 emissions by fuel share

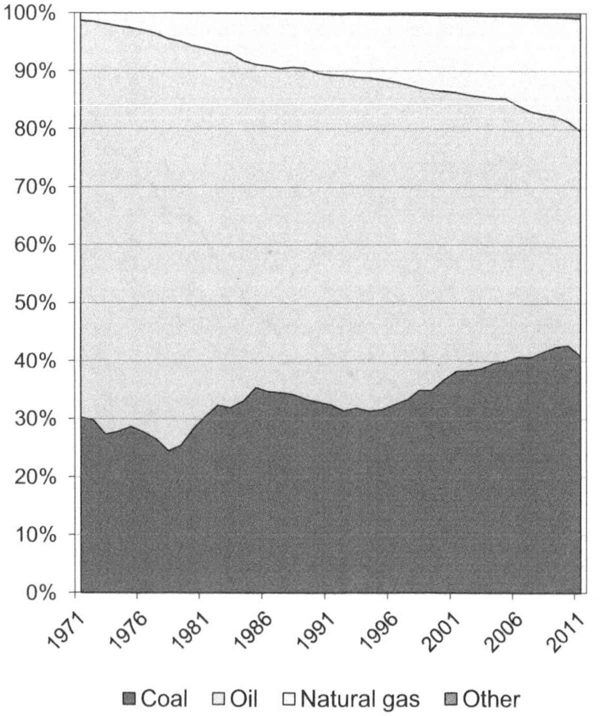

*Please refer to notes and definitions in Part I. Peat is included in coal for all figures.

Source: *IEA/OECD Electricity Information, IEA/OECD CO2 Emissions from Fuel Combustion*

IEA ASIA OCEANIA

1. Total primary energy supply (TPES) by fuel*

	1973	1980	1990	2000	2010	2011	2012e	Average annual percent change 73-90	90-11
TPES (Mtce)	**581.41**	**663.32**	**902.15**	**1188.99**	**1271.22**	**1232.80**	**1238.21**	**2.62**	**1.50**
Coal and peat	128.34	144.83	197.63	268.68	342.50	338.96	344.60	2.57	2.60
Oil	418.56	420.59	478.43	562.98	491.35	496.16	506.81	0.79	0.17
Natural Gas	12.48	42.36	93.63	152.83	220.82	245.65	269.52	12.59	4.70
Biofuels and waste	5.04	5.90	14.92	19.13	26.74	27.88	28.34	6.59	3.02
Nuclear	3.61	32.04	94.99	160.46	162.63	95.49	61.85	21.20	0.02
Hydro	11.52	15.01	16.34	16.22	15.25	15.93	14.52	2.08	-0.12
Geothermal	1.86	2.56	4.36	7.21	8.74	9.04	8.85	5.15	3.53
Solar, wind, tide	-	0.03	1.80	1.42	3.02	3.52	3.54	-	3.25
Net electricity trade(1)	-	-	-	-	-	-	-	-	-
Heat(2)	-	-	0.05	0.05	0.18	0.18	0.18	-	5.82

(1) Net trade between OECD and non-OECD countries.

(2) Ambient heat from heat pumps used in the transformation sector.

Source: *IEA/OECD Energy Balances of OECD Countries*

2. Energy supply, GDP and population*

	1973	1980	1990	2000	2010	2011	2012e	Average annual percent change 73-90	90-11
GDP (2005 bil. USD)	2355	2982	4732	5720	6656	6697	6783	4.19	1.67
TPES/GDP(1)	0.25	0.22	0.19	0.21	0.19	0.18	0.18	-1.51	-0.17
Population (millions)	159.6	173.1	187.0	197.1	204.3	204.8	204.8	0.94	0.43
TPES/population(1)	3.64	3.83	4.82	6.03	6.22	6.02	6.05	1.66	1.06
TPES/GDP(2)	126.2	113.7	97.5	106.3	97.7	94.1	93.3	-1.51	-0.17
Coal and peat supply/GDP(2)	112.8	100.5	86.4	97.2	106.5	104.8	105.1	-1.55	0.92
Elec. consumption/GDP(2)	84.8	87.0	85.2	99.5	102.6	99.7	..	0.02	0.76
Elec. generation (TWh)	563	728	1127	1587	1902	1860	1851	4.17	2.41
Industrial production(2)	..	..	..	..	..	..	..	..	..

(1) TPES/GDP in units of tce/2005 thousand USD, TPES/population in units of tce/capita.

(2) As index, 2005 (not shown) = 100

Source: *IEA/OECD Energy Balances of OECD Countries* , *OECD Main Economic Indicators*

3. Coal and peat production by type*

	1978(1)	1990	2000	2005	2010	2011	2012e	Average annual percent change 78-90	90-11
Mtce:									
Coking coal	44.87	63.09	102.24	127.32	160.81	144.83	145.01	2.88	4.04
Steam coal	42.82	93.04	122.17	143.26	158.51	154.74	168.99	6.68	2.45
Lignite	9.73	14.69	21.06	23.71	24.25	23.90	24.75	3.49	2.35
Peat	-	-	-	-	-	-	-	-	-
Mt:									
Coking coal	46.34	65.21	105.06	130.80	165.27	148.83	149.02	2.89	4.01
Steam coal	60.93	120.98	148.88	177.11	194.16	189.11	204.87	5.88	2.15
Lignite	30.66	46.15	67.51	70.78	72.39	71.32	73.86	3.47	2.09
Peat	-	-	-	-	-	-	-	-	-

(1) Earliest year for which split by coal type is available.

* Please refer to notes and definitions in Part I.

Source: *IEA/OECD Energy Balances of OECD Countries, IEA/OECD Energy Statistics of OECD Countries*

IEA ASIA OCEANIA

4. Final consumption of energy by fuel*

(Mtce)

	1973	1980	1990	2000	2009	2010	2011	Average annual percent change 73-90	90-11
Total final consumption(1)	**424.17**	**452.67**	**615.15**	**789.83**	**783.34**	**814.20**	**808.60**	**2.21**	**1.31**
Coal and peat	52.37	57.60	69.33	55.71	52.17	57.85	57.20	1.66	-0.91
Oil	298.16	294.04	372.40	470.65	422.62	425.61	421.70	1.32	0.59
Natural Gas	7.63	16.02	37.65	69.17	91.98	99.50	102.32	9.84	4.88
Biofuels and wastes	4.99	5.63	10.33	13.20	12.93	14.07	14.63	4.37	1.67
Geothermal	-	-	0.35	0.59	0.67	0.62	0.64	-	2.87
Solar, wind, tide	-	0.03	1.80	1.33	0.97	0.99	1.01	-	-2.70
Electricity	60.98	79.21	123.03	173.70	194.88	208.52	203.89	4.22	2.43
Heat	0.04	0.15	0.27	5.47	7.14	7.04	7.20	11.86	17.01
of which:									
Total industry	**183.62**	**173.62**	**206.12**	**234.88**	**211.50**	**228.82**	**227.12**	**0.68**	**0.46**
Coal and peat	35.20	39.16	54.42	53.20	48.71	54.39	53.90	2.60	-0.05
Oil	101.61	74.98	72.46	70.95	48.00	47.61	46.79	-1.97	-2.06
Natural Gas	4.26	8.31	15.09	23.49	30.05	33.26	35.98	7.72	4.23
Biofuels and wastes	2.13	3.17	6.80	9.92	9.33	10.16	10.43	7.06	2.05
Geothermal	-	-	0.16	0.19	0.22	0.21	0.21	-	1.44
Solar, wind, tide	-	-	-	-	0.00	0.00	0.00	-	-
Electricity	40.43	48.00	57.19	74.09	71.33	79.85	76.32	2.06	1.38
Heat	-	-	-	3.04	3.86	3.34	3.47	-	-
Total transport	**83.03**	**111.24**	**157.70**	**205.60**	**198.69**	**200.91**	**199.65**	**3.85**	**1.13**
Coal and peat	0.33	0.01	0.11	0.13	0.25	0.25	0.17	-6.54	2.44
Oil	80.97	109.19	155.08	202.21	192.77	194.66	193.33	3.90	1.06
Natural Gas	-	0.01	0.10	0.44	1.91	2.10	2.19	-	16.10
Biofuels and wastes	-	-	-	-	0.65	0.85	0.90	-	-
Electricity	1.73	2.03	2.42	2.83	3.11	3.07	3.05	2.00	1.11
Residential	**47.24**	**60.07**	**85.18**	**105.82**	**110.22**	**116.11**	**113.60**	**3.53**	**1.38**
Coal and peat	15.17	16.84	12.64	0.87	1.31	1.25	1.23	-1.07	-10.51
Oil	13.84	15.69	22.55	29.81	22.99	24.44	23.53	2.91	0.20
Natural Gas	2.24	5.62	13.97	25.60	29.25	30.48	30.53	11.37	3.79
Biofuels and wastes	2.86	2.43	3.45	3.04	2.26	2.24	2.27	1.11	-1.98
Geothermal	-	-	0.01	0.01	0.01	0.01	0.02	-	2.23
Solar, wind, tide	-	0.03	1.73	1.29	0.93	0.94	0.96	-	-2.77
Electricity	13.13	19.46	30.79	43.61	51.23	54.20	52.45	5.14	2.57
Heat	-	-	0.04	1.57	2.25	2.56	2.62	-	21.51
Comm & public services	**32.48**	**37.14**	**84.25**	**121.50**	**130.30**	**132.88**	**133.12**	**5.77**	**2.20**
Coal and peat	1.56	1.59	1.58	1.00	0.85	0.84	0.81	0.06	-3.10
Oil	24.66	24.45	43.47	50.54	31.74	29.49	29.13	3.39	-1.89
Natural Gas	0.85	1.61	6.93	16.90	28.60	31.26	31.29	13.16	7.45
Biofuels and waste	-	0.03	0.08	0.24	0.70	0.83	1.00	-	13.10
Geothermal	-	-	0.13	0.24	0.29	0.27	0.27	-	3.68
Solar, wind, tide	-	-	0.07	0.04	0.04	0.05	0.05	-	-1.32
Electricity	5.38	9.32	31.78	51.67	67.05	68.99	69.46	11.02	3.79
Heat	0.04	0.15	0.22	0.86	1.03	1.14	1.11	10.68	7.94
Non-energy use	**58.27**	**49.41**	**65.49**	**103.72**	**117.96**	**120.14**	**119.45**	**0.69**	**2.90**
Coal and peat	-	-	0.57	0.50	1.01	1.07	1.00	-	2.68
Oil	58.02	48.96	63.42	100.57	114.86	116.73	116.20	0.53	2.93
Natural Gas	0.26	0.45	1.50	2.66	2.10	2.34	2.26	10.96	1.96

(1) Total Final Consumption (TFC) excludes use in transformation processes and energy industry own use.

* Please refer to notes and definitions in Part I. Agriculture, Forestry, Fishing and "Non-specified other" sectors are not reported here individually, but are accounted for in Total final consumption.

Source: *IEA/OECD Energy Balances of OECD Countries*

IEA ASIA OCEANIA

5. Coal balance*

(Mtce)

	1973	1980	1990	2000	2005	2010	2011	2012e	Average annual percent change 73-90	90-11
Production	94.4	103.1	170.8	245.5	294.3	343.6	323.5	338.8	3.5	3.1
Imports	59.6	74.9	127.4	194.9	227.1	269.2	268.3	273.3	4.6	3.6
Exports	-26.0	-41.9	-98.3	-177.7	-219.8	-274.8	-266.2	-281.8	8.1	4.9
Stock changes	0.3	8.7	-2.2	6.0	2.2	4.5	13.3	14.4		
Primary supply	**128.3**	**144.8**	**197.6**	**268.7**	**303.9**	**342.5**	**339.0**	**344.6**	**2.6**	**2.6**
Statistical differences	**-2.5**	**-2.3**	**3.2**	**1.7**	**10.5**	**-4.2**	**-1.3**	..		
Total transformation	**-68.7**	**-79.1**	**-125.3**	**-207.9**	**-251.5**	**-271.7**	**-271.4**	..	**3.6**	**3.7**
Electricity and heat gen.	-37.8	-45.2	-86.3	-168.3	-211.3	-230.7	-227.6	..	5.0	4.7
Main activity producers (1)	*-37.8*	*-40.9*	*-75.3*	*-152.1*	*-193.0*	*-210.7*	*-206.9*	..	*4.1*	*4.9*
Autoproducers	*-0.1*	*-4.3*	*-11.0*	*-16.3*	*-18.2*	*-20.0*	*-20.7*	..	*36.1*	*3.1*
Gas works	3.7	5.2	-0.3	-0.2	0.1	0.0	0.0	..	-	-
Coal transformation(2)	-34.7	-39.1	-38.6	-39.4	-40.3	-41.0	-43.8	..	0.6	0.6
BKB plants	*-0.2*	*-0.2*	*-0.1*	*-0.0*	*-0.1*	*-0.0*	*-0.0*	..	*-5.3*	*-3.7*
Blast furnaces	*-27.1*	*-23.8*	*-32.2*	*-37.8*	*-36.0*	*-39.6*	*-40.9*	..	*1.0*	*1.2*
Coke ovens	*-5.8*	*-14.5*	*-1.6*	*-0.6*	*-4.3*	*-1.4*	*-2.9*	..	*-7.3*	*2.8*
Patent fuel plants	*-1.5*	*-0.6*	*-4.8*	*-0.9*	*0.1*	-	-	..	*6.9*	-
Other transformation(3)	-	-	-	-	-	-	-	..	-	-
Energy ind. own use	**-4.6**	**-5.5**	**-6.1**	**-6.8**	**-5.5**	**-8.8**	**-9.0**	..	**1.7**	**1.9**
Losses	**-0.1**	**-0.3**	**-0.2**	**-0.0**	**-0.0**	**-0.0**	**-0.0**	..		
Final consumption(4)	**52.4**	**57.6**	**69.3**	**55.7**	**57.3**	**57.9**	**57.2**	..	**1.7**	**-0.9**
Industry(5)	35.2	39.2	54.4	53.2	54.1	54.4	53.9	..	2.6	-0.0
Iron and steel	*26.9*	*27.3*	*19.8*	*18.7*	*21.6*	*23.2*	*24.1*	..	*-1.8*	*0.9*
Chemical	*0.4*	*0.8*	*4.3*	*4.4*	*4.5*	*4.7*	*5.0*	..	*15.3*	*0.6*
Non-metallic minerals	*0.9*	*5.9*	*11.9*	*12.1*	*11.1*	*10.0*	*10.2*	..	*16.4*	*-0.7*
Paper, pulp and print	*0.3*	*0.5*	*1.9*	*2.3*	*2.3*	*2.2*	*2.2*	..	*10.8*	*0.6*
Other industry (6)	*6.7*	*4.7*	*16.5*	*15.7*	*14.6*	*14.3*	*12.4*	..	*5.5*	*-1.3*
Transport(7)	0.3	0.0	0.1	0.1	0.3	0.2	0.2	..	-6.5	2.4
Other	16.8	18.4	14.2	1.9	2.4	2.2	2.1	..	-1.0	-8.6
Comm. and pub. services	*1.6*	*1.6*	*1.6*	*1.0*	*0.9*	*0.8*	*0.8*	..	*0.1*	*-3.1*
Residential	*15.2*	*16.8*	*12.6*	*0.9*	*1.4*	*1.2*	*1.2*	..	*-1.1*	*-10.5*
Other sectors (8)	*0.1*	-	*0.0*	*0.0*	*0.0*	*0.1*	*0.1*	..	*-11.4*	*9.5*
Non-energy use	-	-	0.6	0.5	0.5	1.1	1.0	..	-	2.7

(1) Main activity electricity and heat generation includes district heating.

(2) Coal transformation refers to the transformation of coal and peat to secondary and tertiary products (mainly cokes, briquettes, coal tar and off-gases). Gas works and Liquefaction are shown separately to match the *Energy Balances of OECD Countries.*

(3) Other transformation includes Liquefaction and Non-specified transformations.

(4) Final Consumption includes non-energy use and energy use (Industry, Transport and Other).

(5) Please refer to notes and definitions in Part I for detailed explanation.

(6) Other industry includes Non-ferrous metals, Transport equipment, Machinery, Mining and quarrying, Food and tobacco, Wood and wood products, Construction, Textile and leather, and Non-specified industry.

(7) Transport includes Rail and Inland waterways.

(8) Other sectors includes Agriculture/Forestry, Fishing and Non-specified other.

* "Coal" refers to all types of coal, primary (anthracite, coking coal, other bituminous coal, sub-bituminous coal, lignite/brown coal) and derived fuels (including patent fuel, BKB, coke oven coke, gas coke, coal tar, coke oven gas, gas works gas, blast furnace gas and other recovered gases), as well as peat, peat briquettes and oil shale. Quantities have been converted to Mtce using calorific values reported by the respective countries.
Please refer to notes and definitions in Part I.

Source: *IEA/OECD Energy Balances of OECD Countries*

IEA ASIA OCEANIA

6. Use of coal for selected end-uses*

(million tonnes)

	1978	1990	2000	2005	2009	2010	2011	Average annual percent change 78-90	90-11
Steam coal	**57.24**	**128.28**	**206.41**	**250.36**	**268.14**	**284.74**	**279.48**	**6.96**	**3.78**
Total electricity and heat	30.57	77.21	155.44	206.60	228.57	230.86	227.32	8.03	5.28
Main activity producers	*30.21*	*72.22*	*147.71*	*195.31*	*216.08*	*217.57*	*213.07*	*7.53*	*5.29*
Autoproducers	*0.36*	*4.99*	*7.72*	*11.29*	*12.49*	*13.30*	*14.25*	*24.49*	*5.12*
Patent fuel/BKB plants	18.64	20.81	2.41	2.01	1.94	1.86	1.82	0.92	-10.95
Coke ovens/Liquefaction(1)	0.13	6.53	12.13	12.41	10.20	13.36	14.36	38.86	3.83
Blast furnace inputs	-	-	-	0.80	0.55	0.70	0.62	-	-
Gas manufacture	0.19	-	-	-	-	-	-	-	-
Industry	7.37	25.15	32.43	28.52	26.01	26.96	28.00	10.78	0.51
Iron and steel	*1.46*	*2.05*	*3.42*	*4.81*	*4.05*	*4.59*	*5.57*	*2.86*	*4.88*
Chemical	*0.34*	*2.63*	*3.92*	*4.21*	*4.19*	*4.54*	*4.86*	*18.50*	*2.97*
Non-metallic minerals	*1.88*	*12.63*	*13.30*	*12.29*	*10.90*	*11.02*	*11.44*	*17.23*	*-0.47*
Paper, pulp and print	*0.35*	*2.25*	*2.64*	*2.83*	*2.71*	*2.66*	*2.62*	*16.65*	*0.73*
Other industry	*3.34*	*5.60*	*9.15*	*4.37*	*4.16*	*4.16*	*3.52*	*4.42*	*-2.19*
Other sectors(2)	1.49	1.04	1.05	1.01	0.94	0.94	0.97	-2.95	-0.37
Non-energy use	-	0.02	-	-	-	-	-	-	-
Coking coal	**67.95**	**82.84**	**81.30**	**81.99**	**77.00**	**89.10**	**90.82**	**1.67**	**0.44**
Total electricity and heat	-	-	-	-	-	-	-	-	-
Main activity producers	-	-	-	-	-	-	-	-	-
Autoproducers	-	-	-	-	-	-	-	-	-
Patent fuel/BKB plants	-	-	-	-	-	-	-	-	-
Coke ovens/Liquefaction(1)	64.35	77.00	67.15	67.67	62.33	69.01	68.56	1.51	-0.55
Blast furnace inputs	-	5.25	13.97	15.92	14.75	19.19	21.90	-	7.04
Gas manufacture	4.45	-	-	-	-	-	-	-	-
Industry	0.12	0.00	0.00	0.00	0.32	0.23	0.12	-32.71	25.61
Iron and steel	*0.06*	*0.00*	*0.00*	*0.00*	*0.26*	*0.15*	*0.11*	*-29.20*	*24.92*
Chemical	*0.01*	-	-	-	-	-	-	-	-
Non-metallic minerals	-	-	-	-	*0.00*	*0.00*	-	-	-
Paper, pulp and print	-	-	-	-	-	-	-	-	-
Other industry	*0.04*	-	-	-	*0.06*	*0.07*	*0.01*	-	-
Other sectors(2)	0.01	-	-	-	-	0.01	-	-	-
Non-energy use	-	-	-	-	-	-	-	-	-
Lignite	**30.64**	**46.15**	**67.51**	**70.78**	**72.30**	**72.38**	**71.32**	**3.47**	**2.09**
Total electricity and heat	27.32	44.15	66.35	68.11	72.21	71.58	70.17	4.08	2.23
Main activity producers	*26.32*	*43.07*	*65.25*	*68.10*	*72.19*	*71.56*	*70.16*	*4.19*	*2.35*
Autoproducers	*0.99*	*1.08*	*1.10*	*0.01*	*0.02*	*0.02*	*0.02*	*0.72*	*-18.18*
Patent fuel/BKB plants	2.85	1.78	0.96	1.00	0.60	0.48	0.54	-3.84	-5.55
Coke ovens/Liquefaction(1)	-	-	-	-	-	-	-	-	-
Gas manufacture	-	-	-	-	-	-	-	-	-
Industry	0.42	0.21	0.17	0.21	0.21	0.20	0.24	-5.76	0.68
Iron and steel	-	-	-	-	-	-	-	-	-
Chemical	*0.03*	*0.03*	-	-	-	-	-	*-1.23*	-
Non-metallic minerals	*0.00*	*0.00*	-	*0.00*	*0.00*	*0.00*	*0.00*	*-3.32*	*0.00*
Paper, pulp and print	*0.30*	-	-	-	-	-	-	-	-
Other industry	*0.10*	*0.18*	*0.17*	*0.21*	*0.21*	*0.20*	*0.24*	*5.52*	*1.31*
Other sectors(2)	0.05	0.06	0.08	0.06	0.03	0.06	0.07	0.90	0.46
Non-energy use	-	-	-	-	-	-	-	-	-

(1) Liquefaction primarily refers to direct distillation processes. Indirect liquefaction may be reported here or under gas manufacture.

(2) Other sectors are Residential, Commercial and public services, Agriculture/Forestry, Fishing and Non-specified other.

* Coal totals may not match the sum of the individual rows due to statistical differences. Please refer to notes and definitions in Part I.

Source: *IEA/OECD Energy Statistics of OECD Countries*

IEA ASIA OCEANIA

7. Fuel prices to end users*

	1978[1]	1990	1995	2000	2005	2009	2010	2011	2012
				(US dollars / tce)					
For electricity generation									
Steam coal	..	..	..	..	..	..	..	..	..
Heavy fuel oil	..	..	..	..	..	..	..	..	..
Natural gas	..	..	..	..	..	..	..	..	..
For industry									
Steam coal	..	..	..	..	..	..	..	..	..
Coking coal	..	..	..	..	..	..	..	..	..
High sulphur fuel oil	..	..	..	..	..	..	..	..	..
Low sulphur fuel oil	..	..	..	..	..	..	..	..	..
Natural gas	..	..	..	..	..	..	..	..	..
				(US dollars / unit)[2]					
For electricity generation									
Steam coal	..	..	..	..	..	..	..	..	..
Heavy fuel oil	..	..	..	..	..	..	..	..	..
Natural gas	..	..	..	..	..	..	..	..	..
For industry									
Steam coal	..	..	..	..	..	..	..	..	..
Coking coal	..	..	..	..	..	..	..	..	..
High sulphur fuel oil	..	..	..	..	..	..	..	..	..
Low sulphur fuel oil	..	..	..	..	..	..	..	..	..
Natural gas	..	..	..	..	..	..	..	..	..

(1) Earliest year for which split by coal type is available.

(2) For steam coal, coking coal and heavy fuel oil per metric tonne; for natural gas per 10^7 kilocalories GCV.

* Please refer to notes and definitions in Part I. Steam coal prices relate to historical definitions.

Source: *IEA/OECD Energy Prices & Taxes*

8. Coal and peat trade by type of coal*

(Mtce)

	1978	1990	1995	2000	2005	2009	2010	2011	2012e
Total imports	**54.88**	**127.35**	**161.69**	**194.93**	**227.12**	**235.58**	**269.19**	**268.31**	**273.26**
Bituminous coal[1]	1.59	46.73	80.35	115.64	147.54	161.74	181.24	181.20	189.61
Coking coal	53.19	79.50	79.14	73.02	73.71	70.15	82.52	82.74	80.24
Sub-bituminous coal	-	0.60	1.59	3.67	2.78	3.00	3.78	3.11	2.02
Lignite	-	-	-	-	-	-	-	-	-
Peat	-	-	-	-	-	-	-	-	-
Coal products[2]	0.10	0.51	0.61	2.60	3.10	0.69	1.65	1.27	1.39
Total exports	**35.25**	**98.34**	**131.47**	**177.72**	**219.77**	**244.68**	**274.75**	**266.15**	**281.80**
Bituminous coal[1]	4.25	39.96	54.64	77.02	94.23	119.71	118.82	126.41	139.59
Coking coal	29.83	55.92	72.93	98.02	123.85	123.85	155.26	138.74	140.69
Sub-bituminous coal	-	-	-	-	-	0.05	-	0.01	-
Lignite	-	-	-	-	-	-	-	-	-
Peat	-	-	-	-	-	-	-	-	-
Coal products[2]	1.17	2.46	3.91	2.68	1.68	1.07	0.67	1.00	1.52

(1) Bituminous coal includes anthracite.

(2) Coal products includes products derived from coal, for example: coke, coal tar, briquettes, patent fuels.

* Please refer to notes and definitions in Part I.

Source: *IEA/OECD Energy Balances of OECD Countries, IEA/OECD Energy Statistics of OECD Countries*

IEA ASIA OCEANIA

9. Total coal imports by origin*

(thousand tonnes)

	1978[1]	1985	1990	1995	2000	2005	2010	2011	2012e
Total coal[2]	**55041**	**110580**	**130648**	**173185**	**215251**	**254817**	**304236**	**303463**	**309307**
Coking coal	**52885**	**77367**	**76223**	**76950**	**76656**	**77155**	**85840**	**86054**	**83456**
Australia	25299	33741	39558	39880	45398	40884	48994	45837	44454
Canada	11304	19321	13316	13007	9389	8923	11261	11891	11532
Czech Republic	-	-	-	-	-	-	-	-	-
Germany	398	-	-	-	-	-	-	-	-
Poland	429	-	-	-	-	-	-	-	-
United Kingdom	-	-	-	-	-	-	-	-	-
United States	10406	14729	10275	7541	2841	1838	5859	9802	9505
Other OECD	11	401	225	298	367	530	384	364	354
China, People's Rep.	420	1211	2802	5327	9555	7031	2615	2437	2364
Colombia	-	-	153	52	-	-	24	24	23
Indonesia	-	103	668	4629	5886	15374	13157	11497	11151
South Africa	2360	5086	3055	3437	603	-	-	-	-
Former Soviet Union[3]	2244	2775	x	x	x	x	x	x	x
Russian Federation [4]	*x*	*x*	*6150*	*2741*	*2498*	*2534*	*3499*	*4184*	*4057*
Other FSU [4]	*x*	*x*	-	-	-	-	-	-	-
Venezuela	-	-	-	-	-	-	-	-	-
Vietnam	-	-	-	-	27	-	-	-	-
Non-specified/other	14	-	21	38	92	41	47	18	1
Steam coal	**2156**	**33213**	**54425**	**96235**	**138595**	**177662**	**218396**	**217409**	**225851**
Australia	668	19238	24852	43431	67182	96282	111677	107820	114091
Canada	105	2249	8170	11604	9944	1194	7610	10521	10706
Czech Republic	-	-	-	-	-	-	-	-	-
Germany	-	-	-	1	-	-	-	-	-
Poland	-	-	-	-	-	-	-	-	-
United Kingdom	-	-	-	-	-	-	-	-	-
United States	1	1980	5751	7780	2396	365	492	1521	1615
Other OECD	-	14	138	319	637	56	-	-	-
China, People's Rep.	513	2427	3504	13065	33751	38226	11592	7134	7008
Colombia	-	-	94	56	-	-	138	470	483
Indonesia	-	306	806	8323	14185	26174	61492	62376	63051
South Africa	157	5508	7057	6493	3448	97	2691	3724	3758
Former Soviet Union[3]	149	1020	x	x	x	x	x	x	x
Russian Federation [4]	*x*	*x*	*3402*	*3189*	*5497*	*12369*	*17071*	*20384*	*21832*
Other FSU [4]	*x*	*x*	-	-	-	*21*	-	-	-
Venezuela	-	-	-	-	-	15	-	-	-
Vietnam	-	157	150	1496	1079	2401	3454	2815	2678
Non-specified/other	563	314	501	478	476	462	2179	644	629
Lignite	-	-	-	-	-	-	-	-	-

(1) Earliest year for which split by coal type is available.

(2) Total coal consists of steam coal, coking coal and lignite.

(3) For years prior to 1990, or if no details for individual Former Soviet Union republics.

(4) Data for former republics are not available separately for some OECD member countries.

* Please refer to notes and definitions in Part I.

Source: IEA/OECD Coal Statistics

IEA ASIA OCEANIA

10. Coking coal exports by destination*

(thousand tonnes)

	1978(1)	1985	1990	1995	2000	2005	2010	2011	2012e
World	**31085**	**49698**	**58086**	**75625**	**100712**	**127246**	**159566**	**142568**	**144573**
Total OECD	**29188**	**42067**	**44287**	**54979**	**74911**	**85006**	**82157**	**77956**	**77181**
Australia	-	-	-	123	122	49	-	-	-
Austria	-	-	-	-	-	-	-	-	-
Belgium	374	16	1054	1524	1833	1881	1034	1314	1276
Canada	-	-	30	-	-	-	-	-	-
Chile	32	64	200	466	776	822	446	413	543
Czech Republic	-	-	-	-	-	-	-	-	-
Denmark	-	-	-	-	-	-	-	-	-
Estonia	-	-	-	-	-	-	-	-	-
Finland	-	-	-	-	-	392	76	68	314
France	1173	2106	1917	2310	3739	3895	2506	2560	2453
Germany	-	-	25	204	2619	1746	1160	1275	908
Greece	191	-	-	-	-	-	-	-	-
Hungary	-	-	-	-	-	-	-	-	-
Iceland	-	-	-	-	-	-	-	-	-
Ireland	-	-	-	-	-	14	-	-	-
Israel	-	-	-	-	-	129	55	-	-
Italy	1319	2055	1181	2752	2914	2527	1559	1621	1751
Japan	22651	29559	28822	31291	39861	45780	48934	43038	40668
Korea	1330	3064	4944	9039	10358	12457	15864	16443	16398
Luxembourg	-	-	-	-	-	-	-	-	-
Mexico	10	-	-	-	666	588	336	254	423
Netherlands	992	1380	725	703	2279	5652	3342	3837	5029
New Zealand	-	-	-	-	-	-	-	-	-
Norway	-	-	-	-	-	-	-	-	-
Poland	-	-	-	-	-	-	-	-	40
Portugal	-	-	-	129	-	-	-	-	-
Slovak Republic	-	-	-	-	-	-	-	-	-
Slovenia	-	-	-	-	-	-	-	-	72
Spain	593	1186	694	896	1767	2470	1069	1721	1486
Sweden	-	-	702	535	992	1226	1486	1223	980
Switzerland	-	-	-	-	46	-	-	-	-
Turkey	-	516	912	1608	1451	720	909	736	1514
United Kingdom	469	2121	3081	3360	5488	4451	3381	3453	3326
United States	54	-	-	39	-	207	-	-	-
Total non-OECD	**1897**	**7631**	**13799**	**20415**	**25569**	**38109**	**72894**	**61243**	**65438**
Brazil	164	958	1291	3448	4988	3218	4234	2882	2338
China(2)	-	192	560	317	265	4330	27500	16040	21638
Chinese Taipei	981	1990	2798	3851	6273	7091	5357	8050	8585
Egypt	-	312	323	459	-	353	324	-	-
India	32	2042	4895	9831	10773	18039	32450	32057	30425
Romania	675	595	2256	710	-	46	-	-	-
Oth. Africa & Mid. East	-	688	454	597	1554	3108	1610	1180	1322
Oth. non-OECD Americas	-	35	300	384	595	784	603	608	570
Other Asia & Oceania	45	688	734	818	1051	737	671	356	560
Other non-OECD Europe and Eurasia	-	131	188	-	70	403	145	70	-
Non-specified/Other	**-**	**-**	**-**	**231**	**232**	**4131**	**4515**	**3369**	**1954**

(1) Earliest year for which split by coal type is available.
(2) China includes the People's Republic of China and Hong Kong, China.
* Please refer to notes and definitions in Part I.
Source: IEA/OECD Coal Statistics

IEA ASIA OCEANIA

11. Steam coal exports by destination*

(thousand tonnes)

	1978(1)	1985	1990	1995	2000	2005	2010	2011	2012e
World	**5533**	**39339**	**45647**	**62413**	**87804**	**107418**	**135479**	**144129**	**159155**
Total OECD	**4809**	**29593**	**38304**	**51291**	**70501**	**85460**	**95692**	**98650**	**103280**
Australia	-	-	-	-	-	-	-	-	-
Austria	-	-	-	-	-	-	-	-	-
Belgium	-	771	129	1116	428	17	-	-	-
Canada	-	-	-	-	-	-	-	-	-
Chile	-	99	120	1023	1301	412	309	691	388
Czech Republic	-	-	-	-	-	-	-	-	-
Denmark	150	2557	1149	1076	142	-	-	-	-
Estonia	-	-	-	-	-	-	-	-	-
Finland	-	-	-	-	-	-	-	-	-
France	682	508	934	276	434	469	66	-	-
Germany	458	591	125	61	72	115	-	-	-
Greece	-	125	-	-	110	-	-	-	-
Hungary	-	-	-	-	-	-	-	-	-
Iceland	-	-	-	-	-	-	-	-	-
Ireland	-	-	133	121	284	568	-	-	-
Israel	-	577	528	695	2623	1165	516	501	494
Italy	-	226	-	366	428	141	-	-	-
Japan	1489	15326	26569	34556	47449	57574	66413	66961	69817
Korea	-	5453	3633	9662	11455	17970	24840	28273	28925
Luxembourg	-	-	-	-	-	-	-	-	-
Mexico	-	-	-	-	-	4238	3109	1703	3595
Netherlands	320	2464	4236	1562	2550	760	127	-	-
New Zealand	-	-	-	-	16	56	59	58	22
Norway	-	-	-	-	-	-	-	-	-
Poland	-	-	-	-	-	-	-	-	-
Portugal	-	-	-	264	-	-	-	-	-
Slovak Republic	-	-	-	-	-	-	-	-	-
Slovenia	-	-	-	-	-	-	-	-	-
Spain	-	455	205	110	1445	671	-	-	-
Sweden	-	-	155	44	83	164	73	145	39
Switzerland	-	-	29	-	-	-	-	-	-
Turkey	-	-	-	25	55	45	-	-	-
United Kingdom	932	441	328	142	1499	993	13	-	-
United States	778	-	31	192	127	102	167	318	-
Total non-OECD	**204**	**8082**	**7343**	**11122**	**17303**	**21958**	**39668**	**45432**	**55875**
Brazil	-	-	158	-	-	33	20	41	-
China(2)	-	2189	2443	3420	1429	2121	15192	17867	29968
Chinese Taipei	76	3761	3046	5920	10034	14329	19553	20124	17516
Egypt	-	-	-	-	-	-	-	-	-
India	-	93	47	-	2469	1461	610	495	1165
Romania	-	-	33	110	-	-	-	-	-
Oth. Africa & Mid. East	-	-	-	-	-	-	-	2	-
Oth. non-OECD Americas	-	-	-	-	72	-	-	-	-
Other Asia & Oceania	128	2039	1616	1672	3299	4014	4293	6903	7226
Other non-OECD Europe and Eurasia	-	-	-	-	-	-	-	-	-
Non-specified/Other	**520**	**1664**	**-**	**-**	**-**	**-**	**119**	**47**	**-**

(1) Earliest year for which split by coal type is available.

(2) China includes the People's Republic of China and Hong Kong, China.

* Please refer to notes and definitions in Part I. Steam coal includes all sub-bituminous coal.

Source: IEA/OECD Coal Statistics

IEA EUROPE*

Figure 1: TPES by fuel (Mtce)

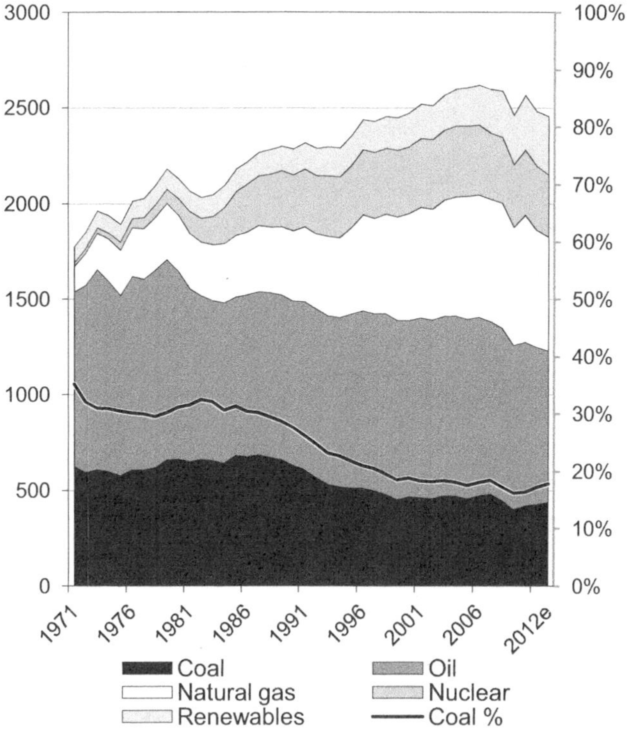

Figure 2: Coal supply indicators (1971 = 100)

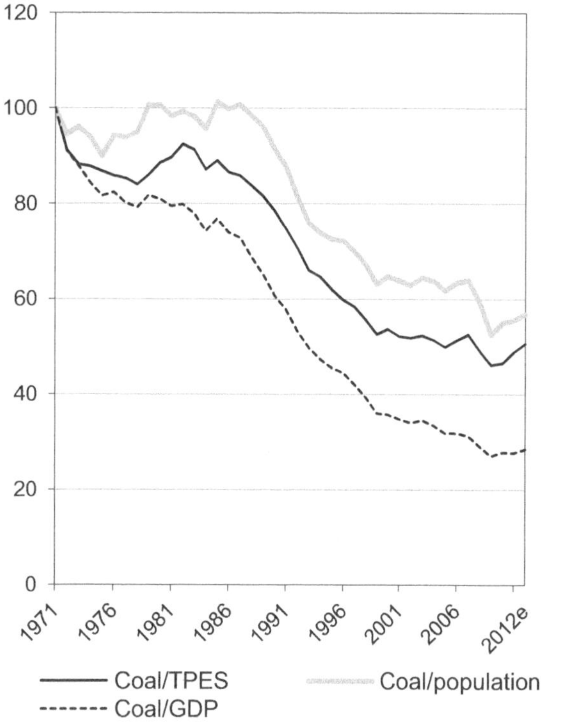

Figure 3: Primary coal supply (Mtce)

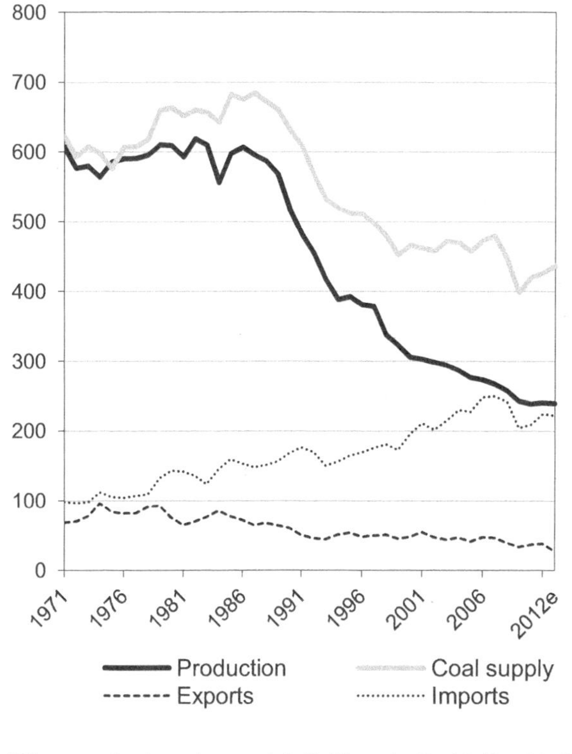

Figure 4: Coal consumption (Mtce)

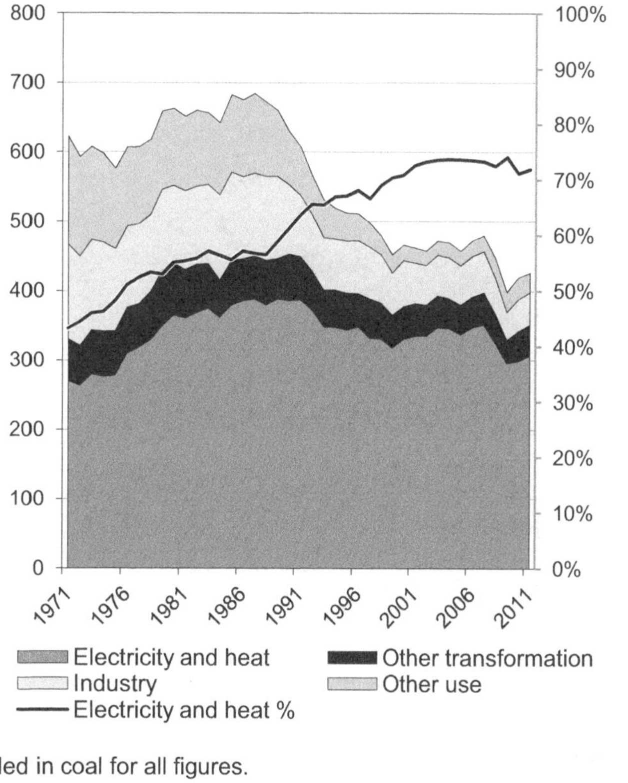

*Please refer to notes and definitions in Part I. Peat is included in coal for all figures.

Source: *IEA/OECD Coal Information, IEA/OECD Energy Balances of OECD Countries, OECD Main Economic Indicators*

IEA EUROPE*

Figure 5: Electricity generation by fuel (TWh)

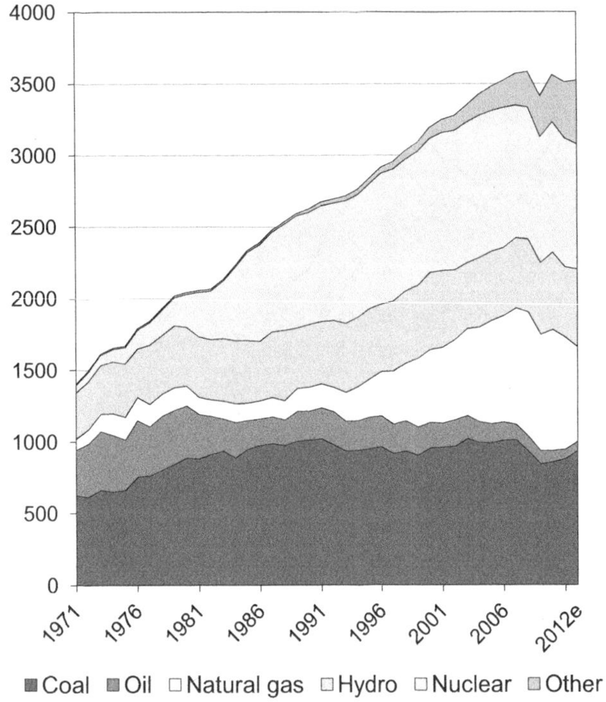

Figure 6: CO_2 emissions by fuel (Mt CO_2)

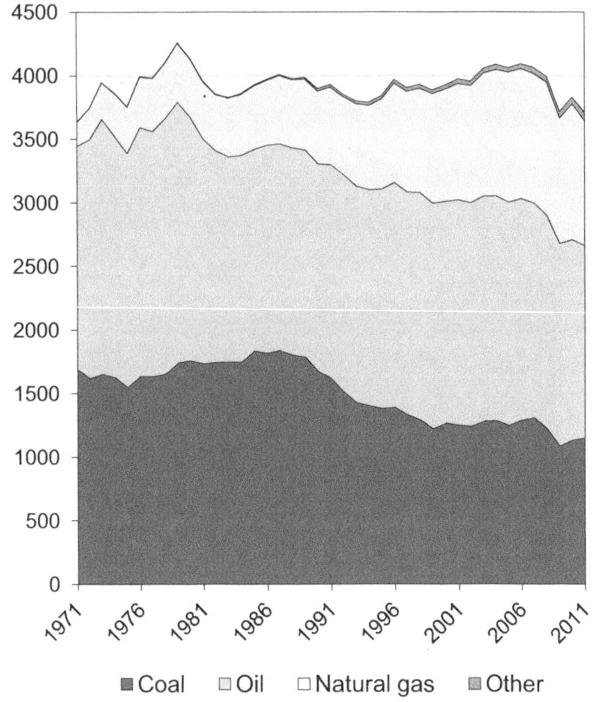

Figure 7: Electricity generation by fuel share

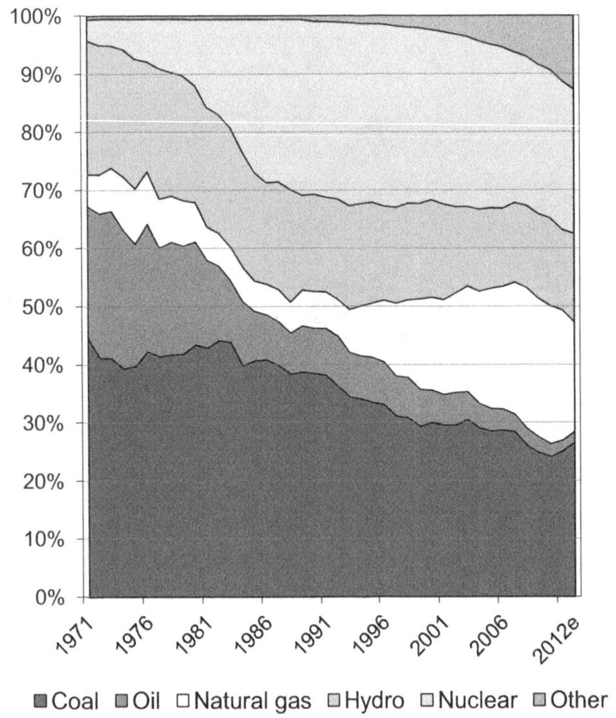

Figure 8: CO_2 emissions by fuel share

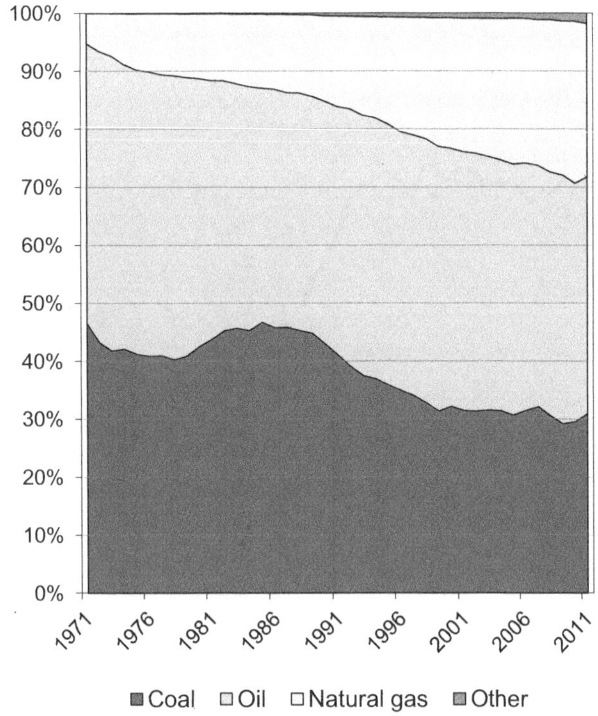

*Please refer to notes and definitions in Part I. Peat is included in coal for all figures.

Source: *IEA/OECD Electricity Information, IEA/OECD CO2 Emissions from Fuel Combustion*

IEA EUROPE

1. Total primary energy supply (TPES) by fuel*

	1973	1980	1990	2000	2010	2011	2012e	Average annual percent change 73-90	90-11
TPES (Mtce)	**1963.32**	**2132.49**	**2288.26**	**2475.08**	**2569.55**	**2481.71**	**2454.98**	**0.90**	**0.39**
Coal and peat	607.73	662.58	630.78	466.17	419.59	426.32	436.67	0.22	-1.85
Oil	1045.48	982.64	858.07	923.37	855.56	822.27	794.33	-1.16	-0.20
Natural Gas	192.47	294.44	368.49	559.36	665.43	609.80	593.11	3.89	2.43
Biofuels and waste	44.27	50.97	77.03	99.51	184.48	182.57	189.33	3.31	4.19
Nuclear	27.69	85.79	291.38	348.01	339.13	334.96	325.71	14.85	0.67
Hydro	41.79	50.69	53.96	65.89	66.10	59.91	66.17	1.52	0.50
Geothermal	3.12	3.38	5.28	7.84	11.47	12.02	12.21	3.15	4.00
Solar, wind, tide	0.07	0.07	0.41	3.80	24.55	31.91	37.02	11.02	23.10
Net electricity trade(1)	0.71	1.92	2.85	0.48	2.46	1.22	-0.34	8.55	-3.93
Heat(2)	-	-	0.03	0.65	0.78	0.72	0.76	-	16.82

(1) Net trade between OECD and non-OECD countries.

(2) Ambient heat from heat pumps used in the transformation sector.

Source: *IEA/OECD Energy Balances of OECD Countries*

2. Energy supply, GDP and population*

	1973	1980	1990	2000	2010	2011	2012e	Average annual percent change 73-90	90-11
GDP (2005 bil. USD)	7074	8373	10634	13359	15439	15724	15704	2.43	1.88
TPES/GDP(1)	0.28	0.25	0.22	0.19	0.17	0.16	0.16	-1.49	-1.47
Population (millions)	454.8	473.6	496.2	518.0	548.7	551.2	552.7	0.51	0.50
TPES/population(1)	4.32	4.50	4.61	4.78	4.68	4.50	4.44	0.39	-0.11
TPES/GDP(2)	156.3	143.4	121.2	104.4	93.7	88.9	88.1	-1.49	-1.47
Coal and peat supply/GDP(2)	276.0	254.3	190.6	112.1	87.3	87.1	89.3	-2.16	-3.66
Elec. consumption/GDP(2)	93.9	101.0	103.1	99.8	98.2	94.7	..	0.55	-0.40
Elec. generation (TWh)	1616	2046	2628	3194	3560	3512	3523	2.90	1.39
Industrial production(2)	..	..	..	..	..	..	..	..	..

(1) TPES/GDP in units of tce/2005 thousand USD, TPES/population in units of tce/capita.

(2) As index, 2005 (not shown) = 100

Source: *IEA/OECD Energy Balances of OECD Countries , OECD Main Economic Indicators*

3. Coal and peat production by type*

	1978(1)	1990	2000	2005	2010	2011	2012e	Average annual percent change 78-90	90-11
Mtce:									
Coking coal	136.40	92.69	44.80	37.08	25.90	25.24	24.35	-3.17	-6.01
Steam coal	278.59	224.43	129.41	109.50	84.51	82.43	83.80	-1.79	-4.66
Lignite	177.75	194.47	128.07	125.23	123.66	128.59	128.86	0.75	-1.95
Peat	2.30	4.96	3.21	4.56	4.32	3.81	2.18	6.61	-1.24
Mt:									
Coking coal	141.07	93.45	45.21	37.30	26.19	25.49	24.57	-3.37	-6.00
Steam coal	347.05	283.08	163.72	139.62	107.72	104.73	106.33	-1.68	-4.62
Lignite	585.45	644.55	429.99	432.66	407.40	429.13	432.57	0.80	-1.92
Peat	8.05	14.68	9.96	13.57	13.21	11.38	6.35	5.13	-1.20

(1) Earliest year for which split by coal type is available.

* Please refer to notes and definitions in Part I.

Source: *IEA/OECD Energy Balances of OECD Countries, IEA/OECD Energy Statistics of OECD Countries*

IEA EUROPE

4. Final consumption of energy by fuel*

(Mtce)

	1973	1980	1990	2000	2009	2010	2011	Average annual percent change 73-90	90-11
Total final consumption[(1)]	**1458.56**	**1541.90**	**1588.66**	**1738.05**	**1721.97**	**1805.92**	**1730.78**	**0.50**	**0.41**
Coal and peat	264.40	222.67	177.90	88.41	68.74	76.94	74.85	-2.30	-4.04
Oil	823.83	786.98	734.25	806.60	759.60	759.63	733.23	-0.67	-0.01
Natural Gas	135.85	229.74	285.98	381.86	366.23	404.93	370.39	4.48	1.24
Biofuels and wastes	40.36	43.56	64.36	73.03	99.65	110.90	110.83	2.78	2.62
Geothermal	0.07	0.10	1.21	1.83	3.32	3.57	3.81	18.12	5.61
Solar, wind, tide	-	0.01	0.24	0.94	2.21	2.49	2.96	-	12.75
Electricity	164.64	209.58	271.50	330.21	358.07	375.56	369.02	2.99	1.47
Heat	29.41	49.26	53.22	55.17	64.16	71.90	65.69	3.55	1.01
of which:									
Total industry	**531.97**	**508.68**	**454.11**	**459.42**	**384.94**	**414.75**	**415.05**	**-0.93**	**-0.43**
Coal and peat	130.29	112.12	100.47	65.96	40.08	45.64	46.77	-1.52	-3.58
Oil	214.57	166.86	81.18	75.94	52.20	50.83	47.20	-5.56	-2.55
Natural Gas	74.30	95.66	109.85	139.97	111.51	123.58	124.16	2.33	0.58
Biofuels and wastes	7.03	8.39	20.58	24.53	30.32	33.64	34.81	6.52	2.53
Geothermal	-	-	0.01	0.01	0.02	0.03	0.03	-	7.39
Solar, wind, tide	-	-	0.01	0.14	0.00	0.01	0.02	-	1.48
Electricity	90.40	105.03	123.24	139.22	129.41	138.42	139.81	1.84	0.60
Heat	15.38	20.62	18.78	13.65	21.39	22.60	22.25	1.18	0.81
Total transport	**248.29**	**297.71**	**376.60**	**449.30**	**474.60**	**473.04**	**468.70**	**2.48**	**1.05**
Coal and peat	9.79	3.63	0.30	0.01	0.02	0.02	0.01	-18.56	-13.86
Oil	233.42	287.37	368.26	438.36	445.58	441.56	436.33	2.72	0.81
Natural Gas	0.22	0.81	0.45	0.93	3.90	4.09	4.24	4.25	11.23
Biofuels and wastes	0.00	0.00	0.01	1.01	16.69	18.77	19.56	5.41	44.36
Electricity	4.86	5.89	7.58	8.99	8.42	8.61	8.55	2.65	0.58
Residential	**345.30**	**366.05**	**397.40**	**427.55**	**433.08**	**466.00**	**418.73**	**0.83**	**0.25**
Coal and peat	89.20	71.99	52.62	16.84	21.52	24.46	21.51	-3.06	-4.17
Oil	144.38	116.38	93.54	89.01	66.90	66.35	57.51	-2.52	-2.29
Natural Gas	33.12	71.78	105.64	155.35	159.51	176.93	152.72	7.06	1.77
Biofuels and wastes	26.32	28.45	39.85	43.86	47.61	53.11	49.45	2.47	1.03
Geothermal	0.07	0.10	0.76	1.36	2.72	3.07	3.22	14.94	7.12
Solar, wind, tide	-	0.01	0.18	0.64	1.87	2.07	2.48	-	13.23
Electricity	41.14	57.95	77.99	93.67	107.82	111.39	106.47	3.83	1.49
Heat	11.07	19.38	26.82	26.82	25.13	28.62	25.39	5.34	-0.26
Comm & public services	**136.40**	**149.66**	**156.91**	**178.09**	**218.26**	**229.21**	**211.08**	**0.83**	**1.42**
Coal and peat	22.02	24.44	18.02	2.09	3.68	2.73	2.76	-1.17	-8.55
Oil	82.12	72.12	39.80	33.46	29.33	29.16	25.96	-4.17	-2.01
Natural Gas	5.07	13.02	37.04	50.58	65.72	71.03	61.64	12.41	2.45
Biofuels and waste	1.16	1.50	0.43	1.87	2.74	2.87	2.98	-5.67	9.65
Geothermal	-	-	0.14	0.33	0.41	0.34	0.36	-	4.65
Solar, wind, tide	-	-	0.02	0.07	0.23	0.26	0.30	-	12.93
Electricity	24.87	35.55	56.55	81.41	104.62	109.18	106.08	4.95	3.04
Heat	1.17	3.03	4.90	8.28	11.53	13.64	11.01	8.81	3.93
Non-energy use	**115.93**	**128.81**	**142.19**	**159.57**	**153.60**	**161.95**	**156.31**	**1.21**	**0.45**
Coal and peat	4.42	3.08	2.44	1.52	1.45	1.86	1.89	-3.45	-1.19
Oil	104.59	107.94	119.74	138.10	136.09	142.09	136.62	0.80	0.63
Natural Gas	6.92	17.79	20.02	19.94	16.06	18.01	17.79	6.45	-0.56

(1) Total Final Consumption (TFC) excludes use in transformation processes and energy industry own use.

* Please refer to notes and definitions in Part I. Agriculture, Forestry, Fishing and "Non-specified other" sectors are not reported here individually, but are accounted for in Total final consumption.

Source: *IEA/OECD Energy Balances of OECD Countries*

IEA EUROPE

5. Coal balance*

(Mtce)

	1973	1980	1990	2000	2005	2010	2011	2012e	Average annual percent change 73-90	90-11
Production	579.5	608.9	516.5	305.5	276.4	238.4	240.1	239.2	-0.7	-3.6
Imports	97.9	143.1	168.1	195.2	227.0	208.6	223.6	222.0	3.2	1.4
Exports	-77.5	-75.1	-60.5	-48.0	-41.6	-36.7	-38.3	-27.7	-1.4	-2.1
Stock changes	7.9	-14.3	6.6	13.5	-4.4	9.2	1.0	3.2		
Primary supply	**607.7**	**662.6**	**630.8**	**466.2**	**457.4**	**419.6**	**426.3**	**436.7**	**0.2**	**-1.8**
Statistical differences	**5.1**	**-7.5**	**-11.2**	**0.8**	**-0.8**	**-3.7**	**-5.5**	..		
Total transformation	**-321.5**	**-411.9**	**-427.3**	**-367.2**	**-368.8**	**-328.8**	**-336.4**	..	**1.7**	**-1.1**
Electricity and heat gen.	-279.3	-365.1	-386.0	-330.2	-336.9	-298.6	-306.6	..	1.9	-1.1
Main activity producers (1)	*-258.3*	*-305.8*	*-336.0*	*-310.4*	*-318.3*	*-282.2*	*-290.6*	..	*1.6*	*-0.7*
Autoproducers	*-21.1*	*-59.3*	*-50.1*	*-19.8*	*-18.6*	*-16.4*	*-16.0*	..	*5.2*	*-5.3*
Gas works	11.5	2.5	2.4	0.1	-0.1	-0.1	-0.1	..	-8.7	-
Coal transformation(2)	-53.7	-49.4	-43.7	-37.1	-31.9	-30.2	-29.8	..	-1.2	-1.8
BKB plants	*2.4*	*1.6*	*-1.4*	*-0.2*	*-0.2*	*0.0*	*-0.0*	..	*-*	*-15.3*
Blast furnaces	*-38.6*	*-37.7*	*-33.0*	*-30.1*	*-27.6*	*-25.6*	*-24.9*	..	*-0.9*	*-1.3*
Coke ovens	*-18.2*	*-14.0*	*-9.9*	*-7.0*	*-4.2*	*-4.6*	*-4.9*	..	*-3.5*	*-3.3*
Patent fuel plants	*0.8*	*0.7*	*0.7*	*0.2*	*0.1*	*0.0*	*0.0*	..	*-1.0*	*-14.0*
Other transformation(3)	-	-	-	0.0	0.0	0.0	0.0	..	-	-
Energy ind. own use	**-21.7**	**-18.5**	**-13.2**	**-10.3**	**-9.3**	**-8.8**	**-8.2**	..	**-2.9**	**-2.2**
Losses	**-5.2**	**-2.0**	**-1.2**	**-1.1**	**-1.3**	**-1.3**	**-1.3**	..		
Final consumption(4)	**264.4**	**222.7**	**177.9**	**88.4**	**77.2**	**76.9**	**74.8**	..	**-2.3**	**-4.0**
Industry(5)	130.3	112.1	100.5	66.0	56.3	45.6	46.8	..	-1.5	-3.6
Iron and steel	*52.7*	*42.8*	*39.1*	*27.1*	*25.2*	*20.5*	*21.9*	..	*-1.8*	*-2.7*
Chemical	*14.3*	*12.9*	*11.4*	*4.3*	*4.7*	*5.3*	*4.9*	..	*-1.3*	*-3.9*
Non-metallic minerals	*14.3*	*16.3*	*17.9*	*11.1*	*7.7*	*6.8*	*7.7*	..	*1.3*	*-3.9*
Paper, pulp and print	*4.2*	*3.4*	*3.8*	*2.2*	*1.9*	*1.6*	*1.6*	..	*-0.6*	*-4.1*
Other industry (6)	*44.9*	*36.8*	*28.4*	*21.2*	*16.7*	*11.4*	*10.7*	..	*-2.7*	*-4.5*
Transport(7)	9.8	3.6	0.3	0.0	0.0	0.0	0.0	..	-18.6	-13.9
Other	119.9	103.8	74.7	20.9	18.9	29.4	26.2	..	-2.7	-4.9
Comm. and pub. services	*22.0*	*24.4*	*18.0*	*2.1*	*1.8*	*2.7*	*2.8*	..	*-1.2*	*-8.6*
Residential	*89.2*	*72.0*	*52.6*	*16.8*	*15.4*	*24.5*	*21.5*	..	*-3.1*	*-4.2*
Other sectors (8)	*8.7*	*7.4*	*4.1*	*2.0*	*1.7*	*2.2*	*1.9*	..	*-4.4*	*-3.5*
Non-energy use	4.4	3.1	2.4	1.5	2.0	1.9	1.9	..	-3.4	-1.2

(1) Main activity electricity and heat generation includes district heating.

(2) Coal transformation refers to the transformation of coal and peat to secondary and tertiary products (mainly cokes, briquettes, coal tar and off-gases). Gas works and Liquefaction are shown separately to match the *Energy Balances of OECD Countries.*

(3) Other transformation includes Liquefaction and Non-specified transformations.

(4) Final Consumption includes non-energy use and energy use (Industry, Transport and Other).

(5) Please refer to notes and definitions in Part I for detailed explanation.

(6) Other industry includes Non-ferrous metals, Transport equipment, Machinery, Mining and quarrying, Food and tobacco, Wood and wood products, Construction, Textile and leather, and Non-specified industry.

(7) Transport includes Rail and Inland waterways.

(8) Other sectors includes Agriculture/Forestry, Fishing and Non-specified other.

* "Coal" refers to all types of coal, primary (anthracite, coking coal, other bituminous coal, sub-bituminous coal, lignite/brown coal) and derived fuels (including patent fuel, BKB, coke oven coke, gas coke, coal tar, coke oven gas, gas works gas, blast furnace gas and other recovered gases), as well as peat, peat briquettes and oil shale. Quantities have been converted to Mtce using calorific values reported by the respective countries.
Please refer to notes and definitions in Part I.

Source: *IEA/OECD Energy Balances of OECD Countries*

IEA EUROPE

6. Use of coal for selected end-uses*

(million tonnes)

	1978	1990	2000	2005	2009	2010	2011	Average annual percent change 78-90	90-11
Steam coal	**366.47**	**365.71**	**282.83**	**291.89**	**246.90**	**259.97**	**264.95**	**-0.02**	**-1.52**
Total electricity and heat	256.93	283.88	228.93	237.42	198.25	204.33	209.93	0.83	-1.43
Main activity producers	*207.55*	*246.13*	*219.07*	*228.39*	*191.78*	*197.07*	*202.55*	*1.43*	*-0.92*
Autoproducers	*49.38*	*37.75*	*9.86*	*9.03*	*6.47*	*7.26*	*7.38*	*-2.21*	*-7.48*
Patent fuel/BKB plants	6.90	2.94	0.84	0.42	0.35	0.35	0.34	-6.86	-9.83
Coke ovens/Liquefaction(1)	2.28	0.09	-	-	-	0.25	0.12	-23.82	1.38
Blast furnace inputs	0.01	3.58	7.32	7.69	4.48	8.25	7.94	72.95	3.87
Gas manufacture	1.63	0.41	-	-	-	-	-	-10.93	-
Industry	40.00	40.42	32.01	29.54	19.20	21.25	21.72	0.09	-2.91
Iron and steel	*4.59*	*4.25*	*4.01*	*5.10*	*2.82*	*3.56*	*4.59*	*-0.64*	*0.36*
Chemical	*3.18*	*5.71*	*3.06*	*3.29*	*3.37*	*4.15*	*3.63*	*5.00*	*-2.13*
Non-metallic minerals	*13.49*	*16.54*	*10.00*	*6.29*	*5.32*	*5.45*	*6.34*	*1.71*	*-4.46*
Paper, pulp and print	*0.84*	*2.21*	*1.79*	*1.63*	*1.29*	*1.23*	*1.14*	*8.39*	*-3.11*
Other industry	*17.90*	*11.71*	*13.14*	*13.23*	*6.41*	*6.87*	*6.02*	*-3.47*	*-3.12*
Other sectors(2)	50.74	31.24	15.48	15.28	22.25	24.46	21.34	-3.96	-1.80
Non-energy use	0.08	0.05	0.12	0.24	0.14	0.22	0.23	-3.32	7.34
Coking coal	**145.14**	**129.44**	**92.80**	**81.79**	**63.66**	**72.25**	**73.65**	**-0.95**	**-2.65**
Total electricity and heat	7.14	18.23	13.37	12.70	7.70	5.14	6.72	8.12	-4.64
Main activity producers	*6.19*	*14.32*	*10.17*	*10.73*	*6.71*	*4.21*	*5.60*	*7.23*	*-4.37*
Autoproducers	*0.95*	*3.91*	*3.20*	*1.97*	*0.98*	*0.93*	*1.12*	*12.52*	*-5.77*
Patent fuel/BKB plants	-	-	-	-	-	-	-	-	-
Coke ovens/Liquefaction(1)	135.41	107.92	73.13	64.90	52.53	63.40	62.64	-1.87	-2.56
Blast furnace inputs	-	1.45	1.78	2.81	2.01	2.30	2.66	-	2.93
Gas manufacture	2.46	0.26	-	-	-	-	-	-17.08	-
Industry	0.14	1.97	2.64	0.79	1.54	1.52	2.16	24.42	0.43
Iron and steel	*0.01*	*1.63*	*0.24*	*0.23*	*0.64*	*0.97*	*1.22*	*57.52*	*-1.38*
Chemical	*0.00*	-	*0.00*	-	*0.02*	-	-	-	-
Non-metallic minerals	*0.00*	-	*0.00*	-	-	-	-	-	-
Paper, pulp and print	-	-	-	-	-	-	-	-	-
Other industry	*0.13*	*0.34*	*2.39*	*0.55*	*0.88*	*0.55*	*0.94*	*8.07*	*5.01*
Other sectors(2)	0.27	0.14	0.10	0.13	0.28	0.18	0.24	-5.56	2.75
Non-energy use	-	-	0.23	0.22	0.14	0.18	0.22	-	-
Lignite	**585.58**	**654.20**	**436.90**	**435.76**	**424.54**	**409.19**	**430.74**	**0.93**	**-1.97**
Total electricity and heat	355.04	458.83	399.95	405.09	390.09	374.24	393.79	2.16	-0.73
Main activity producers	*339.04*	*432.38*	*393.73*	*399.07*	*384.69*	*367.29*	*387.29*	*2.05*	*-0.52*
Autoproducers	*16.00*	*26.44*	*6.22*	*6.02*	*5.41*	*6.96*	*6.51*	*4.28*	*-6.46*
Patent fuel/BKB plants	134.01	108.37	12.15	12.42	12.22	13.34	14.44	-1.75	-9.15
Coke ovens/Liquefaction(1)	-	-	-	-	-	-	-	-	-
Gas manufacture	6.40	2.83	1.35	1.49	1.51	1.56	1.65	-6.57	-2.55
Industry	47.82	40.21	13.76	8.58	9.91	11.63	10.47	-1.44	-6.21
Iron and steel	*0.93*	*1.35*	*0.22*	*0.09*	*0.06*	*0.20*	*0.16*	*3.17*	*-9.70*
Chemical	*14.32*	*10.67*	*3.13*	*2.99*	*2.86*	*3.23*	*2.88*	*-2.42*	*-6.05*
Non-metallic minerals	*1.57*	*1.63*	*0.50*	*0.28*	*0.43*	*0.57*	*0.46*	*0.30*	*-5.80*
Paper, pulp and print	*4.04*	*1.93*	*0.47*	*0.43*	*0.42*	*0.26*	*0.37*	*-5.97*	*-7.58*
Other industry	*26.97*	*24.63*	*9.45*	*4.80*	*6.14*	*7.38*	*6.60*	*-0.75*	*-6.08*
Other sectors(2)	32.65	40.74	8.91	7.20	8.63	8.55	9.44	1.86	-6.72
Non-energy use	0.83	0.58	0.02	0.04	0.03	0.03	0.03	-2.94	-12.76

(1) Liquefaction primarily refers to direct distillation processes. Indirect liquefaction may be reported here or under gas manufacture.

(2) Other sectors are Residential, Commercial and public services, Agriculture/Forestry, Fishing and Non-specified other.

* Coal totals may not match the sum of the individual rows due to statistical differences. Please refer to notes and definitions in Part I.

Source: *IEA/OECD Energy Statistics of OECD Countries*

IEA EUROPE

7. Fuel prices to end users*

	1978(1)	1990	1995	2000	2005	2009	2010	2011	2012
				(US dollars / tce)					
For electricity generation									
Steam coal	..	..	..	..	..	..	..	..	..
Heavy fuel oil	..	..	..	..	..	..	..	..	..
Natural gas	..	..	..	..	..	..	..	..	..
For industry									
Steam coal	..	..	..	..	..	..	..	..	..
Coking coal	..	..	..	..	..	..	..	..	..
High sulphur fuel oil	..	..	..	..	..	..	..	..	..
Low sulphur fuel oil	..	..	..	..	..	..	..	..	..
Natural gas	..	..	..	..	..	..	..	..	..
				(US dollars / unit)(2)					
For electricity generation									
Steam coal	..	..	..	..	..	..	..	..	..
Heavy fuel oil	..	..	..	..	..	..	..	..	..
Natural gas	..	..	..	..	..	..	..	..	..
For industry									
Steam coal	..	..	..	..	..	..	..	..	..
Coking coal	..	..	..	..	..	..	..	..	..
High sulphur fuel oil	..	..	..	..	..	..	..	..	..
Low sulphur fuel oil	..	..	..	..	..	..	..	..	..
Natural gas	..	..	..	..	..	..	..	..	..

(1) Earliest year for which split by coal type is available.

(2) For steam coal, coking coal and heavy fuel oil per metric tonne; for natural gas per 10^7 kilocalories GCV.

* Please refer to notes and definitions in Part I. Steam coal prices relate to historical definitions.

Source: *IEA/OECD Energy Prices & Taxes*

8. Coal and peat trade by type of coal*

(Mtce)

	1978	1990	1995	2000	2005	2009	2010	2011	2012e
Total imports	**108.62**	**168.08**	**164.69**	**195.18**	**227.04**	**203.95**	**208.65**	**223.59**	**222.04**
Bituminous coal(1)	46.33	93.68	95.85	122.30	159.48	153.62	142.92	159.65	160.81
Coking coal	40.54	56.93	53.59	56.67	53.92	40.70	52.97	52.66	50.70
Sub-bituminous coal	0.60	0.65	0.52	0.36	0.48	0.64	0.48	0.38	0.39
Lignite	4.69	4.43	2.80	1.52	0.52	0.63	0.54	0.61	0.82
Peat	-	0.10	0.13	0.10	0.14	0.23	0.16	0.20	0.19
Coal products(2)	16.47	12.28	11.80	14.24	12.48	8.14	11.58	10.09	9.12
Total exports	**90.65**	**60.46**	**53.74**	**47.97**	**41.63**	**33.72**	**36.66**	**38.34**	**27.75**
Bituminous coal(1)	30.61	22.45	23.74	28.49	25.58	18.02	19.24	22.19	11.07
Coking coal	35.17	19.56	18.19	9.76	6.47	5.88	5.44	4.70	4.69
Sub-bituminous coal	-	0.04	0.00	0.04	0.02	0.00	0.00	0.00	0.00
Lignite	4.33	4.28	2.95	1.31	0.65	0.69	0.63	0.72	0.97
Peat	0.01	0.12	0.07	0.06	0.05	0.02	0.01	0.00	0.00
Coal products(2)	20.53	14.00	8.78	8.32	8.86	9.11	11.35	10.73	11.02

(1) Bituminous coal includes anthracite.

(2) Coal products includes products derived from coal, for example: coke, coal tar, briquettes, patent fuels.

* Please refer to notes and definitions in Part I.

Source: *IEA/OECD Energy Balances of OECD Countries, IEA/OECD Energy Statistics of OECD Countries*

IEA EUROPE

9. Total coal imports by origin*

(thousand tonnes)

	1978[1]	1985	1990	1995	2000	2005	2010	2011	2012e
Total coal[2]	**106042**	**158117**	**170722**	**167955**	**199932**	**237834**	**218086**	**238962**	**238288**
Coking coal	**40078**	**52204**	**55532**	**52716**	**56057**	**52927**	**52878**	**52433**	**50595**
Australia	4497	9730	10043	13043	21757	20330	17145	15479	15021
Canada	294	1107	2812	4176	7072	6743	3732	4002	3005
Czech Republic	909	695	774	2985	3388	3366	3704	2990	2698
Germany	10550	5219	3141	37	2	289	1	15	46
Poland	6190	5025	2570	6573	3118	3246	1821	2168	2384
United Kingdom	79	71	52	-	-	6	1	-	-
United States	8745	22793	28433	21067	16901	13914	21519	21143	20988
Other OECD	105	285	80	384	24	180	92	136	370
China, People's Rep.	-	-	1	12	2	279	3	-	-
Colombia	-	50	24	85	140	193	766	921	685
Indonesia	-	-	46	214	441	-	-	201	32
South Africa	206	256	158	206	388	295	574	515	452
Former Soviet Union[3]	3005	1441	x	x	x	x	x	x	x
Russian Federation [4]	*x*	*x*	*2419*	*526*	*2327*	*2668*	*2711*	*3684*	*3227*
Other FSU [4]	*x*	*x*	*-*	*-*	*26*	*520*	*80*	*661*	*97*
Venezuela	-	-	-	990	418	872	187	318	413
Vietnam	-	-	-	-	53	-	37	-	-
Non-specified/other	5498	5532	4979	2418	-	26	505	200	1177
Steam coal	**53202**	**92197**	**105117**	**108847**	**140645**	**183721**	**164037**	**185118**	**185918**
Australia	2955	11160	9790	7478	9416	7428	3504	3315	2425
Canada	816	1435	832	307	12	807	859	1257	1084
Czech Republic	243	462	327	1685	2435	1394	2880	3929	2406
Germany	6716	3523	2123	834	464	601	611	530	293
Poland	15647	15429	13084	15821	18884	15475	10510	5746	4864
United Kingdom	2285	2394	2423	713	589	297	303	427	265
United States	499	16499	20452	20423	4676	2322	11535	17599	29565
Other OECD	986	2698	3515	1706	3654	2788	3769	3119	3461
China, People's Rep.	21	236	2783	2307	1887	1894	279	134	59
Colombia	-	1542	9060	11231	22490	26594	37713	51963	50102
Indonesia	-	-	194	2890	8230	14312	9724	9885	9458
South Africa	10814	26339	26099	32967	42839	52875	18829	18184	14580
Former Soviet Union[3]	2863	1895	x	x	x	x	x	x	x
Russian Federation [4]	*x*	*x*	*6091*	*7289*	*15392*	*49113*	*53880*	*55744*	*57898*
Other FSU [4]	*x*	*x*	*-*	*647*	*845*	*2857*	*2160*	*3461*	*2059*
Venezuela	-	-	1475	1831	3081	1142	811	951	702
Vietnam	-	-	-	364	560	213	23	18	-
Non-specified/other	9357	8585	6869	354	5191	3609	6647	8856	6697
Lignite	**12762**	**13716**	**10073**	**6392**	**3230**	**1186**	**1171**	**1411**	**1775**

(1) Earliest year for which split by coal type is available.

(2) Total coal consists of steam coal, coking coal and lignite.

(3) For years prior to 1990, or if no details for individual Former Soviet Union republics.

(4) Data for former republics are not available separately for some OECD member countries.

* Please refer to notes and definitions in Part I.

Source: IEA/OECD Coal Statistics

IEA EUROPE

10. Coking coal exports by destination*

(thousand tonnes)

	1978(1)	1985	1990	1995	2000	2005	2010	2011	2012e
World	**35537**	**25596**	**19781**	**18218**	**9747**	**6663**	**5491**	**4867**	**4851**
Total OECD	**25095**	**15893**	**11025**	**11983**	**8106**	**6172**	**5361**	**4687**	**4471**
Australia	-	-	-	2	-	-	-	-	-
Austria	1283	1582	1351	1897	1844	1514	1312	972	1041
Belgium	2383	1572	858	559	1	20	16	31	132
Canada	-	-	-	-	-	-	-	-	-
Chile	-	-	-	-	-	-	-	-	-
Czech Republic	-	-	-	871	214	523	720	1017	848
Denmark	-	-	-	9	-	-	-	-	-
Estonia	-	-	-	-	-	-	-	-	-
Finland	-	-	203	534	717	13	-	-	20
France	6331	2680	1697	634	108	30	16	2	-
Germany	154	241	144	362	1061	397	152	150	173
Greece	54	-	-	1	-	-	-	-	-
Hungary	-	-	-	1287	1010	449	502	495	391
Iceland	-	-	-	3	-	-	-	-	-
Ireland	-	8	3	3	4	1	-	2	9
Israel	-	-	-	-	-	-	-	-	-
Italy	4005	2439	1017	18	-	295	-	-	-
Japan	804	-	-	-	-	-	-	-	-
Korea	-	-	-	-	-	-	-	-	-
Luxembourg	286	-	-	10	-	-	-	-	-
Mexico	-	-	-	-	-	-	-	-	-
Netherlands	1426	851	573	995	323	48	-	-	-
New Zealand	-	-	-	-	-	-	-	-	-
Norway	125	134	-	156	53	-	-	-	-
Poland	-	-	-	896	538	592	733	686	519
Portugal	122	125	52	117	-	-	-	-	-
Slovak Republic	5126	4183	3681	2657	1570	1909	1832	1332	1335
Slovenia	-	-	-	-	-	-	-	-	-
Spain	1826	637	649	58	65	99	-	-	2
Sweden	387	254	-	310	494	-	-	-	-
Switzerland	18	5	3	3	-	-	-	-	1
Turkey	89	-	-	178	100	212	78	-	-
United Kingdom	573	1182	794	421	4	-	-	-	-
United States	103	-	-	2	-	70	-	-	-
Total non-OECD	**2491**	**3862**	**4676**	**6157**	**1641**	**489**	**130**	**180**	**378**
Brazil	15	861	1249	790	143	-	-	-	-
China(2)	-	-	-	-	-	-	-	-	-
Chinese Taipei	-	-	-	-	-	-	-	-	-
Egypt	-	-	100	48	529	355	-	-	-
India	-	-	284	-	-	-	-	-	-
Romania	-	-	100	217	62	-	-	-	-
Oth. Africa & Mid. East	520	4	-	-	2	-	-	-	-
Oth. non-OECD Americas	5	861	1249	-	2	-	-	-	-
Other Asia & Oceania	-	-	-	-	-	-	1	1	1
Other non-OECD Europe and Eurasia	1951	2136	1694	5102	903	134	129	179	377
Non-specified/Other	**7951**	**5841**	**4080**	**78**	**-**	**2**	**-**	**-**	**2**

(1) Earliest year for which split by coal type is available.

(2) China includes the People's Republic of China and Hong Kong, China.

* Please refer to notes and definitions in Part I.

Source: IEA/OECD Coal Statistics

IEA EUROPE

11. Steam coal exports by destination*

(thousand tonnes)

	1978(1)	1985	1990	1995	2000	2005	2010	2011	2012e
World	**33590**	**31963**	**25277**	**28036**	**32165**	**29452**	**21706**	**25329**	**12215**
Total OECD	**25347**	**23616**	**20776**	**22161**	**31217**	**28922**	**20982**	**17766**	**11849**
Australia	-	1	1	12	-	-	-	23	6
Austria	275	1001	1216	1050	1643	2182	1447	1634	1321
Belgium	2633	1373	1471	1414	610	864	607	258	173
Canada	-	-	-	-	-	19	46	45	2
Chile	-	-	-	-	-	-	-	-	-
Czech Republic	274	1378	2282	1874	864	719	685	844	676
Denmark	4161	3454	1387	3214	2382	1154	800	258	273
Estonia	-	-	-	-	3	-	-	-	-
Finland	4095	2461	2815	2559	1228	963	190	52	129
France	6035	1975	1167	1152	2415	3115	1909	906	434
Germany	3171	4042	4942	5655	15884	13364	8845	7764	3777
Greece	1	15	-	1	2	144	-	66	32
Hungary	-	-	-	374	322	321	158	202	38
Iceland	-	21	18	20	7	44	56	54	6
Ireland	540	1287	793	686	468	596	604	477	394
Israel	-	-	-	-	-	5	-	-	-
Italy	1530	1211	600	307	923	245	1172	846	1540
Japan	-	-	61	-	-	-	-	-	-
Korea	-	-	-	-	-	-	-	-	-
Luxembourg	52	107	3	50	164	38	50	109	-
Mexico	-	-	-	-	-	-	-	-	-
Netherlands	938	1282	1341	617	363	651	370	163	58
New Zealand	-	-	-	-	-	-	-	-	2
Norway	167	286	339	311	603	495	461	423	280
Poland	-	-	1	304	117	127	1386	1371	1015
Portugal	15	179	186	151	5	437	245	190	31
Slovak Republic	237	79	198	404	1217	722	449	602	491
Slovenia	-	-	-	2	12	10	4	-	2
Spain	21	389	275	129	441	58	52	93	351
Sweden	244	1666	864	809	172	208	311	173	240
Switzerland	90	287	52	11	20	8	32	7	30
Turkey	74	-	-	-	-	66	218	17	148
United Kingdom	149	1122	764	1054	1351	2320	885	1164	400
United States	645	-	-	1	1	47	-	25	-
Total non-OECD	**85**	**1636**	**722**	**4980**	**908**	**526**	**316**	**394**	**332**
Brazil	-	13	10	37	-	-	-	44	-
China(2)	-	-	3	-	2	-	-	1	-
Chinese Taipei	-	-	-	-	-	-	-	-	-
Egypt	-	-	1	-	2	4	22	35	22
India	-	-	1	1	-	1	2	2	4
Romania	-	-	16	-	-	-	-	-	26
Oth. Africa & Mid. East	31	77	375	-	3	436	31	75	27
Oth. non-OECD Americas	-	3	-	-	16	-	15	46	46
Other Asia & Oceania	-	188	-	-	4	5	4	8	40
Other non-OECD Europe and Eurasia	54	1355	316	4942	881	80	242	183	167
Non-specified/Other	**8158**	**6711**	**3779**	**895**	**40**	**4**	**408**	**7169**	**34**

(1) Earliest year for which split by coal type is available.

(2) China includes the People's Republic of China and Hong Kong, China.

* Please refer to notes and definitions in Part I. Steam coal includes all sub-bituminous coal.

Source: IEA/OECD Coal Statistics

AUSTRALIA*

Figure 1: TPES by fuel (Mtce)

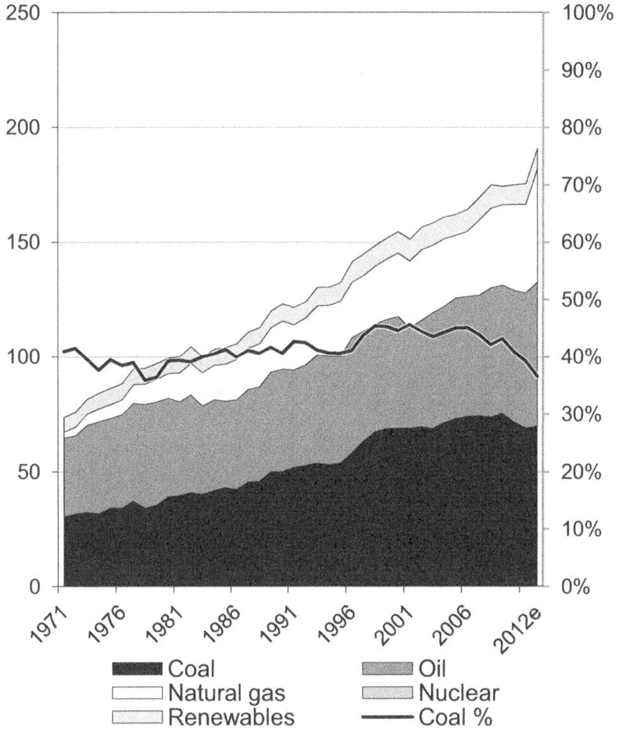

Figure 2: Coal supply indicators (1971 = 100)

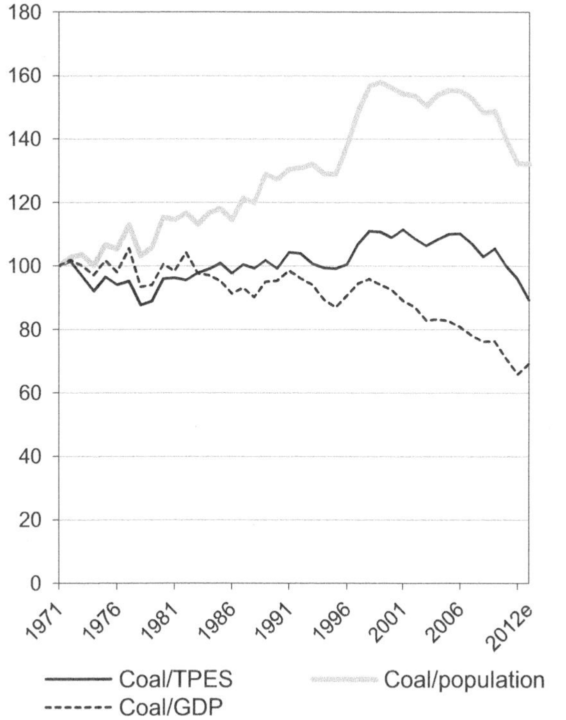

Figure 3: Primary coal supply (Mtce)

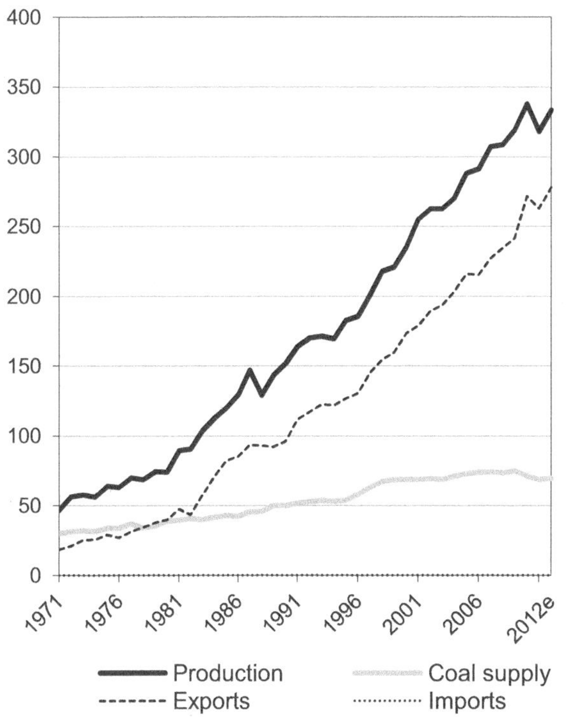

Figure 4: Coal consumption (Mtce)

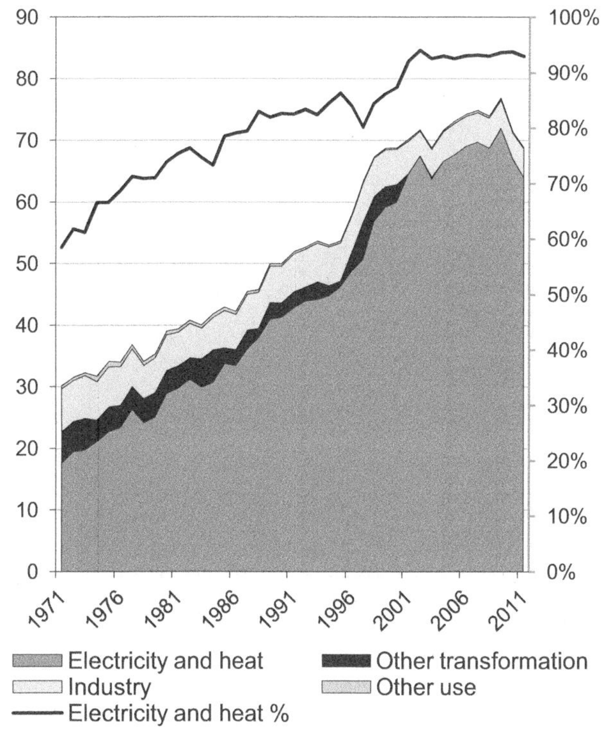

*Please refer to notes and definitions in Part I. Peat is included in coal for all figures.

Source: *IEA/OECD Coal Information, IEA/OECD Energy Balances of OECD Countries, OECD Main Economic Indicators*

AUSTRALIA*

Figure 5: Electricity generation by fuel (TWh)

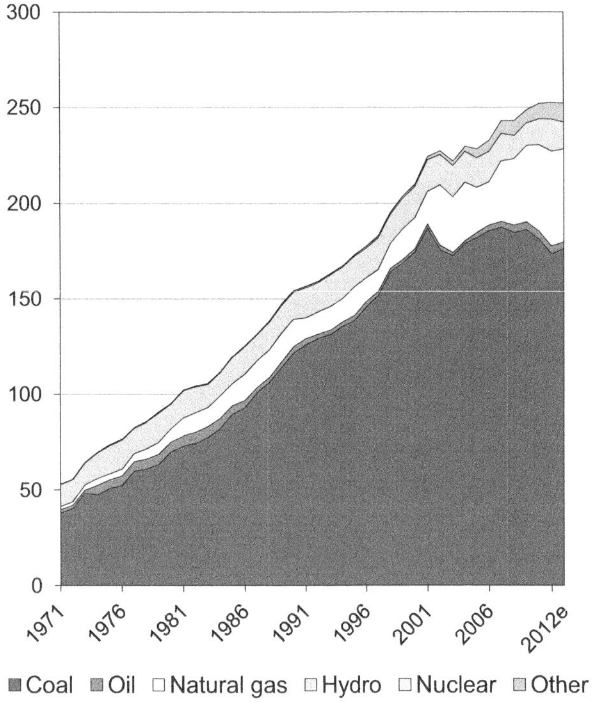

Figure 6: CO_2 emissions by fuel (Mt CO_2)

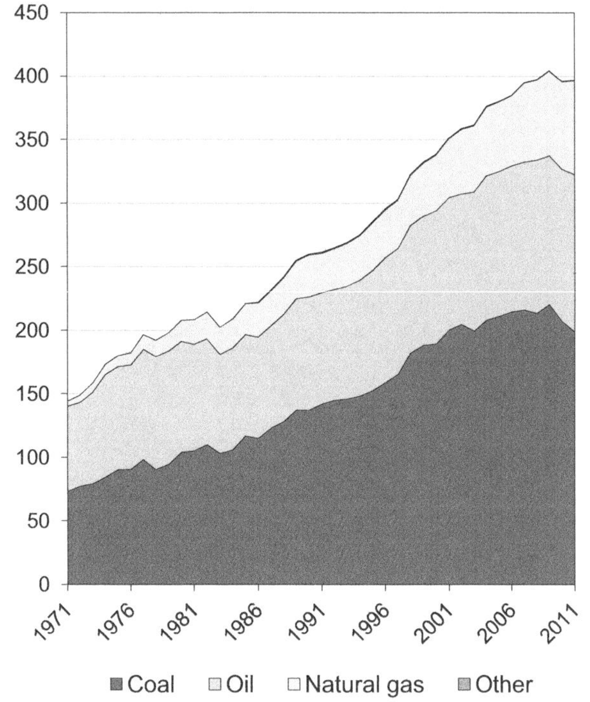

Figure 7: Electricity generation by fuel share

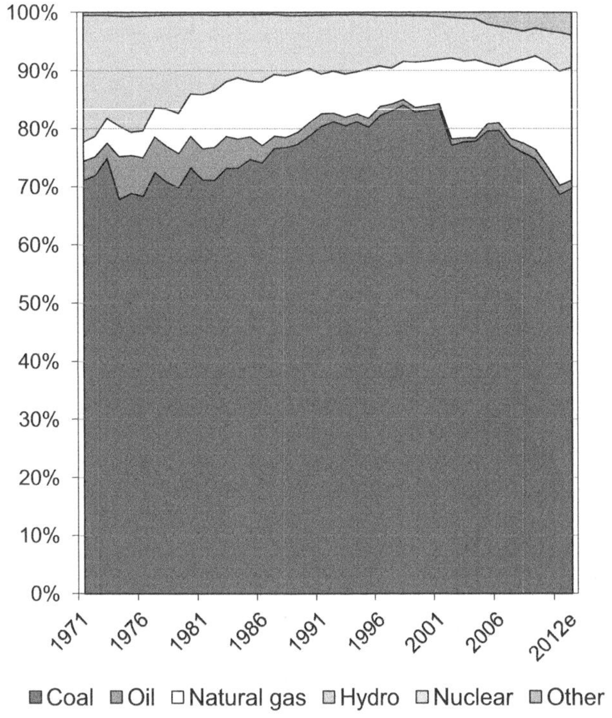

Figure 8: CO_2 emissions by fuel share

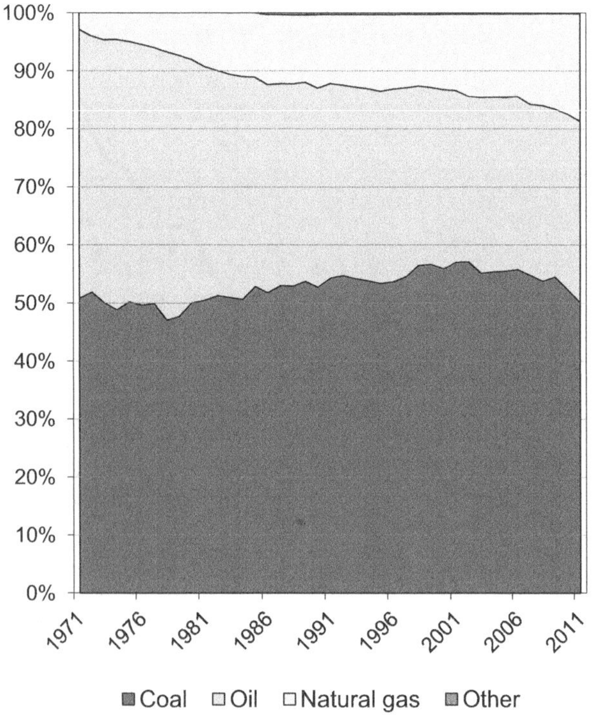

*Please refer to notes and definitions in Part I. Peat is included in coal for all figures.

Source: *IEA/OECD Electricity Information, IEA/OECD CO2 Emissions from Fuel Combustion*

AUSTRALIA

1. Total primary energy supply (TPES) by fuel*

	1973	1980	1990	2000	2010	2011	2012e	Average annual percent change 73-90	90-11
TPES (Mtce)	**81.51**	**99.43**	**123.18**	**154.44**	**175.02**	**175.55**	**190.97**	**2.46**	**1.70**
Coal and peat	32.26	39.03	49.97	68.79	71.50	68.83	69.76	2.61	1.54
Oil	37.98	42.96	44.57	48.79	57.46	59.08	62.81	0.95	1.35
Natural Gas	4.83	10.66	21.12	27.53	37.44	38.60	49.55	9.07	2.91
Biofuels and waste	5.04	5.16	5.66	7.19	5.96	5.79	5.81	0.68	0.11
Nuclear	-	-	-	-	-	-	-	-	-
Hydro	1.40	1.59	1.74	2.01	1.66	2.06	1.72	1.27	0.81
Geothermal	-	-	-	-	0.00	0.00	0.00	-	-
Solar, wind, tide	-	0.03	0.12	0.13	1.00	1.19	1.32	-	11.74
Net electricity trade(1)	-	-	-	-	-	-	-	-	-
Heat(2)	-	-	-	-	-	-	-	-	-

(1) Net trade between OECD and non-OECD countries.

(2) Ambient heat from heat pumps used in the transformation sector.

Source: *IEA/OECD Energy Balances of OECD Countries*

2. Energy supply, GDP and population*

	1973	1980	1990	2000	2010	2011	2012e	Average annual percent change 73-90	90-11
GDP (2005 bil. USD)	277	334	451	640	870	899	868	2.90	3.34
TPES/GDP(1)	0.29	0.30	0.27	0.24	0.20	0.20	0.22	-0.43	-1.58
Population (millions)	13.6	14.8	17.2	19.3	22.4	22.8	23.1	1.37	1.35
TPES/population(1)	5.99	6.72	7.17	8.01	7.80	7.71	8.27	1.07	0.35
TPES/GDP(2)	137.7	139.4	128.0	113.1	94.3	91.5	103.1	-0.43	-1.58
Coal and peat supply/GDP(2)	121.2	121.7	115.4	111.9	85.6	79.7	83.7	-0.29	-1.74
Elec. consumption/GDP(2)	75.9	95.2	115.0	108.4	96.5	94.2	..	2.47	-0.95
Elec. generation (TWh)	64	95	154	210	252	253	252	5.27	2.37
Industrial production(2)	..	53.5	73.1	94.1	112.3	111.2	115.9	..	2.02

(1) TPES/GDP in units of tce/2005 thousand USD, TPES/population in units of tce/capita.

(2) As index, 2005 (not shown) = 100

Source: *IEA/OECD Energy Balances of OECD Countries , OECD Main Economic Indicators*

3. Coal and peat production by type*

	1978(1)	1990	2000	2005	2010	2011	2012e	Average annual percent change 78-90	90-11
Mtce:									
Coking coal	36.14	62.50	100.89	124.82	158.44	142.67	142.89	4.67	4.01
Steam coal	22.71	74.47	113.27 e	139.56	155.18	151.55	165.79	10.40	3.44
Lignite	9.64	14.61	20.95	23.59	24.11	23.74	24.59	3.52	2.34
Peat	-	-	-	-	-	-	-	-	-
Mt:									
Coking coal	37.67	64.63	103.75	128.36	162.93	146.71	146.94	4.60	3.98
Steam coal	30.88	93.94	135.68 e	171.70	189.38	184.53	200.26	9.72	3.27
Lignite	30.47	45.99	67.29	70.53	72.09	71.00	73.54	3.49	2.09
Peat	-	-	-	-	-	-	-	-	-

(1) Earliest year for which split by coal type is available.

* Please refer to notes and definitions in Part I.

Source: *IEA/OECD Energy Balances of OECD Countries, IEA/OECD Energy Statistics of OECD Countries*

AUSTRALIA

4. Final consumption of energy by fuel*

(Mtce)

	1973	1980	1990	2000	2009	2010	2011	Average annual percent change 73-90	90-11
Total final consumption[(1)]	**56.54**	**66.84**	**80.78**	**99.38**	**107.30**	**109.02**	**111.21**	**2.12**	**1.53**
Coal and peat	7.42	6.44	6.37	5.98	4.92	4.45	4.88	-0.89	-1.27
Oil	34.66	38.46	41.43	49.59	54.25	55.31	56.54	1.05	1.49
Natural Gas	3.02	7.18	12.35	16.27	17.68	18.32	18.58	8.64	1.96
Biofuels and wastes	4.99	5.00	4.65	6.20	4.79	4.90	4.93	-0.42	0.28
Geothermal	-	-	-	-	-	-	-	-	-
Solar, wind, tide	-	0.03	0.12	0.12	0.28	0.35	0.37	-	5.74
Electricity	6.45	9.73	15.87	21.22	25.37	25.69	25.91	5.45	2.36
Heat	-	-	-	-	-	-	-	-	-
of which:									
Total industry	**21.81**	**25.12**	**27.44**	**33.96**	**31.74**	**32.35**	**33.03**	**1.36**	**0.89**
Coal and peat	6.99	5.85	5.98	5.75	4.61	4.15	4.65	-0.92	-1.18
Oil	7.96	7.69	4.13	5.07	4.65	5.15	5.01	-3.79	0.93
Natural Gas	1.88	4.89	7.95	10.16	10.39	10.72	10.86	8.86	1.50
Biofuels and wastes	2.13	2.71	2.12	3.51	2.43	2.50	2.46	-0.04	0.72
Geothermal	-	-	-	-	-	-	-	-	-
Solar, wind, tide	-	-	-	-	-	-	-	-	-
Electricity	2.85	3.99	7.27	9.46	9.67	9.82	10.04	5.67	1.55
Heat	-	-	-	-	-	-	-	-	-
Total transport	**18.47**	**24.04**	**30.16**	**36.65**	**41.41**	**41.84**	**42.76**	**2.93**	**1.68**
Coal and peat	0.03	0.01	0.10	0.13	0.25	0.24	0.17	7.63	2.50
Oil	18.36	23.92	29.82	35.81	39.67	40.06	40.96	2.89	1.52
Natural Gas	-	-	0.01	0.43	0.63	0.64	0.66	-	20.06
Biofuels and wastes	-	-	-	-	0.34	0.41	0.49	-	-
Electricity	0.08	0.11	0.22	0.29	0.52	0.48	0.48	6.13	3.77
Residential	**7.21**	**8.63**	**10.70**	**12.94**	**14.39**	**14.67**	**14.91**	**2.35**	**1.59**
Coal and peat	0.27	0.28	0.08	0.03 e	0.01	0.01	0.00	-6.76	-12.47
Oil	1.13	1.12	0.54	0.50	0.41	0.40	0.42	-4.23	-1.21
Natural Gas	0.67	1.38	2.70	3.64	4.33	4.42	4.55	8.58	2.51
Biofuels and wastes	2.86	2.26	2.53	2.67	1.99	1.96	1.96	-0.72	-1.20
Geothermal	-	-	-	-	-	-	-	-	-
Solar, wind, tide	-	0.03	0.12	0.12	0.27	0.33	0.36	-	5.56
Electricity	2.29	3.56	4.74	5.99	7.38	7.55	7.61	4.35	2.29
Heat	-	-	-	-	-	-	-	-	-
Comm & public services	**1.61**	**3.43**	**4.97**	**7.20**	**9.94**	**10.02**	**9.98**	**6.85**	**3.37**
Coal and peat	0.13	0.31	0.21	0.07 e	0.05	0.05	0.04	2.67	-7.35
Oil	0.16	0.72	0.37	0.45	0.89	0.89	0.89	5.00	4.34
Natural Gas	0.22	0.46	1.04	1.54	1.45	1.49	1.51	9.62	1.78
Biofuels and waste	-	0.03	-	0.01	0.03	0.03	0.02	-	-
Geothermal	-	-	-	-	-	-	-	-	-
Solar, wind, tide	-	-	-	-	0.01	0.01	0.01	-	-
Electricity	1.10	1.90	3.35	5.13	7.51	7.54	7.50	6.79	3.91
Heat	-	-	-	-	-	-	-	-	-
Non-energy use	**3.63**	**4.12**	**5.65**	**6.32**	**6.76**	**7.02**	**7.32**	**2.63**	**1.24**
Coal and peat	-	-	-	-	-	-	-	-	-
Oil	3.38	3.67	5.01	5.84	5.88	5.97	6.32	2.35	1.11
Natural Gas	0.26	0.45	0.64	0.49	0.88	1.05	1.00	5.54	2.15

(1) Total Final Consumption (TFC) excludes use in transformation processes and energy industry own use.

* Please refer to notes and definitions in Part I. Agriculture, Forestry, Fishing and "Non-specified other" sectors are not reported here individually, but are accounted for in Total final consumption.

Source: *IEA/OECD Energy Balances of OECD Countries*

AUSTRALIA

5. Coal balance*

(Mtce)

	1973	1980	1990	2000	2005	2010	2011	2012e	Average annual percent change 73-90	90-11
Production	57.5	74.1	151.6	235.1 e	288.0	337.7	318.0	333.3	5.9	3.6
Imports	-	0.0	-	-	0.0	0.0	0.0	0.0	-	-
Exports	-25.2	-39.7	-96.1	-173.5	-215.7	-271.6	-263.0	-278.0	8.2	4.9
Stock changes	-0.0	4.6	-5.5	7.1	0.7	5.4	13.8	14.5		
Primary supply	**32.3**	**39.0**	**50.0**	**68.8**	**72.9**	**71.5**	**68.8**	**69.8**	**2.6**	**1.5**
Statistical differences	**0.0**	**0.0**	**0.1**	**-0.4 e**	**2.9**	**3.3**	**3.0**	..		
Total transformation	**-24.8 e**	**-31.9 e**	**-43.2 e**	**-62.0 e**	**-69.4 e**	**-69.4**	**-66.1 e**	..	**3.3**	**2.0**
Electricity and heat gen.	-19.7	-28.8	-41.3	-60.0 e	-67.7	-67.0	-64.0	..	4.4	2.1
Main activity producers (1)	*-19.7*	*-28.1*	*-40.4*	*-59.4 e*	*-67.5*	*-66.9*	*-63.9*	..	*4.3*	*2.2*
Autoproducers	-	*-0.7*	*-0.9*	*-0.7 e*	*-0.1*	*-0.1*	*-0.1*	..	-	*-9.7*
Gas works	0.3	0.5	0.3	0.1	0.2	0.0	0.0	..	-0.4	-15.5
Coal transformation(2)	-5.4 e	-3.6 e	-2.2 e	-2.1 e	-1.9 e	-2.4	-2.2 e	..	-5.1	-0.0
BKB plants	*-0.2*	*-0.2*	*-0.1*	*-0.0 e*	*-0.1*	*-0.0*	*-0.0*	..	*-5.3*	*-3.7*
Blast furnaces	*-2.4 e*	*-2.2 e*	*-1.4 e*	*-1.5*	*-1.6 e*	*-1.6*	*-1.6 e*	..	*-3.3*	*0.6*
Coke ovens	*-2.8*	*-1.2*	*-0.8*	*-0.5*	*-0.2*	*-0.8*	*-0.6*	..	*-7.3*	*-1.2*
Patent fuel plants	-	-	-	-	-	-	-	..	-	-
Other transformation(3)	-	-	-	-	-	-	-	..	-	-
Energy ind. own use	**-0.0**	**-0.7**	**-0.5**	**-0.5 e**	**-0.9**	**-1.0**	**-0.8**	..	**19.7**	**2.4**
Losses	-	-	-	-	-	-	-	..		
Final consumption(4)	**7.4**	**6.4**	**6.4**	**6.0**	**5.4**	**4.5**	**4.9**	..	**-0.9**	**-1.3**
Industry(5)	7.0	5.8	6.0	5.8	5.0	4.2	4.7	..	-0.9	-1.2
Iron and steel	*3.0 e*	*2.9 e*	*2.1 e*	*1.8 e*	*1.4 e*	*1.0*	*1.7 e*	..	*-2.1*	*-1.1*
Chemical	*0.4*	*0.4*	*0.4*	*0.3 e*	*0.4*	*0.2*	*0.2*	..	*0.2*	*-2.4*
Non-metallic minerals	*0.9*	*0.8*	*0.8*	*0.7 e*	*0.9*	*0.8*	*0.8*	..	*-0.7*	*0.1*
Paper, pulp and print	*0.3*	*0.3*	*0.3*	*0.3*	*0.2*	*0.1*	*0.1*	..	*-0.4*	*-5.7*
Other industry (6)	*2.3*	*1.5*	*2.3*	*2.7 e*	*2.1*	*1.9*	*1.8*	..	*0.0*	*-1.1*
Transport(7)	0.0	0.0	0.1	0.1	0.3	0.2	0.2	..	7.6	2.5
Other	0.4	0.6	0.3	0.1	0.1	0.1	0.0	..	-1.9	-8.3
Comm. and pub. services	*0.1*	*0.3*	*0.2*	*0.1 e*	*0.1*	*0.0*	*0.0*	..	*2.7*	*-7.3*
Residential	*0.3*	*0.3*	*0.1*	*0.0 e*	*0.0*	*0.0*	*0.0*	..	*-6.8*	*-12.5*
Other sectors (8)	-	-	-	-	-	-	-	..	-	-
Non-energy use	-	-	-	-	-	-	-	..	-	-

(1) Main activity electricity and heat generation includes district heating.

(2) Coal transformation refers to the transformation of coal and peat to secondary and tertiary products (mainly cokes, briquettes, coal tar and off-gases). Gas works and Liquefaction are shown separately to match the *Energy Balances of OECD Countries.*

(3) Other transformation includes Liquefaction and Non-specified transformations.

(4) Final Consumption includes non-energy use and energy use (Industry, Transport and Other).

(5) Please refer to notes and definitions in Part I for detailed explanation.

(6) Other industry includes Non-ferrous metals, Transport equipment, Machinery, Mining and quarrying, Food and tobacco, Wood and wood products, Construction, Textile and leather, and Non-specified industry.

(7) Transport includes Rail and Inland waterways.

(8) Other sectors includes Agriculture/Forestry, Fishing and Non-specified other.

* "Coal" refers to all types of coal, primary (anthracite, coking coal, other bituminous coal, sub-bituminous coal, lignite/brown coal) and derived fuels (including patent fuel, BKB, coke oven coke, gas coke, coal tar, coke oven gas, gas works gas, blast furnace gas and other recovered gases), as well as peat, peat briquettes and oil shale. Quantities have been converted to Mtce using calorific values reported by the respective countries. Please refer to notes and definitions in Part I.

Source: *IEA/OECD Energy Balances of OECD Countries*

AUSTRALIA

6. Use of coal for selected end-uses*

(million tonnes)

	1978	1990	2000	2005	2009	2010	2011	Average annual percent change 78-90	90-11
Steam coal	**24.24**	**43.37**	**56.03**	**64.61**	**68.35**	**61.91**	**58.23**	**4.97**	**1.41**
Total electricity and heat	20.89	38.20	51.10 e	59.17	63.26	57.00	53.58	5.16	1.62
Main activity producers	*20.53*	*37.75*	*50.91 e*	*59.17*	*63.26*	*57.00*	*53.58*	*5.21*	*1.68*
Autoproducers	*0.36*	*0.45*	*0.19 e*	-	-	-	-	*1.82*	-
Patent fuel/BKB plants	-	-	-	-	-	-	-	-	-
Coke ovens/Liquefaction(1)	-	-	-	-	-	-	-	-	-
Blast furnace inputs	-	-	-	0.80 e	0.55 e	0.70	0.62 e	-	-
Gas manufacture	0.04	-	-	-	-	-	-	-	-
Industry	3.17	4.85	4.69	4.17	4.12	3.80	3.74	3.59	-1.23
Iron and steel	*0.37*	*0.34*	*0.10 e*	*0.12 e*	*0.12 e*	*0.02*	*0.17 e*	*-0.73*	*-3.33*
Chemical	*0.15*	*0.13*	*0.09*	*0.19*	*0.23*	*0.23*	*0.22*	*-1.34*	*2.45*
Non-metallic minerals	*0.88*	*0.97*	*0.79 e*	*1.08*	*0.96*	*0.99*	*0.95*	*0.83*	*-0.13*
Paper, pulp and print	*0.28*	*0.35*	*0.33*	*0.34*	*0.22*	*0.20*	*0.14*	*1.89*	*-4.24*
Other industry	*1.49*	*3.06*	*3.39 e*	*2.43 e*	*2.58 e*	*2.36*	*2.27 e*	*6.15*	*-1.41*
Other sectors(2)	0.25	0.15	0.04	0.05	0.02	0.02	0.02	-3.93	-10.22
Non-energy use	-	-	-	-	-	-	-	-	-
Coking coal	**7.22**	**5.93**	**4.80**	**4.46**	**3.48**	**4.15**	**4.42**	**-1.62**	**-1.39**
Total electricity and heat	-	-	-	-	-	-	-	-	-
Main activity producers	-	-	-	-	-	-	-	-	-
Autoproducers	-	-	-	-	-	-	-	-	-
Patent fuel/BKB plants	-	-	-	-	-	-	-	-	-
Coke ovens/Liquefaction(1)	7.22	5.93	4.80	4.46	3.48	4.15	4.42	-1.62	-1.39
Blast furnace inputs	-	-	-	-	-	-	-	-	-
Gas manufacture	-	-	-	-	-	-	-	-	-
Industry	-	-	-	-	-	-	-	-	-
Iron and steel	-	-	-	-	-	-	-	-	-
Chemical	-	-	-	-	-	-	-	-	-
Non-metallic minerals	-	-	-	-	-	-	-	-	-
Paper, pulp and print	-	-	-	-	-	-	-	-	-
Other industry	-	-	-	-	-	-	-	-	-
Other sectors(2)	-	-	-	-	-	-	-	-	-
Non-energy use	-	-	-	-	-	-	-	-	-
Lignite	**30.47**	**45.99**	**67.29**	**70.53**	**72.04**	**72.09**	**71.00**	**3.49**	**2.09**
Total electricity and heat	27.32	44.14	66.34 e	68.10	72.19	71.56	70.16	4.08	2.23
Main activity producers	*26.32*	*43.07*	*65.25 e*	*68.10*	*72.19*	*71.56*	*70.16*	*4.19*	*2.35*
Autoproducers	*0.99*	*1.07*	*1.09 e*	-	-	-	-	*0.62*	-
Patent fuel/BKB plants	2.85	1.78	0.96 e	1.00	0.60	0.48	0.54	-3.84	-5.55
Coke ovens/Liquefaction(1)	-	-	-	-	-	-	-	-	-
Gas manufacture	-	-	-	-	-	-	-	-	-
Industry	0.31	0.07	-	0.00	0.00	0.00	0.00	-11.23	-15.74
Iron and steel	-	-	-	-	-	-	-	-	-
Chemical	-	*0.03*	-	-	-	-	-	-	-
Non-metallic minerals	-	*0.00*	-	*0.00*	*0.00*	*0.00*	*0.00*	-	*0.00*
Paper, pulp and print	*0.26*	-	-	-	-	-	-	-	-
Other industry	*0.04*	*0.05*	-	-	-	-	-	*0.96*	-
Other sectors(2)	0.00	-	-	-	-	-	-	-	-
Non-energy use	-	-	-	-	-	-	-	-	-

(1) Liquefaction primarily refers to direct distillation processes. Indirect liquefaction may be reported here or under gas manufacture.

(2) Other sectors are Residential, Commercial and public services, Agriculture/Forestry, Fishing and Non-specified other.

* Coal totals may not match the sum of the individual rows due to statistical differences. Please refer to notes and definitions in Part I.

Source: *IEA/OECD Energy Statistics of OECD Countries*

AUSTRALIA

7. Fuel prices to end users*

	1978[1]	1990	1995	2000	2005	2009	2010	2011	2012
			(US dollars / tce)						
For electricity generation									
Steam coal	10.83	29.05	..	..	..	..	..	..	..
Heavy fuel oil	67.03	..	..	..	..	..	..	..	..
Natural gas	..	..	..	..	..	..	..	..	..
For industry									
Steam coal	..	..	..	..	..	..	..	..	..
Coking coal	29.69	..	..	..	..	..	..	..	..
High sulphur fuel oil	67.03	..	..	..	..	..	..	..	..
Low sulphur fuel oil	..	..	..	..	..	..	..	..	..
Natural gas	34.33	95.78	103.03	..	..	..	..	..	..
			(Australian dollars / unit)[2]						
For electricity generation									
Steam coal	8.91	35.11	..	..	..	..	..	..	..
Heavy fuel oil	80.22	..	..	..	..	..	..	..	..
Natural gas	..	..	..	..	..	..	..	..	..
For industry									
Steam coal	..	..	..	..	..	..	..	..	..
Coking coal	25.19	..	..	..	..	..	..	..	..
High sulphur fuel oil	80.22	..	..	..	..	..	..	..	..
Low sulphur fuel oil	..	..	..	..	..	..	..	..	..
Natural gas	38.52	157.84	178.78	..	..	..	..	..	..

(1) Earliest year for which split by coal type is available.

(2) For steam coal, coking coal and heavy fuel oil per metric tonne; for natural gas per 10^7 kilocalories GCV.

* Please refer to notes and definitions in Part I. Steam coal prices relate to historical definitions.

Source: *IEA/OECD Energy Prices & Taxes*

8. Coal and peat trade by type of coal*

(Mtce)

	1978	1990	1995	2000	2005	2009	2010	2011	2012e
Total imports	**0.01**	**-**	**-**	**-**	**0.02**	**0.01**	**0.01**	**0.05**	**0.03**
Bituminous coal[1]	0.01	-	-	-	-	-	-	-	-
Coking coal	-	-	-	-	-	-	-	-	-
Sub-bituminous coal	-	-	-	-	-	-	-	-	-
Lignite	-	-	-	-	-	-	-	-	-
Peat	-	-	-	-	-	-	-	-	-
Coal products[2]	-	-	-	-	0.02	0.01	0.01	0.05	0.03
Total exports	**34.17**	**96.07**	**126.55**	**173.46**	**215.70**	**241.50**	**271.64**	**262.96**	**278.02**
Bituminous coal[1]	4.24	39.96	54.63	77.02	94.23	119.71	118.71	126.37	139.58
Coking coal	29.77	55.58	71.55	96.43	121.47	121.79	152.93	136.58	138.44
Sub-bituminous coal	-	-	-	-	-	-	-	-	-
Lignite	-	-	-	-	-	-	-	-	-
Peat	-	-	-	-	-	-	-	-	-
Coal products[2]	0.16	0.52	0.37	0.02	c	c	c	c	-

(1) Bituminous coal includes anthracite.

(2) Coal products includes products derived from coal, for example: coke, coal tar, briquettes, patent fuels.

* Please refer to notes and definitions in Part I.

Source: *IEA/OECD Energy Balances of OECD Countries, IEA/OECD Energy Statistics of OECD Countries*

AUSTRALIA

10. Coking coal exports by destination*

(thousand tonnes)

	1978[1]	1985	1990	1995	2000	2005	2010	2011	2012e
World	**31029**	**49382**	**57750**	**74291**	**99161**	**124915**	**157265**	**140455**	**142363**
Total OECD	**29177**	**41759**	**44002**	**54258**	**73658**	**83834**	**81628**	**77522**	**76636**
Australia	-	-	-	-	-	-	-	-	-
Austria	-	-	-	-	-	-	-	-	-
Belgium	374	16	1054	1524	1833	1881	1034	1314	1276
Canada	-	-	-	-	-	-	-	-	-
Chile	32	64	200	466	463	555	389	413	543
Czech Republic	-	-	-	-	-	-	-	-	-
Denmark	-	-	-	-	-	-	-	-	-
Estonia	-	-	-	-	-	-	-	-	-
Finland	-	-	-	-	-	392	76	68	314
France	1173	2106	1917	2310	3739	3895	2506	2560	2453
Germany	-	-	25	204	2619	1746	1160	1275	908
Greece	191	-	-	-	-	-	-	-	-
Hungary	-	-	-	-	-	-	-	-	-
Iceland	-	-	-	-	-	-	-	-	-
Ireland	-	-	-	-	-	14	-	-	-
Israel	-	-	-	-	-	129	55	-	-
Italy	1319	2055	1181	2752	2914	2527	1559	1621	1751
Japan	22640	29251	28579	30774	39174	44962	48462	42604	40123
Korea	1330	3064	4944	9032	10358	12457	15864	16443	16398
Luxembourg	-	-	-	-	-	-	-	-	-
Mexico	10	-	-	-	666	588	336	254	423
Netherlands	992	1380	713	703	2194	5652	3342	3837	5029
New Zealand	-	-	-	-	-	-	-	-	-
Norway	-	-	-	-	-	-	-	-	-
Poland	-	-	-	-	-	-	-	-	40
Portugal	-	-	-	129	-	-	-	-	-
Slovak Republic	-	-	-	-	-	-	-	-	-
Slovenia	-	-	-	-	-	-	-	-	72
Spain	593	1186	694	896	1767	2470	1069	1721	1486
Sweden	-	-	702	535	992	1226	1486	1223	980
Switzerland	-	-	-	-	-	-	-	-	-
Turkey	-	516	912	1608	1451	720	909	736	1514
United Kingdom	469	2121	3081	3325	5488	4451	3381	3453	3326
United States	54	-	-	-	-	169	-	-	-
Total non-OECD	**1852**	**7623**	**13748**	**20033**	**25293**	**36950**	**71427**	**59564**	**63773**
Brazil	164	958	1291	3448	4988	3094	4234	2882	2338
China[2]	-	192	560	211	174	4193	27282	15716	21317
Chinese Taipei	981	1990	2798	3851	6273	7091	5357	8050	8585
Egypt	-	312	323	459	-	353	324	-	-
India	32	2042	4844	9555	10588	17442	31377	30915	29292
Romania	675	595	2256	710	-	46	-	-	-
Oth. Africa & Mid. East	-	688	454	597	1554	2807	1434	967	1111
Oth. non-OECD Americas	-	35	300	384	595	784	603	608	570
Other Asia & Oceania	-	680	734	818	1051	737	671	356	560
Other non-OECD Europe and Eurasia	-	131	188	-	70	403	145	70	-
Non-specified/Other	**-**	**-**	**-**	**-**	**210**	**4131**	**4210**	**3369**	**1954**

(1) Earliest year for which split by coal type is available.

(2) China includes the People's Republic of China and Hong Kong, China.

* Please refer to notes and definitions in Part I.

Source: IEA/OECD Coal Statistics

AUSTRALIA

11. Steam coal exports by destination*

(thousand tonnes)

	1978(1)	1985	1990	1995	2000	2005	2010	2011	2012e
World	**5522**	**39259**	**45646**	**62411**	**87801**	**107415**	**135356**	**144081**	**159152**
Total OECD	**4809**	**29523**	**38304**	**51291**	**70498**	**85460**	**95692**	**98650**	**103280**
Australia	-	-	-	-	-	-	-	-	-
Austria	-	-	-	-	-	-	-	-	-
Belgium	-	771	129	1116	428	17	-	-	-
Canada	-	-	-	-	-	-	-	-	-
Chile	-	99	120	1023	1301	412	309	691	388
Czech Republic	-	-	-	-	-	-	-	-	-
Denmark	150	2557	1149	1076	142	-	-	-	-
Estonia	-	-	-	-	-	-	-	-	-
Finland	-	-	-	-	-	-	-	-	-
France	682	508	934	276	434	469	66	-	-
Germany	458	591	125	61	72	115	-	-	-
Greece	-	125	-	-	110	-	-	-	-
Hungary	-	-	-	-	-	-	-	-	-
Iceland	-	-	-	-	-	-	-	-	-
Ireland	-	-	133	121	284	568	-	-	-
Israel	-	577	528	695	2623	1165	516	501	494
Italy	-	226	-	366	428	141	-	-	-
Japan	1489	15256	26569	34556	47449	57574	66413	66961	69817
Korea	-	5453	3633	9662	11452	17970	24840	28273	28925
Luxembourg	-	-	-	-	-	-	-	-	-
Mexico	-	-	-	-	-	4238	3109	1703	3595
Netherlands	320	2464	4236	1562	2550	760	127	-	-
New Zealand	-	-	-	-	16	56	59	58	22
Norway	-	-	-	-	-	-	-	-	-
Poland	-	-	-	-	-	-	-	-	-
Portugal	-	-	-	264	-	-	-	-	-
Slovak Republic	-	-	-	-	-	-	-	-	-
Slovenia	-	-	-	-	-	-	-	-	-
Spain	-	455	205	110	1445	671	-	-	-
Sweden	-	-	155	44	83	164	73	145	39
Switzerland	-	-	29	-	-	-	-	-	-
Turkey	-	-	-	25	55	45	-	-	-
United Kingdom	932	441	328	142	1499	993	13	-	-
United States	778	-	31	192	127	102	167	318	-
Total non-OECD	**193**	**8072**	**7342**	**11120**	**17303**	**21955**	**39664**	**45430**	**55872**
Brazil	-	-	158	-	-	33	20	41	-
China(2)	-	2189	2443	3420	1429	2121	15192	17867	29968
Chinese Taipei	76	3761	3046	5920	10034	14329	19553	20124	17516
Egypt	-	-	-	-	-	-	-	-	-
India	-	93	47	-	2469	1461	610	495	1165
Romania	-	-	33	110	-	-	-	-	-
Oth. Africa & Mid. East	-	-	-	-	-	-	-	2	-
Oth. non-OECD Americas	-	-	-	-	72	-	-	-	-
Other Asia & Oceania	117	2029	1615	1670	3299	4011	4289	6901	7223
Other non-OECD Europe and Eurasia	-	-	-	-	-	-	-	-	-
Non-specified/Other	**520**	**1664**	**-**	**-**	**-**	**-**	**-**	**1**	**-**

(1) Earliest year for which split by coal type is available.

(2) China includes the People's Republic of China and Hong Kong, China.

* Please refer to notes and definitions in Part I. Steam coal includes all sub-bituminous coal.

Source: IEA/OECD Coal Statistics

AUSTRALIA

13. Coal export values by destination*

(average unit value, FOB, USD/tonne)

	1990	1995	2000	2005	2007	2008	2009	2010	2011/12(1)
Coking coal(2)	**47.72**	**44.23**	**32.85**	**88.94**	**84.16**	**194.87**	**143.83**	**171.76**	..
Exports to:									
Belgium	51.66	44.65	31.01	104.61	89.48	193.95	133.96	198.00	..
Canada	x	x	x	x	x	x	x	x	..
Denmark	x	x	x	x	x	x	x	x	..
Finland	x	x	x	125.25	104.79	303.82	151.00	..	..
France	47.00	43.07	33.37	96.94	89.99	193.92	193.51	220.90	..
Germany	46.30	43.41	30.46	99.12	102.27	240.18	165.59	184.82	..
Greece	x	x	x	..	66.59	..	..	..	..
Ireland	x	x	x	..	..	..	..	..	..
Italy	46.76	42.06	34.97	102.14	99.64	229.97	174.51	208.24	..
Japan	46.98	43.37	31.64	83.74	79.50	193.79	165.07	154.53	..
Netherlands	50.32	44.75	34.10	98.18	92.15	213.20	196.55	197.71	..
Norway	x	x	x	..	..	..	..	..	..
Portugal	x	49.00	x	..	..	..	..	..	..
Spain	48.27	48.65	34.81	98.95	95.80	231.74	160.60	188.97	..
Sweden	50.15	48.95	38.71	117.03	96.65	275.66	146.05	198.20	..
Switzerland	x	x	x	..	..	..	..	..	..
Turkey	50.25	44.26	32.55	108.84	77.14	237.21	161.37	160.25	..
United Kingdom	47.25	45.13	33.80	100.80	91.38	200.63	149.63	163.56	..
Other OECD	48.21	44.96	37.46	89.68	92.28	208.03	127.84	162.39	..
Non-OECD	48.23	46.32	34.47	88.25	88.74	186.29	131.20	181.42	..
Steam coal(2,3)	**38.34**	**34.99**	**24.27**	**48.86**	**51.11**	**92.23**	**80.03**	**85.82**	..
Exports to:									
Belgium	38.42	33.43	20.72	x	x	153.74	x	x	..
Canada	x	x	x	x	x	x	x	x	..
Denmark	32.29	24.73	17.01	44.09	x	x	87.04	x	..
Finland	x	x	x	..	..	..	..	x	..
France	33.82	32.97	19.14	45.88	44.93	75.81	67.68	x	..
Germany	37.74	29.61	x	x	x	x	x	x	..
Greece	x	x	24.07	..	..	..	..	x	..
Ireland	x	34.14	21.21	29.92	75.25	129.29	..	x	..
Italy	x	37.75	21.30	x	x	x	x	x	..
Japan	40.03	36.85	24.72	49.42	54.61	103.23	88.54	89.03	..
Netherlands	36.31	33.03	20.65	60.71	49.60	86.94	78.94	x	..
Norway	x	x	x	..	..	..	..	..	..
Portugal	x	31.12	x	..	..	..	..	x	..
Spain	32.00	x	21.16	42.82	71.08	122.95	44.47	x	..
Sweden	x	37.85	29.22	67.75	60.32	109.33	90.88	120.56	..
Switzerland	x	x	x	..	..	..	..	..	..
Turkey	x	x	27.11	x	x	x	x	x	..
United Kingdom	35.45	x	20.98	54.27	x	125.83	65.52	x	..
Other OECD	38.42	34.96	25.67	49.30	47.51	73.30	76.03	83.10	..
Non-OECD	35.56	31.43	23.50	46.65	45.14	82.28	71.55	82.42	..

Note: On occasion, shipment of extremely small quantities of high valued coal results in high export costs.

(1) Import data for steam and coking coals are currently unavailable for 2011 and 2012 due to resource constraints.

(2) Weighted average of individual countries using import volumes as weights.

(3) Bituminous steam coal only.

* Please refer to notes and definitions in Part I.

Source: IEA/OECD Coal Statistics

AUSTRIA*

Figure 1: TPES by fuel (Mtce)

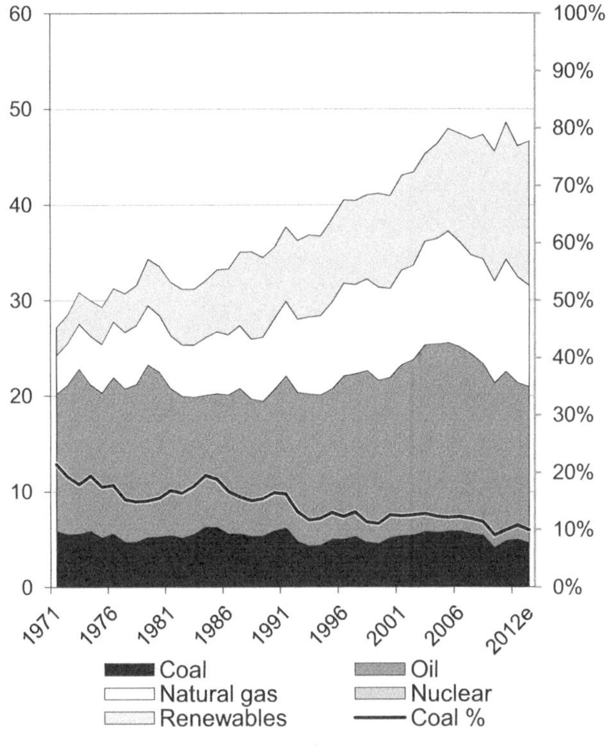

Figure 2: Coal supply indicators (1971 = 100)

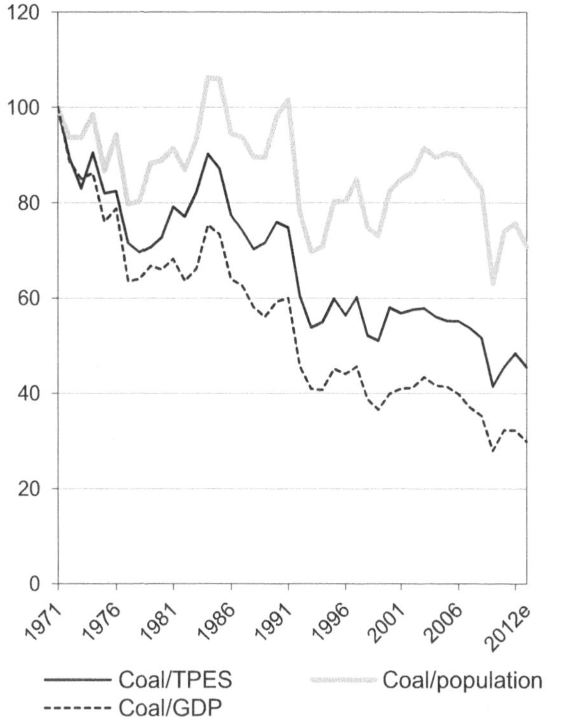

Figure 3: Primary coal supply (Mtce)

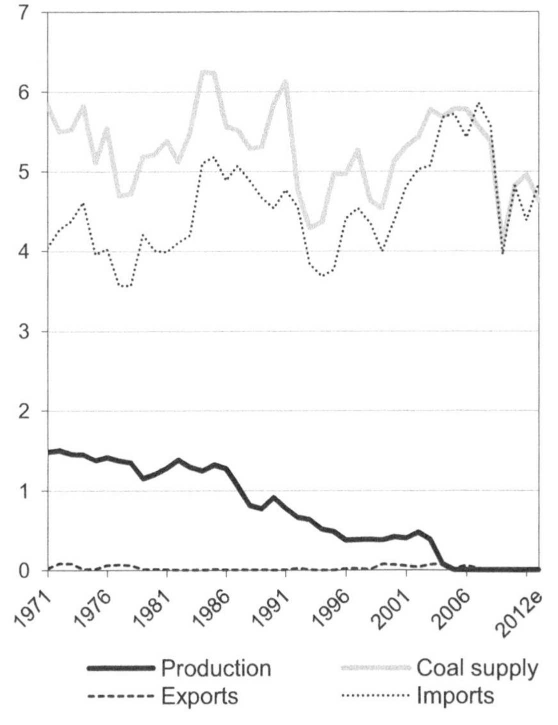

Figure 4: Coal consumption (Mtce)

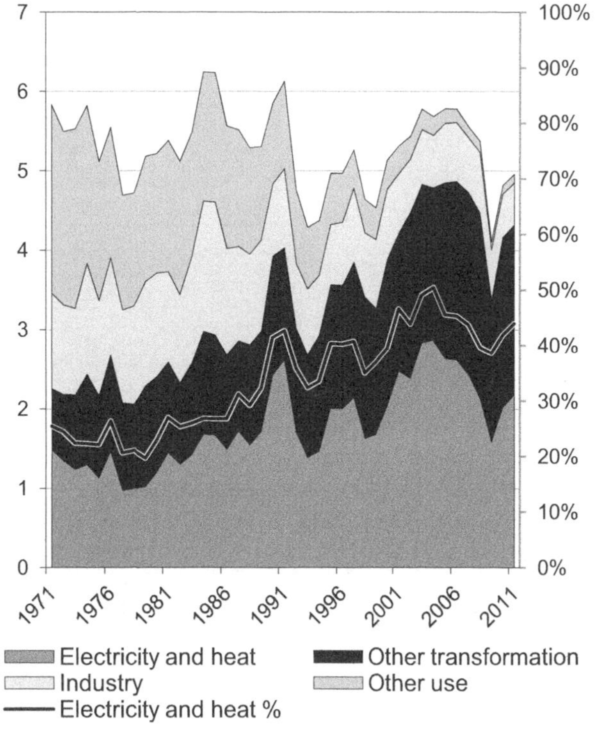

*Please refer to notes and definitions in Part I. Peat is included in coal for all figures.

Source: *IEA/OECD Coal Information, IEA/OECD Energy Balances of OECD Countries, OECD Main Economic Indicators*

AUSTRIA*

Figure 5: Electricity generation by fuel (TWh)

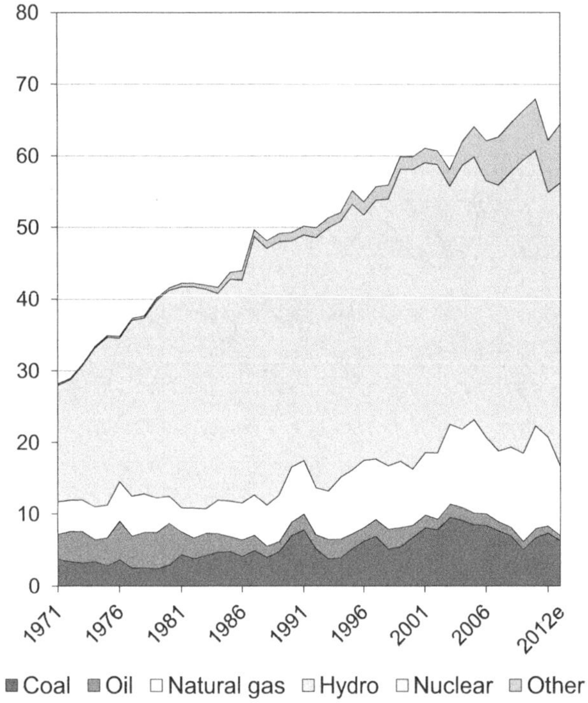

Figure 6: CO_2 emissions by fuel (Mt CO_2)

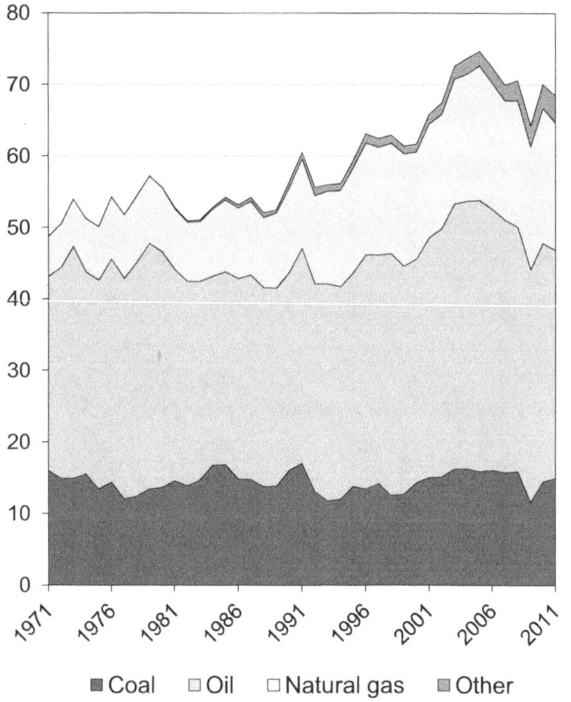

Figure 7: Electricity generation by fuel share

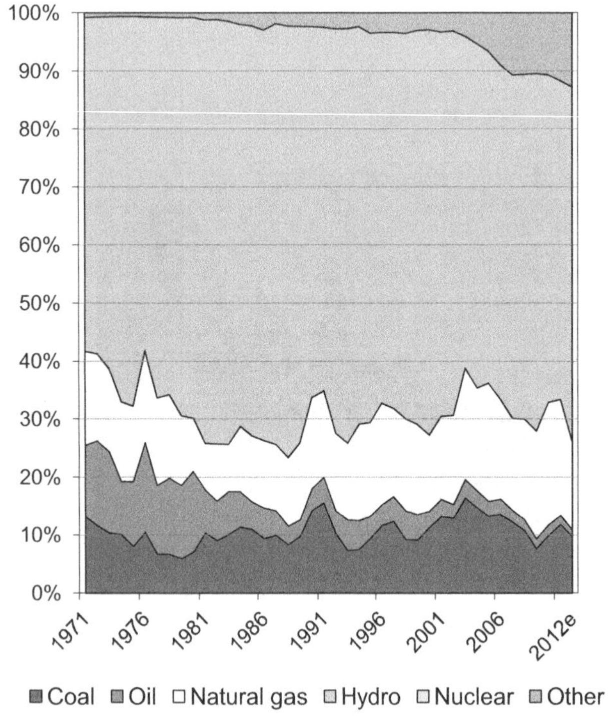

Figure 8: CO_2 emissions by fuel share

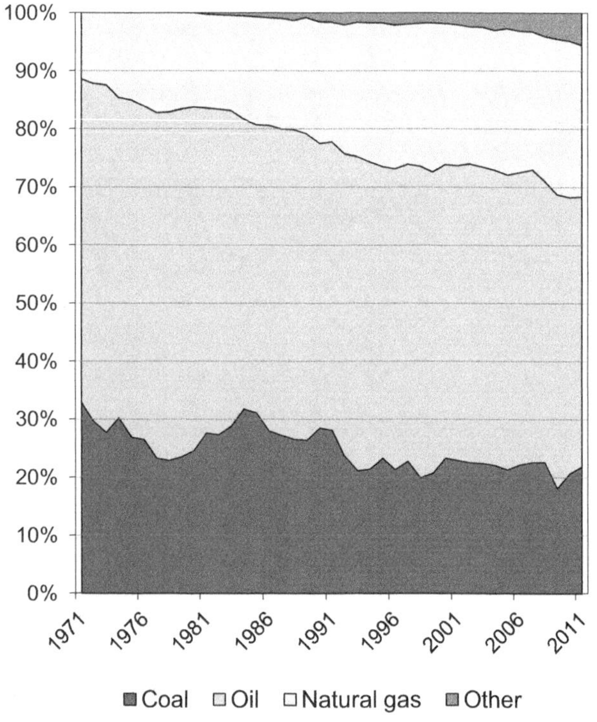

*Please refer to notes and definitions in Part I. Peat is included in coal for all figures.

Source: *IEA/OECD Electricity Information, IEA/OECD CO2 Emissions from Fuel Combustion*

AUSTRIA

1. Total primary energy supply (TPES) by fuel*

	1973	1980	1990	2000	2010	2011	2012e	Average annual percent change 73-90	90-11
TPES (Mtce)	**30.68**	**33.08**	**35.49**	**40.80**	**48.90**	**47.17**	**46.99**	**0.86**	**1.36**
Coal and peat	5.53	5.22	5.85	5.13	4.82	4.96	4.63	0.33	-0.78
Oil	17.30	17.26	14.82	16.76	17.70	16.39	16.30	-0.91	0.48
Natural Gas	4.72	5.93	7.40	9.31	11.73	11.08	10.60	2.68	1.94
Biofuels and waste	1.02	1.62	3.57	4.48	9.09	8.99	9.64	7.63	4.49
Nuclear	-	-	-	-	-	-	-	-	-
Hydro	2.30	3.53	3.87	5.14	4.72	4.20	4.86	3.10	0.39
Geothermal	-	-	0.01	0.04	0.05	0.05	0.04	-	11.21
Solar, wind, tide	-	-	0.02	0.10	0.49	0.50	0.58	-	16.20
Net electricity trade(1)	-0.19	-0.49	-0.06	-0.17	0.29	1.01	0.35	-6.90	x
Heat(2)	-	-	-	0.01	0.01	0.01	0.01	-	-

(1) Net trade between OECD and non-OECD countries.

(2) Ambient heat from heat pumps used in the transformation sector.

Source: *IEA/OECD Energy Balances of OECD Countries*

2. Energy supply, GDP and population*

	1973	1980	1990	2000	2010	2011	2012e	Average annual percent change 73-90	90-11
GDP (2005 bil. USD)	142	172	215	281	327	335	338	2.49	2.13
TPES/GDP(1)	0.22	0.19	0.16	0.15	0.15	0.14	0.14	-1.59	-0.75
Population (millions)	7.6	7.5	7.7	8.0	8.4	8.4	8.4	0.07	0.44
TPES/population(1)	4.04	4.38	4.62	5.09	5.83	5.60	5.59	0.79	0.92
TPES/GDP(2)	136.9	121.2	104.1	91.8	94.6	88.9	87.8	-1.59	-0.75
Coal and peat supply/GDP(2)	205.9	159.6	143.3	96.5	77.9	78.0	72.3	-2.11	-2.86
Elec. consumption/GDP(2)	93.5	100.2	103.9	96.1	98.2	96.0	..	0.63	-0.38
Elec. generation (TWh)	31	42	49	60	68	62	64	2.78	1.11
Industrial production(2)	36.6	44.8	58.8	85.3	110.2	116.8	119.0	2.83	3.32

(1) TPES/GDP in units of tce/2005 thousand USD, TPES/population in units of tce/capita.

(2) As index, 2005 (not shown) = 100

Source: *IEA/OECD Energy Balances of OECD Countries , OECD Main Economic Indicators*

3. Coal and peat production by type*

	1978(1)	1990	2000	2005	2010	2011	2012e	Average annual percent change 78-90	90-11
Mtce:									
Coking coal	-	-	-	-	-	-	-	-	-
Steam coal	-	-	-	-	-	-	-	-	-
Lignite	1.35	0.91	0.42	-	-	-	-	-3.25	-
Peat	-	0.00	0.00	0.00	0.00	0.00	0.00	-	0.00
Mt:									
Coking coal	-	-	-	-	-	-	-	-	-
Steam coal	-	-	-	-	-	-	-	-	-
Lignite	3.08	2.45	1.25	-	-	-	-	-1.88	-
Peat	-	0.00	0.00	0.00	0.00	0.00	0.00	-	0.00

(1) Earliest year for which split by coal type is available.

* Please refer to notes and definitions in Part I.

Source: *IEA/OECD Energy Balances of OECD Countries, IEA/OECD Energy Statistics of OECD Countries*

AUSTRIA

4. Final consumption of energy by fuel*

(Mtce)

	1973	1980	1990	2000	2009	2010	2011	Average annual percent change 73-90	90-11
Total final consumption(1)	**23.73**	**26.63**	**28.16**	**33.63**	**37.74**	**40.16**	**38.29**	**1.01**	**1.47**
Coal and peat	3.35	2.81	1.93	1.25	0.74	0.66	0.64	-3.21	-5.09
Oil	14.21	13.94	12.63	14.83	15.59	15.94	14.95	-0.69	0.81
Natural Gas	2.07	4.04	4.34	6.10	6.56	7.22	6.74	4.45	2.13
Biofuels and wastes	0.99	1.51	3.12	3.55	5.25	5.86	5.65	7.01	2.87
Geothermal	-	-	0.01	0.01	0.01	0.01	0.01	-	3.66
Solar, wind, tide	-	-	0.02	0.09	0.22	0.23	0.23	-	12.14
Electricity	3.11	4.06	5.25	6.33	7.25	7.53	7.56	3.14	1.75
Heat	-	0.27	0.87	1.46	2.13	2.71	2.50	-	5.12
of which:									
Total industry	**7.46**	**7.04**	**7.40**	**8.60**	**10.61**	**10.80**	**10.60**	**-0.05**	**1.73**
Coal and peat	1.09	1.31	0.91	0.89	0.64	0.55	0.53	-1.05	-2.49
Oil	3.19	1.42	0.96	0.81	1.06	0.93	0.88	-6.80	-0.41
Natural Gas	1.65	2.27	2.33	2.98	3.27	3.33	3.14	2.07	1.43
Biofuels and wastes	0.05	0.30	0.87	1.20	2.01	2.32	2.40	18.50	4.95
Geothermal	-	-	-	-	-	-	-	-	-
Solar, wind, tide	-	-	-	-	-	-	-	-	-
Electricity	1.49	1.74	2.21	2.54	3.28	3.30	3.29	2.35	1.92
Heat	-	-	0.11	0.17	0.35	0.36	0.34	-	5.39
Total transport	**5.53**	**5.84**	**6.93**	**9.27**	**11.31**	**11.62**	**11.34**	**1.34**	**2.37**
Coal and peat	0.16	0.02	0.00	0.00	-	-	-	-21.12	-
Oil	5.12	5.49	6.44	8.62	9.95	10.32	10.02	1.36	2.13
Natural Gas	0.03	0.05	0.14	0.21	0.26	0.20	0.25	9.94	2.82
Biofuels and wastes	0.00	0.00	0.01	0.02	0.69	0.68	0.69	5.41	23.06
Electricity	0.22	0.28	0.34	0.43	0.41	0.42	0.38	2.73	0.58
Residential	**8.17**	**7.28**	**8.34**	**9.03**	**8.92**	**10.07**	**9.21**	**0.12**	**0.47**
Coal and peat	2.03	1.33	0.94	0.30	0.07	0.08	0.07	-4.39	-11.81
Oil	4.33	3.11	2.44	2.45	1.77	1.96	1.70	-3.33	-1.70
Natural Gas	0.13	0.55	1.09	1.60	1.74	1.91	1.70	13.14	2.13
Biofuels and wastes	0.93	1.21	1.99	2.05	2.16	2.41	2.16	4.57	0.40
Geothermal	-	-	-	-	-	-	-	-	-
Solar, wind, tide	-	-	0.01	0.05	0.13	0.15	0.14	-	12.41
Electricity	0.75	1.08	1.46	1.84	2.18	2.22	2.19	4.03	1.95
Heat	-	-	0.40	0.73	0.87	1.34	1.25	-	5.51
Comm & public services	**0.71**	**2.62**	**2.44**	**3.63**	**3.67**	**4.32**	**4.19**	**7.56**	**2.60**
Coal and peat	0.06	0.15	0.04	0.04	0.01	0.01	0.01	-3.05	-8.56
Oil	0.02	1.20	0.56	0.62	0.52	0.40	0.33	21.96	-2.52
Natural Gas	0.07	0.43	0.26	0.84	0.78	1.22	1.16	8.17	7.39
Biofuels and waste	-	-	0.11	0.11	0.10	0.12	0.10	-	-0.24
Geothermal	-	-	0.01	0.01	0.01	0.01	0.01	-	3.66
Solar, wind, tide	-	-	0.01	0.04	0.08	0.08	0.09	-	11.77
Electricity	0.56	0.83	1.12	1.42	1.29	1.49	1.59	4.15	1.71
Heat	-	-	0.35	0.55	0.90	1.00	0.90	-	4.55
Non-energy use	**1.76**	**2.30**	**2.22**	**2.35**	**2.46**	**2.53**	**2.18**	**1.39**	**-0.09**
Coal and peat	0.00	0.00	0.01	0.01	0.02	0.02	0.03	12.38	4.15
Oil	1.56	1.56	1.70	1.89	1.96	1.97	1.67	0.50	-0.09
Natural Gas	0.19	0.73	0.50	0.45	0.49	0.54	0.47	5.85	-0.30

(1) Total Final Consumption (TFC) excludes use in transformation processes and energy industry own use.

* Please refer to notes and definitions in Part I. Agriculture, Forestry, Fishing and "Non-specified other" sectors are not reported here individually, but are accounted for in Total final consumption.

Source: *IEA/OECD Energy Balances of OECD Countries*

AUSTRIA

5. Coal balance*

(Mtce)

	1973	1980	1990	2000	2005	2010	2011	2012e	Average annual percent change 73-90	90-11
Production	1.5	1.2	0.9	0.4	0.0	0.0	0.0	0.0	-2.7	-31.7
Imports	4.4	4.0	4.5	4.4	5.7	4.8	4.4	4.8	0.2	-0.2
Exports	-0.1	-0.0	-0.0	-0.1	-0.0	-0.0	-0.0	-0.0	-17.0	-0.4
Stock changes	-0.2	0.0	0.4	0.4	0.1	0.0	0.6	-0.2		
Primary supply	**5.5**	**5.2**	**5.9**	**5.1**	**5.8**	**4.8**	**5.0**	**4.6**	**0.3**	**-0.8**
Statistical differences	**0.0**	**0.0**	**-0.0**	**-0.1**	**-0.0**	**0.0**	**0.0**	..		
Total transformation	**-1.8 e**	**-2.2 e**	**-3.3 e**	**-3.1 e**	**-3.9 e**	**-3.2**	**-3.3**	..	**3.6**	**0.1**
Electricity and heat gen.	-1.2	-1.2	-2.4	-2.0	-2.6	-2.0	-2.2	..	4.0	-0.5
Main activity producers (1)	*-1.0*	*-1.0*	*-2.1*	*-1.6*	*-2.1*	*-1.4*	*-1.6*	..	*4.2*	*-1.4*
Autoproducers	*-0.2*	*-0.2*	*-0.3*	*-0.4*	*-0.5*	*-0.6*	*-0.6*	..	*3.3*	*3.0*
Gas works	0.5	-0.0	0.0	-	-	-	-	..	-15.7	-
Coal transformation(2)	-1.1 e	-1.0 e	-0.9 e	-1.1 e	-1.3 e	-1.1	-1.1	..	-1.1	1.3
BKB plants	*0.1*	-	-	-	-	-	-	..	-	-
Blast furnaces	*-1.0* e	*-0.9* e	*-0.7* e	*-1.0* e	*-1.2* e	*-1.1*	*-1.1*	..	*-2.1*	*1.8*
Coke ovens	*-0.1*	*-0.1*	*-0.1*	*-0.1*	*-0.1*	*-0.1*	*-0.1*	..	*3.0*	*-3.2*
Patent fuel plants	-	-	-	-	-	-	-	..	-	-
Other transformation(3)	-	-	-	-	-	-	-	..	-	-
Energy ind. own use	**-0.3**	**-0.1**	**-0.7**	**-0.7**	**-0.8**	**-1.0**	**-1.0**	..	**5.4**	**1.8**
Losses	**-0.1**	**-0.0**	**-0.0**	**-**	**-0.0**	**-0.0**	**-0.1**	..		
Final consumption(4)	**3.4**	**2.8**	**1.9**	**1.3**	**0.9**	**0.7**	**0.6**	..	**-3.2**	**-5.1**
Industry(5)	1.1	1.3	0.9	0.9	0.7	0.5	0.5	..	-1.1	-2.5
Iron and steel	*0.7* e	*1.1* e	*0.6* e	*0.5* e	*0.4* e	*0.3*	*0.3*	..	*-1.6*	*-2.5*
Chemical	*0.0*	*0.0*	*0.0*	*0.1*	*0.0*	*0.0*	*0.0*	..	*-1.6*	*-10.7*
Non-metallic minerals	*0.1*	*0.0*	*0.2*	*0.2*	*0.2*	*0.1*	*0.1*	..	*7.5*	*-2.6*
Paper, pulp and print	*0.0*	*0.0*	*0.1*	*0.1*	*0.1*	*0.1*	*0.1*	..	*4.4*	*-1.4*
Other industry (6)	*0.2*	*0.1*	*0.0*	*0.0*	*0.0*	*0.0*	*0.0*	..	*-13.0*	*-3.7*
Transport(7)	0.2	0.0	0.0	0.0	-	-	-	..	-21.1	-
Other	2.1	1.5	1.0	0.3	0.2	0.1	0.1	..	-4.3	-11.6
Comm. and pub. services	*0.1*	*0.2*	*0.0*	*0.0*	*0.0*	*0.0*	*0.0*	..	*-3.1*	*-8.6*
Residential	*2.0*	*1.3*	*0.9*	*0.3*	*0.1*	*0.1*	*0.1*	..	*-4.4*	*-11.8*
Other sectors (8)	*0.0*	*0.0*	*0.0*	*0.0*	*0.0*	*0.0*	*0.0*	..	*3.1*	*-10.4*
Non-energy use	0.0	0.0	0.0	0.0	0.0	0.0	0.0	..	12.4	4.1

(1) Main activity electricity and heat generation includes district heating.

(2) Coal transformation refers to the transformation of coal and peat to secondary and tertiary products (mainly cokes, briquettes, coal tar and off-gases). Gas works and Liquefaction are shown separately to match the *Energy Balances of OECD Countries.*

(3) Other transformation includes Liquefaction and Non-specified transformations.

(4) Final Consumption includes non-energy use and energy use (Industry, Transport and Other).

(5) Please refer to notes and definitions in Part I for detailed explanation.

(6) Other industry includes Non-ferrous metals, Transport equipment, Machinery, Mining and quarrying, Food and tobacco, Wood and wood products, Construction, Textile and leather, and Non-specified industry.

(7) Transport includes Rail and Inland waterways.

(8) Other sectors includes Agriculture/Forestry, Fishing and Non-specified other.

* "Coal" refers to all types of coal, primary (anthracite, coking coal, other bituminous coal, sub-bituminous coal, lignite/brown coal) and derived fuels (including patent fuel, BKB, coke oven coke, gas coke, coal tar, coke oven gas, gas works gas, blast furnace gas and other recovered gases), as well as peat, peat briquettes and oil shale. Quantities have been converted to Mtce using calorific values reported by the respective countries. Please refer to notes and definitions in Part I.

Source: *IEA/OECD Energy Balances of OECD Countries*

AUSTRIA

6. Use of coal for selected end-uses*

(million tonnes)

	1978	1990	2000	2005	2009	2010	2011	Average annual percent change 78-90	90-11
Steam coal	**0.37**	**1.82**	**1.89**	**2.22**	**1.64**	**1.97**	**2.07**	**14.21**	**0.60**
Total electricity and heat	0.00	1.42	1.42	1.89	1.20	1.58	1.71	83.11	0.88
Main activity producers	*0.00*	*1.37*	*1.36*	*1.84*	*1.16*	*1.54*	*1.67*	*82.59*	*0.94*
Autoproducers	-	*0.05*	*0.06*	*0.04*	*0.04*	*0.04*	*0.04*	-	*-1.36*
Patent fuel/BKB plants	-	-	-	-	-	-	-	-	-
Coke ovens/Liquefaction(1)	-	-	-	-	-	-	-	-	-
Blast furnace inputs	-	-	-	-	-	-	-	-	-
Gas manufacture	-	-	-	-	-	-	-	-	-
Industry	0.04	0.21	0.35	0.28	0.33	0.24	0.18	15.22	-0.58
Iron and steel	*0.01*	-	-	*0.00*	*0.00*	*0.00*	*0.00*	-	-
Chemical	-	*0.01*	*0.06*	*0.04*	*0.02*	*0.02*	*0.00*	-	*-5.79*
Non-metallic minerals	*0.02*	*0.20*	*0.21*	*0.15*	*0.22*	*0.13*	*0.12*	*21.10*	*-2.46*
Paper, pulp and print	-	*0.00*	*0.09*	*0.09*	*0.09*	*0.09*	*0.06*	-	*17.86*
Other industry	*0.00*	-	-	-	*0.00*	*0.00*	-	-	-
Other sectors(2)	0.32	0.19	0.08	0.05	-	0.00	0.02	-4.29	-10.81
Non-energy use	-	0.00	0.00	0.00	0.00	0.00	-	-	-
Coking coal	**2.01**	**2.34**	**1.88**	**1.90**	**1.69**	**1.84**	**1.79**	**1.28**	**-1.27**
Total electricity and heat	-	-	-	-	-	-	-	-	-
Main activity producers	-	-	-	-	-	-	-	-	-
Autoproducers	-	-	-	-	-	-	-	-	-
Patent fuel/BKB plants	-	-	-	-	-	-	-	-	-
Coke ovens/Liquefaction(1)	2.01	2.34	1.88	1.90	1.69	1.84	1.79	1.28	-1.27
Blast furnace inputs	-	-	-	-	-	-	-	-	-
Gas manufacture	-	-	-	-	-	-	-	-	-
Industry	-	-	-	-	-	-	-	-	-
Iron and steel	-	-	-	-	-	-	-	-	-
Chemical	-	-	-	-	-	-	-	-	-
Non-metallic minerals	-	-	-	-	-	-	-	-	-
Paper, pulp and print	-	-	-	-	-	-	-	-	-
Other industry	-	-	-	-	-	-	-	-	-
Other sectors(2)	-	-	-	-	-	-	-	-	-
Non-energy use	-	-	-	-	-	-	-	-	-
Lignite	**2.99**	**2.50**	**1.34**	**1.21**	**0.01**	**0.01**	**0.02**	**-1.47**	**-21.16**
Total electricity and heat	2.03	2.13	1.23	1.14	-	-	-	0.43	-
Main activity producers	*1.99*	*2.08*	*1.19*	*1.12*	-	-	-	*0.37*	-
Autoproducers	*0.04*	*0.05*	*0.04*	*0.02*	-	-	-	*2.75*	-
Patent fuel/BKB plants	-	-	-	-	-	-	-	-	-
Coke ovens/Liquefaction(1)	-	-	-	-	-	-	-	-	-
Gas manufacture	0.03	-	-	-	-	-	-	-	-
Industry	0.21	0.15	0.07	0.06	0.00	0.01	0.01	-2.72	-10.62
Iron and steel	*0.01*	-	-	-	-	-	-	-	-
Chemical	*0.00*	-	-	-	-	-	-	-	-
Non-metallic minerals	*0.01*	*0.01*	*0.00*	*0.01*	*0.00*	*0.01*	*0.01*	*2.69*	*1.16*
Paper, pulp and print	*0.15*	*0.13*	*0.07*	*0.06*	-	-	-	*-1.05*	-
Other industry	*0.03*	*0.00*	-	-	-	-	-	*-15.69*	-
Other sectors(2)	0.64	0.22	0.05	0.01	0.00	0.01	0.00	-8.63	-17.32
Non-energy use	-	-	-	-	-	-	-	-	-

(1) Liquefaction primarily refers to direct distillation processes. Indirect liquefaction may be reported here or under gas manufacture.

(2) Other sectors are Residential, Commercial and public services, Agriculture/Forestry, Fishing and Non-specified other.

* Coal totals may not match the sum of the individual rows due to statistical differences. Please refer to notes and definitions in Part I.

Source: *IEA/OECD Energy Statistics of OECD Countries*

AUSTRIA

7. Fuel prices to end users*

	1978[1]	1990	1995	2000	2005	2009	2010	2011	2012
				(US dollars / tce)					
For electricity generation									
Steam coal	173.19	80.98	c	56.94	92.70	141.37	115.56	135.53	126.89
Heavy fuel oil	69.15	92.64	c	51.28	132.84	153.71	270.09	332.53	341.08
Natural gas	74.24	115.57	..	..	..	..	..	..	..
For industry									
Steam coal	90.20	78.84	82.29	54.73	178.69	243.64	208.76	249.91	233.09
Coking coal	..	..	..	..	..	..	..	..	..
High sulphur fuel oil	74.62	87.74	98.87	..	..	..	..	..	..
Low sulphur fuel oil	..	..	135.70	128.14	258.08	355.95	438.29	580.23	601.19
Natural gas	73.76	121.26	133.63	..	..	..	..	..	..
				(Euro / unit)[2]					
For electricity generation									
Steam coal	172.67	63.15	c	58.33	70.45	96.07	82.37	91.96	93.20
Heavy fuel oil	100.14	104.94	c	76.31	146.66	151.74	279.65	327.76	363.92
Natural gas	100.80	122.74	..	..	..	..	..	..	..
For industry									
Steam coal	93.75	64.10	59.37	58.45	141.58	172.62	155.14	176.79	178.49
Coking coal	..	..	..	..	..	..	..	..	..
High sulphur fuel oil	108.06	99.39	99.39	..	..	..	..	..	..
Low sulphur fuel oil	..	..	136.41	190.67	284.92	351.38	453.82	571.90	641.45
Natural gas	100.14	128.78	125.94	..	..	..	..	..	..

(1) Earliest year for which split by coal type is available.

(2) For steam coal, coking coal and heavy fuel oil per metric tonne; for natural gas per 10^7 kilocalories GCV.

* Please refer to notes and definitions in Part I. Steam coal prices relate to historical definitions.

Source: *IEA/OECD Energy Prices & Taxes*

8. Coal and peat trade by type of coal*

(Mtce)

	1978	1990	1995	2000	2005	2009	2010	2011	2012e
Total imports	**3.57**	**4.54**	**3.76**	**4.38**	**5.72**	**3.95**	**4.82**	**4.38**	**4.85**
Bituminous coal[1]	0.30	1.18	1.16	1.60	2.18	1.39	1.59	1.25	1.65
Coking coal	1.99	2.36	1.76	1.72	2.05	1.65	1.89	1.73	1.77
Sub-bituminous coal	-	-	-	0.03	0.05	0.05	0.05	0.05	0.05
Lignite	0.11	0.01	0.01	0.01	0.01	0.01	0.01	0.02	0.00
Peat	-	-	-	-	-	-	-	-	-
Coal products[2]	1.17	0.99	0.83	1.03	1.42	0.85	1.28	1.34	1.37
Total exports	**0.06**	**0.00**	**0.00**	**0.07**	**0.02**	**0.02**	**0.01**	**0.00**	**0.01**
Bituminous coal[1]	-	-	0.00	-	0.00	-	0.00	0.00	0.00
Coking coal	-	-	-	-	-	-	-	-	-
Sub-bituminous coal	-	-	-	-	-	-	-	-	-
Lignite	0.00	0.00	-	-	-	0.00	0.00	0.00	-
Peat	-	-	-	-	-	-	-	-	-
Coal products[2]	0.06	0.00	0.00	0.07	0.02	0.02	0.00	0.00	0.01

(1) Bituminous coal includes anthracite.

(2) Coal products includes products derived from coal, for example: coke, coal tar, briquettes, patent fuels.

* Please refer to notes and definitions in Part I.

Source: *IEA/OECD Energy Balances of OECD Countries, IEA/OECD Energy Statistics of OECD Countries*

AUSTRIA

9. Total coal imports by origin*

(thousand tonnes)

	1978(1)	1985	1990	1995	2000	2005	2010	2011	2012e
Total coal(2)	**2553**	**3983**	**3645**	**3023**	**3463**	**4448**	**3725**	**3211**	**3641**
Coking coal	**2006**	**2483**	**2376**	**1778**	**1738**	**2063**	**1907**	**1742**	**1786**
Australia	-	-	-	-	-	-	-	-	-
Canada	-	-	-	-	-	-	-	-	5
Czech Republic	662	652	746	-	1187	992	986	711	807
Germany	205	147	-	-	-	9	-	-	-
Poland	470	784	566	1778	551	519	366	135	289
United Kingdom	-	-	-	-	-	-	-	-	-
United States	-	503	456	-	-	501	539	872	685
Other OECD	-	-	-	-	-	-	-	-	-
China, People's Rep.	-	-	-	-	-	-	-	-	-
Colombia	-	-	-	-	-	-	16	24	-
Indonesia	-	-	-	-	-	-	-	-	-
South Africa	-	-	-	-	-	-	-	-	-
Former Soviet Union(3)	669	397	x	x	x	x	x	x	x
Russian Federation (4)	*x*	*x*	*608*	-	-	*31*	-	-	-
Other FSU (4)	*x*	*x*	-	-	-	*11*	-	-	-
Venezuela	-	-	-	-	-	-	-	-	-
Vietnam	-	-	-	-	-	-	-	-	-
Non-specified/other	-	-	-	-	-	-	-	-	-
Steam coal	**301**	**1116**	**1233**	**1216**	**1709**	**2337**	**1796**	**1413**	**1854**
Australia	-	66	-	-	-	-	-	78	-
Canada	-	-	-	-	-	-	-	-	-
Czech Republic	-	-	-	15	307	922	837	1021	863
Germany	8	57	29	10	44	72	74	77	86
Poland	260	949	1189	1087	1358	1327	300	235	226
United Kingdom	-	-	-	-	-	-	-	-	-
United States	-	19	-	-	-	-	540	-	-
Other OECD	-	-	9	9	-	1	-	-	408
China, People's Rep.	-	-	-	-	-	-	-	-	-
Colombia	-	-	-	-	-	-	-	-	34
Indonesia	-	-	-	-	-	-	1	-	-
South Africa	-	-	6	95	-	10	-	-	-
Former Soviet Union(3)	33	-	x	x	x	x	x	x	x
Russian Federation (4)	*x*	*x*	-	-	-	-	-	-	*203*
Other FSU (4)	*x*	*x*	-	-	-	*5*	*24*	*2*	*1*
Venezuela	-	-	-	-	-	-	20	-	13
Vietnam	-	-	-	-	-	-	-	-	-
Non-specified/other	-	25	-	-	-	-	-	-	20
Lignite	**246**	**384**	**36**	**29**	**16**	**48**	**22**	**56**	**1**

(1) Earliest year for which split by coal type is available.

(2) Total coal consists of steam coal, coking coal and lignite.

(3) For years prior to 1990, or if no details for individual Former Soviet Union republics.

(4) Data for former republics are not available separately for some OECD member countries.

* Please refer to notes and definitions in Part I.

Source: IEA/OECD Coal Statistics

AUSTRIA

12. Coal import values by origin*

(average unit value, CIF, USD/tonne)

	1990	1995	2000	2005	2007	2008	2009	2010	2011/12[1]
Coking coal[2]	..	**214.20**	**68.06**	**85.39**	..	**307.83**	**295.92**	**289.94**	..
Imports from:									
Australia	..	..	..	..	..	..	..	..	..
Canada	..	..	..	..	..	..	..	..	..
Czech Republic	..	..	..	242.07	..	..	..	..	..
Poland	..	..	..	..	..	..	..	..	..
United States	..	..	..	..	..	..	..	..	..
China	..	..	..	..	..	..	..	..	..
Colombia	..	..	..	..	..	..	..	..	..
Indonesia	..	..	..	..	..	..	..	..	..
South Africa	..	..	..	..	..	..	..	..	..
Former Soviet Union[3]	..	..	..	..	..	..	..	..	..
Other bituminous coal[4]	..	**66.80**	**44.57**	**116.18**	**114.81**	**170.22**	**148.12**	**161.66**	..
Imports from:									
Australia	..	..	..	..	135.44	..	..	..	..
Canada	..	..	..	406.70	..	321.27	..	..	..
Czech Republic	..	63.72	43.50	129.52	116.79	165.34	144.57	152.93	..
Poland	..	73.62	45.11	100.29	106.74	165.52	133.63	138.27	..
United States	..	..	..	155.09	159.43	185.60	173.85	226.78	..
China	..	..	..	..	472.80	505.58	..	..	..
Colombia	..	..	..	..	..	..	..	241.64	..
Indonesia	..	..	..	..	..	..	..	..	..
South Africa	..	..	..	133.90	117.62	..	..	..	..
Former Soviet Union[3]	..	..	..	70.33	118.57	177.48	175.05	126.39	..

Note: On occasion, shipment of extremely small quantities of high valued coal results in high import costs.

(1) Import data for steam and coking coals are currently unavailable for 2011 and 2012 due to resource constraints.

(2) Weighted average of individual countries using import volumes as weights. Prices exclude intra-EU trade.

(3) Former Soviet Union until 1991, Russian Federation starting in 1992.

(4) Bituminous steam coal only. Weighted average of trades. Prices exclude intra-EU trade.

* Please refer to notes and definitions in Part I.

Source: IEA/OECD Coal Statistics

BELGIUM*

Figure 1: TPES by fuel (Mtce)

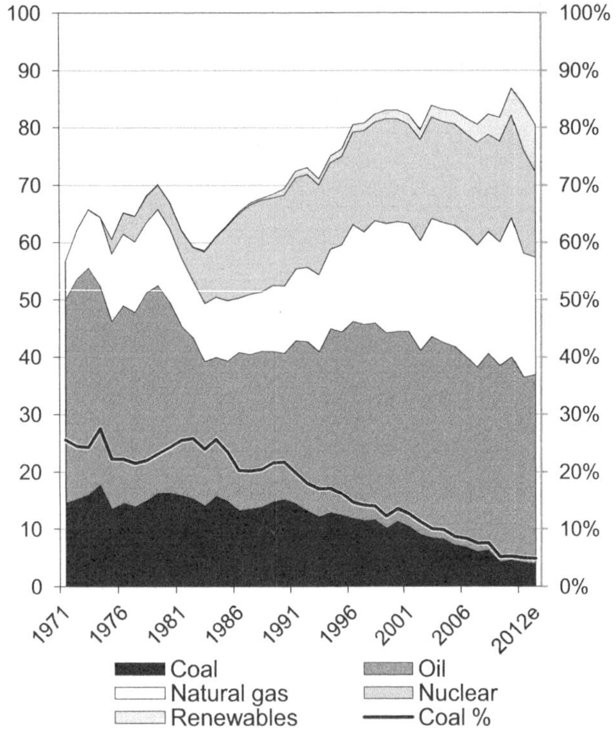

Figure 2: Coal supply indicators (1971 = 100)

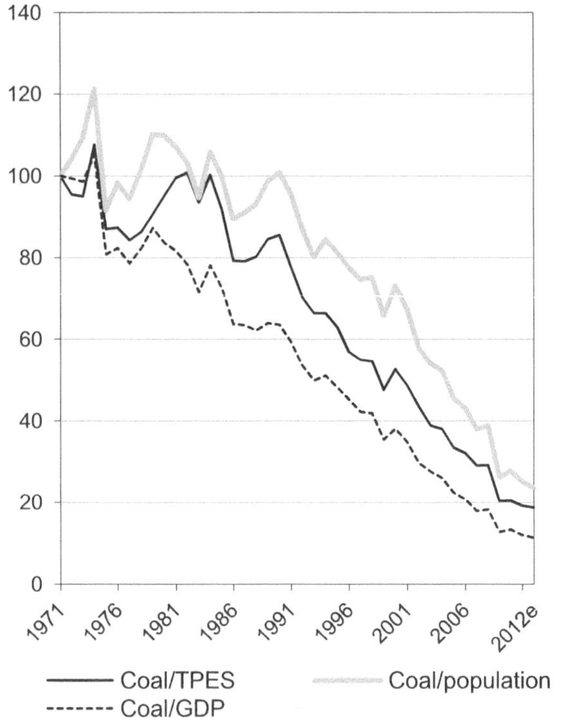

Figure 3: Primary coal supply (Mtce)

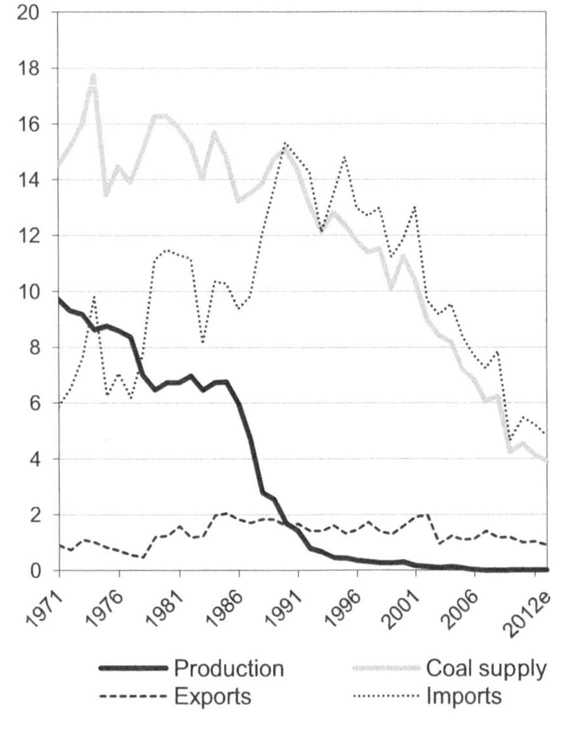

Figure 4: Coal consumption (Mtce)

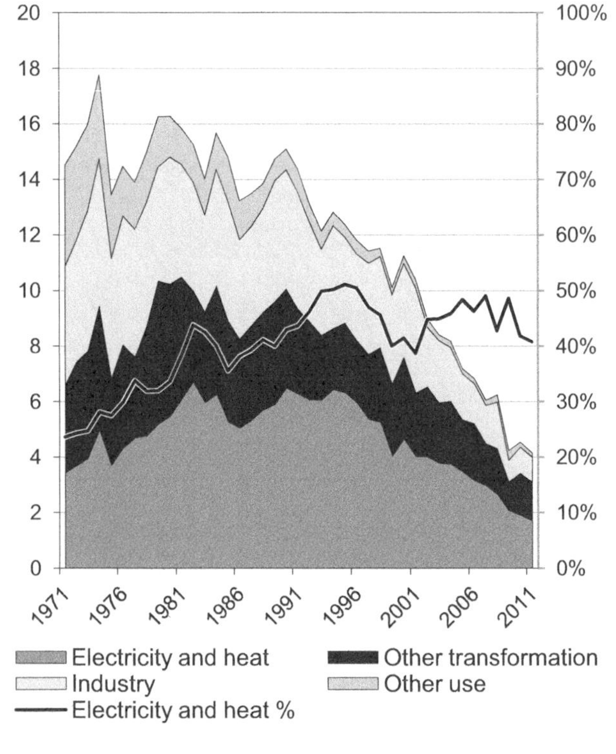

*Please refer to notes and definitions in Part I. Peat is included in coal for all figures.

Source: *IEA/OECD Coal Information, IEA/OECD Energy Balances of OECD Countries, OECD Main Economic Indicators*

BELGIUM*

Figure 5: Electricity generation by fuel (TWh)

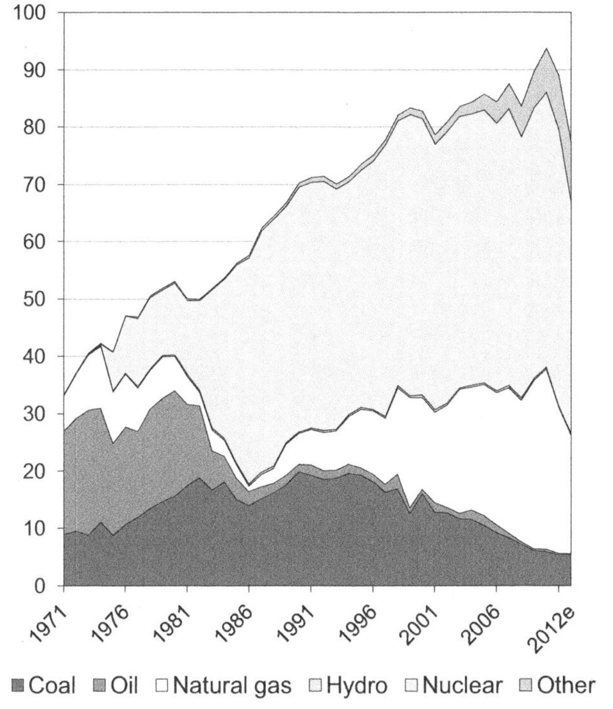

Figure 6: CO_2 emissions by fuel (Mt CO_2)

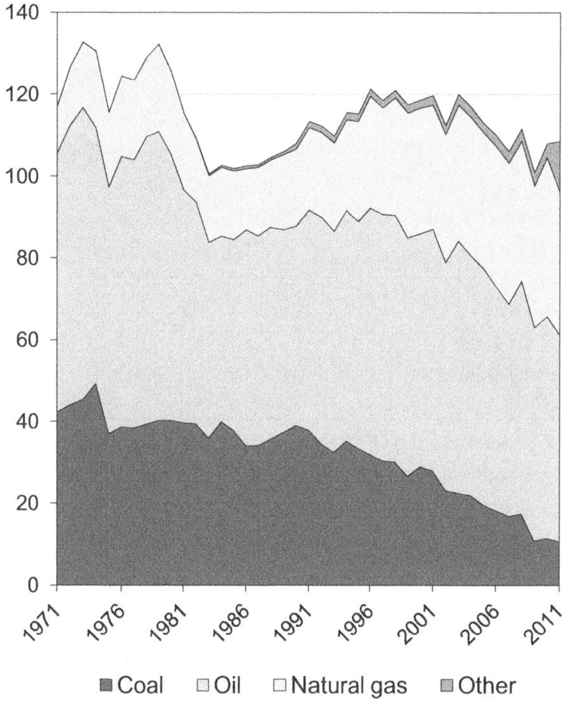

Figure 7: Electricity generation by fuel share

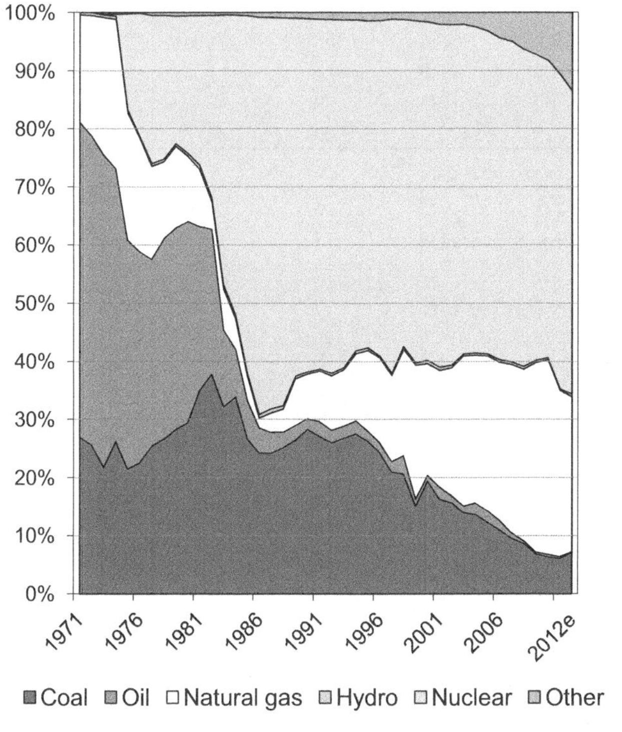

Figure 8: CO_2 emissions by fuel share

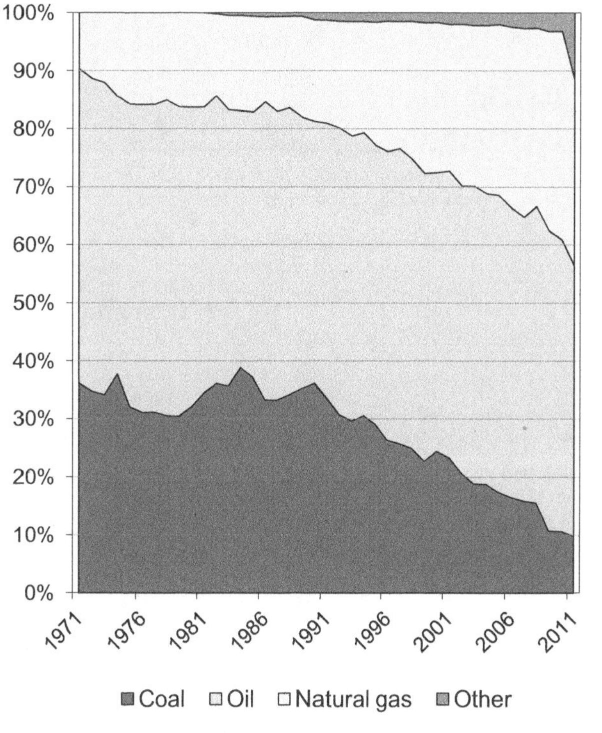

*Please refer to notes and definitions in Part I. Peat is included in coal for all figures.

Source: *IEA/OECD Electricity Information, IEA/OECD CO2 Emissions from Fuel Combustion*

BELGIUM

1. Total primary energy supply (TPES) by fuel*

	1973	1980	1990	2000	2010	2011	2012e	Average annual percent change 73-90	90-11
TPES (Mtce)	**65.70**	**66.81**	**68.98**	**83.58**	**86.99**	**84.42**	**81.84**	**0.29**	**0.97**
Coal and peat	15.97	16.28	15.10	11.25	4.55	4.15	3.92	-0.33	-5.96
Oil	39.55	33.34	25.65	33.23	35.42	32.29	32.99	-2.52	1.10
Natural Gas	10.20	12.73	11.67	19.09	24.28	21.60	20.46	0.79	2.98
Biofuels and waste	0.01	0.08	1.08	1.48	4.44	7.53	7.51	30.09	9.70
Nuclear	0.03	4.67	15.91	17.93	17.85	17.96	15.00	45.13	0.58
Hydro	0.02	0.03	0.03	0.06	0.04	0.02	0.05	2.63	-1.44
Geothermal	-	-	0.00	0.00	0.01	0.01	0.01	-	2.98
Solar, wind, tide	-	-	0.00	0.00	0.24	0.45	0.57	-	29.23
Net electricity trade(1)	-0.09	-0.32	-0.46	0.53	0.07	0.31	1.22	9.84	x
Heat(2)	-	-	-	-	0.09	0.10	0.12	-	-

(1) Net trade between OECD and non-OECD countries.

(2) Ambient heat from heat pumps used in the transformation sector.

Source: *IEA/OECD Energy Balances of OECD Countries*

2. Energy supply, GDP and population*

	1973	1980	1990	2000	2010	2011	2012e	Average annual percent change 73-90	90-11
GDP (2005 bil. USD)	191	229	280	349	401	408	407	2.28	1.81
TPES/GDP(1)	0.34	0.29	0.25	0.24	0.22	0.21	0.20	-1.95	-0.83
Population (millions)	9.7	9.9	10.0	10.2	10.9	11.0	11.0	0.15	0.46
TPES/population(1)	6.76	6.78	6.92	8.16	7.99	7.69	7.41	0.14	0.50
TPES/GDP(2)	155.0	131.1	110.9	107.9	97.7	93.1	90.5	-1.95	-0.83
Coal and peat supply/GDP(2)	440.1	373.1	283.6	169.7	59.6	53.5	50.5	-2.55	-7.63
Elec. consumption/GDP(2)	84.3	88.9	97.5	104.6	97.8	92.4	..	0.86	-0.26
Elec. generation (TWh)	41	53	70	83	94	89	77	3.28	1.13
Industrial production(2)	59.1	63.5	76.8	87.0	105.0	108.9	105.7	1.55	1.68

(1) TPES/GDP in units of tce/2005 thousand USD, TPES/population in units of tce/capita.

(2) As index, 2005 (not shown) = 100

Source: *IEA/OECD Energy Balances of OECD Countries* , *OECD Main Economic Indicators*

3. Coal and peat production by type*

	1978(1)	1990	2000	2005	2010	2011	2012e	Average annual percent change 78-90	90-11
Mtce:									
Coking coal	3.81	-	-	-	-	-	-	-	-
Steam coal	3.19	1.69	0.29	0.08	-	-	-	-5.18	-
Lignite	-	-	-	-	-	-	-	-	-
Peat	-	-	-	-	-	-	-	-	-
Mt:									
Coking coal	3.81	-	-	-	-	-	-	-	-
Steam coal	4.90	2.36	0.38	0.11	-	-	-	-5.92	-
Lignite	-	-	-	-	-	-	-	-	-
Peat	-	-	-	-	-	-	-	-	-

(1) Earliest year for which split by coal type is available.

* Please refer to notes and definitions in Part I.

Source: *IEA/OECD Energy Balances of OECD Countries, IEA/OECD Energy Statistics of OECD Countries*

BELGIUM

4. Final consumption of energy by fuel*

(Mtce)

	1973	1980	1990	2000	2009	2010	2011	Average annual percent change 73-90	90-11
Total final consumption(1)	**48.19**	**46.12**	**45.81**	**59.12**	**56.09**	**60.28**	**60.91**	**-0.30**	**1.37**
Coal and peat	8.15	6.05	5.05	3.68	1.13	1.13	1.05	-2.78	-7.22
Oil	28.80	24.08	23.04	29.93	28.47	29.38	28.68	-1.30	1.05
Natural Gas	6.57	10.11	9.74	14.52	14.45	16.57	15.64	2.34	2.28
Biofuels and wastes	-	-	0.54	0.76	1.72	2.00	4.74	-	10.92
Geothermal	-	-	-	-	-	-	0.00	-	-
Solar, wind, tide	-	-	0.00	0.00	0.02	0.02	0.02	-	14.50
Electricity	4.20	5.33	7.12	9.53	9.49	10.24	9.84	3.15	1.55
Heat	0.46	0.56	0.32	0.70	0.82	0.95	0.95	-2.14	5.38
of which:									
Total industry	**19.41**	**16.00**	**15.02**	**18.32**	**13.03**	**14.95**	**17.96**	**-1.50**	**0.86**
Coal and peat	5.06	4.57	4.30	3.40	0.77	0.94	0.90	-0.95	-7.20
Oil	7.46	3.65	2.31	2.29	0.93	0.87	0.59	-6.67	-6.28
Natural Gas	3.68	4.33	4.12	6.58	5.71	6.60	7.32	0.67	2.78
Biofuels and wastes	-	-	0.27	0.54	0.93	1.06	3.83	-	13.40
Geothermal	-	-	-	-	-	-	-	-	-
Solar, wind, tide	-	-	-	-	-	-	-	-	-
Electricity	2.75	2.94	3.75	4.90	4.01	4.69	4.58	1.83	0.95
Heat	0.46	0.51	0.27	0.61	0.67	0.78	0.74	-3.08	4.98
Total transport	**6.31**	**7.72**	**9.69**	**11.69**	**13.06**	**13.14**	**13.09**	**2.55**	**1.44**
Coal and peat	0.02	0.00	-	-	-	-	-	-	-
Oil	6.20	7.60	9.53	11.52	12.44	12.41	12.38	2.57	1.25
Natural Gas	-	-	-	-	-	-	0.01	-	-
Biofuels and wastes	-	-	-	-	0.41	0.52	0.50	-	-
Electricity	0.10	0.12	0.15	0.18	0.22	0.21	0.20	2.53	1.28
Residential	**15.78**	**13.70**	**11.82**	**13.53**	**11.83**	**12.81**	**10.58**	**-1.68**	**-0.53**
Coal and peat	3.07	1.48	0.75	0.28	0.33	0.17	0.15	-7.99	-7.41
Oil	10.21	7.00	4.98	5.39	3.93	4.30	3.70	-4.14	-1.40
Natural Gas	1.56	3.58	3.55	4.70	4.73	5.46	3.98	4.95	0.56
Biofuels and wastes	-	-	0.26	0.22	0.32	0.36	0.35	-	1.32
Geothermal	-	-	-	-	-	-	0.00	-	-
Solar, wind, tide	-	-	-	0.00	0.01	0.01	0.01	-	-
Electricity	0.93	1.61	2.26	2.92	2.48	2.49	2.37	5.34	0.21
Heat	-	0.02	0.03	0.02	0.02	0.02	0.02	-	-1.46
Comm & public services	**1.52**	**4.25**	**4.11**	**4.92**	**6.57**	**7.19**	**6.29**	**6.02**	**2.05**
Coal and peat	0.00	0.00	-	-	-	-	-	-	-
Oil	0.59	2.23	1.65	1.19	1.36	1.55	1.28	6.20	-1.21
Natural Gas	0.51	1.34	1.48	2.21	2.46	2.79	2.20	6.42	1.92
Biofuels and waste	-	-	-	0.00	0.01	0.01	0.02	-	-
Geothermal	-	-	-	-	-	-	0.00	-	-
Solar, wind, tide	-	-	-	0.00	0.00	0.00	0.00	-	-
Electricity	0.41	0.65	0.96	1.50	2.64	2.73	2.66	5.08	4.98
Heat	-	0.03	0.02	0.02	0.10	0.11	0.13	-	9.05
Non-energy use	**4.52**	**3.73**	**4.48**	**9.63**	**10.36**	**10.85**	**10.94**	**-0.06**	**4.34**
Coal and peat	-	-	-	-	0.01	-	-	-	-
Oil	3.69	2.87	3.88	8.60	9.13	9.61	10.18	0.29	4.70
Natural Gas	0.83	0.86	0.60	1.03	1.21	1.23	0.76	-1.87	1.13

(1) Total Final Consumption (TFC) excludes use in transformation processes and energy industry own use.

* Please refer to notes and definitions in Part I. Agriculture, Forestry, Fishing and "Non-specified other" sectors are not reported here individually, but are accounted for in Total final consumption.

Source: *IEA/OECD Energy Balances of OECD Countries*

BELGIUM

5. Coal balance*

(Mtce)

	1973	1980	1990	2000	2005	2010	2011	2012e	Average annual percent change 73-90	90-11
Production	9.2	6.7	1.7	0.3	0.1	-	-	-	-9.5	-
Imports	7.6	11.5	15.3	11.8	8.4	5.5	5.2	4.8	4.2	-5.0
Exports	-1.1	-1.2	-1.6	-1.6	-1.1	-1.0	-1.0	-0.9	2.2	-2.1
Stock changes	0.3	-0.7	-0.3	0.7	-0.2	0.1	-0.0	0.0		
Primary supply	**16.0**	**16.3**	**15.1**	**11.3**	**7.2**	**4.5**	**4.2**	**3.9**	**-0.3**	**-6.0**
Statistical differences	**1.5**	**-0.4**	**-0.0**	**-0.5**	**0.1**	**-0.0**	**0.0**	**..**		
Total transformation	**-8.2**	**-9.1 e**	**-9.4 e**	**-6.7 e**	**-5.1 e**	**-3.2**	**-2.9**	**..**	**0.8**	**-5.4**
Electricity and heat gen.	-3.9	-5.5	-6.5	-4.7	-3.5	-1.9	-1.7	..	3.0	-6.2
Main activity producers (1)	*-3.9*	*-4.8*	*-6.0*	*-4.5*	*-3.3*	*-1.8*	*-1.6*	..	*2.5*	*-6.1*
Autoproducers	-	*-0.7*	*-0.5*	*-0.2*	*-0.2*	*-0.1*	*-0.1*	..	-	*-7.8*
Gas works	-0.0	0.0	-	-	-	-	-	..	-	-
Coal transformation(2)	-4.3	-3.6 e	-2.9 e	-2.0 e	-1.6 e	-1.3	-1.2	..	-2.3	-3.9
BKB plants	-	-	-	-	-	-	-	..	-	-
Blast furnaces	*-3.8*	*-2.9 e*	*-2.3 e*	*-2.0 e*	*-1.7 e*	*-1.1*	*-1.1*	..	*-2.7*	*-3.7*
Coke ovens	*-0.6*	*-0.8*	*-0.5*	*-0.0*	*0.1*	*-0.2*	*-0.2*	..	*-0.6*	*-5.1*
Patent fuel plants	*0.1*	*0.0*	*0.0 e*	*0.0*	*0.0*	-	-	..	*-34.9*	-
Other transformation(3)	-	-	-	-	-	-	-	..	-	-
Energy ind. own use	**-1.1**	**-0.8**	**-0.7**	**-0.3**	**-0.3**	**-0.2**	**-0.2**	**..**	**-2.6**	**-5.6**
Losses	**-0.0**	**-0.0**	**-**	**-**	**-**	**-**	**-**	**..**		
Final consumption(4)	**8.2**	**6.0**	**5.1**	**3.7**	**1.8**	**1.1**	**1.0**	**..**	**-2.8**	**-7.2**
Industry(5)	5.1	4.6	4.3	3.4	1.7	0.9	0.9	..	-1.0	-7.2
Iron and steel	*3.5*	*2.9 e*	*2.9 e*	*2.8 e*	*1.3 e*	*0.5*	*0.4*	..	*-1.1*	*-8.6*
Chemical	*0.1*	*0.1*	*0.2*	*0.0*	*0.0*	-	-	..	*2.2*	-
Non-metallic minerals	*1.0*	*1.5*	*0.8*	*0.4*	*0.2*	*0.4*	*0.3*	..	*-1.0*	*-4.1*
Paper, pulp and print	-	*0.0*	*0.0*	*0.0*	*0.0*	*0.0*	*0.0*	..	-	*-2.1*
Other industry (6)	*0.4*	*0.2*	*0.3*	*0.2*	*0.1*	*0.1*	*0.1*	..	*-2.0*	*-6.1*
Transport(7)	0.0	0.0	-	-	-	-	-	..	-	-
Other	3.1	1.5	0.7	0.3	0.2	0.2	0.2	..	-8.0	-7.3
Comm. and pub. services	*0.0*	*0.0*	-	-	-	-	-	..	-	-
Residential	*3.1*	*1.5*	*0.7*	*0.3*	*0.2*	*0.2*	*0.1*	..	*-8.0*	*-7.4*
Other sectors (8)	-	-	-	-	-	*0.0*	*0.0*	..	-	-
Non-energy use	-	-	-	-	0.0	-	-	..	-	-

(1) Main activity electricity and heat generation includes district heating.

(2) Coal transformation refers to the transformation of coal and peat to secondary and tertiary products (mainly cokes, briquettes, coal tar and off-gases). Gas works and Liquefaction are shown separately to match the *Energy Balances of OECD Countries.*

(3) Other transformation includes Liquefaction and Non-specified transformations.

(4) Final Consumption includes non-energy use and energy use (Industry, Transport and Other).

(5) Please refer to notes and definitions in Part I for detailed explanation.

(6) Other industry includes Non-ferrous metals, Transport equipment, Machinery, Mining and quarrying, Food and tobacco, Wood and wood products, Construction, Textile and leather, and Non-specified industry.

(7) Transport includes Rail and Inland waterways.

(8) Other sectors includes Agriculture/Forestry, Fishing and Non-specified other.

* "Coal" refers to all types of coal, primary (anthracite, coking coal, other bituminous coal, sub-bituminous coal, lignite/brown coal) and derived fuels (including patent fuel, BKB, coke oven coke, gas coke, coal tar, coke oven gas, gas works gas, blast furnace gas and other recovered gases), as well as peat, peat briquettes and oil shale. Quantities have been converted to Mtce using calorific values reported by the respective countries.
Please refer to notes and definitions in Part I.

Source: *IEA/OECD Energy Balances of OECD Countries*

BELGIUM

6. Use of coal for selected end-uses*

(million tonnes)

	1978	1990	2000	2005	2009	2010	2011	Average annual percent change 78-90	90-11
Steam coal	**8.76**	**9.00**	**7.00**	**4.29**	**2.72**	**2.29**	**1.82**	**0.22**	**-7.33**
Total electricity and heat	4.44	6.64	4.37	3.05	1.74	1.52	1.27	3.41	-7.58
Main activity producers	*4.18*	*6.34*	*4.31*	*2.98*	*1.66*	*1.47*	*1.23*	*3.53*	*-7.50*
Autoproducers	*0.26*	*0.30*	*0.07*	*0.06*	*0.08*	*0.05*	*0.03*	*1.22*	*-9.79*
Patent fuel/BKB plants	0.12	0.00 e	0.01	0.01	-	-	-	-32.76	-
Coke ovens/Liquefaction(1)	0.03	-	-	-	-	-	-	-	-
Blast furnace inputs	-	0.41 e	0.98 e	0.46 e	0.04 e	0.17	0.15	-	-4.81
Gas manufacture	-	-	-	-	-	-	-	-	-
Industry	2.66	1.37	0.98	0.38	0.30	0.36	0.33	-5.39	-6.53
Iron and steel	*0.17*	*0.10 e*	*0.48 e*	*0.14 e*	*0.00 e*	-	-	*-4.36*	-
Chemical	*0.03*	*0.14*	-	-	-	-	-	*13.74*	-
Non-metallic minerals	*2.44*	*0.87*	*0.36*	*0.12*	*0.17*	*0.24*	*0.22*	*-8.25*	*-6.32*
Paper, pulp and print	-	*0.05*	*0.04*	*0.04*	*0.05*	*0.04*	*0.03*	-	*-1.53*
Other industry	*0.02*	*0.22 e*	*0.10 e*	*0.08 e*	*0.09 e*	*0.08*	*0.08*	*20.19*	*-4.83*
Other sectors(2)	1.63	0.70	0.30	0.19	0.37	0.20	0.15	-6.77	-7.01
Non-energy use	-	-	-	0.01	0.02	-	-	-	-
Coking coal	**7.52**	**7.16**	**4.05**	**3.26**	**1.96**	**2.63**	**2.58**	**-0.42**	**-4.74**
Total electricity and heat	0.20	-	-	-	-	-	-	-	-
Main activity producers	*0.18*	-	-	-	-	-	-	-	-
Autoproducers	*0.02*	-	-	-	-	-	-	-	-
Patent fuel/BKB plants	-	-	-	-	-	-	-	-	-
Coke ovens/Liquefaction(1)	7.19	7.16	3.86	3.37	2.05	2.59	2.58	-0.03	-4.75
Blast furnace inputs	-	-	-	-	-	-	-	-	-
Gas manufacture	-	-	-	-	-	-	-	-	-
Industry	0.00	-	-	-	-	-	-	-	-
Iron and steel	-	-	-	-	-	-	-	-	-
Chemical	-	-	-	-	-	-	-	-	-
Non-metallic minerals	*0.00*	-	-	-	-	-	-	-	-
Paper, pulp and print	-	-	-	-	-	-	-	-	-
Other industry	-	-	-	-	-	-	-	-	-
Other sectors(2)	0.00	-	-	-	-	-	-	-	-
Non-energy use	-	-	-	-	-	-	-	-	-
Lignite	**-**	**0.28**	**0.22**	**0.21**	**0.17**	**0.24**	**0.29**	**-**	**0.27**
Total electricity and heat	-	-	-	-	-	-	-	-	-
Main activity producers	-	-	-	-	-	-	-	-	-
Autoproducers	-	-	-	-	-	-	-	-	-
Patent fuel/BKB plants	-	-	-	-	-	-	-	-	-
Coke ovens/Liquefaction(1)	-	-	-	-	-	-	-	-	-
Gas manufacture	-	-	-	-	-	-	-	-	-
Industry	-	0.28	0.22	0.21	0.15	0.24	0.26	-	-0.34
Iron and steel	-	-	-	-	-	-	-	-	-
Chemical	-	-	-	*0.01*	*0.00*	-	-	-	-
Non-metallic minerals	-	*0.28*	*0.22*	*0.20*	*0.14*	*0.24*	*0.26*	-	*-0.34*
Paper, pulp and print	-	-	-	-	-	-	-	-	-
Other industry	-	-	-	-	*0.00*	-	-	-	-
Other sectors(2)	-	-	-	-	-	0.01	-	-	-
Non-energy use	-	-	-	-	-	-	-	-	-

(1) Liquefaction primarily refers to direct distillation processes. Indirect liquefaction may be reported here or under gas manufacture.

(2) Other sectors are Residential, Commercial and public services, Agriculture/Forestry, Fishing and Non-specified other.

* Coal totals may not match the sum of the individual rows due to statistical differences. Please refer to notes and definitions in Part I.

Source: *IEA/OECD Energy Statistics of OECD Countries*

BELGIUM

7. Fuel prices to end users*

	1978[1]	1990	1995	2000	2005	2009	2010	2011	2012
				(US dollars / tce)					
For electricity generation									
Steam coal	53.08	59.60	57.28	42.95	105.36	100.77	..	..	..
Heavy fuel oil	66.67	91.13	112.16	129.79	x	x	x	x	x
Natural gas	65.42	96.10	85.35	c	c	c	c	c	c
For industry									
Steam coal	29.62	55.41	..	..	..	..	..	..	..
Coking coal	63.67	61.72	57.55	47.21	..	..	..	..	..
High sulphur fuel oil	66.67	91.13	112.16	129.79	x	x	x	x	x
Low sulphur fuel oil	..	102.02	105.16	117.76	220.57	300.97	384.64	516.81	538.90
Natural gas	71.79	112.22	110.12	c	c	324.61	276.04	329.12	325.55
				(Euro / unit)[2]					
For electricity generation									
Steam coal	31.61	37.68	31.93	35.54	64.68	55.31	..	..	..
Heavy fuel oil	72.53	105.23	114.20	196.15	x	x	x	x	x
Natural gas	65.69	102.43	80.22	c	c	c	c	c	c
For industry									
Steam coal	18.34	36.42	..	..	..	..	..	..	..
Coking coal	49.73	51.17	42.07	51.22	..	..	..	..	..
High sulphur fuel oil	72.53	105.23	114.20	196.15	x	x	x	x	x
Low sulphur fuel oil	..	117.80	107.07	177.97	247.31	301.74	404.49	517.35	583.98
Natural gas	72.09	119.61	103.50	c	c	300.41	267.96	304.12	325.64

(1) Earliest year for which split by coal type is available.

(2) For steam coal, coking coal and heavy fuel oil per metric tonne; for natural gas per 10^7 kilocalories GCV.

* Please refer to notes and definitions in Part I. Steam coal prices relate to historical definitions.

Source: *IEA/OECD Energy Prices & Taxes*

8. Coal and peat trade by type of coal*

(Mtce)

	1978	1990	1995	2000	2005	2009	2010	2011	2012e
Total imports	**7.72**	**15.33**	**14.81**	**11.84**	**8.38**	**4.65**	**5.46**	**5.23**	**4.81**
Bituminous coal[1]	3.51	7.24	8.21	6.84	4.61	2.74	2.38	2.17	1.67
Coking coal	3.49	7.13	5.27	3.82	3.53	1.67	2.80	2.70	2.60
Sub-bituminous coal	-	-	-	-	-	-	-	-	-
Lignite	-	0.08	0.06	0.06	0.06	0.05	0.07	0.08	0.09
Peat	-	-	-	-	-	-	-	-	-
Coal products[2]	0.71	0.88	1.27	1.12	0.18	0.20	0.21	0.27	0.46
Total exports	**0.47**	**1.60**	**1.33**	**1.59**	**1.10**	**1.18**	**0.99**	**1.03**	**0.89**
Bituminous coal[1]	0.12	0.65	0.76	1.16	1.02	0.76	0.47	0.56	0.09
Coking coal	0.10	-	-	0.09	0.03	0.01	0.06	0.03	-
Sub-bituminous coal	-	0.03	0.00	0.04	0.02	-	-	-	-
Lignite	-	-	-	-	-	-	-	-	0.00
Peat	-	-	-	-	-	-	-	-	-
Coal products[2]	0.24	0.92	0.57	0.30	0.03	0.41	0.47	0.44	0.80

(1) Bituminous coal includes anthracite.

(2) Coal products includes products derived from coal, for example: coke, coal tar, briquettes, patent fuels.

* Please refer to notes and definitions in Part I.

Source: *IEA/OECD Energy Balances of OECD Countries, IEA/OECD Energy Statistics of OECD Countries*

BELGIUM

9. Total coal imports by origin*

(thousand tonnes)

	1978(1)	1985	1990	1995	2000	2005	2010	2011	2012e
Total coal(2)	**7007**	**9598**	**15037**	**14294**	**11566**	**9010**	**5772**	**5483**	**4804**
Coking coal	**3490**	**4663**	**7132**	**5267**	**3818**	**3533**	**2801**	**2704**	**2600**
Australia	209	16	1015	1614	1109	2003	1149	1368	1278
Canada	148	94	236	205	678	227	-	-	-
Czech Republic	59	-	-	-	-	-	-	-	-
Germany	1751	1028	690	20	-	-	-	-	-
Poland	392	540	105	66	74	-	-	-	-
United Kingdom	22	-	32	-	-	-	-	-	-
United States	833	2951	4897	3171	1898	1303	1562	1211	1192
Other OECD	15	-	-	15	-	-	90	79	130
China, People's Rep.	-	-	-	-	-	-	-	-	-
Colombia	-	-	-	-	-	-	-	-	-
Indonesia	-	-	-	-	-	-	-	-	-
South Africa	15	34	157	176	59	-	-	-	-
Former Soviet Union(3)	46	-	x	x	x	x	x	x	x
Russian Federation (4)	*x*	*x*	-	-	-	-	-	*12*	-
Other FSU (4)	*x*	*x*	-	-	-	-	-	-	-
Venezuela	-	-	-	-	-	-	-	-	-
Vietnam	-	-	-	-	-	-	-	-	-
Non-specified/other	-	-	-	-	-	-	-	34	-
Steam coal	**3517**	**4657**	**7629**	**8832**	**7529**	**5271**	**2726**	**2488**	**1904**
Australia	-	205	388	1283	2354	947	200	167	-
Canada	26	28	66	20	12	43	19	4	-
Czech Republic	-	-	-	-	-	-	-	-	-
Germany	2316	1067	721	301	118	155	145	147	13
Poland	105	59	242	318	225	436	23	2	78
United Kingdom	129	39	67	66	76	62	46	35	51
United States	-	671	981	983	309	411	636	580	812
Other OECD	72	246	60	75	48	29	58	32	30
China, People's Rep.	-	112	292	299	167	58	-	-	-
Colombia	-	51	170	269	431	5	166	395	75
Indonesia	-	-	6	-	11	-	-	-	-
South Africa	606	2078	4365	4855	3028	2081	1120	694	-
Former Soviet Union(3)	233	73	x	x	x	x	x	x	x
Russian Federation (4)	*x*	*x*	*234*	*154*	*705*	*944*	*125*	*58*	*800*
Other FSU (4)	*x*	*x*	-	*111*	*19*	-	-	-	-
Venezuela	-	-	1	17	3	-	-	-	45
Vietnam	-	-	-	81	23	98	-	-	-
Non-specified/other	30	28	36	-	-	2	188	374	-
Lignite	**-**	**278**	**276**	**195**	**219**	**206**	**245**	**291**	**300**

(1) Earliest year for which split by coal type is available.

(2) Total coal consists of steam coal, coking coal and lignite.

(3) For years prior to 1990, or if no details for individual Former Soviet Union republics.

(4) Data for former republics are not available separately for some OECD member countries.

* Please refer to notes and definitions in Part I.

Source: IEA/OECD Coal Statistics

BELGIUM

11. Steam coal exports by destination*

(thousand tonnes)

	1978[1]	1985	1990	1995	2000	2005	2010	2011	2012e
World	**122**	**1075**	**724**	**818**	**1340**	**1199**	**545**	**640**	**105**
Total OECD	**122**	**1021**	**724**	**818**	**1340**	**1199**	**535**	**637**	**105**
Australia	-	-	-	-	-	-	-	-	-
Austria	-	-	-	-	-	-	-	1	-
Belgium	-	-	-	-	-	-	-	-	-
Canada	-	-	-	-	-	-	-	-	-
Chile	-	-	-	-	-	-	-	-	-
Czech Republic	-	-	-	-	-	-	-	-	-
Denmark	-	2	1	-	-	-	-	1	-
Estonia	-	-	-	-	-	-	-	-	-
Finland	-	-	-	-	2	-	-	-	-
France	61	317	360	352	609	377	257	208	-
Germany	1	345	67	204	495	334	159	169	29
Greece	-	-	-	-	-	2	-	-	-
Hungary	-	-	-	-	-	-	-	-	-
Iceland	-	-	-	-	-	-	-	-	-
Ireland	-	-	-	-	7	-	-	-	-
Israel	-	-	-	-	-	-	-	-	-
Italy	-	14	5	-	-	-	-	-	-
Japan	-	-	61	-	-	-	-	-	-
Korea	-	-	-	-	-	-	-	-	-
Luxembourg	21	2	-	50	40	37	32	102	-
Mexico	-	-	-	-	-	-	-	-	-
Netherlands	1	79	179	182	136	423	6	155	48
New Zealand	-	-	-	-	-	-	-	-	-
Norway	13	87	-	25	-	-	-	-	-
Poland	-	-	-	-	-	-	-	-	-
Portugal	6	21	4	-	-	-	-	-	-
Slovak Republic	-	-	-	-	-	-	-	-	-
Slovenia	-	-	-	-	-	-	-	-	-
Spain	18	7	43	3	6	-	-	-	-
Sweden	-	28	-	-	15	17	81	1	20
Switzerland	1	18	-	-	15	7	-	-	-
Turkey	-	-	-	-	-	-	-	-	-
United Kingdom	-	101	4	2	15	2	-	-	8
United States	-	-	-	-	-	-	-	-	-
Total non-OECD	**-**	**54**	**-**	**-**	**-**	**-**	**-**	**-**	**-**
Brazil	-	-	-	-	-	-	-	-	-
China[2]	-	-	-	-	-	-	-	-	-
Chinese Taipei	-	-	-	-	-	-	-	-	-
Egypt	-	-	-	-	-	-	-	-	-
India	-	-	-	-	-	-	-	-	-
Romania	-	-	-	-	-	-	-	-	-
Oth. Africa & Mid. East	-	-	-	-	-	-	-	-	-
Oth. non-OECD Americas	-	2	-	-	-	-	-	-	-
Other Asia & Oceania	-	-	-	-	-	-	-	-	-
Other non-OECD Europe and Eurasia	-	52	-	-	-	-	-	-	-
Non-specified/Other	**-**	**-**	**-**	**-**	**-**	**-**	**10**	**3**	**-**

(1) Earliest year for which split by coal type is available.

(2) China includes the People's Republic of China and Hong Kong, China.

* Please refer to notes and definitions in Part I. Steam coal includes all sub-bituminous coal.

Source: IEA/OECD Coal Statistics

BELGIUM

12. Coal import values by origin*

(average unit value, CIF, USD/tonne)

	1990	1995	2000	2005	2007	2008	2009	2010	2011/12[1]
Coking coal[2]	**63.93**	**57.74**	**44.05**	**108.68**	**120.51**	**163.26**	**146.61**	**180.40**	..
Imports from:									
Australia	63.54	58.30	38.58	125.12	135.93	184.60	314.05	215.58	..
Canada	63.80	56.91	43.97	148.24	136.78	..	166.36	109.55	..
Czech Republic	254.00	..	..	..	..	..	..	..	..
Poland	60.08	57.96	..	133.14	..	..	..	..	..
United States	62.61	58.28	52.54	96.51	104.33	141.00	140.30	186.51	..
China	..	63.59	..	285.09	163.87	241.46	..	..	..
Colombia	..	..	..	..	..	..	..	..	..
Indonesia	..	..	..	..	..	..	..	..	..
South Africa	54.18	49.03	41.49	67.23	78.76	139.94	79.53	91.21	..
Former Soviet Union[3]	..	..	..	75.13	91.44	172.22	..	139.54	..
Other bituminous coal[4]	**45.57**	**47.94**	**39.88**	**70.83**	**82.81**	**150.58**	**109.92**	**114.02**	..
Imports from:									
Australia	50.57	45.93	40.41	97.75	90.32	199.82	272.28	160.77	..
Canada	57.14	66.64	42.87	102.49	72.46	..	95.33	111.61	..
Czech Republic	..	..	..	..	..	..	..	..	..
Poland	48.40	46.66	31.66	63.56	158.69	270.57	72.53	89.28	..
United States	51.01	49.77	48.91	..	90.95	140.60	139.10	141.31	..
China	48.49	45.65	58.92	..	653.12	..	x	x	..
Colombia	51.92	53.17	35.18	73.98	84.14	..	152.05	96.91	..
Indonesia	70.09	..	25.33	..	71.07	163.57	273.22	..	..
South Africa	44.88	43.64	36.80	68.15	80.05	128.43	97.91	97.65	..
Former Soviet Union[3]	44.83	43.65	37.03	67.19	82.65	139.75	89.68	106.33	..

Note: On occasion, shipment of extremely small quantities of high valued coal results in high import costs.

(1) Import data for steam and coking coals are currently unavailable for 2011 and 2012 due to resource constraints.

(2) Weighted average of individual countries using import volumes as weights. Prices exclude intra-EU trade.

(3) Former Soviet Union until 1991, Russian Federation starting in 1992.

(4) Bituminous steam coal only. Weighted average of trades. Prices exclude intra-EU trade.

* Please refer to notes and definitions in Part I.

Source: IEA/OECD Coal Statistics

CANADA*

Figure 1: TPES by fuel (Mtce)

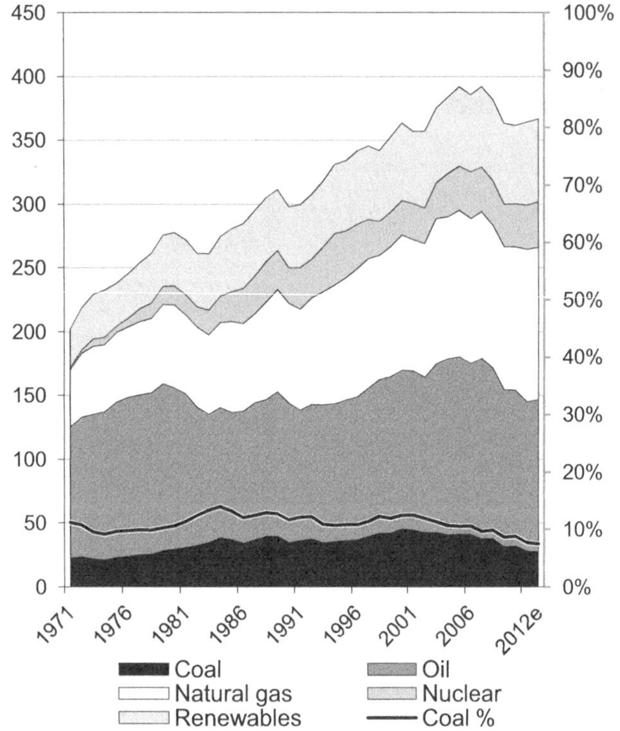

Figure 2: Coal supply indicators (1971 = 100)

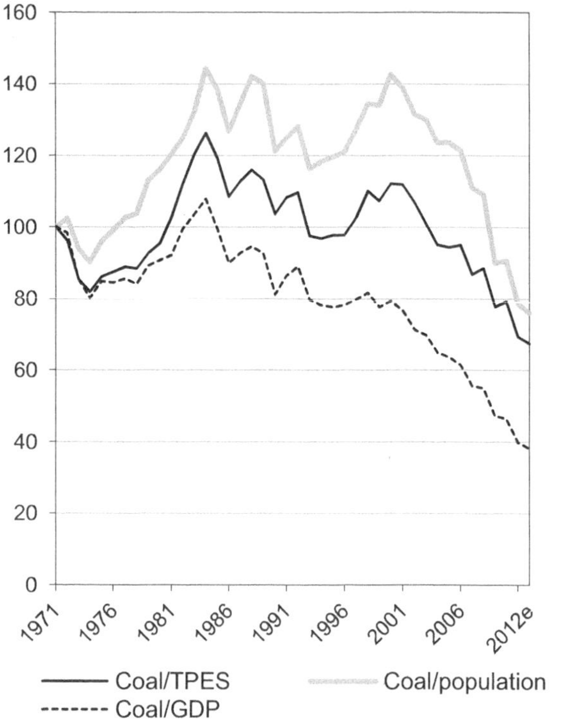

Figure 3: Primary coal supply (Mtce)

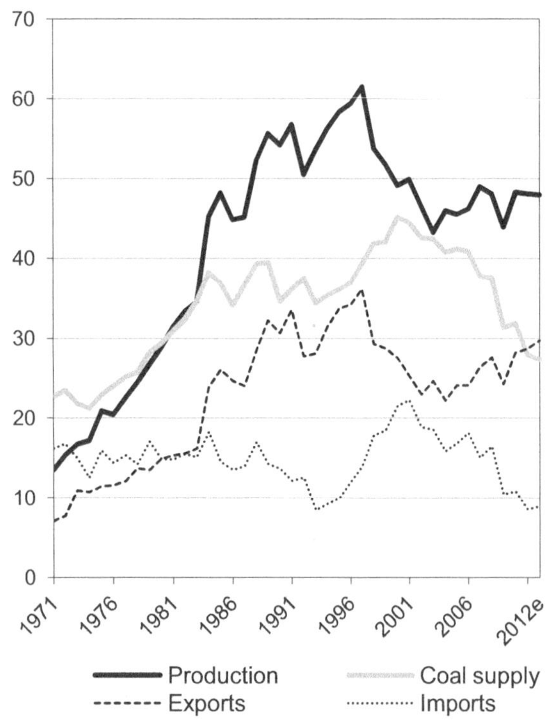

Figure 4: Coal consumption (Mtce)

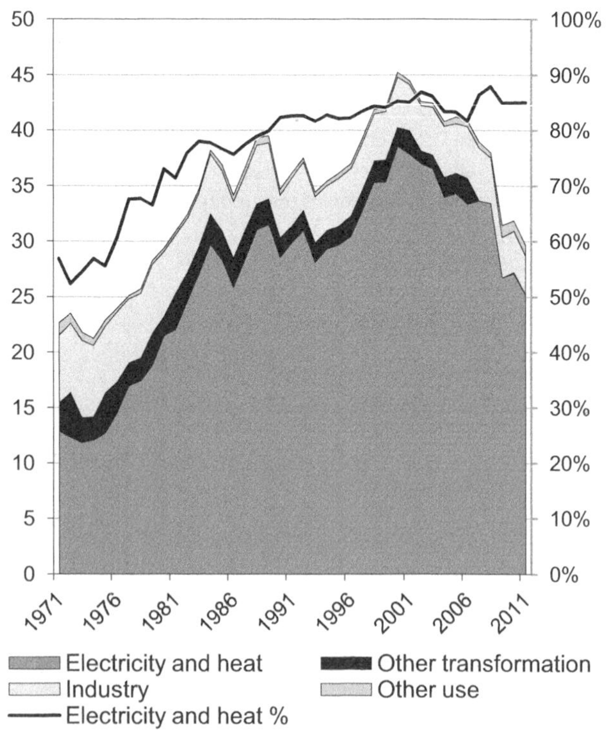

*Please refer to notes and definitions in Part I. Peat is included in coal for all figures.

Source: *IEA/OECD Coal Information, IEA/OECD Energy Balances of OECD Countries, OECD Main Economic Indicators*

CANADA*

Figure 5: Electricity generation by fuel (TWh)

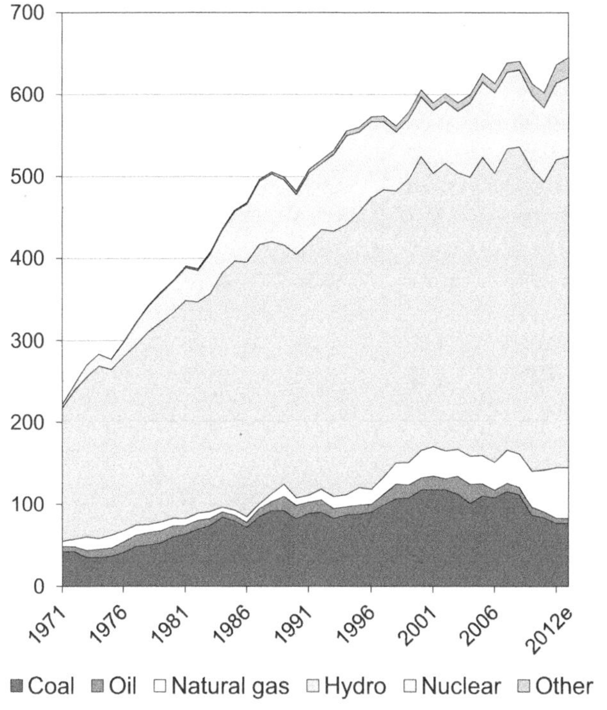

Figure 6: CO_2 emissions by fuel (Mt CO_2)

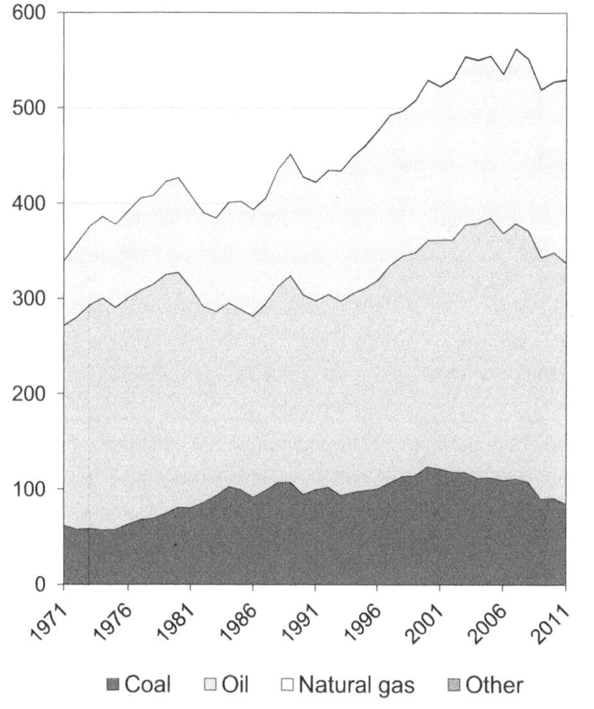

Figure 7: Electricity generation by fuel share

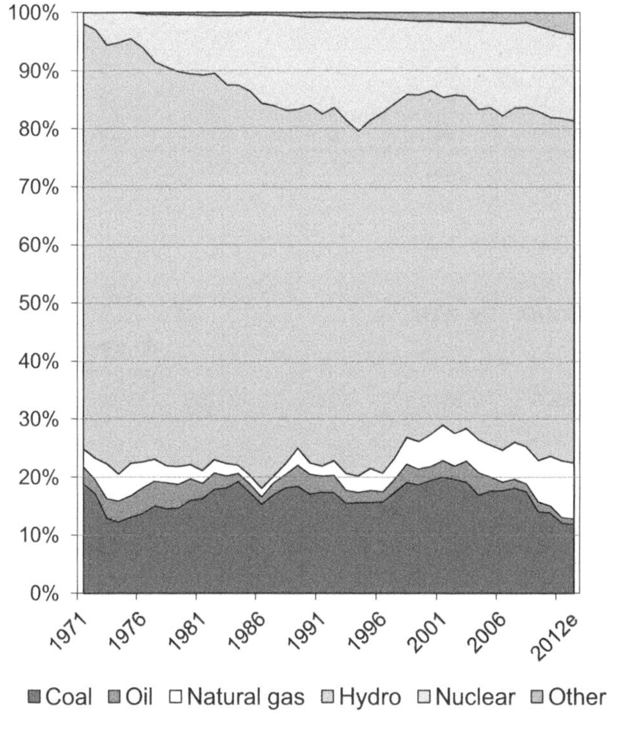

Figure 8: CO_2 emissions by fuel share

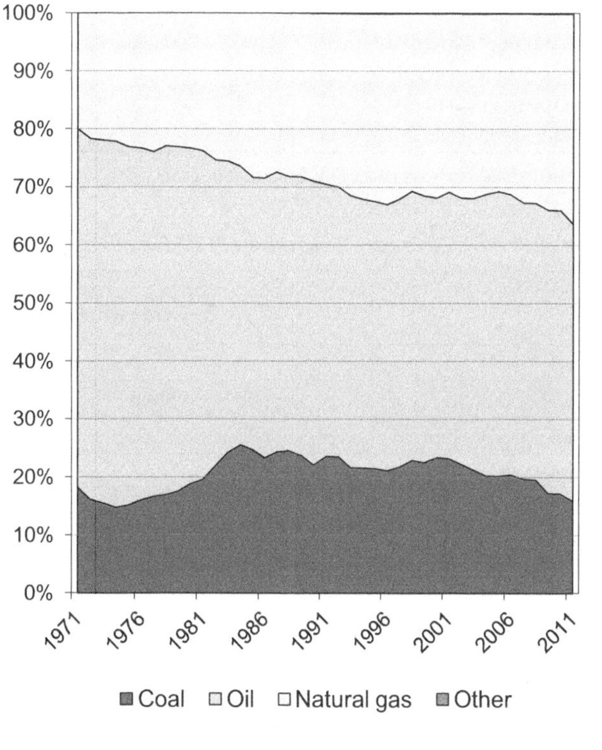

*Please refer to notes and definitions in Part I. Peat is included in coal for all figures.

Source: *IEA/OECD Electricity Information, IEA/OECD CO2 Emissions from Fuel Combustion*

CANADA

1. Total primary energy supply (TPES) by fuel*

	1973	1980	1990	2000	2010	2011	2012e	Average annual percent change 73-90	90-11
TPES (Mtce)	**227.64**	**274.22**	**297.95**	**359.21**	**358.56**	**359.78**	**360.93**	**1.60**	**0.90**
Coal and peat	21.81	29.40	34.68	45.23	31.88	28.00	27.32	2.77	-1.01
Oil	113.41	126.46	109.31	124.43	122.22	117.03	119.45	-0.22	0.33
Natural Gas	53.25	65.07	78.18	106.05	112.40	119.38	119.22	2.28	2.04
Biofuels and waste	11.16	10.92	11.65	16.70	17.21	17.56	16.62	0.25	1.97
Nuclear	5.82	14.86	27.71	27.10	33.75	34.84	35.91	9.62	1.10
Hydro	23.91	30.86	36.46	44.05	43.17	46.16	46.69	2.51	1.13
Geothermal	-	-	-	-	-	-	-	-	-
Solar, wind, tide	-	-	0.00	0.04 e	1.09	1.29	1.48	-	33.06
Net electricity trade(1)	-1.72	-3.35	-0.04	-4.38	-3.16	-4.49	-5.75	-19.53	24.79
Heat(2)	-	-	-	-	-	-	-	-	-

(1) Net trade between OECD and non-OECD countries.

(2) Ambient heat from heat pumps used in the transformation sector.

Source: *IEA/OECD Energy Balances of OECD Countries*

2. Energy supply, GDP and population*

	1973	1980	1990	2000	2010	2011	2012e	Average annual percent change 73-90	90-11
GDP (2005 bil. USD)	449	568	750	1000	1204	1235	1257	3.07	2.40
TPES/GDP(1)	0.51	0.48	0.40	0.36	0.30	0.29	0.29	-1.43	-1.47
Population (millions)	22.5	24.5	27.7	30.7	34.1	34.5	34.8	1.23	1.05
TPES/population(1)	10.12	11.19	10.76	11.71	10.51	10.43	10.37	0.36	-0.15
TPES/GDP(2)	148.0	140.7	115.8	104.7	86.8	85.0	83.7	-1.43	-1.47
Coal and peat supply/GDP(2)	133.8	142.4	127.3	124.5	72.9	62.4	59.8	-0.29	-3.34
Elec. consumption/GDP(2)	108.6	118.1	123.4	106.6	93.0	93.0	..	0.76	-1.34
Elec. generation (TWh)	270	373	482	606	602	637	646	3.47	1.34
Industrial production(2)	50.7	55.6	68.3	98.9	89.6	92.8	93.9	1.77	1.47

(1) TPES/GDP in units of tce/2005 thousand USD, TPES/population in units of tce/capita.

(2) As index, 2005 (not shown) = 100

Source: *IEA/OECD Energy Balances of OECD Countries , OECD Main Economic Indicators*

3. Coal and peat production by type*

	1978(1)	1990	2000	2005	2010	2011	2012e	Average annual percent change 78-90	90-11
Mtce:									
Coking coal	13.39	27.17	23.92	22.60 e	23.81	24.90	26.28	6.07	-0.41
Steam coal	8.55	22.44	19.79	17.48 e	19.48	18.44	17.04	8.37	-0.93
Lignite	2.52	4.57	5.44	5.44 e	5.00	4.74	4.63	5.10	0.17
Peat	-	-	-	-	-	-	-	-	-
Mt:									
Coking coal	13.78	27.66	28.16	26.61 e	28.16	29.45	31.09	5.98	0.30
Steam coal	11.64	31.27	29.81	27.72 e	29.48	27.93	25.98	8.58	-0.54
Lignite	5.07	9.41	11.19	11.02 e	10.26	9.73	9.50	5.29	0.16
Peat	-	-	-	-	-	-	-	-	-

(1) Earliest year for which split by coal type is available.

* Please refer to notes and definitions in Part I.

Source: *IEA/OECD Energy Balances of OECD Countries, IEA/OECD Energy Statistics of OECD Countries*

CANADA

4. Final consumption of energy by fuel*

(Mtce)

	1973	1980	1990	2000	2009	2010	2011	Average annual percent change 73-90	90-11
Total final consumption(1)	**187.75**	**221.52**	**227.07**	**270.90**	**273.67**	**284.50**	**291.39**	**1.12**	**1.19**
Coal and peat	7.73	6.18	4.41	5.02	4.75	4.69	4.45	-3.25	0.04
Oil	108.07	114.27	98.27	115.40	121.99	128.28	128.58	-0.56	1.29
Natural Gas	33.89	51.75	61.86	76.30	74.15	74.52	79.87	3.60	1.22
Biofuels and wastes	10.88	10.60	10.26	13.85	14.36	14.15	13.95	-0.35	1.48
Geothermal	-	-	-	-	-	-	-	-	-
Solar, wind, tide	-	-	-	-	-	-	-	-	-
Electricity	27.04	37.25	51.37	59.17	57.15	62.16	63.75	3.85	1.03
Heat	0.14	1.47	0.91	1.16	1.26	0.70	0.78	11.78	-0.72
of which:									
Total industry	**65.29**	**71.52**	**67.30**	**78.56**	**72.97**	**80.52**	**80.68**	**0.18**	**0.87**
Coal and peat	6.98	5.85	3.87	4.59	3.64	3.72	3.50	-3.41	-0.48
Oil	20.06	16.56	9.70	9.22	8.87	9.65	8.67	-4.19	-0.53
Natural Gas	16.95	23.14	24.07	27.59	29.38	32.35	34.11	2.08	1.67
Biofuels and wastes	8.15	7.87	8.13	11.03	9.91	9.20	8.34	-0.02	0.13
Geothermal	-	-	-	-	-	-	-	-	-
Solar, wind, tide	-	-	-	-	-	-	-	-	-
Electricity	13.01	16.67	20.63	24.98	19.91	24.91	25.28	2.75	0.97
Heat	0.14	1.43	0.90	1.15	1.26	0.70	0.78	11.73	-0.68
Total transport	**48.00**	**63.31**	**61.60**	**74.49**	**82.13**	**84.03**	**84.01**	**1.48**	**1.49**
Coal and peat	0.18	-	-	-	-	-	-	-	-
Oil	47.42	60.70	57.06	66.95	76.64	78.29	77.72	1.09	1.48
Natural Gas	-	2.33	4.14	6.79	3.89	3.52	3.48	-	-0.82
Biofuels and wastes	-	-	-	0.19	1.12	1.76	2.34	-	-
Electricity	0.40	0.28	0.40	0.56	0.47	0.46	0.47	0.06	0.78
Residential	**36.34**	**40.43**	**41.03**	**44.10**	**45.95**	**44.01**	**46.83**	**0.72**	**0.63**
Coal and peat	0.57	0.11	0.07	0.05	0.04	0.05	0.05	-11.87	-1.57
Oil	17.92	14.67	6.66	4.64	4.10	3.85	3.78	-5.66	-2.66
Natural Gas	8.73	12.49	16.22	19.80	20.27	18.89	20.94	3.71	1.22
Biofuels and wastes	2.73	2.73 e	2.13	2.63	3.32	3.18	3.26	-1.44	2.04
Geothermal	-	-	-	-	-	-	-	-	-
Solar, wind, tide	-	-	-	-	-	-	-	-	-
Electricity	6.38	10.42	15.95	16.98	18.22	18.04	18.80	5.54	0.79
Heat	-	-	0.00	-	0.00	-	-	-	-
Comm & public services	**24.13**	**26.13**	**31.04**	**39.17**	**36.63**	**35.92**	**37.81**	**1.49**	**0.94**
Coal and peat	0.00	0.04	0.01	0.00	-	-	-	6.38	-
Oil	8.93	7.08	5.82	8.25	3.62	3.61	4.30	-2.48	-1.44
Natural Gas	8.20	10.05	11.89	15.45	15.62	14.69	15.46	2.21	1.26
Biofuels and waste	-	-	-	-	0.01 e	0.01 e	0.01 e	-	-
Geothermal	-	-	-	-	-	-	-	-	-
Solar, wind, tide	-	-	-	-	-	-	-	-	-
Electricity	7.00	8.92	13.31	15.46	17.39	17.60	18.03	3.86	1.46
Heat	-	0.04	0.01	0.01	0.00	0.00	0.00	-	-12.94
Non-energy use	**10.44**	**16.65**	**21.46**	**28.56**	**30.68**	**33.99**	**35.18**	**4.33**	**2.38**
Coal and peat	-	0.18	0.47	0.38	1.07	0.93	0.91	-	3.22
Oil	10.44	13.14	16.16	22.35	25.35	28.80	29.43	2.61	2.90
Natural Gas	-	3.33	4.83	5.84	4.26	4.26	4.85	-	0.02

(1) Total Final Consumption (TFC) excludes use in transformation processes and energy industry own use.

* Please refer to notes and definitions in Part I. Agriculture, Forestry, Fishing and "Non-specified other" sectors are not reported here individually, but are accounted for in Total final consumption.

Source: *IEA/OECD Energy Balances of OECD Countries*

CANADA

5. Coal balance*

(Mtce)

	1973	1980	1990	2000	2005	2010	2011	2012e	Average annual percent change 73-90	90-11
Production	16.7	28.9	54.2	49.2	45.5 e	48.3	48.1	48.0	7.2	-0.6
Imports	15.0	14.8	13.6	21.5	16.8 e	10.8	8.5	8.9	-0.5	-2.2
Exports	-10.9	-14.9	-30.6	-27.5	-24.1 e	-28.2	-28.7	-29.7	6.3	-0.3
Stock changes	1.0	0.5	-2.5	2.1	2.9 e	1.0	0.1	0.2		
Primary supply	**21.8**	**29.4**	**34.7**	**45.2**	**41.2**	**31.9**	**28.0**	**27.3**	**2.8**	**-1.0**
Statistical differences	**0.5**	**1.1**	**0.2**	**0.0**	**-0.1**	**1.7**	**3.3**	..		
Total transformation	**-14.5 e**	**-24.3 e**	**-30.3 e**	**-40.1 e**	**-35.9 e**	**-28.8 e**	**-26.9 e**	..	**4.4**	**-0.6**
Electricity and heat gen.	-11.9	-21.5	-28.5	-38.5	-34.3 e	-27.1	-25.2	..	5.3	-0.6
Main activity producers (1)	*-11.9*	*-21.4*	*-28.5*	*-38.4*	*-34.1 e*	*-27.1*	*-25.2*	..	*5.3*	*-0.6*
Autoproducers	-	*-0.0*	*-0.0*	*-0.1*	*-0.2 e*	*-0.0*	*-0.0*	..	-	*-16.4*
Gas works	-	-	-	-	-	-	-	..	-	-
Coal transformation(2)	-2.7 e	-2.8 e	-1.8 e	-1.6 e	-1.6 e	-1.7 e	-1.7 e	..	-2.3	-0.3
BKB plants	-	-	-	-	-	-	-	..	-	-
Blast furnaces	*-1.9 e*	*-2.1 e*	*-1.5 e*	*-1.4 e*	*-1.5 e*	*-1.1 e*	*-0.9 e*	..	*-1.2*	*-2.5*
Coke ovens	*-0.7*	*-0.7*	*-0.3*	*-0.2*	*-0.2 e*	*-0.7*	*-0.8*	..	*-6.2*	*5.5*
Patent fuel plants	-	-	-	-	-	-	-	..	-	-
Other transformation(3)	-	-	-	-	-	-	-	..	-	-
Energy ind. own use	**-0.1**	**-0.0**	**-0.1**	**-0.1**	**-0.1 e**	**-0.0**	**-0.0**	..	**2.6**	**-14.0**
Losses	-	-	-	-	-	-	-	..		
Final consumption(4)	**7.7**	**6.2**	**4.4**	**5.0**	**5.1**	**4.7**	**4.5**	..	**-3.2**	**0.0**
Industry(5)	7.0	5.9	3.9	4.6	4.4	3.7	3.5	..	-3.4	-0.5
Iron and steel	*3.9 e*	*4.1 e*	*2.3 e*	*2.6 e*	*2.5 e*	*2.1 e*	*1.8 e*	..	*-3.1*	*-1.0*
Chemical	-	*0.0*	-	-	-	-	-	..	-	-
Non-metallic minerals	*0.7*	*0.3*	*0.7*	*1.0*	*1.1 e*	*0.7*	*0.7*	..	*-0.0*	*0.5*
Paper, pulp and print	*0.3*	*0.3*	*0.1*	*0.1*	*0.0 e*	-	-	..	*-4.4*	-
Other industry (6)	*2.2*	*1.1*	*0.8*	*1.0*	*0.9 e*	*0.9*	*0.9*	..	*-5.6*	*0.7*
Transport(7)	0.2	-	-	-	-	-	-	..	-	-
Other	0.6	0.1	0.1	0.1	0.1	0.0	0.0	..	-11.4	-2.0
Comm. and pub. services	*0.0*	*0.0*	*0.0*	*0.0*	-	-	-	..	*6.4*	-
Residential	*0.6*	*0.1*	*0.1*	*0.1*	*0.1 e*	*0.0*	*0.0*	..	*-11.9*	*-1.6*
Other sectors (8)	-	-	-	-	-	-	-	..	-	-
Non-energy use	-	0.2	0.5	0.4	0.6	0.9	0.9	..	-	3.2

(1) Main activity electricity and heat generation includes district heating.

(2) Coal transformation refers to the transformation of coal and peat to secondary and tertiary products (mainly cokes, briquettes, coal tar and off-gases). Gas works and Liquefaction are shown separately to match the *Energy Balances of OECD Countries.*

(3) Other transformation includes Liquefaction and Non-specified transformations.

(4) Final Consumption includes non-energy use and energy use (Industry, Transport and Other).

(5) Please refer to notes and definitions in Part I for detailed explanation.

(6) Other industry includes Non-ferrous metals, Transport equipment, Machinery, Mining and quarrying, Food and tobacco, Wood and wood products, Construction, Textile and leather, and Non-specified industry.

(7) Transport includes Rail and Inland waterways.

(8) Other sectors includes Agriculture/Forestry, Fishing and Non-specified other.

* "Coal" refers to all types of coal, primary (anthracite, coking coal, other bituminous coal, sub-bituminous coal, lignite/brown coal) and derived fuels (including patent fuel, BKB, coke oven coke, gas coke, coal tar, coke oven gas, gas works gas, blast furnace gas and other recovered gases), as well as peat, peat briquettes and oil shale. Quantities have been converted to Mtce using calorific values reported by the respective countries. Please refer to notes and definitions in Part I.

Source: *IEA/OECD Energy Balances of OECD Countries*

CANADA

6. Use of coal for selected end-uses*

(million tonnes)

	1978	1990	2000	2005	2009	2010	2011	Average annual percent change 78-90	90-11
Steam coal	**19.71**	**34.77**	**47.30**	**44.78**	**35.16**	**34.32**	**27.86**	**4.84**	**-1.05**
Total electricity and heat	18.31	33.07	44.97	41.87 e	33.13	33.65	31.71	5.05	-0.20
Main activity producers	*18.30*	*33.05*	*44.97*	*41.86 e*	*33.12*	*33.64*	*31.71*	*5.05*	*-0.20*
Autoproducers	*0.01*	*0.02*	*0.01*	*0.01 e*	*0.00*	*0.00*	*0.00*	*2.12*	*-12.86*
Patent fuel/BKB plants	-	-	-	-	-	-	-	-	-
Coke ovens/Liquefaction(1)	-	-	-	-	-	-	-	-	-
Blast furnace inputs	-	-	-	-	-	-	-	-	-
Gas manufacture	-	-	-	-	-	-	-	-	-
Industry	1.42	1.22	1.74	1.86	1.91	1.79	1.78	-1.26	1.81
Iron and steel	*0.07*	-	-	-	-	-	-	-	-
Chemical	*0.26*	-	-	-	-	-	-	-	-
Non-metallic minerals	*0.36*	*0.69*	*1.03*	*1.20 e*	*0.95*	*0.89*	*0.88*	*5.41*	*1.21*
Paper, pulp and print	*0.28*	*0.09*	*0.03*	*0.02 e*	-	-	-	*-8.64*	-
Other industry	*0.45*	*0.44*	*0.67*	*0.64 e*	*0.96*	*0.90*	*0.90*	*-0.17*	*3.44*
Other sectors(2)	0.18	0.09	0.05	0.03	-	-	-	-5.61	-
Non-energy use	-	0.26	0.33	0.53	1.22	1.07	1.05	-	6.90
Coking coal	**6.78**	**5.02**	**4.46**	**4.29**	**3.67**	**3.72**	**5.56**	**-2.46**	**0.48**
Total electricity and heat	-	-	-	-	-	-	-	-	-
Main activity producers	-	-	-	-	-	-	-	-	-
Autoproducers	-	-	-	-	-	-	-	-	-
Patent fuel/BKB plants	-	-	-	-	-	-	-	-	-
Coke ovens/Liquefaction(1)	6.98	5.00	4.23	4.29 e	3.04	4.02	4.19	-2.74	-0.83
Blast furnace inputs	-	-	-	-	-	-	-	-	-
Gas manufacture	-	-	-	-	-	-	-	-	-
Industry	-	-	-	-	-	-	-	-	-
Iron and steel	-	-	-	-	-	-	-	-	-
Chemical	-	-	-	-	-	-	-	-	-
Non-metallic minerals	-	-	-	-	-	-	-	-	-
Paper, pulp and print	-	-	-	-	-	-	-	-	-
Other industry	-	-	-	-	-	-	-	-	-
Other sectors(2)	-	-	-	-	-	-	-	-	-
Non-energy use	-	-	-	-	-	-	-	-	-
Lignite	**4.98**	**9.36**	**11.21**	**11.14**	**10.42**	**10.14**	**9.61**	**5.40**	**0.13**
Total electricity and heat	4.60	9.08	10.86	10.61 e	10.08	10.10	9.35	5.83	0.13
Main activity producers	*4.60*	*9.08*	*10.86*	*10.61 e*	*10.08*	*10.10*	*9.35*	*5.83*	*0.13*
Autoproducers	-	-	-	-	-	-	-	-	-
Patent fuel/BKB plants	-	-	-	-	-	-	-	-	-
Coke ovens/Liquefaction(1)	-	-	-	-	-	-	-	-	-
Gas manufacture	-	-	-	-	-	-	-	-	-
Industry	0.28	0.17	0.16	0.10	-	-	-	-3.97	-
Iron and steel	-	-	-	-	-	-	-	-	-
Chemical	-	-	-	-	-	-	-	-	-
Non-metallic minerals	-	-	-	-	-	-	-	-	-
Paper, pulp and print	*0.17*	*0.10*	*0.10*	*0.06 e*	-	-	-	*-4.60*	-
Other industry	*0.11*	*0.08*	*0.07*	*0.04 e*	-	-	-	*-3.07*	-
Other sectors(2)	0.11	0.01	0.04	0.05	0.09	0.09	0.10	-18.83	12.04
Non-energy use	-	0.09	0.14	0.16	-	-	-	-	-

(1) Liquefaction primarily refers to direct distillation processes. Indirect liquefaction may be reported here or under gas manufacture.

(2) Other sectors are Residential, Commercial and public services, Agriculture/Forestry, Fishing and Non-specified other.

* Coal totals may not match the sum of the individual rows due to statistical differences. Please refer to notes and definitions in Part I.

Source: *IEA/OECD Energy Statistics of OECD Countries*

CANADA

7. Fuel prices to end users*

	1978(1)	1990	1995	2000	2005	2009	2010	2011	2012
				(US dollars / tce)					
For electricity generation									
Steam coal	37.88	61.02	..	20.54	28.16	28.67	35.65	33.13	..
Heavy fuel oil	51.11	..	..	87.53	144.70	214.77	311.23	393.28	..
Natural gas	34.43	48.90	..	104.82	173.64	153.35	159.01	143.60	..
For industry									
Steam coal	43.98	..	..	..	..	..	..	..	..
Coking coal	62.70	..	..	..	..	..	..	..	..
High sulphur fuel oil	51.91	93.40	87.58	148.04	243.30	315.10	402.25	540.64	568.58
Low sulphur fuel oil	..	..	..	..	..	..	..	..	..
Natural gas	43.30	65.09	54.28	69.84	226.41	134.26	124.55	139.38	107.62
				(Canadian dollars / unit)(2)					
For electricity generation									
Steam coal	37.62	62.00	..	26.55	29.70	28.49	31.97	28.54	..
Heavy fuel oil	79.95	..	..	178.28	240.46	336.19	439.72	533.74	..
Natural gas	50.49	73.37	..	200.15	270.51	225.04	210.62	182.71	..
For industry									
Steam coal	51.08	..	..	..	..	..	..	..	..
Coking coal	73.20	..	..	..	..	..	..	..	..
High sulphur fuel oil	81.21	149.50	164.86	301.51	404.30	493.24	568.31	733.74	779.22
Low sulphur fuel oil	..	..	..	..	..	..	..	..	..
Natural gas	63.50	97.67	95.79	133.36	352.73	197.03	164.97	177.34	138.27

(1) Earliest year for which split by coal type is available.

(2) For steam coal, coking coal and heavy fuel oil per metric tonne; for natural gas per 10^7 kilocalories GCV.

* Please refer to notes and definitions in Part I. Steam coal prices relate to historical definitions.

Source: *IEA/OECD Energy Prices & Taxes*

8. Coal and peat trade by type of coal*

(Mtce)

	1978	1990	1995	2000	2005	2009	2010	2011	2012e
Total imports	**14.21**	**13.64**	**9.92**	**21.50**	**16.83 e**	**10.41**	**10.81**	**8.51**	**8.90**
Bituminous coal(1)	8.51	9.10	5.27	14.00	8.62 e	5.11	4.75	3.65	3.74
Coking coal	5.30	4.22	3.88	4.15	4.09 e	2.15	3.00	3.65	4.25
Sub-bituminous coal	-	-	-	2.76	4.03 e	2.83	2.36	0.82	0.59
Lignite	-	-	-	-	0.00 e	0.00	0.00	0.00	0.01
Peat	-	-	-	-	-	-	-	-	-
Coal products(2)	0.40	0.32	0.76	0.59	0.08	0.32	0.70	0.38	0.32
Total exports	**13.68**	**30.63**	**33.70**	**27.53**	**24.08 e**	**24.21**	**28.17**	**28.68**	**29.70**
Bituminous coal(1)	0.94	4.08	5.33	3.14	1.18 e	5.72	4.75	4.86	3.24
Coking coal	12.65	26.37	28.05	24.11	22.68 e	18.21	23.28	23.39	25.98
Sub-bituminous coal	-	-	-	-	0.01 e	0.01	0.00	0.03	0.03
Lignite	0.00	0.00	-	-	0.06 e	0.06	0.06	0.06	0.05
Peat	-	-	-	-	-	-	-	-	-
Coal products(2)	0.08	0.18	0.31	0.28	0.14	0.21	0.08	0.33	0.39

(1) Bituminous coal includes anthracite.

(2) Coal products includes products derived from coal, for example: coke, coal tar, briquettes, patent fuels.

* Please refer to notes and definitions in Part I.

Source: *IEA/OECD Energy Balances of OECD Countries, IEA/OECD Energy Statistics of OECD Countries*

CANADA

9. Total coal imports by origin*

(thousand tonnes)

	1978(1)	1985	1990	1995	2000	2005	2010	2011	2012e
Total coal(2)	**14250**	**14579**	**14169**	**9735**	**23231**	**20745 e**	**12611**	**9472**	**9817**
Coking coal	**5454**	**6188**	**4491**	**4129**	**4296**	**4228 e**	**3092**	**3770**	**4382**
Australia	-	-	-	-	-	-	-	-	-
Canada	-	-	-	-	-	-	-	-	-
Czech Republic	-	-	-	-	-	-	-	-	-
Germany	-	-	-	-	-	-	-	-	-
Poland	-	-	-	-	-	-	-	-	-
United Kingdom	-	-	-	-	-	-	-	-	-
United States	5454	6188	4491	4129	4296	4160 e	3092	3770	4382
Other OECD	-	-	-	-	-	-	-	-	-
China, People's Rep.	-	-	-	-	-	-	-	-	-
Colombia	-	-	-	-	-	9 e	-	-	-
Indonesia	-	-	-	-	-	-	-	-	-
South Africa	-	-	-	-	-	-	-	-	-
Former Soviet Union(3)	-	-	x	x	x	x	x	x	x
Russian Federation (4)	*x*	*x*	-	-	-	-	-	-	-
Other FSU (4)	*x*	*x*	-	-	-	-	-	-	-
Venezuela	-	-	-	-	-	-	-	-	-
Vietnam	-	-	-	-	-	-	-	-	-
Non-specified/other	-	-	-	-	-	59 e	-	-	-
Steam coal	**8796**	**8391**	**9678**	**5606**	**18935**	**16516 e**	**9515**	**5695**	**5424**
Australia	-	-	-	-	-	-	-	-	-
Canada	-	-	-	-	-	-	-	-	-
Czech Republic	-	-	-	-	-	-	-	-	-
Germany	-	-	-	-	-	-	-	3	-
Poland	-	-	-	-	-	-	-	-	-
United Kingdom	-	-	-	-	-	21 e	1	302	-
United States	8796	8391	9639	5095	18531	12855 e	7247	3266	3777
Other OECD	-	-	-	-	-	-	26	7	-
China, People's Rep.	-	-	-	-	155	51 e	-	-	-
Colombia	-	-	-	511	-	2086	2044	1803	1344
Indonesia	-	-	-	-	-	-	-	-	-
South Africa	-	-	-	-	61	-	-	-	-
Former Soviet Union(3)	-	-	x	x	x	x	x	x	x
Russian Federation (4)	*x*	*x*	-	-	*26*	*350 e*	*12*	*145*	*83*
Other FSU (4)	*x*	*x*	-	-	-	*62 e*	*153*	*108*	*220*
Venezuela	-	-	39	-	91	625 e	32	61	-
Vietnam	-	-	-	-	-	-	-	-	-
Non-specified/other	-	-	-	-	71	466	-	-	-
Lignite	**-**	**-**	**-**	**-**	**-**	**1 e**	**4**	**7**	**11**

(1) Earliest year for which split by coal type is available.

(2) Total coal consists of steam coal, coking coal and lignite.

(3) For years prior to 1990, or if no details for individual Former Soviet Union republics.

(4) Data for former republics are not available separately for some OECD member countries.

* Please refer to notes and definitions in Part I.

Source: IEA/OECD Coal Statistics

CANADA

10. Coking coal exports by destination*

(thousand tonnes)

	1978(1)	1985	1990	1995	2000	2005	2010	2011	2012e
World	**13017**	**22483**	**26851**	**28564**	**28386**	**26710 e**	**27528**	**27666**	**30725**
Total OECD	**12217**	**20715**	**19908**	**25594**	**22752**	**22087 e**	**20735**	**20792**	**17460**
Australia	-	-	-	-	-	-	-	-	-
Austria	-	-	-	-	-	-	-	-	-
Belgium	147	62	6	261	374	58 e	48	55	-
Canada	-	-	-	-	-	-	-	-	-
Chile	-	130	224	267	312	368 e	215	217	253
Czech Republic	-	-	-	-	-	-	-	-	-
Denmark	-	-	-	-	-	-	-	-	-
Estonia	-	-	-	-	-	-	-	-	-
Finland	-	-	41	58	-	516 e	416	422	303
France	-	352	379	370	585	494 e	166	208	55
Germany	-	123	72	104	792	1757 e	1250	640	471
Greece	-	-	-	-	-	-	-	-	-
Hungary	-	-	-	-	-	-	-	-	-
Iceland	-	-	-	-	-	-	-	-	-
Ireland	-	-	-	-	-	-	-	-	-
Israel	-	-	-	-	-	-	-	-	-
Italy	164	33	159	968	1170	1469 e	1014	1000	767
Japan	10934	17026	16569	15495	12085	6794 e	8691	7307	7482
Korea	668	2041	-	4280	3851	4770 e	5294	6514	5051
Luxembourg	-	-	-	-	-	-	-	-	-
Mexico	-	195	-	517	385	406 e	302	193	87
Netherlands	-	-	369	459	408	807 e	707	1263	1457
New Zealand	-	-	-	-	-	-	-	-	-
Norway	-	-	-	-	-	-	-	-	-
Poland	-	-	-	-	-	-	-	-	-
Portugal	-	-	519	161	-	-	-	-	-
Slovak Republic	-	-	-	-	-	-	59	221	-
Slovenia	-	-	-	-	-	-	-	-	-
Spain	150	-	-	337	338	344 e	69	118	-
Sweden	154	313	102	155	-	-	-	-	60
Switzerland	-	-	-	-	-	-	-	-	-
Turkey	-	-	51	257	819	1025 e	838	849	500
United Kingdom	-	330	645	1171	1093	1677 e	283	430	99
United States	-	110	772	734	540	1602 e	1383	1355	875
Total non-OECD	**800**	**1768**	**1745**	**2586**	**2795**	**4623 e**	**6793**	**6874**	**13265**
Brazil	600	899	1108	1078	1471	1718 e	1637	2281	1813
China(2)	-	-	300	-	-	956 e	4341	3074	9565
Chinese Taipei	-	496	-	1264	1324	1274 e	637	1069	1004
Egypt	-	-	-	46	-	426 e	-	60	-
India	200	38	-	-	-	-	-	280	828
Romania	-	-	-	-	-	-	-	-	-
Oth. Africa & Mid. East	-	31	129	50	-	-	69	-	-
Oth. non-OECD Americas	-	-	-	-	-	-	-	-	-
Other Asia & Oceania	-	235	208	148	-	104 e	109	110	55
Other non-OECD Europe and Eurasia	-	69	-	-	-	145 e	-	-	-
Non-specified/Other	**-**	**-**	**5198**	**384**	**2839**	**-**	**-**	**-**	**-**

(1) Earliest year for which split by coal type is available.

(2) China includes the People's Republic of China and Hong Kong, China.

* Please refer to notes and definitions in Part I.

Source: IEA/OECD Coal Statistics

CANADA

11. Steam coal exports by destination*

(thousand tonnes)

	1978(1)	1985	1990	1995	2000	2005	2010	2011	2012e
World	**971**	**4895**	**4149**	**5429**	**3696**	**1366 e**	**5751**	**5933**	**3971**
Total OECD	**971**	**4389**	**2826**	**5289**	**3419**	**1319 e**	**4202**	**4625**	**3658**
Australia	-	-	-	-	-	-	-	-	-
Austria	-	-	-	-	-	-	-	-	-
Belgium	27	78	-	13	151	-	-	-	-
Canada	-	-	-	-	-	-	-	-	-
Chile	-	-	120	207	-	180 e	44	-	-
Czech Republic	-	-	-	-	-	-	-	-	-
Denmark	309	328	479	-	-	-	-	-	-
Estonia	-	-	-	-	-	-	-	-	-
Finland	-	56	-	-	-	-	-	-	-
France	-	384	18	164	-	-	-	-	-
Germany	492	203	64	86	55	-	-	-	-
Greece	-	26	-	-	-	-	-	-	-
Hungary	-	-	-	-	-	-	-	-	-
Iceland	-	-	-	-	-	-	-	-	-
Ireland	-	-	-	-	-	-	-	-	-
Israel	-	-	-	-	-	-	-	-	-
Italy	-	34	-	255	14	-	1	1	-
Japan	83	1516	1933	2483	1244	726 e	2005	2040	2088
Korea	56	1469	-	1719	1767	169 e	1670	2229	1419
Luxembourg	-	-	-	-	-	-	-	-	-
Mexico	-	-	-	-	-	-	395	207	127
Netherlands	-	28	-	-	-	-	-	99	-
New Zealand	-	-	-	-	-	-	-	-	-
Norway	-	-	-	-	-	-	-	-	-
Poland	-	-	-	-	-	-	-	-	-
Portugal	-	-	-	-	-	-	-	-	-
Slovak Republic	-	-	-	-	-	-	-	-	-
Slovenia	-	-	-	-	-	-	-	-	-
Spain	-	112	-	-	-	-	1	2	1
Sweden	-	-	-	-	-	-	-	-	-
Switzerland	-	-	-	-	-	-	-	-	-
Turkey	-	-	-	-	-	-	-	-	-
United Kingdom	-	-	18	315	80	-	-	-	-
United States	4	155	194	47	108	244 e	86	47	23
Total non-OECD	**-**	**506**	**98**	**140**	**-**	**2 e**	**1549**	**1307**	**313**
Brazil	-	-	98	140	-	-	55	-	-
China(2)	-	506	-	-	-	-	1493	1306	312
Chinese Taipei	-	-	-	-	-	2 e	1	-	1
Egypt	-	-	-	-	-	-	-	-	-
India	-	-	-	-	-	-	-	1	-
Romania	-	-	-	-	-	-	-	-	-
Oth. Africa & Mid. East	-	-	-	-	-	-	-	-	-
Oth. non-OECD Americas	-	-	-	-	-	-	-	-	-
Other Asia & Oceania	-	-	-	-	-	-	-	-	-
Other non-OECD Europe and Eurasia	-	-	-	-	-	-	-	-	-
Non-specified/Other	**-**	**-**	**1225**	**-**	**277**	**45 e**	**-**	**1**	**-**

(1) Earliest year for which split by coal type is available.

(2) China includes the People's Republic of China and Hong Kong, China.

* Please refer to notes and definitions in Part I. Steam coal includes all sub-bituminous coal.

Source: IEA/OECD Coal Statistics

CHILE*

Figure 1: TPES by fuel (Mtce)

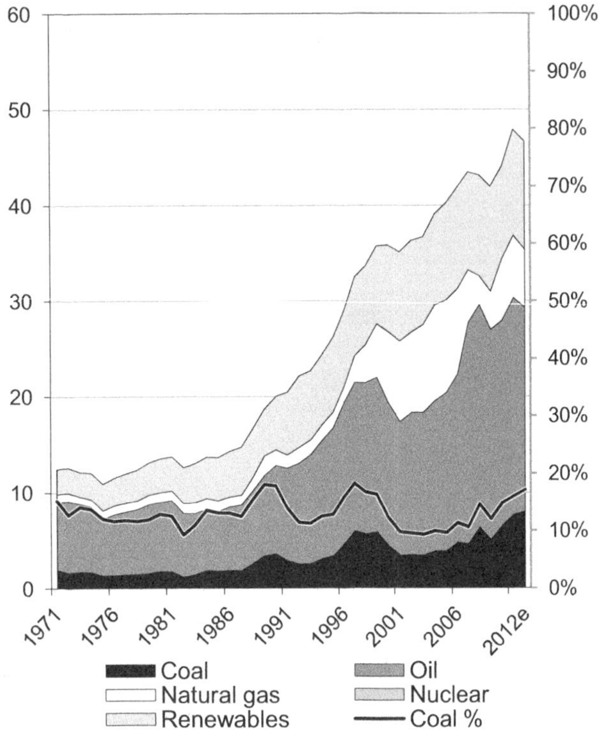

Figure 2: Coal supply indicators (1971 = 100)

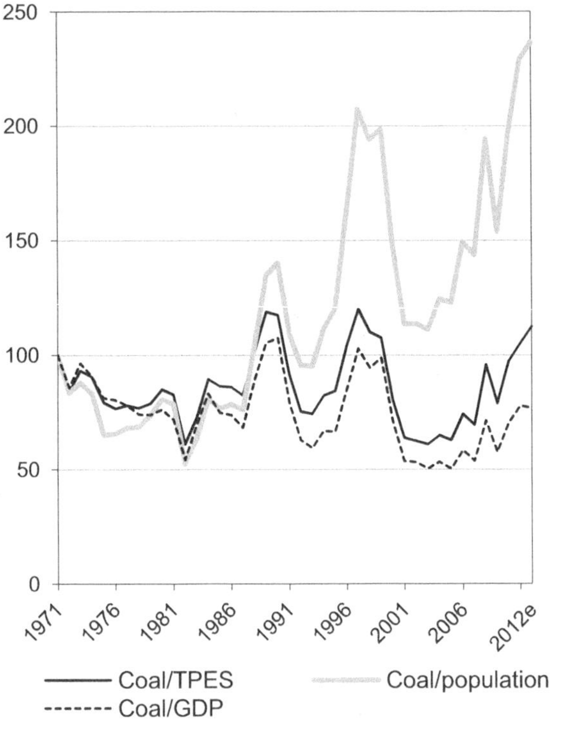

Figure 3: Primary coal supply (Mtce)

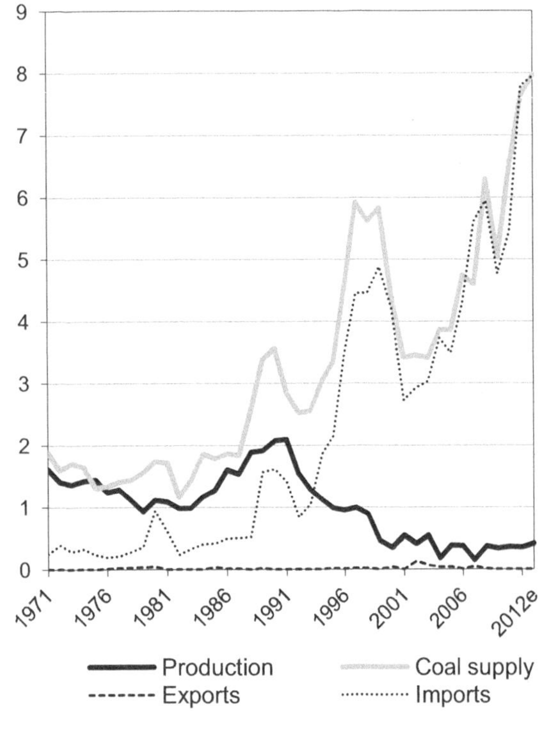

Figure 4: Coal consumption (Mtce)

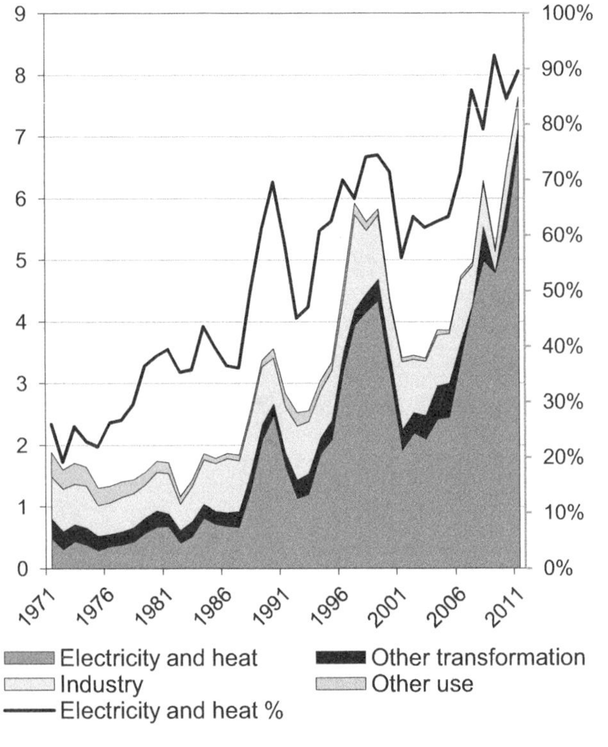

*Please refer to notes and definitions in Part I. Peat is included in coal for all figures.

Source: *IEA/OECD Coal Information, IEA/OECD Energy Balances of OECD Countries, OECD Main Economic Indicators*

CHILE*

Figure 5: Electricity generation by fuel (TWh)

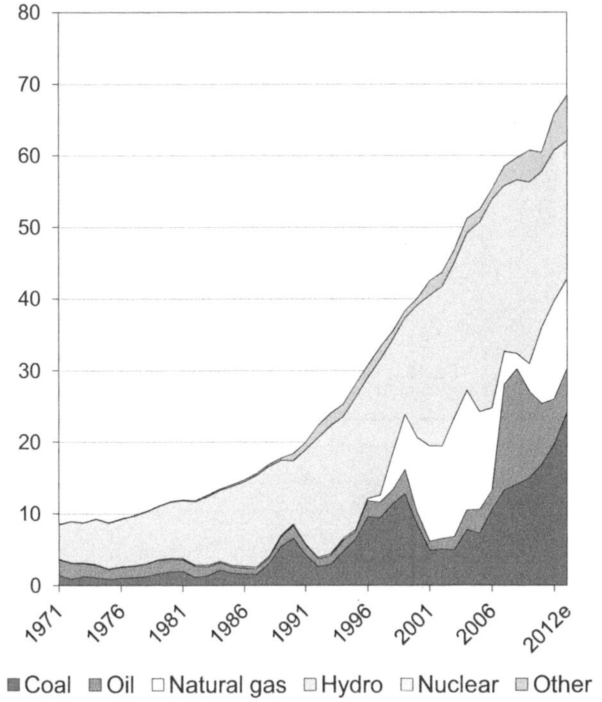

Figure 6: CO_2 emissions by fuel (Mt CO_2)

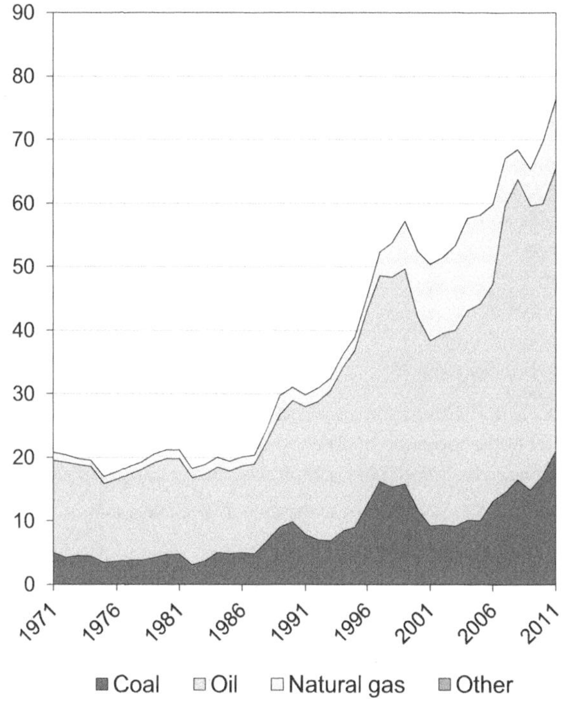

Figure 7: Electricity generation by fuel share

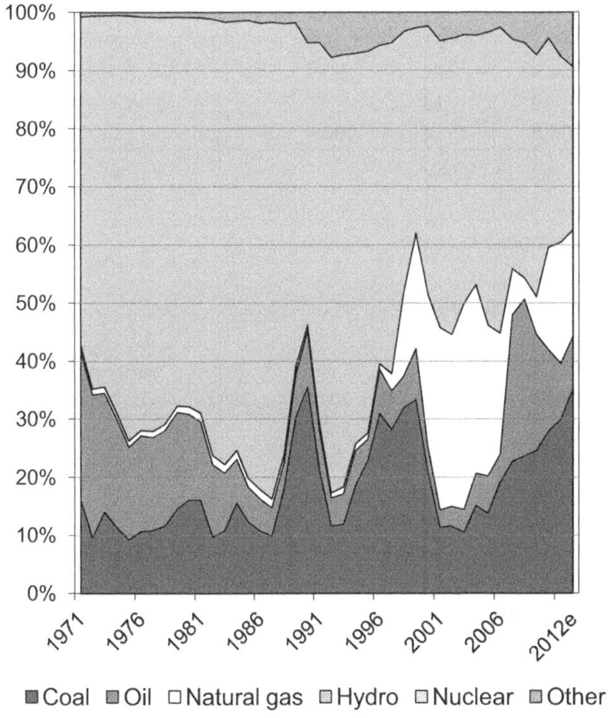

Figure 8: CO_2 emissions by fuel share

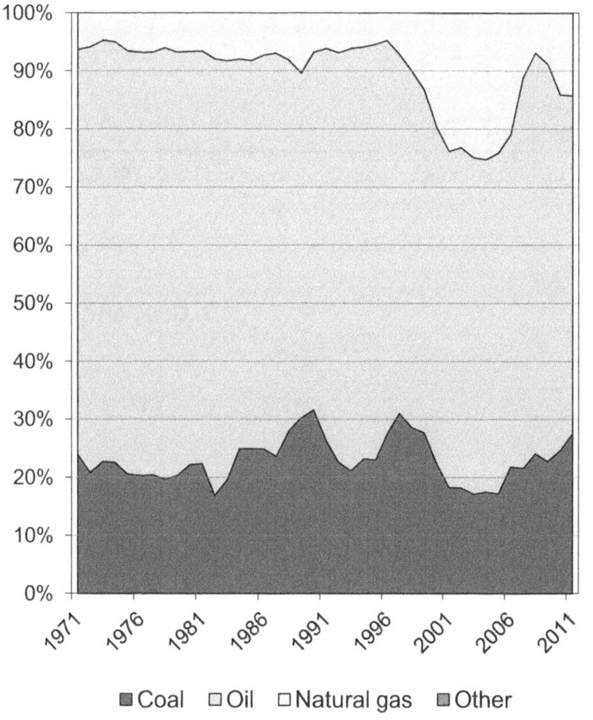

*Please refer to notes and definitions in Part I. Peat is included in coal for all figures.

Source: *IEA/OECD Electricity Information, IEA/OECD CO2 Emissions from Fuel Combustion*

CHILE

1. Total primary energy supply (TPES) by fuel*

	1973	1980	1990	2000	2010	2011	2012e	Average annual percent change 73-90	90-11
TPES (Mtce)	**12.14**	**13.54**	**20.01**	**35.96**	**44.17**	**47.96**	**46.74**	**2.98**	**4.25**
Coal and peat	1.71	1.74	3.57	4.39	6.52	7.64	7.97	4.41	3.69
Oil	7.10	7.24	9.24	14.97	21.44	22.64	21.33	1.56	4.36
Natural Gas	0.75	1.03	1.63	7.44	6.38	6.52	6.05	4.67	6.82
Biofuels and waste	1.89	2.56	4.48	6.75	7.01	8.45	8.89	5.19	3.07
Nuclear	-	-	-	-	-	-	-	-	-
Hydro	0.69	0.97	1.10	2.27	2.67	2.58 e	2.36	2.79	4.16
Geothermal	-	-	-	-	-	-	-	-	-
Solar, wind, tide	-	-	-	-	0.04	0.04	0.05	-	-
Net electricity trade(1)	0.00	-	-	0.15	0.12	0.09	0.09	-	-
Heat(2)	-	-	-	-	-	-	-	-	-

(1) Net trade between OECD and non-OECD countries.

(2) Ambient heat from heat pumps used in the transformation sector.

Source: *IEA/OECD Energy Balances of OECD Countries*

2. Energy supply, GDP and population*

	1973	1980	1990	2000	2010	2011	2012e	Average annual percent change 73-90	90-11
GDP (2005 bil. USD)	28 e	37 e	53	99	148	157	166	3.74	5.31
TPES/GDP(1)	0.43 e	0.37 e	0.38	0.36	0.30	0.31	0.28	-0.73	-1.00
Population (millions)	10.1 e	11.2 e	13.2	15.4	17.1	17.3	17.4	1.57	1.30
TPES/population(1)	1.20 e	1.21 e	1.52	2.34	2.58	2.78	2.68	1.39	2.92
TPES/GDP(2)	129.8 e	112.2 e	114.6	110.7	90.5	92.7	85.6	-0.73	-1.00
Coal and peat supply/GDP(2)	191.9 e	151.5 e	214.0	141.7	140.0	154.8	153.0	0.64	-1.53
Elec. consumption/GDP(2)	65.4 e	67.8 e	74.3	94.9	94.0	93.9	..	0.76	1.12
Elec. generation (TWh)	9	12	18	40	60	66	68	4.45	6.26
Industrial production(2)	..	..	..	80.0	101.0	106.5	114.5	..	..

(1) TPES/GDP in units of tce/2005 thousand USD, TPES/population in units of tce/capita.

(2) As index, 2005 (not shown) = 100

Source: *IEA/OECD Energy Balances of OECD Countries , OECD Main Economic Indicators*

3. Coal and peat production by type*

	1978(1)	1990	2000	2005	2010	2011	2012e	Average annual percent change 78-90	90-11
Mtce:									
Coking coal	-	-	-	-	-	-	-	-	-
Steam coal	1.10	2.07	0.35	0.38	0.36	0.35	0.41	5.46	-8.11
Lignite	0.02	-	-	-	-	-	-	-	-
Peat	-	-	-	-	-	-	-	-	-
Mt:									
Coking coal	-	-	-	-	-	-	-	-	-
Steam coal	1.13	2.18	0.37	0.54	0.62	0.65	0.72	5.65	-5.58
Lignite	0.03	-	-	-	-	-	-	-	-
Peat	-	-	-	-	-	-	-	-	-

(1) Earliest year for which split by coal type is available.

* Please refer to notes and definitions in Part I.

Source: *IEA/OECD Energy Balances of OECD Countries, IEA/OECD Energy Statistics of OECD Countries*

CHILE

4. Final consumption of energy by fuel*

(Mtce)

	1973	1980	1990	2000	2009	2010	2011	Average annual percent change 73-90	90-11
Total final consumption[1]	**9.32**	**10.42**	**15.86**	**29.12**	**32.17**	**34.06**	**35.97**	**3.18**	**3.98**
Coal and peat	1.01	0.81	0.90	0.91	0.39	0.60	0.51	-0.66	-2.70
Oil	5.49	5.76	7.84	13.12	16.65	17.29	18.40	2.12	4.15
Natural Gas	0.06	0.14	1.29	4.70	2.07	3.36	2.79	20.19	3.76
Biofuels and wastes	1.87	2.51	3.92	5.88	6.43	6.08	7.16	4.47	2.90
Geothermal	-	-	-	-	-	-	-	-	-
Solar, wind, tide	-	-	-	-	-	-	-	-	-
Electricity	0.90	1.20	1.90	4.52	6.64	6.72	7.12	4.53	6.48
Heat	-	-	-	-	-	-	-	-	-
of which:									
Total industry	**3.30**	**3.97**	**4.87**	**9.50**	**11.73**	**12.46**	**13.82**	**2.31**	**5.09**
Coal and peat	0.66	0.63	0.75	0.84	0.35	0.57	0.48	0.72	-2.09
Oil	1.72	1.78	1.91	2.93	4.53	4.35	4.82	0.63	4.50
Natural Gas	0.00	0.01	0.00	0.96	0.22	1.57	1.42	3.08	32.08
Biofuels and wastes	0.34	0.75	0.96	1.62	2.14	1.56	2.38	6.35	4.43
Geothermal	-	-	-	-	-	-	-	-	-
Solar, wind, tide	-	-	-	-	-	-	-	-	-
Electricity	0.58	0.79	1.25	3.16	4.50	4.40	4.72	4.56	6.54
Heat	-	-	-	-	-	-	-	-	-
Total transport	**2.62**	**2.99**	**4.35**	**8.09**	**9.85**	**10.20**	**10.39**	**3.03**	**4.23**
Coal and peat	0.18	0.07	-	-	-	-	-	-	-
Oil	2.42	2.89	4.32	8.06	9.77	10.13	10.31	3.47	4.23
Natural Gas	-	-	0.01	0.01	0.03	0.03	0.02	-	5.44
Biofuels and wastes	-	-	-	-	-	-	-	-	-
Electricity	0.02	0.02	0.03	0.03	0.05	0.05	0.06	0.55	3.95
Residential	**3.02**	**3.12**	**4.46**	**6.95**	**7.29**	**7.57**	**7.94**	**2.31**	**2.78**
Coal and peat	0.06	0.06	0.08	0.03	0.01	0.01	0.01	2.15	-9.32
Oil	1.25	0.98	1.08	1.55	1.37	1.33	1.43	-0.86	1.35
Natural Gas	0.04	0.11	0.18	0.35	0.53	0.56	0.56	8.79	5.40
Biofuels and wastes	1.53	1.76	2.58 e	4.25 e	4.29 e	4.52 e	4.77 e	3.13	2.96
Geothermal	-	-	-	-	-	-	-	-	-
Solar, wind, tide	-	-	-	-	-	-	-	-	-
Electricity	0.14	0.22	0.53	0.76	1.09	1.15	1.17	7.96	3.84
Heat	-	-	-	-	-	-	-	-	-
Comm & public services	**0.15**	**0.21**	**0.68**	**0.89**	**1.65**	**1.95**	**2.27**	**9.17**	**5.88**
Coal and peat	0.02	0.02	0.01	0.03	0.02	0.01	0.01	-2.72	-0.43
Oil	-	-	0.18	0.23	0.47	0.66	0.93	-	8.29
Natural Gas	0.01	0.02	0.03	0.08	0.19	0.17	0.17	6.78	8.53
Biofuels and waste	-	-	0.38	0.00	-	-	0.00	-	-23.48
Geothermal	-	-	-	-	-	-	-	-	-
Solar, wind, tide	-	-	-	-	-	-	-	-	-
Electricity	0.12	0.17	0.09	0.55	0.98	1.11	1.15	-2.01	13.12
Heat	-	-	-	-	-	-	-	-	-
Non-energy use	**0.01**	**0.01**	**1.31**	**3.42**	**1.41**	**1.46**	**1.02**	**30.79**	**-1.18**
Coal and peat	-	-	-	-	-	-	-	-	-
Oil	0.01	0.01	0.25	0.12	0.31	0.43	0.40	18.65	2.22
Natural Gas	-	-	1.06	3.30	1.10	1.03	0.62	-	-2.49

(1) Total Final Consumption (TFC) excludes use in transformation processes and energy industry own use.

* Please refer to notes and definitions in Part I. Agriculture, Forestry, Fishing and "Non-specified other" sectors are not reported here individually, but are accounted for in Total final consumption.

Source: *IEA/OECD Energy Balances of OECD Countries*

CHILE

5. Coal balance*

(Mtce)

	1973	1980	1990	2000	2005	2010	2011	2012e	Average annual percent change 73-90	90-11
Production	1.4	1.1	2.1	0.3	0.4	0.4	0.4	0.4	2.5	-8.1
Imports	0.3	0.9	1.6	4.2	3.5	5.4	7.8	7.9	10.7	7.8
Exports	-0.0	-0.0	-	-0.0	-0.0	-	-	-	-	-
Stock changes	0.1	-0.3	-0.1	-0.1	0.0	0.7	-0.5	-0.4		
Primary supply	**1.7**	**1.7**	**3.6**	**4.4**	**3.9**	**6.5**	**7.6**	**8.0**	**4.4**	**3.7**
Statistical differences	**-**	**0.0**	**0.1**	**0.1**	**-0.1**	**-0.1**	**0.2**	..		
Total transformation	**-0.6**	**-0.8**	**-2.7 e**	**-3.4 e**	**-2.7 e**	**-5.7 e**	**-7.1 e**	..	**9.0**	**4.8**
Electricity and heat gen.	-0.4	-0.7	-2.5	-3.1	-2.4	-5.5	-6.8	..	10.8	4.9
Main activity producers (1)	*-0.4*	*-0.7*	*-1.7*	*-3.1*	*-2.4*	*-5.5*	*-6.8*	..	*8.4*	*6.9*
Autoproducers	*-0.0*	*-0.0*	*-0.8*	*-0.0*	*-0.0*	-	-	..	*33.5*	-
Gas works	0.0	0.0	0.1	0.1	0.0	0.0	0.0	..	5.2	-4.5
Coal transformation(2)	-0.2	-0.2	-0.2 e	-0.4 e	-0.3 e	-0.2 e	-0.3 e	..	1.0	0.8
BKB plants	-	-	-	-	-	-	-	..	-	-
Blast furnaces	*-0.1*	*-0.2*	*-0.2 e*	*-0.3 e*	*-0.3 e*	*-0.2 e*	*-0.3 e*	..	*4.2*	*1.7*
Coke ovens	*-0.1*	*-0.0*	*-0.0*	*-0.0*	*-0.0*	*-0.0*	*0.0*	..	*-8.4*	-
Patent fuel plants	-	-	-	-	-	-	-	..	-	-
Other transformation(3)	-	-	-	-	-	-	-	..	-	-
Energy ind. own use	**-0.0**	**-0.0**	**-0.1**	**-0.1**	**-0.2**	**-0.1**	**-0.2**	..	**4.0**	**2.8**
Losses	**-0.0**	**-0.1**	**-0.0**	**-0.1**	**-0.0**	**-0.0**	**-0.0**	..		
Final consumption(4)	**1.0**	**0.8**	**0.9**	**0.9**	**0.9**	**0.6**	**0.5**	..	**-0.7**	**-2.7**
Industry(5)	0.7	0.6	0.7	0.8	0.8	0.6	0.5	..	0.7	-2.1
Iron and steel	*0.2*	*0.2*	*0.1 e*	*0.1 e*	*0.2 e*	*0.1 e*	*0.2 e*	..	*-2.6*	*2.4*
Chemical	-	-	-	-	-	-	-	..	-	-
Non-metallic minerals	*0.2*	*0.2*	*0.2*	*0.2*	*0.2*	*0.1*	*0.0*	..	*-0.0*	*-22.2*
Paper, pulp and print	*0.0*	*0.0*	*0.0*	*0.0*	-	-	-	..	*-0.1*	-
Other industry (6)	*0.3*	*0.2*	*0.5*	*0.4*	*0.4*	*0.4*	*0.3*	..	*2.2*	*-1.7*
Transport(7)	0.2	0.1	-	-	-	-	-	..	-	-
Other	0.2	0.1	0.2	0.1	0.1	0.0	0.0	..	-0.4	-7.9
Comm. and pub. services	*0.0*	*0.0*	*0.0*	*0.0*	*0.0*	*0.0*	*0.0*	..	*-2.7*	*-0.4*
Residential	*0.1*	*0.1*	*0.1*	*0.0*	*0.0*	*0.0*	*0.0*	..	*2.1*	*-9.3*
Other sectors (8)	*0.1*	*0.0*	*0.1*	*0.0*	*0.0*	*0.0*	*0.0*	..	*-2.2*	*-12.0*
Non-energy use	-	-	-	-	-	-	-	..	-	-

(1) Main activity electricity and heat generation includes district heating.

(2) Coal transformation refers to the transformation of coal and peat to secondary and tertiary products (mainly cokes, briquettes, coal tar and off-gases). Gas works and Liquefaction are shown separately to match the *Energy Balances of OECD Countries.*

(3) Other transformation includes Liquefaction and Non-specified transformations.

(4) Final Consumption includes non-energy use and energy use (Industry, Transport and Other).

(5) Please refer to notes and definitions in Part I for detailed explanation.

(6) Other industry includes Non-ferrous metals, Transport equipment, Machinery, Mining and quarrying, Food and tobacco, Wood and wood products, Construction, Textile and leather, and Non-specified industry.

(7) Transport includes Rail and Inland waterways.

(8) Other sectors includes Agriculture/Forestry, Fishing and Non-specified other.

* "Coal" refers to all types of coal, primary (anthracite, coking coal, other bituminous coal, sub-bituminous coal, lignite/brown coal) and derived fuels (including patent fuel, BKB, coke oven coke, gas coke, coal tar, coke oven gas, gas works gas, blast furnace gas and other recovered gases), as well as peat, peat briquettes and oil shale. Quantities have been converted to Mtce using calorific values reported by the respective countries. Please refer to notes and definitions in Part I.

Source: *IEA/OECD Energy Balances of OECD Countries*

CHILE

6. Use of coal for selected end-uses*

(million tonnes)

	1978	1990	2000	2005	2009	2010	2011	Average annual percent change 78-90	90-11
Steam coal	**0.99**	**3.23**	**3.88**	**3.87**	**6.23**	**7.84**	**9.15**	**10.37**	**5.08**
Total electricity and heat	0.42	2.48	3.13	3.20	5.85	7.42	8.56	15.91	6.08
Main activity producers	*0.41*	*1.69*	*3.12*	*3.19*	*5.85*	*7.42*	*8.56*	*12.52*	*8.02*
Autoproducers	*0.01*	*0.79*	*0.01*	*0.01*	-	-	-	*42.76*	-
Patent fuel/BKB plants	-	-	-	-	-	-	-	-	-
Coke ovens/Liquefaction(1)	-	-	-	-	-	-	-	-	-
Blast furnace inputs	-	-	-	-	-	-	-	-	-
Gas manufacture	0.03	-	-	-	-	-	-	-	-
Industry	0.38	0.66	0.72	0.64	0.19	0.50	0.31	4.78	-3.49
Iron and steel	-	-	-	-	-	-	-	-	-
Chemical	-	-	-	-	-	-	-	-	-
Non-metallic minerals	*0.13*	*0.19*	*0.24*	*0.20*	*0.05*	*0.08*	*0.00*	*3.46*	*-22.19*
Paper, pulp and print	*0.00*	*0.01*	*0.04*	-	*0.01*	-	-	*7.32*	-
Other industry	*0.25*	*0.46*	*0.45*	*0.44*	*0.13*	*0.42*	*0.31*	*5.37*	*-1.83*
Other sectors(2)	0.06	0.09	0.02	0.03	0.02	0.01	0.00	3.64	-13.59
Non-energy use	-	-	-	-	-	-	-	-	-
Coking coal	**0.39**	**0.49**	**0.71**	**0.70**	**0.59**	**0.51**	**0.64**	**2.06**	**1.28**
Total electricity and heat	-	-	-	-	-	-	-	-	-
Main activity producers	-	-	-	-	-	-	-	-	-
Autoproducers	-	-	-	-	-	-	-	-	-
Patent fuel/BKB plants	-	-	-	-	-	-	-	-	-
Coke ovens/Liquefaction(1)	0.39	0.49	0.71	0.70	0.59	0.51	0.68	2.06	1.54
Blast furnace inputs	-	-	-	-	-	-	-	-	-
Gas manufacture	-	-	-	-	-	-	-	-	-
Industry	-	-	-	-	-	-	-	-	-
Iron and steel	-	-	-	-	-	-	-	-	-
Chemical	-	-	-	-	-	-	-	-	-
Non-metallic minerals	-	-	-	-	-	-	-	-	-
Paper, pulp and print	-	-	-	-	-	-	-	-	-
Other industry	-	-	-	-	-	-	-	-	-
Other sectors(2)	-	-	-	-	-	-	-	-	-
Non-energy use	-	-	-	-	-	-	-	-	-
Lignite	**0.03**	**-**	**-**	**-**	**-**	**-**	**-**	**-**	**-**
Total electricity and heat	0.03	-	-	-	-	-	-	-	-
Main activity producers	*0.03*	-	-	-	-	-	-	-	-
Autoproducers	-	-	-	-	-	-	-	-	-
Patent fuel/BKB plants	-	-	-	-	-	-	-	-	-
Coke ovens/Liquefaction(1)	-	-	-	-	-	-	-	-	-
Gas manufacture	-	-	-	-	-	-	-	-	-
Industry	-	-	-	-	-	-	-	-	-
Iron and steel	-	-	-	-	-	-	-	-	-
Chemical	-	-	-	-	-	-	-	-	-
Non-metallic minerals	-	-	-	-	-	-	-	-	-
Paper, pulp and print	-	-	-	-	-	-	-	-	-
Other industry	-	-	-	-	-	-	-	-	-
Other sectors(2)	-	-	-	-	-	-	-	-	-
Non-energy use	-	-	-	-	-	-	-	-	-

(1) Liquefaction primarily refers to direct distillation processes. Indirect liquefaction may be reported here or under gas manufacture.

(2) Other sectors are Residential, Commercial and public services, Agriculture/Forestry, Fishing and Non-specified other.

* Coal totals may not match the sum of the individual rows due to statistical differences. Please refer to notes and definitions in Part I.

Source: *IEA/OECD Energy Statistics of OECD Countries*

CHILE

7. Fuel prices to end users*

	1978[(1)]	1990	1995	2000	2005	2009	2010	2011	2012
				(US dollars / tce)					
For electricity generation									
Steam coal	..	..	..	32.41	58.26	95.43	83.62	106.83	96.33
Heavy fuel oil	..	..	..	..	..	..	..	..	..
Natural gas	..	..	..	..	..	..	..	..	..
For industry									
Steam coal	..	..	..	..	..	..	..	..	..
Coking coal	..	..	..	..	99.06	196.62	130.57	269.88	222.89
High sulphur fuel oil	..	..	..	..	..	..	..	..	..
Low sulphur fuel oil	..	..	..	..	..	..	..	..	..
Natural gas	..	..	..	..	..	..	..	..	..
				(1 000 Chilean pesos / unit)[(2)]					
For electricity generation									
Steam coal	..	..	..	17.49	32.62	53.34	42.65	51.66	46.82
Heavy fuel oil	..	..	..	..	..	..	..	..	..
Natural gas	..	..	..	..	..	..	..	..	..
For industry									
Steam coal	..	..	..	..	..	..	..	..	..
Coking coal	..	..	..	..	55.45	109.90	66.59	130.50	108.32
High sulphur fuel oil	..	..	..	..	..	..	..	..	..
Low sulphur fuel oil	..	..	..	..	..	..	..	..	..
Natural gas	..	..	..	..	..	..	..	..	..

(1) Earliest year for which split by coal type is available.

(2) For steam coal, coking coal and heavy fuel oil per metric tonne; for natural gas per 10^7 kilocalories GCV.

* Please refer to notes and definitions in Part I. Steam coal prices relate to historical definitions.

Source: *IEA/OECD Energy Prices & Taxes*

8. Coal and peat trade by type of coal*

(Mtce)

	1978	1990	1995	2000	2005	2009	2010	2011	2012e
Total imports	**0.28**	**1.62**	**2.14**	**4.20**	**3.49**	**4.76**	**5.44**	**7.79**	**7.94**
Bituminous coal[(1)]	-	1.11	1.42	3.50	2.75	4.22	5.00	7.17	6.69
Coking coal	0.18	0.47	0.68	0.68	0.68	0.55	0.44	0.59	0.67
Sub-bituminous coal	-	-	-	-	-	-	-	-	0.59
Lignite	-	-	-	-	-	-	-	-	-
Peat	-	-	-	-	-	-	-	-	-
Coal products[(2)]	0.10	0.04	0.05	0.03	0.06	-	0.00	0.03	-
Total exports	**0.03**	**-**	**0.02**	**0.04**	**0.04**	**-**	**-**	**-**	**-**
Bituminous coal[(1)]	-	-	-	-	-	-	-	-	-
Coking coal	-	-	-	-	-	-	-	-	-
Sub-bituminous coal	-	-	-	-	-	-	-	-	-
Lignite	-	-	-	-	-	-	-	-	-
Peat	-	-	-	-	-	-	-	-	-
Coal products[(2)]	0.03	-	0.02	0.04	0.04	-	-	-	-

(1) Bituminous coal includes anthracite.

(2) Coal products includes products derived from coal, for example: coke, coal tar, briquettes, patent fuels.

* Please refer to notes and definitions in Part I.

Source: *IEA/OECD Energy Balances of OECD Countries, IEA/OECD Energy Statistics of OECD Countries*

CHILE

9. Total coal imports by origin*
(thousand tonnes)

	1978(1)	1985	1990	1995	2000	2005	2010	2011	2012e
Total coal(2)	**185**	**372**	**1660**	**2209**	**4402**	**4010**	**6933**	**9665**	**10139**
Coking coal	**185**	**372**	**492**	**715**	**714**	**692**	**450**	**602**	**685**
Australia	-	-	-	-	-	411	296	360	421
Canada	-	-	-	-	-	281	154	242	264
Czech Republic	-	-	-	-	-	-	-	-	-
Germany	-	-	-	-	-	-	-	-	-
Poland	-	-	-	-	-	-	-	-	-
United Kingdom	-	-	-	-	-	-	-	-	-
United States	-	-	-	-	-	-	-	-	-
Other OECD	-	-	-	-	-	-	-	-	-
China, People's Rep.	-	-	-	-	-	-	-	-	-
Colombia	-	-	-	-	-	-	-	-	-
Indonesia	-	-	-	-	-	-	-	-	-
South Africa	-	-	-	-	-	-	-	-	-
Former Soviet Union(3)	-	-	x	x	x	x	x	x	x
Russian Federation (4)	x	x	-	-	-	-	-	-	-
Other FSU (4)	x	x	-	-	-	-	-	-	-
Venezuela	-	-	-	-	-	-	-	-	-
Vietnam	-	-	-	-	-	-	-	-	-
Non-specified/other	185	372	492	715	714	-	-	-	-
Steam coal	**-**	**-**	**1168**	**1494**	**3688**	**3318**	**6483**	**9063**	**9454**
Australia	-	-	-	-	1330	437	647	698	203
Canada	-	-	-	-	805	260	-	-	-
Czech Republic	-	-	-	-	-	-	-	-	-
Germany	-	-	-	-	-	-	-	-	-
Poland	-	-	-	-	-	-	-	-	-
United Kingdom	-	-	-	-	-	-	-	-	-
United States	-	-	-	-	-	523	1591	3588	3914
Other OECD	-	-	-	-	158	254	57	-	-
China, People's Rep.	-	-	-	-	62	-	-	-	-
Colombia	-	-	-	-	-	-	3427	4184	5100
Indonesia	-	-	-	-	631	1266	746	590	160
South Africa	-	-	-	-	121	-	-	-	-
Former Soviet Union(3)	-	-	x	x	x	x	x	x	x
Russian Federation (4)	x	x	-	-	-	-	-	-	-
Other FSU (4)	x	x	-	-	-	-	-	-	-
Venezuela	-	-	-	-	111	-	-	-	-
Vietnam	-	-	-	-	-	-	-	-	-
Non-specified/other	-	-	1168	1494	470	578	15	3	77
Lignite	**-**	**-**	**-**	**-**	**-**	**-**	**-**	**-**	**-**

(1) Earliest year for which split by coal type is available.

(2) Total coal consists of steam coal, coking coal and lignite.

(3) For years prior to 1990, or if no details for individual Former Soviet Union republics.

(4) Data for former republics are not available separately for some OECD member countries.

* Please refer to notes and definitions in Part I.

Source: IEA/OECD Coal Statistics

CHILE

12. Coal import values by origin*

(average unit value, CIF, USD/tonne)

	1990	1995	2000	2005	2007	2008	2009	2010	2011/12[1]
Coking coal[2]	..	..	..	**107.89**	**115.82**	**229.08**	**220.52**	**198.18**	..
Imports from:									
Australia	..	..	..	107.43	114.51	197.90	218.94	192.86	..
Canada	..	..	..	108.56	118.06	280.70	236.39	208.56	..
Czech Republic	..	..	..	..	..	..	..	..	..
Poland	..	..	..	..	..	..	..	..	..
United States	..	..	..	..	..	903.61	156.11	..	..
China	..	..	..	..	..	..	..	..	..
Colombia	..	..	..	2551.00	..	..	2738.47	..	..
Indonesia	..	..	..	..	..	..	..	..	..
South Africa	..	..	..	..	..	..	..	..	..
Former Soviet Union[3]	..	..	..	..	..	..	..	..	..
Other bituminous coal[4]	..	..	..	**57.96**	**69.83**	**116.10**	**93.65**	**84.48**	..
Imports from:									
Australia	..	..	..	70.59	75.37	133.55	93.89	81.90	..
Canada	..	..	..	63.95	70.19	148.33	97.93	122.44	..
Czech Republic	..	..	..	..	..	..	..	..	..
Poland	..	..	..	..	..	..	..	..	..
United States	..	..	..	47.31	43.23	91.35	55.63	63.59	..
China	..	..	..	..	..	..	..	..	..
Colombia	..	..	..	78.19	84.44	..	92.14	94.19	..
Indonesia	..	..	..	45.50	51.52	96.33	138.21	82.57	..
South Africa	..	..	..	..	..	..	..	..	..
Former Soviet Union[3]	..	..	..	..	..	..	..	..	..

Note: On occasion, shipment of extremely small quantities of high valued coal results in high import costs.

(1) Import data for steam and coking coals are currently unavailable for 2011 and 2012 due to resource constraints.

(2) Weighted average of individual countries using import volumes as weights. Prices exclude intra-EU trade.

(3) Former Soviet Union until 1991, Russian Federation starting in 1992.

(4) Bituminous steam coal only. Weighted average of trades. Prices exclude intra-EU trade.

* Please refer to notes and definitions in Part I.

Source: IEA/OECD Coal Statistics

CZECH REPUBLIC*

Figure 1: TPES by fuel (Mtce)

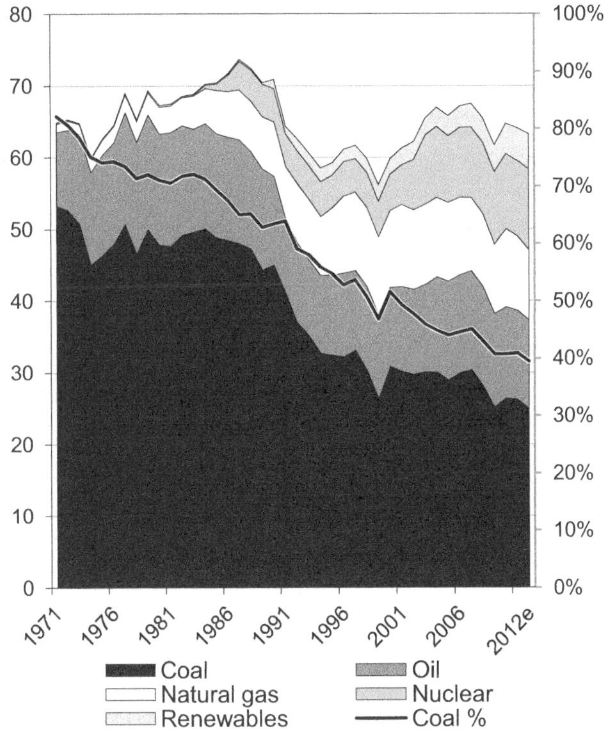

Figure 2: Coal supply indicators (1971 = 100)

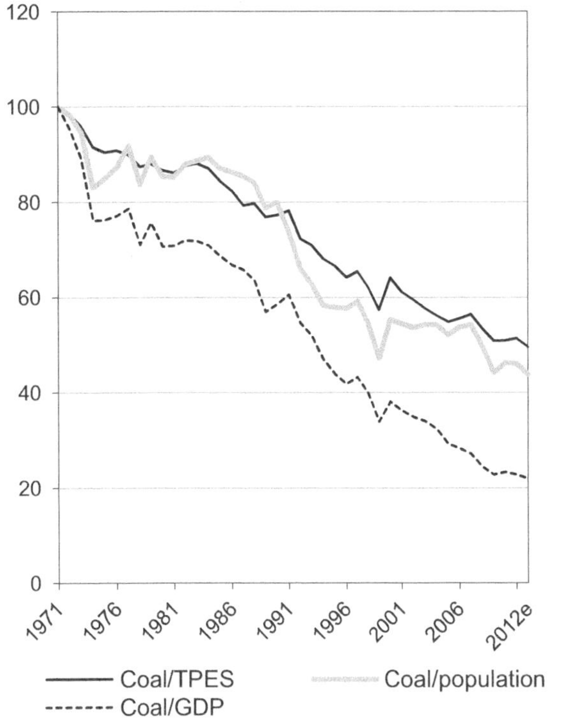

Figure 3: Primary coal supply (Mtce)

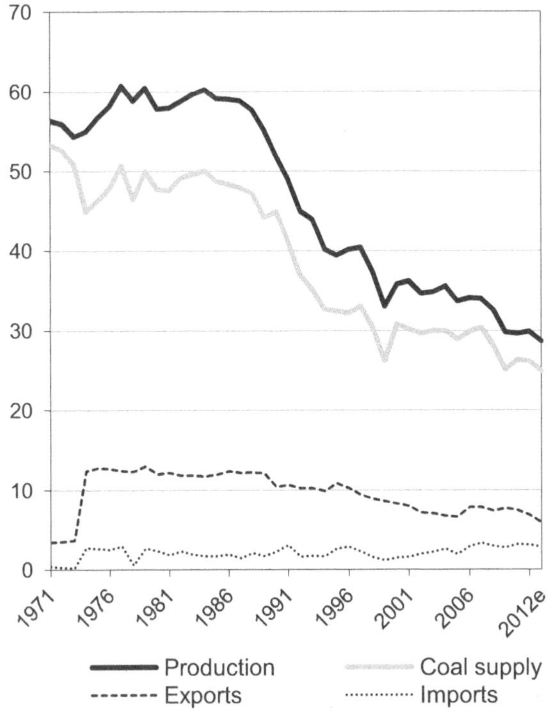

Figure 4: Coal consumption (Mtce)

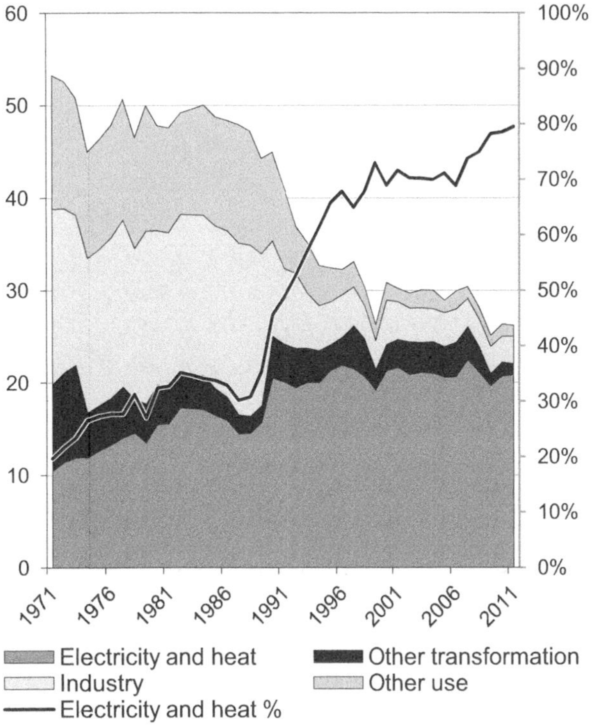

*Please refer to notes and definitions in Part I. Peat is included in coal for all figures.

Source: *IEA/OECD Coal Information, IEA/OECD Energy Balances of OECD Countries, OECD Main Economic Indicators*

CZECH REPUBLIC*

Figure 5: Electricity generation by fuel (TWh)

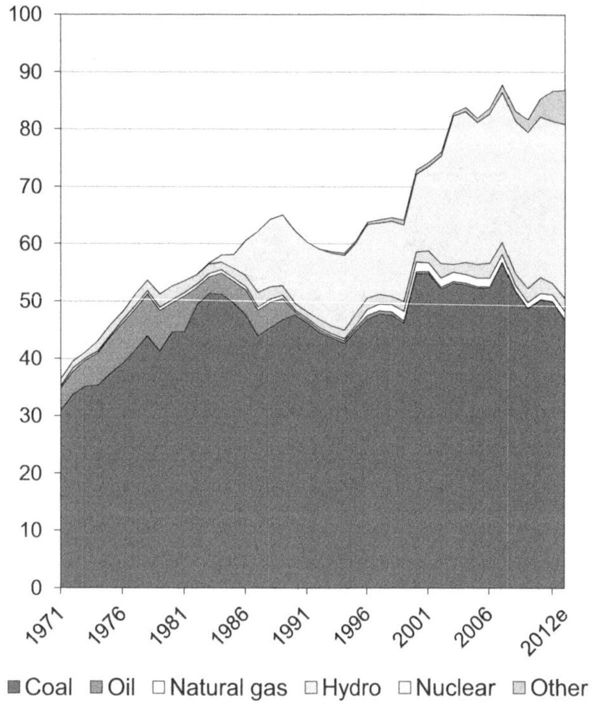

Figure 6: CO_2 emissions by fuel (Mt CO_2)

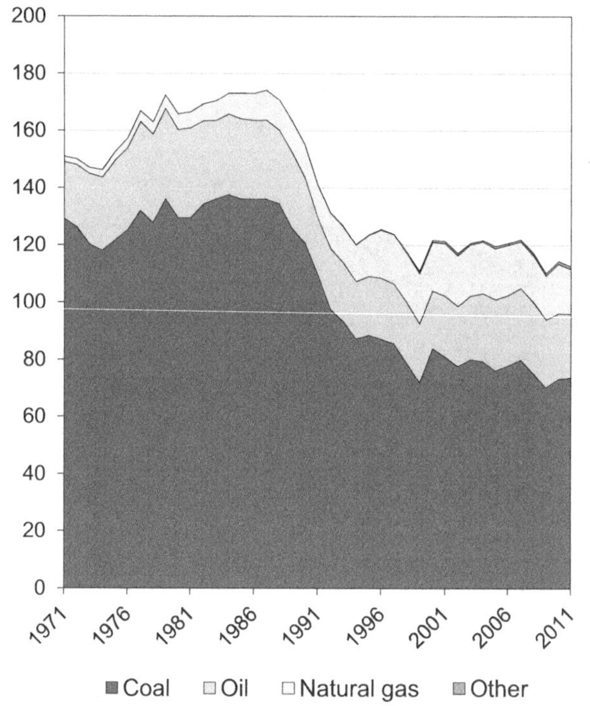

Figure 7: Electricity generation by fuel share

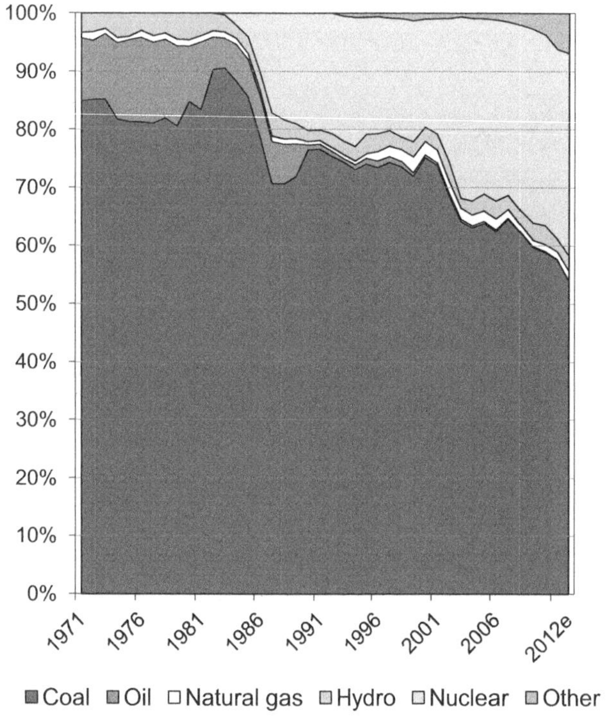

Figure 8: CO_2 emissions by fuel share

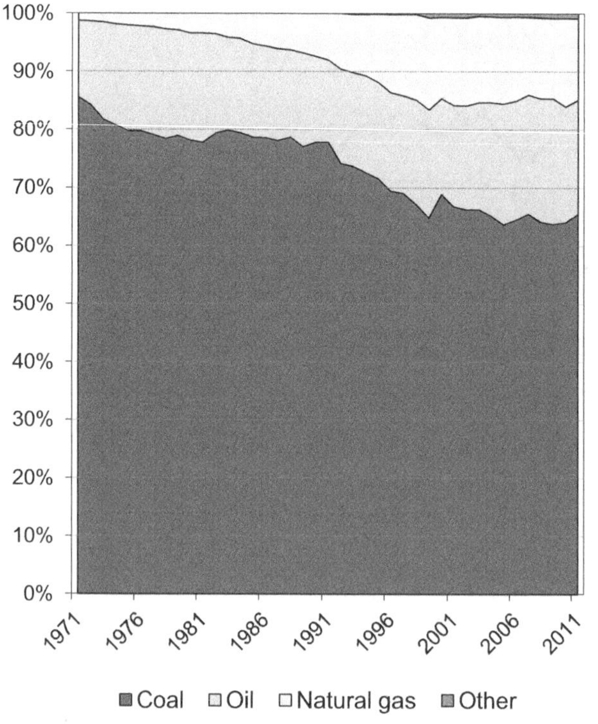

*Please refer to notes and definitions in Part I. Peat is included in coal for all figures.

Source: *IEA/OECD Electricity Information, IEA/OECD CO2 Emissions from Fuel Combustion*

CZECH REPUBLIC

1. Total primary energy supply (TPES) by fuel*

	1973	1980	1990	2000	2010	2011	2012e	Average annual percent change 73-90	90-11
TPES (Mtce)	**64.51**	**67.07**	**70.81**	**58.56**	**62.92**	**62.04**	**61.18**	**0.55**	**-0.63**
Coal and peat	50.83	47.78	44.95	30.83	26.34	26.20	24.91	-0.72	-2.54
Oil	12.37	15.48	12.47	11.03	12.81	12.45	12.38	0.05	-0.01
Natural Gas	1.46	3.70	7.49	10.71	10.85	10.30	9.73	10.11	1.52
Biofuels and waste	-	-	1.15	1.95	3.79	4.01	4.28	-	6.11
Nuclear	-	-	4.69	5.06	10.46	10.56	11.32	-	3.95
Hydro	0.13	0.29	0.14	0.22	0.34	0.24	0.28	0.42	2.53
Geothermal	-	-	-	-	-	-	-	-	-
Solar, wind, tide	-	-	-	-	0.13	0.33	0.34	-	-
Net electricity trade(1)	-0.27	-0.19	-0.09	-1.23	-1.84	-2.09	-2.10	-6.60	16.48
Heat(2)	-	-	-	-0.00	0.04	0.05	0.05	-	-

(1) Net trade between OECD and non-OECD countries.

(2) Ambient heat from heat pumps used in the transformation sector.

Source: *IEA/OECD Energy Balances of OECD Countries*

2. Energy supply, GDP and population*

	1973	1980	1990	2000	2010	2011	2012e	Average annual percent change 73-90	90-11
GDP (2005 bil. USD)	75 e	89 e	101	106	148	151	149	1.77	1.94
TPES/GDP(1)	0.86 e	0.75 e	0.70	0.55	0.42	0.41	0.41	-1.20	-2.52
Population (millions)	9.9	10.3	10.4	10.3	10.5	10.5	10.5	0.26	0.06
TPES/population(1)	6.50	6.49	6.83	5.70	5.98	5.91	5.82	0.29	-0.69
TPES/GDP(2)	174.3 e	152.8 e	142.0	111.5	85.8	83.1	83.0	-1.20	-2.52
Coal and peat supply/GDP(2)	305.0 e	241.8 e	200.2	130.3	79.8	78.0	75.1	-2.45	-4.39
Elec. consumption/GDP(2)	92.6 e	100.2 e	112.2	109.1	90.6	88.2	..	1.14	-1.14
Elec. generation (TWh)	41	53	62	73	85	87	87	2.46	1.59
Industrial production(2)	..	..	92.6	77.3	110.4	116.9	116.0	..	1.12

(1) TPES/GDP in units of tce/2005 thousand USD, TPES/population in units of tce/capita.

(2) As index, 2005 (not shown) = 100

Source: *IEA/OECD Energy Balances of OECD Countries* , *OECD Main Economic Indicators*

3. Coal and peat production by type*

	1978(1)	1990	2000	2005	2010	2011	2012e	Average annual percent change 78-90	90-11
Mtce:									
Coking coal	14.81	13.78	7.92	6.92	5.87	5.06	4.93	-0.60	-4.66
Steam coal	6.57	4.98	5.54	5.09	4.71	5.20	5.46	-2.28	0.20
Lignite	37.25	33.12	22.32	21.66	19.03	19.59	18.29	-0.98	-2.47
Peat	0.17	-	-	-	-	-	-	-	-
Mt:									
Coking coal	18.55	14.38	8.14	7.14	6.02	5.18	5.05	-2.10	-4.74
Steam coal	10.60	8.03	6.72	6.12	5.41	6.08	6.39	-2.28	-1.32
Lignite	88.84	78.98	50.31	48.77	43.77	46.64	43.53	-0.98	-2.48
Peat	0.60	-	-	-	-	-	-	-	-

(1) Earliest year for which split by coal type is available.

* Please refer to notes and definitions in Part I.

Source: *IEA/OECD Energy Balances of OECD Countries, IEA/OECD Energy Statistics of OECD Countries*

CZECH REPUBLIC

4. Final consumption of energy by fuel*

(Mtce)

	1973	1980	1990	2000	2009	2010	2011	Average annual percent change 73-90	90-11
Total final consumption(1)	**44.78**	**49.51**	**48.99**	**36.70**	**36.89**	**38.62**	**37.09**	**0.53**	**-1.32**
Coal and peat	28.93	28.04	19.92	6.65	4.21	4.14	4.17	-2.17	-7.17
Oil	11.07	13.18	11.71	10.42	12.52	12.24	11.90	0.33	0.08
Natural Gas	1.16	1.68	6.06	8.45	7.80	9.07	8.11	10.21	1.40
Biofuels and wastes	-	-	1.15	1.36	2.64	2.91	2.93	-	4.54
Geothermal	-	-	-	-	-	-	-	-	-
Solar, wind, tide	-	-	-	-	0.01	0.01	0.02	-	-
Electricity	3.62	4.65	5.92	6.07	6.75	7.03	6.96	2.93	0.77
Heat	-	1.95	4.23	3.75	2.96	3.21	3.00	-	-1.62
of which:									
Total industry	**25.18**	**26.80**	**22.73**	**13.03**	**10.59**	**11.20**	**11.09**	**-0.60**	**-3.36**
Coal and peat	16.34	16.70	10.29	4.75	3.04	2.84	3.00	-2.68	-5.70
Oil	5.88	6.96	4.13	0.93	0.61	0.67	0.61	-2.06	-8.74
Natural Gas	0.66	0.40	3.45	3.72	2.74	3.28	3.01	10.26	-0.65
Biofuels and wastes	-	-	-	0.20	0.76	0.77	0.75	-	-
Geothermal	-	-	-	-	-	-	-	-	-
Solar, wind, tide	-	-	-	-	-	-	-	-	-
Electricity	2.31	2.73	3.31	2.33	2.68	2.78	2.84	2.14	-0.72
Heat	-	-	1.54	1.11	0.76	0.86	0.88	-	-2.61
Total transport	**3.10**	**3.28**	**3.70**	**6.05**	**8.89**	**8.46**	**8.50**	**1.05**	**4.04**
Coal and peat	0.17	0.14	-	-	0.00	0.00	0.00	-	-
Oil	2.70	2.86	3.31	5.63	8.25	7.75	7.69	1.22	4.09
Natural Gas	-	-	-	0.04	0.10	0.11	0.10	-	-
Biofuels and wastes	-	-	-	0.09	0.28	0.33	0.43	-	-
Electricity	0.23	0.28	0.39	0.29	0.25	0.27	0.28	3.08	-1.61
Residential	**10.05**	**10.65**	**12.49**	**8.62**	**8.60**	**9.44**	**8.41**	**1.28**	**-1.87**
Coal and peat	9.29 e	8.04 e	6.94 e	1.12	0.77	0.81	0.72	-1.70	-10.21
Oil	0.07	0.12	0.12	0.11	0.01	0.01	0.01	3.64	-13.24
Natural Gas	0.22	0.55	1.31	2.93	2.94	3.40	2.86	11.22	3.78
Biofuels and wastes	-	-	1.15	1.04	1.48	1.65	1.58	-	1.50
Geothermal	-	-	-	-	-	-	-	-	-
Solar, wind, tide	-	-	-	-	0.01	0.01	0.01	-	-
Electricity	0.48	0.76	1.18	1.70	1.80	1.85	1.74	5.42	1.87
Heat	-	1.18	1.78	1.73	1.59	1.71	1.48	-	-0.88
Comm & public services	**0.89**	**2.26**	**4.32**	**4.26**	**4.20**	**4.46**	**4.32**	**9.76**	**-0.01**
Coal and peat	0.39 e	0.33 e	2.09 e	0.34	0.03	0.04	0.04	10.30	-17.00
Oil	-	0.13	0.07	0.01	0.02	0.02	0.02	-	-6.25
Natural Gas	0.25	0.63	1.04	1.58	1.79	1.97	1.84	8.86	2.73
Biofuels and waste	-	-	-	0.04	0.08	0.10	0.10	-	-
Geothermal	-	-	-	-	-	-	-	-	-
Solar, wind, tide	-	-	-	-	0.00	0.00	0.00	-	-
Electricity	0.25	0.41	0.45	1.42	1.68	1.72	1.70	3.56	6.55
Heat	-	0.77	0.67	0.88	0.60	0.62	0.62	-	-0.41
Non-energy use	**1.64**	**1.78**	**2.47**	**3.05**	**3.56**	**3.86**	**3.62**	**2.45**	**1.83**
Coal and peat	-	-	-	-	0.34	0.42	0.39	-	-
Oil	1.64	1.78	2.47	3.05	3.13	3.30	3.08	2.45	1.06
Natural Gas	-	-	-	-	0.08	0.14	0.14	-	-

(1) Total Final Consumption (TFC) excludes use in transformation processes and energy industry own use.

* Please refer to notes and definitions in Part I. Agriculture, Forestry, Fishing and "Non-specified other" sectors are not reported here individually, but are accounted for in Total final consumption.

Source: *IEA/OECD Energy Balances of OECD Countries*

CZECH REPUBLIC

5. Coal balance*

(Mtce)

	1973	1980	1990	2000	2005	2010	2011	2012e	Average annual percent change 73-90	90-11
Production	54.3 e	57.8	51.9	35.8	33.7	29.6	29.9	28.7	-0.3	-2.6
Imports	0.2 e	2.3	2.2	1.5	1.9	3.1	3.1	2.8	15.0	1.6
Exports	-3.7 e	-12.0	-10.4	-8.3	-6.6	-7.4	-6.8	-5.9	6.3	-2.0
Stock changes	-0.0 e	-0.3	1.2	1.8	-0.1	1.0	0.0	-0.7		
Primary supply	**50.8**	**47.8**	**45.0**	**30.8**	**28.9**	**26.3**	**26.2**	**24.9**	**-0.7**	**-2.5**
Statistical differences	**1.8**	**2.0**	**-1.3**	**-0.6**	**-1.1**	**0.3**	**0.5**	..		
Total transformation	**-22.7 e**	**-20.7 e**	**-22.6 e**	**-23.1**	**-22.1**	**-22.0**	**-22.0**	..	**-0.0**	**-0.1**
Electricity and heat gen.	-11.9 e	-15.5 e	-20.5 e	-21.3	-20.6	-20.7	-20.8	..	3.3	0.1
Main activity producers (1)	*-11.9 e*	*-15.5 e*	*-17.4 e*	*-18.9*	*-18.3*	*-18.4*	*-18.9*	..	*2.3*	*0.4*
Autoproducers	-	-	*-3.1 e*	*-2.4*	*-2.2*	*-2.3*	*-1.9*	..	-	*-2.3*
Gas works	-0.9 e	-0.9	-0.4	-0.2	-0.2	-0.2	-0.1	..	-5.1	-4.0
Coal transformation(2)	-10.0 e	-4.3 e	-1.8 e	-1.7	-1.3	-1.1	-1.0	..	-9.7	-2.5
BKB plants	*0.0 e*	*0.0*	*-0.0*	*0.0*	*-0.0*	*-0.0*	-	..	-	-
Blast furnaces	*-2.6 e*	*-2.6 e*	*-1.7 e*	*-1.3*	*-1.2*	*-1.1*	*-1.0*	..	*-2.3*	*-2.4*
Coke ovens	*-7.4 e*	*-1.8 e*	*-0.0 e*	*-0.4*	*-0.1*	*0.0*	*-0.0*	..	*-26.5*	*-11.1*
Patent fuel plants	-	-	-	-	-	-	-	..	-	-
Other transformation(3)	-	-	-	-	-	-	-	..	-	-
Energy ind. own use	**-1.0 e**	**-1.0 e**	**-1.1 e**	**-0.5**	**-0.6**	**-0.5**	**-0.4**	..	**0.8**	**-4.2**
Losses	**-0.1 e**	**-0.1 e**	**-0.1 e**	**-0.0**	**-0.1**	**-0.1**	**-0.1**	..		
Final consumption(4)	**28.9**	**28.0**	**19.9**	**6.7**	**5.0**	**4.1**	**4.2**	..	**-2.2**	**-7.2**
Industry(5)	16.3	16.7	10.3	4.7	3.6	2.8	3.0	..	-2.7	-5.7
Iron and steel	*3.4 e*	*3.9 e*	*5.0 e*	*2.0*	*1.8*	*1.2*	*1.5*	..	*2.3*	*-5.5*
Chemical	*1.1 e*	*0.9 e*	*0.5*	*1.2*	*1.3*	*1.2*	*1.0*	..	*-4.6*	*3.4*
Non-metallic minerals	*0.5 e*	*0.7 e*	*0.5*	*0.3*	*0.2*	*0.2*	*0.2*	..	*-0.3*	*-3.5*
Paper, pulp and print	*1.3 e*	*1.1 e*	*0.3*	*0.1*	*0.1*	*0.1*	*0.1*	..	*-7.7*	*-5.6*
Other industry (6)	*9.9 e*	*10.1 e*	*3.9 e*	*1.1*	*0.3*	*0.1*	*0.1*	..	*-5.3*	*-15.8*
Transport(7)	0.2	0.1	-	-	0.0	0.0	0.0	..	-	-
Other	12.4	11.2	9.6	1.9	0.9	0.9	0.8	..	-1.5	-11.3
Comm. and pub. services	*0.4 e*	*0.3 e*	*2.1 e*	*0.3*	*0.1*	*0.0*	*0.0*	..	*10.3*	*-17.0*
Residential	*9.3 e*	*8.0 e*	*6.9 e*	*1.1*	*0.8*	*0.8*	*0.7*	..	*-1.7*	*-10.2*
Other sectors (8)	*2.7 e*	*2.8 e*	*0.6 e*	*0.4*	*0.0*	*0.0*	*0.0*	..	*-8.5*	*-15.3*
Non-energy use	-	-	-	-	0.4	0.4	0.4	..	-	-

(1) Main activity electricity and heat generation includes district heating.

(2) Coal transformation refers to the transformation of coal and peat to secondary and tertiary products (mainly cokes, briquettes, coal tar and off-gases). Gas works and Liquefaction are shown separately to match the *Energy Balances of OECD Countries.*

(3) Other transformation includes Liquefaction and Non-specified transformations.

(4) Final Consumption includes non-energy use and energy use (Industry, Transport and Other).

(5) Please refer to notes and definitions in Part I for detailed explanation.

(6) Other industry includes Non-ferrous metals, Transport equipment, Machinery, Mining and quarrying, Food and tobacco, Wood and wood products, Construction, Textile and leather, and Non-specified industry.

(7) Transport includes Rail and Inland waterways.

(8) Other sectors includes Agriculture/Forestry, Fishing and Non-specified other.

* "Coal" refers to all types of coal, primary (anthracite, coking coal, other bituminous coal, sub-bituminous coal, lignite/brown coal) and derived fuels (including patent fuel, BKB, coke oven coke, gas coke, coal tar, coke oven gas, gas works gas, blast furnace gas and other recovered gases), as well as peat, peat briquettes and oil shale. Quantities have been converted to Mtce using calorific values reported by the respective countries. Please refer to notes and definitions in Part I.

Source: *IEA/OECD Energy Balances of OECD Countries*

CZECH REPUBLIC

6. Use of coal for selected end-uses*

(million tonnes)

	1978	1990	2000	2005	2009	2010	2011	Average annual percent change 78-90	90-11
Steam coal	**9.81**	**10.12**	**5.75**	**4.89**	**3.83**	**4.44**	**4.67**	**0.26**	**-3.61**
Total electricity and heat	-	4.94	3.90	3.27	3.54	3.88	3.72	-	-1.34
Main activity producers	-	*3.75*	*3.79*	*3.23*	*3.16*	*3.65*	*3.50*	-	*-0.33*
Autoproducers	-	*1.19*	*0.11*	*0.04*	*0.38*	*0.23*	*0.22*	-	*-7.66*
Patent fuel/BKB plants	-	-	-	-	-	-	-	-	-
Coke ovens/Liquefaction[1]	-	-	-	-	-	-	-	-	-
Blast furnace inputs	-	-	-	-	-	-	-	-	-
Gas manufacture	-	0.03	-	-	-	-	-	-	-
Industry	7.85	3.05	1.46	1.31	0.96	0.68	0.81	-7.58	-6.10
Iron and steel	*1.10 e*	*0.85*	*0.79*	*0.83*	*0.59*	*0.29*	*0.40*	*-2.11*	*-3.49*
Chemical	*0.05 e*	*0.04*	*0.15*	*0.20*	*0.16*	*0.18*	*0.17*	*-2.32*	*7.04*
Non-metallic minerals	*0.60 e*	*0.47*	*0.25*	*0.21*	*0.16*	*0.18*	*0.21*	*-2.01*	*-3.78*
Paper, pulp and print	*0.04 e*	*0.03*	*0.01*	*0.02*	*0.02*	*0.02*	*0.01*	*-2.51*	*-3.95*
Other industry	*6.05 e*	*1.66*	*0.26*	*0.04*	*0.02*	*0.02*	*0.02*	*-10.23*	*-18.97*
Other sectors[2]	1.96	2.11	0.75	0.21	0.08	0.11	0.11	0.60	-13.19
Non-energy use	-	-	-	-	-	-	-	-	-
Coking coal	**12.57**	**9.94**	**4.97**	**4.33**	**2.93**	**3.37**	**3.35**	**-1.93**	**-5.05**
Total electricity and heat	-	-	-	-	-	-	-	-	-
Main activity producers	-	-	-	-	-	-	-	-	-
Autoproducers	-	-	-	-	-	-	-	-	-
Patent fuel/BKB plants	-	-	-	-	-	-	-	-	-
Coke ovens/Liquefaction[1]	12.57 e	8.54 e	4.56	4.34	2.95	3.24	3.29	-3.17	-4.44
Blast furnace inputs	-	-	-	-	-	-	-	-	-
Gas manufacture	-	-	-	-	-	-	-	-	-
Industry	-	1.40	0.04	-	-	-	-	-	-
Iron and steel	-	*1.40 e*	*0.04*	-	-	-	-	-	-
Chemical	-	-	*0.00*	-	-	-	-	-	-
Non-metallic minerals	-	-	-	-	-	-	-	-	-
Paper, pulp and print	-	-	-	-	-	-	-	-	-
Other industry	-	-	-	-	-	-	-	-	-
Other sectors[2]	-	-	-	-	-	-	-	-	-
Non-energy use	-	-	-	-	-	-	-	-	-
Lignite	**81.32**	**71.77**	**50.37**	**47.62**	**44.24**	**43.73**	**44.10**	**-1.04**	**-2.29**
Total electricity and heat	37.97 e	43.71	41.15	40.95	37.56	38.88	39.26	1.18	-0.51
Main activity producers	*37.97 e*	*38.58*	*37.36*	*37.13*	*34.23*	*35.53*	*36.28*	*0.13*	*-0.29*
Autoproducers	-	*5.13*	*3.79*	*3.82*	*3.33*	*3.35*	*2.98*	-	*-2.55*
Patent fuel/BKB plants	1.96	1.94	0.44	0.54	0.34	0.29	-	-0.10	-
Coke ovens/Liquefaction[1]	-	-	-	-	-	-	-	-	-
Gas manufacture	6.35	2.67	1.35	1.49	1.51	1.56	1.65	-6.97	-2.27
Industry	16.58	7.12	4.94	3.32	2.85	2.85	2.57	-6.80	-4.73
Iron and steel	*0.54*	*0.35*	*0.21*	*0.09*	*0.05*	*0.06*	*0.05*	*-3.52*	*-8.70*
Chemical	*2.49*	*1.20*	*2.41*	*2.60*	*2.35*	*2.34*	*2.10*	*-5.93*	*2.71*
Non-metallic minerals	*0.36*	*0.48*	*0.20*	*0.03*	*0.04*	*0.04*	*0.04*	*2.43*	*-10.91*
Paper, pulp and print	*2.95*	*0.79*	*0.23*	*0.21*	*0.22*	*0.24*	*0.22*	*-10.39*	*-5.97*
Other industry	*10.24*	*4.31*	*1.90*	*0.40*	*0.19*	*0.17*	*0.16*	*-6.96*	*-14.44*
Other sectors[2]	18.87	16.19	2.47	1.31	1.20	1.30	1.31	-1.27	-11.28
Non-energy use	-	-	-	-	-	-	-	-	-

(1) Liquefaction primarily refers to direct distillation processes. Indirect liquefaction may be reported here or under gas manufacture.

(2) Other sectors are Residential, Commercial and public services, Agriculture/Forestry, Fishing and Non-specified other.

* Coal totals may not match the sum of the individual rows due to statistical differences. Please refer to notes and definitions in Part I.

Source: *IEA/OECD Energy Statistics of OECD Countries*

CZECH REPUBLIC

7. Fuel prices to end users*

	1978(1)	1990	1995	2000	2005	2009	2010	2011	2012
			(US dollars / tce)						
For electricity generation									
Steam coal	10.29	20.49	26.48	21.88	c	c	c	c	c
Heavy fuel oil	52.02	111.20	69.18	81.86	152.18	c	c	c	c
Natural gas	..	94.07	122.50	113.74	..	..	..	..	..
For industry									
Steam coal	..	24.44	31.30	26.96	c	c	c	c	c
Coking coal	..	52.62	64.92	50.73	c	c	c	c	c
High sulphur fuel oil	52.02	111.20	69.18	81.86	152.18	c	c	c	c
Low sulphur fuel oil	..	..	..	..	180.93	260.90	309.60	388.75	398.01
Natural gas	..	94.07	122.50	114.82	227.47	410.76	412.63	459.68	441.61
			(1 000 Czech crowns / unit)(2)						
For electricity generation									
Steam coal	0.05	0.13	0.26	0.31	c	c	c	c	c
Heavy fuel oil	1.02	2.72	2.50	4.31	4.96	c	c	c	c
Natural gas	..	2.17	4.18	5.65	..	..	..	..	..
For industry									
Steam coal	..	0.24	0.45	0.56	c	c	c	c	c
Coking coal	..	0.88	1.60	1.82	c	c	c	c	c
High sulphur fuel oil	1.02	2.72	2.50	4.31	4.96	c	c	c	c
Low sulphur fuel oil	..	..	..	..	5.90	6.77	8.04	9.35	10.59
Natural gas	..	2.17	4.18	5.70	7.01	10.06	10.12	10.43	11.09

(1) Earliest year for which split by coal type is available.

(2) For steam coal, coking coal and heavy fuel oil per metric tonne; for natural gas per 10^7 kilocalories GCV.

* Please refer to notes and definitions in Part I. Steam coal prices relate to historical definitions.

Source: *IEA/OECD Energy Prices & Taxes*

8. Coal and peat trade by type of coal*

(Mtce)

	1978	1990	1995	2000	2005	2009	2010	2011	2012e
Total imports	**0.52**	**2.25**	**2.61**	**1.48**	**1.91**	**2.74**	**3.14**	**3.12**	**2.79**
Bituminous coal(1)	0.25	2.09	1.46	0.68	0.66	0.90	0.92	0.98	0.70
Coking coal	-	-	0.86	0.22	0.42	0.85	0.92	1.12	1.01
Sub-bituminous coal	-	-	-	-	-	-	-	-	-
Lignite	-	-	-	0.00	0.00	0.05	0.03	0.05	0.22
Peat	-	-	-	-	-	-	-	-	-
Coal products(2)	0.27	0.15	0.29	0.59	0.83	0.94	1.28	0.97	0.87
Total exports	**12.32**	**10.38**	**10.84**	**8.26**	**6.60**	**7.69**	**7.43**	**6.81**	**5.87**
Bituminous coal(1)	0.37	0.47	1.35	2.51	2.06	2.48	2.50	2.82	1.86
Coking coal	5.56	4.22	5.05	3.42	2.99	3.90	3.43	2.80	2.78
Sub-bituminous coal	-	-	-	-	-	-	-	-	-
Lignite	3.31	4.03	2.82	1.29	0.54	0.64	0.59	0.67	0.76
Peat	-	-	-	-	-	-	-	-	-
Coal products(2)	3.09	1.66	1.62	1.03	1.02	0.66	0.92	0.53	0.47

(1) Bituminous coal includes anthracite.

(2) Coal products includes products derived from coal, for example: coke, coal tar, briquettes, patent fuels.

* Please refer to notes and definitions in Part I.

Source: *IEA/OECD Energy Balances of OECD Countries, IEA/OECD Energy Statistics of OECD Countries*

CZECH REPUBLIC

9. Total coal imports by origin*

(thousand tonnes)

	1978[1]	1985	1990	1995	2000	2005	2010	2011	2012e
Total coal[2]	**274 e**	**1378 e**	**2282**	**2690 e**	**1095**	**1264**	**2067**	**2386**	**2201**
Coking coal	**-**	**-**	**-**	**890 e**	**217**	**492**	**909**	**1181**	**1060**
Australia	-	-	-	-	-	-	10	-	-
Canada	-	-	-	-	-	-	-	-	-
Czech Republic	-	-	-	-	-	-	-	-	-
Germany	-	-	-	-	-	-	-	-	-
Poland	-	-	-	890	217	492	803	1100	919
United Kingdom	-	-	-	-	-	-	-	-	-
United States	-	-	-	-	-	-	94	81	125
Other OECD	-	-	-	-	-	-	-	-	-
China, People's Rep.	-	-	-	-	-	-	-	-	-
Colombia	-	-	-	-	-	-	-	-	16
Indonesia	-	-	-	-	-	-	-	-	-
South Africa	-	-	-	-	-	-	-	-	-
Former Soviet Union[3]	-	-	x	x	x	x	x	x	x
Russian Federation[4]	*x*	*x*	-	-	-	-	*2*	-	-
Other FSU[4]	*x*	*x*	-	-	-	-	-	-	-
Venezuela	-	-	-	-	-	-	-	-	-
Vietnam	-	-	-	-	-	-	-	-	-
Non-specified/other	-	-	-	-	-	-	-	-	-
Steam coal	**274 e**	**1378 e**	**2282**	**1800**	**877**	**771**	**1100**	**1129**	**791**
Australia	-	-	-	-	-	-	-	-	-
Canada	-	-	-	-	-	-	-	-	-
Czech Republic	-	-	-	-	-	-	-	-	-
Germany	-	-	-	-	-	2	2	-	5
Poland	274 e	1378 e	2282	1800	869	733	699	708	547
United Kingdom	-	-	-	-	-	-	-	-	-
United States	-	-	-	-	-	-	-	-	-
Other OECD	-	-	-	-	-	1	8	-	5
China, People's Rep.	-	-	-	-	-	-	-	-	-
Colombia	-	-	-	-	-	-	-	-	-
Indonesia	-	-	-	-	-	-	-	-	-
South Africa	-	-	-	-	-	-	-	-	-
Former Soviet Union[3]	-	-	x	x	x	x	x	x	x
Russian Federation[4]	*x*	*x*	-	-	*6*	*32*	*390*	*421*	*152*
Other FSU[4]	*x*	*x*	-	-	*2*	*3*	*1*	-	*82*
Venezuela	-	-	-	-	-	-	-	-	-
Vietnam	-	-	-	-	-	-	-	-	-
Non-specified/other	-	-	-	-	-	-	-	-	-
Lignite	**-**	**-**	**-**	**-**	**1**	**1**	**58**	**76**	**350**

(1) Earliest year for which split by coal type is available.

(2) Total coal consists of steam coal, coking coal and lignite.

(3) For years prior to 1990, or if no details for individual Former Soviet Union republics.

(4) Data for former republics are not available separately for some OECD member countries.

* Please refer to notes and definitions in Part I.

Source: IEA/OECD Coal Statistics

CZECH REPUBLIC

10. Coking coal exports by destination*

(thousand tonnes)

	1978(1)	1985	1990	1995	2000	2005	2010	2011	2012e
World	**5934 e**	**5086**	**4504**	**5180 e**	**3427**	**3210**	**3499**	**2972**	**2946**
Total OECD	**5934 e**	**4874 e**	**4504 e**	**5041**	**3427**	**3103**	**3388**	**2872**	**2887**
Australia	-	-	-	-	-	-	-	-	-
Austria	600 e	652 e	785 e	1650	1244	994	945	837	850
Belgium	59 e	-	-	-	-	-	-	-	-
Canada	-	-	-	-	-	-	-	-	-
Chile	-	-	-	-	-	-	-	-	-
Czech Republic	-	-	-	-	-	-	-	-	-
Denmark	-	-	-	-	-	-	-	-	-
Estonia	-	-	-	-	-	-	-	-	-
Finland	-	-	-	-	-	-	-	-	-
France	-	-	-	-	-	-	-	-	-
Germany	-	-	28 e	50	-	-	-	-	-
Greece	-	-	-	1	-	-	-	-	-
Hungary	-	-	-	850	744	230	383	410	319
Iceland	-	-	-	-	-	-	-	-	-
Ireland	-	-	-	-	-	-	-	-	-
Israel	-	-	-	-	-	-	-	-	-
Italy	-	-	-	-	-	-	-	-	-
Japan	-	-	-	-	-	-	-	-	-
Korea	-	-	-	-	-	-	-	-	-
Luxembourg	-	-	-	-	-	-	-	-	-
Mexico	-	-	-	-	-	-	-	-	-
Netherlands	69 e	6 e	10 e	-	-	-	-	-	-
New Zealand	-	-	-	-	-	-	-	-	-
Norway	-	-	-	-	-	-	-	-	-
Poland	-	-	-	896	538	592	727	675	514
Portugal	-	-	-	-	-	-	-	-	-
Slovak Republic	5126 e	4183 e	3681 e	1594	901	1287	1333	950	1204
Slovenia	-	-	-	-	-	-	-	-	-
Spain	-	-	-	-	-	-	-	-	-
Sweden	80 e	30 e	-	-	-	-	-	-	-
Switzerland	-	-	-	-	-	-	-	-	-
Turkey	-	-	-	-	-	-	-	-	-
United Kingdom	-	3 e	-	-	-	-	-	-	-
United States	-	-	-	-	-	-	-	-	-
Total non-OECD	**-**	**-**	**-**	**61**	**-**	**107**	**111**	**100**	**59**
Brazil	-	-	-	-	-	-	-	-	-
China(2)	-	-	-	-	-	-	-	-	-
Chinese Taipei	-	-	-	-	-	-	-	-	-
Egypt	-	-	-	-	-	-	-	-	-
India	-	-	-	-	-	-	-	-	-
Romania	-	-	-	61	-	-	-	-	-
Oth. Africa & Mid. East	-	-	-	-	-	-	-	-	-
Oth. non-OECD Americas	-	-	-	-	-	-	-	-	-
Other Asia & Oceania	-	-	-	-	-	-	-	-	-
Other non-OECD Europe and Eurasia	-	-	-	-	-	107	111	100	59
Non-specified/Other	**-**	**212 e**	**-**	**78**	**-**	**-**	**-**	**-**	**-**

(1) Earliest year for which split by coal type is available.
(2) China includes the People's Republic of China and Hong Kong, China.
* Please refer to notes and definitions in Part I.
Source: IEA/OECD Coal Statistics

CZECH REPUBLIC

11. Steam coal exports by destination*

(thousand tonnes)

	1978[1]	1985	1990	1995	2000	2005	2010	2011	2012e
World	**394 e**	**319**	**498**	**1835**	**2459**	**2051**	**2773**	**3307**	**2185**
Total OECD	**394 e**	**284 e**	**498**	**1000**	**2442**	**2042**	**2772**	**3307**	**2185**
Australia	-	-	-	-	-	-	-	-	-
Austria	-	-	-	458	331	981	1000	1335	727
Belgium	-	-	-	-	-	-	-	-	-
Canada	-	-	-	-	-	-	-	-	-
Chile	-	-	-	-	-	-	-	-	-
Czech Republic	-	-	-	-	-	-	-	-	-
Denmark	6 e	38 e	40	-	-	-	-	-	-
Estonia	-	-	-	-	-	-	-	-	-
Finland	-	-	-	-	-	-	-	-	-
France	-	-	-	-	-	-	-	-	-
Germany	100 e	89 e	220	-	1033	526	27	-	17
Greece	1 e	2 e	-	1	-	-	-	-	-
Hungary	-	-	-	59	43	21	103	150	15
Iceland	-	-	-	-	-	-	-	-	-
Ireland	-	2 e	-	-	-	-	-	-	-
Israel	-	-	-	-	-	-	-	-	-
Italy	-	-	-	-	-	-	-	-	-
Japan	-	-	-	-	-	-	-	-	-
Korea	-	-	-	-	-	-	-	-	-
Luxembourg	-	-	-	-	-	-	-	-	-
Mexico	-	-	-	-	-	-	-	-	-
Netherlands	36 e	64 e	-	-	-	-	-	-	-
New Zealand	-	-	-	-	-	-	-	-	-
Norway	-	-	32	-	-	-	-	-	-
Poland	-	-	-	304	102	45	1257	1357	994
Portugal	-	-	-	-	-	-	-	-	-
Slovak Republic	237 e	79 e	198	178	931	469	315	400	317
Slovenia	-	-	-	-	-	-	1	-	2
Spain	-	-	-	-	-	-	-	-	-
Sweden	6 e	-	6	-	2	-	69	65	113
Switzerland	8 e	-	-	-	-	-	-	-	-
Turkey	-	-	-	-	-	-	-	-	-
United Kingdom	-	10 e	2	-	-	-	-	-	-
United States	-	-	-	-	-	-	-	-	-
Total non-OECD	**-**	**-**	**-**	**-**	**17**	**9**	**1**	**-**	**-**
Brazil	-	-	-	-	-	-	-	-	-
China[2]	-	-	-	-	-	-	-	-	-
Chinese Taipei	-	-	-	-	-	-	-	-	-
Egypt	-	-	-	-	-	-	-	-	-
India	-	-	-	-	-	-	-	-	-
Romania	-	-	-	-	-	-	-	-	-
Oth. Africa & Mid. East	-	-	-	-	-	-	-	-	-
Oth. non-OECD Americas	-	-	-	-	-	-	-	-	-
Other Asia & Oceania	-	-	-	-	-	-	-	-	-
Other non-OECD Europe and Eurasia	-	-	-	-	17	9	1	-	-
Non-specified/Other	**-**	**35 e**	**-**	**835**	**-**	**-**	**-**	**-**	**-**

(1) Earliest year for which split by coal type is available.

(2) China includes the People's Republic of China and Hong Kong, China.

* Please refer to notes and definitions in Part I. Steam coal includes all sub-bituminous coal.

Source: IEA/OECD Coal Statistics

DENMARK*

Figure 1: TPES by fuel (Mtce)

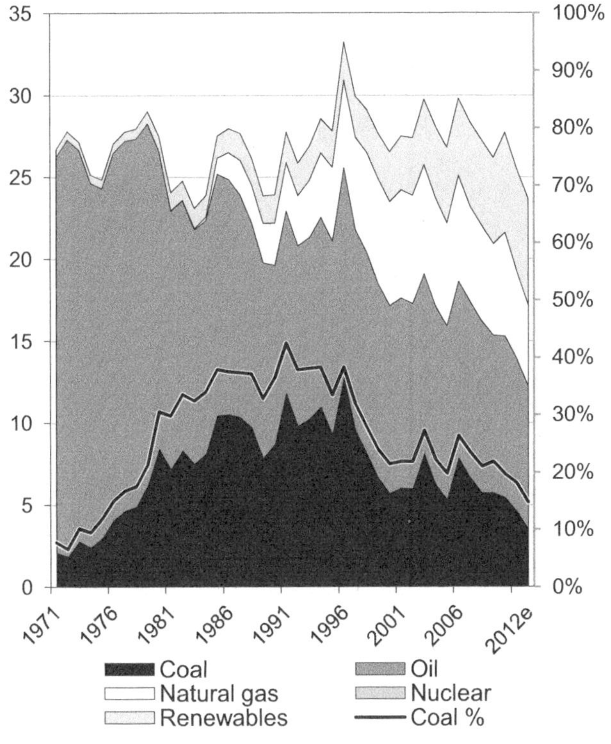

Figure 2: Coal supply indicators (1971 = 100)

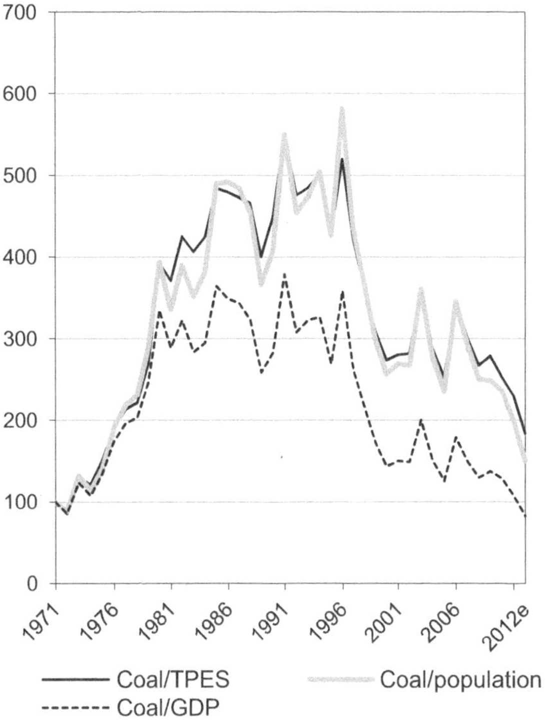

Figure 3: Primary coal supply (Mtce)

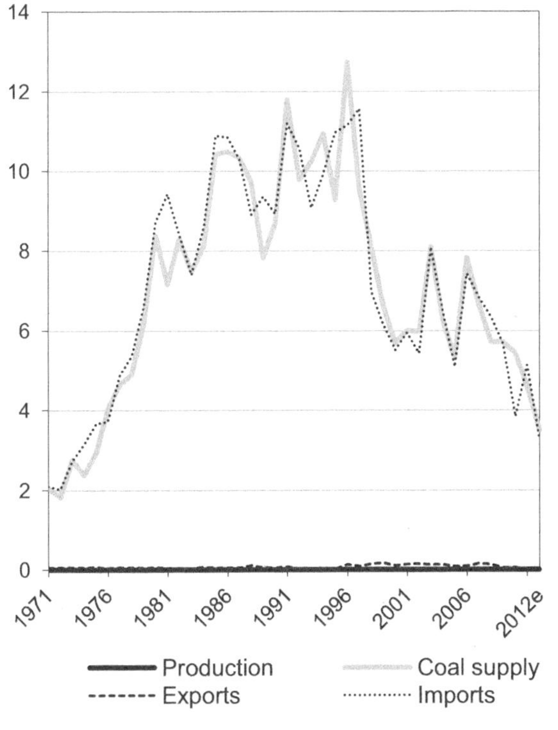

Figure 4: Coal consumption (Mtce)

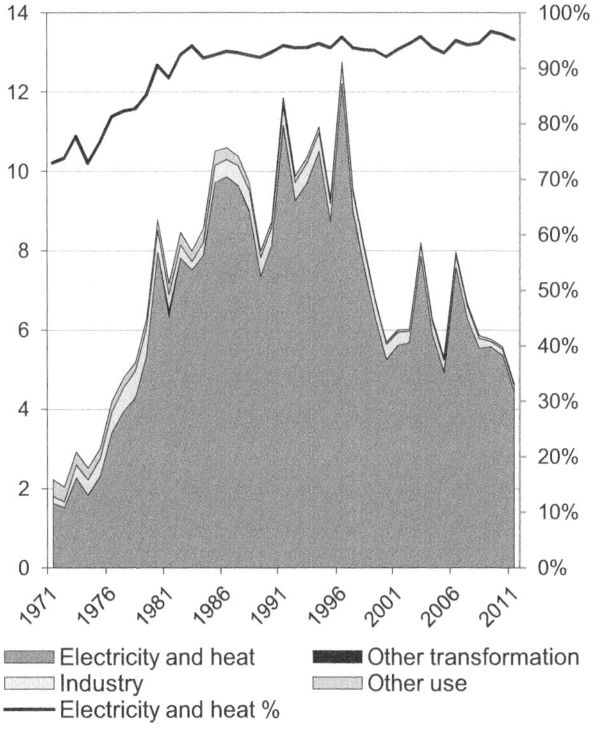

*Please refer to notes and definitions in Part I. Peat is included in coal for all figures.

Source: *IEA/OECD Coal Information, IEA/OECD Energy Balances of OECD Countries, OECD Main Economic Indicators*

DENMARK*

Figure 5: Electricity generation by fuel (TWh)

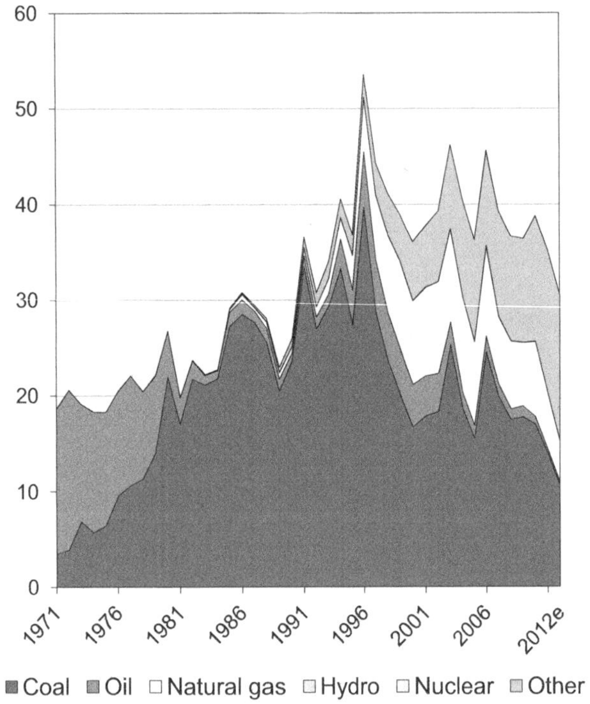

Figure 6: CO_2 emissions by fuel (Mt CO_2)

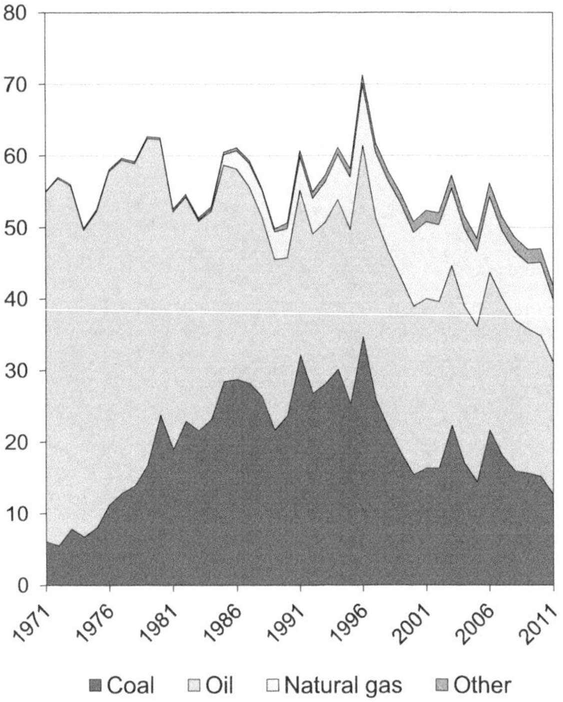

Figure 7: Electricity generation by fuel share

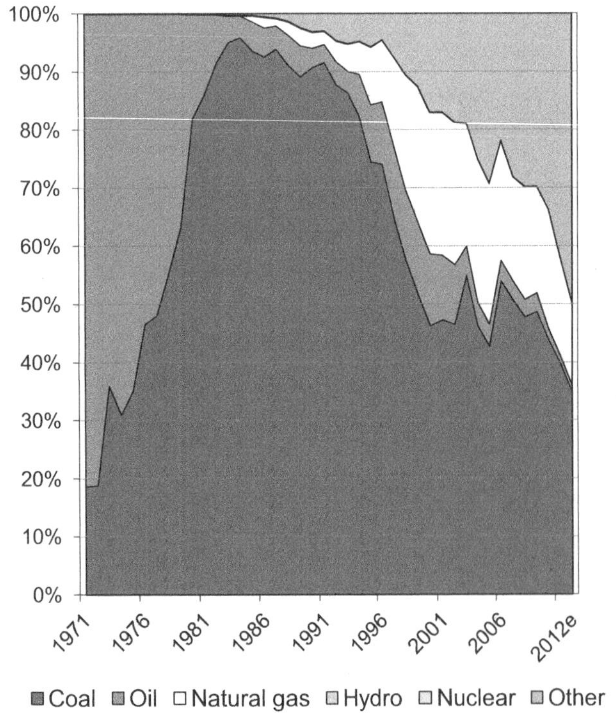

Figure 8: CO_2 emissions by fuel share

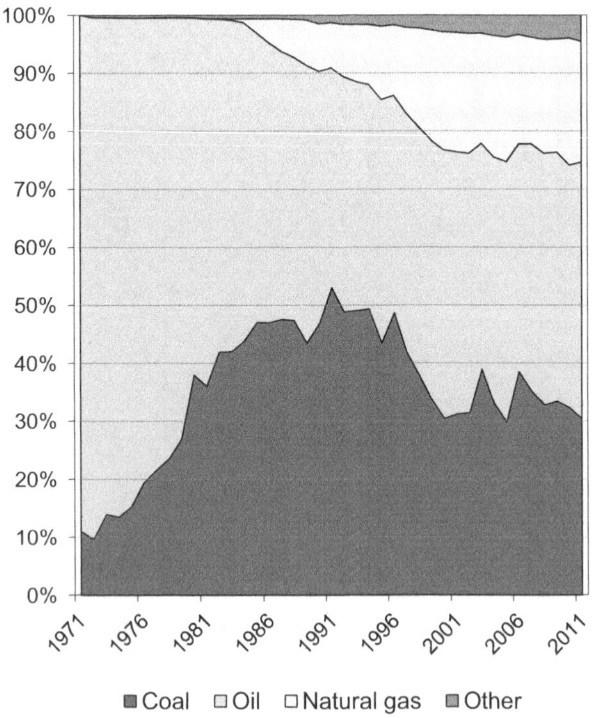

*Please refer to notes and definitions in Part I. Peat is included in coal for all figures.

Source: *IEA/OECD Electricity Information, IEA/OECD CO2 Emissions from Fuel Combustion*

DENMARK

1. Total primary energy supply (TPES) by fuel*

	1973	1980	1990	2000	2010	2011	2012e	Average annual percent change 73-90	90-11
TPES (Mtce)	**27.13**	**27.34**	**24.80**	**26.62**	**27.58**	**25.71**	**24.34**	**-0.53**	**0.17**
Coal and peat	2.76	8.39	8.70	5.69	5.44	4.62	3.50	6.99	-2.97
Oil	23.89	18.17	10.93	11.45	9.85	9.32	8.72	-4.50	-0.76
Natural Gas	0.00	0.00	2.60	6.35	6.32	5.31	4.96	65.00	3.47
Biofuels and waste	0.51	0.92	1.63	2.49	5.11	5.05	5.19	7.11	5.54
Nuclear	-	-	-	-	-	-	-	-	-
Hydro	0.00	0.00	0.00	0.00	0.00	0.00	0.00	0.91	-2.35
Geothermal	-	-	0.00	0.00	0.01	0.01	0.02	-	6.07
Solar, wind, tide	-	0.00	0.08	0.53	0.98	1.23	1.31	-	14.01
Net electricity trade(1)	-0.03	-0.15	0.87	0.08	-0.14	0.16	0.64	x	-7.67
Heat(2)	-	-	0.00	0.00	0.01	0.01	0.00	-	1.13

(1) Net trade between OECD and non-OECD countries.

(2) Ambient heat from heat pumps used in the transformation sector.

Source: *IEA/OECD Energy Balances of OECD Countries*

2. Energy supply, GDP and population*

	1973	1980	1990	2000	2010	2011	2012e	Average annual percent change 73-90	90-11
GDP (2005 bil. USD)	136	153	187	242	257	260	259	1.90	1.58
TPES/GDP(1)	0.20	0.18	0.13	0.11	0.11	0.10	0.09	-2.38	-1.38
Population (millions)	5.0	5.1	5.1	5.3	5.5	5.6	5.6	0.14	0.38
TPES/population(1)	5.40	5.33	4.83	4.99	4.97	4.62	4.36	-0.66	-0.21
TPES/GDP(2)	190.4	171.1	126.4	105.0	102.4	94.4	89.8	-2.38	-1.38
Coal and peat supply/GDP(2)	98.5	267.2	225.4	114.2	102.7	86.2	65.6	4.99	-4.47
Elec. consumption/GDP(2)	91.1	109.2	116.6	103.2	96.1	92.9	..	1.46	-1.07
Elec. generation (TWh)	19	27	26	36	39	35	30	1.82	1.45
Industrial production(2)	..	55.0	72.5	96.5	87.3	88.9	88.3	..	0.98

(1) TPES/GDP in units of tce/2005 thousand USD, TPES/population in units of tce/capita.

(2) As index, 2005 (not shown) = 100

Source: *IEA/OECD Energy Balances of OECD Countries , OECD Main Economic Indicators*

3. Coal and peat production by type*

	1978(1)	1990	2000	2005	2010	2011	2012e	Average annual percent change 78-90	90-11
Mtce:									
Coking coal	-	-	-	-	-	-	-	-	-
Steam coal	-	-	-	-	-	-	-	-	-
Lignite	-	-	-	-	-	-	-	-	-
Peat	-	-	-	-	-	-	-	-	-
Mt:									
Coking coal	-	-	-	-	-	-	-	-	-
Steam coal	-	-	-	-	-	-	-	-	-
Lignite	-	-	-	-	-	-	-	-	-
Peat	-	-	-	-	-	-	-	-	-

(1) Earliest year for which split by coal type is available.

* Please refer to notes and definitions in Part I.

Source: *IEA/OECD Energy Balances of OECD Countries, IEA/OECD Energy Statistics of OECD Countries*

DENMARK

4. Final consumption of energy by fuel*

(Mtce)

	1973	1980	1990	2000	2009	2010	2011	Average annual percent change 73-90	90-11
Total final consumption[(1)]	**21.87**	**21.05**	**18.81**	**20.32**	**20.38**	**21.28**	**20.13**	**-0.88**	**0.32**
Coal and peat	0.65	0.83	0.62	0.44	0.19	0.21	0.22	-0.32	-4.71
Oil	19.02	16.17	9.79	9.38	8.85	8.77	8.36	-3.83	-0.75
Natural Gas	-	-	1.60	2.36	2.24	2.48	2.26	-	1.65
Biofuels and wastes	0.23	0.55	0.80	0.93	1.70	1.80	1.83	7.62	4.01
Geothermal	-	-	-	-	-	-	-	-	-
Solar, wind, tide	-	0.00	0.00	0.01	0.02	0.02	0.02	-	8.98
Electricity	1.98	2.66	3.48	3.99	3.86	3.95	3.86	3.39	0.48
Heat	-	0.84	2.51	3.22	3.50	4.05	3.57	-	1.70
of which:									
Total industry	**4.79**	**4.51**	**3.85**	**4.19**	**3.34**	**3.45**	**3.41**	**-1.28**	**-0.57**
Coal and peat	0.33	0.56	0.46	0.38	0.14	0.15	0.17	2.07	-4.75
Oil	3.81	3.02	1.31	1.08	0.77	0.78	0.74	-6.07	-2.69
Natural Gas	-	-	0.76	1.11	0.95	1.02	1.01	-	1.34
Biofuels and wastes	0.08	0.11	0.16	0.16	0.28	0.26	0.27	4.01	2.47
Geothermal	-	-	-	-	-	-	-	-	-
Solar, wind, tide	-	-	-	-	-	-	-	-	-
Electricity	0.57	0.71	1.04	1.23	1.04	1.06	1.04	3.64	0.02
Heat	-	0.11	0.11	0.23	0.17	0.17	0.17	-	2.42
Total transport	**3.86**	**4.32**	**4.93**	**5.76**	**6.29**	**6.18**	**6.08**	**1.44**	**1.01**
Coal and peat	0.00	-	-	-	-	-	-	-	-
Oil	3.85	4.31	4.90	5.71	6.23	6.09	5.84	1.43	0.84
Natural Gas	-	-	-	-	-	-	-	-	-
Biofuels and wastes	-	-	-	-	0.01	0.04	0.19	-	-
Electricity	0.01	0.02	0.03	0.04	0.05	0.05	0.05	3.87	3.22
Residential	**9.90**	**7.97**	**5.72**	**5.94**	**6.32**	**7.00**	**6.24**	**-3.18**	**0.42**
Coal and peat	0.31	0.25	0.07	0.02	0.01	0.02	0.02	-8.54	-6.69
Oil	8.66	5.96	1.81	1.11	0.65	0.68	0.57	-8.79	-5.34
Natural Gas	-	-	0.55	0.92	0.90	1.05	0.88	-	2.29
Biofuels and wastes	0.12	0.35	0.48	0.61	1.27	1.36	1.22	8.64	4.57
Geothermal	-	-	-	-	-	-	-	-	-
Solar, wind, tide	-	0.00	0.00	0.01	0.01	0.02	0.02	-	9.06
Electricity	0.81	0.91	1.19	1.25	1.24	1.28	1.24	2.27	0.21
Heat	-	0.49	1.62	2.03	2.23	2.60	2.29	-	1.65
Comm & public services	**0.46**	**1.75**	**2.45**	**2.61**	**2.83**	**3.03**	**2.81**	**10.37**	**0.64**
Coal and peat	0.01	0.01	0.01	0.00	0.00	0.00	0.00	-3.42	-7.14
Oil	-	0.72	0.46	0.19	0.11	0.10	0.09	-	-7.73
Natural Gas	-	-	0.20	0.23	0.33	0.33	0.30	-	1.92
Biofuels and waste	0.02	0.02	0.04	0.08	0.06	0.06	0.07	4.47	2.28
Geothermal	-	-	-	-	-	-	-	-	-
Solar, wind, tide	-	-	0.00	0.00	0.00	0.00	0.00	-	8.53
Electricity	0.43	0.75	1.03	1.22	1.31	1.32	1.30	5.28	1.12
Heat	-	0.24	0.72	0.90	1.03	1.20	1.06	-	1.84
Non-energy use	**1.02**	**0.59**	**0.43**	**0.42**	**0.35**	**0.37**	**0.42**	**-4.88**	**-0.22**
Coal and peat	-	-	-	-	-	-	-	-	-
Oil	1.02	0.59	0.43	0.42	0.35	0.37	0.42	-4.88	-0.22
Natural Gas	-	-	-	-	-	-	-	-	-

(1) Total Final Consumption (TFC) excludes use in transformation processes and energy industry own use.

* Please refer to notes and definitions in Part I. Agriculture, Forestry, Fishing and "Non-specified other" sectors are not reported here individually, but are accounted for in Total final consumption.

Source: *IEA/OECD Energy Balances of OECD Countries*

DENMARK

5. Coal balance*

(Mtce)

	1973	1980	1990	2000	2005	2010	2011	2012e	Average annual percent change 73-90	90-11
Production	-	-	-	-	-	-	-	-	-	-
Imports	2.7	8.7	8.9	5.5	5.1	3.8	5.1	3.3	7.2	-2.6
Exports	-0.1	-0.1	-0.0	-0.1	-0.1	-0.1	-	-	-1.0	-
Stock changes	0.1	-0.3	-0.2	0.3	0.3	1.7	-0.5	0.2		
Primary supply	**2.8**	**8.4**	**8.7**	**5.7**	**5.3**	**5.4**	**4.6**	**3.5**	**7.0**	**-3.0**
Statistical differences	**0.0**	**0.2**	**-0.0**	**-0.0**	**-0.0**	**0.1**	**0.0**	..		
Total transformation	**-2.1**	**-7.8**	**-8.1**	**-5.2**	**-4.9**	**-5.4**	**-4.4**	..	**8.2**	**-2.8**
Electricity and heat gen.	-2.3	-8.0	-8.1	-5.2	-4.9	-5.4	-4.4	..	7.8	-2.8
Main activity producers (1)	*-2.3*	*-7.9*	*-8.1*	*-5.2*	*-4.9*	*-5.4*	*-4.4*	..	*7.7*	*-2.8*
Autoproducers	-	*-0.1*	*-0.1*	*-0.0*	*-0.0*	*-0.0*	*-0.0*	..	-	*-11.1*
Gas works	0.2	0.2	0.1	0.0	0.0	0.0	0.0	..	-6.4	-4.9
Coal transformation (2)	-	-	-	-	-	-	-	..	-	-
BKB plants	-	-	-	-	-	-	-	..	-	-
Blast furnaces	-	-	-	-	-	-	-	..	-	-
Coke ovens	-	-	-	-	-	-	-	..	-	-
Patent fuel plants	-	-	-	-	-	-	-	..	-	-
Other transformation (3)	-	-	-	-	-	-	-	..	-	-
Energy ind. own use	**-0.0**	**-0.0**	**-**	**-**	**-**	**-**	**-**	..	**-**	**-**
Losses	**-0.0**	**-0.0**	**-0.0**	**-0.0**	**-0.0**	**-0.0**	**-0.0**	..		
Final consumption (4)	**0.7**	**0.8**	**0.6**	**0.4**	**0.4**	**0.2**	**0.2**	..	**-0.3**	**-4.7**
Industry (5)	0.3	0.6	0.5	0.4	0.3	0.2	0.2	..	2.1	-4.8
Iron and steel	*0.0*	*0.0*	*0.0*	*0.0*	*0.0*	-	-	..	*-15.0*	-
Chemical	*0.0*	*0.0*	*0.0*	*0.0*	*0.0*	-	-	..	*15.1*	-
Non-metallic minerals	*0.1*	*0.4*	*0.2*	*0.2*	*0.2*	*0.1*	*0.1*	..	*2.3*	*-3.8*
Paper, pulp and print	*0.0*	*0.1*	*0.0*	*0.0*	-	-	-	..	*15.2*	-
Other industry (6)	*0.2*	*0.1*	*0.2*	*0.1*	*0.1*	*0.1*	*0.1*	..	*1.3*	*-4.8*
Transport (7)	0.0	-	-	-	-	-	-	..	-	-
Other	0.3	0.3	0.2	0.1	0.1	0.1	0.1	..	-4.2	-4.6
Comm. and pub. services	*0.0*	*0.0*	*0.0*	*0.0*	*0.0*	*0.0*	*0.0*	..	*-3.4*	*-7.1*
Residential	*0.3*	*0.2*	*0.1*	*0.0*	*0.0*	*0.0*	*0.0*	..	*-8.5*	*-6.7*
Other sectors (8)	-	*0.0*	*0.1*	*0.0*	*0.1*	*0.0*	*0.0*	..	-	*-3.2*
Non-energy use	-	-	-	-	-	-	-	..	-	-

(1) Main activity electricity and heat generation includes district heating.

(2) Coal transformation refers to the transformation of coal and peat to secondary and tertiary products (mainly cokes, briquettes, coal tar and off-gases). Gas works and Liquefaction are shown separately to match the *Energy Balances of OECD Countries.*

(3) Other transformation includes Liquefaction and Non-specified transformations.

(4) Final Consumption includes non-energy use and energy use (Industry, Transport and Other).

(5) Please refer to notes and definitions in Part I for detailed explanation.

(6) Other industry includes Non-ferrous metals, Transport equipment, Machinery, Mining and quarrying, Food and tobacco, Wood and wood products, Construction, Textile and leather, and Non-specified industry.

(7) Transport includes Rail and Inland waterways.

(8) Other sectors includes Agriculture/Forestry, Fishing and Non-specified other.

* "Coal" refers to all types of coal, primary (anthracite, coking coal, other bituminous coal, sub-bituminous coal, lignite/brown coal) and derived fuels (including patent fuel, BKB, coke oven coke, gas coke, coal tar, coke oven gas, gas works gas, blast furnace gas and other recovered gases), as well as peat, peat briquettes and oil shale. Quantities have been converted to Mtce using calorific values reported by the respective countries.
Please refer to notes and definitions in Part I.

Source: *IEA/OECD Energy Balances of OECD Countries*

DENMARK

6. Use of coal for selected end-uses*

(million tonnes)

	1978	1990	2000	2005	2009	2010	2011	Average annual percent change 78-90	90-11
Steam coal	**5.65**	**9.99**	**6.64**	**6.29**	**6.78**	**6.50**	**5.52**	**4.86**	**-2.79**
Total electricity and heat	4.99	9.40	6.19	5.91	6.64	6.44	5.33	5.43	-2.66
Main activity producers	*4.93*	*9.34*	*6.18*	*5.90*	*6.63*	*6.44*	*5.33*	*5.47*	*-2.64*
Autoproducers	*0.06*	*0.06*	*0.02*	*0.01*	*0.01*	*0.01*	*0.01*	*0.41*	*-10.52*
Patent fuel/BKB plants	-	-	-	-	-	-	-	-	-
Coke ovens/Liquefaction(1)	-	-	-	-	-	-	-	-	-
Blast furnace inputs	-	-	-	-	-	-	-	-	-
Gas manufacture	0.09	-	-	-	-	-	-	-	-
Industry	0.69	0.47	0.38	0.30	0.12	0.14	0.16	-3.18	-5.15
Iron and steel	-	-	-	-	-	-	-	-	-
Chemical	-	*0.00*	*0.02*	*0.02*	-	-	-	-	-
Non-metallic minerals	*0.55*	*0.20*	*0.23*	*0.21*	*0.06*	*0.07*	*0.08*	*-8.07*	*-4.25*
Paper, pulp and print	*0.06*	*0.05*	-	-	-	-	-	*-1.35*	-
Other industry	*0.09*	*0.22*	*0.13*	*0.07*	*0.06*	*0.07*	*0.08*	*7.85*	*-4.95*
Other sectors(2)	0.01	0.12	0.04	0.07	0.05	0.05	0.05	20.35	-4.18
Non-energy use	-	-	-	-	-	-	-	-	-
Coking coal	**-**	**-**	**-**	**-**	**-**	**-**	**-**	**-**	**-**
Total electricity and heat	-	-	-	-	-	-	-	-	-
Main activity producers	-	-	-	-	-	-	-	-	-
Autoproducers	-	-	-	-	-	-	-	-	-
Patent fuel/BKB plants	-	-	-	-	-	-	-	-	-
Coke ovens/Liquefaction(1)	-	-	-	-	-	-	-	-	-
Blast furnace inputs	-	-	-	-	-	-	-	-	-
Gas manufacture	-	-	-	-	-	-	-	-	-
Industry	-	-	-	-	-	-	-	-	-
Iron and steel	-	-	-	-	-	-	-	-	-
Chemical	-	-	-	-	-	-	-	-	-
Non-metallic minerals	-	-	-	-	-	-	-	-	-
Paper, pulp and print	-	-	-	-	-	-	-	-	-
Other industry	-	-	-	-	-	-	-	-	-
Other sectors(2)	-	-	-	-	-	-	-	-	-
Non-energy use	-	-	-	-	-	-	-	-	-
Lignite	**-**	**-**	**-**	**-**	**-**	**-**	**-**	**-**	**-**
Total electricity and heat	-	-	-	-	-	-	-	-	-
Main activity producers	-	-	-	-	-	-	-	-	-
Autoproducers	-	-	-	-	-	-	-	-	-
Patent fuel/BKB plants	-	-	-	-	-	-	-	-	-
Coke ovens/Liquefaction(1)	-	-	-	-	-	-	-	-	-
Gas manufacture	-	-	-	-	-	-	-	-	-
Industry	-	-	-	-	-	-	-	-	-
Iron and steel	-	-	-	-	-	-	-	-	-
Chemical	-	-	-	-	-	-	-	-	-
Non-metallic minerals	-	-	-	-	-	-	-	-	-
Paper, pulp and print	-	-	-	-	-	-	-	-	-
Other industry	-	-	-	-	-	-	-	-	-
Other sectors(2)	-	-	-	-	-	-	-	-	-
Non-energy use	-	-	-	-	-	-	-	-	-

(1) Liquefaction primarily refers to direct distillation processes. Indirect liquefaction may be reported here or under gas manufacture.

(2) Other sectors are Residential, Commercial and public services, Agriculture/Forestry, Fishing and Non-specified other.

* Coal totals may not match the sum of the individual rows due to statistical differences. Please refer to notes and definitions in Part I.

Source: *IEA/OECD Energy Statistics of OECD Countries*

DENMARK

7. Fuel prices to end users*

	1978[1]	1990	1995	2000	2005	2009	2010	2011	2012
				(US dollars / tce)					
For electricity generation									
Steam coal	..	..	..	..	..	..	..	..	..
Heavy fuel oil	..	..	..	..	..	..	..	..	..
Natural gas	c	c	c	c	c	c	c	c	c
For industry									
Steam coal	76.39	119.94	84.94	..	..	..	..	..	..
Coking coal	x	x	x	x	x	x	x	x	x
High sulphur fuel oil	69.50	..	..	..	..	..	..	..	..
Low sulphur fuel oil	..	108.11	112.49	162.96	280.27	360.54	477.71	588.87	676.41
Natural gas	258.29	c	c	c	c	585.00	..	..	..
				(Danish crowns / unit)[2]					
For electricity generation									
Steam coal	..	..	..	..	..	..	..	..	..
Heavy fuel oil	..	..	..	..	..	..	..	..	..
Natural gas	c	c	c	c	c	c	c	c	c
For industry									
Steam coal	421	742	476	..	..	..	..	..	..
Coking coal	x	x	x	x	x	x	x	x	x
High sulphur fuel oil	528	..	..	..	..	..	..	..	..
Low sulphur fuel oil	..	922	869	1817	2317	2664	3702	4349	5399
Natural gas	1830	c	c	c	c	4031	..	..	..

(1) Earliest year for which split by coal type is available.

(2) For steam coal, coking coal and heavy fuel oil per metric tonne; for natural gas per 10^7 kilocalories GCV.

* Please refer to notes and definitions in Part I. Steam coal prices relate to historical definitions.

Source: *IEA/OECD Energy Prices & Taxes*

8. Coal and peat trade by type of coal*

(Mtce)

	1978	1990	1995	2000	2005	2009	2010	2011	2012e
Total imports	**5.36**	**8.93**	**10.97**	**5.51**	**5.09**	**5.66**	**3.83**	**5.13**	**3.27**
Bituminous coal[1]	5.24	8.88	10.92	5.46	5.05	5.64	3.81	5.10	3.25
Coking coal	-	-	-	-	-	-	-	-	-
Sub-bituminous coal	-	-	-	-	-	-	-	-	-
Lignite	-	-	-	-	-	-	-	-	-
Peat	-	-	-	-	-	-	-	-	-
Coal products[2]	0.12	0.05	0.05	0.05	0.04	0.02	0.02	0.02	0.02
Total exports	**0.06**	**0.05**	**0.02**	**0.10**	**0.08**	**0.05**	**0.06**	**-**	**-**
Bituminous coal[1]	-	0.05	0.02	0.10	0.08	0.05	0.06	-	-
Coking coal	-	-	-	-	-	-	-	-	-
Sub-bituminous coal	-	-	-	-	-	-	-	-	-
Lignite	-	-	-	-	-	-	-	-	-
Peat	-	-	-	-	-	-	-	-	-
Coal products[2]	0.06	-	0.00	0.00	-	-	-	-	-

(1) Bituminous coal includes anthracite.

(2) Coal products includes products derived from coal, for example: coke, coal tar, briquettes, patent fuels.

* Please refer to notes and definitions in Part I.

Source: *IEA/OECD Energy Balances of OECD Countries, IEA/OECD Energy Statistics of OECD Countries*

DENMARK

9. Total coal imports by origin*

(thousand tonnes)

	1978[1]	1985	1990	1995	2000	2005	2010	2011	2012e
Total coal[2]	**6110**	**12457**	**10255**	**13009**	**6416**	**6031**	**4570**	**6133**	**3906**
Coking coal	**-**	**6**	**-**	**-**	**-**	**-**	**-**	**-**	**-**
Australia	-	-	-	-	-	-	-	-	-
Canada	-	-	-	-	-	-	-	-	-
Czech Republic	-	4	-	-	-	-	-	-	-
Germany	-	2	-	-	-	-	-	-	-
Poland	-	-	-	-	-	-	-	-	-
United Kingdom	-	-	-	-	-	-	-	-	-
United States	-	-	-	-	-	-	-	-	-
Other OECD	-	-	-	-	-	-	-	-	-
China, People's Rep.	-	-	-	-	-	-	-	-	-
Colombia	-	-	-	-	-	-	-	-	-
Indonesia	-	-	-	-	-	-	-	-	-
South Africa	-	-	-	-	-	-	-	-	-
Former Soviet Union[3]	-	-	x	x	x	x	x	x	x
Russian Federation[4]	*x*	*x*	-	-	-	-	-	-	-
Other FSU[4]	*x*	*x*	-	-	-	-	-	-	-
Venezuela	-	-	-	-	-	-	-	-	-
Vietnam	-	-	-	-	-	-	-	-	-
Non-specified/other	-	-	-	-	-	-	-	-	-
Steam coal	**6110**	**12451**	**10255**	**13009**	**6416**	**6031**	**4570**	**6133**	**3906**
Australia	177	2484	1127	917	143	129	-	-	-
Canada	307	328	646	-	-	-	-	-	-
Czech Republic	6	38	39	3	-	-	37	22	-
Germany	941	14	52	1	-	-	-	-	-
Poland	3078	2056	972	3476	2311	830	459	61	57
United Kingdom	145	1311	592	30	3	-	-	-	-
United States	2	2271	3223	2423	-	66	375	80	-
Other OECD	8	30	8	26	177	369	255	144	205
China, People's Rep.	-	29	57	-	-	-	-	-	-
Colombia	-	332	2057	1116	812	1254	1338	2622	1508
Indonesia	-	-	-	382	-	-	-	-	-
South Africa	868	3496	-	3493	1672	1852	735	1269	808
Former Soviet Union[3]	528	-	x	x	x	x	x	x	x
Russian Federation[4]	*x*	*x*	-	*1078*	*1295*	*1531*	*1371*	*1905*	*1328*
Other FSU[4]	*x*	*x*	-	*59*	-	-	-	*30*	-
Venezuela	-	-	-	-	-	-	-	-	-
Vietnam	-	-	-	2	3	-	-	-	-
Non-specified/other	50	62 e	1482	3	-	-	-	-	-
Lignite	**-**	**-**	**-**	**-**	**-**	**-**	**-**	**-**	**-**

(1) Earliest year for which split by coal type is available.

(2) Total coal consists of steam coal, coking coal and lignite.

(3) For years prior to 1990, or if no details for individual Former Soviet Union republics.

(4) Data for former republics are not available separately for some OECD member countries.

* Please refer to notes and definitions in Part I.

Source: IEA/OECD Coal Statistics

DENMARK

11. Steam coal exports by destination*

(thousand tonnes)

	1978(1)	1985	1990	1995	2000	2005	2010	2011	2012e
World	**-**	**57**	**54**	**25**	**123**	**96**	**71**	**-**	**-**
Total OECD	**-**	**57**	**54**	**25**	**123**	**96**	**71**	**-**	**-**
Australia	-	-	-	-	-	-	-	-	-
Austria	-	-	-	-	-	-	-	-	-
Belgium	-	-	-	-	-	-	-	-	-
Canada	-	-	-	-	-	-	-	-	-
Chile	-	-	-	-	-	-	-	-	-
Czech Republic	-	-	-	-	-	-	-	-	-
Denmark	-	-	-	-	-	-	-	-	-
Estonia	-	-	-	-	-	-	-	-	-
Finland	-	-	-	-	-	-	-	-	-
France	-	-	-	-	-	-	-	-	-
Germany	-	-	9	10	76	-	70	-	-
Greece	-	-	-	-	-	-	-	-	-
Hungary	-	-	-	-	-	-	-	-	-
Iceland	-	-	-	-	-	-	-	-	-
Ireland	-	-	-	-	-	-	-	-	-
Israel	-	-	-	-	-	-	-	-	-
Italy	-	-	-	-	-	-	-	-	-
Japan	-	-	-	-	-	-	-	-	-
Korea	-	-	-	-	-	-	-	-	-
Luxembourg	-	-	-	-	-	-	-	-	-
Mexico	-	-	-	-	-	-	-	-	-
Netherlands	-	-	-	-	-	-	-	-	-
New Zealand	-	-	-	-	-	-	-	-	-
Norway	-	-	14	7	34	95	-	-	-
Poland	-	-	-	-	13	-	1	-	-
Portugal	-	-	-	-	-	-	-	-	-
Slovak Republic	-	-	-	-	-	-	-	-	-
Slovenia	-	-	-	-	-	-	-	-	-
Spain	-	-	-	-	-	-	-	-	-
Sweden	-	54	31	8	-	1	-	-	-
Switzerland	-	-	-	-	-	-	-	-	-
Turkey	-	-	-	-	-	-	-	-	-
United Kingdom	-	3	-	-	-	-	-	-	-
United States	-	-	-	-	-	-	-	-	-
Total non-OECD	**-**	**-**	**-**	**-**	**-**	**-**	**-**	**-**	**-**
Brazil	-	-	-	-	-	-	-	-	-
China(2)	-	-	-	-	-	-	-	-	-
Chinese Taipei	-	-	-	-	-	-	-	-	-
Egypt	-	-	-	-	-	-	-	-	-
India	-	-	-	-	-	-	-	-	-
Romania	-	-	-	-	-	-	-	-	-
Oth. Africa & Mid. East	-	-	-	-	-	-	-	-	-
Oth. non-OECD Americas	-	-	-	-	-	-	-	-	-
Other Asia & Oceania	-	-	-	-	-	-	-	-	-
Other non-OECD Europe and Eurasia	-	-	-	-	-	-	-	-	-
Non-specified/Other	**-**	**-**	**-**	**-**	**-**	**-**	**-**	**-**	**-**

(1) Earliest year for which split by coal type is available.

(2) China includes the People's Republic of China and Hong Kong, China.

* Please refer to notes and definitions in Part I. Steam coal includes all sub-bituminous coal.

Source: IEA/OECD Coal Statistics

DENMARK

12. Coal import values by origin*

(average unit value, CIF, USD/tonne)

	1990	1995	2000	2005	2007	2008	2009	2010	2011/12[1]
Coking coal[2]	**93.62**	**535.51**	**390.83**	**625.03**	**154.99**	**434.34**	..	..	..
Imports from:									
Australia	..	..	..	..	..	..	..	..	..
Canada	..	..	..	..	..	..	..	..	..
Czech Republic	..	..	..	..	..	..	..	..	..
Poland	..	..	390.83	..	..	..	..	..	..
United States	..	..	..	..	..	..	..	..	..
China	..	..	..	..	..	..	..	..	..
Colombia	..	..	..	..	..	..	..	..	..
Indonesia	..	..	..	..	..	..	..	..	..
South Africa	..	..	..	..	..	..	..	..	..
Former Soviet Union[3]	..	..	..	..	..	..	..	..	..
Other bituminous coal[4]	**46.02**	**38.53**	**31.67**	**61.84**	**75.20**	**113.34**	**88.22**	**86.27**	..
Imports from:									
Australia	43.67	35.66	25.20	56.06	67.34	..	75.00	..	..
Canada	50.24	39.52	..	..	..	..	..	..	..
Czech Republic	72.83	..	..	..	..	..	..	..	..
Poland	45.29	36.22	31.78	52.10	89.54	153.20	96.54	83.70	..
United States	46.31	42.57	36.07	62.81	..	138.51	74.51	..	..
China	43.94	..	..	..	..	..	..	..	..
Colombia	47.11	36.57	30.52	62.06	78.38	..	97.37	84.83	..
Indonesia	..	31.89	..	..	70.23	..	..	..	..
South Africa	..	39.01	31.93	67.27	70.30	136.65	76.04	82.40	..
Former Soviet Union[3]	41.65	40.89	32.65	59.88	73.66	109.46	90.53	92.94	..

Note: On occasion, shipment of extremely small quantities of high valued coal results in high import costs.

(1) Import data for steam and coking coals are currently unavailable for 2011 and 2012 due to resource constraints.

(2) Weighted average of individual countries using import volumes as weights. Prices exclude intra-EU trade.

(3) Former Soviet Union until 1991, Russian Federation starting in 1992.

(4) Bituminous steam coal only. Weighted average of trades. Prices exclude intra-EU trade.

* Please refer to notes and definitions in Part I.

Source: IEA/OECD Coal Statistics

ESTONIA*

Figure 1: TPES by fuel (Mtce)

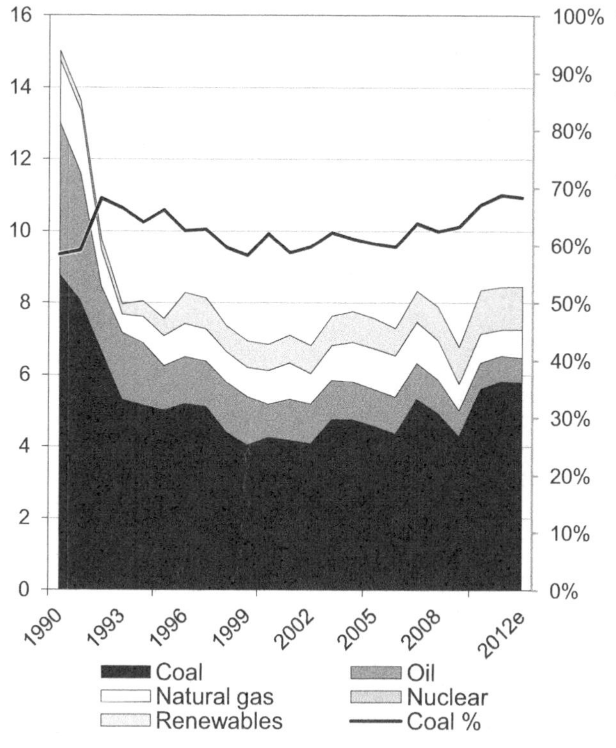

Figure 2: Coal supply indicators (1971 = 100)

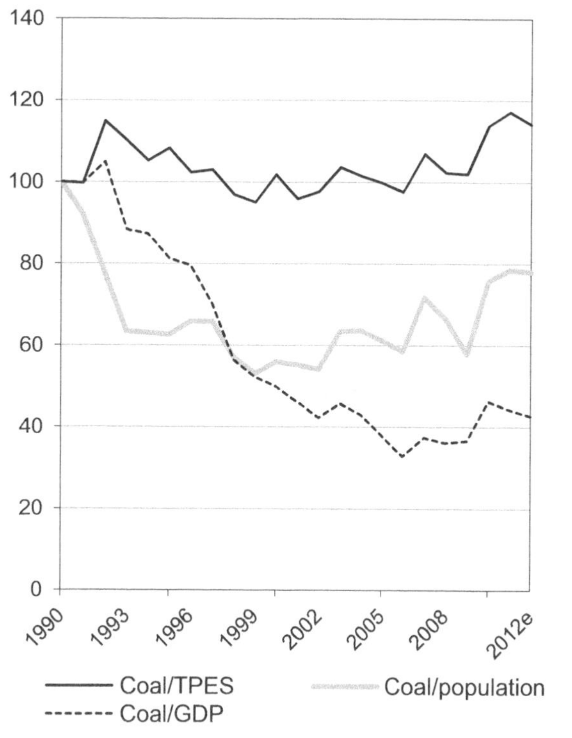

Figure 3: Primary coal supply (Mtce)

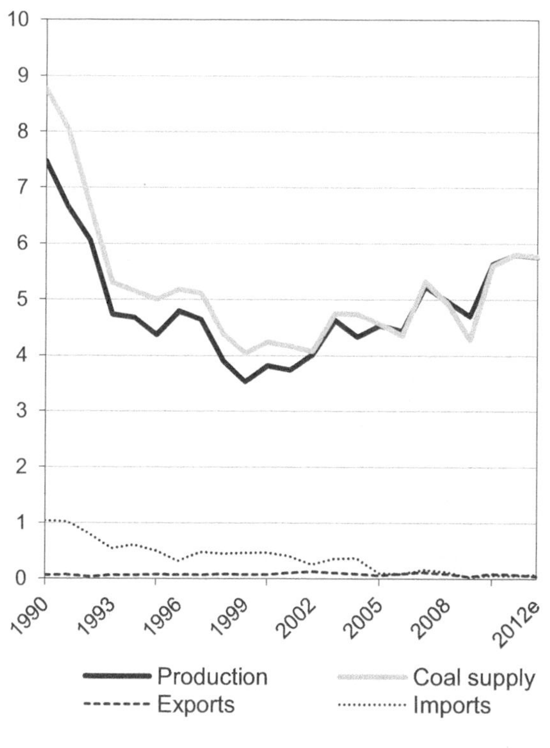

Figure 4: Coal consumption (Mtce)

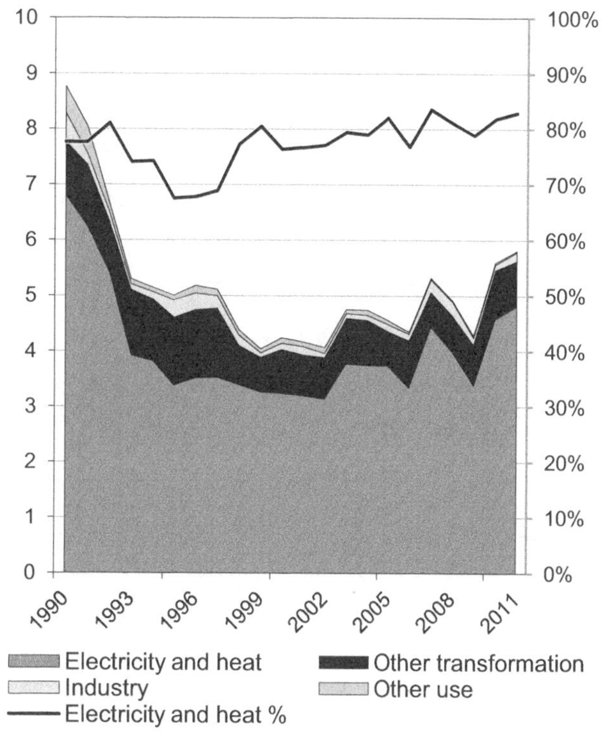

*Please refer to notes and definitions in Part I. Peat is included in coal for all figures.

Source: *IEA/OECD Coal Information, IEA/OECD Energy Balances of OECD Countries, OECD Main Economic Indicators*

ESTONIA*

Figure 5: Electricity generation by fuel (TWh)

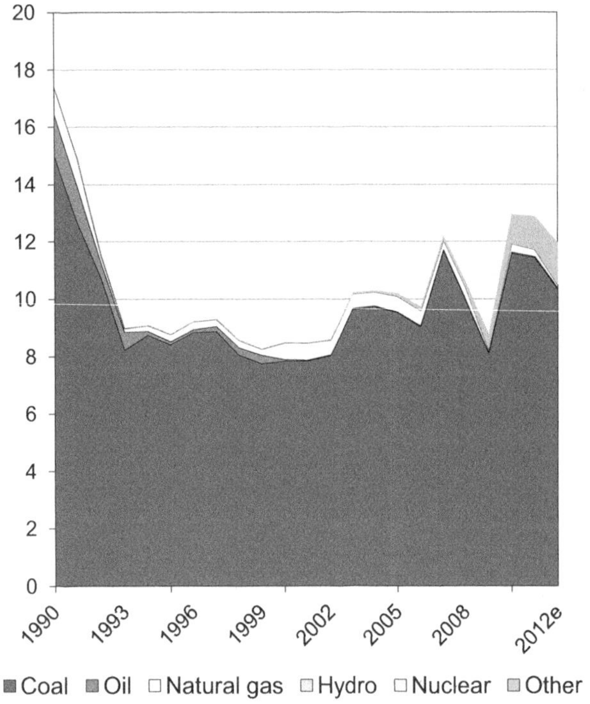

Figure 6: CO_2 emissions by fuel (Mt CO_2)

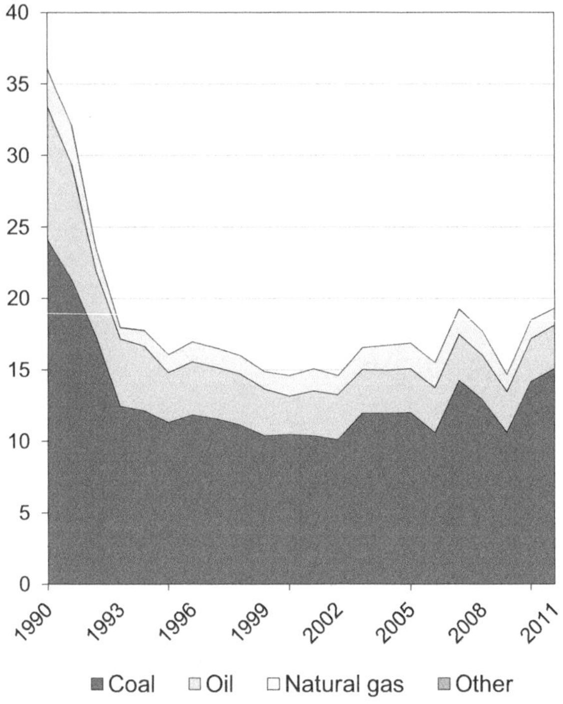

Figure 7: Electricity generation by fuel share

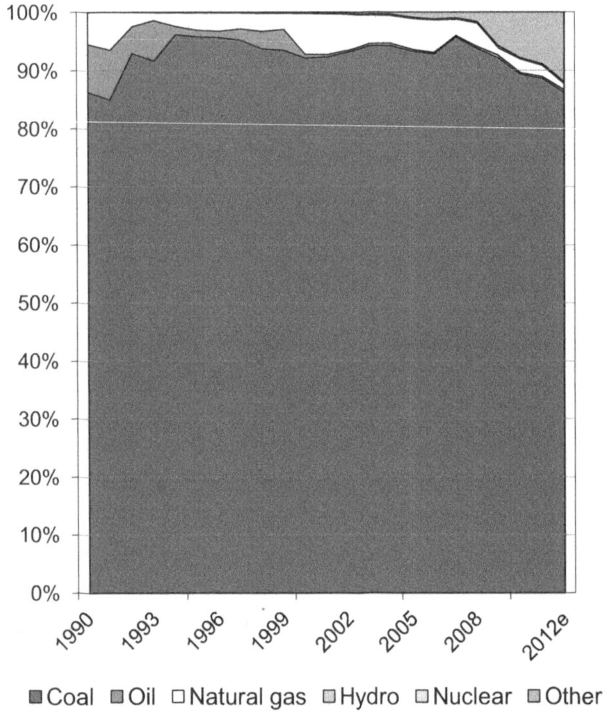

Figure 8: CO_2 emissions by fuel share

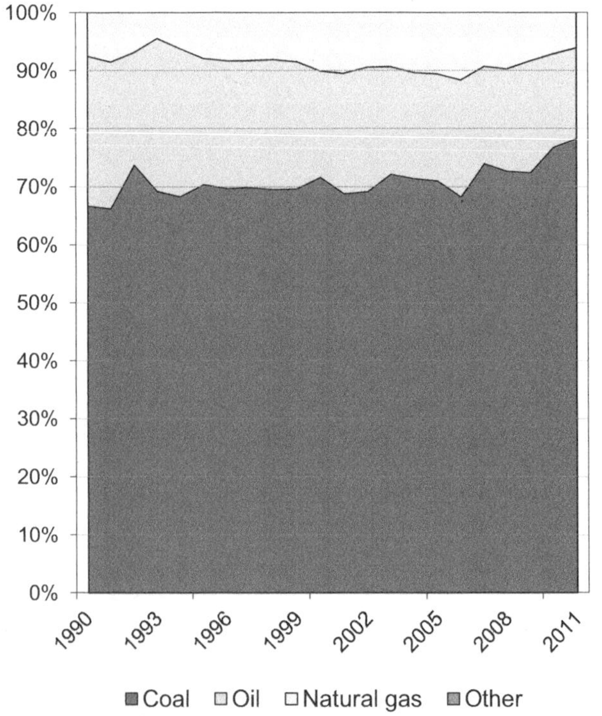

*Please refer to notes and definitions in Part I. Peat is included in coal for all figures.

Source: *IEA/OECD Electricity Information, IEA/OECD CO2 Emissions from Fuel Combustion*

ESTONIA

1. Total primary energy supply (TPES) by fuel*

	1973	1980	1990	2000	2010	2011	2012e	Average annual percent change 73-90	90-11
TPES (Mtce)	**x**	**x**	**14.16**	**6.74**	**7.95**	**8.00**	**8.17**	**x**	**-2.68**
Coal and peat	x	x	8.77	4.25	5.60	5.81	5.77	x	-1.94
Oil	x	x	4.24	0.93	0.74	0.73	0.70	x	-8.03
Natural Gas	x	x	1.75	0.95	0.80	0.72	0.78	x	-4.14
Biofuels and waste	x	x	0.27	0.73	1.17	1.14	1.13	x	7.13
Nuclear	x	x	-	-	-	-	-	x	-
Hydro	x	x	-	0.00	0.00	0.00	0.01	x	-
Geothermal	x	x	-	-	-	-	-	x	-
Solar, wind, tide	x	x	-	-	0.03	0.05	0.05	x	-
Net electricity trade(1)	x	x	-0.86	-0.11	-0.40	-0.44	-0.28	x	-3.17
Heat(2)	x	x	-	-	-	-	-	x	-

(1) Net trade between OECD and non-OECD countries.

(2) Ambient heat from heat pumps used in the transformation sector.

Source: *IEA/OECD Energy Balances of OECD Countries*

2. Energy supply, GDP and population*

	1973	1980	1990	2000	2010	2011	2012e	Average annual percent change 73-90	90-11
GDP (2005 bil. USD)	x	x	10 e	10	14	15	16	x	1.94
TPES/GDP(1)	x	x	1.40 e	0.68	0.57	0.53	0.52	x	-4.53
Population (millions)	x	x	1.6 e	1.4	1.3	1.3	1.3	x	-0.81
TPES/population(1)	x	x	8.92 e	4.91	5.94	5.97	6.10	x	-1.89
TPES/GDP(2)	x	x	263.3 e	129.0	107.0	99.5	98.4	x	-4.53
Coal and peat supply/GDP(2)	x	x	263.7 e	131.5	121.9	116.7	112.4	x	-3.81
Elec. consumption/GDP(2)	x	x	159.4 e	116.7	113.6	100.6	..	x	-2.17
Elec. generation (TWh)	x	x	17	9	13	13	12	x	-1.42
Industrial production(2)	x	x	..	62.3	104.2	124.9	124.8	x	..

(1) TPES/GDP in units of tce/2005 thousand USD, TPES/population in units of tce/capita.

(2) As index, 2005 (not shown) = 100

Source: *IEA/OECD Energy Balances of OECD Countries , OECD Main Economic Indicators*

3. Coal and peat production by type*

	1978(1)	1990	2000	2005	2010	2011	2012e	Average annual percent change 78-90	90-11
Mtce:									
Coking coal	x	-	-	-	-	-	-	x	-
Steam coal	x	-	-	-	-	-	-	x	-
Lignite	x	6.91	3.70	4.42	5.51	5.69	5.71	x	-0.92
Peat	x	0.56	0.12	0.12	0.13	0.11	0.06	x	-7.35
Mt:									
Coking coal	x	-	-	-	-	-	-	x	-
Steam coal	x	-	-	-	-	-	-	x	-
Lignite	x	22.49	11.73	14.59	17.93	18.73	18.80	x	-0.87
Peat	x	1.73	0.35	0.38	0.36	0.32	0.17	x	-7.69

(1) Earliest year for which split by coal type is available.

* Please refer to notes and definitions in Part I.

Source: *IEA/OECD Energy Balances of OECD Countries, IEA/OECD Energy Statistics of OECD Countries*

ESTONIA

4. Final consumption of energy by fuel*

(Mtce)

	1973	1980	1990	2000	2009	2010	2011	Average annual percent change 73-90	90-11
Total final consumption[(1)]	**x**	**x**	**8.57**	**3.69**	**3.97**	**4.16**	**4.07**	**x**	**-3.49**
Coal and peat	x	x	1.01	0.21	0.16	0.14	0.18	x	-7.87
Oil	x	x	2.83	1.13	1.27	1.33	1.40	x	-3.30
Natural Gas	x	x	0.63	0.39	0.28	0.30	0.29	x	-3.60
Biofuels and wastes	x	x	0.27	0.61	0.74	0.79	0.71	x	4.75
Geothermal	x	x	-	-	-	-	-	x	-
Solar, wind, tide	x	x	-	-	-	-	-	x	-
Electricity	x	x	0.86	0.61	0.82	0.85	0.81	x	-0.27
Heat	x	x	2.98	0.73	0.71	0.76	0.68	x	-6.81
of which:									
Total industry	**x**	**x**	**3.63**	**0.82**	**0.77**	**0.82**	**0.87**	**x**	**-6.59**
Coal and peat	x	x	0.53	0.11	0.13	0.10	0.16	x	-5.58
Oil	x	x	1.00	0.13	0.08	0.09	0.09	x	-11.00
Natural Gas	x	x	0.27	0.17	0.14	0.16	0.16	x	-2.45
Biofuels and wastes	x	x	0.01	0.11	0.12	0.15	0.16	x	14.56
Geothermal	x	x	-	-	-	-	-	x	-
Solar, wind, tide	x	x	-	-	-	-	-	x	-
Electricity	x	x	0.36	0.22	0.24	0.26	0.25	x	-1.74
Heat	x	x	1.46	0.06	0.06	0.06	0.05	x	-14.61
Total transport	**x**	**x**	**1.15**	**0.80**	**1.02**	**1.07**	**1.07**	**x**	**-0.36**
Coal and peat	x	x	0.01	-	-	-	-	x	-
Oil	x	x	1.13	0.79	1.01	1.06	1.06	x	-0.27
Natural Gas	x	x	-	-	-	-	-	x	-
Biofuels and wastes	x	x	-	-	-	-	-	x	-
Electricity	x	x	0.02	0.01	0.01	0.01	0.01	x	-5.68
Residential	**x**	**x**	**1.43**	**1.32**	**1.39**	**1.47**	**1.34**	**x**	**-0.33**
Coal and peat	x	x	0.25	0.04	0.01	0.01	0.02	x	-12.45
Oil	x	x	0.08	0.03	0.01	0.01	0.01	x	-8.26
Natural Gas	x	x	0.08	0.06	0.07	0.08	0.07	x	-0.46
Biofuels and wastes	x	x	0.17	0.47	0.59	0.60	0.52	x	5.37
Geothermal	x	x	-	-	-	-	-	x	-
Solar, wind, tide	x	x	-	-	-	-	-	x	-
Electricity	x	x	0.11	0.18	0.23	0.25	0.24	x	3.55
Heat	x	x	0.73	0.54	0.47	0.51	0.48	x	-2.02
Comm & public services	**x**	**x**	**0.48**	**0.41**	**0.60**	**0.61**	**0.58**	**x**	**0.87**
Coal and peat	x	x	0.07	0.01	0.00	0.00	0.00	x	-15.06
Oil	x	x	0.06	0.08	0.05	0.04	0.06	x	-0.15
Natural Gas	x	x	0.01	0.02	0.04	0.04	0.04	x	6.67
Biofuels and waste	x	x	0.08	0.01	0.02	0.03	0.03	x	-4.58
Geothermal	x	x	-	-	-	-	-	x	-
Solar, wind, tide	x	x	-	-	-	-	-	x	-
Electricity	x	x	0.03	0.17	0.32	0.31	0.30	x	12.29
Heat	x	x	0.23	0.13	0.17	0.18	0.15	x	-2.18
Non-energy use	**x**	**x**	**0.37**	**0.26**	**0.07**	**0.05**	**0.06**	**x**	**-8.18**
Coal and peat	x	x	-	0.05	0.02	0.02	0.00	x	-
Oil	x	x	0.11	0.07	0.03	0.03	0.06	x	-3.05
Natural Gas	x	x	0.26	0.14	0.02	-	-	x	-

(1) Total Final Consumption (TFC) excludes use in transformation processes and energy industry own use.

* Please refer to notes and definitions in Part I. Agriculture, Forestry, Fishing and "Non-specified other" sectors are not reported here individually, but are accounted for in Total final consumption.

Source: *IEA/OECD Energy Balances of OECD Countries*

ESTONIA

5. Coal balance*

(Mtce)

	1973	1980	1990	2000	2005	2010	2011	2012e	Average annual percent change 73-90	90-11
Production	x	x	7.5	3.8	4.5	5.6	5.8	5.8	x	-1.2
Imports	x	x	1.0	0.5	0.1	0.1	0.1	0.1	x	-12.5
Exports	x	x	-0.1	-0.1	-0.1	-0.1	-0.1	-0.1	x	0.9
Stock changes	x	x	0.3	0.0	-0.0	-0.0	0.0	-0.0		
Primary supply	**x**	**x**	**8.8**	**4.2**	**4.6**	**5.6**	**5.8**	**5.8**	**x**	**-1.9**
Statistical differences	**x**	**x**	**-0.4**	**-0.1**	**0.3**	**0.4**	**0.5**	..		
Total transformation	**x**	**x**	**-7.2**	**-3.9**	**-4.6**	**-5.9**	**-6.1**	..	**x**	**-0.8**
Electricity and heat gen.	x	x	-6.8	-3.2	-3.7	-4.6	-4.8	..	x	-1.6
Main activity producers (1)	*x*	*x*	*-6.8*	*-3.2*	*-3.7*	*-4.6*	*-4.8*	..	*x*	*-1.7*
Autoproducers	*x*	*x*	-	*-0.0*	*-0.0*	*-0.0*	*-0.0*	..	*x*	-
Gas works	x	x	-0.1	-0.1	-0.1	-0.2	-0.2	..	x	3.2
Coal transformation(2)	x	x	-0.1	-0.0	-0.0	-0.0	-0.0	..	x	-11.4
BKB plants	*x*	*x*	*-0.1*	*-0.0*	*-0.0*	*-0.0*	*-0.0*	..	*x*	*-15.8*
Blast furnaces	*x*	*x*	-	-	-	-	-	..	*x*	-
Coke ovens	*x*	*x*	*-0.0*	*-0.0*	*-0.0*	*-0.0*	*-0.0*	..	*x*	*-7.7*
Patent fuel plants	*x*	*x*	-	-	-	-	-	..	*x*	-
Other transformation(3)	x	x	-0.3	-0.5	-0.7	-1.1	-1.1	..	x	7.1
Energy ind. own use	**x**	**x**	**-0.1**	**-0.0**	**-0.0**	**-0.0**	**-0.0**	..	**x**	**-5.2**
Losses	**x**	**x**	**-0.0**	**-0.0**	-	-	-	..		
Final consumption(4)	**x**	**x**	**1.0**	**0.2**	**0.2**	**0.1**	**0.2**	..	**x**	**-7.9**
Industry(5)	x	x	0.5	0.1	0.1	0.1	0.2	..	x	-5.6
Iron and steel	*x*	*x*	-	-	-	-	-	..	*x*	-
Chemical	*x*	*x*	*0.0*	*0.0*	-	-	-	..	*x*	-
Non-metallic minerals	*x*	*x*	*0.0*	*0.1*	*0.1*	*0.1*	*0.2*	..	*x*	*24.2*
Paper, pulp and print	*x*	*x*	-	*0.0*	-	-	*0.0*	..	*x*	-
Other industry (6)	*x*	*x*	*0.5*	*0.0*	*0.0*	*0.0*	*0.0*	..	*x*	*-22.6*
Transport(7)	x	x	0.0	-	-	-	-	..	x	-
Other	x	x	0.5	0.1	0.0	0.0	0.0	..	x	-14.5
Comm. and pub. services	*x*	*x*	*0.1*	*0.0*	*0.0*	*0.0*	*0.0*	..	*x*	*-15.1*
Residential	*x*	*x*	*0.2*	*0.0*	*0.0*	*0.0*	*0.0*	..	*x*	*-12.4*
Other sectors (8)	*x*	*x*	*0.1*	*0.0*	*0.0*	-	-	..	*x*	-
Non-energy use	x	x	-	0.0	0.0	0.0	0.0	..	x	-

(1) Main activity electricity and heat generation includes district heating.

(2) Coal transformation refers to the transformation of coal and peat to secondary and tertiary products (mainly cokes, briquettes, coal tar and off-gases). Gas works and Liquefaction are shown separately to match the *Energy Balances of OECD Countries.*

(3) Other transformation includes Liquefaction and Non-specified transformations.

(4) Final Consumption includes non-energy use and energy use (Industry, Transport and Other).

(5) Please refer to notes and definitions in Part I for detailed explanation.

(6) Other industry includes Non-ferrous metals, Transport equipment, Machinery, Mining and quarrying, Food and tobacco, Wood and wood products, Construction, Textile and leather, and Non-specified industry.

(7) Transport includes Rail and Inland waterways.

(8) Other sectors includes Agriculture/Forestry, Fishing and Non-specified other.

* "Coal" refers to all types of coal, primary (anthracite, coking coal, other bituminous coal, sub-bituminous coal, lignite/brown coal) and derived fuels (including patent fuel, BKB, coke oven coke, gas coke, coal tar, coke oven gas, gas works gas, blast furnace gas and other recovered gases), as well as peat, peat briquettes and oil shale. Quantities have been converted to Mtce using calorific values reported by the respective countries. Please refer to notes and definitions in Part I.

Source: *IEA/OECD Energy Balances of OECD Countries*

ESTONIA

6. Use of coal for selected end-uses*

(million tonnes)

	1978	1990	2000	2005	2009	2010	2011	Average annual percent change 78-90	90-11
Steam coal	**x**	**0.38**	**0.09**	**0.06**	**0.09**	**0.06**	**0.07**	**x**	**-7.76**
Total electricity and heat	x	0.17	0.01	0.00	0.00	0.00	0.00	x	-21.72
Main activity producers	*x*	*0.17*	*0.00*	*0.00*	*0.00*	*0.00*	*0.00*	*x*	*-21.72*
Autoproducers	*x*	-	*0.01*	*0.00*	*0.00*	*0.00*	-	*x*	-
Patent fuel/BKB plants	x	-	-	-	-	-	-	x	-
Coke ovens/Liquefaction(1)	x	-	-	-	-	-	-	x	-
Blast furnace inputs	x	-	-	-	-	-	-	x	-
Gas manufacture	x	-	-	-	-	-	-	x	-
Industry	x	0.09	0.04	0.01	0.08	0.05	0.06	x	-1.94
Iron and steel	*x*	-	-	-	-	-	-	*x*	-
Chemical	*x*	*0.00*	-	-	-	-	-	*x*	-
Non-metallic minerals	*x*	-	*0.04*	*0.01*	*0.07*	*0.05*	*0.06*	*x*	-
Paper, pulp and print	*x*	-	-	-	-	-	-	*x*	-
Other industry	*x*	*0.08*	*0.00*	*0.00*	*0.00*	*0.00*	*0.00*	*x*	*-16.30*
Other sectors(2)	x	0.12	0.04	0.04	0.01	0.01	0.01	x	-10.31
Non-energy use	x	-	-	-	-	-	-	x	-
Coking coal	**x**	**-**	**-**	**-**	**-**	**-**	**-**	**x**	**-**
Total electricity and heat	x	-	-	-	-	-	-	x	-
Main activity producers	*x*	-	-	-	-	-	-	*x*	-
Autoproducers	*x*	-	-	-	-	-	-	*x*	-
Patent fuel/BKB plants	x	-	-	-	-	-	-	x	-
Coke ovens/Liquefaction(1)	x	-	-	-	-	-	-	x	-
Blast furnace inputs	x	-	-	-	-	-	-	x	-
Gas manufacture	x	-	-	-	-	-	-	x	-
Industry	x	-	-	-	-	-	-	x	-
Iron and steel	*x*	-	-	-	-	-	-	*x*	-
Chemical	*x*	-	-	-	-	-	-	*x*	-
Non-metallic minerals	*x*	-	-	-	-	-	-	*x*	-
Paper, pulp and print	*x*	-	-	-	-	-	-	*x*	-
Other industry	*x*	-	-	-	-	-	-	*x*	-
Other sectors(2)	x	-	-	-	-	-	-	x	-
Non-energy use	x	-	-	-	-	-	-	x	-
Lignite	**x**	**25.95**	**13.23**	**14.80**	**13.77**	**17.89**	**18.74**	**x**	**-1.54**
Total electricity and heat	x	22.57	10.84	11.63	9.85	13.55	14.39	x	-2.12
Main activity producers	*x*	22.57	*10.82*	*11.62*	*9.82*	*13.53*	*14.38*	*x*	*-2.12*
Autoproducers	*x*	-	*0.02*	*0.01*	*0.03*	*0.02*	*0.01*	*x*	-
Patent fuel/BKB plants	x	-	-	-	-	-	-	x	-
Coke ovens/Liquefaction(1)	x	0.88	1.39	1.98	2.79	3.09	3.08	x	6.16
Gas manufacture	x	0.65	0.61	0.83	0.90	1.03	0.98	x	1.97
Industry	x	1.39	0.22	0.27	0.16	0.16	0.26	x	-7.65
Iron and steel	*x*	-	-	-	-	-	-	*x*	-
Chemical	*x*	-	-	-	-	-	-	*x*	-
Non-metallic minerals	*x*	-	*0.22*	*0.27*	*0.16*	*0.16*	*0.26*	*x*	-
Paper, pulp and print	*x*	-	-	-	-	-	-	*x*	-
Other industry	*x*	*1.39*	*0.00*	-	-	-	-	*x*	-
Other sectors(2)	x	-	0.00	-	-	-	-	x	-
Non-energy use	x	-	0.15	0.08	0.05	0.06	0.01	x	-

(1) Liquefaction primarily refers to direct distillation processes. Indirect liquefaction may be reported here or under gas manufacture.

(2) Other sectors are Residential, Commercial and public services, Agriculture/Forestry, Fishing and Non-specified other.

* Coal totals may not match the sum of the individual rows due to statistical differences. Please refer to notes and definitions in Part I.

Source: *IEA/OECD Energy Statistics of OECD Countries*

ESTONIA

7. Fuel prices to end users*

	1978(1)	1990	1995	2000	2005	2009	2010	2011	2012
				(US dollars / tce)					
For electricity generation									
Steam coal	x	..	..	..	..	..	..	..	..
Heavy fuel oil	x	..	..	..	..	..	..	..	..
Natural gas	x	..	..	..	..	..	..	..	..
For industry									
Steam coal	x	..	..	..	..	..	..	..	..
Coking coal	x	..	..	..	..	..	..	..	..
High sulphur fuel oil	x	..	..	..	..	..	..	..	..
Low sulphur fuel oil	x	..	..	..	..	..	..	..	..
Natural gas	x	..	..	..	..	305.01	329.68	358.78	412.52
				(Euro / unit)(2)					
For electricity generation									
Steam coal	x	..	..	..	..	..	..	..	..
Heavy fuel oil	x	..	..	..	..	..	..	..	..
Natural gas	x	..	..	..	..	..	..	..	..
For industry									
Steam coal	x	..	..	..	..	..	..	..	..
Coking coal	x	..	..	..	..	..	..	..	..
High sulphur fuel oil	x	..	..	..	..	..	..	..	..
Low sulphur fuel oil	x	..	..	..	..	..	..	..	..
Natural gas	x	..	..	..	..	282.16	319.86	331.53	412.64

(1) Earliest year for which split by coal type is available.

(2) For steam coal, coking coal and heavy fuel oil per metric tonne; for natural gas per 10^7 kilocalories GCV.

* Please refer to notes and definitions in Part I. Steam coal prices relate to historical definitions.

Source: *IEA/OECD Energy Prices & Taxes*

8. Coal and peat trade by type of coal*

(Mtce)

	1978	1990	1995	2000	2005	2009	2010	2011	2012e
Total imports	**x**	**1.04**	**0.50**	**0.47**	**0.10**	**0.03**	**0.07**	**0.06**	**0.08**
Bituminous coal(1)	x	0.32	0.08	0.09	0.05	0.03	0.07	0.06	0.08
Coking coal	x	-	-	-	-	-	-	-	-
Sub-bituminous coal	x	-	-	-	-	-	-	-	-
Lignite	x	0.70	0.43	0.37	0.05	0.01	-	-	-
Peat	x	-	-	-	-	-	-	-	0.00
Coal products(2)	x	0.01	-	0.00	-	-	-	-	-
Total exports	**x**	**0.06**	**0.07**	**0.08**	**0.06**	**0.04**	**0.09**	**0.07**	**0.05**
Bituminous coal(1)	x	-	-	-	-	-	-	-	-
Coking coal	x	-	-	-	-	-	-	-	-
Sub-bituminous coal	x	-	-	-	-	-	-	-	-
Lignite	x	0.03	0.00	0.00	0.00	-	-	-	-
Peat	x	-	0.00	0.02	0.00	0.01	0.03	0.01	0.01
Coal products(2)	x	0.03	0.07	0.05	0.06	0.04	0.06	0.07	0.04

(1) Bituminous coal includes anthracite.

(2) Coal products includes products derived from coal, for example: coke, coal tar, briquettes, patent fuels.

* Please refer to notes and definitions in Part I.

Source: *IEA/OECD Energy Balances of OECD Countries, IEA/OECD Energy Statistics of OECD Countries*

ESTONIA

9. Total coal imports by origin*

(thousand tonnes)

	1978(1)	1985	1990	1995	2000	2005	2010	2011	2012e
Total coal(2)	**x**	**x**	**2876**	**1602**	**1469**	**234**	**71**	**67**	**82**
Coking coal	**x**	**x**	**-**	**-**	**-**	**-**	**-**	**-**	**-**
Australia	x	x	-	-	-	-	-	-	-
Canada	x	x	-	-	-	-	-	-	-
Czech Republic	x	x	-	-	-	-	-	-	-
Germany	x	x	-	-	-	-	-	-	-
Poland	x	x	-	-	-	-	-	-	-
United Kingdom	x	x	-	-	-	-	-	-	-
United States	x	x	-	-	-	-	-	-	-
Other OECD	x	x	-	-	-	-	-	-	-
China, People's Rep.	x	x	-	-	-	-	-	-	-
Colombia	x	x	-	-	-	-	-	-	-
Indonesia	x	x	-	-	-	-	-	-	-
South Africa	x	x	-	-	-	-	-	-	-
Former Soviet Union(3)	x	x	x	x	x	x	x	x	x
Russian Federation (4)	*x*	*x*	*-*	*-*	*-*	*-*	*-*	*-*	*-*
Other FSU (4)	*x*	*x*	*-*	*-*	*-*	*-*	*-*	*-*	*-*
Venezuela	x	x	-	-	-	-	-	-	-
Vietnam	x	x	-	-	-	-	-	-	-
Non-specified/other	x	x	-	-	-	-	-	-	-
Steam coal	**x**	**x**	**373**	**87**	**101**	**54**	**71**	**67**	**82**
Australia	x	x	-	-	-	-	-	-	-
Canada	x	x	-	-	-	-	-	-	-
Czech Republic	x	x	-	-	-	-	-	-	-
Germany	x	x	-	-	-	-	-	-	-
Poland	x	x	-	-	3	-	-	-	-
United Kingdom	x	x	-	-	-	-	-	-	-
United States	x	x	-	-	-	-	-	-	-
Other OECD	x	x	-	-	-	-	-	-	-
China, People's Rep.	x	x	-	-	-	-	-	-	-
Colombia	x	x	-	-	-	-	-	-	-
Indonesia	x	x	-	-	-	-	-	-	-
South Africa	x	x	-	-	-	-	-	-	-
Former Soviet Union(3)	x	x	x	x	x	x	x	x	x
Russian Federation (4)	*x*	*x*	*373*	*87*	*98*	*54*	*71*	*67*	*79*
Other FSU (4)	*x*	*x*	*-*	*-*	*-*	*-*	*-*	*-*	*3*
Venezuela	x	x	-	-	-	-	-	-	-
Vietnam	x	x	-	-	-	-	-	-	-
Non-specified/other	x	x	-	-	-	-	-	-	-
Lignite	**x**	**x**	**2503**	**1515**	**1368**	**180**	**-**	**-**	**-**

(1) Earliest year for which split by coal type is available.

(2) Total coal consists of steam coal, coking coal and lignite.

(3) For years prior to 1990, or if no details for individual Former Soviet Union republics.

(4) Data for former republics are not available separately for some OECD member countries.

* Please refer to notes and definitions in Part I.

Source: IEA/OECD Coal Statistics

ESTONIA

12. Coal import values by origin*

(average unit value, CIF, USD/tonne)

	1990	1995	2000	2005	2007	2008	2009	2010	2011/12[1]
Coking coal[2]	..	..	..	..	..	..	..	**909.41**	..
Imports from:									
Australia	..	..	..	..	..	..	..	..	..
Canada	..	..	..	..	..	..	..	..	..
Czech Republic	..	..	..	..	..	..	..	..	..
Poland	..	..	..	..	..	..	..	..	..
United States	..	..	..	..	..	..	..	..	..
China	..	..	..	..	..	..	..	..	..
Colombia	..	..	..	..	..	..	..	..	..
Indonesia	..	..	..	..	..	..	..	..	..
South Africa	..	..	..	..	..	..	..	..	..
Former Soviet Union[3]	..	..	..	..	..	..	..	909.41	..
Other bituminous coal[4]	..	..	..	..	**72.73**	**115.06**	**93.40**	**78.63**	..
Imports from:									
Australia	..	..	..	..	..	..	..	..	..
Canada	..	..	..	..	..	..	..	..	..
Czech Republic	..	..	..	..	..	..	..	..	..
Poland	..	..	..	..	..	..	..	..	..
United States	..	..	..	..	..	..	..	..	..
China	..	..	..	..	..	..	..	..	..
Colombia	..	..	..	..	..	..	..	..	..
Indonesia	..	..	..	..	..	..	..	..	..
South Africa	..	..	..	..	..	..	..	..	..
Former Soviet Union[3]	..	..	..	..	72.59	109.74	91.69	78.60	..

Note: On occasion, shipment of extremely small quantities of high valued coal results in high import costs.

(1) Import data for steam and coking coals are currently unavailable for 2011 and 2012 due to resource constraints.

(2) Weighted average of individual countries using import volumes as weights. Prices exclude intra-EU trade.

(3) Former Soviet Union until 1991, Russian Federation starting in 1992.

(4) Bituminous steam coal only. Weighted average of trades. Prices exclude intra-EU trade.

* Please refer to notes and definitions in Part I.

Source: IEA/OECD Coal Statistics

FINLAND*

Figure 1: TPES by fuel (Mtce)

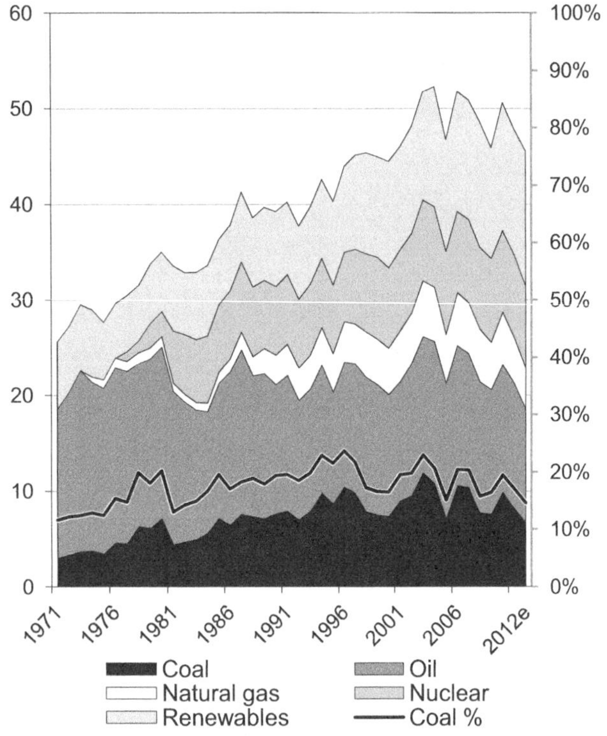

Figure 2: Coal supply indicators (1971 = 100)

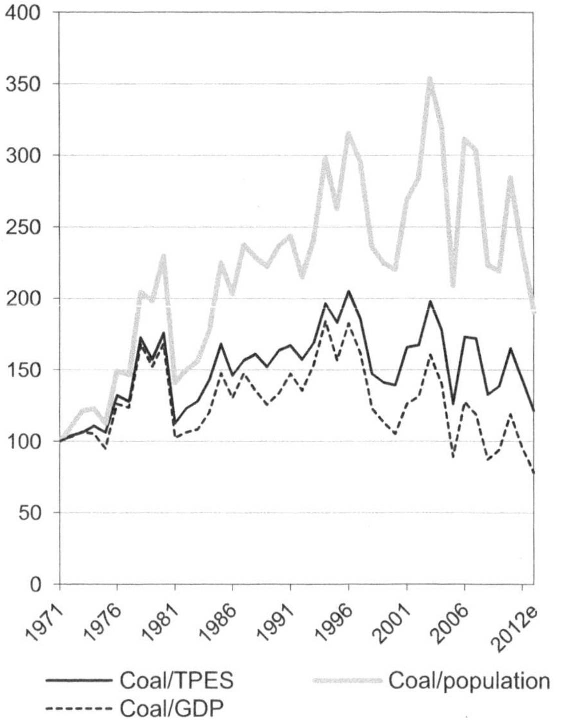

Figure 3: Primary coal supply (Mtce)

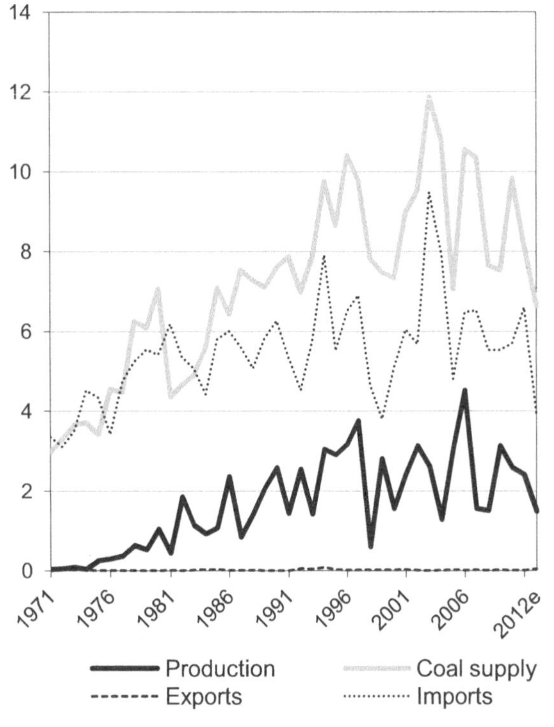

Figure 4: Coal consumption (Mtce)

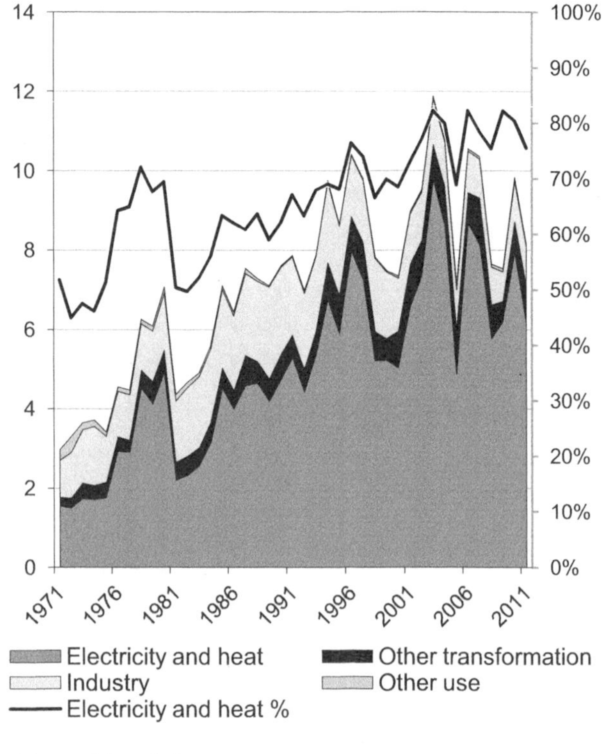

*Please refer to notes and definitions in Part I. Peat is included in coal for all figures.

Source: *IEA/OECD Coal Information, IEA/OECD Energy Balances of OECD Countries, OECD Main Economic Indicators*

FINLAND*

Figure 5: Electricity generation by fuel (TWh)

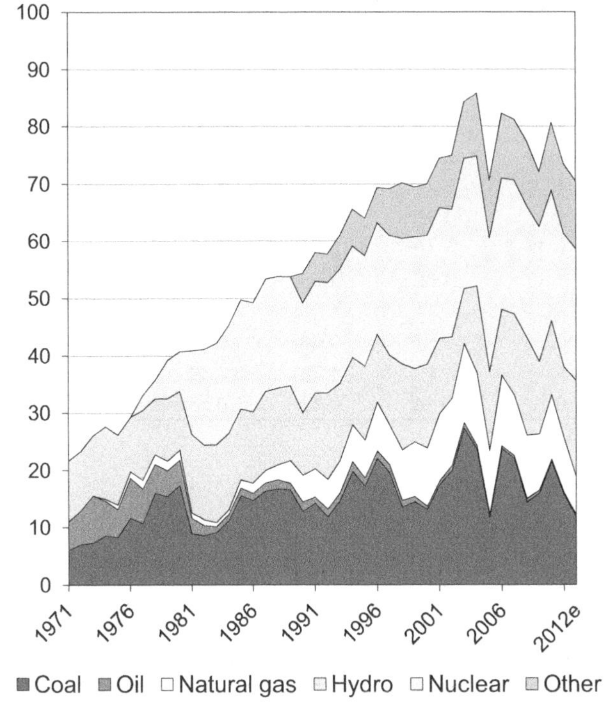

Figure 6: CO_2 emissions by fuel (Mt CO_2)

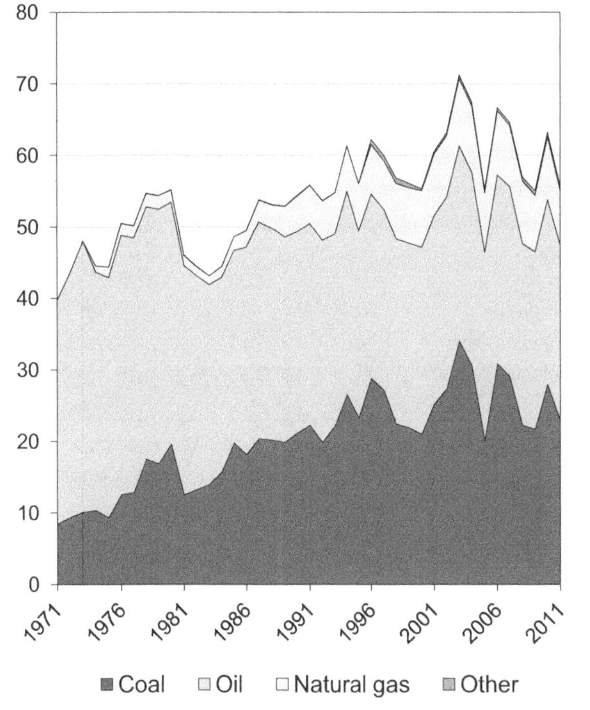

Figure 7: Electricity generation by fuel share

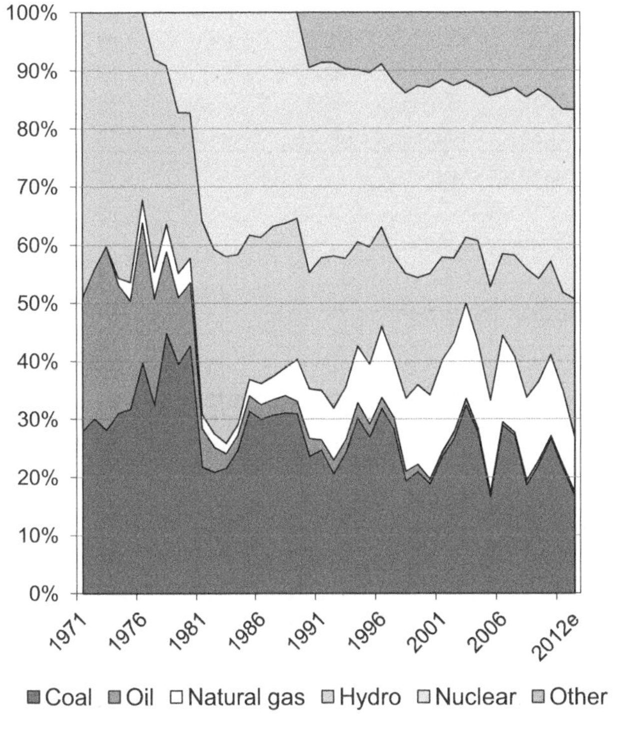

Figure 8: CO_2 emissions by fuel share

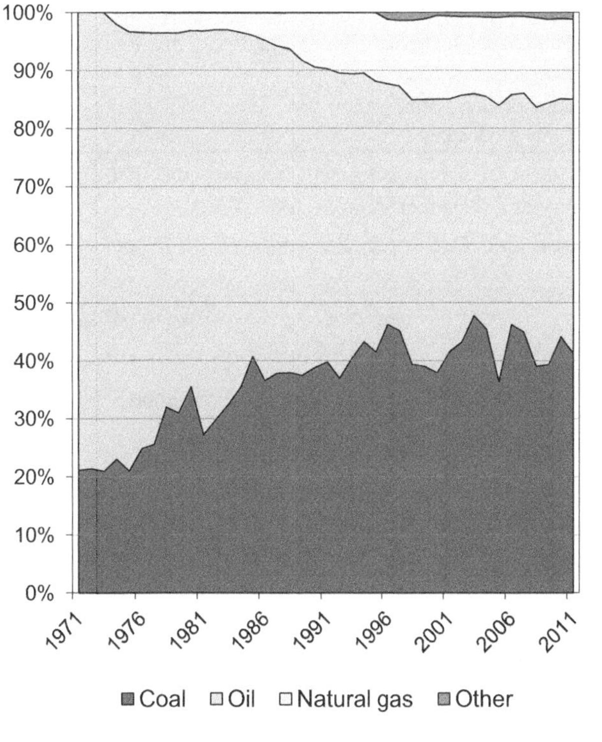

*Please refer to notes and definitions in Part I. Peat is included in coal for all figures.

Source: *IEA/OECD Electricity Information, IEA/OECD CO2 Emissions from Fuel Combustion*

FINLAND

1. Total primary energy supply (TPES) by fuel*

	1973	1980	1990	2000	2010	2011	2012e	Average annual percent change 73-90	90-11
TPES (Mtce)	**30.05**	**35.14**	**40.54**	**46.05**	**52.04**	**49.64**	**47.82**	**1.78**	**0.97**
Coal and peat	3.65	7.07	7.60	7.34	9.84	8.14	6.65	4.41	0.33
Oil	18.94	18.00	13.52	12.73	13.38	13.10	11.91	-1.97	-0.15
Natural Gas	-	1.10	3.12	4.89	5.48	4.80	4.28	-	2.08
Biofuels and waste	5.63	4.96	6.52	9.35	11.80	11.55	12.03	0.86	2.76
Nuclear	-	2.61	7.15	8.37	8.49	8.63	8.55	-	0.90
Hydro	1.29	1.26	1.33	1.80	1.59	1.53	2.06	0.19	0.65
Geothermal	-	-	-	-	-	-	-	-	-
Solar, wind, tide	-	-	0.00	0.01	0.04	0.06	0.06	-	25.20
Net electricity trade(1)	0.53	0.15	1.31	1.46	1.29	1.70	2.14	5.45	1.26
Heat(2)	-	-	-	0.10	0.14	0.13	0.13	-	-

(1) Net trade between OECD and non-OECD countries.

(2) Ambient heat from heat pumps used in the transformation sector.

Source: *IEA/OECD Energy Balances of OECD Countries*

2. Energy supply, GDP and population*

	1973	1980	1990	2000	2010	2011	2012e	Average annual percent change 73-90	90-11
GDP (2005 bil. USD)	84	104	140	172	204	210	209	3.03	1.94
TPES/GDP(1)	0.36	0.34	0.29	0.27	0.26	0.24	0.23	-1.21	-0.95
Population (millions)	4.7	4.8	5.0	5.2	5.4	5.4	5.4	0.39	0.37
TPES/population(1)	6.44	7.35	8.13	8.90	9.70	9.22	8.84	1.38	0.60
TPES/GDP(2)	142.3	135.5	115.6	107.1	102.0	94.7	91.4	-1.21	-0.95
Coal and peat supply/GDP(2)	119.8	188.9	150.2	118.3	133.6	107.6	88.1	1.34	-1.58
Elec. consumption/GDP(2)	77.4	86.9	101.9	106.7	99.2	92.8	..	1.63	-0.45
Elec. generation (TWh)	26	41	54	70	81	73	70	4.41	1.44
Industrial production(2)	34.8	44.6	59.0	91.2	100.0	101.7	99.6	3.15	2.63

(1) TPES/GDP in units of tce/2005 thousand USD, TPES/population in units of tce/capita.

(2) As index, 2005 (not shown) = 100

Source: *IEA/OECD Energy Balances of OECD Countries , OECD Main Economic Indicators*

3. Coal and peat production by type*

	1978(1)	1990	2000	2005	2010	2011	2012e	Average annual percent change 78-90	90-11
Mtce:									
Coking coal	-	-	-	-	-	-	-	-	-
Steam coal	-	-	-	-	-	-	-	-	-
Lignite	-	-	-	-	-	-	-	-	-
Peat	0.63	2.59	1.55	3.05	2.58	2.41	1.48	12.47	-0.34
Mt:									
Coking coal	-	-	-	-	-	-	-	-	-
Steam coal	-	-	-	-	-	-	-	-	-
Lignite	-	-	-	-	-	-	-	-	-
Peat	2.21	7.15	4.47	8.78	7.42	6.94	4.27	10.28	-0.15

(1) Earliest year for which split by coal type is available.

* Please refer to notes and definitions in Part I.

Source: *IEA/OECD Energy Balances of OECD Countries, IEA/OECD Energy Statistics of OECD Countries*

FINLAND

4. Final consumption of energy by fuel*

(Mtce)

	1973	1980	1990	2000	2009	2010	2011	Average annual percent change 73-90	90-11
Total final consumption(1)	**27.41**	**27.63**	**31.77**	**35.21**	**34.92**	**38.30**	**35.99**	**0.87**	**0.60**
Coal and peat	1.52	1.59	2.23	1.38	0.83	1.12	1.04	2.28	-3.59
Oil	16.08	14.29	13.13	11.99	11.21	11.84	11.08	-1.19	-0.80
Natural Gas	-	0.62	1.40	1.35	1.40	1.51	1.43	-	0.11
Biofuels and wastes	5.63	4.93	5.03	6.43	6.12	6.91	6.75	-0.66	1.41
Geothermal	-	-	-	-	-	-	-	-	-
Solar, wind, tide	-	-	0.00	0.00	0.00	0.00	0.00	-	4.82
Electricity	3.31	4.57	7.24	9.30	9.47	10.26	9.86	4.71	1.48
Heat	0.86	1.64	2.74	4.76	5.90	6.66	5.83	7.02	3.67
of which:									
Total industry	**9.31**	**8.55**	**12.89**	**16.57**	**13.53**	**15.53**	**15.14**	**1.93**	**0.77**
Coal and peat	1.35	1.44	2.20	1.33	0.74	1.02	0.95	2.93	-3.90
Oil	5.63	3.57	1.63	2.07	1.57	1.78	1.68	-7.04	0.14
Natural Gas	-	0.57	1.31	1.18	0.88	0.96	0.95	-	-1.51
Biofuels and wastes	-	-	3.50	4.86	3.75	4.30	4.40	-	1.09
Geothermal	-	-	-	-	-	-	-	-	-
Solar, wind, tide	-	-	-	-	-	-	-	-	-
Electricity	2.22	2.80	4.00	5.27	4.44	4.96	4.86	3.52	0.94
Heat	0.11	0.17	0.25	1.86	2.13	2.51	2.30	4.83	11.17
Total transport	**3.45**	**4.00**	**5.64**	**5.71**	**6.00**	**6.24**	**6.20**	**2.94**	**0.45**
Coal and peat	0.02	-	-	-	-	-	-	-	-
Oil	3.42	3.97	5.59	5.62	5.72	5.95	5.83	2.93	0.20
Natural Gas	-	-	-	0.02	0.02	0.02	0.02	-	-
Biofuels and wastes	-	-	-	-	0.18	0.17	0.26	-	-
Electricity	0.01	0.03	0.05	0.07	0.09	0.09	0.09	12.32	2.61
Residential	**6.13**	**6.39**	**7.62**	**6.48**	**7.70**	**8.38**	**7.20**	**1.29**	**-0.27**
Coal and peat	0.16	0.15	0.03	0.02	0.01	0.02	0.01	-8.59	-4.59
Oil	4.80	4.23	2.91	1.04	0.73	0.79	0.59	-2.91	-7.29
Natural Gas	-	0.05	0.04	0.03	0.06	0.07	0.05	-	0.74
Biofuels and wastes	-	-	1.30	1.34	1.84	2.04	1.75	-	1.40
Geothermal	-	-	-	-	-	-	-	-	-
Solar, wind, tide	-	-	0.00	0.00	0.00	0.00	0.00	-	4.82
Electricity	0.62	0.99	1.79	2.23	2.71	2.90	2.68	6.48	1.93
Heat	0.55	0.96	1.54	1.83	2.34	2.57	2.13	6.28	1.56
Comm & public services	**0.43**	**0.68**	**1.28**	**2.22**	**2.66**	**2.79**	**2.64**	**6.58**	**3.52**
Coal and peat	-	-	-	0.00	0.00	0.00	0.00	-	-
Oil	-	-	-	0.47	0.39	0.43	0.37	-	-
Natural Gas	-	-	-	0.04	0.04	0.04	0.05	-	-
Biofuels and waste	-	-	-	0.08	0.11	0.13	0.11	-	-
Geothermal	-	-	-	-	-	-	-	-	-
Solar, wind, tide	-	-	-	-	-	-	-	-	-
Electricity	0.43	0.68	1.28	1.63	2.12	2.19	2.11	6.58	2.43
Heat	-	-	-	-	-	-	-	-	-
Non-energy use	**1.51**	**1.76**	**2.08**	**1.53**	**2.01**	**2.10**	**1.94**	**1.90**	**-0.32**
Coal and peat	-	-	-	-	-	-	-	-	-
Oil	1.51	1.76	2.05	1.48	1.62	1.69	1.58	1.80	-1.22
Natural Gas	-	-	0.03	0.05	0.38	0.41	0.36	-	11.99

(1) Total Final Consumption (TFC) excludes use in transformation processes and energy industry own use.

* Please refer to notes and definitions in Part I. Agriculture, Forestry, Fishing and "Non-specified other" sectors are not reported here individually, but are accounted for in Total final consumption.

Source: *IEA/OECD Energy Balances of OECD Countries*

FINLAND

5. Coal balance*

(Mtce)

	1973	1980	1990	2000	2005	2010	2011	2012e	Average annual percent change 73-90	90-11
Production	0.1	1.0	2.6	1.6	3.1	2.6	2.4	1.5	22.6	-0.3
Imports	3.5	5.4	6.3	5.1	4.8	5.7	6.6	3.9	3.5	0.2
Exports	-0.0	-0.0	-0.0	-0.0	-0.0	-0.0	-0.0	-0.0	-17.1	10.3
Stock changes	0.1	0.6	-1.3	0.7	-0.8	1.6	-0.9	1.3		
Primary supply	**3.6**	**7.1**	**7.6**	**7.3**	**7.1**	**9.8**	**8.1**	**6.7**	**4.4**	**0.3**
Statistical differences	**-0.0**	**-0.2**	**-0.1**	**0.0**	**-0.1**	**0.0**	**-0.1**	..		
Total transformation	**-2.1 e**	**-5.3 e**	**-5.3 e**	**-5.7 e**	**-5.5 e**	**-8.5 e**	**-6.7 e**	..	**5.5**	**1.1**
Electricity and heat gen.	-1.7	-4.9	-4.7	-5.0	-4.9	-7.9	-6.1	..	6.1	1.3
Main activity producers (1)	*-1.7*	*-4.5*	*-4.4*	*-4.6*	*-4.4*	*-7.5*	*-5.8*	..	*5.7*	*1.3*
Autoproducers	-	*-0.5*	*-0.3*	*-0.4*	*-0.4*	*-0.4*	*-0.4*	..	-	*1.2*
Gas works	0.0	0.0	0.0	-	-	-	-	..	1.1	-
Coal transformation(2)	-0.4 e	-0.4 e	-0.5 e	-0.7 e	-0.6 e	-0.6 e	-0.5 e	..	2.2	-0.1
BKB plants	-	-	-	-	-	-	-	..	-	-
Blast furnaces	*-0.4 e*	*-0.4 e*	*-0.5 e*	*-0.6 e*	*-0.5 e*	*-0.5 e*	*-0.5 e*	..	*2.0*	*-0.7*
Coke ovens	-	-	*-0.0*	*-0.1*	*-0.1*	*-0.1*	*-0.1*	..	-	*8.4*
Patent fuel plants	-	-	-	-	-	-	-	..	-	-
Other transformation(3)	-	-	-	-	-	-	-	..	-	-
Energy ind. own use	**-0.0**	**-**	**-**	**-0.2**	**-0.3**	**-0.2**	**-0.2**	..	**-**	**-**
Losses	**-0.0**	**-0.0**	**-0.0**	**-0.0**	**-0.1**	**-0.1**	**-0.1**	..		
Final consumption(4)	**1.5**	**1.6**	**2.2**	**1.4**	**1.1**	**1.1**	**1.0**	..	**2.3**	**-3.6**
Industry(5)	1.3	1.4	2.2	1.3	1.1	1.0	1.0	..	2.9	-3.9
Iron and steel	*0.4 e*	*0.5 e*	*0.8 e*	*0.6 e*	*0.5 e*	*0.5 e*	*0.5 e*	..	*4.3*	*-2.5*
Chemical	*0.0*	*0.0*	*0.1*	*0.1*	*0.1*	-	-	..	*17.2*	-
Non-metallic minerals	-	*0.5*	*0.7*	*0.2*	*0.2*	*0.1*	*0.1*	..	-	*-7.9*
Paper, pulp and print	*0.0*	*0.3*	*0.5*	*0.4*	*0.2*	*0.3*	*0.3*	..	*20.2*	*-2.9*
Other industry (6)	*0.9*	*0.2*	*0.1*	*0.0*	*0.0*	*0.1*	*0.1*	..	*-15.0*	*1.6*
Transport(7)	0.0	-	-	-	-	-	-	..	-	-
Other	0.2	0.2	0.0	0.1	0.1	0.1	0.1	..	-8.6	4.2
Comm. and pub. services	-	-	-	*0.0*	*0.0*	*0.0*	*0.0*	..	-	-
Residential	*0.2*	*0.2*	*0.0*	*0.0*	*0.0*	*0.0*	*0.0*	..	*-8.6*	*-4.6*
Other sectors (8)	-	-	-	*0.0*	*0.0*	*0.1*	*0.1*	..	-	-
Non-energy use	-	-	-	-	-	-	-	..	-	-

(1) Main activity electricity and heat generation includes district heating.

(2) Coal transformation refers to the transformation of coal and peat to secondary and tertiary products (mainly cokes, briquettes, coal tar and off-gases). Gas works and Liquefaction are shown separately to match the *Energy Balances of OECD Countries*.

(3) Other transformation includes Liquefaction and Non-specified transformations.

(4) Final Consumption includes non-energy use and energy use (Industry, Transport and Other).

(5) Please refer to notes and definitions in Part I for detailed explanation.

(6) Other industry includes Non-ferrous metals, Transport equipment, Machinery, Mining and quarrying, Food and tobacco, Wood and wood products, Construction, Textile and leather, and Non-specified industry.

(7) Transport includes Rail and Inland waterways.

(8) Other sectors includes Agriculture/Forestry, Fishing and Non-specified other.

* "Coal" refers to all types of coal, primary (anthracite, coking coal, other bituminous coal, sub-bituminous coal, lignite/brown coal) and derived fuels (including patent fuel, BKB, coke oven coke, gas coke, coal tar, coke oven gas, gas works gas, blast furnace gas and other recovered gases), as well as peat, peat briquettes and oil shale. Quantities have been converted to Mtce using calorific values reported by the respective countries. Please refer to notes and definitions in Part I.

Source: *IEA/OECD Energy Balances of OECD Countries*

FINLAND

6. Use of coal for selected end-uses*

(million tonnes)

	1978	1990	2000	2005	2009	2010	2011	Average annual percent change 78-90	90-11
Steam coal	**5.31**	**4.94**	**3.91**	**3.20**	**4.50**	**5.78**	**4.20**	**-0.60**	**-0.77**
Total electricity and heat	4.57	3.88	3.55	2.92	4.45	5.60	3.98	-1.34	0.12
Main activity producers	*4.52*	*3.77*	*3.51*	*2.90*	*4.41*	*5.55*	*3.97*	*-1.49*	*0.24*
Autoproducers	*0.05*	*0.11*	*0.04*	*0.02*	*0.05*	*0.05*	*0.02*	*7.24*	*-9.09*
Patent fuel/BKB plants	-	-	-	-	-	-	-	-	-
Coke ovens/Liquefaction(1)	-	-	-	-	-	-	-	-	-
Blast furnace inputs	-	-	-	-	-	-	-	-	-
Gas manufacture	-	-	-	-	-	-	-	-	-
Industry	0.67	1.04	0.36	0.27	0.12	0.16	0.16	3.82	-8.46
Iron and steel	*0.09*	*0.08*	-	-	-	-	-	*-0.61*	-
Chemical	*0.01*	*0.08*	*0.08*	*0.08*	-	-	-	*16.10*	-
Non-metallic minerals	*0.41*	*0.76*	*0.18*	*0.17*	*0.08*	*0.11*	*0.13*	*5.22*	*-8.25*
Paper, pulp and print	-	*0.09*	*0.09*	*0.02*	*0.01*	*0.02*	*0.01*	-	*-10.57*
Other industry	*0.15*	*0.02*	*0.01*	-	*0.03*	*0.03*	*0.03*	*-14.26*	*0.91*
Other sectors(2)	0.08	0.01	0.01	0.01	0.01	0.01	0.01	-16.20	-0.56
Non-energy use	-	-	-	-	-	-	-	-	-
Coking coal	**-**	**0.71**	**1.28**	**1.40**	**0.92**	**1.21**	**1.28**	**-**	**2.84**
Total electricity and heat	-	-	-	-	-	-	-	-	-
Main activity producers	-	-	-	-	-	-	-	-	-
Autoproducers	-	-	-	-	-	-	-	-	-
Patent fuel/BKB plants	-	-	-	-	-	-	-	-	-
Coke ovens/Liquefaction(1)	-	0.71	1.28	1.28	1.07	1.20	1.22	-	2.58
Blast furnace inputs	-	-	-	-	-	-	-	-	-
Gas manufacture	-	-	-	-	-	-	-	-	-
Industry	-	-	-	-	-	-	-	-	-
Iron and steel	-	-	-	-	-	-	-	-	-
Chemical	-	-	-	-	-	-	-	-	-
Non-metallic minerals	-	-	-	-	-	-	-	-	-
Paper, pulp and print	-	-	-	-	-	-	-	-	-
Other industry	-	-	-	-	-	-	-	-	-
Other sectors(2)	-	-	-	-	-	-	-	-	-
Non-energy use	-	-	-	-	-	-	-	-	-
Lignite	**-**	**-**	**-**	**-**	**-**	**-**	**-**	**-**	**-**
Total electricity and heat	-	-	-	-	-	-	-	-	-
Main activity producers	-	-	-	-	-	-	-	-	-
Autoproducers	-	-	-	-	-	-	-	-	-
Patent fuel/BKB plants	-	-	-	-	-	-	-	-	-
Coke ovens/Liquefaction(1)	-	-	-	-	-	-	-	-	-
Gas manufacture	-	-	-	-	-	-	-	-	-
Industry	-	-	-	-	-	-	-	-	-
Iron and steel	-	-	-	-	-	-	-	-	-
Chemical	-	-	-	-	-	-	-	-	-
Non-metallic minerals	-	-	-	-	-	-	-	-	-
Paper, pulp and print	-	-	-	-	-	-	-	-	-
Other industry	-	-	-	-	-	-	-	-	-
Other sectors(2)	-	-	-	-	-	-	-	-	-
Non-energy use	-	-	-	-	-	-	-	-	-

(1) Liquefaction primarily refers to direct distillation processes. Indirect liquefaction may be reported here or under gas manufacture.

(2) Other sectors are Residential, Commercial and public services, Agriculture/Forestry, Fishing and Non-specified other.

* Coal totals may not match the sum of the individual rows due to statistical differences. Please refer to notes and definitions in Part I.

Source: *IEA/OECD Energy Statistics of OECD Countries*

FINLAND

7. Fuel prices to end users*

	1978(1)	1990	1995	2000	2005	2009	2010	2011	2012
				(US dollars / tce)					
For electricity generation									
Steam coal	51.74	72.35	96.93	44.40	82.82	111.56	116.89	157.56	133.45
Heavy fuel oil	71.34	130.84	150.64	133.67	232.79	307.65	390.31	..	..
Natural gas	78.45	96.13	113.53	88.04	128.66	235.04	249.85	306.44	319.32
For industry									
Steam coal	51.74	72.35	96.93	89.47	146.63	192.19	193.76	362.38	322.66
Coking coal	..	..	101.01	99.78	196.65	287.47	288.81	479.99	398.61
High sulphur fuel oil	71.34	130.84	..	..	..	..	..	..	..
Low sulphur fuel oil	..	..	150.64	172.12	286.78	375.52	454.89	..	..
Natural gas	78.45	96.13	113.53	101.68	147.93	258.81	272.51	408.76	413.84
				(Euro / unit)(2)					
For electricity generation									
Steam coal	31.11	40.48	61.91	41.92	58.01	69.88	76.80	98.53	90.34
Heavy fuel oil	67.61	115.38	151.64	198.90	257.00	303.70	404.13	..	..
Natural gas	69.70	79.47	107.14	122.82	133.16	217.51	242.53	283.16	319.41
For industry									
Steam coal	31.11	40.48	61.91	84.47	102.71	120.38	127.30	226.62	218.43
Coking coal	..	..	74.12	108.23	158.25	206.86	217.99	344.87	310.03
High sulphur fuel oil	67.61	115.38	..	..	..	..	..	..	..
Low sulphur fuel oil	..	..	151.64	256.12	316.60	370.70	471.00	..	..
Natural gas	69.70	79.47	107.14	141.85	153.11	239.52	264.53	377.71	413.96

(1) Earliest year for which split by coal type is available.

(2) For steam coal, coking coal and heavy fuel oil per metric tonne; for natural gas per 10^7 kilocalories GCV.

* Please refer to notes and definitions in Part I. Steam coal prices relate to historical definitions.

Source: *IEA/OECD Energy Prices & Taxes*

8. Coal and peat trade by type of coal*

(Mtce)

	1978	1990	1995	2000	2005	2009	2010	2011	2012e
Total imports	**5.23**	**6.27**	**5.51**	**5.08**	**4.80**	**5.53**	**5.70**	**6.60**	**3.88**
Bituminous coal(1)	4.31	4.85	3.75	3.32	2.89	4.29	3.93	4.85	2.37
Coking coal	-	0.64	1.49	1.26	1.40	0.96	1.33	1.27	1.17
Sub-bituminous coal	-	-	-	-	-	-	-	-	-
Lignite	-	-	-	-	-	-	-	-	-
Peat	-	-	0.00	-	0.01	0.04	0.00	0.04	0.03
Coal products(2)	0.92	0.77	0.27	0.50	0.50	0.25	0.44	0.44	0.32
Total exports	**-**	**0.00**	**0.03**	**0.02**	**0.02**	**0.02**	**0.01**	**0.01**	**0.05**
Bituminous coal(1)	-	-	-	-	-	-	-	-	-
Coking coal	-	-	0.00	-	-	-	-	-	-
Sub-bituminous coal	-	-	-	-	-	-	-	-	-
Lignite	-	-	-	-	-	-	-	-	-
Peat	-	-	0.03	0.02	0.02	0.02	0.01	0.00	0.00
Coal products(2)	-	0.00	0.00	-	-	0.00	0.00	0.01	0.05

(1) Bituminous coal includes anthracite.

(2) Coal products includes products derived from coal, for example: coke, coal tar, briquettes, patent fuels.

* Please refer to notes and definitions in Part I.

Source: *IEA/OECD Energy Balances of OECD Countries, IEA/OECD Energy Statistics of OECD Countries*

FINLAND

9. Total coal imports by origin*

(thousand tonnes)

	1978[(1)]	1985	1990	1995	2000	2005	2010	2011	2012e
Total coal[(2)]	**4789**	**5089**	**6101**	**5821**	**5072**	**4723**	**5920**	**6970**	**3954**
Coking coal	**-**	**-**	**711**	**1651**	**1258**	**1401**	**1327**	**1269**	**1169**
Australia	-	-	-	-	-	487	406	293	241
Canada	-	-	-	-	100	519	416	482	364
Czech Republic	-	-	-	-	-	-	-	-	-
Germany	-	-	-	-	2	-	-	-	-
Poland	-	-	203	602	705	13	-	-	20
United Kingdom	-	-	-	-	-	-	-	-	-
United States	-	-	45	695	360	382	437	334	391
Other OECD	-	-	-	49	-	-	1	24	87
China, People's Rep.	-	-	-	-	-	-	3	-	-
Colombia	-	-	-	-	-	-	64	87	-
Indonesia	-	-	-	-	-	-	-	-	-
South Africa	-	-	-	-	-	-	-	-	-
Former Soviet Union[(3)]	-	-	x	x	x	x	x	x	x
Russian Federation [(4)]	*x*	*x*	*463*	*305*	*91*	-	-	*49*	*66*
Other FSU [(4)]	*x*	*x*	-	-	-	-	-	-	-
Venezuela	-	-	-	-	-	-	-	-	-
Vietnam	-	-	-	-	-	-	-	-	-
Non-specified/other	-	-	-	-	-	-	-	-	-
Steam coal	**4789**	**5089**	**5390**	**4170**	**3814**	**3322**	**4593**	**5701**	**2785**
Australia	-	404	-	51	-	-	-	-	-
Canada	-	56	-	-	-	-	-	-	-
Czech Republic	-	-	-	-	-	-	7	31	20
Germany	-	-	-	-	-	-	6	21	-
Poland	4089	2407	2609	1899	1313	551	211	29	133
United Kingdom	3	1	253	88	-	3	-	6	-
United States	-	1018	41	460	-	-	166	173	-
Other OECD	-	4	-	-	51	1	14	51	11
China, People's Rep.	-	-	100	-	-	-	-	-	-
Colombia	-	347	334	203	-	-	413	33	17
Indonesia	-	-	-	24	-	-	-	-	-
South Africa	-	-	-	-	-	-	-	-	-
Former Soviet Union[(3)]	697	852	x	x	x	x	x	x	x
Russian Federation [(4)]	*x*	*x*	*1905*	*1240*	*2448*	*2736*	*3680*	*5199*	*2556*
Other FSU [(4)]	*x*	*x*	-	*176*	*2*	*31*	*55*	*152*	*48*
Venezuela	-	-	148	29	-	-	41	6	-
Vietnam	-	-	-	-	-	-	-	-	-
Non-specified/other	-	-	-	-	-	-	-	-	-
Lignite	**-**	**-**	**-**	**-**	**-**	**-**	**-**	**-**	**-**

(1) Earliest year for which split by coal type is available.

(2) Total coal consists of steam coal, coking coal and lignite.

(3) For years prior to 1990, or if no details for individual Former Soviet Union republics.

(4) Data for former republics are not available separately for some OECD member countries.

* Please refer to notes and definitions in Part I.

Source: IEA/OECD Coal Statistics

FINLAND

12. Coal import values by origin*
(average unit value, CIF, USD/tonne)

	1990	1995	2000	2005	2007	2008	2009	2010	2011/12(1)
Coking coal(2)	..	**60.49**	**53.94**	**131.60**	**152.40**	**326.47**	**192.34**	**221.52**	..
Imports from:									
Australia	..	..	..	148.54	150.14	392.82	166.94	221.06	..
Canada	..	..	58.26	153.33	164.99	350.70	216.54	225.40	..
Czech Republic	..	..	..	..	..	..	..	..	..
Poland	..	60.94	52.76	147.80	..	..	..	..	..
United States	..	60.37	55.42	121.25	135.83	258.32	200.58	220.62	..
China	..	..	..	..	..	..	..	490.92	..
Colombia	..	..	..	..	..	..	..	192.27	..
Indonesia	..	..	..	..	..	..	..	..	..
South Africa	..	..	..	..	..	..	..	..	..
Former Soviet Union(3)	..	59.06	51.74	57.46	129.54	..	129.25	..	..
Other bituminous coal(4)	**49.71**	**44.30**	**32.08**	**62.80**	**72.64**	**134.21**	**86.94**	**92.80**	..
Imports from:									
Australia	..	46.59	..	..	..	..	..	..	..
Canada	..	..	..	..	..	..	..	..	..
Czech Republic	..	..	..	..	..	..	..	..	..
Poland	..	44.28	34.83	68.11	79.86	164.75	79.19	80.73	..
United States	..	53.46	..	..	72.38	136.19	69.73	114.19	..
China	..	..	..	..	..	..	..	..	..
Colombia	..	52.73	..	..	77.53	..	82.32	91.88	..
Indonesia	..	38.94	..	..	..	238.93	145.71	..	..
South Africa	..	..	..	..	..	195.76	..	..	..
Former Soviet Union(3)	..	40.34	30.46	61.59	71.89	124.32	87.26	91.84	..

Note: On occasion, shipment of extremely small quantities of high valued coal results in high import costs.

(1) Import data for steam and coking coals are currently unavailable for 2011 and 2012 due to resource constraints.

(2) Weighted average of individual countries using import volumes as weights. Prices exclude intra-EU trade.

(3) Former Soviet Union until 1991, Russian Federation starting in 1992.

(4) Bituminous steam coal only. Weighted average of trades. Prices exclude intra-EU trade.

* Please refer to notes and definitions in Part I.

Source: IEA/OECD Coal Statistics

FRANCE*

Figure 1: TPES by fuel (Mtce)

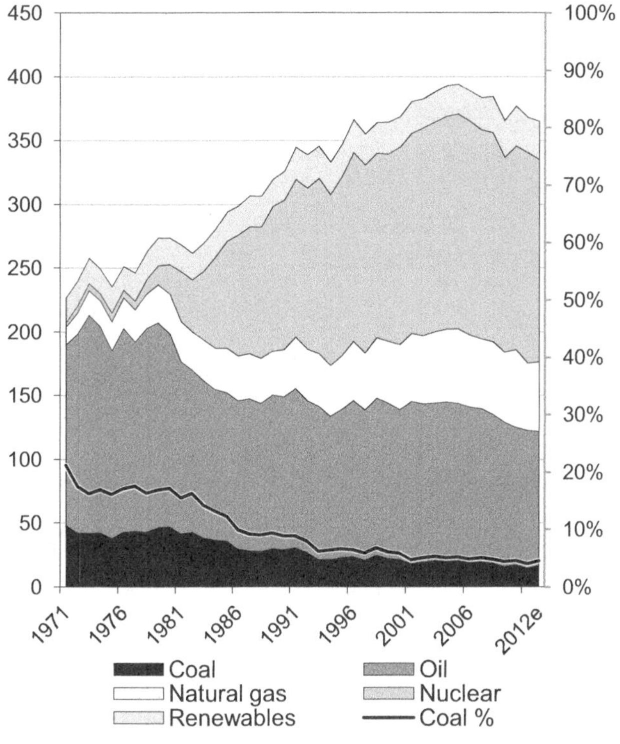

Figure 2: Coal supply indicators (1971 = 100)

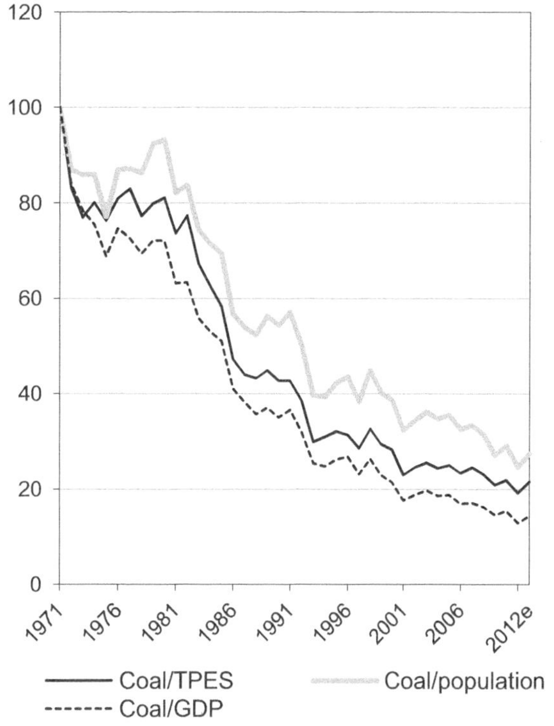

Figure 3: Primary coal supply (Mtce)

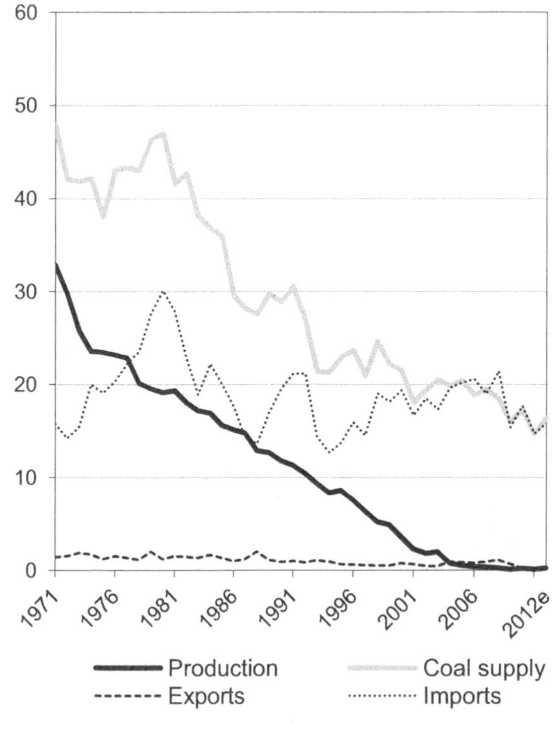

Figure 4: Coal consumption (Mtce)

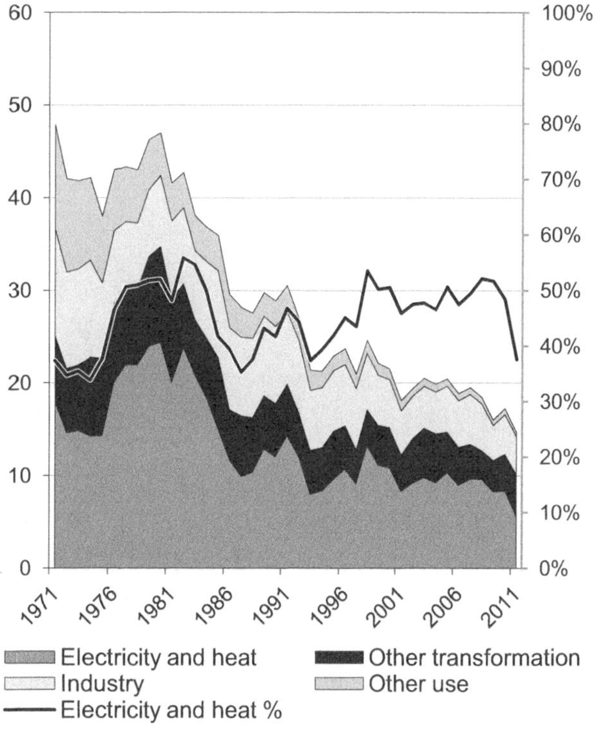

*Please refer to notes and definitions in Part I. Peat is included in coal for all figures.

Source: *IEA/OECD Coal Information, IEA/OECD Energy Balances of OECD Countries, OECD Main Economic Indicators*

FRANCE*

Figure 5: Electricity generation by fuel (TWh)

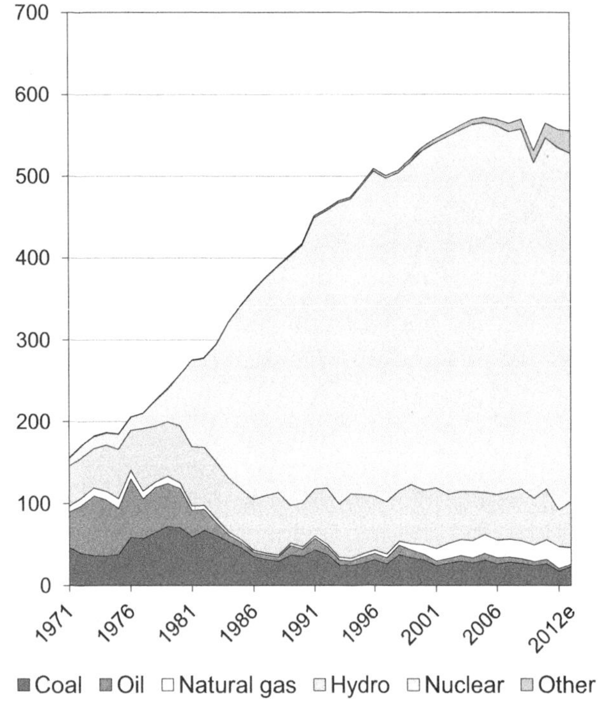

Figure 6: CO_2 emissions by fuel (Mt CO_2)

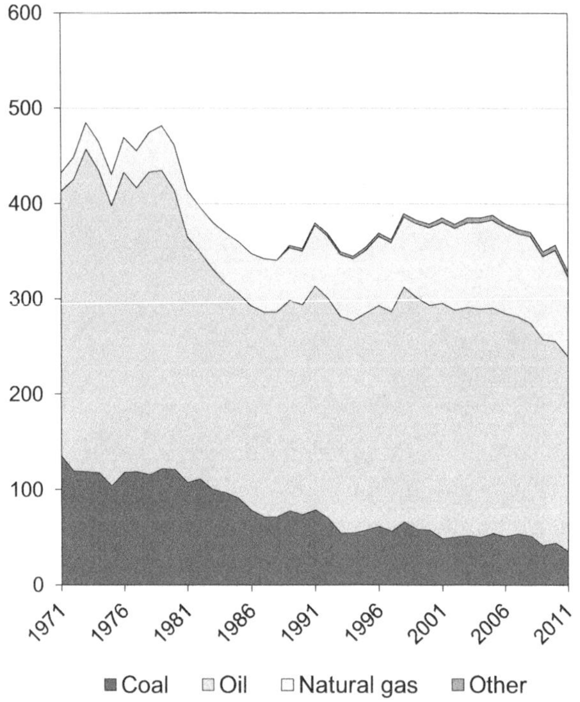

Figure 7: Electricity generation by fuel share

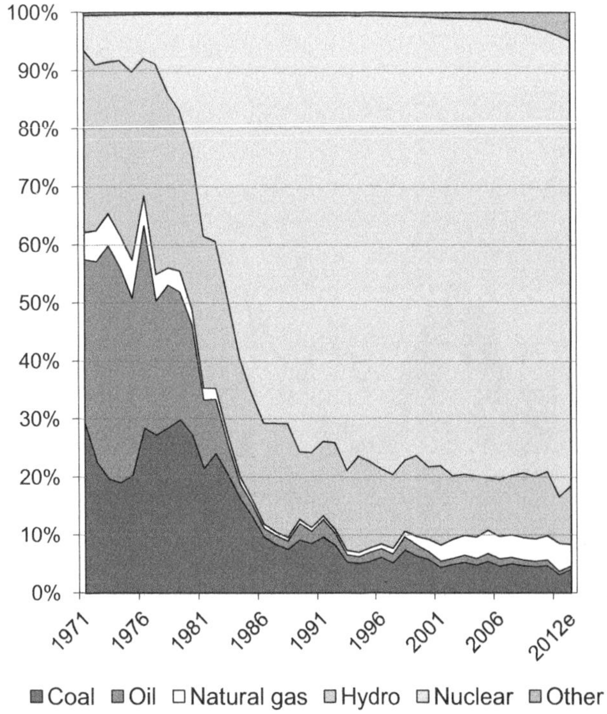

Figure 8: CO_2 emissions by fuel share

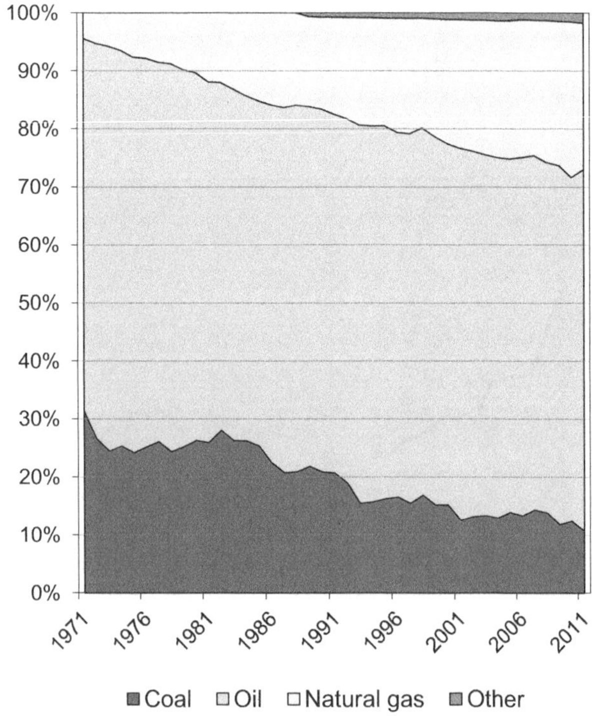

*Please refer to notes and definitions in Part I. Peat is included in coal for all figures.

Source: *IEA/OECD Electricity Information, IEA/OECD CO2 Emissions from Fuel Combustion*

FRANCE

1. Total primary energy supply (TPES) by fuel*

	1973	1980	1990	2000	2010	2011	2012e	Average annual percent change 73-90	90-11
TPES (Mtce)	**257.34**	**273.96**	**320.00**	**359.97**	**373.08**	**361.18**	**359.58**	**1.29**	**0.58**
Coal and peat	41.86	46.99	28.87	21.49	17.24	14.68	16.38	-2.16	-3.17
Oil	171.15	151.88	120.04	117.46	107.98	107.94	105.59	-2.06	-0.50
Natural Gas	19.28	30.91	37.18	51.08	60.75	52.90	54.61	3.94	1.69
Biofuels and waste	13.98	12.35	15.69	15.39	22.11	20.36	20.56	0.68	1.25
Nuclear	5.49	22.80	116.93	154.56	159.54	164.70	158.36	19.71	1.64
Hydro	5.86	8.54	6.61	8.25	7.66	5.51	6.93	0.71	-0.87
Geothermal	0.00	0.02	0.16	0.18	0.13	0.12	0.12	28.87	-1.33
Solar, wind, tide	0.07	0.07	0.10	0.10	1.44	1.91	2.49	2.11	15.20
Net electricity trade(1)	-0.35	0.40	-5.58	-8.54	-3.77	-6.93	-5.47	17.60	1.04
Heat(2)	-	-	-	-	-	-	-	-	-

(1) Net trade between OECD and non-OECD countries.

(2) Ambient heat from heat pumps used in the transformation sector.

Source: *IEA/OECD Energy Balances of OECD Countries*

2. Energy supply, GDP and population*

	1973	1980	1990	2000	2010	2011	2012e	Average annual percent change 73-90	90-11
GDP (2005 bil. USD)	1050	1284	1624	1973	2204	2249	2249	2.60	1.56
TPES/GDP(1)	0.25	0.21	0.20	0.18	0.17	0.16	0.16	-1.27	-0.97
Population (millions)	53.3	55.1	58.1	60.7	64.8	65.1	65.3	0.51	0.54
TPES/population(1)	4.83	4.97	5.50	5.93	5.76	5.55	5.51	0.77	0.04
TPES/GDP(2)	135.4	117.9	108.9	100.8	93.5	88.7	88.3	-1.27	-0.97
Coal and peat supply/GDP(2)	416.8	382.8	185.9	113.9	81.8	68.3	76.2	-4.64	-4.66
Elec. consumption/GDP(2)	71.5	82.3	94.1	98.6	101.8	94.3	..	1.62	0.01
Elec. generation (TWh)	183	257	417	536	564	557	555	4.98	1.38
Industrial production(2)	70.0	76.9	91.9	100.8	88.9	90.8	88.3	1.61	-0.06

(1) TPES/GDP in units of tce/2005 thousand USD, TPES/population in units of tce/capita.

(2) As index, 2005 (not shown) = 100

Source: *IEA/OECD Energy Balances of OECD Countries , OECD Main Economic Indicators*

3. Coal and peat production by type*

	1978(1)	1990	2000	2005	2010	2011	2012e	Average annual percent change 78-90	90-11
Mtce:									
Coking coal	4.80	1.80	-	-	-	-	-	-7.86	-
Steam coal	14.14	8.55	3.37	0.55	0.23	0.13	0.26	-4.11	-18.01
Lignite	1.17	1.43	0.17	-	-	-	-	1.68	-
Peat	-	-	-	-	-	-	-	-	-
Mt:									
Coking coal	4.66	1.82	-	-	-	-	-	-7.54	-
Steam coal	16.50	9.38	3.80	0.62	0.26	0.15	0.29	-4.60	-17.90
Lignite	2.73	2.33	0.30	-	-	-	-	-1.30	-
Peat	-	-	-	-	-	-	-	-	-

(1) Earliest year for which split by coal type is available.

* Please refer to notes and definitions in Part I.

Source: *IEA/OECD Energy Balances of OECD Countries, IEA/OECD Energy Statistics of OECD Countries*

FRANCE

4. Final consumption of energy by fuel*

(Mtce)

	1973	1980	1990	2000	2009	2010	2011	Average annual percent change 73-90	90-11
Total final consumption[(1)]	**203.17**	**201.84**	**204.40**	**233.16**	**225.53**	**232.43**	**217.43**	**0.04**	**0.29**
Coal and peat	19.95	12.30	11.11	6.33	4.36	4.95	4.56	-3.38	-4.15
Oil	137.19	124.80	107.32	115.98	103.96	102.19	99.79	-1.43	-0.35
Natural Gas	14.68	27.52	34.17	45.91	44.39	47.13	39.44	5.10	0.68
Biofuels and wastes	12.73	11.16	13.80	12.83	17.03	18.36	16.65	0.47	0.90
Geothermal	0.00	0.02	0.16	0.18	0.12	0.13	0.12	28.87	-1.33
Solar, wind, tide	-	0.01	0.03	0.02	0.07	0.08	0.09	-	5.82
Electricity	18.26	25.69	37.13	47.29	51.35	54.56	51.57	4.26	1.58
Heat	0.37	0.34	0.68	4.62	4.25	5.03	5.22	3.69	10.17
of which:									
Total industry	**59.55**	**58.87**	**47.09**	**49.63**	**39.14**	**41.50**	**40.75**	**-1.37**	**-0.69**
Coal and peat	10.40	7.71	8.37	5.20	3.76	4.30	4.05	-1.27	-3.39
Oil	31.22	27.53	9.37	8.02	7.38	7.15	7.15	-6.83	-1.28
Natural Gas	5.93	10.30	13.12	17.64	11.63	12.98	12.26	4.78	-0.32
Biofuels and wastes	1.67	1.61	2.14	2.23	2.64	2.64	2.80	1.44	1.30
Geothermal	-	-	-	-	-	-	-	-	-
Solar, wind, tide	-	-	-	-	-	-	-	-	-
Electricity	10.32	11.72	14.09	16.54	13.73	14.43	14.48	1.85	0.13
Heat	-	-	-	-	-	-	-	-	-
Total transport	**35.29**	**43.30**	**55.05**	**64.34**	**62.58**	**62.60**	**62.40**	**2.65**	**0.60**
Coal and peat	0.09	0.03	-	-	-	-	-	-	-
Oil	34.39	42.42	53.96	62.43	57.44	57.53	57.33	2.68	0.29
Natural Gas	0.02	0.01	0.00	0.00	0.09	0.07	0.07	-23.40	31.79
Biofuels and wastes	-	-	-	0.47	3.52	3.46	3.48	-	-
Electricity	0.79	0.85	1.09	1.44	1.54	1.54	1.52	1.95	1.60
Residential	**30.57**	**32.28**	**51.06**	**58.20**	**60.94**	**62.08**	**52.78**	**3.06**	**0.16**
Coal and peat	9.17	4.56	2.37 e	0.90	0.29	0.32	0.25	-7.64	-10.19
Oil	2.61	2.99	15.65	13.66	11.77	10.62	8.86	11.12	-2.67
Natural Gas	4.17	7.74	9.41	18.08 e	20.27	19.81	16.11	4.91	2.59
Biofuels and wastes	10.92	9.41	11.54	9.54	9.83	11.22	9.16	0.32	-1.09
Geothermal	0.00	0.01	0.15	0.17	0.07	0.08	0.07	28.81	-3.55
Solar, wind, tide	-	0.01	0.03	0.02	0.06	0.07	0.08	-	5.00
Electricity	3.71	7.56	11.91	15.81	18.64	19.96	18.26	7.11	2.06
Heat	-	-	-	-	-	-	-	-	-
Comm & public services	**52.32**	**43.15**	**25.82**	**25.91**	**31.86**	**33.67**	**29.92**	**-4.07**	**0.71**
Coal and peat	0.28	0.00	-	-	0.15	0.19	0.14	-	-
Oil	46.32	31.67	7.28	6.15	5.51	5.02	4.25	-10.31	-2.53
Natural Gas	2.42	6.12	8.72	6.45 e	8.48	9.56	8.02	7.84	-0.40
Biofuels and waste	-	-	0.07	0.53	0.98	0.98	1.14	-	14.35
Geothermal	-	-	-	-	0.04	0.04	0.04	-	-
Solar, wind, tide	-	-	-	-	0.01	0.01	0.01	-	-
Electricity	3.30	5.36	9.75	12.78	16.69	17.87	16.32	6.58	2.48
Heat	-	-	-	-	-	-	-	-	-
Non-energy use	**20.86**	**19.16**	**19.09**	**23.09**	**17.41**	**17.34**	**17.60**	**-0.52**	**-0.39**
Coal and peat	-	-	0.37	0.22	0.09	0.09	0.08	-	-7.18
Oil	18.72	15.99	16.00	19.56	15.65	15.35	15.56	-0.92	-0.13
Natural Gas	2.14	3.17	2.72	3.32	1.66	1.91	1.96	1.43	-1.55

(1) Total Final Consumption (TFC) excludes use in transformation processes and energy industry own use.

* Please refer to notes and definitions in Part I. Agriculture, Forestry, Fishing and "Non-specified other" sectors are not reported here individually, but are accounted for in Total final consumption.

Source: *IEA/OECD Energy Balances of OECD Countries*

FRANCE

5. Coal balance*

(Mtce)

	1973	1980	1990	2000	2005	2010	2011	2012e	Average annual percent change 73-90	90-11
Production	25.8	19.1	11.8	3.5	0.5	0.2	0.1	0.3	-4.5	-19.2
Imports	15.4	30.1	19.5	19.3	20.2	17.7	14.7	15.8	1.4	-1.3
Exports	-1.9	-1.2	-0.9	-0.8	-0.9	-0.3	-0.1	-0.2	-4.1	-8.4
Stock changes	2.5	-1.0	-1.5	-0.6	0.6	-0.4	0.0	0.6		
Primary supply	**41.9**	**47.0**	**28.9**	**21.5**	**20.4**	**17.2**	**14.7**	**16.4**	**-2.2**	**-3.2**
Statistical differences	**2.0**	**-0.0**	**-0.3**	**0.4**	**0.2**	**-0.2**	**-0.9**	..		
Total transformation	**-20.8**	**-31.8 e**	**-16.4 e**	**-14.7 e**	**-14.2 e**	**-11.7 e**	**-8.7 e**	..	**-1.4**	**-3.0**
Electricity and heat gen.	-14.8	-24.4	-12.0	-10.8	-10.3	-8.3	-5.5	..	-1.2	-3.7
Main activity producers (1)	*-14.8*	*-16.9*	*-6.9*	*-9.4*	*-8.9*	*-7.4*	*-4.6*	..	*-4.4*	*-1.9*
Autoproducers	-	*-7.5*	*-5.1*	*-1.4*	*-1.4*	*-0.9*	*-0.9*	..	-	*-8.1*
Gas works	1.1	0.0	-	-	-	-	-	..	-	-
Coal transformation(2)	-7.1	-7.4 e	-4.4 e	-3.9 e	-3.9 e	-3.3 e	-3.2 e	..	-2.8	-1.4
BKB plants	-	-	-	-	-	-	-	..	-	-
Blast furnaces	*-6.3*	*-6.3 e*	*-3.8 e*	*-3.6 e*	*-3.2 e*	*-2.7 e*	*-2.6 e*	..	*-2.8*	*-1.9*
Coke ovens	*-1.3*	*-1.5*	*-0.8*	*-0.3*	*-0.7*	*-0.7*	*-0.7*	..	*-3.0*	*-0.6*
Patent fuel plants	*0.4*	*0.3*	*0.2*	*0.0*	*0.0*	-	-	..	*-3.9*	-
Other transformation(3)	-	-	-	-	-	-	-	..	-	-
Energy ind. own use	**-2.5**	**-2.7**	**-0.9**	**-0.8**	**-0.6**	**-0.5**	**-0.5**	..	**-6.1**	**-2.9**
Losses	**-0.6**	**-0.2**	**-0.2**	**-0.1**	-	-	-	..		
Final consumption(4)	**19.9**	**12.3**	**11.1**	**6.3**	**5.7**	**4.9**	**4.6**	..	**-3.4**	**-4.2**
Industry(5)	10.4	7.7	8.4	5.2	5.0	4.3	4.1	..	-1.3	-3.4
Iron and steel	*6.3*	*4.6 e*	*4.8 e*	*2.8 e*	*3.0 e*	*2.7 e*	*2.4 e*	..	*-1.6*	*-3.1*
Chemical	*1.2*	*0.8*	*1.0 e*	*c*	*0.5*	*0.4*	*0.4*	..	*-0.9*	*-4.7*
Non-metallic minerals	*0.5*	*0.6*	*1.0 e*	*0.2 e*	*0.3*	*0.6*	*0.7*	..	*4.1*	*-1.8*
Paper, pulp and print	*0.2*	*0.1*	*0.4 e*	*0.1 e*	*0.2*	*0.0*	*0.0*	..	*5.6*	*-12.1*
Other industry (6)	*2.3*	*1.6*	*1.2 e*	*2.1*	*0.9*	*0.6*	*0.6*	..	*-3.7*	*-3.5*
Transport(7)	0.1	0.0	-	-	-	-	-	..	-	-
Other	9.5	4.6	2.4	0.9	0.6	0.6	0.4	..	-7.8	-7.8
Comm. and pub. services	*0.3*	*0.0*	-	-	-	*0.2*	*0.1*	..	-	-
Residential	*9.2*	*4.6*	*2.4 e*	*0.9*	*0.5*	*0.3*	*0.2*	..	*-7.6*	*-10.2*
Other sectors (8)	-	-	-	-	*0.1 e*	*0.1*	*0.0*	..	-	-
Non-energy use	-	-	0.4	0.2	0.2	0.1	0.1	..	-	-7.2

(1) Main activity electricity and heat generation includes district heating.

(2) Coal transformation refers to the transformation of coal and peat to secondary and tertiary products (mainly cokes, briquettes, coal tar and off-gases). Gas works and Liquefaction are shown separately to match the *Energy Balances of OECD Countries.*

(3) Other transformation includes Liquefaction and Non-specified transformations.

(4) Final Consumption includes non-energy use and energy use (Industry, Transport and Other).

(5) Please refer to notes and definitions in Part I for detailed explanation.

(6) Other industry includes Non-ferrous metals, Transport equipment, Machinery, Mining and quarrying, Food and tobacco, Wood and wood products, Construction, Textile and leather, and Non-specified industry.

(7) Transport includes Rail and Inland waterways.

(8) Other sectors includes Agriculture/Forestry, Fishing and Non-specified other.

* "Coal" refers to all types of coal, primary (anthracite, coking coal, other bituminous coal, sub-bituminous coal, lignite/brown coal) and derived fuels (including patent fuel, BKB, coke oven coke, gas coke, coal tar, coke oven gas, gas works gas, blast furnace gas and other recovered gases), as well as peat, peat briquettes and oil shale. Quantities have been converted to Mtce using calorific values reported by the respective countries. Please refer to notes and definitions in Part I.

Source: *IEA/OECD Energy Balances of OECD Countries*

FRANCE

6. Use of coal for selected end-uses*

(million tonnes)

	1978	1990	2000	2005	2009	2010	2011	Average annual percent change 78-90	90-11
Steam coal	**31.60**	**19.12**	**15.26**	**14.92**	**12.82**	**12.83**	**10.79**	**-4.10**	**-2.69**
Total electricity and heat	22.16	11.03	10.42	10.31	8.11	7.60	5.27	-5.65	-3.46
Main activity producers	*16.75*	*7.00*	*9.49*	*9.35*	*7.45*	*6.97*	*4.65*	*-7.01*	*-1.93*
Autoproducers	*5.41*	*4.03*	*0.94*	*0.97*	*0.66*	*0.63*	*0.62*	*-2.43*	*-8.53*
Patent fuel/BKB plants	2.09	0.38	0.11	0.03	0.00	-	-	-13.19	-
Coke ovens/Liquefaction[1]	-	-	-	-	-	-	-	-	-
Blast furnace inputs	-	1.09 e	2.03 e	2.37 e	1.05 e	1.88 e	1.81 e	-	2.46
Gas manufacture	-	-	-	-	-	-	-	-	-
Industry	2.91	4.68	2.46	2.43	1.76	2.16	2.08	4.04	-3.79
Iron and steel	*1.47*	*1.63 e*	*0.43 e*	*0.59 e*	*0.32 e*	*0.56 e*	*0.52 e*	*0.88*	*-5.30*
Chemical	*0.38*	*0.81 e*	*..*	*0.43*	*0.30*	*0.44*	*0.39*	*6.50*	*-3.40*
Non-metallic minerals	*0.15*	*0.99 e*	*0.20 e*	*0.32*	*0.53*	*0.56*	*0.63*	*17.02*	*-2.12*
Paper, pulp and print	*0.04*	*0.49 e*	*0.16 e*	*0.19*	*0.04*	*0.04*	*0.03*	*22.74*	*-11.94*
Other industry	*0.87*	*0.76 e*	*1.67 e*	*0.91 e*	*0.57 e*	*0.56 e*	*0.50 e*	*-1.14*	*-1.91*
Other sectors[2]	3.09	1.65	0.70	0.48	0.43	0.51	0.39	-5.10	-6.68
Non-energy use	-	-	-	-	-	-	-	-	-
Coking coal	**12.98**	**9.67**	**6.54**	**6.22**	**3.50**	**4.50**	**3.75**	**-2.42**	**-4.41**
Total electricity and heat	-	-	-	-	-	-	-	-	-
Main activity producers	-	-	-	-	-	-	-	-	-
Autoproducers	-	-	-	-	-	-	-	-	-
Patent fuel/BKB plants	-	-	-	-	-	-	-	-	-
Coke ovens/Liquefaction[1]	12.98	9.52	6.54	5.85	4.35	4.33	4.08	-2.55	-3.96
Blast furnace inputs	-	-	-	-	-	-	-	-	-
Gas manufacture	-	-	-	-	-	-	-	-	-
Industry	-	-	-	-	-	-	-	-	-
Iron and steel	-	-	-	-	-	-	-	-	-
Chemical	-	-	-	-	-	-	-	-	-
Non-metallic minerals	-	-	-	-	-	-	-	-	-
Paper, pulp and print	-	-	-	-	-	-	-	-	-
Other industry	-	-	-	-	-	-	-	-	-
Other sectors[2]	-	-	-	-	-	-	-	-	-
Non-energy use	-	-	-	-	-	-	-	-	-
Lignite	**2.62**	**2.09**	**0.36**	**0.04**	**0.05**	**0.05**	**0.08**	**-1.85**	**-14.30**
Total electricity and heat	2.34	1.84	0.27	-	-	-	-	-1.97	-
Main activity producers	*1.17*	*0.59*	*0.27*	-	-	-	-	*-5.51*	-
Autoproducers	*1.17*	*1.25*	-	-	-	-	-	*0.55*	-
Patent fuel/BKB plants	-	-	-	-	-	-	-	-	-
Coke ovens/Liquefaction[1]	-	-	-	-	-	-	-	-	-
Gas manufacture	-	-	-	-	-	-	-	-	-
Industry	0.20	0.23	0.05	0.04	0.05	0.05	0.08	1.17	-4.79
Iron and steel	-	-	-	-	-	-	-	-	-
Chemical	*0.01*	*0.09 e*	-	-	-	-	-	*20.31*	-
Non-metallic minerals	*0.00*	-	-	-	-	-	-	-	-
Paper, pulp and print	-	-	-	-	-	-	-	-	-
Other industry	*0.19*	*0.14 e*	*0.05*	*0.04*	*0.05*	*0.05*	*0.08*	*-2.59*	*-2.45*
Other sectors[2]	0.07	0.08	0.04	-	-	-	-	0.80	-
Non-energy use	-	-	-	-	-	-	-	-	-

(1) Liquefaction primarily refers to direct distillation processes. Indirect liquefaction may be reported here or under gas manufacture.

(2) Other sectors are Residential, Commercial and public services, Agriculture/Forestry, Fishing and Non-specified other.

* Coal totals may not match the sum of the individual rows due to statistical differences. Please refer to notes and definitions in Part I.

Source: *IEA/OECD Energy Statistics of OECD Countries*

FRANCE

7. Fuel prices to end users*

	1978(1)	1990	1995	2000	2005	2009	2010	2011	2012
				(US dollars / tce)					
For electricity generation									
Steam coal	..	57.35	55.34	41.98	84.41	128.44	122.05	144.71	..
Heavy fuel oil	..	..	..	..	..	..	..	..	..
Natural gas	..	..	..	..	..	..	..	..	..
For industry									
Steam coal	61.82	128.20	145.13	105.48	..	..	..	..	..
Coking coal	51.84	62.62	59.55	45.43	106.19	197.67	209.31	278.39	..
High sulphur fuel oil	66.80	97.93	104.13	128.55	217.27	308.22	..	..	..
Low sulphur fuel oil	..	..	118.22	140.36	232.51	315.88	390.84	521.86	554.96
Natural gas	76.57	128.49	135.93	138.50	256.81	341.47	376.91	466.02	462.57
				(Euro / unit)(2)					
For electricity generation									
Steam coal	..	42.23	37.36	40.41	60.28	82.02	81.75	92.26	..
Heavy fuel oil	..	..	..	..	..	..	..	..	..
Natural gas	..	..	..	..	..	..	..	..	..
For industry									
Steam coal	37.73	94.40	97.98	101.53	..	..	..	..	..
Coking coal	37.12	54.09	47.16	51.29	88.96	148.08	164.46	208.22	..
High sulphur fuel oil	64.01	113.21	110.37	194.27	243.61	309.01	..	..	..
Low sulphur fuel oil	..	..	125.31	212.12	260.70	316.70	411.01	522.40	601.37
Natural gas	67.73	137.11	133.00	193.21	265.79	316.01	365.87	430.62	462.70

(1) Earliest year for which split by coal type is available.

(2) For steam coal, coking coal and heavy fuel oil per metric tonne; for natural gas per 10^7 kilocalories GCV.

* Please refer to notes and definitions in Part I. Steam coal prices relate to historical definitions.

Source: *IEA/OECD Energy Prices & Taxes*

8. Coal and peat trade by type of coal*

(Mtce)

	1978	1990	1995	2000	2005	2009	2010	2011	2012e
Total imports	**23.52**	**19.50**	**13.72**	**19.34**	**20.18**	**15.37**	**17.66**	**14.67**	**15.83**
Bituminous coal(1)	12.96	10.05	5.13	11.03	12.06	10.49	11.50	9.43	9.86
Coking coal	8.55	8.17	7.60	6.81	6.51	3.73	4.80	3.95	4.94
Sub-bituminous coal	-	-	-	-	-	-	-	-	-
Lignite	0.00	0.04	0.03	0.03	0.02	0.03	0.03	0.05	0.04
Peat	-	-	-	-	-	-	-	-	-
Coal products(2)	2.00	1.24	0.96	1.47	1.59	1.12	1.33	1.24	0.99
Total exports	**1.16**	**0.92**	**0.66**	**0.77**	**0.88**	**0.69**	**0.25**	**0.15**	**0.25**
Bituminous coal(1)	0.38	0.53	0.35	0.08	0.23	0.02	0.02	0.01	0.02
Coking coal	-	-	-	-	0.03	0.08	0.12	0.05	0.16
Sub-bituminous coal	-	-	-	-	-	-	-	-	-
Lignite	0.00	-	-	-	-	-	-	-	-
Peat	-	-	-	-	-	-	-	-	-
Coal products(2)	0.77	0.39	0.31	0.69	0.61	0.60	0.12	0.08	0.07

(1) Bituminous coal includes anthracite.

(2) Coal products includes products derived from coal, for example: coke, coal tar, briquettes, patent fuels.

* Please refer to notes and definitions in Part I.

Source: *IEA/OECD Energy Balances of OECD Countries, IEA/OECD Energy Statistics of OECD Countries*

FRANCE

9. Total coal imports by origin*

(thousand tonnes)

	1978(1)	1985	1990	1995	2000	2005	2010	2011	2012e
Total coal(2)	**23451**	**18873**	**19458**	**13243**	**19032**	**19887**	**17631**	**14509**	**15932**
Coking coal	**8316**	**7773**	**7848**	**7300**	**6543**	**6255**	**4615**	**3799**	**4744**
Australia	987	1949	2071	2310	2818	3800	2811	2245	3033
Canada	-	351	637	51	577	421	8	14	-
Czech Republic	-	-	-	-	-	-	-	-	-
Germany	4364	1853	817	-	-	263	-	15	46
Poland	1311	185	254	900	182	3	-	-	-
United Kingdom	1	38	-	-	-	-	-	-	-
United States	1462	3307	4019	3891	2667	1627	1636	1391	1496
Other OECD	-	-	50	87	1	80	-	1	8
China, People's Rep.	-	-	-	-	-	-	-	-	-
Colombia	-	1	-	30	130	61	-	-	-
Indonesia	-	-	-	-	-	-	-	-	-
South Africa	191	-	-	30	154	-	-	-	-
Former Soviet Union(3)	-	89	x	x	x	x	x	x	x
Russian Federation (4)	*x*	*x*	-	-	-	-	*107*	*133*	*100*
Other FSU (4)	*x*	*x*	-	-	-	-	-	-	*5*
Venezuela	-	-	-	1	14	-	-	-	25
Vietnam	-	-	-	-	-	-	37	-	-
Non-specified/other	-	-	-	-	-	-	16	-	31
Steam coal	**15125**	**11062**	**11541**	**5890**	**12437**	**13596**	**12964**	**10628**	**11112**
Australia	785	850	1480	276	1022	1509	670	410	352
Canada	-	410	50	-	-	70	163	186	52
Czech Republic	-	-	-	-	-	-	2	4	-
Germany	2217	1162	718	221	98	55	159	77	39
Poland	3441	883	141	159	881	1451	1349	125	241
United Kingdom	891	204	311	77	61	35	7	20	21
United States	36	678	2586	53	425	301	1648	2105	2574
Other OECD	98	280	222	5	1125	1149	460	130	93
China, People's Rep.	-	46	1776	900	524	15	6	17	34
Colombia	-	2	2033	900	1043	2455	3024	2909	3145
Indonesia	-	-	-	-	6	245	-	-	-
South Africa	6643	6444	863	2100	5694	4225	2407	1929	1861
Former Soviet Union(3)	853	75	x	x	x	x	x	x	x
Russian Federation (4)	*x*	*x*	*777*	*302*	*383*	*905*	*2753*	*2066*	*2426*
Other FSU (4)	*x*	*x*	-	*21*	-	-	*27*	*201*	*137*
Venezuela	-	-	560	707	935	542	262	348	74
Vietnam	-	-	-	156	140	100	23	1	-
Non-specified/other	161	28	24	13	100	539	4	100	63
Lignite	**10**	**38**	**69**	**53**	**52**	**36**	**52**	**82**	**76**

(1) Earliest year for which split by coal type is available.

(2) Total coal consists of steam coal, coking coal and lignite.

(3) For years prior to 1990, or if no details for individual Former Soviet Union republics.

(4) Data for former republics are not available separately for some OECD member countries.

* Please refer to notes and definitions in Part I.

Source: IEA/OECD Coal Statistics

FRANCE

11. Steam coal exports by destination*

(thousand tonnes)

	1978(1)	1985	1990	1995	2000	2005	2010	2011	2012e
World	**449**	**976**	**585**	**392**	**89**	**262**	**20**	**15**	**25**
Total OECD	**443**	**700**	**573**	**392**	**88**	**262**	**20**	**15**	**25**
Australia	-	-	-	-	-	-	-	-	-
Austria	-	1	8	1	-	-	-	-	-
Belgium	51	25	15	1	5	16	4	11	8
Canada	-	-	-	-	-	-	-	-	-
Chile	-	-	-	-	-	-	-	-	-
Czech Republic	-	-	-	-	-	-	-	-	-
Denmark	-	4	-	-	-	-	-	-	-
Estonia	-	-	-	-	-	-	-	-	-
Finland	-	-	-	-	-	-	-	-	-
France	-	-	-	-	-	-	-	-	-
Germany	264	412	343	250	75	1	5	-	16
Greece	-	-	-	-	-	-	-	-	-
Hungary	-	-	-	-	-	-	-	-	-
Iceland	-	-	-	20	-	-	-	-	-
Ireland	-	46	36	24	-	-	-	-	-
Israel	-	-	-	-	-	-	-	-	-
Italy	8	22	56	5	-	-	-	-	-
Japan	-	-	-	-	-	-	-	-	-
Korea	-	-	-	-	-	-	-	-	-
Luxembourg	17	2	2	-	-	-	2	4	-
Mexico	-	-	-	-	-	-	-	-	-
Netherlands	-	15	-	-	1	-	-	-	1
New Zealand	-	-	-	-	-	-	-	-	-
Norway	53	105	95	70	7	-	-	-	-
Poland	-	-	-	-	-	-	-	-	-
Portugal	-	-	1	-	-	-	-	-	-
Slovak Republic	-	-	-	-	-	-	-	-	-
Slovenia	-	-	-	-	-	-	-	-	-
Spain	-	-	10	6	-	-	2	-	-
Sweden	-	-	-	-	-	-	-	-	-
Switzerland	8	20	6	-	-	-	-	-	-
Turkey	42	-	-	-	-	-	-	-	-
United Kingdom	-	48	1	15	-	245	7	-	-
United States	-	-	-	-	-	-	-	-	-
Total non-OECD	**1**	**266**	**11**	**-**	**-**	**-**	**-**	**-**	**-**
Brazil	-	-	-	-	-	-	-	-	-
China(2)	-	-	-	-	-	-	-	-	-
Chinese Taipei	-	-	-	-	-	-	-	-	-
Egypt	-	-	-	-	-	-	-	-	-
India	-	-	-	-	-	-	-	-	-
Romania	-	-	-	-	-	-	-	-	-
Oth. Africa & Mid. East	-	-	11	-	-	-	-	-	-
Oth. non-OECD Americas	-	-	-	-	-	-	-	-	-
Other Asia & Oceania	-	-	-	-	-	-	-	-	-
Other non-OECD Europe and Eurasia	1	266	-	-	-	-	-	-	-
Non-specified/Other	**5**	**10**	**1**	**-**	**1**	**-**	**-**	**-**	**-**

(1) Earliest year for which split by coal type is available.

(2) China includes the People's Republic of China and Hong Kong, China.

* Please refer to notes and definitions in Part I. Steam coal includes all sub-bituminous coal.

Source: IEA/OECD Coal Statistics

FRANCE

12. Coal import values by origin*

(average unit value, CIF, USD/tonne)

	1990	1995	2000	2005	2007	2008	2009	2010	2011/12[1]
Coking coal[2]	**65.14**	**61.82**	**47.26**	**109.69**	**124.63**	**212.51**	**204.75**	**217.32**	**..**
Imports from:									
Australia	65.23	60.65	45.74	115.61	125.44	230.00	239.82	226.06	..
Canada	64.81	62.14	46.36	117.56	134.95	186.97	289.34	182.99	..
Czech Republic	..	..	..	..	..	..	..	..	..
Poland	65.24	67.65	47.64	..	..	..	..	..	..
United States	64.95	62.68	50.47	107.04	116.61	169.04	164.69	207.96	..
China	64.53	52.07	..	..	..	258.16	..	..	..
Colombia	..	..	34.84	..	..	..	..	..	..
Indonesia	..	..	..	..	..	..	..	..	..
South Africa	66.18	53.98	32.99	..	..	..	157.57	..	..
Former Soviet Union[3]	..	..	..	..	118.36	205.53	215.17	148.73	..
Other bituminous coal[4]	**50.42**	**49.29**	**37.21**	**75.23**	**84.49**	**135.53**	**113.60**	**108.32**	**..**
Imports from:									
Australia	47.89	50.95	38.29	79.28	72.97	118.16	101.22	151.45	..
Canada	62.44	42.97	..	121.47	146.11	121.26	165.48	263.60	..
Czech Republic	..	..	..	..	..	..	..	..	..
Poland	46.61	34.06	34.55	83.83	163.83	..	80.58	89.65	..
United States	48.77	50.21	34.84	75.04	92.02	125.30	146.07	111.45	..
China	48.79	36.52	30.35	..	..	230.24	..	..	..
Colombia	49.99	46.98	37.94	77.69	84.63	..	122.16	104.16	..
Indonesia	55.24	54.12	36.19	50.11	56.80	..	36.66	..	..
South Africa	..	45.55	33.28	70.02	76.34	139.33	109.51	95.16	..
Former Soviet Union[3]	44.72	47.86	36.55	71.20	77.92	142.92	92.95	110.63	..

Note: On occasion, shipment of extremely small quantities of high valued coal results in high import costs.

(1) Import data for steam and coking coals are currently unavailable for 2011 and 2012 due to resource constraints.

(2) Weighted average of individual countries using import volumes as weights. Prices exclude intra-EU trade.

(3) Former Soviet Union until 1991, Russian Federation starting in 1992.

(4) Bituminous steam coal only. Weighted average of trades. Prices exclude intra-EU trade.

* Please refer to notes and definitions in Part I.

Source: IEA/OECD Coal Statistics

GERMANY*

Figure 1: TPES by fuel (Mtce)

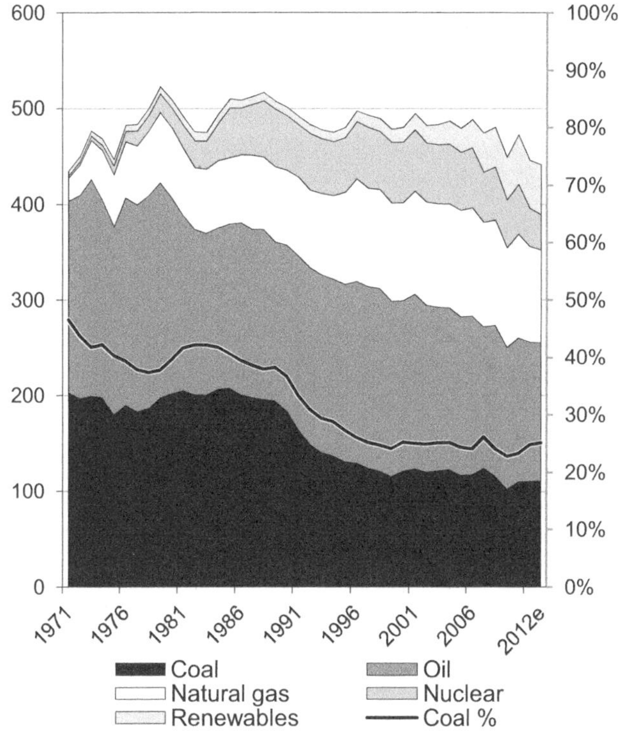

Figure 2: Coal supply indicators (1971 = 100)

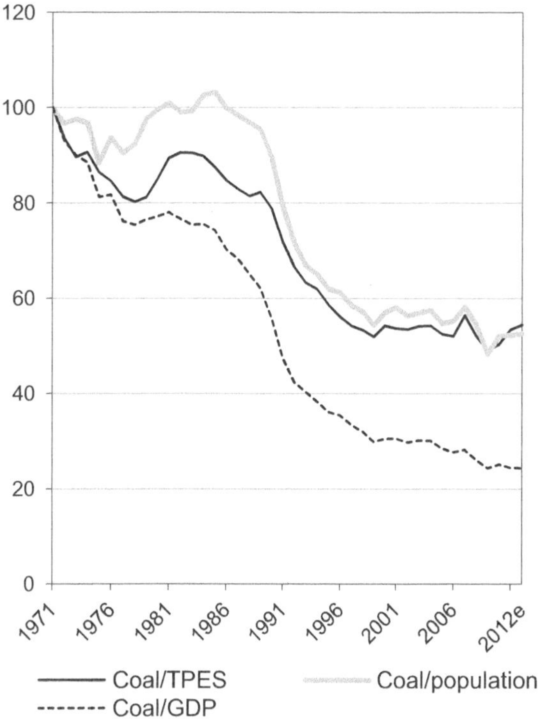

Figure 3: Primary coal supply (Mtce)

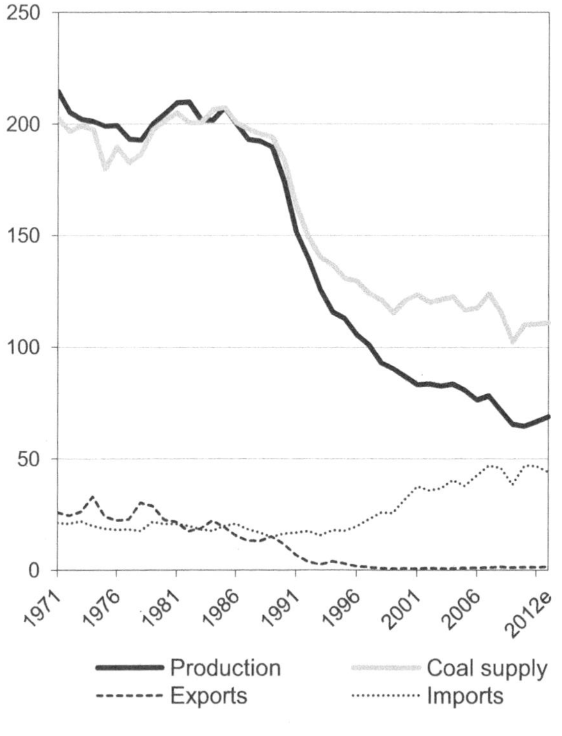

Figure 4: Coal consumption (Mtce)

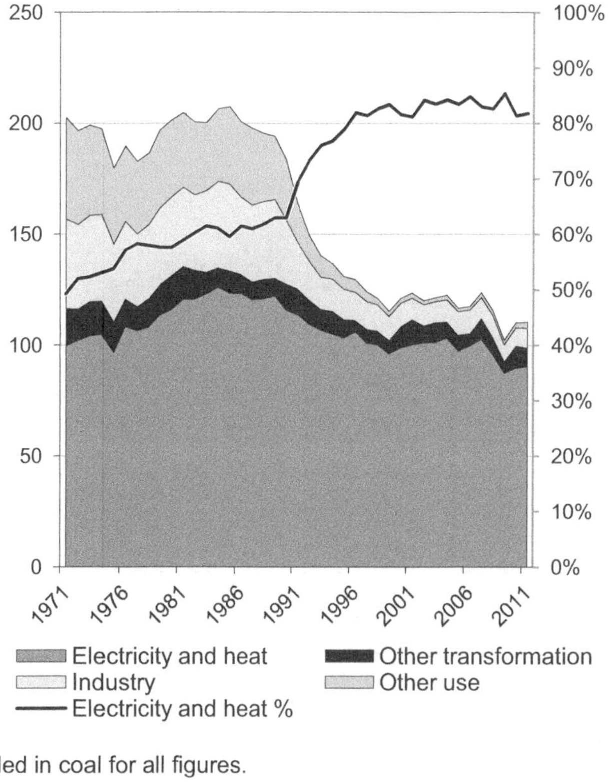

*Please refer to notes and definitions in Part I. Peat is included in coal for all figures.

Source: *IEA/OECD Coal Information, IEA/OECD Energy Balances of OECD Countries, OECD Main Economic Indicators*

GERMANY*

Figure 5: Electricity generation by fuel (TWh)

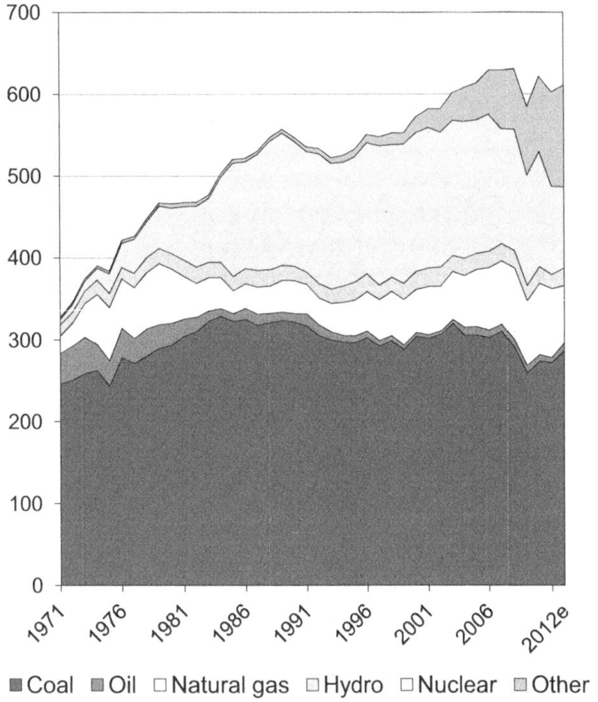

Figure 6: CO_2 emissions by fuel (Mt CO_2)

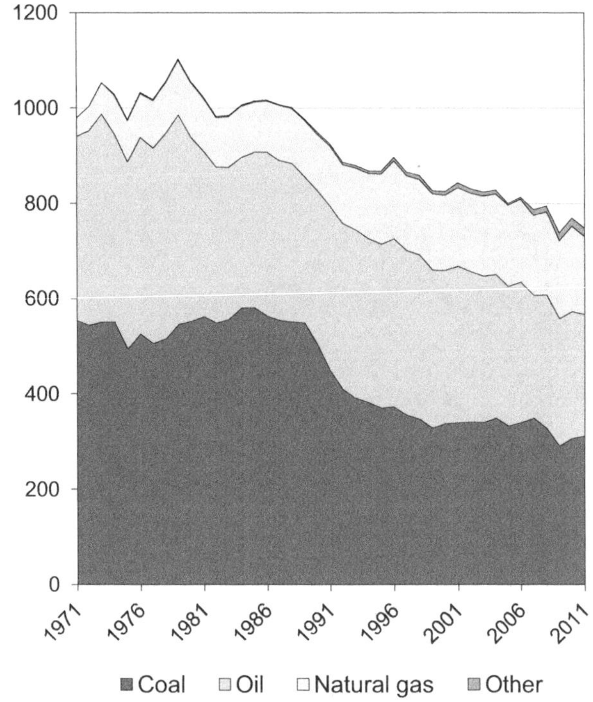

Figure 7: Electricity generation by fuel share

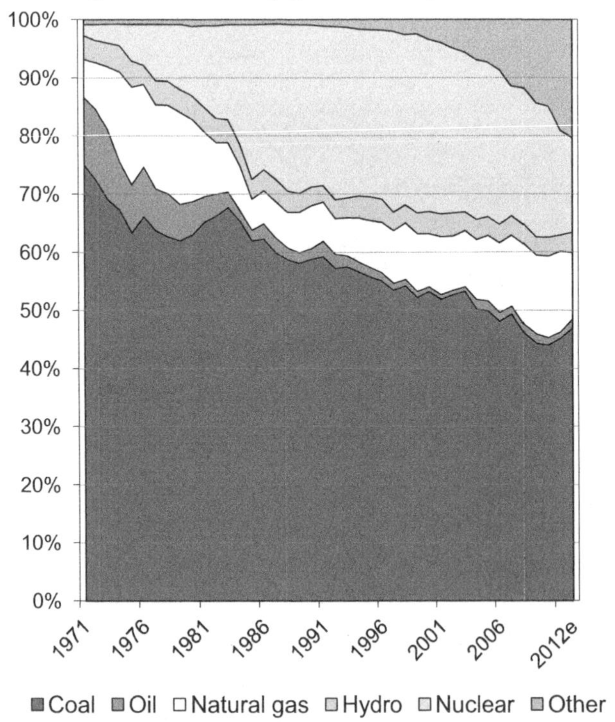

Figure 8: CO_2 emissions by fuel share

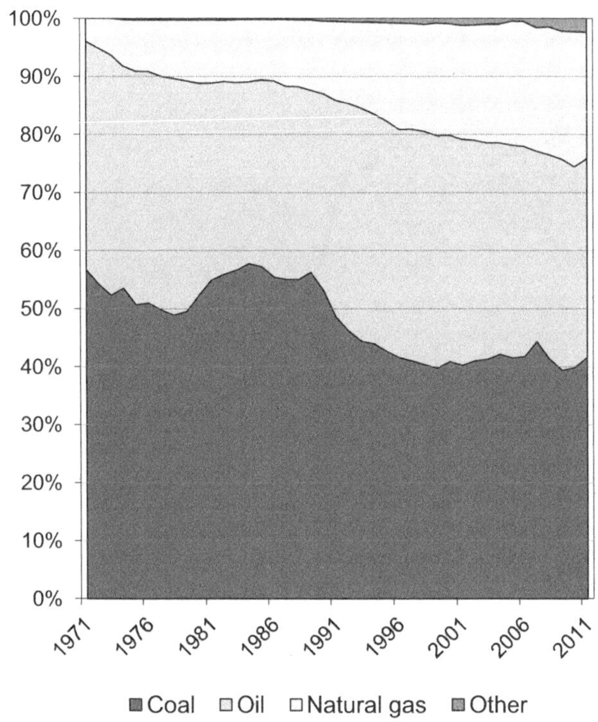

*Please refer to notes and definitions in Part I. Peat is included in coal for all figures.

Source: *IEA/OECD Electricity Information, IEA/OECD CO2 Emissions from Fuel Combustion*

GERMANY

1. Total primary energy supply (TPES) by fuel*

	1973	1980	1990	2000	2010	2011	2012e	Average annual percent change 73-90	90-11
TPES (Mtce)	**478.14**	**510.25**	**501.64**	**480.83**	**471.10**	**445.39**	**439.12**	**0.28**	**-0.56**
Coal and peat	199.14	201.45	183.66	121.18	110.17	110.50	110.99	-0.47	-2.39
Oil	226.71	205.52	173.41	178.13	150.23	145.60	144.31	-1.56	-0.83
Natural Gas	40.92	73.13	78.52	102.62	108.22	99.39	96.82	3.91	1.13
Biofuels and waste	3.57	6.22	6.85	11.23	42.01	38.00	39.01	3.90	8.50
Nuclear	4.51	20.71	56.91	63.14	52.33	40.23	37.03	16.09	-1.64
Hydro	1.87	2.34	2.14	2.67	2.51	2.12	2.63	0.79	-0.04
Geothermal	-	-	0.01	0.18	0.76	0.83	0.95	-	23.47
Solar, wind, tide	-	-	0.02	1.31	6.72	9.18	9.91	-	32.54
Net electricity trade(1)	1.42	0.88	0.11	0.38	-1.84	-0.46	-2.52	-13.78	x
Heat(2)	-	-	-0.00	-0.00	-0.01	-0.01	-0.01	-	3.23

(1) Net trade between OECD and non-OECD countries.

(2) Ambient heat from heat pumps used in the transformation sector.

Source: *IEA/OECD Energy Balances of OECD Countries*

2. Energy supply, GDP and population*

	1973	1980	1990	2000	2010	2011	2012e	Average annual percent change 73-90	90-11
GDP (2005 bil. USD)	1492	1761	2216	2685	2959	3049	3069	2.36	1.53
TPES/GDP(1)	0.32	0.29	0.23	0.18	0.16	0.15	0.14	-2.03	-2.06
Population (millions)	79.0	78.3	79.4	82.2	81.8	81.8	81.6	0.03	0.14
TPES/population(1)	6.06	6.52	6.32	5.85	5.76	5.45	5.38	0.25	-0.71
TPES/GDP(2)	185.2	167.4	130.8	103.4	92.0	84.4	82.7	-2.03	-2.06
Coal and peat supply/GDP(2)	316.4	271.2	196.4	107.0	88.3	85.9	85.7	-2.77	-3.86
Elec. consumption/GDP(2)	111.4	118.2	109.0	95.6	94.9	90.8	..	-0.13	-0.87
Elec. generation (TWh)	374	466	548	572	622	602	611	2.26	0.45
Industrial production(2)	65.4	70.6	85.1	93.9	105.0	113.5	112.8	1.56	1.38

(1) TPES/GDP in units of tce/2005 thousand USD, TPES/population in units of tce/capita.

(2) As index, 2005 (not shown) = 100

Source: *IEA/OECD Energy Balances of OECD Countries , OECD Main Economic Indicators*

3. Coal and peat production by type*

	1978(1)	1990	2000	2005	2010	2011	2012e	Average annual percent change 78-90	90-11
Mtce:									
Coking coal	51.64	44.06	18.66	15.01	7.07	7.23	6.26	-1.32	-8.25
Steam coal	34.97	27.23	15.86	10.75	5.03	4.81	5.30	-2.06	-7.92
Lignite	106.08	102.55	52.05	54.89	52.36	54.43	57.18	-0.28	-2.97
Peat	-	0.12	0.04	0.04	-	-	-	-	-
Mt:									
Coking coal	52.24	44.58	18.86	15.17	7.15	7.30	6.33	-1.31	-8.25
Steam coal	37.95	31.98	18.51	12.85	5.75	4.76	5.23	-1.42	-8.68
Lignite	377.89	357.47	167.69	177.91	169.40	176.50	185.43	-0.46	-3.30
Peat	-	0.43	0.15	0.13	-	-	-	-	-

(1) Earliest year for which split by coal type is available.

* Please refer to notes and definitions in Part I.

Source: *IEA/OECD Energy Balances of OECD Countries, IEA/OECD Energy Statistics of OECD Countries*

GERMANY

4. Final consumption of energy by fuel*

(Mtce)

	1973	1980	1990	2000	2009	2010	2011	Average annual percent change 73-90	90-11
Total final consumption(1)	**345.30**	**355.23**	**343.90**	**330.52**	**309.23**	**328.44**	**315.75**	**-0.02**	**-0.41**
Coal and peat	79.56	70.29	56.06	12.80	9.75	10.52	11.76	-2.04	-7.17
Oil	190.42	175.26	158.79	162.80	134.29	135.33	131.38	-1.06	-0.90
Natural Gas	26.54	47.83	55.79	78.71	72.80	80.85	73.10	4.47	1.29
Biofuels and wastes	2.50	2.72	4.25	6.72	15.73	19.34	19.50	3.17	7.53
Geothermal	-	-	0.01	0.18	0.57	0.65	0.81	-	23.28
Solar, wind, tide	-	-	0.02	0.16	0.58	0.64	0.79	-	20.47
Electricity	38.44	48.14	55.91	59.40	60.88	64.99	64.07	2.23	0.65
Heat	7.84	10.99	13.07	9.76	14.62	16.13	14.33	3.05	0.44
of which:									
Total industry	**126.06**	**112.62**	**94.52**	**73.33**	**70.82**	**78.14**	**78.50**	**-1.68**	**-0.88**
Coal and peat	38.94	35.60	28.91	10.56	7.67	8.22	8.94	-1.74	-5.44
Oil	46.15	25.59	10.10	7.35	4.23	3.76	2.99	-8.55	-5.63
Natural Gas	16.74	24.05	24.35	27.50	24.18	27.33	27.12	2.23	0.51
Biofuels and wastes	0.04	0.04	1.13	0.65	4.73	6.17	5.34	21.74	7.66
Geothermal	-	-	-	-	-	-	-	-	-
Solar, wind, tide	-	-	-	-	-	-	-	-	-
Electricity	21.91	24.52	26.60	26.00	24.82	27.69	28.34	1.15	0.30
Heat	2.27	2.82	3.44	1.29	5.18	4.97	5.78	2.47	2.50
Total transport	**51.61**	**63.14**	**77.71**	**84.80**	**76.35**	**76.67**	**77.38**	**2.44**	**-0.02**
Coal and peat	2.46	0.36	0.02	0.01	-	-	-	-24.22	-
Oil	47.94	61.23	76.01	82.51	69.54	69.25	70.20	2.75	-0.38
Natural Gas	-	0.09	-	-	1.00	1.14	0.93	-	-
Biofuels and wastes	-	-	-	0.32	3.85	4.23	4.20	-	-
Electricity	1.21	1.47	1.68	1.95	1.95	2.05	2.04	1.94	0.93
Residential	**86.85**	**78.39**	**89.96**	**93.14**	**84.37**	**90.40**	**80.09**	**0.21**	**-0.55**
Coal and peat	22.03	19.01	15.37	1.33 e	1.37	1.58	1.80	-2.10	-9.71
Oil	48.19	35.76	25.84	27.83	20.03	20.17	17.34	-3.60	-1.88
Natural Gas	0.00	0.05	19.15	33.46	31.59	34.60	28.76	69.17	1.96
Biofuels and wastes	1.41	1.28	3.11	5.75 e	7.13	8.93	8.30	4.77	4.78
Geothermal	-	-	0.01	0.18	0.57	0.65	0.74	-	22.74
Solar, wind, tide	-	-	0.01 e	0.08 e	0.56	0.62	0.77	-	22.79
Electricity	9.65	14.13	16.84	16.03	17.10	17.41	16.78	3.33	-0.02
Heat	5.57	8.17	9.63	8.47	6.02	6.44	5.61	3.27	-2.54
Comm & public services	**39.17**	**40.70**	**40.51**	**35.82**	**44.38**	**47.75**	**42.91**	**0.20**	**0.28**
Coal and peat	10.22	10.76	8.87	0.44 e	0.34	0.25	0.52	-0.83	-12.60
Oil	22.98	21.38	15.04	12.60	11.36	11.65	10.63	-2.46	-1.64
Natural Gas	-	-	6.69	8.29	13.26	14.37	12.98	-	3.21
Biofuels and waste	1.05	1.41	-	-	0.02	0.02	-	-	-
Geothermal	-	-	-	-	-	-	-	-	-
Solar, wind, tide	-	-	-	-	0.02	0.02	0.02	-	-
Electricity	4.92	7.15	9.91	14.49	15.95	16.73	15.81	4.20	2.25
Heat	-	-	-	-	3.43	4.71	2.95	-	-
Non-energy use	**24.61**	**32.80**	**32.78**	**35.81**	**32.25**	**34.37**	**34.03**	**1.70**	**0.18**
Coal and peat	3.21	2.22	1.20	0.39	0.36	0.47	0.50	-5.62	-4.09
Oil	20.27	26.76	28.36	32.35	29.12	30.49	30.22	2.00	0.30
Natural Gas	1.14	3.82	3.22	3.07	2.77	3.42	3.31	6.32	0.13

(1) Total Final Consumption (TFC) excludes use in transformation processes and energy industry own use.

* Please refer to notes and definitions in Part I. Agriculture, Forestry, Fishing and "Non-specified other" sectors are not reported here individually, but are accounted for in Total final consumption.

Source: *IEA/OECD Energy Balances of OECD Countries*

GERMANY

5. Coal balance*

(Mtce)

	1973	1980	1990	2000	2005	2010	2011	2012e	Average annual percent change 73-90	90-11
Production	202.0	204.5	174.0	86.6	80.7	64.5	66.5	68.7	-0.9	-4.5
Imports	21.8	20.6	16.4	31.7	37.6	46.8	46.7	44.1	-1.6	5.1
Exports	-26.1	-22.5	-11.7	-0.8	-0.9	-1.4	-1.4	-1.5	-4.6	-9.7
Stock changes	1.5	-1.1	4.9	3.6	-0.7	0.2	-1.2	-0.4		
Primary supply	**199.1**	**201.5**	**183.7**	**121.2**	**116.7**	**110.2**	**110.5**	**111.0**	**-0.5**	**-2.4**
Statistical differences	**0.4**	**0.2**	**0.6**	**0.4**	**1.1**	**-0.5**	**0.9**	..		
Total transformation	**-111.3 e**	**-125.3 e**	**-124.1 e**	**-106.6 e**	**-103.6**	**-97.3**	**-97.6**	..	**0.6**	**-1.1**
Electricity and heat gen.	-104.2	-115.9	-115.6 e	-98.8 e	-97.3	-89.6	-90.3	..	0.6	-1.2
Main activity producers (1)	*-99.8*	*-94.2*	*-98.7 e*	*-91.5 e*	*-91.2*	*-83.8*	*-84.3*	..	*-0.1*	*-0.7*
Autoproducers	*-4.3*	*-21.6*	*-16.9 e*	*-7.3*	*-6.1*	*-5.8*	*-6.0*	..	*8.4*	*-4.8*
Gas works	0.8	1.8	1.9	0.0	0.0	-	-	..	5.0	-
Coal transformation(2)	-8.0 e	-11.2 e	-10.5 e	-7.8 e	-6.3	-7.7	-7.3	..	1.6	-1.7
BKB plants	*1.9*	*1.2*	*-1.4*	*-0.2*	*-0.1*	*0.0*	*-0.0*	..	*-*	*-17.2*
Blast furnaces	*-9.2 e*	*-10.1 e*	*-8.0 e*	*-7.9 e*	*-6.3*	*-7.3*	*-6.8*	..	*-0.8*	*-0.8*
Coke ovens	*-0.9*	*-2.5*	*-1.2*	*0.2*	*0.1*	*-0.4*	*-0.4*	..	*1.5*	*-4.7*
Patent fuel plants	*0.3*	*0.2*	*0.2*	*0.0*	*0.0*	-	-	..	*-3.6*	-
Other transformation(3)	-	-	-	-	-	-	-	..	-	-
Energy ind. own use	**-7.0**	**-5.1**	**-3.5**	**-1.7 e**	**-1.5**	**-1.3**	**-1.4**	..	**-4.0**	**-4.1**
Losses	**-1.7**	**-1.0**	**-0.6**	**-0.5 e**	**-0.5**	**-0.6**	**-0.5**	..		
Final consumption(4)	**79.6**	**70.3**	**56.1**	**12.8**	**12.2**	**10.5**	**11.8**	..	**-2.0**	**-7.2**
Industry(5)	38.9	35.6	28.9	10.6	10.6	8.2	8.9	..	-1.7	-5.4
Iron and steel	*13.7 e*	*10.6 e*	*7.8 e*	*5.6 e*	*6.0*	*4.5*	*5.2*	..	*-3.2*	*-1.9*
Chemical	*9.3*	*8.3*	*6.4*	*0.6 e*	*0.9*	*1.0*	*0.7*	..	*-2.2*	*-9.9*
Non-metallic minerals	*1.8*	*3.0*	*3.5*	*2.8 e*	*1.9*	*1.8*	*2.0*	..	*3.8*	*-2.6*
Paper, pulp and print	*1.2*	*1.3*	*1.3*	*0.5 e*	*0.5*	*0.5*	*0.5*	..	*0.2*	*-4.6*
Other industry (6)	*12.9*	*12.4*	*9.9*	*1.0 e*	*1.1*	*0.5*	*0.5*	..	*-1.5*	*-12.9*
Transport(7)	2.5	0.4	0.0	0.0	-	-	-	..	-24.2	-
Other	34.9	32.1	25.9	1.8	1.3	1.8	2.3	..	-1.7	-10.9
Comm. and pub. services	*10.2*	*10.8*	*8.9*	*0.4 e*	*0.6*	*0.3*	*0.5*	..	*-0.8*	*-12.6*
Residential	*22.0*	*19.0*	*15.4*	*1.3 e*	*0.7*	*1.6*	*1.8*	..	*-2.1*	*-9.7*
Other sectors (8)	*2.7*	*2.3*	*1.7*	*0.1 e*	*0.0*	-	-	..	*-2.7*	-
Non-energy use	3.2	2.2	1.2	0.4	0.3	0.5	0.5	..	-5.6	-4.1

(1) Main activity electricity and heat generation includes district heating.

(2) Coal transformation refers to the transformation of coal and peat to secondary and tertiary products (mainly cokes, briquettes, coal tar and off-gases). Gas works and Liquefaction are shown separately to match the *Energy Balances of OECD Countries.*

(3) Other transformation includes Liquefaction and Non-specified transformations.

(4) Final Consumption includes non-energy use and energy use (Industry, Transport and Other).

(5) Please refer to notes and definitions in Part I for detailed explanation.

(6) Other industry includes Non-ferrous metals, Transport equipment, Machinery, Mining and quarrying, Food and tobacco, Wood and wood products, Construction, Textile and leather, and Non-specified industry.

(7) Transport includes Rail and Inland waterways.

(8) Other sectors includes Agriculture/Forestry, Fishing and Non-specified other.

* "Coal" refers to all types of coal, primary (anthracite, coking coal, other bituminous coal, sub-bituminous coal, lignite/brown coal) and derived fuels (including patent fuel, BKB, coke oven coke, gas coke, coal tar, coke oven gas, gas works gas, blast furnace gas and other recovered gases), as well as peat, peat briquettes and oil shale. Quantities have been converted to Mtce using calorific values reported by the respective countries. Please refer to notes and definitions in Part I.

Source: *IEA/OECD Energy Balances of OECD Countries*

GERMANY

6. Use of coal for selected end-uses*

(million tonnes)

	1978	1990	2000	2005	2009	2010	2011	Average annual percent change 78-90	90-11
Steam coal	**45.08**	**44.75**	**44.50**	**41.76**	**36.54**	**43.48**	**41.54**	**-0.06**	**-0.35**
Total electricity and heat	34.59	33.78 e	38.77 e	35.38	32.60	38.88	36.10	-0.20	0.32
Main activity producers	*21.05*	*26.06 e*	*35.67 e*	*32.74*	*31.03*	*36.21*	*33.25*	*1.79*	*1.17*
Autoproducers	*13.55*	*7.73 e*	*3.10*	*2.65*	*1.57*	*2.67*	*2.85*	*-4.57*	*-4.63*
Patent fuel/BKB plants	1.46	0.78	0.15	0.09	-	-	-	-5.14	-
Coke ovens/Liquefaction(1)	-	-	-	-	-	-	-	-	-
Blast furnace inputs	-	1.73	2.53	2.77	1.83	3.96	3.77	-	3.78
Gas manufacture	-	-	-	-	-	-	-	-	-
Industry	5.15	6.53	2.44	3.34	2.39	2.64	2.81	2.00	-3.95
Iron and steel	*0.25*	*-*	*0.21*	*1.41*	*0.74*	*0.82*	*1.53*	*-*	*-*
Chemical	*1.88*	*2.57*	*0.27*	*0.70*	*0.45*	*0.73*	*0.26*	*2.66*	*-10.39*
Non-metallic minerals	*0.48*	*1.64*	*1.35*	*0.34*	*0.59*	*0.49*	*0.51*	*10.75*	*-5.37*
Paper, pulp and print	*0.28*	*0.56*	*0.43*	*0.47*	*0.48*	*0.44*	*0.37*	*5.92*	*-1.97*
Other industry	*2.26*	*1.76*	*0.19*	*0.43*	*0.14*	*0.17*	*0.13*	*-2.05*	*-11.54*
Other sectors(2)	2.87	0.99	0.36	0.61	1.03	0.81	1.44	-8.47	1.80
Non-energy use	-	0.01	-	0.06	0.02	0.06	0.05	-	9.42
Coking coal	**41.88**	**42.22**	**24.46**	**22.19**	**15.47**	**15.11**	**16.48**	**0.07**	**-4.38**
Total electricity and heat	6.58	18.12	13.04	12.02	7.05	4.51	6.08	8.81	-5.07
Main activity producers	*5.70*	*14.24*	*10.17*	*10.73*	*6.71*	*4.21*	*5.60*	*7.93*	*-4.35*
Autoproducers	*0.88*	*3.88*	*2.87*	*1.28*	*0.34*	*0.31*	*0.48*	*13.17*	*-9.46*
Patent fuel/BKB plants	-	-	-	-	-	-	-	-	-
Coke ovens/Liquefaction(1)	34.30	24.09	11.42	10.18	8.42	10.60	10.40	-2.90	-3.92
Blast furnace inputs	-	-	-	-	-	-	-	-	-
Gas manufacture	1.01	-	-	-	-	-	-	-	-
Industry	-	-	-	-	-	-	-	-	-
Iron and steel	-	-	-	-	-	-	-	-	-
Chemical	-	-	-	-	-	-	-	-	-
Non-metallic minerals	-	-	-	-	-	-	-	-	-
Paper, pulp and print	-	-	-	-	-	-	-	-	-
Other industry	-	-	-	-	-	-	-	-	-
Other sectors(2)	-	-	-	-	-	-	-	-	-
Non-energy use	-	-	-	-	-	-	-	-	-
Lignite	**383.17**	**364.05**	**169.94**	**177.89**	**169.85**	**169.52**	**176.68**	**-0.43**	**-3.38**
Total electricity and heat	224.46	227.90 e	157.39 e	165.28	156.80	154.93	160.21	0.13	-1.66
Main activity producers	*210.89*	*209.11 e*	*155.39 e*	*163.32*	*155.02*	*151.80*	*157.25*	*-0.07*	*-1.35*
Autoproducers	*13.57*	*18.79*	*2.00*	*1.95*	*1.78*	*3.13*	*2.97*	*2.75*	*-8.42*
Patent fuel/BKB plants	130.03	104.92	11.36	11.31	11.88	13.05	14.44	-1.77	-9.01
Coke ovens/Liquefaction(1)	-	-	-	-	-	-	-	-	-
Gas manufacture	0.02	0.14	-	-	-	-	-	18.25	-
Industry	20.55	20.36	0.45	0.37	0.35	0.57	0.60	-0.08	-15.44
Iron and steel	*0.24*	*0.76*	*-*	*-*	*0.01*	*0.00*	*-*	*10.07*	*-*
Chemical	*10.63*	*7.55*	*0.31*	*0.18*	*0.23*	*0.41*	*0.42*	*-2.81*	*-12.89*
Non-metallic minerals	*0.13*	*0.49*	*0.00*	*0.01*	*-*	*0.02*	*0.02*	*11.48*	*-13.93*
Paper, pulp and print	*0.88*	*0.60*	*0.00*	*0.03*	*-*	*-*	*-*	*-3.20*	*-*
Other industry	*8.66*	*10.96*	*0.13*	*0.15*	*0.12*	*0.14*	*0.17*	*1.98*	*-18.11*
Other sectors(2)	2.38	8.38	0.05	0.02	-	-	-	11.06	-
Non-energy use	-	-	0.02	0.04	0.03	0.03	0.03	-	-

(1) Liquefaction primarily refers to direct distillation processes. Indirect liquefaction may be reported here or under gas manufacture.

(2) Other sectors are Residential, Commercial and public services, Agriculture/Forestry, Fishing and Non-specified other.

* Coal totals may not match the sum of the individual rows due to statistical differences. Please refer to notes and definitions in Part I.

Source: *IEA/OECD Energy Statistics of OECD Countries*

GERMANY

7. Fuel prices to end users*

	1978(1)	1990	1995	2000	2005	2009	2010	2011	2012
				(US dollars / tce)					
For electricity generation									
Steam coal	83.97	141.95	163.77	42.94	80.73	111.47	119.27	155.16	128.02
Heavy fuel oil	74.14	104.43	100.64	119.49	205.66	300.16	359.41	..	c
Natural gas	60.70	122.96	135.75	119.34	..	..	..	..	..
For industry									
Steam coal	98.52	185.76	..	..	..	..	..	..	..
Coking coal	62.91	63.02	60.34	c	c	c	c	c	c
High sulphur fuel oil	74.07	102.27	..	..	..	..	..	..	..
Low sulphur fuel oil	..	..	105.62	123.70	216.35	312.12	372.28	509.53	520.87
Natural gas	83.25	145.88	161.02	146.17	309.84	432.01	421.14	491.80	461.66
				(Euro / unit)(2)					
For electricity generation									
Steam coal	85.18	115.81	118.57	46.02	64.19	79.25	88.94	110.14	98.38
Heavy fuel oil	106.60	120.76	103.28	181.50	231.78	302.48	379.90	..	c
Natural gas	80.15	130.58	127.93	166.48	..	..	..	..	..
For industry									
Steam coal	101.18	153.44	..	..	..	..	..	..	..
Coking coal	64.15	51.68	43.91	c	c	c	c	c	c
High sulphur fuel oil	106.50	118.26	..	..	..	..	..	..	..
Low sulphur fuel oil	..	..	108.39	187.90	243.83	314.53	393.50	512.68	567.33
Natural gas	109.93	154.92	151.75	203.90	320.68	399.80	408.81	454.44	461.79

(1) Earliest year for which split by coal type is available.

(2) For steam coal, coking coal and heavy fuel oil per metric tonne; for natural gas per 10^7 kilocalories GCV.

* Please refer to notes and definitions in Part I. Steam coal prices relate to historical definitions.

Source: *IEA/OECD Energy Prices & Taxes*

8. Coal and peat trade by type of coal*

(Mtce)

	1978	1990	1995	2000	2005	2009	2010	2011	2012e
Total imports	**17.48**	**16.43**	**17.52**	**31.74**	**37.64**	**38.23**	**46.84**	**46.65**	**44.06**
Bituminous coal(1)	9.61	10.75	12.16	20.28	26.50	28.61	34.47	34.28	31.54
Coking coal	2.41	1.69	1.41	4.56	7.36	6.38	7.71	8.69	9.16
Sub-bituminous coal	-	-	-	-	-	-	-	-	-
Lignite	1.34	1.06	1.01	0.91	0.00	0.01	-	-	-
Peat	-	-	-	-	-	-	-	-	-
Coal products(2)	4.12	2.94	2.94	5.99	3.78	3.24	4.66	3.68	3.36
Total exports	**30.12**	**11.65**	**2.87**	**0.80**	**0.91**	**1.25**	**1.36**	**1.38**	**1.45**
Bituminous coal(1)	5.29	1.63	1.18	0.28	0.26	0.21	0.25	0.20	0.11
Coking coal	13.20	3.96	0.72	0.00	-	0.11	0.01	0.01	0.01
Sub-bituminous coal	-	-	-	-	-	-	-	-	-
Lignite	0.00	0.05	0.00	0.00	0.00	0.01	-	0.00	0.12
Peat	-	0.12	0.05	0.04	0.03	-	-	-	-
Coal products(2)	11.63	5.89	0.93	0.47	0.61	0.92	1.10	1.16	1.22

(1) Bituminous coal includes anthracite.

(2) Coal products includes products derived from coal, for example: coke, coal tar, briquettes, patent fuels.

* Please refer to notes and definitions in Part I.

Source: *IEA/OECD Energy Balances of OECD Countries, IEA/OECD Energy Statistics of OECD Countries*

GERMANY

9. Total coal imports by origin*

(thousand tonnes)

	1978(1)	1985	1990	1995	2000	2005	2010	2011	2012e
Total coal(2)	**17656**	**18199**	**15663**	**17184**	**29744**	**37114**	**45725**	**47845**	**45195**
Coking coal	**2435**	**2235**	**1706**	**1427**	**4608**	**7152**	**7793**	**8778**	**9256**
Australia	-	4	58	434	3414	3403	2577	2551	2709
Canada	-	-	25	20	865	1485	557	830	783
Czech Republic	-	-	28	1	-	-	-	-	-
Germany	-	-	-	-	-	-	-	-	-
Poland	-	82	116	147	50	832	-	492	774
United Kingdom	48	20	-	-	-	-	-	-	-
United States	553	26	48	613	257	1135	2393	2815	2965
Other OECD	89	102	2	210	19	-	-	-	-
China, People's Rep.	-	-	-	1	2	-	-	-	-
Colombia	-	-	-	1	-	132	312	519	527
Indonesia	-	-	-	-	-	-	-	-	-
South Africa	-	-	-	-	1	90	574	515	452
Former Soviet Union(3)	20	363	x	x	x	x	x	x	x
Russian Federation (4)	*x*	*x*	*177*	-	-	*49*	*1030*	*987*	*989*
Other FSU (4)	*x*	*x*	-	-	-	-	-	-	-
Venezuela	-	-	-	-	-	-	-	-	-
Vietnam	-	-	-	-	-	-	-	-	-
Non-specified/other	1725	1638	1252	-	-	26	350	69	57
Steam coal	**10430**	**13312**	**11874**	**13625**	**23340**	**29953**	**37932**	**39067**	**35939**
Australia	763	720	1094	505	301	768	1529	1559	1640
Canada	428	235	45	83	-	172	587	893	719
Czech Republic	152	346	248	922	1061	12	15	39	-
Germany	-	-	-	-	-	-	-	-	-
Poland	2041	2709	2583	3116	6744	7924	5835	3229	2721
United Kingdom	554	180	284	36	37	-	1	-	-
United States	399	1072	689	2066	432	132	3321	4865	7454
Other OECD	637	918	637	988	1236	64	137	179	174
China, People's Rep.	21	9	8	84	67	-	-	17	-
Colombia	-	19	128	995	2719	2937	7548	9919	8008
Indonesia	-	-	38	194	149	-	-	-	-
South Africa	1108	3196	4512	4228	4577	8215	2714	2152	1670
Former Soviet Union(3)	96	166	x	x	x	x	x	x	x
Russian Federation (4)	*x*	*x*	*157*	*199*	*937*	*7497*	*10104*	*8700*	*9004*
Other FSU (4)	*x*	*x*	-	*13*	-	-	-	*201*	-
Venezuela	-	-	-	193	341	-	-	-	-
Vietnam	-	-	-	3	114	-	-	17	-
Non-specified/other	4231	3742	1451	-	4625	2232	6141	7297	4549
Lignite	**4791**	**2652**	**2083**	**2132**	**1796**	**9**	**-**	**-**	**-**

(1) Earliest year for which split by coal type is available.

(2) Total coal consists of steam coal, coking coal and lignite.

(3) For years prior to 1990, or if no details for individual Former Soviet Union republics.

(4) Data for former republics are not available separately for some OECD member countries.

* Please refer to notes and definitions in Part I.

Source: IEA/OECD Coal Statistics

GERMANY

10. Coking coal exports by destination*

(thousand tonnes)

	1978(1)	1985	1990	1995	2000	2005	2010	2011	2012e
World	**13354**	**5701**	**4002**	**730**	**3**	**-**	**6**	**11**	**6**
Total OECD	**12335**	**5640**	**3902**	**730**	**3**	**-**	**6**	**11**	**6**
Australia	-	-	-	1	-	-	-	-	-
Austria	213	146	-	-	1	-	-	-	-
Belgium	1897	960	717	481	1	-	-	-	-
Canada	-	-	-	-	-	-	-	-	-
Chile	-	-	-	-	-	-	-	-	-
Czech Republic	-	-	-	-	-	-	-	-	-
Denmark	-	-	-	-	-	-	-	-	-
Estonia	-	-	-	-	-	-	-	-	-
Finland	-	-	-	-	-	-	-	-	-
France	5020	2446	1443	1	1	-	-	-	-
Germany	-	-	-	-	-	-	-	-	-
Greece	-	-	-	-	-	-	-	-	-
Hungary	-	-	-	-	-	-	-	-	-
Iceland	-	-	-	-	-	-	-	-	-
Ireland	-	-	-	-	-	-	-	-	-
Israel	-	-	-	-	-	-	-	-	-
Italy	2480	1600	859	1	-	-	-	-	-
Japan	375	-	-	-	-	-	-	-	-
Korea	-	-	-	-	-	-	-	-	-
Luxembourg	286	-	-	-	-	-	-	-	-
Mexico	-	-	-	-	-	-	-	-	-
Netherlands	975	486	465	238	-	-	-	-	-
New Zealand	-	-	-	-	-	-	-	-	-
Norway	-	-	-	-	-	-	-	-	-
Poland	-	-	-	-	-	-	6	11	5
Portugal	-	-	-	7	-	-	-	-	-
Slovak Republic	-	-	-	-	-	-	-	-	-
Slovenia	-	-	-	-	-	-	-	-	-
Spain	447	-	415	-	-	-	-	-	-
Sweden	234	-	-	-	-	-	-	-	-
Switzerland	18	2	3	1	-	-	-	-	1
Turkey	89	-	-	-	-	-	-	-	-
United Kingdom	198	-	-	-	-	-	-	-	-
United States	103	-	-	-	-	-	-	-	-
Total non-OECD	**1019**	**61**	**100**	**-**	**-**	**-**	**-**	**-**	**-**
Brazil	15	-	-	-	-	-	-	-	-
China(2)	-	-	-	-	-	-	-	-	-
Chinese Taipei	-	-	-	-	-	-	-	-	-
Egypt	-	-	100	-	-	-	-	-	-
India	-	-	-	-	-	-	-	-	-
Romania	-	-	-	-	-	-	-	-	-
Oth. Africa & Mid. East	520	-	-	-	-	-	-	-	-
Oth. non-OECD Americas	5	-	-	-	-	-	-	-	-
Other Asia & Oceania	-	-	-	-	-	-	-	-	-
Other non-OECD Europe and Eurasia	479	61	-	-	-	-	-	-	-
Non-specified/Other	**-**	**-**	**-**	**-**	**-**	**-**	**-**	**-**	**-**

(1) Earliest year for which split by coal type is available.
(2) China includes the People's Republic of China and Hong Kong, China.
* Please refer to notes and definitions in Part I.
Source: IEA/OECD Coal Statistics

GERMANY

11. Steam coal exports by destination*

(thousand tonnes)

	1978[1]	1985	1990	1995	2000	2005	2010	2011	2012e
World	**5737**	**3539**	**1510**	**1120**	**276**	**255**	**247**	**193**	**102**
Total OECD	**5381**	**2293**	**1186**	**1120**	**275**	**252**	**41**	**36**	**71**
Australia	-	-	-	12	-	-	-	-	-
Austria	15	51	6	-	98	9	3	1	-
Belgium	2123	1022	678	853	82	78	5	5	13
Canada	-	-	-	-	-	-	-	-	-
Chile	-	-	-	-	-	-	-	-	-
Czech Republic	-	-	-	-	-	2	1	4	2
Denmark	944	3	3	-	-	1	1	-	3
Estonia	-	-	-	-	-	-	-	-	-
Finland	-	1	-	2	-	-	-	-	-
France	1574	525	81	144	75	74	-	15	12
Germany	-	-	-	-	-	-	-	-	-
Greece	-	-	-	-	1	-	-	-	-
Hungary	-	-	-	-	-	-	1	1	-
Iceland	-	-	-	-	-	-	-	-	-
Ireland	7	9	5	-	-	-	-	-	-
Israel	-	-	-	-	-	5	-	-	-
Italy	32	25	28	6	6	-	-	-	-
Japan	-	-	-	-	-	-	-	-	-
Korea	-	-	-	-	-	-	-	-	-
Luxembourg	14	7	1	-	-	-	1	-	-
Mexico	-	-	-	-	-	-	-	-	-
Netherlands	467	157	48	-	3	3	2	-	2
New Zealand	-	-	-	-	-	-	-	-	-
Norway	18	71	45	32	-	-	-	-	-
Poland	-	-	1	-	-	75	2	1	6
Portugal	6	1	-	-	-	-	-	-	-
Slovak Republic	-	-	-	-	-	2	1	1	-
Slovenia	-	-	-	-	-	1	2	-	-
Spain	-	-	42	-	3	-	-	-	-
Sweden	38	-	5	-	1	-	3	-	3
Switzerland	60	67	38	11	5	1	10	7	30
Turkey	32	-	-	-	-	-	4	-	-
United Kingdom	51	354	205	60	1	1	5	1	-
United States	-	-	-	-	-	-	-	-	-
Total non-OECD	**53**	**963**	**324**	**-**	**-**	**1**	**1**	**1**	**-**
Brazil	-	13	5	-	-	-	-	-	-
China[2]	-	-	-	-	-	-	-	-	-
Chinese Taipei	-	-	-	-	-	-	-	-	-
Egypt	-	-	-	-	-	-	-	-	-
India	-	-	-	-	-	-	-	-	-
Romania	-	-	-	-	-	-	-	-	-
Oth. Africa & Mid. East	-	6	3	-	-	-	-	-	-
Oth. non-OECD Americas	-	1	-	-	-	-	-	-	-
Other Asia & Oceania	-	-	-	-	-	1	1	-	-
Other non-OECD Europe and Eurasia	53	943	316	-	-	-	-	1	-
Non-specified/Other	**303**	**283**	**-**	**-**	**1**	**2**	**205**	**156**	**31**

(1) Earliest year for which split by coal type is available.

(2) China includes the People's Republic of China and Hong Kong, China.

* Please refer to notes and definitions in Part I. Steam coal includes all sub-bituminous coal.

Source: IEA/OECD Coal Statistics

GERMANY

12. Coal import values by origin*

(average unit value, CIF, USD/tonne)

	1990	1995	2000	2005	2007	2008	2009	2010	2011/12(1)
Coking coal(2)	**63.76**	**67.24**	**46.10**	**113.48**	**133.45**	**182.72**	**240.22**	**191.06**	**..**
Imports from:									
Australia	55.97	67.69	43.04	104.77	132.39	217.74	282.87	206.44	..
Canada	60.33	62.38	40.03	110.62	137.52	180.05	223.30	220.05	..
Czech Republic	66.67	78.37	..	..	..	..	..	189.56	..
Poland	66.68	57.59	44.16	157.75	147.19	430.07	158.89	292.74	..
United States	60.93	73.55	43.16	120.72	131.66	151.09	151.81	201.28	..
China	..	..	53.44	538.07	..	172.05	..	..	..
Colombia	96.62	..	..	..	..	..	193.09	211.34	..
Indonesia	..	..	..	..	..	..	..	..	..
South Africa	..	..	..	..	..	..	..	..	..
Former Soviet Union(3)	65.10	..	..	89.56	81.15	136.07	130.86	116.19	..
Other bituminous coal(4)	**57.87**	**53.53**	**33.45**	**72.48**	**81.49**	**138.84**	**100.87**	**101.29**	**..**
Imports from:									
Australia	62.06	51.00	32.19	..	64.73	129.56	152.06	174.90	..
Canada	66.66	88.38	..	..	116.92	..	..	..	..
Czech Republic	63.93	64.04	37.25	85.54	90.41	152.10	199.57	151.99	..
Poland	62.25	46.86	31.39	73.74	91.16	154.54	104.83	93.05	..
United States	55.69	50.00	35.85	73.70	98.66	135.78	102.27	104.33	..
China	73.16	33.06	37.57	85.99	228.55	..	..	..	..
Colombia	61.65	45.24	34.09	69.49	76.67	..	87.91	87.27	..
Indonesia	62.96	51.35	35.41	76.10	71.34	138.95	..	..	..
South Africa	47.20	46.38	33.42	67.69	75.05	144.76	82.97	113.77	..
Former Soviet Union(3)	47.72	41.35	34.88	68.13	76.21	139.38	93.84	105.75	..

Note: On occasion, shipment of extremely small quantities of high valued coal results in high import costs.

(1) Import data for steam and coking coals are currently unavailable for 2011 and 2012 due to resource constraints.

(2) Weighted average of individual countries using import volumes as weights. Prices exclude intra-EU trade.

(3) Former Soviet Union until 1991, Russian Federation starting in 1992.

(4) Bituminous steam coal only. Weighted average of trades. Prices exclude intra-EU trade.

* Please refer to notes and definitions in Part I.

Source: IEA/OECD Coal Statistics

GREECE*

Figure 1: TPES by fuel (Mtce)

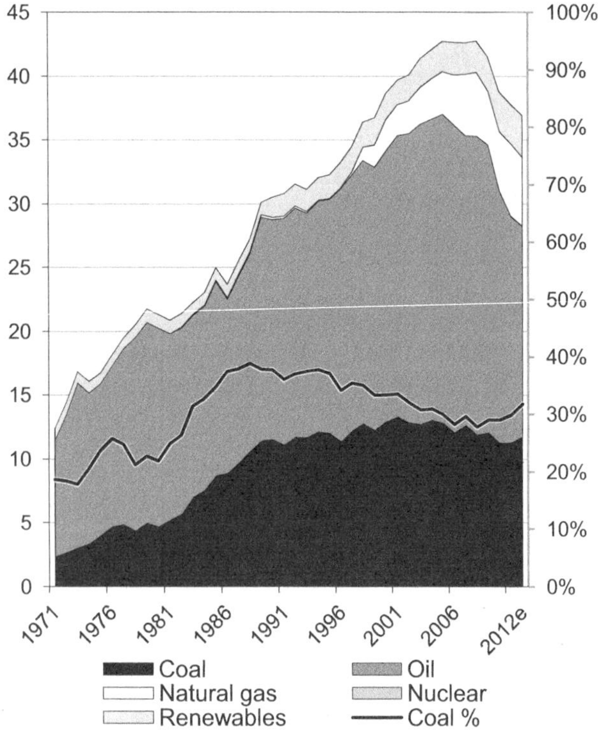

Figure 2: Coal supply indicators (1971 = 100)

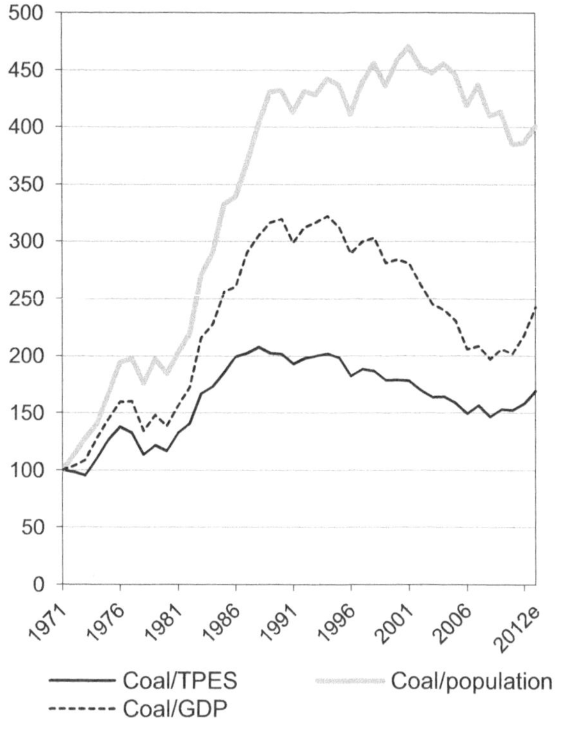

Figure 3: Primary coal supply (Mtce)

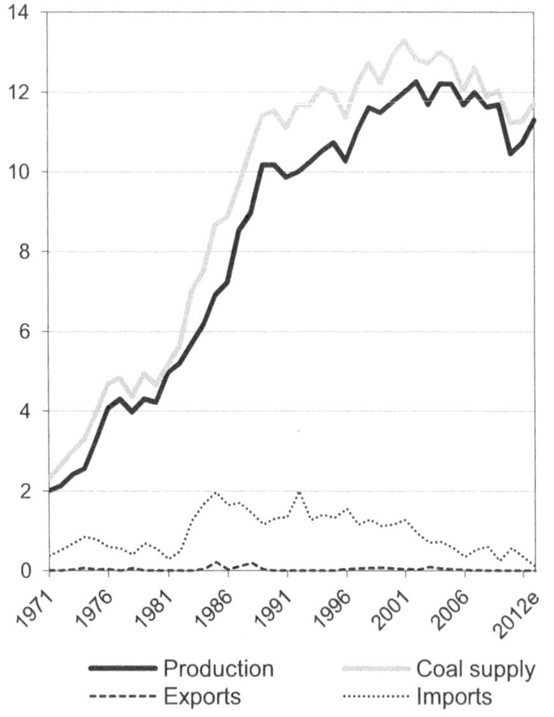

Figure 4: Coal consumption (Mtce)

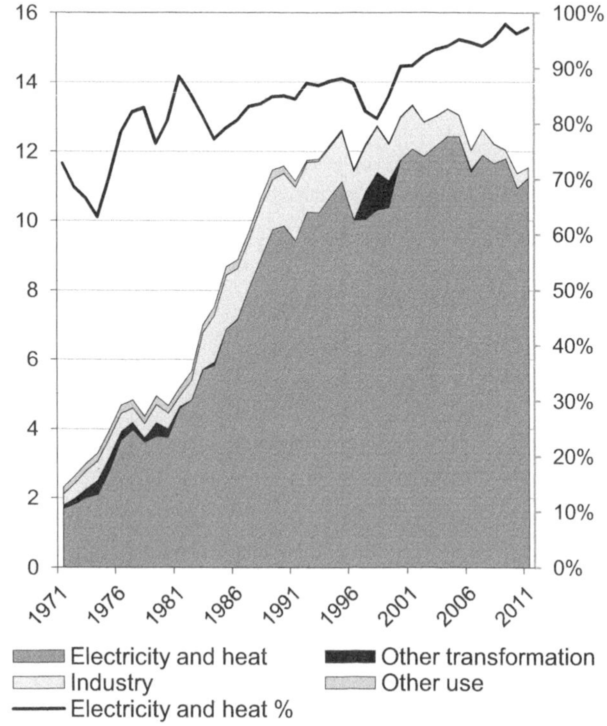

*Please refer to notes and definitions in Part I. Peat is included in coal for all figures.

Source: *IEA/OECD Coal Information, IEA/OECD Energy Balances of OECD Countries, OECD Main Economic Indicators*

GREECE*

Figure 5: Electricity generation by fuel (TWh)

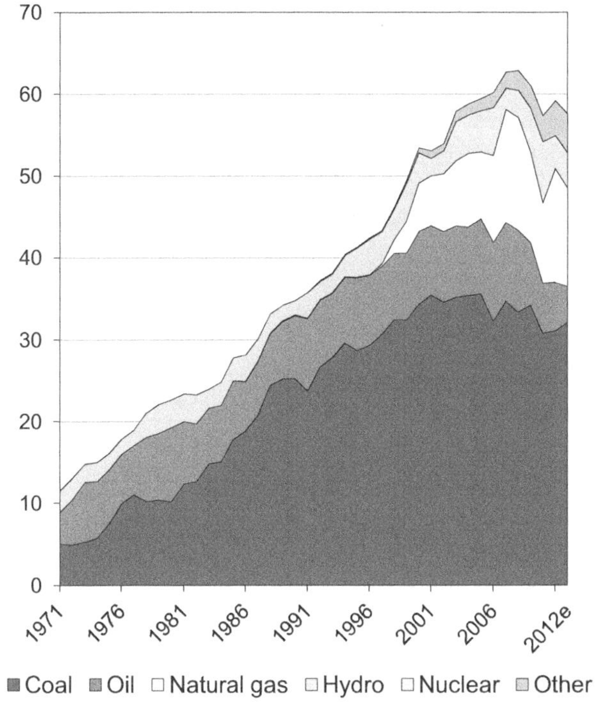

Figure 6: CO_2 emissions by fuel (Mt CO_2)

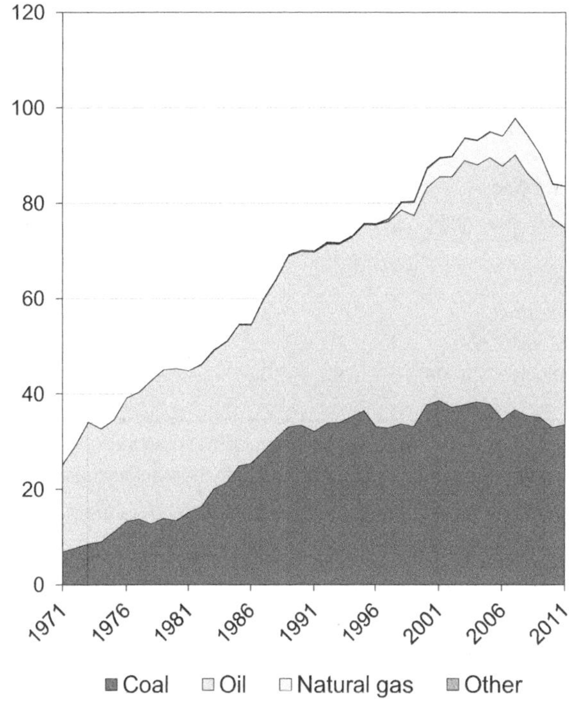

Figure 7: Electricity generation by fuel share

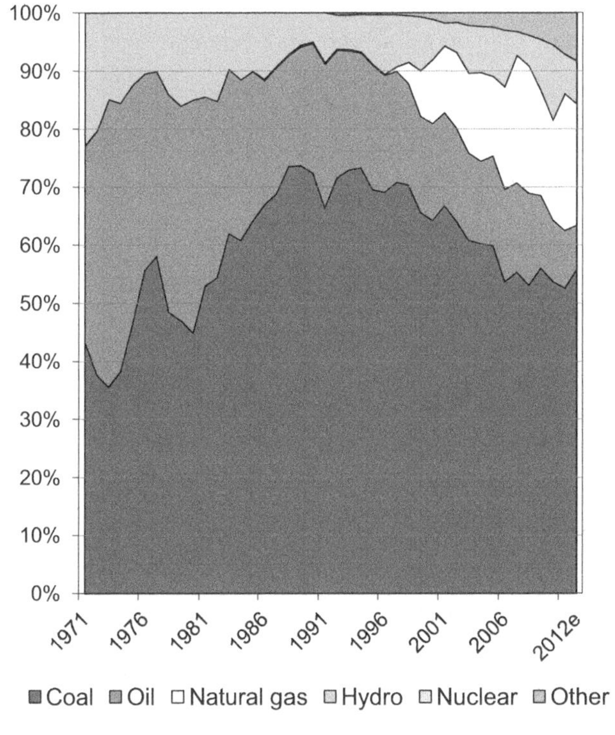

Figure 8: CO_2 emissions by fuel share

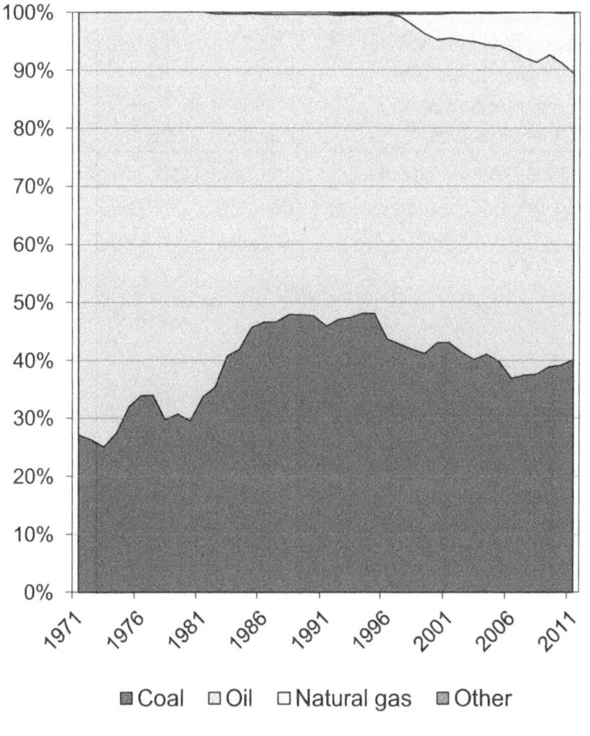

*Please refer to notes and definitions in Part I. Peat is included in coal for all figures.

Source: *IEA/OECD Electricity Information, IEA/OECD CO2 Emissions from Fuel Combustion*

GREECE

1. Total primary energy supply (TPES) by fuel*

	1973	1980	1990	2000	2010	2011	2012e	Average annual percent change 73-90	90-11
TPES (Mtce)	**16.87**	**21.40**	**30.63**	**38.69**	**39.45**	**38.18**	**37.13**	**3.57**	**1.05**
Coal and peat	3.01	4.66	11.52	12.91	11.23	11.27	11.72	8.23	-0.11
Oil	12.95	15.60	17.24	21.26	19.79	17.76	16.53	1.70	0.14
Natural Gas	-	-	0.20	2.43	4.62	5.67	5.38	-	17.34
Biofuels and waste	0.64	0.64	1.28	1.44	1.54	1.80	1.91	4.12	1.65
Nuclear	-	-	-	-	-	-	-	-	-
Hydro	0.27	0.42	0.22	0.45	0.92	0.49	0.53	-1.33	3.98
Geothermal	-	-	0.00	0.00	0.04	0.04	0.04	-	11.73
Solar, wind, tide	-	-	0.08	0.20	0.61	0.74	0.80	-	11.14
Net electricity trade(1)	0.01	0.08	0.09	-0.00	0.70	0.40	0.22	17.63	7.48
Heat(2)	-	-	-	-	-	-	-	-	-

(1) Net trade between OECD and non-OECD countries.

(2) Ambient heat from heat pumps used in the transformation sector.

Source: *IEA/OECD Energy Balances of OECD Countries*

2. Energy supply, GDP and population*

	1973	1980	1990	2000	2010	2011	2012e	Average annual percent change 73-90	90-11
GDP (2005 bil. USD)	120	146	156	197	241	224	210	1.59	1.73
TPES/GDP(1)	0.14	0.15	0.20	0.20	0.16	0.17	0.18	1.95	-0.66
Population (millions)	9.1	9.8	10.3	10.9	11.3	11.3	11.3	0.76	0.43
TPES/population(1)	1.86	2.18	2.96	3.54	3.49	3.38	3.28	2.78	0.62
TPES/GDP(2)	78.4	81.5	108.9	109.2	91.0	94.8	98.4	1.95	-0.66
Coal and peat supply/GDP(2)	47.2	60.0	138.4	123.1	87.5	94.5	105.0	6.53	-1.80
Elec. consumption/GDP(2)	50.1	64.3	85.9	103.3	104.0	109.1	..	3.22	1.14
Elec. generation (TWh)	15	23	35	53	57	59	58	5.15	2.56
Industrial production(2)	56.9	76.1	83.7	100.6	84.7	78.1	75.5	2.30	-0.33

(1) TPES/GDP in units of tce/2005 thousand USD, TPES/population in units of tce/capita.

(2) As index, 2005 (not shown) = 100

Source: *IEA/OECD Energy Balances of OECD Countries* , *OECD Main Economic Indicators*

3. Coal and peat production by type*

	1978(1)	1990	2000	2005	2010	2011	2012e	Average annual percent change 78-90	90-11
Mtce:									
Coking coal	-	-	-	-	-	-	-	-	-
Steam coal	-	-	-	-	-	-	-	-	-
Lignite	3.97	10.17	11.75	12.20	10.45	10.72	11.29	8.16	0.25
Peat	-	-	-	-	-	-	-	-	-
Mt:									
Coking coal	-	-	-	-	-	-	-	-	-
Steam coal	-	-	-	-	-	-	-	-	-
Lignite	21.82	51.90	63.89	69.40	56.52	58.67	61.79	7.49	0.59
Peat	-	-	-	-	-	-	-	-	-

(1) Earliest year for which split by coal type is available.

* Please refer to notes and definitions in Part I.

Source: *IEA/OECD Energy Balances of OECD Countries, IEA/OECD Energy Statistics of OECD Countries*

GREECE

4. Final consumption of energy by fuel*

(Mtce)

	1973	1980	1990	2000	2009	2010	2011	Average annual percent change 73-90	90-11
Total final consumption[(1)]	**12.18**	**15.29**	**20.71**	**26.36**	**29.41**	**27.77**	**27.14**	**3.17**	**1.30**
Coal and peat	0.74	0.67	1.74	1.25	0.25	0.43	0.31	5.13	-7.89
Oil	9.24	11.53	13.97	17.73	19.23	17.40	16.32	2.46	0.74
Natural Gas	-	-	0.14	0.54	1.53	1.62	2.10	-	13.83
Biofuels and wastes	0.64	0.64	1.28	1.35	1.32	1.42	1.68	4.12	1.31
Geothermal	-	-	0.00	0.00	0.03	0.04	0.04	-	11.73
Solar, wind, tide	-	-	0.08	0.14	0.26	0.26	0.26	-	5.76
Electricity	1.56	2.45	3.50	5.30	6.72	6.53	6.36	4.86	2.89
Heat	-	-	-	0.04	0.07	0.07	0.08	-	-
of which:									
Total industry	**4.19**	**5.52**	**5.69**	**6.34**	**4.94**	**4.96**	**4.75**	**1.81**	**-0.86**
Coal and peat	0.53	0.45	1.53	1.22	0.24	0.43	0.30	6.39	-7.43
Oil	2.75	3.78	2.40	2.78	2.05	1.91	1.48	-0.80	-2.29
Natural Gas	-	-	-	0.35	0.58	0.53	0.79	-	-
Biofuels and wastes	-	-	0.27	0.33	0.35	0.35	0.37	-	1.50
Geothermal	-	-	-	-	-	-	-	-	-
Solar, wind, tide	-	-	-	-	-	-	0.00	-	-
Electricity	0.91	1.29	1.49	1.66	1.73	1.74	1.80	2.97	0.91
Heat	-	-	-	-	-	-	-	-	-
Total transport	**2.95**	**4.55**	**7.20**	**9.14**	**11.94**	**10.69**	**9.59**	**5.38**	**1.37**
Coal and peat	0.02	0.00	0.00	-	-	-	-	-17.43	-
Oil	2.92	4.54	7.18	9.11	11.78	10.46	9.39	5.43	1.28
Natural Gas	-	-	-	-	0.02	0.02	0.02	-	-
Biofuels and wastes	-	-	-	-	0.11	0.18	0.15	-	-
Electricity	0.01	0.01	0.02	0.03	0.03	0.02	0.03	5.92	2.38
Residential	**2.80**	**2.84**	**4.35**	**6.40**	**6.93**	**6.62**	**7.78**	**2.63**	**2.81**
Coal and peat	0.06	0.06	0.03	0.03	0.01	0.00	0.01	-3.78	-7.40
Oil	1.72	1.44	2.12	3.44	3.14	2.81	3.69	1.25	2.67
Natural Gas	-	-	-	0.01	0.37	0.36	0.50	-	-
Biofuels and wastes	0.64	0.64	1.00	1.00	0.84	0.86	1.09	2.65	0.40
Geothermal	-	-	..	..	0.02	0.02	0.01	..	..
Solar, wind, tide	-	-	0.08	0.14	0.26	0.26	0.25	-	5.57
Electricity	0.38	0.69	1.11	1.75	2.23	2.23	2.17	6.58	3.21
Heat	-	-	-	0.04	0.07	0.07	0.08	-	-
Comm & public services	**0.25**	**0.40**	**0.93**	**1.87**	**3.06**	**2.78**	**2.68**	**7.96**	**5.17**
Coal and peat	0.00	0.00	0.01	-	-	-	-	11.79	-
Oil	-	-	0.23	0.35	0.42	0.36	0.33	-	1.72
Natural Gas	-	-	-	0.01	0.21	0.20	0.24	-	-
Biofuels and waste	-	-	-	0.00	0.00	0.00	0.02	-	-
Geothermal	-	-	-	-	..	..	0.02	-	-
Solar, wind, tide	-	-	0.00	0.00	0.01	0.01	0.01	-	14.94
Electricity	0.25	0.40	0.69	1.51	2.43	2.21	2.07	6.12	5.37
Heat	-	-	-	-	-	-	-	-	-
Non-energy use	**0.76**	**0.71**	**1.00**	**1.03**	**1.29**	**1.58**	**1.26**	**1.63**	**1.11**
Coal and peat	0.12	0.15	0.17	-	-	-	-	1.77	-
Oil	0.64	0.56	0.70	0.85	0.94	1.08	0.71	0.52	0.10
Natural Gas	-	-	0.14	0.17	0.35	0.51	0.55	-	6.81

(1) Total Final Consumption (TFC) excludes use in transformation processes and energy industry own use.

* Please refer to notes and definitions in Part I. Agriculture, Forestry, Fishing and "Non-specified other" sectors are not reported here individually, but are accounted for in Total final consumption.

Source: *IEA/OECD Energy Balances of OECD Countries*

GREECE

5. Coal balance*

(Mtce)

	1973	1980	1990	2000	2005	2010	2011	2012e	Average annual percent change 73-90	90-11
Production	2.4	4.2	10.2	11.7	12.2	10.5	10.7	11.3	8.8	0.3
Imports	0.7	0.5	1.3	1.2	0.6	0.6	0.3	0.1	4.0	-6.3
Exports	-0.0	-	-	-0.1	-0.0	-	-0.0	-0.0	-	-
Stock changes	-0.1	-0.1	0.0	0.1	0.1	0.2	0.2	0.3		
Primary supply	**3.0**	**4.7**	**11.5**	**12.9**	**12.8**	**11.2**	**11.3**	**11.7**	**8.2**	**-0.1**
Statistical differences	**-0.0**	**-0.0**	**0.1**	**0.1**	**0.3**	**0.1**	**0.3**	..		
Total transformation	**-2.1 e**	**-3.9 e**	**-9.9**	**-11.8**	**-12.5**	**-10.9**	**-11.2**	..	**9.5**	**0.6**
Electricity and heat gen.	-2.0	-3.8	-9.8	-11.8	-12.4	-10.9	-11.2	..	9.8	0.6
Main activity producers (1)	*-2.0*	*-3.8*	*-9.8*	*-11.8*	*-12.4*	*-10.9*	*-11.2*	..	*9.8*	*0.6*
Autoproducers	-	-	-	-	-	-	-	..	-	-
Gas works	-0.0	-0.0	0.0	-	-	-	-	..	-	-
Coal transformation(2)	-0.1 e	-0.1 e	-0.0	-0.0	-0.0	-	-	..	-6.3	-
BKB plants	*0.0*	*0.0*	*-0.0*	*-0.0*	*-0.0*	-	-	..	-	-
Blast furnaces	*-0.1* e	*-0.1* e	-	-	-	-	-	..	-	-
Coke ovens	*0.0*	*-0.1*	-	-	-	-	-	..	-	-
Patent fuel plants	-	-	-	-	-	-	-	..	-	-
Other transformation(3)	-	-	-	-	-	-	-	..	-	-
Energy ind. own use	**-0.1**	**-0.1**	**-**	**-**	**-**	**-**	**-**	..	**-**	**-**
Losses	**-0.0**	**-0.0**	**-0.0**	**-**	**-**	**-**	**-**	..		
Final consumption(4)	**0.7**	**0.7**	**1.7**	**1.3**	**0.6**	**0.4**	**0.3**	..	**5.1**	**-7.9**
Industry(5)	0.5	0.5	1.5	1.2	0.6	0.4	0.3	..	6.4	-7.4
Iron and steel	*0.2*	*0.2*	-	-	-	-	-	..	-	-
Chemical	*0.0*	*0.0*	*0.1*	-	-	-	-	..	*5.6*	-
Non-metallic minerals	*0.0*	*0.0*	*1.2*	*1.0*	*0.4*	*0.2*	*0.1*	..	*36.6*	*-11.4*
Paper, pulp and print	-	-	*0.0*	-	-	-	-	..	-	-
Other industry (6)	*0.3*	*0.3*	*0.3*	*0.2*	*0.2*	*0.2*	*0.2*	..	*-0.1*	*-1.1*
Transport(7)	0.0	0.0	0.0	-	-	-	-	..	-17.4	-
Other	0.1	0.1	0.0	0.0	0.0	0.0	0.0	..	-1.8	-8.1
Comm. and pub. services	*0.0*	*0.0*	*0.0*	-	-	-	-	..	*11.8*	-
Residential	*0.1*	*0.1*	*0.0*	*0.0*	*0.0*	*0.0*	*0.0*	..	*-3.8*	*-7.4*
Other sectors (8)	-	-	*0.0*	*0.0*	*0.0*	-	*0.0*	..	-	*-3.8*
Non-energy use	0.1	0.1	0.2	-	-	-	-	..	1.8	-

(1) Main activity electricity and heat generation includes district heating.

(2) Coal transformation refers to the transformation of coal and peat to secondary and tertiary products (mainly cokes, briquettes, coal tar and off-gases). Gas works and Liquefaction are shown separately to match the *Energy Balances of OECD Countries.*

(3) Other transformation includes Liquefaction and Non-specified transformations.

(4) Final Consumption includes non-energy use and energy use (Industry, Transport and Other).

(5) Please refer to notes and definitions in Part I for detailed explanation.

(6) Other industry includes Non-ferrous metals, Transport equipment, Machinery, Mining and quarrying, Food and tobacco, Wood and wood products, Construction, Textile and leather, and Non-specified industry.

(7) Transport includes Rail and Inland waterways.

(8) Other sectors includes Agriculture/Forestry, Fishing and Non-specified other.

* "Coal" refers to all types of coal, primary (anthracite, coking coal, other bituminous coal, sub-bituminous coal, lignite/brown coal) and derived fuels (including patent fuel, BKB, coke oven coke, gas coke, coal tar, coke oven gas, gas works gas, blast furnace gas and other recovered gases), as well as peat, peat briquettes and oil shale. Quantities have been converted to Mtce using calorific values reported by the respective countries. Please refer to notes and definitions in Part I.

Source: *IEA/OECD Energy Balances of OECD Countries*

GREECE

6. Use of coal for selected end-uses*

(million tonnes)

	1978	1990	2000	2005	2009	2010	2011	Average annual percent change 78-90	90-11
Steam coal	**0.15**	**1.38**	**1.12**	**0.56**	**0.34**	**0.61**	**0.38**	**20.18**	**-5.96**
Total electricity and heat	-	-	0.01	-	0.02	0.16	0.03	-	-
Main activity producers	-	-	*0.01*	-	*0.02*	*0.16*	*0.03*	-	-
Autoproducers	-	-	-	-	-	-	-	-	-
Patent fuel/BKB plants	-	-	-	-	-	-	-	-	-
Coke ovens/Liquefaction(1)	-	-	-	-	-	-	-	-	-
Blast furnace inputs	-	-	-	-	-	-	-	-	-
Gas manufacture	0.02	-	-	-	-	-	-	-	-
Industry	0.13	1.38	1.12	0.56	0.26	0.46	0.35	21.66	-6.37
Iron and steel	-	-	-	-	-	-	-	-	-
Chemical	-	-	-	-	-	-	-	-	-
Non-metallic minerals	-	*1.29*	*1.05*	*0.43*	*0.13*	*0.27*	*0.12*	-	*-10.85*
Paper, pulp and print	-	-	-	-	-	-	-	-	-
Other industry	*0.13*	*0.09*	*0.07*	*0.13*	*0.13*	*0.19*	*0.23*	*-3.54*	*4.85*
Other sectors(2)	0.00	0.00	-	0.00	0.00	-	-	-8.75	-
Non-energy use	-	-	-	-	-	-	-	-	-
Coking coal	**0.21**	**-**	**-**	**-**	**-**	**-**	**-**	**-**	**-**
Total electricity and heat	-	-	-	-	-	-	-	-	-
Main activity producers	-	-	-	-	-	-	-	-	-
Autoproducers	-	-	-	-	-	-	-	-	-
Patent fuel/BKB plants	-	-	-	-	-	-	-	-	-
Coke ovens/Liquefaction(1)	0.21	-	-	-	-	-	-	-	-
Blast furnace inputs	-	-	-	-	-	-	-	-	-
Gas manufacture	-	-	-	-	-	-	-	-	-
Industry	-	-	-	-	-	-	-	-	-
Iron and steel	-	-	-	-	-	-	-	-	-
Chemical	-	-	-	-	-	-	-	-	-
Non-metallic minerals	-	-	-	-	-	-	-	-	-
Paper, pulp and print	-	-	-	-	-	-	-	-	-
Other industry	-	-	-	-	-	-	-	-	-
Other sectors(2)	-	-	-	-	-	-	-	-	-
Non-energy use	-	-	-	-	-	-	-	-	-
Lignite	**21.74**	**52.05**	**64.56**	**70.10**	**65.21**	**57.70**	**59.98**	**7.55**	**0.68**
Total electricity and heat	19.83	50.53	63.86	69.43	65.17	57.66	59.87	8.11	0.81
Main activity producers	*19.83*	*50.53*	*63.86*	*69.43*	*65.17*	*57.66*	*59.87*	*8.11*	*0.81*
Autoproducers	-	-	-	-	-	-	-	-	-
Patent fuel/BKB plants	0.66	0.35	0.24	0.41	-	-	-	-5.14	-
Coke ovens/Liquefaction(1)	-	-	-	-	-	-	-	-	-
Gas manufacture	-	-	-	-	-	-	-	-	-
Industry	0.36	0.52	0.38	0.22	0.03	0.02	0.07	2.96	-9.32
Iron and steel	-	-	-	-	-	-	-	-	-
Chemical	*0.21*	*0.20*	-	-	-	-	-	*-0.25*	-
Non-metallic minerals	-	*0.02*	*0.00*	-	-	-	-	-	-
Paper, pulp and print	-	-	-	-	-	-	-	-	-
Other industry	*0.16*	*0.30*	*0.38*	*0.22*	*0.03*	*0.02*	*0.07*	*5.46*	*-6.94*
Other sectors(2)	-	0.08	0.08	0.03	0.02	0.03	0.04	-	-2.69
Non-energy use	0.83	0.58	-	-	-	-	-	-2.96	-

(1) Liquefaction primarily refers to direct distillation processes. Indirect liquefaction may be reported here or under gas manufacture.

(2) Other sectors are Residential, Commercial and public services, Agriculture/Forestry, Fishing and Non-specified other.

* Coal totals may not match the sum of the individual rows due to statistical differences. Please refer to notes and definitions in Part I.

Source: *IEA/OECD Energy Statistics of OECD Countries*

GREECE

7. Fuel prices to end users*

	1978(1)	1990	1995	2000	2005	2009	2010	2011	2012
				(US dollars / tce)					
For electricity generation									
Steam coal	..	..	..	..	..	..	..	..	..
Heavy fuel oil	49.62	116.53	141.39	..	..	..	..	..	..
Natural gas	..	..	..	c	c	c	c	c	c
For industry									
Steam coal	93.42	..	..	..	..	..	..	..	..
Coking coal	..	..	..	..	..	..	..	..	..
High sulphur fuel oil	58.87	116.53	..	..	240.10	..	..	..	..
Low sulphur fuel oil	..	..	141.39	170.14	261.75	355.68	431.57	583.17	613.50
Natural gas	x	x	x	153.52	243.00	343.12	402.55	506.54	603.86
				(Euro / unit)(2)					
For electricity generation									
Steam coal	..	..	..	..	..	..	..	..	..
Heavy fuel oil	7.35	74.15	131.86	..	..	..	..	..	..
Natural gas	..	..	..	c	c	c	c	c	c
For industry									
Steam coal	8.82	..	..	..	..	..	..	..	..
Coking coal	..	..	..	..	..	..	..	..	..
High sulphur fuel oil	8.72	74.15	..	..	265.07	..	..	..	..
Low sulphur fuel oil	..	..	131.86	249.44	288.97	351.11	446.86	574.80	654.59
Natural gas	x	x	x	211.00	251.50	317.54	390.76	468.06	604.04

(1) Earliest year for which split by coal type is available.

(2) For steam coal, coking coal and heavy fuel oil per metric tonne; for natural gas per 10^7 kilocalories GCV.

* Please refer to notes and definitions in Part I. Steam coal prices relate to historical definitions.

Source: *IEA/OECD Energy Prices & Taxes*

8. Coal and peat trade by type of coal*

(Mtce)

	1978	1990	1995	2000	2005	2009	2010	2011	2012e
Total imports	**0.40**	**1.31**	**1.32**	**1.16**	**0.57**	**0.24**	**0.57**	**0.34**	**0.10**
Bituminous coal(1)	0.12	1.28	1.31	1.16	0.56	0.24	0.57	0.32	0.08
Coking coal	0.21	-	-	-	-	-	-	-	-
Sub-bituminous coal	-	-	-	-	-	-	-	-	-
Lignite	-	-	-	-	-	0.00	0.01	0.02	0.02
Peat	-	-	-	-	-	-	-	-	-
Coal products(2)	0.07	0.03	0.01	0.00	0.00	0.00	0.00	-	-
Total exports	**0.06**	**-**	**-**	**0.06**	**0.04**	**0.00**	**-**	**0.01**	**0.01**
Bituminous coal(1)	-	-	-	0.05	0.01	0.00	-	0.01	0.01
Coking coal	-	-	-	-	-	-	-	-	-
Sub-bituminous coal	-	-	-	-	-	-	-	-	-
Lignite	-	-	-	0.00	-	-	-	0.00	-
Peat	-	-	-	-	-	-	-	-	-
Coal products(2)	0.06	-	-	-	0.03	-	-	-	-

(1) Bituminous coal includes anthracite.

(2) Coal products includes products derived from coal, for example: coke, coal tar, briquettes, patent fuels.

* Please refer to notes and definitions in Part I.

Source: *IEA/OECD Energy Balances of OECD Countries, IEA/OECD Energy Statistics of OECD Countries*

GREECE

9. Total coal imports by origin*

(thousand tonnes)

	1978(1)	1985	1990	1995	2000	2005	2010	2011	2012e
Total coal(2)	**349**	**2040**	**1380**	**1409**	**1245**	**646**	**651**	**506**	**240**
Coking coal	**213**	**-**	**-**	**-**	**-**	**-**	**-**	**-**	**-**
Australia	159	-	-	-	-	-	-	-	-
Canada	-	-	-	-	-	-	-	-	-
Czech Republic	-	-	-	-	-	-	-	-	-
Germany	-	-	-	-	-	-	-	-	-
Poland	54	-	-	-	-	-	-	-	-
United Kingdom	-	-	-	-	-	-	-	-	-
United States	-	-	-	-	-	-	-	-	-
Other OECD	-	-	-	-	-	-	-	-	-
China, People's Rep.	-	-	-	-	-	-	-	-	-
Colombia	-	-	-	-	-	-	-	-	-
Indonesia	-	-	-	-	-	-	-	-	-
South Africa	-	-	-	-	-	-	-	-	-
Former Soviet Union(3)	-	-	x	x	x	x	x	x	x
Russian Federation (4)	*x*	*x*	-	-	-	-	-	-	-
Other FSU (4)	*x*	*x*	-	-	-	-	-	-	-
Venezuela	-	-	-	-	-	-	-	-	-
Vietnam	-	-	-	-	-	-	-	-	-
Non-specified/other	-	-	-	-	-	-	-	-	-
Steam coal	**136**	**2040**	**1380**	**1409**	**1245**	**646**	**617**	**395**	**103**
Australia	48	588	-	-	110	-	-	-	-
Canada	-	-	-	-	-	-	-	-	-
Czech Republic	1	2	-	-	-	-	-	-	-
Germany	-	-	-	-	-	-	-	-	-
Poland	75	17	-	-	-	-	-	-	-
United Kingdom	-	61	-	-	-	-	-	-	-
United States	-	1136	35	36	-	-	47	-	-
Other OECD	-	-	4	-	-	71	51	-	-
China, People's Rep.	-	-	-	-	119	-	-	-	-
Colombia	-	-	-	-	-	-	76	230	-
Indonesia	-	-	-	-	205	63	-	-	-
South Africa	-	207	1017	1123	447	132	45	-	-
Former Soviet Union(3)	12	29	x	x	x	x	x	x	x
Russian Federation (4)	*x*	*x*	*324*	*250*	*321*	*380*	*398*	*84*	*49*
Other FSU (4)	*x*	*x*	-	-	*43*	-	-	*81*	-
Venezuela	-	-	-	-	-	-	-	-	-
Vietnam	-	-	-	-	-	-	-	-	-
Non-specified/other	-	-	-	-	-	-	-	-	54
Lignite	**-**	**-**	**-**	**-**	**-**	**-**	**34**	**111**	**137**

(1) Earliest year for which split by coal type is available.

(2) Total coal consists of steam coal, coking coal and lignite.

(3) For years prior to 1990, or if no details for individual Former Soviet Union republics.

(4) Data for former republics are not available separately for some OECD member countries.

* Please refer to notes and definitions in Part I.

Source: IEA/OECD Coal Statistics

GREECE

12. Coal import values by origin*
(average unit value, CIF, USD/tonne)

	1990	1995	2000	2005	2007	2008	2009	2010	2011/12[1]
Coking coal[2]	**40.74**	**288.28**	**402.73**	**304.84**	**310.94**	**380.10**	**255.93**	**354.06**	**..**
Imports from:									
Australia	40.88	..	..	..	..	..	..	..	..
Canada	..	..	..	..	..	..	..	..	..
Czech Republic	201.72	288.28	..	..	..	..	..	..	..
Poland	..	..	..	..	..	..	..	..	..
United States	..	..	..	..	..	..	..	..	..
China	..	..	..	..	..	..	..	..	..
Colombia	..	..	..	..	..	..	..	..	..
Indonesia	..	..	..	..	..	..	..	..	..
South Africa	..	..	..	..	..	..	..	..	..
Former Soviet Union[3]	..	..	..	..	..	..	..	..	..
Other bituminous coal[4]	**43.47**	**42.95**	**32.02**	**54.72**	**78.44**	**117.37**	**133.42**	**105.45**	**..**
Imports from:									
Australia	..	..	..	..	..	..	..	..	..
Canada	..	..	..	..	..	..	..	..	..
Czech Republic	131.94	179.85	112.43	198.06	252.12	341.67	346.31	306.04	..
Poland	..	..	..	..	..	..	..	..	..
United States	52.03	..	24.40	..	..	120.31	..	94.73	..
China	..	..	31.20	..	..	..	..	..	..
Colombia	29.06	45.00	33.32	..	79.58	..	99.14	112.67	..
Indonesia	..	..	31.02	76.16	..	..	..	..	..
South Africa	..	40.52	29.73	69.89	..	..	195.66	187.04	..
Former Soviet Union[3]	42.36	54.16	37.42	46.37	75.73	97.49	54.30	95.73	..

Note: On occasion, shipment of extremely small quantities of high valued coal results in high import costs.

(1) Import data for steam and coking coals are currently unavailable for 2011 and 2012 due to resource constraints.

(2) Weighted average of individual countries using import volumes as weights. Prices exclude intra-EU trade.

(3) Former Soviet Union until 1991, Russian Federation starting in 1992.

(4) Bituminous steam coal only. Weighted average of trades. Prices exclude intra-EU trade.

* Please refer to notes and definitions in Part I.

Source: IEA/OECD Coal Statistics

HUNGARY*

Figure 1: TPES by fuel (Mtce)

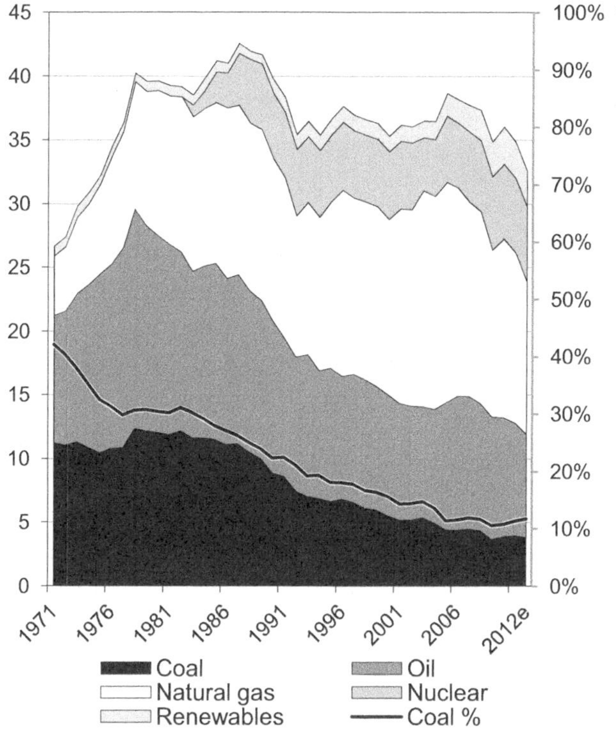

Figure 2: Coal supply indicators (1971 = 100)

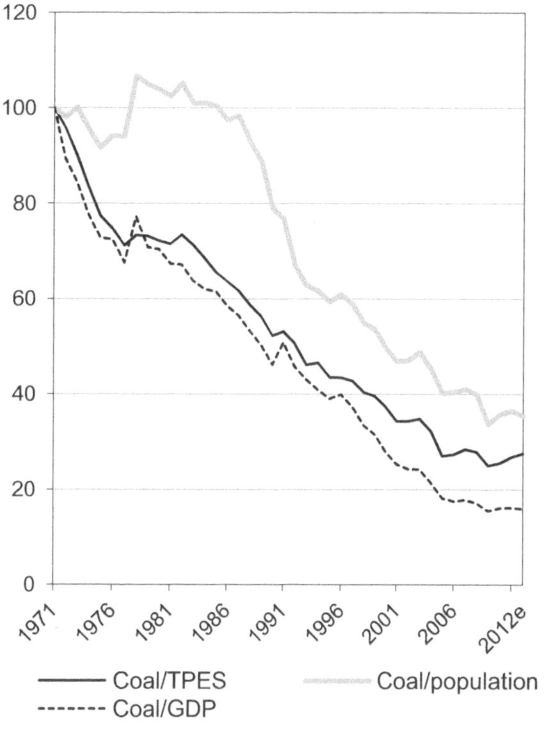

Figure 3: Primary coal supply (Mtce)

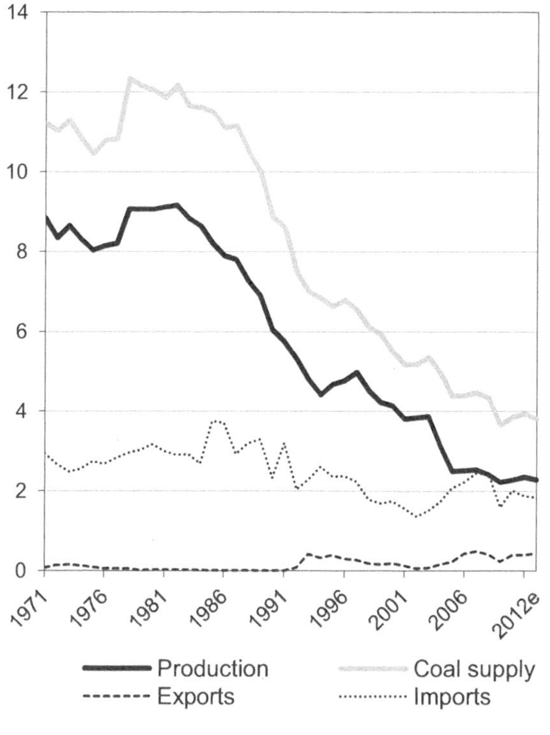

Figure 4: Coal consumption (Mtce)

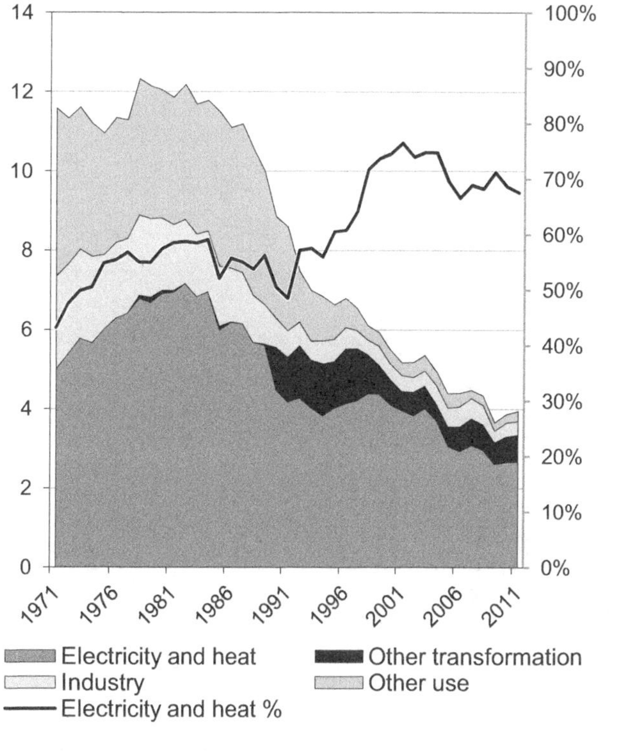

*Please refer to notes and definitions in Part I. Peat is included in coal for all figures.

Source: *IEA/OECD Coal Information, IEA/OECD Energy Balances of OECD Countries, OECD Main Economic Indicators*

HUNGARY*

Figure 5: Electricity generation by fuel (TWh)

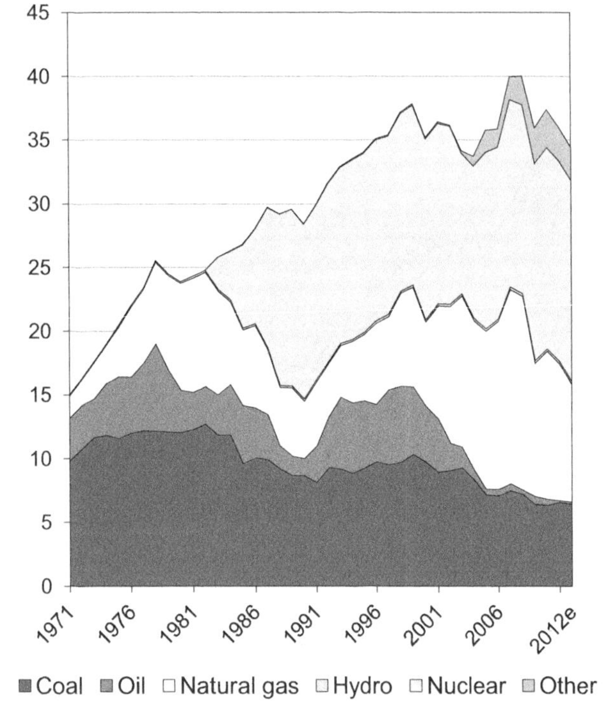

Figure 6: CO_2 emissions by fuel (Mt CO_2)

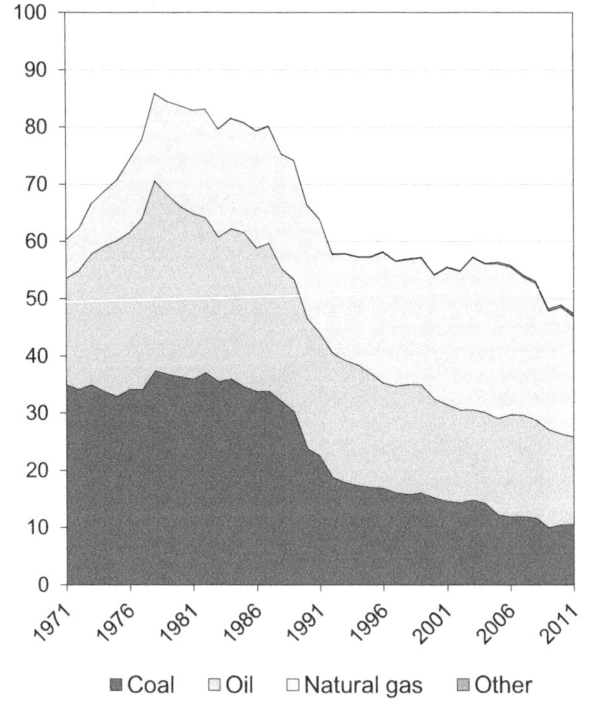

Figure 7: Electricity generation by fuel share

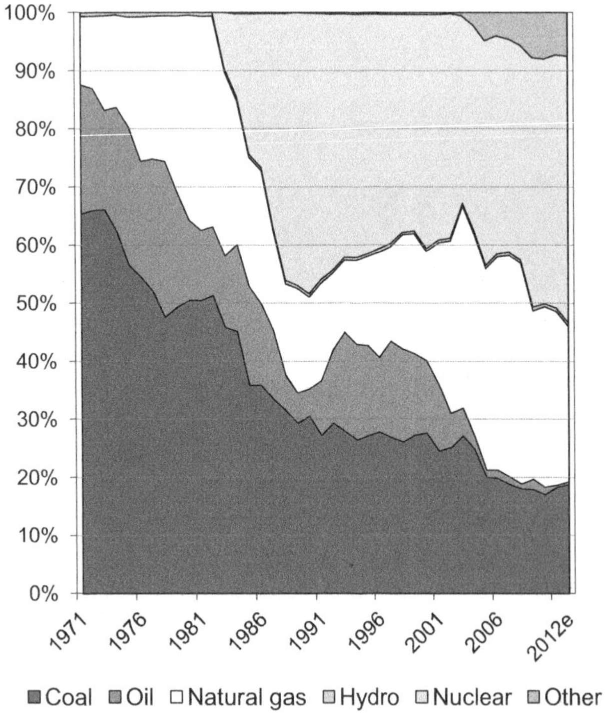

Figure 8: CO_2 emissions by fuel share

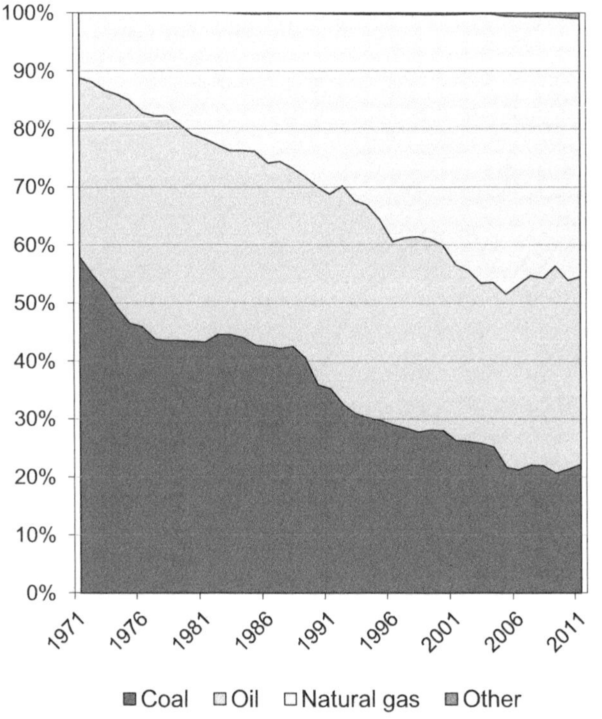

*Please refer to notes and definitions in Part I. Peat is included in coal for all figures.

Source: *IEA/OECD Electricity Information, IEA/OECD CO2 Emissions from Fuel Combustion*

HUNGARY

1. Total primary energy supply (TPES) by fuel*

	1973	1980	1990	2000	2010	2011	2012e	Average annual percent change 73-90	90-11
TPES (Mtce)	**30.39**	**40.50**	**41.08**	**35.71**	**36.67**	**35.66**	**33.57**	**1.79**	**-0.67**
Coal and peat	11.29	12.04	8.86	5.50	3.86	3.94	3.81	-1.42	-3.79
Oil	11.64	15.42	11.93	9.47	9.35	8.85	8.10	0.14	-1.41
Natural Gas	5.96	11.39	12.73	13.79	14.02	13.36	11.97	4.57	0.23
Biofuels and waste	0.91	0.74	0.94	1.08	2.68	2.58	2.54	0.15	4.93
Nuclear	-	-	5.11	5.30	5.88	5.86	5.90	-	0.65
Hydro	0.01	0.01	0.02	0.02	0.02	0.03	0.03	3.39	1.06
Geothermal	-	-	0.12	0.12	0.14	0.15	0.15	-	0.93
Solar, wind, tide	-	-	-	-	0.07	0.09	0.10	-	-
Net electricity trade(1)	0.57	0.91	1.37	0.42	0.64	0.82	0.98	5.26	-2.43
Heat(2)	-	-	-	-	-	-	-	-	-

(1) Net trade between OECD and non-OECD countries.

(2) Ambient heat from heat pumps used in the transformation sector.

Source: *IEA/OECD Energy Balances of OECD Countries*

2. Energy supply, GDP and population*

	1973	1980	1990	2000	2010	2011	2012e	Average annual percent change 73-90	90-11
GDP (2005 bil. USD)	61 e	78 e	88 e	90	109	111	109	2.14	1.14
TPES/GDP(1)	0.50 e	0.52 e	0.47 e	0.40	0.34	0.32	0.31	-0.35	-1.79
Population (millions)	10.4	10.7	10.4	10.2	10.0	10.0	9.9	-0.03	-0.18
TPES/population(1)	2.92	3.78	3.96	3.50	3.67	3.58	3.38	1.82	-0.49
TPES/GDP(2)	139.2 e	144.8 e	131.2 e	111.1	93.9	89.8	86.0	-0.35	-1.79
Coal and peat supply/GDP(2)	464.7 e	386.9 e	254.2 e	153.7	88.8	89.2	87.7	-3.49	-4.87
Elec. consumption/GDP(2)	98.0 e	111.4 e	122.9 e	111.6	106.7	106.0	..	1.34	-0.70
Elec. generation (TWh)	18	24	28	35	37	36	34	2.85	1.13
Industrial production(2)	..	53.2	51.9	76.0	107.9	113.9	111.9	..	3.81

(1) TPES/GDP in units of tce/2005 thousand USD, TPES/population in units of tce/capita.

(2) As index, 2005 (not shown) = 100

Source: *IEA/OECD Energy Balances of OECD Countries , OECD Main Economic Indicators*

3. Coal and peat production by type*

	1978(1)	1990	2000	2005	2010	2011	2012e	Average annual percent change 78-90	90-11
Mtce:									
Coking coal	0.87	0.17	-	-	-	-	-	-12.69	-
Steam coal	1.08	0.14	-	-	-	-	-	-15.56	-
Lignite	7.12	5.72	4.13	2.50	2.28	2.35	2.29	-1.80	-4.15
Peat	-	-	-	-	-	-	-	-	-
Mt:									
Coking coal	0.86	0.17	-	-	-	-	-	-12.69	-
Steam coal	2.50	0.33	-	-	-	-	-	-15.56	-
Lignite	22.74	17.33	14.03	9.57	9.11	9.56	9.29	-2.24	-2.80
Peat	-	-	-	-	-	-	-	-	-

(1) Earliest year for which split by coal type is available.

* Please refer to notes and definitions in Part I.

Source: *IEA/OECD Energy Balances of OECD Countries, IEA/OECD Energy Statistics of OECD Countries*

HUNGARY

4. Final consumption of energy by fuel*

(Mtce)

	1973	1980	1990	2000	2009	2010	2011	Average annual percent change 73-90	90-11
Total final consumption(1)	**23.61**	**30.81**	**29.49**	**24.55**	**25.48**	**25.89**	**25.29**	**1.32**	**-0.73**
Coal and peat	5.83	5.06	3.30	0.77	0.51	0.56	0.60	-3.28	-7.81
Oil	9.23	12.86	10.17	7.43	9.09	8.69	8.35	0.57	-0.93
Natural Gas	4.01	6.59	8.85	9.56	8.70	9.18	8.86	4.78	0.00
Biofuels and wastes	0.88	0.71	0.88	0.99	1.43	1.55	1.64	0.01	3.00
Geothermal	-	-	0.12	0.11	0.13	0.13	0.14	-	0.60
Solar, wind, tide	-	-	-	-	0.01	0.01	0.01	-	-
Electricity	2.16	3.14	3.88	3.62	4.07	4.20	4.24	3.52	0.43
Heat	1.51	2.46	2.27	2.07	1.56	1.56	1.44	2.43	-2.13
of which:									
Total industry	**8.76**	**10.59**	**8.62**	**4.66**	**3.47**	**3.73**	**3.65**	**-0.09**	**-4.01**
Coal and peat	2.24	1.84	0.74	0.42	0.30	0.35	0.35	-6.30	-3.52
Oil	2.18	2.43	1.28	0.38	0.22	0.20	0.12	-3.09	-10.78
Natural Gas	2.32	3.70	4.58	1.96	1.27	1.38	1.39	4.07	-5.50
Biofuels and wastes	0.03	0.02	0.00	0.09	0.16	0.17	0.17	-10.48	19.46
Geothermal	-	-	-	-	0.00	0.00	0.00	-	-
Solar, wind, tide	-	-	-	-	-	-	-	-	-
Electricity	1.32	1.70	1.69	1.08	1.05	1.20	1.21	1.46	-1.56
Heat	0.66	0.91	0.33	0.74	0.47	0.43	0.40	-4.03	0.93
Total transport	**3.26**	**4.11**	**4.19**	**4.34**	**6.44**	**5.88**	**5.70**	**1.49**	**1.48**
Coal and peat	0.54	0.18	0.00	0.00	-	-	-	-30.33	-
Oil	2.62	3.80	4.04	4.21	6.05	5.49	5.33	2.58	1.32
Natural Gas	-	-	0.00	0.00	0.00	0.00	0.00	-	9.45
Biofuels and wastes	-	-	-	-	0.24	0.25	0.24	-	-
Electricity	0.10	0.13	0.15	0.12	0.15	0.14	0.14	2.40	-0.29
Residential	**5.42**	**7.36**	**9.58**	**7.99**	**7.88**	**8.19**	**7.83**	**3.41**	**-0.95**
Coal and peat	2.22	2.66	2.40	0.28	0.20	0.21	0.25	0.47	-10.29
Oil	1.41	1.75	1.67	0.44	0.15	0.19	0.16	0.98	-10.44
Natural Gas	0.26	0.69	2.35	4.32	4.54	4.66	4.24	13.85	2.85
Biofuels and wastes	0.80	0.65	0.85	0.79	0.83	0.94	1.03	0.31	0.95
Geothermal	-	-	-	-	-	-	-	-	-
Solar, wind, tide	-	-	-	-	0.01	0.01	0.01	-	-
Electricity	0.31	0.62	1.13	1.20	1.38	1.38	1.39	7.83	0.99
Heat	0.41	0.98	1.19	0.95	0.77	0.81	0.76	6.43	-2.13
Comm & public services	**1.60**	**2.00**	**2.88**	**4.32**	**4.27**	**4.48**	**4.49**	**3.51**	**2.13**
Coal and peat	0.67	0.32	0.13	0.03	0.00	0.00	0.00	-9.18	-14.95
Oil	0.24	0.64	0.41	0.10	0.03	0.03	0.03	3.15	-12.42
Natural Gas	0.13	0.28	0.88	2.53	2.24	2.45	2.46	11.98	5.04
Biofuels and waste	-	0.00	0.01	0.08	0.17	0.17	0.18	-	12.57
Geothermal	-	-	0.12	0.11	0.11	0.11	0.12	-	-0.15
Solar, wind, tide	-	-	-	-	0.00	0.00	0.00	-	-
Electricity	0.28	0.33	0.68	1.09	1.40	1.40	1.41	5.28	3.53
Heat	0.28	0.43	0.65	0.37	0.32	0.31	0.29	5.08	-3.77
Non-energy use	**1.84**	**3.51**	**2.48**	**2.27**	**2.79**	**2.90**	**2.92**	**1.79**	**0.77**
Coal and peat	-	-	-	-	-	-	-	-	-
Oil	0.98	2.20	1.69	1.80	2.33	2.39	2.33	3.27	1.54
Natural Gas	0.86	1.30	0.79	0.47	0.46	0.52	0.58	-0.48	-1.42

(1) Total Final Consumption (TFC) excludes use in transformation processes and energy industry own use.

* Please refer to notes and definitions in Part I. Agriculture, Forestry, Fishing and "Non-specified other" sectors are not reported here individually, but are accounted for in Total final consumption.

Source: *IEA/OECD Energy Balances of OECD Countries*

HUNGARY

5. Coal balance*

(Mtce)

	1973	1980	1990	2000	2005	2010	2011	2012e	Average annual percent change 73-90	90-11
Production	8.6	9.1	6.0	4.1	2.5	2.3	2.4	2.3	-2.1	-4.4
Imports	2.5	3.2	2.3	1.7	2.1	2.0	1.9	1.8	-0.4	-1.0
Exports	-0.2	-0.0	-0.0	-0.2	-0.2	-0.4	-0.4	-0.4	-28.6	37.3
Stock changes	0.3	-0.2	0.5	-0.2	0.0	-0.0	0.1	0.1		
Primary supply	**11.3**	**12.0**	**8.9**	**5.5**	**4.4**	**3.9**	**3.9**	**3.8**	**-1.4**	**-3.8**
Statistical differences	**1.2**	**0.6**	**-0.7**	**0.0**	**-0.0**	**-0.0**	**0.0**	..		
Total transformation	**-6.5 e**	**-7.5 e**	**-4.7 e**	**-4.6 e**	**-3.4 e**	**-3.1 e**	**-3.2 e**	..	**-1.9**	**-1.8**
Electricity and heat gen.	-5.8	-6.9	-4.5	-4.1	-3.1	-2.7	-2.7	..	-1.5	-2.4
Main activity producers (1)	*-5.4*	*-6.5*	*-4.2*	*-4.1*	*-3.1*	*-2.6*	*-2.7*	..	*-1.4*	*-2.2*
Autoproducers	*-0.4*	*-0.5*	*-0.2*	-	-	*-0.0*	*-0.0*	..	*-3.5*	*-20.4*
Gas works	0.3	0.2	-	-	-	-	-	..	-	-
Coal transformation(2)	-1.0 e	-0.8 e	-0.3 e	-0.5 e	-0.4 e	-0.5 e	-0.5 e	..	-7.8	3.7
BKB plants	*0.4*	*0.4*	-	-	-	-	-	..	-	-
Blast furnaces	*-0.9 e*	*-0.9 e*	*-0.6 e*	*-0.4 e*	*-0.3 e*	*-0.4 e*	*-0.4 e*	..	*-2.6*	*-1.8*
Coke ovens	*-0.4*	*-0.3*	*-0.1*	*-0.2*	*-0.1*	*-0.1*	*-0.1*	..	*-5.3*	*0.1*
Patent fuel plants	*-0.1*	*0.0*	*0.5*	*0.0*	*0.0*	-	*0.0*	..	-	*-28.1*
Other transformation(3)	-	-	-	-	-	-	-	..	-	-
Energy ind. own use	**-0.1**	**-0.1**	**-0.1**	**-0.1**	**-0.1**	**-0.1**	**-0.1**	..	**-0.2**	**1.3**
Losses	**-0.0**	**-0.0**	**-0.0**	-	**-0.0**	**-0.0**	**-0.0**	..		
Final consumption(4)	**5.8**	**5.1**	**3.3**	**0.8**	**0.8**	**0.6**	**0.6**	..	**-3.3**	**-7.8**
Industry(5)	2.2	1.8	0.7	0.4	0.5	0.3	0.3	..	-6.3	-3.5
Iron and steel	*0.9 e*	*0.9 e*	*0.5 e*	*0.3 e*	*0.3 e*	*0.3 e*	*0.3 e*	..	*-3.6*	*-2.6*
Chemical	*0.2*	*0.1*	*0.0*	-	-	*0.0*	*0.0*	..	*-9.1*	*-14.3*
Non-metallic minerals	*0.5*	*0.3*	*0.1*	*0.1*	*0.2*	*0.1*	*0.1*	..	*-9.8*	*-1.5*
Paper, pulp and print	*0.0*	*0.0*	*0.0*	*0.0*	-	-	-	..	*-14.5*	-
Other industry (6)	*0.7*	*0.5*	*0.1*	*0.0*	*0.0*	*0.0*	*0.0*	..	*-8.9*	*-11.7*
Transport(7)	0.5	0.2	0.0	0.0	0.0	-	-	..	-30.3	-
Other	3.0	3.0	2.6	0.4	0.4	0.2	0.3	..	-1.0	-10.5
Comm. and pub. services	*0.7*	*0.3*	*0.1*	*0.0*	*0.0*	*0.0*	*0.0*	..	*-9.2*	*-14.9*
Residential	*2.2*	*2.7*	*2.4*	*0.3*	*0.3*	*0.2*	*0.2*	..	*0.5*	*-10.3*
Other sectors (8)	*0.2*	*0.1*	*0.0*	*0.0*	*0.0*	*0.0*	*0.0*	..	*-8.9*	*-16.2*
Non-energy use	-	-	-	-	-	-	-	..	-	-

(1) Main activity electricity and heat generation includes district heating.

(2) Coal transformation refers to the transformation of coal and peat to secondary and tertiary products (mainly cokes, briquettes, coal tar and off-gases). Gas works and Liquefaction are shown separately to match the *Energy Balances of OECD Countries.*

(3) Other transformation includes Liquefaction and Non-specified transformations.

(4) Final Consumption includes non-energy use and energy use (Industry, Transport and Other).

(5) Please refer to notes and definitions in Part I for detailed explanation.

(6) Other industry includes Non-ferrous metals, Transport equipment, Machinery, Mining and quarrying, Food and tobacco, Wood and wood products, Construction, Textile and leather, and Non-specified industry.

(7) Transport includes Rail and Inland waterways.

(8) Other sectors includes Agriculture/Forestry, Fishing and Non-specified other.

* "Coal" refers to all types of coal, primary (anthracite, coking coal, other bituminous coal, sub-bituminous coal, lignite/brown coal) and derived fuels (including patent fuel, BKB, coke oven coke, gas coke, coal tar, coke oven gas, gas works gas, blast furnace gas and other recovered gases), as well as peat, peat briquettes and oil shale. Quantities have been converted to Mtce using calorific values reported by the respective countries. Please refer to notes and definitions in Part I.

Source: *IEA/OECD Energy Balances of OECD Countries*

HUNGARY

6. Use of coal for selected end-uses*

(million tonnes)

	1978	1990	2000	2005	2009	2010	2011	Average annual percent change 78-90	90-11
Steam coal	**3.34**	**1.44**	**0.39**	**1.27**	**0.76**	**0.65**	**0.49**	**-6.81**	**-5.02**
Total electricity and heat	1.62	-	0.04	0.66	0.43	0.37	0.20	-	-
Main activity producers	*1.56*	-	*0.04*	*0.66*	*0.42*	*0.37*	*0.20*	-	-
Autoproducers	*0.06*	-	-	-	*0.00*	*0.00*	*0.00*	-	-
Patent fuel/BKB plants	0.33	0.93	0.03	0.02	0.01	-	0.00	8.97	-22.83
Coke ovens/Liquefaction(1)	-	-	-	-	-	-	-	-	-
Blast furnace inputs	-	-	-	-	-	-	-	-	-
Gas manufacture	-	-	-	-	-	-	-	-	-
Industry	0.38	0.10	0.11	0.19	0.10	0.06	0.07	-10.29	-2.14
Iron and steel	*0.19*	*0.05*	-	-	-	-	-	*-9.95*	-
Chemical	*0.02*	*0.01*	-	-	-	-	-	*-7.85*	-
Non-metallic minerals	*0.07*	*0.01*	*0.11*	*0.19*	*0.09*	*0.06*	*0.06*	*-13.19*	*7.47*
Paper, pulp and print	-	-	-	-	-	-	-	-	-
Other industry	*0.10*	*0.03*	-	*0.00*	*0.01*	*0.01*	*0.01*	*-9.91*	*-6.39*
Other sectors(2)	0.87	0.40	0.21	0.37	0.22	0.22	0.22	-6.23	-2.83
Non-energy use	-	-	-	-	-	-	-	-	-
Coking coal	**1.74**	**0.97**	**1.28**	**0.81**	**1.02**	**1.42**	**1.47**	**-4.74**	**1.98**
Total electricity and heat	-	-	-	-	-	-	-	-	-
Main activity producers	-	-	-	-	-	-	-	-	-
Autoproducers	-	-	-	-	-	-	-	-	-
Patent fuel/BKB plants	-	-	-	-	-	-	-	-	-
Coke ovens/Liquefaction(1)	1.38	0.96	1.28	0.81	1.02	1.41	1.46	-3.04	2.06
Blast furnace inputs	-	-	-	-	-	-	-	-	-
Gas manufacture	-	-	-	-	-	-	-	-	-
Industry	0.01	0.00	-	-	-	-	-	-10.91	-
Iron and steel	*0.01*	*0.00*	-	-	-	-	-	*-9.91*	-
Chemical	*0.00*	-	-	-	-	-	-	-	-
Non-metallic minerals	-	-	-	-	-	-	-	-	-
Paper, pulp and print	-	-	-	-	-	-	-	-	-
Other industry	-	-	-	-	-	-	-	-	-
Other sectors(2)	0.16	0.01	-	-	-	-	-	-20.46	-
Non-energy use	-	-	-	-	-	-	-	-	-
Lignite	**22.88**	**17.90**	**13.50**	**9.51**	**8.90**	**8.92**	**9.66**	**-2.02**	**-2.89**
Total electricity and heat	16.68	14.53	12.97	9.37	8.76	8.74	9.39	-1.14	-2.06
Main activity producers	*16.06*	*14.34*	*12.97*	*9.37*	*8.76*	*8.74*	*9.39*	*-0.94*	*-1.99*
Autoproducers	*0.62*	*0.20*	-	-	-	-	-	*-9.07*	-
Patent fuel/BKB plants	1.09	0.90	0.07	0.00	-	-	-	-1.55	-
Coke ovens/Liquefaction(1)	-	-	-	-	-	-	-	-	-
Gas manufacture	-	-	-	-	-	-	-	-	-
Industry	2.28	0.54	0.03	0.00	0.01	0.01	0.01	-11.31	-17.72
Iron and steel	*0.08*	*0.03*	-	-	-	-	-	*-8.93*	-
Chemical	*0.25*	*0.08*	-	-	-	-	-	*-8.81*	-
Non-metallic minerals	*0.99*	*0.19*	*0.02*	-	*0.01*	*0.01*	*0.01*	*-13.02*	*-13.41*
Paper, pulp and print	*0.06*	*0.01*	*0.00*	-	-	-	-	*-17.57*	-
Other industry	*0.91*	*0.24*	*0.01*	*0.00*	-	-	-	*-10.44*	-
Other sectors(2)	2.48	1.88	0.43	0.10	0.13	0.17	0.26	-2.31	-8.96
Non-energy use	-	-	-	-	-	-	-	-	-

(1) Liquefaction primarily refers to direct distillation processes. Indirect liquefaction may be reported here or under gas manufacture.

(2) Other sectors are Residential, Commercial and public services, Agriculture/Forestry, Fishing and Non-specified other.

* Coal totals may not match the sum of the individual rows due to statistical differences. Please refer to notes and definitions in Part I.

Source: *IEA/OECD Energy Statistics of OECD Countries*

HUNGARY

7. Fuel prices to end users*

	1978(1)	1990	1995	2000	2005	2009	2010	2011	2012
				(US dollars / tce)					
For electricity generation									
Steam coal	..	90.60	c	c	c	c	c	c	c
Heavy fuel oil	..	64.33	58.61	105.44	..	..	..	..	..
Natural gas	..	76.50	68.34	77.67	222.34	399.65	295.07	379.58	397.63
For industry									
Steam coal	..	100.92	..	..	..	..	..	..	..
Coking coal	c	c	c	c	c	c	c	c	c
High sulphur fuel oil	..	70.64	71.29	95.91	..	..	..	..	..
Low sulphur fuel oil	..	..	81.57	99.70	243.38	324.26	399.70	544.68	574.82
Natural gas	..	102.01	82.13	85.27	268.82	475.56	332.40	394.69	432.83
				(1 000 Hungarian forints / unit)(2)					
For electricity generation									
Steam coal	..	1.70	c	c	c	c	c	c	c
Heavy fuel oil	..	5.64	10.21	41.25	..	..	..	..	..
Natural gas	..	6.22	11.05	28.19	57.04	103.82	78.82	97.88	114.94
For industry									
Steam coal	..	3.70	..	..	..	..	..	..	..
Coking coal	c	c	c	c	c	c	c	c	c
High sulphur fuel oil	..	6.06	12.16	36.74	..	..	..	..	..
Low sulphur fuel oil	..	..	13.92	38.19	65.91	88.92	112.70	148.26	175.38
Natural gas	..	8.29	13.28	30.95	68.96	123.55	88.79	101.78	125.11

(1) Earliest year for which split by coal type is available.

(2) For steam coal, coking coal and heavy fuel oil per metric tonne; for natural gas per 10^7 kilocalories GCV.

* Please refer to notes and definitions in Part I. Steam coal prices relate to historical definitions.

Source: *IEA/OECD Energy Prices & Taxes*

8. Coal and peat trade by type of coal*

(Mtce)

	1978	1990	1995	2000	2005	2009	2010	2011	2012e
Total imports	**2.96**	**2.33**	**2.37**	**1.73**	**2.08**	**1.58**	**2.02**	**1.87**	**1.83**
Bituminous coal(1)	-	-	-	-	0.47	0.31	0.26	0.11	0.04
Coking coal	0.86	0.64	1.47	1.26	0.96	1.04	1.57	1.55	1.60
Sub-bituminous coal	0.60	0.65	0.28	0.33	0.43	0.21	0.17	0.19	0.18
Lignite	-	0.14	0.24	0.13	-	-	-	-	-
Peat	-	-	-	-	-	-	-	-	-
Coal products(2)	1.51	0.90	0.37	0.01	0.22	0.02	0.02	0.02	0.01
Total exports	**0.05**	**0.00**	**0.39**	**0.19**	**0.22**	**0.23**	**0.40**	**0.40**	**0.43**
Bituminous coal(1)	-	-	-	-	0.00	0.00	0.01	-	-
Coking coal	-	-	-	-	-	-	-	-	-
Sub-bituminous coal	-	0.00	-	-	0.00	0.00	0.00	0.00	0.00
Lignite	0.02	-	-	0.00	0.11	0.01	0.00	0.00	0.04
Peat	-	-	-	-	-	-	-	-	-
Coal products(2)	0.03	-	0.39	0.19	0.10	0.22	0.39	0.40	0.39

(1) Bituminous coal includes anthracite.

(2) Coal products includes products derived from coal, for example: coke, coal tar, briquettes, patent fuels.

* Please refer to notes and definitions in Part I.

Source: *IEA/OECD Energy Balances of OECD Countries, IEA/OECD Energy Statistics of OECD Countries*

HUNGARY

9. Total coal imports by origin*

(thousand tonnes)

	1978[(1)]	1985	1990	1995	2000	2005	2010	2011	2012e
Total coal[(2)]	**1630**	**2413**	**1763**	**2209 e**	**1891**	**2153**	**2059**	**1877**	**1822**
Coking coal	**817**	**478**	**610**	**1418 e**	**1234**	**898**	**1460**	**1430**	**1471**
Australia	-	-	-	-	-	115	-	-	-
Canada	-	-	-	-	-	-	-	-	-
Czech Republic	-	-	-	-	800	230	380	215	346
Germany	-	-	-	-	-	-	-	-	-
Poland	-	-	-	-	251	217	129	86	81
United Kingdom	-	-	-	-	-	-	-	-	-
United States	-	-	-	-	51	52	840	1017	870
Other OECD	-	-	-	-	-	-	-	-	-
China, People's Rep.	-	-	-	-	-	-	-	-	-
Colombia	-	-	-	-	-	-	46	72	142
Indonesia	-	-	-	-	-	-	-	40	32
South Africa	-	-	-	-	-	-	-	-	-
Former Soviet Union[(3)]	817 e	478 e	x	x	x	x	x	x	x
Russian Federation [(4)]	*x*	*x*	*610 e*	-	*132*	*284*	*65*	-	-
Other FSU [(4)]	*x*	*x*	-	-	-	-	-	-	-
Venezuela	-	-	-	-	-	-	-	-	-
Vietnam	-	-	-	-	-	-	-	-	-
Non-specified/other	-	-	-	1418	-	-	-	-	-
Steam coal	**813**	**1733**	**890**	**384 e**	**430**	**1255**	**599**	**447**	**351**
Australia	-	-	-	-	-	-	-	-	-
Canada	-	-	-	-	-	-	-	-	-
Czech Republic	-	-	-	168 e	42	409	292	274	252
Germany	-	-	-	-	-	-	17	16	15
Poland	-	-	-	212 e	294	263	59	49	26
United Kingdom	-	-	-	-	-	-	-	-	-
United States	-	-	-	-	-	-	-	-	-
Other OECD	-	-	-	-	-	1	-	-	-
China, People's Rep.	-	-	-	-	-	-	-	-	-
Colombia	-	-	-	-	-	-	-	-	-
Indonesia	-	-	-	-	-	-	-	-	-
South Africa	-	-	-	-	-	-	-	-	-
Former Soviet Union[(3)]	-	-	x	x	x	x	x	x	x
Russian Federation [(4)]	*x*	*x*	-	*4 e*	*94*	*582*	*220*	*108*	*58*
Other FSU [(4)]	*x*	*x*	-	-	-	-	*11*	-	-
Venezuela	-	-	-	-	-	-	-	-	-
Vietnam	-	-	-	-	-	-	-	-	-
Non-specified/other	813	1733	890	-	-	-	-	-	-
Lignite	**-**	**202**	**263**	**407**	**227**	**-**	**-**	**-**	**-**

(1) Earliest year for which split by coal type is available.

(2) Total coal consists of steam coal, coking coal and lignite.

(3) For years prior to 1990, or if no details for individual Former Soviet Union republics.

(4) Data for former republics are not available separately for some OECD member countries.

* Please refer to notes and definitions in Part I.

Source: IEA/OECD Coal Statistics

ICELAND*

Figure 1: TPES by fuel (Mtce)

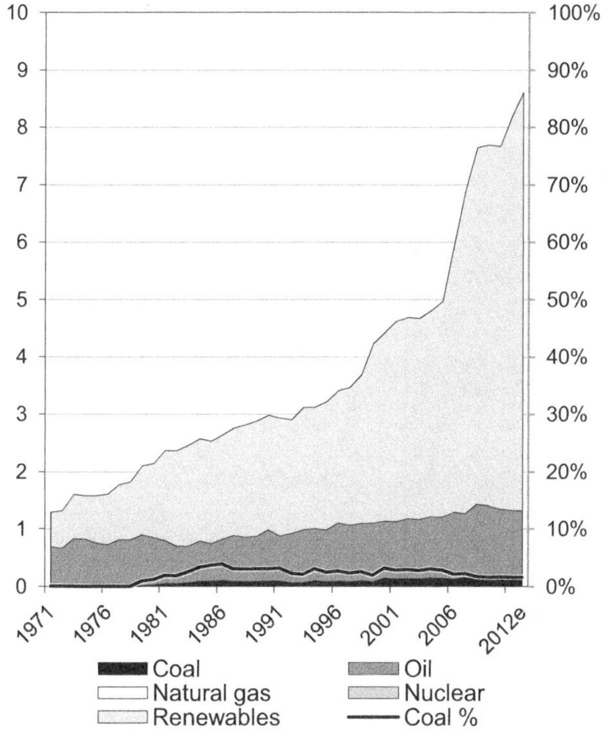

Figure 2: Coal supply indicators (1971 = 100)

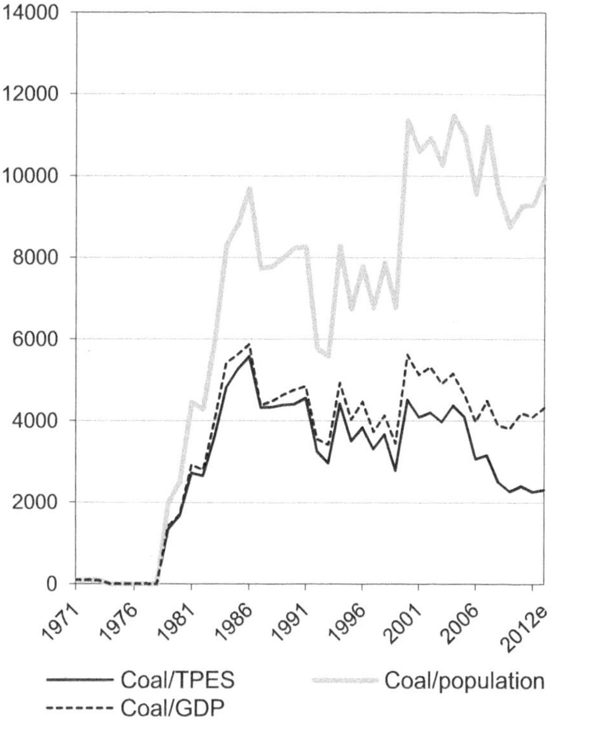

Figure 3: Primary coal supply (Mtce)

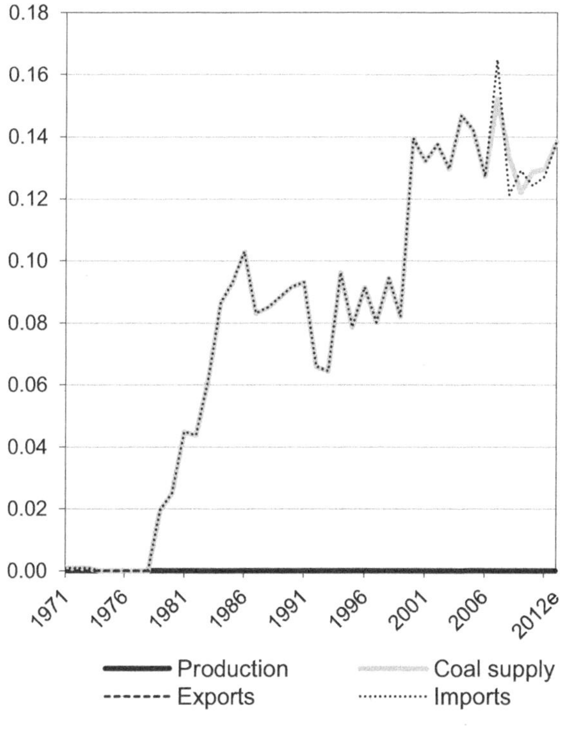

Figure 4: Coal consumption (Mtce)

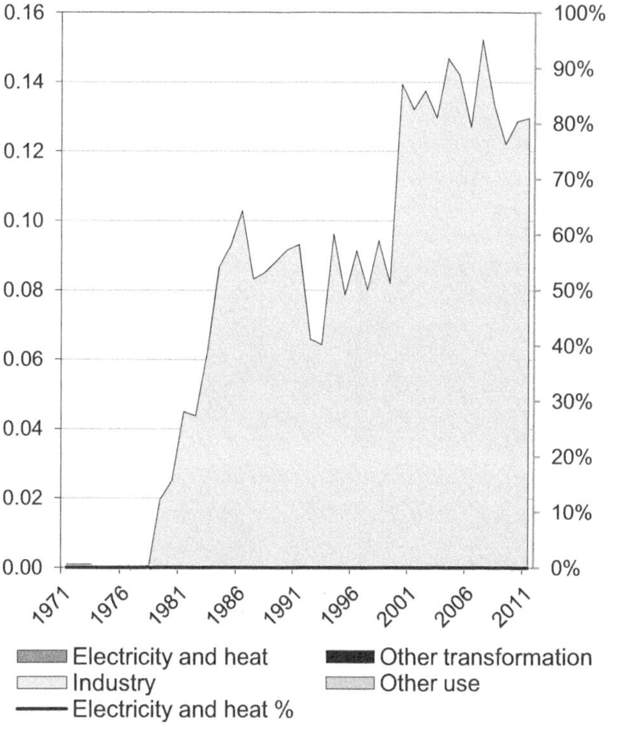

*Please refer to notes and definitions in Part I. Peat is included in coal for all figures.

Source: *IEA/OECD Coal Information, IEA/OECD Energy Balances of OECD Countries, OECD Main Economic Indicators*

ICELAND*

Figure 5: Electricity generation by fuel (TWh)

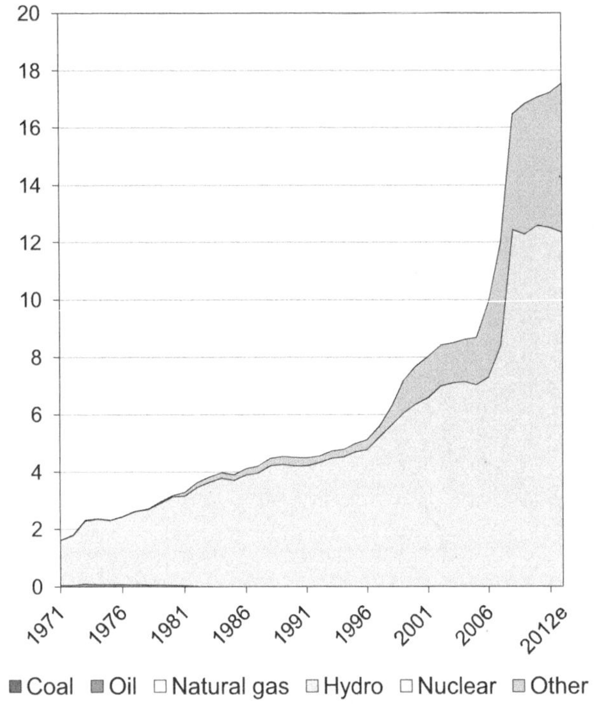

Figure 6: CO_2 emissions by fuel (Mt CO_2)

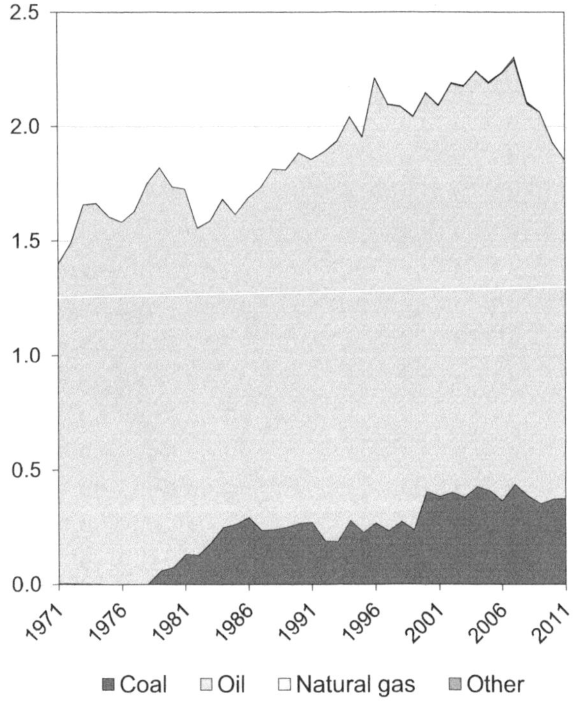

Figure 7: Electricity generation by fuel share

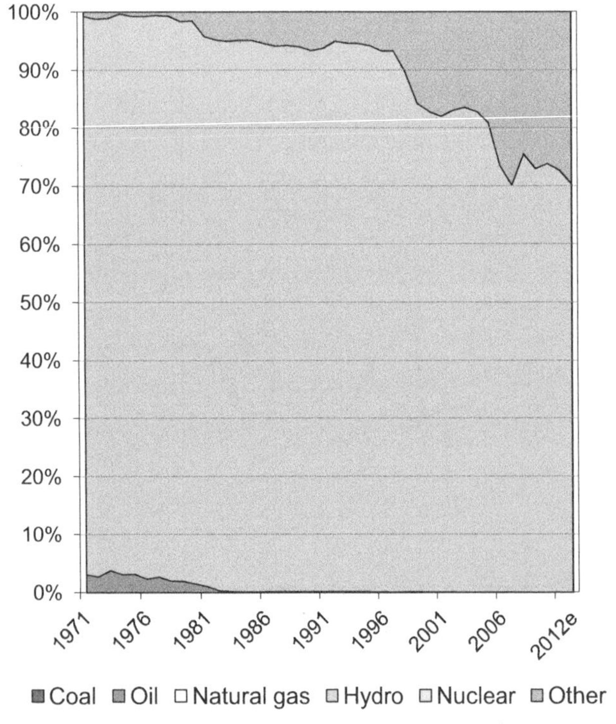

Figure 8: CO_2 emissions by fuel share

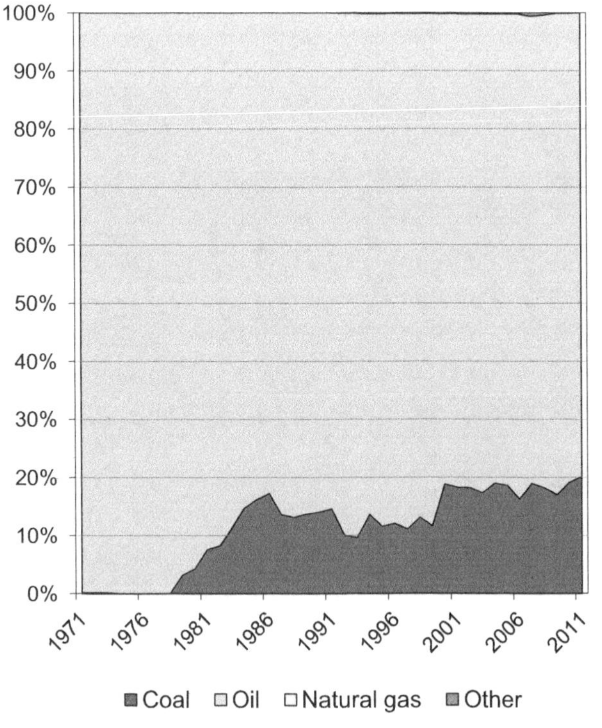

*Please refer to notes and definitions in Part I. Peat is included in coal for all figures.

Source: *IEA/OECD Electricity Information, IEA/OECD CO2 Emissions from Fuel Combustion*

ICELAND

1. Total primary energy supply (TPES) by fuel*

	1973	1980	1990	2000	2010	2011	2012e	Average annual percent change 73-90	90-11
TPES (Mtce)	**1.60**	**2.14**	**2.98**	**4.43**	**7.67**	**8.19**	**8.60**	**3.74**	**4.92**
Coal and peat	0.00	0.03	0.09	0.14	0.13	0.13	0.14	31.25	1.66
Oil	0.82	0.82	0.89	1.00	1.21	1.19	1.18	0.47	1.40
Natural Gas	-	-	-	-	-	-	-	-	-
Biofuels and waste	-	-	-	0.00	0.00	0.00	0.00	-	-
Nuclear	-	-	-	-	-	-	-	-	-
Hydro	0.27	0.38	0.52	0.78	1.55	1.54	1.52	3.86	5.33
Geothermal	0.50	0.91	1.48	2.51	4.78	5.33	5.77	6.58	6.27
Solar, wind, tide	-	-	-	-	-	-	-	-	-
Net electricity trade[1]	-	-	-	-	-	-	-	-	-
Heat[2]	-	-	-	-	-	-	-	-	-

(1) Net trade between OECD and non-OECD countries.

(2) Ambient heat from heat pumps used in the transformation sector.

Source: *IEA/OECD Energy Balances of OECD Countries*

2. Energy supply, GDP and population*

	1973	1980	1990	2000	2010	2011	2012e	Average annual percent change 73-90	90-11
GDP (2005 bil. USD)	5	8	10	13	16	17	17	3.81	2.39
TPES/GDP[1]	0.29	0.27	0.29	0.34	0.47	0.49	0.50	-0.07	2.48
Population (millions)	0.2	0.2	0.3	0.3	0.3	0.3	0.3	1.09	1.07
TPES/population[1]	7.54	9.38	11.70	15.76	24.12	25.66	26.97	2.62	3.81
TPES/GDP[2]	96.3	89.3	95.1	109.8	153.3	159.0	164.4	-0.07	2.48
Coal and peat supply/GDP[2]	1.9	36.6	102.3	121.0	90.1	88.2	92.8	26.43	-0.71
Elec. consumption/GDP[2]	79.5	76.2	79.5	109.2	200.1	198.3	..	-0.01	4.45
Elec. generation (TWh)	2	3	5	8	17	17	18	3.99	6.59
Industrial production[2]	..	..	..	..	..	..	..	..	..

(1) TPES/GDP in units of tce/2005 thousand USD, TPES/population in units of tce/capita.

(2) As index, 2005 (not shown) = 100

Source: *IEA/OECD Energy Balances of OECD Countries , OECD Main Economic Indicators*

3. Coal and peat production by type*

	1978[1]	1990	2000	2005	2010	2011	2012e	Average annual percent change 78-90	90-11
Mtce:									
Coking coal	-	-	-	-	-	-	-	-	-
Steam coal	-	-	-	-	-	-	-	-	-
Lignite	-	-	-	-	-	-	-	-	-
Peat	-	-	-	-	-	-	-	-	-
Mt:									
Coking coal	-	-	-	-	-	-	-	-	-
Steam coal	-	-	-	-	-	-	-	-	-
Lignite	-	-	-	-	-	-	-	-	-
Peat	-	-	-	-	-	-	-	-	-

(1) Earliest year for which split by coal type is available.

* Please refer to notes and definitions in Part I.

Source: *IEA/OECD Energy Balances of OECD Countries, IEA/OECD Energy Statistics of OECD Countries*

ICELAND

4. Final consumption of energy by fuel*

(Mtce)

	1973	1980	1990	2000	2009	2010	2011	Average annual percent change 73-90	90-11
Total final consumption(1)	**1.46**	**1.83**	**2.34**	**3.01**	**4.27**	**4.20**	**4.23**	**2.82**	**2.85**
Coal and peat	0.00	0.03	0.09	0.14	0.12	0.13	0.13	31.25	1.66
Oil	0.77	0.78	0.85	0.99	1.30	1.21	1.19	0.56	1.64
Natural Gas	-	-	-	-	-	-	-	-	-
Biofuels and wastes	-	-	-	-	0.00	0.00	0.00	-	-
Geothermal	0.44	0.63	0.78	0.78	0.59	0.59	0.60	3.52	-1.28
Solar, wind, tide	-	-	-	-	-	-	-	-	-
Electricity	0.25	0.35	0.48	0.85	1.92	1.93	1.97	3.81	6.94
Heat	-	0.05	0.14	0.25	0.34	0.34	0.34	-	4.35
of which:									
Total industry	**0.38**	**0.50**	**0.60**	**1.03**	**1.87**	**1.86**	**1.89**	**2.62**	**5.66**
Coal and peat	-	0.03	0.09	0.14	0.12	0.13	0.13	-	1.66
Oil	0.16	0.20	0.13	0.16	0.07	0.05	0.04	-1.31	-5.07
Natural Gas	-	-	-	-	-	-	-	-	-
Biofuels and wastes	-	-	-	-	-	-	-	-	-
Geothermal	0.04	0.04	0.06	0.08	0.02	0.02	0.02	2.53	-5.73
Solar, wind, tide	-	-	-	-	-	-	-	-	-
Electricity	0.18	0.24	0.31	0.64	1.66	1.66	1.70	3.19	8.38
Heat	-	-	-	-	-	-	-	-	-
Total transport	**0.18**	**0.23**	**0.30**	**0.30**	**0.42**	**0.40**	**0.39**	**3.02**	**1.25**
Coal and peat	-	-	-	-	-	-	-	-	-
Oil	0.18	0.23	0.30	0.30	0.42	0.40	0.39	3.02	1.23
Natural Gas	-	-	-	-	-	-	-	-	-
Biofuels and wastes	-	-	-	-	0.00	0.00	0.00	-	-
Electricity	-	-	-	-	-	-	-	-	-
Residential	**0.64**	**0.76**	**0.83**	**0.86**	**0.75**	**0.76**	**0.76**	**1.55**	**-0.40**
Coal and peat	0.00	-	-	-	-	-	-	-	-
Oil	0.25	0.10	0.02	0.01	0.00	0.00	0.00	-13.32	-7.23
Natural Gas	-	-	-	-	-	-	-	-	-
Biofuels and wastes	-	-	-	-	-	-	-	-	-
Geothermal	0.35	0.54	0.63	0.56	0.41	0.41	0.41	3.58	-1.97
Solar, wind, tide	-	-	-	-	-	-	-	-	-
Electricity	0.04	0.08	0.07	0.07	0.11	0.11	0.11	2.88	1.93
Heat	-	0.04	0.11	0.21	0.23	0.23	0.24	-	3.89
Comm & public services	**0.03**	**0.04**	**0.10**	**0.17**	**0.30**	**0.30**	**0.30**	**7.17**	**5.35**
Coal and peat	-	-	-	-	-	-	-	-	-
Oil	-	-	-	-	0.00	0.00	0.00	-	-
Natural Gas	-	-	-	-	-	-	-	-	-
Biofuels and waste	-	-	-	-	-	-	-	-	-
Geothermal	0.02	0.02	0.03	0.06	0.11	0.11	0.11	3.13	6.23
Solar, wind, tide	-	-	-	-	-	-	-	-	-
Electricity	0.01	0.02	0.06	0.09	0.13	0.13	0.13	9.56	3.59
Heat	-	0.00	0.01	0.01	0.06	0.06	0.06	-	9.35
Non-energy use	**0.03**	**0.02**	**0.09**	**0.18**	**0.49**	**0.48**	**0.50**	**6.57**	**8.31**
Coal and peat	-	-	-	-	-	-	-	-	-
Oil	0.03	0.02	0.09	0.18	0.49	0.48	0.50	6.57	8.31
Natural Gas	-	-	-	-	-	-	-	-	-

(1) Total Final Consumption (TFC) excludes use in transformation processes and energy industry own use.

* Please refer to notes and definitions in Part I. Agriculture, Forestry, Fishing and "Non-specified other" sectors are not reported here individually, but are accounted for in Total final consumption.

Source: *IEA/OECD Energy Balances of OECD Countries*

ICELAND

5. Coal balance*

(Mtce)

	1973	1980	1990	2000	2005	2010	2011	2012e	Average annual percent change 73-90	90-11
Production	-	-	-	-	-	-	-	-	-	-
Imports	0.0	0.0	0.1	0.1	0.1	0.1	0.1	0.1	31.3	1.6
Exports	-	-	-	-	-	-	-	-	-	-
Stock changes	-	-	-	-	-	0.0	0.0	-		
Primary supply	**0.0**	**0.0**	**0.1**	**0.1**	**0.1**	**0.1**	**0.1**	**0.1**	**31.3**	**1.7**
Statistical differences	-	-	-	-	-	-	-	..		
Total transformation	-	-	-	-	-	-	-	..	-	-
Electricity and heat gen.	-	-	-	-	-	-	-	..	-	-
Main activity producers (1)	-	-	-	-	-	-	-	..	-	-
Autoproducers	-	-	-	-	-	-	-	..	-	-
Gas works	-	-	-	-	-	-	-	..	-	-
Coal transformation(2)	-	-	-	-	-	-	-	..	-	-
BKB plants	-	-	-	-	-	-	-	..	-	-
Blast furnaces	-	-	-	-	-	-	-	..	-	-
Coke ovens	-	-	-	-	-	-	-	..	-	-
Patent fuel plants	-	-	-	-	-	-	-	..	-	-
Other transformation(3)	-	-	-	-	-	-	-	..	-	-
Energy ind. own use	-	-	-	-	-	-	-	..	-	-
Losses	-	-	-	-	-	-	-	..		
Final consumption(4)	**0.0**	**0.0**	**0.1**	**0.1**	**0.1**	**0.1**	**0.1**	..	**31.3**	**1.7**
Industry(5)	-	0.0	0.1	0.1	0.1	0.1	0.1	..	-	1.7
Iron and steel	-	*0.0*	*0.1*	*0.1*	*0.1*	*0.1*	*0.1*	..	-	*2.1*
Chemical	-	-	-	-	-	-	-	..	-	-
Non-metallic minerals	-	-	*0.0*	*0.0*	*0.0*	*0.0*	*0.0*	..	-	*-2.4*
Paper, pulp and print	-	-	-	-	-	-	-	..	-	-
Other industry (6)	-	-	-	-	-	-	-	..	-	-
Transport(7)	-	-	-	-	-	-	-	..	-	-
Other	0.0	-	-	-	-	-	-	..	-	-
Comm. and pub. services	-	-	-	-	-	-	-	..	-	-
Residential	*0.0*	-	-	-	-	-	-	..	-	-
Other sectors (8)	-	-	-	-	-	-	-	..	-	-
Non-energy use	-	-	-	-	-	-	-	..	-	-

(1) Main activity electricity and heat generation includes district heating.

(2) Coal transformation refers to the transformation of coal and peat to secondary and tertiary products (mainly cokes, briquettes, coal tar and off-gases). Gas works and Liquefaction are shown separately to match the *Energy Balances of OECD Countries*.

(3) Other transformation includes Liquefaction and Non-specified transformations.

(4) Final Consumption includes non-energy use and energy use (Industry, Transport and Other).

(5) Please refer to notes and definitions in Part I for detailed explanation.

(6) Other industry includes Non-ferrous metals, Transport equipment, Machinery, Mining and quarrying, Food and tobacco, Wood and wood products, Construction, Textile and leather, and Non-specified industry.

(7) Transport includes Rail and Inland waterways.

(8) Other sectors includes Agriculture/Forestry, Fishing and Non-specified other.

* "Coal" refers to all types of coal, primary (anthracite, coking coal, other bituminous coal, sub-bituminous coal, lignite/brown coal) and derived fuels (including patent fuel, BKB, coke oven coke, gas coke, coal tar, coke oven gas, gas works gas, blast furnace gas and other recovered gases), as well as peat, peat briquettes and oil shale. Quantities have been converted to Mtce using calorific values reported by the respective countries.
Please refer to notes and definitions in Part I.

Source: *IEA/OECD Energy Balances of OECD Countries*

ICELAND

6. Use of coal for selected end-uses*

(million tonnes)

	1978	1990	2000	2005	2009	2010	2011	Average annual percent change 78-90	90-11
Steam coal	**-**	**0.07**	**0.10**	**0.12**	**0.10**	**0.11**	**0.11**	**-**	**2.31**
Total electricity and heat	-	-	-	-	-	-	-	-	-
Main activity producers	-	-	-	-	-	-	-	-	-
Autoproducers	-	-	-	-	-	-	-	-	-
Patent fuel/BKB plants	-	-	-	-	-	-	-	-	-
Coke ovens/Liquefaction(1)	-	-	-	-	-	-	-	-	-
Blast furnace inputs	-	-	-	-	-	-	-	-	-
Gas manufacture	-	-	-	-	-	-	-	-	-
Industry	-	0.07	0.10	0.12	0.10	0.11	0.11	-	2.31
Iron and steel	-	*0.05*	*0.09*	*0.10*	*0.09*	*0.10*	*0.10*	-	*3.01*
Chemical	-	-	-	-	-	-	-	-	-
Non-metallic minerals	-	*0.01*	*0.01*	*0.01*	*0.01*	*0.01*	*0.01*	-	*-2.29*
Paper, pulp and print	-	-	-	-	-	-	-	-	-
Other industry	-	-	*0.00*	*0.00*	*0.00*	-	-	-	-
Other sectors(2)	-	-	-	-	-	-	-	-	-
Non-energy use	-	-	-	-	-	-	-	-	-
Coking coal	**-**	**-**	**-**	**-**	**-**	**-**	**-**	**-**	**-**
Total electricity and heat	-	-	-	-	-	-	-	-	-
Main activity producers	-	-	-	-	-	-	-	-	-
Autoproducers	-	-	-	-	-	-	-	-	-
Patent fuel/BKB plants	-	-	-	-	-	-	-	-	-
Coke ovens/Liquefaction(1)	-	-	-	-	-	-	-	-	-
Blast furnace inputs	-	-	-	-	-	-	-	-	-
Gas manufacture	-	-	-	-	-	-	-	-	-
Industry	-	-	-	-	-	-	-	-	-
Iron and steel	-	-	-	-	-	-	-	-	-
Chemical	-	-	-	-	-	-	-	-	-
Non-metallic minerals	-	-	-	-	-	-	-	-	-
Paper, pulp and print	-	-	-	-	-	-	-	-	-
Other industry	-	-	-	-	-	-	-	-	-
Other sectors(2)	-	-	-	-	-	-	-	-	-
Non-energy use	-	-	-	-	-	-	-	-	-
Lignite	**-**	**-**	**-**	**-**	**-**	**-**	**-**	**-**	**-**
Total electricity and heat	-	-	-	-	-	-	-	-	-
Main activity producers	-	-	-	-	-	-	-	-	-
Autoproducers	-	-	-	-	-	-	-	-	-
Patent fuel/BKB plants	-	-	-	-	-	-	-	-	-
Coke ovens/Liquefaction(1)	-	-	-	-	-	-	-	-	-
Gas manufacture	-	-	-	-	-	-	-	-	-
Industry	-	-	-	-	-	-	-	-	-
Iron and steel	-	-	-	-	-	-	-	-	-
Chemical	-	-	-	-	-	-	-	-	-
Non-metallic minerals	-	-	-	-	-	-	-	-	-
Paper, pulp and print	-	-	-	-	-	-	-	-	-
Other industry	-	-	-	-	-	-	-	-	-
Other sectors(2)	-	-	-	-	-	-	-	-	-
Non-energy use	-	-	-	-	-	-	-	-	-

(1) Liquefaction primarily refers to direct distillation processes. Indirect liquefaction may be reported here or under gas manufacture.

(2) Other sectors are Residential, Commercial and public services, Agriculture/Forestry, Fishing and Non-specified other.

* Coal totals may not match the sum of the individual rows due to statistical differences. Please refer to notes and definitions in Part I.

Source: *IEA/OECD Energy Statistics of OECD Countries*

ICELAND

7. Fuel prices to end users*

	1978(1)	1990	1995	2000	2005	2009	2010	2011	2012
				(US dollars / tce)					
For electricity generation									
Steam coal	..	..	..	..	..	..	..	..	..
Heavy fuel oil	..	..	..	..	..	..	..	..	..
Natural gas	..	..	..	..	..	..	..	..	..
For industry									
Steam coal	..	..	..	..	..	..	..	..	..
Coking coal	..	..	..	..	..	..	..	..	..
High sulphur fuel oil	..	..	..	..	..	..	..	..	..
Low sulphur fuel oil	..	..	..	..	..	..	..	..	..
Natural gas	..	..	..	..	..	..	..	..	..
				(Icelandic krona / unit)(2)					
For electricity generation									
Steam coal	..	..	..	..	..	..	..	..	..
Heavy fuel oil	..	..	..	..	..	..	..	..	..
Natural gas	..	..	..	..	..	..	..	..	..
For industry									
Steam coal	..	..	..	..	..	..	..	..	..
Coking coal	..	..	..	..	..	..	..	..	..
High sulphur fuel oil	..	..	..	..	..	..	..	..	..
Low sulphur fuel oil	..	..	..	..	..	..	..	..	..
Natural gas	..	..	..	..	..	..	..	..	..

(1) Earliest year for which split by coal type is available.

(2) For steam coal, coking coal and heavy fuel oil per metric tonne; for natural gas per 10^7 kilocalories GCV.

* Please refer to notes and definitions in Part I. Steam coal prices relate to historical definitions.

Source: *IEA/OECD Energy Prices & Taxes*

8. Coal and peat trade by type of coal*

(Mtce)

	1978	1990	1995	2000	2005	2009	2010	2011	2012e
Total imports	**-**	**0.09**	**0.08**	**0.14**	**0.14**	**0.13**	**0.12**	**0.13**	**0.14**
Bituminous coal(1)	-	0.06	0.06	0.10	0.11	0.11	0.10	0.11	0.10
Coking coal	-	-	-	-	-	-	-	-	-
Sub-bituminous coal	-	-	-	-	-	-	-	-	-
Lignite	-	-	-	-	-	-	-	-	-
Peat	-	-	-	-	-	-	-	-	-
Coal products(2)	-	0.03	0.02	0.04	0.03	0.02	0.02	0.02	0.04
Total exports	**-**	**-**	**-**	**-**	**-**	**-**	**-**	**-**	**-**
Bituminous coal(1)	-	-	-	-	-	-	-	-	-
Coking coal	-	-	-	-	-	-	-	-	-
Sub-bituminous coal	-	-	-	-	-	-	-	-	-
Lignite	-	-	-	-	-	-	-	-	-
Peat	-	-	-	-	-	-	-	-	-
Coal products(2)	-	-	-	-	-	-	-	-	-

(1) Bituminous coal includes anthracite.

(2) Coal products includes products derived from coal, for example: coke, coal tar, briquettes, patent fuels.

* Please refer to notes and definitions in Part I.

Source: *IEA/OECD Energy Balances of OECD Countries, IEA/OECD Energy Statistics of OECD Countries*

ICELAND

9. Total coal imports by origin*

(thousand tonnes)

	1978[1]	1985	1990	1995	2000	2005	2010	2011	2012e
Total coal[2]	-	**69**	**65**	**65**	**101**	**117**	**106**	**111**	**104**
Coking coal	-	**20**	-	-	-	-	-	-	-
Australia	-	4	-	-	-	-	-	-	-
Canada	-	-	-	-	-	-	-	-	-
Czech Republic	-	-	-	-	-	-	-	-	-
Germany	-	-	-	-	-	-	-	-	-
Poland	-	-	-	-	-	-	-	-	-
United Kingdom	-	16	-	-	-	-	-	-	-
United States	-	-	-	-	-	-	-	-	-
Other OECD	-	-	-	-	-	-	-	-	-
China, People's Rep.	-	-	-	-	-	-	-	-	-
Colombia	-	-	-	-	-	-	-	-	-
Indonesia	-	-	-	-	-	-	-	-	-
South Africa	-	-	-	-	-	-	-	-	-
Former Soviet Union[3]	-	-	x	x	x	x	x	x	x
Russian Federation[4]	x	x	-	-	-	-	-	-	-
Other FSU[4]	x	x	-	-	-	-	-	-	-
Venezuela	-	-	-	-	-	-	-	-	-
Vietnam	-	-	-	-	-	-	-	-	-
Non-specified/other	-	-	-	-	-	-	-	-	-
Steam coal	-	**49**	**65**	**65**	**101**	**117**	**106**	**111**	**104**
Australia	-	-	-	-	-	-	-	-	-
Canada	-	-	-	-	-	-	-	-	-
Czech Republic	-	-	-	-	-	-	-	-	-
Germany	-	3	-	-	6	39	14	30	33
Poland	-	-	-	11	7	8	-	-	-
United Kingdom	-	-	13	-	4	-	4	-	-
United States	-	42	52	37	48	9	20	14	11
Other OECD	-	2	-	17	26	61	68	60	60
China, People's Rep.	-	-	-	-	-	-	-	-	-
Colombia	-	2	-	-	10	-	-	4	-
Indonesia	-	-	-	-	-	-	-	-	-
South Africa	-	-	-	-	-	-	-	-	-
Former Soviet Union[3]	-	-	x	x	x	x	x	x	x
Russian Federation[4]	x	x	-	-	-	-	-	-	-
Other FSU[4]	x	x	-	-	-	-	-	-	-
Venezuela	-	-	-	-	-	-	-	-	-
Vietnam	-	-	-	-	-	-	-	-	-
Non-specified/other	-	-	-	-	-	-	-	3	-
Lignite	-	-	-	-	-	-	-	-	-

(1) Earliest year for which split by coal type is available.

(2) Total coal consists of steam coal, coking coal and lignite.

(3) For years prior to 1990, or if no details for individual Former Soviet Union republics.

(4) Data for former republics are not available separately for some OECD member countries.

* Please refer to notes and definitions in Part I.

Source: IEA/OECD Coal Statistics

IRELAND*

Figure 1: TPES by fuel (Mtce)

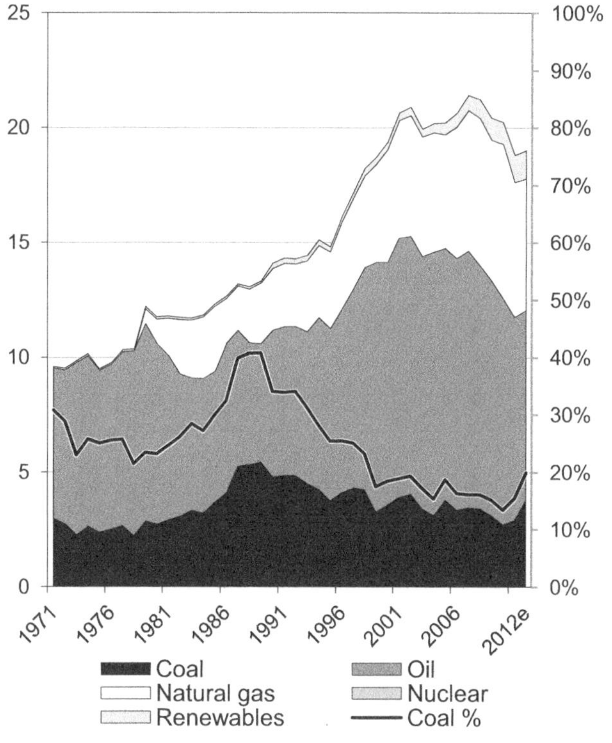

Figure 2: Coal supply indicators (1971 = 100)

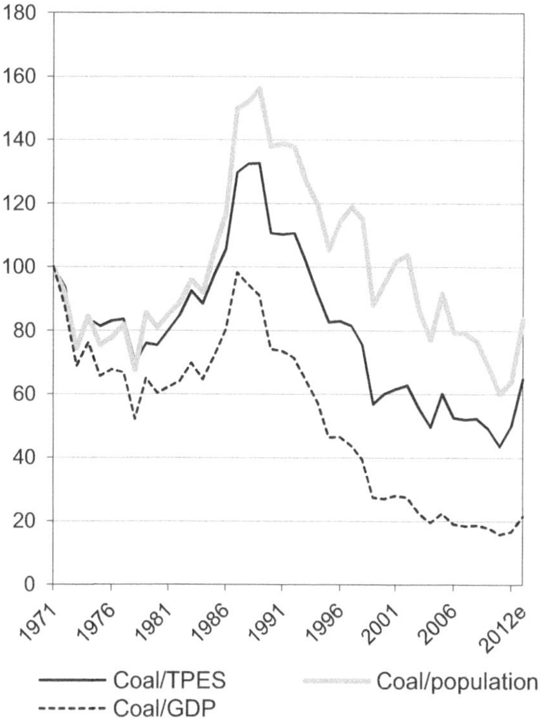

Figure 3: Primary coal supply (Mtce)

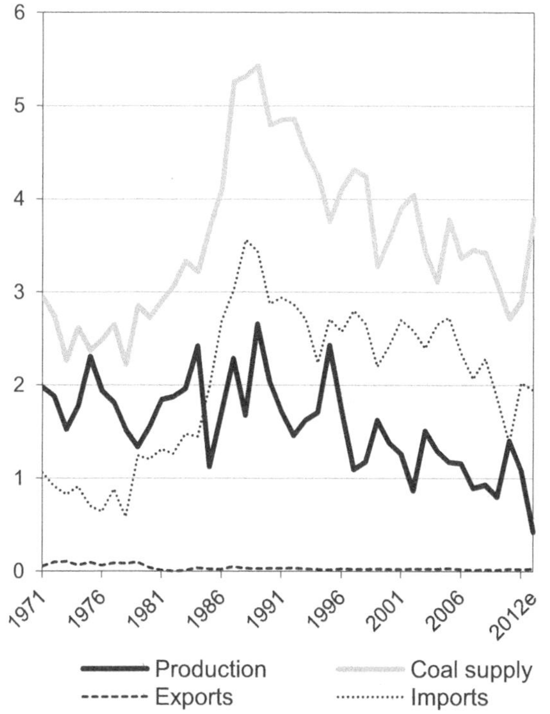

Figure 4: Coal consumption (Mtce)

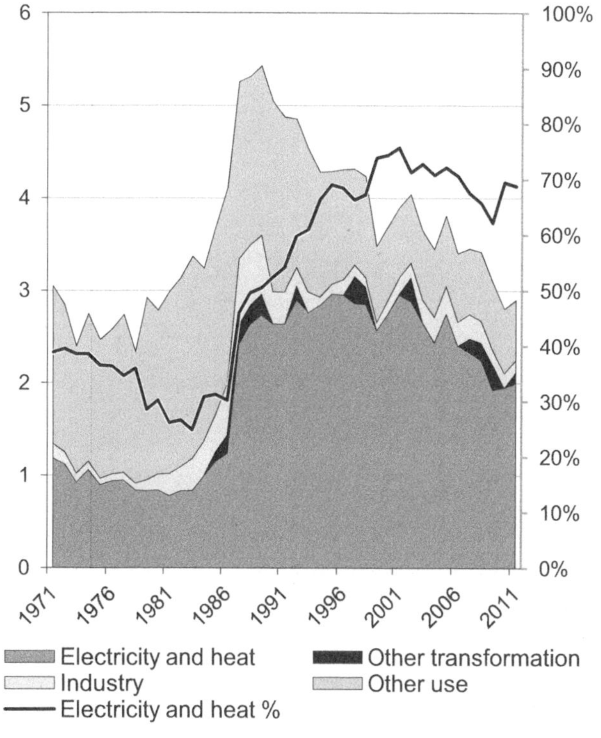

*Please refer to notes and definitions in Part I. Peat is included in coal for all figures.

Source: *IEA/OECD Coal Information, IEA/OECD Energy Balances of OECD Countries, OECD Main Economic Indicators*

IRELAND*

Figure 5: Electricity generation by fuel (TWh)

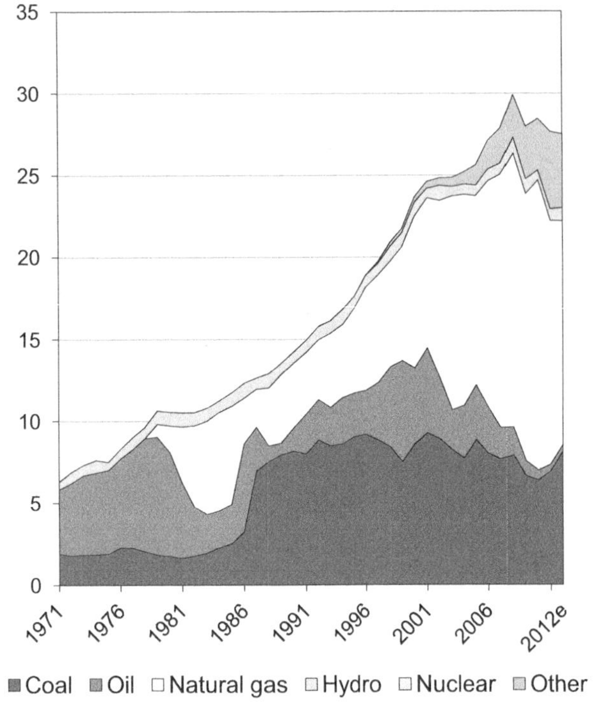

Figure 6: CO_2 emissions by fuel (Mt CO_2)

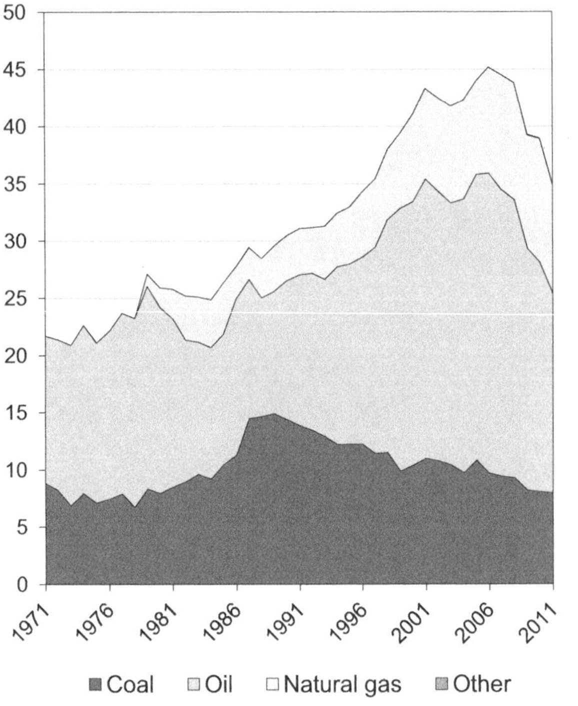

Figure 7: Electricity generation by fuel share

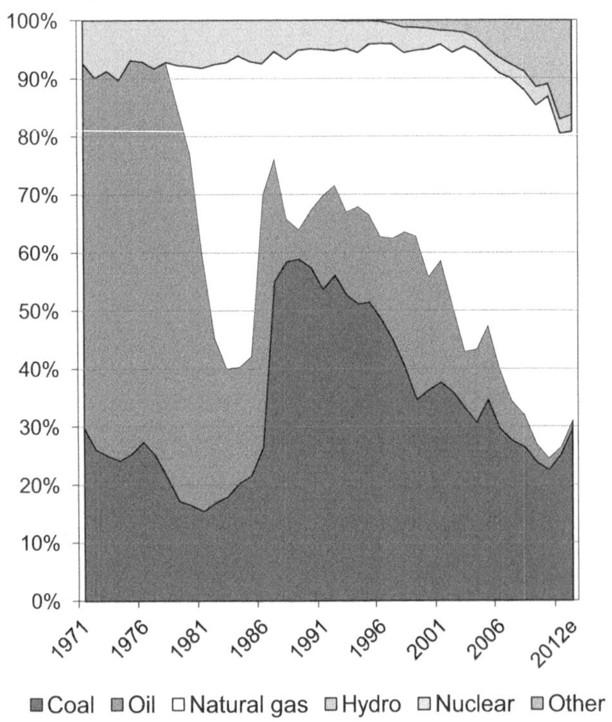

Figure 8: CO_2 emissions by fuel share

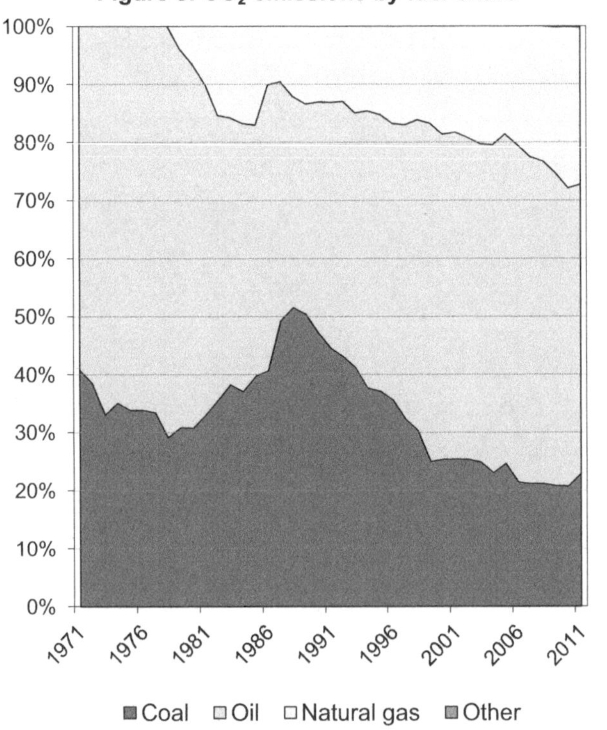

*Please refer to notes and definitions in Part I. Peat is included in coal for all figures.

Source: *IEA/OECD Electricity Information, IEA/OECD CO2 Emissions from Fuel Combustion*

IRELAND

1. Total primary energy supply (TPES) by fuel*

	1973	1980	1990	2000	2010	2011	2012e	Average annual percent change 73-90	90-11
TPES (Mtce)	**9.87**	**11.77**	**14.10**	**19.39**	**20.31**	**18.88**	**19.07**	**2.12**	**1.40**
Coal and peat	2.26	2.73	4.79	3.57	2.71	2.90	3.80	4.51	-2.36
Oil	7.52	7.88	6.39	10.56	9.87	8.85	8.27	-0.95	1.56
Natural Gas	-	1.05	2.67	4.91	6.71	5.88	5.73	-	3.82
Biofuels and waste	-	-	0.15	0.20	0.54	0.55	0.62	-	6.27
Nuclear	-	-	-	-	-	-	-	-	-
Hydro	0.08	0.10	0.09	0.10	0.07	0.09	0.10	0.47	0.07
Geothermal	-	-	-	-	-	-	-	-	-
Solar, wind, tide	-	-	0.00	0.03	0.36	0.55	0.51	-	53.43
Net electricity trade(1)	0.01	-	-	0.01	0.06	0.06	0.05	-	-
Heat(2)	-	-	-	-	-	-	-	-	-

(1) Net trade between OECD and non-OECD countries.

(2) Ambient heat from heat pumps used in the transformation sector.

Source: *IEA/OECD Energy Balances of OECD Countries*

2. Energy supply, GDP and population*

	1973	1980	1990	2000	2010	2011	2012e	Average annual percent change 73-90	90-11
GDP (2005 bil. USD)	40	54	78	159	207	210	212	4.05	4.85
TPES/GDP(1)	0.25	0.22	0.18	0.12	0.10	0.09	0.09	-1.86	-3.29
Population (millions)	3.1	3.4	3.5	3.8	4.6	4.6	4.6	0.78	1.28
TPES/population(1)	3.21	3.46	4.02	5.10	4.45	4.13	4.16	1.33	0.12
TPES/GDP(2)	247.3	214.2	179.9	120.9	97.2	89.1	89.1	-1.86	-3.29
Coal and peat supply/GDP(2)	307.3	268.7	331.1	120.7	70.3	74.1	96.1	0.44	-6.88
Elec. consumption/GDP(2)	129.6	131.6	127.3	106.4	102.3	98.7	..	-0.11	-1.20
Elec. generation (TWh)	7	11	14	24	28	28	28	3.96	3.22
Industrial production(2)	..	12.8	23.6	74.9	109.0	109.0	107.2	..	7.56

(1) TPES/GDP in units of tce/2005 thousand USD, TPES/population in units of tce/capita.

(2) As index, 2005 (not shown) = 100

Source: *IEA/OECD Energy Balances of OECD Countries , OECD Main Economic Indicators*

3. Coal and peat production by type*

	1978(1)	1990	2000	2005	2010	2011	2012e	Average annual percent change 78-90	90-11
Mtce:									
Coking coal	-	-	-	-	-	-	-	-	-
Steam coal	0.02	0.02	-	-	-	-	-	0.45	-
Lignite	-	-	-	-	-	-	-	-	-
Peat	1.50	2.02	1.38	1.17	1.40	1.09	0.43	2.50	-2.90
Mt:									
Coking coal	-	-	-	-	-	-	-	-	-
Steam coal	0.02	0.03	-	-	-	-	-	1.46	-
Lignite	-	-	-	-	-	-	-	-	-
Peat	5.24	6.52	4.81	3.96	4.99	3.71	1.45	1.82	-2.65

(1) Earliest year for which split by coal type is available.

* Please refer to notes and definitions in Part I.

Source: *IEA/OECD Energy Balances of OECD Countries, IEA/OECD Energy Statistics of OECD Countries*

IRELAND

4. Final consumption of energy by fuel*

(Mtce)

	1973	1980	1990	2000	2009	2010	2011	Average annual percent change 73-90	90-11
Total final consumption[(1)]	**7.30**	**9.06**	**10.78**	**15.18**	**16.21**	**16.20**	**14.91**	**2.31**	**1.56**
Coal and peat	1.47	1.94	2.41	0.94	0.90	0.86	0.78	2.93	-5.19
Oil	5.08	5.57	5.33	9.32	9.70	9.48	8.40	0.29	2.19
Natural Gas	-	0.49	1.43	2.26	2.11	2.30	2.22	-	2.13
Biofuels and wastes	-	-	0.15	0.17	0.39	0.42	0.44	-	5.08
Geothermal	-	-	-	-	-	-	-	-	-
Solar, wind, tide	-	-	0.00	0.00	0.01	0.01	0.01	-	27.64
Electricity	0.76	1.06	1.46	2.49	3.10	3.12	3.06	3.94	3.59
Heat	-	-	-	-	-	-	-	-	-
of which:									
Total industry	**2.43**	**3.10**	**2.47**	**3.56**	**2.97**	**3.04**	**3.22**	**0.10**	**1.28**
Coal and peat	0.10	0.17	0.35	0.15	0.16	0.15	0.14	7.70	-4.32
Oil	2.06	2.04	0.97	1.64	0.93	0.87	0.80	-4.33	-0.92
Natural Gas	-	0.49	0.51	0.67	0.61	0.67	0.89	-	2.68
Biofuels and wastes	-	-	0.09	0.14	0.21	0.23	0.23	-	4.65
Geothermal	-	-	-	-	-	-	-	-	-
Solar, wind, tide	-	-	-	-	-	-	-	-	-
Electricity	0.27	0.39	0.55	0.95	1.06	1.12	1.17	4.28	3.63
Heat	-	-	-	-	-	-	-	-	-
Total transport	**1.61**	**2.20**	**2.34**	**4.91**	**5.88**	**5.59**	**5.15**	**2.22**	**3.82**
Coal and peat	-	-	-	-	-	-	-	-	-
Oil	1.61	2.20	2.34	4.90	5.76	5.46	5.01	2.22	3.68
Natural Gas	-	-	-	-	-	-	-	-	-
Biofuels and wastes	-	-	-	-	0.11	0.13	0.14	-	-
Electricity	-	-	0.00	0.00	0.01	0.01	0.01	-	5.05
Residential	**2.58**	**2.60**	**3.32**	**3.55**	**4.45**	**4.63**	**3.89**	**1.50**	**0.76**
Coal and peat	1.38	1.66	2.04	0.79	0.74	0.70	0.65	2.34	-5.31
Oil	0.89	0.50	0.55	1.32	1.77	1.82	1.37	-2.81	4.49
Natural Gas	-	-	0.17	0.63	0.89	1.01	0.81	-	7.81
Biofuels and wastes	-	-	0.06	0.02	0.04	0.04	0.03	-	-3.21
Geothermal	-	-	-	-	-	-	-	-	-
Solar, wind, tide	-	-	0.00	0.00	0.01	0.01	0.01	-	27.51
Electricity	0.32	0.44	0.51	0.78	1.00	1.05	1.02	2.80	3.36
Heat	-	-	-	-	-	-	-	-	-
Comm & public services	**0.36**	**0.83**	**1.40**	**1.94**	**2.30**	**2.17**	**1.89**	**8.37**	**1.45**
Coal and peat	-	0.11	0.02	0.01	-	-	-	-	-
Oil	0.19	0.50	0.90	0.83	0.68	0.63	0.54	9.60	-2.36
Natural Gas	-	-	0.13	0.42	0.61	0.63	0.52	-	6.67
Biofuels and waste	-	-	-	-	0.03	0.02	0.03	-	-
Geothermal	-	-	-	-	-	-	-	-	-
Solar, wind, tide	-	-	-	-	0.00	0.00	0.00	-	-
Electricity	0.17	0.22	0.34	0.69	0.98	0.88	0.80	4.32	4.11
Heat	-	-	-	-	-	-	-	-	-
Non-energy use	**0.30**	**0.29**	**0.89**	**0.79**	**0.23**	**0.38**	**0.39**	**6.71**	**-3.82**
Coal and peat	-	-	-	-	-	-	-	-	-
Oil	0.30	0.29	0.28	0.24	0.23	0.38	0.39	-0.37	1.67
Natural Gas	-	-	0.61	0.55	-	-	-	-	-

(1) Total Final Consumption (TFC) excludes use in transformation processes and energy industry own use.

* Please refer to notes and definitions in Part I. Agriculture, Forestry, Fishing and "Non-specified other" sectors are not reported here individually, but are accounted for in Total final consumption.

Source: *IEA/OECD Energy Balances of OECD Countries*

IRELAND

5. Coal balance*

(Mtce)

	1973	1980	1990	2000	2005	2010	2011	2012e	Average annual percent change 73-90	90-11
Production	1.5	1.5	2.0	1.4	1.2	1.4	1.1	0.4	1.7	-3.0
Imports	0.8	1.2	2.9	2.4	2.7	1.4	2.0	1.9	7.6	-1.7
Exports	-0.1	-0.0	-0.0	-0.0	-0.0	-0.0	-0.0	-0.0	-6.6	-1.5
Stock changes	0.0	0.0	-0.1	-0.2	-0.1	-0.0	-0.2	1.4		
Primary supply	**2.3**	**2.7**	**4.8**	**3.6**	**3.8**	**2.7**	**2.9**	**3.8**	**4.5**	**-2.4**
Statistical differences	**-**	**-0.0**	**0.3**	**0.1**	**0.1**	**0.1**	**-0.1**	..		
Total transformation	**-0.8**	**-0.7**	**-2.7**	**-2.8**	**-2.8**	**-2.0**	**-2.0**	..	**7.5**	**-1.3**
Electricity and heat gen.	-0.9	-0.8	-2.6	-2.7	-2.7	-1.9	-2.0	..	6.4	-1.3
Main activity producers (1)	*-0.9*	*-0.8*	*-2.6*	*-2.7*	*-2.7*	*-1.9*	*-2.0*	..	*6.3*	*-1.3*
Autoproducers	-	*-0.0*	*-0.0*	*-0.0*	*-0.0*	*-0.0*	*-0.0*	..	-	*-1.4*
Gas works	0.2	0.1	-	-	-	-	-	..	-	-
Coal transformation(2)	-	0.0	-0.0	-0.0	-0.0	-0.0	-0.0	..	-	-0.5
BKB plants	-	*0.0*	*-0.0*	*-0.0*	*-0.0*	*-0.0*	*-0.0*	..	-	*-0.5*
Blast furnaces	-	-	-	-	-	-	-	..	-	-
Coke ovens	-	-	-	-	-	-	-	..	-	-
Patent fuel plants	-	-	*c*	*c*	*c*	*c*	*c*	..	*c*	*c*
Other transformation(3)	-	-	-	-	-	-	-	..	-	-
Energy ind. own use	**-0.0**	**-0.0**	**-0.0**	**-0.0**	**-0.0**	**-0.0**	**-0.0**	..	**4.4**	**4.1**
Losses	**-0.0**	**-0.0**	**-**	**-**	**-**	**-**	**-**	..		
Final consumption(4)	**1.5**	**1.9**	**2.4**	**0.9**	**1.1**	**0.9**	**0.8**	..	**2.9**	**-5.2**
Industry(5)	0.1	0.2	0.3	0.1	0.3	0.2	0.1	..	7.7	-4.3
Iron and steel	*0.0*	*0.0*	*0.0*	-	-	-	-	..	*8.0*	-
Chemical	-	-	*0.0*	-	-	-	-	..	-	-
Non-metallic minerals	-	-	*0.2*	*0.1*	*0.2*	*0.1*	*0.1*	..	-	*-1.9*
Paper, pulp and print	-	-	-	-	-	-	-	..	-	-
Other industry (6)	*0.1*	*0.2*	*0.1*	*0.1*	*0.1*	*0.0*	*0.0*	..	*1.1*	*-6.9*
Transport(7)	-	-	-	-	-	-	-	..	-	-
Other	1.4	1.8	2.1	0.8	0.8	0.7	0.6	..	2.4	-5.4
Comm. and pub. services	-	*0.1*	*0.0*	*0.0*	*0.0*	-	-	..	-	-
Residential	*1.4*	*1.7*	*2.0*	*0.8*	*0.7*	*0.7*	*0.6*	..	*2.3*	*-5.3*
Other sectors (8)	-	-	-	-	-	-	-	..	-	-
Non-energy use	-	-	-	-	-	-	-	..	-	-

(1) Main activity electricity and heat generation includes district heating.

(2) Coal transformation refers to the transformation of coal and peat to secondary and tertiary products (mainly cokes, briquettes, coal tar and off-gases). Gas works and Liquefaction are shown separately to match the *Energy Balances of OECD Countries*.

(3) Other transformation includes Liquefaction and Non-specified transformations.

(4) Final Consumption includes non-energy use and energy use (Industry, Transport and Other).

(5) Please refer to notes and definitions in Part I for detailed explanation.

(6) Other industry includes Non-ferrous metals, Transport equipment, Machinery, Mining and quarrying, Food and tobacco, Wood and wood products, Construction, Textile and leather, and Non-specified industry.

(7) Transport includes Rail and Inland waterways.

(8) Other sectors includes Agriculture/Forestry, Fishing and Non-specified other.

* "Coal" refers to all types of coal, primary (anthracite, coking coal, other bituminous coal, sub-bituminous coal, lignite/brown coal) and derived fuels (including patent fuel, BKB, coke oven coke, gas coke, coal tar, coke oven gas, gas works gas, blast furnace gas and other recovered gases), as well as peat, peat briquettes and oil shale. Quantities have been converted to Mtce using calorific values reported by the respective countries.
Please refer to notes and definitions in Part I.

Source: *IEA/OECD Energy Balances of OECD Countries*

IRELAND

6. Use of coal for selected end-uses*

(million tonnes)

	1978	1990	2000	2005	2009	2010	2011	Average annual percent change 78-90	90-11
Steam coal	**0.54**	**3.18**	**2.94**	**2.99**	**1.85**	**2.00**	**2.05**	**15.88**	**-2.07**
Total electricity and heat	0.02	1.97	2.35	2.34	1.31	1.47	1.55	45.40	-1.13
Main activity producers	*0.02*	*1.96*	*2.34*	*2.33*	*1.31*	*1.47*	*1.55*	*45.37*	*-1.11*
Autoproducers	-	*0.01*	*0.01*	*0.01*	-	-	-	-	-
Patent fuel/BKB plants	-	c	c	c	c	c	c	c	c
Coke ovens/Liquefaction(1)	-	-	-	-	-	-	-	-	-
Blast furnace inputs	-	-	-	-	-	-	-	-	-
Gas manufacture	0.03	-	-	-	-	-	-	-	-
Industry	0.04	0.33	0.17	0.32	0.17	0.16	0.14	19.33	-3.80
Iron and steel	-	-	-	-	-	-	-	-	-
Chemical	-	*0.03*	-	-	-	-	-	-	-
Non-metallic minerals	-	*0.19*	*0.11*	*0.22*	*0.13*	*0.14*	*0.12*	-	*-2.07*
Paper, pulp and print	-	-	-	-	*0.00*	-	-	-	-
Other industry	*0.04*	*0.11*	*0.06*	*0.09*	*0.04*	*0.02*	*0.02*	*9.19*	*-7.07*
Other sectors(2)	0.46	1.05	0.38	0.35	0.35	0.34	0.30	7.25	-5.76
Non-energy use	-	-	-	-	-	-	-	-	-
Coking coal	**-**	**0.01**	**-**	**-**	**-**	**-**	**-**	**-**	**-**
Total electricity and heat	-	-	-	-	-	-	-	-	-
Main activity producers	-	-	-	-	-	-	-	-	-
Autoproducers	-	-	-	-	-	-	-	-	-
Patent fuel/BKB plants	-	-	-	-	-	-	-	-	-
Coke ovens/Liquefaction(1)	-	-	-	-	-	-	-	-	-
Blast furnace inputs	-	-	-	-	-	-	-	-	-
Gas manufacture	-	-	-	-	-	-	-	-	-
Industry	-	0.01	-	-	-	-	-	-	-
Iron and steel	-	*0.01*	-	-	-	-	-	-	-
Chemical	-	-	-	-	-	-	-	-	-
Non-metallic minerals	-	-	-	-	-	-	-	-	-
Paper, pulp and print	-	-	-	-	-	-	-	-	-
Other industry	-	-	-	-	-	-	-	-	-
Other sectors(2)	-	-	-	-	-	-	-	-	-
Non-energy use	-	-	-	-	-	-	-	-	-
Lignite	**-**	**0.04**	**0.04**	**0.05**	**0.02**	**0.02**	**0.02**	**-**	**-2.57**
Total electricity and heat	-	-	-	-	-	-	-	-	-
Main activity producers	-	-	-	-	-	-	-	-	-
Autoproducers	-	-	-	-	-	-	-	-	-
Patent fuel/BKB plants	-	-	-	-	-	-	-	-	-
Coke ovens/Liquefaction(1)	-	-	-	-	-	-	-	-	-
Gas manufacture	-	-	-	-	-	-	-	-	-
Industry	-	-	-	-	-	-	-	-	-
Iron and steel	-	-	-	-	-	-	-	-	-
Chemical	-	-	-	-	-	-	-	-	-
Non-metallic minerals	-	-	-	-	-	-	-	-	-
Paper, pulp and print	-	-	-	-	-	-	-	-	-
Other industry	-	-	-	-	-	-	-	-	-
Other sectors(2)	-	0.04	0.04	0.05	0.03	0.02	0.02	-	-2.57
Non-energy use	-	-	-	-	-	-	-	-	-

(1) Liquefaction primarily refers to direct distillation processes. Indirect liquefaction may be reported here or under gas manufacture.

(2) Other sectors are Residential, Commercial and public services, Agriculture/Forestry, Fishing and Non-specified other.

* Coal totals may not match the sum of the individual rows due to statistical differences. Please refer to notes and definitions in Part I.

Source: *IEA/OECD Energy Statistics of OECD Countries*

IRELAND

7. Fuel prices to end users*

	1978(1)	1990	1995	2000	2005	2009	2010	2011	2012
				(US dollars / tce)					
For electricity generation									
Steam coal	..	63.11	53.27	34.26	79.21	128.53	94.21	141.76	99.60
Heavy fuel oil	..	82.37	90.43	103.73	210.16	..	..	..	..
Natural gas	..	90.97	92.51	77.04	c	c	c	c	c
For industry									
Steam coal	..	..	..	..	..	..	..	..	..
Coking coal	x	x	x	x	x	x	x	x	x
High sulphur fuel oil	102.26	115.42	126.31	168.24	309.06	540.65	625.70	818.99	838.05
Low sulphur fuel oil	..	..	..	..	..	..	..	..	..
Natural gas	189.24	257.44	247.33	112.77	290.90	376.03	335.39	397.16	412.32
				(Euro / unit)(2)					
For electricity generation									
Steam coal	..	42.89	37.38	32.89	56.42	81.87	62.94	90.15	68.57
Heavy fuel oil	..	85.67	97.12	152.42	229.12	..	..	..	..
Natural gas	..	89.83	94.32	107.47	c	c	c	c	c
For industry									
Steam coal	..	..	..	..	..	..	..	..	..
Coking coal	x	x	x	x	x	x	x	x	x
High sulphur fuel oil	91.68	120.05	135.65	247.21	336.94	527.04	639.76	797.14	883.00
Low sulphur fuel oil	..	..	..	..	..	..	..	..	..
Natural gas	161.07	254.20	252.17	157.32	301.08	348.00	325.57	366.99	412.44

(1) Earliest year for which split by coal type is available.

(2) For steam coal, coking coal and heavy fuel oil per metric tonne; for natural gas per 10^7 kilocalories GCV.

* Please refer to notes and definitions in Part I. Steam coal prices relate to historical definitions.

Source: *IEA/OECD Energy Prices & Taxes*

8. Coal and peat trade by type of coal*

(Mtce)

	1978	1990	1995	2000	2005	2009	2010	2011	2012e
Total imports	**0.58**	**2.87**	**2.71**	**2.42**	**2.72**	**1.85**	**1.38**	**2.02**	**1.95**
Bituminous coal(1)	0.57	2.80	2.67	2.40	2.69	1.83	1.36	2.01	1.93
Coking coal	-	0.01	-	-	-	-	-	-	-
Sub-bituminous coal	-	-	-	-	-	-	-	-	-
Lignite	-	0.03	0.03	0.03	0.04	0.02	0.01	0.01	0.02
Peat	-	-	-	-	-	-	-	-	-
Coal products(2)	0.01	0.03	0.01	c	c	c	c	c	-
Total exports	**0.08**	**0.03**	**0.02**	**0.02**	**0.03**	**0.02**	**0.03**	**0.02**	**0.03**
Bituminous coal(1)	0.05	0.03	0.00	0.01	0.01	0.01	0.01	0.01	0.01
Coking coal	-	-	-	-	-	-	-	-	-
Sub-bituminous coal	-	-	-	-	-	-	-	-	-
Lignite	-	0.00	0.00	-	-	0.00	0.00	0.00	-
Peat	0.01	-	-	-	-	-	-	-	-
Coal products(2)	0.02	0.01	0.01	0.01	0.01	0.01	0.01	0.01	0.01

(1) Bituminous coal includes anthracite.

(2) Coal products includes products derived from coal, for example: coke, coal tar, briquettes, patent fuels.

* Please refer to notes and definitions in Part I.

Source: *IEA/OECD Energy Balances of OECD Countries, IEA/OECD Energy Statistics of OECD Countries*

IRELAND

9. Total coal imports by origin*

(thousand tonnes)

	1978(1)	1985	1990	1995	2000	2005	2010	2011	2012e
Total coal(2)	**565**	**1918**	**3121**	**2910**	**2784**	**3076**	**1619**	**2311**	**2224**
Coking coal	**-**	**7**	**14**	**-**	**-**	**-**	**-**	**-**	**-**
Australia	-	-	-	-	-	-	-	-	-
Canada	-	-	-	-	-	-	-	-	-
Czech Republic	-	-	-	-	-	-	-	-	-
Germany	-	-	2	-	-	-	-	-	-
Poland	-	-	2	-	-	-	-	-	-
United Kingdom	-	6	-	-	-	-	-	-	-
United States	-	-	2	-	-	-	-	-	-
Other OECD	-	-	5	-	-	-	-	-	-
China, People's Rep.	-	-	-	-	-	-	-	-	-
Colombia	-	-	3	-	-	-	-	-	-
Indonesia	-	-	-	-	-	-	-	-	-
South Africa	-	-	-	-	-	-	-	-	-
Former Soviet Union(3)	-	-	x	x	x	x	x	x	x
Russian Federation (4)	*x*	*x*	-	-	-	-	-	-	-
Other FSU (4)	*x*	*x*	-	-	-	-	-	-	-
Venezuela	-	-	-	-	-	-	-	-	-
Vietnam	-	-	-	-	-	-	-	-	-
Non-specified/other	-	1	-	-	-	-	-	-	-
Steam coal	**565**	**1911**	**3067**	**2860**	**2747**	**3024**	**1597**	**2289**	**2199**
Australia	6	76	-	118	292	299	-	-	-
Canada	-	-	-	-	-	-	-	-	-
Czech Republic	-	2	-	-	-	-	-	-	-
Germany	-	65	21	3	-	5	-	-	-
Poland	331	735	430	300	176	288	278	189	156
United Kingdom	214	390	302	101	123	19	38	58	30
United States	-	329	1277	1031	452	-	-	-	147
Other OECD	-	122	115	10	-	36	-	-	-
China, People's Rep.	-	5	7	-	-	9	-	-	-
Colombia	-	-	667	955	904	1013	1184	1867	-
Indonesia	-	-	17	43	327	602	-	-	-
South Africa	11	70	71	226	418	634	38	3	-
Former Soviet Union(3)	-	-	x	x	x	x	x	x	x
Russian Federation (4)	*x*	*x*	*15*	-	-	-	-	-	-
Other FSU (4)	*x*	*x*	-	-	-	*9*	-	-	-
Venezuela	-	-	5	-	-	-	-	-	-
Vietnam	-	-	-	-	-	-	-	-	-
Non-specified/other	3	117	140	73	55	110	59	172	1866
Lignite	**-**	**-**	**40**	**50**	**37**	**52**	**22**	**22**	**25**

(1) Earliest year for which split by coal type is available.

(2) Total coal consists of steam coal, coking coal and lignite.

(3) For years prior to 1990, or if no details for individual Former Soviet Union republics.

(4) Data for former republics are not available separately for some OECD member countries.

* Please refer to notes and definitions in Part I.

Source: IEA/OECD Coal Statistics

IRELAND

12. Coal import values by origin*

(average unit value, CIF, USD/tonne)

	1990	1995	2000	2005	2007	2008	2009	2010	2011/12(1)
Coking coal(2)	**107.07**	**101.41**	**104.63**	**412.43**	**86.14**	**102.25**	**794.41**	**571.13**	..
Imports from:									
Australia	..	..	..	..	..	..	..	..	..
Canada	..	..	..	..	..	..	..	..	..
Czech Republic	..	..	..	..	..	..	..	..	..
Poland	100.40	..	..	..	..	..	..	..	..
United States	..	98.08	97.32	..	..	..	..	..	..
China	..	..	..	..	..	..	..	..	..
Colombia	106.09	..	..	..	..	..	..	..	..
Indonesia	..	..	..	..	..	..	..	..	..
South Africa	..	..	..	..	..	96.99	..	..	..
Former Soviet Union(3)	..	..	..	..	77.79	95.68	..	..	..
Other bituminous coal(4)	**70.76**	**69.55**	**42.85**	**74.91**	**125.15**	**143.08**	**148.41**	**116.63**	..
Imports from:									
Australia	..	41.76	36.34	84.49	618.20	142.53	..	..	..
Canada	..	..	..	..	..	..	..	..	..
Czech Republic	..	..	..	..	..	..	..	..	..
Poland	98.25	99.09	83.30	114.10	177.54	256.72	234.09	211.92	..
United States	62.06	57.93	55.28	..	..	..	69.93	74.57	..
China	82.52	..	..	..	..	123.41	..	..	..
Colombia	72.99	36.07	..	..	..	..	233.72	..	..
Indonesia	69.73	80.28	30.19	68.00	89.61	131.11	96.31	..	..
South Africa	85.34	84.69	41.55	64.62	85.73	126.72	100.80	108.11	..
Former Soviet Union(3)	81.51	70.14	..	713.45	..	..	..	..	..

Note: On occasion, shipment of extremely small quantities of high valued coal results in high import costs.

(1) Import data for steam and coking coals are currently unavailable for 2011 and 2012 due to resource constraints.

(2) Weighted average of individual countries using import volumes as weights. Prices exclude intra-EU trade.

(3) Former Soviet Union until 1991, Russian Federation starting in 1992.

(4) Bituminous steam coal only. Weighted average of trades. Prices exclude intra-EU trade.

* Please refer to notes and definitions in Part I.

Source: IEA/OECD Coal Statistics

ISRAEL*

Figure 1: TPES by fuel (Mtce)

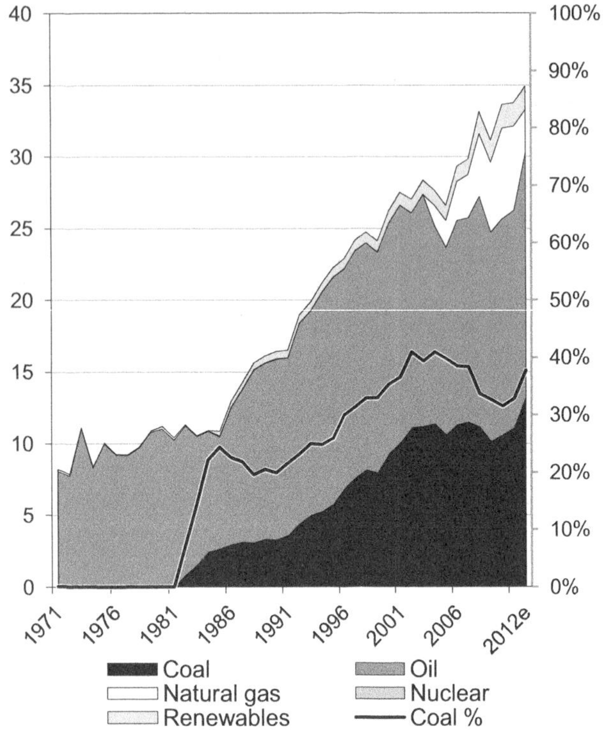

Figure 2: Coal supply indicators (1971 = 100)

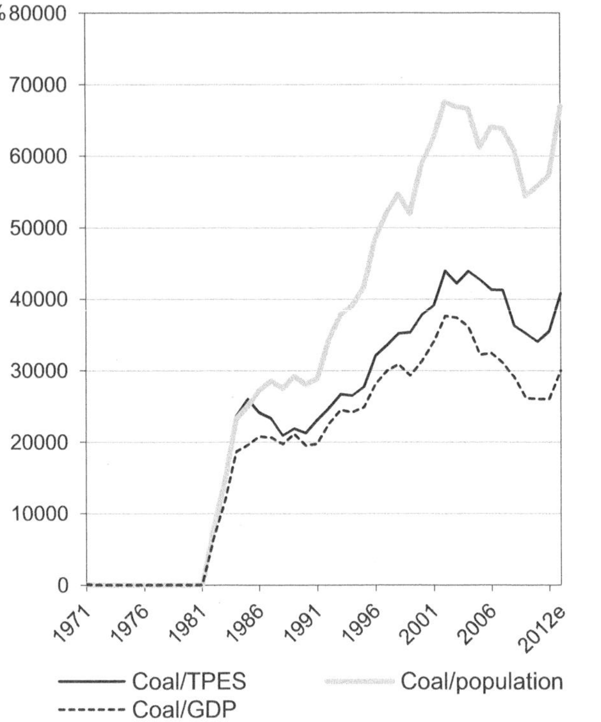

Figure 3: Primary coal supply (Mtce)

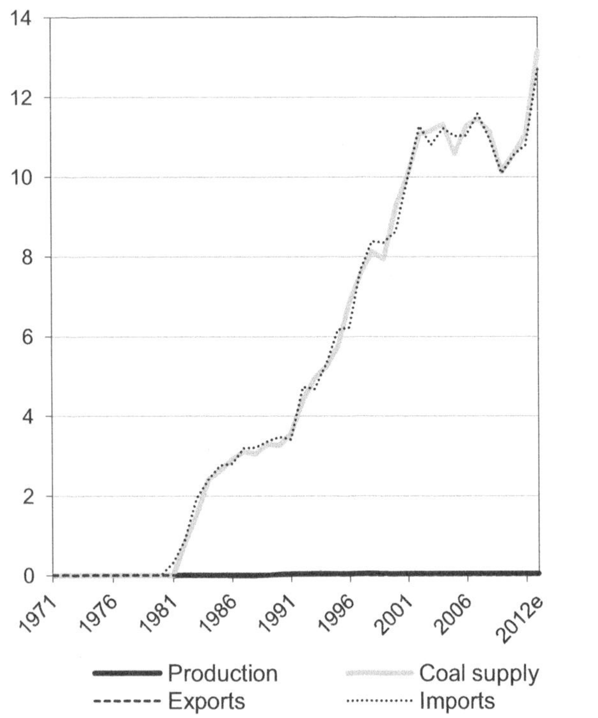

Figure 4: Coal consumption (Mtce)

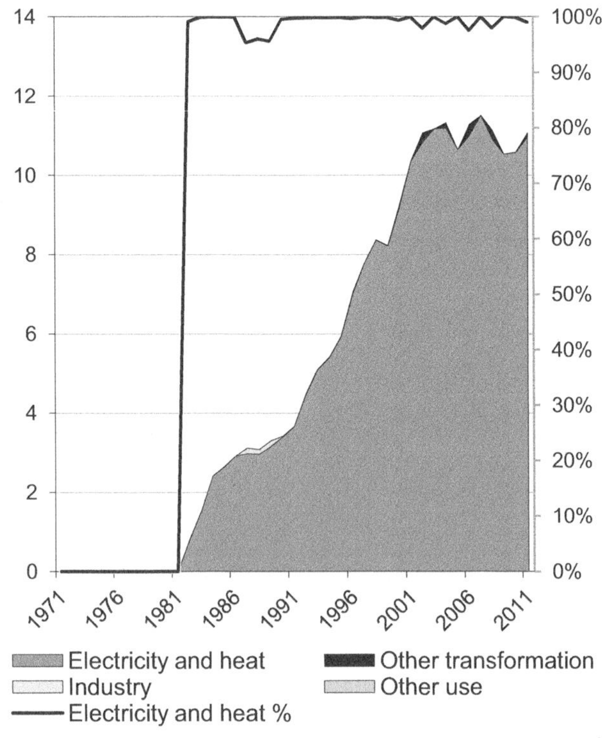

*Please refer to notes and definitions in Part I. Peat is included in coal for all figures.

Source: *IEA/OECD Coal Information, IEA/OECD Energy Balances of OECD Countries, OECD Main Economic Indicators*

ISRAEL*

Figure 5: Electricity generation by fuel (TWh)

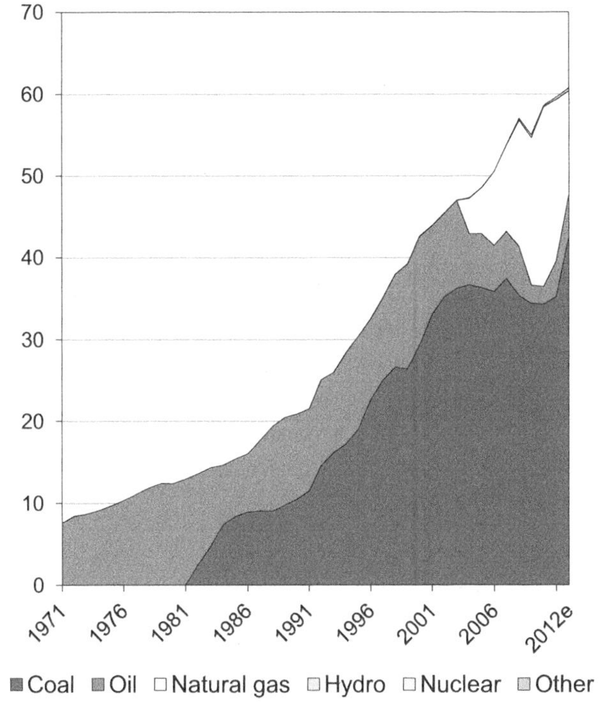

Figure 6: CO_2 emissions by fuel (Mt CO_2)

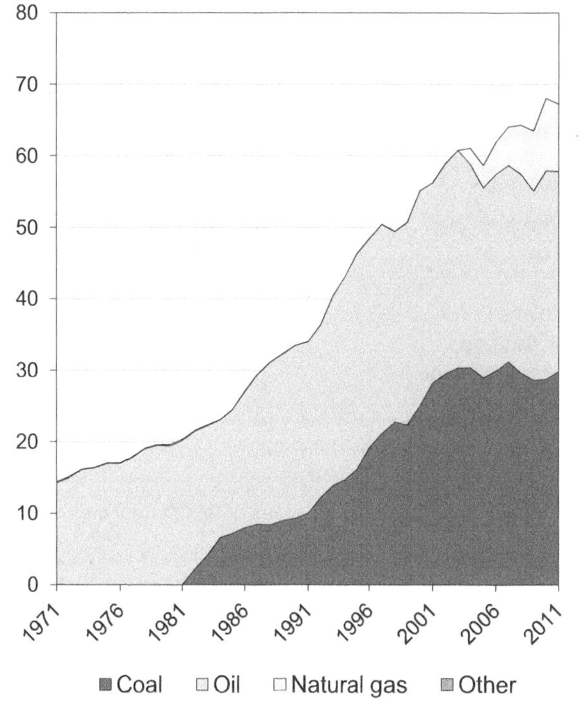

Figure 7: Electricity generation by fuel share

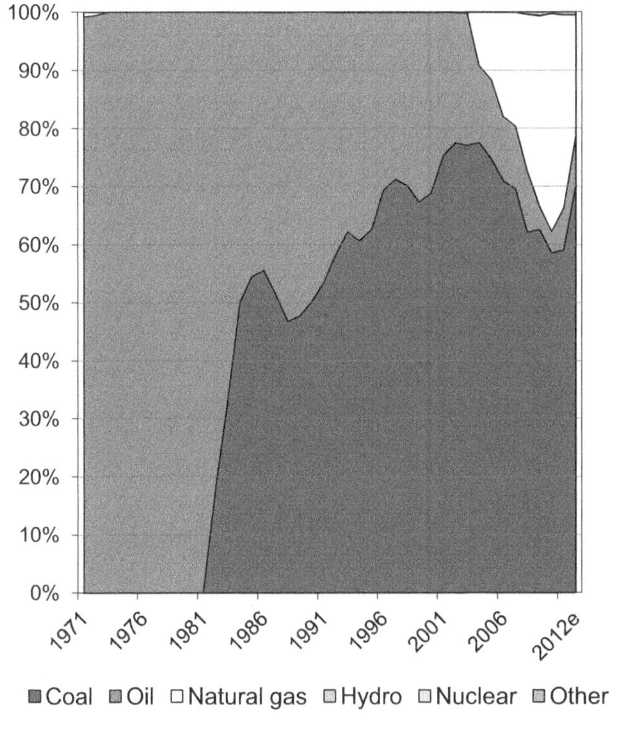

Figure 8: CO_2 emissions by fuel share

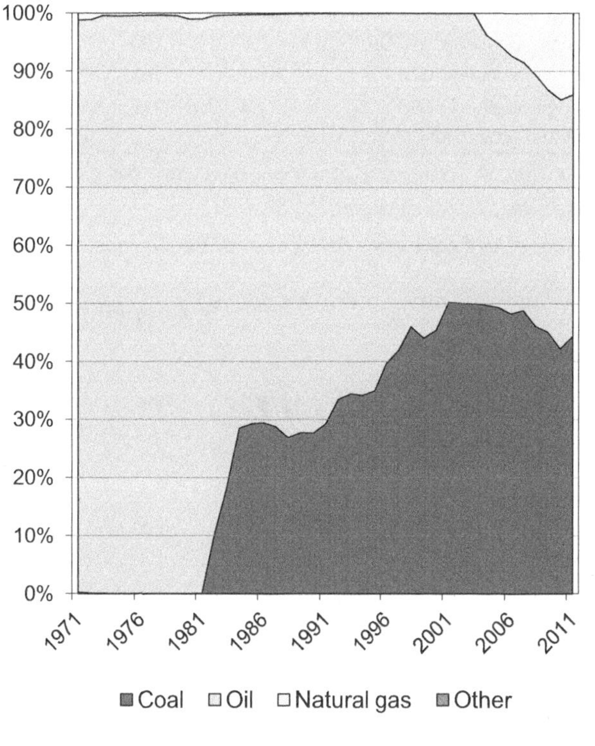

*Please refer to notes and definitions in Part I. Peat is included in coal for all figures.

Source: *IEA/OECD Electricity Information, IEA/OECD CO2 Emissions from Fuel Combustion*

ISRAEL

1. Total primary energy supply (TPES) by fuel*

	1973	1980	1990	2000	2010	2011	2012e	Average annual percent change 73-90	90-11
TPES (Mtce)	**11.09**	**11.18**	**16.38**	**26.05**	**33.14**	**33.22**	**34.40**	**2.32**	**3.42**
Coal and peat	0.00	0.00	3.27	9.25	10.58	11.07	13.17	54.87	5.99
Oil	11.03	11.00	12.61	16.10	15.10	15.18	17.09	0.79	0.89
Natural Gas	0.06	0.19	0.04	0.01	6.29	5.86	3.00	-2.70	26.74
Biofuels and waste	0.00	0.00	0.00	0.01	0.03	0.03	0.03	0.00	8.80
Nuclear	-	-	-	-	-	-	-	-	-
Hydro	-	-	0.00	0.00	0.00	0.00	0.00	-	11.22
Geothermal	-	-	-	-	-	-	-	-	-
Solar, wind, tide	-	-	0.51	0.85	1.61	1.60	1.63	-	5.57
Net electricity trade(1)	-0.01	-0.02	-0.06	-0.18	-0.49	-0.52	-0.52	13.01	11.18
Heat(2)	-	-	-	-	-	-	-	-	-

(1) Net trade between OECD and non-OECD countries.

(2) Ambient heat from heat pumps used in the transformation sector.

Source: *IEA/OECD Energy Balances of OECD Countries*

2. Energy supply, GDP and population*

	1973	1980	1990	2000	2010	2011	2012e	Average annual percent change 73-90	90-11
GDP (2005 bil. USD)	37	47	68	120	166	173	179	3.64	4.54
TPES/GDP(1)	0.30	0.24	0.24	0.22	0.20	0.19	0.19	-1.28	-1.07
Population (millions)	3.3 e	3.9 e	4.7 e	6.3	7.6	7.8	7.9	2.07	2.44
TPES/population(1)	3.36 e	2.87 e	3.50 e	4.13	4.35	4.28	4.35	0.24	0.96
TPES/GDP(2)	151.6	120.3	121.8	109.9	101.4	97.2	97.6	-1.28	-1.07
Coal and peat supply/GDP(2)	0.1	0.0	60.6	97.4	80.8	80.8	93.2	49.43	1.38
Elec. consumption/GDP(2)	64.6	73.3	84.0	101.2	92.6	89.0	..	1.56	0.27
Elec. generation (TWh)	9	12	21	43	59	60	61	5.28	5.12
Industrial production(2)	..	..	54.1	95.7	125.1	127.9	133.9	..	4.18

(1) TPES/GDP in units of tce/2005 thousand USD, TPES/population in units of tce/capita.

(2) As index, 2005 (not shown) = 100

Source: *IEA/OECD Energy Balances of OECD Countries* , *OECD Main Economic Indicators*

3. Coal and peat production by type*

	1978(1)	1990	2000	2005	2010	2011	2012e	Average annual percent change 78-90	90-11
Mtce:									
Coking coal	-	-	-	-	-	-	-	-	-
Steam coal	-	-	-	-	-	-	-	-	-
Lignite	-	0.03	0.04	0.04	0.04	0.04	0.04	-	1.64
Peat	-	-	-	-	-	-	-	-	-
Mt:									
Coking coal	-	-	-	-	-	-	-	-	-
Steam coal	-	-	-	-	-	-	-	-	-
Lignite	-	0.30	0.39	0.43	0.43	0.42	0.40	-	1.52
Peat	-	-	-	-	-	-	-	-	-

(1) Earliest year for which split by coal type is available.

* Please refer to notes and definitions in Part I.

Source: *IEA/OECD Energy Balances of OECD Countries, IEA/OECD Energy Statistics of OECD Countries*

ISRAEL

4. Final consumption of energy by fuel*

(Mtce)

	1973	1980	1990	2000	2009	2010	2011	Average annual percent change 73-90	90-11
Total final consumption[(1)]	**5.16**	**6.45**	**9.95**	**17.11**	**19.88**	**21.20**	**20.11**	**3.94**	**3.41**
Coal and peat	0.00	0.00	0.02	0.03	-	-	-	10.95	-
Oil	4.15	4.91	7.15	11.48	12.56	13.50	12.34	3.24	2.64
Natural Gas	0.06	0.19	0.04	0.01	0.24	0.09	0.17	-2.70	7.04
Biofuels and wastes	0.00	0.00	0.00	0.01	0.02	0.02	0.01	0.00	5.19
Geothermal	-	-	-	-	-	-	-	-	-
Solar, wind, tide	-	-	0.51	0.85	1.48	1.60	1.57	-	5.49
Electricity	0.93	1.35	2.23	4.74	5.58	5.99	6.01	5.26	4.83
Heat	-	-	-	-	-	-	-	-	-
of which:									
Total industry	**1.56**	**1.73**	**2.05**	**1.94**	**2.28**	**2.95**	**1.90**	**1.63**	**-0.36**
Coal and peat	0.00	0.00	0.02	0.03	-	-	-	10.95	-
Oil	1.27	1.30	1.38	0.63	0.74	1.32	0.26	0.50	-7.69
Natural Gas	-	-	0.00	-	0.24	0.09	0.17	-	21.37
Biofuels and wastes	-	-	-	-	-	-	-	-	-
Geothermal	-	-	-	-	-	-	-	-	-
Solar, wind, tide	-	-	-	-	-	-	-	-	-
Electricity	0.28	0.43	0.65	1.28	1.29	1.53	1.48	5.00	3.98
Heat	-	-	-	-	-	-	-	-	-
Total transport	**1.65**	**1.98**	**3.16**	**4.82**	**5.43**	**5.79**	**5.41**	**3.90**	**2.59**
Coal and peat	-	-	-	-	-	-	-	-	-
Oil	1.65	1.98	3.16	4.82	5.43	5.79	5.41	3.90	2.59
Natural Gas	-	-	-	-	-	-	-	-	-
Biofuels and wastes	-	-	-	-	-	-	-	-	-
Electricity	-	-	-	-	-	-	-	-	-
Residential	**0.41**	**0.56**	**1.89**	**3.47**	**4.75**	**4.85**	**4.96**	**9.32**	**4.71**
Coal and peat	-	-	-	-	-	-	-	-	-
Oil	0.14	0.19	0.72	1.19	1.40	1.36	1.42	9.98	3.32
Natural Gas	-	0.00	0.00	-	-	-	-	-	-
Biofuels and wastes	0.00	0.00	0.00	0.01	0.01	0.01	0.01	0.00	1.75
Geothermal	-	-	-	-	-	-	-	-	-
Solar, wind, tide	-	-	0.51	0.85	1.48	1.60	1.57	-	5.49
Electricity	0.27	0.37	0.65	1.42	1.86	1.88	1.96	5.38	5.36
Heat	-	-	-	-	-	-	-	-	-
Comm & public services	**-**	**0.38**	**0.63**	**1.44**	**1.92**	**2.09**	**2.10**	**-**	**5.89**
Coal and peat	-	-	-	-	-	-	-	-	-
Oil	-	-	-	-	-	-	-	-	-
Natural Gas	-	-	-	-	-	-	-	-	-
Biofuels and waste	-	-	-	-	-	-	-	-	-
Geothermal	-	-	-	-	-	-	-	-	-
Solar, wind, tide	-	-	-	-	-	-	-	-	-
Electricity	-	0.38	0.63	1.44	1.92	2.09	2.10	-	5.89
Heat	-	-	-	-	-	-	-	-	-
Non-energy use	**0.40**	**0.93**	**1.05**	**2.59**	**1.92**	**1.99**	**2.22**	**5.92**	**3.61**
Coal and peat	-	-	-	-	-	-	-	-	-
Oil	0.33	0.75	1.02	2.58	1.92	1.99	2.22	6.81	3.78
Natural Gas	0.06	0.19	0.04	0.01	-	-	-	-3.16	-

(1) Total Final Consumption (TFC) excludes use in transformation processes and energy industry own use.

* Please refer to notes and definitions in Part I. Agriculture, Forestry, Fishing and "Non-specified other" sectors are not reported here individually, but are accounted for in Total final consumption.

Source: *IEA/OECD Energy Balances of OECD Countries*

ISRAEL

5. Coal balance*

(Mtce)

	1973	1980	1990	2000	2005	2010	2011	2012e	Average annual percent change 73-90	90-11
Production	-	-	0.0	0.0	0.0	0.0	0.0	0.0	-	1.6
Imports	0.0	0.0	3.5	8.6	11.0	10.5	10.8	12.7	55.4	5.5
Exports	-	-	-	-	-	-	-	-	-	-
Stock changes	-	-0.0	-0.2	0.6	-0.5	-	0.2	0.4		
Primary supply	**0.0**	**0.0**	**3.3**	**9.2**	**10.6**	**10.6**	**11.1**	**13.2**	**54.9**	**6.0**
Statistical differences	**-**	**0.0**	**0.1**	**-0.0**	**0.1**	**-0.0**	**-0.1**	**..**		
Total transformation	**0.0**	**0.0**	**-3.4**	**-9.2**	**-10.6**	**-10.6**	**-11.0**	**..**	**-**	**5.7**
Electricity and heat gen.	-	-	-3.4	-9.2	-10.6	-10.6	-11.0	..	-	5.7
Main activity producers (1)	-	-	*-3.4*	*-9.2*	*-10.6*	*-10.5*	*-10.9*	..	-	*5.7*
Autoproducers	-	-	-	*-0.0*	*-0.0*	*-0.0*	*-0.0*	..	-	-
Gas works	-	-	-	-	-	-	-	..	-	-
Coal transformation(2)	0.0	0.0	-	-	-	-	-	..	-	-
BKB plants	-	-	-	-	-	-	-	..	-	-
Blast furnaces	*0.0*	*0.0*	-	-	-	-	-	..	-	-
Coke ovens	-	-	-	-	-	-	-	..	-	-
Patent fuel plants	-	-	-	-	-	-	-	..	-	-
Other transformation(3)	-	-	-	-	-	-	-	..	-	-
Energy ind. own use	**-**	**-**	**-**	**-**	**-**	**-**	**-**	**..**	**-**	**-**
Losses	**-**	**-**	**-**	**-**	**-**	**-**	**-**	**..**		
Final consumption(4)	**0.0**	**0.0**	**0.0**	**0.0**	**-**	**-**	**-**	**..**	**11.0**	**-**
Industry(5)	0.0	0.0	0.0	0.0	-	-	-	..	11.0	-
Iron and steel	*0.0*	*0.0*	-	-	-	-	-	..	-	-
Chemical	-	-	-	-	-	-	-	..	-	-
Non-metallic minerals	-	-	-	-	-	-	-	..	-	-
Paper, pulp and print	-	-	-	-	-	-	-	..	-	-
Other industry (6)	-	-	*0.0*	*0.0*	-	-	-	..	-	-
Transport(7)	-	-	-	-	-	-	-	..	-	-
Other	-	-	-	-	-	-	-	..	-	-
Comm. and pub. services	-	-	-	-	-	-	-	..	-	-
Residential	-	-	-	-	-	-	-	..	-	-
Other sectors (8)	-	-	-	-	-	-	-	..	-	-
Non-energy use	-	-	-	-	-	-	-	..	-	-

(1) Main activity electricity and heat generation includes district heating.

(2) Coal transformation refers to the transformation of coal and peat to secondary and tertiary products (mainly cokes, briquettes, coal tar and off-gases). Gas works and Liquefaction are shown separately to match the *Energy Balances of OECD Countries.*

(3) Other transformation includes Liquefaction and Non-specified transformations.

(4) Final Consumption includes non-energy use and energy use (Industry, Transport and Other).

(5) Please refer to notes and definitions in Part I for detailed explanation.

(6) Other industry includes Non-ferrous metals, Transport equipment, Machinery, Mining and quarrying, Food and tobacco, Wood and wood products, Construction, Textile and leather, and Non-specified industry.

(7) Transport includes Rail and Inland waterways.

(8) Other sectors includes Agriculture/Forestry, Fishing and Non-specified other.

* "Coal" refers to all types of coal, primary (anthracite, coking coal, other bituminous coal, sub-bituminous coal, lignite/brown coal) and derived fuels (including patent fuel, BKB, coke oven coke, gas coke, coal tar, coke oven gas, gas works gas, blast furnace gas and other recovered gases), as well as peat, peat briquettes and oil shale. Quantities have been converted to Mtce using calorific values reported by the respective countries.
Please refer to notes and definitions in Part I.

Source: *IEA/OECD Energy Balances of OECD Countries*

ISRAEL

6. Use of coal for selected end-uses*

(million tonnes)

	1978	1990	2000	2005	2009	2010	2011	Average annual percent change 78-90	90-11
Steam coal	**-**	**3.72**	**10.59**	**12.12**	**11.85**	**12.31**	**12.75**	**-**	**6.04**
Total electricity and heat	-	3.70	10.22	12.69	12.31	12.30	12.62	-	6.01
Main activity producers	-	*3.70*	*10.22*	*12.69*	*12.31*	*12.30*	*12.62*	-	*6.01*
Autoproducers	-	-	-	-	-	-	-	-	-
Patent fuel/BKB plants	-	-	-	-	-	-	-	-	-
Coke ovens/Liquefaction(1)	-	-	-	-	-	-	-	-	-
Blast furnace inputs	-	-	-	-	-	-	-	-	-
Gas manufacture	-	-	-	-	-	-	-	-	-
Industry	-	0.02	0.03	-	-	-	-	-	-
Iron and steel	-	-	-	-	-	-	-	-	-
Chemical	-	-	-	-	-	-	-	-	-
Non-metallic minerals	-	-	-	-	-	-	-	-	-
Paper, pulp and print	-	-	-	-	-	-	-	-	-
Other industry	-	*0.02*	*0.03*	-	-	-	-	-	-
Other sectors(2)	-	-	-	-	-	-	-	-	-
Non-energy use	-	-	-	-	-	-	-	-	-
Coking coal	**-**	**-**	**-**	**-**	**-**	**-**	**-**	**-**	**-**
Total electricity and heat	-	-	-	-	-	-	-	-	-
Main activity producers	-	-	-	-	-	-	-	-	-
Autoproducers	-	-	-	-	-	-	-	-	-
Patent fuel/BKB plants	-	-	-	-	-	-	-	-	-
Coke ovens/Liquefaction(1)	-	-	-	-	-	-	-	-	-
Blast furnace inputs	-	-	-	-	-	-	-	-	-
Gas manufacture	-	-	-	-	-	-	-	-	-
Industry	-	-	-	-	-	-	-	-	-
Iron and steel	-	-	-	-	-	-	-	-	-
Chemical	-	-	-	-	-	-	-	-	-
Non-metallic minerals	-	-	-	-	-	-	-	-	-
Paper, pulp and print	-	-	-	-	-	-	-	-	-
Other industry	-	-	-	-	-	-	-	-	-
Other sectors(2)	-	-	-	-	-	-	-	-	-
Non-energy use	-	-	-	-	-	-	-	-	-
Lignite	**-**	**0.30**	**0.39**	**0.43**	**0.44**	**0.43**	**0.42**	**-**	**1.52**
Total electricity and heat	-	0.30	0.46	0.43	0.44	0.43	0.42	-	1.52
Main activity producers	-	*0.30*	*0.27*	*0.26*	-	-	-	-	-
Autoproducers	-	-	*0.18*	*0.17*	*0.44*	*0.43*	*0.42*	-	-
Patent fuel/BKB plants	-	-	-	-	-	-	-	-	-
Coke ovens/Liquefaction(1)	-	-	-	-	-	-	-	-	-
Gas manufacture	-	-	-	-	-	-	-	-	-
Industry	-	-	-	-	-	-	-	-	-
Iron and steel	-	-	-	-	-	-	-	-	-
Chemical	-	-	-	-	-	-	-	-	-
Non-metallic minerals	-	-	-	-	-	-	-	-	-
Paper, pulp and print	-	-	-	-	-	-	-	-	-
Other industry	-	-	-	-	-	-	-	-	-
Other sectors(2)	-	-	-	-	-	-	-	-	-
Non-energy use	-	-	-	-	-	-	-	-	-

(1) Liquefaction primarily refers to direct distillation processes. Indirect liquefaction may be reported here or under gas manufacture.

(2) Other sectors are Residential, Commercial and public services, Agriculture/Forestry, Fishing and Non-specified other.

* Coal totals may not match the sum of the individual rows due to statistical differences. Please refer to notes and definitions in Part I.

Source: *IEA/OECD Energy Statistics of OECD Countries*

ISRAEL

7. Fuel prices to end users*

	1978(1)	1990	1995	2000	2005	2009	2010	2011	2012
				(US dollars / tce)					
For electricity generation									
Steam coal	..	..	..	..	..	165.17	130.46	189.72	179.45
Heavy fuel oil	..	..	..	..	..	297.57	361.89	560.87	587.59
Natural gas	x	x	x	x	..	255.20	339.46	378.82	384.02
For industry									
Steam coal	x	x	x	x	x	x	x	x	x
Coking coal	x	x	x	x	x	x	x	x	x
High sulphur fuel oil	..	..	..	..	c	c	c	c	c
Low sulphur fuel oil	..	..	..	..	c	c	c	c	c
Natural gas	x	x	x	x	c	c	c	c	c
				(New Israeli sheqel / unit)(2)					
For electricity generation									
Steam coal	..	..	..	..	..	481.81	361.56	503.72	513.25
Heavy fuel oil	..	..	..	..	..	1602.52	1851.54	2749.27	3102.54
Natural gas	x	x	x	x	..	1288.44	1628.20	1740.84	1900.94
For industry									
Steam coal	x	x	x	x	x	x	x	x	x
Coking coal	x	x	x	x	x	x	x	x	x
High sulphur fuel oil	..	..	..	..	c	c	c	c	c
Low sulphur fuel oil	..	..	..	..	c	c	c	c	c
Natural gas	x	x	x	x	c	c	c	c	c

(1) Earliest year for which split by coal type is available.

(2) For steam coal, coking coal and heavy fuel oil per metric tonne; for natural gas per 10^7 kilocalories GCV.

* Please refer to notes and definitions in Part I. Steam coal prices relate to historical definitions.

Source: *IEA/OECD Energy Prices & Taxes*

8. Coal and peat trade by type of coal*

(Mtce)

	1978	1990	1995	2000	2005	2009	2010	2011	2012e
Total imports	**0.00**	**3.48**	**6.17**	**8.62**	**11.03**	**10.09**	**10.54**	**10.79**	**12.70**
Bituminous coal(1)	-	3.48	6.17	8.62	11.03	10.09	10.54	10.68	12.58
Coking coal	-	-	-	-	-	-	-	-	-
Sub-bituminous coal	-	-	-	-	-	-	-	0.11	0.12
Lignite	-	-	-	-	-	-	-	-	-
Peat	-	-	-	-	-	-	-	-	-
Coal products(2)	0.00	-	-	-	-	-	-	-	-
Total exports	**-**	**-**	**-**	**-**	**-**	**-**	**-**	**-**	**-**
Bituminous coal(1)	-	-	-	-	-	-	-	-	-
Coking coal	-	-	-	-	-	-	-	-	-
Sub-bituminous coal	-	-	-	-	-	-	-	-	-
Lignite	-	-	-	-	-	-	-	-	-
Peat	-	-	-	-	-	-	-	-	-
Coal products(2)	-	-	-	-	-	-	-	-	-

(1) Bituminous coal includes anthracite.

(2) Coal products includes products derived from coal, for example: coke, coal tar, briquettes, patent fuels.

* Please refer to notes and definitions in Part I.

Source: *IEA/OECD Energy Balances of OECD Countries, IEA/OECD Energy Statistics of OECD Countries*

ISRAEL

9. Total coal imports by origin*

(thousand tonnes)

	1978(1)	1985	1990	1995	2000	2005	2010	2011	2012e
Total coal(2)	**-**	**3063**	**3998**	**7090**	**9917**	**12685**	**12310**	**12475**	**14668**
Coking coal	**-**	**-**	**-**	**-**	**-**	**-**	**-**	**-**	**-**
Australia	-	-	-	-	-	-	-	-	-
Canada	-	-	-	-	-	-	-	-	-
Czech Republic	-	-	-	-	-	-	-	-	-
Germany	-	-	-	-	-	-	-	-	-
Poland	-	-	-	-	-	-	-	-	-
United Kingdom	-	-	-	-	-	-	-	-	-
United States	-	-	-	-	-	-	-	-	-
Other OECD	-	-	-	-	-	-	-	-	-
China, People's Rep.	-	-	-	-	-	-	-	-	-
Colombia	-	-	-	-	-	-	-	-	-
Indonesia	-	-	-	-	-	-	-	-	-
South Africa	-	-	-	-	-	-	-	-	-
Former Soviet Union(3)	-	-	x	x	x	x	x	x	x
Russian Federation (4)	*x*	*x*	-	-	-	-	-	-	-
Other FSU (4)	*x*	*x*	-	-	-	-	-	-	-
Venezuela	-	-	-	-	-	-	-	-	-
Vietnam	-	-	-	-	-	-	-	-	-
Non-specified/other	-	-	-	-	-	-	-	-	-
Steam coal	**-**	**3063**	**3998**	**7090**	**9917**	**12685**	**12310**	**12475**	**14668**
Australia	-	-	-	-	-	-	-	-	494
Canada	-	-	-	-	-	-	-	-	-
Czech Republic	-	-	-	-	-	-	-	-	-
Germany	-	-	-	-	-	-	-	-	-
Poland	-	-	-	-	-	-	-	-	-
United Kingdom	-	-	-	-	-	-	-	-	-
United States	-	-	-	-	-	-	-	-	-
Other OECD	-	-	-	-	-	-	-	-	-
China, People's Rep.	-	-	-	-	-	-	-	-	-
Colombia	-	-	-	-	-	-	4234	5591	5712
Indonesia	-	-	-	-	-	-	-	-	-
South Africa	-	-	-	-	-	-	3145	3337	4752
Former Soviet Union(3)	-	-	x	x	x	x	x	x	x
Russian Federation (4)	*x*	*x*	-	-	-	-	*1931*	*1711*	*2404*
Other FSU (4)	*x*	*x*	-	-	-	-	-	-	-
Venezuela	-	-	-	-	-	-	-	-	-
Vietnam	-	-	-	-	-	-	-	-	-
Non-specified/other	-	3063	3998	7090	9917	12685	3000	1836	1306
Lignite	**-**	**-**	**-**	**-**	**-**	**-**	**-**	**-**	**-**

(1) Earliest year for which split by coal type is available.

(2) Total coal consists of steam coal, coking coal and lignite.

(3) For years prior to 1990, or if no details for individual Former Soviet Union republics.

(4) Data for former republics are not available separately for some OECD member countries.

* Please refer to notes and definitions in Part I.

Source: IEA/OECD Coal Statistics

ITALY*

Figure 1: TPES by fuel (Mtce)

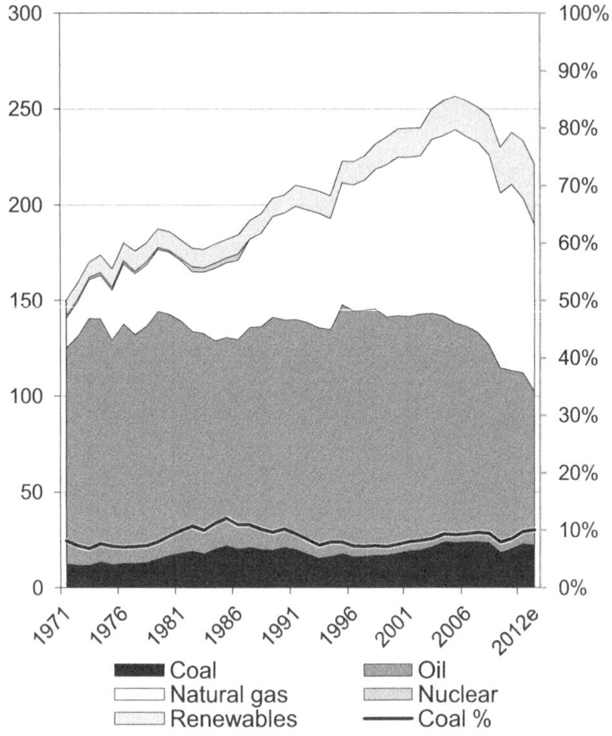

Figure 2: Coal supply indicators (1971 = 100)

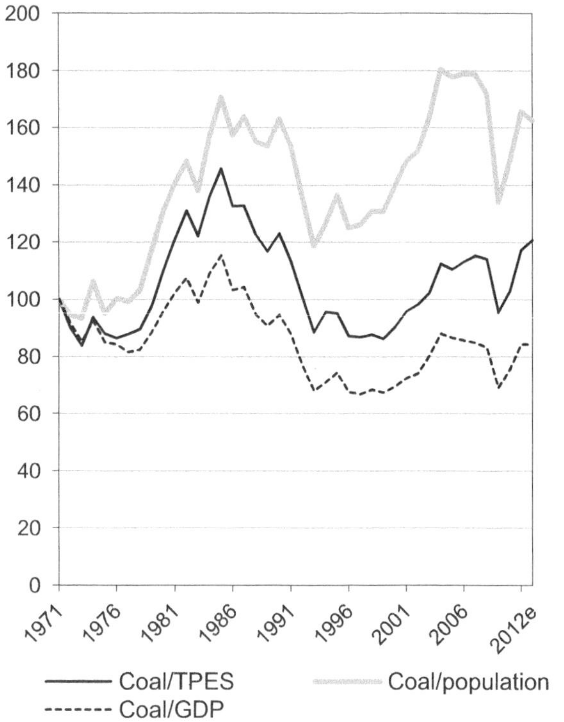

Figure 3: Primary coal supply (Mtce)

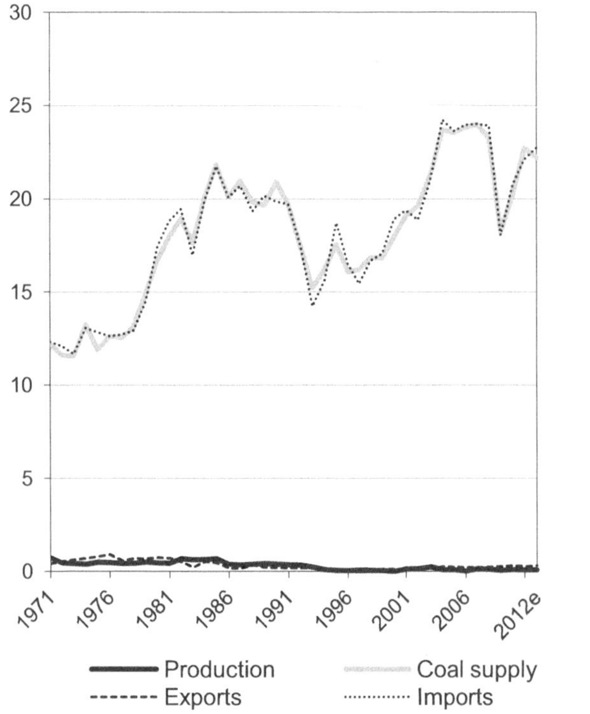

Figure 4: Coal consumption (Mtce)

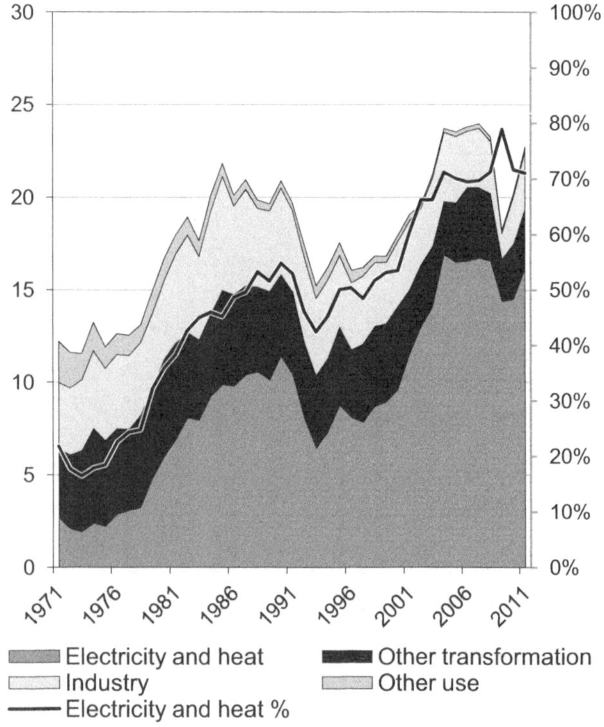

*Please refer to notes and definitions in Part I. Peat is included in coal for all figures.

Source: *IEA/OECD Coal Information, IEA/OECD Energy Balances of OECD Countries, OECD Main Economic Indicators*

ITALY*

Figure 5: Electricity generation by fuel (TWh)

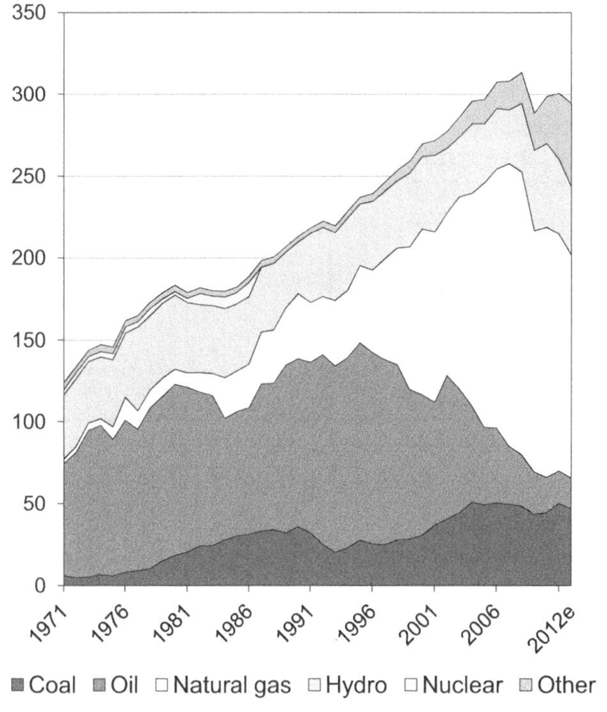

Figure 6: CO_2 emissions by fuel (Mt CO_2)

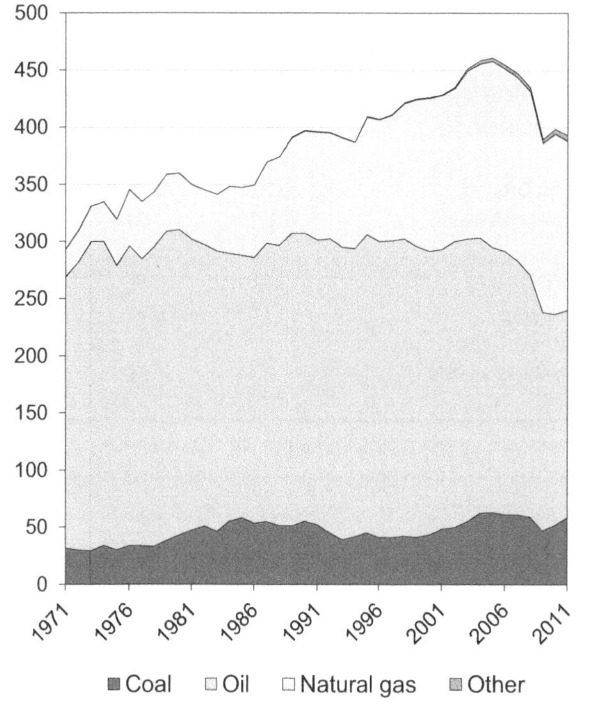

Figure 7: Electricity generation by fuel share

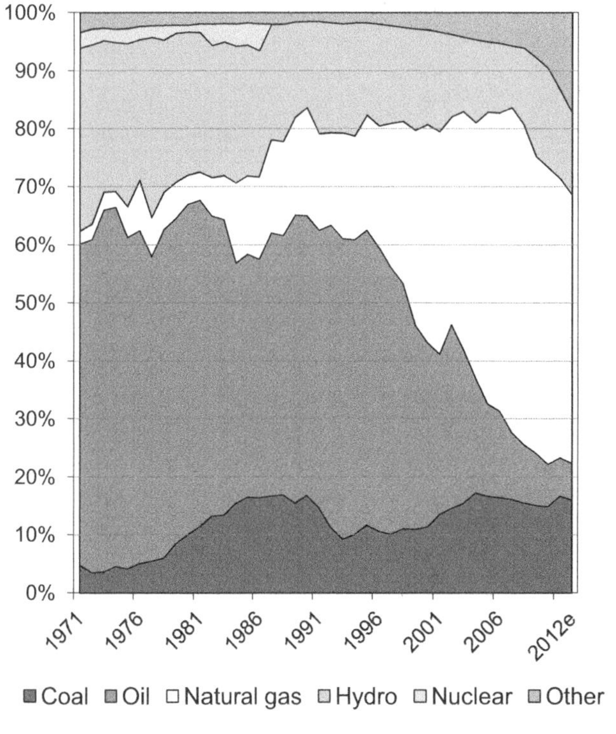

Figure 8: CO_2 emissions by fuel share

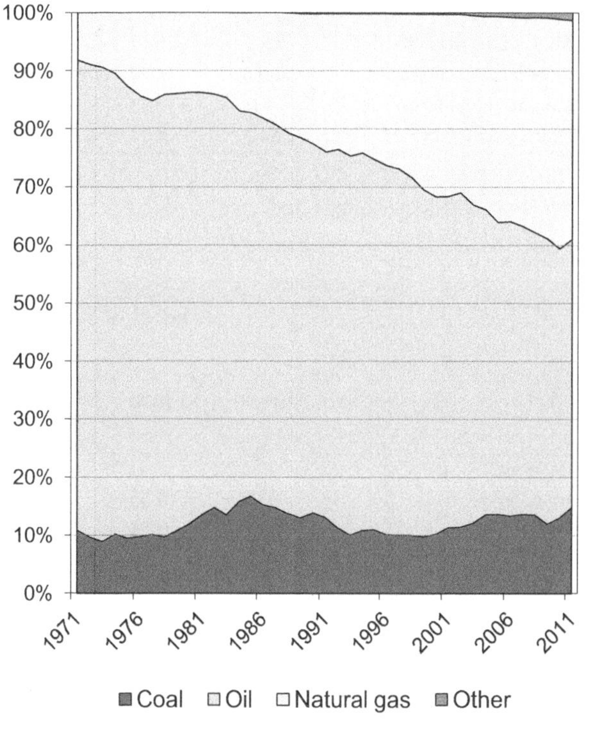

*Please refer to notes and definitions in Part I. Peat is included in coal for all figures.

Source: *IEA/OECD Electricity Information, IEA/OECD CO2 Emissions from Fuel Combustion*

ITALY

1. Total primary energy supply (TPES) by fuel*

	1973	1980	1990	2000	2010	2011	2012e	Average annual percent change 73-90	90-11
TPES (Mtce)	**170.17**	**186.91**	**209.37**	**245.03**	**243.20**	**239.17**	**226.59**	**1.23**	**0.64**
Coal and peat	11.57	16.69	20.90	17.94	20.24	22.73	22.19	3.54	0.40
Oil	129.00	126.05	119.03	124.08	93.21	89.58	79.90	-0.47	-1.34
Natural Gas	20.32	32.46	55.70	82.75	97.20	91.14	87.63	6.11	2.37
Biofuels and waste	0.35	1.31	1.34	3.22	12.48	14.57	15.22	8.27	12.02
Nuclear	1.17	0.82	-	-	-	-	-	-	-
Hydro	4.61	5.56	3.89	5.43	6.28	5.63	5.15	-1.00	1.78
Geothermal	3.05	3.28	4.24	6.08	6.82	7.16	7.06	1.97	2.52
Solar, wind, tide	-	-	0.01	0.09 e	1.55	2.74	4.16	-	32.33
Net electricity trade(1)	0.11	0.75	4.26	5.45	5.43	5.62	5.29	24.13	1.33
Heat(2)	-	-	-	-	-	-	-	-	-

(1) Net trade between OECD and non-OECD countries.

(2) Ambient heat from heat pumps used in the transformation sector.

Source: *IEA/OECD Energy Balances of OECD Countries*

2. Energy supply, GDP and population*

	1973	1980	1990	2000	2010	2011	2012e	Average annual percent change 73-90	90-11
GDP (2005 bil. USD)	891	1144	1451	1701	1764	1770	1729	2.91	0.95
TPES/GDP(1)	0.19	0.16	0.14	0.14	0.14	0.14	0.13	-1.63	-0.31
Population (millions)	54.8	56.4	56.7	56.9	60.5	60.7	60.4	0.21	0.33
TPES/population(1)	3.11	3.31	3.69	4.30	4.02	3.94	3.75	1.02	0.31
TPES/GDP(2)	129.8	111.1	98.1	98.0	93.8	91.9	89.1	-1.63	-0.31
Coal and peat supply/GDP(2)	98.5	110.7	109.4	80.1	87.1	97.5	97.5	0.62	-0.55
Elec. consumption/GDP(2)	82.0	82.9	87.8	95.3	100.7	101.2	..	0.41	0.68
Elec. generation (TWh)	144	183	213	270	299	301	294	2.34	1.65
Industrial production(2)	64.8	80.1	90.4	104.1	89.0	89.3	83.8	1.98	-0.06

(1) TPES/GDP in units of tce/2005 thousand USD, TPES/population in units of tce/capita.

(2) As index, 2005 (not shown) = 100

Source: *IEA/OECD Energy Balances of OECD Countries , OECD Main Economic Indicators*

3. Coal and peat production by type*

	1978(1)	1990	2000	2005	2010	2011	2012e	Average annual percent change 78-90	90-11
Mtce:									
Coking coal	-	-	-	-	-	-	-	-	-
Steam coal	0.01	0.05	-	0.09	0.09	0.08	0.07	20.81	2.30
Lignite	0.43	0.34	0.00	-	-	-	-	-1.89	-
Peat	-	-	-	-	-	-	-	-	-
Mt:									
Coking coal	-	-	-	-	-	-	-	-	-
Steam coal	0.01	0.06	-	0.10	0.10	0.09	0.08	20.81	2.22
Lignite	1.20	0.96	0.01	-	-	-	-	-1.89	-
Peat	-	-	-	-	-	-	-	-	-

(1) Earliest year for which split by coal type is available.

* Please refer to notes and definitions in Part I.

Source: *IEA/OECD Energy Balances of OECD Countries, IEA/OECD Energy Statistics of OECD Countries*

ITALY

4. Final consumption of energy by fuel*

(Mtce)

	1973	1980	1990	2000	2009	2010	2011	Average annual percent change 73-90	90-11
Total final consumption(1)	**137.94**	**146.04**	**164.20**	**184.05**	**179.39**	**185.38**	**181.07**	**1.03**	**0.47**
Coal and peat	5.26	5.45	5.10	3.83	1.58	2.76	3.34	-0.18	-2.00
Oil	99.91	91.71	87.79	88.99	79.55	77.61	76.58	-0.76	-0.65
Natural Gas	17.65	28.18	43.42	55.11	52.32	55.79	51.36	5.44	0.80
Biofuels and wastes	-	1.06	1.23	2.25	5.47	7.32	7.77	-	9.16
Geothermal	-	-	0.29	0.30	0.30	0.18	0.18	-	-2.22
Solar, wind, tide	-	-	0.01	0.02	0.12	0.19	0.20	-	17.41
Electricity	15.12	19.63	26.37	33.54	35.63	36.77	37.08	3.33	1.64
Heat	-	-	-	-	4.42	4.76	4.56	-	-
of which:									
Total industry	**55.80**	**51.20**	**48.70**	**54.64**	**41.32**	**42.80**	**41.27**	**-0.80**	**-0.79**
Coal and peat	3.80	4.26	4.70	3.49	1.43	2.57	3.10	1.25	-1.96
Oil	30.18	22.30	11.54	9.57	6.18	4.97	4.99	-5.50	-3.92
Natural Gas	12.34	12.92	18.52	23.74	14.13	14.78	13.29	2.42	-1.57
Biofuels and wastes	-	0.17	0.31	0.41	0.56	0.28	0.65	-	3.50
Geothermal	-	-	-	-	-	0.00	0.00	-	-
Solar, wind, tide	-	-	-	-	-	0.01	0.01	-	-
Electricity	9.48	11.55	13.63	17.43	14.82	15.71	15.74	2.16	0.69
Heat	-	-	-	-	4.20	4.47	3.49	-	-
Total transport	**27.09**	**34.79**	**46.72**	**56.70**	**55.84**	**55.01**	**55.05**	**3.26**	**0.78**
Coal and peat	0.21	0.00	-	-	-	-	-	-	-
Oil	26.24	33.83	45.60	55.19	52.00	50.61	50.49	3.30	0.49
Natural Gas	0.18	0.37	0.30	0.47	0.86	0.99	1.22	3.13	6.94
Biofuels and wastes	-	-	-	-	1.69	2.09	2.02	-	-
Electricity	0.46	0.59	0.83	1.05	1.29	1.31	1.33	3.44	2.28
Residential	**38.26**	**41.38**	**37.23**	**39.41**	**41.28**	**45.23**	**44.74**	**-0.16**	**0.88**
Coal and peat	1.25	1.19	0.40	0.10	0.01	0.01	0.01	-6.45	-18.53
Oil	28.65	22.71	13.31	8.58	5.41	4.75	4.40	-4.41	-5.13
Natural Gas	5.13	11.94	16.16	21.38	24.02	26.70	25.69	6.98	2.23
Biofuels and wastes	-	0.90 e	0.88	1.83	3.20	4.91	5.06	-	8.71
Geothermal	-	-	-	0.00	0.00	0.00	0.00	-	-
Solar, wind, tide	-	-	0.01	0.01	0.09	0.14	0.15	-	15.74
Electricity	3.23	4.65	6.48	7.51	8.47	8.54	8.62	4.18	1.37
Heat	-	-	-	-	0.07	0.18	0.81	-	-
Comm & public services	**2.22**	**3.10**	**11.68**	**16.49**	**24.17**	**24.25**	**22.50**	**10.25**	**3.17**
Coal and peat	-	-	-	-	-	-	-	-	-
Oil	0.44	0.58	0.75	1.37	1.19	1.18	1.08	3.20	1.79
Natural Gas	-	-	6.01	7.96	12.30	12.30	10.36	-	2.62
Biofuels and waste	-	-	-	0.01	0.02	0.03	0.03	-	-
Geothermal	-	-	-	0.19	0.19	0.11	0.11	-	-
Solar, wind, tide	-	-	-	0.00	0.03	0.04	0.04	-	-
Electricity	1.79	2.52	4.92	6.95	10.36	10.52	10.68	6.14	3.76
Heat	-	-	-	-	0.09	0.07	0.20	-	-
Non-energy use	**11.83**	**12.42**	**14.89**	**12.04**	**12.13**	**13.66**	**13.12**	**1.37**	**-0.60**
Coal and peat	-	-	-	0.24	0.14	0.19	0.23	-	-
Oil	11.83	9.48	12.50	10.41	11.17	12.66	12.28	0.32	-0.08
Natural Gas	-	2.94	2.40	1.40	0.81	0.81	0.61	-	-6.31

(1) Total Final Consumption (TFC) excludes use in transformation processes and energy industry own use.

* Please refer to notes and definitions in Part I. Agriculture, Forestry, Fishing and "Non-specified other" sectors are not reported here individually, but are accounted for in Total final consumption.

Source: *IEA/OECD Energy Balances of OECD Countries*

ITALY

5. Coal balance*

(Mtce)

	1973	1980	1990	2000	2005	2010	2011	2012e	Average annual percent change 73-90	90-11
Production	0.4	0.5	0.4	0.0	0.1	0.1	0.1	0.1	-0.5	-7.1
Imports	11.7	17.4	19.8	18.9	23.6	20.7	22.1	22.7	3.2	0.5
Exports	-0.6	-0.7	-0.2	-0.1	-0.2	-0.3	-0.3	-0.3	-6.5	1.5
Stock changes	0.1	-0.4	0.9	-0.8	0.0	-0.3	0.8	-0.3		
Primary supply	**11.6**	**16.7**	**20.9**	**17.9**	**23.5**	**20.2**	**22.7**	**22.2**	**3.5**	**0.4**
Statistical differences	**0.1**	**0.0**	**0.2**	**-0.5**	**-0.2**	**-0.6**	**-0.7**	..		
Total transformation	**-4.2 e**	**-9.1 e**	**-14.7**	**-13.2 e**	**-19.1 e**	**-16.8 e**	**-18.6 e**	..	**7.6**	**1.1**
Electricity and heat gen.	-1.9	-6.0	-11.4	-9.6	-16.5	-14.5	-16.2	..	11.1	1.7
Main activity producers (1)	*-1.9*	*-4.9*	*-10.2*	*-9.6*	*-16.2*	*-14.5*	*-16.1*	..	*10.4*	*2.2*
Autoproducers	-	*-1.1*	*-1.3*	..	*-0.3*	*-0.0*	*-0.0*	..	-	*-18.5*
Gas works	0.7	0.7	0.3	0.0	-	-	-	..	-4.9	-
Coal transformation(2)	-3.0 e	-3.8 e	-3.5	-3.7 e	-2.6 e	-2.3 e	-2.4 e	..	1.0	-1.8
BKB plants	-	-	-	-	-	-	-	..	-	-
Blast furnaces	*-2.2 e*	*-2.2 e*	*-2.3*	*-2.1 e*	*-2.2 e*	*-1.6 e*	*-1.8 e*	..	*0.3*	*-1.0*
Coke ovens	*-0.8*	*-1.6*	*-1.2*	*-1.6*	*-0.4*	*-0.6*	*-0.6*	..	*2.4*	*-3.6*
Patent fuel plants	-	*0.0*	-	-	-	-	-	..	-	-
Other transformation(3)	-	-	-	-	-	-	-	..	-	-
Energy ind. own use	**-1.7**	**-2.0**	**-1.2**	**-0.4**	**-0.4**	**-0.1**	**-0.1**	..	**-1.9**	**-12.5**
Losses	**-0.5**	**-0.2**	**-0.1**	**-0.0**	-	-	-	..		
Final consumption(4)	**5.3**	**5.5**	**5.1**	**3.8**	**3.8**	**2.8**	**3.3**	..	**-0.2**	**-2.0**
Industry(5)	3.8	4.3	4.7	3.5	3.6	2.6	3.1	..	1.3	-2.0
Iron and steel	*2.8*	*2.9*	*2.9*	*2.7 e*	*3.1 e*	*2.5 e*	*2.6 e*	..	*0.4*	*-0.6*
Chemical	*0.3*	*0.2*	*0.2*	*0.0*	*0.0*	*0.0*	*0.0*	..	*-1.0*	*-16.6*
Non-metallic minerals	*0.2*	*0.5*	*1.3*	*0.5*	*0.5*	*0.1*	*0.5*	..	*11.3*	*-4.5*
Paper, pulp and print	-	*0.0*	*0.0*	*0.0*	-	-	-	..	-	-
Other industry (6)	*0.6*	*0.5*	*0.2*	*0.3*	*0.0*	*0.0*	*0.0*	..	*-5.7*	*-11.3*
Transport(7)	0.2	0.0	-	-	-	-	-	..	-	-
Other	1.2	1.2	0.4	0.1	0.0	0.0	0.0	..	-6.4	-18.5
Comm. and pub. services	-	-	-	-	-	-	-	..	-	-
Residential	*1.2*	*1.2*	*0.4*	*0.1*	*0.0*	*0.0*	*0.0*	..	*-6.4*	*-18.5*
Other sectors (8)	-	-	-	-	-	-	-	..	-	-
Non-energy use	-	-	-	0.2	0.2	0.2	0.2	..	-	-

(1) Main activity electricity and heat generation includes district heating.

(2) Coal transformation refers to the transformation of coal and peat to secondary and tertiary products (mainly cokes, briquettes, coal tar and off-gases). Gas works and Liquefaction are shown separately to match the *Energy Balances of OECD Countries.*

(3) Other transformation includes Liquefaction and Non-specified transformations.

(4) Final Consumption includes non-energy use and energy use (Industry, Transport and Other).

(5) Please refer to notes and definitions in Part I for detailed explanation.

(6) Other industry includes Non-ferrous metals, Transport equipment, Machinery, Mining and quarrying, Food and tobacco, Wood and wood products, Construction, Textile and leather, and Non-specified industry.

(7) Transport includes Rail and Inland waterways.

(8) Other sectors includes Agriculture/Forestry, Fishing and Non-specified other.

* "Coal" refers to all types of coal, primary (anthracite, coking coal, other bituminous coal, sub-bituminous coal, lignite/brown coal) and derived fuels (including patent fuel, BKB, coke oven coke, gas coke, coal tar, coke oven gas, gas works gas, blast furnace gas and other recovered gases), as well as peat, peat briquettes and oil shale. Quantities have been converted to Mtce using calorific values reported by the respective countries. Please refer to notes and definitions in Part I.

Source: *IEA/OECD Energy Balances of OECD Countries*

ITALY

6. Use of coal for selected end-uses*

(million tonnes)

	1978	1990	2000	2005	2009	2010	2011	Average annual percent change 78-90	90-11
Steam coal	**2.51**	**12.69**	**11.36**	**18.61**	**16.36**	**16.62**	**18.57**	**14.46**	**1.83**
Total electricity and heat	2.07	10.78	9.53	16.38	15.23	15.05	16.68	14.77	2.10
Main activity producers	*2.02*	*10.77*	*9.53*	*16.22*	*15.22*	*15.03*	*16.66*	*14.97*	*2.10*
Autoproducers	*0.05*	*0.02*	*..*	*0.16*	*0.01*	*0.02*	*0.02*	*-8.42*	*0.56*
Patent fuel/BKB plants	0.01	-	-	-	-	-	-	-	-
Coke ovens/Liquefaction(1)	-	-	-	-	-	-	-	-	-
Blast furnace inputs	-	0.17	0.94 e	1.15 e	0.73 e	0.91 e	0.86 e	-	8.04
Gas manufacture	-	-	-	-	-	-	-	-	-
Industry	0.31	1.68	0.88	1.07	0.40	0.66	1.02	15.08	-2.35
Iron and steel	*0.03*	*0.22*	*0.26 e*	*0.52 e*	*0.22 e*	*0.55 e*	*0.47 e*	*19.30*	*3.73*
Chemical	*0.01*	*0.00*	*0.00*	*0.00*	*0.00*	*0.00*	*0.00*	*-9.91*	*-3.25*
Non-metallic minerals	*0.20*	*1.40*	*0.54*	*0.54*	*0.17*	*0.10*	*0.55*	*17.56*	*-4.36*
Paper, pulp and print	-	-	*0.00*	-	-	-	-	-	-
Other industry	*0.07*	*0.06*	*0.08 e*	*0.00 e*	*0.01 e*	*0.01 e*	*0.01 e*	*-1.12*	*-11.30*
Other sectors(2)	0.07	0.06	0.00	0.01	0.01	0.01	0.01	-1.87	-10.31
Non-energy use	-	-	-	-	-	-	-	-	-
Coking coal	**9.91**	**8.63**	**6.66**	**5.63**	**3.59**	**5.15**	**5.87**	**-1.14**	**-1.82**
Total electricity and heat	-	-	-	-	-	-	-	-	-
Main activity producers	-	-	-	-	-	-	-	-	-
Autoproducers	-	-	-	-	-	-	-	-	-
Patent fuel/BKB plants	-	-	-	-	-	-	-	-	-
Coke ovens/Liquefaction(1)	9.90	8.61	6.38	5.06	3.39	4.92	5.59	-1.15	-2.04
Blast furnace inputs	-	-	-	-	-	-	-	-	-
Gas manufacture	-	-	-	-	-	-	-	-	-
Industry	-	-	-	-	-	-	-	-	-
Iron and steel	-	-	-	-	-	-	-	-	-
Chemical	-	-	-	-	-	-	-	-	-
Non-metallic minerals	-	-	-	-	-	-	-	-	-
Paper, pulp and print	-	-	-	-	-	-	-	-	-
Other industry	-	-	-	-	-	-	-	-	-
Other sectors(2)	-	-	-	-	-	-	-	-	-
Non-energy use	-	-	0.23	0.22	0.14	0.18	0.22	-	-
Lignite	**1.27**	**1.09**	**0.03**	**0.01**	**0.01**	**0.01**	**0.01**	**-1.27**	**-22.61**
Total electricity and heat	1.20	1.06	0.00	-	-	-	-	-1.07	-
Main activity producers	*1.20*	*1.06*	*0.00*	-	-	-	-	*-1.07*	-
Autoproducers	-	-	*..*	-	-	-	-	-	-
Patent fuel/BKB plants	-	-	-	-	-	-	-	-	-
Coke ovens/Liquefaction(1)	-	-	-	-	-	-	-	-	-
Gas manufacture	-	-	-	-	-	-	-	-	-
Industry	0.04	0.03	0.03	0.01	0.01	0.01	0.01	-0.49	-8.59
Iron and steel	*0.03*	*0.02*	-	-	-	-	-	*-2.55*	-
Chemical	-	-	-	-	-	-	-	-	-
Non-metallic minerals	*0.01*	*0.01*	*0.01*	*0.01*	*0.01*	*0.01*	*0.01*	*5.02*	*-2.76*
Paper, pulp and print	-	-	-	-	-	-	-	-	-
Other industry	*0.00*	*0.00*	*0.02*	-	-	-	-	*1503.17*	-
Other sectors(2)	0.03	-	-	-	-	-	-	-	-
Non-energy use	-	-	-	-	-	-	-	-	-

(1) Liquefaction primarily refers to direct distillation processes. Indirect liquefaction may be reported here or under gas manufacture.
(2) Other sectors are Residential, Commercial and public services, Agriculture/Forestry, Fishing and Non-specified other.
* Coal totals may not match the sum of the individual rows due to statistical differences. Please refer to notes and definitions in Part I.

Source: *IEA/OECD Energy Statistics of OECD Countries*

ITALY

7. Fuel prices to end users*

	1978(1)	1990	1995	2000	2005	2009	2010	2011	2012
				(US dollars / tce)					
For electricity generation									
Steam coal	41.32	67.72	65.27	c	83.45	122.09	127.81	..	..
Heavy fuel oil	62.53	98.15	98.18	c	c	c	c	c	c
Natural gas	63.55	90.90	101.30	c	c	c	c	c	c
For industry									
Steam coal	70.05	58.84	57.35	43.46	83.23	129.51	127.19	159.34	..
Coking coal	69.79	64.88	60.70	53.26	108.96	152.00	185.72	254.80	..
High sulphur fuel oil	63.44	136.94	127.10	161.05	..	..	..	..	..
Low sulphur fuel oil	..	..	122.24	155.19	278.73	346.16	425.43	570.17	631.02
Natural gas	64.94	122.20	135.14	c	273.65	433.75	375.67	..	..
				(Euro / unit)(2)					
For electricity generation									
Steam coal	16.03	37.13	48.62	c	58.54	74.07	81.33	..	..
Heavy fuel oil	37.56	83.32	113.24	c	c	c	c	c	c
Natural gas	35.79	72.34	109.53	c	c	c	c	c	c
For industry									
Steam coal	27.01	32.06	42.46	41.51	58.98	82.06	84.53	100.81	..
Coking coal	29.04	38.15	48.50	54.90	83.33	103.94	133.21	173.97	..
High sulphur fuel oil	38.11	116.25	146.59	239.64	..	..	..	..	..
Low sulphur fuel oil	..	..	140.99	230.93	307.72	341.72	440.50	561.99	673.28
Natural gas	36.57	97.25	146.12	c	283.23	401.42	364.67	..	..

(1) Earliest year for which split by coal type is available.

(2) For steam coal, coking coal and heavy fuel oil per metric tonne; for natural gas per 10^7 kilocalories GCV.

* Please refer to notes and definitions in Part I. Steam coal prices relate to historical definitions.

Source: *IEA/OECD Energy Prices & Taxes*

8. Coal and peat trade by type of coal*

(Mtce)

	1978	1990	1995	2000	2005	2009	2010	2011	2012e
Total imports	**12.92**	**19.83**	**18.72**	**18.90**	**23.62**	**18.01**	**20.73**	**22.11**	**22.72**
Bituminous coal(1)	2.19	10.53	10.51	10.78	16.66	14.25	15.10	16.02	17.75
Coking coal	10.58	9.14	7.46	7.61	6.11	3.39	5.36	5.93	4.83
Sub-bituminous coal	-	-	-	-	-	0.37	0.25	0.13	0.13
Lignite	0.02	0.05	0.00	0.00	0.00	0.00	0.00	0.00	0.00
Peat	-	-	-	-	-	-	-	-	-
Coal products(2)	0.13	0.11	0.75	0.51	0.84	-	0.02	0.03	0.00
Total exports	**0.70**	**0.20**	**0.13**	**0.12**	**0.23**	**0.27**	**0.30**	**0.27**	**0.30**
Bituminous coal(1)	-	-	-	0.00	-	-	0.00	0.00	0.05
Coking coal	-	-	-	-	-	-	-	-	-
Sub-bituminous coal	-	-	-	-	-	-	-	-	-
Lignite	-	-	-	-	-	-	-	-	-
Peat	-	-	-	-	-	-	-	-	-
Coal products(2)	0.70	0.20	0.13	0.12	0.23	0.27	0.30	0.27	0.25

(1) Bituminous coal includes anthracite.

(2) Coal products includes products derived from coal, for example: coke, coal tar, briquettes, patent fuels.

* Please refer to notes and definitions in Part I.

Source: *IEA/OECD Energy Balances of OECD Countries, IEA/OECD Energy Statistics of OECD Countries*

ITALY

9. Total coal imports by origin*

(thousand tonnes)

	1978(1)	1985	1990	1995	2000	2005	2010	2011	2012e
Total coal(2)	**12526**	**22362**	**20578**	**18499**	**19027**	**24158**	**22112**	**23471**	**24348**
Coking coal	**10007**	**10248**	**8648**	**7052**	**7198**	**5784**	**5066**	**5606**	**4571**
Australia	1347	1973	1045	2002	2463	2109	1896	1394	1359
Canada	-	12	212	1090	1280	637	865	906	724
Czech Republic	-	-	-	-	-	-	-	-	-
Germany	2537	1734	804	-	-	-	-	-	-
Poland	1525	839	158	17	-	283	-	-	-
United Kingdom	-	-	-	-	-	-	-	-	-
United States	3027	5541	6236	3868	3281	2431	2305	3128	2455
Other OECD	-	-	12	-	-	-	-	-	-
China, People's Rep.	-	-	-	11	-	-	-	-	-
Colombia	-	-	15	-	-	-	-	-	-
Indonesia	-	-	-	-	-	-	-	-	-
South Africa	-	-	-	-	174	205	-	-	-
Former Soviet Union(3)	1036	112	x	x	x	x	x	x	x
Russian Federation (4)	*x*	*x*	*131*	*64*	*-*	*14*	*-*	*-*	*-*
Other FSU (4)	*x*	*x*	*-*	*-*	*-*	*-*	*-*	*-*	*-*
Venezuela	-	-	-	-	-	105	-	104	-
Vietnam	-	-	-	-	-	-	-	-	-
Non-specified/other	535	37	35	-	-	-	-	74	33
Steam coal	**2451**	**11965**	**11797**	**11433**	**11817**	**18366**	**17040**	**17860**	**19772**
Australia	-	260	-	1120	1141	682	598	338	107
Canada	-	-	-	110	-	506	-	-	69
Czech Republic	-	-	-	-	-	-	-	-	-
Germany	-	-	109	1	1	1	-	-	-
Poland	1437	1150	507	360	912	182	-	-	-
United Kingdom	52	-	-	-	-	-	-	-	-
United States	-	3518	4800	4681	4	204	460	1731	5637
Other OECD	-	26	148	3	-	1	995	724	1029
China, People's Rep.	-	-	310	479	369	-	-	-	-
Colombia	-	180	290	148	1759	2997	1762	1593	2756
Indonesia	-	-	-	31	1919	6800	7027	6573	3782
South Africa	960	6641	4884	3487	3548	4395	3919	3917	3055
Former Soviet Union(3)	-	150	x	x	x	x	x	x	x
Russian Federation (4)	*x*	*x*	*609*	*689*	*993*	*1082*	*2092*	*2763*	*2431*
Other FSU (4)	*x*	*x*	*-*	*22*	*56*	*1025*	*89*	*120*	*729*
Venezuela	-	-	140	302	1115	445	98	101	135
Vietnam	-	-	-	-	-	-	-	-	-
Non-specified/other	2	40	-	-	-	46	-	-	42
Lignite	**68**	**149**	**133**	**14**	**12**	**8**	**6**	**5**	**5**

(1) Earliest year for which split by coal type is available.

(2) Total coal consists of steam coal, coking coal and lignite.

(3) For years prior to 1990, or if no details for individual Former Soviet Union republics.

(4) Data for former republics are not available separately for some OECD member countries.

* Please refer to notes and definitions in Part I.

Source: IEA/OECD Coal Statistics

ITALY

12. Coal import values by origin*

(average unit value, CIF, USD/tonne)

	1990	1995	2000	2005	2007	2008	2009	2010	2011/12[1]
Coking coal[2]	**61.39**	**57.69**	**50.55**	**103.02**	**118.05**	**188.97**	**144.59**	**176.04**	..
Imports from:									
Australia	57.32	52.91	46.44	111.81	126.24	240.02	242.48	226.56	..
Canada	57.65	57.67	48.02	119.76	131.04	232.32	216.35	198.18	..
Czech Republic	..	..	..	..	488.51	752.71	..	..	..
Poland	64.68	62.38	..	128.50	476.51	..	..	..	..
United States	62.37	62.18	56.50	113.75	132.19	165.11	148.53	191.47	..
China	..	..	..	..	..	..	..	..	..
Colombia	68.24	..	46.42	87.23	87.00	..	77.81	95.05	..
Indonesia	..	..	..	67.23	81.67	141.10	76.85	109.86	..
South Africa	..	49.38	40.62	74.08	101.09	159.46	..	113.18	..
Former Soviet Union[3]	54.53	45.76	..	77.71	86.60	150.61	88.18	106.58	..
Other bituminous coal[4]	**51.84**	**50.48**	**37.92**	**73.20**	**86.00**	**143.68**	**113.67**	**112.42**	..
Imports from:									
Australia	..	45.90	37.05	..	..	163.91	216.33	154.63	..
Canada	..	2798.99	..	1441.74	427.07	311.61	481.07	478.93	..
Czech Republic	..	..	..	..	..	..	..	..	..
Poland	58.63	51.68	35.99	125.35	111.14	440.45	565.39	528.54	..
United States	56.62	54.54	47.06	81.23	..	158.43	88.49	111.33	..
China	47.34	51.34	31.19	..	..	..	..	..	..
Colombia	56.10	53.06	40.67	72.04	82.16	..	131.32	111.83	..
Indonesia	48.46	49.59	36.48	69.52	81.77	132.94	100.80	106.53	..
South Africa	44.39	44.85	37.26	74.49	92.44	153.17	115.16	104.65	..
Former Soviet Union[3]	42.54	45.77	36.92	74.31	93.36	160.60	103.87	160.05	..

Note: On occasion, shipment of extremely small quantities of high valued coal results in high import costs.

(1) Import data for steam and coking coals are currently unavailable for 2011 and 2012 due to resource constraints.

(2) Weighted average of individual countries using import volumes as weights. Prices exclude intra-EU trade.

(3) Former Soviet Union until 1991, Russian Federation starting in 1992.

(4) Bituminous steam coal only. Weighted average of trades. Prices exclude intra-EU trade.

* Please refer to notes and definitions in Part I.

Source: IEA/OECD Coal Statistics

JAPAN*

Figure 1: TPES by fuel (Mtce)

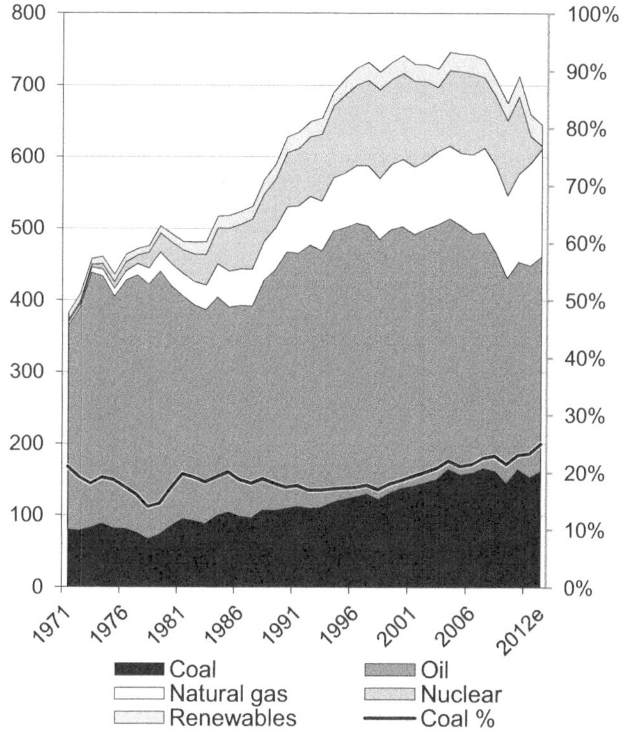

Figure 2: Coal supply indicators (1971 = 100)

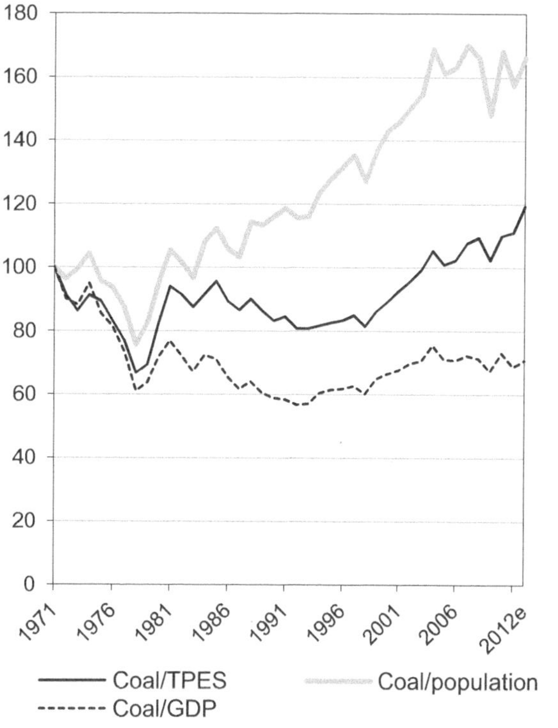

Figure 3: Primary coal supply (Mtce)

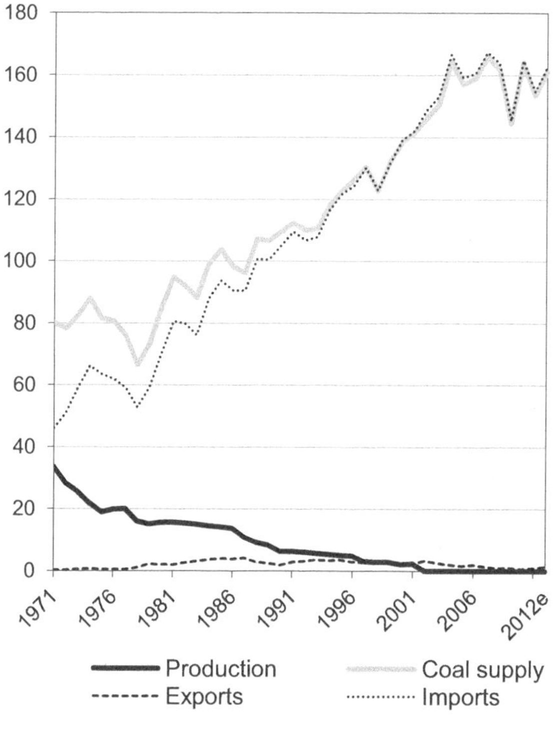

Figure 4: Coal consumption (Mtce)

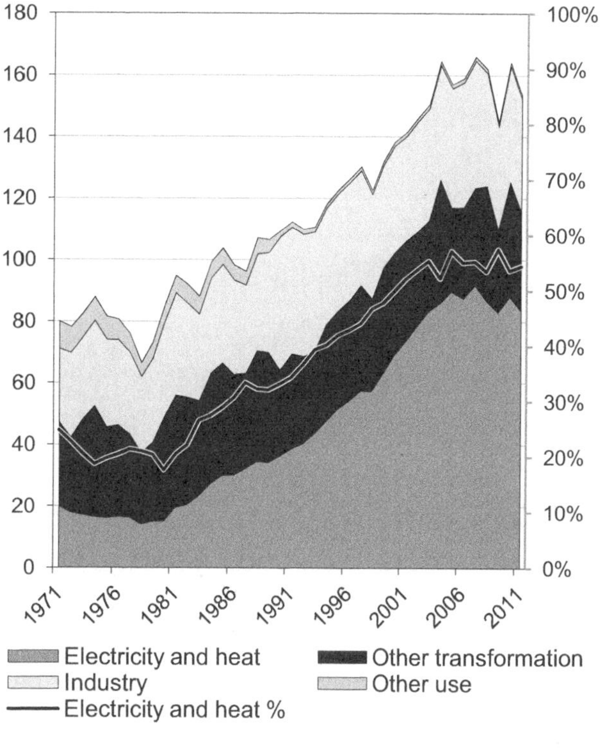

*Please refer to notes and definitions in Part I. Peat is included in coal for all figures.

Source: *IEA/OECD Coal Information, IEA/OECD Energy Balances of OECD Countries, OECD Main Economic Indicators*

JAPAN*

Figure 5: Electricity generation by fuel (TWh)

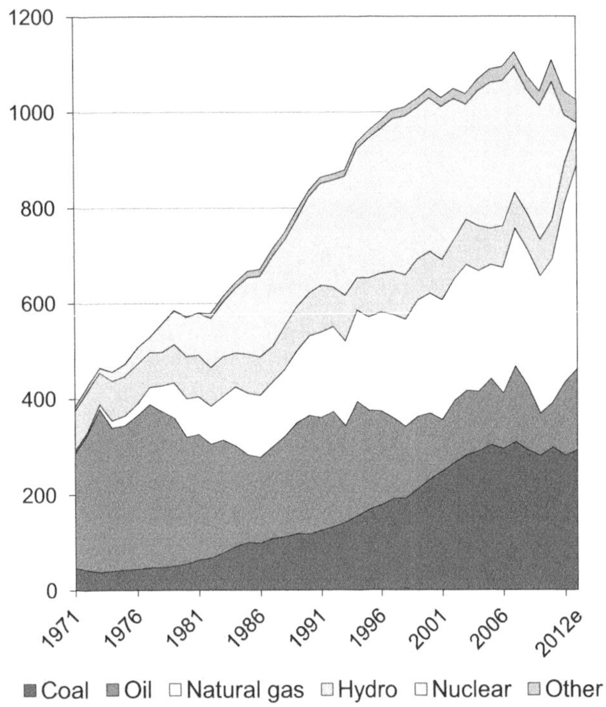

Figure 6: CO_2 emissions by fuel (Mt CO_2)

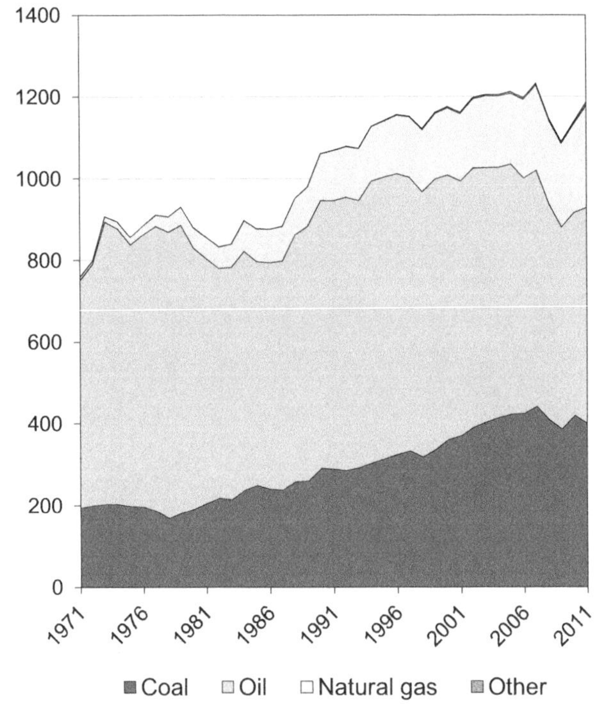

Figure 7: Electricity generation by fuel share

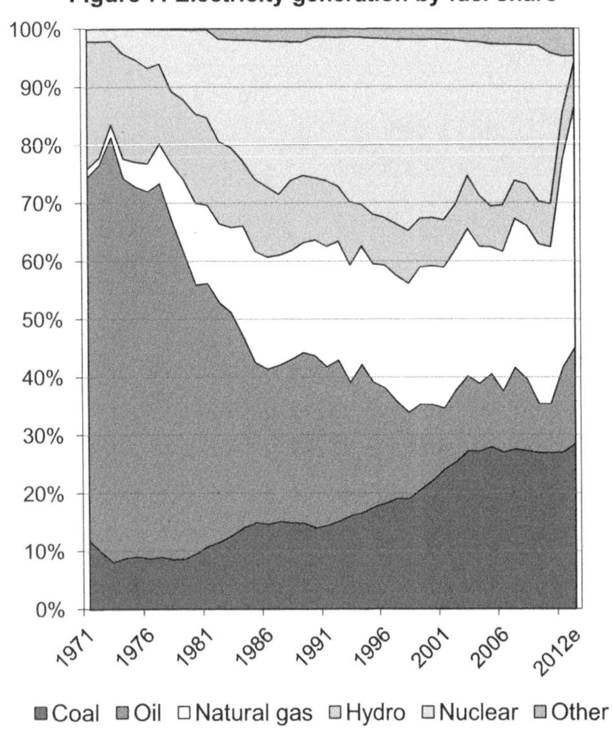

Figure 8: CO_2 emissions by fuel share

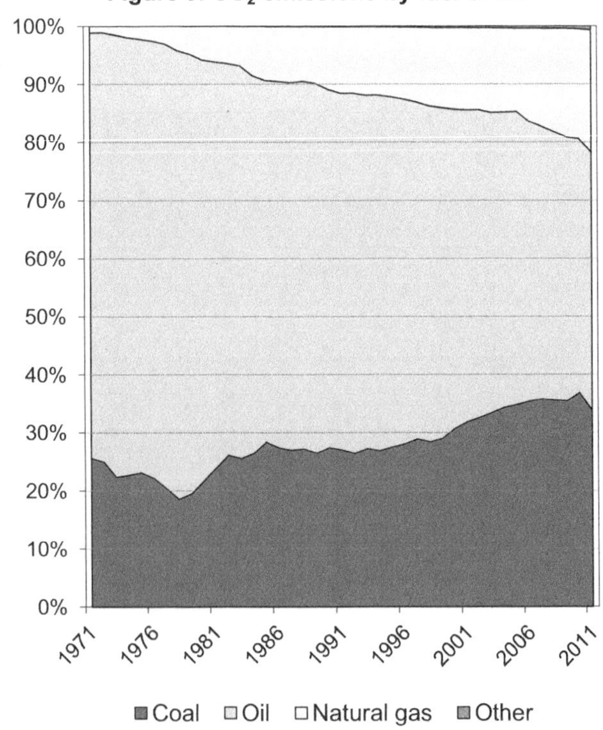

*Please refer to notes and definitions in Part I. Peat is included in coal for all figures.

Source: *IEA/OECD Electricity Information, IEA/OECD CO2 Emissions from Fuel Combustion*

JAPAN

1. Total primary energy supply (TPES) by fuel*

	1973	1980	1990	2000	2010	2011	2012e	Average annual percent change 73-90	90-11
TPES (Mtce)	**457.67**	**492.18**	**627.61**	**741.38**	**712.99**	**659.24**	**645.00**	**1.87**	**0.23**
Coal and peat	82.66	85.08	109.45	138.37	164.24	153.46	161.29	1.67	1.62
Oil	355.61	333.83	357.74	364.59	289.29	294.71	299.85	0.04	-0.92
Natural Gas	7.25	30.57	63.09	93.79	122.88	142.79	150.24	13.57	3.97
Biofuels and waste	-	-	7.14	8.38	14.16	14.85	14.85	-	3.55
Nuclear	3.61	30.75	75.30	119.90	107.31	37.89	4.18	19.56	-3.22
Hydro	8.20	10.85	10.97	10.72	10.10	10.22	9.55	1.72	-0.34
Geothermal	0.33	1.11	2.25	4.43	3.49	3.54	3.36	11.95	2.18
Solar, wind, tide	-	-	1.67	1.21 e	1.54 e	1.78 e	1.67	-	0.31
Net electricity trade(1)	-	-	-	-	-	-	-	-	-
Heat(2)	-	-	-	-	-	-	-	-	-

(1) Net trade between OECD and non-OECD countries.

(2) Ambient heat from heat pumps used in the transformation sector.

Source: *IEA/OECD Energy Balances of OECD Countries*

2. Energy supply, GDP and population*

	1973	1980	1990	2000	2010	2011	2012e	Average annual percent change 73-90	90-11
GDP (2005 bil. USD)	1940	2448	3851	4308	4648	4622	4714	4.12	0.87
TPES/GDP(1)	0.24	0.20	0.16	0.17	0.15	0.14	0.14	-2.15	-0.63
Population (millions)	108.9	117.1	123.6	126.9	128.0	127.8	127.3	0.75	0.16
TPES/population(1)	4.20	4.20	5.08	5.84	5.57	5.16	5.07	1.12	0.07
TPES/GDP(2)	145.0	123.6	100.2	105.8	94.3	87.7	84.1	-2.15	-0.63
Coal and peat supply/GDP(2)	124.1	101.2	82.7	93.5	102.9	96.7	99.6	-2.35	0.74
Elec. consumption/GDP(2)	100.0	98.0	91.0	102.4	100.5	95.0	..	-0.56	0.21
Elec. generation (TWh)	465	573	836	1049	1109	1043	1026	3.50	1.06
Industrial production(2)	55.6	65.7	96.9	98.3	94.8	92.4	92.1	3.32	-0.23

(1) TPES/GDP in units of tce/2005 thousand USD, TPES/population in units of tce/capita.

(2) As index, 2005 (not shown) = 100

Source: *IEA/OECD Energy Balances of OECD Countries* , *OECD Main Economic Indicators*

3. Coal and peat production by type*

	1978(1)	1990	2000	2005	2010	2011	2012e	Average annual percent change 78-90	90-11
Mtce:									
Coking coal	8.72	-	-	-	-	-	-	-	-
Steam coal	7.23	6.38	2.20	-	-	-	-	-1.04	-
Lignite	0.02	-	-	-	-	-	-	-	-
Peat	-	-	-	-	-	-	-	-	-
Mt:									
Coking coal	8.66	-	-	-	-	-	-	-	-
Steam coal	9.99	7.98	2.96	-	-	-	-	-1.85	-
Lignite	0.04	-	-	-	-	-	-	-	-
Peat	-	-	-	-	-	-	-	-	-

(1) Earliest year for which split by coal type is available.

* Please refer to notes and definitions in Part I.

Source: *IEA/OECD Energy Balances of OECD Countries, IEA/OECD Energy Statistics of OECD Countries*

JAPAN

4. Final consumption of energy by fuel*

(Mtce)

	1973	1980	1990	2000	2009	2010	2011	Average annual percent change 73-90	90-11
Total final consumption(1)	**334.26**	**331.27**	**427.76**	**490.39**	**447.08**	**461.65**	**449.25**	**1.46**	**0.23**
Coal and peat	34.39	36.07	45.25	36.04	34.77	38.90	37.80	1.63	-0.85
Oil	244.37	223.65	262.85	299.36	246.20	245.00	239.62	0.43	-0.44
Natural Gas	4.45	8.34	21.77	33.00	46.03	49.30	50.50	9.79	4.09
Biofuels and wastes	-	-	3.75	3.81	3.35	3.93	4.29	-	0.64
Geothermal	-	-	0.11	0.32	0.30	0.25	0.26	-	4.00
Solar, wind, tide	-	-	1.67	1.15	0.63	0.59	0.59	-	-4.86
Electricity	51.00	63.06	92.09	115.94	114.99	122.82	115.41	3.54	1.08
Heat	0.04	0.15	0.27	0.77	0.80	0.85	0.77	11.86	5.22
of which:									
Total industry	**149.84**	**130.33**	**146.02**	**139.87**	**117.38**	**126.53**	**121.04**	**-0.15**	**-0.89**
Coal and peat	26.64	30.60	43.31	34.67	33.55	37.66	36.55	2.90	-0.81
Oil	85.06	56.41	52.38	49.73	34.47	33.54	33.89	-2.81	-2.05
Natural Gas	2.34	3.05	5.28	7.30	10.49	11.16	12.02	4.91	4.00
Biofuels and wastes	-	-	3.61	3.74	3.32	3.90	4.26	-	0.79
Geothermal	-	-	-	-	-	-	-	-	-
Solar, wind, tide	-	-	-	-	-	-	-	-	-
Electricity	35.80	40.27	41.44	44.43	35.55	40.27	34.32	0.86	-0.89
Heat	-	-	-	-	-	-	-	-	-
Total transport	**58.22**	**77.11**	**102.50**	**125.62**	**108.93**	**109.82**	**108.32**	**3.38**	**0.26**
Coal and peat	0.29	-	-	-	-	-	-	-	-
Oil	56.31	75.24	100.44	123.34	106.61	107.51	106.03	3.46	0.26
Natural Gas	-	-	-	-	-	-	-	-	-
Biofuels and wastes	-	-	-	-	-	-	-	-	-
Electricity	1.63	1.87	2.07	2.28	2.31	2.31	2.28	1.42	0.47
Residential	**29.89**	**36.76**	**54.27**	**69.46**	**67.05**	**70.93**	**67.81**	**3.57**	**1.07**
Coal and peat	6.12	4.51	0.09	-	-	-	-	-21.76	-
Oil	12.50	13.82	19.26	23.70	18.31	19.68	18.43	2.57	-0.21
Natural Gas	1.54	4.18	10.50	12.85	12.91	13.11	13.09	11.96	1.05
Biofuels and wastes	-	-	0.14	0.07	0.03	0.03	0.03	-	-6.80
Geothermal	-	-	-	-	-	-	-	-	-
Solar, wind, tide	-	-	1.61	1.12	0.61	0.56	0.56	-	-4.86
Electricity	9.73	14.26	22.62	31.68	35.14	37.50	35.65	5.09	2.19
Heat	-	-	0.04	0.04	0.04	0.04	0.04	-	-0.24
Comm & public services	**28.86**	**29.06**	**65.71**	**88.38**	**91.29**	**91.90**	**92.73**	**4.96**	**1.65**
Coal and peat	1.28	0.97	1.28	0.87	0.75	0.75	0.73	-0.02	-2.63
Oil	23.29	20.36	32.77	36.67	25.86	23.49	23.80	2.03	-1.51
Natural Gas	0.55	1.08	5.56	12.58	22.18	24.52	24.88	14.63	7.40
Biofuels and waste	-	-	-	-	-	-	-	-	-
Geothermal	-	-	0.06	0.18	0.17	0.14	0.15	-	4.02
Solar, wind, tide	-	-	0.06	0.03	0.02	0.02	0.02	-	-4.78
Electricity	3.70	6.51	25.76	37.31	41.54	42.17	42.42	12.09	2.40
Heat	0.04	0.15	0.22	0.73	0.76	0.81	0.73	10.68	5.85
Non-energy use	**52.68**	**40.61**	**49.35**	**59.30**	**57.14**	**57.03**	**53.70**	**-0.38**	**0.40**
Coal and peat	-	-	0.57	0.50	0.47	0.49	0.53	-	-0.39
Oil	52.68	40.61	48.35	58.54	56.23	56.03	52.66	-0.50	0.41
Natural Gas	-	-	0.43	0.26	0.44	0.51	0.51	-	0.77

(1) Total Final Consumption (TFC) excludes use in transformation processes and energy industry own use.

* Please refer to notes and definitions in Part I. Agriculture, Forestry, Fishing and "Non-specified other" sectors are not reported here individually, but are accounted for in Total final consumption.

Source: *IEA/OECD Energy Balances of OECD Countries*

JAPAN

5. Coal balance*

(Mtce)

	1973	1980	1990	2000	2005	2010	2011	2012e	Average annual percent change 73-90	90-11
Production	25.6	15.6	6.4	2.2	-	-	-	-	-7.8	-
Imports	59.0	70.0	104.9	139.0	159.3	164.8	154.6	162.7	3.4	1.9
Exports	-0.6	-2.1	-1.9	-2.7	-1.7	-0.7	-1.0	-1.5	7.3	-3.1
Stock changes	-1.3	1.6	0.1	-0.2	-0.6	0.1	-0.1	0.1		
Primary supply	**82.7**	**85.1**	**109.5**	**138.4**	**157.0**	**164.2**	**153.5**	**161.3**	**1.7**	**1.6**
Statistical differences	**-2.5**	**-1.9**	**2.0**	**-1.9**	**2.8**	**-6.9**	**-2.6**	..		
Total transformation	**-41.2 e**	**-43.4 e**	**-62.8 e**	**-97.2 e**	**-116.5 e**	**-114.1 e**	**-108.8 e**	..	**2.5**	**2.7**
Electricity and heat gen.	-17.1	-15.0	-36.3	-69.1	-89.7	-88.1	-83.5	..	4.5	4.0
Main activity producers (1)	*-17.1*	*-11.5*	*-29.2*	*-57.5*	*-77.6*	*-76.0*	*-71.9*	..	*3.2*	*4.4*
Autoproducers	-	*-3.6*	*-7.1*	*-11.6*	*-12.1*	*-12.0*	*-11.6*	..	-	*2.4*
Gas works	3.4	4.7	-0.6	-0.3	-0.1	-	-	..	-	-
Coal transformation(2)	-27.5 e	-33.1 e	-25.9 e	-27.8 e	-26.7 e	-26.0 e	-25.3 e	..	-0.4	-0.1
BKB plants	-	-	-	-	-	-	-	..	-	-
Blast furnaces	*-24.5 e*	*-20.1 e*	*-25.6 e*	*-28.3 e*	*-26.3 e*	*-26.7 e*	*-25.3 e*	..	*0.3*	*-0.1*
Coke ovens	*-3.0*	*-13.1*	*-0.3*	*0.5*	*-0.4*	*0.7*	*-0.0*	..	*-12.6*	*-15.8*
Patent fuel plants	*-0.0*	*0.1*	*-0.0*	-	-	-	-	..	*-9.7*	-
Other transformation(3)	-	-	-	-	-	-	-	..	-	-
Energy ind. own use	**-4.4**	**-3.4**	**-3.4**	**-3.2**	**-3.3**	**-4.4**	**-4.2**	..	**-1.5**	**1.0**
Losses	**-0.1**	**-0.3**	**-**	**-**	**-**	**-**	**-**	..		
Final consumption(4)	**34.4**	**36.1**	**45.3**	**36.0**	**40.0**	**38.9**	**37.8**	..	**1.6**	**-0.9**
Industry(5)	26.6	30.6	43.3	34.7	38.7	37.7	36.5	..	2.9	-0.8
Iron and steel	*23.2 e*	*23.0 e*	*16.9 e*	*13.2 e*	*15.7 e*	*15.9 e*	*16.1 e*	..	*-1.8*	*-0.2*
Chemical	-	*0.4*	*3.9*	*3.9*	*4.0*	*4.3*	*4.2*	..	-	*0.4*
Non-metallic minerals	-	*4.3*	*8.2*	*7.0*	*6.1*	*5.2*	*5.1*	..	-	*-2.3*
Paper, pulp and print	-	*0.1*	*1.6*	*2.0*	*2.1*	*2.0*	*2.1*	..	-	*1.2*
Other industry (6)	*3.4*	*2.6*	*12.7*	*8.6*	*10.8*	*10.2*	*9.1*	..	*8.1*	*-1.6*
Transport(7)	0.3	-	-	-	-	-	-	..	-	-
Other	7.5	5.5	1.4	0.9	0.8	0.8	0.7	..	-9.5	-3.0
Comm. and pub. services	*1.3*	*1.0*	*1.3*	*0.9*	*0.8*	*0.8*	*0.7*	..	*-0.0*	*-2.6*
Residential	*6.1*	*4.5*	*0.1*	-	-	-	-	..	*-21.8*	-
Other sectors (8)	*0.1*	-	*0.0*	-	-	-	-	..	*-21.5*	-
Non-energy use	-	-	0.6	0.5	0.5	0.5	0.5	..	-	-0.4

(1) Main activity electricity and heat generation includes district heating.

(2) Coal transformation refers to the transformation of coal and peat to secondary and tertiary products (mainly cokes, briquettes, coal tar and off-gases). Gas works and Liquefaction are shown separately to match the *Energy Balances of OECD Countries.*

(3) Other transformation includes Liquefaction and Non-specified transformations.

(4) Final Consumption includes non-energy use and energy use (Industry, Transport and Other).

(5) Please refer to notes and definitions in Part I for detailed explanation.

(6) Other industry includes Non-ferrous metals, Transport equipment, Machinery, Mining and quarrying, Food and tobacco, Wood and wood products, Construction, Textile and leather, and Non-specified industry.

(7) Transport includes Rail and Inland waterways.

(8) Other sectors includes Agriculture/Forestry, Fishing and Non-specified other.

* "Coal" refers to all types of coal, primary (anthracite, coking coal, other bituminous coal, sub-bituminous coal, lignite/brown coal) and derived fuels (including patent fuel, BKB, coke oven coke, gas coke, coal tar, coke oven gas, gas works gas, blast furnace gas and other recovered gases), as well as peat, peat briquettes and oil shale. Quantities have been converted to Mtce using calorific values reported by the respective countries.
Please refer to notes and definitions in Part I.

Source: *IEA/OECD Energy Balances of OECD Countries*

JAPAN

6. Use of coal for selected end-uses*

(million tonnes)

	1978	1990	2000	2005	2009	2010	2011	Average annual percent change 78-90	90-11
Steam coal	**11.08**	**50.03**	**96.11**	**120.44**	**111.38**	**127.71**	**120.32**	**13.38**	**4.27**
Total electricity and heat	8.57	31.05	64.88	92.29	86.36	90.05	86.20	11.32	4.98
Main activity producers	*8.57*	*26.54*	*57.37*	*83.53*	*77.45*	*81.12*	*77.21*	*9.87*	*5.22*
Autoproducers	-	*4.52*	*7.52*	*8.76*	*8.91*	*8.93*	*8.99*	-	*3.33*
Patent fuel/BKB plants	0.34	0.11	-	-	-	-	-	-9.34	-
Coke ovens/Liquefaction(1)	0.08	5.89	11.46	11.65	9.44	12.52	13.53	43.53	4.04
Blast furnace inputs	-	-	-	-	-	-	-	-	-
Gas manufacture	0.07	-	-	-	-	-	-	-	-
Industry	2.52	14.83	16.13	16.67	14.68	15.28	15.35	15.91	0.16
Iron and steel	*0.23*	*1.54*	*2.58*	*3.50*	*2.77*	*3.19*	*3.41*	*17.29*	*3.86*
Chemical	*0.19*	*2.42*	*3.57*	*3.89*	*3.94*	*4.14*	*4.03*	*23.68*	*2.46*
Non-metallic minerals	*0.71*	*8.12*	*7.20*	*6.45*	*5.42*	*5.42*	*5.37*	*22.52*	*-1.95*
Paper, pulp and print	*0.05*	*1.90*	*2.32*	*2.45*	*2.45*	*2.42*	*2.44*	*36.33*	*1.20*
Other industry	*1.35*	*0.86*	*0.46*	*0.38*	*0.10*	*0.11*	*0.11*	*-3.72*	*-9.39*
Other sectors(2)	0.58	0.67	0.92	0.83	0.81	0.80	0.78	1.32	0.67
Non-energy use	-	0.02	-	-	-	-	-	-	-
Coking coal	**58.72**	**64.93**	**57.08**	**56.53**	**52.33**	**57.68**	**53.82**	**0.84**	**-0.89**
Total electricity and heat	-	-	-	-	-	-	-	-	-
Main activity producers	-	-	-	-	-	-	-	-	-
Autoproducers	-	-	-	-	-	-	-	-	-
Patent fuel/BKB plants	-	-	-	-	-	-	-	-	-
Coke ovens/Liquefaction(1)	55.12	59.33	45.97	47.80	44.04	45.35	41.26	0.61	-1.71
Blast furnace inputs	-	5.25 e	10.93 e	10.44 e	8.62 e	11.64 e	12.59 e	-	4.25
Gas manufacture	4.45	-	-	-	-	-	-	-	-
Industry	0.12	0.00	0.00	0.00	0.00	0.00	0.00	-32.71	5.37
Iron and steel	*0.06*	*0.00 e*	*0.00 e*	*0.00 e*	*0.00 e*	*0.00 e*	*0.00 e*	*-29.20*	*5.37*
Chemical	*0.01*	-	-	-	-	-	-	-	-
Non-metallic minerals	-	-	-	-	-	-	-	-	-
Paper, pulp and print	-	-	-	-	-	-	-	-	-
Other industry	*0.04*	-	-	-	-	-	-	-	-
Other sectors(2)	0.01	-	-	-	-	-	-	-	-
Non-energy use	-	-	-	-	-	-	-	-	-
Lignite	**0.04**	**-**	**-**	**-**	**-**	**-**	**-**	**-**	**-**
Total electricity and heat	-	-	-	-	-	-	-	-	-
Main activity producers	-	-	-	-	-	-	-	-	-
Autoproducers	-	-	-	-	-	-	-	-	-
Patent fuel/BKB plants	-	-	-	-	-	-	-	-	-
Coke ovens/Liquefaction(1)	-	-	-	-	-	-	-	-	-
Gas manufacture	-	-	-	-	-	-	-	-	-
Industry	0.03	-	-	-	-	-	-	-	-
Iron and steel	-	-	-	-	-	-	-	-	-
Chemical	*0.03*	-	-	-	-	-	-	-	-
Non-metallic minerals	-	-	-	-	-	-	-	-	-
Paper, pulp and print	-	-	-	-	-	-	-	-	-
Other industry	*0.00*	-	-	-	-	-	-	-	-
Other sectors(2)	0.01	-	-	-	-	-	-	-	-
Non-energy use	-	-	-	-	-	-	-	-	-

(1) Liquefaction primarily refers to direct distillation processes. Indirect liquefaction may be reported here or under gas manufacture.

(2) Other sectors are Residential, Commercial and public services, Agriculture/Forestry, Fishing and Non-specified other.

* Coal totals may not match the sum of the individual rows due to statistical differences. Please refer to notes and definitions in Part I.

Source: *IEA/OECD Energy Statistics of OECD Countries*

JAPAN

7. Fuel prices to end users*

	1978(1)	1990	1995	2000	2005	2009	2010	2011	2012
				(US dollars / tce)					
For electricity generation									
Steam coal	90.58	101.84	92.27	51.42	..	..	..	..	..
Heavy fuel oil	79.83	151.02	153.70	..	..	..	..	..	..
Natural gas	73.11	129.99	122.57	..	..	..	..	..	..
For industry									
Steam coal	71.23	70.53	62.00	45.68	81.25	152.90	151.91	193.10	189.16
Coking coal	72.47	68.03	61.13	45.29	104.02	203.29	186.84	251.65	212.40
High sulphur fuel oil	82.50	136.92	132.36	163.49	255.46	..	..	..	..
Low sulphur fuel oil	..	..	205.13	219.96	342.15	426.31	541.35	717.29	744.78
Natural gas	302.48	320.89	381.57	351.99	312.30	439.76	493.61	635.70	..
				(1 000 Yen / unit)(2)					
For electricity generation									
Steam coal	15.17	11.74	6.91	4.41	..	..	..	..	..
Heavy fuel oil	23.04	29.99	19.83	..	..	..	..	..	..
Natural gas	19.78	24.20	14.82	..	..	..	..	..	..
For industry									
Steam coal	11.93	8.13	4.64	3.92	7.12	11.39	10.61	12.27	12.02
Coking coal	14.29	9.23	5.39	4.58	10.73	17.82	15.36	19.89	16.78
High sulphur fuel oil	23.81	27.19	17.08	24.18	38.57	..	..	..	..
Low sulphur fuel oil	..	..	26.46	32.53	51.66	54.71	65.15	78.54	81.52
Natural gas	81.84	59.74	46.15	48.80	44.21	52.90	55.70	65.26	..

(1) Earliest year for which split by coal type is available.

(2) For steam coal, coking coal and heavy fuel oil per metric tonne; for natural gas per 10^7 kilocalories GCV.

* Please refer to notes and definitions in Part I. Steam coal prices relate to historical definitions.

Source: *IEA/OECD Energy Prices & Taxes*

8. Coal and peat trade by type of coal*

(Mtce)

	1978	1990	1995	2000	2005	2009	2010	2011	2012e
Total imports	**52.67**	**104.88**	**121.47**	**139.00**	**159.26**	**145.17**	**164.78**	**154.59**	**162.73**
Bituminous coal(1)	1.43	36.00	57.90	81.81	102.24	94.51	108.40	102.12	111.64
Coking coal	51.24	68.37	62.97	54.60	54.26	50.23	55.36	51.65	50.10
Sub-bituminous coal	-	-	-	-	-	-	-	-	-
Lignite	-	-	-	-	-	-	-	-	-
Peat	-	-	-	-	-	-	-	-	-
Coal products(2)	-	0.51	0.61	2.60	2.76	0.43	1.01	0.81	0.99
Total exports	**1.05**	**1.93**	**3.54**	**2.67**	**1.68**	**1.07**	**0.67**	**1.00**	**1.53**
Bituminous coal(1)	0.01	0.00	0.00	0.00	0.00	0.00	0.00	0.00	0.00
Coking coal	0.05	-	-	-	-	-	-	-	-
Sub-bituminous coal	-	-	-	-	-	-	-	-	-
Lignite	-	-	-	-	-	-	-	-	-
Peat	-	-	-	-	-	-	-	-	-
Coal products(2)	1.00	1.93	3.54	2.66	1.68	1.07	0.67	1.00	1.52

(1) Bituminous coal includes anthracite.

(2) Coal products includes products derived from coal, for example: coke, coal tar, briquettes, patent fuels.

* Please refer to notes and definitions in Part I.

Source: *IEA/OECD Energy Balances of OECD Countries, IEA/OECD Energy Statistics of OECD Countries*

JAPAN

9. Total coal imports by origin*

(thousand tonnes)

	1978(1)	1985	1990	1995	2000	2005	2010	2011	2012e
Total coal(2)	**52858**	**93449**	**106918**	**127354**	**150340**	**176975**	**185394**	**174141**	**183771**
Coking coal	**50876**	**69242**	**64935**	**59799**	**57081**	**56528**	**57679**	**53817**	**52199**
Australia	24149	30116	34505	31470	34757	29220	32549	29556	28667
Canada	10895	17221	11290	8756	5292	4608	6317	5331	5171
Czech Republic	-	-	-	-	-	-	-	-	-
Germany	398	-	-	-	-	-	-	-	-
Poland	429	-	-	-	-	-	-	-	-
United Kingdom	-	-	-	-	-	-	-	-	-
United States	9956	12829	7367	5217	1422	937	2621	4589	4450
Other OECD	11	401	225	291	367	530	384	364	354
China, People's Rep.	420	1211	2802	4255	6774	3773	679	515	500
Colombia	-	-	153	52	-	-	24	24	23
Indonesia	-	103	668	4585	5677	15374	13113	11497	11151
South Africa	2360	4586	2955	2741	603	-	-	-	-
Former Soviet Union(3)	2244	2775	x	x	x	x	x	x	x
Russian Federation (4)	*x*	*x*	*4950*	*2394*	*2104*	*2045*	*1946*	*1926*	*1868*
Other FSU (4)	*x*	*x*	-	-	-	-	-	-	-
Venezuela	-	-	-	-	-	-	-	-	-
Vietnam	-	-	-	-	27	-	-	-	-
Non-specified/other	14	-	20	38	58	41	46	15	-
Steam coal	**1982**	**24207**	**41983**	**67555**	**93259**	**120447**	**127715**	**120324**	**131572**
Australia	668	14138	21346	33967	55147	77013	85114	79319	86752
Canada	105	780	6920	9448	8297	1190	2630	2577	2828
Czech Republic	-	-	-	-	-	-	-	-	-
Germany	-	-	-	-	-	-	-	-	-
Poland	-	-	-	-	-	-	-	-	-
United Kingdom	-	-	-	-	-	-	-	-	-
United States	-	960	4516	5630	2230	-	426	667	754
Other OECD	-	14	138	315	575	-	-	-	-
China, People's Rep.	513	2427	2504	5856	12470	20646	6236	3880	3841
Colombia	-	-	94	56	-	-	138	164	180
Indonesia	-	306	409	4948	8908	9770	21167	21954	24088
South Africa	157	4091	2003	2958	945	97	345	694	762
Former Soviet Union(3)	149	1020	x	x	x	x	x	x	x
Russian Federation (4)	*x*	*x*	*3402*	*2656*	*3300*	*9351*	*10067*	*9910*	*11245*
Other FSU (4)	*x*	*x*	-	-	-	*21*	-	-	-
Venezuela	-	-	-	-	-	15	-	-	-
Vietnam	-	157	150	1273	944	2043	1591	1158	1121
Non-specified/other	390	314	501	448	443	301	1	1	1
Lignite	-	-	-	-	-	-	-	-	-

(1) Earliest year for which split by coal type is available.

(2) Total coal consists of steam coal, coking coal and lignite.

(3) For years prior to 1990, or if no details for individual Former Soviet Union republics.

(4) Data for former republics are not available separately for some OECD member countries.

* Please refer to notes and definitions in Part I.

Source: IEA/OECD Coal Statistics

JAPAN

12. Coal import values by origin*

(average unit value, CIF, USD/tonne)

	1990	1995	2000	2005	2007	2008	2009	2010	2011/12[1]
Coking coal[2]	**60.72**	**55.03**	**39.46**	**88.80**	**88.43**	**184.13**	**163.82**	**151.45**	**..**
Imports from:									
Australia	55.27	51.15	39.01	96.44	96.03	206.71	191.58	170.10	..
Canada	71.27	64.49	45.46	106.05	112.61	234.34	220.94	191.97	..
Czech Republic	x	x	x	..	..	..	x	x	..
Poland	x	x	x	..	..	..	x	x	..
United States	66.90	61.37	52.69	159.01	516.32	308.56	241.65	208.24	..
China	54.38	49.49	37.12	100.45	100.91	256.42	138.99	154.71	..
Colombia	59.34	48.26	37.31	x	x	..	120.75	x	..
Indonesia	50.84	48.29	32.36	53.51	60.43	104.80	97.14	96.60	..
South Africa	50.11	49.54	39.99	x	x	x	x	x	..
Former Soviet Union[3]	57.45	54.81	43.62	114.96	105.45	250.90	190.54	185.00	..
Other bituminous coal[4]	**50.97**	**47.85**	**34.59**	**62.73**	**70.92**	**125.42**	**111.12**	**110.40**	**..**
Imports from:									
Australia	52.23	48.87	34.59	61.90	71.03	127.23	111.87	111.12	..
Canada	48.26	44.20	34.72	63.79	66.04	125.52	103.84	107.80	..
Czech Republic	x	x	190.61	..	..	..	x	x	..
Poland	x	x	x	..	..	..	x	x	..
United States	53.17	52.65	45.49	x	553.99	151.26	47.12	84.43	..
China	47.59	44.48	33.69	65.65	73.60	124.10	118.18	112.29	..
Colombia	x	x	x	x	100.34	x	x	118.01	..
Indonesia	54.97	43.79	31.85	59.65	66.45	116.08	112.99	108.20	..
South Africa	47.95	48.27	35.82	80.64	77.46	105.49	87.62	107.07	..
Former Soviet Union[3]	46.54	43.45	30.68	66.10	72.28	121.34	101.62	106.42	..

Note: On occasion, shipment of extremely small quantities of high valued coal results in high import costs.

(1) Import data for steam and coking coals are currently unavailable for 2011 and 2012 due to resource constraints.

(2) Weighted average of individual countries using import volumes as weights. Prices exclude intra-EU trade.

(3) Former Soviet Union until 1991, Russian Federation starting in 1992.

(4) Bituminous steam coal only. Weighted average of trades. Prices exclude intra-EU trade.

* Please refer to notes and definitions in Part I.

Source: IEA/OECD Coal Statistics

KOREA*

Figure 1: TPES by fuel (Mtce)

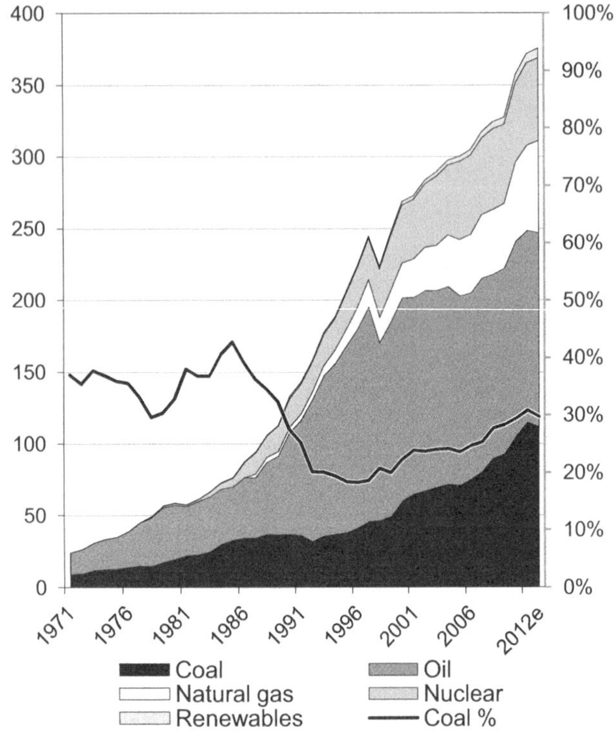

Figure 2: Coal supply indicators (1971 = 100)

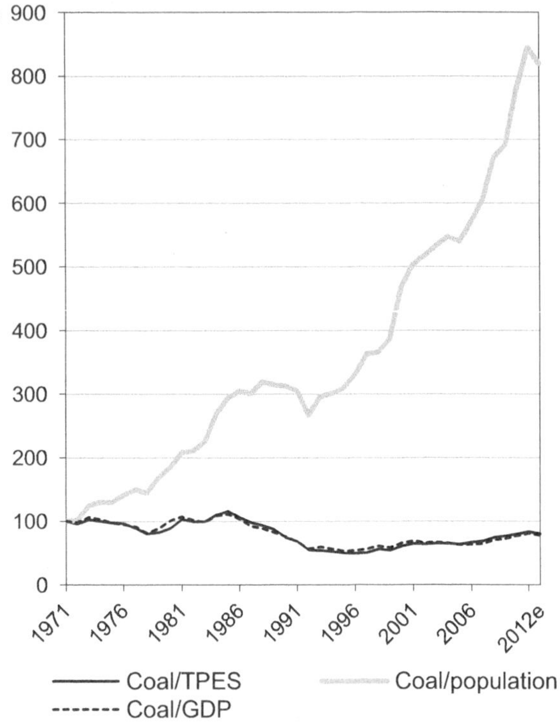

Figure 3: Primary coal supply (Mtce)

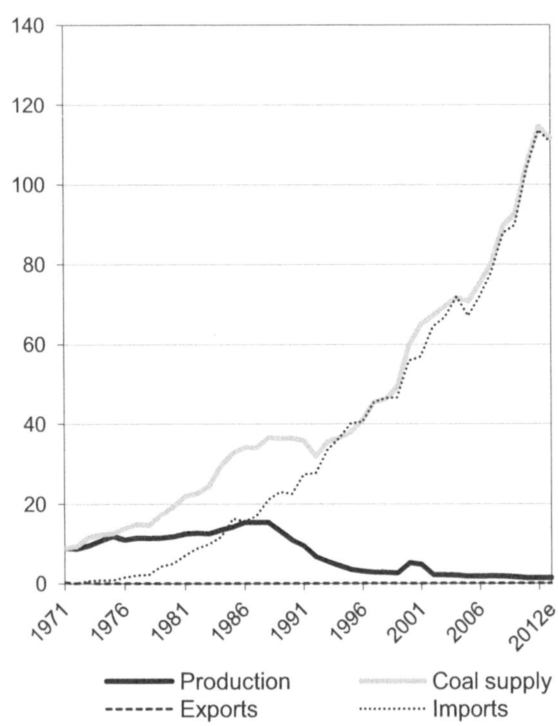

Figure 4: Coal consumption (Mtce)

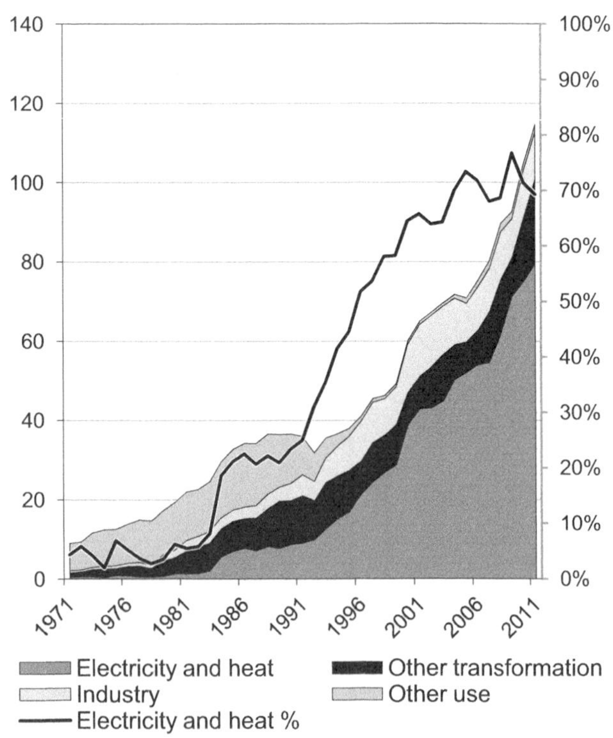

*Please refer to notes and definitions in Part I. Peat is included in coal for all figures.

Source: *IEA/OECD Coal Information, IEA/OECD Energy Balances of OECD Countries, OECD Main Economic Indicators*

KOREA*

Figure 5: Electricity generation by fuel (TWh)

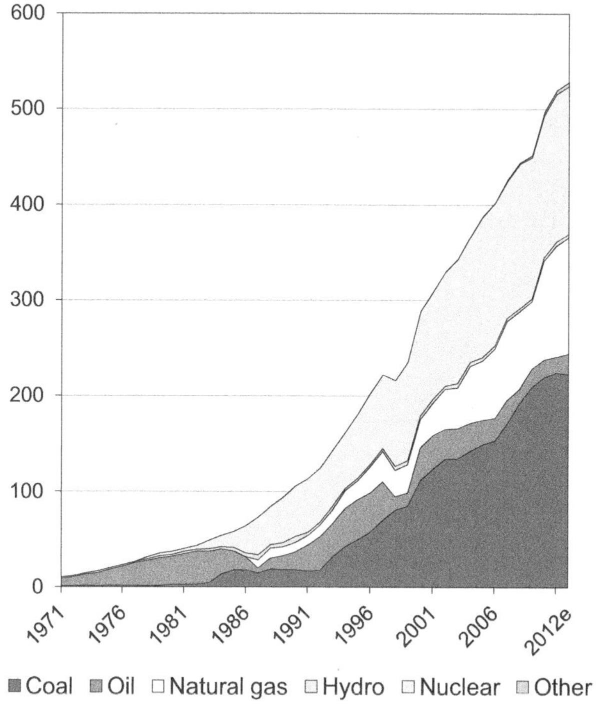

Figure 6: CO_2 emissions by fuel (Mt CO_2)

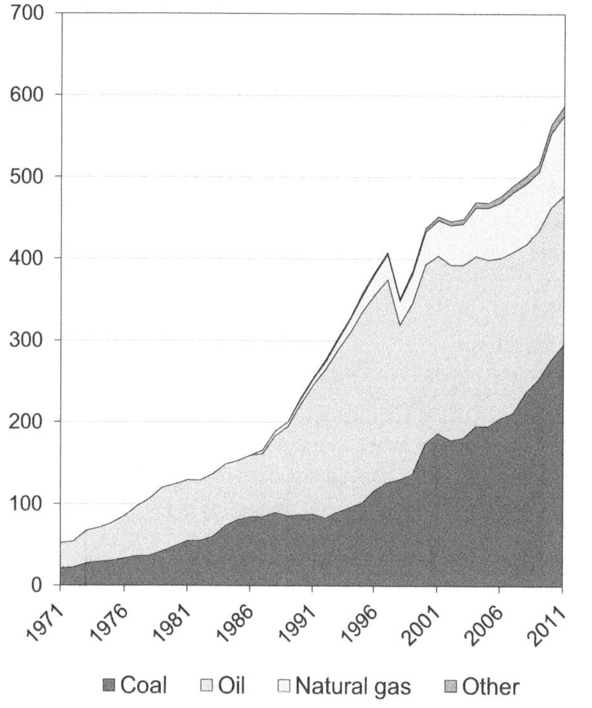

Figure 7: Electricity generation by fuel share

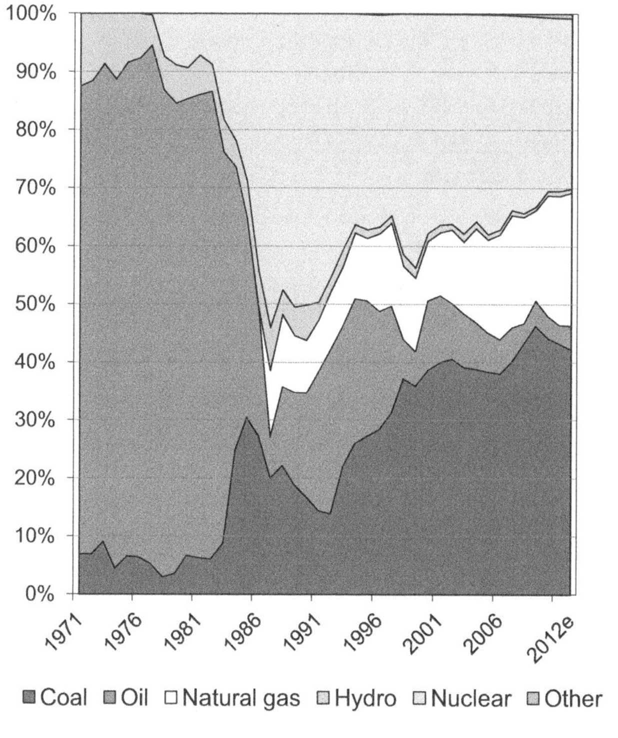

Figure 8: CO_2 emissions by fuel share

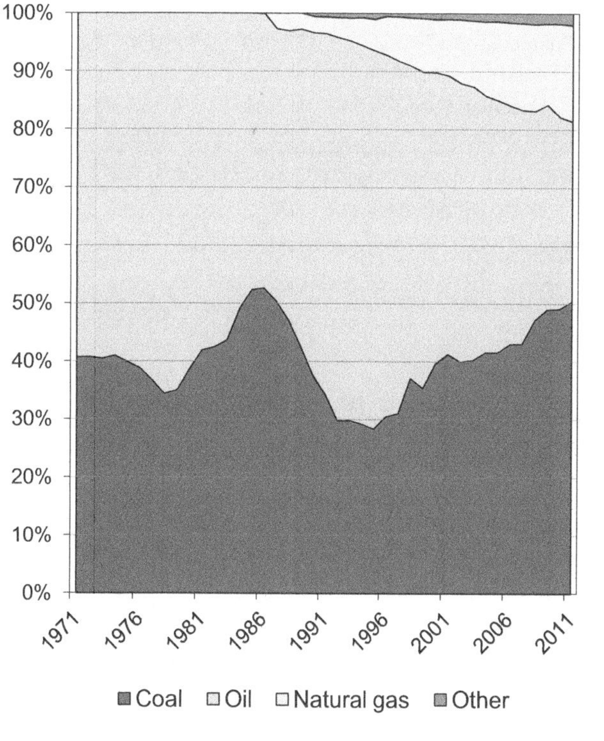

*Please refer to notes and definitions in Part I. Peat is included in coal for all figures.

Source: *IEA/OECD Electricity Information, IEA/OECD CO2 Emissions from Fuel Combustion*

KOREA

1. Total primary energy supply (TPES) by fuel*

	1973	1980	1990	2000	2010	2011	2012e	Average annual percent change 73-90	90-11
TPES (Mtce)	**30.78**	**58.87**	**132.98**	**268.80**	**357.09**	**372.06**	**375.72**	**8.99**	**5.02**
Coal and peat	11.61	19.26	36.51	59.93	104.89	114.62	111.55	6.97	5.60
Oil	19.01	38.07	71.05	141.49	135.88	133.87	135.33	8.06	3.06
Natural Gas	-	-	3.89	24.29	55.18	59.39	64.24	-	13.86
Biofuels and waste	-	-	1.04	1.97	4.93	5.51	5.96	-	8.24
Nuclear	-	1.29	19.69	40.57	55.32	57.60	57.67	-	5.24
Hydro	0.16	0.24	0.78	0.49	0.45	0.56	0.45	9.87	-1.53
Geothermal	-	-	-	-	0.05	0.07	0.10	-	-
Solar, wind, tide	-	-	0.01	0.06 e	0.26	0.29	0.29	-	15.48
Net electricity trade(1)	-	-	-	-	-	-	-	-	-
Heat(2)	-	-	-	-	0.13	0.13	0.13	-	-

(1) Net trade between OECD and non-OECD countries.

(2) Ambient heat from heat pumps used in the transformation sector.

Source: *IEA/OECD Energy Balances of OECD Countries*

2. Energy supply, GDP and population*

	1973	1980	1990	2000	2010	2011	2012e	Average annual percent change 73-90	90-11
GDP (2005 bil. USD)	82	142	360	678	1019	1056	1078	9.13	5.25
TPES/GDP(1)	0.38	0.41	0.37	0.40	0.35	0.35	0.35	-0.13	-0.22
Population (millions)	34.1	38.1	42.9	47.0	49.4	49.8	50.0	1.35	0.71
TPES/population(1)	0.90	1.54	3.10	5.72	7.23	7.47	7.52	7.53	4.28
TPES/GDP(2)	106.2	116.3	103.9	111.5	98.6	99.1	98.1	-0.13	-0.22
Coal and peat supply/GDP(2)	170.0	161.4	121.0	105.5	122.9	129.5	123.6	-1.98	0.33
Elec. consumption/GDP(2)	37.2	54.3	61.9	91.6	104.2	105.3	..	3.05	2.56
Elec. generation (TWh)	15	37	105	289	497	520	528	12.23	7.90
Industrial production(2)	..	10.5	31.7	74.3	139.2	147.4	148.7	..	7.59

(1) TPES/GDP in units of tce/2005 thousand USD, TPES/population in units of tce/capita.

(2) As index, 2005 (not shown) = 100

Source: *IEA/OECD Energy Balances of OECD Countries , OECD Main Economic Indicators*

3. Coal and peat production by type*

	1978(1)	1990	2000	2005	2010	2011	2012e	Average annual percent change 78-90	90-11
Mtce:									
Coking coal	-	-	-	-	-	-	-	-	-
Steam coal	11.35	10.82	5.20	1.80	1.37	1.37	1.37	-0.39	-9.37
Lignite	-	-	-	-	-	-	-	-	-
Peat	-	-	-	-	-	-	-	-	-
Mt:									
Coking coal	-	-	-	-	-	-	-	-	-
Steam coal	18.05	17.22	8.30	2.83	2.08	2.08	2.09	-0.39	-9.57
Lignite	-	-	-	-	-	-	-	-	-
Peat	-	-	-	-	-	-	-	-	-

(1) Earliest year for which split by coal type is available.

* Please refer to notes and definitions in Part I.

Source: *IEA/OECD Energy Balances of OECD Countries, IEA/OECD Energy Statistics of OECD Countries*

KOREA

4. Final consumption of energy by fuel*

(Mtce)

	1973	1980	1990	2000	2009	2010	2011	Average annual percent change 73-90	90-11
Total final consumption[(1)]	**24.98**	**44.70**	**92.72**	**181.58**	**211.17**	**225.22**	**230.06**	**8.02**	**4.42**
Coal and peat	9.26	13.92	16.74	12.96	11.69	13.63	13.73	3.54	-0.94
Oil	14.14	26.76	62.37	114.11	113.85	116.95	117.26	9.12	3.05
Natural Gas	-	-	0.96	15.61	25.88	29.40	30.81	-	17.95
Biofuels and wastes	-	-	1.04	1.82	3.49	3.76	3.91	-	6.50
Geothermal	-	-	-	-	0.03	0.05	0.07	-	-
Solar, wind, tide	-	-	0.01	0.06	0.04	0.04	0.04	-	4.94
Electricity	1.58	4.02	11.60	32.33	49.84	55.21	57.82	12.46	7.95
Heat	-	-	-	4.70	6.34	6.19	6.42	-	-
of which:									
Total industry	**9.17**	**14.73**	**27.52**	**54.96**	**57.08**	**64.25**	**67.43**	**6.68**	**4.36**
Coal and peat	0.56	1.92	4.36	12.15	9.87	11.83	12.06	12.88	4.97
Oil	7.52	10.01	15.55	15.68	8.33	8.43	7.42	4.36	-3.46
Natural Gas	-	-	0.10	4.11	8.05	10.13	11.82	-	25.27
Biofuels and wastes	-	-	0.41	1.51	2.49	2.49	2.42	-	8.87
Geothermal	-	-	-	-	0.00	0.00	0.00	-	-
Solar, wind, tide	-	-	-	-	0.00	0.00	0.00	-	-
Electricity	1.09	2.79	7.10	18.48	24.49	28.03	30.23	11.67	7.14
Heat	-	-	-	3.04	3.86	3.34	3.47	-	-
Total transport	**3.57**	**6.83**	**20.82**	**37.52**	**41.82**	**42.73**	**42.03**	**10.93**	**3.40**
Coal and peat	0.02	0.00	-	-	-	-	-	-	-
Oil	3.54	6.78	20.69	37.27	39.96	40.57	39.82	10.95	3.17
Natural Gas	-	-	-	-	1.28	1.46	1.53	-	-
Biofuels and wastes	-	-	-	-	0.31	0.43	0.40	-	-
Electricity	0.02	0.05	0.12	0.25	0.27	0.27	0.28	12.93	3.87
Residential	**8.85**	**13.23**	**18.49**	**21.48**	**26.63**	**28.37**	**28.78**	**4.43**	**2.13**
Coal and peat	8.55 e	11.85 e	12.34 e	0.81 e	1.28	1.22	1.20	2.18	-10.51
Oil	0.10	0.72	2.74	5.54	4.19	4.26	4.58	21.34	2.48
Natural Gas	-	-	0.66 e	8.90	11.81	12.77	12.72	-	15.10
Biofuels and wastes	-	-	0.56	0.09	0.03	0.05	0.07	-	-9.18
Geothermal	-	-	-	-	0.00	0.00	0.01	-	-
Solar, wind, tide	-	-	0.01	0.05	0.03	0.03	0.02	-	4.90
Electricity	0.19	0.65	2.18	4.56	7.09	7.53	7.60	15.43	6.13
Heat	-	-	-	1.53	2.20	2.51	2.58	-	-
Comm & public services	**1.27**	**3.64**	**12.42**	**24.59**	**27.44**	**29.29**	**28.80**	**14.33**	**4.09**
Coal and peat	0.14	0.14 e	0.04 e	-	-	-	-	-7.29	-
Oil	0.86	3.00	10.09	13.25	4.85	4.93	4.24	15.56	-4.05
Natural Gas	-	-	0.19 e	2.58	4.75	5.04	4.74	-	16.43
Biofuels and waste	-	-	0.07	0.22	0.65	0.79	0.97	-	13.06
Geothermal	-	-	-	-	0.03	0.04	0.05	-	-
Solar, wind, tide	-	-	0.01	0.01	0.01	0.01	0.02	-	4.74
Electricity	0.27	0.50	2.01	8.39	16.88	18.15	18.41	12.47	11.11
Heat	-	-	-	0.14	0.28	0.34	0.37	-	-
Non-energy use	**1.62**	**4.37**	**9.61**	**35.76**	**52.87**	**54.80**	**57.29**	**11.04**	**8.87**
Coal and peat	-	-	-	-	0.54	0.58	0.47	-	-
Oil	1.62	4.37	9.61	35.76	52.33	54.22	56.82	11.04	8.83
Natural Gas	-	-	-	-	-	-	-	-	-

(1) Total Final Consumption (TFC) excludes use in transformation processes and energy industry own use.

* Please refer to notes and definitions in Part I. Agriculture, Forestry, Fishing and "Non-specified other" sectors are not reported here individually, but are accounted for in Total final consumption.

Source: *IEA/OECD Energy Balances of OECD Countries*

KOREA

5. Coal balance*

(Mtce)

	1973	1980	1990	2000	2005	2010	2011	2012e	Average annual percent change 73-90	90-11
Production	9.5 e	11.7	10.8	5.2	1.8	1.4	1.4	1.4	0.8	-9.4
Imports	0.6 e	5.0 e	22.5 e	55.9 e	67.0	104.2	113.5	110.5	23.2	8.0
Exports	-0.2 e	-	-	-	-	-	-	-	-	-
Stock changes	1.6 e	2.6 e	3.2	-1.2	1.9	-0.7	-0.3	-0.3		
Primary supply	**11.6**	**19.3**	**36.5**	**59.9**	**70.8**	**104.9**	**114.6**	**111.6**	**7.0**	**5.6**
Statistical differences	**-0.0**	**-0.4**	**1.2**	**4.1**	**4.9**	**-0.6**	**-1.5**	..		
Total transformation	**-2.2 e**	**-3.5 e**	**-18.8 e**	**-48.0 e**	**-63.4**	**-87.3 e**	**-95.5 e**	..	**13.5**	**8.0**
Electricity and heat gen.	-0.5 e	-1.2 e	-8.5	-38.7	-51.9	-74.9	-79.4	..	18.5	11.2
Main activity producers (1)	*-0.4*	*-1.2*	*-5.6*	*-34.9*	*-46.2*	*-67.3*	*-70.6*	..	*16.4*	*12.8*
Autoproducers	*-0.1 e*	*-0.0 e*	*-2.9*	*-3.7*	*-5.7*	*-7.6*	*-8.8*	..	*25.9*	*5.4*
Gas works	-	-	-	-	-	-	-	..	-	-
Coal transformation(2)	-1.7 e	-2.3 e	-10.3 e	-9.4 e	-11.5	-12.4 e	-16.1 e	..	11.1	2.1
BKB plants	-	-	-	-	-	-	-	..	-	-
Blast furnaces	*-0.2 e*	*-1.4 e*	*-4.9*	*-7.8*	*-7.9*	*-11.0 e*	*-13.8 e*	..	*21.7*	*5.0*
Coke ovens	*-0.0 e*	*-0.2 e*	*-0.6*	*-0.6*	*-3.7*	*-1.4*	*-2.3 e*	..	*28.5*	*7.0*
Patent fuel plants	*-1.5 e*	*-0.7 e*	*-4.8 e*	*-0.9 e*	*0.1*	-	-	..	*7.0*	-
Other transformation(3)	-	-	-	-	-	-	-	..	-	-
Energy ind. own use	**-0.1 e**	**-1.4 e**	**-2.1**	**-3.0**	**-1.2**	**-3.4**	**-3.9**	..	**17.8**	**3.1**
Losses	-	-	-	-	-	-	-	..		
Final consumption(4)	**9.3**	**13.9**	**16.7**	**13.0**	**11.1**	**13.6**	**13.7**	..	**3.5**	**-0.9**
Industry(5)	0.6	1.9	4.4	12.1	9.7	11.8	12.1	..	12.9	5.0
Iron and steel	*0.6 e*	*1.3 e*	*0.7*	*3.7*	*4.4*	*6.2 e*	*6.3 e*	..	*1.4*	*11.0*
Chemical	-	-	*0.1*	*0.2*	*0.1*	*0.1*	*0.5*	..	-	*10.5*
Non-metallic minerals	-	*0.7*	*2.9*	*4.4*	*4.0*	*3.8*	*4.3*	..	-	*1.8*
Paper, pulp and print	-	-	*0.0*	-	*0.0*	*0.0*	*0.0*	..	-	*13.5*
Other industry (6)	-	-	*0.7*	*3.9*	*1.1*	*1.7*	*1.0*	..	-	*1.8*
Transport(7)	0.0	0.0	-	-	-	-	-	..	-	-
Other	8.7	12.0	12.4	0.8	1.4	1.2	1.2	..	2.1	-10.5
Comm. and pub. services	*0.1*	*0.1 e*	*0.0 e*	-	-	-	-	..	*-7.3*	-
Residential	*8.6 e*	*11.9 e*	*12.3 e*	*0.8 e*	*1.4*	*1.2*	*1.2*	..	2.2	-10.5
Other sectors (8)	-	-	-	-	-	-	-	..	-	-
Non-energy use	-	-	-	-	-	0.6	0.5	..	-	-

(1) Main activity electricity and heat generation includes district heating.

(2) Coal transformation refers to the transformation of coal and peat to secondary and tertiary products (mainly cokes, briquettes, coal tar and off-gases). Gas works and Liquefaction are shown separately to match the *Energy Balances of OECD Countries.*

(3) Other transformation includes Liquefaction and Non-specified transformations.

(4) Final Consumption includes non-energy use and energy use (Industry, Transport and Other).

(5) Please refer to notes and definitions in Part I for detailed explanation.

(6) Other industry includes Non-ferrous metals, Transport equipment, Machinery, Mining and quarrying, Food and tobacco, Wood and wood products, Construction, Textile and leather, and Non-specified industry.

(7) Transport includes Rail and Inland waterways.

(8) Other sectors includes Agriculture/Forestry, Fishing and Non-specified other.

* "Coal" refers to all types of coal, primary (anthracite, coking coal, other bituminous coal, sub-bituminous coal, lignite/brown coal) and derived fuels (including patent fuel, BKB, coke oven coke, gas coke, coal tar, coke oven gas, gas works gas, blast furnace gas and other recovered gases), as well as peat, peat briquettes and oil shale. Quantities have been converted to Mtce using calorific values reported by the respective countries.
Please refer to notes and definitions in Part I.

Source: *IEA/OECD Energy Balances of OECD Countries*

KOREA

6. Use of coal for selected end-uses*

(million tonnes)

	1978	1990	2000	2005	2009	2010	2011	Average annual percent change 78-90	90-11
Steam coal	**19.88**	**33.04**	**52.38**	**61.39**	**85.68**	**92.84**	**98.32**	**4.32**	**5.33**
Total electricity and heat	0.52	7.71	39.01	52.72	77.68	83.17	86.77	25.23	12.22
Main activity producers	*0.52*	*7.71*	*39.01*	*50.21*	*74.10*	*78.81*	*81.51*	*25.23*	*11.89*
Autoproducers	-	-	-	*2.51*	*3.58*	*4.36*	*5.26*	-	-
Patent fuel/BKB plants	18.29 e	20.70	2.41	2.01	1.94	1.86	1.82	1.04	-10.93
Coke ovens/Liquefaction(1)	-	-	-	-	-	-	-	-	-
Blast furnace inputs	-	-	-	-	-	-	-	-	-
Gas manufacture	-	-	-	-	-	-	-	-	-
Industry	0.84	4.59	11.02	7.03	6.60	7.23	8.31	15.15	2.87
Iron and steel	*0.68*	*0.17*	*0.73*	*1.19*	*1.16*	*1.37*	*2.00*	*-10.91*	*12.41*
Chemical	-	*0.08*	*0.27*	*0.13*	*0.02*	*0.17*	*0.61*	-	*10.35*
Non-metallic minerals	*0.16*	*3.53*	*5.31*	*4.75*	*4.40*	*4.50*	*5.05*	*29.42*	*1.71*
Paper, pulp and print	-	*0.00*	-	*0.04*	*0.04*	*0.04*	*0.04*	-	*13.39*
Other industry	-	*0.80*	*4.71*	*0.91*	*0.97*	*1.15*	*0.62*	-	*-1.22*
Other sectors(2)	0.22	0.05	-	-	-	-	-	-12.19	-
Non-energy use	-	-	-	-	-	-	-	-	-
Coking coal	**2.01**	**11.74**	**19.42**	**20.88**	**21.18**	**27.21**	**32.58**	**15.84**	**4.98**
Total electricity and heat	-	-	-	-	-	-	-	-	-
Main activity producers	-	-	-	-	-	-	-	-	-
Autoproducers	-	-	-	-	-	-	-	-	-
Patent fuel/BKB plants	-	-	-	-	-	-	-	-	-
Coke ovens/Liquefaction(1)	2.01	11.74	16.38	15.40	14.80	19.52	22.88	15.84	3.23
Blast furnace inputs	-	-	3.03	5.48	6.12 e	7.55 e	9.31 e	-	-
Gas manufacture	-	-	-	-	-	-	-	-	-
Industry	-	-	-	-	0.26	0.15	0.10	-	-
Iron and steel	-	-	-	-	*0.26 e*	*0.15 e*	*0.10 e*	-	-
Chemical	-	-	-	-	-	-	-	-	-
Non-metallic minerals	-	-	-	-	-	-	-	-	-
Paper, pulp and print	-	-	-	-	-	-	-	-	-
Other industry	-	-	-	-	-	-	-	-	-
Other sectors(2)	-	-	-	-	-	-	-	-	-
Non-energy use	-	-	-	-	-	-	-	-	-
Lignite	**-**	**-**	**-**	**-**	**-**	**-**	**-**	**-**	**-**
Total electricity and heat	-	-	-	-	-	-	-	-	-
Main activity producers	-	-	-	-	-	-	-	-	-
Autoproducers	-	-	-	-	-	-	-	-	-
Patent fuel/BKB plants	-	-	-	-	-	-	-	-	-
Coke ovens/Liquefaction(1)	-	-	-	-	-	-	-	-	-
Gas manufacture	-	-	-	-	-	-	-	-	-
Industry	-	-	-	-	-	-	-	-	-
Iron and steel	-	-	-	-	-	-	-	-	-
Chemical	-	-	-	-	-	-	-	-	-
Non-metallic minerals	-	-	-	-	-	-	-	-	-
Paper, pulp and print	-	-	-	-	-	-	-	-	-
Other industry	-	-	-	-	-	-	-	-	-
Other sectors(2)	-	-	-	-	-	-	-	-	-
Non-energy use	-	-	-	-	-	-	-	-	-

(1) Liquefaction primarily refers to direct distillation processes. Indirect liquefaction may be reported here or under gas manufacture.

(2) Other sectors are Residential, Commercial and public services, Agriculture/Forestry, Fishing and Non-specified other.

* Coal totals may not match the sum of the individual rows due to statistical differences. Please refer to notes and definitions in Part I.

Source: *IEA/OECD Energy Statistics of OECD Countries*

KOREA

7. Fuel prices to end users*

	1978(1)	1990	1995	2000	2005	2009	2010	2011	2012
				(US dollars / tce)					
For electricity generation									
Steam coal	..	..	..	..	58.44	87.54	..	..	..
Heavy fuel oil	..	99.47	110.35	209.68	..	..	..	..	..
Natural gas	..	..	..	..	285.40	352.07	..	..	..
For industry									
Steam coal	..	61.03	56.01	58.36	66.77	94.12	..	..	..
Coking coal	..	..	..	..	116.53	169.43	..	..	..
High sulphur fuel oil	..	99.47	110.35	209.68	330.54	365.32	475.63	644.49	689.02
Low sulphur fuel oil	..	..	..	216.62	326.78	400.04	510.18	706.19	761.45
Natural gas	..	..	..	..	305.16	372.93	474.66	544.63	..
				(1 000 Korean wons / unit)(2)					
For electricity generation									
Steam coal	..	..	..	..	56.43	105.23	..	..	..
Heavy fuel oil	..	92.59	111.92	311.68	..	..	..	..	..
Natural gas	..	..	..	..	375.83	577.12	..	..	..
For industry									
Steam coal	..	40.74	40.74	62.21	64.48	113.14	..	..	..
Coking coal	..	..	..	..	112.53	203.67	..	..	..
High sulphur fuel oil	..	92.59	111.92	311.68	445.09	612.35	722.52	938.67	1019.95
Low sulphur fuel oil	..	..	..	322.00	440.03	670.55	774.99	1028.53	1127.16
Natural gas	..	..	..	..	401.85	611.32	705.13	775.73	..

(1) Earliest year for which split by coal type is available.

(2) For steam coal, coking coal and heavy fuel oil per metric tonne; for natural gas per 10^7 kilocalories GCV.

* Please refer to notes and definitions in Part I. Steam coal prices relate to historical definitions.

Source: *IEA/OECD Energy Prices & Taxes*

8. Coal and peat trade by type of coal*

(Mtce)

	1978	1990	1995	2000	2005	2009	2010	2011	2012e
Total imports	**2.20 e**	**22.46 e**	**40.22 e**	**55.91 e**	**67.04**	**89.89**	**104.21**	**113.55**	**110.50**
Bituminous coal(1)	0.15 e	10.73 e	22.45 e	33.82 e	45.24	67.17	72.79	79.05	77.97
Coking coal	1.95 e	11.13 e	16.17	18.42	19.45	19.92	27.15	31.08	30.14
Sub-bituminous coal	-	0.60	1.59	3.67	2.04	2.54	3.64	3.01	2.02
Lignite	-	-	-	-	-	-	-	-	-
Peat	-	-	-	-	-	-	-	-	-
Coal products(2)	0.10	-	-	-	0.32	0.26	0.63	0.41	0.37
Total exports	**0.00**	**-**	**-**	**-**	**-**	**-**	**-**	**-**	**-**
Bituminous coal(1)	-	-	-	-	-	-	-	-	-
Coking coal	-	-	-	-	-	-	-	-	-
Sub-bituminous coal	-	-	-	-	-	-	-	-	-
Lignite	-	-	-	-	-	-	-	-	-
Peat	-	-	-	-	-	-	-	-	-
Coal products(2)	0.00	-	-	-	-	-	-	-	-

(1) Bituminous coal includes anthracite.

(2) Coal products includes products derived from coal, for example: coke, coal tar, briquettes, patent fuels.

* Please refer to notes and definitions in Part I.

Source: *IEA/OECD Energy Balances of OECD Countries, IEA/OECD Energy Statistics of OECD Countries*

KOREA

9. Total coal imports by origin*

(thousand tonnes)

	1978(1)	1985	1990	1995	2000	2005	2010	2011	2012e
Total coal(2)	**2169**	**17131**	**23729 e**	**45831 e**	**64895 e**	**76758**	**118591**	**129150**	**125535**
Coking coal	**2009**	**8125**	**11287 e**	**17151 e**	**19575 e**	**20627**	**28160**	**32234**	**31256**
Australia	1150	3625	5053 e	8410 e	10641 e	11664	16445	16281	15787
Canada	409	2100	2026 e	4251 e	4097 e	4315	4944	6560	6361
Czech Republic	-	-	-	-	-	-	-	-	-
Germany	-	-	-	-	-	-	-	-	-
Poland	-	-	-	-	-	-	-	-	-
United Kingdom	-	-	-	-	-	-	-	-	-
United States	450	1900	2908 e	2324 e	1419 e	901	3238	5213	5055
Other OECD	-	-	-	7 e	-	-	-	-	-
China, People's Rep.	-	-	-	1072 e	2781 e	3258	1936	1922	1864
Colombia	-	-	-	-	-	-	-	-	-
Indonesia	-	-	-	44 e	209 e	-	44	-	-
South Africa	-	500	100 e	696 e	-	-	-	-	-
Former Soviet Union(3)	-	-	x	x	x	x	x	x	x
Russian Federation (4)	*x*	*x*	*1200*	*347 e*	*394 e*	*489*	*1553*	*2258*	*2189*
Other FSU (4)	*x*	*x*	-	-	-	-	-	-	-
Venezuela	-	-	-	-	-	-	-	-	-
Vietnam	-	-	-	-	-	-	-	-	-
Non-specified/other	-	-	-	-	34 e	-	-	-	-
Steam coal	**160**	**9006**	**12442 e**	**28680 e**	**45320 e**	**56131**	**90431**	**96916**	**94279**
Australia	-	5100	3506 e	9464 e	12019 e	19207	26512	28470	27339
Canada	-	1469	1250 e	2156 e	1647 e	4	4980	7944	7878
Czech Republic	-	-	-	-	-	-	-	-	-
Germany	-	-	-	1 e	-	-	-	-	-
Poland	-	-	-	-	-	-	-	-	-
United Kingdom	-	-	-	-	-	-	-	-	-
United States	-	1020	1235 e	2150 e	166 e	365	66	854	861
Other OECD	-	-	-	4 e	62 e	56	-	-	-
China, People's Rep.	-	-	1000 e	7209 e	21281 e	17580	5356	3254	3167
Colombia	-	-	-	-	-	-	-	306	303
Indonesia	-	-	397 e	3375 e	5277 e	15382	40126	40284	38963
South Africa	-	1417	5054 e	3535 e	2503 e	-	2346	3030	2996
Former Soviet Union(3)	-	-	x	x	x	x	x	x	x
Russian Federation (4)	*x*	*x*	-	*533 e*	*2197 e*	*3018*	*7004*	*10474*	*10587*
Other FSU (4)	*x*	*x*	-	-	-	-	-	-	-
Venezuela	-	-	-	-	-	-	-	-	-
Vietnam	-	-	-	223 e	135 e	358	1863	1657	1557
Non-specified/other	160	-	-	30 e	33	161	2178	643	628
Lignite	-	-	-	-	-	-	-	-	-

(1) Earliest year for which split by coal type is available.

(2) Total coal consists of steam coal, coking coal and lignite.

(3) For years prior to 1990, or if no details for individual Former Soviet Union republics.

(4) Data for former republics are not available separately for some OECD member countries.

* Please refer to notes and definitions in Part I.

Source: IEA/OECD Coal Statistics

KOREA

12. Coal import values by origin*

(average unit value, CIF, USD/tonne)

	1990	1995	2000	2005	2007	2008	2009	2010	2011/12(1)
Coking coal(2)	..	..	**42.41**	**110.75**	**105.43**	**198.33**	**135.92**	**181.64**	..
Imports from:									
Australia	..	..	40.71	106.89	103.16	212.42	158.33	178.96	..
Canada	..	..	44.36	106.32	109.12	225.52	168.02	182.70	..
Czech Republic	..	..	..	..	..	..	..	..	..
Poland	..	..	..	..	..	..	..	..	..
United States	..	..	55.50	138.42	123.05	301.67	175.06	193.11	..
China	..	..	39.97	121.09	107.69	275.89	161.91	198.24	..
Colombia	..	..	..	..	..	..	..	170.02	..
Indonesia	..	..	33.76	..	..	266.61	279.45	256.09	..
South Africa	..	..	..	..	..	..	..	..	..
Former Soviet Union(3)	..	..	41.71	124.34	102.14	280.59	121.84	168.55	..
Other bituminous coal(4)	..	..	**28.67**	**55.76**	**61.19**	**88.66**	**79.76**	**86.26**	..
Imports from:									
Australia	..	..	32.70	58.85	60.28	93.33	91.04	98.63	..
Canada	..	..	32.59	95.90	67.51	83.70	80.70	87.56	..
Czech Republic	..	..	..	..	..	..	..	..	..
Poland	..	..	..	..	..	..	..	..	..
United States	..	..	23.75	31.96	29.67	95.92	47.11	45.66	..
China	..	..	28.72	59.44	69.56	113.59	106.18	102.71	..
Colombia	..	..	..	..	..	..	..	93.74	..
Indonesia	..	..	25.90	46.50	53.97	78.62	68.47	74.17	..
South Africa	..	..	32.05	..	71.15	76.36	104.72	96.98	..
Former Soviet Union(3)	..	..	28.59	58.14	71.32	116.34	89.02	94.94	..

Note: On occasion, shipment of extremely small quantities of high valued coal results in high import costs.

(1) Import data for steam and coking coals are currently unavailable for 2011 and 2012 due to resource constraints.

(2) Weighted average of individual countries using import volumes as weights. Prices exclude intra-EU trade.

(3) Former Soviet Union until 1991, Russian Federation starting in 1992.

(4) Bituminous steam coal only. Weighted average of trades. Prices exclude intra-EU trade.

* Please refer to notes and definitions in Part I.

Source: IEA/OECD Coal Statistics

LUXEMBOURG*

Figure 1: TPES by fuel (Mtce)

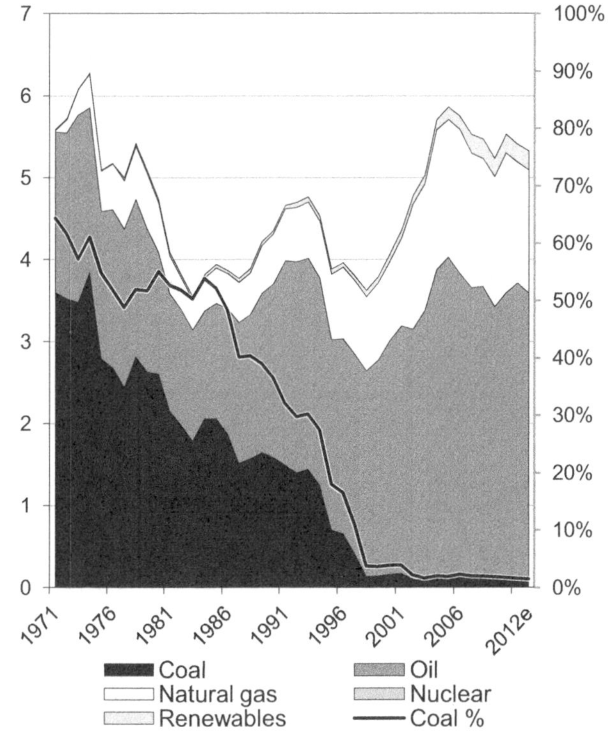

Figure 2: Coal supply indicators (1971 = 100)

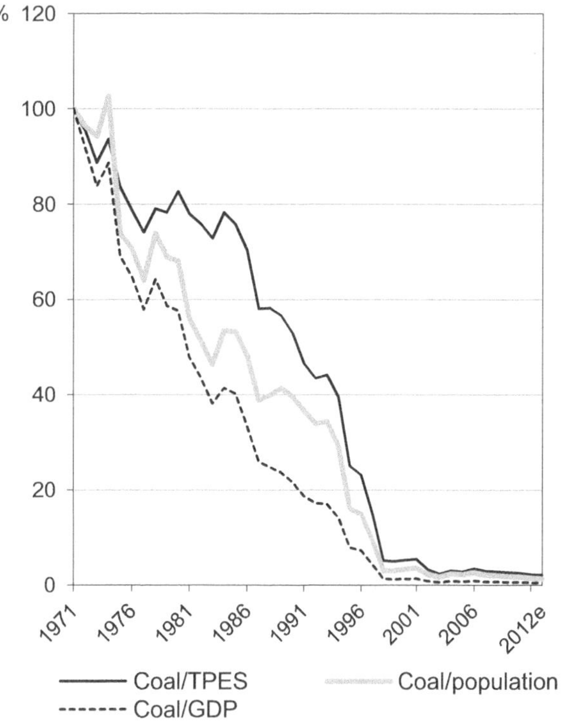

Figure 3: Primary coal supply (Mtce)

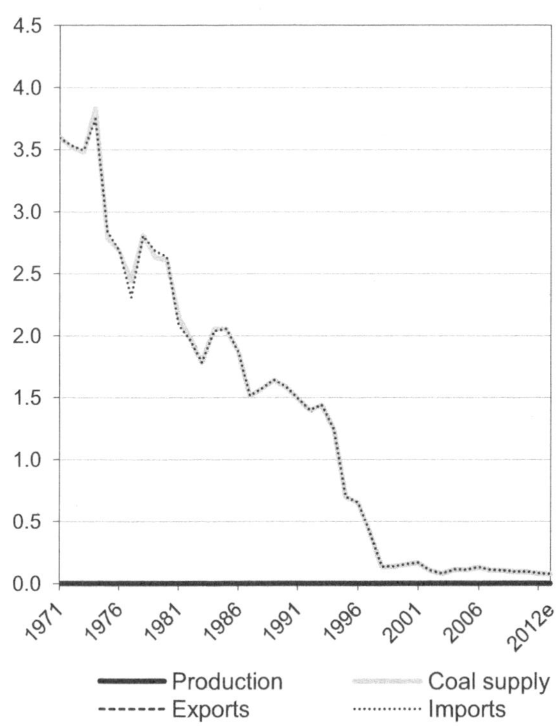

Figure 4: Coal consumption (Mtce)

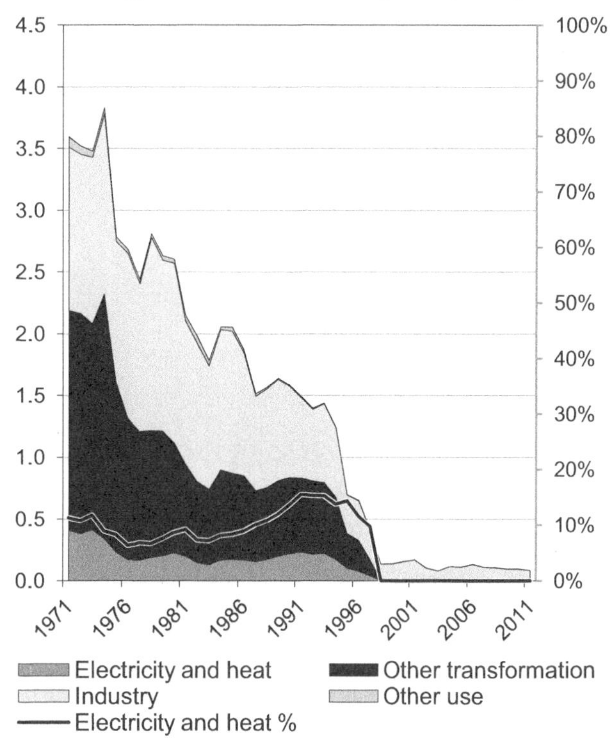

*Please refer to notes and definitions in Part I. Peat is included in coal for all figures.

Source: *IEA/OECD Coal Information, IEA/OECD Energy Balances of OECD Countries, OECD Main Economic Indicators*

LUXEMBOURG*

Figure 5: Electricity generation by fuel (TWh)

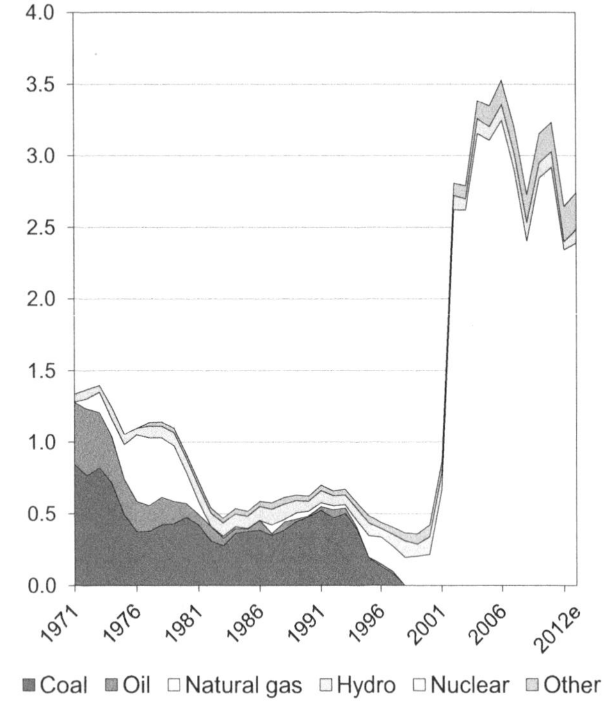

Figure 6: CO_2 emissions by fuel (Mt CO_2)

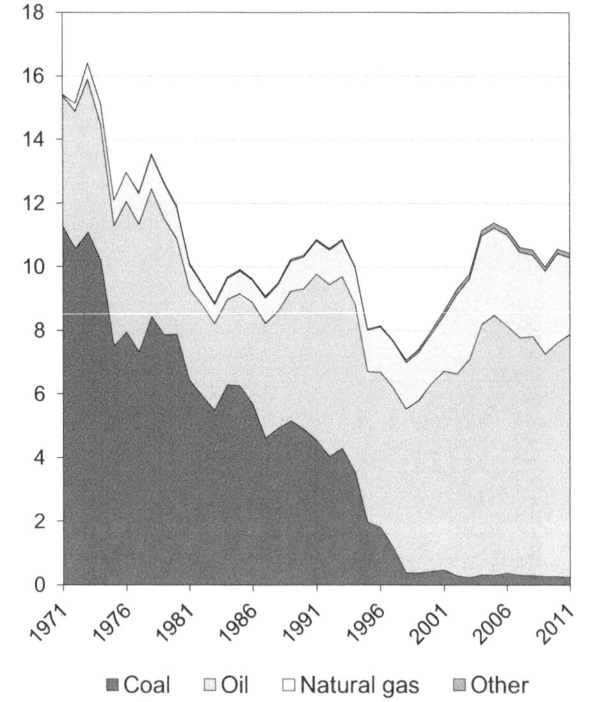

Figure 7: Electricity generation by fuel share

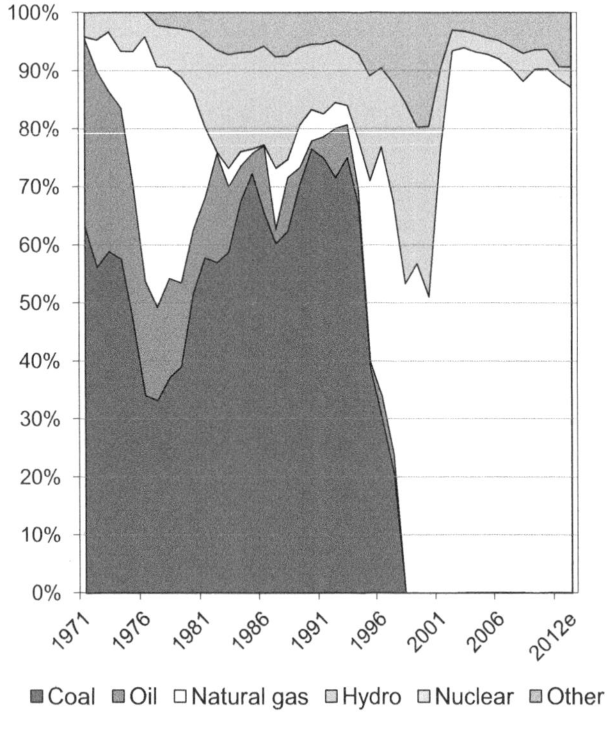

Figure 8: CO_2 emissions by fuel share

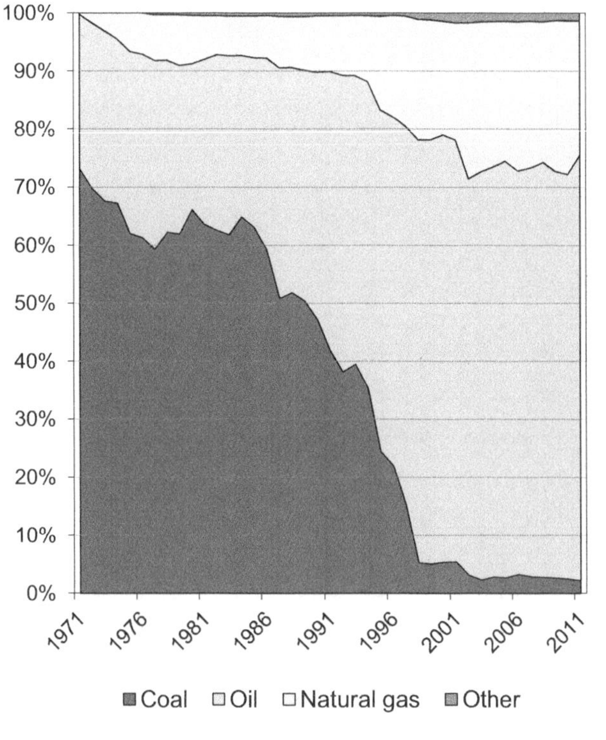

*Please refer to notes and definitions in Part I. Peat is included in coal for all figures.

Source: *IEA/OECD Electricity Information, IEA/OECD CO2 Emissions from Fuel Combustion*

LUXEMBOURG

1. Total primary energy supply (TPES) by fuel*

	1973	1980	1990	2000	2010	2011	2012e	Average annual percent change 73-90	90-11
TPES (Mtce)	**6.33**	**5.08**	**4.84**	**4.76**	**6.03**	**5.96**	**5.83**	**-1.57**	**0.99**
Coal and peat	3.48	2.61	1.59	0.16	0.10	0.08	0.08	-4.51	-13.10
Oil	2.29	1.48	2.12	2.86	3.50	3.63	3.51	-0.45	2.60
Natural Gas	0.31	0.61	0.61	0.96	1.71	1.47	1.50	4.05	4.26
Biofuels and waste	-	0.03	0.03	0.07	0.20	0.20	0.21	-	9.10
Nuclear	-	-	-	-	-	-	-	-	-
Hydro	0.01	0.01	0.01	0.02	0.01	0.01	0.01	2.37	-0.89
Geothermal	-	-	-	-	-	-	-	-	-
Solar, wind, tide	-	-	-	0.00	0.01	0.01	0.01	-	-
Net electricity trade(1)	0.25	0.35	0.48	0.70	0.50	0.55	0.50	3.95	0.63
Heat(2)	-	-	-	-	-	-	-	-	-

(1) Net trade between OECD and non-OECD countries.

(2) Ambient heat from heat pumps used in the transformation sector.

Source: *IEA/OECD Energy Balances of OECD Countries*

2. Energy supply, GDP and population*

	1973	1980	1990	2000	2010	2011	2012e	Average annual percent change 73-90	90-11
GDP (2005 bil. USD)	11	12	19	32	41	42	42	3.40	3.76
TPES/GDP(1)	0.58	0.43	0.25	0.15	0.15	0.14	0.14	-4.81	-2.67
Population (millions)	0.4	0.4	0.4	0.4	0.5	0.5	0.5	0.50	1.47
TPES/population(1)	18.04	13.97	12.67	10.90	11.87	11.48	11.00	-2.06	-0.47
TPES/GDP(2)	347.9	256.5	150.6	90.7	87.9	85.4	83.3	-4.81	-2.67
Coal and peat supply/GDP(2)	10768.2	7405.7	2780.8	167.0	78.4	67.2	62.6	-7.65	-16.25
Elec. consumption/GDP(2)	165.9	183.6	131.4	111.9	97.6	94.7	..	-1.36	-1.55
Elec. generation (TWh)	1	1	1	0	3	3	3	-4.62	7.12
Industrial production(2)	53.5	46.5	66.6	83.8	88.8	86.7	82.2	1.30	1.26

(1) TPES/GDP in units of tce/2005 thousand USD, TPES/population in units of tce/capita.

(2) As index, 2005 (not shown) = 100

Source: *IEA/OECD Energy Balances of OECD Countries , OECD Main Economic Indicators*

3. Coal and peat production by type*

	1978(1)	1990	2000	2005	2010	2011	2012e	Average annual percent change 78-90	90-11
Mtce:									
Coking coal	-	-	-	-	-	-	-	-	-
Steam coal	-	-	-	-	-	-	-	-	-
Lignite	-	-	-	-	-	-	-	-	-
Peat	-	-	-	-	-	-	-	-	-
Mt:									
Coking coal	-	-	-	-	-	-	-	-	-
Steam coal	-	-	-	-	-	-	-	-	-
Lignite	-	-	-	-	-	-	-	-	-
Peat	-	-	-	-	-	-	-	-	-

(1) Earliest year for which split by coal type is available.

* Please refer to notes and definitions in Part I.

Source: *IEA/OECD Energy Balances of OECD Countries, IEA/OECD Energy Statistics of OECD Countries*

LUXEMBOURG

4. Final consumption of energy by fuel*

(Mtce)

	1973	1980	1990	2000	2009	2010	2011	Average annual percent change 73-90	90-11
Total final consumption[(1)]	**4.10**	**3.87**	**3.97**	**4.62**	**5.23**	**5.56**	**5.56**	**-0.20**	**1.62**
Coal and peat	1.40	1.49	0.75	0.16	0.10	0.10	0.08	-3.59	-9.94
Oil	2.09	1.44	2.11	2.85	3.32	3.49	3.63	0.05	2.61
Natural Gas	0.25	0.51	0.60	0.86	0.88	0.97	0.87	5.21	1.76
Biofuels and wastes	-	-	-	0.03	0.14	0.15	0.15	-	-
Geothermal	-	-	-	-	-	-	-	-	-
Solar, wind, tide	-	-	-	-	0.00	0.00	0.00	-	-
Electricity	0.36	0.44	0.51	0.71	0.75	0.81	0.80	1.99	2.15
Heat	-	-	-	0.01	0.04	0.04	0.04	-	-
of which:									
Total industry	**2.94**	**2.34**	**1.87**	**1.02**	**0.93**	**1.04**	**0.95**	**-2.62**	**-3.18**
Coal and peat	1.35	1.45	0.74	0.15	0.09	0.09	0.08	-3.46	-9.93
Oil	1.10	0.24	0.39	0.06	0.01	0.02	0.02	-5.94	-13.78
Natural Gas	0.20	0.35	0.40	0.40	0.39	0.42	0.39	4.17	-0.10
Biofuels and wastes	-	-	-	0.01	0.05	0.06	0.06	-	-
Geothermal	-	-	-	-	-	-	-	-	-
Solar, wind, tide	-	-	-	-	-	-	-	-	-
Electricity	0.29	0.30	0.35	0.40	0.38	0.44	0.40	1.02	0.70
Heat	-	-	-	-	0.01	0.01	0.00	-	-
Total transport	**0.33**	**0.62**	**1.26**	**2.30**	**2.96**	**3.12**	**3.32**	**8.12**	**4.73**
Coal and peat	-	-	-	-	-	-	-	-	-
Oil	0.33	0.61	1.25	2.29	2.88	3.05	3.24	8.16	4.64
Natural Gas	-	-	-	-	-	-	-	-	-
Biofuels and wastes	-	-	-	-	0.06	0.06	0.06	-	-
Electricity	0.00	0.01	0.01	0.01	0.01	0.01	0.01	3.05	3.78
Residential	**0.73**	**0.76**	**0.71**	**0.67**	**0.74**	**0.73**	**0.61**	**-0.10**	**-0.75**
Coal and peat	0.05	0.03	0.01	0.00	0.00	0.00	0.00	-9.35	-11.20
Oil	0.60	0.51	0.43	0.32	0.32	0.27	0.22	-1.86	-3.27
Natural Gas	0.05	0.16	0.20	0.22	0.28	0.33	0.27	8.04	1.39
Biofuels and wastes	-	-	-	0.02	0.02	0.03	0.02	-	-
Geothermal	-	-	-	-	-	-	-	-	-
Solar, wind, tide	-	-	-	-	0.00	0.00	0.00	-	-
Electricity	0.03	0.06	0.07	0.10	0.11	0.10	0.10	5.87	1.84
Heat	-	-	-	-	-	-	-	-	-
Comm & public services	**-**	**0.06**	**0.08**	**0.54**	**0.52**	**0.58**	**0.61**	**-**	**10.24**
Coal and peat	-	-	-	-	-	-	-	-	-
Oil	-	-	-	0.09	0.04	0.09	0.09	-	-
Natural Gas	-	-	-	0.24	0.21	0.22	0.21	-	-
Biofuels and waste	-	-	-	0.00	0.00	0.00	0.00	-	-
Geothermal	-	-	-	-	-	-	-	-	-
Solar, wind, tide	-	-	-	-	-	-	-	-	-
Electricity	-	0.06	0.08	0.20	0.24	0.24	0.27	-	6.16
Heat	-	-	-	0.01	0.03	0.03	0.03	-	-
Non-energy use	**0.04**	**0.06**	**0.03**	**0.06**	**0.03**	**0.04**	**0.04**	**-1.93**	**1.19**
Coal and peat	-	-	-	-	-	-	-	-	-
Oil	0.04	0.06	0.03	0.06	0.03	0.04	0.04	-1.93	1.19
Natural Gas	-	-	-	-	-	-	-	-	-

(1) Total Final Consumption (TFC) excludes use in transformation processes and energy industry own use.

* Please refer to notes and definitions in Part I. Agriculture, Forestry, Fishing and "Non-specified other" sectors are not reported here individually, but are accounted for in Total final consumption.

Source: *IEA/OECD Energy Balances of OECD Countries*

LUXEMBOURG

5. Coal balance*

(Mtce)

	1973	1980	1990	2000	2005	2010	2011	2012e	Average annual percent change 73-90	90-11
Production	-	-	-	-	-	-	-	-	-	-
Imports	3.5	2.6	1.6	0.2	0.1	0.1	0.1	0.1	-4.5	-13.1
Exports	-	-	-	-	-	-	-	-	-	-
Stock changes	-0.0	-0.0	-	-	-	-	-	-		
Primary supply	**3.5**	**2.6**	**1.6**	**0.2**	**0.1**	**0.1**	**0.1**	**0.1**	**-4.5**	**-13.1**
Statistical differences	**0.0**	**-**	**-**	**-**	**-**	**-**	**-**	**..**		
Total transformation	**-1.9 e**	**-1.1 e**	**-0.8**	**-**	**-**	**-**	**-**	**..**	**-5.1**	**-**
Electricity and heat gen.	-0.4	-0.2	-0.2	-	-	-	-	..	-3.7	-
Main activity producers (1)	*-0.4*	-	-	-	-	-	-	..	-	-
Autoproducers	-	*-0.2*	*-0.2*	-	-	-	-	..	-	-
Gas works	-	-	-	-	-	-	-	..	-	-
Coal transformation(2)	-1.5 e	-0.8 e	-0.6	-	-	-	-	..	-5.5	-
BKB plants	-	-	-	-	-	-	-	..	-	-
Blast furnaces	*-1.5 e*	*-0.8 e*	*-0.6*	-	-	-	-	..	*-5.5*	-
Coke ovens	-	-	-	-	-	-	-	..	-	-
Patent fuel plants	-	-	-	-	-	-	-	..	-	-
Other transformation(3)	-	-	-	-	-	-	-	..	-	-
Energy ind. own use	**-**	**-**	**-**	**-**	**-**	**-**	**-**	**..**	**-**	**-**
Losses	**-0.1**	**-0.1**	**-0.0**	**-**	**-**	**-**	**-**	**..**		
Final consumption(4)	**1.4**	**1.5**	**0.7**	**0.2**	**0.1**	**0.1**	**0.1**	**..**	**-3.6**	**-9.9**
Industry(5)	1.3	1.5	0.7	0.2	0.1	0.1	0.1	..	-3.5	-9.9
Iron and steel	*1.3*	*1.3 e*	*0.6*	*0.0*	*0.0*	*0.0*	*0.0*	..	*-4.5*	*-16.0*
Chemical	-	-	-	-	-	-	-	..	-	-
Non-metallic minerals	-	*0.1*	*0.1*	*0.1*	*0.1*	*0.1*	*0.1*	..	-	*-3.6*
Paper, pulp and print	-	-	-	-	-	-	-	..	-	-
Other industry (6)	*0.0*	-	*0.0*	*0.0*	*0.0*	*0.0*	*0.0*	..	*-7.5*	*-0.6*
Transport(7)	-	-	-	-	-	-	-	..	-	-
Other	0.0	0.0	0.0	0.0	0.0	0.0	0.0	..	-9.3	-11.2
Comm. and pub. services	-	-	-	-	-	-	-	..	-	-
Residential	*0.0*	*0.0*	*0.0*	*0.0*	*0.0*	*0.0*	*0.0*	..	*-9.3*	*-11.2*
Other sectors (8)	-	-	-	-	-	-	-	..	-	-
Non-energy use	-	-	-	-	-	-	-	..	-	-

(1) Main activity electricity and heat generation includes district heating.

(2) Coal transformation refers to the transformation of coal and peat to secondary and tertiary products (mainly cokes, briquettes, coal tar and off-gases). Gas works and Liquefaction are shown separately to match the *Energy Balances of OECD Countries.*

(3) Other transformation includes Liquefaction and Non-specified transformations.

(4) Final Consumption includes non-energy use and energy use (Industry, Transport and Other).

(5) Please refer to notes and definitions in Part I for detailed explanation.

(6) Other industry includes Non-ferrous metals, Transport equipment, Machinery, Mining and quarrying, Food and tobacco, Wood and wood products, Construction, Textile and leather, and Non-specified industry.

(7) Transport includes Rail and Inland waterways.

(8) Other sectors includes Agriculture/Forestry, Fishing and Non-specified other.

* "Coal" refers to all types of coal, primary (anthracite, coking coal, other bituminous coal, sub-bituminous coal, lignite/brown coal) and derived fuels (including patent fuel, BKB, coke oven coke, gas coke, coal tar, coke oven gas, gas works gas, blast furnace gas and other recovered gases), as well as peat, peat briquettes and oil shale. Quantities have been converted to Mtce using calorific values reported by the respective countries.
Please refer to notes and definitions in Part I.

Source: *IEA/OECD Energy Balances of OECD Countries*

LUXEMBOURG

6. Use of coal for selected end-uses*

(million tonnes)

	1978	1990	2000	2005	2009	2010	2011	Average annual percent change 78-90	90-11
Steam coal	**0.50**	**0.20**	**0.17**	**0.12**	**0.10**	**0.10**	**0.09**	**-7.48**	**-3.71**
Total electricity and heat	-	-	-	-	-	-	-	-	-
Main activity producers	-	-	-	-	-	-	-	-	-
Autoproducers	-	-	-	-	-	-	-	-	-
Patent fuel/BKB plants	-	-	-	-	-	-	-	-	-
Coke ovens/Liquefaction(1)	-	-	-	-	-	-	-	-	-
Blast furnace inputs	-	-	-	-	-	-	-	-	-
Gas manufacture	-	-	-	-	-	-	-	-	-
Industry	0.50	0.20	0.17	0.12	0.10	0.10	0.09	-7.45	-3.69
Iron and steel	*0.50*	*0.04*	*0.04*	*0.03*	*0.03*	*0.03*	*0.02*	*-19.27*	*-4.04*
Chemical	-	-	-	-	-	-	-	-	-
Non-metallic minerals	-	*0.16*	*0.13*	*0.09*	*0.08*	*0.08*	*0.07*	-	*-3.61*
Paper, pulp and print	-	-	-	-	-	-	-	-	-
Other industry	-	-	-	-	-	-	-	-	-
Other sectors(2)	0.01	0.00	-	-	-	-	-	-12.55	-
Non-energy use	-	-	-	-	-	-	-	-	-
Coking coal	**-**	**-**	**-**	**-**	**-**	**-**	**-**	**-**	**-**
Total electricity and heat	-	-	-	-	-	-	-	-	-
Main activity producers	-	-	-	-	-	-	-	-	-
Autoproducers	-	-	-	-	-	-	-	-	-
Patent fuel/BKB plants	-	-	-	-	-	-	-	-	-
Coke ovens/Liquefaction(1)	-	-	-	-	-	-	-	-	-
Blast furnace inputs	-	-	-	-	-	-	-	-	-
Gas manufacture	-	-	-	-	-	-	-	-	-
Industry	-	-	-	-	-	-	-	-	-
Iron and steel	-	-	-	-	-	-	-	-	-
Chemical	-	-	-	-	-	-	-	-	-
Non-metallic minerals	-	-	-	-	-	-	-	-	-
Paper, pulp and print	-	-	-	-	-	-	-	-	-
Other industry	-	-	-	-	-	-	-	-	-
Other sectors(2)	-	-	-	-	-	-	-	-	-
Non-energy use	-	-	-	-	-	-	-	-	-
Lignite	**0.02**	**-**	**-**	**-**	**-**	**-**	**-**	**-**	**-**
Total electricity and heat	-	-	-	-	-	-	-	-	-
Main activity producers	-	-	-	-	-	-	-	-	-
Autoproducers	-	-	-	-	-	-	-	-	-
Patent fuel/BKB plants	-	-	-	-	-	-	-	-	-
Coke ovens/Liquefaction(1)	-	-	-	-	-	-	-	-	-
Gas manufacture	-	-	-	-	-	-	-	-	-
Industry	0.02	-	-	-	-	-	-	-	-
Iron and steel	*0.02*	-	-	-	-	-	-	-	-
Chemical	-	-	-	-	-	-	-	-	-
Non-metallic minerals	-	-	-	-	-	-	-	-	-
Paper, pulp and print	-	-	-	-	-	-	-	-	-
Other industry	-	-	-	-	-	-	-	-	-
Other sectors(2)	-	-	-	-	-	-	-	-	-
Non-energy use	-	-	-	-	-	-	-	-	-

(1) Liquefaction primarily refers to direct distillation processes. Indirect liquefaction may be reported here or under gas manufacture.

(2) Other sectors are Residential, Commercial and public services, Agriculture/Forestry, Fishing and Non-specified other.

* Coal totals may not match the sum of the individual rows due to statistical differences. Please refer to notes and definitions in Part I.

Source: *IEA/OECD Energy Statistics of OECD Countries*

LUXEMBOURG

7. Fuel prices to end users*

	1978(1)	1990	1995	2000	2005	2009	2010	2011	2012
				(US dollars / tce)					
For electricity generation									
Steam coal	..	..	..	..	..	..	..	..	..
Heavy fuel oil	..	..	..	..	..	..	..	..	..
Natural gas	..	..	..	..	..	..	..	..	..
For industry									
Steam coal	..	..	..	..	..	..	..	..	..
Coking coal	x	x	x	x	x	x	x	x	x
High sulphur fuel oil	..	94.73	x	x	x	x	x	x	x
Low sulphur fuel oil	..	..	108.68	130.60	..	..	..	..	..
Natural gas	..	..	..	..	..	361.19	349.74	452.54	457.09
				(Euro / unit)(2)					
For electricity generation									
Steam coal	..	..	..	..	..	..	..	..	..
Heavy fuel oil	..	..	..	..	..	..	..	..	..
Natural gas	..	..	..	..	..	..	..	..	..
For industry									
Steam coal	..	..	..	..	..	..	..	..	..
Coking coal	x	x	x	x	x	x	x	x	x
High sulphur fuel oil	..	108.26	x	x	x	x	x	x	x
Low sulphur fuel oil	..	..	109.52	195.34	..	..	..	..	..
Natural gas	..	..	..	..	..	334.27	339.50	418.17	457.22

(1) Earliest year for which split by coal type is available.

(2) For steam coal, coking coal and heavy fuel oil per metric tonne; for natural gas per 10^7 kilocalories GCV.

* Please refer to notes and definitions in Part I. Steam coal prices relate to historical definitions.

Source: *IEA/OECD Energy Prices & Taxes*

8. Coal and peat trade by type of coal*

(Mtce)

	1978	1990	1995	2000	2005	2009	2010	2011	2012e
Total imports	**2.80**	**1.59**	**0.70**	**0.16**	**0.11**	**0.10**	**0.10**	**0.08**	**0.08**
Bituminous coal(1)	0.48	0.16	0.18	0.14	0.10	0.09	0.09	0.08	0.07
Coking coal	-	-	-	-	-	-	-	-	-
Sub-bituminous coal	-	-	-	-	-	-	-	-	-
Lignite	0.01	-	-	-	-	-	-	-	-
Peat	-	-	-	-	-	-	-	-	-
Coal products(2)	2.31	1.42	0.52	0.01	0.01	0.01	0.01	0.01	0.01
Total exports	**-**	**-**	**-**	**-**	**-**	**-**	**-**	**-**	**-**
Bituminous coal(1)	-	-	-	-	-	-	-	-	-
Coking coal	-	-	-	-	-	-	-	-	-
Sub-bituminous coal	-	-	-	-	-	-	-	-	-
Lignite	-	-	-	-	-	-	-	-	-
Peat	-	-	-	-	-	-	-	-	-
Coal products(2)	-	-	-	-	-	-	-	-	-

(1) Bituminous coal includes anthracite.

(2) Coal products includes products derived from coal, for example: coke, coal tar, briquettes, patent fuels.

* Please refer to notes and definitions in Part I.

Source: *IEA/OECD Energy Balances of OECD Countries, IEA/OECD Energy Statistics of OECD Countries*

LUXEMBOURG

9. Total coal imports by origin*

(thousand tonnes)

	1978[1]	1985	1990	1995	2000	2005	2010	2011	2012e
Total coal[2]	**512**	**199**	**197**	**217**	**172**	**122**	**102**	**89**	**82**
Coking coal	**-**	**-**	**-**	**-**	**-**	**-**	**-**	**-**	**-**
Australia	-	-	-	-	-	-	-	-	-
Canada	-	-	-	-	-	-	-	-	-
Czech Republic	-	-	-	-	-	-	-	-	-
Germany	-	-	-	-	-	-	-	-	-
Poland	-	-	-	-	-	-	-	-	-
United Kingdom	-	-	-	-	-	-	-	-	-
United States	-	-	-	-	-	-	-	-	-
Other OECD	-	-	-	-	-	-	-	-	-
China, People's Rep.	-	-	-	-	-	-	-	-	-
Colombia	-	-	-	-	-	-	-	-	-
Indonesia	-	-	-	-	-	-	-	-	-
South Africa	-	-	-	-	-	-	-	-	-
Former Soviet Union[3]	-	-	x	x	x	x	x	x	x
Russian Federation [4]	x	x	-	-	-	-	-	-	-
Other FSU [4]	x	x	-	-	-	-	-	-	-
Venezuela	-	-	-	-	-	-	-	-	-
Vietnam	-	-	-	-	-	-	-	-	-
Non-specified/other	-	-	-	-	-	-	-	-	-
Steam coal	**495**	**199**	**197**	**217**	**172**	**122**	**102**	**89**	**82**
Australia	-	52	-	-	-	-	-	-	-
Canada	-	-	-	-	-	-	-	-	-
Czech Republic	-	-	-	-	-	-	-	-	-
Germany	299	33	-	-	-	-	-	-	-
Poland	-	-	-	-	-	-	-	-	-
United Kingdom	12	-	-	-	-	-	-	-	-
United States	1	-	-	-	-	-	-	-	-
Other OECD	6	32	-	-	-	-	-	-	12
China, People's Rep.	-	-	-	-	-	-	-	-	-
Colombia	-	-	-	-	-	-	-	-	-
Indonesia	-	-	-	-	-	-	-	-	-
South Africa	125	72	138	107	125	76	76	73	70
Former Soviet Union[3]	-	10	x	x	x	x	x	x	x
Russian Federation [4]	x	x	-	-	-	-	-	-	-
Other FSU [4]	x	x	-	-	-	-	-	-	-
Venezuela	-	-	-	-	-	-	-	-	-
Vietnam	-	-	-	-	-	-	-	-	-
Non-specified/other	52	-	59	110	47	46	26	16	-
Lignite	**17**	**-**	**-**	**-**	**-**	**-**	**-**	**-**	**-**

(1) Earliest year for which split by coal type is available.

(2) Total coal consists of steam coal, coking coal and lignite.

(3) For years prior to 1990, or if no details for individual Former Soviet Union republics.

(4) Data for former republics are not available separately for some OECD member countries.

* Please refer to notes and definitions in Part I.

Source: IEA/OECD Coal Statistics

MEXICO*

Figure 1: TPES by fuel (Mtce)

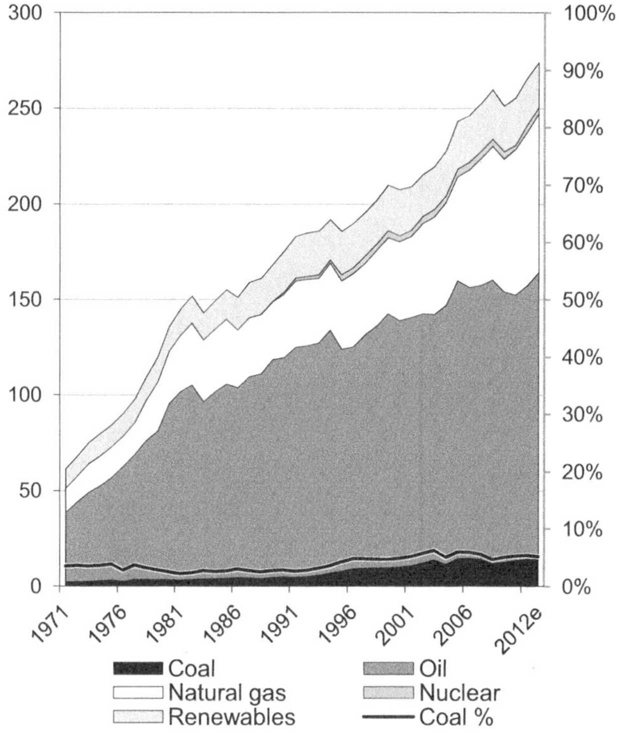

Figure 2: Coal supply indicators (1971 = 100)

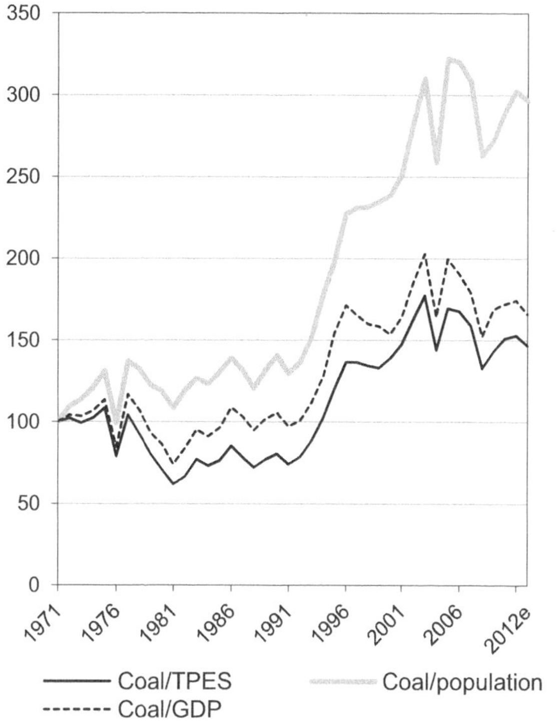

Figure 3: Primary coal supply (Mtce)

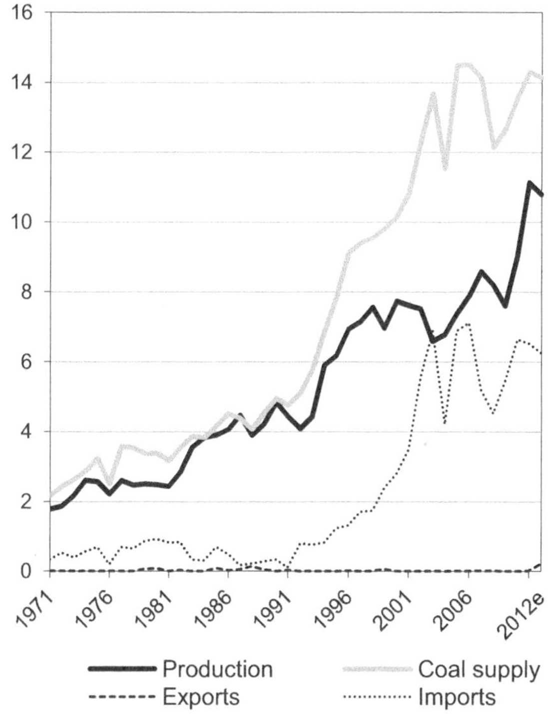

Figure 4: Coal consumption (Mtce)

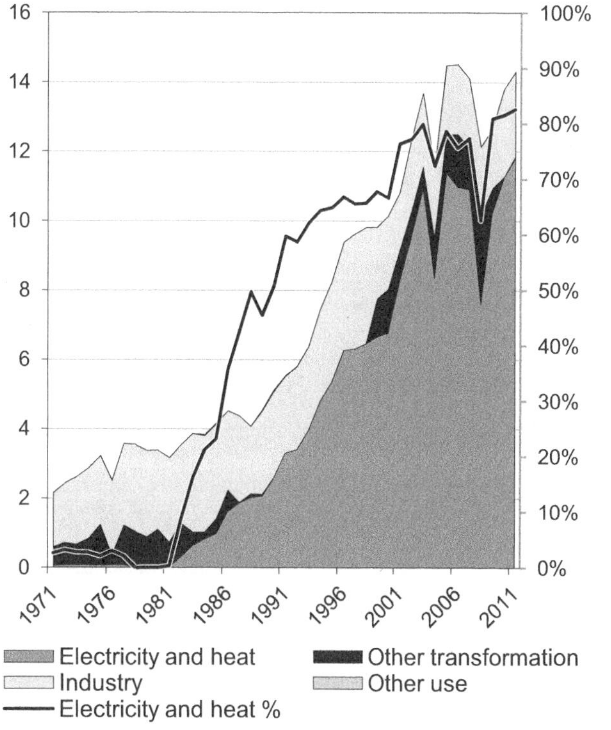

*Please refer to notes and definitions in Part I. Peat is included in coal for all figures.

Source: *IEA/OECD Coal Information, IEA/OECD Energy Balances of OECD Countries, OECD Main Economic Indicators*

MEXICO*

Figure 5: Electricity generation by fuel (TWh)

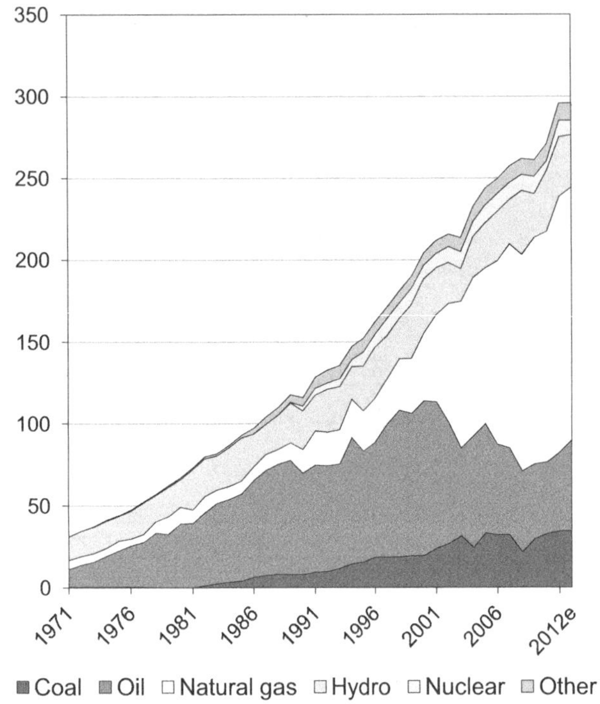

Figure 6: CO_2 emissions by fuel (Mt CO_2)

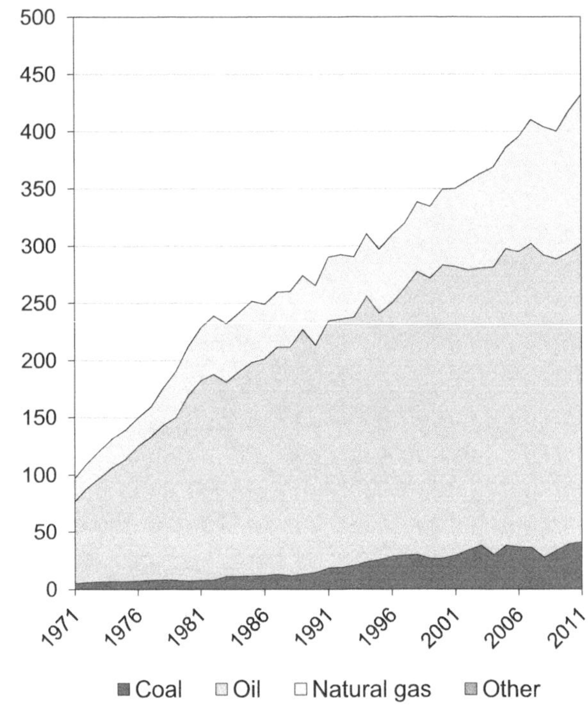

Figure 7: Electricity generation by fuel share

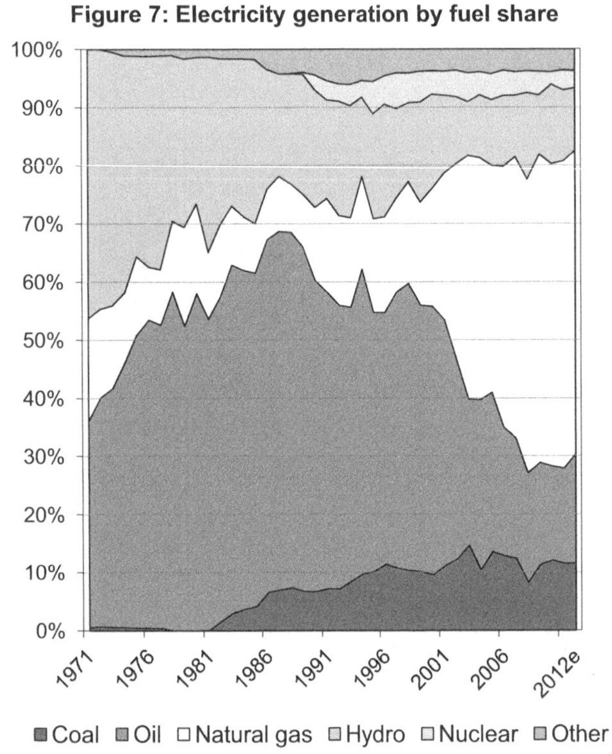

Figure 8: CO_2 emissions by fuel share

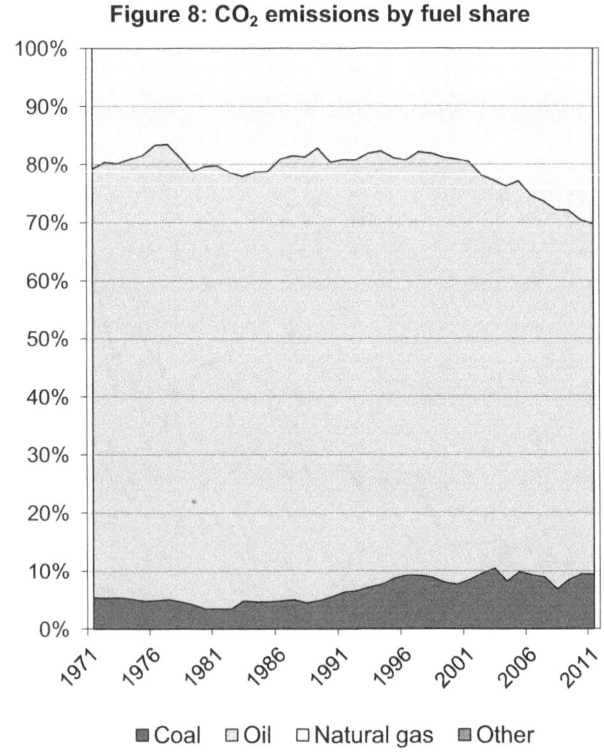

*Please refer to notes and definitions in Part I. Peat is included in coal for all figures.

Source: *IEA/OECD Electricity Information, IEA/OECD CO2 Emissions from Fuel Combustion*

MEXICO

1. Total primary energy supply (TPES) by fuel*

	1973	1980	1990	2000	2010	2011	2012e	Average annual percent change 73-90	90-11
TPES (Mtce)	**75.10**	**135.88**	**174.99**	**207.69**	**255.61**	**265.96**	**274.18**	**5.10**	**2.01**
Coal and peat	2.62	3.38	4.95	10.15	13.58	14.31	14.14	3.81	5.18
Oil	46.39	92.07	114.66	128.90	138.91	142.99	149.89	5.47	1.06
Natural Gas	14.99	27.33	33.03	41.32	76.09	80.31	83.01	4.76	4.32
Biofuels and waste	8.87	9.82	12.22	12.77	11.95	11.81	12.06	1.90	-0.16
Nuclear	-	-	1.09	3.06	2.19	3.76	3.27	-	6.05
Hydro	1.99	2.07	2.88	4.07	4.56	4.46	3.91	2.21	2.09
Geothermal	0.20	1.12	6.29	7.25	8.13	7.99	7.14	22.57	1.14
Solar, wind, tide	-	-	0.03	0.07	0.32	0.41	0.63	-	14.19
Net electricity trade(1)	0.04	0.08	-0.17	0.11	-0.12	-0.07	0.11	x	-4.10
Heat(2)	-	-	-	-	-	-	-	-	-

(1) Net trade between OECD and non-OECD countries.

(2) Ambient heat from heat pumps used in the transformation sector.

Source: *IEA/OECD Energy Balances of OECD Countries*

2. Energy supply, GDP and population*

	1973	1980	1990	2000	2010	2011	2012e	Average annual percent change 73-90	90-11
GDP (2005 bil. USD)	296	458	548	771	921	957	994	3.68	2.69
TPES/GDP(1)	0.25	0.30	0.32	0.27	0.28	0.28	0.28	1.37	-0.66
Population (millions)	53.3	65.7	81.3	98.3	108.3	109.2	110.2	2.51	1.42
TPES/population(1)	1.41	2.07	2.15	2.11	2.36	2.44	2.49	2.52	0.59
TPES/GDP(2)	88.2	103.2	111.1	93.7	96.6	96.7	95.9	1.37	-0.66
Coal and peat supply/GDP(2)	51.7	43.1	52.8	76.9	86.1	87.3	83.1	0.12	2.42
Elec. consumption/GDP(2)	48.2	56.4	82.7	95.2	102.1	106.7	..	3.23	1.22
Elec. generation (TWh)	37	67	116	204	271	296	296	6.93	4.57
Industrial production(2)	..	55.5	66.1	98.1	105.5	109.7	113.6	..	2.44

(1) TPES/GDP in units of tce/2005 thousand USD, TPES/population in units of tce/capita.

(2) As index, 2005 (not shown) = 100

Source: *IEA/OECD Energy Balances of OECD Countries* , *OECD Main Economic Indicators*

3. Coal and peat production by type*

	1978(1)	1990	2000	2005	2010	2011	2012e	Average annual percent change 78-90	90-11
Mtce:									
Coking coal	2.47	2.37	1.77	1.44	1.59	2.04	2.16	-0.34	-0.71
Steam coal	-	2.46	5.96	5.93	7.45	9.08	8.63	-	6.41
Lignite	-	-	-	-	-	-	-	-	-
Peat	-	-	-	-	-	-	-	-	-
Mt:									
Coking coal	3.09	2.96	2.21	1.79	1.59	2.04	2.16	-0.34	-1.75
Steam coal	-	3.97	9.13	8.96	11.25	13.72	13.03	-	6.08
Lignite	-	-	-	-	-	-	-	-	-
Peat	-	-	-	-	-	-	-	-	-

(1) Earliest year for which split by coal type is available.

* Please refer to notes and definitions in Part I.

Source: *IEA/OECD Energy Balances of OECD Countries, IEA/OECD Energy Statistics of OECD Countries*

MEXICO

4. Final consumption of energy by fuel*

(Mtce)

	1973	1980	1990	2000	2009	2010	2011	Average annual percent change 73-90	90-11
Total final consumption[(1)]	**56.77**	**94.17**	**120.45**	**140.41**	**157.46**	**162.31**	**165.81**	**4.52**	**1.53**
Coal and peat	1.96	2.29	2.51	2.14	1.72	2.55	2.46	1.48	-0.10
Oil	31.69	56.70	73.16	88.09	103.58	105.51	105.63	5.04	1.76
Natural Gas	10.38	18.34	20.23	18.50	16.86	18.20	19.49	4.00	-0.18
Biofuels and wastes	8.87	9.82	12.22	11.69	10.36	10.33	10.27	1.90	-0.82
Geothermal	-	-	-	-	-	-	-	-	-
Solar, wind, tide	-	-	0.02	0.06	0.23	0.17	0.20	-	10.45
Electricity	3.88	7.02	12.31	19.93	24.71	25.55	27.76	7.03	3.95
Heat	-	-	-	-	-	-	-	-	-
of which:									
Total industry	**20.16**	**31.75**	**37.93**	**39.88**	**37.08**	**39.38**	**41.69**	**3.79**	**0.45**
Coal and peat	1.96	2.29	2.48	2.14	1.72	2.55	2.46	1.40	-0.04
Oil	5.00	8.61	10.64	10.68	8.83	8.35	8.43	4.54	-1.10
Natural Gas	9.09	14.93	15.46	12.86	11.78	12.88	14.13	3.17	-0.43
Biofuels and wastes	1.89	2.20	2.79	1.97	1.47	1.46	1.42	2.30	-3.15
Geothermal	-	-	-	-	-	-	-	-	-
Solar, wind, tide	-	-	0.00	0.00	0.01	0.01	0.01	-	10.31
Electricity	2.22	3.71	6.56	12.23	13.27	14.14	15.24	6.57	4.10
Heat	-	-	-	-	-	-	-	-	-
Total transport	**17.72**	**32.57**	**40.76**	**51.66**	**72.11**	**74.07**	**74.36**	**5.02**	**2.90**
Coal and peat	-	-	-	-	-	-	-	-	-
Oil	17.68	32.52	40.66	51.51	71.95	73.90	74.21	5.02	2.91
Natural Gas	-	-	-	0.01	0.02	0.02	0.02	-	-
Biofuels and wastes	-	-	-	-	-	-	-	-	-
Electricity	0.04	0.05	0.10	0.14	0.14	0.15	0.14	4.91	1.60
Residential	**12.27**	**15.94**	**22.50**	**25.85**	**25.27**	**25.49**	**25.74**	**3.63**	**0.64**
Coal and peat	-	-	-	-	-	-	-	-	-
Oil	3.68	6.42	9.39	10.91	9.21	9.41	9.31	5.67	-0.04
Natural Gas	0.56	0.67	1.16	0.74	0.99	1.05	1.02	4.40	-0.59
Biofuels and wastes	6.98	7.62	9.43	9.72	8.89	8.86	8.85	1.79	-0.30
Geothermal	-	-	-	-	-	-	-	-	-
Solar, wind, tide	-	-	0.01	0.04	0.13	0.10	0.11	-	10.47
Electricity	1.06	1.23	2.51	4.44	6.05	6.07	6.45	5.17	4.61
Heat	-	-	-	-	-	-	-	-	-
Comm & public services	**1.00**	**2.80**	**2.12**	**4.90**	**5.09**	**5.11**	**5.34**	**4.48**	**4.50**
Coal and peat	-	-	-	-	-	-	-	-	-
Oil	1.00	1.84	0.77	2.47	2.08	2.15	2.15	-1.54	4.99
Natural Gas	-	-	-	0.24	0.29	0.33	0.32	-	-
Biofuels and waste	-	-	-	-	-	-	-	-	-
Geothermal	-	-	-	-	-	-	-	-	-
Solar, wind, tide	-	-	0.01	0.02	0.09	0.06	0.08	-	10.44
Electricity	-	0.97	1.34	2.16	2.63	2.56	2.80	-	3.59
Heat	-	-	-	-	-	-	-	-	-
Non-energy use	**3.36**	**7.12**	**12.97**	**14.07**	**11.29**	**11.66**	**11.44**	**8.27**	**-0.59**
Coal and peat	-	-	0.03	-	-	-	-	-	-
Oil	2.63	4.39	9.32	9.43	7.51	7.73	7.44	7.73	-1.07
Natural Gas	0.73	2.73	3.61	4.64	3.78	3.93	4.00	9.89	0.50

(1) Total Final Consumption (TFC) excludes use in transformation processes and energy industry own use.

* Please refer to notes and definitions in Part I. Agriculture, Forestry, Fishing and "Non-specified other" sectors are not reported here individually, but are accounted for in Total final consumption.

Source: *IEA/OECD Energy Balances of OECD Countries*

MEXICO

5. Coal balance*

(Mtce)

	1973	1980	1990	2000	2005	2010	2011	2012e	Average annual percent change 73-90	90-11
Production	2.2	2.5	4.8	7.7	7.4	9.0	11.1	10.8	4.8	4.0
Imports	0.4	0.9	0.3	2.8	6.9	6.6	6.5	6.3	-0.9	15.2
Exports	-	-0.1	-0.0	-0.0	-0.0	-0.0	-0.0	-0.2	-	0.3
Stock changes	0.1	0.1	-0.2	-0.4	0.2	-2.1	-3.3	-2.7		
Primary supply	**2.6**	**3.4**	**5.0**	**10.2**	**14.5**	**13.6**	**14.3**	**14.1**	**3.8**	**5.2**
Statistical differences	**-0.1**	**-0.2**	**0.1**	**-0.3**	**-0.6**	**1.0**	**1.0**	..		
Total transformation	**-0.5**	**-0.9**	**-2.5 e**	**-7.7 e**	**-11.8 e**	**-11.9 e**	**-12.7 e**	..	**9.7**	**8.1**
Electricity and heat gen.	-0.1	-	-2.6	-6.8	-11.4	-11.3	-11.8	..	23.5	7.5
Main activity producers (1)	*-0.1*	-	*-2.6*	*-6.2*	*-11.2*	*-11.0*	*-11.6*	..	*23.5*	*7.4*
Autoproducers	-	-	-	*-0.5*	*-0.2*	*-0.2*	*-0.2*	..	-	-
Gas works	-	-	-	-	-	-	-	..	-	-
Coal transformation(2)	-0.4	-0.9	0.1 e	-0.9 e	-0.4 e	-0.7 e	-0.9 e	..	-	-
BKB plants	-	-	-	-	-	-	-	..	-	-
Blast furnaces	-	-	-	*-0.6*	*-0.3*	*-0.2*	*-0.2*	..	-	-
Coke ovens	*-0.4*	*-0.9*	*0.1* e	*-0.3* e	*-0.2* e	*-0.4* e	*-0.7* e	..	-	-
Patent fuel plants	-	-	-	-	-	-	-	..	-	-
Other transformation(3)	-	-	-	-	-	-	-	..	-	-
Energy ind. own use	**-0.1**	**-0.0**	**-0.1**	**-0.0**	**-0.0**	**-0.1**	**-0.1**	..	**-1.6**	**3.9**
Losses	**-0.0**	**-0.0**	-	-	-	-	-	..		
Final consumption(4)	**2.0**	**2.3**	**2.5**	**2.1**	**2.0**	**2.5**	**2.5**	..	**1.5**	**-0.1**
Industry(5)	2.0	2.3	2.5	2.1	2.0	2.5	2.5	..	1.4	-0.0
Iron and steel	*2.0*	*2.3*	*2.2* e	*1.9* e	*1.8* e	*2.4* e	*2.3* e	..	*0.7*	*0.3*
Chemical	-	-	-	-	-	-	-	..	-	-
Non-metallic minerals	-	-	*0.0*	*0.0*	*0.1*	*0.1*	*0.1*	..	-	*5.8*
Paper, pulp and print	-	-	-	-	-	-	-	..	-	-
Other industry (6)	-	-	*0.2*	*0.2*	*0.0*	*0.0*	-	..	-	-
Transport(7)	-	-	-	-	-	-	-	..	-	-
Other	-	-	-	-	-	-	-	..	-	-
Comm. and pub. services	-	-	-	-	-	-	-	..	-	-
Residential	-	-	-	-	-	-	-	..	-	-
Other sectors (8)	-	-	-	-	-	-	-	..	-	-
Non-energy use	-	-	0.0	-	-	-	-	..	-	-

(1) Main activity electricity and heat generation includes district heating.

(2) Coal transformation refers to the transformation of coal and peat to secondary and tertiary products (mainly cokes, briquettes, coal tar and off-gases). Gas works and Liquefaction are shown separately to match the *Energy Balances of OECD Countries.*

(3) Other transformation includes Liquefaction and Non-specified transformations.

(4) Final Consumption includes non-energy use and energy use (Industry, Transport and Other).

(5) Please refer to notes and definitions in Part I for detailed explanation.

(6) Other industry includes Non-ferrous metals, Transport equipment, Machinery, Mining and quarrying, Food and tobacco, Wood and wood products, Construction, Textile and leather, and Non-specified industry.

(7) Transport includes Rail and Inland waterways.

(8) Other sectors includes Agriculture/Forestry, Fishing and Non-specified other.

* "Coal" refers to all types of coal, primary (anthracite, coking coal, other bituminous coal, sub-bituminous coal, lignite/brown coal) and derived fuels (including patent fuel, BKB, coke oven coke, gas coke, coal tar, coke oven gas, gas works gas, blast furnace gas and other recovered gases), as well as peat, peat briquettes and oil shale. Quantities have been converted to Mtce using calorific values reported by the respective countries.
Please refer to notes and definitions in Part I.

Source: *IEA/OECD Energy Balances of OECD Countries*

MEXICO

6. Use of coal for selected end-uses*

(million tonnes)

	1978	1990	2000	2005	2009	2010	2011	Average annual percent change 78-90	90-11
Steam coal	-	**3.97**	**9.57**	**15.11**	**13.92**	**14.99**	**15.76**	-	**6.79**
Total electricity and heat	-	3.97	9.57	14.92	13.71	14.77	15.59	-	6.73
Main activity producers	-	*3.97*	*9.57*	*14.92*	*13.68*	*14.69*	*15.52*	-	*6.71*
Autoproducers	-	-	-	-	*0.03*	*0.08*	*0.07*	-	-
Patent fuel/BKB plants	-	-	-	-	-	-	-	-	-
Coke ovens/Liquefaction(1)	-	-	-	-	-	-	-	-	-
Blast furnace inputs	-	-	-	-	-	-	-	-	-
Gas manufacture	-	-	-	-	-	-	-	-	-
Industry	-	-	-	0.19	0.21	0.22	0.17	-	-
Iron and steel	-	-	-	-	-	-	-	-	-
Chemical	-	-	-	-	-	-	-	-	-
Non-metallic minerals	-	-	-	*0.19*	*0.21*	*0.22*	*0.17*	-	-
Paper, pulp and print	-	-	-	-	-	-	-	-	-
Other industry	-	-	-	-	-	-	-	-	-
Other sectors(2)	-	-	-	-	-	-	-	-	-
Non-energy use	-	-	-	-	-	-	-	-	-
Coking coal	**4.05**	**2.91**	**2.72**	**2.45**	**2.44**	**2.70**	**2.86**	**-2.72**	**-0.08**
Total electricity and heat	-	-	-	-	-	-	-	-	-
Main activity producers	-	-	-	-	-	-	-	-	-
Autoproducers	-	-	-	-	-	-	-	-	-
Patent fuel/BKB plants	-	-	-	-	-	-	-	-	-
Coke ovens/Liquefaction(1)	3.97	2.89	2.72	2.45	2.44	2.70	2.86	-2.61	-0.04
Blast furnace inputs	-	-	-	-	-	-	-	-	-
Gas manufacture	-	-	-	-	-	-	-	-	-
Industry	-	-	-	-	-	-	-	-	-
Iron and steel	-	-	-	-	-	-	-	-	-
Chemical	-	-	-	-	-	-	-	-	-
Non-metallic minerals	-	-	-	-	-	-	-	-	-
Paper, pulp and print	-	-	-	-	-	-	-	-	-
Other industry	-	-	-	-	-	-	-	-	-
Other sectors(2)	-	-	-	-	-	-	-	-	-
Non-energy use	-	-	-	-	-	-	-	-	-
Lignite	-	-	**0.00**	**0.00**	**0.00**	**0.00**	**0.00**	-	-
Total electricity and heat	-	-	-	-	-	-	-	-	-
Main activity producers	-	-	-	-	-	-	-	-	-
Autoproducers	-	-	-	-	-	-	-	-	-
Patent fuel/BKB plants	-	-	-	-	-	-	-	-	-
Coke ovens/Liquefaction(1)	-	-	0.00	0.00	0.00	0.00	0.00	-	-
Gas manufacture	-	-	-	-	-	-	-	-	-
Industry	-	-	-	-	-	-	-	-	-
Iron and steel	-	-	-	-	-	-	-	-	-
Chemical	-	-	-	-	-	-	-	-	-
Non-metallic minerals	-	-	-	-	-	-	-	-	-
Paper, pulp and print	-	-	-	-	-	-	-	-	-
Other industry	-	-	-	-	-	-	-	-	-
Other sectors(2)	-	-	-	-	-	-	-	-	-
Non-energy use	-	-	-	-	-	-	-	-	-

(1) Liquefaction primarily refers to direct distillation processes. Indirect liquefaction may be reported here or under gas manufacture.

(2) Other sectors are Residential, Commercial and public services, Agriculture/Forestry, Fishing and Non-specified other.

* Coal totals may not match the sum of the individual rows due to statistical differences. Please refer to notes and definitions in Part I.

Source: *IEA/OECD Energy Statistics of OECD Countries*

MEXICO

7. Fuel prices to end users*

	1978(1)	1990	1995	2000	2005	2009	2010	2011	2012
				(US dollars / tce)					
For electricity generation									
Steam coal	..	48.20	39.17	48.69	62.69	78.20	83.49	85.90	87.11
Heavy fuel oil	7.91	38.05	38.46	86.20	122.84	209.10	284.68	371.33	416.15
Natural gas	10.62	68.39	47.89	116.69	282.94	154.07	181.24	170.31	127.13
For industry									
Steam coal	x	x	x	x	x	x	x	x	x
Coking coal	x	x	x	x	x	x	x	x	x
High sulphur fuel oil	7.91	38.05	43.48	86.29	124.96	209.12	284.20	372.95	414.83
Low sulphur fuel oil	..	..	..	..	..	..	..	..	..
Natural gas	10.53	68.39	47.89	116.69	277.77	..	..	..	..
				(1 000 Mexican pesos / unit)(2)					
For electricity generation									
Steam coal	..	0.09	0.16	0.30	0.45	0.69	0.69	0.70	0.75
Heavy fuel oil	0.00	0.16	0.36	1.18	1.94	4.09	5.21	6.68	7.93
Natural gas	0.00	0.25	0.40	1.42	3.96	2.68	2.94	2.72	2.15
For industry									
Steam coal	x	x	x	x	x	x	x	x	x
Coking coal	x	x	x	x	x	x	x	x	x
High sulphur fuel oil	0.00	0.16	0.40	1.18	1.97	4.09	5.20	6.71	7.90
Low sulphur fuel oil	..	..	..	..	..	..	..	..	..
Natural gas	0.00	0.25	0.40	1.42	3.89	..	..	..	..

(1) Earliest year for which split by coal type is available.

(2) For steam coal, coking coal and heavy fuel oil per metric tonne; for natural gas per 10^7 kilocalories GCV.

* Please refer to notes and definitions in Part I. Steam coal prices relate to historical definitions.

Source: *IEA/OECD Energy Prices & Taxes*

8. Coal and peat trade by type of coal*

(Mtce)

	1978	1990	1995	2000	2005	2009	2010	2011	2012e
Total imports	**0.64**	**0.34**	**1.22**	**2.80**	**6.88**	**5.45**	**6.64**	**6.52**	**6.25**
Bituminous coal(1)	-	-	-	-	-	-	-	-	-
Coking coal	0.56	0.22	0.83	1.81	1.79	0.55	1.54	1.62	0.74
Sub-bituminous coal	-	-	-	0.42	4.74	4.71	4.74	4.60	5.16
Lignite	-	-	0.00	0.00	0.00	0.00	0.00	0.00	0.00
Peat	-	-	-	-	-	-	-	-	-
Coal products(2)	0.08	0.11	0.40	0.57	0.35	0.19	0.35	0.30	0.35
Total exports	**-**	**0.01**	**0.00**	**0.01**	**0.01**	**0.01**	**0.01**	**0.01**	**0.22**
Bituminous coal(1)	-	-	-	-	-	-	-	-	-
Coking coal	-	0.00	-	0.00	0.00	0.01	0.01	0.01	0.22
Sub-bituminous coal	-	-	-	-	-	-	-	-	-
Lignite	-	-	-	-	-	-	-	-	-
Peat	-	-	-	-	-	-	-	-	-
Coal products(2)	-	0.00	0.00	0.00	0.00	-	0.00	-	-

(1) Bituminous coal includes anthracite.

(2) Coal products includes products derived from coal, for example: coke, coal tar, briquettes, patent fuels.

* Please refer to notes and definitions in Part I.

Source: *IEA/OECD Energy Balances of OECD Countries, IEA/OECD Energy Statistics of OECD Countries*

MEXICO

9. Total coal imports by origin*

(thousand tonnes)

	1978[1]	1985	1990	1995	2000	2005	2010	2011	2012e
Total coal[2]	**570**	**590**	**228**	**820**	**2436**	**7265**	**7701**	**7585**	**7440**
Coking coal	**570**	**590**	**228**	**818**	**1796**	**1773**	**1531**	**1603**	**731**
Australia	-	-	-	-	1074	393	405	693	206
Canada	-	195	122	818	538	507	361	103	113
Czech Republic	-	-	-	-	-	-	-	-	-
Germany	-	-	-	-	-	-	-	-	-
Poland	-	-	-	-	-	-	-	-	-
United Kingdom	-	-	-	-	-	-	-	-	-
United States	470	303	6	-	61	537	670	718	203
Other OECD	-	-	-	-	-	222	-	-	-
China, People's Rep.	-	-	-	-	1	3	3	-	4
Colombia	-	-	-	-	-	111	92	89	35
Indonesia	-	-	-	-	-	-	-	-	-
South Africa	-	-	-	-	-	-	-	-	146
Former Soviet Union[3]	-	-	x	x	x	x	x	x	x
Russian Federation[4]	*x*	*x*	-	-	*66*	-	-	-	-
Other FSU[4]	*x*	*x*	-	-	-	-	-	-	*24*
Venezuela	-	-	-	-	56	-	-	-	-
Vietnam	-	-	-	-	-	-	-	-	-
Non-specified/other	100	92	100	-	-	-	-	-	-
Steam coal	**-**	**-**	**-**	**-**	**636**	**5489**	**6166**	**5979**	**6706**
Australia	-	-	-	-	-	4933	2421	2683	3399
Canada	-	-	-	-	-	72	275	207	162
Czech Republic	-	-	-	-	-	-	-	-	-
Germany	-	-	-	-	-	-	-	-	-
Poland	-	-	-	-	-	-	-	-	-
United Kingdom	-	-	-	-	-	-	-	-	-
United States	-	-	-	-	636	484	1185	1520	2884
Other OECD	-	-	-	-	-	-	-	-	-
China, People's Rep.	-	-	-	-	-	-	-	-	-
Colombia	-	-	-	-	-	-	1010	354	67
Indonesia	-	-	-	-	-	-	-	84	-
South Africa	-	-	-	-	-	-	1275	1131	139
Former Soviet Union[3]	-	-	x	x	x	x	x	x	x
Russian Federation[4]	*x*	*x*	-	-	-	-	-	-	-
Other FSU[4]	*x*	*x*	-	-	-	-	-	-	-
Venezuela	-	-	-	-	-	-	-	-	55
Vietnam	-	-	-	-	-	-	-	-	-
Non-specified/other	-	-	-	-	-	-	-	-	-
Lignite	**-**	**-**	**-**	**2**	**4**	**3**	**4**	**3**	**3**

(1) Earliest year for which split by coal type is available.

(2) Total coal consists of steam coal, coking coal and lignite.

(3) For years prior to 1990, or if no details for individual Former Soviet Union republics.

(4) Data for former republics are not available separately for some OECD member countries.

* Please refer to notes and definitions in Part I.

Source: IEA/OECD Coal Statistics

NETHERLANDS*

Figure 1: TPES by fuel (Mtce)

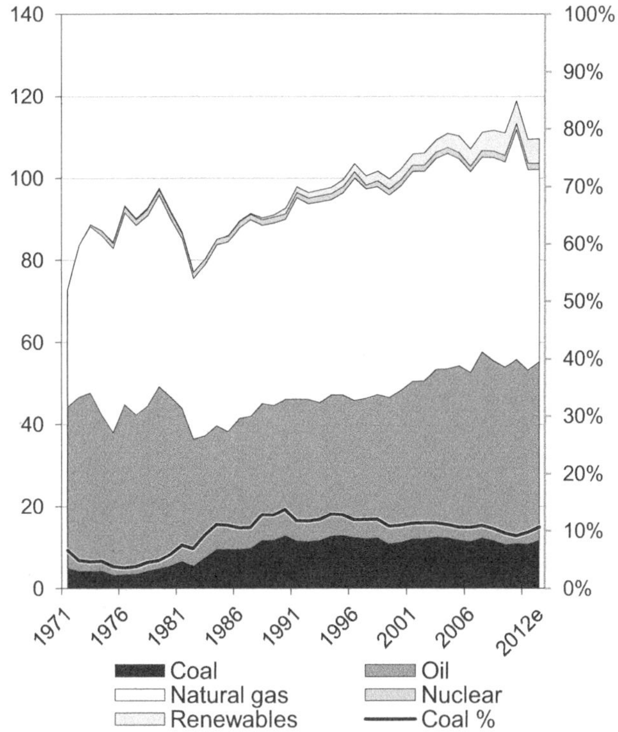

Figure 2: Coal supply indicators (1971 = 100)

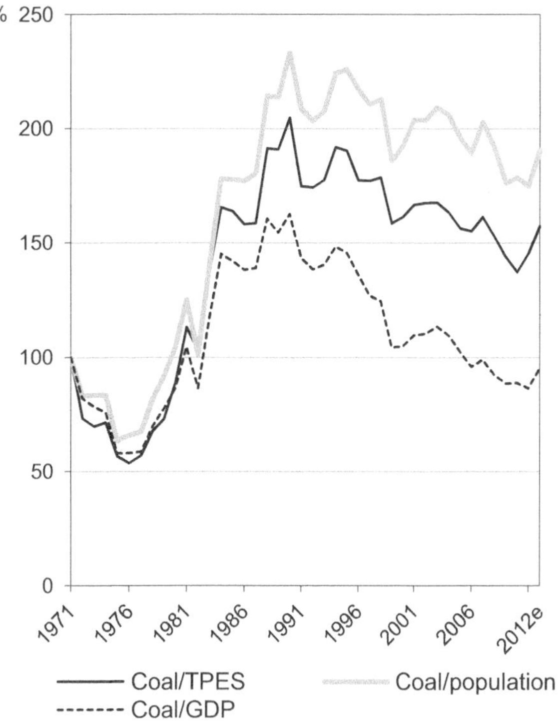

Figure 3: Primary coal supply (Mtce)

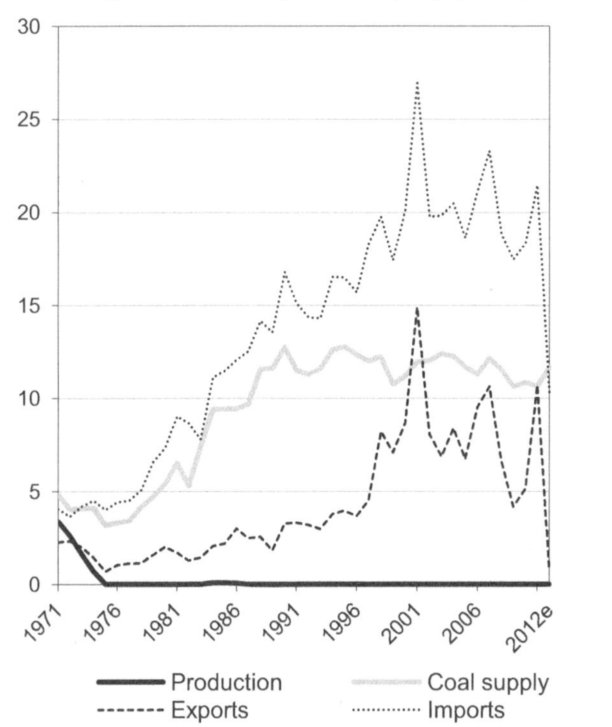

Figure 4: Coal consumption (Mtce)

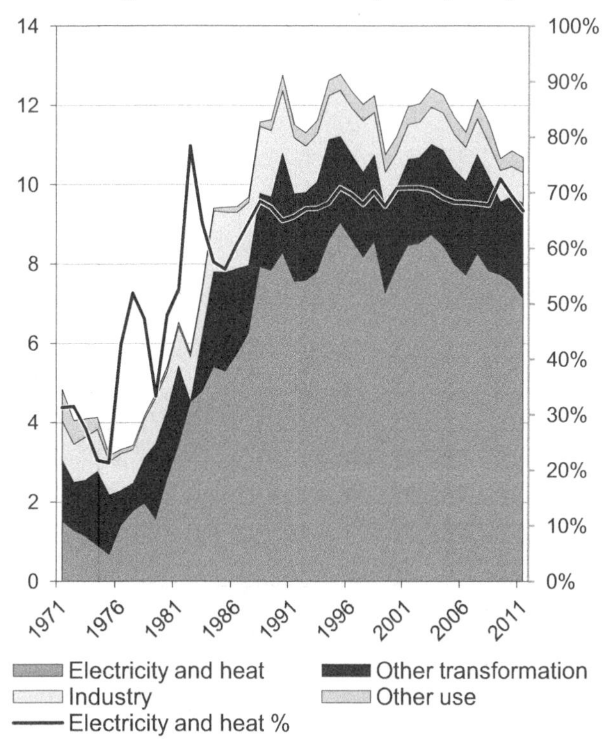

*Please refer to notes and definitions in Part I. Peat is included in coal for all figures.

Source: *IEA/OECD Coal Information, IEA/OECD Energy Balances of OECD Countries, OECD Main Economic Indicators*

NETHERLANDS*

Figure 5: Electricity generation by fuel (TWh)

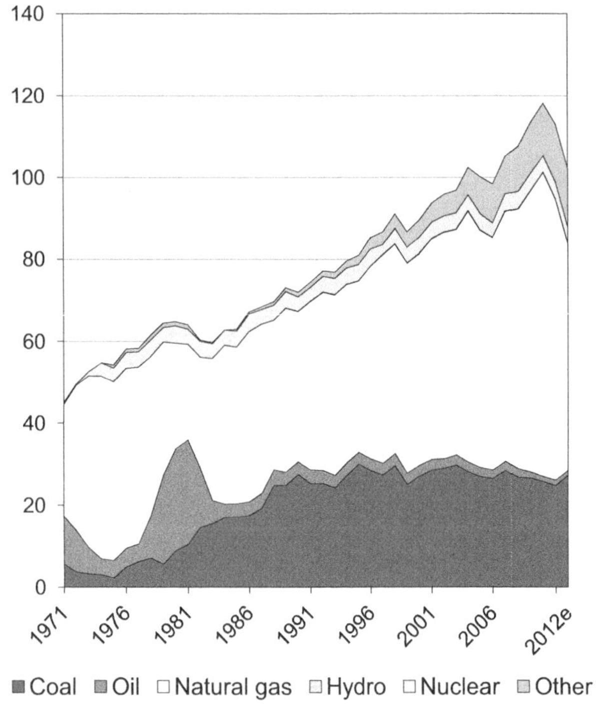

Figure 6: CO_2 emissions by fuel (Mt CO_2)

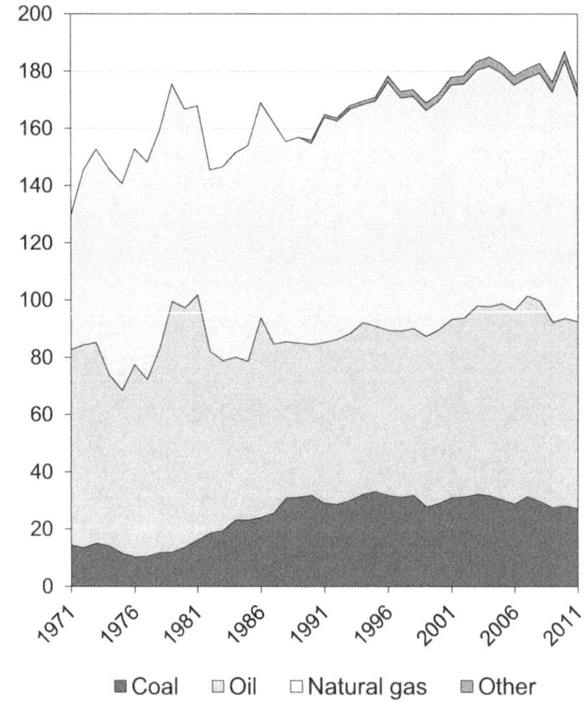

Figure 7: Electricity generation by fuel share

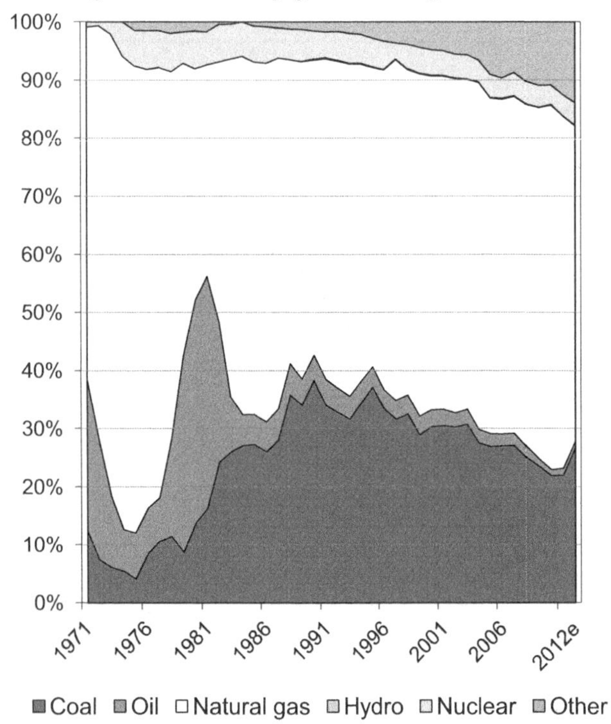

Figure 8: CO_2 emissions by fuel share

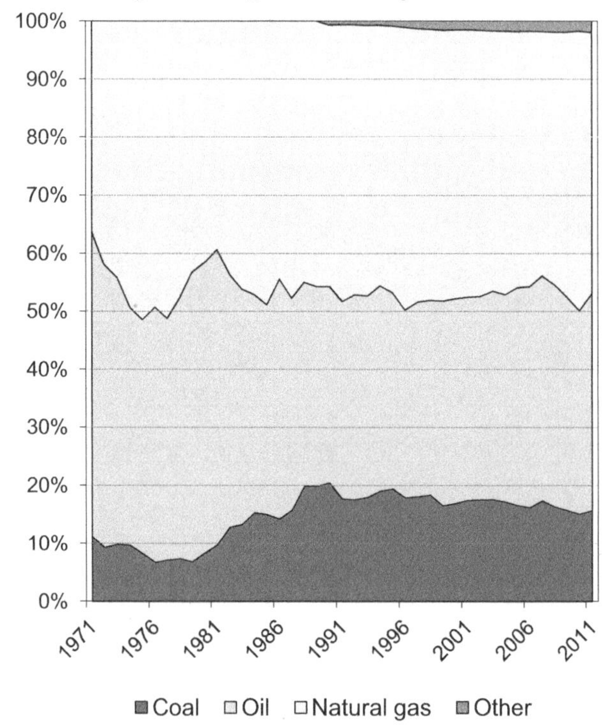

*Please refer to notes and definitions in Part I. Peat is included in coal for all figures.

Source: *IEA/OECD Electricity Information, IEA/OECD CO2 Emissions from Fuel Combustion*

NETHERLANDS

1. Total primary energy supply (TPES) by fuel*

	1973	1980	1990	2000	2010	2011	2012e	Average annual percent change 73-90	90-11
TPES (Mtce)	**88.57**	**91.95**	**93.84**	**104.60**	**119.18**	**110.60**	**111.74**	**0.34**	**0.79**
Coal and peat	4.10	5.42	12.76	11.21	10.86	10.68	11.68	6.91	-0.85
Oil	43.52	41.22	33.25	36.99	44.98	42.58	43.72	-1.57	1.18
Natural Gas	40.71	43.46	44.00	49.98	55.99	48.85	46.78	0.46	0.50
Biofuels and waste	-	0.32	1.36	2.47	4.95	5.12	5.26	-	6.51
Nuclear	0.41	1.56	1.30	1.46	1.48	1.54	1.47	7.00	0.80
Hydro	-	-	0.01	0.02	0.01	0.01	0.01	-	-1.88
Geothermal	-	-	-	-	0.01	0.01	0.02	-	-
Solar, wind, tide	-	-	0.01	0.15	0.55	0.69	0.69	-	22.45
Net electricity trade(1)	-0.17	-0.04	1.13	2.32	0.34	1.12	2.10	x	-0.06
Heat(2)	-	-	-	-	-	-	-	-	-

(1) Net trade between OECD and non-OECD countries.

(2) Ambient heat from heat pumps used in the transformation sector.

Source: *IEA/OECD Energy Balances of OECD Countries*

2. Energy supply, GDP and population*

	1973	1980	1990	2000	2010	2011	2012e	Average annual percent change 73-90	90-11
GDP (2005 bil. USD)	292	351	438	598	684	691	684	2.41	2.19
TPES/GDP(1)	0.30	0.26	0.21	0.17	0.17	0.16	0.16	-2.02	-1.38
Population (millions)	13.4	14.1	14.9	15.9	16.6	16.7	16.8	0.63	0.53
TPES/population(1)	6.59	6.50	6.28	6.57	7.17	6.63	6.67	-0.29	0.26
TPES/GDP(2)	171.8	148.4	121.5	99.2	98.8	90.8	92.6	-2.02	-1.38
Coal and peat supply/GDP(2)	76.4	84.1	159.0	102.3	86.6	84.3	93.2	4.40	-2.97
Elec. consumption/GDP(2)	92.7	99.9	102.6	99.9	95.5	95.1	..	0.60	-0.36
Elec. generation (TWh)	53	65	72	90	118	113	102	1.86	2.17
Industrial production(2)	66.0	73.7	78.4	95.7	106.7	106.2	105.7	1.02	1.46

(1) TPES/GDP in units of tce/2005 thousand USD, TPES/population in units of tce/capita.

(2) As index, 2005 (not shown) = 100

Source: *IEA/OECD Energy Balances of OECD Countries , OECD Main Economic Indicators*

3. Coal and peat production by type*

	1978(1)	1990	2000	2005	2010	2011	2012e	Average annual percent change 78-90	90-11
Mtce:									
Coking coal	-	-	-	-	-	-	-	-	-
Steam coal	-	-	-	-	-	-	-	-	-
Lignite	-	-	-	-	-	-	-	-	-
Peat	-	-	-	-	-	-	-	-	-
Mt:									
Coking coal	-	-	-	-	-	-	-	-	-
Steam coal	-	-	-	-	-	-	-	-	-
Lignite	-	-	-	-	-	-	-	-	-
Peat	-	-	-	-	-	-	-	-	-

(1) Earliest year for which split by coal type is available.

* Please refer to notes and definitions in Part I.

Source: *IEA/OECD Energy Balances of OECD Countries, IEA/OECD Energy Statistics of OECD Countries*

NETHERLANDS

4. Final consumption of energy by fuel*

(Mtce)

	1973	1980	1990	2000	2009	2010	2011	Average annual percent change 73-90	90-11
Total final consumption[(1)]	**68.07**	**77.59**	**70.20**	**80.99**	**85.86**	**92.53**	**85.18**	**0.18**	**0.93**
Coal and peat	1.55	1.11	1.97	1.30	1.09	1.15	1.15	1.44	-2.52
Oil	33.52	34.79	25.82	29.70	37.68	39.56	37.36	-1.52	1.77
Natural Gas	27.56	34.64	32.40	33.34	30.05	34.57	29.28	0.96	-0.48
Biofuels and wastes	-	-	0.54	0.50	1.29	1.07	1.19	-	3.87
Geothermal	-	-	-	-	0.00	0.01	0.01	-	-
Solar, wind, tide	-	-	0.00	0.02	0.03	0.03	0.04	-	12.54
Electricity	5.45	7.05	9.03	12.01	12.77	13.13	13.20	3.02	1.82
Heat	-	-	0.44	4.13	2.94	3.01	2.95	-	9.52
of which:									
Total industry	**19.30**	**17.59**	**16.15**	**19.74**	**16.93**	**18.75**	**18.39**	**-1.04**	**0.62**
Coal and peat	1.09	0.99	1.58	0.87	0.79	0.76	0.78	2.21	-3.30
Oil	3.79	3.63	1.12	2.29	3.19	3.95	3.87	-6.90	6.06
Natural Gas	11.63	9.52	9.30	8.95	6.83	7.51	7.15	-1.31	-1.25
Biofuels and wastes	-	-	0.06	0.09	0.18	0.15	0.13	-	3.88
Geothermal	-	-	-	-	-	-	-	-	-
Solar, wind, tide	-	-	-	-	-	-	-	-	-
Electricity	2.79	3.45	4.08	5.01	4.46	4.80	4.79	2.26	0.77
Heat	-	-	-	2.53	1.48	1.58	1.67	-	-
Total transport	**9.33**	**10.85**	**12.70**	**15.67**	**16.43**	**16.47**	**16.64**	**1.83**	**1.29**
Coal and peat	-	-	-	-	-	-	-	-	-
Oil	9.22	10.73	12.54	15.47	15.68	15.91	15.95	1.82	1.15
Natural Gas	-	-	-	-	0.01	0.02	0.02	-	-
Biofuels and wastes	-	-	-	-	0.53	0.33	0.46	-	-
Electricity	0.11	0.12	0.16	0.20	0.20	0.21	0.21	2.09	1.47
Residential	**18.66**	**20.25**	**14.21**	**14.71**	**14.56**	**16.45**	**13.93**	**-1.59**	**-0.10**
Coal and peat	0.34	0.07	0.03	0.01	0.01	0.01	0.01	-12.77	-7.81
Oil	6.77	2.01	0.35	0.08	0.09	0.15	0.11	-16.03	-5.52
Natural Gas	10.21	16.31	11.23	11.38	10.66	12.34	10.04	0.56	-0.53
Biofuels and wastes	-	-	0.42	0.33	0.43	0.43	0.44	-	0.13
Geothermal	-	-	-	-	-	-	-	-	-
Solar, wind, tide	-	-	0.00	0.01	0.03	0.03	0.03	-	17.58
Electricity	1.35	1.86	2.03	2.68	2.97	3.03	2.91	2.44	1.74
Heat	-	-	0.15	0.21	0.38	0.46	0.40	-	4.70
Comm & public services	**2.74**	**2.52**	**8.25**	**10.32**	**12.69**	**13.90**	**11.67**	**6.71**	**1.67**
Coal and peat	-	-	0.04	0.03	0.00	0.00	0.01	-	-6.59
Oil	1.53	1.07	0.90	0.41	0.60	0.56	0.41	-3.07	-3.67
Natural Gas	-	0.04	4.65	5.31	6.93	8.20	6.07	-	1.27
Biofuels and waste	-	-	0.05	0.07	0.06	0.07	0.07	-	1.29
Geothermal	-	-	-	-	-	-	-	-	-
Solar, wind, tide	-	-	0.00	0.00	0.00	0.00	0.00	-	4.41
Electricity	1.20	1.41	2.47	3.61	4.21	4.29	4.38	4.34	2.76
Heat	-	-	0.13	0.88	0.89	0.78	0.73	-	8.55
Non-energy use	**10.80**	**18.81**	**14.03**	**15.01**	**20.44**	**22.12**	**19.91**	**1.55**	**1.68**
Coal and peat	-	-	0.32	0.39	0.29	0.39	0.36	-	0.57
Oil	10.80	16.32	10.47	10.72	17.33	18.46	16.46	-0.19	2.18
Natural Gas	-	2.49	3.25	3.91	2.83	3.27	3.10	-	-0.23

(1) Total Final Consumption (TFC) excludes use in transformation processes and energy industry own use.

* Please refer to notes and definitions in Part I. Agriculture, Forestry, Fishing and "Non-specified other" sectors are not reported here individually, but are accounted for in Total final consumption.

Source: *IEA/OECD Energy Balances of OECD Countries*

NETHERLANDS

5. Coal balance*

(Mtce)

	1973	1980	1990	2000	2005	2010	2011	2012e	Average annual percent change 73-90	90-11
Production	1.6	-	-	-	-	-	-	-	-	-
Imports	4.2	7.3	16.8	20.1	18.6	18.3	21.5	10.3	8.5	1.2
Exports	-2.0	-2.0	-3.3	-8.7	-6.7	-5.1	-10.7	-0.7	2.9	5.8
Stock changes	0.3	0.1	-0.7	-0.2	-0.2	-2.3	-0.1	2.1		
Primary supply	**4.1**	**5.4**	**12.8**	**11.2**	**11.7**	**10.9**	**10.7**	**11.7**	**6.9**	**-0.8**
Statistical differences	**0.1**	**0.1**	**-0.6**	**-**	**-**	**-0.0**	**-0.0**	..		
Total transformation	**-1.8 e**	**-4.1**	**-9.9 e**	**-9.7 e**	**-10.1 e**	**-9.4**	**-9.3 e**	..	**10.7**	**-0.3**
Electricity and heat gen.	-1.1	-2.6	-8.3	-7.9	-8.0	-7.5	-7.1	..	12.5	-0.7
Main activity producers [(1)]	*-1.1*	*-2.4*	*-8.2*	*-7.9*	*-8.0*	*-7.5*	*-7.1*	..	*12.4*	*-0.7*
Autoproducers	-	*-0.2*	*-0.1*	*-0.0*	*-0.0*	*-0.0*	*-0.0*	..	-	*-8.4*
Gas works	-	-	-	-	-	-	-	..	-	-
Coal transformation[(2)]	-0.6 e	-1.6	-1.6 e	-1.7 e	-2.1 e	-1.9	-2.1 e	..	5.4	1.5
BKB plants	-	-	-	-	-	-	-	..	-	-
Blast furnaces	*-0.8* e	*-1.2*	*-1.5* e	*-1.6* e	*-1.9* e	*-1.7*	*-1.9* e	..	*3.3*	*1.3*
Coke ovens	*0.2*	*-0.4*	*-0.1*	*-0.2*	*-0.2*	*-0.2*	*-0.2*	..	-	*4.1*
Patent fuel plants	*0.0*	-	-	-	-	-	-	..	-	-
Other transformation[(3)]	-	-	-	-	-	-	-	..	-	-
Energy ind. own use	**-0.9**	**-0.3**	**-0.3**	**-0.3**	**-0.3**	**-0.3**	**-0.3**	..	**-5.5**	**-1.3**
Losses	**-0.1**	**-**	**-**	**-**	**-**	**-**	**-**	..		
Final consumption[(4)]	**1.5**	**1.1**	**2.0**	**1.3**	**1.3**	**1.2**	**1.2**	..	**1.4**	**-2.5**
Industry[(5)]	1.1	1.0	1.6	0.9	0.9	0.8	0.8	..	2.2	-3.3
Iron and steel	*0.7*	*0.6*	*0.9* e	*0.7* e	*0.8* e	*0.7*	*0.7* e	..	*1.6*	*-1.3*
Chemical	*0.2*	*0.2*	*0.4*	*0.0*	-	-	-	..	*2.8*	-
Non-metallic minerals	*0.0*	*0.2*	*0.1*	*0.1*	*0.0*	*0.1*	*0.0*	..	*10.5*	*-4.0*
Paper, pulp and print	-	-	*0.0*	-	-	-	-	..	-	-
Other industry [(6)]	*0.1*	*0.1*	*0.2* e	*0.0*	*0.0*	*0.0*	*0.0*	..	*1.2*	*-7.2*
Transport[(7)]	-	-	-	-	-	-	-	..	-	-
Other	0.5	0.1	0.1	0.0	0.0	0.0	0.0	..	-10.1	-7.4
Comm. and pub. services	-	-	*0.0*	*0.0*	*0.0*	*0.0*	*0.0*	..	-	*-6.6*
Residential	*0.3*	*0.1*	*0.0*	*0.0*	*0.0*	*0.0*	*0.0*	..	*-12.8*	*-7.8*
Other sectors [(8)]	*0.1*	*0.1*	*0.0*	-	-	-	-	..	*-18.0*	-
Non-energy use	-	-	0.3	0.4	0.4	0.4	0.4	..	-	0.6

(1) Main activity electricity and heat generation includes district heating.

(2) Coal transformation refers to the transformation of coal and peat to secondary and tertiary products (mainly cokes, briquettes, coal tar and off-gases). Gas works and Liquefaction are shown separately to match the *Energy Balances of OECD Countries*.

(3) Other transformation includes Liquefaction and Non-specified transformations.

(4) Final Consumption includes non-energy use and energy use (Industry, Transport and Other).

(5) Please refer to notes and definitions in Part I for detailed explanation.

(6) Other industry includes Non-ferrous metals, Transport equipment, Machinery, Mining and quarrying, Food and tobacco, Wood and wood products, Construction, Textile and leather, and Non-specified industry.

(7) Transport includes Rail and Inland waterways.

(8) Other sectors includes Agriculture/Forestry, Fishing and Non-specified other.

* "Coal" refers to all types of coal, primary (anthracite, coking coal, other bituminous coal, sub-bituminous coal, lignite/brown coal) and derived fuels (including patent fuel, BKB, coke oven coke, gas coke, coal tar, coke oven gas, gas works gas, blast furnace gas and other recovered gases), as well as peat, peat briquettes and oil shale. Quantities have been converted to Mtce using calorific values reported by the respective countries.
Please refer to notes and definitions in Part I.

Source: *IEA/OECD Energy Balances of OECD Countries*

NETHERLANDS

6. Use of coal for selected end-uses*

(million tonnes)

	1978	1990	2000	2005	2009	2010	2011	Average annual percent change 78-90	90-11
Steam coal	**1.85**	**9.64**	**8.69**	**8.33**	**8.45**	**7.92**	**7.43**	**14.74**	**-1.23**
Total electricity and heat	1.70	8.72	8.59	8.27	8.41	7.88	7.40	14.59	-0.78
Main activity producers	*1.59*	*8.69*	*8.59*	*8.27*	*8.41*	*7.88*	*7.40*	*15.21*	*-0.76*
Autoproducers	*0.11*	*0.03*	-	-	-	-	-	*-9.75*	-
Patent fuel/BKB plants	-	-	-	-	-	-	-	-	-
Coke ovens/Liquefaction(1)	-	-	-	-	-	-	-	-	-
Blast furnace inputs	0.01 e	-	-	-	-	-	-	-	-
Gas manufacture	-	-	-	-	-	-	-	-	-
Industry	0.06	0.48	0.07	0.03	0.04	0.04	0.02	19.17	-13.50
Iron and steel	-	-	*0.01*	*0.01*	-	-	-	-	-
Chemical	*0.05*	*0.36*	*0.01*	-	-	-	-	*17.55*	-
Non-metallic minerals	-	*0.03*	*0.02*	*0.00*	-	*0.00*	-	-	-
Paper, pulp and print	-	*0.00*	-	-	-	-	-	-	-
Other industry	*0.01*	*0.09*	*0.04*	*0.02*	*0.04*	*0.04*	*0.02*	*22.79*	*-6.48*
Other sectors(2)	0.09	0.06	0.03	0.03	0.01	0.01	0.01	-3.15	-10.66
Non-energy use	-	0.01	-	-	-	-	-	-	-
Coking coal	**2.95**	**4.39**	**4.05**	**4.68**	**3.42**	**3.95**	**4.25**	**3.36**	**-0.16**
Total electricity and heat	-	-	-	-	-	-	-	-	-
Main activity producers	-	-	-	-	-	-	-	-	-
Autoproducers	-	-	-	-	-	-	-	-	-
Patent fuel/BKB plants	-	-	-	-	-	-	-	-	-
Coke ovens/Liquefaction(1)	3.28	3.85	3.01	3.16	2.48	2.93	2.96	1.33	-1.24
Blast furnace inputs	-	0.62 e	0.98 e	1.47 e	0.90 e	1.02	1.27 e	-	3.47
Gas manufacture	-	-	-	-	-	-	-	-	-
Industry	-	0.06	0.06	0.05	0.04	-	0.02	-	-5.65
Iron and steel	-	*0.06 e*	*0.06 e*	*0.05 e*	*0.04 e*	-	*0.02 e*	-	*-5.65*
Chemical	-	-	-	-	-	-	-	-	-
Non-metallic minerals	-	-	-	-	-	-	-	-	-
Paper, pulp and print	-	-	-	-	-	-	-	-	-
Other industry	-	-	-	-	-	-	-	-	-
Other sectors(2)	-	-	-	-	-	-	-	-	-
Non-energy use	-	-	0.01	0.00	0.00	0.00	0.00	-	-
Lignite	**0.07**	**0.07**	**0.05**	**0.05**	**0.03**	**0.03**	**0.04**	**0.12**	**-2.80**
Total electricity and heat	-	-	-	-	-	-	-	-	-
Main activity producers	-	-	-	-	-	-	-	-	-
Autoproducers	-	-	-	-	-	-	-	-	-
Patent fuel/BKB plants	-	-	-	-	-	-	-	-	-
Coke ovens/Liquefaction(1)	-	-	-	-	-	-	-	-	-
Gas manufacture	-	-	-	-	-	-	-	-	-
Industry	0.07	0.06	0.03	0.03	0.02	0.03	0.03	-0.95	-3.93
Iron and steel	-	-	-	-	-	-	-	-	-
Chemical	-	-	-	-	-	-	-	-	-
Non-metallic minerals	*0.07*	*0.06*	*0.03*	*0.02*	*0.02*	*0.02*	*0.02*	*-1.23*	*-5.52*
Paper, pulp and print	-	-	-	-	-	-	-	-	-
Other industry	-	*0.00*	*0.00*	*0.00*	*0.01*	*0.01*	*0.01*	-	*6.82*
Other sectors(2)	0.00	0.01	0.02	0.02	0.01	0.00	0.01	5.95	3.75
Non-energy use	-	-	-	-	-	-	-	-	-

(1) Liquefaction primarily refers to direct distillation processes. Indirect liquefaction may be reported here or under gas manufacture.

(2) Other sectors are Residential, Commercial and public services, Agriculture/Forestry, Fishing and Non-specified other.

* Coal totals may not match the sum of the individual rows due to statistical differences. Please refer to notes and definitions in Part I.

Source: *IEA/OECD Energy Statistics of OECD Countries*

NETHERLANDS

7. Fuel prices to end users*

	1978(1)	1990	1995	2000	2005	2009	2010	2011	2012
				(US dollars / tce)					
For electricity generation									
Steam coal	40.99	71.37	..	..	..	..	..	..	..
Heavy fuel oil	65.34	..	..	..	..	..	..	..	..
Natural gas	61.49	103.23	112.23	..	..	..	..	..	..
For industry									
Steam coal	57.76	62.74	..	..	..	..	..	..	..
Coking coal	105.55	154.34	..	..	..	..	..	..	..
High sulphur fuel oil	71.09	131.18	138.43	x	x	x	x	x	x
Low sulphur fuel oil	..	..	..	156.50	255.43	279.91	362.79	493.29	515.14
Natural gas	68.25	97.42	123.52	129.53	..	383.29	311.50	348.63	349.38
				(Euro / unit)(2)					
For electricity generation									
Steam coal	40.25	58.95	..	..	..	..	..	..	..
Heavy fuel oil	89.46	..	..	..	..	..	..	..	..
Natural gas	77.63	109.63	105.05	..	..	..	..	..	..
For industry									
Steam coal	56.72	51.82	..	..	..	..	..	..	..
Coking coal	100.69	123.84	..	..	..	..	..	..	..
High sulphur fuel oil	97.33	151.08	140.51	x	x	x	x	x	x
Low sulphur fuel oil	..	..	..	236.75	286.69	280.92	381.91	494.31	558.80
Natural gas	86.17	103.46	115.61	180.69	..	354.72	302.38	322.15	349.48

(1) Earliest year for which split by coal type is available.

(2) For steam coal, coking coal and heavy fuel oil per metric tonne; for natural gas per 10^7 kilocalories GCV.

* Please refer to notes and definitions in Part I. Steam coal prices relate to historical definitions.

Source: *IEA/OECD Energy Prices & Taxes*

8. Coal and peat trade by type of coal*

(Mtce)

	1978	1990	1995	2000	2005	2009	2010	2011	2012e
Total imports	**5.05**	**16.78**	**16.47**	**20.09**	**18.61**	**17.46**	**18.33**	**21.46**	**10.26**
Bituminous coal(1)	1.88	11.68	10.88	14.71	13.24	14.26	13.45	16.93	6.00
Coking coal	2.64	4.51	4.83	4.76	4.88	2.98	4.47	4.25	4.01
Sub-bituminous coal	-	-	-	-	-	-	-	-	-
Lignite	0.05	0.05	0.04	0.03	0.03	0.03	0.02	0.01	0.02
Peat	-	-	-	-	-	-	-	-	-
Coal products(2)	0.48	0.54	0.72	0.59	0.46	0.18	0.40	0.27	0.24
Total exports	**1.13**	**3.26**	**3.96**	**8.66**	**6.73**	**4.18**	**5.14**	**10.68**	**0.72**
Bituminous coal(1)	0.41	2.31	2.88	7.18	6.14	3.98	4.97	10.52	0.48
Coking coal	0.01	-	0.01	0.92	0.23	0.03	-	0.12	0.14
Sub-bituminous coal	-	-	-	-	-	-	-	-	-
Lignite	0.00	0.00	0.02	0.00	-	-	0.00	0.00	-
Peat	-	-	-	-	-	-	-	-	-
Coal products(2)	0.71	0.95	1.05	0.56	0.36	0.17	0.16	0.04	0.10

(1) Bituminous coal includes anthracite.

(2) Coal products includes products derived from coal, for example: coke, coal tar, briquettes, patent fuels.

* Please refer to notes and definitions in Part I.

Source: *IEA/OECD Energy Balances of OECD Countries, IEA/OECD Energy Statistics of OECD Countries*

NETHERLANDS

9. Total coal imports by origin*

(thousand tonnes)

	1978(1)	1985	1990	1995	2000	2005	2010	2011	2012e
Total coal(2)	**5108**	**11713**	**17513**	**17193**	**22586**	**20465**	**20469**	**24481**	**11259**
Coking coal	**2853**	**4169**	**4603**	**4936**	**4864**	**4987**	**4569**	**4343**	**4098**
Australia	961	1027	449	432	1902	1346	555	700	584
Canada	-	35	426	514	366	902	503	566	481
Czech Republic	89	6	-	-	-	-	-	-	-
Germany	782	455	462	15	-	-	1	-	-
Poland	376	315	98	368	319	-	-	-	-
United Kingdom	-	7	-	-	-	-	-	-	-
United States	644	2255	3110	2378	1621	1364	2859	2421	2181
Other OECD	1	48	4	10	1	99	-	29	33
China, People's Rep.	-	-	1	-	-	-	-	-	-
Colombia	-	-	6	16	10	-	-	-	-
Indonesia	-	-	46	214	441	-	-	-	-
South Africa	-	21	1	-	-	-	-	-	-
Former Soviet Union(3)	-	-	x	x	x	x	x	x	x
Russian Federation (4)	*x*	*x*	-	-	-	*359*	*554*	*527*	*400*
Other FSU (4)	*x*	*x*	-	-	-	*204*	-	-	-
Venezuela	-	-	-	989	204	713	97	100	388
Vietnam	-	-	-	-	-	-	-	-	-
Non-specified/other	-	-	-	-	-	-	-	-	31
Steam coal	**2185**	**7438**	**12837**	**12198**	**17675**	**15428**	**15871**	**20120**	**7138**
Australia	529	1898	5176	2293	1110	439	-	-	-
Canada	55	141	-	-	-	-	-	71	25
Czech Republic	62	64	-	-	-	-	-	-	-
Germany	703	155	72	34	19	43	17	45	16
Poland	264	949	1043	795	1246	97	73	-	-
United Kingdom	166	2	-	9	9	68	20	9	2
United States	14	3105	3023	2731	1242	290	819	2038	723
Other OECD	10	68	237	76	365	239	797	721	250
China, People's Rep.	-	1	152	462	271	83	23	21	3
Colombia	-	162	1514	2450	4989	5703	8656	12480	4456
Indonesia	-	-	133	1034	2644	992	14	-	-
South Africa	309	878	1436	2167	5287	6084	2015	1491	533
Former Soviet Union(3)	57	10	x	x	x	x	x	x	x
Russian Federation (4)	*x*	*x*	*32*	*41*	*174*	*1250*	*3401*	*2944*	*1032*
Other FSU (4)	*x*	*x*	-	-	-	*33*	-	-	-
Venezuela	-	-	19	79	90	-	36	300	98
Vietnam	-	-	-	20	157	-	-	-	-
Non-specified/other	16	5	-	7	72	107	-	-	-
Lignite	**70**	**106**	**73**	**59**	**47**	**50**	**29**	**18**	**23**

(1) Earliest year for which split by coal type is available.

(2) Total coal consists of steam coal, coking coal and lignite.

(3) For years prior to 1990, or if no details for individual Former Soviet Union republics.

(4) Data for former republics are not available separately for some OECD member countries.

* Please refer to notes and definitions in Part I.

Source: IEA/OECD Coal Statistics

NETHERLANDS

10. Coking coal exports by destination*

(thousand tonnes)

	1978(1)	1985	1990	1995	2000	2005	2010	2011	2012e
World	**13**	**103**	**-**	**8**	**938**	**236**	**-**	**123**	**146**
Total OECD	**13**	**103**	**-**	**8**	**938**	**236**	**-**	**123**	**146**
Australia	-	-	-	-	-	-	-	-	-
Austria	-	-	-	-	-	-	-	-	-
Belgium	13	72	-	-	-	-	-	-	-
Canada	-	-	-	-	-	-	-	-	-
Chile	-	-	-	-	-	-	-	-	-
Czech Republic	-	-	-	-	-	-	-	-	-
Denmark	-	-	-	-	-	-	-	-	-
Estonia	-	-	-	-	-	-	-	-	-
Finland	-	-	-	-	-	-	-	-	-
France	-	-	-	-	10	-	-	-	-
Germany	-	9	-	8	928	236	-	123	146
Greece	-	-	-	-	-	-	-	-	-
Hungary	-	-	-	-	-	-	-	-	-
Iceland	-	-	-	-	-	-	-	-	-
Ireland	-	-	-	-	-	-	-	-	-
Israel	-	-	-	-	-	-	-	-	-
Italy	-	-	-	-	-	-	-	-	-
Japan	-	-	-	-	-	-	-	-	-
Korea	-	-	-	-	-	-	-	-	-
Luxembourg	-	-	-	-	-	-	-	-	-
Mexico	-	-	-	-	-	-	-	-	-
Netherlands	-	-	-	-	-	-	-	-	-
New Zealand	-	-	-	-	-	-	-	-	-
Norway	-	-	-	-	-	-	-	-	-
Poland	-	-	-	-	-	-	-	-	-
Portugal	-	-	-	-	-	-	-	-	-
Slovak Republic	-	-	-	-	-	-	-	-	-
Slovenia	-	-	-	-	-	-	-	-	-
Spain	-	-	-	-	-	-	-	-	-
Sweden	-	-	-	-	-	-	-	-	-
Switzerland	-	3	-	-	-	-	-	-	-
Turkey	-	-	-	-	-	-	-	-	-
United Kingdom	-	19	-	-	-	-	-	-	-
United States	-	-	-	-	-	-	-	-	-
Total non-OECD	**-**	**-**	**-**	**-**	**-**	**-**	**-**	**-**	**-**
Brazil	-	-	-	-	-	-	-	-	-
China(2)	-	-	-	-	-	-	-	-	-
Chinese Taipei	-	-	-	-	-	-	-	-	-
Egypt	-	-	-	-	-	-	-	-	-
India	-	-	-	-	-	-	-	-	-
Romania	-	-	-	-	-	-	-	-	-
Oth. Africa & Mid. East	-	-	-	-	-	-	-	-	-
Oth. non-OECD Americas	-	-	-	-	-	-	-	-	-
Other Asia & Oceania	-	-	-	-	-	-	-	-	-
Other non-OECD Europe and Eurasia	-	-	-	-	-	-	-	-	-
Non-specified/Other	**-**	**-**	**-**	**-**	**-**	**-**	**-**	**-**	**-**

(1) Earliest year for which split by coal type is available.
(2) China includes the People's Republic of China and Hong Kong, China.
* Please refer to notes and definitions in Part I.
Source: IEA/OECD Coal Statistics

NETHERLANDS

11. Steam coal exports by destination*

(thousand tonnes)

	1978(1)	1985	1990	1995	2000	2005	2010	2011	2012e
World	**475**	**1336**	**2512**	**3218**	**8621**	**7138**	**5866**	**12485**	**569**
Total OECD	**475**	**1296**	**2448**	**3199**	**8590**	**7114**	**5719**	**5377**	**568**
Australia	-	-	-	-	-	-	-	23	-
Austria	-	-	-	-	1	4	-	-	-
Belgium	231	222	469	168	148	38	216	172	18
Canada	-	-	-	-	-	18	23	43	-
Chile	-	-	-	-	-	-	-	-	-
Czech Republic	-	-	-	-	-	-	-	-	-
Denmark	-	23	2	-	-	67	176	46	5
Estonia	-	-	-	-	-	-	-	-	-
Finland	-	45	2	-	-	309	-	9	1
France	9	67	261	333	318	1291	1057	628	70
Germany	228	216	1410	2466	7466	4599	3752	3901	417
Greece	-	-	-	-	-	71	-	7	1
Hungary	-	-	-	-	-	60	-	-	-
Iceland	-	-	-	-	-	33	56	51	6
Ireland	-	41	48	5	-	-	-	-	-
Israel	-	-	-	-	-	-	-	-	-
Italy	-	-	-	8	3	-	-	-	-
Japan	-	-	-	-	-	-	-	-	-
Korea	-	-	-	-	-	-	-	-	-
Luxembourg	-	96	-	-	53	1	15	3	-
Mexico	-	-	-	-	-	-	-	-	-
Netherlands	-	-	-	-	-	-	-	-	-
New Zealand	-	-	-	-	-	-	-	-	-
Norway	-	10	7	22	315	172	239	204	22
Poland	-	-	-	-	-	6	16	-	-
Portugal	-	-	3	5	-	-	-	-	-
Slovak Republic	-	-	-	-	-	-	-	-	-
Slovenia	-	-	-	-	-	-	-	-	-
Spain	-	-	26	60	19	17	4	4	-
Sweden	-	123	-	2	-	13	1	-	-
Switzerland	5	158	7	-	-	-	7	-	-
Turkey	-	-	-	-	-	-	-	-	-
United Kingdom	2	295	213	130	267	415	157	264	28
United States	-	-	-	-	-	-	-	22	-
Total non-OECD	**-**	**40**	**64**	**-**	**-**	**24**	**147**	**108**	**1**
Brazil	-	-	-	-	-	-	-	44	-
China(2)	-	-	3	-	-	-	-	-	-
Chinese Taipei	-	-	-	-	-	-	-	-	-
Egypt	-	-	-	-	-	3	7	-	-
India	-	-	-	-	-	-	-	-	-
Romania	-	-	-	-	-	-	-	-	-
Oth. Africa & Mid. East	-	-	61	-	-	-	15	16	-
Oth. non-OECD Americas	-	-	-	-	-	-	15	31	-
Other Asia & Oceania	-	-	-	-	-	-	1	2	-
Other non-OECD Europe and Eurasia	-	40	-	-	-	21	109	15	1
Non-specified/Other	**-**	**-**	**-**	**19**	**31**	**-**	**-**	**7000**	**-**

(1) Earliest year for which split by coal type is available.

(2) China includes the People's Republic of China and Hong Kong, China.

* Please refer to notes and definitions in Part I. Steam coal includes all sub-bituminous coal.

Source: IEA/OECD Coal Statistics

NETHERLANDS

12. Coal import values by origin*

(average unit value, CIF, USD/tonne)

	1990	1995	2000	2005	2007	2008	2009	2010	2011/12(1)
Coking coal(2)	**60.87**	**53.77**	**46.87**	**104.06**	**125.84**	**187.06**	**182.14**	**182.24**	..
Imports from:									
Australia	60.19	56.87	48.55	101.15	127.31	191.72	235.19	199.46	..
Canada	59.41	54.78	45.10	100.14	131.67	220.98	214.60	205.14	..
Czech Republic	..	..	..	..	..	..	..	..	..
Poland	60.71	55.46	45.45	..	..	..	..	..	..
United States	51.33	57.65	50.32	108.36	128.72	159.55	152.33	166.55	..
China	46.96	39.24	43.08	352.78	269.39	242.43	..	244.12	..
Colombia	50.24	..	32.01	130.59	..	..	..	..	..
Indonesia	65.10	49.63	34.74	..	..	..	..	..	..
South Africa	47.62	..	..	..	..	..	..	..	..
Former Soviet Union(3)	..	33.59	..	95.25	91.41	155.18	128.98	175.13	..
Other bituminous coal(4)	**52.25**	**43.89**	**29.54**	**68.86**	**79.12**	**133.50**	**98.82**	**95.59**	..
Imports from:									
Australia	52.28	42.39	31.26	106.55	102.57	147.93	198.55	237.68	..
Canada	..	..	29.97	144.61	94.31	139.51	..	164.66	..
Czech Republic	..	..	..	..	..	..	132.46	..	..
Poland	55.66	46.60	30.39	304.89	75.83	161.93	120.11	85.97	..
United States	54.91	49.37	31.13	76.01	88.56	141.08	121.93	110.73	..
China	51.46	42.61	30.20	77.09	..	..	x	x	..
Colombia	51.72	40.75	28.38	65.86	78.24	..	94.29	87.09	..
Indonesia	54.88	41.50	26.45	63.41	69.92	129.35	103.55	66.58	..
South Africa	43.71	41.63	30.08	65.31	75.31	130.98	94.06	79.35	..
Former Soviet Union(3)	47.99	58.39	30.10	67.64	79.08	116.75	93.91	94.11	..

Note: On occasion, shipment of extremely small quantities of high valued coal results in high import costs.

(1) Import data for steam and coking coals are currently unavailable for 2011 and 2012 due to resource constraints.

(2) Weighted average of individual countries using import volumes as weights. Prices exclude intra-EU trade.

(3) Former Soviet Union until 1991, Russian Federation starting in 1992.

(4) Bituminous steam coal only. Weighted average of trades. Prices exclude intra-EU trade.

* Please refer to notes and definitions in Part I.

Source: IEA/OECD Coal Statistics

NEW ZEALAND*

Figure 1: TPES by fuel (Mtce)

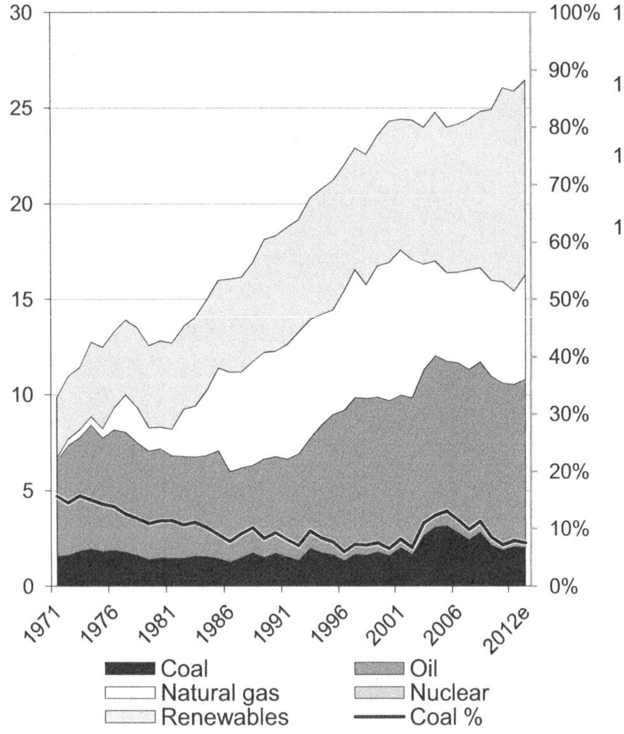

Figure 2: Coal supply indicators (1971 = 100)

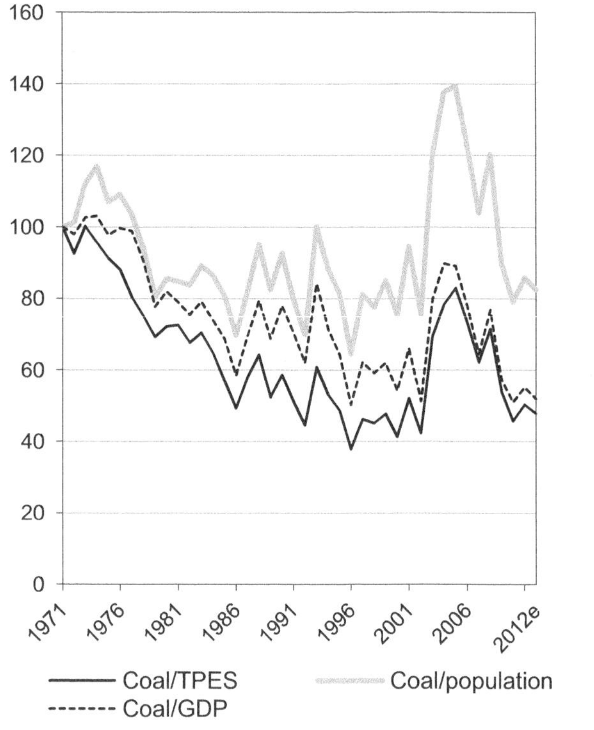

Figure 3: Primary coal supply (Mtce)

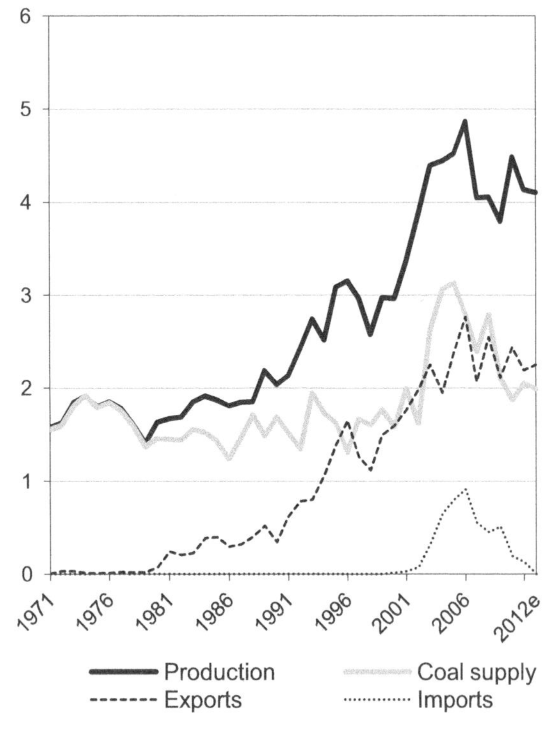

Figure 4: Coal consumption (Mtce)

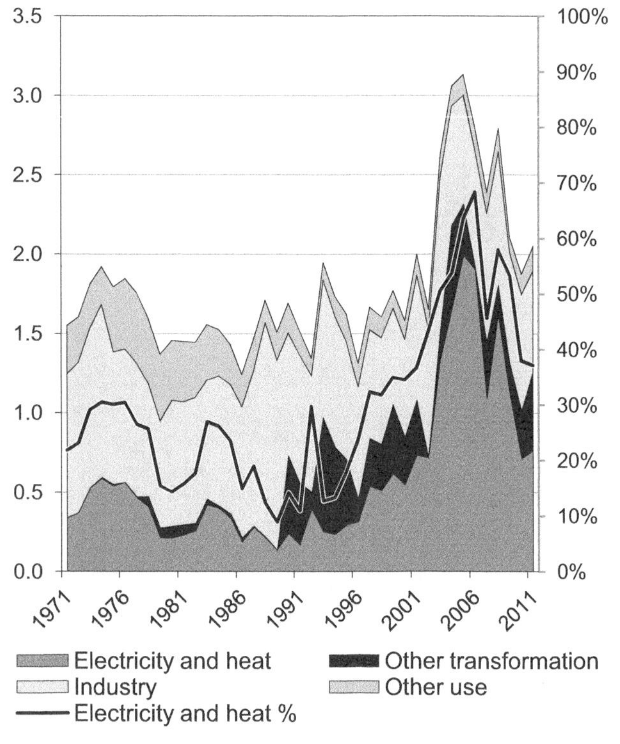

*Please refer to notes and definitions in Part I. Peat is included in coal for all figures.

Source: *IEA/OECD Coal Information, IEA/OECD Energy Balances of OECD Countries, OECD Main Economic Indicators*

NEW ZEALAND*

Figure 5: Electricity generation by fuel (TWh)

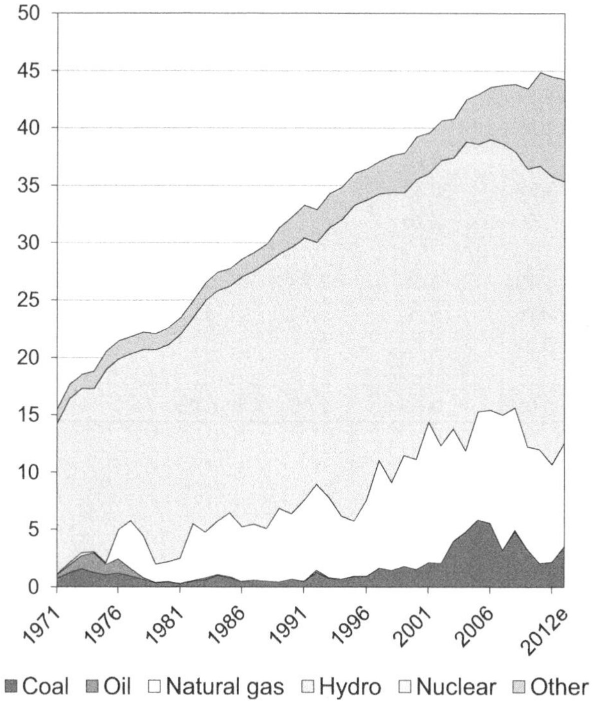

Figure 6: CO_2 emissions by fuel (Mt CO_2)

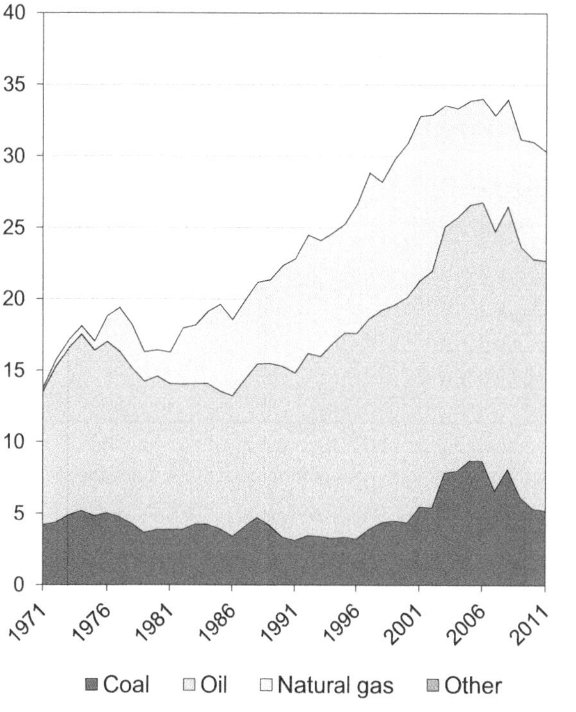

Figure 7: Electricity generation by fuel share

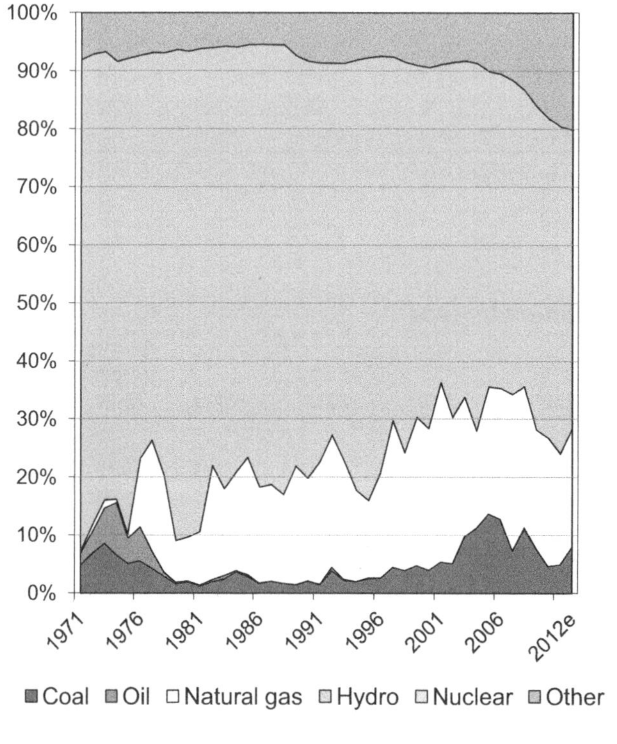

Figure 8: CO_2 emissions by fuel share

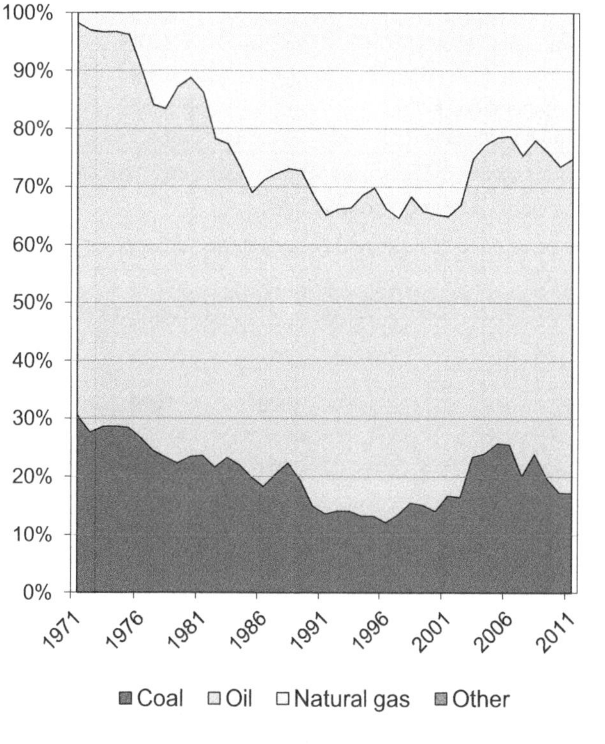

*Please refer to notes and definitions in Part I. Peat is included in coal for all figures.

Source: *IEA/OECD Electricity Information, IEA/OECD CO2 Emissions from Fuel Combustion*

NEW ZEALAND

1. Total primary energy supply (TPES) by fuel*

	1973	1980	1990	2000	2010	2011	2012e	Average annual percent change 73-90	90-11
TPES (Mtce)	**11.46**	**12.84**	**18.38**	**24.37**	**26.12**	**25.95**	**26.52**	**2.82**	**1.66**
Coal and peat	1.80	1.46	1.69	1.58	1.87	2.05	1.99	-0.38	0.92
Oil	5.96	5.73	5.07	8.11	8.72	8.50	8.82	-0.95	2.49
Natural Gas	0.40	1.13	5.53	7.22	5.32	4.87	5.49	16.65	-0.60
Biofuels and waste	-	0.74	1.08	1.59	1.70	1.72	1.73	-	2.26
Nuclear	-	-	-	-	-	-	-	-	-
Hydro	1.76	2.33	2.85	3.00	3.04	3.08	2.80	2.88	0.37
Geothermal	1.53	1.46	2.11	2.78	5.20	5.43	5.39	1.92	4.60
Solar, wind, tide	-	-	-	0.01	0.21	0.25	0.26	-	-
Net electricity trade(1)	-	-	-	-	-	-	-	-	-
Heat(2)	-	-	0.05	0.05	0.05	0.05	0.05	-	-0.28

(1) Net trade between OECD and non-OECD countries.

(2) Ambient heat from heat pumps used in the transformation sector.

Source: *IEA/OECD Energy Balances of OECD Countries*

2. Energy supply, GDP and population*

	1973	1980	1990	2000	2010	2011	2012e	Average annual percent change 73-90	90-11
GDP (2005 bil. USD)	56	57	70	94	118	120	123	1.25	2.60
TPES/GDP(1)	0.20	0.22	0.26	0.26	0.22	0.22	0.22	1.55	-0.92
Population (millions)	3.0	3.1	3.4	3.9	4.4	4.4	4.5	0.75	1.29
TPES/population(1)	3.86	4.08	5.45	6.30	5.96	5.88	5.96	2.06	0.36
TPES/GDP(2)	95.3	105.5	123.8	122.1	103.8	102.0	101.2	1.55	-0.92
Coal and peat supply/GDP(2)	115.3	92.0	87.5	60.8	57.2	61.9	58.3	-1.61	-1.63
Elec. consumption/GDP(2)	83.9	101.6	120.3	108.7	98.4	96.1	..	2.14	-1.06
Elec. generation (TWh)	19	23	32	39	45	44	44	3.32	1.54
Industrial production(2)	..	64.0	73.7	89.1	96.7	95.3	95.8	..	1.23

(1) TPES/GDP in units of tce/2005 thousand USD, TPES/population in units of tce/capita.

(2) As index, 2005 (not shown) = 100

Source: *IEA/OECD Energy Balances of OECD Countries , OECD Main Economic Indicators*

3. Coal and peat production by type*

	1978(1)	1990	2000	2005	2010	2011	2012e	Average annual percent change 78-90	90-11
Mtce:									
Coking coal	0.01	0.59	1.35	2.50	2.37	2.16	2.11	39.10	6.39
Steam coal	1.53	1.37	1.51	1.89	1.97	1.82	1.83	-0.94	1.37
Lignite	0.07	0.08	0.11	0.13	0.15	0.16	0.16	1.07	3.37
Peat	-	-	-	-	-	-	-	-	-
Mt:									
Coking coal	0.01	0.58	1.31	2.45	2.34	2.12	2.08	39.12	6.38
Steam coal	2.01	1.84	1.94	2.58	2.70	2.51	2.53	-0.72	1.48
Lignite	0.15	0.16	0.21	0.25	0.30	0.32	0.33	0.43	3.39
Peat	-	-	-	-	-	-	-	-	-

(1) Earliest year for which split by coal type is available.

* Please refer to notes and definitions in Part I.

Source: *IEA/OECD Energy Balances of OECD Countries, IEA/OECD Energy Statistics of OECD Countries*

NEW ZEALAND

4. Final consumption of energy by fuel*

(Mtce)

	1973	1980	1990	2000	2009	2010	2011	Average annual percent change 73-90	90-11
Total final consumption[(1)]	**8.39**	**9.87**	**13.88**	**18.48**	**17.80**	**18.30**	**18.08**	**3.00**	**1.27**
Coal and peat	1.28	1.17	0.96	0.74	0.78	0.86	0.79	-1.67	-0.92
Oil	4.99	5.17	5.75	7.58	8.31	8.34	8.28	0.84	1.75
Natural Gas	0.17	0.50	2.57	4.30	2.38	2.48	2.43	17.52	-0.26
Biofuels and wastes	-	0.63	0.89	1.37	1.30	1.48	1.50	-	2.53
Geothermal	-	-	0.24	0.28	0.34	0.32	0.32	-	1.29
Solar, wind, tide	-	-	-	-	0.01	0.01	0.01	-	-
Electricity	1.96	2.40	3.47	4.21	4.67	4.80	4.74	3.42	1.51
Heat	-	-	-	-	-	-	-	-	-
of which:									
Total industry	**2.81**	**3.44**	**5.14**	**6.09**	**5.29**	**5.69**	**5.62**	**3.62**	**0.42**
Coal and peat	1.01	0.79	0.77	0.62	0.69	0.74	0.64	-1.54	-0.92
Oil	1.06	0.87	0.41	0.48	0.56	0.50	0.47	-5.49	0.72
Natural Gas	0.05	0.37	1.75	1.91	1.12	1.25	1.29	23.84	-1.47
Biofuels and wastes	-	0.46	0.67	1.15	1.09	1.26	1.28	-	3.15
Geothermal	-	-	0.16	0.19	0.22	0.21	0.21	-	1.40
Solar, wind, tide	-	-	-	-	-	-	-	-	-
Electricity	0.69	0.94	1.38	1.72	1.62	1.73	1.72	4.13	1.08
Heat	-	-	-	-	-	-	-	-	-
Total transport	**2.77**	**3.27**	**4.22**	**5.80**	**6.53**	**6.52**	**6.54**	**2.51**	**2.10**
Coal and peat	-	0.00	0.00	0.00	0.00	0.00	0.00	-	-2.08
Oil	2.77	3.26	4.13	5.79	6.52	6.51	6.52	2.39	2.19
Natural Gas	-	0.01	0.08	0.00	0.00	0.00	0.00	-	-16.85
Biofuels and wastes	-	-	-	-	0.00	0.01	0.01	-	-
Electricity	0.00	0.00	0.01	0.01	0.01	0.01	0.01	2.15	0.25
Residential	**1.29**	**1.45**	**1.73**	**1.94**	**2.16**	**2.14**	**2.10**	**1.74**	**0.94**
Coal and peat	0.23	0.20	0.12	0.04	0.03	0.02	0.02	-3.48	-7.54
Oil	0.11	0.04	0.02	0.07	0.09	0.10	0.10	-10.39	8.83
Natural Gas	0.04	0.06	0.11	0.22	0.20	0.18	0.18	6.77	2.36
Biofuels and wastes	-	0.17 e	0.22	0.22	0.20	0.20	0.20	-	-0.41
Geothermal	-	-	0.01	0.01	0.01	0.01	0.01	-	-0.05
Solar, wind, tide	-	-	-	-	0.01	0.01	0.01	-	-
Electricity	0.91	0.99	1.25	1.38	1.62	1.62	1.58	1.87	1.12
Heat	-	-	-	-	-	-	-	-	-
Comm & public services	**0.73**	**1.01**	**1.14**	**1.33**	**1.62**	**1.66**	**1.61**	**2.64**	**1.64**
Coal and peat	0.01	0.18	0.05	0.06	0.04	0.05	0.04	12.56	-0.86
Oil	0.34	0.36	0.24	0.17	0.14	0.18	0.20	-2.03	-0.83
Natural Gas	0.08	0.07	0.13	0.19	0.22	0.21	0.17	2.86	1.11
Biofuels and waste	-	0.00	0.00	0.00	0.01	0.01	0.01	-	8.93
Geothermal	-	-	0.06	0.06	0.09	0.08	0.07	-	0.62
Solar, wind, tide	-	-	-	-	-	-	-	-	-
Electricity	0.30	0.41	0.65	0.84	1.12	1.13	1.12	4.63	2.58
Heat	-	-	-	-	-	-	-	-	-
Non-energy use	**0.34**	**0.31**	**0.88**	**2.35**	**1.20**	**1.30**	**1.15**	**5.68**	**1.28**
Coal and peat	-	-	-	-	-	-	-	-	-
Oil	0.34	0.31	0.45	0.44	0.42	0.51	0.40	1.61	-0.57
Natural Gas	-	-	0.43	1.91	0.78	0.78	0.75	-	2.69

(1) Total Final Consumption (TFC) excludes use in transformation processes and energy industry own use.

* Please refer to notes and definitions in Part I. Agriculture, Forestry, Fishing and "Non-specified other" sectors are not reported here individually, but are accounted for in Total final consumption.

Source: *IEA/OECD Energy Balances of OECD Countries*

NEW ZEALAND

5. Coal balance*

(Mtce)

	1973	1980	1990	2000	2005	2010	2011	2012e	Average annual percent change 73-90	90-11
Production	1.8	1.6	2.0	3.0	4.5	4.5	4.1	4.1	0.6	3.4
Imports	-	-	0.0	0.0	0.8	0.2	0.1	0.0	-	26.0
Exports	-0.0	-0.1	-0.3	-1.6	-2.4	-2.4	-2.2	-2.3	15.1	9.3
Stock changes	-0.0	-0.1	-	0.2	0.2	-0.4	-0.0	0.1		
Primary supply	**1.8**	**1.5**	**1.7**	**1.6**	**3.1**	**1.9**	**2.1**	**2.0**	**-0.4**	**0.9**
Statistical differences	**-**	**-0.0**	**-0.1**	**-0.1**	**-0.1**	**-0.0**	**-0.2**	..		
Total transformation	**-0.5**	**-0.2**	**-0.4 e**	**-0.7**	**-2.2**	**-0.9**	**-1.0**	..	**-0.9**	**4.0**
Electricity and heat gen.	-0.5	-0.2	-0.2	-0.5	-2.0	-0.7	-0.8	..	-4.5	5.7
Main activity producers (1)	*-0.5*	*-0.2*	*-0.2*	*-0.3*	*-1.7*	*-0.4*	*-0.5*	..	*-6.5*	*5.7*
Autoproducers	-	-	*-0.1*	*-0.2*	*-0.3*	*-0.3*	*-0.2*	..	-	*5.6*
Gas works	0.0	-0.0	-	-	-	-	-	..	-	-
Coal transformation (2)	-0.0	-0.0	-0.2 e	-0.1	-0.2	-0.2	-0.2	..	18.4	0.9
BKB plants	-	-	-	-	-	-	-	..	-	-
Blast furnaces	-	-	*-0.2 e*	*-0.2*	*-0.2*	*-0.2*	*-0.3*	..	-	*0.9*
Coke ovens	*-0.0*	*-0.0*	*0.0*	*0.0*	*0.0*	*0.0*	*0.0*	..	-	*1.5*
Patent fuel plants	*-0.0*	-	-	-	-	-	-	..	-	-
Other transformation (3)	-	-	-	-	-	-	-	..	-	-
Energy ind. own use	**-**	**-0.0**	**-0.1**	**-0.1**	**-0.1**	**-0.1**	**-0.1**	..	**-**	**1.1**
Losses	**-0.0**	**-0.0**	**-0.2 e**	**-0.0**	**-0.0**	**-0.0**	**-0.0**	..		
Final consumption (4)	**1.3**	**1.2**	**1.0**	**0.7**	**0.8**	**0.9**	**0.8**	..	**-1.7**	**-0.9**
Industry (5)	1.0	0.8	0.8	0.6	0.7	0.7	0.6	..	-1.5	-0.9
Iron and steel	*0.1*	*0.1*	*0.0*	*0.0*	*0.1*	*0.1*	*0.1*	..	*-8.7*	*5.7*
Chemical	-	-	-	-	-	-	-	..	-	-
Non-metallic minerals	-	*0.1*	-	-	-	*0.1*	*0.1*	..	-	-
Paper, pulp and print	-	*0.1*	-	-	-	-	*0.0*	..	-	-
Other industry (6)	*0.9*	*0.5*	*0.8*	*0.6*	*0.6*	*0.6*	*0.5*	..	*-1.2*	*-1.9*
Transport (7)	-	0.0	0.0	0.0	0.0	0.0	0.0	..	-	-2.1
Other	0.3	0.4	0.2	0.1	0.1	0.1	0.2	..	-2.3	-0.9
Comm. and pub. services	*0.0*	*0.2*	*0.1*	*0.1*	*0.1*	*0.0*	*0.0*	..	*12.6*	*-0.9*
Residential	*0.2*	*0.2*	*0.1*	*0.0*	*0.0*	*0.0*	*0.0*	..	*-3.5*	*-7.5*
Other sectors (8)	*0.0*	-	*0.0*	*0.0*	*0.0*	*0.1*	*0.1*	..	*-7.0*	*9.9*
Non-energy use	-	-	-	-	-	-	-	..	-	-

(1) Main activity electricity and heat generation includes district heating.

(2) Coal transformation refers to the transformation of coal and peat to secondary and tertiary products (mainly cokes, briquettes, coal tar and off-gases). Gas works and Liquefaction are shown separately to match the *Energy Balances of OECD Countries.*

(3) Other transformation includes Liquefaction and Non-specified transformations.

(4) Final Consumption includes non-energy use and energy use (Industry, Transport and Other).

(5) Please refer to notes and definitions in Part I for detailed explanation.

(6) Other industry includes Non-ferrous metals, Transport equipment, Machinery, Mining and quarrying, Food and tobacco, Wood and wood products, Construction, Textile and leather, and Non-specified industry.

(7) Transport includes Rail and Inland waterways.

(8) Other sectors includes Agriculture/Forestry, Fishing and Non-specified other.

* "Coal" refers to all types of coal, primary (anthracite, coking coal, other bituminous coal, sub-bituminous coal, lignite/brown coal) and derived fuels (including patent fuel, BKB, coke oven coke, gas coke, coal tar, coke oven gas, gas works gas, blast furnace gas and other recovered gases), as well as peat, peat briquettes and oil shale. Quantities have been converted to Mtce using calorific values reported by the respective countries.
Please refer to notes and definitions in Part I.

Source: *IEA/OECD Energy Balances of OECD Countries*

NEW ZEALAND

6. Use of coal for selected end-uses*

(million tonnes)

	1978	1990	2000	2005	2009	2010	2011	Average annual percent change 78-90	90-11
Steam coal	**2.04**	**1.84**	**1.88**	**3.93**	**2.73**	**2.29**	**2.61**	**-0.83**	**1.68**
Total electricity and heat	0.59	0.26	0.45	2.42	1.27	0.64	0.76	-6.61	5.26
Main activity producers	*0.59*	*0.23*	*0.43*	*2.41*	*1.27*	*0.64*	*0.76*	*-7.50*	*5.82*
Autoproducers	-	*0.03*	*0.02*	*0.02*	*0.00*	*0.00*	*0.00*	-	*-11.81*
Patent fuel/BKB plants	-	-	-	-	-	-	-	-	-
Coke ovens/Liquefaction(1)	0.05	0.64	0.67	0.76	0.75	0.84	0.84	23.70	1.27
Blast furnace inputs	-	-	-	-	-	-	-	-	-
Gas manufacture	0.08	-	-	-	-	-	-	-	-
Industry	0.83	0.89	0.59	0.65	0.62	0.65	0.60	0.58	-1.80
Iron and steel	*0.18*	-	-	-	-	-	-	-	-
Chemical	-	-	-	-	-	-	-	-	-
Non-metallic minerals	*0.12*	-	-	-	*0.11*	*0.12*	*0.08*	-	-
Paper, pulp and print	*0.03*	-	-	-	-	-	*0.00*	-	-
Other industry	*0.49*	*0.89*	*0.59*	*0.65*	*0.51*	*0.53*	*0.52*	*5.01*	*-2.46*
Other sectors(2)	0.45	0.17	0.09	0.13	0.11	0.12	0.17	-7.81	0.11
Non-energy use	-	-	-	-	-	-	-	-	-
Coking coal	**-**	**0.24**	**0.00**	**0.12**	**-**	**0.07**	**0.00**	**-**	**-23.02**
Total electricity and heat	-	-	-	-	-	-	-	-	-
Main activity producers	-	-	-	-	-	-	-	-	-
Autoproducers	-	-	-	-	-	-	-	-	-
Patent fuel/BKB plants	-	-	-	-	-	-	-	-	-
Coke ovens/Liquefaction(1)	-	-	-	-	-	-	-	-	-
Blast furnace inputs	-	-	-	-	-	-	-	-	-
Gas manufacture	-	-	-	-	-	-	-	-	-
Industry	-	-	-	-	0.06	0.08	0.01	-	-
Iron and steel	-	-	-	-	-	-	-	-	-
Chemical	-	-	-	-	-	-	-	-	-
Non-metallic minerals	-	-	-	-	*0.00*	*0.00*	-	-	-
Paper, pulp and print	-	-	-	-	-	-	-	-	-
Other industry	-	-	-	-	*0.06*	*0.07*	*0.01*	-	-
Other sectors(2)	-	-	-	-	-	0.01	-	-	-
Non-energy use	-	-	-	-	-	-	-	-	-
Lignite	**0.13**	**0.16**	**0.21**	**0.25**	**0.26**	**0.29**	**0.32**	**1.63**	**3.37**
Total electricity and heat	-	0.01	0.01	0.01	0.02	0.02	0.02	-	0.99
Main activity producers	-	-	-	-	-	-	-	-	-
Autoproducers	-	*0.01*	*0.01*	*0.01*	*0.02*	*0.02*	*0.02*	-	*0.99*
Patent fuel/BKB plants	-	-	-	-	-	-	-	-	-
Coke ovens/Liquefaction(1)	-	-	-	-	-	-	-	-	-
Gas manufacture	-	-	-	-	-	-	-	-	-
Industry	0.09	0.14	0.17	0.21	0.21	0.20	0.24	3.63	2.74
Iron and steel	-	-	-	-	-	-	-	-	-
Chemical	-	-	-	-	-	-	-	-	-
Non-metallic minerals	*0.00*	-	-	-	-	-	-	-	-
Paper, pulp and print	*0.03*	-	-	-	-	-	-	-	-
Other industry	*0.05*	*0.14*	*0.17*	*0.21*	*0.21*	*0.20*	*0.24*	*8.27*	*2.74*
Other sectors(2)	0.04	0.06	0.08	0.06	0.03	0.06	0.07	2.87	0.46
Non-energy use	-	-	-	-	-	-	-	-	-

(1) Liquefaction primarily refers to direct distillation processes. Indirect liquefaction may be reported here or under gas manufacture.

(2) Other sectors are Residential, Commercial and public services, Agriculture/Forestry, Fishing and Non-specified other.

* Coal totals may not match the sum of the individual rows due to statistical differences. Please refer to notes and definitions in Part I.

Source: *IEA/OECD Energy Statistics of OECD Countries*

NEW ZEALAND

7. Fuel prices to end users*

	1978[(1)]	1990	1995	2000	2005	2009	2010	2011	2012
				(US dollars / tce)					
For electricity generation									
Steam coal	c	c	c	c	c	c	c	c	c
Heavy fuel oil	c	c	c	c	c	c	c	c	c
Natural gas	c	c	c	c	c	c	c	c	c
For industry									
Steam coal	36.23	c	c	c	c	c	c	c	c
Coking coal	c	c	c	c	c	c	c	c	c
High sulphur fuel oil	90.62	157.65	159.59	155.06	290.24	295.50	409.64	461.26	481.77
Low sulphur fuel oil	..	..	..	..	..	..	..	..	..
Natural gas	..	83.90	98.96	66.32	173.19	181.46	188.65	201.56	195.54
				(New Zealand dollars / unit)[(2)]					
For electricity generation									
Steam coal	c	c	c	c	c	c	c	c	c
Heavy fuel oil	c	c	c	c	c	c	c	c	c
Natural gas	c	c	c	c	c	c	c	c	c
For industry									
Steam coal	34.00	c	c	c	c	c	c	c	c
Coking coal	c	c	c	c	c	c	c	c	c
High sulphur fuel oil	126.00	382.76	351.94	494.68	596.71	684.13	822.50	846.58	860.88
Low sulphur fuel oil	..	..	..	..	..	..	..	..	..
Natural gas	..	181.01	193.91	187.99	316.37	373.29	336.56	328.71	310.47

(1) Earliest year for which split by coal type is available.

(2) For steam coal, coking coal and heavy fuel oil per metric tonne; for natural gas per 10^7 kilocalories GCV.

* Please refer to notes and definitions in Part I. Steam coal prices relate to historical definitions.

Source: *IEA/OECD Energy Prices & Taxes*

8. Coal and peat trade by type of coal*

(Mtce)

	1978	1990	1995	2000	2005	2009	2010	2011	2012e
Total imports	**0.00**	**0.00**	**-**	**0.02**	**0.80**	**0.51**	**0.19**	**0.13**	**0.00**
Bituminous coal[(1)]	0.00	-	-	0.02	0.06	0.06	0.05	0.03	-
Coking coal	-	0.00	-	-	-	-	0.00	0.00	0.00
Sub-bituminous coal	-	-	-	-	0.74	0.45	0.14	0.10	-
Lignite	-	-	-	-	-	-	-	-	-
Peat	-	-	-	-	-	-	-	-	-
Coal products[(2)]	0.00	-	-	-	-	-	-	-	-
Total exports	**0.02**	**0.34**	**1.38**	**1.59**	**2.38**	**2.11**	**2.44**	**2.19**	**2.25**
Bituminous coal[(1)]	-	-	-	-	-	-	0.11	0.04	-
Coking coal	0.01	0.34	1.38	1.59	2.38	2.06	2.33	2.15	2.25
Sub-bituminous coal	-	-	-	-	-	0.05	-	0.01	-
Lignite	-	-	-	-	-	-	-	-	-
Peat	-	-	-	-	-	-	-	-	-
Coal products[(2)]	0.01	-	-	-	-	-	-	-	-

(1) Bituminous coal includes anthracite.

(2) Coal products includes products derived from coal, for example: coke, coal tar, briquettes, patent fuels.

* Please refer to notes and definitions in Part I.

Source: *IEA/OECD Energy Balances of OECD Countries, IEA/OECD Energy Statistics of OECD Countries*

NEW ZEALAND

9. Total coal imports by origin*

(thousand tonnes)

	1978(1)	1985	1990	1995	2000	2005	2010	2011	2012e
Total coal(2)	**1**	**-**	**1**	**-**	**16**	**1084**	**251**	**172**	**1**
Coking coal	**-**	**-**	**1**	**-**	**-**	**-**	**1**	**3**	**1**
Australia	-	-	-	-	-	-	-	-	-
Canada	-	-	-	-	-	-	-	-	-
Czech Republic	-	-	-	-	-	-	-	-	-
Germany	-	-	-	-	-	-	-	-	-
Poland	-	-	-	-	-	-	-	-	-
United Kingdom	-	-	-	-	-	-	-	-	-
United States	-	-	-	-	-	-	-	-	-
Other OECD	-	-	-	-	-	-	-	-	-
China, People's Rep.	-	-	-	-	-	-	-	-	-
Colombia	-	-	-	-	-	-	-	-	-
Indonesia	-	-	-	-	-	-	-	-	-
South Africa	-	-	-	-	-	-	-	-	-
Former Soviet Union(3)	-	-	x	x	x	x	x	x	x
Russian Federation (4)	x	x	-	-	-	-	-	-	-
Other FSU (4)	x	x	-	-	-	-	-	-	-
Venezuela	-	-	-	-	-	-	-	-	-
Vietnam	-	-	-	-	-	-	-	-	-
Non-specified/other	-	-	1	-	-	-	1	3	1
Steam coal	**1**	**-**	**-**	**-**	**16**	**1084**	**250**	**169**	**-**
Australia	-	-	-	-	16	62	51	31	-
Canada	-	-	-	-	-	-	-	-	-
Czech Republic	-	-	-	-	-	-	-	-	-
Germany	-	-	-	-	-	-	-	-	-
Poland	-	-	-	-	-	-	-	-	-
United Kingdom	-	-	-	-	-	-	-	-	-
United States	1	-	-	-	-	-	-	-	-
Other OECD	-	-	-	-	-	-	-	-	-
China, People's Rep.	-	-	-	-	-	-	-	-	-
Colombia	-	-	-	-	-	-	-	-	-
Indonesia	-	-	-	-	-	1022	199	138	-
South Africa	-	-	-	-	-	-	-	-	-
Former Soviet Union(3)	-	-	x	x	x	x	x	x	x
Russian Federation (4)	x	x	-	-	-	-	-	-	-
Other FSU (4)	x	x	-	-	-	-	-	-	-
Venezuela	-	-	-	-	-	-	-	-	-
Vietnam	-	-	-	-	-	-	-	-	-
Non-specified/other	-	-	-	-	-	-	-	-	-
Lignite	**-**	**-**	**-**	**-**	**-**	**-**	**-**	**-**	**-**

(1) Earliest year for which split by coal type is available.

(2) Total coal consists of steam coal, coking coal and lignite.

(3) For years prior to 1990, or if no details for individual Former Soviet Union republics.

(4) Data for former republics are not available separately for some OECD member countries.

* Please refer to notes and definitions in Part I.

Source: IEA/OECD Coal Statistics

NEW ZEALAND

10. Coking coal exports by destination*

(thousand tonnes)

	1978(1)	1985	1990	1995	2000	2005	2010	2011	2012e
World	**11**	**315**	**336**	**1334**	**1551**	**2331**	**2301**	**2113**	**2210**
Total OECD	**11**	**308**	**285**	**721**	**1253**	**1172 e**	**529 e**	**434 e**	**545**
Australia	-	-	-	123	122	49 e	-	-	-
Austria	-	-	-	-	-	-	-	-	-
Belgium	-	-	-	-	-	-	-	-	-
Canada	-	-	30	-	-	-	-	-	-
Chile	-	-	-	-	313	267 e	57 e	-	-
Czech Republic	-	-	-	-	-	-	-	-	-
Denmark	-	-	-	-	-	-	-	-	-
Estonia	-	-	-	-	-	-	-	-	-
Finland	-	-	-	-	-	-	-	-	-
France	-	-	-	-	-	-	-	-	-
Germany	-	-	-	-	-	-	-	-	-
Greece	-	-	-	-	-	-	-	-	-
Hungary	-	-	-	-	-	-	-	-	-
Iceland	-	-	-	-	-	-	-	-	-
Ireland	-	-	-	-	-	-	-	-	-
Israel	-	-	-	-	-	-	-	-	-
Italy	-	-	-	-	-	-	-	-	-
Japan	11	308	243	517	687	818 e	472 e	434 e	545
Korea	-	-	-	7	-	-	-	-	-
Luxembourg	-	-	-	-	-	-	-	-	-
Mexico	-	-	-	-	-	-	-	-	-
Netherlands	-	-	12	-	85	-	-	-	-
New Zealand	-	-	-	-	-	-	-	-	-
Norway	-	-	-	-	-	-	-	-	-
Poland	-	-	-	-	-	-	-	-	-
Portugal	-	-	-	-	-	-	-	-	-
Slovak Republic	-	-	-	-	-	-	-	-	-
Slovenia	-	-	-	-	-	-	-	-	-
Spain	-	-	-	-	-	-	-	-	-
Sweden	-	-	-	-	-	-	-	-	-
Switzerland	-	-	-	-	46	-	-	-	-
Turkey	-	-	-	-	-	-	-	-	-
United Kingdom	-	-	-	35	-	-	-	-	-
United States	-	-	-	39	-	38 e	-	-	-
Total non-OECD	**-**	**7**	**51**	**382**	**276**	**1159 e**	**1467 e**	**1679 e**	**1665**
Brazil	-	-	-	-	-	124 e	-	-	-
China(2)	-	-	-	106	91	137 e	218 e	324 e	321
Chinese Taipei	-	-	-	-	-	-	-	-	-
Egypt	-	-	-	-	-	-	-	-	-
India	-	-	51	276	185	597 e	1073 e	1142 e	1133
Romania	-	-	-	-	-	-	-	-	-
Oth. Africa & Mid. East	-	-	-	-	-	301 e	176 e	213 e	211
Oth. non-OECD Americas	-	-	-	-	-	-	-	-	-
Other Asia & Oceania	-	7	-	-	-	-	-	-	-
Other non-OECD Europe and Eurasia	-	-	-	-	-	-	-	-	-
Non-specified/Other	**-**	**-**	**-**	**231**	**22**	**-**	**305 e**	**-**	**-**

(1) Earliest year for which split by coal type is available.

(2) China includes the People's Republic of China and Hong Kong, China.

* Please refer to notes and definitions in Part I.

Source: IEA/OECD Coal Statistics

NEW ZEALAND

11. Steam coal exports by destination*

(thousand tonnes)

	1978(1)	1985	1990	1995	2000	2005	2010	2011	2012e
World	-	**80**	-	-	-	-	**119**	**46**	-
Total OECD	-	**70**	-	-	-	-	-	-	-
Australia	-	-	-	-	-	-	-	-	-
Austria	-	-	-	-	-	-	-	-	-
Belgium	-	-	-	-	-	-	-	-	-
Canada	-	-	-	-	-	-	-	-	-
Chile	-	-	-	-	-	-	-	-	-
Czech Republic	-	-	-	-	-	-	-	-	-
Denmark	-	-	-	-	-	-	-	-	-
Estonia	-	-	-	-	-	-	-	-	-
Finland	-	-	-	-	-	-	-	-	-
France	-	-	-	-	-	-	-	-	-
Germany	-	-	-	-	-	-	-	-	-
Greece	-	-	-	-	-	-	-	-	-
Hungary	-	-	-	-	-	-	-	-	-
Iceland	-	-	-	-	-	-	-	-	-
Ireland	-	-	-	-	-	-	-	-	-
Israel	-	-	-	-	-	-	-	-	-
Italy	-	-	-	-	-	-	-	-	-
Japan	-	70	-	-	-	-	-	-	-
Korea	-	-	-	-	-	-	-	-	-
Luxembourg	-	-	-	-	-	-	-	-	-
Mexico	-	-	-	-	-	-	-	-	-
Netherlands	-	-	-	-	-	-	-	-	-
New Zealand	-	-	-	-	-	-	-	-	-
Norway	-	-	-	-	-	-	-	-	-
Poland	-	-	-	-	-	-	-	-	-
Portugal	-	-	-	-	-	-	-	-	-
Slovak Republic	-	-	-	-	-	-	-	-	-
Slovenia	-	-	-	-	-	-	-	-	-
Spain	-	-	-	-	-	-	-	-	-
Sweden	-	-	-	-	-	-	-	-	-
Switzerland	-	-	-	-	-	-	-	-	-
Turkey	-	-	-	-	-	-	-	-	-
United Kingdom	-	-	-	-	-	-	-	-	-
United States	-	-	-	-	-	-	-	-	-
Total non-OECD	-	**10**	-	-	-	-	-	-	-
Brazil	-	-	-	-	-	-	-	-	-
China(2)	-	-	-	-	-	-	-	-	-
Chinese Taipei	-	-	-	-	-	-	-	-	-
Egypt	-	-	-	-	-	-	-	-	-
India	-	-	-	-	-	-	-	-	-
Romania	-	-	-	-	-	-	-	-	-
Oth. Africa & Mid. East	-	-	-	-	-	-	-	-	-
Oth. non-OECD Americas	-	-	-	-	-	-	-	-	-
Other Asia & Oceania	-	10	-	-	-	-	-	-	-
Other non-OECD Europe and Eurasia	-	-	-	-	-	-	-	-	-
Non-specified/Other	-	-	-	-	-	-	**119**	**46**	-

(1) Earliest year for which split by coal type is available.

(2) China includes the People's Republic of China and Hong Kong, China.

* Please refer to notes and definitions in Part I. Steam coal includes all sub-bituminous coal.

Source: IEA/OECD Coal Statistics

NORWAY*

Figure 1: TPES by fuel (Mtce)

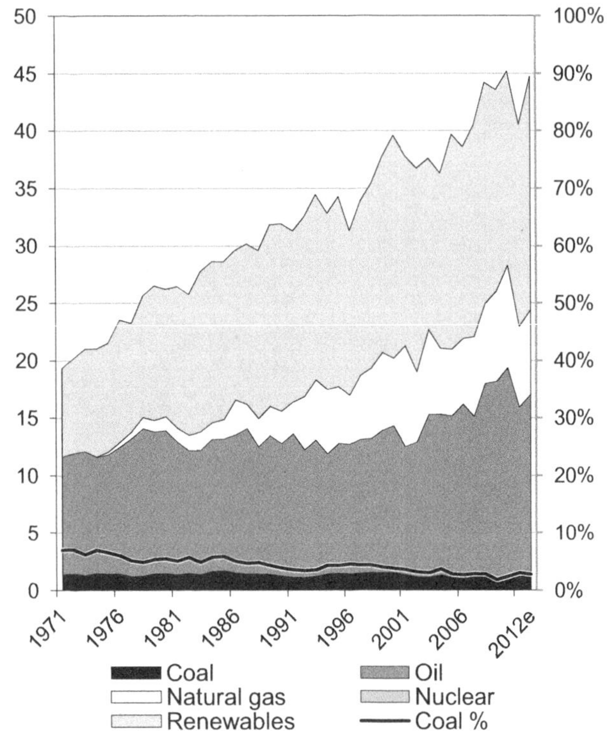

Figure 2: Coal supply indicators (1971 = 100)

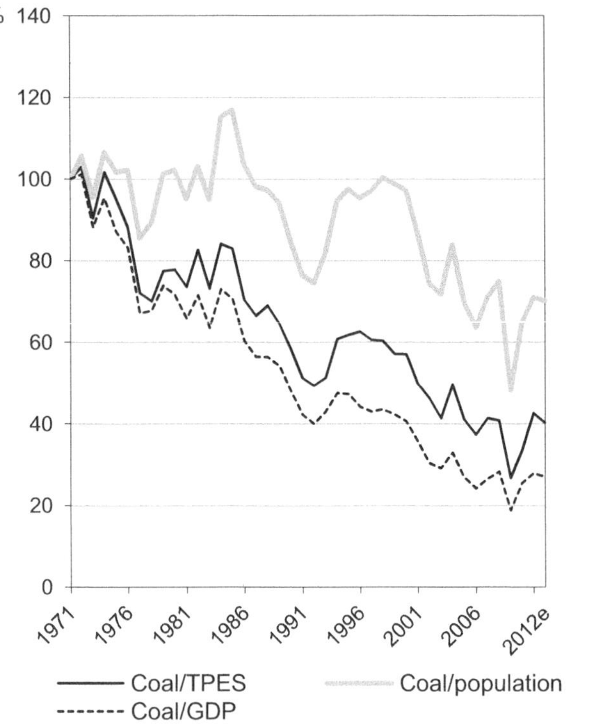

Figure 3: Primary coal supply (Mtce)

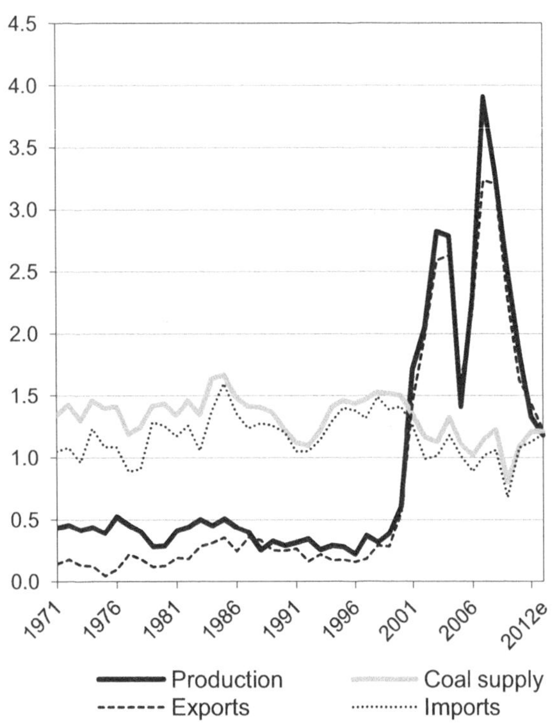

Figure 4: Coal consumption (Mtce)

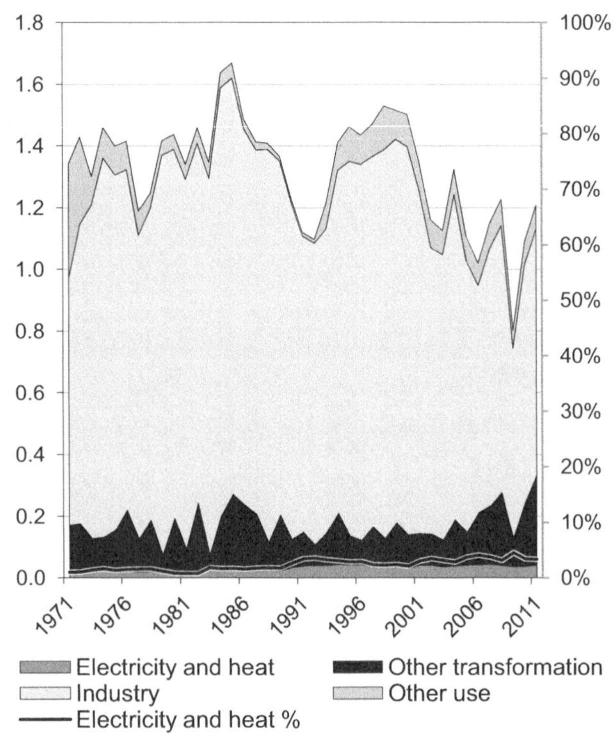

*Please refer to notes and definitions in Part I. Peat is included in coal for all figures.

Source: *IEA/OECD Coal Information, IEA/OECD Energy Balances of OECD Countries, OECD Main Economic Indicators*

NORWAY*

Figure 5: Electricity generation by fuel (TWh)

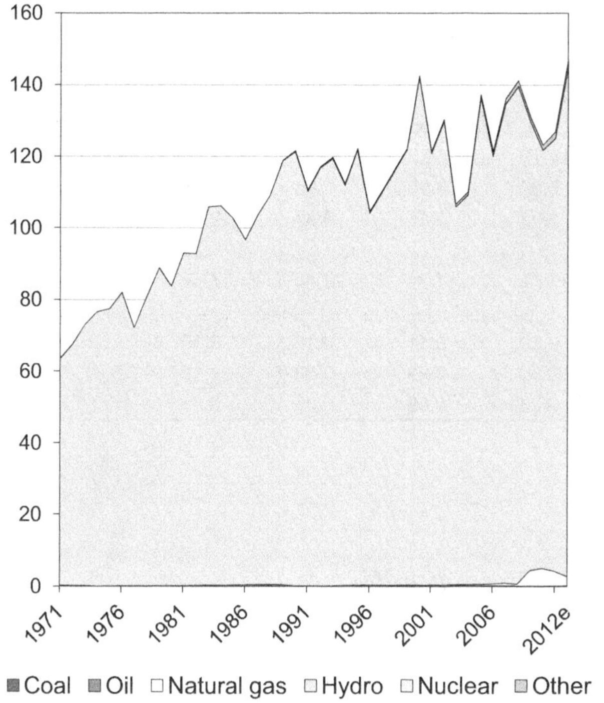

Figure 6: CO_2 emissions by fuel (Mt CO_2)

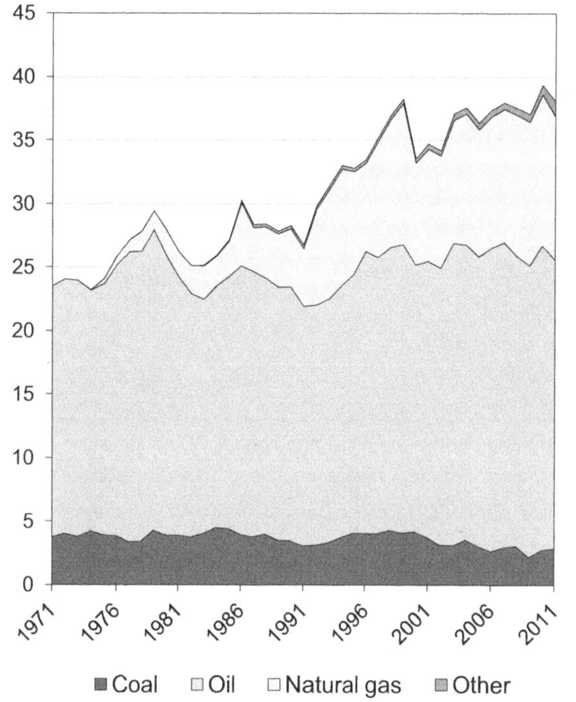

Figure 7: Electricity generation by fuel share

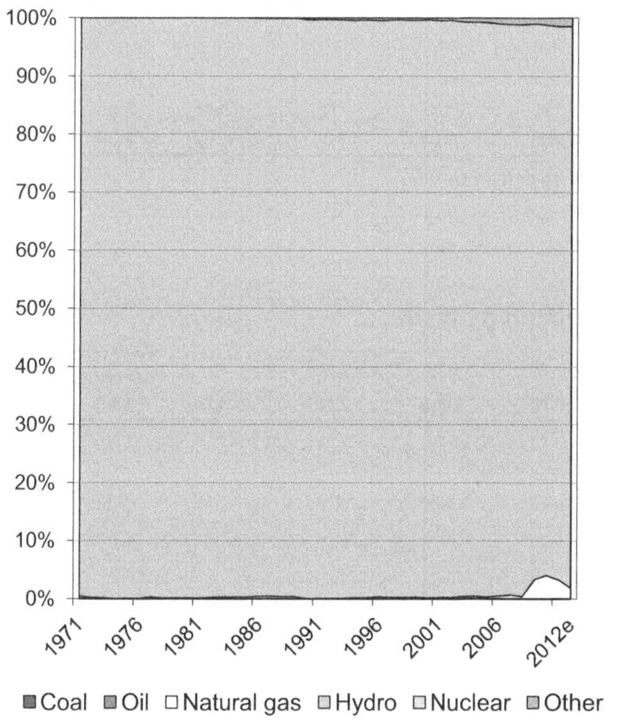

Figure 8: CO_2 emissions by fuel share

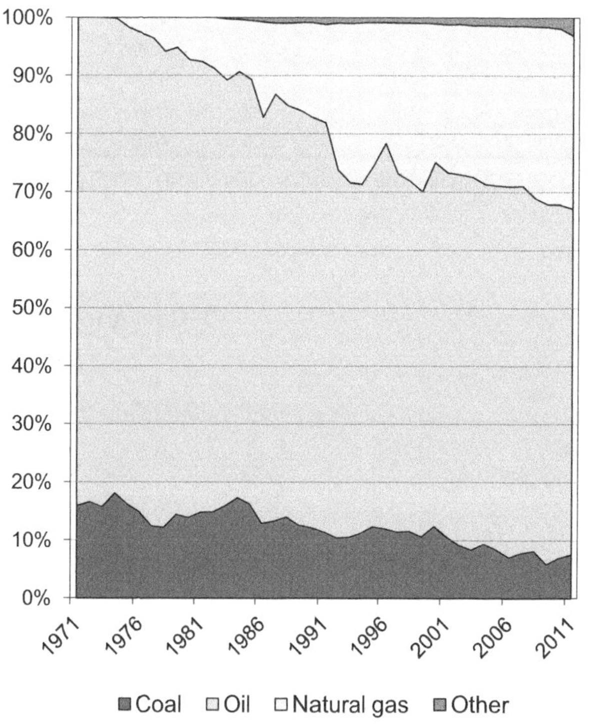

*Please refer to notes and definitions in Part I. Peat is included in coal for all figures.

Source: *IEA/OECD Electricity Information, IEA/OECD CO2 Emissions from Fuel Combustion*

NORWAY

1. Total primary energy supply (TPES) by fuel*

	1973	1980	1990	2000	2010	2011	2012e	Average annual percent change 73-90	90-11
TPES (Mtce)	**20.37**	**26.17**	**30.00**	**37.27**	**46.20**	**40.20**	**42.60**	**2.30**	**1.40**
Coal and peat	1.30	1.44	1.23	1.50	1.09	1.21	1.21	-0.31	-0.09
Oil	10.76	12.44	11.52	12.79	18.28	14.70	15.80	0.40	1.17
Natural Gas	-	1.24	2.82	5.92	8.90	7.05	7.39	-	4.46
Biofuels and waste	-	0.84	1.47	1.95	2.47	2.57	2.73	-	2.68
Nuclear	-	-	-	-	-	-	-	-	-
Hydro	8.95	10.27	14.88	17.42	14.34	14.84	17.42	3.04	-0.01
Geothermal	-	-	-	-	-	-	-	-	-
Solar, wind, tide	-	-	-	0.00	0.11	0.16	0.19	-	-
Net electricity trade(1)	-0.64	-0.06	-1.95	-2.34	0.93	-0.38	-2.19	6.81	-7.53
Heat(2)	-	-	0.03	0.03	0.09	0.05	0.05	-	2.73

(1) Net trade between OECD and non-OECD countries.

(2) Ambient heat from heat pumps used in the transformation sector.

Source: *IEA/OECD Energy Balances of OECD Countries*

2. Energy supply, GDP and population*

	1973	1980	1990	2000	2010	2011	2012e	Average annual percent change 73-90	90-11
GDP (2005 bil. USD)	109	148	190	273	316	320	330	3.33	2.52
TPES/GDP(1)	0.19	0.18	0.16	0.14	0.15	0.13	0.13	-0.99	-1.09
Population (millions)	4.0	4.1	4.2	4.5	4.9	5.0	5.0	0.40	0.74
TPES/population(1)	5.14	6.40	7.07	8.30	9.45	8.12	8.49	1.89	0.66
TPES/GDP(2)	149.0	140.7	125.8	108.7	116.3	100.0	102.7	-0.99	-1.09
Coal and peat supply/GDP(2)	328.3	266.6	178.4	151.0	94.7	103.7	100.8	-3.52	-2.55
Elec. consumption/GDP(2)	153.6	139.0	140.2	110.3	98.7	90.5	..	-0.53	-2.06
Elec. generation (TWh)	73	84	122	143	123	127	147	3.05	0.20
Industrial production(2)	33.5	48.0	74.2	104.5	88.1	83.7	85.2	4.79	0.58

(1) TPES/GDP in units of tce/2005 thousand USD, TPES/population in units of tce/capita.

(2) As index, 2005 (not shown) = 100

Source: *IEA/OECD Energy Balances of OECD Countries* , *OECD Main Economic Indicators*

3. Coal and peat production by type*

	1978(1)	1990	2000	2005	2010	2011	2012e	Average annual percent change 78-90	90-11
Mtce:									
Coking coal	0.23	-	-	-	-	-	-	-	-
Steam coal	0.18	0.29	0.61	1.41	1.86	1.33	1.18	4.31	7.51
Lignite	-	-	-	-	-	-	-	-	-
Peat	-	-	-	-	-	-	-	-	-
Mt:									
Coking coal	0.23	-	-	-	-	-	-	-	-
Steam coal	0.18	0.30	0.63	1.47	1.94	1.39	1.23	4.68	7.51
Lignite	-	-	-	-	-	-	-	-	-
Peat	-	-	-	-	-	-	-	-	-

(1) Earliest year for which split by coal type is available.

* Please refer to notes and definitions in Part I.

Source: *IEA/OECD Energy Balances of OECD Countries, IEA/OECD Energy Statistics of OECD Countries*

NORWAY

4. Final consumption of energy by fuel*

(Mtce)

	1973	1980	1990	2000	2009	2010	2011	Average annual percent change 73-90	90-11
Total final consumption[(1)]	**19.08**	**22.83**	**24.91**	**28.29**	**28.62**	**30.50**	**28.99**	**1.58**	**0.73**
Coal and peat	1.17	1.24	1.11	1.36	0.67	0.85	0.88	-0.34	-1.11
Oil	10.44	11.56	10.52	10.72	11.73	12.21	11.68	0.04	0.50
Natural Gas	-	-	-	0.85	1.00	1.10	1.22	-	-
Biofuels and wastes	-	0.84	1.28	1.71	1.61	1.86	1.79	-	1.61
Geothermal	-	-	-	-	-	-	-	-	-
Solar, wind, tide	-	-	-	-	-	-	-	-	-
Electricity	7.47	9.19	11.89	13.46	13.18	13.94	12.95	2.78	0.41
Heat	-	-	0.11	0.18	0.42	0.55	0.48	-	7.43
of which:									
Total industry	**8.34**	**9.19**	**8.62**	**9.82**	**7.84**	**8.69**	**8.64**	**0.19**	**0.02**
Coal and peat	1.08	1.20	1.09	1.26	0.62	0.77	0.80	0.06	-1.49
Oil	2.69	2.83	1.33	1.09	1.29	1.41	1.36	-4.08	0.12
Natural Gas	-	-	-	0.25	0.30	0.36	0.34	-	-
Biofuels and wastes	-	0.26	0.55	0.85	0.52	0.62	0.68	-	1.07
Geothermal	-	-	-	-	-	-	-	-	-
Solar, wind, tide	-	-	-	-	-	-	-	-	-
Electricity	4.57	4.91	5.63	6.34	5.07	5.47	5.41	1.23	-0.19
Heat	-	-	0.02	0.03	0.04	0.06	0.05	-	3.57
Total transport	**3.28**	**4.13**	**4.87**	**5.80**	**6.64**	**6.93**	**6.70**	**2.35**	**1.53**
Coal and peat	-	-	-	-	-	-	-	-	-
Oil	3.21	4.05	4.79	5.72	6.35	6.61	6.36	2.37	1.36
Natural Gas	-	-	-	0.00	0.07	0.08	0.09	-	-
Biofuels and wastes	-	-	-	-	0.14	0.17	0.16	-	-
Electricity	0.06	0.08	0.08	0.08	0.08	0.08	0.08	1.33	0.23
Residential	**2.68**	**4.55**	**5.15**	**5.46**	**5.71**	**6.32**	**5.57**	**3.92**	**0.37**
Coal and peat	0.09	0.05	0.01	0.00	-	-	-	-12.87	-
Oil	0.63	1.17	0.64	0.35	0.24	0.28	0.21	0.15	-5.09
Natural Gas	-	-	-	0.00	0.00	0.01	0.00	-	-
Biofuels and wastes	-	0.57	0.74	0.82	0.91	1.02	0.90	-	0.94
Geothermal	-	-	-	-	-	-	-	-	-
Solar, wind, tide	-	-	-	-	-	-	-	-	-
Electricity	1.96	2.77	3.72	4.26	4.46	4.88	4.35	3.84	0.75
Heat	-	-	0.04	0.03	0.09	0.13	0.10	-	4.94
Comm & public services	**0.82**	**2.33**	**2.92**	**3.03**	**4.00**	**4.01**	**3.56**	**7.72**	**0.95**
Coal and peat	0.00	-	-	-	-	-	0.00	-	-
Oil	0.01	0.98	0.49	0.34	0.34	0.34	0.30	22.95	-2.26
Natural Gas	-	-	-	0.00	0.03	0.03	0.03	-	-
Biofuels and waste	-	-	-	0.04	0.04	0.05	0.04	-	-
Geothermal	-	-	-	-	-	-	-	-	-
Solar, wind, tide	-	-	-	-	-	-	-	-	-
Electricity	0.81	1.35	2.38	2.53	3.30	3.23	2.84	6.55	0.85
Heat	-	-	0.05	0.13	0.29	0.36	0.33	-	9.74
Non-energy use	**1.58**	**2.24**	**2.63**	**3.08**	**3.16**	**3.25**	**3.26**	**3.03**	**1.03**
Coal and peat	-	-	-	0.10	0.06	0.07	0.07	-	-
Oil	1.58	2.24	2.63	2.38	2.54	2.58	2.47	3.03	-0.30
Natural Gas	-	-	-	0.60	0.57	0.60	0.72	-	-

(1) Total Final Consumption (TFC) excludes use in transformation processes and energy industry own use.

* Please refer to notes and definitions in Part I. Agriculture, Forestry, Fishing and "Non-specified other" sectors are not reported here individually, but are accounted for in Total final consumption.

Source: *IEA/OECD Energy Balances of OECD Countries*

NORWAY

5. Coal balance*

(Mtce)

	1973	1980	1990	2000	2005	2010	2011	2012e	Average annual percent change 73-90	90-11
Production	0.4	0.3	0.3	0.6	1.4	1.9	1.3	1.2	-2.1	7.5
Imports	1.0	1.3	1.2	1.4	1.0	1.1	1.1	1.2	1.3	-0.3
Exports	-0.1	-0.1	-0.2	-0.6	-1.6	-1.6	-1.4	-1.2	3.9	8.7
Stock changes	0.1	0.0	-0.0	0.0	0.3	-0.2	0.2	0.1		
Primary supply	**1.3**	**1.4**	**1.2**	**1.5**	**1.1**	**1.1**	**1.2**	**1.2**	**-0.3**	**-0.1**
Statistical differences	**0.0**	**-0.0**	**-0.0**	**0.0**	**0.0**	**-0.1**	**-0.1**	..		
Total transformation	**-0.1 e**	**-0.1 e**	**-0.1**	**-0.1**	**-0.1**	**-0.2**	**-0.2**	..	**-0.5**	**1.9**
Electricity and heat gen.	-0.0	-0.0	-0.0	-0.0	-0.0	-0.0	-0.0	..	3.5	1.0
Main activity producers (1)	*-0.0*	*-0.0*	*-0.0*	*-0.0*	*-0.0*	*-0.0*	*-0.0*	..	*3.5*	*1.0*
Autoproducers	-	-	-	-	-	-	-	..	-	-
Gas works	0.0	0.0	-	-	-	-	-	..	-	-
Coal transformation(2)	-0.1 e	-0.1 e	-0.1	-0.1	-0.1	-0.1	-0.1	..	-2.3	2.2
BKB plants	-	-	-	-	-	-	-	..	-	-
Blast furnaces	*-0.1* e	*-0.1* e	*-0.1*	*-0.1*	*-0.1*	*-0.1*	*-0.1*	..	*-1.4*	*2.2*
Coke ovens	*-0.0*	*0.0*	-	-	-	-	-	..	-	-
Patent fuel plants	-	-	-	-	-	-	-	..	-	-
Other transformation(3)	-	-	-	-	-	-	-	..	-	-
Energy ind. own use	**-0.0**	**-0.0**	**-**	**-**	**-**	**-**	**-**	..	**-**	**-**
Losses	**-0.0**	**-0.0**	**-0.0**	**-0.0**	**-0.0**	**-0.0**	**-0.0**	..		
Final consumption(4)	**1.2**	**1.2**	**1.1**	**1.4**	**1.0**	**0.8**	**0.9**	..	**-0.3**	**-1.1**
Industry(5)	1.1	1.2	1.1	1.3	0.9	0.8	0.8	..	0.1	-1.5
Iron and steel	*0.8* e	*0.9* e	*0.8*	*0.8*	*0.5*	*0.3*	*0.4*	..	*-0.6*	*-3.5*
Chemical	*0.1*	*0.2*	*0.2*	*0.3*	*0.2*	*0.3*	*0.3*	..	*3.4*	*2.8*
Non-metallic minerals	*0.1*	*0.1*	*0.1*	*0.2*	*0.1*	*0.1*	*0.1*	..	*1.5*	*-0.4*
Paper, pulp and print	-	-	*0.0*	-	-	-	-	..	-	-
Other industry (6)	*0.0*	*0.0*	*0.0*	*0.0*	-	-	-	..	*-3.9*	-
Transport(7)	-	-	-	-	-	-	-	..	-	-
Other	0.1	0.0	0.0	0.0	0.0	-	0.0	..	-10.7	-7.0
Comm. and pub. services	*0.0*	-	-	-	-	-	*0.0*	..	-	-
Residential	*0.1*	*0.0*	*0.0*	*0.0*	*0.0*	-	-	..	*-12.9*	-
Other sectors (8)	-	-	*0.0*	-	-	-	-	..	-	-
Non-energy use	-	-	-	0.1	0.1	0.1	0.1	..	-	-

(1) Main activity electricity and heat generation includes district heating.

(2) Coal transformation refers to the transformation of coal and peat to secondary and tertiary products (mainly cokes, briquettes, coal tar and off-gases). Gas works and Liquefaction are shown separately to match the *Energy Balances of OECD Countries.*

(3) Other transformation includes Liquefaction and Non-specified transformations.

(4) Final Consumption includes non-energy use and energy use (Industry, Transport and Other).

(5) Please refer to notes and definitions in Part I for detailed explanation.

(6) Other industry includes Non-ferrous metals, Transport equipment, Machinery, Mining and quarrying, Food and tobacco, Wood and wood products, Construction, Textile and leather, and Non-specified industry.

(7) Transport includes Rail and Inland waterways.

(8) Other sectors includes Agriculture/Forestry, Fishing and Non-specified other.

* "Coal" refers to all types of coal, primary (anthracite, coking coal, other bituminous coal, sub-bituminous coal, lignite/brown coal) and derived fuels (including patent fuel, BKB, coke oven coke, gas coke, coal tar, coke oven gas, gas works gas, blast furnace gas and other recovered gases), as well as peat, peat briquettes and oil shale. Quantities have been converted to Mtce using calorific values reported by the respective countries.
Please refer to notes and definitions in Part I.

Source: *IEA/OECD Energy Balances of OECD Countries*

NORWAY

6. Use of coal for selected end-uses*

(million tonnes)

	1978	1990	2000	2005	2009	2010	2011	Average annual percent change 78-90	90-11
Steam coal	**0.42**	**0.75**	**1.00**	**0.80**	**0.54**	**0.71**	**0.78**	**5.00**	**0.19**
Total electricity and heat	0.02	0.02	0.02	0.03	0.03	0.03	0.03	0.84	0.83
Main activity producers	*0.02*	*0.02*	*0.02*	*0.03*	*0.03*	*0.03*	*0.03*	*0.84*	*0.83*
Autoproducers	-	-	-	-	-	-	-	-	-
Patent fuel/BKB plants	-	-	-	-	-	-	-	-	-
Coke ovens/Liquefaction(1)	-	-	-	-	-	-	-	-	-
Blast furnace inputs	-	-	-	-	-	-	-	-	-
Gas manufacture	-	-	-	-	-	-	-	-	-
Industry	0.35	0.74	0.86	0.61	0.44	0.58	0.61	6.54	-0.93
Iron and steel	*0.25*	*0.48*	*0.47*	*0.31*	*0.18*	*0.25*	*0.27*	*5.75*	*-2.70*
Chemical	*0.10*	*0.11*	*0.22*	*0.19*	*0.14*	*0.21*	*0.22*	*1.12*	*3.36*
Non-metallic minerals	-	*0.14*	*0.17*	*0.11*	*0.12*	*0.12*	*0.11*	-	*-0.91*
Paper, pulp and print	-	*0.01*	-	-	-	-	-	-	-
Other industry	*0.00*	*0.00*	*0.00*	-	-	-	-	*3.44*	-
Other sectors(2)	0.01	0.01	0.00	-	-	-	-	1.53	-
Non-energy use	-	-	0.11	0.08	0.06	0.08	0.08	-	-
Coking coal	**0.41**	**-**	**-**	**-**	**-**	**-**	**-**	**-**	**-**
Total electricity and heat	-	-	-	-	-	-	-	-	-
Main activity producers	-	-	-	-	-	-	-	-	-
Autoproducers	-	-	-	-	-	-	-	-	-
Patent fuel/BKB plants	-	-	-	-	-	-	-	-	-
Coke ovens/Liquefaction(1)	0.41	-	-	-	-	-	-	-	-
Blast furnace inputs	-	-	-	-	-	-	-	-	-
Gas manufacture	-	-	-	-	-	-	-	-	-
Industry	-	-	-	-	-	-	-	-	-
Iron and steel	-	-	-	-	-	-	-	-	-
Chemical	-	-	-	-	-	-	-	-	-
Non-metallic minerals	-	-	-	-	-	-	-	-	-
Paper, pulp and print	-	-	-	-	-	-	-	-	-
Other industry	-	-	-	-	-	-	-	-	-
Other sectors(2)	-	-	-	-	-	-	-	-	-
Non-energy use	-	-	-	-	-	-	-	-	-
Lignite	**-**	**-**	**-**	**-**	**-**	**-**	**-**	**-**	**-**
Total electricity and heat	-	-	-	-	-	-	-	-	-
Main activity producers	-	-	-	-	-	-	-	-	-
Autoproducers	-	-	-	-	-	-	-	-	-
Patent fuel/BKB plants	-	-	-	-	-	-	-	-	-
Coke ovens/Liquefaction(1)	-	-	-	-	-	-	-	-	-
Gas manufacture	-	-	-	-	-	-	-	-	-
Industry	-	-	-	-	-	-	-	-	-
Iron and steel	-	-	-	-	-	-	-	-	-
Chemical	-	-	-	-	-	-	-	-	-
Non-metallic minerals	-	-	-	-	-	-	-	-	-
Paper, pulp and print	-	-	-	-	-	-	-	-	-
Other industry	-	-	-	-	-	-	-	-	-
Other sectors(2)	-	-	-	-	-	-	-	-	-
Non-energy use	-	-	-	-	-	-	-	-	-

(1) Liquefaction primarily refers to direct distillation processes. Indirect liquefaction may be reported here or under gas manufacture.

(2) Other sectors are Residential, Commercial and public services, Agriculture/Forestry, Fishing and Non-specified other.

* Coal totals may not match the sum of the individual rows due to statistical differences. Please refer to notes and definitions in Part I.

Source: *IEA/OECD Energy Statistics of OECD Countries*

NORWAY

7. Fuel prices to end users*

	1978(1)	1990	1995	2000	2005	2009	2010	2011	2012
				(US dollars / tce)					
For electricity generation									
Steam coal	..	..	..	..	..	..	..	..	..
Heavy fuel oil	x	x	x	x	x	x	x	x	x
Natural gas	..	..	..	..	..	..	..	..	..
For industry									
Steam coal	39.79	53.13	..	..	..	..	..	..	..
Coking coal	..	93.32	89.69	x	x	x	x	x	x
High sulphur fuel oil	70.70	217.89	x	x	x	x	x	x	x
Low sulphur fuel oil	..	..	252.17	249.16	..	..	..	..	..
Natural gas	x	x	x	x	x	x	x	x	x
				(Norwegian crowns / unit)(2)					
For electricity generation									
Steam coal	..	..	..	..	..	..	..	..	..
Heavy fuel oil	x	x	x	x	x	x	x	x	x
Natural gas	..	..	..	..	..	..	..	..	..
For industry									
Steam coal	200	319	..	..	..	..	..	..	..
Coking coal	..	560	545	x	x	x	x	x	x
High sulphur fuel oil	528	1943	x	x	x	x	x	x	x
Low sulphur fuel oil	..	..	2277	3123	..	..	..	..	..
Natural gas	x	x	x	x	x	x	x	x	x

(1) Earliest year for which split by coal type is available.

(2) For steam coal, coking coal and heavy fuel oil per metric tonne; for natural gas per 10^7 kilocalories GCV.

* Please refer to notes and definitions in Part I. Steam coal prices relate to historical definitions.

Source: *IEA/OECD Energy Prices & Taxes*

8. Coal and peat trade by type of coal*

(Mtce)

	1978	1990	1995	2000	2005	2009	2010	2011	2012e
Total imports	**0.91**	**1.20**	**1.40**	**1.41**	**1.01**	**0.68**	**1.08**	**1.12**	**1.19**
Bituminous coal(1)	0.26	0.68	0.89	0.88	0.64	0.41	0.66	0.68	0.73
Coking coal	0.19	-	-	-	-	-	-	-	-
Sub-bituminous coal	-	-	-	-	-	-	-	-	-
Lignite	-	-	-	-	-	-	-	-	-
Peat	-	-	-	-	-	-	-	-	-
Coal products(2)	0.46	0.52	0.51	0.53	0.37	0.27	0.42	0.44	0.46
Total exports	**0.18**	**0.25**	**0.17**	**0.55**	**1.60**	**2.30**	**1.63**	**1.44**	**1.22**
Bituminous coal(1)	0.08	0.24	0.17	0.55	1.60	2.30	1.62	1.44	1.22
Coking coal	-	-	-	-	-	-	-	-	-
Sub-bituminous coal	-	-	-	-	-	-	-	-	-
Lignite	-	-	-	-	-	-	-	-	-
Peat	-	-	-	-	-	-	-	-	-
Coal products(2)	0.10	0.00	-	0.00	0.00	-	0.00	-	-

(1) Bituminous coal includes anthracite.

(2) Coal products includes products derived from coal, for example: coke, coal tar, briquettes, patent fuels.

* Please refer to notes and definitions in Part I.

Source: *IEA/OECD Energy Balances of OECD Countries, IEA/OECD Energy Statistics of OECD Countries*

NORWAY

9. Total coal imports by origin*

(thousand tonnes)

	1978(1)	1985	1990	1995	2000	2005	2010	2011	2012e
Total coal(2)	**450**	**917**	**713**	**930**	**919**	**667**	**684**	**711**	**764**
Coking coal	**187**	**215**	**-**	**-**	**-**	**-**	**-**	**-**	**-**
Australia	-	-	-	-	-	-	-	-	-
Canada	-	-	-	-	-	-	-	-	-
Czech Republic	-	-	-	-	-	-	-	-	-
Germany	-	-	-	-	-	-	-	-	-
Poland	125	134	-	-	-	-	-	-	-
United Kingdom	-	-	-	-	-	-	-	-	-
United States	62	32	-	-	-	-	-	-	-
Other OECD	-	-	-	-	-	-	-	-	-
China, People's Rep.	-	-	-	-	-	-	-	-	-
Colombia	-	49	-	-	-	-	-	-	-
Indonesia	-	-	-	-	-	-	-	-	-
South Africa	-	-	-	-	-	-	-	-	-
Former Soviet Union(3)	-	-	x	x	x	x	x	x	x
Russian Federation (4)	*x*	*x*	*-*	*-*	*-*	*-*	*-*	*-*	*-*
Other FSU (4)	*x*	*x*	*-*	*-*	*-*	*-*	*-*	*-*	*-*
Venezuela	-	-	-	-	-	-	-	-	-
Vietnam	-	-	-	-	-	-	-	-	-
Non-specified/other	-	-	-	-	-	-	-	-	-
Steam coal	**263**	**702**	**713**	**930**	**919**	**667**	**684**	**711**	**764**
Australia	2	2	-	49	-	-	22	14	2
Canada	-	-	-	-	-	-	-	-	-
Czech Republic	-	-	32	-	-	-	-	-	-
Germany	55	99	48	59	64	71	60	20	21
Poland	29	141	87	195	151	159	83	96	128
United Kingdom	77	19	181	153	169	103	91	87	77
United States	46	191	151	103	21	17	51	68	74
Other OECD	38	191	99	143	294	148	123	180	223
China, People's Rep.	-	-	4	-	36	30	6	-	-
Colombia	-	43	9	155	102	97	111	127	147
Indonesia	-	-	-	8	-	-	-	-	-
South Africa	16	14	-	9	-	-	11	-	-
Former Soviet Union(3)	-	2	x	x	x	x	x	x	x
Russian Federation (4)	*x*	*x*	*97*	*52*	*82*	*39*	*122*	*105*	*84*
Other FSU (4)	*x*	*x*	*-*	*-*	*-*	*-*	*-*	*14*	*8*
Venezuela	-	-	5	-	-	-	-	-	-
Vietnam	-	-	-	1	-	-	-	-	-
Non-specified/other	-	-	-	3	-	3	4	-	-
Lignite	**-**	**-**	**-**	**-**	**-**	**-**	**-**	**-**	**-**

(1) Earliest year for which split by coal type is available.

(2) Total coal consists of steam coal, coking coal and lignite.

(3) For years prior to 1990, or if no details for individual Former Soviet Union republics.

(4) Data for former republics are not available separately for some OECD member countries.

* Please refer to notes and definitions in Part I.

Source: IEA/OECD Coal Statistics

NORWAY

11. Steam coal exports by destination*

(thousand tonnes)

	1978[(1)]	1985	1990	1995	2000	2005	2010	2011	2012e
World	**77**	**235**	**254**	**180**	**574**	**1666**	**1691**	**1504**	**1273**
Total OECD	**76**	**235**	**253**	**179**	**570**	**1666**	**1691**	**1504**	**1273**
Australia	-	-	-	-	-	-	-	-	-
Austria	-	-	-	-	-	-	-	-	-
Belgium	-	1	-	-	-	-	-	-	1
Canada	-	-	-	-	-	-	-	-	-
Chile	-	-	-	-	-	-	-	-	-
Czech Republic	-	-	-	-	-	-	-	-	-
Denmark	-	44	2	22	156	245	182	144	205
Estonia	-	-	-	-	-	-	-	-	-
Finland	-	-	-	-	50	-	-	-	-
France	-	-	50	86	-	111	-	-	83
Germany	75	69	87	57	330	994	816	1003	563
Greece	-	-	-	-	-	71	-	-	-
Hungary	-	-	-	-	-	-	-	-	-
Iceland	-	2	-	-	7	-	-	-	-
Ireland	-	21	-	-	-	-	-	-	-
Israel	-	-	-	-	-	-	-	-	-
Italy	-	-	4	3	-	-	-	-	-
Japan	-	-	-	-	-	-	-	-	-
Korea	-	-	-	-	-	-	-	-	-
Luxembourg	-	-	-	-	-	-	-	-	-
Mexico	-	-	-	-	-	-	-	-	-
Netherlands	-	2	46	-	-	-	282	-	-
New Zealand	-	-	-	-	-	-	-	-	-
Norway	-	-	-	-	-	-	-	-	-
Poland	-	-	-	-	2	1	110	-	-
Portugal	-	-	-	-	-	214	217	162	-
Slovak Republic	-	-	-	-	-	-	-	-	-
Slovenia	-	-	-	-	-	-	-	-	-
Spain	-	-	-	-	-	-	-	-	241
Sweden	-	92	3	10	-	-	-	-	-
Switzerland	-	-	-	-	-	-	15	-	-
Turkey	-	-	-	-	-	-	-	-	-
United Kingdom	1	4	61	1	25	30	69	195	180
United States	-	-	-	-	-	-	-	-	-
Total non-OECD	**-**	**-**	**-**	**1**	**4**	**-**	**-**	**-**	**-**
Brazil	-	-	-	-	-	-	-	-	-
China[(2)]	-	-	-	-	-	-	-	-	-
Chinese Taipei	-	-	-	-	-	-	-	-	-
Egypt	-	-	-	-	-	-	-	-	-
India	-	-	-	1	-	-	-	-	-
Romania	-	-	-	-	-	-	-	-	-
Oth. Africa & Mid. East	-	-	-	-	2	-	-	-	-
Oth. non-OECD Americas	-	-	-	-	-	-	-	-	-
Other Asia & Oceania	-	-	-	-	-	-	-	-	-
Other non-OECD Europe and Eurasia	-	-	-	-	2	-	-	-	-
Non-specified/Other	**1**	**-**	**1**	**-**	**-**	**-**	**-**	**-**	**-**

(1) Earliest year for which split by coal type is available.

(2) China includes the People's Republic of China and Hong Kong, China.

* Please refer to notes and definitions in Part I. Steam coal includes all sub-bituminous coal.

Source: IEA/OECD Coal Statistics

POLAND*

Figure 1: TPES by fuel (Mtce)

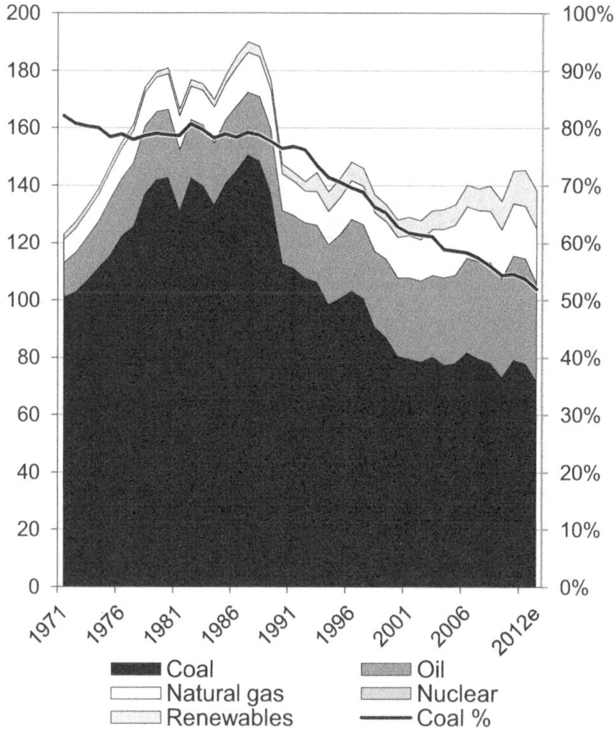

Figure 2: Coal supply indicators (1971 = 100)

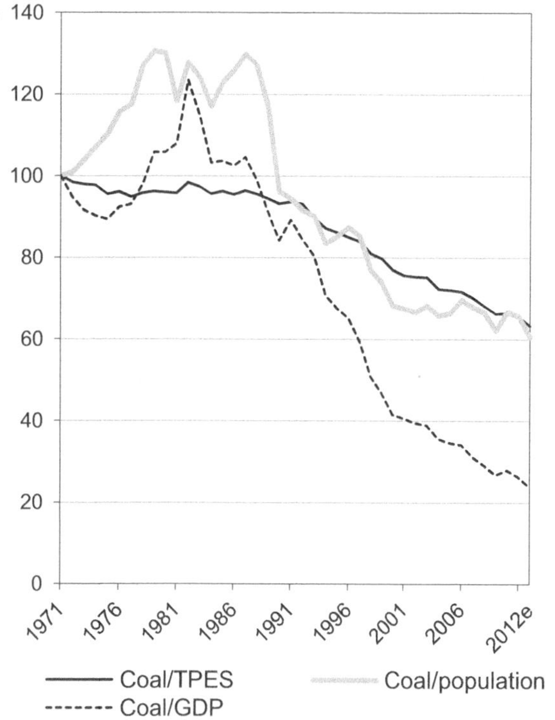

Figure 3: Primary coal supply (Mtce)

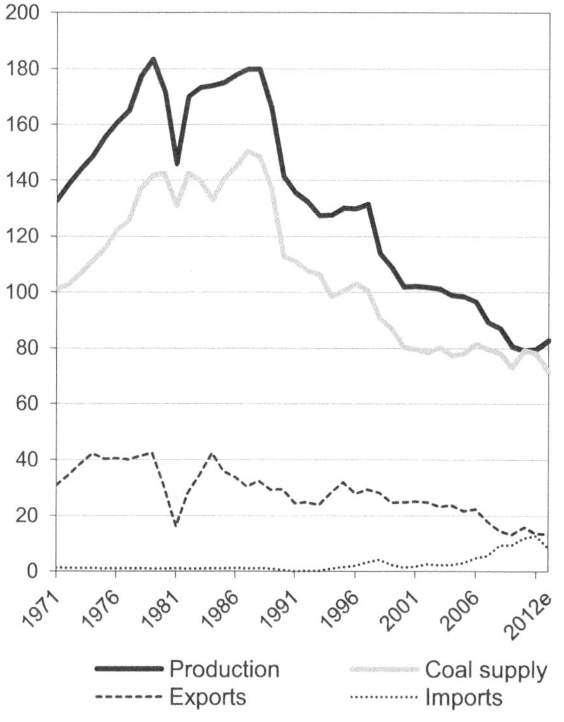

Figure 4: Coal consumption (Mtce)

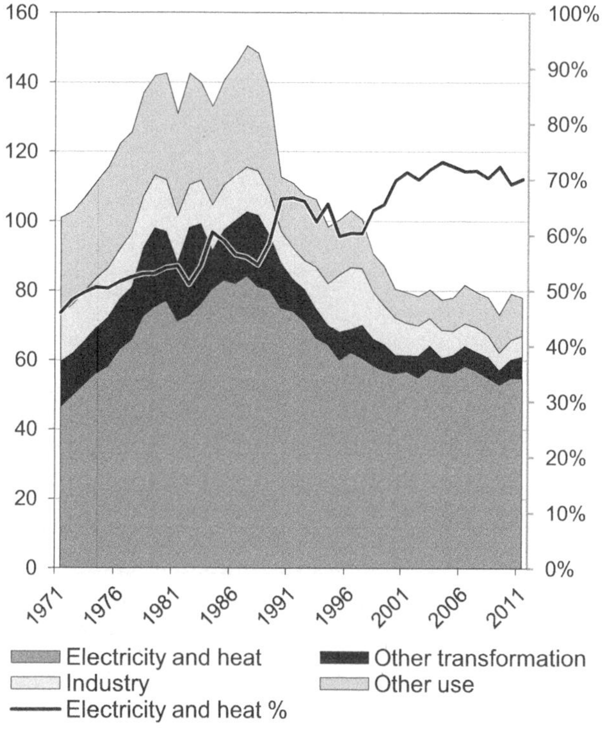

*Please refer to notes and definitions in Part I. Peat is included in coal for all figures.

Source: *IEA/OECD Coal Information, IEA/OECD Energy Balances of OECD Countries, OECD Main Economic Indicators*

POLAND*

Figure 5: Electricity generation by fuel (TWh)

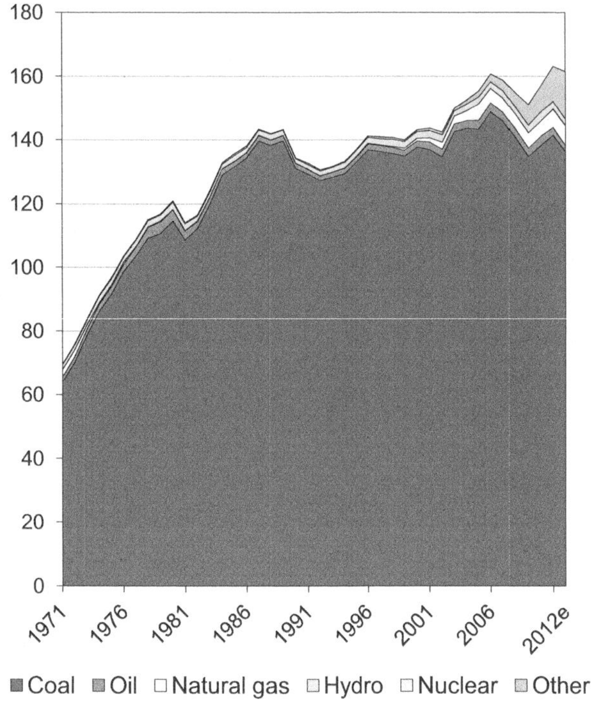

Figure 6: CO_2 emissions by fuel (Mt CO_2)

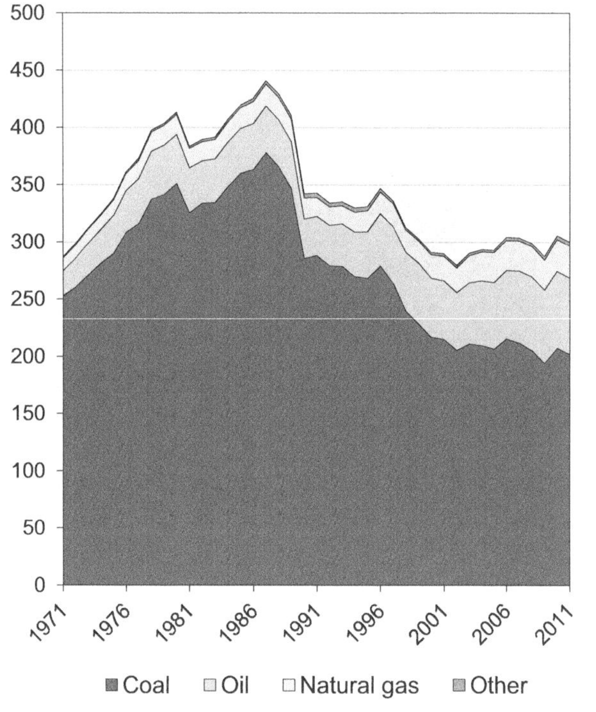

Figure 7: Electricity generation by fuel share

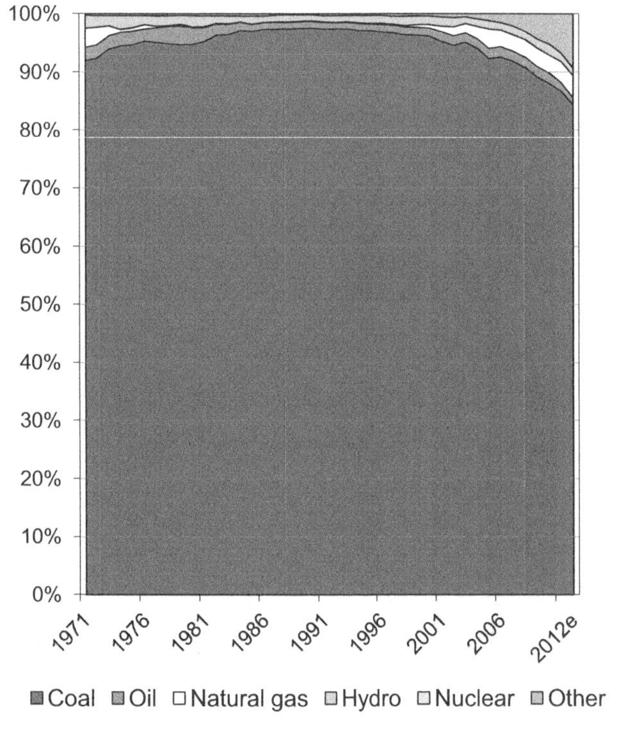

Figure 8: CO_2 emissions by fuel share

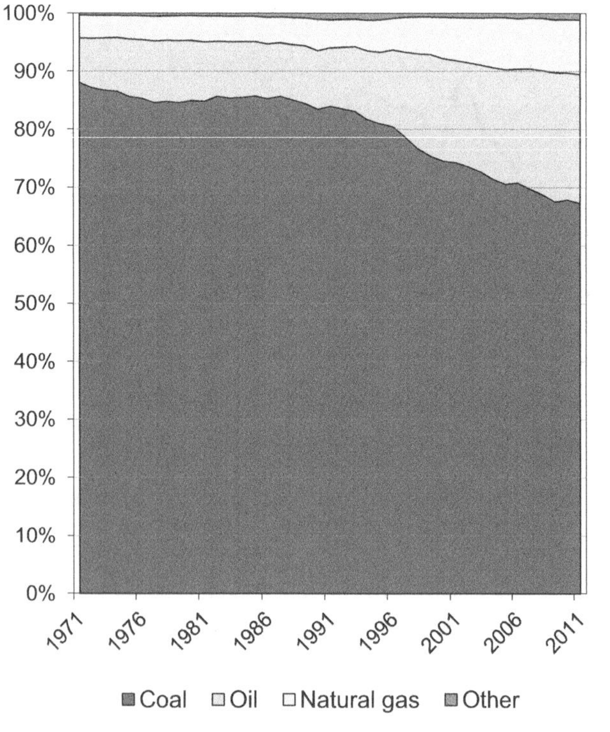

*Please refer to notes and definitions in Part I. Peat is included in coal for all figures.

Source: *IEA/OECD Electricity Information, IEA/OECD CO2 Emissions from Fuel Combustion*

POLAND

1. Total primary energy supply (TPES) by fuel*

	1973	1980	1990	2000	2010	2011	2012e	Average annual percent change 73-90	90-11
TPES (Mtce)	**132.68**	**180.89**	**147.29**	**127.31**	**145.06**	**144.73**	**137.92**	**0.62**	**-0.08**
Coal and peat	106.71	142.57	112.67	80.43	79.14	78.03	71.72	0.32	-1.73
Oil	15.26	23.79	18.62	27.36	36.33	36.55	33.92	1.18	3.26
Natural Gas	8.94	12.53	12.76	14.22	18.29	18.33	19.42	2.12	1.74
Biofuels and waste	1.81	1.74	3.18	5.81	10.85	11.73	12.31	3.37	6.41
Nuclear	-	-	-	-	-	-	-	-	-
Hydro	0.18	0.29	0.17	0.26	0.36	0.29	0.25	-0.16	2.40
Geothermal	-	-	-	0.00	0.02	0.02	0.02	-	-
Solar, wind, tide	-	-	-	0.00	0.22	0.41	0.60	-	-
Net electricity trade(1)	-0.22	-0.03	-0.13	-0.78	-0.17	-0.64	-0.35	-3.02	8.00
Heat(2)	-	-	-	-	0.02	0.02	0.02	-	-

(1) Net trade between OECD and non-OECD countries.

(2) Ambient heat from heat pumps used in the transformation sector.

Source: *IEA/OECD Energy Balances of OECD Countries*

2. Energy supply, GDP and population*

	1973	1980	1990	2000	2010	2011	2012e	Average annual percent change 73-90	90-11
GDP (2005 bil. USD)	157 e	181 e	180	261	383	400	407	0.82	3.87
TPES/GDP(1)	0.85 e	1.00 e	0.82	0.49	0.38	0.36	0.34	-0.20	-3.81
Population (millions)	33.4	35.6	38.0	38.3	38.5	38.5	38.4	0.77	0.06
TPES/population(1)	3.98	5.08	3.87	3.33	3.77	3.76	3.59	-0.15	-0.14
TPES/GDP(2)	195.0 e	229.7 e	188.3	112.3	87.3	83.4	78.0	-0.20	-3.81
Coal and peat supply/GDP(2)	265.2 e	306.2 e	243.7	120.0	80.6	76.0	68.6	-0.50	-5.40
Elec. consumption/GDP(2)	107.1 e	135.1 e	154.0	108.9	89.7	87.9	..	2.16	-2.63
Elec. generation (TWh)	84	121	134	143	157	163	162	2.81	0.93
Industrial production(2)	..	..	47.2	76.8	134.4	143.4	145.2	..	5.43

(1) TPES/GDP in units of tce/2005 thousand USD, TPES/population in units of tce/capita.

(2) As index, 2005 (not shown) = 100

Source: *IEA/OECD Energy Balances of OECD Countries* , *OECD Main Economic Indicators*

3. Coal and peat production by type*

	1978(1)	1990	2000	2005	2010	2011	2012e	Average annual percent change 78-90	90-11
Mtce:									
Coking coal	41.12	28.99	17.24	14.21	11.74	11.54	11.85	-2.87	-4.29
Steam coal	125.04	93.12	67.30	65.96	50.87	50.19	52.55	-2.43	-2.90
Lignite	11.03	19.27	17.32	18.19	16.51	17.93	18.34	4.76	-0.34
Peat	-	-	-	-	-	-	-	-	-
Mt:									
Coking coal	40.85	28.79	17.22	14.07	11.66	11.44	11.74	-2.87	-4.30
Steam coal	151.78	118.94	86.11	83.83	65.07	65.01	68.08	-2.01	-2.84
Lignite	41.01	67.58	59.48	61.64	56.51	62.84	64.28	4.25	-0.35
Peat	-	-	-	-	-	-	-	-	-

(1) Earliest year for which split by coal type is available.

* Please refer to notes and definitions in Part I.

Source: *IEA/OECD Energy Balances of OECD Countries, IEA/OECD Energy Statistics of OECD Countries*

POLAND

4. Final consumption of energy by fuel*

(Mtce)

	1973	1980	1990	2000	2009	2010	2011	Average annual percent change 73-90	90-11
Total final consumption[(1)]	**86.50**	**111.44**	**87.75**	**83.11**	**92.14**	**99.87**	**97.59**	**0.08**	**0.51**
Coal and peat	41.46	45.65	24.77	18.93	15.99	18.97	16.97	-2.98	-1.79
Oil	12.80	18.58	15.61	25.01	31.78	32.95	33.01	1.17	3.63
Natural Gas	6.31	9.95	10.98	11.65	14.04	15.07	14.80	3.31	1.43
Biofuels and wastes	1.16	1.09	2.33	5.55	7.33	8.26	8.64	4.19	6.44
Geothermal	-	-	-	0.00	0.02	0.02	0.02	-	-
Solar, wind, tide	-	-	-	-	0.01	0.01	0.01	-	-
Electricity	7.16	10.45	11.82	12.12	13.84	14.62	14.98	3.00	1.13
Heat	17.62	25.73	22.23	9.84	9.14	9.98	9.16	1.38	-4.13
of which:									
Total industry	**37.49**	**46.00**	**32.86**	**25.01**	**20.06**	**20.91**	**21.90**	**-0.77**	**-1.91**
Coal and peat	14.39	14.89	9.37	10.69	5.17	5.61	6.01	-2.49	-2.09
Oil	2.08	2.95	1.34	2.60	1.82	1.61	1.49	-2.53	0.49
Natural Gas	4.25	5.46	3.58	3.26	4.18	4.43	4.55	-1.00	1.15
Biofuels and wastes	0.24	0.44	0.97	1.41	1.96	2.12	2.26	8.70	4.09
Geothermal	-	-	-	-	-	-	-	-	-
Solar, wind, tide	-	-	-	-	-	-	-	-	-
Electricity	4.68	6.40	5.25	4.97	4.89	5.14	5.48	0.68	0.20
Heat	11.85	15.85	12.34	2.07	2.03	2.01	2.12	0.24	-8.05
Total transport	**12.81**	**13.11**	**10.18**	**13.59**	**22.78**	**24.20**	**24.50**	**-1.34**	**4.27**
Coal and peat	5.23	2.57	0.25	-	-	-	-	-16.42	-
Oil	7.16	9.94	9.26	12.93	21.06	22.22	22.45	1.52	4.31
Natural Gas	-	-	-	0.09	0.38	0.32	0.32	-	-
Biofuels and wastes	-	-	-	-	0.95	1.27	1.33	-	-
Electricity	0.43	0.59	0.67	0.57	0.39	0.40	0.40	2.74	-2.45
Residential	**19.71**	**28.81**	**25.47**	**24.55**	**26.78**	**30.17**	**27.14**	**1.52**	**0.30**
Coal and peat	13.58 e	17.83 e	10.48 e	6.11	8.09	10.07	8.19	-1.52	-1.17
Oil	0.17	0.25	0.06	1.27	1.13	1.00	0.94	-6.21	14.20
Natural Gas	0.44	1.81	4.17	4.35	4.60	5.06	4.62	14.15	0.49
Biofuels and wastes	0.81	0.56	-	3.24	3.50	3.85	3.92	-	-
Geothermal	-	-	-	0.00	0.02	0.02	0.01	-	-
Solar, wind, tide	-	-	-	-	0.01	0.01	0.01	-	-
Electricity	0.67	1.32	2.48	2.58	3.38	3.52	3.47	7.98	1.61
Heat	4.03	7.05	8.28	6.99	6.06	6.65	5.97	4.33	-1.55
Comm & public services	**7.90**	**10.93**	**7.08**	**7.09**	**10.87**	**12.21**	**11.58**	**-0.64**	**2.37**
Coal and peat	6.10 e	8.04 e	3.08 e	0.75	1.14	1.37	1.15	-3.95	-4.60
Oil	-	-	-	0.56	0.86	1.04	0.97	-	-
Natural Gas	0.12	0.41	0.47	1.32	2.58	2.85	2.67	8.40	8.61
Biofuels and waste	0.09	0.06	0.03	0.32	0.27	0.31	0.31	-6.26	11.66
Geothermal	-	-	-	0.00	0.00	0.00	0.00	-	-
Solar, wind, tide	-	-	-	-	0.00	0.00	0.00	-	-
Electricity	0.88	1.18	2.37	3.41	4.98	5.36	5.44	6.01	4.03
Heat	0.71	1.24	1.13	0.74	1.02	1.28	1.05	2.76	-0.36
Non-energy use	**4.60**	**6.50**	**6.01**	**6.22**	**6.60**	**6.97**	**7.25**	**1.59**	**0.90**
Coal and peat	1.03	0.60	0.26	0.10	0.06	0.11	0.13	-7.77	-3.37
Oil	2.10	3.64	3.01	3.49	4.31	4.50	4.54	2.14	1.98
Natural Gas	1.47	2.26	2.74	2.62	2.24	2.36	2.59	3.75	-0.28

(1) Total Final Consumption (TFC) excludes use in transformation processes and energy industry own use.

* Please refer to notes and definitions in Part I. Agriculture, Forestry, Fishing and "Non-specified other" sectors are not reported here individually, but are accounted for in Total final consumption.

Source: *IEA/OECD Energy Balances of OECD Countries*

POLAND

5. Coal balance*

(Mtce)

	1973	1980	1990	2000	2005	2010	2011	2012e	Average annual percent change 73-90	90-11
Production	143.9	171.9	141.4	101.9	98.4	79.1	79.7	82.7	-0.1	-2.7
Imports	1.2	1.0	0.6	1.5	3.1	11.8	12.6	8.2	-4.2	16.0
Exports	-38.5	-30.4	-29.3	-24.8	-21.6	-15.7	-13.4	-13.3	-1.6	-3.6
Stock changes	0.2	0.0	0.0	1.9	-1.8	3.9	-0.9	-6.0		
Primary supply	**106.7**	**142.6**	**112.7**	**80.4**	**78.0**	**79.1**	**78.0**	**71.7**	**0.3**	**-1.7**
Statistical differences	**-4.7**	**-9.8**	**-5.7**	**0.5**	**-1.5**	**-2.2**	**-3.1**	..		
Total transformation	**-57.7 e**	**-84.4 e**	**-80.2 e**	**-59.6 e**	**-58.4 e**	**-56.5**	**-56.6**	..	**2.0**	**-1.6**
Electricity and heat gen.	-52.9 e	-77.0	-74.9	-56.2	-56.4	-54.7	-54.7	..	2.1	-1.5
Main activity producers (1)	*-36.7 e*	*-53.6*	*-56.3*	*-53.4*	*-53.8*	*-53.0*	*-53.0*	..	*2.5*	*-0.3*
Autoproducers	*-16.2 e*	*-23.4*	*-18.7*	*-2.8*	*-2.6*	*-1.7*	*-1.6*	..	*0.8*	*-10.9*
Gas works	-0.5 e	-0.6	-0.0	0.0	0.0	0.0	0.0	..	-14.4	-
Coal transformation(2)	-4.3 e	-6.8 e	-5.2 e	-3.4 e	-2.0 e	-1.8	-1.9	..	1.2	-4.7
BKB plants	*0.0*	*0.0*	*0.0*	*0.0*	-	-	-	..	*-2.9*	-
Blast furnaces	*-3.2 e*	*-4.2 e*	*-3.2 e*	*-1.9 e*	*-1.2 e*	*-1.0*	*-1.1*	..	*0.1*	*-5.0*
Coke ovens	*-1.2*	*-2.7*	*-2.0*	*-1.6*	*-0.8*	*-0.8*	*-0.8*	..	*2.9*	*-4.3*
Patent fuel plants	*0.1*	*0.1*	*0.0*	-	-	-	-	..	*-23.8*	-
Other transformation(3)	-	-	-	-	-	-	-	..	-	-
Energy ind. own use	**-2.8 e**	**-2.7**	**-2.0**	**-2.3**	**-1.5**	**-1.4**	**-1.3**	..	**-2.1**	**-1.9**
Losses	**-0.1 e**	**-0.1**	**-0.0**	**-0.0**	**-0.0**	**-0.1**	**-0.1**	..		
Final consumption(4)	**41.5**	**45.7**	**24.8**	**18.9**	**16.5**	**19.0**	**17.0**	..	**-3.0**	**-1.8**
Industry(5)	14.4	14.9	9.4	10.7	6.9	5.6	6.0	..	-2.5	-2.1
Iron and steel	*3.1 e*	*3.7 e*	*2.6 e*	*3.0 e*	*2.0 e*	*0.9*	*1.1*	..	*-1.0*	*-4.0*
Chemical	*0.7*	*0.7*	*0.4*	*1.8*	*1.1*	*1.7*	*1.7*	..	*-3.7*	*7.7*
Non-metallic minerals	*5.6*	*5.1*	*3.1*	*2.1*	*1.2*	*1.1*	*1.3*	..	*-3.4*	*-4.0*
Paper, pulp and print	*0.1*	*0.1*	*0.1*	*0.5*	*0.5*	*0.3*	*0.4*	..	*-1.0*	*9.1*
Other industry (6)	*5.0*	*5.4*	*3.2*	*3.3*	*2.2*	*1.5*	*1.5*	..	*-2.5*	*-3.7*
Transport(7)	5.2	2.6	0.2	-	-	-	-	..	-16.4	-
Other	20.8	27.6	14.9	8.1	9.4	13.3	10.8	..	-1.9	-1.5
Comm. and pub. services	*6.1 e*	*8.0 e*	*3.1 e*	*0.7*	*1.0*	*1.4*	*1.1*	..	*-3.9*	*-4.6*
Residential	*13.6 e*	*17.8 e*	*10.5 e*	*6.1*	*7.1*	*10.1*	*8.2*	..	*-1.5*	*-1.2*
Other sectors (8)	*1.1*	*1.7*	*1.3*	*1.3*	*1.3*	*1.8*	*1.5*	..	*1.0*	*0.5*
Non-energy use	1.0	0.6	0.3	0.1	0.3	0.1	0.1	..	-7.8	-3.4

(1) Main activity electricity and heat generation includes district heating.

(2) Coal transformation refers to the transformation of coal and peat to secondary and tertiary products (mainly cokes, briquettes, coal tar and off-gases). Gas works and Liquefaction are shown separately to match the *Energy Balances of OECD Countries.*

(3) Other transformation includes Liquefaction and Non-specified transformations.

(4) Final Consumption includes non-energy use and energy use (Industry, Transport and Other).

(5) Please refer to notes and definitions in Part I for detailed explanation.

(6) Other industry includes Non-ferrous metals, Transport equipment, Machinery, Mining and quarrying, Food and tobacco, Wood and wood products, Construction, Textile and leather, and Non-specified industry.

(7) Transport includes Rail and Inland waterways.

(8) Other sectors includes Agriculture/Forestry, Fishing and Non-specified other.

* "Coal" refers to all types of coal, primary (anthracite, coking coal, other bituminous coal, sub-bituminous coal, lignite/brown coal) and derived fuels (including patent fuel, BKB, coke oven coke, gas coke, coal tar, coke oven gas, gas works gas, blast furnace gas and other recovered gases), as well as peat, peat briquettes and oil shale. Quantities have been converted to Mtce using calorific values reported by the respective countries. Please refer to notes and definitions in Part I.

Source: *IEA/OECD Energy Balances of OECD Countries*

POLAND

6. Use of coal for selected end-uses*

(million tonnes)

	1978	1990	2000	2005	2009	2010	2011	Average annual percent change 78-90	90-11
Steam coal	**127.71**	**102.10**	**70.04**	**69.28**	**65.82**	**72.45**	**71.62**	**-1.85**	**-1.67**
Total electricity and heat	84.76	77.55	51.63	50.82	47.01	50.38	48.95	-0.74	-2.17
Main activity producers	*57.61*	*55.18*	*48.60*	*47.98*	*45.59*	*48.91*	*47.51*	*-0.36*	*-0.71*
Autoproducers	*27.15*	*22.38*	*3.03*	*2.84*	*1.42*	*1.47*	*1.45*	*-1.60*	*-12.23*
Patent fuel/BKB plants	1.64	0.08	-	-	-	-	-	-22.50	-
Coke ovens/Liquefaction(1)	-	-	-	-	-	-	-	-	-
Blast furnace inputs	-	-	-	-	-	-	-	-	-
Gas manufacture	1.49	0.38	-	-	-	-	-	-10.73	-
Industry	9.35	6.74	9.70	5.82	5.34	5.60	5.90	-2.69	-0.63
Iron and steel	*0.11*	*0.05*	*0.79*	*0.47*	*0.20*	*0.13*	*0.15*	*-6.00*	*5.47*
Chemical	*0.36*	*0.32*	*2.17*	*1.23*	*1.96*	*2.18*	*2.17*	*-0.96*	*9.59*
Non-metallic minerals	*5.05*	*3.19*	*2.30*	*1.28*	*1.18*	*1.23*	*1.51*	*-3.75*	*-3.49*
Paper, pulp and print	*0.08*	*0.07*	*0.68*	*0.55*	*0.38*	*0.44*	*0.49*	*-1.35*	*9.84*
Other industry	*3.76*	*3.11*	*3.76*	*2.29*	*1.63*	*1.62*	*1.58*	*-1.55*	*-3.19*
Other sectors(2)	26.49	16.18	9.60	10.39	11.81	12.93	11.70	-4.03	-1.53
Non-energy use	0.04	0.01	0.01	0.09	0.04	0.06	0.07	-7.78	8.25
Coking coal	**25.85**	**18.13**	**13.33**	**11.16**	**9.91**	**12.34**	**11.90**	**-2.91**	**-1.98**
Total electricity and heat	..	..	0.05	0.08	0.06	0.08	0.12	..	..
Main activity producers	..	..	*0.00*	-	-	-	-	..	..
Autoproducers	..	..	*0.05*	*0.08*	*0.06*	*0.08*	*0.12*	..	..
Patent fuel/BKB plants	-	-	-	-	-	-	-	-	-
Coke ovens/Liquefaction(1)	25.30	18.21	12.38	11.38	9.37	12.99	12.42	-2.70	-1.81
Blast furnace inputs	-	-	-	-	-	0.03	0.08	-	-
Gas manufacture	1.17	0.16	-	-	-	-	-	-15.08	-
Industry	-	-	0.02	-	0.03	0.01	0.01	-	-
Iron and steel	..	..	-	-	*0.03*	*0.00*	*0.00*	..	..
Chemical	-	-	-	-	-	-	-	-	-
Non-metallic minerals	..	..	*0.00*	-	-	-	-	..	..
Paper, pulp and print	-	-	-	-	-	-	-	-	-
Other industry	..	..	*0.01*	-	*0.00*	*0.01*	*0.01*	..	..
Other sectors(2)	-	-	0.00	-	0.00	0.00	0.00	-	-
Non-energy use	-	-	-	-	-	-	-	-	-
Lignite	**37.71**	**67.39**	**59.49**	**61.59**	**57.08**	**56.59**	**62.71**	**4.96**	**-0.34**
Total electricity and heat	36.91	66.92	59.11	61.08	56.06	55.73	61.80	5.08	-0.38
Main activity producers	*36.45*	*66.46*	*59.10*	*61.07*	*56.06*	*55.73*	*61.80*	*5.13*	*-0.35*
Autoproducers	*0.46*	*0.46*	*0.01*	*0.00*	*0.00*	*0.00*	-	*-0.05*	-
Patent fuel/BKB plants	0.26	0.23	0.04	-	-	-	-	-0.87	-
Coke ovens/Liquefaction(1)	-	-	-	-	-	-	-	-	-
Gas manufacture	0.01	0.02	-	-	-	-	-	10.42	-
Industry	0.08	0.04	0.05	0.01	0.02	0.03	0.06	-4.76	1.80
Iron and steel	-	-	-	-	-	-	-	-	-
Chemical	*0.00*	*0.00*	-	-	-	-	-	*2.43*	-
Non-metallic minerals	*0.02*	*0.02*	*0.01*	-	-	*0.02*	*0.03*	*0.54*	*2.70*
Paper, pulp and print	-	-	-	-	-	-	-	-	-
Other industry	*0.06*	*0.02*	*0.04*	*0.01*	*0.02*	*0.01*	*0.04*	*-7.48*	*1.95*
Other sectors(2)	0.44	0.16	0.25	0.48	0.53	0.86	0.67	-7.84	6.96
Non-energy use	-	0.00	-	-	0.00	0.00	-	-	-

(1) Liquefaction primarily refers to direct distillation processes. Indirect liquefaction may be reported here or under gas manufacture.

(2) Other sectors are Residential, Commercial and public services, Agriculture/Forestry, Fishing and Non-specified other.

* Coal totals may not match the sum of the individual rows due to statistical differences. Please refer to notes and definitions in Part I.

Source: *IEA/OECD Energy Statistics of OECD Countries*

POLAND

7. Fuel prices to end users*

	1978(1)	1990	1995	2000	2005	2009	2010	2011	2012
				(US dollars / tce)					
For electricity generation									
Steam coal	..	17.03	43.52	38.86	65.19	109.84	107.21	116.06	113.53
Heavy fuel oil	..	61.62	69.27	80.55	179.72	280.56	360.13	480.92	510.01
Natural gas	..	..	..	..	160.15	218.11	226.93	248.92	256.45
For industry									
Steam coal	..	..	51.77	48.83	71.39	122.71	125.20	142.48	142.17
Coking coal	..	..	60.05	39.10	114.57	129.55	187.88	253.68	184.15
High sulphur fuel oil	..	61.62	69.27	80.60	181.87	256.43	342.60	469.03	508.98
Low sulphur fuel oil	..	..	..	..	213.05	331.37	418.02	535.69	572.74
Natural gas	..	63.43	100.91	103.47	174.75	336.54	353.71	385.08	397.65
				(1 000 Polish zlotys / unit)(2)					
For electricity generation									
Steam coal	..	0.01	0.08	0.12	0.15	0.25	0.24	0.25	0.27
Heavy fuel oil	..	0.08	0.24	0.49	0.81	1.23	1.52	1.99	2.33
Natural gas	..	..	..	..	0.67	0.87	0.88	0.95	1.07
For industry									
Steam coal	..	..	0.09	0.17	0.20	0.29	0.29	0.32	0.36
Coking coal	..	..	0.15	0.17	0.37	0.41	0.57	0.76	0.60
High sulphur fuel oil	..	0.08	0.24	0.49	0.82	1.11	1.43	1.93	2.30
Low sulphur fuel oil	..	..	..	..	0.96	1.44	1.76	2.21	2.60
Natural gas	..	0.08	0.31	0.58	0.73	1.35	1.37	1.46	1.66

(1) Earliest year for which split by coal type is available.

(2) For steam coal, coking coal and heavy fuel oil per metric tonne; for natural gas per 10^7 kilocalories GCV.

* Please refer to notes and definitions in Part I. Steam coal prices relate to historical definitions.

Source: *IEA/OECD Energy Prices & Taxes*

8. Coal and peat trade by type of coal*

(Mtce)

	1978	1990	1995	2000	2005	2009	2010	2011	2012e
Total imports	**1.04**	**0.56**	**1.54**	**1.46**	**3.08**	**9.34**	**11.81**	**12.65**	**8.25**
Bituminous coal(1)	-	-	0.05	0.17	2.31	6.99	8.48	10.17	6.50
Coking coal	1.04	0.56	1.44	1.27	0.62	2.28	3.18	2.29	1.53
Sub-bituminous coal	-	-	-	-	-	-	-	-	-
Lignite	-	-	0.00	-	-	0.01	0.01	0.02	0.03
Peat	-	-	-	-	-	-	-	-	-
Coal products(2)	-	-	0.04	0.02	0.15	0.06	0.14	0.17	0.18
Total exports	**41.28**	**29.30**	**31.84**	**24.75**	**21.63**	**13.11**	**15.73**	**13.43**	**13.28**
Bituminous coal(1)	22.19	14.41	16.15	15.89	13.60	6.36	7.26	5.02	5.08
Coking coal	16.20	11.34	12.40	5.33	3.19	1.75	1.84	1.68	1.60
Sub-bituminous coal	-	-	-	-	-	-	-	-	-
Lignite	0.90	0.06	0.11	0.00	0.00	0.02	0.03	0.04	0.04
Peat	-	-	-	-	-	-	-	-	-
Coal products(2)	1.98	3.49	3.18	3.53	4.85	4.99	6.60	6.69	6.56

(1) Bituminous coal includes anthracite.

(2) Coal products includes products derived from coal, for example: coke, coal tar, briquettes, patent fuels.

* Please refer to notes and definitions in Part I.

Source: *IEA/OECD Energy Balances of OECD Countries, IEA/OECD Energy Statistics of OECD Countries*

POLAND

9. Total coal imports by origin*

(thousand tonnes)

	1978[1]	1985	1990	1995	2000	2005	2010	2011	2012e
Total coal[2]	**1044**	**1052**	**560**	**1510**	**1452**	**3372**	**13627**	**15031**	**9742**
Coking coal	**1044**	**1052**	**560**	**1433**	**1263**	**610**	**3155**	**2266**	**1516**
Australia	-	-	-	-	-	35	283	137	356
Canada	-	-	-	-	-	-	-	-	-
Czech Republic	-	-	-	1301	500	558	781	772	478
Germany	-	-	-	2	-	17	-	-	-
Poland	-	-	-	-	-	-	-	-	-
United Kingdom	-	-	-	-	-	-	-	-	-
United States	-	-	-	-	-	-	1839	1187	661
Other OECD	-	-	-	-	1	-	-	-	-
China, People's Rep.	-	-	-	-	-	-	-	-	-
Colombia	-	-	-	-	-	-	198	162	-
Indonesia	-	-	-	-	-	-	-	-	-
South Africa	-	-	-	-	-	-	-	-	-
Former Soviet Union[3]	-	-	x	x	x	x	x	x	x
Russian Federation[4]	*x*	*x*	-	*130*	*736*	-	*54*	*4*	*21*
Other FSU[4]	*x*	*x*	-	-	*26*	-	-	*1*	-
Venezuela	-	-	-	-	-	-	-	-	-
Vietnam	-	-	-	-	-	-	-	-	-
Non-specified/other	1044	1052	560	-	-	-	-	3	-
Steam coal	**-**	**-**	**-**	**64**	**189**	**2762**	**10448**	**12689**	**8116**
Australia	-	-	-	-	11	-	-	-	-
Canada	-	-	-	-	-	-	-	-	65
Czech Republic	-	-	-	26	94	29	1408	2156	969
Germany	-	-	-	-	6	4	12	4	2
Poland	-	-	-	-	-	-	-	-	-
United Kingdom	-	-	-	-	-	-	-	2	3
United States	-	-	-	-	1	-	11	132	135
Other OECD	-	-	-	-	2	4	89	3	-
China, People's Rep.	-	-	-	-	-	5	5	6	-
Colombia	-	-	-	-	4	56	146	161	87
Indonesia	-	-	-	-	-	-	-	-	-
South Africa	-	-	-	-	7	-	-	-	-
Former Soviet Union[3]	-	-	x	x	x	x	x	x	x
Russian Federation[4]	*x*	*x*	-	*5*	*13*	*2390*	*8101*	*9306*	*6536*
Other FSU[4]	*x*	*x*	-	*27*	*51*	*274*	*676*	*914*	*287*
Venezuela	-	-	-	-	-	-	-	-	32
Vietnam	-	-	-	-	-	-	-	-	-
Non-specified/other	-	-	-	6	-	-	-	5	-
Lignite	**-**	**-**	**-**	**13**	**-**	**-**	**24**	**76**	**110**

(1) Earliest year for which split by coal type is available.

(2) Total coal consists of steam coal, coking coal and lignite.

(3) For years prior to 1990, or if no details for individual Former Soviet Union republics.

(4) Data for former republics are not available separately for some OECD member countries.

* Please refer to notes and definitions in Part I.

Source: IEA/OECD Coal Statistics

POLAND

10. Coking coal exports by destination*
(thousand tonnes)

	1978(1)	1985	1990	1995	2000	2005	2010	2011	2012e
World	**16042**	**14450**	**11226**	**12296**	**5290**	**3151**	**1815**	**1670**	**1587**
Total OECD	**6619**	**5025**	**2570**	**6200**	**3649**	**2769**	**1797**	**1619**	**1269**
Australia	..	..	..	1	-	-	-	-	-
Austria	470	784	566	247	599	520	367	135	191
Belgium	392	540	105	78	-	-	12	-	2
Canada	..	..	..	-	-	-	-	-	-
Chile	..	..	..	-	-	-	-	-	-
Czech Republic	..	..	..	871	214	523	720	1017	848
Denmark	..	..	..	9	-	-	-	-	-
Estonia	..	..	..	-	-	-	-	-	-
Finland	..	..	203	534	717	13	-	-	20
France	1311	185	254	633	72	-	-	-	-
Germany	..	82	116	303	131	148	2	-	5
Greece	54	..	..	-	-	-	-	-	-
Hungary	..	..	..	437	266	219	119	85	72
Iceland	..	..	..	3	-	-	-	-	-
Ireland	..	..	2	-	-	-	-	-	-
Israel	..	..	..	-	-	-	-	-	-
Italy	1525	839	158	17	-	295	-	-	-
Japan	429	..	..	-	-	-	-	-	-
Korea	..	..	..	-	-	-	-	-	-
Luxembourg	..	..	..	10	-	-	-	-	-
Mexico	..	..	..	-	-	-	-	-	-
Netherlands	376	315	98	757	323	48	-	-	-
New Zealand	..	..	..	-	-	-	-	-	-
Norway	125	134	..	156	53	-	-	-	-
Poland	..	..	..	-	-	-	-	-	-
Portugal	122	125	40	110	-	-	-	-	-
Slovak Republic	..	..	..	1063	669	622	499	382	131
Slovenia	..	..	..	-	-	-	-	-	-
Spain	1369	637	234	58	7	99	-	-	-
Sweden	71	224	..	310	494	-	-	-	-
Switzerland	..	..	..	2	-	-	-	-	-
Turkey	..	..	..	178	100	212	78	-	-
United Kingdom	375	1160	794	421	4	-	-	-	-
United States	..	..	..	2	-	70	-	-	-
Total non-OECD	**1472**	**3797**	**4576**	**6096**	**1641**	**382**	**18**	**51**	**318**
Brazil	..	861	1249	790	143	-	-	-	-
China(2)	..	..	..	-	-	-	-	-	-
Chinese Taipei	..	..	..	-	-	-	-	-	-
Egypt	..	..	..	48	529	355	-	-	-
India	..	..	284	-	-	-	-	-	-
Romania	..	..	100	156	62	-	-	-	-
Oth. Africa & Mid. East	..	..	..	-	2	-	-	-	-
Oth. non-OECD Americas	..	861	1249	-	2	-	-	-	-
Other Asia & Oceania	..	..	..	-	-	-	-	-	-
Other non-OECD Europe and Eurasia	1472	2075	1694	5102	903	27	18	51	318
Non-specified/Other	**7951**	**5628**	**4080**	**-**	**-**	**-**	**-**	**-**	**-**

(1) Earliest year for which split by coal type is available.
(2) China includes the People's Republic of China and Hong Kong, China.
* Please refer to notes and definitions in Part I.
Source: IEA/OECD Coal Statistics

POLAND

11. Steam coal exports by destination*

(thousand tonnes)

	1978(1)	1985	1990	1995	2000	2005	2010	2011	2012e
World	**24064**	**21674**	**16839**	**19572**	**17955**	**16218**	**8150**	**5337**	**5402**
Total OECD	**16188**	**15271**	**13038**	**14569**	**17132**	**15745**	**8053**	**5226**	**5280**
Australia	..	..	..	-	-	-	-	-	-
Austria	260	949	1202	591	1213	1187	443	297	594
Belgium	105	59	242	306	375	649	216	1	78
Canada	..	..	..	-	-	-	-	-	-
Chile	..	..	..	-	-	-	-	-	-
Czech Republic	274	1378	2282	1874	863	717	684	840	674
Denmark	3078	2056	972	3161	2214	837	437	59	60
Estonia	..	..	..	-	3	-	-	-	-
Finland	4089	2407	2609	2556	1175	653	185	37	128
France	3441	883	141	95	1336	1230	583	-	212
Germany	2041	2709	2583	2649	6396	6906	4007	2666	2733
Greece	..	..	..	-	1	-	-	-	-
Hungary	..	..	..	315	279	240	54	50	23
Iceland	..	..	..	-	-	9	-	-	-
Ireland	331	735	458	308	196	286	228	206	134
Israel	..	..	..	-	-	-	-	-	-
Italy	1437	1150	507	285	913	245	-	-	-
Japan	..	..	..	-	-	-	-	-	-
Korea	..	..	..	-	-	-	-	-	-
Luxembourg	..	..	..	-	-	-	-	-	-
Mexico	..	..	..	-	-	-	-	-	-
Netherlands	264	949	1043	408	208	222	73	-	-
New Zealand	..	..	..	-	-	-	-	-	-
Norway	..	..	..	35	88	142	73	76	117
Poland	..	..	..	-	-	-	-	-	-
Portugal	..	43	..	146	5	223	-	-	-
Slovak Republic	..	..	..	226	286	250	133	195	170
Slovenia	..	..	..	2	12	9	-	-	-
Spain	..	379	16	-	382	41	23	60	18
Sweden	174	1292	732	768	146	172	132	105	103
Switzerland	8	24	1	-	-	-	-	-	-
Turkey	..	..	..	-	-	66	214	-	147
United Kingdom	41	258	250	843	1040	1614	568	634	89
United States	645	..	..	1	1	47	-	-	-
Total non-OECD	**29**	**23**	**34**	**4974**	**823**	**473**	**97**	**111**	**122**
Brazil	..	..	..	37	-	-	-	-	-
China(2)	..	..	..	-	-	-	-	-	-
Chinese Taipei	..	..	..	-	-	-	-	-	-
Egypt	..	..	..	-	-	-	-	-	-
India	..	..	..	-	-	-	-	-	-
Romania	..	..	16	-	-	-	-	-	26
Oth. Africa & Mid. East	29	23	18	-	1	436	-	-	-
Oth. non-OECD Americas	..	..	..	-	16	-	-	-	-
Other Asia & Oceania	..	..	..	-	3	-	-	3	-
Other non-OECD Europe and Eurasia	..	..	..	4937	803	37	97	108	96
Non-specified/Other	**7847**	**6380**	**3767**	**29**	**-**	**-**	**-**	**-**	**-**

(1) Earliest year for which split by coal type is available.

(2) China includes the People's Republic of China and Hong Kong, China.

* Please refer to notes and definitions in Part I. Steam coal includes all sub-bituminous coal.

Source: IEA/OECD Coal Statistics

PORTUGAL*

Figure 1: TPES by fuel (Mtce)

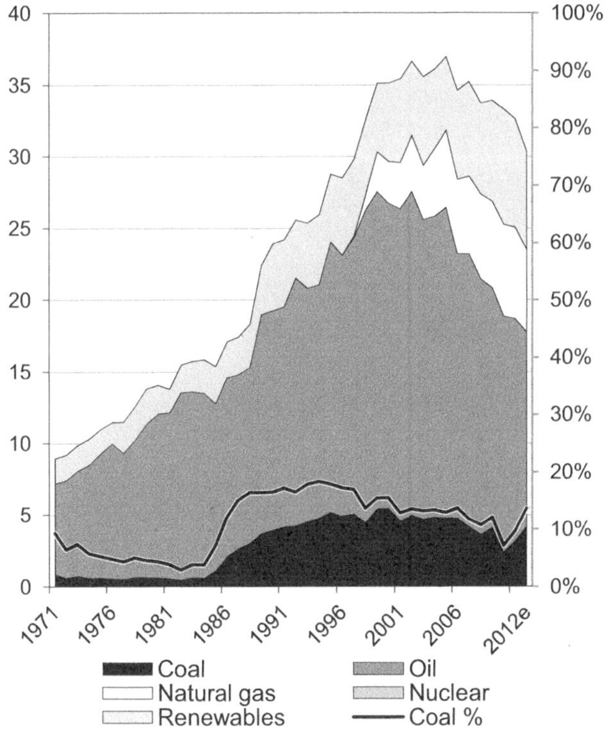

Figure 2: Coal supply indicators (1971 = 100)

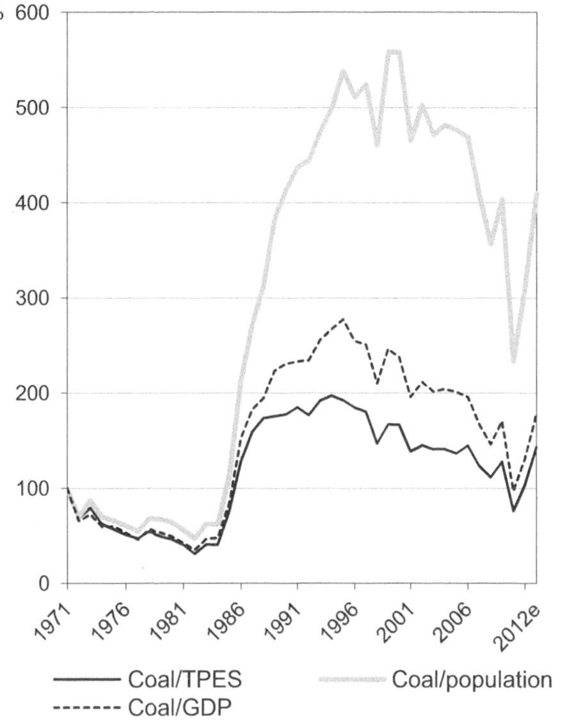

Figure 3: Primary coal supply (Mtce)

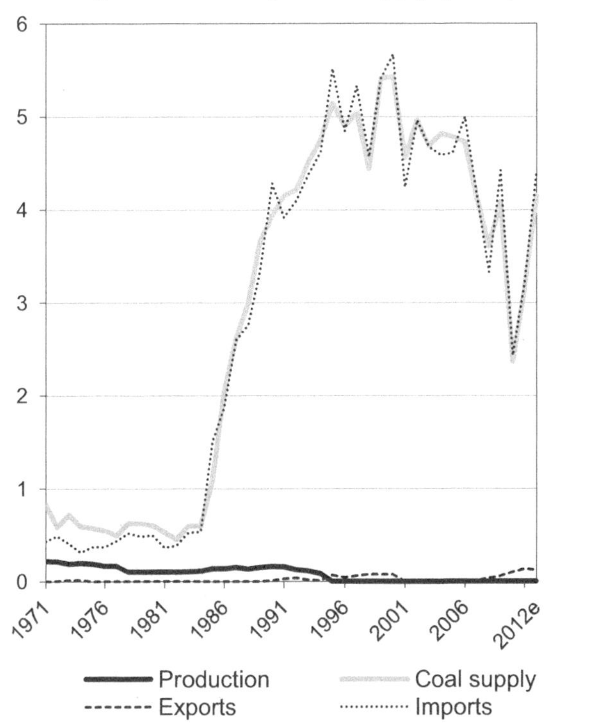

Figure 4: Coal consumption (Mtce)

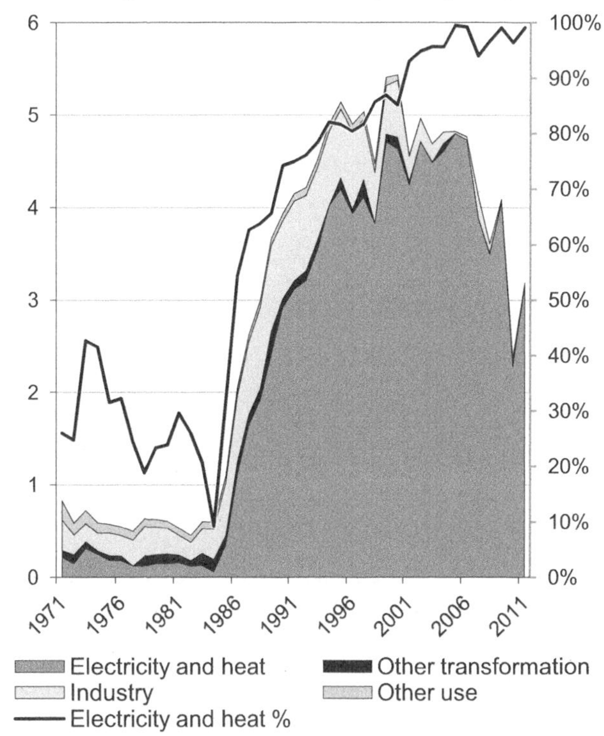

*Please refer to notes and definitions in Part I. Peat is included in coal for all figures.

Source: *IEA/OECD Coal Information, IEA/OECD Energy Balances of OECD Countries, OECD Main Economic Indicators*

PORTUGAL*

Figure 5: Electricity generation by fuel (TWh)

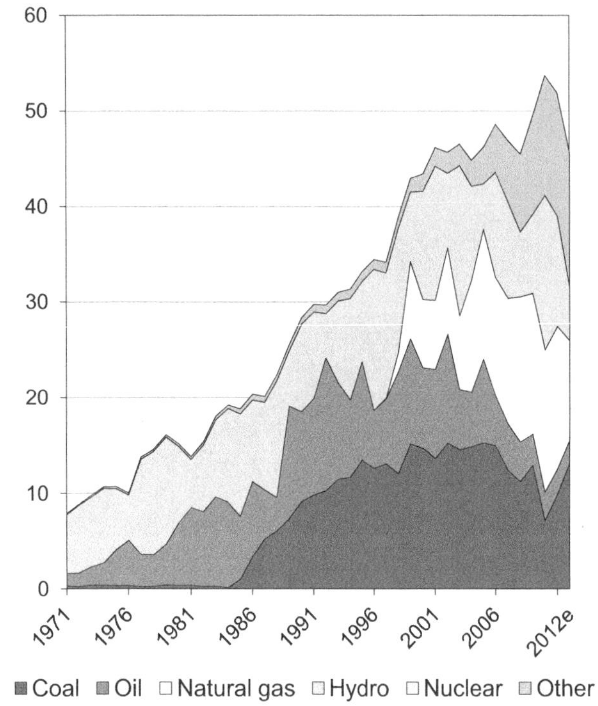

Figure 6: CO_2 emissions by fuel (Mt CO_2)

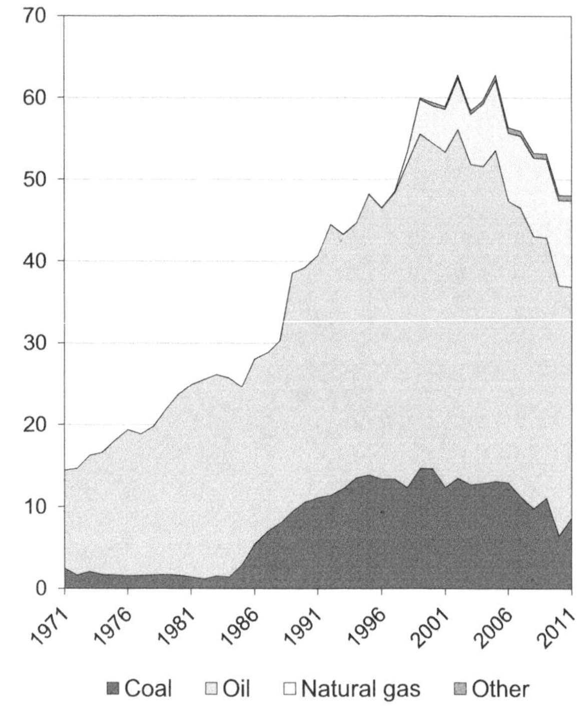

Figure 7: Electricity generation by fuel share

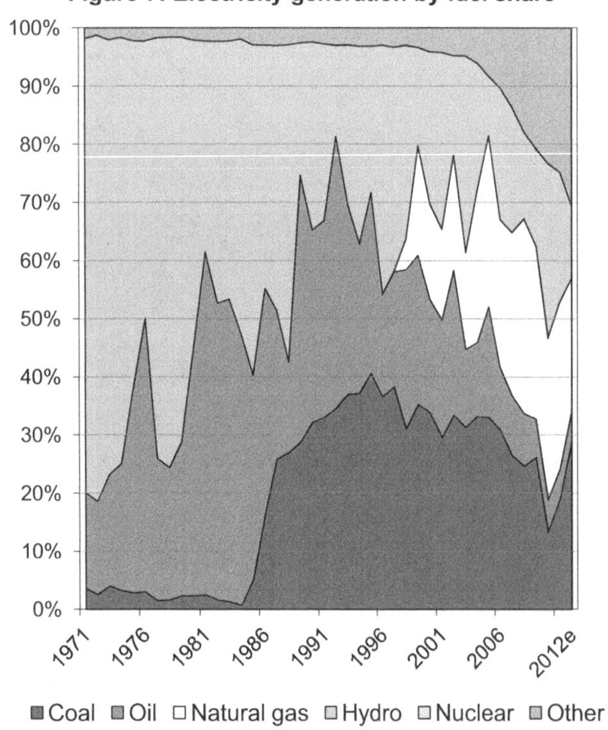

Figure 8: CO_2 emissions by fuel share

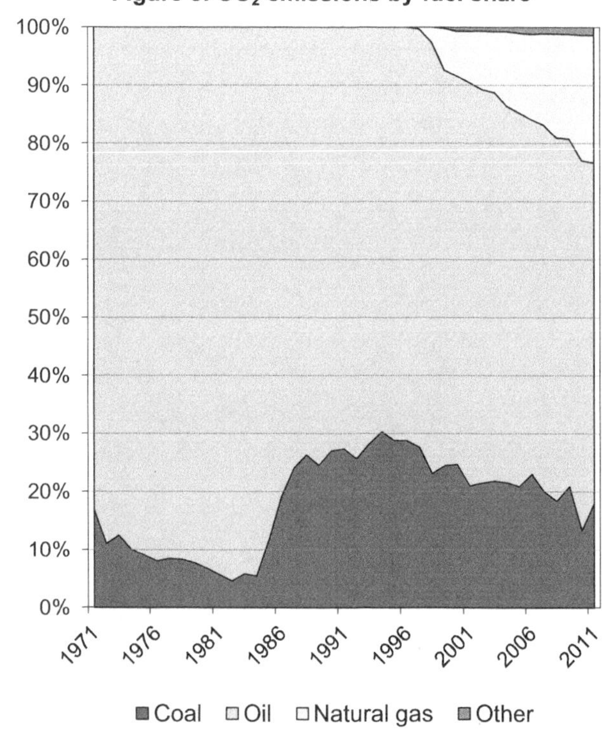

*Please refer to notes and definitions in Part I. Peat is included in coal for all figures.

Source: *IEA/OECD Electricity Information, IEA/OECD CO2 Emissions from Fuel Combustion*

PORTUGAL

1. Total primary energy supply (TPES) by fuel*

	1973	1980	1990	2000	2010	2011	2012e	Average annual percent change 73-90	90-11
TPES (Mtce)	**9.85**	**14.27**	**23.91**	**35.25**	**33.63**	**32.98**	**31.36**	**5.35**	**1.54**
Coal and peat	0.72	0.61	3.94	5.44	2.37	3.16	4.15	10.47	-1.05
Oil	7.32	11.42	15.29	21.30	16.50	15.55	13.59	4.43	0.08
Natural Gas	-	-	-	2.90	6.41	6.37	5.75	-	-
Biofuels and waste	0.91	1.03	3.54	3.96	4.54	4.59	4.62	8.30	1.24
Nuclear	-	-	-	-	-	-	-	-	-
Hydro	0.90	0.98	1.13	1.39	1.98	1.42	0.68	1.32	1.11
Geothermal	-	0.00	0.00	0.10	0.27	0.29	0.19	-	21.82
Solar, wind, tide	-	-	0.02	0.05	1.24	1.26	1.40	-	23.14
Net electricity trade(1)	-0.00	0.22	0.00	0.11	0.32	0.35	0.97	x	22.91
Heat(2)	-	-	-	-	0.00	0.00	0.00	-	-

(1) Net trade between OECD and non-OECD countries.

(2) Ambient heat from heat pumps used in the transformation sector.

Source: *IEA/OECD Energy Balances of OECD Countries*

2. Energy supply, GDP and population*

	1973	1980	1990	2000	2010	2011	2012e	Average annual percent change 73-90	90-11
GDP (2005 bil. USD)	80	100	137	184	197	194	188	3.20	1.66
TPES/GDP(1)	0.12	0.14	0.17	0.19	0.17	0.17	0.17	2.09	-0.11
Population (millions)	8.7	9.9	10.0	10.2	10.6	10.7	10.6	0.81	0.30
TPES/population(1)	1.13	1.45	2.39	3.45	3.16	3.10	2.95	4.51	1.24
TPES/GDP(2)	62.1	72.5	88.3	97.1	86.5	86.2	84.6	2.09	-0.11
Coal and peat supply/GDP(2)	36.1	24.4	114.9	118.4	48.1	65.2	88.6	7.05	-2.66
Elec. consumption/GDP(2)	42.1	59.5	70.9	86.3	104.8	103.2	..	3.12	1.80
Elec. generation (TWh)	10	15	28	43	54	52	46	6.45	2.92
Industrial production(2)	42.5	58.6	93.3	107.1	92.1	90.3	85.8	4.73	-0.16

(1) TPES/GDP in units of tce/2005 thousand USD, TPES/population in units of tce/capita.

(2) As index, 2005 (not shown) = 100

Source: *IEA/OECD Energy Balances of OECD Countries , OECD Main Economic Indicators*

3. Coal and peat production by type*

	1978(1)	1990	2000	2005	2010	2011	2012e	Average annual percent change 78-90	90-11
Mtce:									
Coking coal	-	-	-	-	-	-	-	-	-
Steam coal	0.11	0.16	-	-	-	-	-	3.78	-
Lignite	-	-	-	-	-	-	-	-	-
Peat	-	-	-	-	-	-	-	-	-
Mt:									
Coking coal	-	-	-	-	-	-	-	-	-
Steam coal	0.18	0.28	-	-	-	-	-	3.78	-
Lignite	-	-	-	-	-	-	-	-	-
Peat	-	-	-	-	-	-	-	-	-

(1) Earliest year for which split by coal type is available.

* Please refer to notes and definitions in Part I.

Source: *IEA/OECD Energy Balances of OECD Countries, IEA/OECD Energy Statistics of OECD Countries*

PORTUGAL

4. Final consumption of energy by fuel*

(Mtce)

	1973	1980	1990	2000	2009	2010	2011	Average annual percent change 73-90	90-11
Total final consumption(1)	**8.19**	**11.30**	**19.07**	**27.77**	**27.04**	**27.06**	**25.89**	**5.09**	**1.47**
Coal and peat	0.34	0.36	0.92	0.68	0.03	0.07	0.03	6.00	-15.20
Oil	6.02	8.24	11.87	17.58	14.56	14.45	13.42	4.08	0.59
Natural Gas	-	-	-	1.13	2.05	2.23	2.35	-	-
Biofuels and wastes	0.83	0.91	3.32	3.45	4.01	3.59	3.56	8.52	0.33
Geothermal	-	-	-	0.00	0.01	0.01	0.01	-	-
Solar, wind, tide	-	-	0.02	0.03	0.05	0.08	0.10	-	9.08
Electricity	1.00	1.76	2.89	4.71	5.88	6.13	5.94	6.42	3.49
Heat	-	0.03	0.04	0.19	0.45	0.48	0.48	-	12.46
of which:									
Total industry	**3.32**	**4.76**	**6.65**	**8.89**	**7.48**	**7.80**	**7.62**	**4.17**	**0.65**
Coal and peat	0.20	0.28	0.85	0.62	0.03	0.07	0.03	8.87	-14.85
Oil	2.03	3.00	2.58	3.38	1.64	1.51	1.31	1.40	-3.18
Natural Gas	-	-	-	0.94	1.36	1.50	1.64	-	-
Biofuels and wastes	0.46	0.44	1.68	1.81	2.03	2.11	2.11	7.97	1.08
Geothermal	-	-	-	-	-	-	-	-	-
Solar, wind, tide	-	-	-	-	-	-	-	-	-
Electricity	0.63	1.01	1.50	1.96	1.99	2.15	2.08	5.26	1.57
Heat	-	0.03	0.04	0.18	0.43	0.45	0.45	-	12.15
Total transport	**2.28**	**3.23**	**4.63**	**8.53**	**9.27**	**9.19**	**8.56**	**4.26**	**2.96**
Coal and peat	0.03	0.00	-	-	-	-	-	-	-
Oil	2.23	3.20	4.60	8.49	8.88	8.66	8.06	4.35	2.71
Natural Gas	-	-	-	0.00	0.02	0.02	0.02	-	-
Biofuels and wastes	-	-	-	-	0.32	0.46	0.43	-	-
Electricity	0.03	0.03	0.04	0.04	0.06	0.06	0.05	2.09	1.27
Residential	**1.15**	**1.47**	**3.26**	**4.01**	**4.58**	**4.26**	**3.98**	**6.30**	**0.95**
Coal and peat	0.08	0.04	0.06	0.04	-	-	-	-1.80	-
Oil	0.50	0.56	0.83	0.97	0.76	0.97	0.84	3.08	0.05
Natural Gas	-	-	-	0.10	0.38	0.43	0.37	-	-
Biofuels and wastes	0.37	0.47	1.64	1.64	1.66	1.02	1.02	9.13	-2.26
Geothermal	-	-	-	-	-	-	-	-	-
Solar, wind, tide	-	-	0.01	0.02	0.03	0.05	0.06	-	12.24
Electricity	0.21	0.40	0.73	1.24	1.74	1.78	1.69	7.73	4.10
Heat	-	-	-	0.01	0.01	0.01	0.01	-	-
Comm & public services	**0.36**	**0.62**	**0.86**	**1.99**	**2.93**	**2.71**	**2.64**	**5.26**	**5.50**
Coal and peat	0.02	0.02	0.02	0.02	-	-	-	-0.37	-
Oil	0.20	0.29	0.24	0.48	0.63	0.35	0.26	0.95	0.48
Natural Gas	-	-	-	0.08	0.29	0.28	0.31	-	-
Biofuels and waste	-	-	-	-	-	-	-	-	-
Geothermal	-	-	-	0.00	0.01	0.01	0.01	-	-
Solar, wind, tide	-	-	0.01	0.01	0.02	0.03	0.04	-	6.43
Electricity	0.14	0.31	0.59	1.39	1.96	2.01	2.00	8.92	5.95
Heat	-	-	-	0.00	0.01	0.02	0.02	-	-
Non-energy use	**0.54**	**0.74**	**3.01**	**3.33**	**2.18**	**2.44**	**2.47**	**10.64**	**-0.93**
Coal and peat	-	-	-	-	-	-	-	-	-
Oil	0.54	0.74	3.01	3.33	2.18	2.44	2.47	10.64	-0.93
Natural Gas	-	-	-	-	-	-	-	-	-

(1) Total Final Consumption (TFC) excludes use in transformation processes and energy industry own use.

* Please refer to notes and definitions in Part I. Agriculture, Forestry, Fishing and "Non-specified other" sectors are not reported here individually, but are accounted for in Total final consumption.

Source: *IEA/OECD Energy Balances of OECD Countries*

PORTUGAL

5. Coal balance*

(Mtce)

	1973	1980	1990	2000	2005	2010	2011	2012e	Average annual percent change 73-90	90-11
Production	0.2	0.1	0.2	-	-	-	-	-	-0.8	-
Imports	0.4	0.5	4.3	5.7	4.6	2.4	3.2	4.4	14.9	-1.4
Exports	-0.0	-	-0.0	-0.1	-	-0.1	-0.1	-0.1	-2.2	12.9
Stock changes	0.1	0.0	-0.5	-0.2	0.2	0.0	0.1	-0.1		
Primary supply	**0.7**	**0.6**	**3.9**	**5.4**	**4.8**	**2.4**	**3.2**	**4.2**	**10.5**	**-1.0**
Statistical differences	**0.1**	**0.0**	**-0.0**	**-0.0**	**0.0**	**-0.0**	**0.0**	..		
Total transformation	**-0.4 e**	**-0.2 e**	**-3.0 e**	**-4.7 e**	**-4.8**	**-2.3**	**-3.2**	..	**12.6**	**0.3**
Electricity and heat gen.	-0.3	-0.1	-2.9	-4.6	-4.8	-2.3	-3.2	..	14.1	0.4
Main activity producers (1)	*-0.3*	*-0.1*	*-2.9*	*-4.6*	*-4.8*	*-2.3*	*-3.2*	..	*14.1*	*0.4*
Autoproducers	-	*-0.0*	*-0.0*	*-0.1*	-	-	-	..	-	-
Gas works	0.1	0.1	0.1	0.1	-	-	-	..	1.2	-
Coal transformation(2)	-0.2 e	-0.1 e	-0.1 e	-0.1 e	-	-	-	..	-1.5	-
BKB plants	-	-	-	-	-	-	-	..	-	-
Blast furnaces	*-0.1* e	*-0.1* e	*-0.1* e	*-0.1* e	-	-	-	..	*-2.3*	-
Coke ovens	*-0.0*	*-0.0*	*-0.0*	*-0.0*	-	-	-	..	*-0.1*	-
Patent fuel plants	*0.0*	-	-	-	-	-	-	..	-	-
Other transformation(3)	-	-	-	-	-	-	-	..	-	-
Energy ind. own use	**-0.0**	**-0.0**	**-0.0**	**-0.0**	**-**	**-**	**-**	..	**0.5**	**-**
Losses	**-0.0**	**-0.0**	**-0.0**	**-0.0**	**-**	**-**	**-**	..		
Final consumption(4)	**0.3**	**0.4**	**0.9**	**0.7**	**0.0**	**0.1**	**0.0**	..	**6.0**	**-15.2**
Industry(5)	0.2	0.3	0.8	0.6	0.0	0.1	0.0	..	8.9	-14.9
Iron and steel	*0.1* e	*0.2* e	*0.1* e	*0.1* e	-	*0.0*	*0.0*	..	*0.9*	*-12.8*
Chemical	-	*0.0*	*0.0*	*0.0*	*0.0*	*0.0*	*0.0*	..	-	*0.9*
Non-metallic minerals	*0.0*	*0.0*	*0.7*	*0.4*	*0.0*	*0.0*	*0.0*	..	*26.6*	*-20.9*
Paper, pulp and print	-	-	-	-	-	-	-	..	-	-
Other industry (6)	*0.1*	*0.1*	*0.0*	*0.1*	-	-	-	..	*-9.8*	-
Transport(7)	0.0	0.0	-	-	-	-	-	..	-	-
Other	0.1	0.1	0.1	0.1	-	-	-	..	-2.4	-
Comm. and pub. services	*0.0*	*0.0*	*0.0*	*0.0*	-	-	-	..	*-0.4*	-
Residential	*0.1*	*0.0*	*0.1*	*0.0*	-	-	-	..	*-1.8*	-
Other sectors (8)	*0.0*	*0.0*	-	-	-	-	-	..	-	-
Non-energy use	-	-	-	-	-	-	-	..	-	-

(1) Main activity electricity and heat generation includes district heating.

(2) Coal transformation refers to the transformation of coal and peat to secondary and tertiary products (mainly cokes, briquettes, coal tar and off-gases). Gas works and Liquefaction are shown separately to match the *Energy Balances of OECD Countries.*

(3) Other transformation includes Liquefaction and Non-specified transformations.

(4) Final Consumption includes non-energy use and energy use (Industry, Transport and Other).

(5) Please refer to notes and definitions in Part I for detailed explanation.

(6) Other industry includes Non-ferrous metals, Transport equipment, Machinery, Mining and quarrying, Food and tobacco, Wood and wood products, Construction, Textile and leather, and Non-specified industry.

(7) Transport includes Rail and Inland waterways.

(8) Other sectors includes Agriculture/Forestry, Fishing and Non-specified other.

* "Coal" refers to all types of coal, primary (anthracite, coking coal, other bituminous coal, sub-bituminous coal, lignite/brown coal) and derived fuels (including patent fuel, BKB, coke oven coke, gas coke, coal tar, coke oven gas, gas works gas, blast furnace gas and other recovered gases), as well as peat, peat briquettes and oil shale. Quantities have been converted to Mtce using calorific values reported by the respective countries.
Please refer to notes and definitions in Part I.

Source: *IEA/OECD Energy Balances of OECD Countries*

PORTUGAL

6. Use of coal for selected end-uses*

(million tonnes)

	1978	1990	2000	2005	2009	2010	2011	Average annual percent change 78-90	90-11
Steam coal	**0.21**	**4.08**	**5.66**	**5.48**	**4.68**	**2.70**	**3.70**	**28.21**	**-0.47**
Total electricity and heat	0.17	3.26	5.17	5.44	4.64	2.61	3.69	27.90	0.60
Main activity producers	*0.17*	*3.26*	*5.17*	*5.44*	*4.64*	*2.61*	*3.69*	*27.90*	*0.60*
Autoproducers	-	-	-	-	-	-	-	-	-
Patent fuel/BKB plants	-	-	-	-	-	-	-	-	-
Coke ovens/Liquefaction(1)	-	-	-	-	-	-	-	-	-
Blast furnace inputs	-	-	-	-	-	-	-	-	-
Gas manufacture	-	-	-	-	-	-	-	-	-
Industry	0.04	0.82	0.48	0.02	0.03	0.08	0.03	29.73	-15.15
Iron and steel	*0.01*	-	-	-	*0.01*	*0.01*	*0.01*	-	-
Chemical	*0.00*	*0.01*	-	*0.02*	*0.02*	*0.01*	*0.01*	*6.99*	*2.13*
Non-metallic minerals	*0.03*	*0.81*	*0.48*	*0.00*	*0.01*	*0.06*	*0.01*	*32.70*	*-20.81*
Paper, pulp and print	-	-	-	-	-	-	-	-	-
Other industry	-	*0.00*	-	-	-	-	-	-	-
Other sectors(2)	0.00	0.00	-	-	-	-	-	5.95	-
Non-energy use	-	-	-	-	-	-	-	-	-
Coking coal	**0.42**	**0.31**	**0.50**	**-**	**-**	**-**	**-**	**-2.48**	**-**
Total electricity and heat	0.00	-	-	-	-	-	-	-	-
Main activity producers	*0.00*	-	-	-	-	-	-	-	-
Autoproducers	-	-	-	-	-	-	-	-	-
Patent fuel/BKB plants	-	-	-	-	-	-	-	-	-
Coke ovens/Liquefaction(1)	0.34	0.31	0.49	-	-	-	-	-0.59	-
Blast furnace inputs	-	-	-	-	-	-	-	-	-
Gas manufacture	-	-	-	-	-	-	-	-	-
Industry	0.06	-	-	-	-	-	-	-	-
Iron and steel	-	-	-	-	-	-	-	-	-
Chemical	-	-	-	-	-	-	-	-	-
Non-metallic minerals	-	-	-	-	-	-	-	-	-
Paper, pulp and print	-	-	-	-	-	-	-	-	-
Other industry	*0.06*	-	-	-	-	-	-	-	-
Other sectors(2)	0.01	-	-	-	-	-	-	-	-
Non-energy use	-	-	-	-	-	-	-	-	-
Lignite	**-**	**-**	**-**	**-**	**-**	**-**	**-**	**-**	**-**
Total electricity and heat	-	-	-	-	-	-	-	-	-
Main activity producers	-	-	-	-	-	-	-	-	-
Autoproducers	-	-	-	-	-	-	-	-	-
Patent fuel/BKB plants	-	-	-	-	-	-	-	-	-
Coke ovens/Liquefaction(1)	-	-	-	-	-	-	-	-	-
Gas manufacture	-	-	-	-	-	-	-	-	-
Industry	-	-	-	-	-	-	-	-	-
Iron and steel	-	-	-	-	-	-	-	-	-
Chemical	-	-	-	-	-	-	-	-	-
Non-metallic minerals	-	-	-	-	-	-	-	-	-
Paper, pulp and print	-	-	-	-	-	-	-	-	-
Other industry	-	-	-	-	-	-	-	-	-
Other sectors(2)	-	-	-	-	-	-	-	-	-
Non-energy use	-	-	-	-	-	-	-	-	-

(1) Liquefaction primarily refers to direct distillation processes. Indirect liquefaction may be reported here or under gas manufacture.

(2) Other sectors are Residential, Commercial and public services, Agriculture/Forestry, Fishing and Non-specified other.

* Coal totals may not match the sum of the individual rows due to statistical differences. Please refer to notes and definitions in Part I.

Source: *IEA/OECD Energy Statistics of OECD Countries*

PORTUGAL

7. Fuel prices to end users*

	1978(1)	1990	1995	2000	2005	2009	2010	2011	2012
				(US dollars / tce)					
For electricity generation									
Steam coal	31.26	58.90	50.80	34.67	77.51	93.32	101.20	129.94	103.39
Heavy fuel oil	48.18	76.31	72.25	108.41	158.83	246.45	x	x	x
Natural gas	x	x	x	..	226.98	281.07	287.18	375.00	416.86
For industry									
Steam coal	..	..	..	37.98	..	..	140.85	192.01	189.83
Coking coal	175.90	52.51	43.92	31.67	..	..	321.28	489.93	428.98
High sulphur fuel oil	54.80	131.62	125.85	161.36	x	x	x	x	x
Low sulphur fuel oil	..	..	141.93	170.14	282.61	410.20	477.42	706.86	804.30
Natural gas	x	x	x	..	264.72	376.51	364.47	453.97	476.70
				(Euro / unit)(2)					
For electricity generation									
Steam coal	5.97	36.47	33.14	32.81	54.42	58.58	66.63	81.45	70.15
Heavy fuel oil	14.47	74.30	74.12	161.32	175.35	242.12	x	x	x
Natural gas	x	x	x	..	234.93	260.12	278.77	346.52	416.98
For industry									
Steam coal	..	..	..	35.94	..	..	108.21	168.79	180.64
Coking coal	38.51	37.27	32.84	34.35	..	..	296.69	358.28	339.59
High sulphur fuel oil	16.46	128.16	129.10	240.10	x	x	x	x	x
Low sulphur fuel oil	..	..	145.60	253.17	312.00	402.99	491.96	693.38	854.05
Natural gas	x	x	x	..	273.98	348.44	353.80	419.48	476.84

(1) Earliest year for which split by coal type is available.

(2) For steam coal, coking coal and heavy fuel oil per metric tonne; for natural gas per 10^7 kilocalories GCV.

* Please refer to notes and definitions in Part I. Steam coal prices relate to historical definitions.

Source: *IEA/OECD Energy Prices & Taxes*

8. Coal and peat trade by type of coal*

(Mtce)

	1978	1990	1995	2000	2005	2009	2010	2011	2012e
Total imports	**0.52**	**4.28**	**5.52**	**5.67**	**4.61**	**4.42**	**2.43**	**3.20**	**4.41**
Bituminous coal(1)	0.01	3.89	4.94	5.18	4.60	4.42	2.43	3.20	4.41
Coking coal	0.40	0.38	0.54	0.49	-	-	-	-	-
Sub-bituminous coal	-	-	-	-	-	-	-	-	-
Lignite	-	-	-	-	-	-	-	-	-
Peat	-	-	-	-	-	-	-	-	-
Coal products(2)	0.11	0.01	0.03	-	0.01	0.00	0.00	0.00	0.00
Total exports	**-**	**0.01**	**0.07**	**0.08**	**-**	**0.06**	**0.10**	**0.14**	**0.12**
Bituminous coal(1)	-	-	-	-	-	0.06	0.10	0.14	0.12
Coking coal	-	-	-	-	-	-	-	-	-
Sub-bituminous coal	-	-	-	-	-	-	-	-	-
Lignite	-	-	-	-	-	-	-	-	-
Peat	-	-	-	-	-	-	-	-	-
Coal products(2)	-	0.01	0.07	0.08	-	-	-	-	-

(1) Bituminous coal includes anthracite.

(2) Coal products includes products derived from coal, for example: coke, coal tar, briquettes, patent fuels.

* Please refer to notes and definitions in Part I.

Source: *IEA/OECD Energy Balances of OECD Countries, IEA/OECD Energy Statistics of OECD Countries*

PORTUGAL

9. Total coal imports by origin*

(thousand tonnes)

	1978[1]	1985	1990	1995	2000	2005	2010	2011	2012e
Total coal[2]	**412**	**1395**	**4669**	**6003**	**6367**	**5272**	**2771**	**3753**	**5176**
Coking coal	**404**	**356**	**377**	**540**	**487**	**-**	**-**	**-**	**-**
Australia	-	-	-	191	-	-	-	-	-
Canada	-	-	162	212	289	-	-	-	-
Czech Republic	-	-	-	-	-	-	-	-	-
Germany	8	-	-	-	-	-	-	-	-
Poland	122	125	40	110	-	-	-	-	-
United Kingdom	3	-	20	-	-	-	-	-	-
United States	257	231	155	27	198	-	-	-	-
Other OECD	-	-	-	-	-	-	-	-	-
China, People's Rep.	-	-	-	-	-	-	-	-	-
Colombia	-	-	-	-	-	-	-	-	-
Indonesia	-	-	-	-	-	-	-	-	-
South Africa	-	-	-	-	-	-	-	-	-
Former Soviet Union[3]	-	-	x	x	x	x	x	x	x
Russian Federation [4]	*x*	*x*	-	-	-	-	-	-	-
Other FSU [4]	*x*	*x*	-	-	-	-	-	-	-
Venezuela	-	-	-	-	-	-	-	-	-
Vietnam	-	-	-	-	-	-	-	-	-
Non-specified/other	14	-	-	-	-	-	-	-	-
Steam coal	**8**	**1039**	**4292**	**5463**	**5880**	**5272**	**2771**	**3753**	**5176**
Australia	-	34	-	265	-	-	-	-	-
Canada	-	-	-	-	-	-	-	-	-
Czech Republic	-	-	-	-	-	-	-	-	-
Germany	3	2	-	-	-	-	-	-	-
Poland	-	43	-	-	-	222	-	-	-
United Kingdom	-	111	177	-	-	-	-	-	-
United States	-	669	1555	1431	343	378	609	785	1124
Other OECD	-	68	4	-	19	228	221	89	10
China, People's Rep.	-	-	-	-	-	-	-	-	-
Colombia	-	-	390	727	2846	2314	1405	2857	3721
Indonesia	-	-	-	-	156	144	-	-	-
South Africa	5	104	2112	2654	2426	1985	483	-	295
Former Soviet Union[3]	-	7	x	x	x	x	x	x	x
Russian Federation [4]	*x*	*x*	*54*	*80*	*90*	-	*53*	*10*	*15*
Other FSU [4]	*x*	*x*	-	-	-	*1*	-	*12*	*11*
Venezuela	-	-	-	306	-	-	-	-	-
Vietnam	-	-	-	-	-	-	-	-	-
Non-specified/other	-	1	-	-	-	-	-	-	-
Lignite	**-**	**-**	**-**	**-**	**-**	**-**	**-**	**-**	**-**

(1) Earliest year for which split by coal type is available.

(2) Total coal consists of steam coal, coking coal and lignite.

(3) For years prior to 1990, or if no details for individual Former Soviet Union republics.

(4) Data for former republics are not available separately for some OECD member countries.

* Please refer to notes and definitions in Part I.

Source: IEA/OECD Coal Statistics

PORTUGAL

11. Steam coal exports by destination*

(thousand tonnes)

	1978(1)	1985	1990	1995	2000	2005	2010	2011	2012e
World	-	-	-	-	-	-	**114**	**158**	**141**
Total OECD	-	-	-	-	-	-	**94**	**132**	**136**
Australia	-	-	-	-	-	-	-	-	6
Austria	-	-	-	-	-	-	-	-	-
Belgium	-	-	-	-	-	-	-	-	-
Canada	-	-	-	-	-	-	22	-	-
Chile	-	-	-	-	-	-	-	-	-
Czech Republic	-	-	-	-	-	-	-	-	-
Denmark	-	-	-	-	-	-	-	-	-
Estonia	-	-	-	-	-	-	-	-	-
Finland	-	-	-	-	-	-	-	-	-
France	-	-	-	-	-	-	4	33	28
Germany	-	-	-	-	-	-	-	-	-
Greece	-	-	-	-	-	-	-	4	5
Hungary	-	-	-	-	-	-	-	-	-
Iceland	-	-	-	-	-	-	-	-	-
Ireland	-	-	-	-	-	-	-	2	-
Israel	-	-	-	-	-	-	-	-	-
Italy	-	-	-	-	-	-	-	-	-
Japan	-	-	-	-	-	-	-	-	-
Korea	-	-	-	-	-	-	-	-	-
Luxembourg	-	-	-	-	-	-	-	-	-
Mexico	-	-	-	-	-	-	-	-	-
Netherlands	-	-	-	-	-	-	-	-	-
New Zealand	-	-	-	-	-	-	-	-	-
Norway	-	-	-	-	-	-	58	69	80
Poland	-	-	-	-	-	-	-	-	-
Portugal	-	-	-	-	-	-	-	-	-
Slovak Republic	-	-	-	-	-	-	-	-	-
Slovenia	-	-	-	-	-	-	-	-	-
Spain	-	-	-	-	-	-	10	21	17
Sweden	-	-	-	-	-	-	-	-	-
Switzerland	-	-	-	-	-	-	-	-	-
Turkey	-	-	-	-	-	-	-	-	-
United Kingdom	-	-	-	-	-	-	-	-	-
United States	-	-	-	-	-	-	-	3	-
Total non-OECD	-	-	-	-	-	-	**20**	**26**	**5**
Brazil	-	-	-	-	-	-	-	-	-
China(2)	-	-	-	-	-	-	-	1	-
Chinese Taipei	-	-	-	-	-	-	-	-	-
Egypt	-	-	-	-	-	-	-	-	-
India	-	-	-	-	-	-	-	-	-
Romania	-	-	-	-	-	-	-	-	-
Oth. Africa & Mid. East	-	-	-	-	-	-	-	14	5
Oth. non-OECD Americas	-	-	-	-	-	-	-	-	-
Other Asia & Oceania	-	-	-	-	-	-	-	-	-
Other non-OECD Europe and Eurasia	-	-	-	-	-	-	20	11	-
Non-specified/Other	-	-	-	-	-	-	-	-	-

(1) Earliest year for which split by coal type is available.

(2) China includes the People's Republic of China and Hong Kong, China.

* Please refer to notes and definitions in Part I. Steam coal includes all sub-bituminous coal.

Source: IEA/OECD Coal Statistics

PORTUGAL

12. Coal import values by origin*
(average unit value, CIF, USD/tonne)

	1990	1995	2000	2005	2007	2008	2009	2010	2011/12(1)
Coking coal(2)	**75.65**	**55.35**	**46.39**	**231.71**	**360.89**	**..**	**..**	**..**	**..**
Imports from:									
Australia	..	51.91	..	..	..	..	..	..	..
Canada	73.86	56.95	42.28	..	..	..	..	..	..
Czech Republic	..	..	..	..	..	..	..	..	..
Poland	57.60	57.64	..	..	..	..	..	..	..
United States	63.00	..	50.09	..	..	..	..	..	..
China	..	..	..	..	..	..	..	..	..
Colombia	..	..	..	..	..	..	..	..	..
Indonesia	..	..	..	..	..	..	..	..	..
South Africa	..	..	..	..	..	..	..	..	..
Former Soviet Union(3)	..	..	..	..	..	..	..	..	..
Other bituminous coal(4)	**54.62**	**44.77**	**29.35**	**64.49**	**77.76**	**139.87**	**80.41**	**91.70**	**..**
Imports from:									
Australia	..	44.95	..	..	..	..	..	..	..
Canada	..	..	..	..	..	..	..	..	..
Czech Republic	..	..	..	..	..	..	..	..	..
Poland	..	39.36	47.92	61.53	..	..	..	..	..
United States	59.44	47.42	37.75	62.53	91.17	98.21	85.92	99.26	..
China	..	..	..	..	..	..	..	..	..
Colombia	54.07	42.91	26.97	64.93	76.29	..	75.73	86.34	..
Indonesia	..	..	34.78	65.83	67.70	74.01	..	..	..
South Africa	49.05	44.07	29.80	63.24	76.81	157.27	80.98	96.48	..
Former Soviet Union(3)	45.19	42.86	40.29	..	..	..	69.63	73.90	..

Note: On occasion, shipment of extremely small quantities of high valued coal results in high import costs.

(1) Import data for steam and coking coals are currently unavailable for 2011 and 2012 due to resource constraints.

(2) Weighted average of individual countries using import volumes as weights. Prices exclude intra-EU trade.

(3) Former Soviet Union until 1991, Russian Federation starting in 1992.

(4) Bituminous steam coal only. Weighted average of trades. Prices exclude intra-EU trade.

* Please refer to notes and definitions in Part I.

Source: IEA/OECD Coal Statistics

SLOVAK REPUBLIC*

Figure 1: TPES by fuel (Mtce)

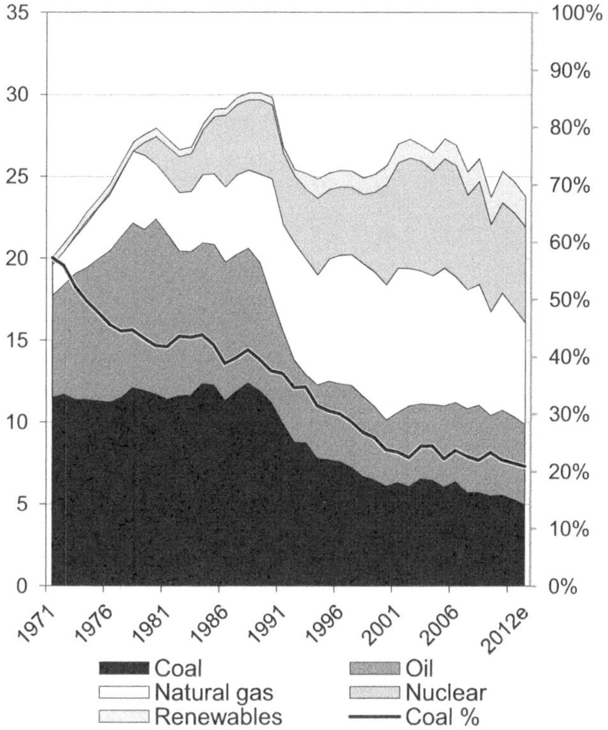

Figure 2: Coal supply indicators (1971 = 100)

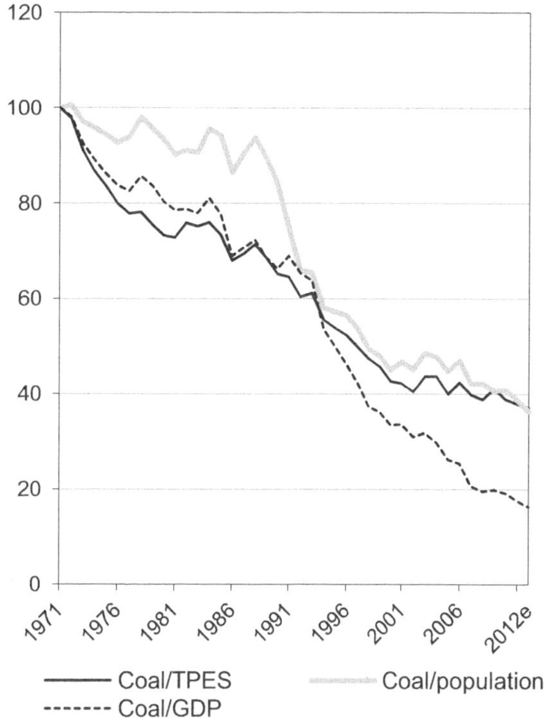

Figure 3: Primary coal supply (Mtce)

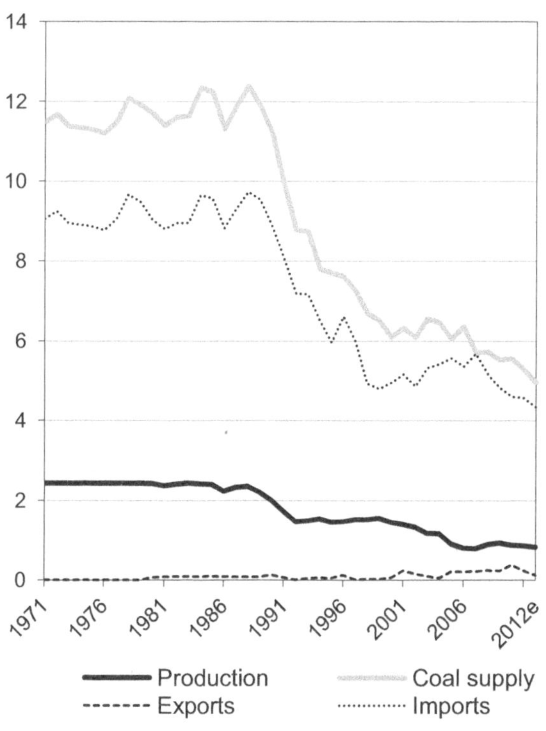

Figure 4: Coal consumption (Mtce)

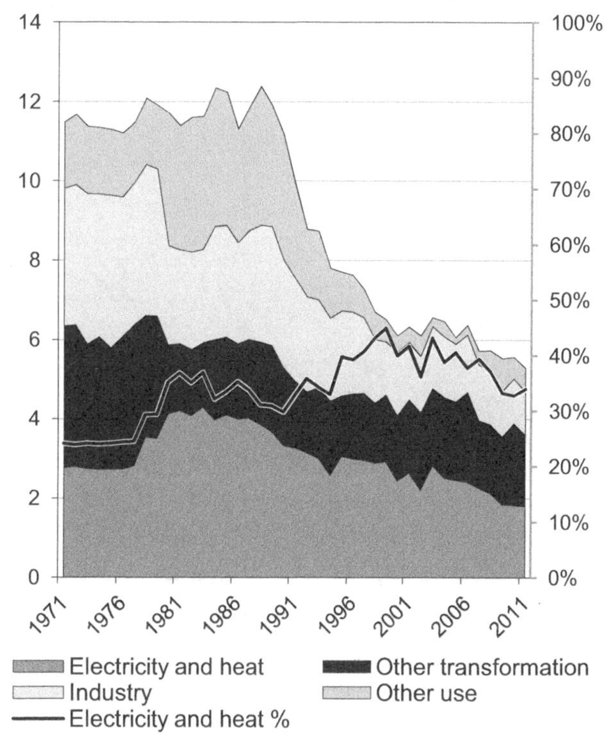

*Please refer to notes and definitions in Part I. Peat is included in coal for all figures.

Source: *IEA/OECD Coal Information, IEA/OECD Energy Balances of OECD Countries, OECD Main Economic Indicators*

SLOVAK REPUBLIC*

Figure 5: Electricity generation by fuel (TWh)

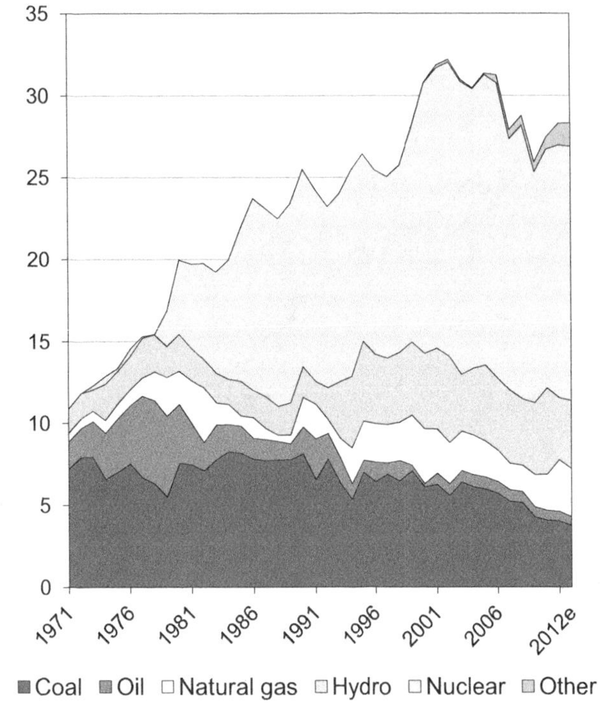

Figure 6: CO_2 emissions by fuel (Mt CO_2)

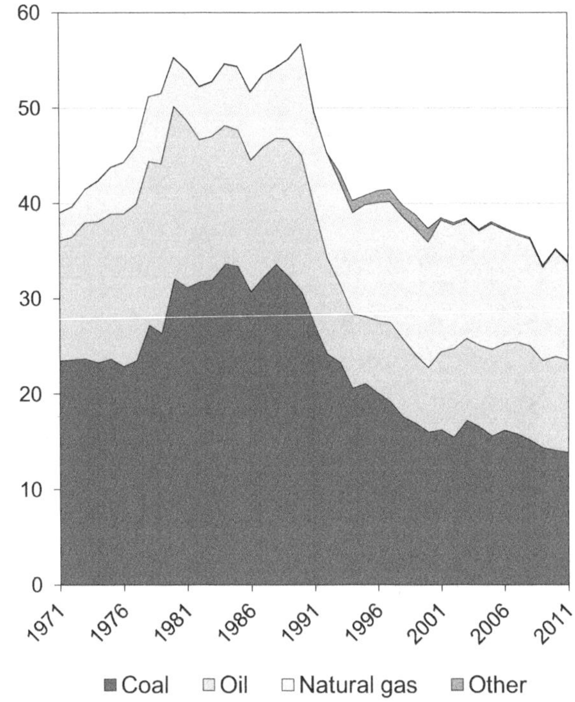

Figure 7: Electricity generation by fuel share

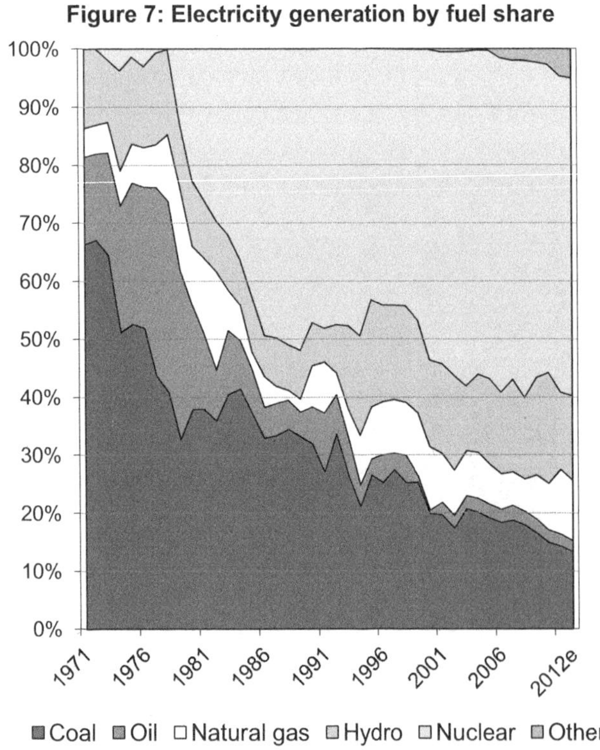

Figure 8: CO_2 emissions by fuel share

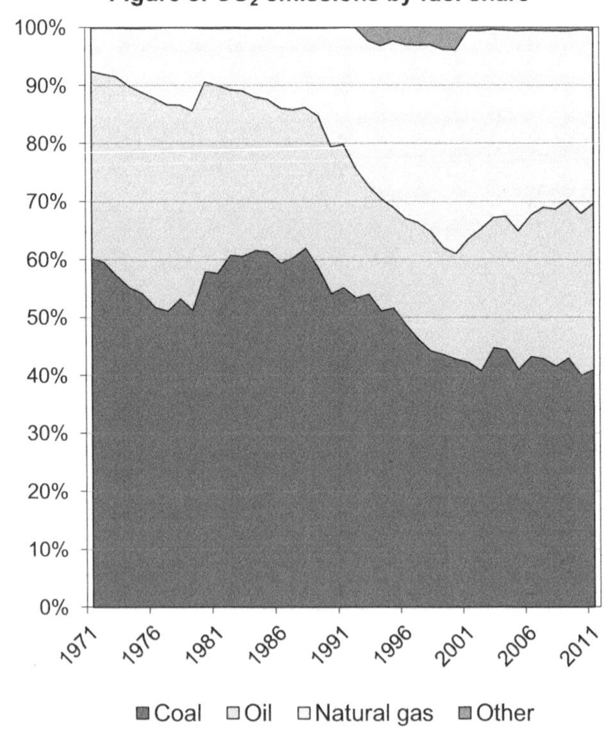

*Please refer to notes and definitions in Part I. Peat is included in coal for all figures.

Source: *IEA/OECD Electricity Information, IEA/OECD CO2 Emissions from Fuel Combustion*

SLOVAK REPUBLIC

1. Total primary energy supply (TPES) by fuel*

	1973	1980	1990	2000	2010	2011	2012e	Average annual percent change 73-90	90-11
TPES (Mtce)	**22.17**	**28.36**	**30.47**	**25.35**	**25.47**	**24.78**	**23.82**	**1.89**	**-0.98**
Coal and peat	11.37	11.71	11.19	6.10	5.57	5.29	4.97	-0.10	-3.50
Oil	7.70	10.71	6.42	4.03	5.17	5.07	4.88	-1.07	-1.12
Natural Gas	2.23	3.31	7.27	8.25	7.15	6.62	6.23	7.20	-0.44
Biofuels and waste	0.27	0.26	0.25	0.59	1.27	1.37	1.26	-0.47	8.48
Nuclear	0.09	1.68	4.48	6.14	5.51	5.81	5.84	26.15	1.25
Hydro	0.16	0.28	0.23	0.57	0.65	0.46	0.51	2.09	3.38
Geothermal	-	-	-	-	0.01	0.01	0.01	-	-
Solar, wind, tide	-	-	-	-	0.01	0.06	0.08	-	-
Net electricity trade(1)	0.35	0.41	0.64	-0.33	0.13	0.09	0.04	3.62	-8.94
Heat(2)	-	-	-	-	-	-	-0.00	-	-

(1) Net trade between OECD and non-OECD countries.

(2) Ambient heat from heat pumps used in the transformation sector.

Source: *IEA/OECD Energy Balances of OECD Countries*

2. Energy supply, GDP and population*

	1973	1980	1990	2000	2010	2011	2012e	Average annual percent change 73-90	90-11
GDP (2005 bil. USD)	25 e	30 e	35 e	38	60	62	63	1.89	2.78
TPES/GDP(1)	0.87 e	0.94 e	0.87 e	0.67	0.42	0.40	0.38	0.00	-3.65
Population (millions)	4.7 e	5.0 e	5.3	5.4	5.4	5.4	5.5	0.75	0.13
TPES/population(1)	4.75 e	5.68 e	5.75	4.69	4.69	4.55	4.37	1.13	-1.10
TPES/GDP(2)	155.2 e	167.3 e	155.3 e	119.7	75.4	71.0	66.9	0.00	-3.65
Coal and peat supply/GDP(2)	353.7 e	307.0 e	253.3 e	128.0	73.2	67.4	62.0	-1.95	-6.11
Elec. consumption/GDP(2)	101.5 e	132.2 e	140.5 e	122.4	84.1	83.7	..	1.93	-2.43
Elec. generation (TWh)	12	20	25	31	27	28	28	4.38	0.50
Industrial production(2)	..	..	86.6	77.2	143.2	152.2	167.2	..	2.72

(1) TPES/GDP in units of tce/2005 thousand USD, TPES/population in units of tce/capita.

(2) As index, 2005 (not shown) = 100

Source: *IEA/OECD Energy Balances of OECD Countries , OECD Main Economic Indicators*

3. Coal and peat production by type*

	1978(1)	1990	2000	2005	2010	2011	2012e	Average annual percent change 78-90	90-11
Mtce:									
Coking coal	-	-	-	-	-	-	-	-	-
Steam coal	-	-	-	-	-	-	-	-	-
Lignite	2.43	2.00	1.45	0.91	0.88	0.86	0.83	-1.62	-3.93
Peat	-	-	-	-	-	-	-	-	-
Mt:									
Coking coal	-	-	-	-	-	-	-	-	-
Steam coal	-	-	-	-	-	-	-	-	-
Lignite	5.80	4.77	3.65	2.51	2.38	2.38	2.29	-1.63	-3.26
Peat	-	-	-	-	-	-	-	-	-

(1) Earliest year for which split by coal type is available.

* Please refer to notes and definitions in Part I.

Source: *IEA/OECD Energy Balances of OECD Countries, IEA/OECD Energy Statistics of OECD Countries*

SLOVAK REPUBLIC

4. Final consumption of energy by fuel*

(Mtce)

	1973	1980	1990	2000	2009	2010	2011	Average annual percent change 73-90	90-11
Total final consumption[(1)]	**15.51**	**18.61**	**22.50**	**16.32**	**15.49**	**16.34**	**15.67**	**2.21**	**-1.71**
Coal and peat	5.49	5.84	5.88	2.02	1.98	1.68	1.69	0.40	-5.76
Oil	5.47	7.20	6.99	4.30	4.14	4.39	4.28	1.44	-2.30
Natural Gas	2.01	2.33	5.59	5.95	4.63	5.29	4.74	6.22	-0.79
Biofuels and wastes	0.27	0.26	0.25	0.46	0.81	0.79	0.81	-0.47	5.77
Geothermal	-	-	-	-	0.00	0.00	0.00	-	-
Solar, wind, tide	-	-	-	-	0.00	0.01	0.01	-	-
Electricity	1.51	2.34	2.88	2.70	2.84	2.97	3.05	3.85	0.28
Heat	0.75	0.64	0.93	0.88	1.10	1.22	1.10	1.22	0.81
of which:									
Total industry	**8.06**	**9.06**	**8.69**	**5.19**	**4.46**	**4.61**	**4.65**	**0.45**	**-2.93**
Coal and peat	3.80	2.50	2.71	1.60	1.15	1.13	1.09	-1.96	-4.24
Oil	2.07	4.03	1.96	0.31	0.19	0.16	0.16	-0.31	-11.38
Natural Gas	0.88	0.65	1.90	1.60	1.08	1.27	1.33	4.67	-1.71
Biofuels and wastes	0.27	0.26	0.25	0.46	0.59	0.56	0.55	-0.47	3.84
Geothermal	-	-	-	-	-	-	-	-	-
Solar, wind, tide	-	-	-	-	-	-	-	-	-
Electricity	1.02	1.58	1.84	1.20	1.32	1.34	1.38	3.53	-1.37
Heat	0.03	0.02	0.03	0.03	0.12	0.15	0.15	-0.84	8.92
Total transport	**2.39**	**2.15**	**2.07**	**2.05**	**3.30**	**3.69**	**3.69**	**-0.86**	**2.81**
Coal and peat	-	-	-	-	-	-	-	-	-
Oil	2.32	1.73	1.92	1.93	2.52	2.92	2.87	-1.09	1.92
Natural Gas	-	0.30	-	-	0.59	0.57	0.62	-	-
Biofuels and wastes	-	-	-	-	0.12	0.14	0.14	-	-
Electricity	0.08	0.12	0.14	0.12	0.06	0.07	0.07	3.65	-3.63
Residential	**1.03**	**1.64**	**3.21**	**3.69**	**3.07**	**3.30**	**3.03**	**6.89**	**-0.27**
Coal and peat	-	0.29	0.62	0.08	0.07	0.08	0.07	-	-10.00
Oil	0.02	0.03	0.05	0.02	0.03	0.02	0.01	4.98	-7.66
Natural Gas	0.30	0.59	1.56	2.35	1.72	1.90	1.67	10.10	0.34
Biofuels and wastes	-	-	-	0.00 e	0.06	0.06	0.06	-	-
Geothermal	-	-	-	-	-	-	-	-	-
Solar, wind, tide	-	-	-	-	-	0.01	0.01	-	-
Electricity	0.19	0.29	0.45	0.67	0.54	0.54	0.55	5.24	0.97
Heat	0.52	0.44	0.53	0.58	0.65	0.70	0.65	0.07	1.05
Comm & public services	**0.77**	**3.95**	**5.22**	**3.14**	**2.78**	**3.01**	**2.29**	**11.89**	**-3.85**
Coal and peat	-	2.65	2.36	0.27	0.70	0.40	0.45	-	-7.61
Oil	-	0.41	0.34	0.02	0.03	0.03	0.00	-	-19.98
Natural Gas	0.47	0.54	1.90	1.94	0.82	1.20	0.51	8.53	-6.06
Biofuels and waste	-	-	-	0.00	0.04	0.03	0.03	-	-
Geothermal	-	-	-	-	0.00	0.00	0.00	-	-
Solar, wind, tide	-	-	-	-	0.00	0.00	0.00	-	-
Electricity	0.13	0.20	0.30	0.65	0.87	0.98	1.01	5.04	6.02
Heat	0.17	0.15	0.33	0.26	0.32	0.36	0.29	3.86	-0.69
Non-energy use	**0.70**	**0.36**	**2.29**	**1.95**	**1.71**	**1.54**	**1.78**	**7.20**	**-1.20**
Coal and peat	-	0.05	0.05	0.05	0.06	0.07	0.08	-	2.61
Oil	0.40	0.11	2.25	1.90	1.27	1.16	1.15	10.64	-3.14
Natural Gas	0.30	0.20	-	-	0.38	0.30	0.55	-	-

(1) Total Final Consumption (TFC) excludes use in transformation processes and energy industry own use.

* Please refer to notes and definitions in Part I. Agriculture, Forestry, Fishing and "Non-specified other" sectors are not reported here individually, but are accounted for in Total final consumption.

Source: *IEA/OECD Energy Balances of OECD Countries*

SLOVAK REPUBLIC

5. Coal balance*

(Mtce)

	1973	1980	1990	2000	2005	2010	2011	2012e	Average annual percent change 73-90	90-11
Production	2.4	2.4	2.0	1.5	0.9	0.9	0.9	0.8	-1.1	-3.9
Imports	8.9	9.0	8.9	5.0	5.6	4.6	4.6	4.3	-0.0	-3.1
Exports	-	-0.1	-0.1	-0.1	-0.2	-0.4	-0.2	-0.1	-	2.8
Stock changes	-	0.3	0.5	-0.2	-0.2	0.5	0.1	-0.1		
Primary supply	**11.4**	**11.7**	**11.2**	**6.1**	**6.1**	**5.6**	**5.3**	**5.0**	**-0.1**	**-3.5**
Statistical differences	**-1.9**	**-0.1**	**-0.0**	**0.0**	**-0.1**	**-0.1**	**-0.1**	**..**		
Total transformation	**-4.0 e**	**-5.1 e**	**-4.6 e**	**-3.6 e**	**-3.7 e**	**-3.2**	**-3.0**	**..**	**0.8**	**-2.1**
Electricity and heat gen.	-2.7	-4.1	-3.3	-2.4	-2.5	-1.8	-1.8	..	1.2	-2.9
Main activity producers (1)	*-2.7*	*-3.5*	*-2.9*	*-2.1*	*-2.2*	*-1.6*	*-1.6*	..	*0.3*	*-2.7*
Autoproducers	-	*-0.6*	*-0.5*	*-0.3*	*-0.3*	*-0.2*	*-0.2*	..	-	*-4.4*
Gas works	-	-	-	-	-	-	-	..	-	-
Coal transformation(2)	-1.3 e	-1.0 e	-1.3 e	-1.2 e	-1.3 e	-1.4	-1.2	..	0.1	-0.4
BKB plants	-	-	-	-	-	-	-	..	-	-
Blast furnaces	*-0.6 e*	*-0.9 e*	*-1.2 e*	*-1.0 e*	*-1.3 e*	*-1.3*	*-1.1*	..	*4.4*	*-0.3*
Coke ovens	*-0.7*	*-0.1*	*-0.1*	*-0.2*	*-0.0*	*-0.1*	*-0.0*	..	*-11.8*	*-3.4*
Patent fuel plants	-	-	-	-	-	-	-	..	-	-
Other transformation(3)	-	-	-	-	-	-	-	..	-	-
Energy ind. own use	**-**	**-0.6**	**-0.6**	**-0.4**	**-0.5**	**-0.5**	**-0.5**	**..**	**-**	**-1.2**
Losses	**-0.0**	**-0.0**	**-0.0**	**-0.0**	**-0.1**	**-0.1**	**-0.1**	**..**		
Final consumption(4)	**5.5**	**5.8**	**5.9**	**2.0**	**1.6**	**1.7**	**1.7**	**..**	**0.4**	**-5.8**
Industry(5)	3.8	2.5	2.7	1.6	1.5	1.1	1.1	..	-2.0	-4.2
Iron and steel	*1.5 e*	*1.0 e*	*1.4 e*	*1.0 e*	*1.0 e*	*0.9*	*0.9*	..	*-0.4*	*-2.1*
Chemical	-	*0.3*	*0.3*	*0.1*	*0.0*	*0.0*	-	..	-	-
Non-metallic minerals	-	*0.2*	*0.2*	*0.2*	*0.2*	*0.2*	*0.2*	..	-	*-0.5*
Paper, pulp and print	-	*0.2*	*0.2*	*0.2*	*0.1*	*0.0*	*0.0*	..	-	*-25.7*
Other industry (6)	*2.3*	*0.8*	*0.6*	*0.1*	*0.1*	*0.0*	*0.0*	..	*-7.2*	*-17.9*
Transport(7)	-	-	-	-	-	-	-	..	-	-
Other	1.7	3.3	3.1	0.4	0.1	0.5	0.5	..	3.6	-8.2
Comm. and pub. services	-	*2.7*	*2.4*	*0.3*	*0.0*	*0.4*	*0.4*	..	-	*-7.6*
Residential	-	*0.3*	*0.6*	*0.1*	*0.1*	*0.1*	*0.1*	..	-	*-10.0*
Other sectors (8)	*1.7*	*0.4*	*0.1*	*0.0*	*0.0*	*0.0*	*0.0*	..	*-13.9*	*-21.8*
Non-energy use	-	0.0	0.0	0.0	0.1	0.1	0.1	..	-	2.6

(1) Main activity electricity and heat generation includes district heating.

(2) Coal transformation refers to the transformation of coal and peat to secondary and tertiary products (mainly cokes, briquettes, coal tar and off-gases). Gas works and Liquefaction are shown separately to match the *Energy Balances of OECD Countries.*

(3) Other transformation includes Liquefaction and Non-specified transformations.

(4) Final Consumption includes non-energy use and energy use (Industry, Transport and Other).

(5) Please refer to notes and definitions in Part I for detailed explanation.

(6) Other industry includes Non-ferrous metals, Transport equipment, Machinery, Mining and quarrying, Food and tobacco, Wood and wood products, Construction, Textile and leather, and Non-specified industry.

(7) Transport includes Rail and Inland waterways.

(8) Other sectors includes Agriculture/Forestry, Fishing and Non-specified other.

* "Coal" refers to all types of coal, primary (anthracite, coking coal, other bituminous coal, sub-bituminous coal, lignite/brown coal) and derived fuels (including patent fuel, BKB, coke oven coke, gas coke, coal tar, coke oven gas, gas works gas, blast furnace gas and other recovered gases), as well as peat, peat briquettes and oil shale. Quantities have been converted to Mtce using calorific values reported by the respective countries. Please refer to notes and definitions in Part I.

Source: *IEA/OECD Energy Balances of OECD Countries*

SLOVAK REPUBLIC

6. Use of coal for selected end-uses*

(million tonnes)

	1978	1990	2000	2005	2009	2010	2011	Average annual percent change 78-90	90-11
Steam coal	**3.97**	**2.74**	**2.06**	**2.25**	**1.94**	**1.67**	**1.56**	**-3.04**	**-2.67**
Total electricity and heat	2.21	1.98	1.33	1.54	0.77	0.71	0.68	-0.91	-4.97
Main activity producers	*2.21*	*1.68*	*1.12*	*1.33*	*0.66*	*0.62*	*0.59*	*-2.25*	*-4.87*
Autoproducers	-	*0.30*	*0.21*	*0.21*	*0.11*	*0.09*	*0.09*	-	*-5.54*
Patent fuel/BKB plants	-	-	-	-	-	-	-	-	-
Coke ovens/Liquefaction(1)	-	-	-	-	-	-	-	-	-
Blast furnace inputs	-	-	-	-	-	-	-	-	-
Gas manufacture	-	-	-	-	-	-	-	-	-
Industry	1.24	0.65	0.68	0.70	0.55	0.50	0.51	-5.27	-1.12
Iron and steel	-	*0.39*	*0.36*	*0.46*	*0.31*	*0.33*	*0.34*	-	*-0.65*
Chemical	-	*0.01*	-	-	-	-	-	-	-
Non-metallic minerals	-	*0.12*	*0.20*	*0.17*	*0.17*	*0.16*	*0.17*	-	*1.60*
Paper, pulp and print	-	*0.03*	*0.11*	*0.06*	*0.07*	*0.01*	-	-	-
Other industry	*1.24*	*0.10*	*0.01*	*0.01*	*0.00*	-	*0.00*	*-19.12*	*-19.57*
Other sectors(2)	0.08	0.12	0.05	0.00	0.61	0.44	0.34	3.99	5.11
Non-energy use	-	-	-	-	0.01	0.02	0.03	-	-
Coking coal	**2.18**	**3.14**	**2.60**	**2.74**	**2.39**	**2.49**	**2.50**	**3.08**	**-1.07**
Total electricity and heat	-	-	-	-	-	-	-	-	-
Main activity producers	-	-	-	-	-	-	-	-	-
Autoproducers	-	-	-	-	-	-	-	-	-
Patent fuel/BKB plants	-	-	-	-	-	-	-	-	-
Coke ovens/Liquefaction(1)	2.15	2.92	2.21	2.35	2.01	2.13	2.08	2.56	-1.59
Blast furnace inputs	-	0.18 e	0.35 e	0.38 e	0.38	0.36	0.42	-	4.18
Gas manufacture	-	-	-	-	-	-	-	-	-
Industry	0.03	0.04	0.04	0.01	-	-	-	3.54	-
Iron and steel	-	*0.04 e*	*0.04 e*	*0.01 e*	-	-	-	-	-
Chemical	-	-	-	-	-	-	-	-	-
Non-metallic minerals	-	-	-	-	-	-	-	-	-
Paper, pulp and print	-	-	-	-	-	-	-	-	-
Other industry	*0.03*	-	-	-	-	-	-	-	-
Other sectors(2)	-	-	-	-	-	-	-	-	-
Non-energy use	-	-	-	-	-	-	-	-	-
Lignite	**13.36**	**12.48**	**4.21**	**3.31**	**3.27**	**3.05**	**3.15**	**-0.57**	**-6.34**
Total electricity and heat	3.98	3.74	3.03	2.86	2.86	2.86	2.99	-0.52	-1.06
Main activity producers	*3.98*	*3.43*	*2.89*	*2.83*	*2.83*	*2.86*	*2.99*	*-1.23*	*-0.66*
Autoproducers	-	*0.31*	*0.14*	*0.03*	*0.04*	*0.00*	*0.00*	-	*-23.89*
Patent fuel/BKB plants	-	-	-	-	-	-	-	-	-
Coke ovens/Liquefaction(1)	-	-	-	-	-	-	-	-	-
Gas manufacture	-	-	-	-	-	-	-	-	-
Industry	3.82	2.41	0.60	0.20	0.21	0.01	0.02	-3.77	-21.48
Iron and steel	-	*0.19*	*0.01*	*0.00*	-	-	*0.00*	-	*-16.73*
Chemical	-	*0.61*	*0.30*	*0.05*	-	-	-	-	-
Non-metallic minerals	-	*0.09*	*0.01*	-	*0.03*	*0.01*	*0.00*	-	*-16.67*
Paper, pulp and print	-	*0.41*	*0.17*	*0.13*	*0.18*	*0.00*	*0.00*	-	*-24.87*
Other industry	*3.82*	*1.11*	*0.11*	*0.02*	*0.00*	*0.00*	*0.01*	*-9.77*	*-20.94*
Other sectors(2)	2.85	6.23	0.57	0.24	0.19	0.17	0.14	6.72	-16.59
Non-energy use	-	-	-	-	-	-	-	-	-

(1) Liquefaction primarily refers to direct distillation processes. Indirect liquefaction may be reported here or under gas manufacture.

(2) Other sectors are Residential, Commercial and public services, Agriculture/Forestry, Fishing and Non-specified other.

* Coal totals may not match the sum of the individual rows due to statistical differences. Please refer to notes and definitions in Part I.

Source: *IEA/OECD Energy Statistics of OECD Countries*

SLOVAK REPUBLIC

7. Fuel prices to end users*

	1978[(1)]	1990	1995	2000	2005	2009	2010	2011	2012
				(US dollars / tce)					
For electricity generation									
Steam coal	4.30	8.61	..	..	..	..	..	..	..
Heavy fuel oil	51.12	88.03	56.14	..	x	x	x	x	x
Natural gas	44.83	77.00	99.39	78.88	223.61	514.06	310.11	401.52	481.86
For industry									
Steam coal	7.88	15.25	..	..	..	..	..	..	..
Coking coal	27.50	56.55	..	..	..	..	..	..	..
High sulphur fuel oil	51.12	88.03	56.14	..	x	x	x	x	x
Low sulphur fuel oil	x	x	x	..	162.63	235.31	321.05	444.82	514.42
Natural gas	44.83	77.00	99.39	78.88	223.61	401.88	415.89	454.23	475.17
				(Euro / unit)[(2)]					
For electricity generation									
Steam coal	1.79	4.48	..	..	..	..	..	..	..
Heavy fuel oil	33.86	72.70	76.78	..	x	x	x	x	x
Natural gas	27.55	59.00	126.13	155.67	296.12	475.74	301.03	371.02	482.00
For industry									
Steam coal	3.29	7.93	..	..	..	..	..	..	..
Coking coal	13.18	33.79	..	..	..	..	..	..	..
High sulphur fuel oil	33.86	72.70	76.78	..	x	x	x	x	x
Low sulphur fuel oil	x	x	x	..	232.09	234.68	335.85	442.96	554.53
Natural gas	27.55	59.00	126.13	155.67	296.12	371.92	403.71	419.73	475.30

(1) Earliest year for which split by coal type is available.

(2) For steam coal, coking coal and heavy fuel oil per metric tonne; for natural gas per 10^7 kilocalories GCV.

* Please refer to notes and definitions in Part I. Steam coal prices relate to historical definitions.

Source: *IEA/OECD Energy Prices & Taxes*

8. Coal and peat trade by type of coal*

(Mtce)

	1978	1990	1995	2000	2005	2009	2010	2011	2012e
Total imports	**9.67**	**8.87**	**5.95**	**4.95**	**5.56**	**4.82**	**4.59**	**4.56**	**4.33**
Bituminous coal[(1)]	3.24	2.33	1.84	1.92	2.23	1.80	1.18	1.31	1.15
Coking coal	2.08	2.99	2.59	2.58	2.74	2.38	2.49	2.49	2.63
Sub-bituminous coal	-	-	-	-	-	-	-	-	-
Lignite	3.15	2.96	1.36	0.31	0.33	0.37	0.31	0.30	0.35
Peat	-	-	-	-	-	-	-	-	-
Coal products[(2)]	1.21	0.60	0.16	0.14	0.26	0.27	0.61	0.47	0.20
Total exports	**-**	**0.13**	**0.05**	**0.06**	**0.21**	**0.22**	**0.38**	**0.24**	**0.12**
Bituminous coal[(1)]	-	-	0.00	-	-	-	-	-	-
Coking coal	-	-	-	-	-	-	-	-	-
Sub-bituminous coal	-	-	-	-	-	-	-	-	-
Lignite	-	0.13	0.00	0.00	-	-	-	-	-
Peat	-	-	-	-	-	-	-	-	-
Coal products[(2)]	-	-	0.05	0.06	0.21	0.22	0.38	0.24	0.12

(1) Bituminous coal includes anthracite.

(2) Coal products includes products derived from coal, for example: coke, coal tar, briquettes, patent fuels.

* Please refer to notes and definitions in Part I.

Source: *IEA/OECD Energy Balances of OECD Countries, IEA/OECD Energy Statistics of OECD Countries*

SLOVAK REPUBLIC

9. Total coal imports by origin*

(thousand tonnes)

	1978(1)	1985	1990	1995	2000	2005	2010	2011	2012e
Total coal(2)	**13711**	**15381**	**12941**	**8355**	**5657**	**6000**	**4411**	**4574**	**4630**
Coking coal	**2180**	**2783**	**3132**	**2746**	**2596**	**2732**	**2472**	**2493**	**2636**
Australia	-	-	-	-	-	-	-	-	-
Canada	-	-	-	-	-	-	-	-	-
Czech Republic	-	-	-	1683 e	901 e	1586	1557	1292	1067
Germany	-	-	-	-	-	-	-	-	-
Poland	-	-	-	1063 e	669 e	579	445	355	301
United Kingdom	-	-	-	-	-	-	-	-	-
United States	-	-	-	-	-	290	309	571	830
Other OECD	-	-	-	-	-	-	-	-	-
China, People's Rep.	-	-	-	-	-	-	-	-	-
Colombia	-	-	-	-	-	-	-	-	-
Indonesia	-	-	-	-	-	-	-	-	-
South Africa	-	-	-	-	-	-	-	-	-
Former Soviet Union(3)	-	-	x	x	x	x	x	x	x
Russian Federation (4)	*x*	*x*	-	-	*1026 e*	*180*	*153*	*189*	*346*
Other FSU (4)	*x*	*x*	-	-	-	*97*	*8*	*86*	*92*
Venezuela	-	-	-	-	-	-	-	-	-
Vietnam	-	-	-	-	-	-	-	-	-
Non-specified/other	2180	2783	3132	-	-	-	-	-	-
Steam coal	**3971**	**2691**	**2734**	**2186**	**2255**	**2531**	**1326**	**1479**	**1299**
Australia	-	-	-	-	-	-	-	-	-
Canada	-	-	-	-	-	-	-	-	-
Czech Republic	-	-	-	547 e	931 e	22	279	373	297
Germany	-	-	-	-	-	-	2	-	-
Poland	-	-	-	226 e	286 e	227	96	115	121
United Kingdom	-	-	-	-	-	-	-	-	-
United States	-	-	-	-	-	-	-	-	-
Other OECD	-	-	-	-	-	2	-	-	-
China, People's Rep.	-	-	-	-	-	-	-	-	-
Colombia	-	-	-	-	-	-	-	-	-
Indonesia	-	-	-	-	-	-	-	-	-
South Africa	-	-	-	-	-	-	-	-	-
Former Soviet Union(3)	-	-	x	x	x	x	x	x	x
Russian Federation (4)	*x*	*x*	-	*1413 e*	*579 e*	*1968*	*863*	*879*	*647*
Other FSU (4)	*x*	*x*	-	-	*459 e*	*163*	*29*	*97*	*155*
Venezuela	-	-	-	-	-	-	-	-	-
Vietnam	-	-	-	-	-	-	-	-	-
Non-specified/other	3971	2691	2734	-	-	149	57	15	79
Lignite	**7560**	**9907**	**7075**	**3423**	**806**	**737**	**613**	**602**	**695**

(1) Earliest year for which split by coal type is available.

(2) Total coal consists of steam coal, coking coal and lignite.

(3) For years prior to 1990, or if no details for individual Former Soviet Union republics.

(4) Data for former republics are not available separately for some OECD member countries.

* Please refer to notes and definitions in Part I.

Source: IEA/OECD Coal Statistics

SLOVENIA*

Figure 1: TPES by fuel (Mtce)

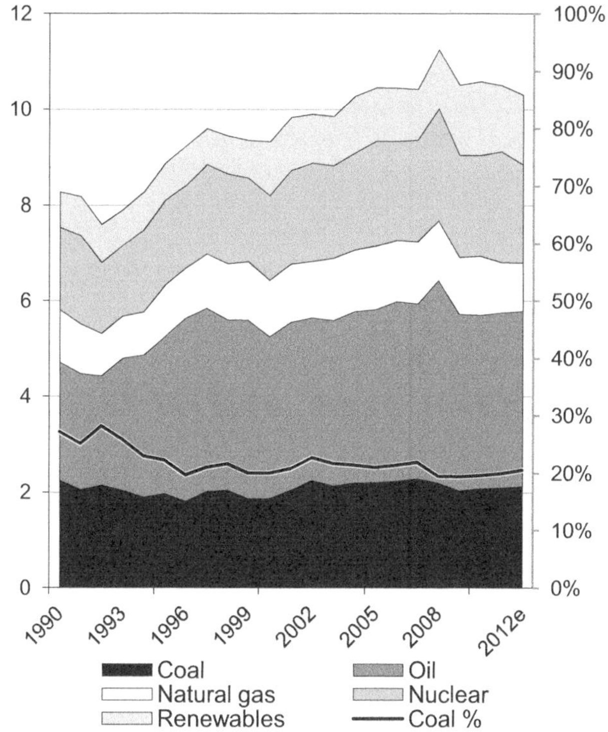

Figure 2: Coal supply indicators (1971 = 100)

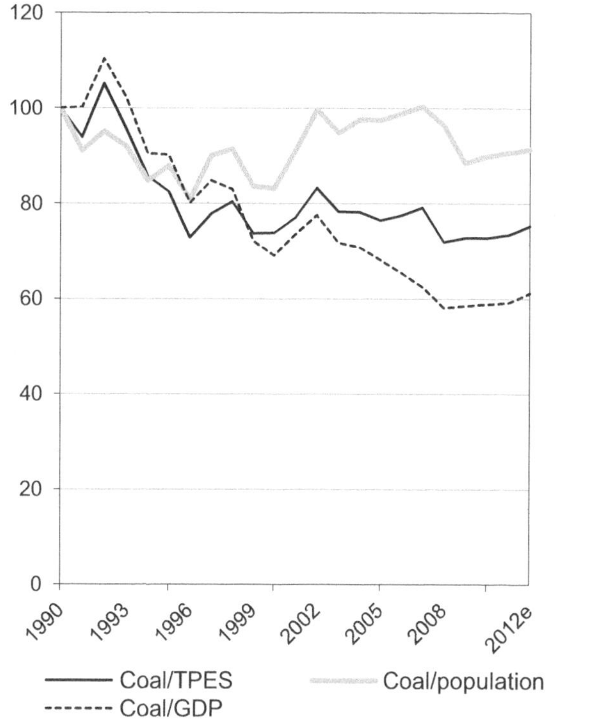

Figure 3: Primary coal supply (Mtce)

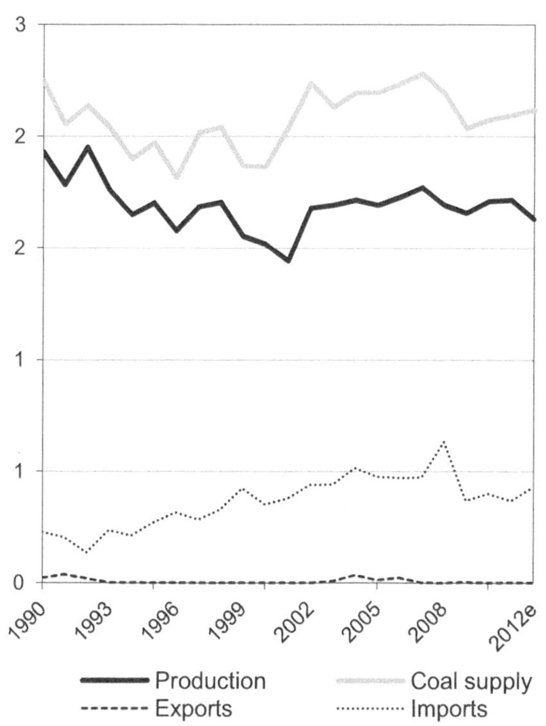

Figure 4: Coal consumption (Mtce)

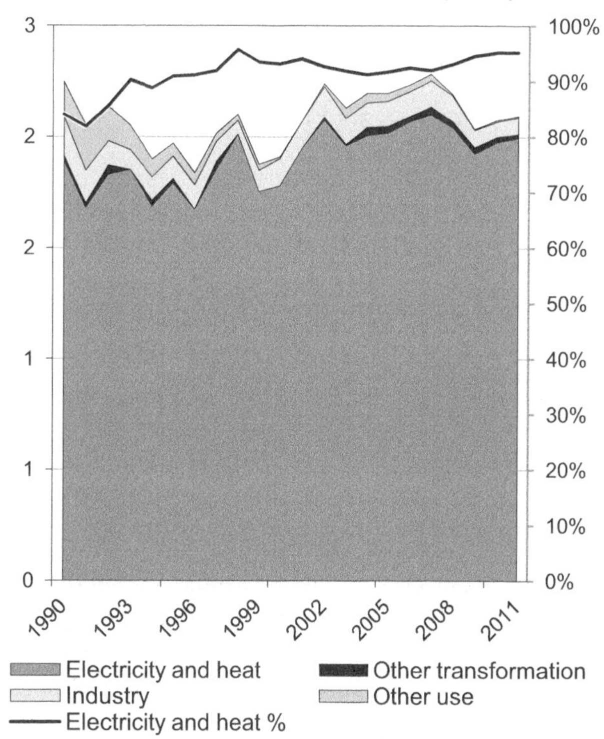

*Please refer to notes and definitions in Part I. Peat is included in coal for all figures.

Source: *IEA/OECD Coal Information, IEA/OECD Energy Balances of OECD Countries, OECD Main Economic Indicators*

SLOVENIA*

Figure 5: Electricity generation by fuel (TWh)

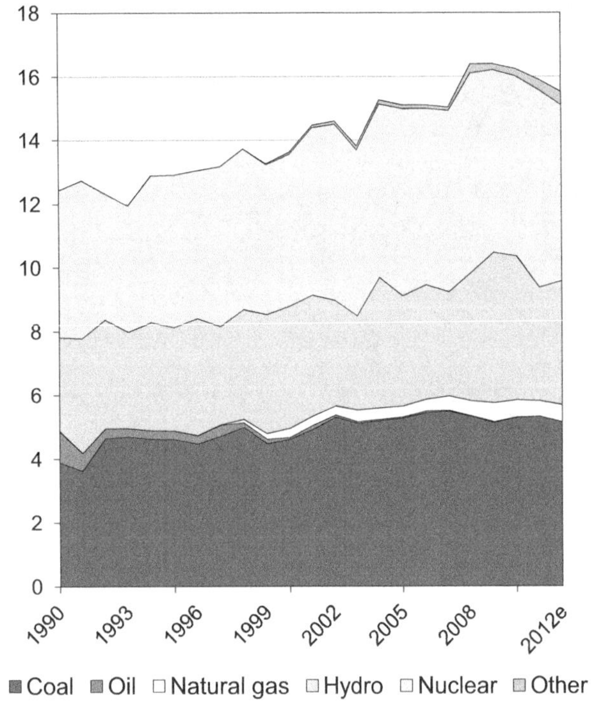

Figure 6: CO_2 emissions by fuel (Mt CO_2)

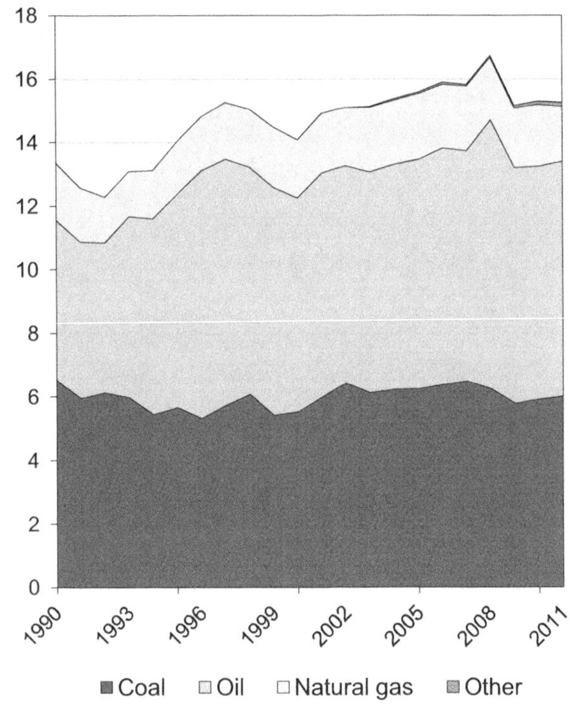

Figure 7: Electricity generation by fuel share

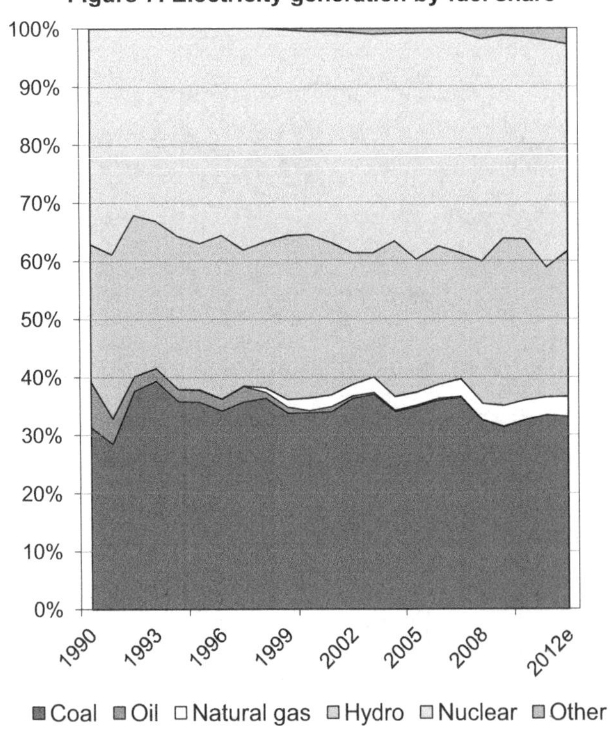

Figure 8: CO_2 emissions by fuel share

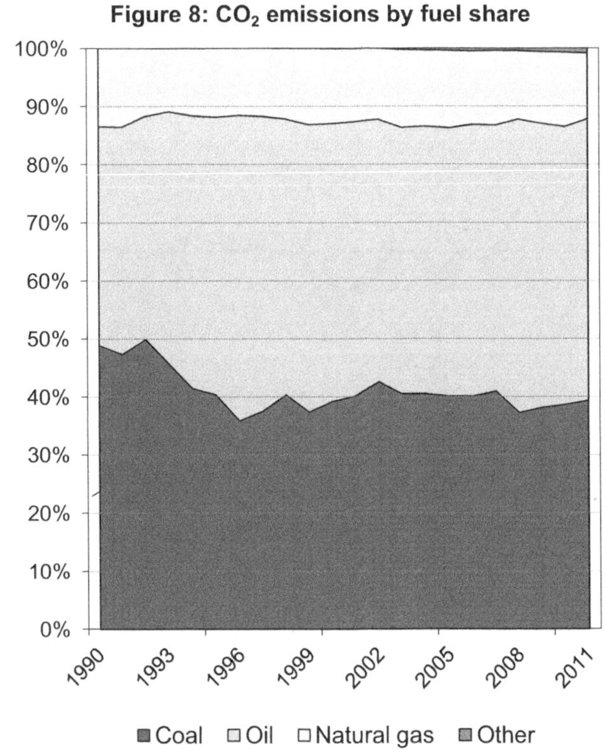

*Please refer to notes and definitions in Part I. Peat is included in coal for all figures.

Source: *IEA/OECD Electricity Information, IEA/OECD CO2 Emissions from Fuel Combustion*

SLOVENIA

1. Total primary energy supply (TPES) by fuel*

	1973	1980	1990	2000	2010	2011	2012e	Average annual percent change 73-90	90-11
TPES (Mtce)	**x**	**x**	**8.16**	**9.16**	**10.33**	**10.36**	**10.20**	**x**	**1.14**
Coal and peat	x	x	2.25	1.87	2.07	2.10	2.12	x	-0.34
Oil	x	x	2.47	3.38	3.63	3.65	3.66	x	1.87
Natural Gas	x	x	1.09	1.18	1.23	1.05	1.01	x	-0.16
Biofuels and waste	x	x	0.38	0.65	0.93	0.88	0.90	x	4.08
Nuclear	x	x	1.72	1.77	2.11	2.31	2.06	x	1.42
Hydro	x	x	0.36	0.47	0.55	0.44	0.48	x	0.90
Geothermal	x	x	-	-	0.05	0.05	0.05	x	-
Solar, wind, tide	x	x	-	-	0.01	0.02	0.03	x	-
Net electricity trade(1)	x	x	-0.12	-0.16	-0.26	-0.16	-0.11	x	1.17
Heat(2)	x	x	-	-	-	-	-	x	-

(1) Net trade between OECD and non-OECD countries.

(2) Ambient heat from heat pumps used in the transformation sector.

Source: *IEA/OECD Energy Balances of OECD Countries*

2. Energy supply, GDP and population*

	1973	1980	1990	2000	2010	2011	2012e	Average annual percent change 73-90	90-11
GDP (2005 bil. USD)	x	x	25	30	39	39	38	x	2.19
TPES/GDP(1)	x	x	0.33	0.31	0.26	0.26	0.27	x	-1.03
Population (millions)	x	x	2.0 e	2.0	2.0	2.1	2.1	x	0.13
TPES/population(1)	x	x	4.08 e	4.61	5.04	5.04	4.96	x	1.01
TPES/GDP(2)	x	x	112.3	105.0	90.8	90.5	91.3	x	-1.03
Coal and peat supply/GDP(2)	x	x	146.9	101.4	86.4	86.8	89.8	x	-2.47
Elec. consumption/GDP(2)	x	x	104.1	98.6	86.0	90.1	..	x	-0.69
Elec. generation (TWh)	x	x	12	14	16	16	16	x	1.18
Industrial production(2)	x	x	..	86.2	102.8	104.1	103.2	x	..

(1) TPES/GDP in units of tce/2005 thousand USD, TPES/population in units of tce/capita.

(2) As index, 2005 (not shown) = 100

Source: *IEA/OECD Energy Balances of OECD Countries , OECD Main Economic Indicators*

3. Coal and peat production by type*

	1978(1)	1990	2000	2005	2010	2011	2012e	Average annual percent change 78-90	90-11
Mtce:									
Coking coal	x	-	-	-	-	-	-	x	-
Steam coal	x	-	-	-	-	-	-	x	-
Lignite	x	1.93	1.52	1.69	1.71	1.72	1.63	x	-0.56
Peat	x	-	-	-	-	-	-	x	-
Mt:									
Coking coal	x	-	-	-	-	-	-	x	-
Steam coal	x	-	-	-	-	-	-	x	-
Lignite	x	5.58	4.48	4.54	4.43	4.50	4.28	x	-1.02
Peat	x	-	-	-	-	-	-	x	-

(1) Earliest year for which split by coal type is available.

* Please refer to notes and definitions in Part I.

Source: *IEA/OECD Energy Balances of OECD Countries, IEA/OECD Energy Statistics of OECD Countries*

SLOVENIA

4. Final consumption of energy by fuel*

(Mtce)

	1973	1980	1990	2000	2009	2010	2011	Average annual percent change 73-90	90-11
Total final consumption[(1)]	**x**	**x**	**5.28**	**6.64**	**7.14**	**7.32**	**7.20**	**x**	**1.49**
Coal and peat	x	x	0.33	0.13	0.08	0.07	0.08	x	-6.48
Oil	x	x	2.14	3.33	3.66	3.62	3.64	x	2.57
Natural Gas	x	x	1.02	0.98	0.95	1.00	0.83	x	-0.96
Biofuels and wastes	x	x	0.38	0.62	0.77	0.83	0.76	x	3.41
Geothermal	x	x	-	-	0.02	0.04	0.05	x	-
Solar, wind, tide	x	x	-	-	0.01	0.01	0.01	x	-
Electricity	x	x	1.14	1.29	1.39	1.47	1.55	x	1.49
Heat	x	x	0.27	0.28	0.26	0.27	0.27	x	-0.02
of which:									
Total industry	**x**	**x**	**2.18**	**2.03**	**1.74**	**1.82**	**1.76**	**x**	**-1.01**
Coal and peat	x	x	0.18	0.12	0.08	0.07	0.08	x	-3.96
Oil	x	x	0.32	0.39	0.22	0.18	0.16	x	-3.21
Natural Gas	x	x	0.81	0.70	0.64	0.69	0.61	x	-1.36
Biofuels and wastes	x	x	0.09	0.11	0.11	0.13	0.12	x	1.45
Geothermal	x	x	-	-	-	-	-	x	-
Solar, wind, tide	x	x	-	-	-	-	-	x	-
Electricity	x	x	0.73	0.68	0.61	0.67	0.72	x	-0.08
Heat	x	x	0.06	0.04	0.08	0.08	0.08	x	1.89
Total transport	**x**	**x**	**1.29**	**1.74**	**2.49**	**2.53**	**2.71**	**x**	**3.59**
Coal and peat	x	x	-	-	-	-	-	x	-
Oil	x	x	1.26	1.70	2.43	2.44	2.64	x	3.57
Natural Gas	x	x	-	-	-	-	-	x	-
Biofuels and wastes	x	x	-	-	0.04	0.06	0.05	x	-
Electricity	x	x	0.03	0.03	0.02	0.02	0.02	x	-1.47
Residential	**x**	**x**	**1.36**	**1.61**	**1.74**	**1.80**	**1.68**	**x**	**1.01**
Coal and peat	x	x	0.16	0.01	-	-	-	x	-
Oil	x	x	0.54	0.55	0.42	0.44	0.36	x	-1.90
Natural Gas	x	x	0.04	0.08	0.15	0.16	0.16	x	7.07
Biofuels and wastes	x	x	0.21	0.51	0.61	0.63	0.59	x	4.96
Geothermal	x	x	-	-	0.02	0.02	0.03	x	-
Solar, wind, tide	x	x	-	-	0.01	0.01	0.01	x	-
Electricity	x	x	0.27	0.32	0.39	0.40	0.39	x	1.74
Heat	x	x	0.13	0.13	0.14	0.14	0.13	x	-0.21
Comm & public services	**x**	**x**	**0.42**	**0.75**	**0.71**	**0.73**	**0.75**	**x**	**2.82**
Coal and peat	x	x	-	-	-	-	-	x	-
Oil	x	x	-	0.35	0.27	0.24	0.20	x	-
Natural Gas	x	x	0.16	0.03	0.03	0.03	0.06	x	-4.82
Biofuels and waste	x	x	0.08	0.00	0.00	0.00	0.00	x	-14.21
Geothermal	x	x	-	-	-	0.01	0.01	x	-
Solar, wind, tide	x	x	-	-	-	-	-	x	-
Electricity	x	x	0.10	0.26	0.37	0.38	0.41	x	6.98
Heat	x	x	0.08	0.11	0.04	0.05	0.06	x	-1.47
Non-energy use	**x**	**x**	**0.01**	**0.34**	**0.33**	**0.30**	**0.17**	**x**	**12.95**
Coal and peat	x	x	-	-	0.00	0.01	0.01	x	-
Oil	x	x	0.01	0.17	0.19	0.18	0.16	x	12.52
Natural Gas	x	x	-	0.17	0.13	0.11	0.01	x	-

(1) Total Final Consumption (TFC) excludes use in transformation processes and energy industry own use.

* Please refer to notes and definitions in Part I. Agriculture, Forestry, Fishing and "Non-specified other" sectors are not reported here individually, but are accounted for in Total final consumption.

Source: *IEA/OECD Energy Balances of OECD Countries*

SLOVENIA

5. Coal balance*

(Mtce)

	1973	1980	1990	2000	2005	2010	2011	2012e	Average annual percent change 73-90	90-11
Production	x	x	1.9	1.5	1.7	1.7	1.7	1.6	x	-0.6
Imports	x	x	0.2	0.4	0.5	0.4	0.4	0.4	x	2.3
Exports	x	x	-0.0	-0.0	-0.0	-	-0.0	-0.0	x	-11.6
Stock changes	x	x	0.1	-0.0	0.0	-0.0	0.0	0.1		
Primary supply	**x**	**x**	**2.3**	**1.9**	**2.2**	**2.1**	**2.1**	**2.1**	**x**	**-0.3**
Statistical differences	**x**	**x**	**-0.0**	**0.0**	**-0.0**	**-0.0**	**-0.0**	**..**		
Total transformation	**x**	**x**	**-1.9 e**	**-1.8**	**-2.0**	**-2.0**	**-2.0**	**..**	**x**	**0.2**
Electricity and heat gen.	x	x	-1.9	-1.8	-2.0	-2.0	-2.0	..	x	0.3
Main activity producers (1)	*x*	*x*	*-1.9*	*-1.8*	*-2.0*	*-2.0*	*-2.0*	..	*x*	*0.3*
Autoproducers	*x*	*x*	*-0.0*	*-0.0*	*-0.0*	*-0.0*	*-0.0*	..	*x*	*-3.2*
Gas works	x	x	0.0	-	-	-	-	..	x	-
Coal transformation(2)	x	x	-0.0 e	-	-	-	-	..	x	-
BKB plants	*x*	*x*	-	-	-	-	-	..	*x*	-
Blast furnaces	*x*	*x*	*-0.0* e	-	-	-	-	..	*x*	-
Coke ovens	*x*	*x*	-	-	-	-	-	..	*x*	-
Patent fuel plants	*x*	*x*	-	-	-	-	-	..	*x*	-
Other transformation(3)	x	x	-	-	-	-	-	..	x	-
Energy ind. own use	**x**	**x**	**-0.0**	**-**	**-**	**-**	**-**	**..**	**x**	**-**
Losses	**x**	**x**	**-0.0**	**-**	**-**	**-**	**-**	**..**		
Final consumption(4)	**x**	**x**	**0.3**	**0.1**	**0.2**	**0.1**	**0.1**	**..**	**x**	**-6.5**
Industry(5)	x	x	0.2	0.1	0.1	0.1	0.1	..	x	-4.0
Iron and steel	*x*	*x*	*0.0* e	*0.0*	*0.0*	*0.0*	*0.0*	..	*x*	*-6.8*
Chemical	*x*	*x*	*0.0*	-	-	-	-	..	*x*	-
Non-metallic minerals	*x*	*x*	*0.0*	*0.0*	*0.1*	*0.0*	*0.0*	..	*x*	*2.6*
Paper, pulp and print	*x*	*x*	*0.0*	*0.0*	*0.1*	*0.0*	*0.0*	..	*x*	*0.8*
Other industry (6)	*x*	*x*	*0.1*	*0.0*	*0.0*	*0.0*	*0.0*	..	*x*	*-17.1*
Transport(7)	x	x	-	-	-	-	-	..	x	-
Other	x	x	0.2	0.0	-	-	-	..	x	-
Comm. and pub. services	*x*	*x*	-	-	-	-	-	..	*x*	-
Residential	*x*	*x*	*0.2*	*0.0*	-	-	-	..	*x*	-
Other sectors (8)	*x*	*x*	-	-	-	-	-	..	*x*	-
Non-energy use	x	x	-	-	0.0	0.0	0.0	..	x	-

(1) Main activity electricity and heat generation includes district heating.

(2) Coal transformation refers to the transformation of coal and peat to secondary and tertiary products (mainly cokes, briquettes, coal tar and off-gases). Gas works and Liquefaction are shown separately to match the *Energy Balances of OECD Countries*.

(3) Other transformation includes Liquefaction and Non-specified transformations.

(4) Final Consumption includes non-energy use and energy use (Industry, Transport and Other).

(5) Please refer to notes and definitions in Part I for detailed explanation.

(6) Other industry includes Non-ferrous metals, Transport equipment, Machinery, Mining and quarrying, Food and tobacco, Wood and wood products, Construction, Textile and leather, and Non-specified industry.

(7) Transport includes Rail and Inland waterways.

(8) Other sectors includes Agriculture/Forestry, Fishing and Non-specified other.

* "Coal" refers to all types of coal, primary (anthracite, coking coal, other bituminous coal, sub-bituminous coal, lignite/brown coal) and derived fuels (including patent fuel, BKB, coke oven coke, gas coke, coal tar, coke oven gas, gas works gas, blast furnace gas and other recovered gases), as well as peat, peat briquettes and oil shale. Quantities have been converted to Mtce using calorific values reported by the respective countries. Please refer to notes and definitions in Part I.

Source: *IEA/OECD Energy Balances of OECD Countries*

SLOVENIA

6. Use of coal for selected end-uses*

(million tonnes)

	1978	1990	2000	2005	2009	2010	2011	Average annual percent change 78-90	90-11
Steam coal	**x**	**0.26**	**0.45**	**0.61**	**0.49**	**0.50**	**0.47**	**x**	**2.84**
Total electricity and heat	x	0.23	0.36	0.49	0.42	0.44	0.42	x	2.84
Main activity producers	*x*	*0.23*	*0.35*	*0.47*	*0.39*	*0.42*	*0.40*	*x*	*2.66*
Autoproducers	*x*	-	*0.02*	*0.02*	*0.02*	*0.02*	*0.02*	*x*	-
Patent fuel/BKB plants	x	-	-	-	-	-	-	x	-
Coke ovens/Liquefaction(1)	x	-	-	-	-	-	-	x	-
Blast furnace inputs	x	-	-	-	-	-	-	x	-
Gas manufacture	x	-	-	-	-	-	-	x	-
Industry	x	0.03	0.08	0.11	0.08	0.06	0.05	x	2.40
Iron and steel	*x*	*0.01*	-	-	-	-	-	*x*	-
Chemical	*x*	-	-	-	-	-	-	*x*	-
Non-metallic minerals	*x*	-	-	*0.04*	*0.03*	*0.01*	*0.01*	*x*	-
Paper, pulp and print	*x*	-	*0.06*	*0.07*	*0.05*	*0.04*	*0.04*	*x*	-
Other industry	*x*	*0.02*	*0.01*	*0.00*	*0.00*	*0.00*	*0.00*	*x*	*-10.38*
Other sectors(2)	x	-	0.01	-	-	-	-	x	-
Non-energy use	x	-	-	0.01	-	-	-	x	-
Coking coal	**x**	**0.00**	**-**	**-**	**-**	**-**	**-**	**x**	**-**
Total electricity and heat	x	-	-	-	-	-	-	x	-
Main activity producers	*x*	-	-	-	-	-	-	*x*	-
Autoproducers	*x*	-	-	-	-	-	-	*x*	-
Patent fuel/BKB plants	x	-	-	-	-	-	-	x	-
Coke ovens/Liquefaction(1)	x	-	-	-	-	-	-	x	-
Blast furnace inputs	x	-	-	-	-	-	-	x	-
Gas manufacture	x	-	-	-	-	-	-	x	-
Industry	x	0.00	-	-	-	-	-	x	-
Iron and steel	*x*	-	-	-	-	-	-	*x*	-
Chemical	*x*	-	-	-	-	-	-	*x*	-
Non-metallic minerals	*x*	-	-	-	-	-	-	*x*	-
Paper, pulp and print	*x*	-	-	-	-	-	-	*x*	-
Other industry	*x*	*0.00*	-	-	-	-	-	*x*	-
Other sectors(2)	x	-	-	-	-	-	-	x	-
Non-energy use	x	-	-	-	-	-	-	x	-
Lignite	**x**	**5.83**	**4.48**	**4.58**	**4.45**	**4.42**	**4.56**	**x**	**-1.16**
Total electricity and heat	x	5.07	4.47	4.58	4.45	4.40	4.53	x	-0.54
Main activity producers	*x*	*4.99*	*4.47*	*4.58*	*4.45*	*4.39*	*4.52*	*x*	*-0.48*
Autoproducers	*x*	*0.08*	-	-	-	*0.01*	*0.01*	*x*	*-8.75*
Patent fuel/BKB plants	x	-	-	-	-	-	-	x	-
Coke ovens/Liquefaction(1)	x	-	-	-	-	-	-	x	-
Gas manufacture	x	-	-	-	-	-	-	x	-
Industry	x	0.28	0.01	-	-	0.02	0.03	x	-10.01
Iron and steel	*x*	*0.02*	-	-	-	-	-	*x*	-
Chemical	*x*	*0.00*	-	-	-	-	-	*x*	-
Non-metallic minerals	*x*	-	-	-	-	-	-	*x*	-
Paper, pulp and print	*x*	*0.08*	-	-	-	*0.02*	*0.03*	*x*	*-4.68*
Other industry	*x*	*0.17*	*0.01*	-	-	-	-	*x*	-
Other sectors(2)	x	0.45	0.00	-	-	-	-	x	-
Non-energy use	x	-	-	-	-	-	-	x	-

(1) Liquefaction primarily refers to direct distillation processes. Indirect liquefaction may be reported here or under gas manufacture.

(2) Other sectors are Residential, Commercial and public services, Agriculture/Forestry, Fishing and Non-specified other.

* Coal totals may not match the sum of the individual rows due to statistical differences. Please refer to notes and definitions in Part I.

Source: *IEA/OECD Energy Statistics of OECD Countries*

SLOVENIA

7. Fuel prices to end users*

	1978(1)	1990	1995	2000	2005	2009	2010	2011	2012
				(US dollars / tce)					
For electricity generation									
Steam coal	x	..	..	..	..	..	..	..	..
Heavy fuel oil	x	..	..	..	..	..	..	..	..
Natural gas	x	..	..	..	..	..	..	..	..
For industry									
Steam coal	x	..	..	..	..	..	..	..	..
Coking coal	x	..	..	..	..	..	..	..	..
High sulphur fuel oil	x	..	..	..	..	..	..	..	..
Low sulphur fuel oil	x	..	..	..	277.70	373.47	464.78	603.75	645.47
Natural gas	x	..	..	..	..	444.01	466.82	527.72	582.38
				(Euro / unit)(2)					
For electricity generation									
Steam coal	x	..	..	..	..	..	..	..	..
Heavy fuel oil	x	..	..	..	..	..	..	..	..
Natural gas	x	..	..	..	..	..	..	..	..
For industry									
Steam coal	x	..	..	..	..	..	..	..	..
Coking coal	x	..	..	..	..	..	..	..	..
High sulphur fuel oil	x	..	..	..	..	..	..	..	..
Low sulphur fuel oil	x	..	..	..	306.24	368.67	481.25	595.09	688.70
Natural gas	x	..	..	..	..	410.91	453.14	487.64	582.54

(1) Earliest year for which split by coal type is available.

(2) For steam coal, coking coal and heavy fuel oil per metric tonne; for natural gas per 10^7 kilocalories GCV.

* Please refer to notes and definitions in Part I. Steam coal prices relate to historical definitions.

Source: *IEA/OECD Energy Prices & Taxes*

8. Coal and peat trade by type of coal*

(Mtce)

	1978	1990	1995	2000	2005	2009	2010	2011	2012e
Total imports	**x**	**0.23**	**0.27**	**0.35**	**0.48**	**0.37**	**0.40**	**0.37**	**0.43**
Bituminous coal(1)	x	0.02	0.02	0.01	0.04	0.03	0.02	0.02	0.02
Coking coal	x	0.00	-	-	-	-	-	-	-
Sub-bituminous coal	x	0.14	0.18	0.27	0.38	0.25	0.31	0.28	0.31
Lignite	x	-	0.01	-	-	0.06	0.04	0.04	0.07
Peat	x	-	-	-	-	-	-	-	-
Coal products(2)	x	0.07	0.06	0.07	0.05	0.03	0.03	0.03	0.03
Total exports	**x**	**0.02**	**0.00**	**0.00**	**0.01**	**0.00**	**-**	**0.00**	**0.00**
Bituminous coal(1)	x	-	-	-	-	0.00	-	0.00	0.00
Coking coal	x	-	-	-	-	-	-	-	-
Sub-bituminous coal	x	-	-	-	0.01	-	-	-	-
Lignite	x	0.02	0.00	0.00	-	-	-	-	-
Peat	x	-	-	-	-	-	-	-	-
Coal products(2)	x	-	-	-	-	-	-	-	-

(1) Bituminous coal includes anthracite.

(2) Coal products includes products derived from coal, for example: coke, coal tar, briquettes, patent fuels.

* Please refer to notes and definitions in Part I.

Source: *IEA/OECD Energy Balances of OECD Countries, IEA/OECD Energy Statistics of OECD Countries*

SLOVENIA

9. Total coal imports by origin*

(thousand tonnes)

	1978(1)	1985	1990	1995	2000	2005	2010	2011	2012e
Total coal(2)	**x**	**x**	**254**	**361**	**448**	**593**	**575**	**527**	**638**
Coking coal	**x**	**x**	**1**	**-**	**-**	**-**	**-**	**-**	**-**
Australia	x	x	-	-	-	-	-	-	-
Canada	x	x	-	-	-	-	-	-	-
Czech Republic	x	x	-	-	-	-	-	-	-
Germany	x	x	-	-	-	-	-	-	-
Poland	x	x	-	-	-	-	-	-	-
United Kingdom	x	x	-	-	-	-	-	-	-
United States	x	x	-	-	-	-	-	-	-
Other OECD	x	x	-	-	-	-	-	-	-
China, People's Rep.	x	x	-	-	-	-	-	-	-
Colombia	x	x	-	-	-	-	-	-	-
Indonesia	x	x	-	-	-	-	-	-	-
South Africa	x	x	-	-	-	-	-	-	-
Former Soviet Union(3)	x	x	x	x	x	x	x	x	x
Russian Federation (4)	*x*	*x*	-	-	-	-	-	-	-
Other FSU (4)	*x*	*x*	-	-	-	-	-	-	-
Venezuela	x	x	-	-	-	-	-	-	-
Vietnam	x	x	-	-	-	-	-	-	-
Non-specified/other	x	x	1	-	-	-	-	-	-
Steam coal	**x**	**x**	**253**	**328**	**448**	**593**	**500**	**442**	**489**
Australia	x	x	-	-	-	-	-	-	-
Canada	x	x	-	-	-	-	-	-	-
Czech Republic	x	x	-	-	8	22	9	29	54
Germany	x	x	-	-	-	1	2	1	1
Poland	x	x	-	-	-	11	-	-	-
United Kingdom	x	x	-	-	-	-	-	-	-
United States	x	x	-	-	-	-	-	-	8
Other OECD	x	x	-	-	-	12	13	7	7
China, People's Rep.	x	x	-	-	-	1	-	-	-
Colombia	x	x	-	-	-	2	-	-	-
Indonesia	x	x	-	307	427	501	434	356	376
South Africa	x	x	-	-	-	-	-	-	-
Former Soviet Union(3)	x	x	x	x	x	x	x	x	x
Russian Federation (4)	*x*	*x*	-	-	-	*41*	*7*	*5*	*6*
Other FSU (4)	*x*	*x*	-	-	-	-	-	-	-
Venezuela	x	x	-	-	-	-	11	9	1
Vietnam	x	x	-	-	-	-	-	-	-
Non-specified/other	x	x	253	21	13	2	24	35	36
Lignite	**x**	**x**	**-**	**33**	**-**	**-**	**75**	**85**	**149**

(1) Earliest year for which split by coal type is available.

(2) Total coal consists of steam coal, coking coal and lignite.

(3) For years prior to 1990, or if no details for individual Former Soviet Union republics.

(4) Data for former republics are not available separately for some OECD member countries.

* Please refer to notes and definitions in Part I.

Source: IEA/OECD Coal Statistics

SLOVENIA

12. Coal import values by origin*

(average unit value, CIF, USD/tonne)

	1990	1995	2000	2005	2007	2008	2009	2010	2011/12(1)
Coking coal(2)	..	..	..	..	..	..	**112.88**	..	..
Imports from:									
Australia	..	..	..	..	..	..	..	..	..
Canada	..	..	..	..	..	..	..	..	..
Czech Republic	..	..	..	..	..	..	..	..	..
Poland	..	..	..	..	..	..	..	..	..
United States	..	..	..	..	..	..	..	..	..
China	..	..	..	..	..	..	..	..	..
Colombia	..	..	..	..	..	..	..	..	..
Indonesia	..	..	..	..	..	..	..	..	..
South Africa	..	..	..	..	..	..	..	..	..
Former Soviet Union(3)	..	..	..	..	..	..	..	..	..
Other bituminous coal(4)	..	..	..	..	**114.24**	**180.69**	**159.08**	**160.43**	..
Imports from:									
Australia	..	..	..	..	..	..	..	..	..
Canada	..	..	..	..	..	..	..	..	..
Czech Republic	..	..	..	..	..	..	..	..	..
Poland	..	..	..	..	94.93	..	..	..	..
United States	..	..	..	..	..	..	..	..	..
China	..	..	..	..	..	..	..	..	..
Colombia	..	..	..	..	131.81	..	..	..	..
Indonesia	..	..	..	..	..	..	..	..	..
South Africa	..	..	..	..	110.18	157.58	..	..	..
Former Soviet Union(3)	..	..	..	..	114.08	187.38	151.37	144.12	..

Note: On occasion, shipment of extremely small quantities of high valued coal results in high import costs.

(1) Import data for steam and coking coals are currently unavailable for 2011 and 2012 due to resource constraints.

(2) Weighted average of individual countries using import volumes as weights. Prices exclude intra-EU trade.

(3) Former Soviet Union until 1991, Russian Federation starting in 1992.

(4) Bituminous steam coal only. Weighted average of trades. Prices exclude intra-EU trade.

* Please refer to notes and definitions in Part I.

Source: IEA/OECD Coal Statistics

SPAIN*

Figure 1: TPES by fuel (Mtce)

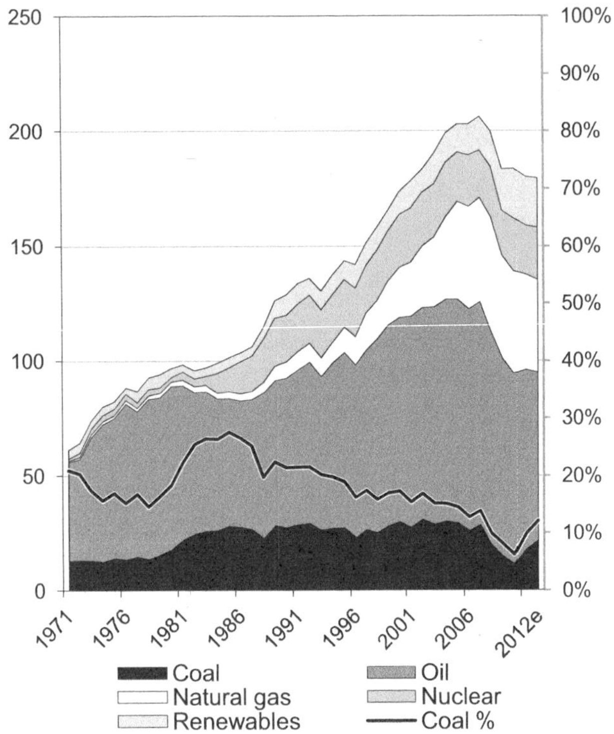

Figure 2: Coal supply indicators (1971 = 100)

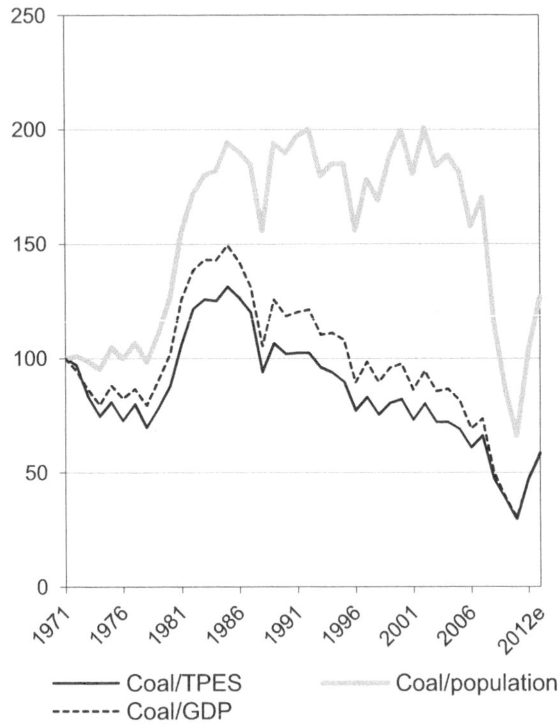

Figure 3: Primary coal supply (Mtce)

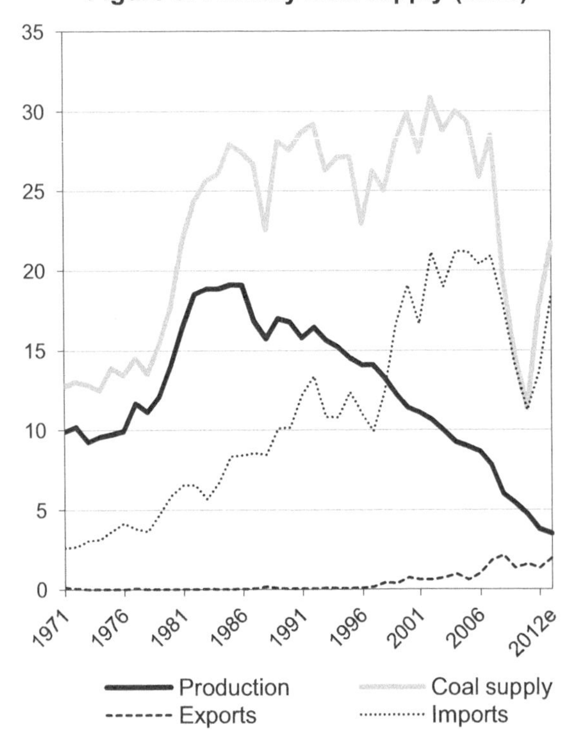

Figure 4: Coal consumption (Mtce)

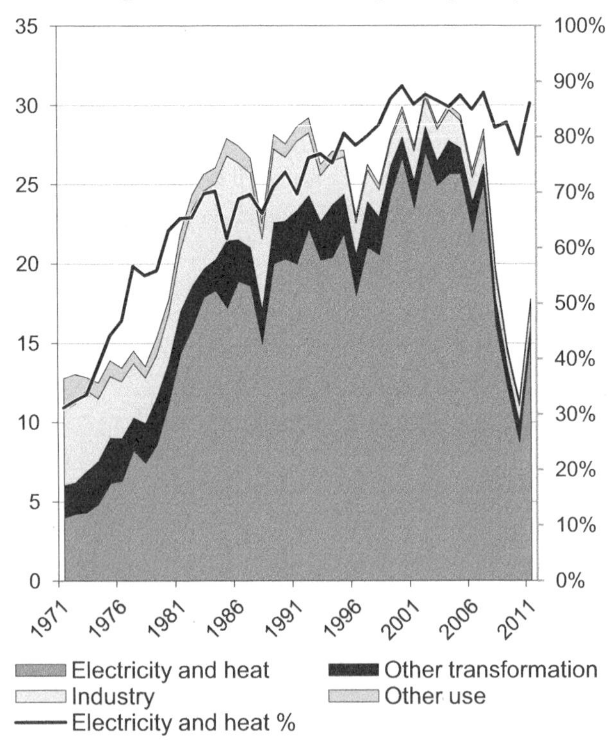

*Please refer to notes and definitions in Part I. Peat is included in coal for all figures.

Source: *IEA/OECD Coal Information, IEA/OECD Energy Balances of OECD Countries, OECD Main Economic Indicators*

SPAIN*

Figure 5: Electricity generation by fuel (TWh)

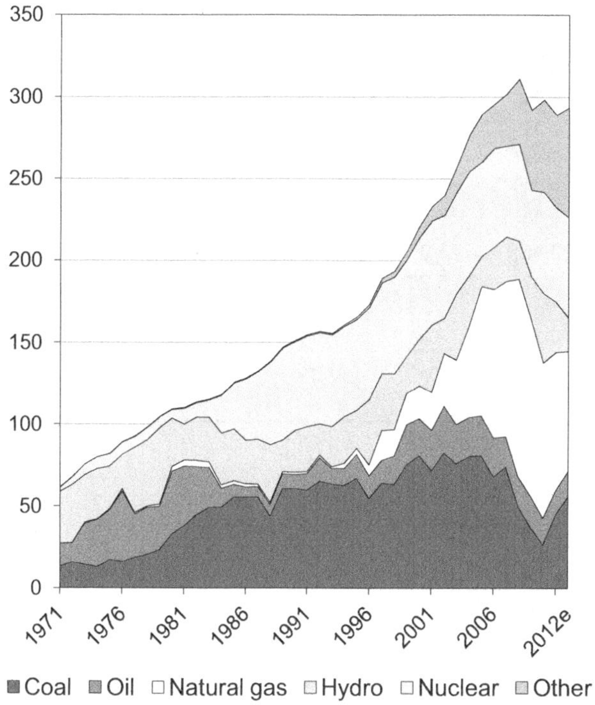

Figure 6: CO_2 emissions by fuel (Mt CO_2)

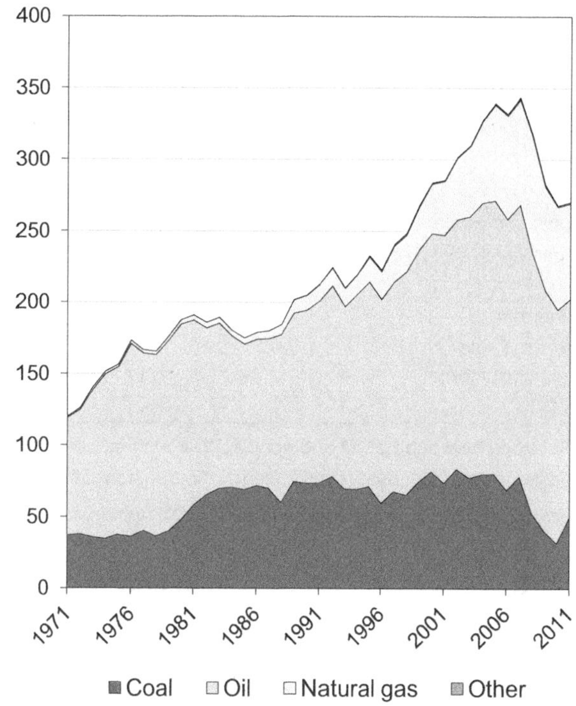

Figure 7: Electricity generation by fuel share

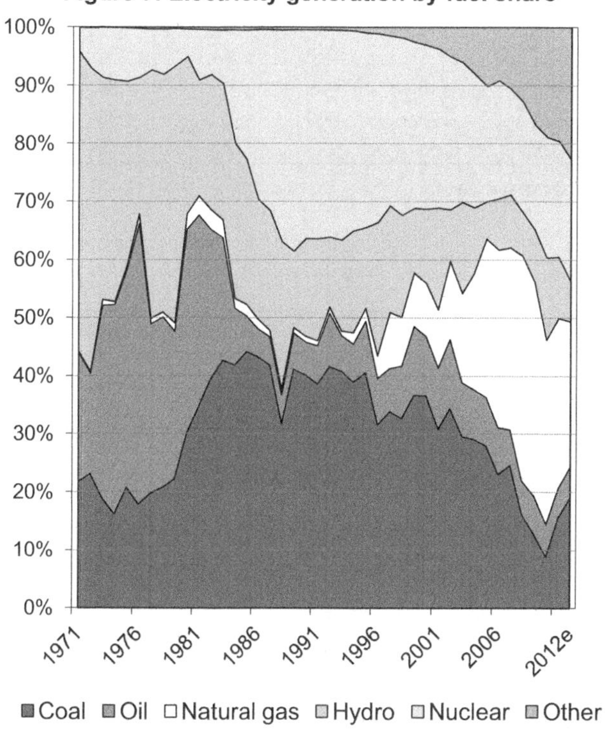

Figure 8: CO_2 emissions by fuel share

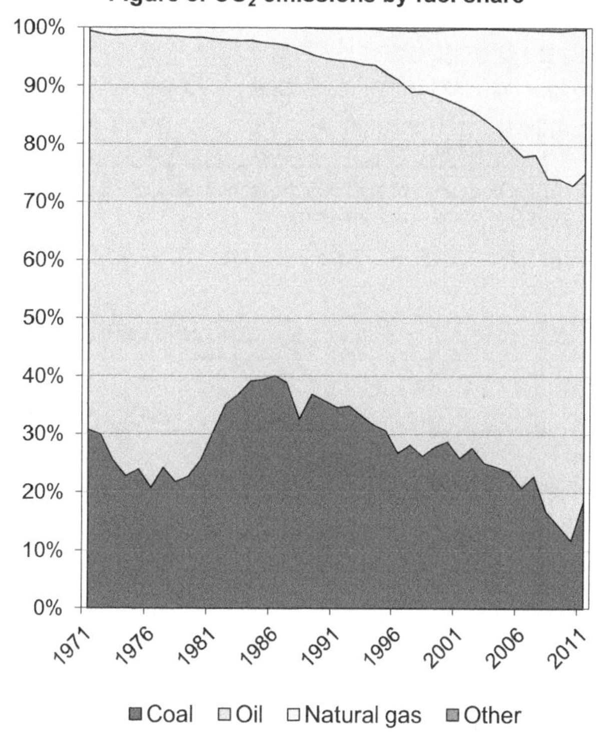

*Please refer to notes and definitions in Part I. Peat is included in coal for all figures.

Source: *IEA/OECD Electricity Information, IEA/OECD CO2 Emissions from Fuel Combustion*

SPAIN

1. Total primary energy supply (TPES) by fuel*

	1973	1980	1990	2000	2010	2011	2012e	Average annual percent change 73-90	90-11
TPES (Mtce)	**73.67**	**96.71**	**128.70**	**174.08**	**182.50**	**179.39**	**178.11**	**3.34**	**1.59**
Coal and peat	12.85	17.75	27.55	29.91	11.34	17.77	21.72	4.59	-2.07
Oil	53.72	71.11	64.96	88.72	83.09	78.31	73.10	1.12	0.89
Natural Gas	1.35	2.08	7.10	21.74	44.46	41.41	40.34	10.26	8.76
Biofuels and waste	0.02	0.38	5.81	5.90	9.42	10.24	10.77	40.22	2.74
Nuclear	2.44	1.93	20.20	23.16	23.08	21.49	22.85	13.25	0.30
Hydro	3.55	3.63	3.13	3.47	5.20	3.76	2.52	-0.74	0.88
Geothermal	-	-	0.01	0.01	0.02	0.02	0.03	-	7.51
Solar, wind, tide	-	-	0.00	0.63	6.92	7.14	8.17	-	46.19
Net electricity trade(1)	-0.25	-0.17	-0.05	0.55	-1.02	-0.75	-1.38	-8.81	13.58
Heat(2)	-	-	-	-	-	-	-	-	-

(1) Net trade between OECD and non-OECD countries.

(2) Ambient heat from heat pumps used in the transformation sector.

Source: *IEA/OECD Energy Balances of OECD Countries*

2. Energy supply, GDP and population*

	1973	1980	1990	2000	2010	2011	2012e	Average annual percent change 73-90	90-11
GDP (2005 bil. USD)	468	547	731	963	1179	1184	1167	2.66	2.32
TPES/GDP(1)	0.16	0.18	0.18	0.18	0.15	0.15	0.15	0.66	-0.71
Population (millions)	35.0	37.7	39.0	40.3	46.1	46.1	46.1	0.65	0.80
TPES/population(1)	2.11	2.57	3.30	4.32	3.96	3.89	3.87	2.67	0.79
TPES/GDP(2)	87.9	98.6	98.2	100.8	86.3	84.5	85.1	0.66	-0.71
Coal and peat supply/GDP(2)	105.9	125.0	145.2	119.7	37.1	57.8	71.7	1.88	-4.29
Elec. consumption/GDP(2)	58.9	76.6	80.4	91.3	96.9	94.6	..	1.84	0.78
Elec. generation (TWh)	76	109	151	221	298	289	294	4.16	3.13
Industrial production(2)	55.6	65.6	79.1	97.8	83.5	82.3	77.3	2.10	0.19

(1) TPES/GDP in units of tce/2005 thousand USD, TPES/population in units of tce/capita.

(2) As index, 2005 (not shown) = 100

Source: *IEA/OECD Energy Balances of OECD Countries , OECD Main Economic Indicators*

3. Coal and peat production by type*

	1978(1)	1990	2000	2005	2010	2011	2012e	Average annual percent change 78-90	90-11
Mtce:									
Coking coal	1.87	0.28	-	-	-	-	-	-14.69	-
Steam coal	7.86	12.12	9.23	7.27	4.71	3.78	3.50	3.68	-5.39
Lignite	1.38	4.38	2.15	1.68	-	-	-	10.09	-
Peat	-	-	-	-	-	-	-	-	-
Mt:									
Coking coal	1.80	0.28	-	-	-	-	-	-14.39	-
Steam coal	12.63	19.03	14.95	11.89	8.43	6.62	6.15	3.48	-4.90
Lignite	5.22	16.37	8.52	7.59	-	-	-	9.99	-
Peat	-	-	-	-	-	-	-	-	-

(1) Earliest year for which split by coal type is available.

* Please refer to notes and definitions in Part I.

Source: *IEA/OECD Energy Balances of OECD Countries, IEA/OECD Energy Statistics of OECD Countries*

SPAIN

4. Final consumption of energy by fuel*

(Mtce)

	1973	1980	1990	2000	2009	2010	2011	Average annual percent change 73-90	90-11
Total final consumption[(1)]	**55.06**	**68.74**	**86.59**	**122.13**	**129.84**	**131.70**	**126.57**	**2.70**	**1.82**
Coal and peat	5.94	3.96	4.85	1.95	1.25	1.40	1.76	-1.19	-4.72
Oil	41.23	52.48	54.50	74.51	72.87	71.41	66.20	1.66	0.93
Natural Gas	0.64	1.03	6.18	17.56	19.11	21.16	20.84	14.27	5.96
Biofuels and wastes	-	0.24	5.60	4.90	6.91	7.38	7.97	-	1.69
Geothermal	-	-	0.01	0.01	0.02	0.02	0.02	-	7.51
Solar, wind, tide	-	-	-	0.04	0.22	0.26	0.29	-	-
Electricity	7.25	11.03	15.46	23.15	29.46	30.08	29.48	4.55	3.12
Heat	-	-	-	-	-	-	-	-	-
of which:									
Total industry	**24.58**	**26.88**	**27.51**	**35.20**	**29.59**	**29.86**	**29.27**	**0.67**	**0.30**
Coal and peat	5.13	3.11	4.02	1.59	0.93	1.10	1.51	-1.42	-4.56
Oil	14.22	16.03	8.23	8.16	6.75	7.01	6.21	-3.17	-1.33
Natural Gas	0.56	0.86	4.85	13.07	10.75	11.09	10.91	13.52	3.93
Biofuels and wastes	-	0.24	2.64	1.86	1.72	1.63	1.79	-	-1.83
Geothermal	-	-	-	-	-	0.00	0.00	-	-
Solar, wind, tide	-	-	-	0.00	0.00	0.00	0.00	-	-
Electricity	4.66	6.63	7.77	10.52	9.43	9.03	8.85	3.05	0.62
Heat	-	-	-	-	-	-	-	-	-
Total transport	**15.50**	**21.54**	**30.40**	**43.15**	**49.23**	**48.41**	**45.79**	**4.04**	**1.97**
Coal and peat	0.02	0.01	-	-	-	-	-	-	-
Oil	15.30	21.29	29.95	42.52	47.21	45.83	42.81	4.03	1.72
Natural Gas	-	-	-	0.01	0.12	0.13	0.13	-	-
Biofuels and wastes	-	-	-	0.10	1.53	2.05	2.46	-	-
Electricity	0.18	0.23	0.45	0.51	0.37	0.40	0.39	5.60	-0.71
Residential	**5.18**	**7.33**	**13.08**	**17.12**	**22.66**	**24.09**	**23.14**	**5.59**	**2.76**
Coal and peat	0.71	0.73	0.75	0.28	0.24	0.19	0.15	0.34	-7.54
Oil	3.07	4.08	5.09	5.78	4.74	4.79	4.13	3.02	-0.99
Natural Gas	0.06	0.12	0.55	2.82	5.24	6.08	5.98	13.52	11.99
Biofuels and wastes	-	-	2.96	2.85	3.49	3.52	3.53	-	0.84
Geothermal	-	-	-	-	0.01	0.01	0.01	-	-
Solar, wind, tide	-	-	-	0.03	0.17	0.21	0.23	-	-
Electricity	1.33	2.40	3.71	5.36	8.77	9.30	9.11	6.21	4.37
Heat	-	-	-	-	-	-	-	-	-
Comm & public services	**1.96**	**3.10**	**4.88**	**9.57**	**13.43**	**13.99**	**13.63**	**5.50**	**5.02**
Coal and peat	0.08	0.11	0.06	0.08 e	0.02	0.04	0.00	-1.08	-12.04
Oil	0.95	1.44	1.49	2.41	2.03	2.02	1.92	2.67	1.21
Natural Gas	0.01	0.04	0.23	0.85	1.28	1.48	1.45	18.34	9.13
Biofuels and waste	-	-	-	0.07	0.08	0.09	0.09	-	-
Geothermal	-	-	0.00	0.00	0.00	0.00	0.00	-	9.89
Solar, wind, tide	-	-	-	0.01	0.05	0.05	0.05	-	-
Electricity	0.92	1.50	3.08	6.15	9.95	10.31	10.10	7.39	5.81
Heat	-	-	-	-	-	-	-	-	-
Non-energy use	**4.80**	**6.58**	**8.34**	**13.42**	**10.22**	**10.05**	**9.67**	**3.30**	**0.71**
Coal and peat	-	-	-	-	-	-	-	-	-
Oil	4.80	6.58	7.81	12.74	9.67	9.38	8.98	2.90	0.67
Natural Gas	-	-	0.53	0.68	0.54	0.67	0.69	-	1.24

(1) Total Final Consumption (TFC) excludes use in transformation processes and energy industry own use.

* Please refer to notes and definitions in Part I. Agriculture, Forestry, Fishing and "Non-specified other" sectors are not reported here individually, but are accounted for in Total final consumption.

Source: *IEA/OECD Energy Balances of OECD Countries*

SPAIN

5. Coal balance*

(Mtce)

	1973	1980	1990	2000	2005	2010	2011	2012e	Average annual percent change 73-90	90-11
Production	9.3	14.0	16.8	11.4	8.9	4.7	3.8	3.5	3.6	-6.8
Imports	3.1	5.9	10.1	19.1	21.2	11.2	13.6	18.5	7.3	1.4
Exports	-0.0	-0.0	-0.0	-0.7	-0.6	-1.6	-1.3	-2.0	9.7	17.7
Stock changes	0.6	-2.2	0.7	0.2	-0.2	-3.0	1.7	1.7		
Primary supply	**12.9**	**17.8**	**27.5**	**29.9**	**29.4**	**11.3**	**17.8**	**21.7**	**4.6**	**-2.1**
Statistical differences	**1.2**	**0.0**	**-0.1**	**0.7**	**0.4**	**0.4**	**1.0**	..		
Total transformation	**-7.4 e**	**-12.9 e**	**-22.0 e**	**-28.3 e**	**-27.0 e**	**-9.5 e**	**-16.5 e**	..	**6.6**	**-1.4**
Electricity and heat gen.	-4.3	-11.2	-20.3	-26.7	-25.7 e	-8.7	-15.3	..	9.5	-1.3
Main activity producers (1)	*-4.3*	*-11.1*	*-20.1*	*-26.4*	*-25.5 e*	*-8.5*	*-15.1*	..	*9.5*	*-1.4*
Autoproducers	-	*-0.1*	*-0.2*	*-0.3*	*-0.2 e*	*-0.2*	*-0.2*	..	-	*0.2*
Gas works	0.4	0.5	0.4	0.1	0.1	0.0	0.0	..	-0.2	-19.1
Coal transformation(2)	-3.5 e	-2.3 e	-2.2 e	-1.7 e	-1.4 e	-0.8 e	-1.2 e	..	-2.9	-2.7
BKB plants	-	-	-	-	-	-	-	..	-	-
Blast furnaces	*-2.3 e*	*-1.6 e*	*-1.4 e*	*-1.0 e*	*-0.9 e*	*-0.9 e*	*-0.9 e*	..	*-2.8*	*-1.9*
Coke ovens	*-1.3*	*-0.7*	*-0.8*	*-0.7*	*-0.4*	*0.0*	*-0.3*	..	*-2.9*	*-4.9*
Patent fuel plants	*-0.0*	*0.0 e*	*0.0*	-	-	-	-	..	-	-
Other transformation(3)	-	-	-	-	-	-	-	..	-	-
Energy ind. own use	**-0.7**	**-0.8**	**-0.5**	**-0.4**	**-0.6**	**-0.8**	**-0.4**	..	**-1.7**	**-0.9**
Losses	**-0.0**	**-0.0**	**-0.0**	**-0.0**	**-0.0**	**-0.1**	**-0.1**	..		
Final consumption(4)	**5.9**	**4.0**	**4.8**	**2.0**	**2.1**	**1.4**	**1.8**	..	**-1.2**	**-4.7**
Industry(5)	5.1	3.1	4.0	1.6	1.7	1.1	1.5	..	-1.4	-4.6
Iron and steel	*3.1 e*	*2.3 e*	*1.9 e*	*1.0 e*	*1.1 e*	*0.8 e*	*0.9 e*	..	*-2.9*	*-3.5*
Chemical	*0.4*	*0.3*	*0.2*	*0.1*	*0.2*	*0.2*	*0.3*	..	*-3.9*	*1.2*
Non-metallic minerals	-	*0.3*	*1.6*	*0.4*	*0.1*	*0.0*	*0.2*	..	-	*-8.8*
Paper, pulp and print	-	*0.0*	*0.0*	-	-	-	-	..	-	-
Other industry (6)	*1.6*	*0.2*	*0.2*	*0.1*	*0.2*	*0.1*	*0.1*	..	*-10.4*	*-3.4*
Transport(7)	0.0	0.0	-	-	-	-	-	..	-	-
Other	0.8	0.8	0.8	0.4	0.4	0.3	0.2	..	0.3	-5.6
Comm. and pub. services	*0.1*	*0.1*	*0.1*	*0.1 e*	*0.0*	*0.0*	*0.0*	..	*-1.1*	*-12.0*
Residential	*0.7*	*0.7*	*0.8*	*0.3*	*0.3*	*0.2*	*0.1*	..	*0.3*	*-7.5*
Other sectors (8)	-	-	*0.0*	-	*0.0*	*0.1*	*0.1*	..	-	*11.1*
Non-energy use	-	-	-	-	-	-	-	..	-	-

(1) Main activity electricity and heat generation includes district heating.

(2) Coal transformation refers to the transformation of coal and peat to secondary and tertiary products (mainly cokes, briquettes, coal tar and off-gases). Gas works and Liquefaction are shown separately to match the *Energy Balances of OECD Countries.*

(3) Other transformation includes Liquefaction and Non-specified transformations.

(4) Final Consumption includes non-energy use and energy use (Industry, Transport and Other).

(5) Please refer to notes and definitions in Part I for detailed explanation.

(6) Other industry includes Non-ferrous metals, Transport equipment, Machinery, Mining and quarrying, Food and tobacco, Wood and wood products, Construction, Textile and leather, and Non-specified industry.

(7) Transport includes Rail and Inland waterways.

(8) Other sectors includes Agriculture/Forestry, Fishing and Non-specified other.

* "Coal" refers to all types of coal, primary (anthracite, coking coal, other bituminous coal, sub-bituminous coal, lignite/brown coal) and derived fuels (including patent fuel, BKB, coke oven coke, gas coke, coal tar, coke oven gas, gas works gas, blast furnace gas and other recovered gases), as well as peat, peat briquettes and oil shale. Quantities have been converted to Mtce using calorific values reported by the respective countries. Please refer to notes and definitions in Part I.

Source: *IEA/OECD Energy Balances of OECD Countries*

SPAIN

6. Use of coal for selected end-uses*

(million tonnes)

	1978	1990	2000	2005	2009	2010	2011	Average annual percent change 78-90	90-11
Steam coal	**11.30**	**25.79**	**33.70**	**33.47**	**17.18**	**12.16**	**21.65**	**7.12**	**-0.83**
Total electricity and heat	9.71	22.92	32.62	31.71 e	15.67	10.97	20.66	7.42	-0.49
Main activity producers	*9.58*	*22.84*	*32.43*	*31.63*	*15.43*	*10.71*	*20.39*	*7.51*	*-0.54*
Autoproducers	*0.13*	*0.08*	*0.19*	*0.08 e*	*0.24*	*0.26*	*0.27*	*-4.23*	*6.11*
Patent fuel/BKB plants	0.06	0.01	-	-	-	-	-	-18.93	-
Coke ovens/Liquefaction(1)	0.30	0.02	-	-	-	-	-	-18.94	-
Blast furnace inputs	-	-	0.57 e	0.51 e	0.25 e	0.56 e	0.64 e	-	-
Gas manufacture	-	-	-	-	-	-	-	-	-
Industry	0.78	2.33	0.65	0.73	0.25	0.38	0.73	9.49	-5.35
Iron and steel	*0.31*	*0.30*	*0.11 e*	*0.20 e*	*0.02 e*	*0.12 e*	*0.24 e*	*-0.16*	*-0.98*
Chemical	*0.23*	*0.14*	*0.06*	*0.25*	*0.20*	*0.22*	*0.20*	*-3.99*	*1.85*
Non-metallic minerals	*0.18*	*1.78*	*0.47*	*0.18*	*0.03*	*0.04*	*0.28*	*21.25*	*-8.38*
Paper, pulp and print	*0.00*	*0.06*	-	-	-	-	-	*33.31*	-
Other industry	*0.08*	*0.05*	-	*0.10 e*	-	-	*0.00 e*	*-3.32*	*-12.54*
Other sectors(2)	0.38	0.58	0.30	0.38	0.38	0.35	0.28	3.53	-3.39
Non-energy use	-	-	-	-	-	-	-	-	-
Coking coal	**4.86**	**4.46**	**3.56**	**3.46**	**2.50**	**2.50**	**2.54**	**-0.73**	**-2.64**
Total electricity and heat	-	-	-	-	-	-	-	-	-
Main activity producers	-	-	-	-	-	-	-	-	-
Autoproducers	-	-	-	-	-	-	-	-	-
Patent fuel/BKB plants	-	-	-	-	-	-	-	-	-
Coke ovens/Liquefaction(1)	4.86	4.46	3.64	3.60	2.33	2.65	2.69	-0.73	-2.37
Blast furnace inputs	-	-	-	-	-	-	-	-	-
Gas manufacture	-	-	-	-	-	-	-	-	-
Industry	-	-	-	-	-	-	-	-	-
Iron and steel	-	-	-	-	-	-	-	-	-
Chemical	-	-	-	-	-	-	-	-	-
Non-metallic minerals	-	-	-	-	-	-	-	-	-
Paper, pulp and print	-	-	-	-	-	-	-	-	-
Other industry	-	-	-	-	-	-	-	-	-
Other sectors(2)	-	-	-	-	-	-	-	-	-
Non-energy use	-	-	-	-	-	-	-	-	-
Lignite	**5.20**	**16.58**	**8.40**	**7.56**	**-**	**-**	**-**	**10.15**	**-**
Total electricity and heat	5.20	16.58	8.40	7.58	-	-	-	10.15	-
Main activity producers	*5.20*	*16.58*	*8.40*	*7.58*	-	-	-	*10.15*	-
Autoproducers	-	-	-	-	-	-	-	-	-
Patent fuel/BKB plants	-	-	-	-	-	-	-	-	-
Coke ovens/Liquefaction(1)	-	-	-	-	-	-	-	-	-
Gas manufacture	-	-	-	-	-	-	-	-	-
Industry	-	-	-	-	-	-	-	-	-
Iron and steel	-	-	-	-	-	-	-	-	-
Chemical	-	-	-	-	-	-	-	-	-
Non-metallic minerals	-	-	-	-	-	-	-	-	-
Paper, pulp and print	-	-	-	-	-	-	-	-	-
Other industry	-	-	-	-	-	-	-	-	-
Other sectors(2)	-	-	-	-	-	-	-	-	-
Non-energy use	-	-	-	-	-	-	-	-	-

(1) Liquefaction primarily refers to direct distillation processes. Indirect liquefaction may be reported here or under gas manufacture.

(2) Other sectors are Residential, Commercial and public services, Agriculture/Forestry, Fishing and Non-specified other.

* Coal totals may not match the sum of the individual rows due to statistical differences. Please refer to notes and definitions in Part I.

Source: *IEA/OECD Energy Statistics of OECD Countries*

SPAIN

7. Fuel prices to end users*

	1978(1)	1990	1995	2000	2005	2009	2010	2011	2012
				(US dollars / tce)					
For electricity generation									
Steam coal	42.60	..	..	..	..	..	..	..	..
Heavy fuel oil	65.54	106.42	130.25	169.30	..	..	..	..	..
Natural gas	60.18	118.39	129.93	128.33	..	..	..	..	..
For industry									
Steam coal	..	..	..	..	..	..	..	..	..
Coking coal	68.57	c	c	c	c	c	c	c	c
High sulphur fuel oil	65.54	106.42	130.25	138.84	..	..	..	..	..
Low sulphur fuel oil	..	..	130.63	169.30	278.37	339.80	416.86	543.46	575.50
Natural gas	100.39	127.20	123.77	136.41	197.69	337.39	303.66	341.20	397.73
				(Euro / unit)(2)					
For electricity generation									
Steam coal	15.43	..	..	..	..	..	..	..	..
Heavy fuel oil	40.57	87.60	131.01	246.67	..	..	..	..	..
Natural gas	35.67	93.31	125.12	179.02	..	..	..	..	..
For industry									
Steam coal	..	..	..	..	..	..	..	..	..
Coking coal	32.38	c	c	c	c	c	c	c	c
High sulphur fuel oil	40.57	87.60	131.01	202.29	..	..	..	..	..
Low sulphur fuel oil	..	..	131.39	246.67	300.92	328.45	422.63	524.50	601.25
Natural gas	59.50	100.25	119.19	190.29	204.61	312.24	294.77	315.28	397.84

(1) Earliest year for which split by coal type is available.

(2) For steam coal, coking coal and heavy fuel oil per metric tonne; for natural gas per 10^7 kilocalories GCV.

* Please refer to notes and definitions in Part I. Steam coal prices relate to historical definitions.

Source: *IEA/OECD Energy Prices & Taxes*

8. Coal and peat trade by type of coal*

(Mtce)

	1978	1990	1995	2000	2005	2009	2010	2011	2012e
Total imports	**3.63**	**10.15**	**12.38**	**19.07**	**21.19**	**14.15**	**11.21**	**13.59**	**18.50**
Bituminous coal(1)	0.24	5.48	8.00	15.09	17.40	12.02	8.26	10.92	16.08
Coking coal	3.14	4.29	3.32	3.84	3.66	2.01	2.75	2.52	2.27
Sub-bituminous coal	0.00	0.00	0.24	-	-	-	-	-	-
Lignite	-	-	-	-	-	-	-	-	-
Peat	-	-	-	-	-	-	-	-	-
Coal products(2)	0.24	0.39	0.82	0.13	0.13	0.12	0.20	0.15	0.15
Total exports	**0.01**	**0.04**	**0.08**	**0.72**	**0.59**	**1.32**	**1.59**	**1.33**	**1.96**
Bituminous coal(1)	0.01	0.00	-	-	-	1.13	1.23	0.98	1.54
Coking coal	-	-	-	-	-	-	-	0.02	-
Sub-bituminous coal	-	-	-	-	-	-	-	-	-
Lignite	-	-	-	-	-	-	-	-	-
Peat	-	-	-	-	-	-	-	-	-
Coal products(2)	-	0.04	0.08	0.72	0.59	0.19	0.36	0.34	0.42

(1) Bituminous coal includes anthracite.

(2) Coal products includes products derived from coal, for example: coke, coal tar, briquettes, patent fuels.

* Please refer to notes and definitions in Part I.

Source: *IEA/OECD Energy Balances of OECD Countries, IEA/OECD Energy Statistics of OECD Countries*

SPAIN

9. Total coal imports by origin*

(thousand tonnes)

	1978[(1)]	1985	1990	1995	2000	2005	2010	2011	2012e
Total coal[(2)]	**3382**	**8418**	**10456**	**13889**	**21649**	**24756**	**12817**	**16168**	**22414**
Coking coal	**3029**	**4090**	**4169**	**3244**	**3755**	**3571**	**2777**	**2505**	**2260**
Australia	454	996	672	879	1388	1682	1272	1463	1202
Canada	68	-	-	389	383	285	60	43	-
Czech Republic	-	-	-	-	-	-	-	-	-
Germany	395	-	364	-	-	-	-	-	-
Poland	1369	637	234	58	-	99	-	-	-
United Kingdom	-	-	-	-	-	-	1	-	-
United States	743	2401	2899	1916	1912	1273	1384	999	950
Other OECD	-	56	-	2	-	-	-	-	-
China, People's Rep.	-	-	-	-	-	178	-	-	-
Colombia	-	-	-	-	-	-	60	-	-
Indonesia	-	-	-	-	-	-	-	-	-
South Africa	-	-	-	-	-	-	-	-	-
Former Soviet Union[(3)]	-	-	x	x	x	x	x	x	x
Russian Federation [(4)]	*x*	*x*	-	-	*72*	-	-	-	*108*
Other FSU [(4)]	*x*	*x*	-	-	-	-	-	-	-
Venezuela	-	-	-	-	-	54	-	-	-
Vietnam	-	-	-	-	-	-	-	-	-
Non-specified/other	-	-	-	-	-	-	-	-	-
Steam coal	**353**	**4328**	**6287**	**10645**	**17894**	**21185**	**10040**	**13663**	**20154**
Australia	-	407	271	110	1627	1433	400	533	149
Canada	-	172	-	-	-	16	1	103	-
Czech Republic	-	-	-	-	-	-	-	-	-
Germany	-	2	43	53	102	169	80	76	49
Poland	-	379	16	-	313	28	12	13	-
United Kingdom	12	5	153	103	27	4	16	19	37
United States	-	855	275	2314	585	227	481	527	1236
Other OECD	27	9	90	136	39	12	18	101	305
China, People's Rep.	-	-	-	-	191	47	7	7	9
Colombia	-	134	404	665	1112	1938	2796	3647	6294
Indonesia	-	-	-	1089	2804	3784	2411	3312	5634
South Africa	114	2274	4667	6037	9180	8736	2584	3065	2818
Former Soviet Union[(3)]	196	91	x	x	x	x	x	x	x
Russian Federation [(4)]	*x*	*x*	*285*	*138*	*1496*	*4235*	*852*	*1852*	*3181*
Other FSU [(4)]	*x*	*x*	-	-	*29*	*450*	*322*	*298*	*307*
Venezuela	-	-	53	-	389	91	54	110	133
Vietnam	-	-	-	-	-	15	-	-	-
Non-specified/other	4	-	30	-	-	-	6	-	2
Lignite	-	-	-	-	-	-	-	-	-

(1) Earliest year for which split by coal type is available.

(2) Total coal consists of steam coal, coking coal and lignite.

(3) For years prior to 1990, or if no details for individual Former Soviet Union republics.

(4) Data for former republics are not available separately for some OECD member countries.

* Please refer to notes and definitions in Part I.

Source: IEA/OECD Coal Statistics

SPAIN

11. Steam coal exports by destination*

(thousand tonnes)

	1978[(1)]	1985	1990	1995	2000	2005	2010	2011	2012e
World	**10**	**5**	**3**	**-**	**-**	**-**	**1488**	**1175**	**1861**
Total OECD	**10**	**5**	**3**	**-**	**-**	**-**	**1288**	**1049**	**1703**
Australia	-	-	-	-	-	-	-	-	-
Austria	-	-	-	-	-	-	-	-	-
Belgium	10	-	-	-	-	-	-	2	4
Canada	-	-	-	-	-	-	-	-	-
Chile	-	-	-	-	-	-	-	-	-
Czech Republic	-	-	-	-	-	-	-	-	-
Denmark	-	2	-	-	-	-	-	-	-
Estonia	-	-	-	-	-	-	-	-	-
Finland	-	-	-	-	-	-	-	-	-
France	-	-	3	-	-	-	1	4	1
Germany	-	-	-	-	-	-	-	-	-
Greece	-	-	-	-	-	-	-	55	19
Hungary	-	-	-	-	-	-	-	-	-
Iceland	-	-	-	-	-	-	-	-	-
Ireland	-	-	-	-	-	-	4	14	5
Israel	-	-	-	-	-	-	-	-	-
Italy	-	-	-	-	-	-	1172	846	1540
Japan	-	-	-	-	-	-	-	-	-
Korea	-	-	-	-	-	-	-	-	-
Luxembourg	-	-	-	-	-	-	-	-	-
Mexico	-	-	-	-	-	-	-	-	-
Netherlands	-	-	-	-	-	-	2	3	3
New Zealand	-	-	-	-	-	-	-	-	2
Norway	-	-	-	-	-	-	11	17	15
Poland	-	-	-	-	-	-	-	2	-
Portugal	-	3	-	-	-	-	28	28	31
Slovak Republic	-	-	-	-	-	-	-	-	-
Slovenia	-	-	-	-	-	-	-	-	-
Spain	-	-	-	-	-	-	-	-	-
Sweden	-	-	-	-	-	-	-	-	-
Switzerland	-	-	-	-	-	-	-	-	-
Turkey	-	-	-	-	-	-	-	17	-
United Kingdom	-	-	-	-	-	-	70	61	83
United States	-	-	-	-	-	-	-	-	-
Total non-OECD	**-**	**-**	**-**	**-**	**-**	**-**	**29**	**126**	**158**
Brazil	-	-	-	-	-	-	-	-	-
China[(2)]	-	-	-	-	-	-	-	-	-
Chinese Taipei	-	-	-	-	-	-	-	-	-
Egypt	-	-	-	-	-	-	14	35	21
India	-	-	-	-	-	-	-	-	-
Romania	-	-	-	-	-	-	-	-	-
Oth. Africa & Mid. East	-	-	-	-	-	-	15	45	20
Oth. non-OECD Americas	-	-	-	-	-	-	-	15	43
Other Asia & Oceania	-	-	-	-	-	-	-	2	37
Other non-OECD Europe and Eurasia	-	-	-	-	-	-	-	29	37
Non-specified/Other	**-**	**-**	**-**	**-**	**-**	**-**	**171**	**-**	**-**

(1) Earliest year for which split by coal type is available.
(2) China includes the People's Republic of China and Hong Kong, China.
* Please refer to notes and definitions in Part I. Steam coal includes all sub-bituminous coal.
Source: IEA/OECD Coal Statistics

SPAIN

12. Coal import values by origin*

(average unit value, CIF, USD/tonne)

	1990	1995	2000	2005	2007	2008	2009	2010	2011/12(1)
Coking coal(2)	**62.38**	**58.59**	**49.11**	**116.50**	**124.87**	**211.23**	**184.13**	**217.85**	..
Imports from:									
Australia	59.85	52.80	46.17	110.07	125.17	238.36	232.74	240.31	..
Canada	..	57.34	48.43	141.45	142.00	259.29	..	178.78	..
Czech Republic	..	..	..	..	..	..	..	..	..
Poland	59.09	51.07	..	135.89	..	..	130.38	..	..
United States	63.21	61.65	53.22	116.42	122.09	160.69	163.34	200.96	..
China	..	..	..	137.51	..	344.75	..	..	..
Colombia	..	..	..	..	..	..	244.67	172.03	..
Indonesia	..	..	..	..	..	..	..	..	..
South Africa	..	..	..	..	..	..	..	..	..
Former Soviet Union(3)	..	..	30.40	..	..	..	..	..	..
Other bituminous coal(4)	**44.74**	**43.24**	**32.03**	**62.94**	**74.06**	**128.55**	**85.30**	**96.14**	..
Imports from:									
Australia	52.89	50.56	33.83	93.53	79.48	158.54	103.14	163.85	..
Canada	..	..	..	64.94	403.91	142.45	462.40	460.06	..
Czech Republic	..	..	..	..	..	..	..	..	..
Poland	102.86	..	38.10	114.05	142.48	214.64	181.26	164.60	..
United States	44.32	38.79	31.62	46.84	126.87	125.44	87.64	108.71	..
China	..	..	36.51	93.42	..	..	..	..	..
Colombia	48.08	46.44	30.43	63.06	72.34	..	90.93	101.33	..
Indonesia	..	39.04	28.63	49.68	57.17	78.56	75.72	68.31	..
South Africa	43.00	44.04	31.91	65.12	75.14	138.56	89.96	99.04	..
Former Soviet Union(3)	42.90	46.33	33.96	65.99	92.84	156.74	83.10	97.41	..

Note: On occasion, shipment of extremely small quantities of high valued coal results in high import costs.

(1) Import data for steam and coking coals are currently unavailable for 2011 and 2012 due to resource constraints.

(2) Weighted average of individual countries using import volumes as weights. Prices exclude intra-EU trade.

(3) Former Soviet Union until 1991, Russian Federation starting in 1992.

(4) Bituminous steam coal only. Weighted average of trades. Prices exclude intra-EU trade.

* Please refer to notes and definitions in Part I.

Source: IEA/OECD Coal Statistics

SWEDEN*

Figure 1: TPES by fuel (Mtce)

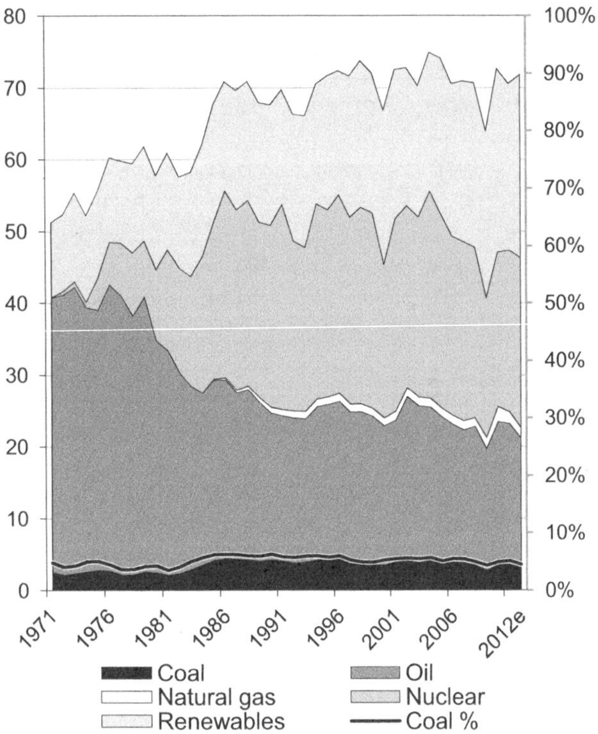

Figure 2: Coal supply indicators (1971 = 100)

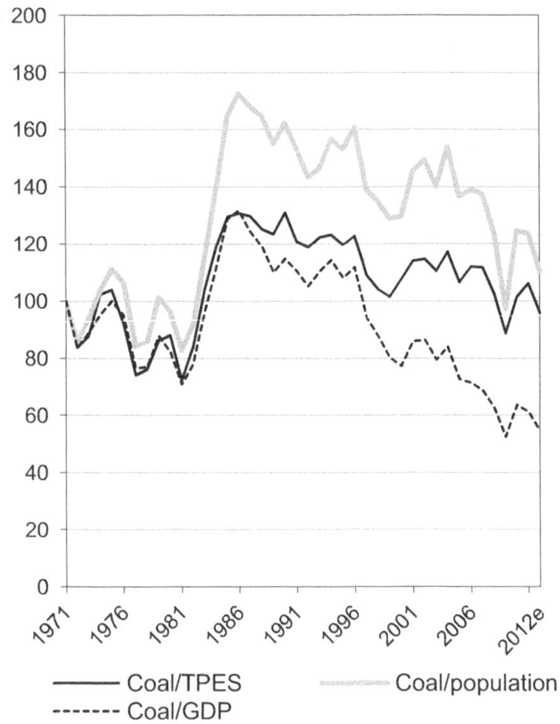

Figure 3: Primary coal supply (Mtce)

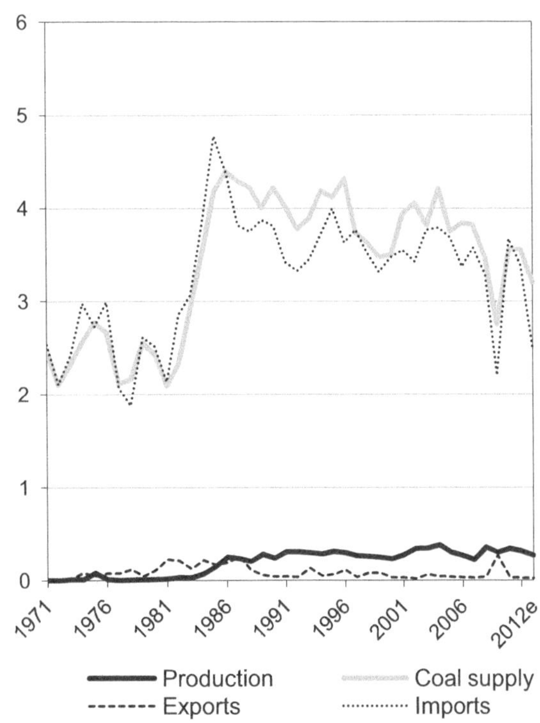

Figure 4: Coal consumption (Mtce)

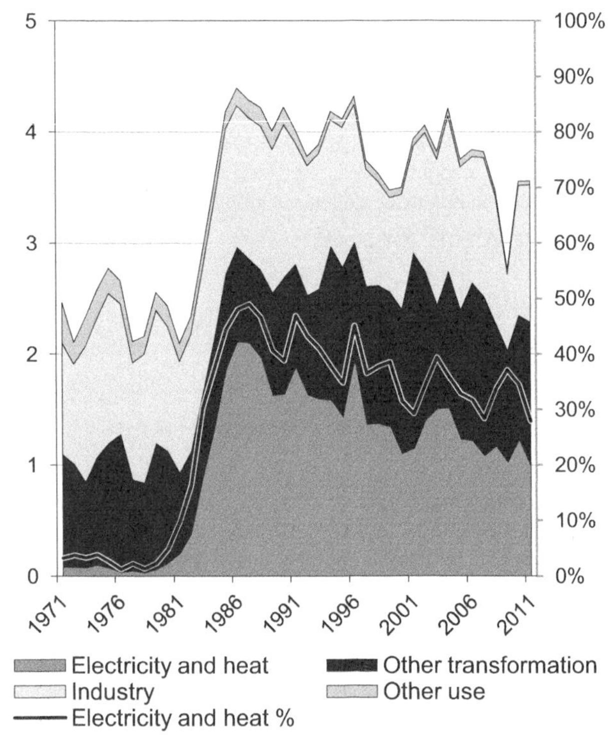

*Please refer to notes and definitions in Part I. Peat is included in coal for all figures.

Source: *IEA/OECD Coal Information, IEA/OECD Energy Balances of OECD Countries, OECD Main Economic Indicators*

SWEDEN*

Figure 5: Electricity generation by fuel (TWh)

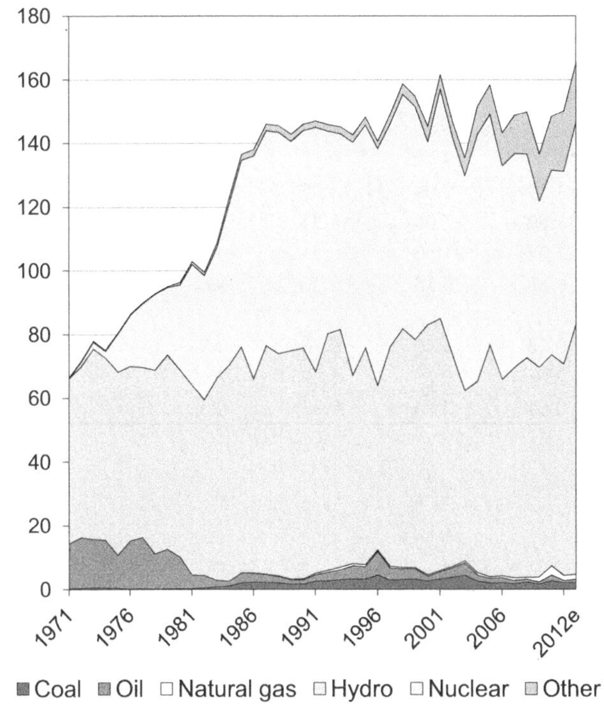

Figure 6: CO_2 emissions by fuel (Mt CO_2)

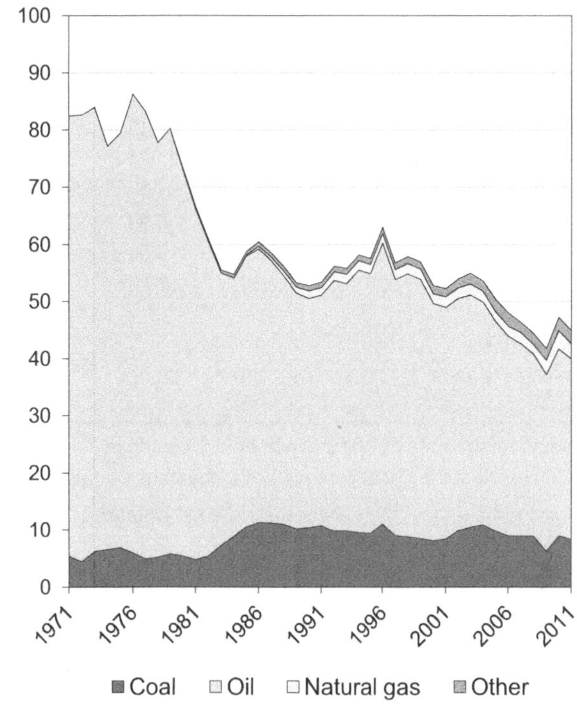

Figure 7: Electricity generation by fuel share

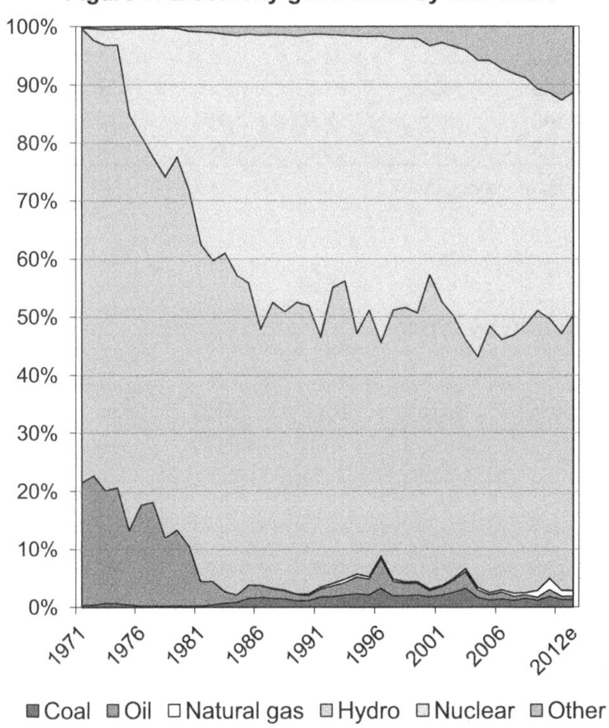

Figure 8: CO_2 emissions by fuel share

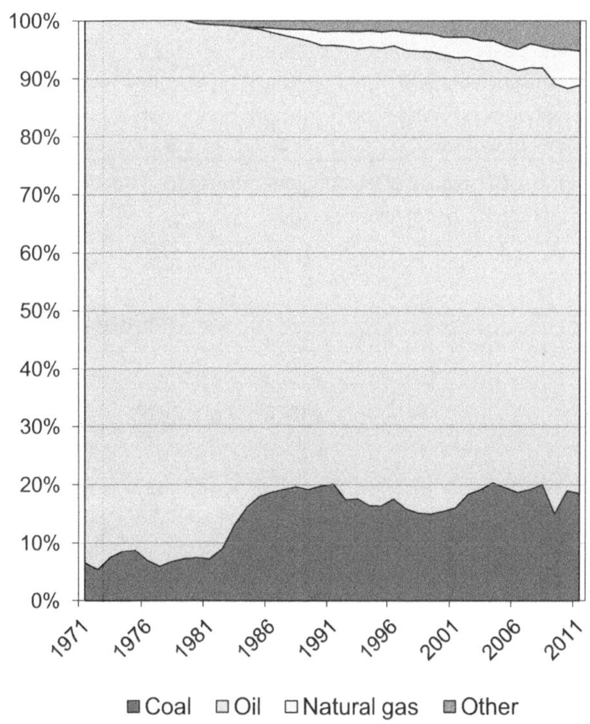

*Please refer to notes and definitions in Part I. Peat is included in coal for all figures.

Source: *IEA/OECD Electricity Information, IEA/OECD CO2 Emissions from Fuel Combustion*

SWEDEN

1. Total primary energy supply (TPES) by fuel*

	1973	1980	1990	2000	2010	2011	2012e	Average annual percent change 73-90	90-11
TPES (Mtce)	**55.49**	**57.84**	**67.43**	**67.94**	**73.31**	**70.06**	**69.82**	**1.15**	**0.18**
Coal and peat	2.33	2.43	4.22	3.50	3.56	3.56	3.20	3.57	-0.81
Oil	39.87	32.34	20.43	19.39	19.88	19.62	17.96	-3.86	-0.19
Natural Gas	-	-	0.82	1.11	2.09	1.65	1.44	-	3.38
Biofuels and waste	5.06	5.90	7.87	11.80	17.00	14.31	15.04	2.63	2.89
Nuclear	0.79	9.86	25.38	21.34	21.53	22.51	23.66	22.68	-0.57
Hydro	7.36	7.23	8.91	9.65	8.16	8.16	9.66	1.13	-0.42
Geothermal	-	-	-	-	-	-	-	-	-
Solar, wind, tide	-	-	0.01	0.06	0.45	0.76	0.90	-	26.73
Net electricity trade(1)	0.09	0.07	-0.22	0.57	0.26	-0.89	-2.40	x	6.94
Heat(2)	-	-	-	0.51	0.39	0.37	0.39	-	-

(1) Net trade between OECD and non-OECD countries.

(2) Ambient heat from heat pumps used in the transformation sector.

Source: *IEA/OECD Energy Balances of OECD Countries*

2. Energy supply, GDP and population*

	1973	1980	1990	2000	2010	2011	2012e	Average annual percent change 73-90	90-11
GDP (2005 bil. USD)	188	212	264	325	402	417	420	2.02	2.20
TPES/GDP(1)	0.30	0.27	0.26	0.21	0.18	0.17	0.17	-0.85	-1.97
Population (millions)	8.1	8.3	8.6	8.9	9.4	9.5	9.5	0.30	0.47
TPES/population(1)	6.82	6.96	7.88	7.66	7.82	7.41	7.33	0.85	-0.29
TPES/GDP(2)	148.6	137.0	128.5	105.3	91.8	84.6	83.6	-0.85	-1.97
Coal and peat supply/GDP(2)	122.4	113.2	158.1	106.5	87.5	84.4	75.3	1.52	-2.94
Elec. consumption/GDP(2)	104.5	113.4	129.3	112.5	92.6	84.8	..	1.26	-1.99
Elec. generation (TWh)	78	96	146	145	148	150	165	3.75	0.14
Industrial production(2)	50.6	50.4	63.7	91.9	93.2	98.6	96.6	1.36	2.10

(1) TPES/GDP in units of tce/2005 thousand USD, TPES/population in units of tce/capita.

(2) As index, 2005 (not shown) = 100

Source: *IEA/OECD Energy Balances of OECD Countries* , *OECD Main Economic Indicators*

3. Coal and peat production by type*

	1978(1)	1990	2000	2005	2010	2011	2012e	Average annual percent change 78-90	90-11
Mtce:									
Coking coal	-	-	-	-	-	-	-	-	-
Steam coal	0.01	0.01	-	-	-	-	-	-3.07	-
Lignite	-	-	-	-	-	-	-	-	-
Peat	-	0.23	0.23	0.30	0.34	0.31	0.27	-	1.43
Mt:									
Coking coal	-	-	-	-	-	-	-	-	-
Steam coal	0.02	0.01	-	-	-	-	-	-3.07	-
Lignite	-	-	-	-	-	-	-	-	-
Peat	-	0.58	0.54	0.71	0.80	0.74	0.63	-	1.13

(1) Earliest year for which split by coal type is available.

* Please refer to notes and definitions in Part I.

Source: *IEA/OECD Energy Balances of OECD Countries, IEA/OECD Energy Statistics of OECD Countries*

SWEDEN

4. Final consumption of energy by fuel*

(Mtce)

	1973	1980	1990	2000	2009	2010	2011	Average annual percent change 73-90	90-11
Total final consumption[(1)]	**49.74**	**49.43**	**45.88**	**50.42**	**45.73**	**50.39**	**46.75**	**-0.47**	**0.09**
Coal and peat	1.48	1.31	1.53	1.10	0.73	1.21	1.27	0.22	-0.89
Oil	34.83	28.81	20.02	20.25	15.16	16.06	15.91	-3.20	-1.09
Natural Gas	-	-	0.48	0.63	0.79	0.95	0.95	-	3.32
Biofuels and wastes	4.94	5.51	6.62	7.55	7.79	8.70	7.39	1.74	0.53
Geothermal	-	-	-	-	-	-	-	-	-
Solar, wind, tide	-	-	0.00	0.01	0.01	0.01	0.02	-	6.10
Electricity	8.50	10.43	14.79	15.81	15.16	16.12	15.31	3.31	0.17
Heat	-	3.37	2.44	5.07	6.08	7.34	5.91	-	4.31
of which:									
Total industry	**18.95**	**17.07**	**16.94**	**19.54**	**15.24**	**17.25**	**15.19**	**-0.66**	**-0.52**
Coal and peat	1.22	1.13	1.37	1.04	0.69	1.17	1.24	0.72	-0.50
Oil	8.68	6.64	3.09	4.39	1.53	1.66	1.58	-5.90	-3.14
Natural Gas	-	-	0.36	0.43	0.41	0.46	0.48	-	1.37
Biofuels and wastes	4.19	4.32	5.25	6.19	5.75	6.63	4.75	1.33	-0.48
Geothermal	-	-	-	-	-	-	-	-	-
Solar, wind, tide	-	-	-	-	-	-	-	-	-
Electricity	4.86	4.98	6.63	7.00	6.32	6.68	6.61	1.84	-0.01
Heat	-	-	0.24	0.49	0.55	0.64	0.54	-	4.00
Total transport	**7.52**	**8.33**	**9.88**	**10.56**	**10.96**	**11.17**	**11.71**	**1.62**	**0.82**
Coal and peat	-	0.00	-	-	-	-	-	-	-
Oil	7.27	8.05	9.57	10.15	10.09	10.25	10.64	1.63	0.51
Natural Gas	-	-	-	0.01	0.03	0.05	0.06	-	-
Biofuels and wastes	-	-	-	-	0.55	0.57	0.69	-	-
Electricity	0.26	0.28	0.30	0.39	0.30	0.30	0.32	1.03	0.31
Residential	**15.35**	**13.93**	**9.35**	**10.42**	**9.93**	**10.79**	**9.94**	**-2.88**	**0.29**
Coal and peat	0.20	0.08	0.02	0.04	0.01	0.01	0.01	-11.78	-6.55
Oil	12.47	9.51	2.20	1.25	0.11	0.10	0.07	-9.71	-14.90
Natural Gas	-	-	0.05	0.11	0.09	0.11	0.10	-	3.45
Biofuels and wastes	0.74	1.18	0.90	0.88	0.98	0.98	1.69	1.15	3.03
Geothermal	-	-	-	-	-	-	-	-	-
Solar, wind, tide	-	-	0.00	0.01	0.01	0.01	0.02	-	6.10
Electricity	1.93	3.16	4.68	5.16	5.03	4.97	4.48	5.34	-0.21
Heat	-	-	1.49	2.97	3.69	4.62	3.58	-	4.26
Comm & public services	**1.41**	**1.93**	**5.72**	**6.28**	**6.05**	**7.18**	**6.48**	**8.59**	**0.59**
Coal and peat	-	0.03	0.01	0.00	0.01	0.01	0.01	-	-2.74
Oil	0.04	0.02	1.93	1.50	0.78	0.86	0.75	24.93	-4.42
Natural Gas	-	-	0.05	0.05	0.07	0.15	0.13	-	4.97
Biofuels and waste	-	0.01	0.02	0.02	0.06	0.06	0.06	-	4.23
Geothermal	-	-	-	-	-	-	-	-	-
Solar, wind, tide	-	-	-	-	-	-	-	-	-
Electricity	1.37	1.87	2.99	3.12	3.29	4.02	3.75	4.72	1.09
Heat	-	-	0.71	1.60	1.84	2.07	1.78	-	4.48
Non-energy use	**3.11**	**2.24**	**2.76**	**2.47**	**2.51**	**3.02**	**2.76**	**-0.71**	**-0.00**
Coal and peat	0.06	0.06	0.06	0.02	0.02	0.02	0.02	0.32	-4.90
Oil	3.06	2.18	2.70	2.45	2.35	2.84	2.60	-0.73	-0.18
Natural Gas	-	-	-	-	0.15	0.15	0.14	-	-

(1) Total Final Consumption (TFC) excludes use in transformation processes and energy industry own use.

* Please refer to notes and definitions in Part I. Agriculture, Forestry, Fishing and "Non-specified other" sectors are not reported here individually, but are accounted for in Total final consumption.

Source: *IEA/OECD Energy Balances of OECD Countries*

SWEDEN

5. Coal balance*

(Mtce)

	1973	1980	1990	2000	2005	2010	2011	2012e	Average annual percent change 73-90	90-11
Production	0.0	0.0	0.2	0.2	0.3	0.3	0.3	0.3	19.5	1.3
Imports	2.4	2.5	3.8	3.5	3.7	3.7	3.4	2.5	2.7	-0.6
Exports	-0.0	-0.1	-0.0	-0.0	-0.0	-0.0	-0.0	-0.0	3.5	-2.4
Stock changes	-0.1	0.0	0.2	-0.2	-0.2	-0.4	-0.1	0.5		
Primary supply	**2.3**	**2.4**	**4.2**	**3.5**	**3.8**	**3.6**	**3.6**	**3.2**	**3.6**	**-0.8**
Statistical differences	**0.0**	**0.0**	**0.0**	**0.1**	**0.1**	**0.3**	**-0.2**	..		
Total transformation	**-0.7 e**	**-0.9 e**	**-2.5 e**	**-2.3 e**	**-2.4 e**	**-2.4 e**	**-1.8 e**	..	**7.5**	**-1.4**
Electricity and heat gen.	-0.1	-0.1	-1.6	-1.1	-1.2	-1.2	-1.0	..	19.8	-2.4
Main activity producers (1)	*-0.1*	*-0.1*	*-1.6*	*-1.0*	*-1.1*	*-1.1*	*-0.9*	..	*19.7*	*-2.6*
Autoproducers	-	*-0.0*	*-0.0*	*-0.1*	*-0.1*	*-0.1*	*-0.0*	..	-	*6.3*
Gas works	0.2	0.1	0.1	0.1	0.1	0.0	0.0	..	-6.0	-6.5
Coal transformation(2)	-0.8 e	-0.9 e	-0.9 e	-1.2 e	-1.2 e	-1.2 e	-0.9 e	..	0.7	-0.2
BKB plants	-	-	-	-	-	-	-	..	-	-
Blast furnaces	*-0.8 e*	*-0.7 e*	*-0.7 e*	*-0.8 e*	*-0.9 e*	*-0.8 e*	*-0.7 e*	..	*-0.8*	*-0.0*
Coke ovens	*-0.0*	*-0.2*	*-0.2*	*-0.4*	*-0.3*	*-0.5*	*-0.2*	..	*11.2*	*-1.0*
Patent fuel plants	-	-	-	-	-	-	-	..	-	-
Other transformation(3)	-	-	-	-	-	-	-	..	-	-
Energy ind. own use	**-0.1 e**	**-0.1**	**-0.1**	**-0.1**	**-0.1**	**-0.1**	**-0.2**	..	**3.6**	**2.0**
Losses	**-0.1 e**	**-0.1**	**-0.1**	**-0.1**	**-0.0**	**-0.1**	**-0.1**	..		
Final consumption(4)	**1.5**	**1.3**	**1.5**	**1.1**	**1.4**	**1.2**	**1.3**	..	**0.2**	**-0.9**
Industry(5)	1.2	1.1	1.4	1.0	1.3	1.2	1.2	..	0.7	-0.5
Iron and steel	*0.7 e*	*0.6 e*	*0.6 e*	*0.5 e*	*0.8 e*	*0.7 e*	*0.7 e*	..	*-1.4*	*1.4*
Chemical	*0.0*	*0.0*	*0.0*	*0.0*	*0.0*	*0.0*	*0.0*	..	*9.6*	*-6.1*
Non-metallic minerals	*0.3*	*0.2*	*0.4*	*0.3*	*0.2*	*0.2*	*0.3*	..	*2.8*	*-2.0*
Paper, pulp and print	*0.0*	*0.0*	*0.1*	*0.0*	*0.0*	*0.0*	*0.0*	..	*16.8*	*-13.0*
Other industry (6)	*0.2*	*0.2*	*0.3*	*0.2*	*0.2*	*0.2*	*0.2*	..	*0.8*	*-1.3*
Transport(7)	-	0.0	-	-	-	-	-	..	-	-
Other	0.2	0.1	0.1	0.0	0.0	0.0	0.0	..	-4.2	-9.4
Comm. and pub. services	-	*0.0*	*0.0*	*0.0*	*0.0*	*0.0*	*0.0*	..	-	*-2.7*
Residential	*0.2*	*0.1*	*0.0*	*0.0*	*0.0*	*0.0*	*0.0*	..	*-11.8*	*-6.5*
Other sectors (8)	-	*0.0*	*0.1*	-	-	-	-	..	-	-
Non-energy use	0.1	0.1	0.1	0.0	0.0	0.0	0.0	..	0.3	-4.9

(1) Main activity electricity and heat generation includes district heating.

(2) Coal transformation refers to the transformation of coal and peat to secondary and tertiary products (mainly cokes, briquettes, coal tar and off-gases). Gas works and Liquefaction are shown separately to match the *Energy Balances of OECD Countries.*

(3) Other transformation includes Liquefaction and Non-specified transformations.

(4) Final Consumption includes non-energy use and energy use (Industry, Transport and Other).

(5) Please refer to notes and definitions in Part I for detailed explanation.

(6) Other industry includes Non-ferrous metals, Transport equipment, Machinery, Mining and quarrying, Food and tobacco, Wood and wood products, Construction, Textile and leather, and Non-specified industry.

(7) Transport includes Rail and Inland waterways.

(8) Other sectors includes Agriculture/Forestry, Fishing and Non-specified other.

* "Coal" refers to all types of coal, primary (anthracite, coking coal, other bituminous coal, sub-bituminous coal, lignite/brown coal) and derived fuels (including patent fuel, BKB, coke oven coke, gas coke, coal tar, coke oven gas, gas works gas, blast furnace gas and other recovered gases), as well as peat, peat briquettes and oil shale. Quantities have been converted to Mtce using calorific values reported by the respective countries. Please refer to notes and definitions in Part I.

Source: *IEA/OECD Energy Balances of OECD Countries*

SWEDEN

6. Use of coal for selected end-uses*

(million tonnes)

	1978	1990	2000	2005	2009	2010	2011	Average annual percent change 78-90	90-11
Steam coal	**0.34**	**2.19**	**1.09**	**1.22**	**0.99**	**0.99**	**1.41**	**16.72**	**-2.08**
Total electricity and heat	0.02	1.19	0.50	0.42	0.34	0.44	0.37	43.16	-5.46
Main activity producers	*0.02*	*1.17*	*0.49*	*0.42*	*0.34*	*0.44*	*0.37*	*43.02*	*-5.40*
Autoproducers	-	*0.01*	*0.01*	*0.00*	-	-	-	-	-
Patent fuel/BKB plants	-	-	-	-	-	-	-	-	-
Coke ovens/Liquefaction(1)	-	-	-	-	-	-	-	-	-
Blast furnace inputs	-	0.18 e	0.27 e	0.42 e	0.20 e	0.31 e	0.32 e	-	2.69
Gas manufacture	-	-	-	-	-	-	-	-	-
Industry	0.30	0.74	0.50	0.57	0.40	0.53	0.62	7.88	-0.84
Iron and steel	*0.02*	*0.05 e*	*0.04 e*	*0.11 e*	*0.03 e*	*0.12 e*	*0.14 e*	*6.32*	*5.23*
Chemical	*0.01*	*0.03*	-	-	-	-	-	*9.89*	-
Non-metallic minerals	*0.19*	*0.39*	*0.25*	*0.25*	*0.21*	*0.23*	*0.27*	*6.43*	*-1.71*
Paper, pulp and print	*0.00*	*0.10*	*0.03*	*0.04*	*0.01*	*0.01*	*0.00*	*33.48*	*-14.04*
Other industry	*0.08*	*0.18 e*	*0.17 e*	*0.18 e*	*0.14 e*	*0.18 e*	*0.21 e*	*7.07*	*0.75*
Other sectors(2)	0.00	0.07	-	-	-	-	-	33.66	-
Non-energy use	0.03	0.02	-	-	-	-	-	-3.02	-
Coking coal	**1.23**	**1.52**	**1.77**	**1.85**	**1.44**	**1.87**	**1.63**	**1.77**	**0.36**
Total electricity and heat	-	-	-	-	-	-	-	-	-
Main activity producers	-	-	-	-	-	-	-	-	-
Autoproducers	-	-	-	-	-	-	-	-	-
Patent fuel/BKB plants	-	-	-	-	-	-	-	-	-
Coke ovens/Liquefaction(1)	1.23	1.52	1.77	1.85	1.55	1.85	1.62	1.77	0.31
Blast furnace inputs	-	-	-	-	-	-	-	-	-
Gas manufacture	-	-	-	-	-	-	-	-	-
Industry	-	-	-	-	-	-	-	-	-
Iron and steel	-	-	-	-	-	-	-	-	-
Chemical	-	-	-	-	-	-	-	-	-
Non-metallic minerals	-	-	-	-	-	-	-	-	-
Paper, pulp and print	-	-	-	-	-	-	-	-	-
Other industry	-	-	-	-	-	-	-	-	-
Other sectors(2)	-	-	-	-	-	-	-	-	-
Non-energy use	-	-	-	-	-	-	-	-	-
Lignite	**-**	**-**	**-**	**-**	**-**	**-**	**-**	**-**	**-**
Total electricity and heat	-	-	-	-	-	-	-	-	-
Main activity producers	-	-	-	-	-	-	-	-	-
Autoproducers	-	-	-	-	-	-	-	-	-
Patent fuel/BKB plants	-	-	-	-	-	-	-	-	-
Coke ovens/Liquefaction(1)	-	-	-	-	-	-	-	-	-
Gas manufacture	-	-	-	-	-	-	-	-	-
Industry	-	-	-	-	-	-	-	-	-
Iron and steel	-	-	-	-	-	-	-	-	-
Chemical	-	-	-	-	-	-	-	-	-
Non-metallic minerals	-	-	-	-	-	-	-	-	-
Paper, pulp and print	-	-	-	-	-	-	-	-	-
Other industry	-	-	-	-	-	-	-	-	-
Other sectors(2)	-	-	-	-	-	-	-	-	-
Non-energy use	-	-	-	-	-	-	-	-	-

(1) Liquefaction primarily refers to direct distillation processes. Indirect liquefaction may be reported here or under gas manufacture.

(2) Other sectors are Residential, Commercial and public services, Agriculture/Forestry, Fishing and Non-specified other.

* Coal totals may not match the sum of the individual rows due to statistical differences. Please refer to notes and definitions in Part I.

Source: *IEA/OECD Energy Statistics of OECD Countries*

SWEDEN

7. Fuel prices to end users*

	1978(1)	1990	1995	2000	2005	2009	2010	2011	2012
				(US dollars / tce)					
For electricity generation									
Steam coal	50.00	..	..	..	..	..	..	..	..
Heavy fuel oil	67.15	77.87	94.34	..	..	..	..	..	..
Natural gas	..	..	..	..	..	..	..	..	..
For industry									
Steam coal	77.55	92.56	..	..	..	..	..	..	..
Coking coal	64.11	..	..	..	..	..	..	..	..
High sulphur fuel oil	80.71	x	x	x	x	x	x	x	x
Low sulphur fuel oil	..	270.45	134.82	320.71	623.71	717.42	842.81	1048.50	1046.28
Natural gas	..	..	..	..	..	442.82	506.66	612.62	572.75
				(Swedish crowns / unit)(2)					
For electricity generation									
Steam coal	215	..	..	..	..	..	..	..	..
Heavy fuel oil	416	632	923	..	..	..	..	..	..
Natural gas	..	..	..	..	..	..	..	..	..
For industry									
Steam coal	321	502	..	..	..	..	..	..	..
Coking coal	267	..	..	..	..	..	..	..	..
High sulphur fuel oil	500	x	x	x	x	x	x	x	x
Low sulphur fuel oil	..	2195	1319	4029	6391	7530	8324	9321	9713
Natural gas	..	..	..	..	..	4357	4692	5106	4985

(1) Earliest year for which split by coal type is available.

(2) For steam coal, coking coal and heavy fuel oil per metric tonne; for natural gas per 10^7 kilocalories GCV.

* Please refer to notes and definitions in Part I. Steam coal prices relate to historical definitions.

Source: *IEA/OECD Energy Prices & Taxes*

8. Coal and peat trade by type of coal*

(Mtce)

	1978	1990	1995	2000	2005	2009	2010	2011	2012e
Total imports	**1.88**	**3.81**	**3.99**	**3.47**	**3.68**	**2.20**	**3.67**	**3.38**	**2.47**
Bituminous coal(1)	0.26	2.05	1.67	1.19	1.17	0.78	0.96	1.36	1.18
Coking coal	1.21	1.35	1.71	1.86	2.00	1.09	2.31	1.65	1.04
Sub-bituminous coal	-	-	-	-	-	-	-	-	-
Lignite	-	0.00	-	-	-	-	-	-	-
Peat	-	0.10	0.13	0.10	0.13	0.19	0.16	0.16	0.16
Coal products(2)	0.41	0.31	0.47	0.32	0.38	0.14	0.24	0.21	0.09
Total exports	**0.12**	**0.04**	**0.07**	**0.03**	**0.04**	**0.27**	**0.03**	**0.02**	**0.02**
Bituminous coal(1)	0.03	0.00	0.00	0.00	0.00	0.01	0.00	0.00	0.00
Coking coal	-	-	-	-	-	-	-	-	-
Sub-bituminous coal	-	-	-	-	-	-	-	-	-
Lignite	-	-	-	-	-	-	-	-	-
Peat	-	-	-	-	-	-	-	-	-
Coal products(2)	0.09	0.04	0.07	0.03	0.04	0.26	0.03	0.02	0.02

(1) Bituminous coal includes anthracite.

(2) Coal products includes products derived from coal, for example: coke, coal tar, briquettes, patent fuels.

* Please refer to notes and definitions in Part I.

Source: *IEA/OECD Energy Balances of OECD Countries, IEA/OECD Energy Statistics of OECD Countries*

SWEDEN

9. Total coal imports by origin*

(thousand tonnes)

	1978(1)	1985	1990	1995	2000	2005	2010	2011	2012e
Total coal(2)	**1545**	**4868**	**3545**	**3499**	**3092**	**3203**	**3285**	**3076**	**2281**
Coking coal	**1239**	**1834**	**1315**	**1674**	**1814**	**1955**	**2258**	**1616**	**1016**
Australia	-	140	449 e	535 e	990	1000	1373	963	492
Canada	78	278	102 e	179 e	54	-	-	-	-
Czech Republic	99	30	-	-	-	-	-	-	-
Germany	231	-	-	-	-	-	-	-	-
Poland	71	224	-	-	-	22	-	-	-
United Kingdom	5	-	-	-	-	-	-	-	-
United States	338	1074	764 e	950 e	570	462	529	653	524
Other OECD	-	65	-	10 e	-	1	-	-	-
China, People's Rep.	-	-	-	-	-	-	-	-	-
Colombia	-	-	-	-	-	-	-	-	-
Indonesia	-	-	-	-	-	-	-	-	-
South Africa	-	-	-	-	-	-	-	-	-
Former Soviet Union(3)	417	2	x	x	x	x	x	x	x
Russian Federation (4)	*x*	*x*	-	-	-	*470*	*356*	-	-
Other FSU (4)	*x*	*x*	-	-	-	-	-	-	-
Venezuela	-	-	-	-	200	-	-	-	-
Vietnam	-	-	-	-	-	-	-	-	-
Non-specified/other	-	21	-	-	-	-	-	-	-
Steam coal	**306**	**3034**	**2228**	**1825**	**1278**	**1248**	**1027**	**1460**	**1265**
Australia	-	614	187 e	44 e	83	370	73	135	175
Canada	-	-	1 e	-	-	-	-	-	-
Czech Republic	8	-	6 e	2 e	-	-	-	-	-
Germany	42	4	7 e	-	-	3	1	14	15
Poland	174	1292	732 e	1000 e	698	127	178	240	196
United Kingdom	30	71	103 e	-	15	2	-	4	44
United States	-	343	117 e	55 e	25	30	24	-	-
Other OECD	-	96	8 e	26 e	56	118	49	44	34
China, People's Rep.	-	-	8 e	6 e	-	-	-	-	-
Colombia	-	183	108 e	-	120	-	-	-	-
Indonesia	-	-	-	20 e	-	-	-	-	-
South Africa	-	19	-	-	-	-	-	-	-
Former Soviet Union(3)	52	407	x	x	x	x	x	x	x
Russian Federation (4)	*x*	*x*	*573*	*455*	*271*	*377*	*300*	*636*	*502*
Other FSU (4)	*x*	*x*	-	*19 e*	*10*	*168*	*336*	*382*	*294*
Venezuela	-	-	375 e	198 e	-	53	66	5	-
Vietnam	-	-	-	-	-	-	-	-	-
Non-specified/other	-	5	3 e	-	-	-	-	-	5
Lignite	**-**	**-**	**2**	**-**	**-**	**-**	**-**	**-**	**-**

(1) Earliest year for which split by coal type is available.
(2) Total coal consists of steam coal, coking coal and lignite.
(3) For years prior to 1990, or if no details for individual Former Soviet Union republics.
(4) Data for former republics are not available separately for some OECD member countries.
* Please refer to notes and definitions in Part I.

Source: IEA/OECD Coal Statistics

SWEDEN

12. Coal import values by origin*

(average unit value, CIF, USD/tonne)

	1990	1995	2000	2005	2007	2008	2009	2010	2011/12[1]
Coking coal[2]	..	**60.87**	**53.95**	**94.31**	**65.83**	**138.31**	..	**88.55**	..
Imports from:									
Australia	..	65.10	52.52	..	..	..	..	..	..
Canada	..	..	..	..	..	..	..	..	..
Czech Republic	..	..	..	..	..	..	..	..	..
Poland	..	55.47	57.84	96.64	..	..	..	..	..
United States	..	61.89	..	..	..	..	..	..	..
China	..	..	..	..	..	..	..	..	..
Colombia	..	..	..	..	..	..	..	..	..
Indonesia	..	..	..	..	..	..	..	..	..
South Africa	..	..	..	..	..	..	..	..	..
Former Soviet Union[3]	..	47.54	31.94	65.37	65.83	138.31	..	88.55	..
Other bituminous coal[4]	..	**55.37**	**49.38**	**111.18**	**124.51**	**227.64**	**157.77**	**189.34**	..
Imports from:									
Australia	..	67.27	55.29	152.75	161.71	328.87	197.49	231.84	..
Canada	..	..	49.62	..	..	..	..	..	..
Czech Republic	..	..	..	..	..	3000.00	..	..	..
Poland	..	48.74	41.89	75.95	84.84	239.79	106.87	115.81	..
United States	..	66.14	58.19	91.28	138.55	181.58	141.83	214.20	..
China	..	..	..	..	..	..	..	..	..
Colombia	..	53.77	45.79	..	..	..	..	..	..
Indonesia	..	..	..	..	..	..	..	..	..
South Africa	..	61.06	..	..	..	..	95.94	..	..
Former Soviet Union[3]	..	43.13	31.04	74.35	85.79	139.75	108.32	134.59	..

Note: On occasion, shipment of extremely small quantities of high valued coal results in high import costs.

(1) Import data for steam and coking coals are currently unavailable for 2011 and 2012 due to resource constraints.

(2) Weighted average of individual countries using import volumes as weights. Prices exclude intra-EU trade.

(3) Former Soviet Union until 1991, Russian Federation starting in 1992.

(4) Bituminous steam coal only. Weighted average of trades. Prices exclude intra-EU trade.

* Please refer to notes and definitions in Part I.

Source: IEA/OECD Coal Statistics

SWITZERLAND*

Figure 1: TPES by fuel (Mtce)

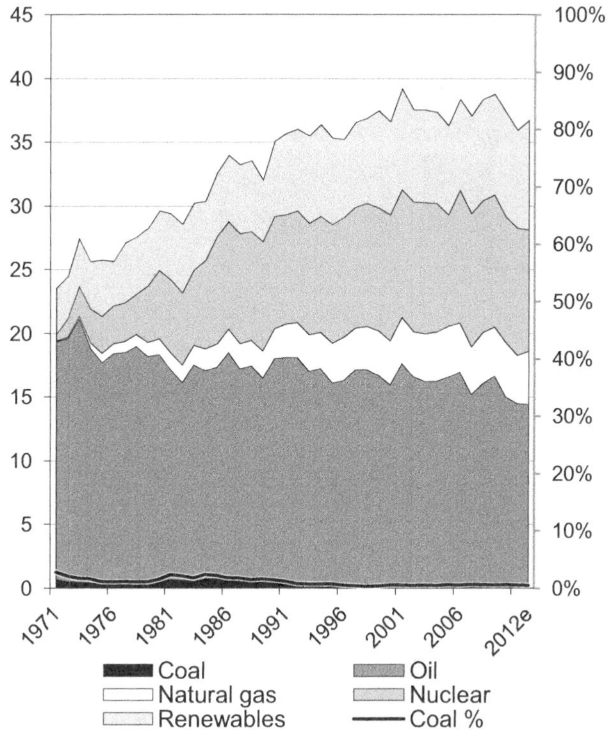

Figure 2: Coal supply indicators (1971 = 100)

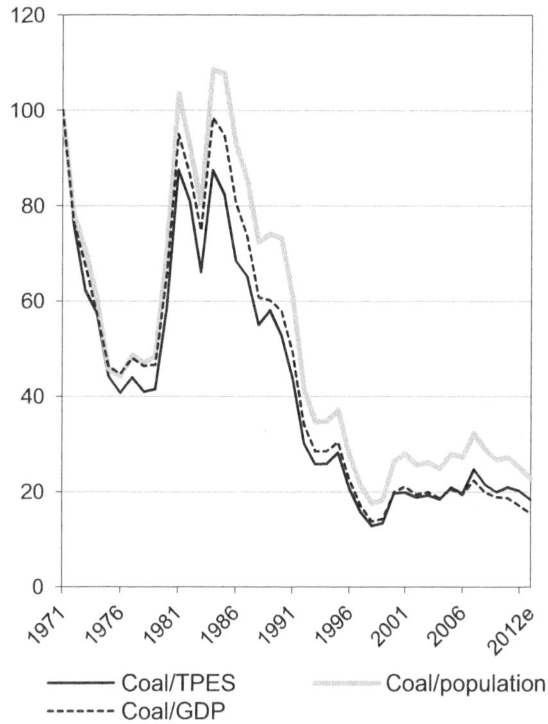

Figure 3: Primary coal supply (Mtce)

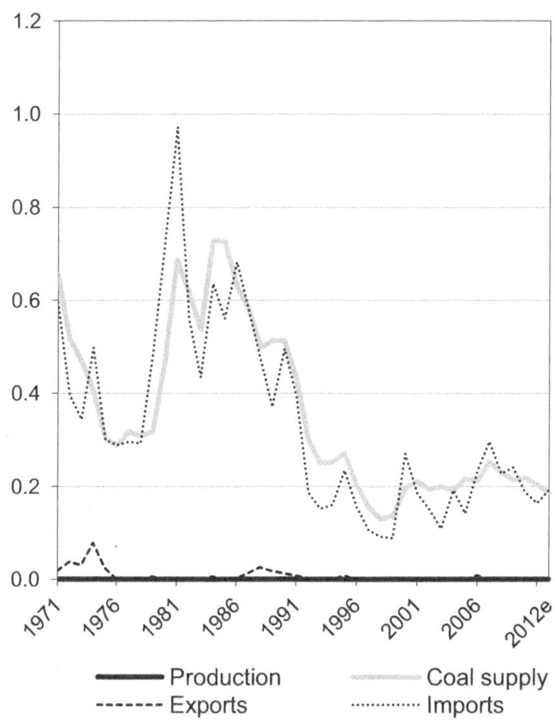

Figure 4: Coal consumption (Mtce)

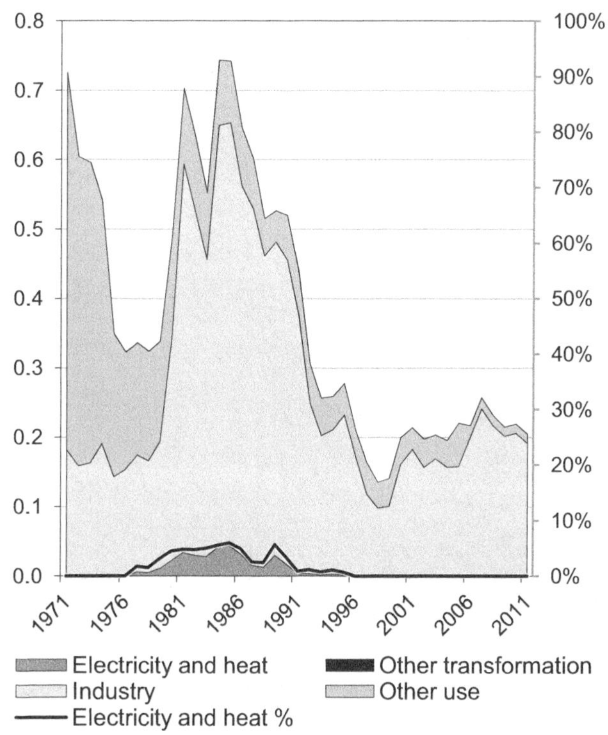

*Please refer to notes and definitions in Part I. Peat is included in coal for all figures.

Source: *IEA/OECD Coal Information, IEA/OECD Energy Balances of OECD Countries, OECD Main Economic Indicators*

SWITZERLAND*

Figure 5: Electricity generation by fuel (TWh)

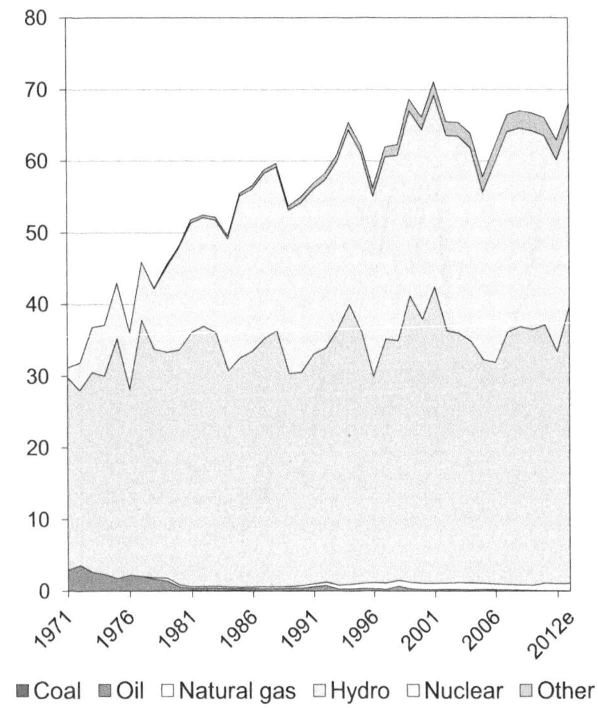

Figure 6: CO_2 emissions by fuel (Mt CO_2)

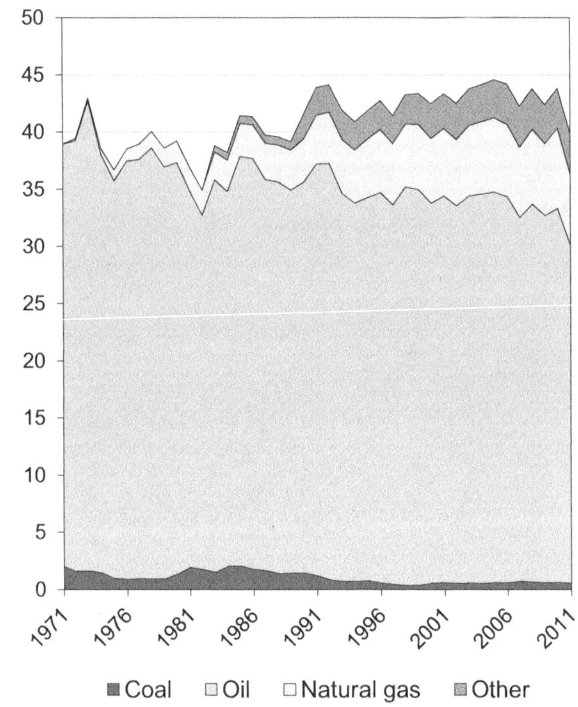

Figure 7: Electricity generation by fuel share

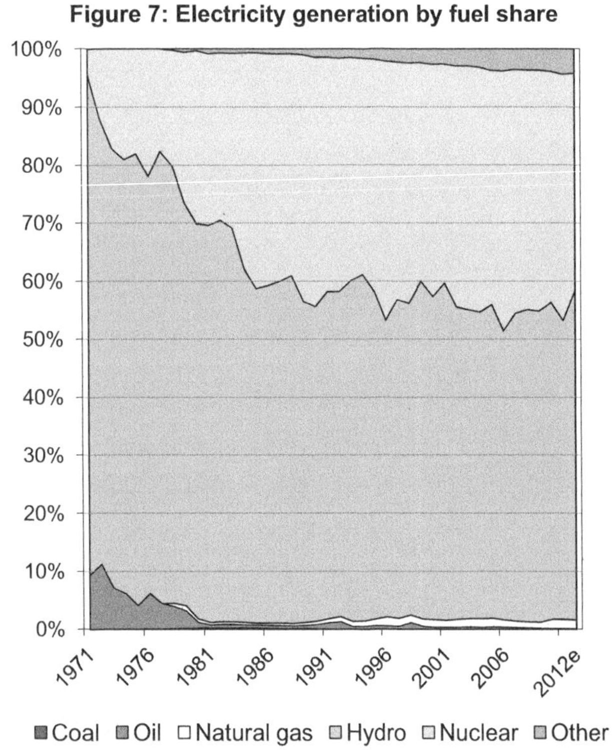

Figure 8: CO_2 emissions by fuel share

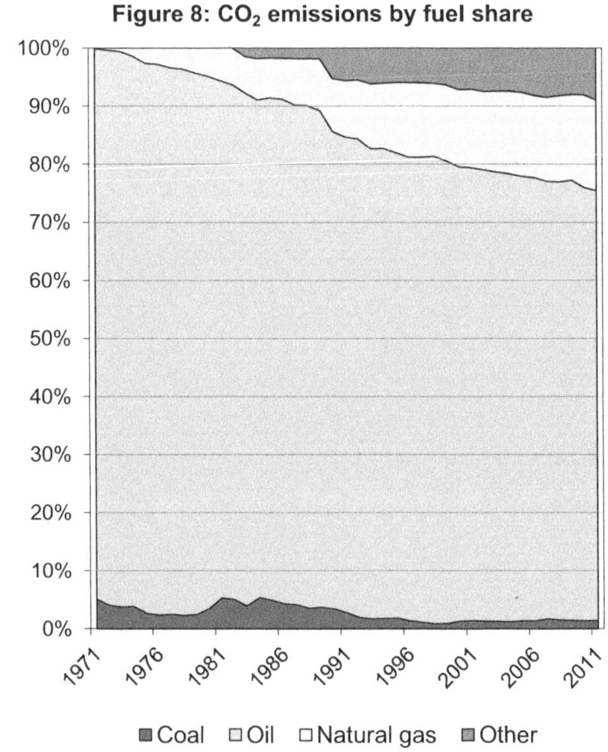

*Please refer to notes and definitions in Part I. Peat is included in coal for all figures.

Source: *IEA/OECD Electricity Information, IEA/OECD CO2 Emissions from Fuel Combustion*

SWITZERLAND

1. Total primary energy supply (TPES) by fuel*

	1973	1980	1990	2000	2010	2011	2012e	Average annual percent change 73-90	90-11
TPES (Mtce)	**27.01**	**28.62**	**34.80**	**35.72**	**37.43**	**36.25**	**36.43**	**1.50**	**0.19**
Coal and peat	0.47	0.47	0.51	0.20	0.22	0.20	0.19	0.52	-4.29
Oil	20.64	17.87	17.51	15.75	14.78	14.26	14.24	-0.96	-0.97
Natural Gas	0.21	1.24	2.33	3.48	4.30	3.81	4.18	15.14	2.37
Biofuels and waste	0.34	0.67	2.12	2.59	3.35	3.25	3.40	11.27	2.06
Nuclear	2.35	5.35	8.83	9.88	9.85	9.99	9.52	8.10	0.59
Hydro	3.43	4.03	3.66	4.53	4.43	3.98	4.74	0.39	0.40
Geothermal	-	-	0.10	0.15	0.37	0.36	0.34	-	6.33
Solar, wind, tide	-	-	0.00	0.02	0.06	0.08	0.09	-	15.93
Net electricity trade(1)	-0.43	-1.01	-0.26	-0.87	0.06	0.32	-0.27	-2.94	x
Heat(2)	-	-	-	0.00	0.00	0.00	0.00	-	-

(1) Net trade between OECD and non-OECD countries.

(2) Ambient heat from heat pumps used in the transformation sector.

Source: *IEA/OECD Energy Balances of OECD Countries*

2. Energy supply, GDP and population*

	1973	1980	1990	2000	2010	2011	2012e	Average annual percent change 73-90	90-11
GDP (2005 bil. USD)	254	260	324	361	428	436	440	1.43	1.43
TPES/GDP(1)	0.11	0.11	0.11	0.10	0.09	0.08	0.08	0.07	-1.22
Population (millions)	6.4	6.4	6.8	7.2	7.8	7.9	7.9	0.32	0.70
TPES/population(1)	4.19	4.48	5.12	4.96	4.81	4.61	4.60	1.18	-0.50
TPES/GDP(2)	110.4	114.4	111.7	102.9	90.8	86.3	85.9	0.07	-1.22
Coal and peat supply/GDP(2)	328.7	320.1	281.9	96.8	90.9	83.3	75.3	-0.90	-5.64
Elec. consumption/GDP(2)	76.7	91.1	96.6	97.5	93.8	90.2	..	1.37	-0.33
Elec. generation (TWh)	37	48	55	66	66	63	68	2.39	0.64
Industrial production(2)	63.6	64.2	78.3	99.0	117.1	118.2	..	1.23	1.98

(1) TPES/GDP in units of tce/2005 thousand USD, TPES/population in units of tce/capita.

(2) As index, 2005 (not shown) = 100

Source: *IEA/OECD Energy Balances of OECD Countries , OECD Main Economic Indicators*

3. Coal and peat production by type*

	1978(1)	1990	2000	2005	2010	2011	2012e	Average annual percent change 78-90	90-11
Mtce:									
Coking coal	-	-	-	-	-	-	-	-	-
Steam coal	-	-	-	-	-	-	-	-	-
Lignite	-	-	-	-	-	-	-	-	-
Peat	-	-	-	-	-	-	-	-	-
Mt:									
Coking coal	-	-	-	-	-	-	-	-	-
Steam coal	-	-	-	-	-	-	-	-	-
Lignite	-	-	-	-	-	-	-	-	-
Peat	-	-	-	-	-	-	-	-	-

(1) Earliest year for which split by coal type is available.

* Please refer to notes and definitions in Part I.

Source: *IEA/OECD Energy Balances of OECD Countries, IEA/OECD Energy Statistics of OECD Countries*

SWITZERLAND

4. Final consumption of energy by fuel*

(Mtce)

	1973	1980	1990	2000	2009	2010	2011	Average annual percent change 73-90	90-11
Total final consumption(1)	**23.81**	**23.74**	**26.67**	**27.96**	**28.78**	**30.06**	**27.79**	**0.67**	**0.20**
Coal and peat	0.60	0.47	0.50	0.20	0.21	0.22	0.20	-0.99	-4.19
Oil	19.15	17.20	16.57	16.17	15.39	15.73	14.29	-0.85	-0.70
Natural Gas	0.16	1.01	2.12	3.12	3.55	3.94	3.51	16.62	2.42
Biofuels and wastes	0.34	0.46	1.30	1.42	1.69	1.81	1.64	8.10	1.12
Geothermal	-	-	0.10	0.15	0.31	0.37	0.36	-	6.33
Solar, wind, tide	-	-	0.00	0.02	0.04	0.05	0.06	-	13.98
Electricity	3.56	4.33	5.72	6.43	7.06	7.34	7.20	2.82	1.10
Heat	-	0.27	0.36	0.45	0.52	0.59	0.54	-	2.02
of which:									
Total industry	**5.71**	**5.55**	**4.81**	**5.48**	**5.36**	**5.63**	**5.35**	**-1.00**	**0.50**
Coal and peat	0.16	0.32	0.44	0.16	0.20	0.21	0.19	5.99	-3.87
Oil	4.16	3.05	1.07	1.28	1.00	0.97	0.76	-7.68	-1.60
Natural Gas	0.02	0.50	0.82	0.99	1.12	1.22	1.17	24.64	1.69
Biofuels and wastes	-	0.17	0.29	0.63	0.57	0.62	0.62	-	3.68
Geothermal	-	-	0.01	0.01	0.02	0.03	0.03	-	6.37
Solar, wind, tide	-	-	0.00	0.00	0.00	0.00	0.00	-	14.20
Electricity	1.36	1.46	2.12	2.22	2.24	2.37	2.36	2.64	0.52
Heat	-	0.04	0.07	0.19	0.21	0.21	0.22	-	5.71
Total transport	**5.05**	**5.28**	**7.37**	**8.34**	**8.67**	**8.64**	**8.53**	**2.24**	**0.70**
Coal and peat	-	-	-	-	-	-	-	-	-
Oil	4.80	5.03	7.03	8.00	8.23	8.19	8.09	2.27	0.67
Natural Gas	-	-	0.02	0.01	0.05	0.05	0.05	-	5.05
Biofuels and wastes	-	-	-	0.00	0.01	0.01	0.01	-	-
Electricity	0.25	0.26	0.32	0.32	0.38	0.39	0.38	1.42	0.83
Residential	**1.81**	**7.71**	**7.48**	**8.01**	**8.55**	**9.23**	**7.91**	**8.70**	**0.27**
Coal and peat	0.43	0.14	0.02	0.01	0.01	0.01	0.01	-16.97	-1.48
Oil	-	5.63	4.75	4.13	3.79	4.03	3.21	-	-1.85
Natural Gas	0.14	0.38	0.85	1.22	1.45	1.65	1.39	11.38	2.36
Biofuels and wastes	0.34	0.25	-	0.42	0.62	0.67	0.56	-	-
Geothermal	-	-	0.08	0.12	0.25	0.30	0.29	-	6.39
Solar, wind, tide	-	-	0.00	0.01	0.03	0.04	0.05	-	14.28
Electricity	0.90	1.24	1.62	1.93	2.20	2.29	2.20	3.55	1.47
Heat	-	0.07	0.15	0.16	0.19	0.24	0.20	-	1.34
Comm & public services	**1.03**	**3.82**	**4.81**	**4.76**	**4.94**	**5.23**	**4.66**	**9.52**	**-0.15**
Coal and peat	-	-	0.00	-	-	-	-	-	-
Oil	-	2.21	2.63	1.77	1.45	1.56	1.24	-	-3.52
Natural Gas	-	0.13	0.43	0.71	0.73	0.82	0.68	-	2.18
Biofuels and waste	-	-	0.04	0.32	0.47	0.48	0.43	-	12.58
Geothermal	-	-	0.01	0.02	0.04	0.04	0.04	-	5.92
Solar, wind, tide	-	-	0.00	0.00	0.01	0.01	0.01	-	12.52
Electricity	1.03	1.33	1.56	1.83	2.12	2.18	2.14	2.49	1.52
Heat	-	0.16	0.14	0.10	0.12	0.14	0.12	-	-0.45
Non-energy use	**1.07**	**0.88**	**0.80**	**0.81**	**0.72**	**0.77**	**0.82**	**-1.72**	**0.11**
Coal and peat	-	0.00	0.00	-	-	-	-	-	-
Oil	1.07	0.88	0.79	0.81	0.72	0.77	0.82	-1.75	0.14
Natural Gas	-	-	-	-	-	-	-	-	-

(1) Total Final Consumption (TFC) excludes use in transformation processes and energy industry own use.

* Please refer to notes and definitions in Part I. Agriculture, Forestry, Fishing and "Non-specified other" sectors are not reported here individually, but are accounted for in Total final consumption.

Source: *IEA/OECD Energy Balances of OECD Countries*

SWITZERLAND

5. Coal balance*

(Mtce)

	1973	1980	1990	2000	2005	2010	2011	2012e	Average annual percent change 73-90	90-11
Production	-	-	-	-	-	-	-	-	-	-
Imports	0.3	0.7	0.5	0.3	0.1	0.2	0.2	0.2	2.2	-5.1
Exports	-0.0	-	-0.0	-	-	-	-	-	-5.2	-
Stock changes	0.2	-0.3	0.0	-0.1	0.1	0.0	0.0	-0.0		
Primary supply	**0.5**	**0.5**	**0.5**	**0.2**	**0.2**	**0.2**	**0.2**	**0.2**	**0.5**	**-4.3**
Statistical differences	**-**	**-**	**-0.0**	**-**	**-**	**-**	**-**	..		
Total transformation	**0.2**	**0.0**	**-0.0**	**0.0**	**0.0**	**-**	**-**	..	**-**	**-**
Electricity and heat gen.	-	-0.0	-0.0	-	-	-	-	..	-	-
Main activity producers (1)	-	*-0.0*	*-0.0*	-	-	-	-	..	-	-
Autoproducers	-	*-0.0*	*-0.0*	-	-	-	-	..	-	-
Gas works	0.2	0.0	0.0	0.0	0.0	-	-	..	-15.7	-
Coal transformation(2)	-	-	-	-	-	-	-	..	-	-
BKB plants	-	-	-	-	-	-	-	..	-	-
Blast furnaces	-	-	-	-	-	-	-	..	-	-
Coke ovens	-	-	-	-	-	-	-	..	-	-
Patent fuel plants	-	-	-	-	-	-	-	..	-	-
Other transformation(3)	-	-	-	-	-	-	-	..	-	-
Energy ind. own use	**-0.0**	**-0.0**	**-**	**-**	**-**	**-**	**-**	..	**-**	**-**
Losses	**-0.0**	**-0.0**	**-0.0**	**-0.0**	**-0.0**	**-**	**-**	..		
Final consumption(4)	**0.6**	**0.5**	**0.5**	**0.2**	**0.2**	**0.2**	**0.2**	..	**-1.0**	**-4.2**
Industry(5)	0.2	0.3	0.4	0.2	0.2	0.2	0.2	..	6.0	-3.9
Iron and steel	*0.0*	-	-	*0.0*	*0.0*	*0.0*	*0.0*	..	-	-
Chemical	*0.0*	*0.0*	*0.0*	-	-	-	-	..	*-11.1*	-
Non-metallic minerals	*0.0*	*0.2*	*0.4*	*0.1*	*0.1*	*0.2*	*0.2*	..	*19.2*	*-3.7*
Paper, pulp and print	*0.0*	*0.0*	*0.0*	-	-	-	-	..	*3.0*	-
Other industry (6)	*0.1*	*0.1*	*0.0*	*0.0*	*0.0*	*0.0*	*0.0*	..	*-6.9*	*-7.8*
Transport(7)	-	-	-	-	-	-	-	..	-	-
Other	0.4	0.1	0.1	0.0	0.1	0.0	0.0	..	-10.9	-6.9
Comm. and pub. services	-	-	*0.0*	-	-	-	-	..	-	-
Residential	*0.4*	*0.1*	*0.0*	*0.0*	*0.0*	*0.0*	*0.0*	..	*-17.0*	*-1.5*
Other sectors (8)	-	-	*0.0*	*0.0*	*0.0*	-	-	..	-	-
Non-energy use	-	0.0	0.0	-	-	-	-	..	-	-

(1) Main activity electricity and heat generation includes district heating.

(2) Coal transformation refers to the transformation of coal and peat to secondary and tertiary products (mainly cokes, briquettes, coal tar and off-gases). Gas works and Liquefaction are shown separately to match the *Energy Balances of OECD Countries.*

(3) Other transformation includes Liquefaction and Non-specified transformations.

(4) Final Consumption includes non-energy use and energy use (Industry, Transport and Other).

(5) Please refer to notes and definitions in Part I for detailed explanation.

(6) Other industry includes Non-ferrous metals, Transport equipment, Machinery, Mining and quarrying, Food and tobacco, Wood and wood products, Construction, Textile and leather, and Non-specified industry.

(7) Transport includes Rail and Inland waterways.

(8) Other sectors includes Agriculture/Forestry, Fishing and Non-specified other.

* "Coal" refers to all types of coal, primary (anthracite, coking coal, other bituminous coal, sub-bituminous coal, lignite/brown coal) and derived fuels (including patent fuel, BKB, coke oven coke, gas coke, coal tar, coke oven gas, gas works gas, blast furnace gas and other recovered gases), as well as peat, peat briquettes and oil shale. Quantities have been converted to Mtce using calorific values reported by the respective countries.
Please refer to notes and definitions in Part I.

Source: *IEA/OECD Energy Balances of OECD Countries*

SWITZERLAND

6. Use of coal for selected end-uses*

(million tonnes)

	1978	1990	2000	2005	2009	2010	2011	Average annual percent change 78-90	90-11
Steam coal	**0.15**	**0.48**	**0.17**	**0.18**	**0.16**	**0.17**	**0.14**	**10.26**	**-5.58**
Total electricity and heat	0.01	0.02	-	-	-	-	-	12.70	-
Main activity producers	*0.00*	*0.02*	-	-	-	-	-	*26.63*	-
Autoproducers	*0.00*	*0.00*	-	-	-	-	-	*0.00*	-
Patent fuel/BKB plants	-	-	-	-	-	-	-	-	-
Coke ovens/Liquefaction(1)	-	-	-	-	-	-	-	-	-
Blast furnace inputs	-	-	-	-	-	-	-	-	-
Gas manufacture	-	-	-	-	-	-	-	-	-
Industry	0.10	0.45	0.17	0.16	0.14	0.15	0.13	13.51	-5.77
Iron and steel	-	-	*0.02*	*0.02*	*0.01*	*0.01*	*0.01*	-	-
Chemical	*0.01*	*0.01*	-	-	-	-	-	*-6.82*	-
Non-metallic minerals	*0.05*	*0.40*	*0.14*	*0.14*	*0.13*	*0.14*	*0.11*	*19.28*	*-5.82*
Paper, pulp and print	*0.02*	*0.04*	-	-	-	-	-	*6.70*	-
Other industry	*0.02*	*0.01*	*0.01*	*0.01*	*0.00*	*0.00*	*0.00*	*-4.17*	*-6.39*
Other sectors(2)	0.03	0.00	0.01	0.01	0.01	0.01	0.01	-15.46	6.15
Non-energy use	0.02	0.00	-	-	-	-	-	-12.55	-
Coking coal	-	-	-	-	-	-	-	-	-
Total electricity and heat	-	-	-	-	-	-	-	-	-
Main activity producers	-	-	-	-	-	-	-	-	-
Autoproducers	-	-	-	-	-	-	-	-	-
Patent fuel/BKB plants	-	-	-	-	-	-	-	-	-
Coke ovens/Liquefaction(1)	-	-	-	-	-	-	-	-	-
Blast furnace inputs	-	-	-	-	-	-	-	-	-
Gas manufacture	-	-	-	-	-	-	-	-	-
Industry	-	-	-	-	-	-	-	-	-
Iron and steel	-	-	-	-	-	-	-	-	-
Chemical	-	-	-	-	-	-	-	-	-
Non-metallic minerals	-	-	-	-	-	-	-	-	-
Paper, pulp and print	-	-	-	-	-	-	-	-	-
Other industry	-	-	-	-	-	-	-	-	-
Other sectors(2)	-	-	-	-	-	-	-	-	-
Non-energy use	-	-	-	-	-	-	-	-	-
Lignite	-	**0.01**	**0.01**	**0.04**	**0.07**	**0.06**	**0.07**	-	**8.27**
Total electricity and heat	-	-	-	-	-	-	-	-	-
Main activity producers	-	-	-	-	-	-	-	-	-
Autoproducers	-	-	-	-	-	-	-	-	-
Patent fuel/BKB plants	-	-	-	-	-	-	-	-	-
Coke ovens/Liquefaction(1)	-	-	-	-	-	-	-	-	-
Gas manufacture	-	-	-	-	-	-	-	-	-
Industry	-	-	-	-	0.07	0.06	0.07	-	-
Iron and steel	-	-	-	-	-	-	-	-	-
Chemical	-	-	-	-	-	-	-	-	-
Non-metallic minerals	-	-	-	-	*0.07*	*0.06*	*0.07*	-	-
Paper, pulp and print	-	-	-	-	-	-	-	-	-
Other industry	-	-	-	-	-	-	-	-	-
Other sectors(2)	-	0.01	0.01	0.04	-	-	-	-	-
Non-energy use	-	-	-	-	-	-	-	-	-

(1) Liquefaction primarily refers to direct distillation processes. Indirect liquefaction may be reported here or under gas manufacture.

(2) Other sectors are Residential, Commercial and public services, Agriculture/Forestry, Fishing and Non-specified other.

* Coal totals may not match the sum of the individual rows due to statistical differences. Please refer to notes and definitions in Part I.

Source: *IEA/OECD Energy Statistics of OECD Countries*

SWITZERLAND

7. Fuel prices to end users*

	1978(1)	1990	1995	2000	2005	2009	2010	2011	2012
				(US dollars / tce)					
For electricity generation									
Steam coal	x	x	x	x	x	x	x	x	x
Heavy fuel oil	..	..	..	..	..	..	..	..	..
Natural gas	..	..	..	..	..	..	..	..	..
For industry									
Steam coal	68.81	72.45	71.06	60.13	110.03	160.86	181.26	233.62	177.96
Coking coal	..	x	x	x	x	x	x	x	x
High sulphur fuel oil	82.08	118.75	x	x	x	x	x	x	x
Low sulphur fuel oil	..	..	115.66	139.45	224.74	327.62	453.85	608.46	612.89
Natural gas	153.66	182.84	230.88	173.40	313.15	547.80	514.87	654.58	648.63
				(Swiss francs / unit)(2)					
For electricity generation									
Steam coal	x	x	x	x	x	x	x	x	x
Heavy fuel oil	..	..	..	..	..	..	..	..	..
Natural gas	..	..	..	..	..	..	..	..	..
For industry									
Steam coal	105.5	86.3	72.0	87.0	117.5	149.8	162.0	178.0	143.0
Coking coal	..	x	x	x	x	x	x	x	x
High sulphur fuel oil	201.3	226.2	x	x	x	x	x	x	x
Low sulphur fuel oil	..	..	187.5	322.8	384.0	488.0	649.0	741.8	788.0
Natural gas	353.3	326.5	350.9	376.3	501.6	765.0	690.2	748.1	781.8

(1) Earliest year for which split by coal type is available.

(2) For steam coal, coking coal and heavy fuel oil per metric tonne; for natural gas per 10^7 kilocalories GCV.

* Please refer to notes and definitions in Part I. Steam coal prices relate to historical definitions.

Source: *IEA/OECD Energy Prices & Taxes*

8. Coal and peat trade by type of coal*

(Mtce)

	1978	1990	1995	2000	2005	2009	2010	2011	2012e
Total imports	**0.29**	**0.50**	**0.23**	**0.27**	**0.14**	**0.24**	**0.19**	**0.16**	**0.19**
Bituminous coal(1)	0.13	0.46	0.20	0.24	0.10	0.17	0.12	0.09	0.14
Coking coal	-	-	-	-	-	-	-	-	-
Sub-bituminous coal	-	-	-	-	-	-	-	-	-
Lignite	-	0.01	0.00	0.00	0.03	0.05	0.05	0.05	0.04
Peat	-	-	-	-	-	-	-	-	-
Coal products(2)	0.15	0.03	0.02	0.03	0.02	0.02	0.02	0.02	0.02
Total exports	**-**	**0.01**	**0.01**	**-**	**-**	**-**	**-**	**-**	**-**
Bituminous coal(1)	-	0.01	0.01	-	-	-	-	-	-
Coking coal	-	-	-	-	-	-	-	-	-
Sub-bituminous coal	-	-	-	-	-	-	-	-	-
Lignite	-	-	-	-	-	-	-	-	-
Peat	-	-	-	-	-	-	-	-	-
Coal products(2)	-	0.00	-	-	-	-	-	-	-

(1) Bituminous coal includes anthracite.

(2) Coal products includes products derived from coal, for example: coke, coal tar, briquettes, patent fuels.

* Please refer to notes and definitions in Part I.

Source: *IEA/OECD Energy Balances of OECD Countries, IEA/OECD Energy Statistics of OECD Countries*

SWITZERLAND

9. Total coal imports by origin*

(thousand tonnes)

	1978(1)	1985	1990	1995	2000	2005	2010	2011	2012e
Total coal(2)	**141**	**470**	**490**	**221**	**256**	**139**	**196**	**171**	**194**
Coking coal	**-**	**-**	**-**	**-**	**-**	**-**	**-**	**-**	**-**
Australia	-	-	-	-	-	-	-	-	-
Canada	-	-	-	-	-	-	-	-	-
Czech Republic	-	-	-	-	-	-	-	-	-
Germany	-	-	-	-	-	-	-	-	-
Poland	-	-	-	-	-	-	-	-	-
United Kingdom	-	-	-	-	-	-	-	-	-
United States	-	-	-	-	-	-	-	-	-
Other OECD	-	-	-	-	-	-	-	-	-
China, People's Rep.	-	-	-	-	-	-	-	-	-
Colombia	-	-	-	-	-	-	-	-	-
Indonesia	-	-	-	-	-	-	-	-	-
South Africa	-	-	-	-	-	-	-	-	-
Former Soviet Union(3)	-	-	x	x	x	x	x	x	x
Russian Federation (4)	*x*	*x*	-	-	-	-	-	-	-
Other FSU (4)	*x*	*x*	-	-	-	-	-	-	-
Venezuela	-	-	-	-	-	-	-	-	-
Vietnam	-	-	-	-	-	-	-	-	-
Non-specified/other	-	-	-	-	-	-	-	-	-
Steam coal	**141**	**470**	**482**	**214**	**250**	**100**	**130**	**99**	**141**
Australia	-	31	-	-	-	-	-	-	-
Canada	-	4	-	-	-	-	-	-	-
Czech Republic	14	-	-	-	-	-	-	-	-
Germany	79	80	55	10	12	4	5	7	12
Poland	8	24	1	-	-	-	-	-	-
United Kingdom	-	-	-	-	-	-	-	-	-
United States	-	126	-	-	-	-	-	-	-
Other OECD	17	83	6	-	2	2	16	-	8
China, People's Rep.	-	-	-	-	-	-	-	-	-
Colombia	-	-	-	-	-	-	-	-	-
Indonesia	-	-	-	-	-	-	-	-	-
South Africa	23	122	420	196	234	91	107	91	121
Former Soviet Union(3)	-	-	x	x	x	x	x	x	x
Russian Federation (4)	*x*	*x*	-	*6*	*2*	*3*	*2*	*1*	-
Other FSU (4)	*x*	*x*	-	-	-	-	-	-	-
Venezuela	-	-	-	-	-	-	-	-	-
Vietnam	-	-	-	-	-	-	-	-	-
Non-specified/other	-	-	-	2	-	-	-	-	-
Lignite	**-**	**-**	**8**	**7**	**6**	**39**	**66**	**72**	**53**

(1) Earliest year for which split by coal type is available.

(2) Total coal consists of steam coal, coking coal and lignite.

(3) For years prior to 1990, or if no details for individual Former Soviet Union republics.

(4) Data for former republics are not available separately for some OECD member countries.

* Please refer to notes and definitions in Part I.

Source: IEA/OECD Coal Statistics

TURKEY*

Figure 1: TPES by fuel (Mtce)

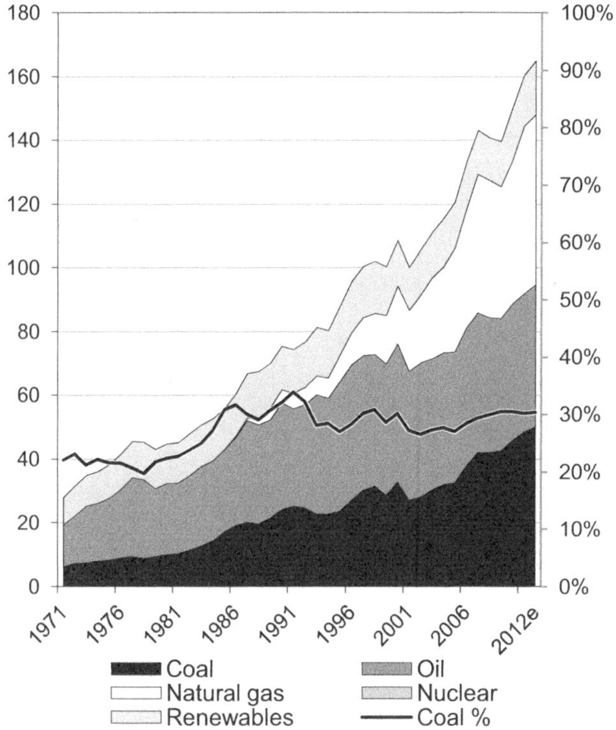

Figure 2: Coal supply indicators (1971 = 100)

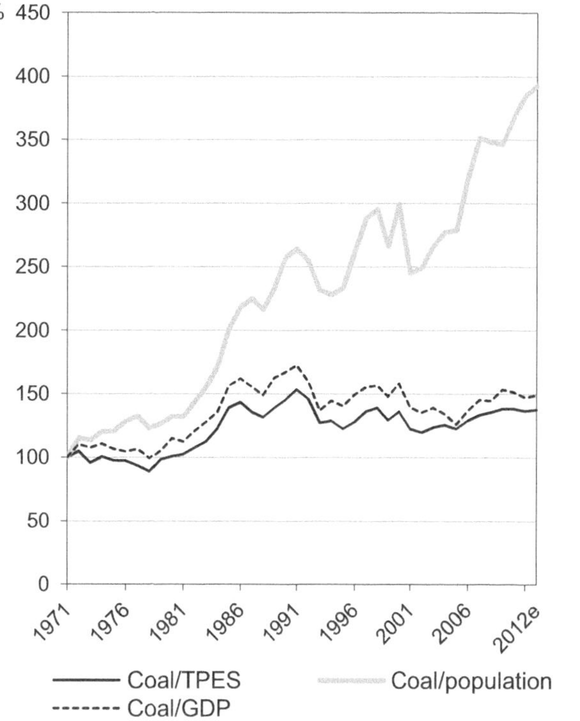

Figure 3: Primary coal supply (Mtce)

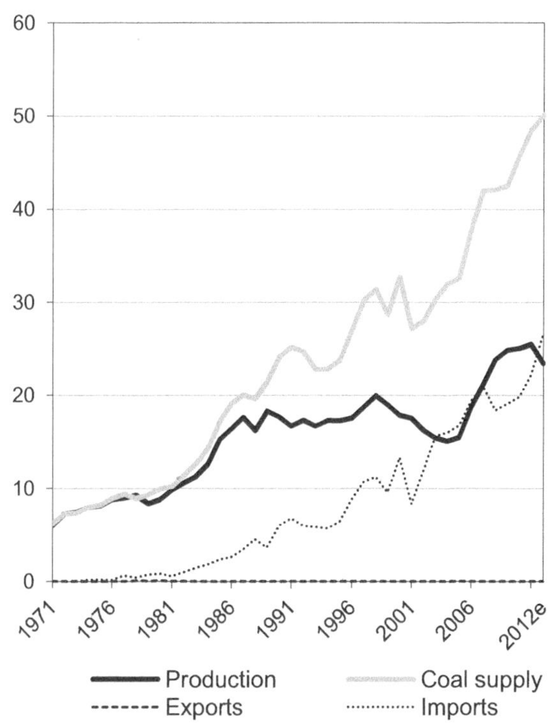

Figure 4: Coal consumption (Mtce)

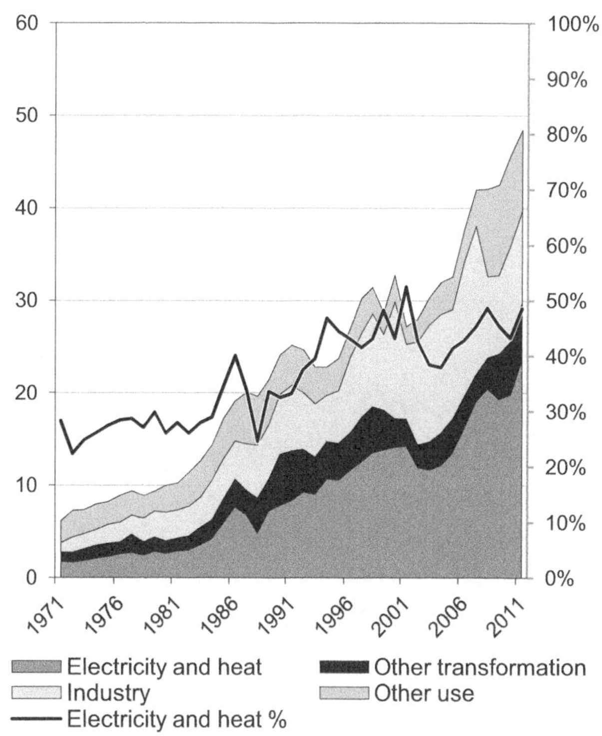

*Please refer to notes and definitions in Part I. Peat is included in coal for all figures.

Source: *IEA/OECD Coal Information, IEA/OECD Energy Balances of OECD Countries, OECD Main Economic Indicators*

TURKEY*

Figure 5: Electricity generation by fuel (TWh)

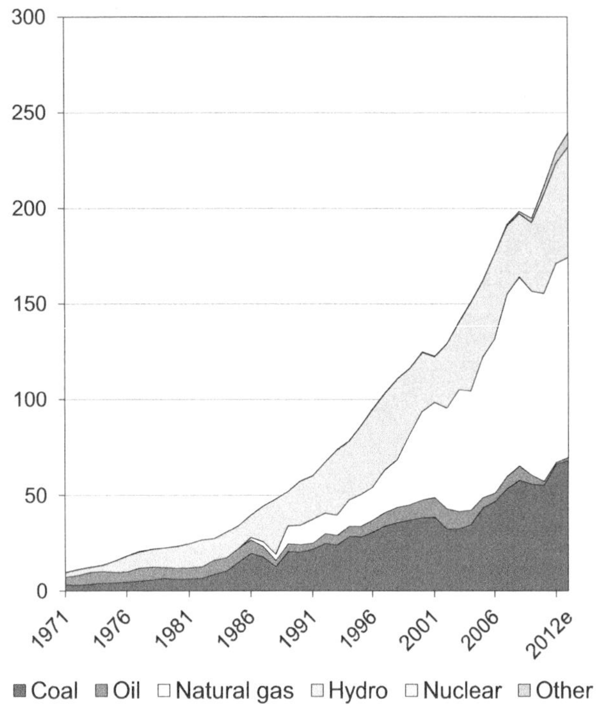

Figure 6: CO_2 emissions by fuel (Mt CO_2)

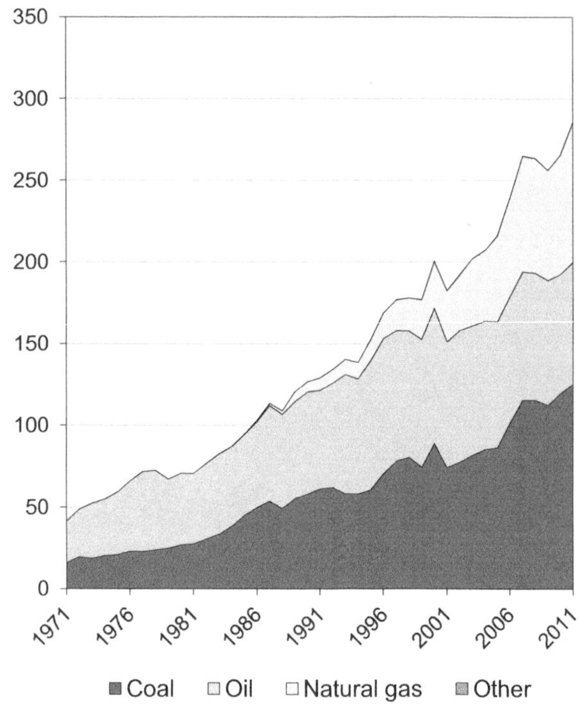

Figure 7: Electricity generation by fuel share

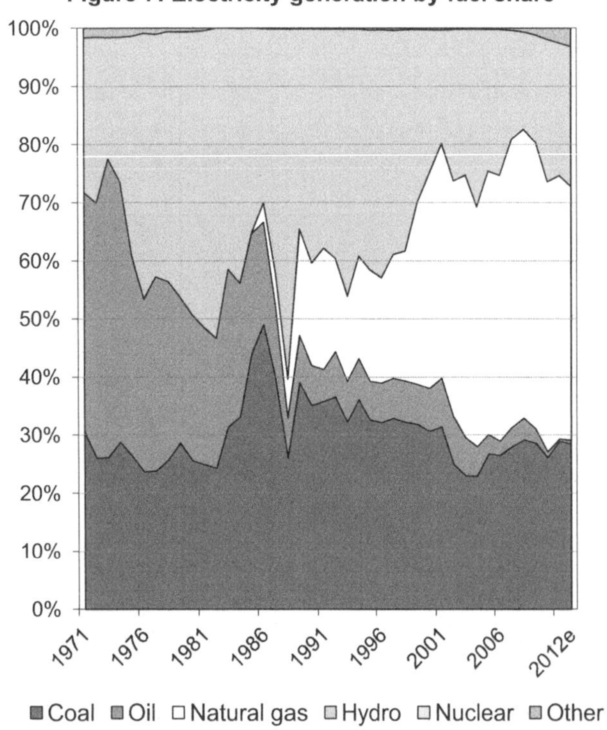

Figure 8: CO_2 emissions by fuel share

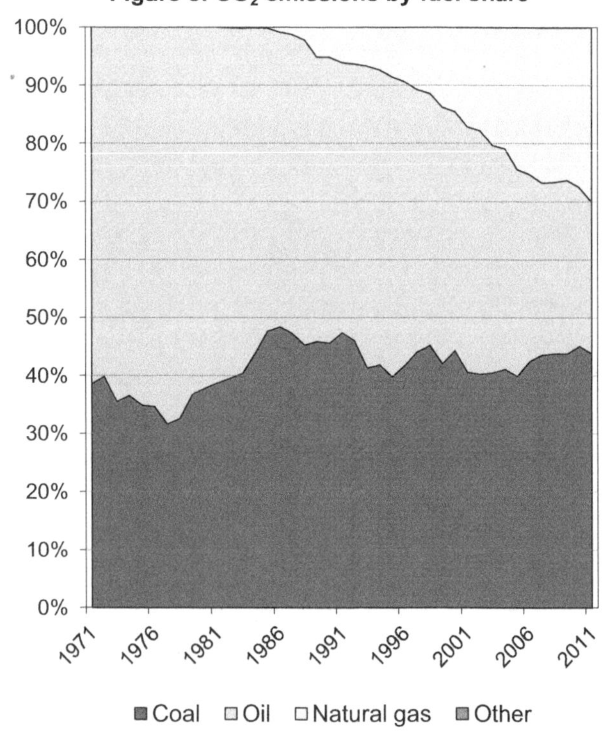

*Please refer to notes and definitions in Part I. Peat is included in coal for all figures.

Source: *IEA/OECD Electricity Information, IEA/OECD CO2 Emissions from Fuel Combustion*

TURKEY

1. Total primary energy supply (TPES) by fuel*

	1973	1980	1990	2000	2010	2011	2012e	Average annual percent change 73-90	90-11
TPES (Mtce)	**34.79**	**44.92**	**75.37**	**109.07**	**150.19**	**160.66**	**165.29**	**4.65**	**3.67**
Coal and peat	7.36	9.98	24.15	32.72	45.76	48.46	50.10	7.24	3.37
Oil	17.83	22.32	33.43	43.43	43.03	43.46	44.64	3.77	1.26
Natural Gas	-	-	4.08	18.05	44.84	52.54	53.22	-	12.94
Biofuels and waste	9.22	10.97	10.29	9.30	6.51	5.23	5.28	0.65	-3.17
Nuclear	-	-	-	-	-	-	-	-	-
Hydro	0.32	1.39	2.84	3.79	6.36	6.43	7.11	13.72	3.96
Geothermal	0.07	0.09	0.62	0.98	2.81	2.94	3.24	13.75	7.71
Solar, wind, tide	-	-	0.04	0.38	0.98	1.48	1.36	-	18.76
Net electricity trade(1)	-	0.16	-0.09	0.41	-0.10	0.11	0.35	-	x
Heat(2)	-	-	-	-	-	-	-	-	-

(1) Net trade between OECD and non-OECD countries.

(2) Ambient heat from heat pumps used in the transformation sector.

Source: *IEA/OECD Energy Balances of OECD Countries*

2. Energy supply, GDP and population*

	1973	1980	1990	2000	2010	2011	2012e	Average annual percent change 73-90	90-11
GDP (2005 bil. USD)	128	162	270	387	565	615	628	4.50	4.00
TPES/GDP(1)	0.27	0.28	0.28	0.28	0.27	0.26	0.26	0.14	-0.32
Population (millions)	38.1	44.4	55.1	64.3	73.0	74.0	75.0	2.20	1.41
TPES/population(1)	0.91	1.01	1.37	1.70	2.06	2.17	2.20	2.40	2.23
TPES/GDP(2)	109.3	110.9	112.0	113.0	106.5	104.7	105.4	0.14	-0.32
Coal and peat supply/GDP(2)	85.5	91.2	132.8	125.5	120.1	116.9	118.2	2.62	-0.60
Elec. consumption/GDP(2)	29.2	45.2	62.6	93.1	112.9	112.3	..	4.59	2.82
Elec. generation (TWh)	12	23	58	125	211	229	239	9.44	6.81
Industrial production(2)	..	..	56.1	79.5	116.1	127.8	131.0	..	4.00

(1) TPES/GDP in units of tce/2005 thousand USD, TPES/population in units of tce/capita.

(2) As index, 2005 (not shown) = 100

Source: *IEA/OECD Energy Balances of OECD Countries , OECD Main Economic Indicators*

3. Coal and peat production by type*

	1978(1)	1990	2000	2005	2010	2011	2012e	Average annual percent change 78-90	90-11
Mtce:									
Coking coal	2.58	2.03	0.71	0.65	0.94	1.01	0.96	-2.00	-3.27
Steam coal	1.16	1.06	0.81	1.59	1.95	1.78	1.83	-0.77	2.51
Lignite	5.55	14.59	16.31	13.20	22.15	22.70	20.63	8.38	2.13
Peat	-	-	-	-	-	-	-	-	-
Mt:									
Coking coal	2.96	1.82	0.74	0.65	1.09	1.18	1.12	-3.97	-2.05
Steam coal	1.33	1.20	1.68	2.41	2.61	2.39	2.44	-0.88	3.34
Lignite	15.12	44.41	60.85	55.28	69.70	72.55	65.95	9.39	2.37
Peat	-	-	-	-	-	-	-	-	-

(1) Earliest year for which split by coal type is available.

* Please refer to notes and definitions in Part I.

Source: *IEA/OECD Energy Balances of OECD Countries, IEA/OECD Energy Statistics of OECD Countries*

TURKEY

4. Final consumption of energy by fuel*

(Mtce)

	1973	1980	1990	2000	2009	2010	2011	Average annual percent change 73-90	Average annual percent change 90-11
Total final consumption(1)	**28.38**	**37.60**	**57.25**	**82.64**	**104.52**	**110.87**	**116.37**	**4.21**	**3.44**
Coal and peat	4.24	6.00	10.76	15.49	18.30	20.17	18.71	5.63	2.67
Oil	13.63	18.14	29.10	37.32	40.86	40.56	40.33	4.56	1.57
Natural Gas	-	-	1.01	7.01	16.07	18.76	25.24	-	16.54
Biofuels and wastes	9.22	10.97	10.29	9.22	6.55	6.34	5.07	0.65	-3.32
Geothermal	0.07	0.09	0.52	0.88	1.78	1.99	2.09	12.60	6.84
Solar, wind, tide	-	-	0.04	0.37	0.43	0.43	0.63	-	14.03
Electricity	1.22	2.40	5.52	11.78	19.02	20.87	22.58	9.29	6.94
Heat	-	-	-	0.55	1.51	1.75	1.74	-	-
of which:									
Total industry	**5.30**	**9.35**	**15.57**	**28.20**	**26.58**	**32.26**	**35.65**	**6.54**	**4.02**
Coal and peat	1.63	3.10	6.43	12.61	8.49	10.41	9.91	8.40	2.08
Oil	2.88	4.76	5.06	6.86	1.90	1.61	2.23	3.37	-3.84
Natural Gas	-	-	0.71	2.38	6.27	9.01	11.25	-	14.10
Biofuels and wastes	-	-	-	-	-	-	-	-	-
Geothermal	-	-	-	-	-	-	-	-	-
Solar, wind, tide	-	-	0.01	0.14	-	-	-	-	-
Electricity	0.79	1.49	3.36	5.66	8.40	9.48	10.52	8.88	5.59
Heat	-	-	-	0.55	1.51	1.75	1.74	-	-
Total transport	**6.26**	**7.84**	**13.18**	**16.80**	**21.34**	**20.90**	**21.21**	**4.48**	**2.29**
Coal and peat	0.75	0.26	0.02	0.00	-	-	-	-18.90	-
Oil	5.50	7.56	13.11	16.65	20.98	20.51	20.80	5.25	2.22
Natural Gas	-	-	-	0.06	0.27	0.31	0.31	-	-
Biofuels and wastes	-	-	-	-	0.01	0.01	0.02	-	-
Electricity	0.01	0.02	0.04	0.09	0.08	0.07	0.08	7.56	3.20
Residential	**14.40**	**17.44**	**20.76**	**25.13**	**30.57**	**32.08**	**33.61**	**2.18**	**2.32**
Coal and peat	1.86	2.64	4.30	2.88	8.43	9.24	8.27	5.07	3.16
Oil	3.09	3.32	4.45	5.13	2.31	2.07	1.82	2.17	-4.16
Natural Gas	-	-	0.06	3.85	6.28	6.92	10.32	-	27.98
Biofuels and wastes	9.22	10.97	10.29	9.22	6.54	6.33	5.04	0.65	-3.34
Geothermal	0.07	0.09	0.52	0.88	1.78	1.99	2.09	12.60	6.84
Solar, wind, tide	-	-	0.03	0.24	0.43	0.43	0.63	-	15.87
Electricity	0.17	0.43	1.11	2.93	4.81	5.09	5.44	11.65	7.85
Heat	-	-	-	-	-	-	-	-	-
Comm & public services	**0.45**	**0.44**	**0.91**	**3.31**	**9.28**	**8.13**	**9.21**	**4.30**	**11.65**
Coal and peat	0.00	0.01	0.00	-	1.19	0.39	0.40	-9.07	33.50
Oil	0.22	-	-	-	-	-	-	-	-
Natural Gas	-	-	-	0.60	2.96	2.20	2.91	-	-
Biofuels and waste	-	-	-	-	-	-	-	-	-
Geothermal	-	-	-	-	-	-	-	-	-
Solar, wind, tide	-	-	-	-	-	-	-	-	-
Electricity	0.22	0.44	0.91	2.71	5.13	5.54	5.90	8.81	9.31
Heat	-	-	-	-	-	-	-	-	-
Non-energy use	**0.81**	**1.20**	**4.01**	**5.03**	**9.79**	**10.29**	**8.85**	**9.83**	**3.84**
Coal and peat	-	-	-	-	-	-	-	-	-
Oil	0.81	1.20	3.76	4.90	9.54	10.01	8.49	9.41	3.96
Natural Gas	-	-	0.25	0.13	0.25	0.28	0.36	-	1.73

(1) Total Final Consumption (TFC) excludes use in transformation processes and energy industry own use.

* Please refer to notes and definitions in Part I. Agriculture, Forestry, Fishing and "Non-specified other" sectors are not reported here individually, but are accounted for in Total final consumption.

Source: *IEA/OECD Energy Balances of OECD Countries*

TURKEY

5. Coal balance*

(Mtce)

	1973	1980	1990	2000	2005	2010	2011	2012e	Average annual percent change 73-90	90-11
Production	7.4	8.8	17.7	17.8	15.4	25.0	25.5	23.4	5.2	1.8
Imports	0.0	0.8	6.0	13.3	16.7	19.8	22.2	26.7	42.3	6.4
Exports	-	-0.1	-	-	-	-	-	-	-	-
Stock changes	-0.1	0.4	0.5	1.6	0.4	0.9	0.8	..		
Primary supply	**7.4**	**10.0**	**24.2**	**32.7**	**32.6**	**45.8**	**48.5**	**50.1**	**7.2**	**3.4**
Statistical differences	**-**	**-0.3**	**-2.0**	**-0.1**	**-0.6**	**-2.5**	**-3.0**	**..**		
Total transformation	**-2.9 e**	**-3.4 e**	**-11.0 e**	**-16.7 e**	**-16.2**	**-22.2 e**	**-26.0 e**	**..**	**8.2**	**4.2**
Electricity and heat gen.	-1.8	-2.6	-7.8	-14.1	-13.5	-19.8	-23.5	..	9.0	5.4
Main activity producers (1)	*-1.8*	*-2.2*	*-7.2*	*-13.4*	*-12.3*	*-18.2*	*-21.9*	..	*8.4*	*5.5*
Autoproducers	-	*-0.4*	*-0.7*	*-0.7*	*-1.1*	*-1.6*	*-1.6*	..	-	*4.3*
Gas works	-0.1	-0.0	-0.0	-	-	-	-	..	-6.8	-
Coal transformation(2)	-0.9 e	-0.7 e	-3.1 e	-2.6 e	-2.7	-2.4 e	-2.5 e	..	7.2	-1.0
BKB plants	*0.0*	*0.0*	*0.0*	*0.0*	*0.0*	-	-	..	*3.1*	-
Blast furnaces	*-0.6 e*	*-0.9 e*	*-1.6 e*	*-1.8 e*	*-1.8*	*-2.0 e*	*-1.8 e*	..	*6.3*	*0.5*
Coke ovens	*-0.4*	*0.1*	*-1.5*	*-0.8*	*-0.9*	*-0.4*	*-0.7*	..	*8.5*	*-3.7*
Patent fuel plants	-	-	-	-	-	-	-	..	-	-
Other transformation(3)	-	-	-	-	-	-	-	..	-	-
Energy ind. own use	**-0.2**	**-0.3**	**-0.4**	**-0.4**	**-0.4**	**-0.9**	**-0.7**	**..**	**3.9**	**2.5**
Losses	**-0.0**	**-0.0**	**-0.0**	**-0.0**	**-0.1**	**-0.0**	**-**	**..**		
Final consumption(4)	**4.2**	**6.0**	**10.8**	**15.5**	**15.3**	**20.2**	**18.7**	**..**	**5.6**	**2.7**
Industry(5)	1.6	3.1	6.4	12.6	11.8	10.4	9.9	..	8.4	2.1
Iron and steel	*0.4 e*	*0.9 e*	*1.0 e*	*1.3 e*	*1.1*	*2.7 e*	*2.9 e*	..	*6.2*	*5.3*
Chemical	*0.3*	*0.2*	*0.4*	*0.0*	*0.1*	*0.3*	*0.4*	..	*2.1*	*-0.5*
Non-metallic minerals	-	-	*0.0*	*0.1*	*0.1*	*0.1*	-	..	-	-
Paper, pulp and print	-	-	*0.0*	*0.0*	*0.0*	*0.0*	*0.1*	..	-	*23.9*
Other industry (6)	*1.0*	*2.0*	*5.0*	*11.1*	*10.5*	*7.3*	*6.5*	..	*10.0*	*1.3*
Transport(7)	0.7	0.3	0.0	0.0	-	-	-	..	-18.9	-
Other	1.9	2.6	4.3	2.9	3.5	9.8	8.8	..	5.1	3.5
Comm. and pub. services	*0.0*	*0.0*	*0.0*	-	-	*0.4*	*0.4*	..	*-9.1*	*33.5*
Residential	*1.9*	*2.6*	*4.3*	*2.9*	*3.5*	*9.2*	*8.3*	..	*5.1*	*3.2*
Other sectors (8)	-	-	-	-	-	*0.1*	*0.1*	..	-	-
Non-energy use	-	-	-	-	-	-	-	..	-	-

(1) Main activity electricity and heat generation includes district heating.

(2) Coal transformation refers to the transformation of coal and peat to secondary and tertiary products (mainly cokes, briquettes, coal tar and off-gases). Gas works and Liquefaction are shown separately to match the *Energy Balances of OECD Countries.*

(3) Other transformation includes Liquefaction and Non-specified transformations.

(4) Final Consumption includes non-energy use and energy use (Industry, Transport and Other).

(5) Please refer to notes and definitions in Part I for detailed explanation.

(6) Other industry includes Non-ferrous metals, Transport equipment, Machinery, Mining and quarrying, Food and tobacco, Wood and wood products, Construction, Textile and leather, and Non-specified industry.

(7) Transport includes Rail and Inland waterways.

(8) Other sectors includes Agriculture/Forestry, Fishing and Non-specified other.

* "Coal" refers to all types of coal, primary (anthracite, coking coal, other bituminous coal, sub-bituminous coal, lignite/brown coal) and derived fuels (including patent fuel, BKB, coke oven coke, gas coke, coal tar, coke oven gas, gas works gas, blast furnace gas and other recovered gases), as well as peat, peat briquettes and oil shale. Quantities have been converted to Mtce using calorific values reported by the respective countries. Please refer to notes and definitions in Part I.

Source: *IEA/OECD Energy Balances of OECD Countries*

TURKEY

6. Use of coal for selected end-uses*

(million tonnes)

	1978	1990	2000	2005	2009	2010	2011	Average annual percent change 78-90	90-11
Steam coal	**1.21**	**3.10**	**8.51**	**14.57**	**15.90**	**18.85**	**19.24**	**8.19**	**9.09**
Total electricity and heat	0.68	0.37	1.66	4.57	5.97	6.81	9.99	-5.01	17.01
Main activity producers	*0.68*	*0.37*	*1.66*	*4.57*	*5.97*	*6.81*	*9.99*	*-5.01*	*17.01*
Autoproducers	-	-	-	-	-	-	-	-	-
Patent fuel/BKB plants	-	-	-	-	-	-	-	-	-
Coke ovens/Liquefaction(1)	0.09	0.06	-	-	-	-	-	-3.28	-
Blast furnace inputs	-	-	-	-	0.38 e	0.46 e	0.40 e	-	-
Gas manufacture	-	-	-	-	-	-	-	-	-
Industry	0.29	1.18	6.16	8.56	3.28	3.85	3.29	12.55	5.02
Iron and steel	-	-	-	-	*0.10 e*	*0.28 e*	*0.43 e*	-	-
Chemical	-	-	-	-	*0.05*	*0.08*	*0.13*	-	-
Non-metallic minerals	-	*0.01*	*0.03*	*0.07*	-	-	-	-	-
Paper, pulp and print	-	*0.00*	*0.02*	*0.02*	*0.02*	-	*0.00*	-	*3.36*
Other industry	*0.29*	*1.17*	*6.12*	*8.47*	*3.11 e*	*3.49 e*	*2.74 e*	*12.48*	*4.13*
Other sectors(2)	0.10	1.49	0.69	1.45	6.13	7.72	5.56	25.02	6.48
Non-energy use	-	-	-	-	-	-	-	-	-
Coking coal	**3.43**	**5.34**	**7.04**	**5.59**	**7.15**	**7.52**	**8.00**	**3.75**	**1.95**
Total electricity and heat	0.36	0.11	0.28	0.60	0.58	0.54	0.52	-9.80	7.96
Main activity producers	*0.31*	*0.08*	-	-	-	-	-	*-11.22*	-
Autoproducers	*0.05*	*0.03*	*0.28*	*0.60*	*0.58*	*0.54*	*0.52*	*-4.01*	*14.59*
Patent fuel/BKB plants	-	-	-	-	-	-	-	-	-
Coke ovens/Liquefaction(1)	2.39	4.66	4.19	4.22	4.90	5.32	5.20	5.71	0.52
Blast furnace inputs	-	-	-	-	0.01 e	-	-	-	-
Gas manufacture	0.29	0.10	-	-	-	-	-	-8.83	-
Industry	0.05	0.34	2.47	0.64	1.34	1.43	2.03	17.37	8.96
Iron and steel	-	-	*0.09*	*0.09*	*0.44 e*	*0.89*	*1.10*	-	-
Chemical	-	-	-	-	*0.02*	-	-	-	-
Non-metallic minerals	-	-	-	-	-	-	-	-	-
Paper, pulp and print	-	-	-	-	-	-	-	-	-
Other industry	*0.05*	*0.34*	*2.38*	*0.55*	*0.88 e*	*0.54*	*0.93*	*17.37*	*4.98*
Other sectors(2)	0.11	0.13	0.10	0.13	0.28	0.18	0.24	1.19	3.10
Non-energy use	-	-	-	-	-	-	-	-	-
Lignite	**13.23**	**45.89**	**64.38**	**56.58**	**75.64**	**69.24**	**73.93**	**10.92**	**2.30**
Total electricity and heat	4.45	29.88	52.54	47.41	62.89	55.44	60.27	17.21	3.40
Main activity producers	*4.30*	*29.63*	*52.29*	*47.21*	*62.64*	*54.96*	*59.71*	*17.45*	*3.39*
Autoproducers	*0.15*	*0.25*	*0.26*	*0.20*	*0.26*	*0.48*	*0.56*	*4.71*	*3.85*
Patent fuel/BKB plants	0.02	0.04	0.00	0.16	-	-	-	6.74	-
Coke ovens/Liquefaction(1)	-	-	-	-	-	-	-	-	-
Gas manufacture	-	-	-	-	-	-	-	-	-
Industry	3.63	8.47	6.92	4.11	6.14	7.75	6.69	7.31	-1.12
Iron and steel	*0.00*	-	-	-	-	*0.13*	*0.10*	-	-
Chemical	*0.73*	*0.93*	*0.11*	*0.16*	*0.28*	*0.49*	*0.36*	*2.10*	*-4.44*
Non-metallic minerals	-	-	-	-	*0.13*	*0.14*	-	-	-
Paper, pulp and print	-	-	-	-	*0.02*	*0.02*	*0.15*	-	-
Other industry	*2.90*	*7.54*	*6.81*	*3.95*	*5.72*	*6.98*	*6.07*	*8.27*	*-1.02*
Other sectors(2)	4.88	7.48	4.93	4.89	6.53	5.98	6.98	3.62	-0.33
Non-energy use	-	-	-	-	-	-	-	-	-

(1) Liquefaction primarily refers to direct distillation processes. Indirect liquefaction may be reported here or under gas manufacture.

(2) Other sectors are Residential, Commercial and public services, Agriculture/Forestry, Fishing and Non-specified other.

* Coal totals may not match the sum of the individual rows due to statistical differences. Please refer to notes and definitions in Part I.

Source: *IEA/OECD Energy Statistics of OECD Countries*

TURKEY

7. Fuel prices to end users*

	1978(1)	1990	1995	2000	2005	2009	2010	2011	2012
				(US dollars / tce)					
For electricity generation									
Steam coal	..	31.85	62.46	50.45	88.29	112.43	127.32	131.08	135.91
Heavy fuel oil	105.18	174.87	126.90	150.29	397.46	585.24	678.67	818.46	888.53
Natural gas	..	110.22	125.64	131.07	234.26	362.67	308.66	301.91	370.57
For industry									
Steam coal	..	79.67	58.81	53.26	78.71	139.04	137.83	142.56	161.45
Coking coal	70.23	82.07	86.05	80.19	148.07	269.73	272.13	254.23	268.97
High sulphur fuel oil	105.18	174.87	126.90	150.29	397.46	585.24	678.67	818.46	888.53
Low sulphur fuel oil	..	..	..	..	..	..	..	..	..
Natural gas	..	98.09	122.15	136.10	236.81	363.70	316.76	306.05	372.21
				(New Turkish lira / unit)(2)					
For electricity generation									
Steam coal	..	0.02	0.82	9.01	33.83	49.69	54.53	62.50	69.57
Heavy fuel oil	0.00	0.63	7.98	128.86	731.11	1241.53	1395.16	1873.31	2183.19
Natural gas	..	0.37	7.41	105.36	403.98	721.28	594.87	647.84	853.61
For industry									
Steam coal	..	0.13	1.64	20.22	64.10	130.59	125.44	144.45	175.62
Coking coal	0.00	0.19	3.44	43.69	173.07	363.59	355.48	369.73	419.93
High sulphur fuel oil	0.00	0.63	7.98	128.86	731.11	1241.53	1395.16	1873.31	2183.19
Low sulphur fuel oil	..	..	..	..	..	..	..	..	..
Natural gas	..	0.33	7.20	109.40	408.38	723.34	610.48	656.71	857.39

(1) Earliest year for which split by coal type is available.

(2) For steam coal, coking coal and heavy fuel oil per metric tonne; for natural gas per 10^7 kilocalories GCV.

* Please refer to notes and definitions in Part I. Steam coal prices relate to historical definitions.

Source: *IEA/OECD Energy Prices & Taxes*

8. Coal and peat trade by type of coal*

(Mtce)

	1978	1990	1995	2000	2005	2009	2010	2011	2012e
Total imports	**0.41**	**6.02**	**6.39**	**13.30**	**16.74**	**19.05**	**19.78**	**22.19**	**26.67**
Bituminous coal(1)	-	1.76	2.14	6.30	11.10	13.69	14.76	15.50	19.33
Coking coal	0.41	4.25	4.07	6.27	5.23	5.10	4.86	6.40	7.14
Sub-bituminous coal	-	-	-	-	-	-	-	-	0.03
Lignite	-	0.01	0.00	0.00	-	-	-	-	-
Peat	-	-	-	-	-	-	-	-	-
Coal products(2)	-	-	0.18	0.72	0.41	0.26	0.16	0.29	0.18
Total exports	**0.09**	**-**	**-**	**-**	**-**	**-**	**-**	**-**	**-**
Bituminous coal(1)	-	-	-	-	-	-	-	-	-
Coking coal	-	-	-	-	-	-	-	-	-
Sub-bituminous coal	-	-	-	-	-	-	-	-	-
Lignite	0.09	-	-	-	-	-	-	-	-
Peat	-	-	-	-	-	-	-	-	-
Coal products(2)	-	-	-	-	-	-	-	-	-

(1) Bituminous coal includes anthracite.

(2) Coal products includes products derived from coal, for example: coke, coal tar, briquettes, patent fuels.

* Please refer to notes and definitions in Part I.

Source: *IEA/OECD Energy Balances of OECD Countries, IEA/OECD Energy Statistics of OECD Countries*

TURKEY

9. Total coal imports by origin*
(thousand tonnes)

	1978[1]	1985	1990	1995	2000	2005	2010	2011	2012e
Total coal[2]	**475**	**2662**	**5572**	**5951**	**13001**	**17360**	**21333**	**23678**	**28682**
Coking coal	**475**	**2662**	**3717**	**3606**	**6202**	**4943**	**5135**	**6793**	**7577**
Australia	-	637	1318	1452	2793	882	1376	985	1514
Canada	-	-	51	210	847	1175	865	798	500
Czech Republic	-	-	-	-	-	-	-	-	-
Germany	70	-	-	-	-	-	-	-	-
Poland	-	-	-	178	100	187	78	-	-
United Kingdom	-	-	-	-	-	6	-	-	-
United States	405	1825	2011	1766	2150	1810	2518	2590	3823
Other OECD	-	-	-	-	-	-	-	-	108
China, People's Rep.	-	-	-	-	-	101	-	-	-
Colombia	-	-	-	-	-	-	-	57	-
Indonesia	-	-	-	-	-	-	-	161	-
South Africa	-	200	-	-	-	-	-	-	-
Former Soviet Union[3]	-	-	x	x	x	x	x	x	x
Russian Federation[4]	*x*	*x*	*337*	-	*259*	*574*	*14*	*1544*	*632*
Other FSU[4]	*x*	*x*	-	-	-	*208*	*55*	*524*	-
Venezuela	-	-	-	-	-	-	90	114	-
Vietnam	-	-	-	-	53	-	-	-	-
Non-specified/other	-	-	-	-	-	-	139	20	1000
Steam coal	**-**	**-**	**1840**	**2335**	**6788**	**12417**	**16198**	**16885**	**21105**
Australia	-	-	-	-	-	24	-	81	-
Canada	-	-	-	36	-	-	89	-	-
Czech Republic	-	-	-	-	-	-	-	-	-
Germany	-	-	-	10	-	-	-	-	-
Poland	-	-	-	-	-	-	292	-	147
United Kingdom	-	-	-	50	69	1	84	187	-
United States	-	-	58	-	-	-	-	54	1015
Other OECD	-	-	-	47	36	-	35	67	-
China, People's Rep.	-	-	-	-	-	1517	215	15	-
Colombia	-	-	-	-	-	2456	2841	5113	7935
Indonesia	-	-	-	-	-	-	-	-	42
South Africa	-	-	1252	877	1440	1448	1812	2853	2795
Former Soviet Union[3]	-	-	x	x	x	x	x	x	x
Russian Federation[4]	*x*	*x*	*530*	*1050*	*5023*	*6198*	*9844*	*6614*	*9171*
Other FSU[4]	*x*	*x*	-	*199*	*174*	*611*	*590*	*957*	-
Venezuela	-	-	-	-	-	11	234	81	-
Vietnam	-	-	-	-	-	-	-	-	-
Non-specified/other	-	-	-	66	46	151	162	863	-
Lignite	**-**	**-**	**15**	**10**	**11**	**-**	**-**	**-**	**-**

(1) Earliest year for which split by coal type is available.

(2) Total coal consists of steam coal, coking coal and lignite.

(3) For years prior to 1990, or if no details for individual Former Soviet Union republics.

(4) Data for former republics are not available separately for some OECD member countries.

* Please refer to notes and definitions in Part I.

Source: IEA/OECD Coal Statistics

UNITED KINGDOM*

Figure 1: TPES by fuel (Mtce)

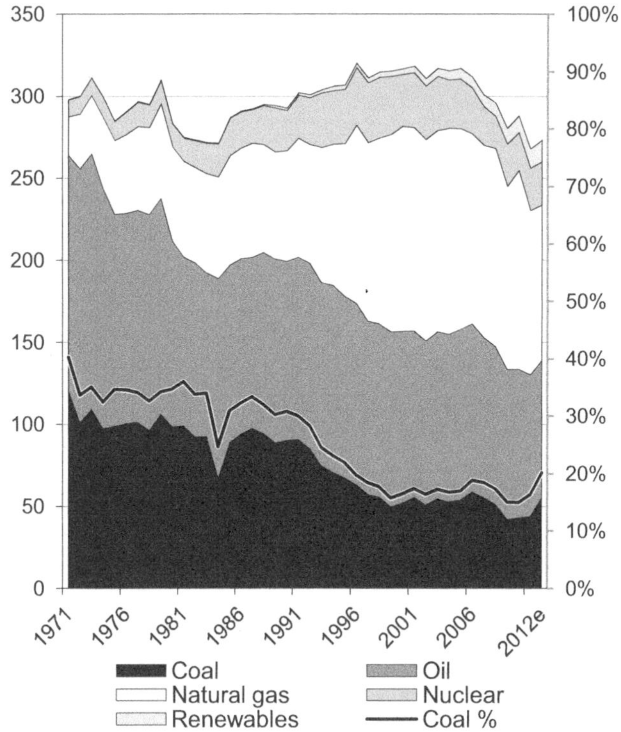

Figure 2: Coal supply indicators (1971 = 100)

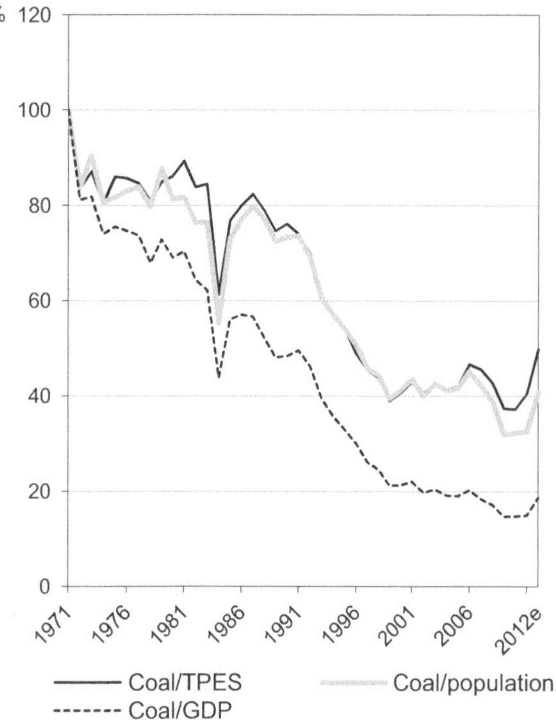

Figure 3: Primary coal supply (Mtce)

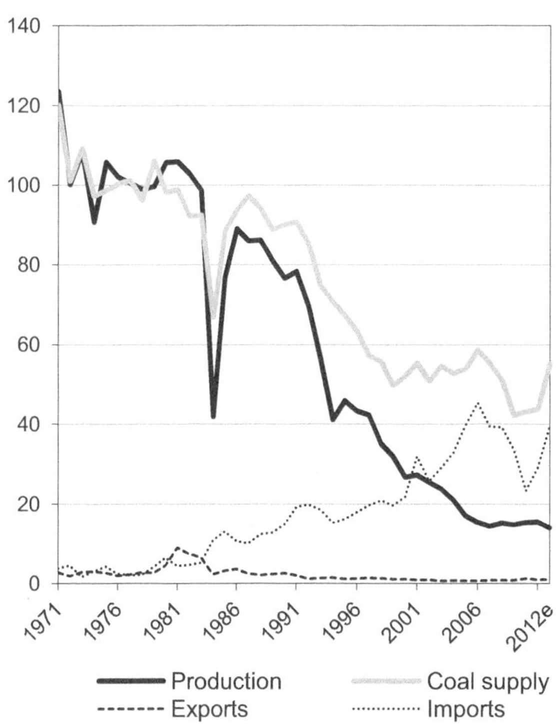

Figure 4: Coal consumption (Mtce)

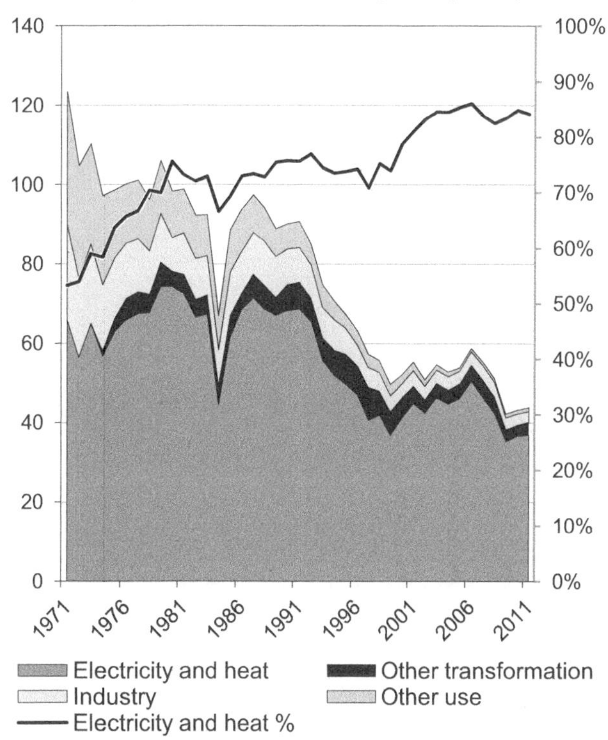

*Please refer to notes and definitions in Part I. Peat is included in coal for all figures.

Source: *IEA/OECD Coal Information, IEA/OECD Energy Balances of OECD Countries, OECD Main Economic Indicators*

UNITED KINGDOM*

Figure 5: Electricity generation by fuel (TWh)

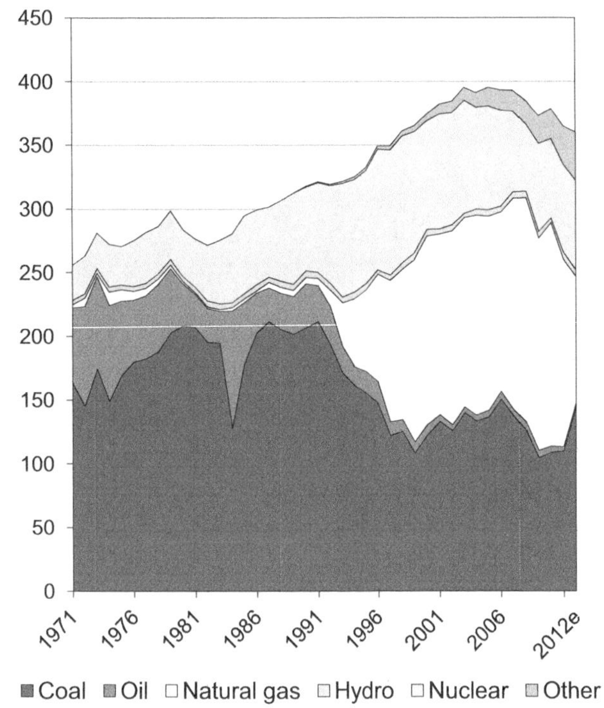

Figure 6: CO_2 emissions by fuel (Mt CO_2)

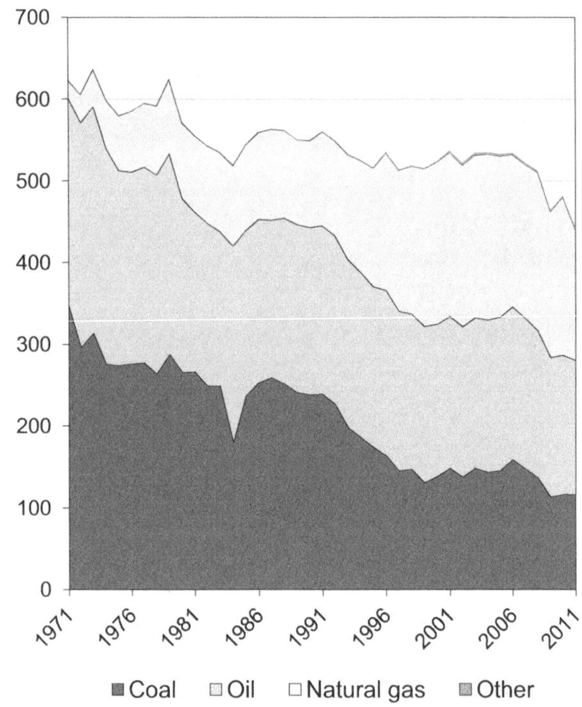

Figure 7: Electricity generation by fuel share

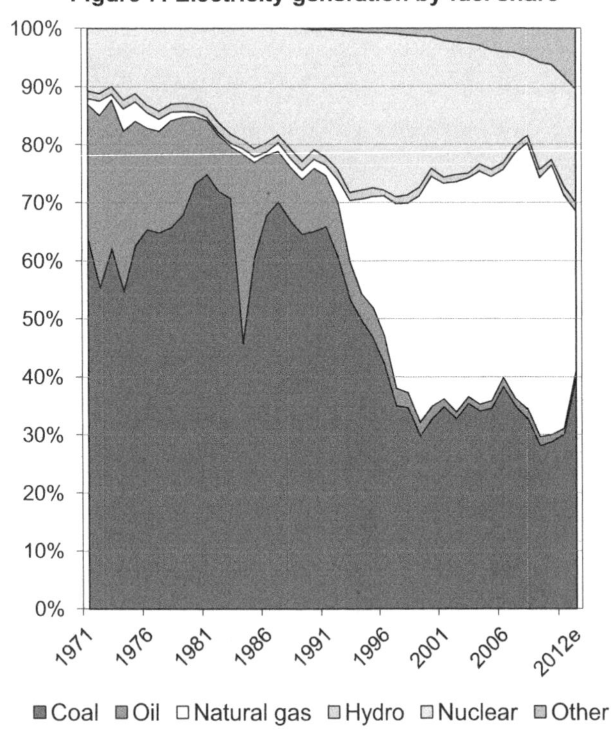

Figure 8: CO_2 emissions by fuel share

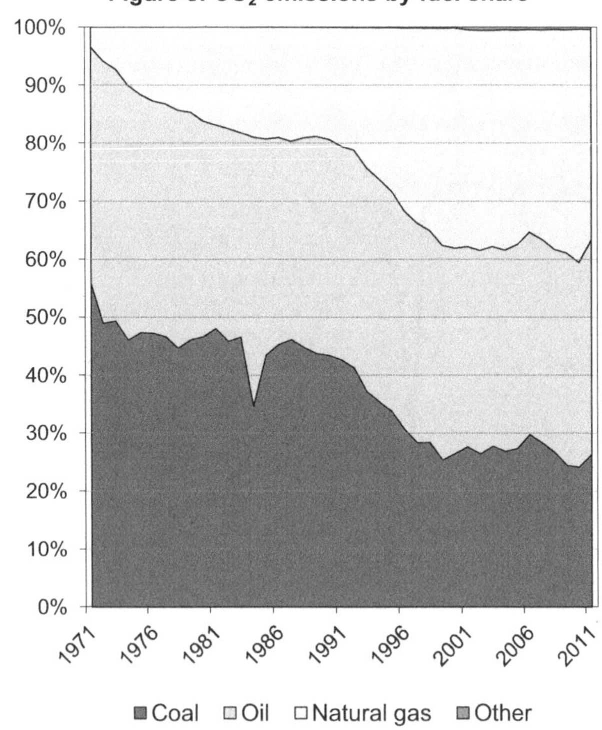

*Please refer to notes and definitions in Part I. Peat is included in coal for all figures.

Source: *IEA/OECD Electricity Information, IEA/OECD CO2 Emissions from Fuel Combustion*

UNITED KINGDOM

1. Total primary energy supply (TPES) by fuel*

	1973	1980	1990	2000	2010	2011	2012e	Average annual percent change 73-90	90-11
TPES (Mtce)	**311.53**	**283.47**	**294.17**	**318.49**	**288.33**	**268.68**	**274.83**	**-0.34**	**-0.43**
Coal and peat	109.18	98.28	90.15	52.15	43.15	43.79	55.16	-1.12	-3.38
Oil	155.58	113.34	109.11	104.60	90.44	86.42	83.98	-2.07	-1.10
Natural Gas	35.87	57.59	67.41	124.82	121.13	100.26	94.70	3.78	1.91
Biofuels and waste	-	-	0.90	2.75	8.32	8.96	9.95	-	11.59
Nuclear	10.42	13.78	24.48	31.67	23.13	25.68	26.21	5.15	0.23
Hydro	0.47	0.48	0.64	0.62	0.45	0.70	0.64	1.78	0.42
Geothermal	-	-	0.00	0.00	0.00	0.00	0.00	-	0.00
Solar, wind, tide	-	-	0.02	0.13	1.38	2.09	2.70	-	26.23
Net electricity trade(1)	0.01	0.00	1.47	1.74	0.33	0.76	1.48	36.14	-3.06
Heat(2)	-	-	-	-	-	-	-	-	-

(1) Net trade between OECD and non-OECD countries.

(2) Ambient heat from heat pumps used in the transformation sector.

Source: *IEA/OECD Energy Balances of OECD Countries*

2. Energy supply, GDP and population*

	1973	1980	1990	2000	2010	2011	2012e	Average annual percent change 73-90	90-11
GDP (2005 bil. USD)	1082	1155	1510	1984	2363	2387	2393	1.98	2.20
TPES/GDP(1)	0.29	0.25	0.19	0.16	0.12	0.11	0.11	-2.28	-2.58
Population (millions)	56.2	56.3	57.2	58.9	62.3	62.7	63.2	0.11	0.44
TPES/population(1)	5.54	5.03	5.14	5.41	4.63	4.28	4.35	-0.44	-0.86
TPES/GDP(2)	207.9	177.1	140.6	115.9	88.1	81.3	82.9	-2.28	-2.58
Coal and peat supply/GDP(2)	430.1	362.5	254.3	112.0	77.8	78.2	98.2	-3.04	-5.46
Elec. consumption/GDP(2)	141.9	133.6	119.6	109.3	91.6	87.7	..	-1.00	-1.47
Elec. generation (TWh)	281	284	318	374	379	365	360	0.72	0.66
Industrial production(2)	74.6	73.4	81.3	103.4	90.9	90.3	88.2	0.51	0.50

(1) TPES/GDP in units of tce/2005 thousand USD, TPES/population in units of tce/capita.

(2) As index, 2005 (not shown) = 100

Source: *IEA/OECD Energy Balances of OECD Countries* , *OECD Main Economic Indicators*

3. Coal and peat production by type*

	1978(1)	1990	2000	2005	2010	2011	2012e	Average annual percent change 78-90	90-11
Mtce:									
Coking coal	14.68	1.60	0.26	0.29	0.28	0.40	0.35	-16.87	-6.38
Steam coal	84.26	74.99	26.40	16.71	15.08	15.13	13.64	-0.97	-7.34
Lignite	-	-	-	-	-	-	-	-	-
Peat	-	-	-	-	-	-	-	-	-
Mt:									
Coking coal	15.11	1.60	0.26	0.27	0.27	0.38	0.34	-17.06	-6.58
Steam coal	108.47	91.16	30.94	20.22	18.15	18.24	16.45	-1.44	-7.37
Lignite	-	-	-	-	-	-	-	-	-
Peat	-	-	-	-	-	-	-	-	-

(1) Earliest year for which split by coal type is available.

* Please refer to notes and definitions in Part I.

Source: *IEA/OECD Energy Balances of OECD Countries, IEA/OECD Energy Statistics of OECD Countries*

UNITED KINGDOM

4. Final consumption of energy by fuel*

(Mtce)

	1973	1980	1990	2000	2009	2010	2011	Average annual percent change 73-90	90-11
Total final consumption[1]	**204.61**	**187.54**	**196.85**	**215.00**	**187.45**	**196.29**	**180.43**	**-0.23**	**-0.41**
Coal and peat	45.32	20.20	15.39	5.88	3.95	3.79	3.63	-6.16	-6.64
Oil	104.41	85.17	87.46	89.39	79.67	79.96	77.34	-1.04	-0.58
Natural Gas	26.25	53.21	59.68	74.89	59.75	67.16	55.36	4.95	-0.36
Biofuels and wastes	-	-	0.59	0.88	2.72	3.05	3.06	-	8.12
Geothermal	-	-	0.00	0.00	0.00	0.00	0.00	-	0.00
Solar, wind, tide	-	-	0.01	0.02	0.10	0.12	0.16	-	11.95
Electricity	28.63	28.79	33.72	40.47	39.53	40.39	39.07	0.97	0.70
Heat	-	0.17	-	3.48	1.72	1.81	1.82	-	-
of which:									
Total industry	**75.06**	**56.11**	**45.47**	**48.46**	**36.70**	**37.87**	**37.10**	**-2.91**	**-0.96**
Coal and peat	20.06	8.52	9.11	3.59	3.01	2.80	2.69	-4.54	-5.64
Oil	30.33	17.41	9.00	8.61	6.94	7.02	6.21	-6.90	-1.75
Natural Gas	13.45	19.28	14.86	20.27	12.87	13.45	13.75	0.59	-0.37
Biofuels and wastes	-	-	0.14	0.40	0.55	0.59	0.66	-	7.73
Geothermal	-	-	-	-	-	-	-	-	-
Solar, wind, tide	-	-	-	-	-	-	-	-	-
Electricity	11.21	10.72	12.36	14.02	12.25	12.84	12.58	0.58	0.08
Heat	-	0.17	-	1.57	1.09	1.17	1.20	-	-
Total transport	**39.47**	**43.39**	**55.97**	**59.81**	**58.45**	**58.21**	**57.58**	**2.08**	**0.14**
Coal and peat	0.08	0.05	0.00	-	0.02	0.02	0.01	-18.29	8.06
Oil	39.07	42.96	55.32	58.76	56.55	56.08	55.58	2.07	0.02
Natural Gas	-	-	-	-	-	-	-	-	-
Biofuels and wastes	-	-	-	-	1.39	1.61	1.49	-	-
Electricity	0.32	0.37	0.65	1.06	0.50	0.50	0.50	4.23	-1.22
Residential	**48.09**	**51.32**	**53.24**	**61.47**	**56.73**	**63.71**	**51.13**	**0.60**	**-0.19**
Coal and peat	20.89	9.71	4.97	2.22	0.87	0.94	0.90	-8.10	-7.83
Oil	5.54	3.76	3.30	4.31	4.01	4.56	3.55	-3.00	0.34
Natural Gas	10.45	27.27	33.20	40.88	36.75	43.06	32.38	7.04	-0.12
Biofuels and wastes	-	-	0.25	0.25	0.47	0.48	0.52	-	3.61
Geothermal	-	-	-	-	-	-	-	-	-
Solar, wind, tide	-	-	-	-	-	-	-	-	-
Electricity	11.22	10.58	11.52	13.74	14.56	14.60	13.71	0.16	0.83
Heat	-	-	-	0.06	0.07	0.07	0.07	-	-
Comm & public services	**19.03**	**18.20**	**18.37**	**24.07**	**20.82**	**21.37**	**20.11**	**-0.21**	**0.43**
Coal and peat	4.18	1.89	1.29	0.06	0.06	0.03	0.03	-6.70	-17.08
Oil	8.36	6.63	4.43	2.01	0.98	0.93	1.07	-3.68	-6.55
Natural Gas	1.09	3.06	3.90	8.93	7.34	7.75	6.53	7.78	2.48
Biofuels and waste	-	-	0.05	0.11	0.13	0.15	0.15	-	5.04
Geothermal	-	-	-	-	-	-	-	-	-
Solar, wind, tide	-	-	-	-	-	-	-	-	-
Electricity	5.40	6.62	8.71	11.12	11.75	11.96	11.79	2.85	1.46
Heat	-	-	-	1.85	0.56	0.56	0.54	-	-
Non-energy use	**17.22**	**10.14**	**15.46**	**16.18**	**11.39**	**11.52**	**11.08**	**-0.63**	**-1.58**
Coal and peat	-	-	-	-	-	-	-	-	-
Oil	17.22	10.14	13.24	14.65	10.50	10.62	10.19	-1.53	-1.24
Natural Gas	-	-	2.22	1.52	0.88	0.90	0.89	-	-4.26

(1) Total Final Consumption (TFC) excludes use in transformation processes and energy industry own use.

* Please refer to notes and definitions in Part I. Agriculture, Forestry, Fishing and "Non-specified other" sectors are not reported here individually, but are accounted for in Total final consumption.

Source: *IEA/OECD Energy Balances of OECD Countries*

UNITED KINGDOM

5. Coal balance*

(Mtce)

	1973	1980	1990	2000	2005	2010	2011	2012e	Average annual percent change 73-90	90-11
Production	108.4	105.7	76.6	26.7	17.0	15.4	15.5	14.0	-2.0	-7.3
Imports	1.6	6.4	14.7	21.8	39.6	23.3	29.1	39.6	13.9	3.3
Exports	-2.8	-4.4	-2.6	-1.1	-0.7	-1.2	-1.0	-1.0	-0.6	-4.5
Stock changes	2.0	-9.4	1.4	4.8	-2.0	5.7	0.2	2.6		
Primary supply	**109.2**	**98.3**	**90.2**	**52.1**	**53.9**	**43.1**	**43.8**	**55.2**	**-1.1**	**-3.4**
Statistical differences	**3.3**	**0.1**	**-1.5**	**0.1**	**0.6**	**0.9**	**0.2**	..		
Total transformation	**-62.1 e**	**-76.3 e**	**-72.1 e**	**-44.6 e**	**-49.0 e**	**-39.1 e**	**-39.2 e**	..	**0.9**	**-2.9**
Electricity and heat gen.	-65.0	-74.3	-68.3	-41.1	-46.0	-36.6	-36.8	..	0.3	-2.9
Main activity producers (1)	*-65.0*	*-72.0*	*-66.3*	*-37.7*	*-42.9*	*-34.1*	*-34.4*	..	*0.1*	*-3.1*
Autoproducers	-	*-2.3*	*-1.9*	*-3.3*	*-3.1*	*-2.5*	*-2.4*	..	-	*1.2*
Gas works	8.4	0.3	0.0	-	-	-	-	..	-37.0	-
Coal transformation(2)	-5.5 e	-2.4 e	-3.9 e	-3.6 e	-3.0 e	-2.6 e	-2.4 e	..	-2.1	-2.3
BKB plants	-	-	-	-	-	-	-	..	-	-
Blast furnaces	*-2.3 e*	*-0.9 e*	*-2.6 e*	*-3.0 e*	*-2.6 e*	*-2.1 e*	*-1.9 e*	..	*0.8*	*-1.4*
Coke ovens	*-3.2*	*-1.4*	*-1.1*	*-0.6*	*-0.4*	*-0.5*	*-0.5*	..	*-6.2*	*-3.8*
Patent fuel plants	*-0.0*	*-0.0*	*-0.2*	*0.1*	*0.0*	*0.0*	*0.0*	..	*9.7*	-
Other transformation(3)	-	-	-	0.0	0.0	0.0	0.0	..	-	-
Energy ind. own use	**-3.2**	**-1.9**	**-1.1**	**-1.5**	**-1.1**	**-0.9**	**-0.9**	..	**-6.1**	**-1.0**
Losses	**-1.8**	**-0.0**	**-0.0**	**-0.2**	**-0.3**	**-0.2**	**-0.2**	..		
Final consumption(4)	**45.3**	**20.2**	**15.4**	**5.9**	**4.0**	**3.8**	**3.6**	..	**-6.2**	**-6.6**
Industry(5)	20.1	8.5	9.1	3.6	3.2	2.8	2.7	..	-4.5	-5.6
Iron and steel	*9.0 e*	*3.6 e*	*3.9 e*	*1.2 e*	*1.3 e*	*1.0 e*	*1.0 e*	..	*-4.8*	*-6.2*
Chemical	*0.3*	*0.2*	*0.9*	*0.0*	*0.1*	*0.1*	*0.1*	..	*6.2*	*-11.8*
Non-metallic minerals	*3.5*	*2.5*	*1.4*	*1.1*	*1.0*	*1.0*	*0.9*	..	*-5.4*	*-1.8*
Paper, pulp and print	*1.2*	*0.1*	*0.6*	*0.1*	*0.1*	*0.1*	*0.1*	..	*-4.5*	*-7.7*
Other industry (6)	*6.0*	*2.0*	*2.4*	*1.1*	*0.7*	*0.6*	*0.6*	..	*-5.3*	*-6.4*
Transport(7)	0.1	0.1	0.0	-	0.0	0.0	0.0	..	-18.3	8.1
Other	25.2	11.6	6.3	2.3	0.9	1.0	0.9	..	-7.8	-8.7
Comm. and pub. services	*4.2*	*1.9*	*1.3*	*0.1*	*0.0*	*0.0*	*0.0*	..	*-6.7*	*-17.1*
Residential	*20.9*	*9.7*	*5.0*	*2.2*	*0.8*	*0.9*	*0.9*	..	*-8.1*	*-7.8*
Other sectors (8)	*0.1*	*0.0*	*0.0*	*0.0*	*0.0*	*0.0*	*0.0*	..	*-8.3*	*-6.2*
Non-energy use	-	-	-	-	-	-	-	..	-	-

(1) Main activity electricity and heat generation includes district heating.

(2) Coal transformation refers to the transformation of coal and peat to secondary and tertiary products (mainly cokes, briquettes, coal tar and off-gases). Gas works and Liquefaction are shown separately to match the *Energy Balances of OECD Countries.*

(3) Other transformation includes Liquefaction and Non-specified transformations.

(4) Final Consumption includes non-energy use and energy use (Industry, Transport and Other).

(5) Please refer to notes and definitions in Part I for detailed explanation.

(6) Other industry includes Non-ferrous metals, Transport equipment, Machinery, Mining and quarrying, Food and tobacco, Wood and wood products, Construction, Textile and leather, and Non-specified industry.

(7) Transport includes Rail and Inland waterways.

(8) Other sectors includes Agriculture/Forestry, Fishing and Non-specified other.

* "Coal" refers to all types of coal, primary (anthracite, coking coal, other bituminous coal, sub-bituminous coal, lignite/brown coal) and derived fuels (including patent fuel, BKB, coke oven coke, gas coke, coal tar, coke oven gas, gas works gas, blast furnace gas and other recovered gases), as well as peat, peat briquettes and oil shale. Quantities have been converted to Mtce using calorific values reported by the respective countries.
Please refer to notes and definitions in Part I.

Source: *IEA/OECD Energy Balances of OECD Countries*

UNITED KINGDOM

6. Use of coal for selected end-uses*

(million tonnes)

	1978	1990	2000	2005	2009	2010	2011	Average annual percent change 78-90	90-11
Steam coal	**105.68**	**96.21**	**51.02**	**55.21**	**43.02**	**45.08**	**45.23**	**-0.78**	**-3.53**
Total electricity and heat	83.21	84.01	46.85	52.52	40.16	41.98	42.33	0.08	-3.21
Main activity producers	*80.64*	*82.56*	*44.76*	*50.58*	*38.26*	*40.23*	*40.57*	*0.20*	*-3.33*
Autoproducers	*2.57*	*1.46*	*2.09*	*1.94*	*1.90*	*1.75*	*1.77*	*-4.59*	*0.92*
Patent fuel/BKB plants	1.19	0.78	0.54	0.27	0.34	0.35	0.33	-3.51	-3.98
Coke ovens/Liquefaction(1)	1.85	-	-	-	-	0.25	0.12	-	-
Blast furnace inputs	-	-	-	-	-	-	-	-	-
Gas manufacture	0.01	-	-	-	-	-	-	-	-
Industry	6.19	5.25	1.88	1.78	1.74	1.72	1.68	-1.36	-5.28
Iron and steel	*0.11*	*0.01*	*0.00*	-	*0.06*	*0.06*	*0.05*	*-16.73*	*7.33*
Chemical	*0.03*	*1.04*	*0.03*	*0.13*	*0.08*	*0.08*	*0.08*	*35.13*	*-11.60*
Non-metallic minerals	*3.09*	*1.50*	*1.24*	*1.12*	*1.08*	*1.06*	*1.06*	*-5.84*	*-1.66*
Paper, pulp and print	*0.32*	*0.63*	*0.12*	*0.14*	*0.12*	*0.12*	*0.12*	*5.93*	*-7.51*
Other industry	*2.65*	*2.07*	*0.48*	*0.39*	*0.40*	*0.39*	*0.37*	*-2.04*	*-7.86*
Other sectors(2)	12.19	5.45	1.97	0.67	0.77	0.76	0.76	-6.50	-8.98
Non-energy use	-	-	-	-	-	-	-	-	-
Coking coal	**14.99**	**10.52**	**8.82**	**6.57**	**5.76**	**6.37**	**6.27**	**-2.91**	**-2.43**
Total electricity and heat	-	-	-	-	-	-	-	-	-
Main activity producers	-	-	-	-	-	-	-	-	-
Autoproducers	-	-	-	-	-	-	-	-	-
Patent fuel/BKB plants	-	-	-	-	-	-	-	-	-
Coke ovens/Liquefaction(1)	14.91	10.09	8.23	5.57	4.94	5.40	5.28	-3.21	-3.03
Blast furnace inputs	-	0.66 e	0.45 e	0.96 e	0.72 e	0.90 e	0.90 e	-	1.49
Gas manufacture	-	-	-	-	-	-	-	-	-
Industry	-	0.11	0.01	0.08	0.13	0.08	0.10	-	-0.50
Iron and steel	-	*0.11 e*	*0.01 e*	*0.08 e*	*0.13 e*	*0.08 e*	*0.10 e*	-	*-0.50*
Chemical	-	-	-	-	-	-	-	-	-
Non-metallic minerals	-	-	-	-	-	-	-	-	-
Paper, pulp and print	-	-	-	-	-	-	-	-	-
Other industry	-	-	-	-	-	-	-	-	-
Other sectors(2)	-	-	-	-	-	-	-	-	-
Non-energy use	-	-	-	-	-	-	-	-	-
Lignite	**-**	**-**	**-**	**-**	**-**	**-**	**-**	**-**	**-**
Total electricity and heat	-	-	-	-	-	-	-	-	-
Main activity producers	-	-	-	-	-	-	-	-	-
Autoproducers	-	-	-	-	-	-	-	-	-
Patent fuel/BKB plants	-	-	-	-	-	-	-	-	-
Coke ovens/Liquefaction(1)	-	-	-	-	-	-	-	-	-
Gas manufacture	-	-	-	-	-	-	-	-	-
Industry	-	-	-	-	-	-	-	-	-
Iron and steel	-	-	-	-	-	-	-	-	-
Chemical	-	-	-	-	-	-	-	-	-
Non-metallic minerals	-	-	-	-	-	-	-	-	-
Paper, pulp and print	-	-	-	-	-	-	-	-	-
Other industry	-	-	-	-	-	-	-	-	-
Other sectors(2)	-	-	-	-	-	-	-	-	-
Non-energy use	-	-	-	-	-	-	-	-	-

(1) Liquefaction primarily refers to direct distillation processes. Indirect liquefaction may be reported here or under gas manufacture.

(2) Other sectors are Residential, Commercial and public services, Agriculture/Forestry, Fishing and Non-specified other.

* Coal totals may not match the sum of the individual rows due to statistical differences. Please refer to notes and definitions in Part I.

Source: *IEA/OECD Energy Statistics of OECD Countries*

UNITED KINGDOM

7. Fuel prices to end users*

	1978(1)	1990	1995	2000	2005	2009	2010	2011	2012
				(US dollars / tce)					
For electricity generation									
Steam coal	52.18	96.27	68.61	55.02	81.19	105.08	119.15	158.99	130.13
Heavy fuel oil	67.63	66.97	90.00	128.62	272.68	287.50	447.83	571.44	607.73
Natural gas	52.91	c	91.78	80.92	166.90	197.91	204.12	277.40	306.03
For industry									
Steam coal	49.74	88.51	65.61	59.32	99.65	111.10	130.90	166.67	165.78
Coking coal	..	..	..	..	..	..	..	..	..
High sulphur fuel oil	72.81	95.94	99.93	135.97	x	x	x	x	x
Low sulphur fuel oil	..	..	..	..	266.21	c	c	c	c
Natural gas	67.78	123.24	98.93	81.44	232.37	274.63	255.74	321.21	347.76
				(Pounds sterling / unit)(2)					
For electricity generation									
Steam coal	21.97	43.77	35.11	29.35	36.07	54.42	62.30	80.14	66.33
Heavy fuel oil	49.72	53.16	80.42	119.80	211.50	259.97	408.85	502.94	540.87
Natural gas	35.47	c	74.78	68.73	118.04	163.18	169.93	222.63	248.36
For industry									
Steam coal	23.25	44.67	37.27	35.13	49.14	63.87	75.98	93.26	93.80
Coking coal	..	..	..	..	..	..	..	..	..
High sulphur fuel oil	53.53	76.16	89.29	126.65	x	x	x	x	x
Low sulphur fuel oil	..	..	..	..	206.48	c	c	c	c
Natural gas	45.44	89.21	80.60	69.17	164.35	226.44	212.91	257.79	282.22

(1) Earliest year for which split by coal type is available.

(2) For steam coal, coking coal and heavy fuel oil per metric tonne; for natural gas per 10^7 kilocalories GCV.

* Please refer to notes and definitions in Part I. Steam coal prices relate to historical definitions.

Source: *IEA/OECD Energy Prices & Taxes*

8. Coal and peat trade by type of coal*

(Mtce)

	1978	1990	1995	2000	2005	2009	2010	2011	2012e
Total imports	**2.18**	**14.75**	**16.11**	**21.76**	**39.59**	**33.67**	**23.30**	**29.07**	**39.58**
Bituminous coal(1)	0.75	5.54	7.76	12.94	32.25	28.30	16.64	22.90	34.37
Coking coal	1.34	8.84	7.77	8.34	6.45	5.19	6.54	6.12	5.01
Sub-bituminous coal	-	-	-	-	-	-	-	-	-
Lignite	-	-	-	-	-	-	-	-	-
Peat	-	-	-	-	-	-	-	-	-
Coal products(2)	0.08	0.37	0.58	0.48	0.88	0.18	0.12	0.05	0.20
Total exports	**2.75**	**2.56**	**1.19**	**1.11**	**0.70**	**0.83**	**1.23**	**0.98**	**1.01**
Bituminous coal(1)	1.69	2.12	0.86	0.66	0.57	0.65	0.72	0.49	0.47
Coking coal	0.09	0.05	0.00	0.00	0.00	0.01	0.00	0.00	0.01
Sub-bituminous coal	-	-	-	-	-	-	-	-	-
Lignite	-	-	-	-	-	-	-	-	-
Peat	-	-	-	-	-	-	-	-	-
Coal products(2)	0.97	0.39	0.32	0.45	0.13	0.17	0.50	0.48	0.52

(1) Bituminous coal includes anthracite.

(2) Coal products includes products derived from coal, for example: coke, coal tar, briquettes, patent fuels.

* Please refer to notes and definitions in Part I.

Source: *IEA/OECD Energy Balances of OECD Countries, IEA/OECD Energy Statistics of OECD Countries*

UNITED KINGDOM

9. Total coal imports by origin*

(thousand tonnes)

	1978(1)	1985	1990	1995	2000	2005	2010	2011	2012e
Total coal(2)	**2352**	**12732**	**14783**	**15896**	**23446**	**43968**	**26540**	**32528**	**44797**
Coking coal	**1383**	**7150**	**8614**	**7754**	**8462**	**6551**	**6634**	**5908**	**4835**
Australia	380	2988	2966	3194	4880	3468	3437	3380	2253
Canada	-	337	961	1306	1633	1092	458	363	148
Czech Republic	-	3	-	-	-	-	-	-	-
Germany	207	-	2	-	-	-	-	-	-
Poland	375	1160	794	396	-	-	-	-	-
United Kingdom	-	-	-	-	-	-	-	-	-
United States	421	2647	3791	1792	1936	1284	2275	1873	1840
Other OECD	-	14	7	1	2	-	1	3	4
China, People's Rep.	-	-	-	-	-	-	-	-	-
Colombia	-	-	-	38	-	-	70	-	-
Indonesia	-	-	-	-	-	-	-	-	-
South Africa	-	1	-	-	-	-	-	-	-
Former Soviet Union(3)	-	-	x	x	x	x	x	x	x
Russian Federation (4)	*x*	*x*	*93*	*27*	*11*	*707*	*376*	*239*	*565*
Other FSU (4)	*x*	*x*	*-*	*-*	*-*	*-*	*17*	*50*	*-*
Venezuela	-	-	-	-	-	-	-	-	-
Vietnam	-	-	-	-	-	-	-	-	-
Non-specified/other	-	-	-	1000	-	-	-	-	25
Steam coal	**969**	**5582**	**6169**	**8142**	**14984**	**37417**	**19906**	**26620**	**39962**
Australia	645	2469	67	447	1222	828	12	-	-
Canada	-	61	24	58	-	-	-	-	154
Czech Republic	-	10	2	2	-	-	3	9	5
Germany	53	783	248	131	-	17	31	26	20
Poland	41	258	250	878	1107	630	563	655	87
United Kingdom	-	-	-	-	-	-	-	-	-
United States	1	498	1641	2056	837	266	2347	4461	8634
Other OECD	73	525	1868	162	204	312	443	654	664
China, People's Rep.	-	34	69	77	143	130	17	51	13
Colombia	-	89	956	2648	5649	3369	6247	8010	11919
Indonesia	-	-	-	65	9	1682	271	-	-
South Africa	26	724	356	1313	4756	12911	763	647	554
Former Soviet Union(3)	106	23	x	x	x	x	x	x	x
Russian Federation (4)	*x*	*x*	*499*	*133*	*480*	*16964*	*9209*	*12093*	*17723*
Other FSU (4)	*x*	*x*	*-*	*-*	*-*	*84*	*-*	*-*	*-*
Venezuela	-	-	169	-	208	-	-	-	172
Vietnam	-	-	-	101	123	-	-	-	-
Non-specified/other	24	108	20	71	246	224	-	14	17
Lignite	**-**	**-**	**-**	**-**	**-**	**-**	**-**	**-**	**-**

(1) Earliest year for which split by coal type is available.

(2) Total coal consists of steam coal, coking coal and lignite.

(3) For years prior to 1990, or if no details for individual Former Soviet Union republics.

(4) Data for former republics are not available separately for some OECD member countries.

* Please refer to notes and definitions in Part I.

Source: IEA/OECD Coal Statistics

UNITED KINGDOM

11. Steam coal exports by destination*

(thousand tonnes)

	1978(1)	1985	1990	1995	2000	2005	2010	2011	2012e
World	**2176**	**2379**	**2258**	**856**	**656**	**533**	**714**	**488**	**466**
Total OECD	**2174**	**2327**	**1960**	**845**	**651**	**525**	**682**	**464**	**427**
Australia	-	1	1	-	-	-	-	-	-
Austria	-	-	-	-	-	-	-	-	-
Belgium	113	44	67	86	-	83	166	67	51
Canada	-	-	-	-	-	1	1	2	2
Chile	-	-	-	-	-	-	-	-	-
Czech Republic	-	-	-	-	-	-	-	-	-
Denmark	125	1278	366	31	12	4	4	8	-
Estonia	-	-	-	-	-	-	-	-	-
Finland	6	2	204	1	-	-	5	6	-
France	950	183	271	142	77	32	5	15	26
Germany	446	191	214	10	13	1	9	25	-
Greece	-	13	-	-	-	-	-	-	7
Hungary	-	-	-	-	-	-	-	-	-
Iceland	-	19	18	-	-	2	-	3	-
Ireland	202	391	246	349	265	310	372	255	255
Israel	-	-	-	-	-	-	-	-	-
Italy	53	-	-	-	1	-	-	-	-
Japan	-	-	-	-	-	-	-	-	-
Korea	-	-	-	-	-	-	-	-	-
Luxembourg	-	-	-	-	71	-	-	-	-
Mexico	-	-	-	-	-	-	-	-	-
Netherlands	168	14	25	27	15	3	5	5	4
New Zealand	-	-	-	-	-	-	-	-	-
Norway	79	12	145	118	158	84	80	57	46
Poland	-	-	-	-	-	-	-	11	15
Portugal	3	111	178	-	-	-	-	-	-
Slovak Republic	-	-	-	-	-	-	-	-	-
Slovenia	-	-	-	-	-	-	-	-	-
Spain	3	3	138	60	31	-	10	8	19
Sweden	26	65	87	21	8	5	25	2	1
Switzerland	-	-	-	-	-	-	-	-	-
Turkey	-	-	-	-	-	-	-	-	1
United Kingdom	-	-	-	-	-	-	-	-	-
United States	-	-	-	-	-	-	-	-	-
Total non-OECD	**2**	**49**	**289**	**5**	**4**	**6**	**11**	**15**	**37**
Brazil	-	-	5	-	-	-	-	-	-
China(2)	-	-	-	-	2	-	-	-	-
Chinese Taipei	-	-	-	-	-	-	-	-	-
Egypt	-	-	1	-	1	1	1	-	1
India	-	-	1	-	-	1	2	2	4
Romania	-	-	-	-	-	-	-	-	-
Oth. Africa & Mid. East	2	48	282	-	-	-	1	-	2
Oth. non-OECD Americas	-	-	-	-	-	-	-	-	3
Other Asia & Oceania	-	-	-	-	1	4	2	1	3
Other non-OECD Europe and Eurasia	-	1	-	5	-	-	5	12	24
Non-specified/Other	**-**	**3**	**9**	**6**	**1**	**2**	**21**	**9**	**2**

(1) Earliest year for which split by coal type is available.

(2) China includes the People's Republic of China and Hong Kong, China.

* Please refer to notes and definitions in Part I. Steam coal includes all sub-bituminous coal.

Source: IEA/OECD Coal Statistics

UNITED KINGDOM

12. Coal import values by origin*

(average unit value, CIF, USD/tonne)

	1990	1995	2000	2005	2007	2008	2009	2010	2011/12[1]
Coking coal[2]	**72.56**	**57.61**	**46.84**	**116.05**	**120.24**	**187.79**	**196.99**	**184.15**	..
Imports from:									
Australia	76.77	56.28	45.23	127.10	124.16	201.79	231.88	190.25	..
Canada	67.89	55.72	45.32	99.81	104.80	192.51	241.64	186.81	..
Czech Republic	..	..	..	..	..	..	..	..	..
Poland	65.40	53.12	..	..	..	..	..	..	..
United States	72.53	60.78	52.11	113.10	131.69	176.47	145.57	181.09	..
China	..	..	..	..	207.67	..	..	..	..
Colombia	..	59.04	..	..	..	..	..	135.86	..
Indonesia	..	..	..	..	..	..	..	..	..
South Africa	..	..	..	..	..	..	..	..	..
Former Soviet Union[3]	46.86	59.92	43.01	91.89	82.35	69.16	98.63	147.54	..
Other bituminous coal[4]	**68.56**	**45.19**	**39.73**	**70.24**	**84.03**	**139.40**	**94.70**	**100.52**	..
Imports from:									
Australia	95.59	41.59	35.25	85.03	91.62	168.33	91.04	..	..
Canada	59.29	..	..	..	..	..	..	2075.56	..
Czech Republic	101.88	..	..	..	..	..	..	..	..
Poland	84.94	51.92	41.04	71.87	112.37	145.69	92.85	97.54	..
United States	69.28	44.64	38.57	83.66	79.53	133.03	85.17	92.99	..
China	73.47	..	..	98.27	71.23	..	x	x	..
Colombia	73.07	43.65	38.30	70.13	78.56	..	89.35	105.53	..
Indonesia	..	46.10	..	53.73	74.02	94.81	114.42	119.57	..
South Africa	..	43.16	41.04	69.27	90.44	155.46	85.90	106.00	..
Former Soviet Union[3]	53.72	50.58	39.22	70.78	82.58	139.84	96.44	94.88	..

Note: On occasion, shipment of extremely small quantities of high valued coal results in high import costs.

(1) Import data for steam and coking coals are currently unavailable for 2011 and 2012 due to resource constraints.

(2) Weighted average of individual countries using import volumes as weights. Prices exclude intra-EU trade.

(3) Former Soviet Union until 1991, Russian Federation starting in 1992.

(4) Bituminous steam coal only. Weighted average of trades. Prices exclude intra-EU trade.

* Please refer to notes and definitions in Part I.

Source: IEA/OECD Coal Statistics

UNITED STATES*

Figure 1: TPES by fuel (Mtce)

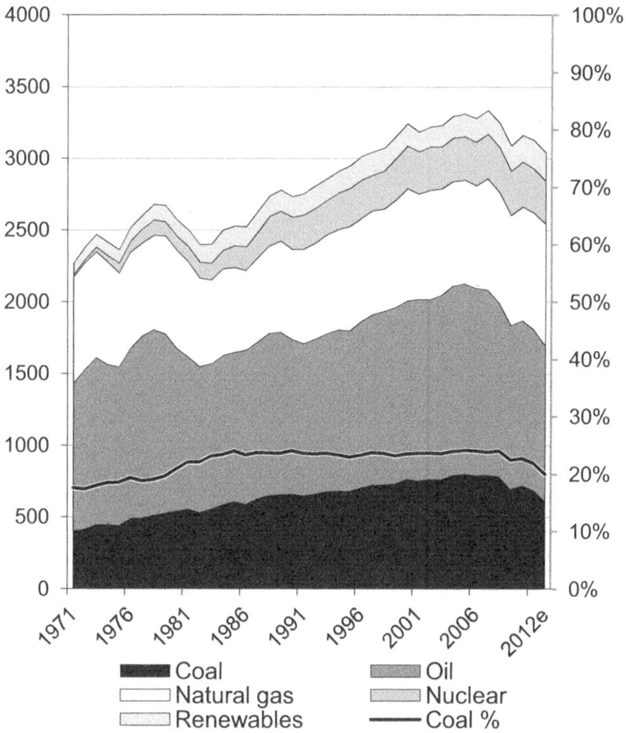

Figure 2: Coal supply indicators (1971 = 100)

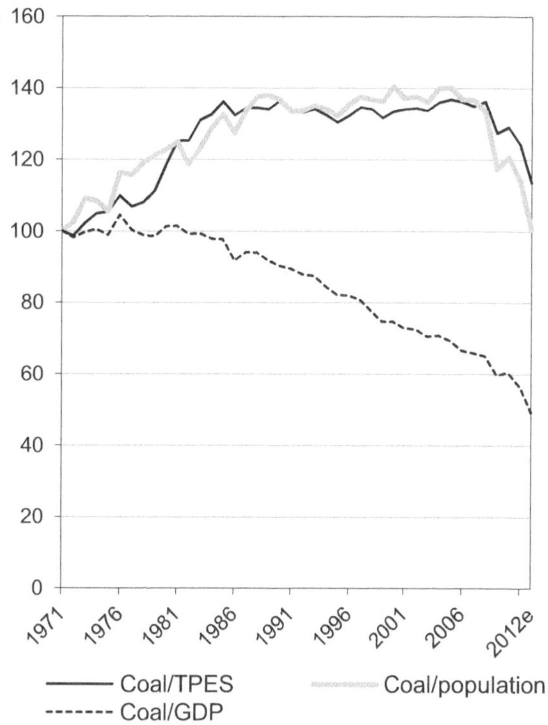

Figure 3: Primary coal supply (Mtce)

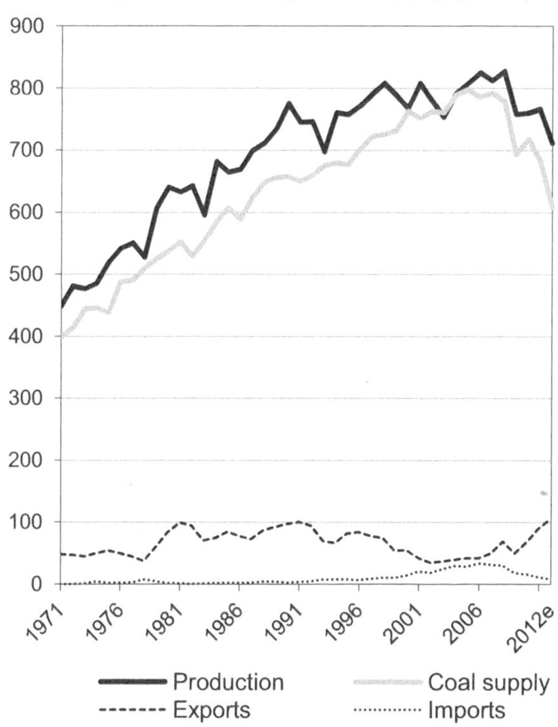

Figure 4: Coal consumption (Mtce)

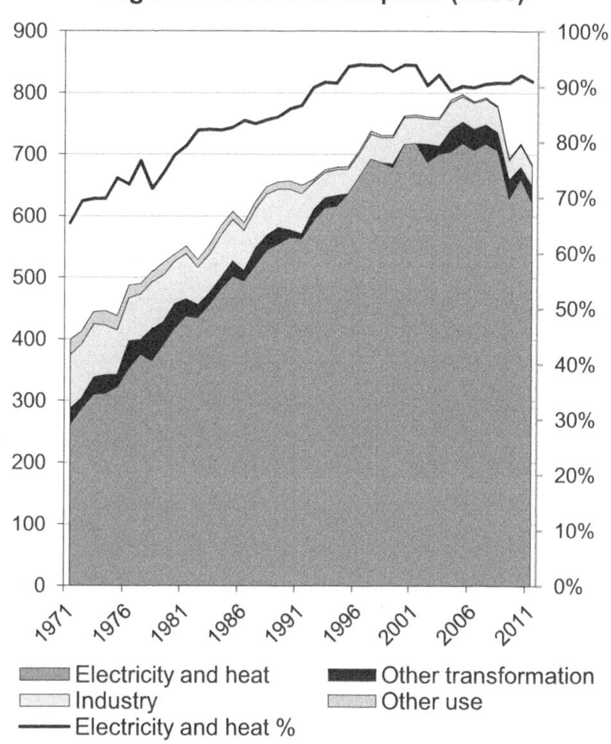

*Please refer to notes and definitions in Part I. Peat is included in coal for all figures.

Source: *IEA/OECD Coal Information, IEA/OECD Energy Balances of OECD Countries, OECD Main Economic Indicators*

UNITED STATES*

Figure 5: Electricity generation by fuel (TWh)

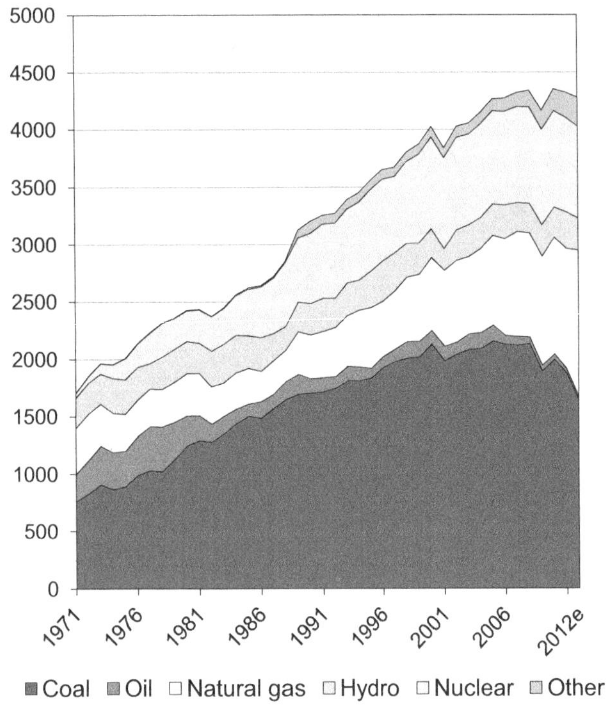

Figure 6: CO_2 emissions by fuel (Mt CO_2)

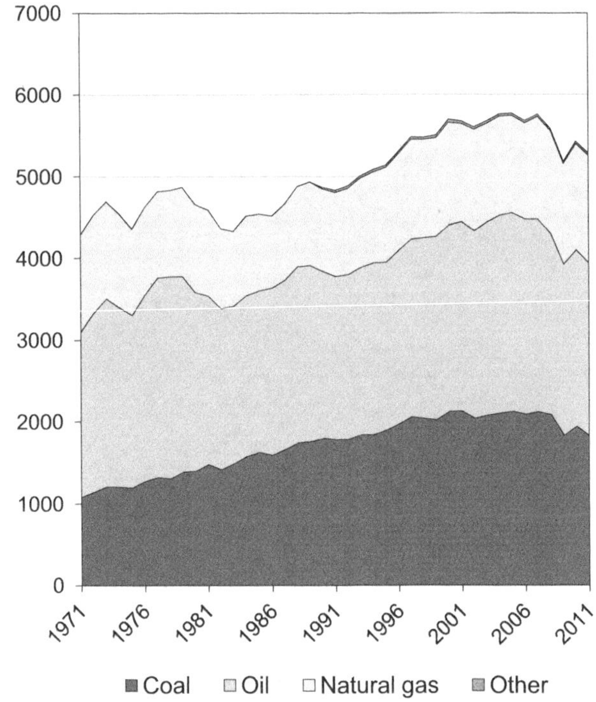

Figure 7: Electricity generation by fuel share

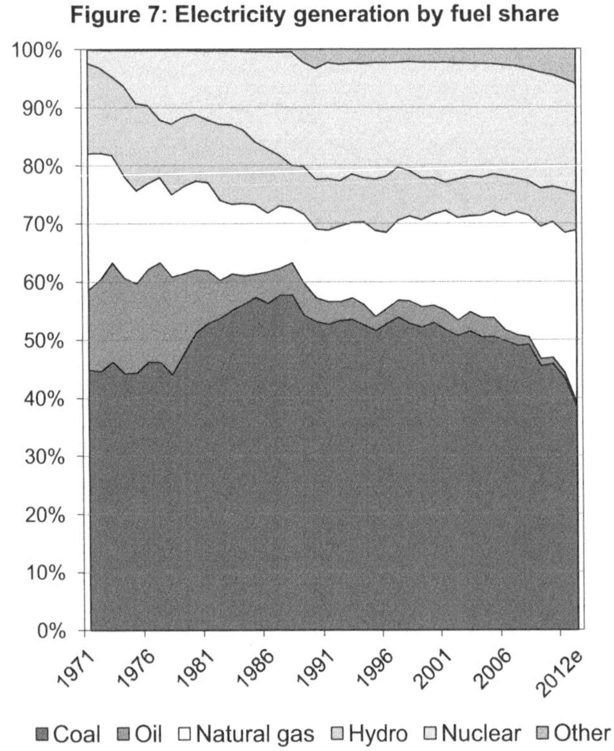

Figure 8: CO_2 emissions by fuel share

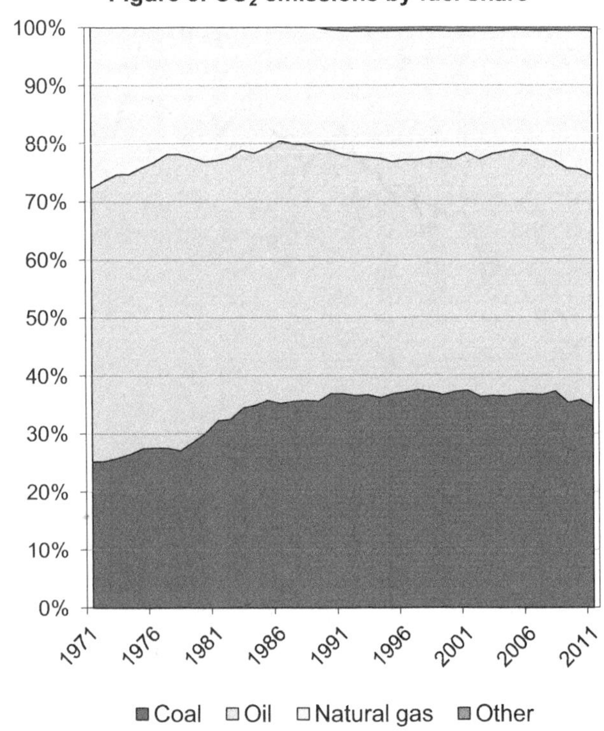

*Please refer to notes and definitions in Part I. Peat is included in coal for all figures.

Source: *IEA/OECD Electricity Information, IEA/OECD CO2 Emissions from Fuel Combustion*

UNITED STATES

1. Total primary energy supply (TPES) by fuel*

	1973	1980	1990	2000	2010	2011	2012e	Average annual percent change 73-90	90-11
TPES (Mtce)	**2471.34**	**2578.11**	**2735.71**	**3247.62**	**3165.01**	**3130.28**	**3046.35**	**0.60**	**0.64**
Coal and peat	444.35	537.48	657.43	762.32	718.06	684.29	608.69	2.33	0.19
Oil	1167.84	1138.47	1081.20	1244.50	1151.12	1122.91	1086.85	-0.45	0.18
Natural Gas	735.02	681.12	626.05	782.26	794.17	812.47	849.69	-0.94	1.25
Biofuels and waste	53.57	77.84	88.94	104.62	127.58	130.65	129.02	3.03	1.85
Nuclear	33.20	99.10	227.69	296.99	312.33	305.80	297.73	11.99	1.41
Hydro	32.60	34.25	33.56	31.11	32.22	39.53	34.40	0.17	0.78
Geothermal	3.01	6.57	20.14	18.70	12.01	12.22	13.70	11.83	-2.35
Solar, wind, tide	-	-	0.46	2.96 e	14.32 e	17.83 e	21.70	-	19.04
Net electricity trade(1)	1.75	3.28	0.24	4.17	3.19	4.58	4.58	-10.97	15.00
Heat(2)	-	-	-	-	-	-	-	-	-

(1) Net trade between OECD and non-OECD countries.

(2) Ambient heat from heat pumps used in the transformation sector.

Source: *IEA/OECD Energy Balances of OECD Countries*

2. Energy supply, GDP and population*

	1973	1980	1990	2000	2010	2011	2012e	Average annual percent change 73-90	90-11
GDP (2005 bil. USD)	4872	5796	7963	11158	12992	13226	13518	2.93	2.45
TPES/GDP(1)	0.51	0.44	0.34	0.29	0.24	0.24	0.23	-2.27	-1.76
Population (millions)	211.9	227.7	250.2	282.4	309.8	312.0	315.1	0.98	1.06
TPES/population(1)	11.66	11.32	10.93	11.50	10.22	10.03	9.67	-0.38	-0.41
TPES/GDP(2)	192.4	168.7	130.3	110.4	92.4	89.8	85.5	-2.27	-1.76
Coal and peat supply/GDP(2)	143.7	146.0	130.0	107.6	87.0	81.5	70.9	-0.58	-2.20
Elec. consumption/GDP(2)	115.7	118.1	111.8	106.0	98.9	96.9	..	-0.20	-0.68
Elec. generation (TWh)	1966	2427	3203	4026	4354	4327	4282	2.91	1.44
Industrial production(2)	47.1	52.5	65.2	96.6	94.9	98.1	101.6	1.93	1.96

(1) TPES/GDP in units of tce/2005 thousand USD, TPES/population in units of tce/capita.

(2) As index, 2005 (not shown) = 100

Source: *IEA/OECD Energy Balances of OECD Countries , OECD Main Economic Indicators*

3. Coal and peat production by type*

	1978(1)	1990	2000	2005	2010	2011	2012e	Average annual percent change 78-90	90-11
Mtce:									
Coking coal	93.39	96.20	56.00	45.21	65.91	78.21	77.87	0.25	-0.98
Steam coal	417.76	641.20	674.42	726.85	660.23	653.01	598.94	3.63	0.09
Lignite	15.55	37.33	36.51	35.48	33.64	35.02	34.09	7.57	-0.30
Peat	-	-	-	-	-	-	-	-	-
Mt:									
Coking coal	92.20	93.26	54.29	46.44	68.65	81.66	81.30	0.10	-0.63
Steam coal	484.60	760.39	839.69	916.00	856.49	850.69	782.02	3.83	0.54
Lignite	31.16	79.91	77.62	76.15	70.97	73.57	71.61	8.16	-0.39
Peat	-	-	-	-	-	-	-	-	-

(1) Earliest year for which split by coal type is available.

* Please refer to notes and definitions in Part I.

Source: *IEA/OECD Energy Balances of OECD Countries, IEA/OECD Energy Statistics of OECD Countries*

UNITED STATES

4. Final consumption of energy by fuel*

(Mtce)

	1973	1980	1990	2000	2009	2010	2011	Average annual percent change 73-90	90-11
Total final consumption(1)	**1879.10**	**1873.27**	**1847.86**	**2208.90**	**2083.81**	**2169.75**	**2148.15**	**-0.10**	**0.72**
Coal and peat	105.84	80.24	79.51	46.54	33.63	38.36	35.11	-1.67	-3.82
Oil	990.71	984.48	976.13	1133.46	1052.56	1086.47	1067.58	-0.09	0.43
Natural Gas	524.24	482.02	432.84	514.12	445.70	467.39	466.84	-1.12	0.36
Biofuels and wastes	53.46	77.69	32.27	74.50	92.46	98.66	101.15	-2.93	5.59
Geothermal	-	-	0.48	0.74	0.31	0.32	0.32	-	-1.96
Solar, wind, tide	-	-	-	2.06	1.96	1.98	2.06	-	-
Electricity	204.84	248.85	323.55	429.93	447.47	467.09	465.61	2.73	1.75
Heat	-	-	3.07	7.54	9.70	9.48	9.49	-	5.51
of which:									
Total industry	**562.57**	**553.39**	**405.33**	**474.65**	**370.23**	**410.64**	**410.01**	**-1.91**	**0.05**
Coal and peat	86.07	68.92	65.74	43.37	31.36	36.21	33.25	-1.57	-3.19
Oil	103.18	127.85	63.22	36.66	39.54	43.24	43.51	-2.84	-1.76
Natural Gas	253.15	216.47	156.98	196.97	150.98	169.55	169.60	-2.77	0.37
Biofuels and wastes	40.82	48.48	12.94	51.22	42.36	45.96	45.60	-6.54	6.18
Geothermal	-	-	-	0.16	0.15	0.15	0.15	-	-
Solar, wind, tide	-	-	-	-	-	-	-	-	-
Electricity	79.34	91.67	106.46	140.32	98.17	108.04	110.40	1.74	0.17
Heat	-	-	-	5.96	7.67	7.49	7.50	-	-
Total transport	**591.81**	**607.53**	**696.52**	**840.34**	**821.54**	**850.48**	**841.96**	**0.96**	**0.91**
Coal and peat	-	-	-	-	-	-	-	-	-
Oil	567.23	586.07	674.00	813.56	768.07	791.57	780.54	1.02	0.70
Natural Gas	24.06	21.07	22.01	21.68	20.83	23.29	23.70	-0.52	0.35
Biofuels and wastes	-	-	-	4.55	31.68	34.67	36.78	-	-
Electricity	0.53	0.38	0.51	0.54	0.96	0.95	0.94	-0.22	3.00
Residential	**339.29**	**307.97**	**299.82**	**377.20**	**374.79**	**383.21**	**376.55**	**-0.72**	**1.09**
Coal and peat	4.54	2.24	2.12	1.90 e	-	-	-	-4.36	-
Oil	102.37	59.79	37.67	45.58	32.06	30.73	28.22	-5.71	-1.37
Natural Gas	161.22	157.78	146.51	166.04	158.65	158.47	156.04	-0.56	0.30
Biofuels and wastes	-	-	-	14.82	14.70	14.36	15.39	-	-
Geothermal	-	-	-	0.31	0.05	0.05	0.05	-	-
Solar, wind, tide	-	-	-	2.06	1.96	1.98	2.06	-	-
Electricity	71.17	88.15	113.52	146.50	167.36	177.62	174.80	2.78	2.08
Heat	-	-	-	-	-	-	-	-	-
Comm & public services	**212.45**	**203.65**	**227.52**	**275.78**	**295.27**	**294.47**	**293.44**	**0.40**	**1.22**
Coal and peat	4.89	2.60	3.45	1.27 e	2.27	2.16	1.86	-2.04	-2.90
Oil	69.72	45.72	30.44	23.79	22.61	21.98	20.33	-4.76	-1.90
Natural Gas	85.81	86.69	87.50	104.53	102.41	101.68	103.09	0.11	0.78
Biofuels and waste	-	-	-	1.77	3.23	3.13	2.89	-	-
Geothermal	-	-	-	0.27	0.12	0.12	0.12	-	-
Solar, wind, tide	-	-	-	-	-	-	-	-	-
Electricity	52.03	68.65	103.06	142.55	162.59	163.41	163.16	4.10	2.21
Heat	-	-	3.07 e	1.58 e	2.04 e	1.99 e	1.99 e	-	-2.05
Non-energy use	**126.74**	**145.63**	**170.06**	**218.61**	**182.97**	**190.92**	**188.87**	**1.74**	**0.50**
Coal and peat	-	-	-	-	-	-	-	-	-
Oil	126.74	145.63	150.23	193.72	170.15	176.52	174.46	1.00	0.71
Natural Gas	-	-	19.84	24.89	12.83	14.40	14.41	-	-1.51

(1) Total Final Consumption (TFC) excludes use in transformation processes and energy industry own use.

* Please refer to notes and definitions in Part I. Agriculture, Forestry, Fishing and "Non-specified other" sectors are not reported here individually, but are accounted for in Total final consumption.

Source: *IEA/OECD Energy Balances of OECD Countries*

UNITED STATES

5. Coal balance*

(Mtce)

	1973	1980	1990	2000	2005	2010	2011	2012e	Average annual percent change 73-90	90-11
Production	476.2	639.9	774.7	766.9	807.5	759.8	766.2	710.9	2.9	-0.1
Imports	1.0	1.5	3.0	14.3	28.0	16.3	11.8	8.4	6.6	6.7
Exports	-44.3	-83.0	-97.1	-54.7	-42.1	-68.9	-90.5	-105.7	4.7	-0.3
Stock changes	11.4	-21.0	-23.2	35.8	4.3	10.9	-3.2	-4.9		
Primary supply	**444.4**	**537.5**	**657.4**	**762.3**	**797.7**	**718.1**	**684.3**	**608.7**	**2.3**	**0.2**
Statistical differences	**18.1**	**-11.1**	**9.0**	**21.7**	**-23.3**	**-4.0**	**-11.1**	..		
Total transformation	**-348.1 e**	**-442.6**	**-585.2 e**	**-735.5 e**	**-726.8 e**	**-673.4 e**	**-635.6 e**	..	**3.1**	**0.4**
Electricity and heat gen.	-309.9	-417.2	-565.7 e	-716.5 e	-718.7	-661.1	-622.6	..	3.6	0.5
Main activity producers (1)	*-309.9*	*-417.2*	*-559.6*	*-699.7 e*	*-710.4*	*-654.8*	*-617.1*	..	*3.5*	*0.5*
Autoproducers	-	-	*-6.2 e*	*-16.8 e*	*-8.3*	*-6.3*	*-5.5*	..	-	*-0.6*
Gas works	0.5	c	-2.6 e	-2.7 e	-2.6	-2.7	-2.8	..	-	0.2
Coal transformation (2)	-38.7 e	-25.3 e	-16.8 e	-16.2 e	-5.5 e	-9.6 e	-10.3 e	..	-4.8	-2.3
BKB plants	-	-	-	-	-	-	-	..	-	-
Blast furnaces	*-25.8 e*	*-16.2 e*	*-11.4 e*	*-10.9 e*	*-8.8 e*	*-6.0 e*	*-7.1 e*	..	*-4.7*	*-2.3*
Coke ovens	*-12.9*	*-9.1*	*-5.4*	*-5.3 e*	*-3.7*	*-3.6*	*-3.3*	..	*-5.0*	*-2.4*
Patent fuel plants	-	-	-	-	*7.0*	-	-	..	-	-
Other transformation (3)	-	-	-	-	-	-	-	..	-	-
Energy ind. own use	**-8.6**	**-3.6**	**-1.8**	**-2.0 e**	**-2.9**	**-2.2**	**-2.5**	..	**-8.9**	**1.6**
Losses	-	-	-	-	-	-	-	..		
Final consumption (4)	**105.8**	**80.2**	**79.5**	**46.5**	**44.8**	**38.4**	**35.1**	..	**-1.7**	**-3.8**
Industry (5)	86.1	68.9	65.7	43.4	41.1	36.2	33.3	..	-1.6	-3.2
Iron and steel	*48.4 e*	*29.6 e*	*18.1 e*	*11.2 e*	*5.9 e*	*7.0 e*	*6.7 e*	..	*-5.6*	*-4.6*
Chemical	*11.9*	*11.2*	*12.5 e*	*9.6 e*	*7.1*	*6.1*	*5.7*	..	*0.3*	*-3.7*
Non-metallic minerals	*5.2*	*10.0*	*11.1*	*11.4*	*10.6*	*7.3*	*7.0*	..	*4.5*	*-2.1*
Paper, pulp and print	*6.8*	*7.1*	*10.4*	*3.7*	*6.5*	*5.8*	*5.3*	..	*2.5*	*-3.1*
Other industry (6)	*13.7*	*11.1*	*13.7 e*	*7.5 e*	*11.1*	*10.0*	*8.6*	..	*0.0*	*-2.2*
Transport (7)	-	-	-	-	-	-	-	..	-	-
Other	19.8	11.3	13.8	3.2	3.6	2.2	1.9	..	-2.1	-9.1
Comm. and pub. services	*4.9*	*2.6*	*3.4*	*1.3 e*	*3.0*	*2.2*	*1.9*	..	*-2.0*	*-2.9*
Residential	*4.5*	*2.2*	*2.1*	*1.9 e*	-	-	-	..	*-4.4*	-
Other sectors (8)	*10.3*	*6.5*	*8.2*	-	*0.6*	-	-	..	*-1.4*	-
Non-energy use	-	-	-	-	-	-	-	..	-	-

(1) Main activity electricity and heat generation includes district heating.

(2) Coal transformation refers to the transformation of coal and peat to secondary and tertiary products (mainly cokes, briquettes, coal tar and off-gases). Gas works and Liquefaction are shown separately to match the *Energy Balances of OECD Countries.*

(3) Other transformation includes Liquefaction and Non-specified transformations.

(4) Final Consumption includes non-energy use and energy use (Industry, Transport and Other).

(5) Please refer to notes and definitions in Part I for detailed explanation.

(6) Other industry includes Non-ferrous metals, Transport equipment, Machinery, Mining and quarrying, Food and tobacco, Wood and wood products, Construction, Textile and leather, and Non-specified industry.

(7) Transport includes Rail and Inland waterways.

(8) Other sectors includes Agriculture/Forestry, Fishing and Non-specified other.

* "Coal" refers to all types of coal, primary (anthracite, coking coal, other bituminous coal, sub-bituminous coal, lignite/brown coal) and derived fuels (including patent fuel, BKB, coke oven coke, gas coke, coal tar, coke oven gas, gas works gas, blast furnace gas and other recovered gases), as well as peat, peat briquettes and oil shale. Quantities have been converted to Mtce using calorific values reported by the respective countries.
Please refer to notes and definitions in Part I.

Source: *IEA/OECD Energy Balances of OECD Countries*

UNITED STATES

6. Use of coal for selected end-uses*

(million tonnes)

	1978	1990	2000	2005	2009	2010	2011	Average annual percent change 78-90	90-11
Steam coal	**484.73**	**701.66**	**866.16**	**932.69**	**839.35**	**862.25**	**826.14**	**3.13**	**0.78**
Total electricity and heat	408.08	637.00	835.53 e	762.86	793.46	830.29	784.23	3.78	1.00
Main activity producers	*408.08*	*630.72*	*820.06 e*	*756.35*	*788.93*	*825.21*	*779.02*	*3.70*	*1.01*
Autoproducers	-	*6.27*	*15.48 e*	*6.50*	*4.53*	*5.08*	*5.21*	-	*-0.88*
Patent fuel/BKB plants	-	-	-	119.79	-	-	-	-	-
Coke ovens/Liquefaction(1)	-	-	-	-	-	-	-	-	-
Blast furnace inputs	-	0.17 e	2.39 e	1.20 e	1.03 e	1.28 e	1.48 e	-	10.73
Gas manufacture	-	-	-	-	-	-	-	-	-
Industry	44.63	52.18	35.88	35.68	29.83	32.42	29.55	1.31	-2.67
Iron and steel	*3.66*	*1.71 e*	*1.46 e*	*0.32 e*	*0.47 e*	*0.68 e*	*0.57 e*	*-6.17*	*-5.11*
Chemical	*10.08*	*14.26 e*	*10.51 e*	*7.23*	*6.68*	*6.84*	*6.48*	*2.93*	*-3.69*
Non-metallic minerals	*11.51*	*11.97*	*12.23*	*11.35*	*7.35*	*8.00*	*7.69*	*0.32*	*-2.09*
Paper, pulp and print	*7.54*	*11.28*	*4.07*	*6.82*	*6.22*	*6.50*	*5.98*	*3.41*	*-2.98*
Other industry	*11.83*	*12.98 e*	*7.62 e*	*9.96 e*	*9.11 e*	*10.41 e*	*8.84 e*	*0.77*	*-1.81*
Other sectors(2)	18.95	15.34	3.68	4.25	2.63	2.51	2.22	-1.75	-8.80
Non-energy use	-	-	-	-	-	-	-	-	-
Coking coal	**68.89**	**35.27**	**25.96**	**20.89**	**14.04**	**19.15**	**19.40**	**-5.43**	**-2.81**
Total electricity and heat	-	-	-	-	-	-	-	-	-
Main activity producers	-	-	-	-	-	-	-	-	-
Autoproducers	-	-	-	-	-	-	-	-	-
Patent fuel/BKB plants	-	-	-	2.81	-	-	-	-	-
Coke ovens/Liquefaction(1)	64.77	35.27	25.96 e	18.45	13.90	19.13	19.45	-4.94	-2.80
Blast furnace inputs	-	-	-	-	-	-	-	-	-
Gas manufacture	-	-	-	-	-	-	-	-	-
Industry	-	-	-	-	-	-	-	-	-
Iron and steel	-	-	-	-	-	-	-	-	-
Chemical	-	-	-	-	-	-	-	-	-
Non-metallic minerals	-	-	-	-	-	-	-	-	-
Paper, pulp and print	-	-	-	-	-	-	-	-	-
Other industry	-	-	-	-	-	-	-	-	-
Other sectors(2)	-	-	-	-	-	-	-	-	-
Non-energy use	-	-	-	-	-	-	-	-	-
Lignite	**30.77**	**79.02**	**74.27**	**76.14**	**65.11**	**68.30**	**74.76**	**8.18**	**-0.26**
Total electricity and heat	28.49	72.04 e	70.01 e	70.49	58.04	61.92	67.22	8.04	-0.33
Main activity producers	*28.49*	*70.94*	*68.76*	*68.44*	*58.04*	*59.34*	*66.91*	*7.90*	*-0.28*
Autoproducers	-	*1.10 e*	*1.25 e*	*2.06*	-	*2.57*	*0.31*	-	*-5.82*
Patent fuel/BKB plants	-	-	-	-	-	-	-	-	-
Coke ovens/Liquefaction(1)	-	-	-	-	-	-	-	-	-
Gas manufacture	-	5.64 e	5.67	5.39	5.43	5.57	5.41	-	-0.20
Industry	2.19	1.24	1.42	0.29	0.29	0.30	0.08	-4.61	-12.19
Iron and steel	-	-	-	-	-	-	-	-	-
Chemical	-	*0.23 e*	*0.24 e*	*0.24*	*0.26*	*0.26*	*0.05*	-	*-7.00*
Non-metallic minerals	-	*0.00*	*0.00*	*0.01*	-	-	-	-	-
Paper, pulp and print	-	-	-	-	-	-	-	-	-
Other industry	*2.19*	*1.01 e*	*1.18 e*	*0.05*	*0.03*	*0.04*	*0.03*	*-6.27*	*-15.40*
Other sectors(2)	0.09	0.11	0.05	0.01	-	-	-	1.52	-
Non-energy use	-	-	-	-	-	-	-	-	-

(1) Liquefaction primarily refers to direct distillation processes. Indirect liquefaction may be reported here or under gas manufacture.

(2) Other sectors are Residential, Commercial and public services, Agriculture/Forestry, Fishing and Non-specified other.

* Coal totals may not match the sum of the individual rows due to statistical differences. Please refer to notes and definitions in Part I.

Source: *IEA/OECD Energy Statistics of OECD Countries*

UNITED STATES

7. Fuel prices to end users*

	1978(1)	1990	1995	2000	2005	2009	2010	2011	2012
			(US dollars / tce)						
For electricity generation									
Steam coal	29.96	38.52	34.58	31.51	40.50	57.91	59.42	62.80	63.12
Heavy fuel oil	62.13	97.68	76.21	125.20	205.30	267.19	370.49	537.14	611.94
Natural gas	43.89	71.64	59.75	134.45	253.47	145.99	156.67	145.92	104.64
For industry									
Steam coal	36.26	40.00	38.59	37.77	56.31	77.28	76.20	85.14	87.90
Coking coal	48.94	52.01	51.52	48.41	91.26	156.19	158.80	197.97	200.07
High sulphur fuel oil	52.81	76.81	76.97	118.29	207.40	260.92	338.09	469.47	509.66
Low sulphur fuel oil	..	..	..	..	..	..	..	..	..
Natural gas	50.94	86.68	78.40	132.99	252.91	159.11	161.27	152.75	115.22
			(US dollars / unit)(2)						
For electricity generation									
Steam coal	26.11	33.57	30.14	27.46	35.30	50.47	51.79	54.74	55.01
Heavy fuel oil	88.65	139.37	108.74	178.64	292.94	381.24	528.64	766.42	873.15
Natural gas	56.43	92.11	76.82	172.86	325.89	187.70	201.43	187.61	134.54
For industry									
Steam coal	33.57	37.03	35.72	34.97	52.12	71.54	70.54	78.81	81.37
Coking coal	49.57	52.68	52.18	49.04	92.44	158.20	160.84	200.52	202.64
High sulphur fuel oil	75.35	109.60	109.82	168.79	295.93	372.29	482.41	669.87	727.21
Low sulphur fuel oil	..	..	..	..	..	..	..	..	..
Natural gas	65.50	111.44	100.80	170.99	325.17	204.57	207.34	196.39	148.14

(1) Earliest year for which split by coal type is available.

(2) For steam coal, coking coal and heavy fuel oil per metric tonne; for natural gas per 10^7 kilocalories GCV.

* Please refer to notes and definitions in Part I. Steam coal prices relate to historical definitions.

Source: *IEA/OECD Energy Prices & Taxes*

8. Coal and peat trade by type of coal*

(Mtce)

	1978	1990	1995	2000	2005	2009	2010	2011	2012e
Total imports	**7.23**	**3.03**	**7.74**	**14.31**	**28.00**	**18.22**	**16.34**	**11.76**	**8.36**
Bituminous coal(1)	2.37	2.34	5.84	9.27	21.96	15.94	12.93	8.77	6.13
Coking coal	-	-	-	1.60	1.54	0.91	1.33	1.39	0.98
Sub-bituminous coal	-	-	0.28	0.03	1.28	1.00	0.94	0.27	0.17
Lignite	-	-	-	0.04	0.06	0.06	0.06	0.06	0.06
Peat	-	-	-	-	-	-	-	-	-
Coal products(2)	4.87	0.68	1.62	3.38	3.15	0.31	1.08	1.27	1.01
Total exports	**37.11**	**97.12**	**81.48**	**54.74**	**42.09**	**49.81**	**68.92**	**90.54**	**105.69**
Bituminous coal(1)	8.69	37.19	30.05	22.24	11.57	13.26	15.23	26.47	39.92
Coking coal	27.79	59.39	48.75	30.72	24.46	31.80	47.88	59.33	59.63
Sub-bituminous coal	-	-	2.01	0.73	4.39	3.47	4.40	3.80	5.25
Lignite	0.04	0.03	-	0.03	0.10	0.11	0.10	0.08	0.02
Peat	-	-	-	-	-	-	-	-	-
Coal products(2)	0.59	0.51	0.67	1.02	1.56	1.17	1.31	0.87	0.87

(1) Bituminous coal includes anthracite.

(2) Coal products includes products derived from coal, for example: coke, coal tar, briquettes, patent fuels.

* Please refer to notes and definitions in Part I.

Source: *IEA/OECD Energy Balances of OECD Countries, IEA/OECD Energy Statistics of OECD Countries*

UNITED STATES

9. Total coal imports by origin*

(thousand tonnes)

	1978(1)	1985	1990	1995	2000	2005	2010	2011	2012e
Total coal(2)	**2679**	**1771**	**2449**	**6533**	**11351**	**27634**	**17556**	**11873**	**8310**
Coking coal	**-**	**-**	**-**	**-**	**1547**	**1603**	**1385**	**1446**	**1015**
Australia	-	-	-	-	-	144	-	-	-
Canada	-	-	-	-	1511	1458	1385	1358	875
Czech Republic	-	-	-	-	-	-	-	-	-
Germany	-	-	-	-	-	-	-	-	-
Poland	-	-	-	-	-	-	-	-	-
United Kingdom	-	-	-	-	-	-	-	-	-
United States	-	-	-	-	-	-	-	-	-
Other OECD	-	-	-	-	-	-	-	-	-
China, People's Rep.	-	-	-	-	-	-	-	-	-
Colombia	-	-	-	-	-	-	-	-	-
Indonesia	-	-	-	-	-	-	-	-	-
South Africa	-	-	-	-	-	-	-	-	-
Former Soviet Union(3)	-	-	x	x	x	x	x	x	x
Russian Federation (4)	*x*	*x*	-	-	-	-	-	-	-
Other FSU (4)	*x*	*x*	-	-	-	-	-	-	-
Venezuela	-	-	-	-	36	-	-	88	140
Vietnam	-	-	-	-	-	-	-	-	-
Non-specified/other	-	-	-	-	-	1	-	-	-
Steam coal	**2679**	**1771**	**2449**	**6533**	**9724**	**25903**	**16036**	**10298**	**7162**
Australia	933	35	22	192	152	66	345	56	-
Canada	49	367	883	1197	155	244	86	46	23
Czech Republic	-	-	-	-	-	-	-	-	-
Germany	-	-	-	-	-	-	-	-	-
Poland	645	-	-	-	-	70	-	-	-
United Kingdom	-	-	5	-	-	1	3	-	-
United States	-	-	-	-	-	-	-	-	-
Other OECD	-	6	-	39	6	39	1	-	-
China, People's Rep.	-	-	-	-	18	18	47	19	59
Colombia	-	539	1296	2483	6928	19247	13230	8619	6306
Indonesia	-	-	-	924	651	2239	1728	777	470
South Africa	996	824	-	-	-	70	-	-	-
Former Soviet Union(3)	-	-	x	x	x	x	x	x	x
Russian Federation (4)	*x*	*x*	-	-	-	*357*	-	-	*13*
Other FSU (4)	*x*	*x*	-	-	-	*79*	*38*	*113*	*170*
Venezuela	-	-	238	1675	1813	3387	528	618	121
Vietnam	-	-	-	-	-	85	-	-	-
Non-specified/other	56	-	5	23	1	1	30	47	-
Lignite	**-**	**-**	**-**	**-**	**80**	**128**	**135**	**129**	**133**

(1) Earliest year for which split by coal type is available.

(2) Total coal consists of steam coal, coking coal and lignite.

(3) For years prior to 1990, or if no details for individual Former Soviet Union republics.

(4) Data for former republics are not available separately for some OECD member countries.

* Please refer to notes and definitions in Part I.

Source: IEA/OECD Coal Statistics

UNITED STATES

10. Coking coal exports by destination*

(thousand tonnes)

	1978(1)	1985	1990	1995	2000	2005	2010	2011	2012e
World	**27433**	**54715**	**57568**	**47255**	**29780**	**26001**	**50906**	**63078**	**63392**
Total OECD	**23813**	**45183**	**47721**	**36210**	**22752**	**19577**	**31644**	**39502**	**39565**
Australia	-	-	-	-	-	-	-	-	-
Austria	-	-	-	-	-	239	412	330	1670
Belgium	956	3381	5532	3546	2343	1470	1689	1316	1392
Canada	5410	6360	3988	4039	3501	4034	3091	3772	4379
Chile	-	151	68	-	-	-	-	394	386
Czech Republic	-	-	-	-	-	-	-	-	-
Denmark	-	-	50	-	-	-	-	-	-
Estonia	-	-	-	-	-	-	-	-	-
Finland	-	-	86	657	288	260	262	307	266
France	1468	3471	4501	2863	2197	1116	1792	1589	2592
Germany	560	586	636	329	419	473	1478	1925	2869
Greece	-	7	-	-	-	-	-	-	-
Hungary	-	-	-	-	65	-	40	26	-
Iceland	-	13	28	36	48	57	59	58	62
Ireland	-	54	-	-	-	-	-	136	-
Israel	-	-	50	127	56	-	-	-	-
Italy	2905	5701	6377	4086	3297	2203	2388	4067	3429
Japan	8991	12613	10019	7193	852	1652	2696	5615	4579
Korea	505	1909	2908	2335	1096	727	2715	4597	4366
Luxembourg	-	-	-	-	-	-	77	-	-
Mexico	-	303	3	-	355	588	700	817	671
Netherlands	929	2975	3606	3609	1735	1548	4929	5889	4253
New Zealand	-	-	-	-	-	-	-	-	-
Norway	68	98	99	83	42	18	75	81	79
Poland	-	-	-	-	-	-	2149	1258	465
Portugal	265	244	234	27	198	-	-	93	134
Slovak Republic	-	-	-	-	-	-	201	249	462
Slovenia	-	-	-	-	-	163	223	663	292
Spain	688	2460	3156	1976	1993	1685	1393	1108	1160
Sweden	299	997	764	1006	642	464	401	480	475
Switzerland	-	-	-	-	-	-	37	-	-
Turkey	409	1716	1906	1638	1584	1642	2076	2248	3323
United Kingdom	360	2144	3710	2660	2041	1238	2761	2484	2261
United States	-	-	-	-	-	-	-	-	-
Total non-OECD	**3620**	**9532**	**9847**	**11045**	**6642**	**6424**	**19193**	**23576**	**23827**
Brazil	1342	5240	5219	5748	4093	3113	7125	7564	7091
China(2)	-	-	-	-	-	-	3808	4191	6526
Chinese Taipei	205	286	357	336	116	-	227	-	-
Egypt	218	457	586	1119	682	280	1042	569	375
India	-	-	-	-	22	1078	2299	3448	4353
Romania	673	1051	1559	1528	443	547	812	937	607
Oth. Africa & Mid. East	1	841	485	300	269	377	161	683	879
Oth. non-OECD Americas	914	588	580	400	184	207	321	306	471
Other Asia & Oceania	24	118	21	400	-	-	-	-	72
Other non-OECD Europe and Eurasia	243	951	1040	1214	833	822	3398	5878	3453
Non-specified/Other	**-**	**-**	**-**	**-**	**386**	**-**	**69**	**-**	**-**

(1) Earliest year for which split by coal type is available.
(2) China includes the People's Republic of China and Hong Kong, China.
* Please refer to notes and definitions in Part I.
Source: IEA/OECD Coal Statistics

UNITED STATES

11. Steam coal exports by destination*

(thousand tonnes)

	1978[1]	1985	1990	1995	2000	2005	2010	2011	2012e
World	**9837**	**29333**	**38344**	**33074**	**23226**	**19094**	**23023**	**34057**	**50629**
Total OECD	**9370**	**26112**	**33473**	**29596**	**22026**	**16990**	**19929**	**28761**	**40134**
Australia	-	1	1	-	-	-	105	302	152
Austria	-	-	-	-	-	-	-	-	-
Belgium	-	634	2178	937	278	411	367	1470	1013
Canada	8782	8508	10083	4541	13524	13625	7245	2431	2161
Chile	-	-	274	-	48	76	1053	1607	1635
Czech Republic	-	-	-	-	-	-	-	-	-
Denmark	-	1958	2842	1905	70	66	73	146	-
Estonia	-	-	-	-	-	-	-	-	-
Finland	-	1152	-	530	-	-	166	145	-
France	38	594	1740	457	564	28	1080	2078	1146
Germany	36	392	320	1562	467	133	935	2394	7402
Greece	-	1112	-	-	-	-	47	-	-
Hungary	-	-	-	-	-	-	-	-	-
Iceland	-	7	15	-	-	-	-	-	-
Ireland	-	335	1322	829	456	-	-	86	208
Israel	-	421	530	562	-	-	-	-	17
Italy	22	3599	4451	3194	65	23	612	1003	4318
Japan	160	1331	2074	3696	3181	236	175	610	590
Korea	300	1020	719	2081	508	580	2523	4882	3884
Luxembourg	-	-	-	-	-	-	-	-	-
Mexico	-	44	188	-	373	341	983	1713	2457
Netherlands	27	2765	3982	3015	643	829	1700	3893	2532
New Zealand	-	-	1	-	-	-	-	-	-
Norway	-	143	62	25	74	-	-	11	17
Poland	-	-	-	-	-	-	65	129	245
Portugal	-	626	1386	1562	343	143	531	798	992
Slovak Republic	-	-	-	-	-	-	-	-	-
Slovenia	-	-	-	-	-	-	182	-	197
Spain	-	674	282	2111	441	-	373	519	815
Sweden	-	230	21	8	-	71	275	153	138
Switzerland	-	8	-	-	-	-	-	169	-
Turkey	5	275	15	191	55	67	220	422	1515
United Kingdom	-	283	987	2390	936	361	1219	3800	8700
United States	-	-	-	-	-	-	-	-	-
Total non-OECD	**95**	**3221**	**4869**	**3476**	**923**	**1946**	**3094**	**5296**	**9251**
Brazil	11	78	79	13	22	693	63	310	125
China[2]	-	-	108	-	9	-	1445	876	2876
Chinese Taipei	-	2575	3820	2033	-	1	-	-	227
Egypt	-	-	-	1	-	-	146	144	-
India	-	-	-	-	-	217	171	634	1829
Romania	-	-	-	271	-	844	-	-	-
Oth. Africa & Mid. East	1	176	682	1099	825	63	1044	2292	2913
Oth. non-OECD Americas	82	110	128	-	1	13	115	463	677
Other Asia & Oceania	1	2	5	-	-	-	77	1	103
Other non-OECD Europe and Eurasia	-	280	47	59	66	115	33	576	501
Non-specified/Other	**372**	**-**	**2**	**2**	**277**	**158**	**-**	**-**	**1244**

(1) Earliest year for which split by coal type is available.

(2) China includes the People's Republic of China and Hong Kong, China.

* Please refer to notes and definitions in Part I. Steam coal includes all sub-bituminous coal.

Source: IEA/OECD Coal Statistics

UNITED STATES

12. Coal import values by origin*

(average unit value, CIF, USD/tonne)

	1990	1995	2000	2005	2007	2008	2009	2010	2011/12(1)
Coking coal(2)	..	..	**43.12**	**111.68**	**105.88**	**129.17**	**127.79**	**198.88**	..
Imports from:									
Australia	..	..	..	61.90	..	..	..	..	..
Canada	..	..	43.15	116.62	107.23	129.17	127.79	198.88	..
Czech Republic	..	..	..	..	..	..	..	..	..
Poland	..	..	..	..	..	..	..	..	..
United States	..	..	..	..	..	..	..	..	..
China	..	..	..	..	..	..	..	..	..
Colombia	..	..	..	..	..	..	..	..	..
Indonesia	..	..	..	..	..	..	..	..	..
South Africa	..	..	..	..	..	..	..	..	..
Former Soviet Union(3)	..	..	..	..	..	..	..	..	..
Other bituminous coal(4)	**36.86**	**36.50**	**31.11**	**47.39**	**49.53**	**61.88**	**68.57**	**67.86**	..
Imports from:									
Australia	45.99	34.16	31.84	35.09	34.93	58.63	63.43	62.55	..
Canada	26.95	35.65	30.82	69.63	59.70	53.24	78.02	76.65	..
Czech Republic	..	..	..	..	..	..	..	..	..
Poland	..	..	..	..	..	..	..	..	..
United States	..	..	..	..	..	..	..	..	..
China	..	..	79.56	103.19	103.58	..	..	65.61	..
Colombia	40.64	34.33	28.92	47.17	51.03	..	67.76	69.79	..
Indonesia	..	38.72	46.17	21.05	28.77	41.54	56.43	49.14	..
South Africa	..	..	..	..	..	107.32	..	..	..
Former Soviet Union(3)	..	..	..	63.72	60.80	..	..	..	..

Note: On occasion, shipment of extremely small quantities of high valued coal results in high import costs.

(1) Import data for steam and coking coals are currently unavailable for 2011 and 2012 due to resource constraints.

(2) Weighted average of individual countries using import volumes as weights. Prices exclude intra-EU trade.

(3) Former Soviet Union until 1991, Russian Federation starting in 1992.

(4) Bituminous steam coal only. Weighted average of trades. Prices exclude intra-EU trade.

* Please refer to notes and definitions in Part I.

Source: IEA/OECD Coal Statistics

UNITED STATES

13. Coal export values by destination*

(average unit value, FOB, USD/tonne)

	1990	1995	2000	2005	2007	2008	2009	2010	2011/12(1)
Coking coal(2)	**51.55**	**48.47**	**42.98**	**89.91**	**98.10**	**148.39**	**129.77**	**160.32**	..
Exports to:									
Belgium	52.98	50.67	44.05	88.93	98.07	127.74	126.93	156.58	..
Canada	41.81	40.31	35.61	71.82	84.83	100.31	99.83	123.73	..
Denmark	30.83	x	x	x	x	61.92	63.50	x	..
Finland	49.30	47.01	43.90	86.04	103.00	241.61	114.55	194.30	..
France	49.93	49.69	44.19	99.69	96.39	126.48	123.78	143.17	..
Germany	51.40	52.38	40.06	111.12	92.86	121.70	109.95	167.39	..
Greece	x	x	x	..	..	147.74	..	..	..
Ireland	x	x	x	..	..	90.00	..	..	..
Italy	52.78	51.38	49.45	86.21	104.34	121.55	128.06	160.27	..
Japan	50.34	45.35	41.99	100.61	x	157.59	181.25	166.81	..
Netherlands	51.94	51.32	45.46	93.04	94.55	129.86	124.48	139.66	..
Norway	66.52	x	74.32	109.84	122.19	144.71	173.02	162.85	..
Portugal	50.47	51.20	38.54	..	..	..	62.88	..	..
Spain	53.97	53.04	43.31	77.98	102.18	131.61	129.74	167.89	..
Sweden	53.76	52.60	48.17	69.27	104.18	117.30	155.54	158.54	..
Switzerland	x	x	x	..	..	116.02	118.09	180.00	..
Turkey	50.88	47.93	38.17	111.52	96.89	224.53	126.34	171.77	..
United Kingdom	53.46	51.83	46.13	92.21	99.41	133.40	117.05	145.13	..
Other OECD	53.05	50.13	45.83	82.48	92.50	158.92	130.89	168.45	..
Non-OECD	51.81	48.65	41.20	96.66	101.87	169.07	140.88	169.67	..
Steam coal(2,3)	**39.99**	**38.94**	**31.84**	**67.09**	**52.61**	**62.76**	**81.33**	**71.63**	..
Exports to:									
Belgium	41.48	38.66	27.48	58.94	66.32	89.15	117.67	78.38	..
Canada	33.62	33.74	31.87	60.16	44.34	38.83	63.73	64.91	..
Denmark	37.89	35.09	27.65	42.00	84.60	96.49	55.70	75.00	..
Finland	x	39.17	x	..	99.11	125.87	68.50	89.00	..
France	40.87	38.72	28.96	40.17	91.66	102.43	152.78	88.90	..
Germany	38.83	36.72	29.11	48.08	63.28	83.25	101.53	72.38	..
Greece	x	40.89	x	..	..	72.00	63.99	75.00	..
Ireland	43.05	39.75	30.17	..	73.21	101.56	..	288.07	..
Italy	44.93	45.97	30.67	89.78	76.87	116.28	107.99	96.09	..
Japan	41.76	38.61	34.90	105.21	198.41	143.02	113.97	83.85	..
Netherlands	43.17	40.53	28.90	76.28	69.83	79.14	82.78	79.70	..
Norway	58.16	x	19.54	..	..	..	..	..	..
Portugal	44.04	39.99	29.44	50.33	59.75	67.00	102.41	65.59	..
Spain	39.25	45.59	23.16	61.22	45.82	72.22	105.24	69.71	..
Sweden	38.57	53.51	x	117.24	..	56.00	..	193.09	..
Switzerland	x	x	189.05	..	..	65.00	108.38	255.54	..
Turkey	23.41	34.12	38.43	104.55	44.99	82.72	81.71	176.85	..
United Kingdom	47.41	37.01	28.13	74.26	51.74	68.99	72.49	69.60	..
Other OECD	43.31	36.81	37.95	65.28	62.58	98.18	76.90	54.52	..
Non-OECD	37.46	38.59	32.19	86.76	54.36	100.39	86.88	75.00	..

Note: On occasion, shipment of extremely small quantities of high valued coal results in high export costs.

(1) Import data for steam and coking coals are currently unavailable for 2011 and 2012 due to resource constraints.

(2) Weighted average of individual countries using import volumes as weights.

(3) Bituminous steam coal only.

* Please refer to notes and definitions in Part I.

Source: IEA/OECD Coal Statistics

PART V

NON-OECD COAL DATA

GENERAL NOTES

In Part V, the statistical tables on world coal supply and end-use, and on coal production, consumption and trade for the major non-OECD coal nations and regions supplement the information presented in Parts II and III of this book[1].

World hard coal supply and end-use statistics

This section provides historical data for world total coal supply and end-use for 1980, 1985, 1990, 1995, 2000, and 2005 to 2011 inclusive. Preliminary supply data for 2012 are also provided. These data summarise information published in the IEA/OECD publications *Energy Statistics of OECD Countries,* 2013 edition, *Energy Balances of Non-OECD Countries,* 2013 edition and *Energy Statistics of Non-OECD Countries*, 2013 edition. In cases where they are available, estimated coal trade statistics are shown for 2012.

Coal balances

These tables show all coal use for selected non-OECD countries in a format whereby coal production, imports, exports and stock changes "balance" transformation uses and final consumption. The transformation of coal into secondary products is clearly accounted for.

Calorific values used to convert primary coal data for OECD countries to million tons of coal equivalent (Mtce) are shown in Part I. Calorific values of coal in non-OECD countries are published annually in *Energy Balances of Non-OECD Countries*.

Coal trade

Trade tables show total[2], steam and coking coal imports by origin, and steam and coking coal exports by destination for major non-OECD coal importers and exporters.

In order to provide a breakdown of trade between steam coal and coking coal, the data have been gathered from a variety of sources, including from official submissions by OECD Member countries to the IEA Secretariat and UNECE Member countries to the UNECE Secretariat (in Geneva), published national sources for each country and from commercial publications. Data reported by exporting countries have been used where no importing country detail is available. Similarly, importing country data have been used in cases where exporting countries do not provide a breakdown of their exports by coal type. Due to classification anomalies and differences in reporting methods and time periods covered, the detailed data in the trade tables may not agree with coal trade data presented elsewhere in *Coal Information.*

Data for 2012 are preliminary and revised in the IEA quarterly publication *Oil, Gas, Coal and Electricity Quarterly Statistics*.

1. Data for Kazakhstan, Russia and Ukraine are available only after 1990. Data for all former Soviet Union republics, as well as data for the total Former Soviet Union before 1990, are available on CD (*Energy Statistics of Non-OECD Countries*).

2. In the 2012 edition, steam coal was revised to include sub-bituminous coal for all countries instead of select countries, so now comprises anthracite, other bituminous coal and sub-bituminous coal in all cases. Total coal comprises steam coal, coking coal and lignite.

World total coal supply and end-use 1975
(million tonnes)[1]

	Production	Imports	Exports	End-use sectors			
				Power and heat plants	Steel industry	Residential	Other Sectors
Total OECD	**1791.6**	**181.4**	**178.8**	**327.7**	**969.4**	**65.7**	**376.9**
Australia[2]	90.0	-	32.4	8.6	41.0	0.0	6.7
Austria	3.2	3.0	0.0	2.2	2.3	0.4	0.9
Belgium	9.9	6.2	0.6	7.6	3.2	2.1	1.9
Canada	25.3	15.8	11.7	7.3	16.6	0.5	1.7
Czech Republic	109.0	2.3	14.4	14.6	32.5	17.6	33.3
Denmark	-	4.1	0.0	-	2.7	0.0	0.5
Finland	-	3.8	-	-	1.8	0.1	0.9
France	26.8	17.4	0.5	16.6	14.6	3.5	6.1
Germany	471.1	17.8	14.9	47.5	242.7	1.3	175.8
Greece	18.1	0.8	-	0.6	15.3	0.0	2.5
Hungary	25.2	1.7	0.1	1.7	16.7	2.7	5.5
Ireland	0.1	0.7	0.1	-	0.0	0.5	0.1
Italy	1.4	12.7	-	11.1	2.1	0.2	0.4
Japan[2]	18.7	62.3	-	64.2	7.2	1.1	9.6
Korea	17.6	0.8	-	1.5	1.3	-	15.4
Mexico	3.1	0.6	-	3.4	0.1	-	0.1
Netherlands	-	4.1	0.1	3.6	0.2	0.1	0.1
New Zealand	2.4	0.0	0.0	0.2	0.7	0.3	1.3
Poland	211.5	1.1	41.9	22.7	103.2	15.8	27.8
Portugal	0.2	0.4	-	0.3	0.2	0.0	0.1
Spain	13.8	4.0	-	7.0	9.2	0.7	1.5
Sweden	0.1	1.6	0.0	1.1	0.0	0.0	0.3
Turkey	14.0	0.2	-	2.3	3.7	3.7	4.3
United Kingdom	128.7	5.1	1.9	21.8	75.5	11.6	13.3
United States	593.9	0.9	60.2	77.6	370.9	3.4	58.7
Other OECD[3]	*7.7*	*14.0*	*0.0*	*4.3*	*5.8*	*0.1*	*8.3*
Non-OECD Europe and Eurasia	**792.2**	**21.6**	**30.6**	**134.6**	**351.6**	**8.3**	**277.9**
Russian Federation	x	x	x	x	x	x	x
Non-OECD Asia	**628.7**	**1.2**	**4.1**	**61.2**	**114.4**	**101.0**	**321.3**
China, People's Republic of	482.0	..	3.0	40.0	79.2	97.2	241.4
Hong Kong, China	-	0.0	0.0	-	-	-	0.0
India[2]	100.5	-	0.4	17.4	28.6	3.7	43.6
Indonesia	0.2	0.0	-	-	-	-	0.2
DPR of Korea	33.0	0.4	-	3.5	4.3	-	25.5
Taipei, Chinese	3.1	0.0	-	0.3	0.7	0.0	2.4
Other Asia	*9.8*	*0.7*	*0.6*	*0.0*	*1.5*	*0.1*	*8.2*
Non-OECD Africa and Middle East	**75.8**	**3.1**	**3.3**	**15.3**	**37.8**	**3.9**	**21.2**
Egypt	-	1.5	-	0.9	-	-	0.0
South Africa	69.4	-	2.7	11.8	36.6	3.8	18.0
Zimbabwe	2.5	0.0	0.3	0.8	0.3	0.0	1.1
Other Africa / Middle East	*3.8*	*1.5*	*0.3*	*1.9*	*0.8*	*0.0*	*2.1*
Non-OECD Americas	**7.1**	**4.2**	**-**	**4.9**	**2.1**	**0.2**	**2.7**
Brazil	2.8	2.8	-	3.2	1.2	-	0.4
Colombia	3.7	-	-	0.7	0.5	0.2	2.0
Other non-OECD Americas	*0.6*	*1.4*	-	*1.0*	*0.4*	-	*0.3*
Total World	**3295.4**	**211.5**	**216.8**	**543.7**	**1475.3**	**179.1**	**999.9**

(1) Total coal comprises anthracite, coking coal, other bituminous coal, sub-bituminous coal and lignite.

(2) Fiscal year. See notes and definitions in Part I.

(3) Chile, Estonia, Iceland, Israel, Luxembourg, Norway, Slovak Republic, Slovenia and Switzerland.

Note: Steel industry consumption includes consumption in coke ovens. Power and heat stations column includes hard coal used in electricity and CHP production by public utilities and autoproducers, and in district heating. Other sectors includes consumption in non-ferrous industries, non-energy use, other transformation, losses and consumption in other sectors.

Source: *IEA/OECD Energy Statistics of OECD Countries* , *Energy Statistics of Non-OECD Countries*

World total coal supply and end-use 1980

(million tonnes)[1]

	Production	Imports	Exports	End-use sectors: Power and heat plants	End-use sectors: Steel industry	End-use sectors: Residential	End-use sectors: Other Sectors
Total OECD	**2035.7**	**232.4**	**207.6**	**302.9**	**1277.6**	**66.7**	**376.6**
Australia[2]	104.5	0.0	42.4	7.4	53.1	0.0	6.9
Austria	2.9	3.2	0.0	2.4	2.4	0.8	0.5
Belgium	8.0	10.2	0.5	8.0	5.5	1.3	2.3
Canada	36.7	15.6	15.3	7.4	28.3	0.2	1.8
Czech Republic	116.8	2.3	14.8	14.0	40.2	17.2	33.1
Denmark	-	10.0	0.0	-	9.3	0.0	0.6
Finland	-	4.7	-	0.1	4.8	0.1	0.7
France	22.8	29.4	0.4	15.7	26.9	2.7	5.4
Germany	484.2	18.3	12.7	39.2	272.8	0.9	174.7
Greece	23.2	0.5	-	0.4	20.6	0.0	2.1
Hungary	26.0	1.6	0.0	1.7	18.6	3.0	4.3
Ireland	0.1	1.2	0.0	-	0.0	0.7	0.3
Italy	1.3	17.3	-	11.3	6.2	0.3	0.6
Japan[2]	18.1	68.6	0.1	66.0	9.6	0.5	11.5
Korea	18.6	5.0	-	4.7	1.9	-	21.2
Mexico	3.1	0.8	-	4.0	-	-	0.0
Netherlands	-	7.3	1.5	3.7	2.4	0.1	0.3
New Zealand	2.1	-	0.1	0.2	0.3	0.2	1.3
Poland	230.0	1.0	31.9	25.4	126.8	19.7	27.5
Portugal	0.2	0.4	-	0.3	0.2	0.0	0.1
Spain	28.3	5.7	0.0	5.9	24.1	0.5	0.8
Sweden	0.0	2.2	0.0	1.7	0.1	0.0	0.4
Turkey	18.6	0.9	0.2	2.6	6.8	6.3	4.7
United Kingdom	130.1	7.3	4.0	13.7	92.0	8.9	8.9
United States	753.0	1.1	83.3	62.9	516.4	2.5	55.4
Other OECD[3]	*7.2*	*17.7*	*0.2*	*4.4*	*8.4*	*0.7*	*11.4*
Non-OECD Europe and Eurasia	**824.1**	**21.8**	**29.8**	**142.1**	**394.9**	**0.9**	**317.6**
Russian Federation	x	x	x	x	x	x	x
Non-OECD Asia	**796.4**	**9.1**	**7.3**	**150.6**	**178.9**	**120.8**	**324.8**
China, People's Republic of	620.2	2.0	6.3	127.3	126.5	115.7	234.2
Hong Kong, China	-	0.0	-	-	-	-	0.0
India[2]	116.1	0.6	0.1	17.4	41.4	4.4	44.6
Indonesia	0.3	0.0	0.1	-	-	-	0.2
DPR of Korea	44.1	0.5	0.1	4.3	5.8	-	34.3
Taipei, Chinese	2.6	4.6	-	1.5	2.4	-	2.0
Other Asia	*13.2*	*1.4*	*0.6*	*0.1*	*2.8*	*0.6*	*9.5*
Non-OECD Africa and Middle East	**120.8**	**2.2**	**28.6**	**17.4**	**51.1**	**2.4**	**26.5**
Egypt	-	0.8	-	0.9	-	-	0.0
South Africa	115.1	-	28.2	13.8	49.9	2.3	23.7
Zimbabwe	2.8	0.1	0.2	1.0	0.4	0.0	1.3
Other Africa / Middle East	*2.9*	*1.3*	*0.2*	*1.8*	*0.8*	*0.0*	*1.4*
Non-OECD Americas	**9.9**	**5.5**	**1.4**	**7.4**	**3.0**	**0.2**	**4.0**
Brazil	5.2	4.5	-	5.8	1.9	-	1.5
Colombia	4.2	-	1.4	0.7	0.7	0.2	2.2
Other non-OECD Americas	*0.5*	*1.0*	-	*0.9*	*0.5*	-	*0.3*
Total World	**3786.9**	**271.0**	**274.6**	**620.3**	**1905.5**	**191.0**	**1049.5**

(1) Total coal comprises anthracite, coking coal, other bituminous coal, sub-bituminous coal and lignite.

(2) Fiscal year. See notes and definitions in Part I.

(3) Chile, Estonia, Iceland, Israel, Luxembourg, Norway, Slovak Republic, Slovenia and Switzerland.

Note: Steel industry consumption includes consumption in coke ovens. Power and heat stations column includes hard coal used in electricity and CHP production by public utilities and autoproducers, and in district heating. Other sectors includes consumption in non-ferrous industries, non-energy use, other transformation, losses and consumption in other sectors.

Source: *IEA/OECD Energy Statistics of OECD Countries , Energy Statistics of Non-OECD Countries*

World total coal supply and end-use 1985

(million tonnes)[1]

				End-use sectors			
	Production	**Imports**	**Exports**	**Power and heat plants**	**Steel industry**	**Residential**	**Other Sectors**
Total OECD	**2236.4**	**289.1**	**269.4**	**262.2**	**1516.8**	**72.7**	**433.9**
Australia[2]	160.7	-	88.6	5.8	65.9	0.0	6.7
Austria	3.1	4.0	0.0	2.4	3.2	0.8	0.6
Belgium	7.7	9.6	1.3	7.8	5.1	1.4	1.5
Canada	60.9	14.6	27.4	6.4	39.6	0.2	1.9
Czech Republic	121.0	1.4	16.3	13.1	42.9	18.2	32.2
Denmark	-	12.5	0.1	-	11.3	0.0	0.6
Finland	-	5.1	0.0	0.1	4.1	0.0	1.1
France	18.9	18.9	1.0	12.0	18.3	2.3	6.0
Germany	522.9	18.2	9.3	32.5	297.7	0.6	195.8
Greece	35.9	2.0	0.2	-	34.7	0.1	3.1
Hungary	24.1	2.4	0.0	1.1	16.3	4.0	4.2
Ireland	0.1	1.9	0.0	0.0	0.1	1.1	0.4
Italy	1.9	22.4	-	10.5	10.8	0.2	2.4
Japan[2]	16.4	93.4	0.0	69.3	23.8	0.1	18.2
Korea	22.5	17.1	-	7.3	6.9	-	28.3
Mexico	5.2	0.6	-	3.8	1.5	-	0.1
Netherlands	0.1	11.7	1.4	4.1	4.9	0.1	0.9
New Zealand	2.5	-	0.4	0.1	0.5	0.2	1.3
Poland	249.4	1.1	36.3	20.5	152.2	20.4	24.3
Portugal	0.2	1.4	-	0.4	0.4	0.0	0.3
Spain	39.7	8.4	0.0	5.2	39.8	0.6	2.9
Sweden	0.0	4.9	0.1	1.6	1.8	-	0.7
Turkey	40.0	2.7	0.0	3.7	20.5	10.4	6.9
United Kingdom	94.1	12.7	2.4	10.3	75.7	8.6	9.4
United States	801.6	1.8	84.1	39.4	627.5	2.6	72.6
Other OECD[3]	*7.6*	*20.5*	*0.4*	*4.7*	*11.4*	*0.7*	*11.7*
Non-OECD Europe and Eurasia	**874.1**	**31.3**	**31.7**	**138.2**	**432.6**	**0.9**	**269.7**
Russian Federation	x	x	x	x	x	x	x
Non-OECD Asia	**1072.6**	**25.7**	**10.1**	**134.8**	**281.3**	**161.4**	**455.5**
China, People's Republic of	837.3	2.3	7.8	107.5	179.0	156.2	338.1
Hong Kong, China	-	5.5	-	-	5.2	-	-
India[2]	158.5	2.0	0.2	18.9	73.2	4.6	59.5
Indonesia	1.9	0.0	1.0	-	0.2	-	1.9
DPR of Korea	52.0	2.5	0.3	5.0	6.8	-	42.4
Taipei, Chinese	1.9	10.1	-	2.7	5.2	-	3.2
Other Asia	*21.0*	*3.2*	*0.8*	*0.7*	*11.8*	*0.5*	*10.4*
Non-OECD Africa and Middle East	**180.1**	**3.7**	**47.8**	**17.6**	**66.0**	**1.9**	**45.1**
Egypt	-	1.2	-	1.2	-	-	0.0
South Africa	173.5	-	47.6	13.0	64.0	1.9	42.1
Zimbabwe	3.1	0.0	0.1	0.8	0.9	0.1	1.2
Other Africa / Middle East	*3.5*	*2.5*	*0.1*	*2.6*	*1.1*	*0.0*	*1.8*
Non-OECD Americas	**17.0**	**9.3**	**4.7**	**11.2**	**4.3**	**0.2**	**7.4**
Brazil	7.7	8.0	-	9.6	2.6	-	4.6
Colombia	8.8	-	4.7	0.8	1.1	0.2	2.3
Other non-OECD Americas	*0.6*	*1.3*	-	*0.9*	*0.5*	*0.0*	*0.5*
Total World	**4380.2**	**359.2**	**363.7**	**564.1**	**2301.0**	**237.1**	**1211.7**

(1) Total coal comprises anthracite, coking coal, other bituminous coal, sub-bituminous coal and lignite.

(2) Fiscal year. See notes and definitions in Part I.

(3) Chile, Estonia, Iceland, Israel, Luxembourg, Norway, Slovak Republic, Slovenia and Switzerland.

Note: Steel industry consumption includes consumption in coke ovens. Power and heat stations column includes hard coal used in electricity and CHP production by public utilities and autoproducers, and in district heating. Other sectors includes consumption in non-ferrous industries, non-energy use, other transformation, losses and consumption in other sectors.

Source: *IEA/OECD Energy Statistics of OECD Countries , Energy Statistics of Non-OECD Countries*

World total coal supply and end-use 1990

(million tonnes)(1)

				End-use sectors			
	Production	Imports	Exports	Power and heat plants	Steel industry	Residential	Other Sectors
Total OECD	**2292.8**	**327.1**	**285.9**	**246.5**	**1672.0**	**49.3**	**355.2**
Australia(2)	204.6	-	103.4	6.3	82.3	0.0	6.7
Austria	2.4	3.6	0.0	2.3	3.6	0.4	0.4
Belgium	2.4	15.0	0.7	7.3	6.6	0.7	2.0
Canada	68.3	14.2	31.0	5.0	42.2	0.1	1.9
Czech Republic	101.4	2.3	14.2	11.1	48.6	13.1	18.9
Denmark	-	10.3	0.1	-	9.4	0.0	0.6
Finland	-	6.1	-	0.8	3.9	0.0	1.0
France	13.5	19.5	0.6	11.1	12.9	1.7	4.9
Germany	434.0	15.7	5.7	24.9	279.8	0.3	145.2
Greece	51.9	1.4	-	-	50.5	0.1	2.8
Hungary	17.8	1.8	0.0	1.0	14.5	2.1	2.7
Ireland	0.0	3.1	0.0	0.0	2.0	1.1	0.3
Italy	1.0	20.6	-	8.9	11.8	0.1	1.7
Japan(2)	8.0	106.9	0.0	66.8	31.1	-	19.5
Korea	17.2	23.7	-	11.9	7.7	-	25.2
Mexico	6.9	0.2	0.0	2.9	4.0	-	-
Netherlands	-	17.5	2.5	3.9	8.7	0.0	1.2
New Zealand	2.6	0.0	0.3	0.6	0.3	0.1	1.1
Poland	215.3	0.6	28.3	18.3	144.5	12.2	12.4
Portugal	0.3	4.7	-	0.3	3.3	0.0	0.8
Spain	35.7	10.5	0.0	4.8	39.5	0.5	2.3
Sweden	0.0	3.5	0.0	1.6	1.2	-	1.0
Turkey	47.4	5.6	-	4.7	30.4	9.1	10.1
United Kingdom	92.8	14.8	2.3	10.2	84.0	4.2	8.0
United States	933.6	2.4	96.0	37.0	709.0	2.3	70.8
Other OECD (3)	*35.6*	*23.2*	*0.7*	*4.8*	*40.3*	*1.2*	*13.8*
Non-OECD Europe and Eurasia	**812.5**	**123.0**	**144.8**	**117.8**	**487.2**	**29.5**	**174.2**
Russian Federation	371.9	53.2	59.2	54.8	225.4	20.2	82.4
Non-OECD Asia	**1350.6**	**44.6**	**23.9**	**169.3**	**472.3**	**173.4**	**537.1**
China, People's Republic of	1039.8	2.0	17.3	134.0	301.9	167.0	411.7
Hong Kong, China	-	8.9	-	-	9.9	-	-
India(2)	225.3	6.0	0.1	24.9	123.2	5.4	67.2
Indonesia	10.2	0.8	4.7	-	4.4	-	3.8
DPR of Korea	46.4	2.6	0.5	5.1	6.2	-	37.1
Taipei, Chinese	0.5	18.5	-	4.2	8.6	-	4.5
Other Asia	*28.5*	*5.7*	*1.3*	*1.1*	*18.2*	*0.9*	*12.8*
Non-OECD Africa and Middle East	**183.3**	**4.1**	**50.1**	**16.4**	**79.2**	**2.4**	**48.8**
Egypt	-	1.3	-	1.3	0.0	-	0.0
South Africa	174.8	-	49.9	12.3	74.3	2.3	45.3
Zimbabwe	5.3	0.1	0.1	0.9	2.7	0.1	1.8
Other Africa / Middle East	*3.1*	*2.7*	*0.1*	*2.0*	*2.2*	*0.0*	*1.7*
Non-OECD Americas	**28.5**	**11.7**	**15.4**	**12.3**	**4.5**	**0.2**	**5.7**
Brazil	4.6	10.1	-	10.3	2.8	-	2.3
Colombia	21.4	-	13.5	0.8	1.5	0.2	2.7
Other non-OECD Americas	*2.6*	*1.5*	*1.9*	*1.2*	*0.2*	*0.0*	*0.7*
Total World	**4667.7**	**510.4**	**520.0**	**562.3**	**2715.2**	**254.8**	**1121.1**

(1) Total coal comprises anthracite, coking coal, other bituminous coal, sub-bituminous coal and lignite.

(2) Fiscal year. See notes and definitions in Part I.

(3) Chile, Estonia, Iceland, Israel, Luxembourg, Norway, Slovak Republic, Slovenia and Switzerland.

Note: Steel industry consumption includes consumption in coke ovens. Power and heat stations column includes hard coal used in electricity and CHP production by public utilities and autoproducers, and in district heating. Other sectors includes consumption in non-ferrous industries, non-energy use, other transformation, losses and consumption in other sectors.

Source: *IEA/OECD Energy Statistics of OECD Countries , Energy Statistics of Non-OECD Countries*

World total coal supply and end-use 1995

(million tonnes)[1]

	Production	Imports	Exports	End-use sectors: Power and heat plants	Steel industry	Residential	Other Sectors
Total OECD	**2050.7**	**369.6**	**305.5**	**213.0**	**1692.6**	**36.8**	**187.5**
Australia[2]	241.8	-	136.7	6.1	89.9	0.0	6.6
Austria	1.3	3.0	0.0	1.9	2.6	0.2	0.4
Belgium	0.6	14.3	0.8	4.9	6.2	0.5	1.6
Canada	75.0	9.7	34.0	4.2	46.0	0.1	2.1
Czech Republic	74.9	2.7	13.5	7.7	45.9	4.1	8.7
Denmark	-	13.0	0.0	-	10.4	0.0	0.5
Finland	-	5.8	0.0	1.7	4.1	0.0	0.8
France	9.9	13.2	0.4	9.4	10.1	1.0	3.6
Germany	251.6	17.2	1.9	13.8	225.4	0.3	31.6
Greece	57.7	1.4	-	-	56.4	0.1	2.0
Hungary	14.8	2.2	-	1.4	13.8	1.0	0.7
Ireland	0.0	2.9	0.0	-	2.3	0.4	0.1
Italy	0.2	18.5	-	7.0	8.4	0.1	2.1
Japan[2]	6.3	127.4	0.0	62.9	47.6	-	23.1
Korea	5.7	45.8	-	14.1	16.7	-	13.6
Mexico	9.3	0.8	-	2.7	7.6	-	-
Netherlands	-	17.2	3.3	4.1	9.3	0.0	0.9
New Zealand	3.6	-	1.3	0.7	0.3	0.1	1.1
Poland	200.7	1.5	32.2	16.6	118.1	14.4	22.6
Portugal	-	6.0	-	0.5	4.6	-	0.6
Spain	28.3	13.9	-	3.3	38.3	0.4	0.6
Sweden	-	3.5	0.0	1.7	0.9	-	0.7
Turkey	55.1	6.0	-	4.2	41.1	7.9	7.9
United Kingdom	53.0	15.9	0.9	8.1	59.6	2.8	6.4
United States	937.1	6.5	80.3	31.4	796.0	2.8	40.8
Other OECD[3]	*23.8*	*21.1*	*0.2*	*4.5*	*31.1*	*0.5*	*8.2*
Non-OECD Europe and Eurasia	**531.7**	**52.4**	**52.2**	**84.5**	**373.0**	**18.2**	**68.6**
Russian Federation	245.7	22.7	28.4	42.7	171.4	13.9	24.9
Non-OECD Asia	**1739.5**	**61.8**	**63.9**	**270.4**	**759.9**	**141.1**	**625.8**
China, People's Republic of	1338.7	1.6	28.6	232.9	500.5	134.8	510.9
Hong Kong, China	-	9.1	-	-	9.7	-	-
India[2]	290.4	12.5	0.7	31.2	200.0	5.4	58.3
Indonesia	41.8	-	31.3	-	6.5	-	5.7
DPR of Korea	31.3	1.0	0.4	1.0	4.6	-	26.3
Taipei, Chinese	0.2	28.8	-	4.2	16.4	-	5.2
Other Asia	*37.0*	*8.8*	*2.9*	*1.1*	*22.1*	*0.9*	*19.4*
Non-OECD Africa and Middle East	**214.2**	**5.4**	**59.9**	**12.0**	**92.7**	**2.6**	**49.9**
Egypt	-	1.5	-	1.5	-	-	0.1
South Africa	206.2	0.4	59.7	7.7	86.4	2.6	47.6
Zimbabwe	4.7	-	0.2	0.7	2.8	0.0	0.9
Other Africa / Middle East	*3.3*	*3.5*	*0.0*	*2.0*	*3.4*	*0.0*	*1.3*
Non-OECD Americas	**35.3**	**13.5**	**22.6**	**13.2**	**6.3**	**0.2**	**5.2**
Brazil	5.2	11.8	-	11.4	3.7	-	2.0
Colombia	25.7	-	18.3	1.1	1.8	0.2	2.6
Other non-OECD Americas	*4.4*	*1.7*	*4.3*	*0.6*	*0.8*	*0.0*	*0.6*
Total World	**4571.4**	**502.7**	**504.1**	**593.0**	**2924.6**	**198.9**	**937.1**

(1) Total coal comprises anthracite, coking coal, other bituminous coal, sub-bituminous coal and lignite.

(2) Fiscal year. See notes and definitions in Part I.

(3) Chile, Estonia, Iceland, Israel, Luxembourg, Norway, Slovak Republic, Slovenia and Switzerland.

Note: Steel industry consumption includes consumption in coke ovens. Power and heat stations column includes hard coal used in electricity and CHP production by public utilities and autoproducers, and in district heating. Other sectors includes consumption in non-ferrous industries, non-energy use, other transformation, losses and consumption in other sectors.

Source: *IEA/OECD Energy Statistics of OECD Countries , Energy Statistics of Non-OECD Countries*

World total coal supply and end-use 2000
(million tonnes)[1]

	Production	Imports	Exports	End-use sectors			
				Power and heat plants	Steel industry	Residential	Other Sectors
Total OECD	**2029.4**	**468.5**	**318.4**	**195.6**	**1864.5**	**22.0**	**174.2**
Australia[2]	306.7	-	187.0	4.9	117.4	-	5.8
Austria	1.2	3.5	-	1.9	2.7	0.1	0.5
Belgium	0.4	11.6	1.4	4.3	4.4	0.3	1.7
Canada	69.2	23.2	32.1	4.2	55.8	0.1	2.5
Czech Republic	65.2	1.1	8.7	5.6	45.1	1.6	8.8
Denmark	-	6.4	0.1	-	6.2	0.0	0.4
Finland	-	5.1	-	1.3	3.5	0.0	0.4
France	4.1	19.0	0.1	7.0	10.7	0.7	4.2
Germany	205.1	29.7	0.3	11.6	209.2	0.1	17.7
Greece	63.9	1.2	0.1	-	63.9	0.0	1.8
Hungary	14.0	1.9	0.0	1.3	13.0	0.5	0.4
Ireland	-	2.8	0.0	-	2.3	0.4	0.2
Italy	0.0	19.0	0.0	6.7	9.5	0.0	1.8
Japan[2]	3.0	150.3	0.0	60.0	64.9	-	25.6
Korea	8.3	64.9	-	17.1	39.0	-	15.7
Mexico	11.3	2.4	0.0	2.7	9.6	-	-
Netherlands	-	22.6	9.6	3.1	8.6	0.0	1.1
New Zealand	3.5	0.0	1.6	0.7	0.5	0.1	0.9
Poland	162.8	1.5	23.3	13.2	110.8	7.6	12.8
Portugal	-	6.4	-	0.5	5.2	-	0.5
Spain	23.5	21.6	-	3.8	41.0	0.3	1.2
Sweden	-	3.1	0.0	1.8	0.5	-	0.7
Turkey	63.3	13.0	-	4.3	54.5	5.7	15.5
United Kingdom	31.2	23.4	0.7	8.2	46.9	1.9	3.0
United States	971.6	11.4	53.1	27.4	905.5	2.2	45.4
Other OECD[3]	*21.2*	*23.3*	*0.6*	*4.0*	*33.9*	*0.3*	*5.7*
Non-OECD Europe and Eurasia	**496.0**	**42.8**	**75.2**	**83.5**	**334.1**	**19.8**	**31.5**
Russian Federation	240.3	25.5	38.3	42.1	163.3	16.0	9.1
Non-OECD Asia	**1839.1**	**91.1**	**117.0**	**254.6**	**1009.3**	**84.0**	**547.0**
China, People's Republic of	1353.8	2.2	55.1	221.5	644.9	77.2	434.4
Hong Kong, China	-	6.1	-	-	6.9	-	-
India[2]	335.7	20.9	1.3	26.3	272.7	5.6	52.4
Indonesia	79.4	0.1	56.8	-	16.0	-	8.8
DPR of Korea	29.7	-	0.4	-	4.5	-	24.9
Taipei, Chinese	0.1	45.4	-	5.9	33.3	-	6.4
Other Asia	*40.4*	*16.4*	*3.5*	*1.0*	*31.0*	*1.2*	*20.1*
Non-OECD Africa and Middle East	**231.6**	**9.5**	**70.4**	**11.9**	**104.9**	**1.6**	**54.1**
Egypt	0.1	1.8	0.1	1.7	-	-	0.1
South Africa	224.2	1.1	69.9	6.8	98.1	1.5	50.7
Zimbabwe	4.3	0.0	0.0	1.0	2.0	0.0	1.2
Other Africa / Middle East	*3.1*	*6.5*	*0.5*	*2.3*	*4.8*	*0.0*	*2.1*
Non-OECD Americas	**53.2**	**15.3**	**43.3**	**14.1**	**8.0**	**0.1**	**5.9**
Brazil	6.8	13.2	-	12.1	6.2	-	2.0
Colombia	38.2	-	35.4	1.2	0.8	0.1	3.0
Other non-OECD Americas	*8.2*	*2.0*	*7.9*	*0.8*	*1.0*	-	*0.9*
Total World	**4649.3**	**627.1**	**624.4**	**559.7**	**3320.8**	**127.4**	**812.7**

(1) Total coal comprises anthracite, coking coal, other bituminous coal, sub-bituminous coal and lignite.

(2) Fiscal year. See notes and definitions in Part I.

(3) Chile, Estonia, Iceland, Israel, Luxembourg, Norway, Slovak Republic, Slovenia and Switzerland.

Note: Steel industry consumption includes consumption in coke ovens. Power and heat stations column includes hard coal used in electricity and CHP production by public utilities and autoproducers, and in district heating. Other sectors includes consumption in non-ferrous industries, non-energy use, other transformation, losses and consumption in other sectors.

Source: *IEA/OECD Energy Statistics of OECD Countries , Energy Statistics of Non-OECD Countries*

World total coal supply and end-use 2005
(million tonnes)[1]

	Production	Imports	Exports	End-use sectors: Power and heat plants	End-use sectors: Steel industry	End-use sectors: Residential	End-use sectors: Other Sectors
Total OECD	**2123.1**	**565.9**	**345.9**	**182.0**	**1863.7**	**19.0**	**285.6**
Australia[2]	370.6	-	232.3	4.6	127.3	0.0	6.3
Austria	-	4.4	0.0	1.9	3.0	0.1	0.3
Belgium	0.1	9.0	1.2	3.5	3.0	0.2	0.9
Canada	65.3	20.7	28.2	4.3	52.5	0.1	2.7
Czech Republic	62.0	1.3	6.5	5.3	44.2	1.3	6.1
Denmark	-	6.0	0.1	-	5.9	-	0.4
Finland	-	4.7	-	1.3	2.9	0.0	0.3
France	0.6	19.9	0.3	6.4	10.3	0.5	4.3
Germany	205.9	37.1	0.3	11.6	212.7	0.2	17.9
Greece	69.4	0.6	0.0	-	69.4	0.0	1.2
Hungary	9.6	2.2	0.4	0.8	10.0	0.5	0.3
Ireland	-	3.1	0.0	-	2.3	0.4	0.4
Italy	0.1	24.2	-	5.9	16.4	0.0	1.9
Japan[2]	-	177.0	0.0	63.0	92.3	-	24.7
Korea	2.8	76.8	-	16.6	52.7	-	13.3
Mexico	10.8	7.3	0.0	2.4	14.9	-	0.2
Netherlands	-	20.5	7.4	3.2	8.3	0.0	1.6
New Zealand	5.3	1.1	2.3	0.8	2.4	0.0	1.0
Poland	159.5	3.4	19.4	11.8	112.0	8.2	8.6
Portugal	-	5.3	-	-	5.4	-	0.0
Spain	19.5	24.8	-	3.8	39.3	0.3	1.1
Sweden	-	3.2	0.0	2.0	0.4	-	0.9
Turkey	58.3	17.4	-	4.3	52.6	6.5	13.4
United Kingdom	20.5	44.0	0.5	5.7	52.5	0.6	3.1
United States	1038.6	27.6	45.3	18.8	833.3	-	169.4
Other OECD[3]	*24.1*	*24.6*	*1.7*	*4.1*	*37.4*	*0.2*	*5.4*
Non-OECD Europe and Eurasia	**548.8**	**43.4**	**115.3**	**88.5**	**333.8**	**7.6**	**44.2**
Russian Federation	282.9	22.6	86.6	45.3	153.2	5.2	11.0
Non-OECD Asia	**3014.5**	**166.9**	**225.5**	**461.2**	**1648.3**	**98.2**	**749.7**
China, People's Republic of	2299.7	26.2	71.7	412.3	1168.1	90.5	597.6
Hong Kong, China	-	10.8	-	-	10.0	-	0.9
India[2]	437.3	38.6	2.0	42.5	349.3	5.6	66.2
Indonesia	170.5	0.1	128.6	-	25.6	-	15.8
DPR of Korea	34.6	-	2.8	-	4.7	-	27.1
Taipei, Chinese	-	60.3	-	5.9	46.0	-	8.2
Other Asia	*72.4*	*30.9*	*20.4*	*0.6*	*44.6*	*2.0*	*34.0*
Non-OECD Africa and Middle East	**252.2**	**11.9**	**72.0**	**12.4**	**118.2**	**4.9**	**56.5**
Egypt	0.0	1.8	0.0	1.7	-	-	0.1
South Africa	245.0	1.9	71.4	7.6	109.2	4.9	53.7
Zimbabwe	3.6	0.1	-	0.8	2.3	0.0	0.5
Other Africa / Middle East	*3.5*	*8.2*	*0.5*	*2.2*	*6.6*	*0.0*	*2.2*
Non-OECD Americas	**72.6**	**17.2**	**60.8**	**14.8**	**7.9**	**0.1**	**5.2**
Brazil	6.3	13.7	0.0	12.9	5.1	-	1.9
Colombia	59.1	0.0	53.6	1.0	1.0	0.1	2.3
Other non-OECD Americas	*7.3*	*3.5*	*7.2*	*0.9*	*1.8*	-	*1.0*
Total World	**6011.1**	**805.3**	**819.5**	**758.9**	**3971.9**	**129.8**	**1141.2**

(1) Total coal comprises anthracite, coking coal, other bituminous coal, sub-bituminous coal and lignite.

(2) Fiscal year. See notes and definitions in Part I.

(3) Chile, Estonia, Iceland, Israel, Luxembourg, Norway, Slovak Republic, Slovenia and Switzerland.

Note: Steel industry consumption includes consumption in coke ovens. Power and heat stations column includes hard coal used in electricity and CHP production by public utilities and autoproducers, and in district heating. Other sectors includes consumption in non-ferrous industries, non-energy use, other transformation, losses and consumption in other sectors.

Source: *IEA/OECD Energy Statistics of OECD Countries , Energy Statistics of Non-OECD Countries*

World total coal supply and end-use 2006

(million tonnes)(1)

				End-use sectors			
	Production	Imports	Exports	Power and heat plants	Steel industry	Residential	Other Sectors
Total OECD	**2148.4**	**600.6**	**347.8**	**187.0**	**1880.5**	**20.8**	**264.4**
Australia(2)	374.7	-	232.5	4.6	130.2	0.0	6.0
Austria	-	4.3	-	1.9	2.7	0.1	0.4
Belgium	0.0	8.3	1.1	3.7	2.6	0.2	1.0
Canada	66.0	20.7	28.2	4.3	50.2	0.1	2.8
Czech Republic	62.9	2.0	7.9	5.2	44.0	1.8	6.4
Denmark	-	8.7	0.1	-	8.9	-	0.4
Finland	-	6.7	-	1.2	6.4	0.0	0.2
France	0.5	20.4	0.1	6.6	8.6	0.5	4.3
Germany	200.1	42.2	0.2	12.0	212.9	0.2	18.6
Greece	64.8	0.4	0.0	-	64.0	0.0	1.1
Hungary	10.0	2.4	0.5	1.3	10.0	0.5	0.3
Ireland	-	2.7	0.0	-	2.0	0.3	0.3
Italy	0.0	24.6	-	6.4	16.7	0.0	1.7
Japan(2)	-	179.1	0.0	63.2	89.5	-	25.1
Korea	2.8	79.7	-	16.7	55.1	-	13.9
Mexico	11.5	7.6	0.0	2.5	14.7	-	0.3
Netherlands	-	22.9	9.9	3.1	8.2	0.0	1.4
New Zealand	5.7	1.2	2.7	0.8	2.3	0.0	1.1
Poland	156.1	5.3	16.7	13.4	113.3	9.3	8.5
Portugal	-	5.8	0.0	-	5.4	-	0.0
Spain	18.4	23.7	-	3.8	34.7	0.3	1.0
Sweden	-	3.1	0.0	2.3	0.6	-	0.9
Turkey	64.3	20.3	-	4.8	55.2	6.8	16.8
United Kingdom	18.5	50.5	0.4	6.1	57.9	0.6	3.1
United States	1067.9	32.9	45.0	19.0	847.0	-	143.1
Other OECD(3)	*24.3*	*25.1*	*2.3*	*4.1*	*37.3*	*0.2*	*5.9*
Non-OECD Europe and Eurasia	**568.8**	**49.1**	**124.7**	**88.7**	**363.1**	**8.8**	**35.8**
Russian Federation	284.6	26.1	91.9	44.7	158.3	5.9	10.9
Non-OECD Asia	**3281.3**	**190.8**	**282.6**	**509.6**	**1838.7**	**99.1**	**782.8**
China, People's Republic of	2475.8	38.1	63.2	456.4	1333.3	91.3	616.5
Hong Kong, China	-	11.4	-	-	9.9	-	1.5
India(2)	462.1	43.1	1.6	46.9	369.7	5.6	70.9
Indonesia	233.3	0.1	183.2	-	28.3	-	21.3
DPR of Korea	35.1	-	2.5	-	4.9	-	27.8
Taipei, Chinese	-	62.3	-	5.8	47.9	-	8.9
Other Asia	*75.0*	*35.8*	*32.2*	*0.5*	*45.0*	*2.1*	*35.8*
Non-OECD Africa and Middle East	**251.8**	**12.6**	**69.3**	**12.1**	**119.2**	**5.6**	**57.8**
Egypt	-	2.0	-	1.9	-	-	0.1
South Africa	244.8	1.9	68.7	7.1	110.7	5.5	54.5
Zimbabwe	3.4	0.0	-	0.8	1.5	0.0	1.1
Other Africa / Middle East	*3.5*	*8.8*	*0.5*	*2.3*	*6.9*	*0.0*	*2.2*
Non-OECD Americas	**79.6**	**16.5**	**69.7**	**14.3**	**8.7**	**0.1**	**4.5**
Brazil	5.9	13.4	-	12.5	5.6	-	1.9
Colombia	65.6	-	62.0	0.9	1.1	0.1	1.9
Other non-OECD Americas	*8.1*	*3.1*	*7.7*	*0.8*	*1.9*	-	*0.7*
Total World	**6329.9**	**869.6**	**894.1**	**811.7**	**4210.2**	**134.3**	**1145.3**

(1) Total coal comprises anthracite, coking coal, other bituminous coal, sub-bituminous coal and lignite.

(2) Fiscal year. See notes and definitions in Part I.

(3) Chile, Estonia, Iceland, Israel, Luxembourg, Norway, Slovak Republic, Slovenia and Switzerland.

Note: Steel industry consumption includes consumption in coke ovens. Power and heat stations column includes hard coal used in electricity and CHP production by public utilities and autoproducers, and in district heating. Other sectors includes consumption in non-ferrous industries, non-energy use, other transformation, losses and consumption in other sectors.

Source: *IEA/OECD Energy Statistics of OECD Countries , Energy Statistics of Non-OECD Countries*

World total coal supply and end-use 2007
(million tonnes)(1)

	Production	Imports	Exports	End-use sectors: Power and heat plants	Steel industry	Residential	Other Sectors
Total OECD	**2160.9**	**617.5**	**370.2**	**187.3**	**1893.3**	**21.5**	**301.8**
Australia(2)	391.0	-	244.4	4.6	131.0	0.0	6.1
Austria	-	4.5	0.0	2.0	2.0	0.0	0.4
Belgium	-	7.7	1.5	3.5	2.2	0.2	1.0
Canada	69.4	18.4	30.8	4.3	53.4	0.1	2.9
Czech Republic	62.6	2.6	8.0	4.9	47.0	1.2	4.8
Denmark	-	8.1	0.2	-	7.5	-	0.4
Finland	-	6.7	-	1.2	5.5	-	0.2
France	0.4	19.0	0.2	6.7	9.3	0.5	4.5
Germany	204.6	46.3	0.3	11.5	220.1	0.8	18.2
Greece	66.3	0.6	0.0	-	65.9	0.0	1.2
Hungary	9.8	2.7	0.4	1.5	10.1	0.3	0.2
Ireland	-	2.3	0.0	-	1.9	0.3	0.3
Italy	0.2	25.0	-	6.2	17.0	0.0	1.9
Japan(2)	-	187.0	0.0	63.8	94.3	-	26.4
Korea	2.9	88.3	-	17.0	60.9	-	14.7
Mexico	12.5	5.5	0.0	2.5	14.7	-	0.3
Netherlands	-	26.1	11.9	3.1	8.7	0.0	1.7
New Zealand	4.8	0.7	2.0	0.8	1.2	0.0	1.2
Poland	145.9	5.9	11.9	14.3	109.8	8.8	9.2
Portugal	-	4.8	0.0	0.0	4.4	-	0.3
Spain	17.2	24.4	1.0	3.9	36.8	0.3	1.4
Sweden	-	3.2	0.0	2.1	0.4	-	0.9
Turkey	75.4	22.9	-	4.5	66.4	8.1	19.1
United Kingdom	17.0	43.4	0.5	6.2	53.0	0.6	3.2
United States	1053.0	33.0	53.7	18.7	829.0	-	175.6
Other OECD (3)	*27.9*	*28.4*	*3.4*	*4.1*	*40.8*	*0.2*	*5.8*
Non-OECD Europe and Eurasia	**575.6**	**51.7**	**129.3**	**91.7**	**356.9**	**8.1**	**44.3**
Russian Federation	289.0	23.7	98.6	46.7	149.9	4.7	10.9
Non-OECD Asia	**3485.6**	**221.8**	**290.6**	**534.0**	**2015.1**	**96.4**	**820.6**
China, People's Republic of	2634.1	51.0	53.1	482.3	1459.4	88.3	639.6
Hong Kong, China	-	12.3	-	-	10.6	-	1.6
India(2)	491.1	49.8	1.6	45.3	406.2	5.8	80.1
Indonesia	248.8	0.1	194.9	-	32.4	-	29.1
DPR of Korea	30.3	0.0	3.7	-	4.0	-	22.7
Taipei, Chinese	-	65.2	-	5.7	49.5	-	10.1
Other Asia	*81.2*	*43.4*	*37.2*	*0.6*	*53.0*	*2.2*	*37.4*
Non-OECD Africa and Middle East	**254.4**	**12.5**	**67.5**	**12.5**	**122.0**	**7.1**	**61.8**
Egypt	-	2.0	-	1.9	-	-	0.1
South Africa	247.7	1.8	67.0	7.5	114.0	7.1	58.0
Zimbabwe	3.3	0.0	-	0.8	1.3	0.0	1.3
Other Africa / Middle East	*3.4*	*8.7*	*0.5*	*2.3*	*6.7*	*0.0*	*2.4*
Non-OECD Americas	**82.2**	**18.4**	**71.7**	**15.6**	**8.5**	**0.1**	**5.0**
Brazil	6.0	14.9	-	13.8	5.1	-	2.0
Colombia	69.9	-	65.3	0.9	1.2	0.1	2.0
Other non-OECD Americas	*6.3*	*3.5*	*6.4*	*1.0*	*2.1*	-	*1.0*
Total World	**6558.6**	**921.9**	**929.2**	**841.2**	**4395.8**	**133.2**	**1233.6**

(1) Total coal comprises anthracite, coking coal, other bituminous coal, sub-bituminous coal and lignite.

(2) Fiscal year. See notes and definitions in Part I.

(3) Chile, Estonia, Iceland, Israel, Luxembourg, Norway, Slovak Republic, Slovenia and Switzerland.

Note: Steel industry consumption includes consumption in coke ovens. Power and heat stations column includes hard coal used in electricity and CHP production by public utilities and autoproducers, and in district heating. Other sectors includes consumption in non-ferrous industries, non-energy use, other transformation, losses and consumption in other sectors.

Source: *IEA/OECD Energy Statistics of OECD Countries , Energy Statistics of Non-OECD Countries*

World total coal supply and end-use 2008
(million tonnes)[1]

	Production	Imports	Exports	End-use sectors: Power and heat plants	Steel industry	Residential	Other Sectors
Total OECD	**2164.3**	**616.0**	**393.4**	**186.3**	**1970.0**	**23.2**	**165.1**
Australia[2]	392.3	-	252.2	4.6	129.9	0.0	6.2
Austria	-	4.2	0.0	2.0	1.7	0.0	0.3
Belgium	-	7.7	1.2	2.9	2.0	0.2	1.0
Canada	67.7	20.5	33.0	4.3	50.7	0.1	2.9
Czech Republic	60.2	2.3	7.6	5.0	43.7	1.2	4.9
Denmark	-	7.6	0.2	-	6.7	-	0.3
Finland	-	5.7	-	1.2	3.6	0.0	0.2
France	0.3	21.4	0.2	6.6	8.3	0.5	4.0
Germany	194.4	45.5	0.5	11.3	208.2	0.7	19.3
Greece	65.7	0.7	0.0	-	64.2	0.0	0.9
Hungary	9.4	2.7	0.2	1.4	9.8	0.4	0.2
Ireland	-	2.6	0.0	-	1.7	0.3	0.3
Italy	0.1	25.1	-	5.9	16.9	0.0	2.1
Japan[2]	-	184.1	0.0	59.9	90.2	-	24.2
Korea	2.8	99.6	-	18.8	68.7	-	15.6
Mexico	11.4	4.6	0.0	2.9	10.9	-	0.2
Netherlands	-	21.2	7.5	3.1	8.2	0.0	1.4
New Zealand	4.8	0.6	2.6	0.7	2.0	0.0	1.3
Poland	144.0	10.4	8.5	13.5	107.6	9.3	9.6
Portugal	-	3.8	0.0	0.0	4.0	-	0.1
Spain	10.2	21.0	1.8	3.6	21.4	0.3	1.1
Sweden	-	2.9	0.0	2.1	0.3	-	0.9
Turkey	79.4	19.5	-	5.9	71.9	9.2	11.6
United Kingdom	18.1	43.9	0.6	6.2	48.3	0.7	3.2
United States	1075.9	31.0	74.0	20.7	948.8	-	46.9
Other OECD[3]	*27.6*	*27.7*	*3.4*	*3.8*	*40.2*	*0.2*	*6.3*
Non-OECD Europe and Eurasia	**616.5**	**58.3**	**136.5**	**85.3**	**380.1**	**11.2**	**55.9**
Russian Federation	305.0	31.3	98.1	42.4	167.9	5.5	13.6
Non-OECD Asia	**3620.0**	**228.2**	**274.7**	**561.5**	**2084.1**	**93.4**	**844.0**
China, People's Republic of	2734.4	40.3	45.3	502.9	1503.8	85.2	651.4
Hong Kong, China	-	11.3	-	-	9.7	-	1.6
India[2]	525.2	65.2	1.7	51.4	431.2	5.9	91.6
Indonesia	248.8	0.1	199.9	-	32.0	-	22.5
DPR of Korea	32.3	0.0	2.6	-	4.4	-	25.3
Taipei, Chinese	-	63.8	-	5.9	47.9	-	9.1
Other Asia	*79.3*	*47.3*	*25.2*	*1.4*	*55.1*	*2.3*	*42.5*
Non-OECD Africa and Middle East	**258.2**	**12.7**	**58.4**	**11.7**	**136.1**	**7.8**	**60.0**
Egypt	-	2.0	-	1.9	-	-	0.1
South Africa	252.2	2.4	57.9	7.5	128.6	7.8	56.5
Zimbabwe	2.5	0.0	-	0.3	1.0	0.0	1.1
Other Africa / Middle East	*3.4*	*8.2*	*0.6*	*1.9*	*6.5*	*0.0*	*2.2*
Non-OECD Americas	**85.3**	**18.7**	**73.3**	**16.0**	**8.0**	**0.1**	**4.8**
Brazil	6.6	15.3	-	13.9	4.8	-	1.9
Colombia	73.5	-	68.5	1.0	1.1	0.1	2.0
Other non-OECD Americas	*5.2*	*3.4*	*4.8*	*1.1*	*2.0*	-	*1.0*
Total World	**6744.3**	**933.9**	**936.4**	**860.9**	**4578.2**	**135.6**	**1129.8**

(1) Total coal comprises anthracite, coking coal, other bituminous coal, sub-bituminous coal and lignite.

(2) Fiscal year. See notes and definitions in Part I.

(3) Chile, Estonia, Iceland, Israel, Luxembourg, Norway, Slovak Republic, Slovenia and Switzerland.

Note: Steel industry consumption includes consumption in coke ovens. Power and heat stations column includes hard coal used in electricity and CHP production by public utilities and autoproducers, and in district heating. Other sectors includes consumption in non-ferrous industries, non-energy use, other transformation, losses and consumption in other sectors.

Source: *IEA/OECD Energy Statistics of OECD Countries , Energy Statistics of Non-OECD Countries*

World total coal supply and end-use 2009
(million tonnes)(1)

	Production	Imports	Exports	End-use sectors			
				Power and heat plants	Steel industry	Residential	Other Sectors
Total OECD	**2057.8**	**545.3**	**372.9**	**153.7**	**1838.6**	**24.9**	**144.9**
Australia(2)	407.9	-	261.7	3.6	135.5	-	5.6
Austria	-	3.2	0.0	1.8	1.2	0.0	0.3
Belgium	-	5.0	0.9	2.1	1.7	0.4	0.5
Canada	62.9	12.9	28.6	3.0	43.2	0.1	3.2
Czech Republic	56.4	2.0	7.7	3.6	41.1	1.2	5.2
Denmark	-	6.7	0.1	-	6.6	-	0.2
Finland	-	5.9	-	1.1	4.5	0.0	0.1
France	0.1	15.5	0.1	4.7	8.1	0.3	2.7
Germany	183.6	38.5	0.3	9.2	196.4	0.6	16.9
Greece	64.9	0.3	0.0	-	65.2	0.0	0.3
Hungary	9.0	1.7	0.0	1.0	9.2	0.3	0.1
Ireland	-	2.1	0.0	-	1.3	0.4	0.2
Italy	0.1	19.5	-	3.7	15.2	0.0	1.1
Japan(2)	-	163.7	0.0	56.3	86.4	-	22.1
Korea	2.5	103.0	-	16.2	77.7	-	13.5
Mexico	10.5	6.0	0.0	2.4	13.7	-	0.2
Netherlands	-	19.9	4.7	2.5	8.4	0.0	1.0
New Zealand	4.6	0.7	2.1	0.8	1.3	0.0	1.0
Poland	135.2	10.8	8.5	9.6	103.1	9.3	8.7
Portugal	-	5.1	0.1	0.0	4.6	-	0.0
Spain	9.4	17.0	1.4	2.4	15.7	0.3	0.6
Sweden	-	1.9	0.0	1.6	0.3	-	0.6
Turkey	79.5	20.4	-	5.4	69.4	11.2	12.6
United Kingdom	17.9	38.2	0.6	5.1	40.2	0.7	2.8
United States	987.6	20.5	53.6	14.4	851.5	-	39.0
Other OECD (3)	*25.7*	*24.7*	*2.4*	*3.3*	*37.0*	*0.2*	*6.3*
Non-OECD Europe and Eurasia	**569.3**	**40.1**	**140.8**	**85.2**	**339.8**	**8.9**	**47.1**
Russian Federation	276.0	24.1	106.4	47.0	142.2	3.5	11.4
Non-OECD Asia	**3873.9**	**338.5**	**296.0**	**606.5**	**2210.3**	**93.9**	**879.0**
China, People's Republic of	2895.3	125.8	22.3	539.3	1593.3	85.4	664.0
Hong Kong, China	-	12.3	-	-	10.2	-	2.2
India(2)	566.1	96.2	2.5	61.3	465.8	6.0	117.0
Indonesia	291.2	0.1	233.4	-	34.0	-	19.8
DPR of Korea	31.6	0.0	3.0	-	4.3	-	24.3
Taipei, Chinese	-	58.6	-	5.2	46.9	-	8.2
Other Asia	*89.7*	*45.5*	*34.7*	*0.7*	*55.8*	*2.5*	*43.5*
Non-OECD Africa and Middle East	**255.0**	**11.1**	**52.5**	**10.2**	**125.8**	**7.5**	**63.2**
Egypt	-	1.8	-	1.7	-	-	0.1
South Africa	249.5	2.0	52.0	6.5	118.7	7.5	59.5
Zimbabwe	2.7	0.0	-	0.3	1.0	0.0	1.3
Other Africa / Middle East	*2.8*	*7.3*	*0.5*	*1.7*	*6.0*	*0.0*	*2.3*
Non-OECD Americas	**82.2**	**16.1**	**69.8**	**14.6**	**7.6**	**0.1**	**5.0**
Brazil	5.7	12.7	-	11.9	4.0	-	1.6
Colombia	72.8	-	66.8	1.6	1.8	0.1	2.3
Other non-OECD Americas	*3.7*	*3.4*	*3.0*	*1.1*	*1.9*	-	*1.2*
Total World	**6838.2**	**951.2**	**932.0**	**870.1**	**4522.0**	**135.4**	**1139.3**

(1) Total coal comprises anthracite, coking coal, other bituminous coal, sub-bituminous coal and lignite.

(2) Fiscal year. See notes and definitions in Part I.

(3) Chile, Estonia, Iceland, Israel, Luxembourg, Norway, Slovak Republic, Slovenia and Switzerland.

Note: Steel industry consumption includes consumption in coke ovens. Power and heat stations column includes hard coal used in electricity and CHP production by public utilities and autoproducers, and in district heating. Other sectors includes consumption in non-ferrous industries, non-energy use, other transformation, losses and consumption in other sectors.

Source: *IEA/OECD Energy Statistics of OECD Countries , Energy Statistics of Non-OECD Countries*

World total coal supply and end-use 2010
(million tonnes)[1]

	Production	Imports	Exports	End-use sectors			
				Power and heat plants	Steel industry	Residential	Other Sectors
Total OECD	**2073.4**	**580.2**	**431.0**	**182.9**	**1875.4**	**27.9**	**159.6**
Australia[2]	424.4	-	292.6	4.2	128.6	-	5.4
Austria	-	3.7	0.0	2.0	1.6	0.0	0.3
Belgium	-	5.8	0.6	2.6	1.5	0.2	0.8
Canada	67.9	12.6	33.4	4.0	43.7	0.1	2.9
Czech Republic	55.2	2.1	7.3	3.6	42.8	1.3	5.2
Denmark	-	4.6	0.1	-	6.4	-	0.2
Finland	-	5.9	-	1.2	5.6	0.0	0.2
France	0.3	17.6	0.1	4.9	7.6	0.3	3.7
Germany	182.3	45.7	0.3	11.4	198.3	0.5	20.8
Greece	56.5	0.7	-	-	57.8	0.0	0.5
Hungary	9.1	2.1	0.0	1.4	9.1	0.4	0.1
Ireland	-	1.6	0.0	-	1.5	0.4	0.2
Italy	0.1	22.1	0.0	5.5	15.1	0.0	1.2
Japan[2]	-	185.4	0.0	61.1	90.0	-	25.4
Korea	2.1	118.6	-	21.0	83.2	-	15.3
Mexico	12.8	7.7	0.0	2.7	14.8	-	0.2
Netherlands	-	20.5	5.9	2.9	7.9	0.0	1.1
New Zealand	5.3	0.3	2.4	0.8	0.7	0.0	1.1
Poland	133.2	13.6	10.1	13.1	106.2	10.4	9.2
Portugal	-	2.8	0.1	0.0	2.6	-	0.1
Spain	8.4	12.8	1.5	2.8	11.0	0.3	0.9
Sweden	-	3.3	0.0	2.0	0.4	-	0.7
Turkey	73.4	21.3	-	6.6	62.8	13.1	13.0
United Kingdom	18.4	26.5	0.7	5.8	42.0	0.7	3.0
United States	996.1	17.6	74.1	19.8	892.2	-	41.7
Other OECD[3]	*27.7*	*25.4*	*1.7*	*3.4*	*42.1*	*0.2*	*6.6*
Non-OECD Europe and Eurasia	**621.6**	**47.2**	**171.1**	**90.1**	**351.2**	**8.2**	**52.4**
Russian Federation	321.7	25.5	132.8	49.8	148.6	2.8	12.2
Non-OECD Asia	**4171.8**	**408.3**	**336.2**	**661.7**	**2340.0**	**94.5**	**887.1**
China, People's Republic of	3140.2	163.1	21.3	580.5	1698.0	85.7	663.5
Hong Kong, China	-	10.3	-	-	8.8	-	1.5
India[2]	570.4	115.7	1.9	73.4	481.4	6.0	116.1
Indonesia	325.0	0.1	267.2	-	35.8	-	24.2
DPR of Korea	32.0	0.1	4.6	-	4.1	-	23.4
Taipei, Chinese	-	63.2	-	7.0	46.2	-	9.6
Other Asia	*104.2*	*55.9*	*41.2*	*0.7*	*65.8*	*2.8*	*48.8*
Non-OECD Africa and Middle East	**260.5**	**12.2**	**67.0**	**14.2**	**133.7**	**3.0**	**55.5**
Egypt	-	2.2	-	2.1	-	-	0.1
South Africa	254.5	2.1	66.4	10.6	126.3	3.0	50.9
Zimbabwe	2.9	0.0	-	0.4	1.1	0.0	1.5
Other Africa / Middle East	*3.1*	*7.9*	*0.6*	*1.2*	*6.2*	*0.0*	*3.1*
Non-OECD Americas	**82.6**	**19.8**	**70.7**	**17.7**	**8.8**	**0.1**	**5.4**
Brazil	5.4	15.9	-	14.1	4.8	-	2.9
Colombia	74.4	-	68.1	2.9	1.6	0.1	1.3
Other non-OECD Americas	*2.9*	*3.9*	*2.6*	*0.8*	*2.5*	-	*1.2*
Total World	**7209.9**	**1067.6**	**1076.1**	**966.6**	**4709.1**	**133.7**	**1159.9**

(1) Total coal comprises anthracite, coking coal, other bituminous coal, sub-bituminous coal and lignite.

(2) Fiscal year. See notes and definitions in Part I.

(3) Chile, Estonia, Iceland, Israel, Luxembourg, Norway, Slovak Republic, Slovenia and Switzerland.

Note: Steel industry consumption includes consumption in coke ovens. Power and heat stations column includes hard coal used in electricity and CHP production by public utilities and autoproducers, and in district heating. Other sectors includes consumption in non-ferrous industries, non-energy use, other transformation, losses and consumption in other sectors.

Source: *IEA/OECD Energy Statistics of OECD Countries , Energy Statistics of Non-OECD Countries*

World total coal supply and end-use 2011
(million tonnes)[1]

	Production	Imports	Exports	End-use sectors			
				Power and heat plants	Steel industry	Residential	Other Sectors
Total OECD	**2081.7**	**594.2**	**449.3**	**185.5**	**1857.0**	**26.0**	**158.2**
Australia[2]	402.2	-	284.5	4.6	123.7	-	5.0
Austria	-	3.2	0.0	1.9	1.7	0.0	0.2
Belgium	-	5.5	0.7	2.6	1.3	0.1	0.7
Canada	67.1	9.5	33.7	4.2	41.1	0.1	2.8
Czech Republic	57.9	2.4	7.4	3.7	43.0	1.3	4.7
Denmark	-	6.1	-	-	5.3	-	0.2
Finland	-	7.0	-	1.2	4.0	0.0	0.2
France	0.1	14.5	0.1	4.6	5.3	0.2	3.6
Germany	188.6	47.8	0.2	11.9	202.4	0.9	21.7
Greece	58.7	0.5	0.0	-	59.9	0.0	0.4
Hungary	9.6	1.9	0.0	1.5	9.6	0.5	0.1
Ireland	-	2.3	0.0	-	1.5	0.3	0.1
Italy	0.1	23.5	0.0	6.1	16.7	0.0	1.6
Japan[2]	-	174.1	0.0	58.2	86.2	-	26.2
Korea	2.1	129.2	-	25.0	86.8	-	17.5
Mexico	15.8	7.6	0.0	2.9	15.6	-	0.2
Netherlands	-	24.5	12.6	3.0	7.4	0.0	1.3
New Zealand	4.9	0.2	2.2	0.8	0.8	0.0	1.1
Poland	139.3	15.0	7.2	12.6	110.9	9.3	9.3
Portugal	-	3.8	0.2	0.0	3.7	-	0.0
Spain	6.6	16.2	1.2	2.9	20.7	0.2	1.2
Sweden	-	3.1	0.0	1.8	0.4	-	0.8
Turkey	76.1	23.7	-	6.8	70.8	12.0	11.5
United Kingdom	18.6	32.5	0.5	5.6	42.3	0.7	2.9
United States	1005.9	11.9	97.3	20.0	851.5	-	38.4
Other OECD[3]	*28.1*	*28.4*	*1.5*	*3.6*	*44.6*	*0.2*	*6.4*
Non-OECD Europe and Eurasia	**657.6**	**50.9**	**162.8**	**97.5**	**382.0**	**11.3**	**54.5**
Russian Federation	321.7	27.6	124.6	53.8	151.7	2.8	13.5
Non-OECD Asia	**4515.1**	**453.5**	**380.4**	**738.0**	**2605.2**	**95.1**	**899.0**
China, People's Republic of	3418.8	182.0	21.7	645.8	1924.1	86.1	677.3
Hong Kong, China	-	12.5	-	-	10.4	-	2.2
India[2]	582.3	132.1	2.0	84.1	506.4	6.1	113.5
Indonesia	360.3	0.0	300.7	-	41.6	-	21.2
DPR of Korea	39.1	-	11.2	-	4.2	-	23.7
Taipei, Chinese	-	66.6	-	7.5	46.8	-	10.5
Other Asia	*114.7*	*60.2*	*44.9*	*0.5*	*71.6*	*2.9*	*50.6*
Non-OECD Africa and Middle East	**259.5**	**13.3**	**70.2**	**13.3**	**127.6**	**3.6**	**56.6**
Egypt	-	2.1	-	2.0	-	-	0.1
South Africa	252.8	2.4	68.8	9.8	119.8	3.6	50.8
Zimbabwe	3.2	0.0	-	0.4	1.3	0.0	1.6
Other Africa / Middle East	*3.5*	*8.8*	*1.4*	*1.1*	*6.5*	*0.0*	*4.0*
Non-OECD Americas	**93.7**	**22.1**	**81.4**	**18.4**	**8.4**	**0.1**	**7.5**
Brazil	5.5	18.0	0.1	14.7	4.6	-	3.6
Colombia	85.8	-	79.3	2.9	0.9	0.1	2.6
Other non-OECD Americas	*2.4*	*4.1*	*2.0*	*0.8*	*2.9*	-	*1.3*
Total World	**7607.6**	**1134.0**	**1144.1**	**1052.7**	**4980.1**	**136.1**	**1175.7**

(1) Total coal comprises anthracite, coking coal, other bituminous coal, sub-bituminous coal and lignite.

(2) Fiscal year. See notes and definitions in Part I.

(3) Chile, Estonia, Iceland, Israel, Luxembourg, Norway, Slovak Republic, Slovenia and Switzerland.

Note: Steel industry consumption includes consumption in coke ovens. Power and heat stations column includes hard coal used in electricity and CHP production by public utilities and autoproducers, and in district heating. Other sectors includes consumption in non-ferrous industries, non-energy use, other transformation, losses and consumption in other sectors.

Source: *IEA/OECD Energy Statistics of OECD Countries , Energy Statistics of Non-OECD Countries*

World total coal supply and end-use 2012e

(million tonnes)[1]

	Production	Imports	Exports	End-use sectors: Power and heat plants	Steel industry	Residential	Other Sectors
Total OECD	**2032.1**	**598.8**	**471.7**	..	..	..	..
Australia[2]	420.7	-	301.5	..	..	..	..
Austria	-	3.6	0.0	..	..	..	..
Belgium	-	4.8	0.1	..	..	..	..
Canada	66.6	9.8	34.8	..	..	..	..
Czech Republic	55.0	2.2	6.4	..	..	..	..
Denmark	-	3.9	-	..	..	..	..
Finland	-	4.0	-	..	..	..	..
France	0.3	15.9	0.2	..	..	..	..
Germany	197.0	45.2	0.4	..	..	..	..
Greece	61.8	0.2	0.0	..	..	..	..
Hungary	9.3	1.8	0.1	..	..	..	..
Ireland	-	2.2	0.0	..	..	..	..
Italy	0.1	24.3	0.1	..	..	..	..
Japan[2]	-	183.8	0.0	..	..	..	..
Korea	2.1	125.5	-	..	..	..	..
Mexico	15.2	7.4	0.2	..	..	..	..
Netherlands	-	11.3	0.7	..	..	..	..
New Zealand	4.9	0.0	2.2	..	..	..	..
Poland	144.1	9.7	7.1	..	..	..	..
Portugal	-	5.2	0.1	..	..	..	..
Spain	6.1	22.4	1.9	..	..	..	..
Sweden	-	2.3	0.0	..	..	..	..
Turkey	69.5	28.7	-	..	..	..	..
United Kingdom	16.8	44.8	0.5	..	..	..	..
United States	934.9	8.3	114.1	..	..	..	..
Other OECD[3]	*27.7*	*31.3*	*1.3*	..	..	..	..
Non-OECD Europe and Eurasia	**692.6**	**54.1**	**171.8**	..	..	..	..
Russian Federation	353.9	31.4	134.2	..	..	..	..
Non-OECD Asia	**4737.9**	**587.1**	**449.2**	..	..	..	..
China, People's Republic of	3549.1	288.8	10.5	..	..	..	..
Hong Kong, China	..	12.6	-	..	..	..	..
India[2]	595.0	159.6	1.5	..	..	..	..
Indonesia	442.8	-	382.6	..	..	..	..
DPR of Korea	39.2	1.1	12.0	..	..	..	..
Taipei, Chinese	..	64.5	-	..	..	..	..
Other Asia	*111.8*	*60.4*	*42.7*	..	..	..	..
Non-OECD Africa and Middle East	**269.0**	**14.2**	**77.6**	..	..	..	..
Egypt	..	2.2	-	..	..	..	..
South Africa	259.3	2.2	74.3	..	..	..	..
Zimbabwe	3.0	0.1	-	..	..	..	..
Other Africa / Middle East	*6.7*	*9.7*	*3.2*	..	..	..	..
Non-OECD Americas	**99.1**	**21.8**	**85.1**	..	..	..	..
Brazil	6.3	18.2	-	..	..	..	..
Colombia	89.5	-	82.2	..	..	..	..
Other non-OECD Americas	*3.3*	*3.7*	*2.9*	..	..	..	..
Total World	**7830.8**	**1276.0**	**1255.3**	..	..	..	..

(1) Total coal comprises anthracite, coking coal, other bituminous coal, sub-bituminous coal and lignite.

(2) Fiscal year. See notes and definitions in Part I.

(3) Chile, Estonia, Iceland, Israel, Luxembourg, Norway, Slovak Republic, Slovenia and Switzerland.

Note: Steel industry consumption includes consumption in coke ovens. Power and heat stations column includes hard coal used in electricity and CHP production by public utilities and autoproducers, and in district heating. Other sectors includes consumption in non-ferrous industries, non-energy use, other transformation, losses and consumption in other sectors.

Source: *IEA/OECD Energy Statistics of OECD Countries , Energy Statistics of Non-OECD Countries*

BRAZIL

Coal balance*

(Mtce)

	1980	1985	1990	1995	2000	2005	2010	2011	Average annual percent change 80-90	90-11
Production	3.6	5.1	2.8	2.9	3.8	3.5	3.0	3.0	-2.5	0.4
Imports	5.3	8.6	11.3	14.0	14.8	15.2	17.3	19.5	7.9	2.6
Exports	-	-	-	-	-	-0.0	-	-0.1	-	-
Stock changes	-0.4	0.7	-0.2	0.0	0.1	-0.1	0.4	-0.5		
Primary supply	**8.5**	**14.4**	**13.8**	**16.9**	**18.6**	**18.6**	**20.6**	**22.0**	**5.0**	**2.3**
Statistical differences	**0.2**	**0.2**	**0.1**	**-0.1**	**0.2**	**-0.0**	**0.0**	**0.0**		
Total transformation	**-4.0**	**-6.1**	**-6.7**	**-8.0**	**-9.3**	**-9.3**	**-9.3**	**-8.8**	**5.2**	**1.3**
Electricity and heat gen.	-1.3	-2.0	-2.1	-2.5	-4.6	-4.1	-4.5	-3.9	5.4	2.9
Main activity producers (1)	*-1.0*	*-1.5*	*-1.4*	*-1.9*	*-3.3*	*-2.7*	*-2.5*	*-2.3*	*3.6*	*2.3*
Autoproducers	*-0.3*	*-0.5*	*-0.7*	*-0.6*	*-1.3*	*-1.5*	*-2.0*	*-1.6*	*10.3*	*3.9*
Gas works	-	-	-	-	-	-	-	-	-	-
Coal transformation(2)	-2.8	-4.2	-4.5	-5.5	-4.7	-5.2	-4.7	-4.9	5.1	0.4
BKB plants	-	-	-	-	-	-	-	-	-	-
Blast furnaces	*-2.2*	*-3.4*	*-3.4*	*-4.6*	*-4.2*	*-4.9*	*-4.9*	*-5.2*	*4.5*	*2.0*
Coke ovens	*-0.6*	*-0.8*	*-1.1*	*-1.0*	*-0.6*	*-0.3*	*0.2*	*0.3*	*7.3*	-
Patent fuel plants	-	-	-	-	-	-	-	-	-	-
Other transformation(3)	-	-	-	-	-	-	-	-	-	-
Energy ind. own use	**-0.4**	**-0.8**	**-0.8**	**-0.8**	**-0.8**	**-0.8**	**-0.7**	**-0.8**	**6.8**	**-0.0**
Losses	**-1.3**	**-2.0**	**-1.2**	**-1.7**	**-0.5**	**-0.5**	**-0.3**	**-0.6**		
Final consumption(4)	**3.0**	**5.6**	**5.2**	**6.4**	**8.2**	**7.9**	**10.4**	**11.9**	**5.9**	**4.0**
Industry(5)	2.8	5.3	5.1	6.3	8.0	7.7	10.2	11.7	6.3	4.0
Iron and steel	*2.1*	*3.2*	*3.5*	*4.8*	*6.5*	*5.9*	*7.8*	*8.6*	*5.0*	*4.4*
Chemical	*0.0*	*0.2*	*0.1*	*0.2*	*0.1*	*0.1*	*0.2*	*0.1*	*47.6*	*0.8*
Non-metallic minerals	*0.4*	*1.4*	*0.9*	*0.6*	*0.3*	*0.2*	*0.1*	*0.3*	*7.2*	*-4.7*
Paper, pulp and print	*0.1*	*0.2*	*0.2*	*0.1*	*0.1*	*0.1*	*0.1*	*0.2*	*8.3*	*-0.9*
Other industry (6)	*0.1*	*0.2*	*0.4*	*0.6*	*1.0*	*1.3*	*2.0*	*2.4*	*14.5*	*8.7*
Transport(7)	0.0	0.0	0.0	-	-	-	-	-	-12.3	-
Other	-	-	-	-	-	-	-	-	-	-
Comm. and pub. services	-	-	-	-	-	-	-	-	-	-
Residential	-	-	-	-	-	-	-	-	-	-
Other sectors (8)	-	-	-	-	-	-	-	-	-	-
Non-energy use	0.2	0.3	0.2	0.1	0.2	0.2	0.2	0.2	-0.9	0.5

(1) Main activity electricity and heat generation includes district heating.

(2) Coal transformation refers to the transformation of coal and peat to secondary and tertiary products (mainly cokes, briquettes, coal tar and off-gases). Gas works and Liquefaction are shown separately to match the *Energy Balances of OECD Countries.*

(3) Other transformation includes Liquefaction and Non-specified transformations.

(4) Final Consumption includes non-energy use and energy use (Industry, Transport and Other).

(5) Please refer to notes and definitions in Part I for detailed explanation.

(6) Other industry includes Non-ferrous metals, Transport equipment, Machinery, Mining and quarrying, Food and tobacco, Wood and wood products, Construction, Textile and leather, and Non-specified industry.

(7) Transport includes Rail and Inland waterways.

(8) Other sectors includes Agriculture/Forestry, Fishing and Non-specified other.

* "Coal" refers to all types of coal, primary (anthracite, coking coal, other bituminous coal, sub-bituminous coal, lignite/brown coal) and derived fuels (including patent fuel, BKB, coke oven coke, gas coke, coal tar, coke oven gas, gas works gas, blast furnace gas and other recovered gases), as well as peat, peat briquettes and oil shale. Quantities have been converted to Mtce using calorific values reported by the respective countries.
Please refer to notes and definitions in Part I.

Source: *IEA/OECD Energy Balances of OECD Countries*

BRAZIL

Total coal imports by origin*

(thousand tonnes)

	1978[1]	1985	1990	1995	2000	2005	2010	2011	2012e
Total coal	**3547**	**8049**	**10146**	**11790**	**14833**	**13699**	**17590**	**20474**	**18157**
Coking coal	**3503**	**7958**	**9801**	**11438**	**10695**	**9396**	**13647**	**12974**	**10852**
Australia	164	958	1291	3448	5247	3501	4627	2337	2514
Canada	600	899	1108	625	1382	1944	1574	2255	846
Czech Republic	-	-	-	-	-	-	-	-	-
Germany	15	-	-	-	-	-	-	-	-
Poland	-	861	1249	790	-	-	-	-	91
United Kingdom	-	-	-	-	-	-	-	-	-
United States	2724	5240	5219	5498	4066	3523	7303	7729	7091
Other OECD	-	-	-	-	-	128	-	-	-
China, People's Rep.	-	-	-	-	-	-	3	-	78
Colombia	-	-	254	85	-	-	-	-	144
Indonesia	-	-	-	-	-	-	-	-	-
South Africa	-	-	670	932	-	300	61	-	-
Former Soviet Union[2]	-	-	x	x	x	x	x	x	x
Russian Federation[3]	*x*	*x*	-	-	-	-	*30*	-	-
Other FSU[3]	*x*	*x*	-	-	-	-	*49*	-	-
Venezuela	-	-	-	-	-	-	-	-	-
Vietnam	-	-	-	39	-	-	-	-	-
Non-specified/other	-	-	10	21	-	-	-	653	88
Steam coal[4]	**44**	**91**	**345**	**352**	**4138**	**4303**	**3943**	**7500**	**7305**
Australia	-	-	158	-	18	32	34	82	609
Canada	-	-	98	140	2	-	134	-	1116
Czech Republic	-	-	-	-	-	-	-	-	-
Germany	-	13	5	-	-	-	-	-	-
Poland	-	-	-	37	-	-	-	-	-
United Kingdom	-	-	5	-	-	-	-	-	22
United States	11	78	79	13	-	660	68	1595	125
Other OECD	-	-	-	-	-	644	39	115	24
China, People's Rep.	-	-	-	-	585	366	6	6	6
Colombia	-	-	-	62	149	426	1601	2512	2422
Indonesia	-	-	-	-	468	211	-	-	-
South Africa	-	-	-	-	1919	249	895	1103	1239
Former Soviet Union[2]	-	-	x	x	x	x	x	x	x
Russian Federation[3]	*x*	*x*	-	-	-	*514*	*387*	*1216*	*1200*
Other FSU[3]	*x*	*x*	-	-	-	*226*	*166*	*174*	*100*
Venezuela	33	-	-	100	997	450	608	420	253
Vietnam	-	-	-	-	-	429	-	-	-
Non-specified/other	-	-	-	-	-	96	5	277	189
Lignite	**-**	**-**	**-**	**-**	**-**	**-**	**-**	**-**	**-**

(1) Earliest year for which split by coal type is available.

(2) For years prior to 1990, or if no details for individual Former Soviet Union republics.

(3) Data for former Republics are not available separately for some countries.

(4) Steam coal now comprises anthracite, other bituminous coal and sub-bituminous coal.

* In these tables coal used for PCI and for blending has been classified by the IEA as steam coal. Accordingly, trade data reported in these tables may differ from those reported in Part III where this coal is shown as coking coal to be consistent with data reported by importing countries and with industry terminology and practice.

Source: IEA/OECD Coal Statistics.

BULGARIA

Coal balance*

(Mtce)

	1980	1985	1990	1995	2000	2005	2010	2011	Average annual percent change 80-90	90-11
Production	7.4	7.5	7.7	7.6	6.2	6.0	7.1	8.9	0.4	0.7
Imports	6.1	7.4	5.0	3.4	3.5	3.7	2.5	2.9	-1.9	-2.5
Exports	-	-	-0.1	-0.0	-0.2	-0.0	-0.1	-0.1	-	0.9
Stock changes	-0.1	-	0.1	-0.2	-0.2	0.3	0.4	-0.1		
Primary supply	**13.4**	**15.0**	**12.7**	**10.8**	**9.3**	**9.9**	**9.9**	**11.6**	**-0.5**	**-0.4**
Statistical differences	**-0.0**	**0.1**	**-0.1**	**-0.1**	**-0.1**	**0.0**	**0.1**	**0.1**		
Total transformation	**-8.3**	**-8.5**	**-10.1**	**-8.9**	**-7.7**	**-8.6**	**-9.3**	**-10.9**	**2.0**	**0.4**
Electricity and heat gen.	-7.7	-7.8	-10.2	-8.1	-7.1	-8.2	-9.2	-10.9	2.9	0.3
Main activity producers (1)	*-7.4*	*-7.6*	*-9.4*	*-7.3*	*-6.8*	*-8.0*	*-9.2*	*-10.9*	*2.5*	*0.7*
Autoproducers	*-0.2*	*-0.3*	*-0.7*	*-0.8*	*-0.2*	*-0.2*	-	-	*11.6*	-
Gas works	-	-	-	-	-	-	-	-	-	-
Coal transformation (2)	-0.6	-0.7	0.0	-0.7	-0.6	-0.4	-0.0	-0.1	-	-
BKB plants	*0.0*	*-0.1*	*-0.0*	*-0.0*	*-0.0*	*0.1*	*-0.0*	*-0.1*	-	*5.0*
Blast furnaces	*-0.5*	*-0.4*	*-0.0*	*-0.5*	*-0.4*	*-0.4*	-	-	*-21.1*	-
Coke ovens	*-0.2*	*-0.2*	*0.1*	*-0.2*	*-0.3*	*-0.1*	-	-	-	-
Patent fuel plants	-	-	-	-	-	-	-	-	-	-
Other transformation (3)	-	-	-	-	-	-	0.0	-	-	-
Energy ind. own use	**-**	**-**	**-0.2**	**-0.2**	**-0.4**	**-0.2**	**-0.0**	**-0.0**	**-**	**-22.4**
Losses	**-0.0**	**-0.0**	**-0.0**	**-**	**-**	**-**	**-0.0**	**-0.0**		
Final consumption (4)	**5.1**	**6.5**	**2.3**	**1.6**	**1.1**	**1.1**	**0.7**	**0.7**	**-7.6**	**-5.4**
Industry (5)	3.6	5.2	1.2	0.9	0.7	0.7	0.3	0.3	-10.2	-6.6
Iron and steel	*0.5*	*0.7*	*0.8*	*0.6*	*0.3*	*0.3*	*0.0*	*0.0*	*3.6*	*-17.2*
Chemical	-	-	*0.2*	*0.1*	*0.2*	*0.1*	*0.1*	*0.1*	-	*-0.9*
Non-metallic minerals	-	-	*0.1*	*0.1*	*0.1*	*0.2*	*0.2*	*0.1*	-	*1.7*
Paper, pulp and print	-	-	-	-	-	-	-	-	-	-
Other industry (6)	*3.1*	*4.5*	*0.2*	*0.1*	*0.1*	*0.1*	*0.0*	*0.0*	*-23.7*	*-15.1*
Transport (7)	-	-	-	0.0	-	-	-	-	-	-
Other	1.5	1.3	1.1	0.7	0.3	0.3	0.3	0.4	-3.1	-5.1
Comm. and pub. services	-	-	*0.0*	*0.0*	-	*0.0*	*0.0*	*0.0*	-	*-5.1*
Residential	*1.2*	*1.1*	*1.0*	*0.7*	*0.3*	*0.3*	*0.3*	*0.3*	*-2.2*	*-4.9*
Other sectors (8)	*0.2*	*0.2*	*0.1*	*0.0*	*0.0*	*0.0*	*0.0*	*0.0*	*-11.9*	*-10.1*
Non-energy use	-	-	-	-	-	-	0.1	0.1	-	-

(1) Main activity electricity and heat generation includes district heating.

(2) Coal transformation refers to the transformation of coal and peat to secondary and tertiary products (mainly cokes, briquettes, coal tar and off-gases). Gas works and Liquefaction are shown separately to match the *Energy Balances of OECD Countries.*

(3) Other transformation includes Liquefaction and Non-specified transformations.

(4) Final Consumption includes non-energy use and energy use (Industry, Transport and Other).

(5) Please refer to notes and definitions in Part I for detailed explanation.

(6) Other industry includes Non-ferrous metals, Transport equipment, Machinery, Mining and quarrying, Food and tobacco, Wood and wood products, Construction, Textile and leather, and Non-specified industry.

(7) Transport includes Rail and Inland waterways.

(8) Other sectors includes Agriculture/Forestry, Fishing and Non-specified other.

* "Coal" refers to all types of coal, primary (anthracite, coking coal, other bituminous coal, sub-bituminous coal, lignite/brown coal) and derived fuels (including patent fuel, BKB, coke oven coke, gas coke, coal tar, coke oven gas, gas works gas, blast furnace gas and other recovered gases), as well as peat, peat briquettes and oil shale. Quantities have been converted to Mtce using calorific values reported by the respective countries. Please refer to notes and definitions in Part I.

Source: *IEA/OECD Energy Balances of OECD Countries*

BULGARIA

Total coal imports by origin*

(thousand tonnes)

	1978(1)	1985	1990	1995	2000	2005	2010	2011	2012e
Total coal	**6201**	**8054**	**5790**	**3453**	**1184**	**1744**	**2939**	**3235**	**2261**
Coking coal	**1921**	**1553**	**1100**	**1734**	**536**	**10**	**-**	**-**	**2**
Australia	-	-	-	-	-	1	-	-	-
Canada	-	-	-	-	-	-	-	-	-
Czech Republic	-	-	-	-	-	-	-	-	-
Germany	-	-	-	-	-	-	-	-	-
Poland	-	-	-	177	140	-	-	-	-
United Kingdom	-	-	-	-	-	-	-	-	-
United States	-	29	44	542	298	8	-	-	-
Other OECD	-	-	-	5	-	-	-	-	2
China, People's Rep.	-	-	-	-	-	-	-	-	-
Colombia	-	-	-	-	-	-	-	-	-
Indonesia	-	-	-	-	-	-	-	-	-
South Africa	-	-	-	100	-	-	-	-	-
Former Soviet Union(2)	1921	1524	x	x	x	x	x	x	x
Russian Federation (3)	*x*	*x*	*156*	*110*	*98*	*1*	*-*	*-*	*-*
Other FSU (3)	*x*	*x*	*900*	*800*	*-*	*-*	*-*	*-*	*-*
Venezuela	-	-	-	-	-	-	-	-	-
Vietnam	-	-	-	-	-	-	-	-	-
Non-specified/other	-	-	-	-	-	-	-	-	-
Steam coal(4)	**4280**	**6501**	**4690**	**1719**	**648**	**1734**	**2939**	**3235**	**2259**
Australia	-	-	-	-	-	66	-	-	-
Canada	-	-	-	-	-	100	-	-	-
Czech Republic	-	-	-	-	-	-	-	-	-
Germany	-	-	-	-	-	-	-	-	-
Poland	-	-	-	-	-	-	-	-	-
United Kingdom	-	-	-	5	-	-	-	-	-
United States	-	-	-	-	-	38	-	-	-
Other OECD	-	-	-	4	-	-	44	2	10
China, People's Rep.	-	-	-	-	-	-	-	-	-
Colombia	-	-	-	-	-	-	-	-	-
Indonesia	-	-	-	-	-	-	-	-	-
South Africa	-	-	-	60	285	-	-	-	-
Former Soviet Union(2)	4280	6501	x	x	x	x	x	x	x
Russian Federation (3)	*x*	*x*	*-*	*1010*	*363*	*858*	*251*	*529*	*632*
Other FSU (3)	*x*	*x*	*4690*	*640*	*-*	*623*	*2599*	*2692*	*1612*
Venezuela	-	-	-	-	-	-	-	-	-
Vietnam	-	-	-	-	-	-	-	-	-
Non-specified/other	-	-	-	-	-	49	45	12	5
Lignite	**-**	**-**	**-**	**-**	**-**	**-**	**-**	**-**	**-**

(1) Earliest year for which split by coal type is available.

(2) For years prior to 1990, or if no details for individual Former Soviet Union republics.

(3) Data for former Republics are not available separately for some countries.

(4) Steam coal now comprises anthracite, other bituminous coal and sub-bituminous coal.

* In these tables coal used for PCI and for blending has been classified by the IEA as steam coal. Accordingly, trade data reported in these tables may differ from those reported in Part III where this coal is shown as coking coal to be consistent with data reported by importing countries and with industry terminology and practice.

Source: IEA/OECD Coal Statistics.

PEOPLE'S REPUBLIC OF CHINA

Coal balance*

(Mtce)

	1980	1985	1990	1995	2000	2005	2010	2011	Average annual percent change 80-90	90-11
Production	443.9	584.4	740.6	962.3	1020.4	1745.4	2383.6	2604.8	5.3	6.2
Imports	1.5	1.7	1.5	1.2	2.1	21.3	128.1	141.2	-0.0	24.2
Exports	-6.0	-7.4	-17.3	-35.0	-65.1	-79.0	-22.9	-23.2	11.1	1.4
Stock changes	7.1	-18.9	28.9	-2.7	29.1	-26.5	-74.4	-67.6		
Primary supply	**446.5**	**559.8**	**753.7**	**925.8**	**986.5**	**1661.2**	**2414.4**	**2655.2**	**5.4**	**6.2**
Statistical differences	**-15.6**	**-15.8**	**-35.5**	**32.4**	**27.5**	**27.4**	**-154.0**	**-166.2**		
Total transformation	**-111.8**	**-143.7**	**-242.5**	**-396.9**	**-525.0**	**-973.9**	**-1426.7**	**-1611.1**	**8.1**	**9.4**
Electricity and heat gen.	-88.7	-122.2	-209.5	-348.7	-472.6	-855.9	-1245.9	-1411.0	9.0	9.5
Main activity producers (1)	*-88.2*	*-121.7*	*-208.5*	*-346.2*	*-468.4*	*-846.9*	*-1230.9*	*-1395.0*	*9.0*	*9.5*
Autoproducers	*-0.5*	*-0.5*	*-1.0*	*-2.4*	*-4.2*	*-9.0*	*-14.9*	*-16.0*	*7.1*	*14.2*
Gas works	-0.5	0.3	-1.6	-0.6	-1.7	-2.5	-1.7	-0.8	12.0	-3.4
Coal transformation(2)	-22.6	-21.8	-31.5	-47.6	-50.7	-115.4	-177.4	-196.5	3.4	9.1
BKB plants	..	..	..	..	..	..	..	-	..	..
Blast furnaces	*-13.9*	*-14.4*	*-22.2*	*-35.5*	*-34.2*	*-92.1*	*-127.8*	*-143.2*	*4.8*	*9.3*
Coke ovens	*-8.7*	*-7.4*	*-9.3*	*-12.0*	*-16.1*	*-20.2*	*-46.8*	*-50.3*	*0.7*	*8.4*
Patent fuel plants	..	..	..	*-0.1*	*-0.4*	*-3.1*	*-2.8*	*-3.0*	..	..
Other transformation(3)	..	..	..	..	..	..	-1.7	-2.8	..	..
Energy ind. own use	**-4.5**	**-8.3**	**-21.2**	**-39.6**	**-54.6**	**-81.1**	**-93.3**	**-90.9**	**16.7**	**7.2**
Losses	**..**	**..**	**..**	**..**	**-0.1**	**..**	**..**	**-**		
Final consumption(4)	**314.5**	**392.1**	**454.4**	**521.6**	**434.3**	**633.6**	**740.4**	**787.0**	**3.7**	**2.7**
Industry(5)	197.8	226.2	258.3	356.4	307.4	473.4	575.9	617.6	2.7	4.2
Iron and steel	*47.4*	*29.9*	*33.5*	*73.8*	*70.6*	*128.0*	*216.7*	*237.3*	*-3.4*	*9.8*
Chemical	*42.1*	*31.1*	*31.6*	*60.4*	*35.4*	*50.4*	*56.5*	*65.9*	*-2.8*	*3.6*
Non-metallic minerals	*32.5*	*59.4*	*70.4*	*95.7*	*102.6*	*145.1*	*171.2*	*182.6*	*8.0*	*4.6*
Paper, pulp and print	*5.5*	*8.5*	*10.0*	*14.0*	*11.3*	*15.3*	*16.7*	*15.9*	*6.2*	*2.2*
Other industry (6)	*70.3*	*97.3*	*112.7*	*112.6*	*87.4*	*134.6*	*114.8*	*115.9*	*4.8*	*0.1*
Transport(7)	13.6	15.8	14.0	9.0	6.1	5.7	4.5	4.6	0.3	-5.2
Other	103.2	138.1	154.6	136.5	94.3	116.5	114.0	116.7	4.1	-1.3
Comm. and pub. services	*3.2*	*4.9*	*8.6*	*8.0*	*10.7*	*13.4*	*14.6*	*16.5*	*10.3*	*3.1*
Residential	*81.1*	*106.9*	*117.0*	*100.2*	*66.5*	*77.8*	*72.2*	*71.7*	*3.7*	*-2.3*
Other sectors (8)	*18.8*	*26.4*	*29.0*	*28.3*	*17.1*	*25.3*	*27.3*	*28.5*	*4.4*	*-0.1*
Non-energy use	-	11.9	27.5	19.7	26.5	37.9	46.0	48.2	-	2.7

(1) Main activity electricity and heat generation includes district heating.

(2) Coal transformation refers to the transformation of coal and peat to secondary and tertiary products (mainly cokes, briquettes, coal tar and off-gases). Gas works and Liquefaction are shown separately to match the *Energy Balances of OECD Countries.*

(3) Other transformation includes Liquefaction and Non-specified transformations.

(4) Final Consumption includes non-energy use and energy use (Industry, Transport and Other).

(5) Please refer to notes and definitions in Part I for detailed explanation.

(6) Other industry includes Non-ferrous metals, Transport equipment, Machinery, Mining and quarrying, Food and tobacco, Wood and wood products, Construction, Textile and leather, and Non-specified industry.

(7) Transport includes Rail and Inland waterways.

(8) Other sectors includes Agriculture/Forestry, Fishing and Non-specified other.

* "Coal" refers to all types of coal, primary (anthracite, coking coal, other bituminous coal, sub-bituminous coal, lignite/brown coal) and derived fuels (including patent fuel, BKB, coke oven coke, gas coke, coal tar, coke oven gas, gas works gas, blast furnace gas and other recovered gases), as well as peat, peat briquettes and oil shale. Quantities have been converted to Mtce using calorific values reported by the respective countries.
Please refer to notes and definitions in Part I.

Source: *IEA/OECD Energy Balances of OECD Countries*

PEOPLE'S REPUBLIC OF CHINA

Total coal imports by origin*

(thousand tonnes)

	1978(1)	1985	1990	1995	2000	2005	2010	2011	2012e
Total coal	**2440**	**2307**	**2003**	**1635**	**2119**	**26173**	**163065**	**181976**	**289135**
Coking coal	**-**	**200**	**900**	**354**	**547**	**7195**	**47082**	**44654**	**70644**
Australia	-	200	600	248	547	4422	24152	13383	30398
Canada	-	-	300	-	-	1239	4018	3074	8937
Czech Republic	-	-	-	-	-	-	-	-	-
Germany	-	-	-	-	-	-	-	-	-
Poland	-	-	-	-	-	-	-	-	-
United Kingdom	-	-	-	-	-	-	-	-	-
United States	-	-	-	-	-	-	2754	4091	6245
Other OECD	-	-	-	106	-	179	138	344	742
China, People's Rep.	-	-	-	-	-	-	-	-	-
Colombia	-	-	-	-	-	-	23	31	84
Indonesia	-	-	-	-	-	-	767	567	-
South Africa	-	-	-	-	-	-	-	-	-
Former Soviet Union(2)	-	-	x	x	x	x	x	x	x
Russian Federation (3)	*x*	*x*	-	-	-	*61*	*1804*	*2975*	*4791*
Other FSU (3)	*x*	*x*	-	-	-	-	-	-	-
Venezuela	-	-	-	-	-	-	39	-	-
Vietnam	-	-	-	-	-	-	-	-	-
Non-specified/other	-	-	-	-	-	1294	13387	20189	19447
Steam coal(4)	**2440**	**2107**	**1103**	**1281**	**1572**	**18978**	**115983**	**137322**	**218143**
Australia	-	-	-	1064	1034	2307	15158	18258	30888
Canada	-	-	-	-	-	-	710	1613	859
Czech Republic	-	-	-	-	-	-	-	-	-
Germany	-	-	-	-	-	-	-	-	-
Poland	-	-	-	-	-	-	-	-	-
United Kingdom	-	-	-	-	-	-	-	-	-
United States	-	-	-	-	-	12	1132	861	5281
Other OECD	-	-	-	-	-	796	26	150	3
China, People's Rep.	-	-	-	-	-	-	-	-	-
Colombia	-	-	-	-	-	-	3756	1279	2789
Indonesia	-	-	-	171	141	2260	56295	65018	114994
South Africa	-	-	-	-	325	-	4183	8402	12952
Former Soviet Union(2)	-	-	x	x	x	x	x	x	x
Russian Federation (3)	*x*	*x*	-	*46*	*72*	*791*	*6801*	*6918*	*17121*
Other FSU (3)	*x*	*x*	-	-	-	-	*1*	*10*	*25*
Venezuela	-	-	-	-	-	-	-	-	-
Vietnam	-	-	-	-	-	11469	18065	23084	17425
Non-specified/other	2440	2107	1103	-	-	1343	9856	11729	15806
Lignite	**-**	**-**	**-**	**-**	**-**	**-**	**-**	**-**	**348**

(1) Earliest year for which split by coal type is available.

(2) For years prior to 1990, or if no details for individual Former Soviet Union republics.

(3) Data for former Republics are not available separately for some countries.

(4) Steam coal now comprises anthracite, other bituminous coal and sub-bituminous coal.

* In these tables coal used for PCI and for blending has been classified by the IEA as steam coal. Accordingly, trade data reported in these tables may differ from those reported in Part III where this coal is shown as coking coal to be consistent with data reported by importing countries and with industry terminology and practice.

Source: IEA/OECD Coal Statistics.

PEOPLE'S REPUBLIC OF CHINA

Coking coal exports by destination

(thousand tonnes)

	1978(1)	1985	1990	1995	2000	2005	2010	2011	2012e
World	**300**	**2500**	**4000**	**4555**	**6470**	**5260**	**1139**	**3594**	**1723**
OECD	**300**	**1211**	**1302**	**3312**	**5989**	**4984**	**932**	**2891**	**1470**
Austria	-	-	-	-	-	-	-	-	-
Belgium	-	-	-	-	-	-	-	-	-
Canada	-	-	-	-	-	-	-	-	-
Czech Republic	-	-	-	-	-	-	-	-	-
Denmark	-	-	-	-	-	-	-	-	-
Finland	-	-	-	-	-	-	-	-	-
France	-	-	-	63	-	-	-	-	-
Germany	-	-	-	63	-	-	-	-	-
Greece	-	-	-	-	-	-	-	-	-
Hungary	-	-	-	-	-	-	-	-	-
Israel	-	-	-	-	-	-	-	-	-
Italy	-	-	-	-	-	-	-	-	-
Japan	300	1211	1301	2201	3631	3279	345	1674	513
Korea	-	-	-	985	2358	1627	587	1103	956
Mexico	-	-	-	-	-	-	-	-	1
Netherlands	-	-	1	-	-	-	-	-	-
Norway	-	-	-	-	-	-	-	-	-
Poland	-	-	-	-	-	-	-	-	-
Portugal	-	-	-	-	-	-	-	-	-
Spain	-	-	-	-	-	78	-	-	-
Sweden	-	-	-	-	-	-	-	-	-
Turkey	-	-	-	-	-	-	-	114	-
United Kingdom	-	-	-	-	-	-	-	-	-
United States	-	-	-	-	-	-	-	-	-
Other OECD(2)	-	-	-	-	-	-	-	-	-
Non-OECD	**-**	**1289**	**2698**	**1243**	**481**	**276**	**207**	**703**	**253**
Brazil	-	-	100	57	-	-	-	-	-
Bulgaria	-	-	-	-	-	-	-	-	-
China, People's Rep.	-	-	-	-	-	-	-	-	-
Chinese Taipei	-	-	-	-	-	-	-	327	81
Hong Kong, China	-	-	-	-	21	-	-	-	-
India	-	-	-	48	360	266	-	173	-
Morocco	-	-	-	-	-	-	-	-	-
Romania	-	400	400	125	-	-	-	-	-
Russian Federation	-	-	-	2	-	-	-	-	-
Ukraine	-	-	-	-	-	-	-	-	-
Other Africa	-	-	-	-	-	-	-	-	-
Other Asia	-	889	2198	1011	100	10	207	203	172
Other Eastern Europe	-	-	-	-	-	-	-	-	-
Other FSU	-	-	-	-	-	-	-	-	-
Other non-OECD Americas	-	-	-	-	-	-	-	-	-
Other Middle East	-	-	-	-	-	-	-	-	-
Non-specified/Other	-	-	-	-	-	-	-	-	-

(1) Earliest year for which split by coal type is available.

(2) Australia, Chile, Estonia, Iceland, Ireland, Luxembourg, New Zealand, Slovak Republic, Slovenia and Switzerland.

Source: IEA/OECD Coal Statistics.

PEOPLE'S REPUBLIC OF CHINA

Steam coal[1] exports by destination

(thousand tonnes)

	1978[2]	1985	1990	1995	2000	2005	2010	2011	2012e
World	**2820**	**5270**	**13280**	**24062**	**48578**	**66413**	**20169**	**18098**	**8742**
OECD	**534**	**2663**	**6879**	**17495**	**33172**	**37207**	**14584**	**13190**	**7143**
Austria	-	-	-	-	-	-	-	-	-
Belgium	-	112	292	406	109	297	114	-	1
Canada	-	-	-	-	114	70	-	-	-
Czech Republic	-	-	-	-	-	-	-	-	-
Denmark	-	29	57	-	-	-	-	-	-
Finland	-	-	100	-	-	-	-	-	-
France	-	46	1776	956	452	8	-	-	30
Germany	21	9	8	57	-	79	-	4	4
Greece	-	-	-	-	228	-	-	-	-
Hungary	-	-	-	-	-	-	-	-	-
Israel	-	-	-	13	-	2	-	-	-
Italy	-	-	310	455	383	-	-	3	3
Japan	513	2427	3258	7717	12585	15947	6877	7008	3546
Korea	-	-	838	7362	18760	18481	7389	6155	3446
Mexico	-	-	-	-	2	8	-	-	-
Netherlands	-	1	152	449	146	148	-	9	11
Norway	-	-	4	-	36	24	6	-	-
Poland	-	-	-	-	-	-	-	-	-
Portugal	-	-	-	-	-	-	-	-	-
Spain	-	-	-	-	199	184	-	-	4
Sweden	-	-	8	-	-	-	-	-	-
Turkey	-	-	-	-	43	1736	189	-	-
United Kingdom	-	34	69	80	107	163	-	-	4
United States	-	-	-	-	8	60	9	11	54
Other OECD[3]	-	5	7	-	-	-	-	-	40
Non-OECD	**2286**	**2607**	**6401**	**6567**	**15406**	**29162**	**5585**	**4908**	**1599**
Brazil	-	-	-	-	585	293	-	-	-
Bulgaria	-	-	-	29	-	-	-	-	-
China, People's Rep.	-	-	-	-	-	-	-	-	-
Chinese Taipei	-	-	-	3988	9076	20992	4989	4889	1581
Hong Kong, China	-	700	1708	1469	1963	993	395	-	-
India	-	-	-	264	1562	3492	-	-	-
Morocco	-	-	-	-	36	138	-	-	-
Romania	-	-	-	-	-	-	-	-	-
Russian Federation	-	-	-	103	-	3	-	-	-
Ukraine	-	-	-	-	-	-	-	-	-
Other Africa	-	-	-	-	-	-	2	-	-
Other Asia	2286	1907	4693	714	2182	3186	199	16	17
Other Eastern Europe	-	-	-	-	-	65	-	-	-
Other FSU	-	-	-	-	2	-	-	-	-
Other non-OECD Americas	-	-	-	-	-	-	-	-	-
Other Middle East	-	-	-	13	-	2	-	3	1
Non-specified/Other	-	-	-	-	-	44	-	-	-

(1) Steam coal now comprises anthracite, other bituminous coal and sub-bituminous coal.

(2) Earliest year for which split by coal type is available.

(3) Australia, Chile, Estonia, Iceland, Ireland, Luxembourg, New Zealand, Slovak Republic, Slovenia and Switzerland.

Source: IEA/OECD Coal Statistics.

CHINESE TAIPEI

Coal balance*

(Mtce)

	1980	1985	1990	1995	2000	2005	2010	2011	Average annual percent change 80-90	90-11
Production	2.3	1.6	0.4	0.2	0.1	-	-	-	-15.6	-
Imports	4.5	9.5	17.5	27.2	41.4	55.2	59.2	58.6	14.7	5.9
Exports	-0.0	-0.0	-0.0	-	-0.0	-0.0	-0.1	-0.1	13.6	6.8
Stock changes	-1.2	-0.8	-1.6	-2.6	1.2	-0.6	0.1	-0.1		
Primary supply	**5.6**	**10.3**	**16.3**	**24.8**	**42.7**	**54.5**	**59.2**	**58.5**	**11.3**	**6.3**
Statistical differences	**-0.0**	**-0.0**	**-0.1**	**-0.9**	**-1.7**	**-0.4**	**-1.6**	**-2.6**		
Total transformation	**-2.4**	**-6.2**	**-10.2**	**-17.5**	**-32.8**	**-44.4**	**-45.0**	**-43.8**	**15.6**	**7.2**
Electricity and heat gen.	-2.2	-5.2	-8.4	-15.5	-29.8	-41.4	-41.8	-40.2	14.3	7.7
Main activity producers (1)	*-2.2*	*-4.8*	*-7.3*	*-12.6*	*-22.7*	*-31.5*	*-32.1*	*-30.6*	*12.6*	*7.1*
Autoproducers	-	*-0.4*	*-1.2*	*-2.9*	*-7.1*	*-9.9*	*-9.7*	*-9.6*	-	*10.6*
Gas works	-	-	-	-	-	-	-	-	-	-
Coal transformation(2)	-0.2	-1.0	-1.8	-1.9	-3.0	-3.0	-3.2	-3.6	25.9	3.3
BKB plants	-	-	-	-	-	-	-	-	-	-
Blast furnaces	-	*-0.9*	*-1.6*	*-1.8*	*-2.7*	*-2.9*	*-3.2*	*-3.5*	-	*3.9*
Coke ovens	*-0.2*	*-0.1*	*-0.2*	*-0.1*	*-0.2*	*-0.1*	*-0.0*	*-0.0*	*2.4*	*-8.1*
Patent fuel plants	-	-	-	-	-	-	-	-	-	-
Other transformation(3)	-	-	-	-	-	-	-	-	-	-
Energy ind. own use	**-0.0**	**-0.3**	**-0.7**	**-0.8**	**-1.1**	**-1.1**	**-1.3**	**-1.3**	**37.5**	**3.0**
Losses	**-**	**-0.2**	**-0.0**	**-0.0**	**-0.0**	**-0.1**	**-0.1**	**-0.1**		
Final consumption(4)	**3.1**	**3.6**	**5.1**	**5.7**	**7.1**	**8.5**	**11.2**	**10.7**	**5.0**	**3.6**
Industry(5)	3.0	3.5	4.9	5.4	6.8	8.2	9.8	10.6	5.0	3.7
Iron and steel	*1.2*	*0.6*	*1.0*	*1.1*	*1.8*	*1.8*	*2.2*	*2.4*	*-1.8*	*4.0*
Chemical	*0.2*	*0.4*	*1.0*	*1.4*	*2.7*	*3.4*	*5.1*	*5.2*	*16.1*	*8.3*
Non-metallic minerals	*1.2*	*2.2*	*2.5*	*2.4*	*2.0*	*2.3*	*2.0*	*2.3*	*7.4*	*-0.5*
Paper, pulp and print	*0.1*	*0.1*	*0.3*	*0.4*	*0.3*	*0.4*	*0.4*	*0.6*	*11.4*	*3.5*
Other industry (6)	*0.2*	*0.1*	*0.1*	*0.2*	*0.2*	*0.3*	*0.2*	*0.2*	*-8.8*	*4.1*
Transport(7)	0.0	-	-	-	-	-	-	-	-	-
Other	0.1	0.1	0.0	-	0.0	-	-	-	-23.3	-
Comm. and pub. services	-	*0.0*	-	-	-	-	-	-	-	-
Residential	*0.0*	-	-	-	-	-	-	-	-	-
Other sectors (8)	*0.1*	*0.1*	*0.0*	-	*0.0*	-	-	-	*-22.5*	-
Non-energy use	-	-	0.2	0.3	0.2	0.3	1.3	0.1	-	-3.3

(1) Main activity electricity and heat generation includes district heating.

(2) Coal transformation refers to the transformation of coal and peat to secondary and tertiary products (mainly cokes, briquettes, coal tar and off-gases). Gas works and Liquefaction are shown separately to match the *Energy Balances of OECD Countries.*

(3) Other transformation includes Liquefaction and Non-specified transformations.

(4) Final Consumption includes non-energy use and energy use (Industry, Transport and Other).

(5) Please refer to notes and definitions in Part I for detailed explanation.

(6) Other industry includes Non-ferrous metals, Transport equipment, Machinery, Mining and quarrying, Food and tobacco, Wood and wood products, Construction, Textile and leather, and Non-specified industry.

(7) Transport includes Rail and Inland waterways.

(8) Other sectors includes Agriculture/Forestry, Fishing and Non-specified other.

* "Coal" refers to all types of coal, primary (anthracite, coking coal, other bituminous coal, sub-bituminous coal, lignite/brown coal) and derived fuels (including patent fuel, BKB, coke oven coke, gas coke, coal tar, coke oven gas, gas works gas, blast furnace gas and other recovered gases), as well as peat, peat briquettes and oil shale. Quantities have been converted to Mtce using calorific values reported by the respective countries.
Please refer to notes and definitions in Part I.

Source: *IEA/OECD Energy Balances of OECD Countries*

CHINESE TAIPEI

Total coal imports by origin*

(thousand tonnes)

	1978(1)	1985	1990	1995	2000	2005	2010	2011	2012e
Total coal	**1386**	**10092**	**18468**	**28681**	**45503**	**60370**	**63951**	**67654**	**64530**
Coking coal	**1386**	**2562**	**4237**	**4580**	**6093**	**5211**	**8490**	**9046**	**8412**
Australia	918	1837	2749	3451	3524	4778	7390	7933	7695
Canada	263	475	1050	693	1232	136	831	965	636
Czech Republic	-	-	-	-	-	-	-	-	-
Germany	-	-	-	-	-	-	-	-	-
Poland	-	-	-	-	-	-	-	-	-
United Kingdom	-	-	-	-	-	-	-	-	-
United States	205	250	438	336	-	47	227	-	-
Other OECD	-	-	-	-	-	-	-	-	-
China, People's Rep.	-	-	-	-	-	3	-	148	81
Colombia	-	-	-	-	-	-	-	-	-
Indonesia	-	-	-	-	-	112	-	-	-
South Africa	-	-	-	-	-	-	-	-	-
Former Soviet Union(2)	-	-	x	x	x	x	x	x	x
Russian Federation (3)	x	x	-	*100*	*1337*	-	*42*	-	-
Other FSU (3)	x	x	-	-	-	-	-	-	-
Venezuela	-	-	-	-	-	-	-	-	-
Vietnam	-	-	-	-	-	17	-	-	-
Non-specified/other	-	-	-	-	-	118	-	-	-
Steam coal(4)	**-**	**7530**	**14231**	**24101**	**39410**	**55159**	**55461**	**58608**	**56118**
Australia	-	3400	3800	6321	12474	15342	21441	19632	17658
Canada	-	-	-	-	-	261	-	90	581
Czech Republic	-	-	-	-	-	-	-	-	-
Germany	-	-	-	-	-	-	-	-	-
Poland	-	-	-	-	-	-	-	-	-
United Kingdom	-	-	-	-	-	-	-	-	-
United States	-	2800	3677	2442	514	-	-	-	227
Other OECD	-	-	-	-	1290	21	9	14	-
China, People's Rep.	-	-	529	3893	8371	18942	4181	2028	1238
Colombia	-	-	-	-	-	-	2099	952	327
Indonesia	-	100	625	6203	13740	18430	23361	27448	27630
South Africa	-	1230	5600	5073	2873	329	2748	4408	4766
Former Soviet Union(2)	-	-	x	x	x	x	x	x	x
Russian Federation (3)	x	x	-	*106*	-	*1226*	*1530*	*3716*	*3329*
Other FSU (3)	x	x	-	-	-	-	-	-	-
Venezuela	-	-	-	-	-	-	-	-	-
Vietnam	-	-	-	63	78	133	50	41	33
Non-specified/other	-	-	-	-	70	475	42	279	329
Lignite	**-**	**-**	**-**	**-**	**-**	**-**	**-**	**-**	**-**

(1) Earliest year for which split by coal type is available.

(2) For years prior to 1990, or if no details for individual Former Soviet Union republics.

(3) Data for former Republics are not available separately for some countries.

(4) Steam coal now comprises anthracite, other bituminous coal and sub-bituminous coal.

* In these tables coal used for PCI and for blending has been classified by the IEA as steam coal. Accordingly, trade data reported in these tables may differ from those reported in Part III where this coal is shown as coking coal to be consistent with data reported by importing countries and with industry terminology and practice.

Source: IEA/OECD Coal Statistics.

COLOMBIA

Coal balance*

(Mtce)

	1980	1985	1990	1995	2000	2005	2010	2011	Average annual percent change 80-90	90-11
Production	3.9	8.1	19.8	23.8	35.5	54.8	69.0	79.7	17.8	6.8
Imports	-	-	-	-	-	0.0	-	-	-	-
Exports	-1.4	-4.3	-12.6	-17.1	-33.0	-49.8	-64.4	-74.6	24.9	8.8
Stock changes	0.1	-0.9	-2.8	-1.6	1.3	-1.2	-	-0.0		
Primary supply	**2.6**	**2.9**	**4.4**	**5.1**	**3.8**	**3.9**	**4.6**	**5.0**	**5.6**	**0.6**
Statistical differences	**1.0**	**1.2**	**0.3**	**0.1**	**0.8**	**0.1**	**-0.4**	**0.0**		
Total transformation	**-1.1**	**-1.6**	**-1.9**	**-2.0**	**-1.0**	**-1.2**	**-1.9**	**-1.3**	**5.7**	**-1.7**
Electricity and heat gen.	-0.7	-1.1	-1.5	-1.7	-0.8	-1.0	-1.5	-0.9	8.2	-2.6
Main activity producers (1)	*-0.4*	*-0.8*	*-1.0*	*-1.2*	*-0.7*	*-0.8*	*-1.4*	*-0.6*	*9.5*	*-2.2*
Autoproducers	*-0.3*	*-0.3*	*-0.5*	*-0.5*	*-0.1*	*-0.2*	*-0.1*	*-0.2*	*6.2*	*-3.4*
Gas works	-	-	-	-	-	-	-	-	-	-
Coal transformation(2)	-0.4	-0.5	-0.4	-0.2	-0.2	-0.2	-0.5	-0.4	-0.3	0.7
BKB plants	-	-	-	-	-	-	-	-	-	-
Blast furnaces	*-0.1*	*-0.1*	*-0.1*	*-0.1*	*-0.1*	*-0.1*	*-0.1*	*-0.1*	*0.5*	*-2.5*
Coke ovens	*-0.3*	*-0.3*	*-0.2*	*-0.1*	*-0.1*	*-0.1*	*-0.4*	*-0.4*	*-0.8*	*2.1*
Patent fuel plants	-	-	-	-	-	-	-	-	-	-
Other transformation(3)	-	-	-	-	-	-	-	-	-	-
Energy ind. own use	**-0.5**	**-0.5**	**-0.5**	**-0.5**	**-0.3**	**-0.3**	**-0.0**	**-0.0**	**0.8**	**-11.0**
Losses	**-0.1**	**-0.1**	**-0.1**	**-0.1**	**-0.1**	**-0.1**	**-0.1**	**-0.1**		
Final consumption(4)	**1.9**	**2.0**	**2.3**	**2.7**	**3.2**	**2.4**	**2.2**	**3.6**	**1.8**	**2.1**
Industry(5)	1.7	1.8	2.1	2.5	3.1	2.4	2.1	3.5	2.3	2.4
Iron and steel	*0.1*	*0.1*	*0.1*	*0.6*	*0.6*	*0.5*	*1.0*	*1.1*	*0.3*	*10.4*
Chemical	*0.1*	*0.1*	*0.1*	*0.1*	*0.2*	*0.1*	*0.1*	*0.1*	*-1.9*	*2.5*
Non-metallic minerals	*0.9*	*0.9*	*1.2*	*0.9*	*1.2*	*0.9*	*0.6*	*1.4*	*2.9*	*0.8*
Paper, pulp and print	*0.2*	*0.3*	*0.3*	*0.4*	*0.5*	*0.4*	*0.2*	*0.4*	*2.8*	*1.3*
Other industry (6)	*0.3*	*0.4*	*0.4*	*0.5*	*0.6*	*0.5*	*0.3*	*0.5*	*2.0*	*0.8*
Transport(7)	0.0	0.0	0.0	0.0	0.0	0.0	0.0	-	-11.8	-
Other	0.2	0.2	0.2	0.2	0.1	0.1	0.1	0.1	-2.6	-2.5
Comm. and pub. services	-	-	-	-	-	-	-	-	-	-
Residential	*0.2*	*0.2*	*0.2*	*0.2*	*0.1*	*0.1*	*0.1*	*0.1*	-2.6	-2.5
Other sectors (8)	-	-	-	-	-	-	-	-	-	-
Non-energy use	-	-	-	-	-	-	-	-	-	-

(1) Main activity electricity and heat generation includes district heating.

(2) Coal transformation refers to the transformation of coal and peat to secondary and tertiary products (mainly cokes, briquettes, coal tar and off-gases). Gas works and Liquefaction are shown separately to match the *Energy Balances of OECD Countries.*

(3) Other transformation includes Liquefaction and Non-specified transformations.

(4) Final Consumption includes non-energy use and energy use (Industry, Transport and Other).

(5) Please refer to notes and definitions in Part I for detailed explanation.

(6) Other industry includes Non-ferrous metals, Transport equipment, Machinery, Mining and quarrying, Food and tobacco, Wood and wood products, Construction, Textile and leather, and Non-specified industry.

(7) Transport includes Rail and Inland waterways.

(8) Other sectors includes Agriculture/Forestry, Fishing and Non-specified other.

* "Coal" refers to all types of coal, primary (anthracite, coking coal, other bituminous coal, sub-bituminous coal, lignite/brown coal) and derived fuels (including patent fuel, BKB, coke oven coke, gas coke, coal tar, coke oven gas, gas works gas, blast furnace gas and other recovered gases), as well as peat, peat briquettes and oil shale. Quantities have been converted to Mtce using calorific values reported by the respective countries.
Please refer to notes and definitions in Part I.

Source: *IEA/OECD Energy Balances of OECD Countries*

COLOMBIA

Steam coal[1] exports by destination

(thousand tonnes)

	1978[2]	1985	1990	1995	2000	2005	2010	2011	2012e
World	**50**	**3073**	**13377**	**17212**	**33565**	**53609**	**65141**	**75413**	**81654**
OECD	**-**	**1587**	**12623**	**15392**	**32284**	**50231**	**54404**	**64363**	**71779**
Austria	-	-	-	-	-	-	-	23	9
Belgium	-	51	170	269	153	499	165	240	331
Canada	-	-	-	213	1590	2086	1731	1487	1204
Czech Republic	-	-	-	-	-	-	-	-	-
Denmark	-	332	2077	1116	819	1225	1358	4405	3069
Finland	-	347	334	203	-	-	411	29	4
France	-	2	2033	1228	2832	2181	1966	1608	1970
Germany	-	19	351	995	899	2873	180	2070	4696
Greece	-	-	-	-	-	-	152	148	-
Hungary	-	-	-	-	-	-	-	-	-
Israel	-	100	548	886	773	4620	3619	5595	5713
Italy	-	180	290	148	1694	2442	1645	1408	2343
Japan	-	-	35	27	-	-	487	231	159
Korea	-	-	-	-	-	-	1115	278	1272
Mexico	-	-	-	-	-	-	854	354	75
Netherlands	-	150	1634	2450	6822	5581	10419	13483	10074
Norway	-	-	-	155	-	-	317	-	-
Poland	-	-	-	-	-	-	194	97	22
Portugal	-	-	390	727	2544	2467	1398	2316	4126
Spain	-	134	404	665	908	1946	2616	3250	5607
Sweden	-	183	108	-	83	-	-	-	-
Turkey	-	-	-	-	-	2525	2738	6046	7455
United Kingdom	-	89	2136	2648	4950	2576	5417	7284	11240
United States	-	-	1305	2483	6412	17260	12973	8292	6306
Other OECD[3]	-	-	808	1179	1805	1950	4649	5719	6104
Non-OECD	**50**	**310**	**702**	**1820**	**1159**	**3051**	**10458**	**11050**	**9835**
Brazil	-	-	-	69	149	279	1312	1631	2014
Bulgaria	-	-	-	-	-	-	-	-	-
China, People's Rep.	-	-	-	-	-	-	3725	1441	2789
Chinese Taipei	-	-	-	-	-	-	2099	660	489
Hong Kong, China	-	100	600	344	-	-	-	-	-
India	-	-	-	-	-	-	289	9	164
Morocco	-	-	29	51	-	-	-	3578	-
Romania	-	-	-	-	69	-	-	30	-
Russian Federation	-	-	-	-	-	7	-	-	-
Ukraine	-	-	-	-	-	-	-	-	-
Other Africa	-	-	-	-	-	-	72	-	152
Other Asia	-	-	-	-	-	-	201	-	319
Other Eastern Europe	-	-	-	-	146	121	412	1139	490
Other FSU	-	-	-	-	-	-	-	-	-
Other non-OECD Americas	50	210	73	1556	1321	3303	6087	6525	8413
Other Middle East	-	100	548	886	773	4620	3619	5595	5713
Non-specified/Other	-	1176	52	-	122	327	279	-	40

(1) Steam coal now comprises anthracite, other bituminous coal and sub-bituminous coal.

(2) Earliest year for which split by coal type is available.

(3) Australia, Chile, Estonia, Iceland, Ireland, Luxembourg, New Zealand, Slovak Republic, Slovenia and Switzerland.

Source: IEA/OECD Coal Statistics.

HONG KONG (CHINA)

Coal balance*

(Mtce)

	1980	1985	1990	1995	2000	2005	2010	2011	Average annual percent change 80-90	90-11
Production	-	-	-	-	-	-	-	-	-	-
Imports	0.0	4.9	7.9	8.0	5.3	9.5	9.1	11.0	98.2	1.6
Exports	-	-	-	-	-	-	-	-	-	-
Stock changes	-	-	-	-	-	-	-	-		
Primary supply	**0.0**	**4.9**	**7.9**	**8.0**	**5.3**	**9.5**	**9.1**	**11.0**	**98.2**	**1.6**
Statistical differences	**-**	**-0.3**	**0.8**	**0.5**	**0.7**	**-**	**-**	**-**		
Total transformation	**0.1**	**-4.3**	**-8.2**	**-7.9**	**-5.3**	**-7.9**	**-6.9**	**-8.3**	**-**	**0.0**
Electricity and heat gen.	-	-4.5	-8.7	-8.6	-6.1	-8.8	-7.7	-9.1	-	0.2
Main activity producers (1)	-	*-4.5*	*-8.7*	*-8.6*	*-6.1*	*-8.8*	*-7.7*	*-9.1*	-	*0.2*
Autoproducers	-	-	-	-	-	-	-	-	-	-
Gas works	0.1	0.2	0.5	0.7	0.8	0.8	0.8	0.9	15.6	3.0
Coal transformation(2)	-0.0	-0.0	-0.0	-0.0	-	-	-	-	-8.9	-
BKB plants	-	-	-	-	-	-	-	-	-	-
Blast furnaces	*-0.0*	*-0.0*	*-0.0*	*-0.0*	-	-	-	-	*-8.9*	-
Coke ovens	-	-	-	-	-	-	-	-	-	-
Patent fuel plants	-	-	-	-	-	-	-	-	-	-
Other transformation(3)	-	-	-	-	-	-	-	-	-	-
Energy ind. own use	**-**	**-**	**-**	**-**	**-**	**-**	**-**	**-**	**-**	**-**
Losses	**-**	**-**	**-**	**-**	**-**	**-**	**-**	**-**		
Final consumption(4)	**0.1**	**0.2**	**0.5**	**0.7**	**0.8**	**1.6**	**2.2**	**2.8**	**15.1**	**8.9**
Industry(5)	0.0	0.0	0.0	0.0	0.0	0.8	1.4	1.9	4.5	24.7
Iron and steel	*0.0*	*0.0*	*0.0*	*0.0*	-	-	-	-	*-12.4*	-
Chemical	-	-	-	-	-	-	-	-	-	-
Non-metallic minerals	-	-	-	-	-	-	-	-	-	-
Paper, pulp and print	-	-	-	-	-	-	-	-	-	-
Other industry (6)	*0.0*	*0.0*	*0.0*	*0.0*	*0.0*	*0.8*	*1.4*	*1.9*	*6.9*	*25.0*
Transport(7)	-	-	-	-	-	-	-	-	-	-
Other	0.1	0.2	0.4	0.6	0.8	0.8	0.8	0.8	15.9	3.0
Comm. and pub. services	*0.0*	*0.1*	*0.2*	*0.3*	*0.3*	*0.3*	*0.3*	*0.4*	*16.0*	*2.5*
Residential	*0.1*	*0.1*	*0.2*	*0.4*	*0.4*	*0.5*	*0.5*	*0.5*	*15.8*	*3.5*
Other sectors (8)	-	-	-	-	-	-	-	-	-	-
Non-energy use	-	-	-	-	-	-	-	-	-	-

(1) Main activity electricity and heat generation includes district heating.

(2) Coal transformation refers to the transformation of coal and peat to secondary and tertiary products (mainly cokes, briquettes, coal tar and off-gases). Gas works and Liquefaction are shown separately to match the *Energy Balances of OECD Countries*.

(3) Other transformation includes Liquefaction and Non-specified transformations.

(4) Final Consumption includes non-energy use and energy use (Industry, Transport and Other).

(5) Please refer to notes and definitions in Part I for detailed explanation.

(6) Other industry includes Non-ferrous metals, Transport equipment, Machinery, Mining and quarrying, Food and tobacco, Wood and wood products, Construction, Textile and leather, and Non-specified industry.

(7) Transport includes Rail and Inland waterways.

(8) Other sectors includes Agriculture/Forestry, Fishing and Non-specified other.

* "Coal" refers to all types of coal, primary (anthracite, coking coal, other bituminous coal, sub-bituminous coal, lignite/brown coal) and derived fuels (including patent fuel, BKB, coke oven coke, gas coke, coal tar, coke oven gas, gas works gas, blast furnace gas and other recovered gases), as well as peat, peat briquettes and oil shale. Quantities have been converted to Mtce using calorific values reported by the respective countries. Please refer to notes and definitions in Part I.

Source: *IEA/OECD Energy Balances of OECD Countries*

HONG KONG (CHINA)

Total coal imports by origin*

(thousand tonnes)

	1978(1)	1985	1990	1995	2000	2005	2010	2011	2012e
Total coal	**8**	**5523**	**8928**	**9109**	**6058**	**10823**	**10324**	**12946**	**12599**
Coking coal	**-**	**-**	**-**	**-**	**43**	**-**	**-**	**-**	**84**
Australia	-	-	-	-	-	-	-	-	-
Canada	-	-	-	-	-	-	-	-	-
Czech Republic	-	-	-	-	-	-	-	-	-
Germany	-	-	-	-	-	-	-	-	-
Poland	-	-	-	-	-	-	-	-	-
United Kingdom	-	-	-	-	-	-	-	-	-
United States	-	-	-	-	-	-	-	-	84
Other OECD	-	-	-	-	-	-	-	-	-
China, People's Rep.	-	-	-	-	43	-	-	-	-
Colombia	-	-	-	-	-	-	-	-	-
Indonesia	-	-	-	-	-	-	-	-	-
South Africa	-	-	-	-	-	-	-	-	-
Former Soviet Union(2)	-	-	x	x	x	x	x	x	x
Russian Federation (3)	*x*	*x*	-	-	-	-	-	-	-
Other FSU (3)	*x*	*x*	-	-	-	-	-	-	-
Venezuela	-	-	-	-	-	-	-	-	-
Vietnam	-	-	-	-	-	-	-	-	-
Non-specified/other	-	-	-	-	-	-	-	-	-
Steam coal(4)	**8**	**5523**	**8928**	**9109**	**6015**	**10823**	**10324**	**12946**	**12515**
Australia	-	1986	3003	2568	276	-	441	1350	666
Canada	-	584	-	-	-	-	-	-	-
Czech Republic	-	-	-	-	-	-	-	-	-
Germany	-	-	-	-	-	-	-	-	-
Poland	-	-	-	-	-	-	-	-	-
United Kingdom	-	-	-	-	-	-	-	-	-
United States	-	-	108	-	-	-	-	-	-
Other OECD	-	-	-	-	-	-	-	-	-
China, People's Rep.	8	709	1708	1233	2265	938	395	-	-
Colombia	-	-	234	344	-	-	-	-	-
Indonesia	-	-	659	2192	2846	9825	9303	11374	11673
South Africa	-	2244	3216	2770	567	-	-	-	-
Former Soviet Union(2)	-	-	x	x	x	x	x	x	x
Russian Federation (3)	*x*	*x*	-	*2*	*61*	*60*	*135*	*222*	*176*
Other FSU (3)	*x*	*x*	-	-	-	-	-	-	-
Venezuela	-	-	-	-	-	-	-	-	-
Vietnam	-	-	-	-	-	-	-	-	-
Non-specified/other	-	-	-	-	-	-	50	-	-
Lignite	**-**	**-**	**-**	**-**	**-**	**-**	**-**	**-**	**-**

(1) Earliest year for which split by coal type is available.

(2) For years prior to 1990, or if no details for individual Former Soviet Union republics.

(3) Data for former Republics are not available separately for some countries.

(4) Steam coal now comprises anthracite, other bituminous coal and sub-bituminous coal.

* In these tables coal used for PCI and for blending has been classified by the IEA as steam coal. Accordingly, trade data reported in these tables may differ from those reported in Part III where this coal is shown as coking coal to be consistent with data reported by importing countries and with industry terminology and practice.

Source: IEA/OECD Coal Statistics.

INDIA

Coal balance*

(Mtce)

	1980	1985	1990	1995	2000	2005	2010	2011	Average annual percent change 80-90	90-11
Production	76.6	105.2	149.6	186.9	208.9	268.2	353.3	360.2	6.9	4.3
Imports	0.5	2.0	6.0	12.5	21.1	37.2	96.5	108.1	27.4	14.8
Exports	-0.1	-0.1	-0.1	-0.5	-1.0	-1.4	-1.9	-1.9	-0.9	17.3
Stock changes	-6.4	-2.6	-7.8	-4.6	1.6	-6.9	-5.0	-1.0		
Primary supply	**70.6**	**104.4**	**147.7**	**194.2**	**230.7**	**297.2**	**443.0**	**465.4**	**7.7**	**5.6**
Statistical differences	**2.2**	**2.3**	**2.1**	**2.2**	**2.1**	**-0.5**	**-12.3**	**-16.9**		
Total transformation	**-32.6**	**-54.5**	**-87.7**	**-139.2**	**-183.2**	**-232.0**	**-311.2**	**-324.5**	**10.4**	**6.4**
Electricity and heat gen.	-26.3	-47.4	-79.4	-128.2	-171.8	-217.3	-296.3	-309.1	11.7	6.7
Main activity producers (1)	*-25.0*	*-44.6*	*-74.8*	*-121.2*	*-161.2*	*-202.0*	*-266.5*	*-277.8*	*11.6*	*6.4*
Autoproducers	*-1.3*	*-2.8*	*-4.6*	*-7.0*	*-10.6*	*-15.3*	*-29.9*	*-31.4*	*13.5*	*9.5*
Gas works	-	-	-	-	-0.0	-0.0	-0.0	-0.0	-	-
Coal transformation(2)	-6.3	-7.2	-8.3	-11.0	-11.4	-14.7	-14.8	-15.4	2.7	3.0
BKB plants	*-0.1*	*-0.2*	*-0.3*	*-0.3*	*-0.2*	*-0.2*	*-0.3*	*-0.3*	*7.9*	*1.3*
Blast furnaces	*-5.1*	*-5.1*	*-5.2*	*-5.7*	*-8.3*	*-8.3*	*-11.3*	*-11.8*	*0.1*	*4.0*
Coke ovens	*-1.1*	*-1.8*	*-2.8*	*-5.0*	*-2.8*	*-6.2*	*-3.2*	*-3.2*	*10.1*	*0.6*
Patent fuel plants	-	-	-	-	-	-	-	-	-	-
Other transformation(3)	-	-	-	-	-	-	-	-	-	-
Energy ind. own use	**-0.9**	**-1.3**	**-2.4**	**-2.4**	**-1.7**	**-1.9**	**-1.6**	**-1.6**	**10.1**	**-2.1**
Losses	-	-	-	-	-	-	-	-		
Final consumption(4)	**39.3**	**50.9**	**59.7**	**54.8**	**47.8**	**62.8**	**118.0**	**122.4**	**4.3**	**3.5**
Industry(5)	21.1	31.5	41.0	38.6	36.0	49.4	101.3	111.1	6.9	4.9
Iron and steel	*8.5*	*9.2*	*12.5*	*15.5*	*13.7*	*21.0*	*41.4*	*48.8*	*3.9*	*6.7*
Chemical	*1.9*	*2.7*	*3.9*	*4.7*	*3.5*	*2.4*	*2.9*	*2.7*	*7.4*	*-1.7*
Non-metallic minerals	*5.1*	*6.2*	*9.5*	*9.9*	*12.7*	*12.7*	*15.2*	*17.4*	*6.3*	*2.9*
Paper, pulp and print	*1.4*	*1.8*	*2.0*	*2.3*	*1.8*	*1.8*	*2.4*	*1.5*	*3.6*	*-1.4*
Other industry (6)	*4.1*	*11.7*	*13.2*	*6.1*	*4.2*	*11.4*	*39.4*	*40.7*	*12.3*	*5.5*
Transport(7)	7.8	6.4	3.5	0.2	-	-	-	-	-7.8	-
Other	10.4	13.0	15.2	16.1	11.9	13.4	16.6	11.3	3.9	-1.4
Comm. and pub. services	*4.4*	*5.0*	*5.3*	*5.6*	*4.1*	*4.2*	*5.3*	*5.6*	*1.7*	*0.3*
Residential	*3.2*	*3.6*	*4.3*	*4.2*	*4.0*	*3.8*	*4.2*	*4.4*	*3.0*	*0.1*
Other sectors (8)	*2.7*	*4.5*	*5.7*	*6.3*	*3.7*	*5.3*	*7.1*	*1.4*	*7.5*	*-6.6*
Non-energy use	-	-	-	-	-	-	-	-	-	-

(1) Main activity electricity and heat generation includes district heating.

(2) Coal transformation refers to the transformation of coal and peat to secondary and tertiary products (mainly cokes, briquettes, coal tar and off-gases). Gas works and Liquefaction are shown separately to match the *Energy Balances of OECD Countries.*

(3) Other transformation includes Liquefaction and Non-specified transformations.

(4) Final Consumption includes non-energy use and energy use (Industry, Transport and Other).

(5) Please refer to notes and definitions in Part I for detailed explanation.

(6) Other industry includes Non-ferrous metals, Transport equipment, Machinery, Mining and quarrying, Food and tobacco, Wood and wood products, Construction, Textile and leather, and Non-specified industry.

(7) Transport includes Rail and Inland waterways.

(8) Other sectors includes Agriculture/Forestry, Fishing and Non-specified other.

* "Coal" refers to all types of coal, primary (anthracite, coking coal, other bituminous coal, sub-bituminous coal, lignite/brown coal) and derived fuels (including patent fuel, BKB, coke oven coke, gas coke, coal tar, coke oven gas, gas works gas, blast furnace gas and other recovered gases), as well as peat, peat briquettes and oil shale. Quantities have been converted to Mtce using calorific values reported by the respective countries. Please refer to notes and definitions in Part I.

Source: *IEA/OECD Energy Balances of OECD Countries*

INDIA

Total coal imports by origin*

(thousand tonnes)

	1978(1)	1985	1990	1995	2000	2005	2010	2011	2012e
Total coal	**220**	**2030**	**5100**	**12510**	**16565**	**38586**	**115717**	**132121**	**159619**
Coking coal	**220**	**1936**	**5000**	**9370**	**8372**	**16892**	**34424**	**33943**	**36577**
Australia	20	1936	4665	8283	7824	13973	30730	28726	28586
Canada	200	-	-	-	-	-	-	280	828
Czech Republic	-	-	-	-	-	-	-	-	-
Germany	-	-	-	-	-	-	-	-	-
Poland	-	-	284	-	-	-	-	-	-
United Kingdom	-	-	-	-	-	-	-	-	-
United States	-	-	-	-	11	837	2299	3448	3922
Other OECD	-	-	51	187	63	92	1073	1142	1132
China, People's Rep.	-	-	-	-	474	603	-	118	-
Colombia	-	-	-	-	-	-	-	-	-
Indonesia	-	-	-	-	-	1365	70	-	2
South Africa	-	-	-	-	-	-	226	-	-
Former Soviet Union(2)	-	-	x	x	x	x	x	x	x
Russian Federation (3)	*x*	*x*	-	-	-	*22*	*6*	*4*	-
Other FSU (3)	*x*	*x*	-	-	-	-	-	-	-
Venezuela	-	-	-	-	-	-	-	-	-
Vietnam	-	-	-	-	-	-	-	-	-
Non-specified/other	-	-	-	900	-	-	20	225	2107
Steam coal(4)	**-**	**94**	**100**	**3140**	**8193**	**21694**	**81293**	**98178**	**123042**
Australia	-	94	100	-	2748	1438	2227	2886	1978
Canada	-	-	-	-	-	-	-	-	-
Czech Republic	-	-	-	-	-	-	-	-	-
Germany	-	-	-	-	-	-	-	-	-
Poland	-	-	-	-	-	-	-	-	-
United Kingdom	-	-	-	-	-	-	1	1	1
United States	-	-	-	-	-	86	773	2557	1828
Other OECD	-	-	-	1	-	-	-	15	-
China, People's Rep.	-	-	-	399	1610	3152	2	-	-
Colombia	-	-	-	-	-	-	260	5	164
Indonesia	-	-	-	-	2256	13889	53677	78291	95840
South Africa	-	-	-	-	1507	3044	24030	13895	22985
Former Soviet Union(2)	-	-	x	x	x	x	x	x	x
Russian Federation (3)	*x*	*x*	-	-	*72*	*33*	*220*	*528*	*220*
Other FSU (3)	*x*	*x*	-	-	-	-	-	-	-
Venezuela	-	-	-	-	-	-	-	-	26
Vietnam	-	-	-	-	-	-	57	-	-
Non-specified/other	-	-	-	2740	-	52	46	-	-
Lignite	**-**	**-**	**-**	**-**	**-**	**-**	**-**	**-**	**-**

(1) Earliest year for which split by coal type is available.

(2) For years prior to 1990, or if no details for individual Former Soviet Union republics.

(3) Data for former Republics are not available separately for some countries.

(4) Steam coal now comprises anthracite, other bituminous coal and sub-bituminous coal.

* In these tables coal used for PCI and for blending has been classified by the IEA as steam coal. Accordingly, trade data reported in these tables may differ from those reported in Part III where this coal is shown as coking coal to be consistent with data reported by importing countries and with industry terminology and practice.

Source: IEA/OECD Coal Statistics.

INDONESIA

Coal balance*

(Mtce)

	1980	1985	1990	1995	2000	2005	2010	2011	Average annual percent change 80-90	90-11
Production	0.2	1.6	8.4	34.1	64.9	140.3	266.2	295.5	42.2	18.5
Imports	0.0	0.0	0.6	-	0.1	0.1	0.1	0.0	32.6	-12.1
Exports	-0.1	-0.8	-3.9	-26.1	-47.9	-108.8	-222.7	-250.6	45.1	21.9
Stock changes	0.0	-	-	1.0	-	-	-	-		
Primary supply	**0.2**	**0.8**	**5.1**	**9.0**	**17.2**	**31.6**	**43.5**	**45.0**	**36.5**	**11.0**
Statistical differences	**-0.0**	**0.9**	**1.3**	**0.4**	**1.5**	**-0.5**	**1.6**	**2.3**		
Total transformation	**-0.0**	**-0.1**	**-3.3**	**-4.9**	**-12.0**	**-19.2**	**-26.9**	**-31.3**	**71.3**	**11.2**
Electricity and heat gen.	-	-0.1	-3.3	-4.9	-12.0	-19.2	-26.9	-31.3	-	11.3
Main activity producers (1)	-	*-0.1*	*-3.3*	*-4.9*	*-9.9*	*-12.7*	*-18.0*	*-20.6*	-	*9.1*
Autoproducers	-	-	-	-	*-2.1*	*-6.5*	*-8.9*	*-10.7*	-	-
Gas works	-	-	-	-	-	-	-	-	-	-
Coal transformation(2)	-0.0	-0.0	-0.0	-0.0	-0.0	-0.0	-0.0	-0.0	3.9	-12.7
BKB plants	-	-	-	*-0.0*	*-0.0*	*-0.0*	*-0.0*	*-0.0*	-	-
Blast furnaces	*-0.0*	*-0.0*	*-0.0*	-	-	-	-	-	*3.9*	-
Coke ovens	-	-	-	-	-	-	-	-	-	-
Patent fuel plants	-	-	-	-	-	-	-	-	-	-
Other transformation(3)	-	-	-	-	-	-	-	-	-	-
Energy ind. own use	**-0.1**	**-0.1**	**-**	**-**	**-**	**-**	**-**	**-**	**-**	**-**
Losses	**-**	**-**	**-**	**-**	**-**	**-**	**-**	**-**		
Final consumption(4)	**0.1**	**1.4**	**3.0**	**4.5**	**6.6**	**11.9**	**18.2**	**16.0**	**37.1**	**8.2**
Industry(5)	0.1	1.4	3.0	4.5	6.6	11.9	18.2	16.0	39.5	8.2
Iron and steel	*0.0*	*0.0*	*0.0*	-	-	-	-	-	*3.3*	-
Chemical	*0.1*	-	-	-	-	-	-	-	-	-
Non-metallic minerals	-	-	-	-	*1.7*	*3.9*	*4.7*	*4.4*	-	-
Paper, pulp and print	-	-	-	-	*0.6*	*0.9*	*1.3*	*1.3*	-	-
Other industry (6)	*0.0*	*1.4*	*3.0*	*4.5*	*4.4*	*7.2*	*12.2*	*10.3*	*57.3*	*6.0*
Transport(7)	0.0	0.0	-	-	-	-	-	-	-	-
Other	-	0.0	-	0.0	0.0	0.0	-	0.0	-	-
Comm. and pub. services	-	-	-	-	-	-	-	-	-	-
Residential	-	-	-	*0.0*	*0.0*	*0.0*	-	*0.0*	-	-
Other sectors (8)	-	*0.0*	-	-	-	-	-	-	-	-
Non-energy use	-	-	-	-	-	-	-	-	-	-

(1) Main activity electricity and heat generation includes district heating.

(2) Coal transformation refers to the transformation of coal and peat to secondary and tertiary products (mainly cokes, briquettes, coal tar and off-gases). Gas works and Liquefaction are shown separately to match the *Energy Balances of OECD Countries.*

(3) Other transformation includes Liquefaction and Non-specified transformations.

(4) Final Consumption includes non-energy use and energy use (Industry, Transport and Other).

(5) Please refer to notes and definitions in Part I for detailed explanation.

(6) Other industry includes Non-ferrous metals, Transport equipment, Machinery, Mining and quarrying, Food and tobacco, Wood and wood products, Construction, Textile and leather, and Non-specified industry.

(7) Transport includes Rail and Inland waterways.

(8) Other sectors includes Agriculture/Forestry, Fishing and Non-specified other.

* "Coal" refers to all types of coal, primary (anthracite, coking coal, other bituminous coal, sub-bituminous coal, lignite/brown coal) and derived fuels (including patent fuel, BKB, coke oven coke, gas coke, coal tar, coke oven gas, gas works gas, blast furnace gas and other recovered gases), as well as peat, peat briquettes and oil shale. Quantities have been converted to Mtce using calorific values reported by the respective countries. Please refer to notes and definitions in Part I.

Source: *IEA/OECD Energy Balances of OECD Countries*

INDONESIA

Steam coal[1] exports by destination

(thousand tonnes)

	1978[2]	1985	1990	1995	2000	2005	2010	2011	2012e
World	**32**	**978**	**4574**	**28908**	**54480**	**118396**	**265000**	**297836**	**382421**
OECD	**-**	**306**	**961**	**11743**	**27791**	**45977**	**94212**	**83726**	**85915**
Austria	-	-	-	-	-	-	-	-	-
Belgium	-	-	6	-	-	6	-	-	-
Canada	-	-	-	-	-	-	-	-	-
Czech Republic	-	-	-	18	-	-	-	-	-
Denmark	-	-	-	285	-	-	-	-	-
Finland	-	-	-	-	-	-	-	-	-
France	-	-	-	65	-	306	-	-	-
Germany	-	-	38	172	105	109	-	33	-
Greece	-	-	-	-	133	80	36	-	-
Hungary	-	-	-	-	-	-	-	-	-
Israel	-	-	-	-	-	-	-	-	-
Italy	-	-	-	31	1637	2422	6835	5768	3836
Japan	-	306	663	5998	13101	19511	35746	32924	33685
Korea	-	-	33	1795	4825	12885	43102	38482	41478
Mexico	-	-	-	-	-	-	-	69	-
Netherlands	-	-	133	855	2700	1076	2804	504	-
Norway	-	-	-	-	-	-	-	-	-
Poland	-	-	-	-	7	-	-	-	-
Portugal	-	-	-	-	70	119	-	-	-
Spain	-	-	-	881	2793	3146	2616	3433	5970
Sweden	-	-	-	-	-	-	-	-	-
Turkey	-	-	-	-	-	-	46	228	53
United Kingdom	-	-	-	86	-	1772	165	-	56
United States	-	-	-	748	627	1884	1583	805	416
Other OECD[3]	-	-	88	809	1793	2661	1279	1480	421
Non-OECD	**32**	**672**	**3613**	**17147**	**26689**	**58073**	**170724**	**214110**	**296506**
Brazil	-	-	-	-	468	146	-	-	-
Bulgaria	-	-	-	123	-	-	-	-	-
China, People's Rep.	-	-	105	336	142	1008	57430	61984	115557
Chinese Taipei	-	-	546	5556	11786	19132	21280	24634	27193
Hong Kong, China	-	-	660	4481	2816	9826	8725	11685	11790
India	-	-	110	1800	3373	11657	44990	71099	96200
Morocco	-	-	-	-	-	-	-	-	-
Romania	-	-	-	-	-	-	-	-	-
Russian Federation	-	-	-	-	-	-	-	-	-
Ukraine	-	-	-	-	-	-	-	75	81
Other Africa	-	-	-	-	-	155	158	85	10
Other Asia	32	672	2192	4582	8032	16004	37194	43619	45296
Other Eastern Europe	-	-	-	269	-	65	770	775	152
Other FSU	-	-	-	-	-	53	-	-	-
Other non-OECD Americas	-	-	70	719	978	919	821	550	190
Other Middle East	-	-	-	-	-	12	101	144	227
Non-specified/Other	-	-	-	18	-	14346	64	-	-

(1) Steam coal now comprises anthracite, other bituminous coal and sub-bituminous coal.

(2) Earliest year for which split by coal type is available.

(3) Australia, Chile, Estonia, Iceland, Ireland, Luxembourg, New Zealand, Slovak Republic, Slovenia and Switzerland.

Source: IEA/OECD Coal Statistics.

KAZAKHSTAN

Coal balance*

(Mtce)

	1980	1985	1990	1995	2000	2005	2010	2011	Average annual percent change 80-90	90-11
Production	x	x	82.9	53.1	48.8	54.7	69.4	72.7	x	-0.6
Imports	x	x	6.8	1.5	1.0	1.4	0.8	0.8	x	-9.9
Exports	x	x	-32.6	-13.3	-21.8	-15.6	-19.7	-19.2	x	-2.5
Stock changes	x	x	-	-	0.3	0.2	-1.2	-0.3		
Primary supply	**x**	**x**	**57.1**	**41.3**	**28.2**	**40.7**	**49.3**	**54.0**	**x**	**-0.3**
Statistical differences	**x**	**x**	**-**	**-**	**1.6**	**-0.0**	**2.9**	**2.0**		
Total transformation	**x**	**x**	**-34.5**	**-24.3**	**-23.5**	**-28.2**	**-28.4**	**-29.9**	**x**	**-0.7**
Electricity and heat gen.	x	x	-34.4	-23.9	-20.9	-25.8	-25.7	-27.1	x	-1.1
Main activity producers (1)	*x*	*x*	*-34.4*	*-23.9*	*-20.9*	*-25.8*	*-25.7*	*-27.1*	*x*	*-1.1*
Autoproducers	*x*	*x*	-	-	-	-	-	-	*x*	-
Gas works	x	x	-	-	-	-	-	-	x	-
Coal transformation (2)	x	x	-0.1	-0.3	-2.6	-2.4	-2.6	-2.8	x	16.4
BKB plants	*x*	*x*	-	-	-	-	-	-	*x*	-
Blast furnaces	*x*	*x*	-	-	*-0.9*	*-0.9*	*-1.1*	*-1.1*	*x*	-
Coke ovens	*x*	*x*	*-0.1*	*-0.3*	*-1.7*	*-1.5*	*-1.5*	*-1.7*	*x*	*13.7*
Patent fuel plants	*x*	*x*	-	-	*0.0*	*0.0*	-	-	*x*	-
Other transformation (3)	x	x	-	-	-	-	-	-	x	-
Energy ind. own use	**x**	**x**	**-**	**-0.4**	**-0.7**	**-0.7**	**-1.2**	**-1.0**	**x**	**-**
Losses	**x**	**x**	**-**	**-**	**-0.1**	**-1.3**	**-1.3**	**-2.1**		
Final consumption (4)	**x**	**x**	**22.5**	**16.7**	**5.5**	**10.5**	**21.4**	**23.0**	**x**	**0.1**
Industry (5)	x	x	22.5	16.7	4.5	9.6	15.6	18.4	x	-1.0
Iron and steel	*x*	*x*	*1.2*	*2.3*	*1.3*	*1.4*	*4.6*	*5.5*	*x*	*7.4*
Chemical	*x*	*x*	-	-	-	-	*0.0*	*0.1*	*x*	-
Non-metallic minerals	*x*	*x*	-	-	-	-	-	-	*x*	-
Paper, pulp and print	*x*	*x*	-	-	-	-	-	-	*x*	-
Other industry (6)	*x*	*x*	*21.3*	*14.4*	*3.2*	*8.2*	*10.9*	*12.8*	*x*	*-2.4*
Transport (7)	x	x	-	-	-	-	-	-	x	-
Other	x	x	-	-	0.0	0.0	5.8	4.6	x	-
Comm. and pub. services	*x*	*x*	-	-	-	-	*1.1*	*1.1*	*x*	-
Residential	*x*	*x*	-	-	*0.0*	*0.0*	*1.7*	*3.3*	*x*	-
Other sectors (8)	*x*	*x*	-	-	-	-	*3.1*	*0.3*	*x*	-
Non-energy use	x	x	-	-	1.0	0.9	-	-	x	-

(1) Main activity electricity and heat generation includes district heating.

(2) Coal transformation refers to the transformation of coal and peat to secondary and tertiary products (mainly cokes, briquettes, coal tar and off-gases). Gas works and Liquefaction are shown separately to match the *Energy Balances of OECD Countries.*

(3) Other transformation includes Liquefaction and Non-specified transformations.

(4) Final Consumption includes non-energy use and energy use (Industry, Transport and Other).

(5) Please refer to notes and definitions in Part I for detailed explanation.

(6) Other industry includes Non-ferrous metals, Transport equipment, Machinery, Mining and quarrying, Food and tobacco, Wood and wood products, Construction, Textile and leather, and Non-specified industry.

(7) Transport includes Rail and Inland waterways.

(8) Other sectors includes Agriculture/Forestry, Fishing and Non-specified other.

* "Coal" refers to all types of coal, primary (anthracite, coking coal, other bituminous coal, sub-bituminous coal, lignite/brown coal) and derived fuels (including patent fuel, BKB, coke oven coke, gas coke, coal tar, coke oven gas, gas works gas, blast furnace gas and other recovered gases), as well as peat, peat briquettes and oil shale. Quantities have been converted to Mtce using calorific values reported by the respective countries. Please refer to notes and definitions in Part I.

Source: *IEA/OECD Energy Balances of OECD Countries*

KAZAKHSTAN

Total coal imports by origin*

(thousand tonnes)

	1978(1)	1985	1990	1995	2000	2005	2010	2011	2012e
Total coal	x	x	3277	1302	668	31	237	159	-
Coking coal	x	x	-	96	-	-	-	-	-
Australia	x	x	-	-	-	-	-	-	-
Canada	x	x	-	-	-	-	-	-	-
Czech Republic	x	x	-	-	-	-	-	-	-
Germany	x	x	-	-	-	-	-	-	-
Poland	x	x	-	-	-	-	-	-	-
United Kingdom	x	x	-	-	-	-	-	-	-
United States	x	x	-	-	-	-	-	-	-
Other OECD	x	x	-	-	-	-	-	-	-
China, People's Rep.	x	x	-	-	-	-	-	-	-
Colombia	x	x	-	-	-	-	-	-	-
Indonesia	x	x	-	-	-	-	-	-	-
South Africa	x	x	-	-	-	-	-	-	-
Former Soviet Union(2)	x	x	x	x	x	x	x	x	x
Russian Federation (3)	*x*	*x*	*-*	*92*	*-*	*-*	*-*	*-*	*-*
Other FSU (3)	*x*	*x*	*-*	*-*	*-*	*-*	*-*	*-*	*-*
Venezuela	x	x	-	-	-	-	-	-	-
Vietnam	x	x	-	-	-	-	-	-	-
Non-specified/other	x	x	-	4	-	-	-	-	-
Steam coal(4)	x	x	3277	1113	668	31	234	158	-
Australia	x	x	-	-	-	-	-	-	-
Canada	x	x	-	-	-	-	-	-	-
Czech Republic	x	x	-	-	-	-	-	-	-
Germany	x	x	-	30	-	-	-	-	-
Poland	x	x	-	-	-	-	-	-	-
United Kingdom	x	x	-	-	-	-	-	-	-
United States	x	x	-	-	-	-	-	-	-
Other OECD	x	x	-	-	-	-	-	-	-
China, People's Rep.	x	x	-	-	2	-	-	-	-
Colombia	x	x	-	-	-	-	-	-	-
Indonesia	x	x	-	-	-	-	-	-	-
South Africa	x	x	-	-	-	-	-	-	-
Former Soviet Union(2)	x	x	x	x	x	x	x	x	x
Russian Federation (3)	*x*	*x*	*3277*	*1037*	*480*	*31*	*234*	*158*	*-*
Other FSU (3)	*x*	*x*	*-*	*46*	*-*	*-*	*-*	*-*	*-*
Venezuela	x	x	-	-	-	-	-	-	-
Vietnam	x	x	-	-	-	-	-	-	-
Non-specified/other	x	x	-	-	186	-	-	-	-
Lignite	x	x	-	93	-	-	3	1	-

(1) Earliest year for which split by coal type is available.

(2) For years prior to 1990, or if no details for individual Former Soviet Union republics.

(3) Data for former Republics are not available separately for some countries.

(4) Steam coal now comprises anthracite, other bituminous coal and sub-bituminous coal.

* In these tables coal used for PCI and for blending has been classified by the IEA as steam coal. Accordingly, trade data reported in these tables may differ from those reported in Part III where this coal is shown as coking coal to be consistent with data reported by importing countries and with industry terminology and practice.

Source: IEA/OECD Coal Statistics.

KAZAKHSTAN

Steam coal(1) exports by destination

(thousand tonnes)

	1978(2)	1985	1990	1995	2000	2005	2010	2011	2012e
World	**x**	**x**	**53210**	**18769**	**25300**	**21308**	**29078**	**29763**	**31608**
OECD	**x**	**x**	**-**	**-**	**-**	**116**	**298**	**499**	**309**
Austria	x	x	-	-	-	-	-	-	-
Belgium	x	x	-	-	-	-	-	-	-
Canada	x	x	-	-	-	-	-	-	-
Czech Republic	x	x	-	-	-	-	-	-	-
Denmark	x	x	-	-	-	-	-	-	-
Finland	x	x	-	-	-	-	49	151	18
France	x	x	-	-	-	-	-	-	-
Germany	x	x	-	-	-	-	2	-	-
Greece	x	x	-	-	-	-	-	-	-
Hungary	x	x	-	-	-	39	9	-	-
Israel	x	x	-	-	-	-	-	-	-
Italy	x	x	-	-	-	4	-	7	-
Japan	x	x	-	-	-	-	-	-	-
Korea	x	x	-	-	-	-	-	-	-
Mexico	x	x	-	-	-	-	-	-	-
Netherlands	x	x	-	-	-	-	-	-	16
Norway	x	x	-	-	-	-	-	-	-
Poland	x	x	-	-	-	73	238	341	275
Portugal	x	x	-	-	-	-	-	-	-
Spain	x	x	-	-	-	-	-	-	-
Sweden	x	x	-	-	-	-	-	-	-
Turkey	x	x	-	-	-	-	-	-	-
United Kingdom	x	x	-	-	-	-	-	-	-
United States	x	x	-	-	-	-	-	-	-
Other OECD(3)	x	x	-	-	-	-	-	-	-
Non-OECD	**x**	**x**	**53210**	**18769**	**25300**	**21192**	**28780**	**29264**	**31299**
Brazil	x	x	-	-	-	-	-	-	-
Bulgaria	x	x	-	-	-	-	-	-	-
China, People's Rep.	x	x	-	-	-	-	-	-	-
Chinese Taipei	x	x	-	-	-	-	-	-	-
Hong Kong, China	x	x	-	-	-	-	-	-	-
India	x	x	-	-	-	-	-	-	-
Morocco	x	x	-	-	-	-	-	-	-
Romania	x	x	-	-	-	-	-	5	23
Russian Federation	x	x	53210	18226	24080	21192	28780	29259	31276
Ukraine	x	x	-	16	1220	-	-	-	-
Other Africa	x	x	-	-	-	-	-	-	-
Other Asia	x	x	-	-	-	-	-	-	-
Other Eastern Europe	x	x	-	-	-	-	-	-	-
Other FSU	x	x	-	527	-	-	-	-	-
Other non-OECD Americas	x	x	-	-	-	-	-	-	-
Other Middle East	x	x	-	-	-	-	-	-	-
Non-specified/Other	x	x	-	-	-	-	-	-	-

(1) Steam coal now comprises anthracite, other bituminous coal and sub-bituminous coal.

(2) Earliest year for which split by coal type is available.

(3) Australia, Chile, Estonia, Iceland, Ireland, Luxembourg, New Zealand, Slovak Republic, Slovenia and Switzerland.

Source: IEA/OECD Coal Statistics.

PEOPLE'S DEMOCRATIC REPUBLIC OF KOREA

Coal balance*

(Mtce)

	1980	1985	1990	1995	2000	2005	2010	2011	Average annual percent change 80-90	90-11
Production	36.3	42.6	38.0	25.4	24.2	28.4	26.4	32.8	0.5	-0.7
Imports	0.7	2.7	2.8	1.3	0.2	0.2	0.3	0.2	14.5	-12.0
Exports	-0.1	-0.3	-0.4	-0.4	-0.3	-2.6	-4.2	-10.2	17.5	16.1
Stock changes	-	-	-	-	-	-	-	-		
Primary supply	**37.0**	**45.0**	**40.4**	**26.3**	**24.0**	**26.0**	**22.5**	**22.9**	**0.9**	**-2.7**
Statistical differences	-	-	-	-	-	-	**-0.0**	**0.0**		
Total transformation	**-7.8**	**-9.0**	**-8.6**	**-4.8**	**-3.8**	**-4.1**	**-3.5**	**-3.6**	**1.0**	**-4.1**
Electricity and heat gen.	-5.0	-5.7	-5.3	-3.9	-3.8	-4.0	-3.4	-3.5	0.6	-2.0
Main activity producers (1)	*-5.0*	*-5.7*	*-5.3*	*-3.9*	*-3.8*	*-4.0*	*-3.4*	*-3.5*	*0.6*	*-2.0*
Autoproducers	-	-	-	-	-	-	-	-	-	-
Gas works	-	-	-	-	-	-	-	-	-	-
Coal transformation(2)	-2.8	-3.2	-3.3	-0.9	-0.1	-0.1	-0.1	-0.1	1.8	-15.4
BKB plants	-	-	-	-	-	-	-	-	-	-
Blast furnaces	*-1.6*	*-1.8*	*-1.8*	*-0.6*	*-0.1*	*-0.1*	*-0.1*	*-0.1*	*1.7*	*-12.9*
Coke ovens	*-1.2*	*-1.4*	*-1.5*	*-0.3*	-	-	-	-	*1.8*	-
Patent fuel plants	-	-	-	-	-	-	-	-	-	-
Other transformation(3)	-	-	-	-	-	-	-	-	-	-
Energy ind. own use	-	-	-	-	-	-	-	-	-	-
Losses	-	-	-	-	-	-	-	-		
Final consumption(4)	**29.2**	**36.1**	**31.8**	**21.6**	**20.2**	**22.0**	**19.0**	**19.3**	**0.8**	**-2.4**
Industry(5)	23.2	28.4	25.8	16.6	15.3	16.8	14.5	14.7	1.1	-2.6
Iron and steel	*1.6*	*1.9*	*1.9*	*0.4*	*0.1*	*0.1*	*0.1*	*0.1*	*1.7*	*-13.4*
Chemical	-	-	-	-	-	-	-	-	-	-
Non-metallic minerals	-	-	-	-	-	-	-	-	-	-
Paper, pulp and print	-	-	-	-	-	-	-	-	-	-
Other industry (6)	*21.6*	*26.5*	*23.9*	*16.2*	*15.2*	*16.7*	*14.4*	*14.6*	*1.0*	*-2.3*
Transport(7)	-	-	-	-	-	-	-	-	-	-
Other	6.0	7.7	5.9	5.0	4.9	5.2	4.5	4.6	-0.2	-1.2
Comm. and pub. services	-	-	-	-	-	-	-	-	-	-
Residential	-	-	-	-	-	-	-	-	-	-
Other sectors (8)	*6.0*	*7.7*	*5.9*	*5.0*	*4.9*	*5.2*	*4.5*	*4.6*	*-0.2*	*-1.2*
Non-energy use	-	-	-	-	-	-	-	-	-	-

(1) Main activity electricity and heat generation includes district heating.

(2) Coal transformation refers to the transformation of coal and peat to secondary and tertiary products (mainly cokes, briquettes, coal tar and off-gases). Gas works and Liquefaction are shown separately to match the *Energy Balances of OECD Countries.*

(3) Other transformation includes Liquefaction and Non-specified transformations.

(4) Final Consumption includes non-energy use and energy use (Industry, Transport and Other).

(5) Please refer to notes and definitions in Part I for detailed explanation.

(6) Other industry includes Non-ferrous metals, Transport equipment, Machinery, Mining and quarrying, Food and tobacco, Wood and wood products, Construction, Textile and leather, and Non-specified industry.

(7) Transport includes Rail and Inland waterways.

(8) Other sectors includes Agriculture/Forestry, Fishing and Non-specified other.

* "Coal" refers to all types of coal, primary (anthracite, coking coal, other bituminous coal, sub-bituminous coal, lignite/brown coal) and derived fuels (including patent fuel, BKB, coke oven coke, gas coke, coal tar, coke oven gas, gas works gas, blast furnace gas and other recovered gases), as well as peat, peat briquettes and oil shale. Quantities have been converted to Mtce using calorific values reported by the respective countries. Please refer to notes and definitions in Part I.

Source: *IEA/OECD Energy Balances of OECD Countries*

MALAYSIA

Coal balance*

(Mtce)

	1980	1985	1990	1995	2000	2005	2010	2011	Average annual percent change 80-90	Average annual percent change 90-11
Production	-	-	0.1	0.1	0.3	0.7	2.2	2.6	-	16.8
Imports	0.1	0.5	2.0	2.3	2.8	9.4	18.7	19.7	39.0	11.4
Exports	-	-	-0.0	-0.1	-0.0	-0.1	-0.1	-0.2	-	8.1
Stock changes	-	-	-0.2	-0.0	0.2	-0.3	0.1	0.1		
Primary supply	**0.1**	**0.5**	**1.9**	**2.3**	**3.3**	**9.8**	**20.9**	**22.3**	**38.3**	**12.3**
Statistical differences	**-**	**-**	**-0.0**	**0.1**	**0.3**	**-0.0**	**0.3**	**-1.2**		
Total transformation	**-**	**-**	**-1.2**	**-1.4**	**-2.1**	**-7.9**	**-18.5**	**-18.6**	**-**	**14.1**
Electricity and heat gen.	-	-	-1.2	-1.4	-2.1	-7.9	-18.5	-18.6	-	14.1
Main activity producers (1)	-	-	*-1.2*	*-1.4*	*-2.1*	*-7.9*	*-18.5*	*-18.6*	-	*14.1*
Autoproducers	-	-	-	-	-	-	-	-	-	-
Gas works	-	-	-	-	-	-	-	-	-	-
Coal transformation(2)	-	-	-	-	-	-	-	-	-	-
BKB plants	-	-	-	-	-	-	-	-	-	-
Blast furnaces	-	-	-	-	-	-	-	-	-	-
Coke ovens	-	-	-	-	-	-	-	-	-	-
Patent fuel plants	-	-	-	-	-	-	-	-	-	-
Other transformation(3)	-	-	-	-	-	-	-	-	-	-
Energy ind. own use	**-**	**-**	**-**	**-**	**-**	**-**	**-**	**-**	**-**	**-**
Losses	**-**	**-**	**-**	**-**	**-**	**-**	**-**	**-**		
Final consumption(4)	**0.1**	**0.5**	**0.7**	**1.0**	**1.4**	**1.9**	**2.6**	**2.5**	**25.5**	**6.0**
Industry(5)	0.1	0.5	0.7	1.0	1.4	1.9	2.6	2.5	25.5	6.0
Iron and steel	-	-	-	-	-	-	-	-	-	-
Chemical	-	-	-	-	-	-	-	-	-	-
Non-metallic minerals	-	-	-	-	-	-	-	-	-	-
Paper, pulp and print	-	-	-	-	-	-	-	-	-	-
Other industry (6)	*0.1*	*0.5*	*0.7*	*1.0*	*1.4*	*1.9*	*2.6*	*2.5*	*25.5*	*6.0*
Transport(7)	-	-	-	-	-	-	-	-	-	-
Other	-	-	-	-	-	-	-	-	-	-
Comm. and pub. services	-	-	-	-	-	-	-	-	-	-
Residential	-	-	-	-	-	-	-	-	-	-
Other sectors (8)	-	-	-	-	-	-	-	-	-	-
Non-energy use	-	-	-	-	-	-	-	-	-	-

(1) Main activity electricity and heat generation includes district heating.

(2) Coal transformation refers to the transformation of coal and peat to secondary and tertiary products (mainly cokes, briquettes, coal tar and off-gases). Gas works and Liquefaction are shown separately to match the *Energy Balances of OECD Countries.*

(3) Other transformation includes Liquefaction and Non-specified transformations.

(4) Final Consumption includes non-energy use and energy use (Industry, Transport and Other).

(5) Please refer to notes and definitions in Part I for detailed explanation.

(6) Other industry includes Non-ferrous metals, Transport equipment, Machinery, Mining and quarrying, Food and tobacco, Wood and wood products, Construction, Textile and leather, and Non-specified industry.

(7) Transport includes Rail and Inland waterways.

(8) Other sectors includes Agriculture/Forestry, Fishing and Non-specified other.

* "Coal" refers to all types of coal, primary (anthracite, coking coal, other bituminous coal, sub-bituminous coal, lignite/brown coal) and derived fuels (including patent fuel, BKB, coke oven coke, gas coke, coal tar, coke oven gas, gas works gas, blast furnace gas and other recovered gases), as well as peat, peat briquettes and oil shale. Quantities have been converted to Mtce using calorific values reported by the respective countries. Please refer to notes and definitions in Part I.

Source: *IEA/OECD Energy Balances of OECD Countries*

MALAYSIA

Total coal imports by origin*

(thousand tonnes)

	1978(1)	1985	1990	1995	2000	2005	2010	2011	2012e
Total coal	**33**	**517**	**972**	**1022**	**802**	**12337**	**26131**	**21491**	**22001**
Coking coal	**-**	**-**	**-**	**-**	**-**	**-**	**-**	**-**	**-**
Australia	-	-	-	-	-	-	-	-	-
Canada	-	-	-	-	-	-	-	-	-
Czech Republic	-	-	-	-	-	-	-	-	-
Germany	-	-	-	-	-	-	-	-	-
Poland	-	-	-	-	-	-	-	-	-
United Kingdom	-	-	-	-	-	-	-	-	-
United States	-	-	-	-	-	-	-	-	-
Other OECD	-	-	-	-	-	-	-	-	-
China, People's Rep.	-	-	-	-	-	-	-	-	-
Colombia	-	-	-	-	-	-	-	-	-
Indonesia	-	-	-	-	-	-	-	-	-
South Africa	-	-	-	-	-	-	-	-	-
Former Soviet Union(2)	-	-	x	x	x	x	x	x	x
Russian Federation (3)	*x*	*x*	-	-	-	-	-	-	-
Other FSU (3)	*x*	*x*	-	-	-	-	-	-	-
Venezuela	-	-	-	-	-	-	-	-	-
Vietnam	-	-	-	-	-	-	-	-	-
Non-specified/other	-	-	-	-	-	-	-	-	-
Steam coal(4)	**33**	**517**	**972**	**1022**	**802**	**12337**	**26131**	**21491**	**22001**
Australia	33	517	697	700	210	1270	3149	2196	3217
Canada	-	-	-	-	-	-	-	-	-
Czech Republic	-	-	-	-	-	-	-	-	-
Germany	-	-	-	-	-	-	-	-	-
Poland	-	-	-	-	-	-	-	-	-
United Kingdom	-	-	-	-	-	-	-	-	-
United States	-	-	-	-	-	-	-	-	-
Other OECD	-	-	-	-	-	66	169	-	-
China, People's Rep.	-	-	-	48	236	58	74	6	-
Colombia	-	-	-	-	-	-	1	-	-
Indonesia	-	-	275	234	144	10739	19984	16267	16034
South Africa	-	-	-	40	200	-	2754	3021	2684
Former Soviet Union(2)	-	-	x	x	x	x	x	x	x
Russian Federation (3)	*x*	*x*	-	-	-	-	-	*1*	*66*
Other FSU (3)	*x*	*x*	-	-	-	-	-	-	-
Venezuela	-	-	-	-	-	-	-	-	-
Vietnam	-	-	-	-	12	125	-	-	-
Non-specified/other	-	-	-	-	-	79	-	-	-
Lignite	**-**	**-**	**-**	**-**	**-**	**-**	**-**	**-**	**-**

(1) Earliest year for which split by coal type is available.

(2) For years prior to 1990, or if no details for individual Former Soviet Union republics.

(3) Data for former Republics are not available separately for some countries.

(4) Steam coal now comprises anthracite, other bituminous coal and sub-bituminous coal.

* In these tables coal used for PCI and for blending has been classified by the IEA as steam coal. Accordingly, trade data reported in these tables may differ from those reported in Part III where this coal is shown as coking coal to be consistent with data reported by importing countries and with industry terminology and practice.

Source: IEA/OECD Coal Statistics.

PHILIPPINES

Coal balance*

(Mtce)

	1980	1985	1990	1995	2000	2005	2010	2011	Average annual percent change 80-90	90-11
Production	0.2	0.9	0.9	1.0	1.0	2.2	5.0	5.2	14.2	8.5
Imports	0.5	1.0	1.1	1.6	6.2	6.0	9.3	9.3	8.4	10.6
Exports	-	-	-	-	-	-	-3.1	-2.1	-	-
Stock changes	-0.0	-0.1	-	-	-	-	-0.1	-0.4		
Primary supply	**0.7**	**1.8**	**2.0**	**2.6**	**7.2**	**8.2**	**11.1**	**12.1**	**10.9**	**8.8**
Statistical differences	**-0.2**	**0.2**	**-0.1**	**-0.0**	**-0.0**	**-0.0**	**-0.2**	**-0.2**		
Total transformation	**-0.2**	**-1.2**	**-0.8**	**-1.2**	**-5.9**	**-6.4**	**-8.0**	**-9.0**	**13.6**	**12.1**
Electricity and heat gen.	-0.1	-1.2	-0.7	-1.1	-5.9	-6.4	-7.9	-8.8	26.4	12.6
Main activity producers (1)	*-0.1*	*-1.2*	*-0.7*	*-1.1*	*-5.9*	*-6.4*	*-7.9*	*-8.8*	*26.4*	*12.6*
Autoproducers	-	-	-	-	-	-	-	-	-	-
Gas works	-	-	-	-	-	-	-	-	-	-
Coal transformation(2)	-0.2	-0.0	-0.1	-0.1	-0.0	-0.0	-0.1	-0.1	-5.7	1.8
BKB plants	-	-	-	-	-	-	-	-	-	-
Blast furnaces	*-0.2*	*-0.0*	*-0.1*	*-0.1*	*-0.0*	*-0.0*	*-0.1*	*-0.1*	*-5.7*	*1.8*
Coke ovens	-	-	-	-	-	-	-	-	-	-
Patent fuel plants	-	-	-	-	-	-	-	-	-	-
Other transformation(3)	-	-	-	-	-	-	-	-	-	-
Energy ind. own use	-	-	-	-	-	-	-	-	-	-
Losses	-	-	-	-	-	-	-	-		
Final consumption(4)	**0.3**	**0.8**	**1.1**	**1.4**	**1.2**	**1.7**	**2.9**	**2.9**	**13.3**	**4.8**
Industry(5)	0.3	0.8	0.8	1.1	1.0	1.6	2.7	2.7	10.7	5.8
Iron and steel	*0.2*	*0.0*	*0.1*	*0.1*	*0.1*	*0.0*	*0.2*	*0.2*	*-4.8*	*3.9*
Chemical	-	-	*0.1*	*0.0*	*0.0*	*0.0*	*0.0*	*0.0*	-	*-3.4*
Non-metallic minerals	*0.1*	*0.5*	*0.7*	*1.0*	*1.0*	*1.5*	*2.3*	*2.4*	*18.7*	*6.3*
Paper, pulp and print	-	-	-	-	-	*0.0*	*0.1*	*0.1*	-	-
Other industry (6)	*0.0*	*0.2*	*0.0*	*0.0*	*0.0*	*0.0*	*0.1*	*0.1*	*4.4*	*1.8*
Transport(7)	-	-	-	-	-	-	-	-	-	-
Other	-	-	-	-	-	-	-	-	-	-
Comm. and pub. services	-	-	-	-	-	-	-	-	-	-
Residential	-	-	-	-	-	-	-	-	-	-
Other sectors (8)	-	-	-	-	-	-	-	-	-	-
Non-energy use	0.0	-	0.2	0.2	0.2	0.2	0.1	0.1	44.7	-2.4

(1) Main activity electricity and heat generation includes district heating.

(2) Coal transformation refers to the transformation of coal and peat to secondary and tertiary products (mainly cokes, briquettes, coal tar and off-gases). Gas works and Liquefaction are shown separately to match the *Energy Balances of OECD Countries.*

(3) Other transformation includes Liquefaction and Non-specified transformations.

(4) Final Consumption includes non-energy use and energy use (Industry, Transport and Other).

(5) Please refer to notes and definitions in Part I for detailed explanation.

(6) Other industry includes Non-ferrous metals, Transport equipment, Machinery, Mining and quarrying, Food and tobacco, Wood and wood products, Construction, Textile and leather, and Non-specified industry.

(7) Transport includes Rail and Inland waterways.

(8) Other sectors includes Agriculture/Forestry, Fishing and Non-specified other.

* "Coal" refers to all types of coal, primary (anthracite, coking coal, other bituminous coal, sub-bituminous coal, lignite/brown coal) and derived fuels (including patent fuel, BKB, coke oven coke, gas coke, coal tar, coke oven gas, gas works gas, blast furnace gas and other recovered gases), as well as peat, peat briquettes and oil shale. Quantities have been converted to Mtce using calorific values reported by the respective countries. Please refer to notes and definitions in Part I.

Source: *IEA/OECD Energy Balances of OECD Countries*

PHILIPPINES

Total coal imports by origin*

(thousand tonnes)

	1978(1)	1985	1990	1995	2000	2005	2010	2011	2012e
Total coal	**20**	**1240**	**1364**	**1096**	**7245**	**7766**	**11181**	**11166**	**12062**
Coking coal	**-**	**-**	**-**	**-**	**-**	**-**	**-**	**-**	**-**
Australia	-	-	-	-	-	-	-	-	-
Canada	-	-	-	-	-	-	-	-	-
Czech Republic	-	-	-	-	-	-	-	-	-
Germany	-	-	-	-	-	-	-	-	-
Poland	-	-	-	-	-	-	-	-	-
United Kingdom	-	-	-	-	-	-	-	-	-
United States	-	-	-	-	-	-	-	-	-
Other OECD	-	-	-	-	-	-	-	-	-
China, People's Rep.	-	-	-	-	-	-	-	-	-
Colombia	-	-	-	-	-	-	-	-	-
Indonesia	-	-	-	-	-	-	-	-	-
South Africa	-	-	-	-	-	-	-	-	-
Former Soviet Union(2)	-	-	x	x	x	x	x	x	x
Russian Federation (3)	*x*	*x*	-	-	-	-	-	-	-
Other FSU (3)	*x*	*x*	-	-	-	-	-	-	-
Venezuela	-	-	-	-	-	-	-	-	-
Vietnam	-	-	-	-	-	-	-	-	-
Non-specified/other	-	-	-	-	-	-	-	-	-
Steam coal(4)	**20**	**1240**	**1364**	**1096**	**7245**	**7766**	**11181**	**11166**	**12062**
Australia	20	827	550	543	1911	646	68	411	441
Canada	-	-	-	-	-	-	-	-	-
Czech Republic	-	-	-	-	-	-	-	-	-
Germany	-	-	-	-	-	-	-	-	-
Poland	-	-	-	-	-	-	-	-	-
United Kingdom	-	-	-	-	-	-	-	-	-
United States	-	-	1	-	-	-	-	-	-
Other OECD	-	-	14	-	-	-	1	-	-
China, People's Rep.	-	413	358	-	1798	1332	1	-	-
Colombia	-	-	-	-	-	-	-	-	-
Indonesia	-	-	238	500	3536	5193	11111	10755	11621
South Africa	-	-	-	-	-	-	-	-	-
Former Soviet Union(2)	-	-	x	x	x	x	x	x	x
Russian Federation (3)	*x*	*x*	-	*53*	-	-	-	-	-
Other FSU (3)	*x*	*x*	-	-	-	-	-	-	-
Venezuela	-	-	-	-	-	-	-	-	-
Vietnam	-	-	203	-	-	583	-	-	-
Non-specified/other	-	-	-	-	-	12	-	-	-
Lignite	**-**	**-**	**-**	**-**	**-**	**-**	**-**	**-**	**-**

(1) Earliest year for which split by coal type is available.

(2) For years prior to 1990, or if no details for individual Former Soviet Union republics.

(3) Data for former Republics are not available separately for some countries.

(4) Steam coal now comprises anthracite, other bituminous coal and sub-bituminous coal.

* In these tables coal used for PCI and for blending has been classified by the IEA as steam coal. Accordingly, trade data reported in these tables may differ from those reported in Part III where this coal is shown as coking coal to be consistent with data reported by importing countries and with industry terminology and practice.

Source: IEA/OECD Coal Statistics.

ROMANIA

Coal balance*

(Mtce)

	1980	1985	1990	1995	2000	2005	2010	2011	Average annual percent change 80-90	Average annual percent change 90-11
Production	11.6	14.7	12.4	11.3	8.0	8.3	8.4	9.5	0.7	-1.2
Imports	6.4	7.3	6.4	4.4	2.7	4.2	1.8	1.6	0.1	-6.5
Exports	-	-	-	-0.3	-0.0	-0.0	-0.1	-0.0	-	-
Stock changes	-	-	-0.3	0.1	-0.1	0.1	-0.2	0.5		
Primary supply	**17.9**	**22.0**	**18.5**	**15.4**	**10.6**	**12.5**	**9.9**	**11.6**	**0.3**	**-2.2**
Statistical differences	**-0.3**	**-0.7**	**-0.1**	**-0.3**	**-0.1**	**0.3**	**0.1**	**0.2**		
Total transformation	**-9.6**	**-13.6**	**-13.6**	**-12.7**	**-9.2**	**-10.6**	**-8.9**	**-10.7**	**3.6**	**-1.2**
Electricity and heat gen.	-6.7	-10.0	-11.3	-10.9	-8.0	-9.1	-8.4	-10.3	5.3	-0.4
Main activity producers (1)	*-6.7*	*-7.5*	*-9.2*	*-10.7*	*-7.7*	*-8.5*	*-7.7*	*-9.4*	*3.2*	*0.1*
Autoproducers	-	*-2.6*	*-2.0*	*-0.2*	*-0.3*	*-0.6*	*-0.6*	*-0.9*	-	*-4.1*
Gas works	-	-	-	-	-	-	-	-	-	-
Coal transformation (2)	-2.9	-3.6	-2.4	-1.8	-1.2	-1.5	-0.5	-0.4	-1.9	-8.4
BKB plants	*-0.5*	*-0.5*	*-0.2*	-	-	-	-	-	*-6.9*	-
Blast furnaces	*-2.6*	*-3.0*	*-1.9*	*-1.3*	*-0.9*	*-1.3*	*-0.5*	*-0.4*	*-3.0*	*-7.5*
Coke ovens	*-0.1*	*-0.5*	*-0.3*	*-0.5*	*-0.4*	*-0.2*	*-0.0*	-	*9.2*	-
Patent fuel plants	*0.4*	*0.4*	*0.1*	-	-	-	-	-	*-9.5*	-
Other transformation (3)	-	-	-	-	-	-	-	-	-	-
Energy ind. own use	**-**	**-**	**-**	**-0.1**	**-0.2**	**-0.4**	**-0.1**	**-0.1**	**-**	**-**
Losses	**-**	**-**	**-0.2**	**-0.2**	**-0.0**	**-0.1**	**-0.0**	**-0.0**		
Final consumption (4)	**8.1**	**7.7**	**4.5**	**2.0**	**1.1**	**1.7**	**1.0**	**1.0**	**-5.7**	**-7.1**
Industry (5)	4.9	4.8	3.1	1.9	1.0	1.6	1.0	0.9	-4.4	-5.7
Iron and steel	*3.8*	*4.0*	*2.5*	*1.8*	*0.8*	*1.4*	*0.7*	*0.7*	*-4.2*	*-5.7*
Chemical	-	-	*0.2*	*0.0*	*0.2*	*0.2*	*0.2*	*0.2*	-	*-0.9*
Non-metallic minerals	-	-	*0.1*	*0.0*	*0.0*	*0.1*	*0.1*	*0.0*	-	*-7.2*
Paper, pulp and print	-	-	-	*0.0*	*0.0*	-	-	-	-	-
Other industry (6)	*1.1*	*0.8*	*0.4*	*0.1*	*0.0*	*0.0*	*0.0*	*0.0*	*-10.9*	*-21.0*
Transport (7)	-	-	0.0	0.0	-	-	-	-	-	-
Other	3.2	2.9	1.3	0.1	0.1	0.0	0.0	0.0	-8.4	-15.3
Comm. and pub. services	-	-	-	-	*0.0*	*0.0*	*0.0*	*0.0*	-	-
Residential	*0.5*	*0.6*	*0.9*	*0.1*	*0.1*	*0.0*	*0.0*	*0.0*	*5.2*	*-15.2*
Other sectors (8)	*2.6*	*2.3*	*0.4*	*0.0*	*0.0*	*0.0*	*0.0*	*0.0*	*-16.4*	*-16.0*
Non-energy use	-	-	-	0.0	-	-	-	-	-	-

(1) Main activity electricity and heat generation includes district heating.

(2) Coal transformation refers to the transformation of coal and peat to secondary and tertiary products (mainly cokes, briquettes, coal tar and off-gases). Gas works and Liquefaction are shown separately to match the *Energy Balances of OECD Countries.*

(3) Other transformation includes Liquefaction and Non-specified transformations.

(4) Final Consumption includes non-energy use and energy use (Industry, Transport and Other).

(5) Please refer to notes and definitions in Part I for detailed explanation.

(6) Other industry includes Non-ferrous metals, Transport equipment, Machinery, Mining and quarrying, Food and tobacco, Wood and wood products, Construction, Textile and leather, and Non-specified industry.

(7) Transport includes Rail and Inland waterways.

(8) Other sectors includes Agriculture/Forestry, Fishing and Non-specified other.

* "Coal" refers to all types of coal, primary (anthracite, coking coal, other bituminous coal, sub-bituminous coal, lignite/brown coal) and derived fuels (including patent fuel, BKB, coke oven coke, gas coke, coal tar, coke oven gas, gas works gas, blast furnace gas and other recovered gases), as well as peat, peat briquettes and oil shale. Quantities have been converted to Mtce using calorific values reported by the respective countries. Please refer to notes and definitions in Part I.

Source: *IEA/OECD Energy Balances of OECD Countries*

ROMANIA

Total coal imports by origin*
(thousand tonnes)

	1978(1)	1985	1990	1995	2000	2005	2010	2011	2012e
Total coal	**4669**	**6048**	**4981**	**4734**	**2522**	**2980**	**1150**	**1613**	**1424**
Coking coal	**3600**	**4000**	**3600**	**4675**	**2370**	**1577**	**233**	**627**	**616**
Australia	675	595	1200	513	-	-	-	-	-
Canada	-	-	-	-	-	-	-	-	-
Czech Republic	-	-	12	57	-	-	-	-	-
Germany	-	-	-	2	-	-	-	-	-
Poland	-	-	100	147	62	-	-	-	1
United Kingdom	-	-	-	98	-	-	-	-	-
United States	673	1051	1188	1437	678	547	233	627	607
Other OECD	-	-	-	165	-	-	-	-	-
China, People's Rep.	-	-	400	124	-	-	-	-	-
Colombia	-	-	-	-	-	-	-	-	-
Indonesia	-	-	-	-	-	-	-	-	-
South Africa	-	-	-	-	-	-	-	-	-
Former Soviet Union(2)	2252	2354	x	x	x	x	x	x	x
Russian Federation (3)	*x*	*x*	*500*	*1741*	*1630*	*1030*	-	-	-
Other FSU (3)	*x*	*x*	*200*	*97*	-	-	-	-	-
Venezuela	-	-	-	-	-	-	-	-	-
Vietnam	-	-	-	-	-	-	-	-	-
Non-specified/other	-	-	-	294	-	-	-	-	8
Steam coal(4)	**1069**	**2048**	**1381**	**30**	**152**	**1140**	**764**	**986**	**805**
Australia	-	-	33	-	-	-	-	-	-
Canada	-	-	-	-	-	-	-	-	-
Czech Republic	-	-	-	-	-	-	-	-	-
Germany	-	-	-	-	-	-	-	-	-
Poland	-	-	16	-	-	-	-	10	23
United Kingdom	-	-	-	-	-	-	-	-	-
United States	-	-	11	2	-	230	-	-	-
Other OECD	-	-	-	-	-	-	-	18	61
China, People's Rep.	-	-	-	-	-	-	-	-	-
Colombia	-	-	-	-	69	-	16	30	-
Indonesia	-	-	-	-	-	-	-	-	-
South Africa	-	-	-	-	-	-	37	148	-
Former Soviet Union(2)	1069	2048	x	x	x	x	x	x	x
Russian Federation (3)	*x*	*x*	*1321*	*3*	*32*	*832*	*448*	*404*	*418*
Other FSU (3)	*x*	*x*	-	*1*	-	*78*	*194*	*376*	*301*
Venezuela	-	-	-	-	51	-	-	-	-
Vietnam	-	-	-	-	-	-	-	-	-
Non-specified/other	-	-	-	24	-	-	69	-	2
Lignite	**-**	**-**	**-**	**29**	**-**	**263**	**153**	**-**	**3**

(1) Earliest year for which split by coal type is available.

(2) For years prior to 1990, or if no details for individual Former Soviet Union republics.

(3) Data for former Republics are not available separately for some countries.

(4) Steam coal now comprises anthracite, other bituminous coal and sub-bituminous coal.

* In these tables coal used for PCI and for blending has been classified by the IEA as steam coal. Accordingly, trade data reported in these tables may differ from those reported in Part III where this coal is shown as coking coal to be consistent with data reported by importing countries and with industry terminology and practice.

Source: IEA/OECD Coal Statistics.

RUSSIAN FEDERATION

Coal balance*

(Mtce)

	1980	1985	1990	1995	2000	2005	2010	2011	Average annual percent change 80-90	90-11
Production	x	x	274.9	186.7	183.6	224.9	257.2	257.2	x	-0.3
Imports	x	x	50.1	20.0	22.0	19.5	21.9	23.3 e	x	-3.6
Exports	x	x	-57.7	-26.8	-36.4	-79.7	-122.8	-115.1	x	3.3
Stock changes	x	x	5.7	4.6	2.2	-3.8	7.8	0.2		
Primary supply	**x**	**x**	**273.0**	**184.5**	**171.4**	**160.9**	**164.0**	**165.6**	**x**	**-2.4**
Statistical differences	**x**	**x**	**6.0**	**5.7**	**-4.0**	**-3.9**	**-12.2**	**-5.9**		
Total transformation	**x**	**x**	**-181.8**	**-145.4**	**-139.3 e**	**-137.3 e**	**-128.3 e**	**-134.2 e**	**x**	**-1.4**
Electricity and heat gen.	x	x	-149.9	-115.6	-114.1	-109.8	-101.2	-104.5	x	-1.7
Main activity producers (1)	*x*	*x*	*-124.2*	*-88.1*	*-80.6*	*-74.6*	*-71.3*	*-72.6*	*x*	*-2.5*
Autoproducers	*x*	*x*	*-25.7*	*-27.5*	*-33.5*	*-35.2*	*-29.9*	*-31.8*	*x*	*1.0*
Gas works	x	x	-	-	-	-	-	-	x	-
Coal transformation(2)	x	x	-31.9	-29.8	-25.1 e	-27.5 e	-27.1 e	-29.7 e	x	-0.3
BKB plants	*x*	*x*	*-1.5*	*-0.6*	*-0.0*	*-0.0*	*-0.0*	*-0.0*	*x*	*-16.1*
Blast furnaces	*x*	*x*	*-19.4*	*-18.9*	*-15.3* e	*-16.7* e	*-20.3* e	*-22.1* e	*x*	*0.6*
Coke ovens	*x*	*x*	*-10.9*	*-10.3*	*-9.8*	*-10.8*	*-6.8* e	*-7.6* e	*x*	*-1.7*
Patent fuel plants	*x*	*x*	-	-	-	-	*0.0*	-	*x*	-
Other transformation(3)	x	x	-	-	-	-	-	-	x	-
Energy ind. own use	**x**	**x**	**-0.8**	**-0.7**	**-0.8**	**-0.8**	**-3.2**	**-3.2**	**x**	**6.9**
Losses	**x**	**x**	**-18.3**	**-9.4**	**-1.6**	**-**	**-**	**-**		
Final consumption(4)	**x**	**x**	**78.2**	**34.7**	**25.7**	**18.9**	**20.4**	**22.4**	**x**	**-5.8**
Industry(5)	x	x	20.8	17.5	10.4	9.8	14.1	16.2	x	-1.2
Iron and steel	*x*	*x*	*15.1*	*13.0*	*6.6* e	*7.6* e	*11.6* e	*13.2* e	*x*	*-0.6*
Chemical	*x*	*x*	*0.6*	*0.5*	*0.1*	*0.0*	*0.3*	*0.4*	*x*	*-2.7*
Non-metallic minerals	*x*	*x*	*0.5*	*0.9*	*0.8*	*1.0*	*1.6*	*2.0*	*x*	*7.1*
Paper, pulp and print	*x*	*x*	-	-	-	*0.0*	*0.1*	*0.1*	*x*	-
Other industry (6)	*x*	*x*	*4.6*	*3.1*	*2.9*	*1.2*	*0.5*	*0.5*	*x*	*-9.9*
Transport(7)	x	x	0.0	0.0	-	-	-	-	x	-
Other	x	x	57.4	17.2	14.2	8.1	5.9	5.8	x	-10.4
Comm. and pub. services	*x*	*x*	*29.4*	*1.2*	*0.6*	*3.8*	*3.5*	*3.5*	*x*	*-9.7*
Residential	*x*	*x*	*17.0*	*11.7*	*12.8*	*4.0*	*2.2*	*2.2*	*x*	*-9.2*
Other sectors (8)	*x*	*x*	*11.0*	*4.3*	*0.9*	*0.2*	*0.1*	*0.1*	*x*	*-20.1*
Non-energy use	x	x	0.0	0.0	1.1	1.0	0.4	0.4	x	13.1

(1) Main activity electricity and heat generation includes district heating.

(2) Coal transformation refers to the transformation of coal and peat to secondary and tertiary products (mainly cokes, briquettes, coal tar and off-gases). Gas works and Liquefaction are shown separately to match the *Energy Balances of OECD Countries*.

(3) Other transformation includes Liquefaction and Non-specified transformations.

(4) Final Consumption includes non-energy use and energy use (Industry, Transport and Other).

(5) Please refer to notes and definitions in Part I for detailed explanation.

(6) Other industry includes Non-ferrous metals, Transport equipment, Machinery, Mining and quarrying, Food and tobacco, Wood and wood products, Construction, Textile and leather, and Non-specified industry.

(7) Transport includes Rail and Inland waterways.

(8) Other sectors includes Agriculture/Forestry, Fishing and Non-specified other.

* "Coal" refers to all types of coal, primary (anthracite, coking coal, other bituminous coal, sub-bituminous coal, lignite/brown coal) and derived fuels (including patent fuel, BKB, coke oven coke, gas coke, coal tar, coke oven gas, gas works gas, blast furnace gas and other recovered gases), as well as peat, peat briquettes and oil shale. Quantities have been converted to Mtce using calorific values reported by the respective countries. Please refer to notes and definitions in Part I.

Source: *IEA/OECD Energy Balances of OECD Countries*

RUSSIAN FEDERATION

Coking coal exports by destination

(thousand tonnes)

	1978[1]	1985	1990	1995	2000	2005	2010	2011	2012e
World	**x**	**x**	**31573**	**8693**	**6704**	**9981**	**16338**	**13916**	**18251**
OECD	**x**	**x**	**10503**	**6133**	**1504**	**4189**	**7534**	**5881**	**6229**
Austria	x	x	608	161	-	31	-	-	-
Belgium	x	x	-	-	-	-	315	-	234
Canada	x	x	-	111	-	-	-	-	-
Czech Republic	x	x	1532	-	-	-	-	-	-
Denmark	x	x	-	67	-	-	7	28	6
Finland	x	x	463	431	91	94	59	100	43
France	x	x	-	84	-	-	143	95	71
Germany	x	x	177	-	-	157	793	519	906
Greece	x	x	-	-	-	-	31	-	-
Hungary	x	x	480	-	132	219	47	-	-
Israel	x	x	-	-	-	-	42	50	24
Italy	x	x	131	857	-	234	414	586	463
Japan	x	x	5482	3505	462	2454	2094	1535	2043
Korea	x	x	1200	571	-	626	1258	1460	1207
Mexico	x	x	-	-	-	-	-	-	-
Netherlands	x	x	-	-	-	223	494	24	311
Norway	x	x	-	-	-	-	-	-	-
Poland	x	x	-	-	736	-	98	18	20
Portugal	x	x	-	-	-	-	-	-	-
Spain	x	x	-	-	72	-	-	-	113
Sweden	x	x	-	156	-	-	19	14	7
Turkey	x	x	337	-	-	35	107	61	-
United Kingdom	x	x	93	158	11	-	512	481	514
United States	x	x	-	32	-	-	-	-	-
Other OECD[2]	x	x	-	-	-	116	1101	910	267
Non-OECD	**x**	**x**	**21070**	**2560**	**5200**	**5792**	**8804**	**8035**	**12022**
Brazil	x	x	-	-	-	-	77	6	96
Bulgaria	x	x	156	822	174	72	23	24	25
China, People's Rep.	x	x	-	-	-	-	2528	1502	4547
Chinese Taipei	x	x	-	-	1337	160	116	66	111
Hong Kong, China	x	x	-	-	-	-	-	-	-
India	x	x	-	-	-	-	9	10	65
Morocco	x	x	-	-	-	-	47	87	28
Romania	x	x	500	876	1525	729	13	-	16
Russian Federation	x	x	-	-	-	-	-	-	-
Ukraine	x	x	19964	-	2164	4606	5579	5633	7076
Other Africa	x	x	150	81	-	16	-	-	-
Other Asia	x	x	300	-	-	7	93	-	39
Other Eastern Europe	x	x	-	781	-	48	-	218	19
Other FSU	x	x	-	-	-	1	319	489	-
Other non-OECD Americas	x	x	-	-	-	-	-	-	-
Other Middle East	x	x	-	-	-	153	42	50	24
Non-specified/Other	x	x	-	-	-	-	-	-	-

(1) Earliest year for which split by coal type is available.

(2) Australia, Chile, Estonia, Iceland, Ireland, Luxembourg, New Zealand, Slovak Republic, Slovenia and Switzerland.

Source: IEA/OECD Coal Statistics.

RUSSIAN FEDERATION

Steam coal[1] exports by destination

(thousand tonnes)

	1978[2]	1985	1990	1995	2000	2005	2010	2011	2012e
World	x	x	**24478**	**17570**	**29738**	**76020**	**114244**	**109358**	**115928**
OECD	x	x	**10225**	**8358**	**25771**	**56879**	**88045**	**80591**	**86679**
Austria	x	x	-	156	-	9	60	61	169
Belgium	x	x	234	871	788	2739	4452	3204	1744
Canada	x	x	-	-	-	453	286	166	154
Czech Republic	x	x	-	242	-	117	431	232	118
Denmark	x	x	1142	-	1295	827	1287	1262	862
Finland	x	x	1905	122	2449	4790	3988	4255	2609
France	x	x	777	-	282	367	2404	1999	2345
Germany	x	x	157	593	928	7333	6875	5947	7564
Greece	x	x	324	215	320	125	412	8	-
Hungary	x	x	-	83	87	132	407	27	7
Israel	x	x	-	-	-	-	962	909	2344
Italy	x	x	609	201	1092	799	1109	1392	2664
Japan	x	x	2845	1936	7041	8119	11839	12121	10740
Korea	x	x	-	200	2565	2446	9970	10717	12147
Mexico	x	x	-	-	-	-	-	-	-
Netherlands	x	x	32	85	209	1400	8773	4691	5878
Norway	x	x	97	52	13	-	173	78	107
Poland	x	x	-	164	14	2042	9610	8738	6190
Portugal	x	x	54	35	-	-	45	-	-
Spain	x	x	285	138	1340	3657	698	1551	3143
Sweden	x	x	573	282	271	453	871	514	565
Turkey	x	x	530	1249	5063	6509	12079	8273	9785
United Kingdom	x	x	499	192	480	12412	10360	13377	16373
United States	x	x	-	-	-	52	-	-	13
Other OECD[3]	x	x	162	1542	1534	2098	954	1069	1158
Non-OECD	x	x	**14253**	**8976**	**3967**	**18845**	**26130**	**28698**	**29124**
Brazil	x	x	-	-	-	538	379	1175	595
Bulgaria	x	x	-	1010	364	1765	888	724	638
China, People's Rep.	x	x	-	81	72	1002	10197	6226	17810
Chinese Taipei	x	x	-	114	-	1286	1519	3613	2944
Hong Kong, China	x	x	-	-	61	-	10	287	103
India	x	x	-	-	36	49	314	1143	683
Morocco	x	x	-	46	-	56	1448	1260	656
Romania	x	x	-	697	5	806	476	319	350
Russian Federation	x	x	-	-	-	-	-	-	-
Ukraine	x	x	6722	5204	2175	2235	9042	8070	3871
Other Africa	x	x	-	-	-	-	61	69	70
Other Asia	x	x	-	261	-	30	98	47	171
Other Eastern Europe	x	x	-	232	-	8828	46	2381	492
Other FSU	x	x	7531	1331	1254	2204	1311	3151	428
Other non-OECD Americas	x	x	-	-	-	-	-	-	-
Other Middle East	x	x	-	-	-	46	1303	1142	2657
Non-specified/Other	x	x	-	236	-	296	69	69	125

(1) Steam coal now comprises anthracite, other bituminous coal and sub-bituminous coal.

(2) Earliest year for which split by coal type is available.

(3) Australia, Chile, Estonia, Iceland, Ireland, Luxembourg, New Zealand, Slovak Republic, Slovenia and Switzerland.

Source: IEA/OECD Coal Statistics.

SERBIA

Coal balance*

(Mtce)

	1980	1985	1990	1995	2000	2005	2010	2011	Average annual percent change 80-90	90-11
Production	..	..	14.5	12.6	11.9	10.7	10.3	11.2	..	-1.2
Imports	..	..	-	0.1	0.4	1.0	1.1	1.1	..	-
Exports	..	..	-	-0.1	-0.0	-0.1	-0.1	-0.0	..	-
Stock changes	..	..	-	-	-	-0.1	-0.3	0.2		
Primary supply	**..**	**..**	**14.5**	**12.6**	**12.3**	**11.5**	**11.1**	**12.5**	**..**	**-0.7**
Statistical differences	**..**	**..**	**-0.0**	**-0.0**	**-0.1**	**0.2**	**-0.0**	**-0.1**		
Total transformation	**..**	**..**	**-13.2**	**-11.9**	**-10.5**	**-10.2**	**-9.7**	**-10.8**	**..**	**-0.9**
Electricity and heat gen.	..	..	-12.9	-11.7	-10.3	-9.9	-9.3	-10.5	..	-1.0
Main activity producers (1)	..	..	*-12.9*	*-11.7*	*-10.3*	*-9.6*	*-9.1*	*-10.2*	..	*-1.1*
Autoproducers	..	..	-	-	-	*-0.3*	*-0.3*	*-0.2*	..	-
Gas works	..	..	-	-	-	-	-	-	..	-
Coal transformation(2)	..	..	-0.3	-0.1	-0.2	-0.3	-0.4	-0.4	..	1.1
BKB plants	..	..	*-0.3*	*-0.1*	*-0.2*	*-0.2*	*-0.1*	*-0.2*	..	*-1.8*
Blast furnaces	..	..	-	-	-	*-0.2*	*-0.3*	*-0.2*	..	-
Coke ovens	..	..	-	-	-	-	-	-	..	-
Patent fuel plants	..	..	-	-	-	-	-	-	..	-
Other transformation(3)	..	..	-	-	-	-	-	-	..	-
Energy ind. own use	**..**	**..**	**-**	**-**	**-**	**-**	**-**	**-**	**..**	**-**
Losses	**..**	**..**	**-**	**-**	**-**	**-0.0**	**-0.1**	**-0.1**		
Final consumption(4)	**..**	**..**	**1.4**	**0.7**	**1.8**	**1.4**	**1.3**	**1.6**	**..**	**0.7**
Industry(5)	..	..	0.5	0.3	1.0	0.6	0.6	0.7	..	1.4
Iron and steel	..	..	*0.1*	*0.0*	*0.3*	*0.4*	*0.3*	*0.3*	..	*7.4*
Chemical	..	..	-	-	-	*0.0*	*0.0*	*0.0*	..	-
Non-metallic minerals	..	..	-	-	-	*0.1*	*0.2*	*0.3*	..	-
Paper, pulp and print	..	..	-	-	-	-	*0.0*	*0.0*	..	-
Other industry (6)	..	..	*0.5*	*0.3*	*0.7*	*0.1*	*0.1*	*0.1*	..	*-6.7*
Transport(7)	..	..	-	-	-	0.0	-	-	..	-
Other	..	..	0.8	0.4	0.8	0.8	0.6	0.8	..	-0.0
Comm. and pub. services	..	..	-	-	-	-	-	-	..	-
Residential	..	..	*0.7*	*0.4*	*0.6*	*0.6*	*0.3*	*0.4*	..	*-2.2*
Other sectors (8)	..	..	*0.1*	*0.1*	*0.2*	*0.1*	*0.3*	*0.4*	..	*4.8*
Non-energy use	..	..	-	-	-	-	0.0	0.0	..	-

(1) Main activity electricity and heat generation includes district heating.

(2) Coal transformation refers to the transformation of coal and peat to secondary and tertiary products (mainly cokes, briquettes, coal tar and off-gases). Gas works and Liquefaction are shown separately to match the *Energy Balances of OECD Countries.*

(3) Other transformation includes Liquefaction and Non-specified transformations.

(4) Final Consumption includes non-energy use and energy use (Industry, Transport and Other).

(5) Please refer to notes and definitions in Part I for detailed explanation.

(6) Other industry includes Non-ferrous metals, Transport equipment, Machinery, Mining and quarrying, Food and tobacco, Wood and wood products, Construction, Textile and leather, and Non-specified industry.

(7) Transport includes Rail and Inland waterways.

(8) Other sectors includes Agriculture/Forestry, Fishing and Non-specified other.

* "Coal" refers to all types of coal, primary (anthracite, coking coal, other bituminous coal, sub-bituminous coal, lignite/brown coal) and derived fuels (including patent fuel, BKB, coke oven coke, gas coke, coal tar, coke oven gas, gas works gas, blast furnace gas and other recovered gases), as well as peat, peat briquettes and oil shale. Quantities have been converted to Mtce using calorific values reported by the respective countries. Please refer to notes and definitions in Part I.

Source: *IEA/OECD Energy Balances of OECD Countries*

SOUTH AFRICA

Coal balance*

(Mtce)

	1980	1985	1990	1995	2000	2005	2010	2011	Average annual percent change 80-90	90-11
Production	95.4	142.5	143.1	168.6	181.3	197.7	205.4	203.9	4.1	1.7
Imports	-	-	-	0.4	1.2	2.0	2.2	2.5	-	-
Exports	-27.2	-46.0	-48.0	-57.6	-67.0	-68.3	-63.4	-65.8	5.8	1.5
Stock changes	-	-	-	0.3	1.6	-	-	-		
Primary supply	**68.1**	**96.5**	**95.1**	**111.6**	**117.1**	**131.3**	**144.2**	**140.7**	**3.4**	**1.9**
Statistical differences	**3.6**	**-3.7**	**6.7**	**-3.5**	**-0.3**	**-0.6**	**-3.4**	**-4.3**		
Total transformation	**-44.7**	**-69.5**	**-78.3**	**-85.7**	**-94.1**	**-106.8**	**-119.0**	**-112.5**	**5.8**	**1.7**
Electricity and heat gen.	-35.0	-44.9	-51.7	-60.2	-68.3	-75.8	-87.5	-83.0	4.0	2.3
Main activity producers (1)	*-32.1*	*-40.8*	*-48.6*	*-56.6*	*-64.4*	*-72.1*	*-84.0*	*-79.6*	*4.2*	*2.4*
Autoproducers	*-2.9*	*-4.1*	*-3.2*	*-3.5*	*-4.0*	*-3.7*	*-3.4*	*-3.4*	*0.9*	*0.4*
Gas works	-2.8	-3.1	-3.2	-3.7	-3.3	-7.3	-6.9	-6.1	1.4	3.1
Coal transformation(2)	-3.8	-3.4	-3.3	-2.0	-1.2	-2.0	-2.5	-2.3	-1.4	-1.8
BKB plants	-	-	-	-	-	-	-	-	-	-
Blast furnaces	*-2.5*	*-2.0*	*-1.7*	*-1.0*	*-0.7*	*-0.8*	*-1.3*	*-1.2*	*-3.8*	*-1.8*
Coke ovens	*-1.3*	*-1.4*	*-1.6*	*-1.0*	*-0.6*	*-1.1*	*-1.3*	*-1.1*	*2.0*	*-1.7*
Patent fuel plants	-	-	-	-	-	-	-	-	-	-
Other transformation(3)	-3.1	-18.2	-20.1	-19.9	-21.2	-21.8	-22.0	-21.2	20.6	0.3
Energy ind. own use	**-0.0**	**-0.0**	**-0.0**	**-**	**-0.0**	**-**	**-**	**-**	**-3.5**	**-**
Losses	**-**	**-**	**-**	**-**	**-**	**-**	**-**	**-**		
Final consumption(4)	**27.0**	**23.2**	**23.4**	**22.4**	**22.8**	**23.9**	**21.9**	**23.8**	**-1.4**	**0.1**
Industry(5)	20.7	16.6	15.8	11.6	13.0	15.1	15.2	15.9	-2.7	0.0
Iron and steel	*10.1*	*9.7*	*9.0*	*6.0*	*5.8*	*6.1*	*8.3*	*7.7*	*-1.1*	*-0.8*
Chemical	*0.1*	*0.1*	*0.1*	*1.5*	*1.5*	*1.4*	*1.4*	*1.2*	*0.3*	*14.4*
Non-metallic minerals	*2.2*	*2.1*	*2.0*	*1.7*	*1.3*	*2.0*	*1.3*	*2.0*	*-1.2*	*0.0*
Paper, pulp and print	-	-	-	*0.0*	*0.1*	*0.1*	*0.1*	*0.1*	-	-
Other industry (6)	*8.3*	*4.7*	*4.7*	*2.4*	*4.3*	*5.5*	*4.0*	*4.8*	*-5.5*	*0.1*
Transport(7)	1.8	0.9	0.1	0.0	-	-	0.0	-	-28.0	-
Other	3.3	3.1	3.5	4.1	2.2	6.8	4.8	6.0	0.5	2.6
Comm. and pub. services	*1.1*	*1.3*	*1.3*	*1.4*	*0.7*	*2.3*	*1.4*	*1.7*	*1.8*	*1.2*
Residential	*2.1*	*1.7*	*2.1*	*2.4*	*1.4*	*4.5*	*2.8*	*3.3*	*-0.1*	*2.1*
Other sectors (8)	*0.1*	*0.0*	*0.0*	*0.3*	*0.1*	*0.0*	*0.6*	*1.1*	*-6.0*	*16.1*
Non-energy use	1.2	2.7	4.0	6.8	7.5	2.0	1.9	2.0	13.1	-3.3

(1) Main activity electricity and heat generation includes district heating.

(2) Coal transformation refers to the transformation of coal and peat to secondary and tertiary products (mainly cokes, briquettes, coal tar and off-gases). Gas works and Liquefaction are shown separately to match the *Energy Balances of OECD Countries.*

(3) Other transformation includes Liquefaction and Non-specified transformations.

(4) Final Consumption includes non-energy use and energy use (Industry, Transport and Other).

(5) Please refer to notes and definitions in Part I for detailed explanation.

(6) Other industry includes Non-ferrous metals, Transport equipment, Machinery, Mining and quarrying, Food and tobacco, Wood and wood products, Construction, Textile and leather, and Non-specified industry.

(7) Transport includes Rail and Inland waterways.

(8) Other sectors includes Agriculture/Forestry, Fishing and Non-specified other.

* "Coal" refers to all types of coal, primary (anthracite, coking coal, other bituminous coal, sub-bituminous coal, lignite/brown coal) and derived fuels (including patent fuel, BKB, coke oven coke, gas coke, coal tar, coke oven gas, gas works gas, blast furnace gas and other recovered gases), as well as peat, peat briquettes and oil shale. Quantities have been converted to Mtce using calorific values reported by the respective countries.
Please refer to notes and definitions in Part I.

Source: *IEA/OECD Energy Balances of OECD Countries*

SOUTH AFRICA

Steam coal[1] exports by destination

(thousand tonnes)

	1978[2]	1985	1990	1995	2000	2005	2010	2011	2012e
World	**12689**	**42488**	**45867**	**53371**	**67001**	**70917**	**65562**	**67368**	**73603**
OECD	**12689**	**36178**	**36477**	**43135**	**54371**	**60360**	**23056**	**25615**	**21880**
Austria	-	-	6	-	-	-	-	-	-
Belgium	606	2078	4365	3641	2504	1757	527	598	59
Canada	-	-	-	-	46	-	-	-	-
Czech Republic	-	-	-	-	-	-	-	-	-
Denmark	868	3496	-	3666	1721	2070	906	1348	630
Finland	-	-	-	325	-	-	-	-	-
France	6643	6444	863	1395	5872	5340	1323	1658	1456
Germany	1108	3196	4512	3670	3979	8812	1149	1283	936
Greece	-	207	1017	630	269	75	70	-	75
Hungary	-	-	-	-	-	-	-	-	-
Israel	-	2200	2583	3900	5523	5329	2826	3319	4753
Italy	960	6641	4884	3653	4173	4939	3236	3773	3432
Japan	157	4091	1427	3039	1661	155	420	846	468
Korea	1861	3700	5733	3850	2385	139	1956	3405	1939
Mexico	-	-	-	-	41	-	1368	388	200
Netherlands	309	812	1304	3503	7564	6527	2723	1778	1447
Norway	-	-	-	30	-	-	5	-	4
Poland	-	-	-	-	265	-	-	-	-
Portugal	3	104	2112	2683	2112	1926	321	-	-
Spain	114	2274	4667	7016	8403	8642	2724	2740	2410
Sweden	-	19	-	135	-	-	-	-	-
Turkey	-	-	1252	315	2547	1324	2080	3615	2873
United Kingdom	26	724	356	735	4503	12144	744	663	767
United States	-	-	-	140	50	135	-	44	50
Other OECD[3]	34	192	1396	809	753	1046	678	157	381
Non-OECD	**-**	**5400**	**9085**	**10236**	**12283**	**10050**	**42506**	**41753**	**51673**
Brazil	-	-	-	-	1919	673	1122	997	1114
Bulgaria	-	-	-	60	107	-	-	-	-
China, People's Rep.	-	-	-	810	522	-	4226	11870	13171
Chinese Taipei	-	1000	5685	5138	2488	522	2566	3862	4767
Hong Kong, China	-	2200	3217	2763	486	-	162	-	-
India	-	-	-	490	3636	3587	23440	16198	20985
Morocco	-	-	-	100	1978	2993	810	65	405
Romania	-	-	-	-	-	-	189	231	316
Russian Federation	-	-	-	-	-	-	-	-	-
Ukraine	-	-	-	-	-	61	-	-	-
Other Africa	-	-	-	145	356	970	2067	1988	3578
Other Asia	-	-	183	40	201	244	4727	3915	4142
Other Eastern Europe	-	-	-	185	-	367	-	-	-
Other FSU	-	-	-	-	-	-	-	-	-
Other non-OECD Americas	-	2200	900	505	639	397	1203	539	430
Other Middle East	-	2200	2583	3900	5559	5565	4820	5407	7518
Non-specified/Other	-	910	305	-	347	507	-	-	50

(1) Steam coal now comprises anthracite, other bituminous coal and sub-bituminous coal.

(2) Earliest year for which split by coal type is available.

(3) Australia, Chile, Estonia, Iceland, Ireland, Luxembourg, New Zealand, Slovak Republic, Slovenia and Switzerland.

Source: IEA/OECD Coal Statistics.

THAILAND

Coal balance*

(Mtce)

	1980	1985	1990	1995	2000	2005	2010	2011	Average annual percent change 80-90	90-11
Production	0.6	2.1	5.1	7.6	7.3	8.6	7.6	8.8	24.2	2.6
Imports	0.1	0.2	0.3	2.2	3.7	7.7	15.3	17.7	13.2	21.5
Exports	-0.0	-0.0	-	-	-	-	-0.0	-0.0	-	-
Stock changes	-	-0.0	0.0	0.0	-0.1	0.1	0.5	-0.4		
Primary supply	**0.7**	**2.4**	**5.5**	**9.8**	**11.0**	**16.4**	**23.4**	**26.1**	**23.3**	**7.7**
Statistical differences	**-0.0**	**-0.1**	**0.1**	**0.4**	**0.1**	**0.2**	**-0.3**	**-0.1**		
Total transformation	**-0.5**	**-1.7**	**-3.7**	**-5.1**	**-6.0**	**-7.0**	**-9.9**	**-12.5**	**21.8**	**6.0**
Electricity and heat gen.	-0.5	-1.7	-3.6	-5.0	-5.9	-6.9	-9.9	-12.5	21.7	6.1
Main activity producers (1)	*-0.5*	*-1.7*	*-3.6*	*-5.0*	*-5.2*	*-6.1*	*-9.2*	*-10.6*	*21.7*	*5.2*
Autoproducers	-	-	-	-	*-0.7*	*-0.8*	*-0.7*	*-1.9*	-	-
Gas works	-	-	-	-	-	-	-	-	-	-
Coal transformation(2)	-	-0.0	-0.0	-0.0	-0.0	-0.0	-0.0	-0.0	-	-2.5
BKB plants	-	-	-	-	-	-	-	-	-	-
Blast furnaces	-	*-0.0*	*-0.0*	*-0.0*	*-0.0*	*-0.0*	*-0.0*	*-0.0*	-	*-2.5*
Coke ovens	-	-	-	-	-	-	-	-	-	-
Patent fuel plants	-	-	-	-	-	-	-	-	-	-
Other transformation(3)	-	-	-	-	-	-	-	-	-	-
Energy ind. own use	-	-	-	-	-	-	-	-	-	-
Losses	-	-	-	-	-	-	-	-		
Final consumption(4)	**0.1**	**0.5**	**1.9**	**5.2**	**5.1**	**9.6**	**13.2**	**13.4**	**30.2**	**9.8**
Industry(5)	0.1	0.5	1.9	5.2	5.1	9.6	13.2	13.4	30.2	9.8
Iron and steel	-	*0.0*	*0.0*	*0.1*	*0.0*	*0.0*	*0.2*	*0.1*	-	*6.6*
Chemical	-	-	-	-	-	-	-	-	-	-
Non-metallic minerals	*0.0*	*0.2*	*1.2*	*3.8*	*4.0*	*7.7*	*10.9*	*11.2*	*94.7*	*11.2*
Paper, pulp and print	-	-	-	-	-	-	-	-	-	-
Other industry (6)	*0.1*	*0.3*	*0.6*	*1.3*	*1.0*	*1.9*	*2.0*	*2.1*	*16.9*	*5.9*
Transport(7)	-	-	-	-	-	-	-	-	-	-
Other	-	-	-	-	-	-	-	-	-	-
Comm. and pub. services	-	-	-	-	-	-	-	-	-	-
Residential	-	-	-	-	-	-	-	-	-	-
Other sectors (8)	-	-	-	-	-	-	-	-	-	-
Non-energy use	-	-	-	-	-	-	-	-	-	-

(1) Main activity electricity and heat generation includes district heating.

(2) Coal transformation refers to the transformation of coal and peat to secondary and tertiary products (mainly cokes, briquettes, coal tar and off-gases). Gas works and Liquefaction are shown separately to match the *Energy Balances of OECD Countries.*

(3) Other transformation includes Liquefaction and Non-specified transformations.

(4) Final Consumption includes non-energy use and energy use (Industry, Transport and Other).

(5) Please refer to notes and definitions in Part I for detailed explanation.

(6) Other industry includes Non-ferrous metals, Transport equipment, Machinery, Mining and quarrying, Food and tobacco, Wood and wood products, Construction, Textile and leather, and Non-specified industry.

(7) Transport includes Rail and Inland waterways.

(8) Other sectors includes Agriculture/Forestry, Fishing and Non-specified other.

* "Coal" refers to all types of coal, primary (anthracite, coking coal, other bituminous coal, sub-bituminous coal, lignite/brown coal) and derived fuels (including patent fuel, BKB, coke oven coke, gas coke, coal tar, coke oven gas, gas works gas, blast furnace gas and other recovered gases), as well as peat, peat briquettes and oil shale. Quantities have been converted to Mtce using calorific values reported by the respective countries. Please refer to notes and definitions in Part I.

Source: *IEA/OECD Energy Balances of OECD Countries*

THAILAND

Total coal imports by origin*

(thousand tonnes)

	1978(1)	1985	1990	1995	2000	2005	2010	2011	2012e
Total coal	**52**	**213**	**250**	**2308**	**2558**	**7989**	**16758**	**16425**	**16797**
Coking coal	**-**	**-**	**-**	**-**	**-**	**-**	**-**	**-**	**4**
Australia	-	-	-	-	-	-	-	-	-
Canada	-	-	-	-	-	-	-	-	-
Czech Republic	-	-	-	-	-	-	-	-	-
Germany	-	-	-	-	-	-	-	-	-
Poland	-	-	-	-	-	-	-	-	-
United Kingdom	-	-	-	-	-	-	-	-	-
United States	-	-	-	-	-	-	-	-	-
Other OECD	-	-	-	-	-	-	-	-	-
China, People's Rep.	-	-	-	-	-	-	-	-	-
Colombia	-	-	-	-	-	-	-	-	-
Indonesia	-	-	-	-	-	-	-	-	4
South Africa	-	-	-	-	-	-	-	-	-
Former Soviet Union(2)	-	-	x	x	x	x	x	x	x
Russian Federation (3)	x	x	-	-	-	-	-	-	-
Other FSU (3)	x	x	-	-	-	-	-	-	-
Venezuela	-	-	-	-	-	-	-	-	-
Vietnam	-	-	-	-	-	-	-	-	-
Non-specified/other	-	-	-	-	-	-	-	-	-
Steam coal(4)	**52**	**213**	**250**	**2308**	**2558**	**7989**	**16758**	**16425**	**16793**
Australia	52	213	-	-	136	-	2488	2946	3300
Canada	-	-	-	-	-	-	-	128	-
Czech Republic	-	-	-	-	-	-	-	-	-
Germany	-	-	-	-	-	-	-	-	-
Poland	-	-	-	-	-	-	-	-	-
United Kingdom	-	-	-	-	-	-	1	-	2
United States	-	-	-	-	-	-	-	-	-
Other OECD	-	-	-	-	-	1	3	2	4
China, People's Rep.	-	-	125	369	66	20	9	2	7
Colombia	-	-	-	-	-	-	144	-	-
Indonesia	-	-	125	1939	2356	6344	12491	13075	12856
South Africa	-	-	-	-	-	-	436	139	286
Former Soviet Union(2)	-	-	x	x	x	x	x	x	x
Russian Federation (3)	x	x	-	-	-	-	-	-	-
Other FSU (3)	x	x	-	-	-	-	-	-	-
Venezuela	-	-	-	-	-	-	-	-	-
Vietnam	-	-	-	-	-	460	150	33	133
Non-specified/other	-	-	-	-	-	1164	1036	100	205
Lignite	**-**	**-**	**-**	**-**	**-**	**-**	**-**	**-**	**-**

(1) Earliest year for which split by coal type is available.

(2) For years prior to 1990, or if no details for individual Former Soviet Union republics.

(3) Data for former Republics are not available separately for some countries.

(4) Steam coal now comprises anthracite, other bituminous coal and sub-bituminous coal.

* In these tables coal used for PCI and for blending has been classified by the IEA as steam coal. Accordingly, trade data reported in these tables may differ from those reported in Part III where this coal is shown as coking coal to be consistent with data reported by importing countries and with industry terminology and practice.

Source: IEA/OECD Coal Statistics.

UKRAINE

Coal balance*

(Mtce)

	1980	1985	1990	1995	2000	2005	2010	2011	Average annual percent change 80-90	90-11
Production	x	x	124.0	61.1	51.9	49.6	48.2	57.6	x	-3.6
Imports	x	x	15.1	13.5	5.8	6.5	11.1	11.9	x	-1.1
Exports	x	x	-21.3	-1.8	-2.6	-2.7	-6.9	-8.0	x	-4.6
Stock changes	x	x	0.8	-	-	0.0	2.2	-2.3		
Primary supply	**x**	**x**	**118.6**	**72.9**	**55.1**	**53.3**	**54.6**	**59.3**	**x**	**-3.2**
Statistical differences	**x**	**x**	**-4.2**	**-0.8**	**-0.3**	**0.4**	**0.0**	**-0.0**		
Total transformation	**x**	**x**	**-72.5**	**-50.4**	**-37.8**	**-33.6**	**-39.8**	**-43.0**	**x**	**-2.5**
Electricity and heat gen.	x	x	-50.7	-33.1	-20.8	-20.4	-29.1	-31.8	x	-2.2
Main activity producers (1)	*x*	*x*	*-48.9*	*-31.7*	*-19.8*	*-17.9*	*-27.4*	*-30.0*	*x*	*-2.3*
Autoproducers	*x*	*x*	*-1.8*	*-1.4*	*-1.1*	*-2.4*	*-1.7*	*-1.8*	*x*	*0.1*
Gas works	x	x	-	-	-	0.1	-0.0	-0.0	x	-
Coal transformation(2)	x	x	-21.8	-17.3	-17.0	-13.2	-10.6	-11.1	x	-3.2
BKB plants	*x*	*x*	*2.9*	*0.7*	*0.3*	*0.0*	*-0.0*	*-0.0*	*x*	*-*
Blast furnaces	*x*	*x*	*-12.6*	*-6.4*	*-6.4*	*-6.1*	*-6.6*	*-6.9*	*x*	*-2.8*
Coke ovens	*x*	*x*	*-12.4*	*-11.3*	*-10.7*	*-7.1*	*-4.0*	*-4.1*	*x*	*-5.1*
Patent fuel plants	*x*	*x*	*0.4*	*-0.3*	*-0.1*	*-0.1*	*-*	*-*	*x*	*-*
Other transformation(3)	x	x	-	-	-	-0.1	-0.1	-0.1	x	-
Energy ind. own use	**x**	**x**	**-5.3**	**-3.5**	**-2.7**	**-3.0**	**-2.6**	**-2.5**	**x**	**-3.4**
Losses	**x**	**x**	**-**	**-**	**-**	**-0.0**	**-0.3**	**-0.3**		
Final consumption(4)	**x**	**x**	**36.6**	**18.2**	**14.2**	**17.1**	**12.0**	**13.4**	**x**	**-4.7**
Industry(5)	x	x	25.7	11.3	10.4	11.9	10.3	11.3	x	-3.8
Iron and steel	*x*	*x*	*15.6*	*8.4*	*8.1*	*10.2*	*8.5*	*9.3*	*x*	*-2.4*
Chemical	*x*	*x*	*0.3*	*-*	*-*	*0.0*	*0.0*	*0.1*	*x*	*-4.6*
Non-metallic minerals	*x*	*x*	*0.0*	*0.0*	*0.0*	*0.2*	*0.9*	*1.2*	*x*	*17.8*
Paper, pulp and print	*x*	*x*	*-*	*-*	*-*	*0.0*	*0.0*	*0.0*	*x*	*-*
Other industry (6)	*x*	*x*	*9.8*	*2.9*	*2.3*	*1.6*	*0.8*	*0.7*	*x*	*-11.6*
Transport(7)	x	x	0.1	-	-	0.1	0.0	0.0	x	-4.5
Other	x	x	10.8	6.9	3.8	3.2	1.0	1.3	x	-9.7
Comm. and pub. services	*x*	*x*	*-*	*-*	*-*	*-*	*0.3*	*0.2*	*x*	*-*
Residential	*x*	*x*	*7.9*	*6.4*	*3.7*	*1.8*	*0.7*	*1.0*	*x*	*-9.3*
Other sectors (8)	*x*	*x*	*2.9*	*0.5*	*0.1*	*1.4*	*0.0*	*0.0*	*x*	*-20.6*
Non-energy use	x	x	-	-	-	1.8	0.7	0.8	x	-

(1) Main activity electricity and heat generation includes district heating.

(2) Coal transformation refers to the transformation of coal and peat to secondary and tertiary products (mainly cokes, briquettes, coal tar and off-gases). Gas works and Liquefaction are shown separately to match the *Energy Balances of OECD Countries.*

(3) Other transformation includes Liquefaction and Non-specified transformations.

(4) Final Consumption includes non-energy use and energy use (Industry, Transport and Other).

(5) Please refer to notes and definitions in Part I for detailed explanation.

(6) Other industry includes Non-ferrous metals, Transport equipment, Machinery, Mining and quarrying, Food and tobacco, Wood and wood products, Construction, Textile and leather, and Non-specified industry.

(7) Transport includes Rail and Inland waterways.

(8) Other sectors includes Agriculture/Forestry, Fishing and Non-specified other.

* "Coal" refers to all types of coal, primary (anthracite, coking coal, other bituminous coal, sub-bituminous coal, lignite/brown coal) and derived fuels (including patent fuel, BKB, coke oven coke, gas coke, coal tar, coke oven gas, gas works gas, blast furnace gas and other recovered gases), as well as peat, peat briquettes and oil shale. Quantities have been converted to Mtce using calorific values reported by the respective countries. Please refer to notes and definitions in Part I.

Source: *IEA/OECD Energy Balances of OECD Countries*

UKRAINE

Total coal imports by origin*

(thousand tonnes)

	1978(1)	1985	1990	1995	2000	2005	2010	2011	2012e
Total coal	**x**	**x**	**26754**	**15360**	**6606**	**7303**	**12181**	**17872**	**13575**
Coking coal	**x**	**x**	**19964**	**5859**	**2759**	**6902**	**7777**	**9565**	**9558**
Australia	x	x	-	-	-	-	-	-	-
Canada	x	x	-	-	-	-	-	-	-
Czech Republic	x	x	-	-	-	-	-	-	-
Germany	x	x	-	-	-	-	-	-	-
Poland	x	x	-	4369	595	31	-	2	58
United Kingdom	x	x	-	-	-	-	-	-	-
United States	x	x	-	-	-	167	2151	4275	2424
Other OECD	x	x	-	-	-	-	-	-	-
China, People's Rep.	x	x	-	-	-	-	-	-	-
Colombia	x	x	-	-	-	-	-	-	-
Indonesia	x	x	-	-	-	-	-	-	-
South Africa	x	x	-	-	-	-	-	-	-
Former Soviet Union(2)	x	x	x	x	x	x	x	x	x
Russian Federation (3)	*x*	*x*	*19964*	-	*2164*	*6704*	*5579*	*5288*	*7076*
Other FSU (3)	*x*	*x*	-	*1490*	-	-	-	-	-
Venezuela	x	x	-	-	-	-	-	-	-
Vietnam	x	x	-	-	-	-	-	-	-
Non-specified/other	x	x	-	-	-	-	47	-	-
Steam coal(4)	**x**	**x**	**6722**	**9437**	**3847**	**401**	**4404**	**8307**	**4017**
Australia	x	x	-	-	-	-	-	-	-
Canada	x	x	-	-	-	-	-	-	-
Czech Republic	x	x	-	-	-	-	-	-	-
Germany	x	x	-	-	-	-	-	-	-
Poland	x	x	-	4722	452	-	33	98	65
United Kingdom	x	x	-	-	-	-	-	-	-
United States	x	x	-	-	-	-	-	64	-
Other OECD	x	x	-	-	-	-	-	-	-
China, People's Rep.	x	x	-	-	-	-	-	-	-
Colombia	x	x	-	-	-	-	-	-	-
Indonesia	x	x	-	-	-	-	-	75	81
South Africa	x	x	-	-	-	11	-	-	-
Former Soviet Union(2)	x	x	x	x	x	x	x	x	x
Russian Federation (3)	*x*	*x*	*6722*	*4259*	*2175*	*390*	*4371*	*8070*	*3871*
Other FSU (3)	*x*	*x*	-	*456*	*1220*	-	-	-	-
Venezuela	x	x	-	-	-	-	-	-	-
Vietnam	x	x	-	-	-	-	-	-	-
Non-specified/other	x	x	-	-	-	-	-	-	-
Lignite	**x**	**x**	**68**	**64**	**-**	**-**	**-**	**-**	**-**

(1) Earliest year for which split by coal type is available.

(2) For years prior to 1990, or if no details for individual Former Soviet Union republics.

(3) Data for former Republics are not available separately for some countries.

(4) Steam coal now comprises anthracite, other bituminous coal and sub-bituminous coal.

* In these tables coal used for PCI and for blending has been classified by the IEA as steam coal. Accordingly, trade data reported in these tables may differ from those reported in Part III where this coal is shown as coking coal to be consistent with data reported by importing countries and with industry terminology and practice.

Source: IEA/OECD Coal Statistics.

UKRAINE

Coking coal exports by destination

(thousand tonnes)

	1978(1)	1985	1990	1995	2000	2005	2010	2011	2012e
World	**x**	**x**	**8114**	**400**	**200**	**509**	**162**	**381**	**192**
OECD	**x**	**x**	**-**	**-**	**-**	**229**	**113**	**381**	**192**
Austria	x	x	-	-	-	-	-	-	-
Belgium	x	x	-	-	-	-	-	-	-
Canada	x	x	-	-	-	-	-	-	-
Czech Republic	x	x	-	-	-	-	-	1	-
Denmark	x	x	-	-	-	-	-	-	-
Finland	x	x	-	-	-	-	-	-	-
France	x	x	-	-	-	-	-	-	-
Germany	x	x	-	-	-	-	-	-	-
Greece	x	x	-	-	-	-	-	-	-
Hungary	x	x	-	-	-	-	-	-	-
Israel	x	x	-	-	-	-	-	-	-
Italy	x	x	-	-	-	-	-	-	-
Japan	x	x	-	-	-	-	-	-	-
Korea	x	x	-	-	-	-	-	150	75
Mexico	x	x	-	-	-	-	-	-	-
Netherlands	x	x	-	-	-	-	-	-	-
Norway	x	x	-	-	-	-	-	-	-
Poland	x	x	-	-	-	-	-	1	-
Portugal	x	x	-	-	-	-	-	-	-
Spain	x	x	-	-	-	-	-	-	-
Sweden	x	x	-	-	-	-	-	-	-
Turkey	x	x	-	-	-	-	100	134	-
United Kingdom	x	x	-	-	-	-	13	-	-
United States	x	x	-	-	-	-	-	-	-
Other OECD(2)	x	x	-	-	-	229	-	95	117
Non-OECD	**x**	**x**	**8114**	**400**	**200**	**280**	**49**	**-**	**-**
Brazil	x	x	-	-	-	-	49	-	-
Bulgaria	x	x	-	200	-	-	-	-	-
China, People's Rep.	x	x	-	-	-	-	-	-	-
Chinese Taipei	x	x	-	-	-	-	-	-	-
Hong Kong, China	x	x	-	-	-	-	-	-	-
India	x	x	-	-	-	-	-	-	-
Morocco	x	x	-	-	-	-	-	-	-
Romania	x	x	-	200	-	-	-	-	-
Russian Federation	x	x	8114	-	200	280	-	-	-
Ukraine	x	x	-	-	-	-	-	-	-
Other Africa	x	x	-	-	-	-	-	-	-
Other Asia	x	x	-	-	-	-	-	-	-
Other Eastern Europe	x	x	-	-	-	-	-	-	-
Other FSU	x	x	-	-	-	-	-	-	-
Other non-OECD Americas	x	x	-	-	-	-	-	-	-
Other Middle East	x	x	-	-	-	-	-	-	-
Non-specified/Other	x	x	-	-	-	-	-	-	-

(1) Earliest year for which split by coal type is available.

(2) Australia, Chile, Estonia, Iceland, Ireland, Luxembourg, New Zealand, Slovak Republic, Slovenia and Switzerland.

Source: IEA/OECD Coal Statistics.

UKRAINE

Steam coal[1] exports by destination

(thousand tonnes)

	1978[2]	1985	1990	1995	2000	2005	2010	2011	2012e
World	**x**	**x**	**16577**	**2000**	**2297**	**3157**	**5918**	**7496**	**5123**
OECD	**x**	**x**	**-**	**550**	**845**	**552**	**2951**	**4256**	**3179**
Austria	x	x	-	-	-	13	1	1	1
Belgium	x	x	-	29	63	11	598	1123	318
Canada	x	x	-	-	-	154	181	107	220
Czech Republic	x	x	-	1	-	7	-	39	78
Denmark	x	x	-	1	-	-	-	-	-
Finland	x	x	-	-	-	-	-	-	-
France	x	x	-	21	-	-	36	165	95
Germany	x	x	-	-	-	5	12	12	20
Greece	x	x	-	-	-	2	30	62	21
Hungary	x	x	-	6	-	-	2	-	-
Israel	x	x	-	-	-	-	-	-	-
Italy	x	x	-	22	83	13	230	280	275
Japan	x	x	-	-	-	52	-	-	-
Korea	x	x	-	-	-	-	-	51	-
Mexico	x	x	-	-	-	-	-	-	-
Netherlands	x	x	-	-	-	82	-	-	33
Norway	x	x	-	-	-	-	-	-	-
Poland	x	x	-	33	-	8	399	532	366
Portugal	x	x	-	-	-	5	4	12	17
Spain	x	x	-	-	52	10	313	226	297
Sweden	x	x	-	-	-	5	3	73	-
Turkey	x	x	-	154	647	40	1083	1301	1079
United Kingdom	x	x	-	-	-	-	-	-	-
United States	x	x	-	-	-	79	38	113	171
Other OECD[3]	x	x	-	283	-	66	21	159	188
Non-OECD	**x**	**x**	**16577**	**1450**	**1452**	**2605**	**2967**	**3240**	**1944**
Brazil	x	x	-	-	-	16	166	174	78
Bulgaria	x	x	-	154	-	2315	2592	2692	1612
China, People's Rep.	x	x	-	-	-	-	-	-	-
Chinese Taipei	x	x	-	-	-	-	-	-	-
Hong Kong, China	x	x	-	-	-	-	-	-	-
India	x	x	-	-	-	-	-	-	-
Morocco	x	x	-	-	-	-	-	-	-
Romania	x	x	-	128	-	192	194	371	251
Russian Federation	x	x	16577	235	1452	10	15	-	-
Ukraine	x	x	-	-	-	-	-	-	-
Other Africa	x	x	-	-	-	-	-	-	-
Other Asia	x	x	-	-	-	-	-	-	-
Other Eastern Europe	x	x	-	27	-	-	-	-	-
Other FSU	x	x	-	906	-	72	-	3	3
Other non-OECD Americas	x	x	-	-	-	-	-	-	-
Other Middle East	x	x	-	-	-	-	-	-	-
Non-specified/Other	x	x	-	-	-	-	-	-	-

(1) Steam coal now comprises anthracite, other bituminous coal and sub-bituminous coal.

(2) Earliest year for which split by coal type is available.

(3) Australia, Chile, Estonia, Iceland, Ireland, Luxembourg, New Zealand, Slovak Republic, Slovenia and Switzerland.

Source: IEA/OECD Coal Statistics.

VENEZUELA

Coal balance*

(Mtce)

	1980	1985	1990	1995	2000	2005	2010	2011	Average annual percent change 80-90	90-11
Production	0.0	0.0	2.3	4.2	8.2	7.5	2.8	2.2	48.5	-0.2
Imports	0.2	0.2	0.3	-	-	-	-	-	4.9	-
Exports	-	-	-1.9	-4.4	-8.3	-7.4	-2.6	-1.9	-	-0.0
Stock changes	-	0.0	-	0.2	0.2	-0.0	-	-		
Primary supply	**0.2**	**0.3**	**0.7**	**0.0**	**0.2**	**0.1**	**0.3**	**0.3**	**11.4**	**-3.8**
Statistical differences	-	-	-	-	-	-	-	-		
Total transformation	**-0.1**	**-0.1**	**-0.2**	-	-	-	-	-	**6.1**	-
Electricity and heat gen.	-	-	-	-	-	-	-	-	-	-
Main activity producers (1)	-	-	-	-	-	-	-	-	-	-
Autoproducers	-	-	-	-	-	-	-	-	-	-
Gas works	-	-	-	-	-	-	-	-	-	-
Coal transformation(2)	-0.1	-0.1	-0.2	-	-	-	-	-	6.1	-
BKB plants	-	-	-	-	-	-	-	-	-	-
Blast furnaces	*-0.1*	*-0.1*	*-0.1*	-	-	-	-	-	*5.4*	-
Coke ovens	-	-	*-0.0*	-	-	-	-	-	-	-
Patent fuel plants	-	-	-	-	-	-	-	-	-	-
Other transformation(3)	-	-	-	-	-	-	-	-	-	-
Energy ind. own use	-	-	-	-	-	-	-	-	-	-
Losses	-	-	-	-	-	-	-	-		
Final consumption(4)	**0.1**	**0.2**	**0.5**	**0.0**	**0.2**	**0.1**	**0.3**	**0.3**	**13.9**	**-2.6**
Industry(5)	0.1	0.2	0.5	0.0	0.2	0.1	0.3	0.3	13.9	-2.6
Iron and steel	*0.1*	*0.1*	*0.2*	-	-	-	-	-	*5.4*	-
Chemical	-	-	-	-	-	-	-	-	-	-
Non-metallic minerals	*0.0*	*0.0*	*0.3*	*0.0*	*0.2*	*0.1*	*0.3*	*0.3*	*23.0*	*-0.8*
Paper, pulp and print	-	-	-	-	-	-	-	-	-	-
Other industry (6)	-	-	-	-	-	-	-	-	-	-
Transport(7)	-	-	-	-	-	-	-	-	-	-
Other	-	-	-	-	-	-	-	-	-	-
Comm. and pub. services	-	-	-	-	-	-	-	-	-	-
Residential	-	-	-	-	-	-	-	-	-	-
Other sectors (8)	-	-	-	-	-	-	-	-	-	-
Non-energy use	-	-	-	-	-	-	-	-	-	-

(1) Main activity electricity and heat generation includes district heating.

(2) Coal transformation refers to the transformation of coal and peat to secondary and tertiary products (mainly cokes, briquettes, coal tar and off-gases). Gas works and Liquefaction are shown separately to match the *Energy Balances of OECD Countries.*

(3) Other transformation includes Liquefaction and Non-specified transformations.

(4) Final Consumption includes non-energy use and energy use (Industry, Transport and Other).

(5) Please refer to notes and definitions in Part I for detailed explanation.

(6) Other industry includes Non-ferrous metals, Transport equipment, Machinery, Mining and quarrying, Food and tobacco, Wood and wood products, Construction, Textile and leather, and Non-specified industry.

(7) Transport includes Rail and Inland waterways.

(8) Other sectors includes Agriculture/Forestry, Fishing and Non-specified other.

* "Coal" refers to all types of coal, primary (anthracite, coking coal, other bituminous coal, sub-bituminous coal, lignite/brown coal) and derived fuels (including patent fuel, BKB, coke oven coke, gas coke, coal tar, coke oven gas, gas works gas, blast furnace gas and other recovered gases), as well as peat, peat briquettes and oil shale. Quantities have been converted to Mtce using calorific values reported by the respective countries. Please refer to notes and definitions in Part I.

Source: *IEA/OECD Energy Balances of OECD Countries*

VENEZUELA

Steam coal[1] exports by destination

(thousand tonnes)

	1978[2]	1985	1990	1995	2000	2005	2010	2011	2012e
World	**33**	**-**	**1834**	**3252**	**7722**	**7141**	**3889**	**3667**	**2740**
OECD	**-**	**-**	**1746**	**3152**	**7227**	**6374**	**2413**	**2248**	**1379**
Austria	-	-	-	-	-	-	3	11	-
Belgium	-	-	1	17	-	135	52	96	45
Canada	-	-	33	-	589	583	93	33	22
Czech Republic	-	-	-	-	-	-	-	-	-
Denmark	-	-	-	-	-	-	-	-	-
Finland	-	-	148	29	-	-	-	-	-
France	-	-	560	550	441	441	227	251	27
Germany	-	-	-	193	476	-	86	91	57
Greece	-	-	-	-	-	-	-	-	-
Hungary	-	-	-	-	-	-	-	-	-
Israel	-	-	-	-	71	-	-	-	-
Italy	-	-	140	302	1067	391	210	220	134
Japan	-	-	-	-	-	-	-	-	-
Korea	-	-	-	-	-	-	-	-	-
Mexico	-	-	-	-	112	-	-	-	54
Netherlands	-	-	19	79	766	452	408	439	410
Norway	-	-	5	-	-	-	-	-	-
Poland	-	-	-	-	-	-	-	-	-
Portugal	-	-	-	306	-	-	-	-	36
Spain	-	-	53	-	378	88	55	110	88
Sweden	-	-	375	198	180	49	112	55	55
Turkey	-	-	-	-	35	48	299	224	-
United Kingdom	-	-	169	-	218	-	43	-	227
United States	-	-	238	1478	2726	4139	825	666	187
Other OECD[3]	-	-	5	-	168	48	-	52	37
Non-OECD	**33**	**-**	**88**	**100**	**495**	**709**	**1476**	**1419**	**1361**
Brazil	33	-	-	100	79	460	972	750	587
Bulgaria	-	-	-	-	-	-	-	-	-
China, People's Rep.	-	-	-	-	-	-	-	-	-
Chinese Taipei	-	-	-	-	-	-	-	-	-
Hong Kong, China	-	-	-	-	-	-	-	-	-
India	-	-	-	-	-	-	-	32	26
Morocco	-	-	-	-	-	-	-	-	-
Romania	-	-	-	-	-	-	-	-	-
Russian Federation	-	-	-	-	-	-	-	-	-
Ukraine	-	-	-	-	-	-	-	-	-
Other Africa	-	-	-	-	-	-	-	-	-
Other Asia	-	-	-	-	-	-	-	-	-
Other Eastern Europe	-	-	-	-	-	-	-	-	-
Other FSU	-	-	-	-	-	-	-	-	-
Other non-OECD Americas	-	-	88	-	584	249	504	683	785
Other Middle East	-	-	-	-	71	-	-	-	-
Non-specified/Other	-	-	-	-	-	58	-	-	-

(1) Steam coal now comprises anthracite, other bituminous coal and sub-bituminous coal.

(2) Earliest year for which split by coal type is available.

(3) Australia, Chile, Estonia, Iceland, Ireland, Luxembourg, New Zealand, Slovak Republic, Slovenia and Switzerland.

Source: IEA/OECD Coal Statistics.

VIETNAM

Coal balance*

(Mtce)

	1980	1985	1990	1995	2000	2005	2010	2011	Average annual percent change 80-90	90-11
Production	4.2	4.5	3.7	6.7	9.3	27.1	35.9	35.6	-1.1	11.4
Imports	0.0	0.0	0.0	0.0	-	0.5	0.8	0.9	2.1	21.9
Exports	-0.5	-0.5	-0.6	-2.3	-2.6	-14.4	-15.9	-14.3	2.1	16.0
Stock changes	-0.4	-	0.1	0.3	-0.4	-1.5	0.1	-		
Primary supply	**3.2**	**4.0**	**3.2**	**4.7**	**6.2**	**11.8**	**20.9**	**22.2**	**-0.2**	**9.7**
Statistical differences	**-**	**-**	**-**	**-**	**-**	**-**	**0.0**	**-**		
Total transformation	**-1.1**	**-1.4**	**-1.3**	**-1.0**	**-1.6**	**-4.3**	**-6.9**	**-7.3**	**1.6**	**8.7**
Electricity and heat gen.	-1.1	-1.4	-1.3	-1.0	-1.6	-4.3	-6.9	-7.3	1.6	8.7
Main activity producers (1)	*-1.1*	*-1.4*	*-1.3*	*-1.0*	*-1.6*	*-3.8*	*-6.7*	*-7.2*	*1.6*	*8.6*
Autoproducers	-	-	-	-	-	*-0.4*	*-0.2*	*-0.2*	-	-
Gas works	-	-	-	-	-	-	-	-	-	-
Coal transformation(2)	-0.0	-0.0	-0.0	-0.0	-	-	-	-	7.1	-
BKB plants	-	-	-	-	-	-	-	-	-	-
Blast furnaces	*-0.0*	*-0.0*	*-0.0*	*-0.0*	-	-	-	-	*7.1*	-
Coke ovens	-	-	-	-	-	-	-	-	-	-
Patent fuel plants	-	-	-	-	-	-	-	-	-	-
Other transformation(3)	-	-	-	-	-	-	-	-	-	-
Energy ind. own use	**-**	**-**	**-**	**-**	**-**	**-**	**-**	**-**	**-**	**-**
Losses	**-**	**-**	**-**	**-**	**-**	**-**	**-**	**-**		
Final consumption(4)	**2.2**	**2.6**	**1.9**	**3.7**	**4.6**	**7.5**	**14.0**	**14.9**	**-1.3**	**10.3**
Industry(5)	1.3	2.1	1.5	2.8	3.3	5.7	11.8	12.5	0.9	10.7
Iron and steel	*0.0*	*0.0*	*0.0*	*0.0*	-	-	-	-	*5.9*	-
Chemical	-	-	-	-	-	-	-	-	-	-
Non-metallic minerals	-	-	-	-	-	-	-	-	-	-
Paper, pulp and print	-	-	-	-	-	-	-	-	-	-
Other industry (6)	*1.3*	*2.1*	*1.5*	*2.8*	*3.3*	*5.7*	*11.8*	*12.5*	*0.9*	*10.8*
Transport(7)	0.1	0.1	0.0	0.0	-	-	-	-	-13.9	-
Other	0.7	0.4	0.4	0.9	1.3	1.9	2.3	2.4	-5.6	8.7
Comm. and pub. services	-	-	*0.0*	*0.2*	*0.4*	*0.5*	*0.5*	*0.6*	-	*15.4*
Residential	*0.5*	*0.4*	*0.3*	*0.6*	*0.8*	*1.4*	*1.7*	*1.8*	*-4.8*	*9.0*
Other sectors (8)	*0.3*	*0.0*	*0.1*	*0.1*	*0.0*	*0.0*	*0.0*	*0.0*	*-9.6*	*-5.3*
Non-energy use	-	-	-	-	-	-	-	-	-	-

(1) Main activity electricity and heat generation includes district heating.

(2) Coal transformation refers to the transformation of coal and peat to secondary and tertiary products (mainly cokes, briquettes, coal tar and off-gases). Gas works and Liquefaction are shown separately to match the *Energy Balances of OECD Countries.*

(3) Other transformation includes Liquefaction and Non-specified transformations.

(4) Final Consumption includes non-energy use and energy use (Industry, Transport and Other).

(5) Please refer to notes and definitions in Part I for detailed explanation.

(6) Other industry includes Non-ferrous metals, Transport equipment, Machinery, Mining and quarrying, Food and tobacco, Wood and wood products, Construction, Textile and leather, and Non-specified industry.

(7) Transport includes Rail and Inland waterways.

(8) Other sectors includes Agriculture/Forestry, Fishing and Non-specified other.

* "Coal" refers to all types of coal, primary (anthracite, coking coal, other bituminous coal, sub-bituminous coal, lignite/brown coal) and derived fuels (including patent fuel, BKB, coke oven coke, gas coke, coal tar, coke oven gas, gas works gas, blast furnace gas and other recovered gases), as well as peat, peat briquettes and oil shale. Quantities have been converted to Mtce using calorific values reported by the respective countries. Please refer to notes and definitions in Part I.

Source: *IEA/OECD Energy Balances of OECD Countries*

VIETNAM

Steam coal[1] exports by destination

(thousand tonnes)

	1978[2]	1985	1990	1995	2000	2005	2010	2011	2012e
World	**1430**	**604**	**745**	**2821**	**3526**	**17987**	**19747**	**21808**	**19074**
OECD	**-**	**100**	**450**	**2308**	**3203**	**6062**	**3401**	**3112**	**1541**
Austria	-	-	-	-	-	-	-	-	-
Belgium	-	-	-	93	240	192	-	-	-
Canada	-	-	-	100	-	-	-	-	-
Czech Republic	-	-	-	-	-	-	-	-	-
Denmark	-	-	-	2	-	-	-	-	-
Finland	-	-	-	-	-	-	-	-	-
France	-	-	-	179	100	155	52	-	-
Germany	-	-	-	3	52	-	-	-	-
Greece	-	-	-	-	-	-	-	-	-
Hungary	-	-	-	-	-	-	-	-	-
Israel	-	-	-	-	-	-	-	-	-
Italy	-	-	-	-	-	-	-	-	-
Japan	-	100	150	1469	2227	4848	1604	1475	761
Korea	-	-	300	400	516	854	1745	1637	780
Mexico	-	-	-	-	-	-	-	-	-
Netherlands	-	-	-	21	20	-	-	-	-
Norway	-	-	-	1	-	-	-	-	-
Poland	-	-	-	-	-	-	-	-	-
Portugal	-	-	-	-	-	-	-	-	-
Spain	-	-	-	-	-	13	-	-	-
Sweden	-	-	-	-	-	-	-	-	-
Turkey	-	-	-	-	-	-	-	-	-
United Kingdom	-	-	-	40	48	-	-	-	-
United States	-	-	-	-	-	-	-	-	-
Other OECD[3]	-	-	-	-	-	-	-	-	-
Non-OECD	**-**	**504**	**295**	**513**	**323**	**11925**	**15946**	**18696**	**17533**
Brazil	-	-	-	100	88	388	-	-	-
Bulgaria	-	-	-	-	-	-	-	-	-
China, People's Rep.	-	504	100	150	27	10532	15626	18655	17425
Chinese Taipei	-	-	-	63	52	118	35	41	33
Hong Kong, China	-	-	-	-	-	-	-	-	-
India	-	-	-	-	-	-	57	-	-
Morocco	-	-	-	-	-	-	-	-	-
Romania	-	-	-	-	-	-	-	-	-
Russian Federation	-	-	-	-	-	-	-	-	-
Ukraine	-	-	-	-	-	-	-	-	-
Other Africa	-	-	-	-	-	-	-	-	-
Other Asia	-	-	195	200	156	887	228	-	75
Other Eastern Europe	-	-	-	-	-	-	-	-	-
Other FSU	-	-	-	-	-	-	-	-	-
Other non-OECD Americas	-	-	-	-	-	-	-	-	-
Other Middle East	-	-	-	-	-	-	-	-	-
Non-specified/Other	1430	-	-	-	-	-	400	-	-

(1) Steam coal now comprises anthracite, other bituminous coal and sub-bituminous coal.

(2) Earliest year for which split by coal type is available.

(3) Australia, Chile, Estonia, Iceland, Ireland, Luxembourg, New Zealand, Slovak Republic, Slovenia and Switzerland.

Source: IEA/OECD Coal Statistics.

AFRICA

Coal balance*

(Mtce)

	1980	1985	1990	1995	2000	2005	2010	2011	Average annual percent change 80-90	90-11
Production	99.6	147.4	150.0	174.8	187.0	202.8	209.9	209.1	4.2	1.6
Imports	1.6	3.6	4.2	4.9	8.6	10.8	11.1	11.3	10.0	4.8
Exports	-27.8	-46.3	-48.4	-58.3	-68.0	-69.4	-64.7	-67.5	5.7	1.6
Stock changes	0.1	-0.3	0.1	0.6	1.6	0.0	-0.0	0.3		
Primary supply	**73.6**	**104.4**	**105.9**	**122.1**	**129.1**	**144.2**	**156.3**	**153.2**	**3.7**	**1.8**
Statistical differences	**3.7**	**-3.7**	**6.5**	**-3.5**	**-0.2**	**-0.5**	**-3.3**	**-4.3**		
Total transformation	**-46.6**	**-72.8**	**-84.0**	**-92.7**	**-102.0**	**-116.7**	**-127.0**	**-120.9**	**6.1**	**1.7**
Electricity and heat gen.	-36.0	-46.7	-56.1	-66.0	-74.6	-84.1	-94.3	-90.2	4.5	2.3
Main activity producers (1)	*-33.1*	*-42.5*	*-52.8*	*-62.2*	*-70.5*	*-80.2*	*-90.8*	*-86.7*	*4.8*	*2.4*
Autoproducers	*-2.9*	*-4.2*	*-3.3*	*-3.7*	*-4.1*	*-3.9*	*-3.5*	*-3.5*	*1.2*	*0.2*
Gas works	-2.8	-3.0	-3.2	-3.7	-3.3	-7.3	-7.0	-6.1	1.5	3.1
Coal transformation(2)	-4.7	-4.9	-4.7	-3.2	-2.9	-3.6	-3.7	-3.5	-0.1	-1.4
BKB plants	-	-	-	-	-	-	-	-	-	-
Blast furnaces	*-3.0*	*-2.9*	*-2.6*	*-1.9*	*-1.7*	*-1.9*	*-2.2*	*-2.0*	*-1.5*	*-1.2*
Coke ovens	*-1.7*	*-2.0*	*-2.1*	*-1.2*	*-1.2*	*-1.7*	*-1.6*	*-1.4*	*2.0*	*-1.7*
Patent fuel plants	-	-	-	-	-	-	-	-	-	-
Other transformation(3)	-3.1	-18.2	-20.1	-19.9	-21.2	-21.8	-22.0	-21.2	20.6	0.3
Energy ind. own use	**-0.1**	**-0.2**	**-0.1**	**-0.0**	**-0.0**	**-0.0**	**-0.0**	**-0.0**	**1.0**	**-3.5**
Losses	**-0.0**	**-0.2**	**-0.1**	**-0.1**	**-0.1**	**-0.2**	**-0.1**	**-0.1**		
Final consumption(4)	**30.6**	**27.4**	**28.1**	**25.8**	**26.8**	**26.8**	**25.8**	**27.9**	**-0.8**	**-0.0**
Industry(5)	23.4	19.8	19.6	14.1	16.0	17.5	17.7	18.6	-1.8	-0.3
Iron and steel	*10.7*	*11.1*	*10.6*	*7.5*	*6.9*	*7.3*	*9.3*	*8.7*	*-0.1*	*-0.9*
Chemical	*0.1*	*0.1*	*0.1*	*1.5*	*1.5*	*1.4*	*1.4*	*1.2*	*-5.8*	*14.4*
Non-metallic minerals	*2.5*	*2.4*	*2.2*	*1.8*	*1.5*	*2.3*	*2.0*	*2.7*	*-1.6*	*1.1*
Paper, pulp and print	-	*0.0*	-	*0.0*	*0.1*	*0.1*	*0.1*	*0.1*	-	-
Other industry (6)	*10.0*	*6.2*	*6.8*	*3.3*	*6.0*	*6.4*	*5.0*	*5.8*	*-3.8*	*-0.7*
Transport(7)	2.0	1.1	0.3	0.1	0.0	0.0	0.0	0.0	-18.4	-12.0
Other	4.0	3.8	4.3	4.8	3.2	7.3	6.2	7.3	0.7	2.6
Comm. and pub. services	*1.3*	*1.5*	*1.6*	*1.7*	*1.0*	*2.4*	*1.7*	*2.0*	*1.9*	*1.1*
Residential	*2.3*	*1.9*	*2.3*	*2.5*	*1.5*	*4.6*	*2.9*	*3.4*	*-0.1*	*1.9*
Other sectors (8)	*0.4*	*0.4*	*0.4*	*0.7*	*0.6*	*0.3*	*1.5*	*1.9*	*0.8*	*7.7*
Non-energy use	1.2	2.7	4.0	6.8	7.5	2.0	1.9	2.0	13.1	-3.3

(1) Main activity electricity and heat generation includes district heating.

(2) Coal transformation refers to the transformation of coal and peat to secondary and tertiary products (mainly cokes, briquettes, coal tar and off-gases). Gas works and Liquefaction are shown separately to match the *Energy Balances of OECD Countries.*

(3) Other transformation includes Liquefaction and Non-specified transformations.

(4) Final Consumption includes non-energy use and energy use (Industry, Transport and Other).

(5) Please refer to notes and definitions in Part I for detailed explanation.

(6) Other industry includes Non-ferrous metals, Transport equipment, Machinery, Mining and quarrying, Food and tobacco, Wood and wood products, Construction, Textile and leather, and Non-specified industry.

(7) Transport includes Rail and Inland waterways.

(8) Other sectors includes Agriculture/Forestry, Fishing and Non-specified other.

* "Coal" refers to all types of coal, primary (anthracite, coking coal, other bituminous coal, sub-bituminous coal, lignite/brown coal) and derived fuels (including patent fuel, BKB, coke oven coke, gas coke, coal tar, coke oven gas, gas works gas, blast furnace gas and other recovered gases), as well as peat, peat briquettes and oil shale. Quantities have been converted to Mtce using calorific values reported by the respective countries. Please refer to notes and definitions in Part I.

Source: *IEA/OECD Energy Balances of OECD Countries*

ASIA EXCLUDING CHINA

Coal balance*

(Mtce)

	1980	1985	1990	1995	2000	2005	2010	2011	Average annual percent change 80-90	90-11
Production	123.8	163.5	211.8	267.3	321.2	484.0	721.4	772.5	5.5	6.4
Imports	7.2	17.0	31.9	48.9	77.5	120.3	206.0	220.8	16.1	9.6
Exports	-0.8	-1.9	-5.4	-29.4	-51.9	-129.4	-264.4	-302.4	21.2	21.2
Stock changes	-8.0	-3.6	-9.1	-5.8	2.5	-9.1	-5.0	-2.6		
Primary supply	**122.2**	**175.1**	**229.2**	**281.0**	**349.3**	**465.8**	**658.0**	**688.2**	**6.5**	**5.4**
Statistical differences	**2.0**	**3.2**	**3.3**	**2.2**	**2.1**	**-1.3**	**-12.6**	**-18.7**		
Total transformation	**-44.7**	**-76.4**	**-119.6**	**-178.0**	**-250.4**	**-328.4**	**-433.7**	**-454.6**	**10.3**	**6.6**
Electricity and heat gen.	-35.2	-64.6	-105.4	-163.5	-235.3	-310.2	-415.2	-435.2	11.6	7.0
Main activity producers (1)	*-33.9*	*-61.4*	*-99.6*	*-153.6*	*-214.7*	*-277.2*	*-365.9*	*-381.5*	*11.4*	*6.6*
Autoproducers	*-1.3*	*-3.2*	*-5.8*	*-9.9*	*-20.6*	*-33.0*	*-49.3*	*-53.8*	*16.1*	*11.2*
Gas works	-	-	-	-	-0.0	-0.0	-0.0	-0.0	-	-
Coal transformation(2)	-9.5	-11.9	-14.2	-14.6	-15.1	-18.2	-18.5	-19.4	4.1	1.5
BKB plants	*-0.1*	*-0.2*	*-0.3*	*-0.3*	*-0.2*	*-0.2*	*-0.3*	*-0.3*	*7.9*	*1.3*
Blast furnaces	*-6.9*	*-8.2*	*-9.1*	*-8.6*	*-11.5*	*-11.5*	*-14.9*	*-15.7*	*2.8*	*2.6*
Coke ovens	*-2.5*	*-3.5*	*-4.8*	*-5.7*	*-3.3*	*-6.4*	*-3.4*	*-3.3*	*6.7*	*-1.7*
Patent fuel plants	*0.0*	*0.0*	*0.0*	-	-	-	-	-	*-0.3*	-
Other transformation(3)	-	-	-	-	-	-	-	-	-	-
Energy ind. own use	**-1.0**	**-1.7**	**-3.2**	**-3.2**	**-2.8**	**-3.0**	**-2.9**	**-2.9**	**12.0**	**-0.3**
Losses	**-**	**-0.2**	**-0.0**	**-0.0**	**-0.0**	**-0.1**	**-0.1**	**-0.1**		
Final consumption(4)	**78.5**	**100.0**	**109.7**	**101.9**	**98.1**	**133.1**	**208.7**	**211.9**	**3.4**	**3.2**
Industry(5)	53.2	71.6	83.5	78.9	79.3	111.8	183.2	192.7	4.6	4.1
Iron and steel	*11.6*	*12.1*	*16.0*	*17.6*	*16.0*	*23.2*	*44.2*	*51.7*	*3.2*	*5.8*
Chemical	*2.2*	*3.1*	*4.9*	*6.1*	*6.2*	*5.8*	*8.0*	*7.9*	*8.5*	*2.3*
Non-metallic minerals	*7.5*	*10.4*	*16.1*	*19.4*	*23.5*	*33.6*	*42.0*	*44.5*	*7.9*	*5.0*
Paper, pulp and print	*1.5*	*1.9*	*2.3*	*2.7*	*2.7*	*3.2*	*4.2*	*3.4*	*4.3*	*1.9*
Other industry (6)	*30.4*	*44.1*	*44.2*	*33.1*	*30.9*	*45.9*	*84.8*	*85.2*	*3.8*	*3.2*
Transport(7)	7.9	6.5	3.5	0.2	0.0	0.0	0.0	0.1	-7.7	-17.6
Other	17.3	21.9	22.2	22.3	18.3	20.8	24.0	18.9	2.5	-0.8
Comm. and pub. services	*4.5*	*5.3*	*5.3*	*5.8*	*4.6*	*4.8*	*5.9*	*6.1*	*1.8*	*0.7*
Residential	*3.7*	*4.0*	*4.9*	*4.9*	*5.0*	*5.4*	*6.2*	*6.5*	*2.8*	*1.4*
Other sectors (8)	*9.2*	*12.6*	*12.0*	*11.6*	*8.7*	*10.6*	*11.9*	*6.2*	*2.8*	*-3.1*
Non-energy use	0.0	-	0.5	0.5	0.4	0.5	1.5	0.3	54.1	-2.8

(1) Main activity electricity and heat generation includes district heating.

(2) Coal transformation refers to the transformation of coal and peat to secondary and tertiary products (mainly cokes, briquettes, coal tar and off-gases). Gas works and Liquefaction are shown separately to match the *Energy Balances of OECD Countries.*

(3) Other transformation includes Liquefaction and Non-specified transformations.

(4) Final Consumption includes non-energy use and energy use (Industry, Transport and Other).

(5) Please refer to notes and definitions in Part I for detailed explanation.

(6) Other industry includes Non-ferrous metals, Transport equipment, Machinery, Mining and quarrying, Food and tobacco, Wood and wood products, Construction, Textile and leather, and Non-specified industry.

(7) Transport includes Rail and Inland waterways.

(8) Other sectors includes Agriculture/Forestry, Fishing and Non-specified other.

* "Coal" refers to all types of coal, primary (anthracite, coking coal, other bituminous coal, sub-bituminous coal, lignite/brown coal) and derived fuels (including patent fuel, BKB, coke oven coke, gas coke, coal tar, coke oven gas, gas works gas, blast furnace gas and other recovered gases), as well as peat, peat briquettes and oil shale. Quantities have been converted to Mtce using calorific values reported by the respective countries. Please refer to notes and definitions in Part I.

Source: *IEA/OECD Energy Balances of OECD Countries*

EUROPEAN UNION - 27

Coal balance*

(Mtce)

	1980	1985	1990	1995	2000	2005	2010	2011	Average annual percent change 80-90	90-11
Production	..	..	528.1	399.7	306.6	280.0	234.4	239.2	..	-3.7
Imports	..	..	175.4	165.7	187.5	218.0	192.8	205.6	..	0.8
Exports	..	..	-60.4	-53.9	-47.7	-40.2	-35.3	-37.1	..	-2.3
Stock changes	..	..	6.4	9.4	11.7	-4.8	8.6	0.4		
Primary supply	..	..	**649.6**	**520.8**	**458.1**	**453.1**	**400.5**	**408.0**	..	**-2.2**
Statistical differences	..	..	**-9.8**	**-2.1**	**0.5**	**0.3**	**-0.5**	**-1.6**		
Total transformation	..	..	**-450.0**	**-396.0**	**-373.0**	**-378.4**	**-332.5**	**-339.9**	..	**-1.3**
Electricity and heat gen.	..	..	-409.2	-357.9	-336.3	-346.5	-303.0	-311.0	..	-1.3
Main activity producers (1)	..	..	*-356.5*	*-326.2*	*-316.6*	*-328.1*	*-287.5*	*-295.7*	..	*-0.9*
Autoproducers	..	..	*-52.7*	*-31.8*	*-19.7*	*-18.3*	*-15.5*	*-15.3*	..	*-5.7*
Gas works	..	..	2.4	0.5	0.1	-0.2	-0.3	-0.3	..	-
Coal transformation(2)	..	..	-42.9	-38.0	-36.3	-31.0	-28.2	-27.6	..	-2.1
BKB plants	..	..	*-1.8*	*-1.2*	*-0.2*	*-0.1*	*-0.0*	*-0.1*	..	*-12.5*
Blast furnaces	..	..	*-33.3*	*-29.2*	*-29.5*	*-27.3*	*-23.9*	*-23.3*	..	*-1.7*
Coke ovens	..	..	*-8.7*	*-7.8*	*-6.8*	*-3.7*	*-4.2*	*-4.2*	..	*-3.4*
Patent fuel plants	..	..	*0.8*	*0.2*	*0.2*	*0.1*	*0.0*	*0.0*	..	*-14.8*
Other transformation(3)	..	..	-0.3	-0.6	-0.5	-0.7	-1.1	-1.1	..	7.0
Energy ind. own use	..	..	**-13.1**	**-12.0**	**-10.4**	**-9.6**	**-8.1**	**-7.7**	..	**-2.5**
Losses	..	..	**-1.4**	**-1.5**	**-1.1**	**-1.2**	**-1.3**	**-1.3**		
Final consumption(4)	..	..	**175.3**	**109.1**	**74.1**	**64.2**	**58.1**	**57.5**	..	**-5.2**
Industry(5)	..	..	97.8	71.0	54.0	46.3	35.9	37.6	..	-4.5
Iron and steel	..	..	*40.6*	*33.2*	*26.1*	*25.2*	*18.2*	*19.3*	..	*-3.5*
Chemical	..	..	*11.2*	*7.5*	*4.4*	*4.7*	*4.9*	*4.5*	..	*-4.2*
Non-metallic minerals	..	..	*17.7*	*13.9*	*11.0*	*8.0*	*7.0*	*8.0*	..	*-3.7*
Paper, pulp and print	..	..	*3.8*	*3.1*	*2.3*	*2.0*	*1.6*	*1.5*	..	*-4.3*
Other industry (6)	..	..	*24.6*	*13.2*	*10.2*	*6.5*	*4.2*	*4.2*	..	*-8.1*
Transport(7)	..	..	0.3	0.0	0.0	0.0	0.0	0.0	..	-13.8
Other	..	..	74.7	36.6	18.6	15.9	20.2	18.0	..	-6.5
Comm. and pub. services	..	..	*18.7*	*5.8*	*2.2*	*2.0*	*2.5*	*2.5*	..	*-9.2*
Residential	..	..	*51.3*	*27.2*	*14.4*	*12.3*	*15.6*	*13.7*	..	*-6.1*
Other sectors (8)	..	..	*4.7*	*3.6*	*2.0*	*1.7*	*2.1*	*1.8*	..	*-4.5*
Non-energy use	..	..	2.4	1.5	1.5	2.0	1.9	1.9	..	-1.2

(1) Main activity electricity and heat generation includes district heating.

(2) Coal transformation refers to the transformation of coal and peat to secondary and tertiary products (mainly cokes, briquettes, coal tar and off-gases). Gas works and Liquefaction are shown separately to match the *Energy Balances of OECD Countries.*

(3) Other transformation includes Liquefaction and Non-specified transformations.

(4) Final Consumption includes non-energy use and energy use (Industry, Transport and Other).

(5) Please refer to notes and definitions in Part I for detailed explanation.

(6) Other industry includes Non-ferrous metals, Transport equipment, Machinery, Mining and quarrying, Food and tobacco, Wood and wood products, Construction, Textile and leather, and Non-specified industry.

(7) Transport includes Rail and Inland waterways.

(8) Other sectors includes Agriculture/Forestry, Fishing and Non-specified other.

* "Coal" refers to all types of coal, primary (anthracite, coking coal, other bituminous coal, sub-bituminous coal, lignite/brown coal) and derived fuels (including patent fuel, BKB, coke oven coke, gas coke, coal tar, coke oven gas, gas works gas, blast furnace gas and other recovered gases), as well as peat, peat briquettes and oil shale. Quantities have been converted to Mtce using calorific values reported by the respective countries. Please refer to notes and definitions in Part I.

Source: *IEA/OECD Energy Balances of OECD Countries*

NON-OECD AMERICAS

Coal balance*

(Mtce)

	1980	1985	1990	1995	2000	2005	2010	2011	Average annual percent change 80-90	90-11
Production	7.8	13.7	25.2	31.3	47.7	66.0	75.0	85.1	12.4	6.0
Imports	6.8	10.3	13.3	15.9	17.0	19.2	21.5	23.9	6.9	2.8
Exports	-1.4	-4.4	-14.6	-21.5	-41.6	-57.5	-67.2	-76.9	26.7	8.2
Stock changes	-0.2	-0.2	-3.1	-1.4	1.7	-1.2	0.6	-0.1		
Primary supply	**13.0**	**19.4**	**20.7**	**24.3**	**24.7**	**26.5**	**29.9**	**32.0**	**4.8**	**2.1**
Statistical differences	**1.2**	**1.4**	**0.4**	**0.0**	**1.1**	**0.0**	**-0.4**	**0.0**		
Total transformation	**-6.2**	**-8.9**	**-9.8**	**-11.3**	**-11.2**	**-12.4**	**-13.5**	**-12.7**	**4.7**	**1.2**
Electricity and heat gen.	-2.4	-3.7	-4.1	-5.3	-6.5	-7.1	-8.5	-7.6	5.5	2.9
Main activity producers (1)	*-1.8*	*-2.7*	*-2.5*	*-3.7*	*-4.8*	*-5.1*	*-6.1*	*-5.4*	*3.6*	*3.7*
Autoproducers	*-0.6*	*-1.0*	*-1.6*	*-1.6*	*-1.7*	*-2.0*	*-2.4*	*-2.2*	*9.8*	*1.5*
Gas works	0.0	0.0	0.0	0.0	0.0	-	-	-	2.6	-
Coal transformation(2)	-3.8	-5.2	-5.7	-6.0	-4.7	-5.4	-5.0	-5.1	4.1	-0.5
BKB plants	-	-	-	-	-	-	-	-	-	-
Blast furnaces	*-2.7*	*-3.9*	*-4.0*	*-5.0*	*-4.8*	*-5.7*	*-5.6*	*-5.8*	*4.3*	*1.8*
Coke ovens	*-1.2*	*-1.3*	*-1.7*	*-1.0*	*0.1*	*0.3*	*0.5*	*0.7*	*3.5*	-
Patent fuel plants	-	-	-	-	-	-	-	-	-	-
Other transformation(3)	-	-	-	-	-	-	-	-	-	-
Energy ind. own use	**-1.0**	**-1.4**	**-1.3**	**-1.4**	**-1.1**	**-1.1**	**-0.7**	**-0.8**	**3.1**	**-2.2**
Losses	**-1.4**	**-2.1**	**-1.3**	**-1.8**	**-0.6**	**-0.6**	**-0.4**	**-0.7**		
Final consumption(4)	**5.6**	**8.5**	**8.8**	**10.0**	**13.0**	**12.4**	**14.9**	**17.7**	**4.5**	**3.4**
Industry(5)	5.2	7.9	8.4	9.7	12.7	12.1	14.6	17.4	5.0	3.6
Iron and steel	*2.8*	*3.9*	*4.1*	*5.7*	*7.8*	*7.3*	*9.6*	*10.5*	*4.1*	*4.5*
Chemical	*0.1*	*0.3*	*0.2*	*0.3*	*0.3*	*0.3*	*0.3*	*0.3*	*7.1*	*1.5*
Non-metallic minerals	*1.4*	*2.3*	*2.5*	*1.6*	*1.7*	*1.2*	*1.0*	*2.1*	*5.9*	*-0.8*
Paper, pulp and print	*0.3*	*0.5*	*0.5*	*0.5*	*0.6*	*0.5*	*0.4*	*0.5*	*4.7*	*0.5*
Other industry (6)	*0.6*	*1.0*	*1.1*	*1.5*	*2.3*	*3.0*	*3.4*	*4.1*	*6.2*	*6.4*
Transport(7)	0.0	0.0	0.0	0.0	0.0	0.0	0.0	-	-12.2	-
Other	0.3	0.2	0.2	0.2	0.1	0.1	0.1	0.1	-2.8	-3.0
Comm. and pub. services	*0.0*	*0.0*	*0.0*	*0.0*	*0.0*	-	-	-	*2.9*	-
Residential	*0.2*	*0.2*	*0.2*	*0.2*	*0.1*	*0.1*	*0.1*	*0.1*	*-2.0*	*-3.0*
Other sectors (8)	*0.0*	*0.0*	*0.0*	*0.0*	*0.0*	*0.0*	*0.0*	*0.0*	*-27.5*	*3.4*
Non-energy use	0.2	0.3	0.2	0.1	0.2	0.2	0.2	0.2	-1.0	0.5

(1) Main activity electricity and heat generation includes district heating.

(2) Coal transformation refers to the transformation of coal and peat to secondary and tertiary products (mainly cokes, briquettes, coal tar and off-gases). Gas works and Liquefaction are shown separately to match the *Energy Balances of OECD Countries.*

(3) Other transformation includes Liquefaction and Non-specified transformations.

(4) Final Consumption includes non-energy use and energy use (Industry, Transport and Other).

(5) Please refer to notes and definitions in Part I for detailed explanation.

(6) Other industry includes Non-ferrous metals, Transport equipment, Machinery, Mining and quarrying, Food and tobacco, Wood and wood products, Construction, Textile and leather, and Non-specified industry.

(7) Transport includes Rail and Inland waterways.

(8) Other sectors includes Agriculture/Forestry, Fishing and Non-specified other.

* "Coal" refers to all types of coal, primary (anthracite, coking coal, other bituminous coal, sub-bituminous coal, lignite/brown coal) and derived fuels (including patent fuel, BKB, coke oven coke, gas coke, coal tar, coke oven gas, gas works gas, blast furnace gas and other recovered gases), as well as peat, peat briquettes and oil shale. Quantities have been converted to Mtce using calorific values reported by the respective countries. Please refer to notes and definitions in Part I.

Source: *IEA/OECD Energy Balances of OECD Countries*

NON-OECD EUROPE AND EURASIA

Coal balance*

(Mtce)

	1980	1985	1990	1995	2000	2005	2010	2011	Average annual percent change 80-90	90-11
Production	518.5	493.0	532.9	337.9	319.3	364.8	413.3	431.4	0.3	-1.0
Imports	21.6	27.4	105.1	46.0	38.1	39.2	43.3	46.1	17.2	-3.8
Exports	-21.4	-20.2	-120.1	-42.5	-61.1	-98.8	-150.5	-143.4	18.8	0.8
Stock changes	-1.4	-3.5	6.7	4.5	2.5	-3.1	8.9	-1.9		
Primary supply	**517.3**	**496.7**	**524.6**	**346.0**	**298.7**	**302.2**	**315.0**	**332.2**	**0.1**	**-2.2**
Statistical differences	**28.4**	**-18.1**	**1.6**	**4.4**	**-3.0**	**-3.0**	**-9.2**	**-3.6**		
Total transformation	**-309.0**	**-285.5**	**-338.5**	**-258.4**	**-236.8**	**-238.6**	**-236.5**	**-252.7**	**0.9**	**-1.4**
Electricity and heat gen.	-223.6	-222.3	-282.0	-208.3	-190.0	-193.1	-194.7	-207.7	2.3	-1.4
Main activity producers (1)	*-211.2*	*-208.9*	*-250.4*	*-177.8*	*-154.5*	*-154.2*	*-161.7*	*-172.6*	*1.7*	*-1.8*
Autoproducers	*-12.4*	*-13.3*	*-31.6*	*-30.5*	*-35.5*	*-38.9*	*-33.0*	*-35.2*	*9.8*	*0.5*
Gas works	-	-	0.0	0.0	0.0	0.1	0.0	-0.0	-	-
Coal transformation(2)	-85.3	-63.2	-56.5	-50.2	-46.8	-45.5	-41.7	-44.8	-4.0	-1.1
BKB plants	*-0.5*	*-0.5*	*0.8*	*-0.0*	*0.0*	*-0.1*	*-0.2*	*-0.3*	-	-
Blast furnaces	*-61.5*	*-56.2*	*-34.0*	*-27.2*	*-23.8*	*-25.5*	*-28.9*	*-30.9*	*-5.7*	*-0.5*
Coke ovens	*-23.8*	*-11.7*	*-23.8*	*-22.7*	*-22.9*	*-19.8*	*-12.6*	*-13.6*	*0.0*	*-2.6*
Patent fuel plants	*0.4*	*5.2*	*0.5*	*-0.3*	*-0.2*	*-0.1*	*0.0*	-	*3.2*	-
Other transformation(3)	-	-	-	-	-	-0.1	-0.1	-0.1	-	-
Energy ind. own use	**-12.9**	**-13.2**	**-6.4**	**-5.0**	**-4.8**	**-5.3**	**-7.3**	**-7.0**	**-6.8**	**0.5**
Losses	**-6.3**	**-6.0**	**-18.6**	**-9.6**	**-1.8**	**-1.5**	**-1.8**	**-2.7**		
Final consumption(4)	**217.5**	**173.8**	**162.7**	**77.4**	**52.2**	**53.8**	**60.3**	**66.2**	**-2.9**	**-4.2**
Industry(5)	97.0	66.2	80.4	49.4	29.3	36.0	43.8	50.2	-1.9	-2.2
Iron and steel	*19.2*	*21.0*	*35.6*	*26.2*	*17.5*	*21.5*	*26.1*	*29.5*	*6.4*	*-0.9*
Chemical	*4.5*	*7.4*	*1.3*	*0.7*	*0.5*	*0.4*	*0.7*	*0.9*	*-11.4*	*-2.0*
Non-metallic minerals	*12.7*	*13.4*	*1.1*	*1.2*	*1.1*	*1.9*	*3.5*	*4.5*	*-21.9*	*7.0*
Paper, pulp and print	-	*0.0*	*0.0*	*0.0*	*0.0*	*0.0*	*0.1*	*0.1*	-	*8.0*
Other industry (6)	*60.6*	*24.4*	*42.4*	*21.4*	*10.3*	*12.2*	*13.4*	*15.3*	*-3.5*	*-4.7*
Transport(7)	8.7	7.0	0.1	0.0	0.0	0.1	0.1	0.1	-33.3	-4.8
Other	102.4	94.6	82.1	27.9	20.8	14.0	15.2	14.6	-2.2	-7.9
Comm. and pub. services	*90.1*	*69.7*	*30.9*	*2.3*	*0.9*	*4.1*	*5.1*	*5.0*	*-10.2*	*-8.3*
Residential	*1.8*	*5.6*	*30.1*	*20.2*	*18.2*	*7.4*	*5.9*	*8.0*	*32.8*	*-6.1*
Other sectors (8)	*10.6*	*19.2*	*21.1*	*5.4*	*1.7*	*2.5*	*4.2*	*1.6*	*7.2*	*-11.6*
Non-energy use	9.4	6.0	0.0	0.0	2.0	3.7	1.2	1.3	-43.3	19.4

(1) Main activity electricity and heat generation includes district heating.

(2) Coal transformation refers to the transformation of coal and peat to secondary and tertiary products (mainly cokes, briquettes, coal tar and off-gases). Gas works and Liquefaction are shown separately to match the *Energy Balances of OECD Countries.*

(3) Other transformation includes Liquefaction and Non-specified transformations.

(4) Final Consumption includes non-energy use and energy use (Industry, Transport and Other).

(5) Please refer to notes and definitions in Part I for detailed explanation.

(6) Other industry includes Non-ferrous metals, Transport equipment, Machinery, Mining and quarrying, Food and tobacco, Wood and wood products, Construction, Textile and leather, and Non-specified industry.

(7) Transport includes Rail and Inland waterways.

(8) Other sectors includes Agriculture/Forestry, Fishing and Non-specified other.

* "Coal" refers to all types of coal, primary (anthracite, coking coal, other bituminous coal, sub-bituminous coal, lignite/brown coal) and derived fuels (including patent fuel, BKB, coke oven coke, gas coke, coal tar, coke oven gas, gas works gas, blast furnace gas and other recovered gases), as well as peat, peat briquettes and oil shale. Quantities have been converted to Mtce using calorific values reported by the respective countries. Please refer to notes and definitions in Part I.

Source: *IEA/OECD Energy Balances of OECD Countries*

NON-OECD TOTAL

Coal balance*

(Mtce)

	1980	1985	1990	1995	2000	2005	2010	2011	Average annual percent change 80-90	Average annual percent change 90-11
Production	1194.6	1403.1	1661.2	1774.7	1896.7	2864.4	3804.2	4103.9	3.4	4.4
Imports	39.5	65.1	164.1	125.6	149.9	221.7	421.5	457.5	15.3	5.0
Exports	-57.3	-80.2	-205.7	-186.6	-287.9	-434.1	-569.7	-613.7	13.6	5.3
Stock changes	-2.4	-26.4	23.4	-4.8	37.3	-39.8	-70.0	-72.0		
Primary supply	**1174.4**	**1361.6**	**1643.0**	**1708.8**	**1796.0**	**2612.2**	**3585.9**	**3875.7**	**3.4**	**4.2**
Statistical differences	**19.3**	**-33.5**	**-23.1**	**35.9**	**27.8**	**22.3**	**-179.9**	**-192.9**		
Total transformation	**-519.0**	**-592.4**	**-803.2**	**-945.9**	**-1131.7**	**-1679.0**	**-2245.3**	**-2461.5**	**4.5**	**5.5**
Electricity and heat gen.	-385.9	-464.1	-665.8	-800.3	-985.3	-1459.4	-1966.4	-2161.1	5.6	5.8
Main activity producers (1)	*-368.1*	*-441.7*	*-622.5*	*-752.1*	*-919.0*	*-1372.4*	*-1863.1*	*-2050.3*	*5.4*	*5.8*
Autoproducers	*-17.8*	*-22.3*	*-43.3*	*-48.1*	*-66.3*	*-87.1*	*-103.3*	*-110.8*	*9.3*	*4.6*
Gas works	-3.2	-2.5	-4.3	-3.6	-4.3	-8.9	-7.9	-6.0	3.1	1.6
Coal transformation (2)	-126.8	-107.7	-113.0	-122.1	-121.0	-188.8	-247.2	-270.2	-1.1	4.2
BKB plants	*-0.6*	*-0.8*	*0.5*	*-0.3*	*-0.2*	*-0.3*	*-0.5*	*-0.6*	*-*	*-*
Blast furnaces	*-88.6*	*-86.1*	*-72.3*	*-78.7*	*-76.8*	*-137.3*	*-180.0*	*-198.3*	*-2.0*	*4.9*
Coke ovens	*-38.1*	*-26.0*	*-41.8*	*-42.8*	*-43.4*	*-48.1*	*-64.0*	*-68.3*	*0.9*	*2.4*
Patent fuel plants	*0.4*	*5.3*	*0.6*	*-0.4*	*-0.6*	*-3.2*	*-2.8*	*-3.0*	*3.0*	*-*
Other transformation (3)	-3.1	-18.2	-20.1	-19.9	-21.2	-21.8	-23.8	-24.1	20.6	0.9
Energy ind. own use	**-19.7**	**-24.9**	**-32.3**	**-49.4**	**-63.6**	**-90.8**	**-104.4**	**-102.0**	**5.1**	**5.6**
Losses	**-7.7**	**-8.5**	**-20.0**	**-11.5**	**-2.7**	**-2.3**	**-2.4**	**-3.6**		
Final consumption (4)	**647.3**	**702.3**	**764.3**	**737.9**	**625.8**	**862.4**	**1053.8**	**1115.9**	**1.7**	**1.8**
Industry (5)	377.0	392.2	450.4	509.1	445.3	652.7	837.9	900.6	1.8	3.4
Iron and steel	*91.9*	*78.0*	*99.8*	*130.8*	*118.9*	*187.3*	*306.0*	*337.6*	*0.8*	*6.0*
Chemical	*49.0*	*42.0*	*38.2*	*69.0*	*43.9*	*58.2*	*66.9*	*76.3*	*-2.5*	*3.4*
Non-metallic minerals	*56.7*	*88.0*	*92.2*	*119.9*	*130.6*	*184.6*	*220.9*	*238.4*	*5.0*	*4.6*
Paper, pulp and print	*7.3*	*10.9*	*12.8*	*17.2*	*14.7*	*19.0*	*21.4*	*20.1*	*5.8*	*2.2*
Other industry (6)	*172.2*	*173.3*	*207.4*	*172.1*	*137.2*	*203.6*	*222.8*	*228.3*	*1.9*	*0.5*
Transport (7)	32.2	30.3	18.0	9.3	6.2	5.9	4.6	4.7	-5.7	-6.2
Other	227.2	258.8	263.8	192.4	137.6	159.5	160.2	158.4	1.5	-2.4
Comm. and pub. services	*99.1*	*81.6*	*46.6*	*18.1*	*17.7*	*25.0*	*27.6*	*30.0*	*-7.3*	*-2.1*
Residential	*89.2*	*118.7*	*154.7*	*128.3*	*91.7*	*95.8*	*87.8*	*90.2*	*5.7*	*-2.5*
Other sectors (8)	*38.9*	*58.5*	*62.6*	*46.0*	*28.2*	*38.7*	*44.9*	*38.2*	*4.9*	*-2.3*
Non-energy use	10.8	20.9	32.1	27.1	36.7	44.3	51.0	52.2	11.5	2.3

(1) Main activity electricity and heat generation includes district heating.

(2) Coal transformation refers to the transformation of coal and peat to secondary and tertiary products (mainly cokes, briquettes, coal tar and off-gases). Gas works and Liquefaction are shown separately to match the *Energy Balances of OECD Countries*.

(3) Other transformation includes Liquefaction and Non-specified transformations.

(4) Final Consumption includes non-energy use and energy use (Industry, Transport and Other).

(5) Please refer to notes and definitions in Part I for detailed explanation.

(6) Other industry includes Non-ferrous metals, Transport equipment, Machinery, Mining and quarrying, Food and tobacco, Wood and wood products, Construction, Textile and leather, and Non-specified industry.

(7) Transport includes Rail and Inland waterways.

(8) Other sectors includes Agriculture/Forestry, Fishing and Non-specified other.

* "Coal" refers to all types of coal, primary (anthracite, coking coal, other bituminous coal, sub-bituminous coal, lignite/brown coal) and derived fuels (including patent fuel, BKB, coke oven coke, gas coke, coal tar, coke oven gas, gas works gas, blast furnace gas and other recovered gases), as well as peat, peat briquettes and oil shale. Quantities have been converted to Mtce using calorific values reported by the respective countries.
Please refer to notes and definitions in Part I.

Source: *IEA/OECD Energy Balances of OECD Countries*

WORLD

Coal balance*

(Mtce)

	1980	1985	1990	1995	2000	2005	2010	2011	Average annual percent change 80-90	90-11
Production	2578.9	2868.3	3193.8	3189.6	3277.2	4302.2	5211.0	5500.8	2.2	2.6
Imports	275.8	354.8	482.9	480.0	592.4	742.8	949.7	995.4	5.8	3.5
Exports	-272.3	-354.1	-492.4	-487.1	-596.0	-761.8	-978.3	-1037.5	6.1	3.6
Stock changes	-28.6	9.1	2.0	-8.1	94.9	-35.0	-45.9	-64.3		
Primary supply	**2553.8**	**2878.1**	**3186.4**	**3174.4**	**3368.4**	**4248.2**	**5136.4**	**5394.4**	**2.2**	**2.5**
Statistical differences	**-0.6**	**-39.8**	**-21.9**	**59.0**	**51.7**	**8.2**	**-188.8**	**-205.8**		
Total transformation	**-1478.5**	**-1681.1**	**-1989.0**	**-2188.0**	**-2508.3**	**-3093.8**	**-3584.1**	**-3770.7**	**3.0**	**3.1**
Electricity and heat gen.	-1235.6	-1448.5	-1749.6	-1944.2	-2263.0	-2790.8	-3217.8	-3379.5	3.5	3.2
Main activity producers (1)	*-1154.2*	*-1362.5*	*-1638.2*	*-1834.6*	*-2143.0*	*-2658.2*	*-3071.4*	*-3226.1*	*3.6*	*3.3*
Autoproducers	*-81.4*	*-86.0*	*-111.4*	*-109.6*	*-120.0*	*-132.6*	*-146.4*	*-153.4*	*3.2*	*1.5*
Gas works	4.6	1.8	-4.8	-6.0	-7.0	-11.5	-10.8	-9.0	-	3.0
Coal transformation(2)	-244.5	-216.3	-214.2	-217.2	-216.6	-269.0	-330.6	-357.1	-1.3	2.5
BKB plants	*0.8*	*-1.2*	*-1.0*	*-1.6*	*-0.4*	*-0.6*	*-0.5*	*-0.7*	*-*	*-1.8*
Blast furnaces	*-168.5*	*-160.1*	*-150.6*	*-155.2*	*-157.9*	*-211.7*	*-252.6*	*-272.6*	*-1.1*	*2.9*
Coke ovens	*-77.2*	*-54.4*	*-58.9*	*-59.8*	*-56.9*	*-60.8*	*-74.8*	*-80.7*	*-2.7*	*1.5*
Patent fuel plants	*0.5*	*-0.6*	*-3.6*	*-0.7*	*-1.4*	*4.0*	*-2.7*	*-3.0*	*-*	*-0.9*
Other transformation(3)	-3.1	-18.2	-20.3	-20.6	-21.7	-22.5	-24.9	-25.2	20.7	1.0
Energy ind. own use	**-47.5**	**-51.9**	**-53.8**	**-69.1**	**-82.9**	**-108.8**	**-124.6**	**-122.0**	**1.3**	**4.0**
Losses	**-10.1**	**-10.5**	**-21.4**	**-13.1**	**-3.9**	**-3.6**	**-3.7**	**-4.9**		
Final consumption(4)	**1017.1**	**1094.8**	**1100.4**	**963.1**	**825.0**	**1050.1**	**1235.2**	**1290.9**	**0.8**	**0.8**
Industry(5)	606.0	637.9	679.0	685.5	615.8	812.0	981.3	1041.3	1.1	2.1
Iron and steel	*198.2*	*171.4*	*181.5*	*199.8*	*180.7*	*244.6*	*361.3*	*394.8*	*-0.9*	*3.8*
Chemical	*73.9*	*73.3*	*66.4*	*86.8*	*62.1*	*74.5*	*83.0*	*91.8*	*-1.1*	*1.6*
Non-metallic minerals	*89.4*	*133.7*	*134.0*	*159.6*	*166.7*	*215.5*	*246.1*	*264.4*	*4.1*	*3.3*
Paper, pulp and print	*18.5*	*26.2*	*29.0*	*26.0*	*23.0*	*29.9*	*31.0*	*29.2*	*4.6*	*0.0*
Other industry (6)	*226.1*	*233.4*	*268.2*	*213.4*	*183.3*	*247.4*	*259.9*	*261.2*	*1.7*	*-0.1*
Transport(7)	35.9	32.4	18.4	9.5	6.3	6.2	4.9	4.9	-6.5	-6.1
Other	361.1	400.3	367.4	238.6	163.8	184.6	194.1	188.6	0.2	-3.1
Comm. and pub. services	*127.8*	*114.7*	*69.7*	*26.2*	*22.1*	*30.8*	*33.3*	*35.5*	*-5.9*	*-3.2*
Residential	*180.4*	*212.9*	*222.6*	*162.9*	*111.5*	*112.7*	*113.6*	*113.0*	*2.1*	*-3.2*
Other sectors (8)	*52.8*	*72.7*	*75.1*	*49.6*	*30.2*	*41.1*	*47.2*	*40.2*	*3.6*	*-2.9*
Non-energy use	14.1	24.2	35.6	29.5	39.2	47.4	54.9	56.0	9.7	2.2

(1) Main activity electricity and heat generation includes district heating.

(2) Coal transformation refers to the transformation of coal and peat to secondary and tertiary products (mainly cokes, briquettes, coal tar and off-gases). Gas works and Liquefaction are shown separately to match the *Energy Balances of OECD Countries.*

(3) Other transformation includes Liquefaction and Non-specified transformations.

(4) Final Consumption includes non-energy use and energy use (Industry, Transport and Other).

(5) Please refer to notes and definitions in Part I for detailed explanation.

(6) Other industry includes Non-ferrous metals, Transport equipment, Machinery, Mining and quarrying, Food and tobacco, Wood and wood products, Construction, Textile and leather, and Non-specified industry.

(7) Transport includes Rail and Inland waterways.

(8) Other sectors includes Agriculture/Forestry, Fishing and Non-specified other.

* "Coal" refers to all types of coal, primary (anthracite, coking coal, other bituminous coal, sub-bituminous coal, lignite/brown coal) and derived fuels (including patent fuel, BKB, coke oven coke, gas coke, coal tar, coke oven gas, gas works gas, blast furnace gas and other recovered gases), as well as peat, peat briquettes and oil shale. Quantities have been converted to Mtce using calorific values reported by the respective countries. Please refer to notes and definitions in Part I.

Source: *IEA/OECD Energy Balances of OECD Countries*

Energy Data Manager / Statistician

Possible Staff Vacancies

International Energy Agency, Paris, France

The IEA

The International Energy Agency, based in Paris, acts as energy policy advisor to 28 member countries in their effort to ensure reliable, affordable and clean energy for their citizens. Founded during the oil crisis of 1973-74, the IEA's initial role was to co-ordinate measures in times of oil supply emergencies. As energy markets have changed, so has the IEA. Its mandate has broadened to incorporate the "Three E's" of balanced energy policy making: energy security, economic development and environmental protection. Current work focuses on climate change policies, market reform, energy technology collaboration and outreach to the rest of the world, especially major consumers and producers of energy like China, India, Russia and the OPEC countries.

The Energy Data Centre, with a staff of around 30 people, provides a dynamic environment for young people just finishing their studies or with one to two years of work experience.

Job description

The data managers/statisticians compile, verify and disseminate information on all aspects of energy including production, transformation and consumption of all fuels, renewables, the emergency reporting system, energy efficiency indicators, CO_2 emissions, and energy prices and taxes. The data managers are responsible for receiving, reviewing and inputting data submissions from member countries and other sources into large computerised databases. They check for completeness, correct calculations, internal consistency, accuracy and consistency with definitions. Often this entails proactively investigating and helping to resolve anomalies in collaboration with national administrations of member and non-member countries. The data managers/statisticians also play a key role in helping to design and implement computer macros used in the preparation of their energy statistics publication(s).

Principal Qualifications

- University degree in a topic relevant to energy, computer programming or statistics. We currently have staff with degrees in Mathematics, Statistics, Information Technology, Economics, Engineering, Physics, Chemistry, Environmental Studies, Hydrology, Public Administration and Business.
- Experience in the basic use of databases and computer software. Good computer programming skills in Visual Basic.
- Ability to work accurately, pay attention to detail and work to deadlines. Ability to deal simultaneously with a wide variety of tasks and to organise work efficiently.
- Good communication skills; ability to work well in a team and in a multicultural environment, particularly in liaising with contacts in national administrations and industry.
- Very good knowledge of one of the two official languages of the Organisation (English or French). Knowledge of other languages would be an advantage.
- Some knowledge of energy industry operations and terminology would also be an advantage, but is not required.

Nationals of any OECD member country are eligible for appointment. Basic salaries start at 3 158 Euros per month. The possibilities for advancement are good for candidates with appropriate qualifications and experience. Tentative enquiries about future vacancies are welcomed from men and women with relevant qualifications and experience. Applications in French or English, accompanied by a curriculum vitae, should be sent to:

Office of Management and Administration
International Energy Agency
9 rue de la Fédération
75739 Paris Cedex 15, France
Email: recruitment@iea.org

On-Line Data Services

Users can instantly access not only all the data published in this book, but also all the time series used for preparing this publication and all the other statistics publications of the IEA. The data are available on-line, either through annual subscription or pay-per-view access. More information on this service can be found on our website: http://data.iea.org

Ten Annual Publications

Energy Statistics of OECD Countries, 2013 Edition

No other publication offers such in-depth statistical coverage. It is intended for anyone involved in analytical or policy work related to energy issues. It contains data on energy supply and consumption in original units for coal, oil, natural gas, biofuels/waste and products derived from these primary fuels, as well as for electricity and heat. Complete data are available for 2010 and 2011 and supply estimates are available for the most recent year (*i.e.* 2012). Historical tables summarise data on production, trade and final consumption. Each issue includes definitions of products and flows and explanatory notes on the individual country data.

Published July 2013 - Price €120

Energy Balances of OECD Countries, 2013 Edition

A companion volume to *Energy Statistics of OECD Countries*, this publication presents standardised energy balances expressed in million tonnes of oil equivalent. Energy supply and consumption data are divided by main fuel: coal, oil, natural gas, nuclear, hydro, geothermal/solar, biofuels/waste, electricity and heat. This allows for easy comparison of the contributions each fuel makes to the economy and their interrelationships through the conversion of one fuel to another. All of this is essential for estimating total energy supply, forecasting, energy conservation, and analysing the potential for interfuel substitution. Complete data are available for 2010 and 2011 and supply estimates are available for the most recent year (*i.e.* 2012). Historical tables summarise key energy and economic indicators as well as data on production, trade and final consumption. Each issue includes definitions of products and flows and explanatory notes on the individual country data as well as conversion factors from original units to tonnes of oil equivalent.

Published July 2013 - Price €120

Energy Statistics of Non-OECD Countries, 2013 Edition

This publication offers the same in-depth statistical coverage as the homonymous publication covering OECD countries. It includes data in original units for more than 100 individual countries and nine main regions. The consistency of OECD and non-OECD countries' detailed statistics provides an accurate picture of the global energy situation for 2010 and 2011. For a description of the content, please see *Energy Statistics of OECD Countries* above.

Published August 2013 - Price €120

Energy Balances of Non-OECD Countries, 2013 Edition

A companion volume to the publication *Energy Statistics of Non-OECD Countries*, this publication presents energy balances in thousand tonnes of oil equivalent and key economic and energy indicators for more than 100 individual countries and nine main regions. It offers the same statistical coverage as the homonymous publication covering OECD countries, and thus provides an accurate picture of the global energy situation for 2010 and 2011. For a description of the content, please see *Energy Balances of OECD Countries* above.

Published August 2013 - Price €120

Electricity Information 2013

This reference document provides essential statistics on electricity and heat for each OECD member country by bringing together information on production, installed capacity, input energy mix to electricity and heat production, input fuel prices, consumption, end-user electricity prices and electricity trades.

Published August 2013 - Price €150

Coal Information 2013

This well-established publication provides detailed information on past and current evolution of the world coal market. It presents country-specific statistics for OECD member countries and selected non-OECD countries on coal production, demand, trade and prices. This publication represents a key reference tool for all those involved in the coal supply or consumption stream, as well as institutions and governments involved in market and policy analysis of the world coal market.

Published August 2013 - Price €165

Natural Gas Information 2013

A detailed reference work on gas supply and demand, covering not only OECD countries but also the rest of the world. Contains essential information on LNG and pipeline trade, gas reserves, storage capacity and prices. The main part of the book, however, concentrates on OECD countries, showing a detailed gas supply and demand balance for each individual country and for the three OECD regions, as well as a breakdown of gas consumption by end-user. Import and export data are reported by source and destination.

Published August 2013 - Price €165

Oil Information 2013

A comprehensive reference book on current developments in oil supply and demand. The first part of this publication contains key data on world production, trade, prices and consumption of major oil product groups, with time series back to the early 1970s. The second part gives a more detailed and comprehensive picture of oil supply, demand, trade, production and consumption by end-user for each OECD country individually and for OECD regions. Trade data are reported extensively by origin and destination.

Published August 2013 - Price €165

Renewables Information 2013

This reference document brings together in one volume essential statistics on renewables and waste energy sources. It presents a detailed and comprehensive picture of developments for renewable and waste energy sources for each of the OECD member countries, encompassing energy indicators, generating capacity, electricity and heat production from renewable and waste sources, as well as production and consumption of renewable and waste products.

Published August 2013 - Price €110

CO_2 Emissions from Fuel Combustion, 2013 Edition

In order for nations to tackle the problem of climate change, they need accurate greenhouse gas emissions data. This publication provides a basis for comparative analysis of CO_2 emissions from fossil fuel combustion, a major source of anthropogenic emissions. The data in this book are designed to assist in understanding the evolution of the emissions of CO_2 from 1971 to 2011 for more than 140 countries and regions by sector and by fuel. Emissions were calculated using IEA energy databases and the default methods and emissions factors from the *Revised 1996 IPCC Guidelines for National Greenhouse Gas Inventories.*

Published November 2013 - Price €165

Two Quarterlies

Oil, Gas, Coal and Electricity, Quarterly Statistics

This publication provides up-to-date, detailed quarterly statistics on oil, coal, natural gas and electricity for OECD countries. Oil statistics cover production, trade, refinery intake and output, stock changes and consumption for crude oil, NGL and nine selected oil product groups. Statistics for electricity, natural gas and coal show supply and trade. Import and export data are reported by origin and destination. Moreover, oil as well as hard coal and brown coal production are reported on a worldwide basis.

Published Quarterly - Price €120, annual subscription €380

Energy Prices and Taxes

This publication responds to the needs of the energy industry and OECD governments for up-to-date information on prices and taxes in national and international energy markets. It contains crude oil import prices by crude stream, industry prices and consumer prices. The end-user prices for OECD member countries cover main petroleum products, gas, coal and electricity. Every issue includes full notes on sources and methods and a description of price mechanisms in each country. Time series availability varies with each data series.

Published Quarterly - Price €120, annual subscription €380

Electronic Editions

CD-ROMs and Online Data Services

To complement its publications, the Energy Data Centre produces CD-ROMs containing the complete databases which are used for preparing the statistics publications. State-of-the-art software allows you to access and manipulate all these data in a very user-friendly manner and includes graphic facilities. These databases are also available on the internet from our online data service.

Annual CD-ROMS / Online Databases

- Energy Statistics of OECD Countries, 1960-2012 — Price: €550 (single user)
- Energy Balances of OECD Countries, 1960-2012 — Price: €550 (single user)
- Energy Statistics of Non-OECD Countries, 1971-2011 — Price: €550 (single user)
- Energy Balances of Non-OECD Countries, 1971-2011 — Price: €550 (single user)
- *Combined subscription of the above four series* — *Price: € 1 400 (single user)*

- Electricity Information 2013 — Price: €550 (single user)
- Coal Information 2013 — Price: €550 (single user)
- Natural Gas Information 2013 — Price: €550 (single user)
- Oil Information 2013 — Price: €550 (single user)
- Renewables Information 2013 — Price: €400 (single user)
- CO_2 Emissions from Fuel Combustion, 1971-2011 — Price: €550 (single user)

Quarterly CD-ROMs / Online Databases

- Energy Prices and Taxes — Price: (four quarters) €900 (single user)

A description of these services is available on our website: **http://data.iea.org**

Other Online Services

The Monthly Oil Data Service

The IEA Monthly Oil Data Service provides the detailed databases of historical and projected information which is used in preparing the IEA's monthly *Oil Market Report* (OMR). The IEA Monthly Oil Data Service comprises three packages available separately or combined as a subscriber service on the Internet. The data are available at the same time as the official release of the Oil Market Report.

The packages include:

- Supply, Demand, Balances and Stocks — Price: €6 000 (single user)
- Trade — Price: €2 000 (single user)
- Field-by-Field Supply — Price: €3 000 (single user)
- *Complete Service* — *Price: €9 000 (single user)*

A description of this service is available on our website: **http://www.iea.org/stats/mods.asp**

The Monthly Gas Data Service

The service provides monthly natural gas data for OECD countries:

- supply balances in terajoules and cubic metres;
- production, trade, stock changes and levels where available, gross inland deliveries, own use and losses;
- highly detailed trade data with about 50 imports origins and exports destinations;
- LNG trade detail available from January 2002.

The databases cover the time period January 1984 to current month with a time lag of two months for the most recent data.

- Monthly Gas Data Service: Natural Gas Balances & Trade
  Historical plus 12 monthly updates Price: €800 (single user)

For more information consult: **http://data.iea.org**

Moreover, the IEA statistics website contains key energy indicators by country, graphs on the world and OECD's energy situation evolution from 1971 to the most recent year available, as well as selected databases for demonstration.

The IEA statistics website can be accessed at www.iea.org/statistics/

IEA Publications, 9, rue de la Fédération, 75739 Paris Cedex 15
Printed in France by Soregraph, August 2013
(61 2013 22 1 P1) ISBN 978-92-64-20310-5

The paper used for this document has received certification from the Programme for the Endorsement of Forest Certification (PEFC) for being produced respecting PEFC's ecological, social and ethical standards. PEFC is an international non-profit, non-governmental organization dedicated to promoting Sustainable Forest Management (SFM) through independent third-party certification.

PEFC
10-31-1510